Numerical Modelling on Metallic Materials

Numerical Modelling on Metallic Materials

Topic Editors

Shuwen Wen
Yongle Sun
Xin Chen

MDPI

Basel • Beijing • Wuhan • Barcelona • Belgrade • Novi Sad • Cluj • Manchester

Topic Editors

Shuwen Wen
AMRC Castings
University of Sheffield
Sheffield
UK

Yongle Sun
Welding and Additive
Manufacturing Centre
Faculty of Engineering and
Applied Sciences
Cranfield University
Cranfield
UK

Xin Chen
Welding and Additive
Manufacturing Centre
Faculty of Engineering and
Applied Sciences
Cranfield University
Cranfield
UK

Editorial Office
MDPI AG
Grosspeteranlage 5
4052 Basel, Switzerland

This is a reprint of the Topic, published open access by the journals *Alloys* (ISSN 2674-063X), *Applied Sciences* (ISSN 2076-3417), *Coatings* (ISSN 2079-6412), *Materials* (ISSN 1996-1944) and *Metals* (ISSN 2075-4701), freely accessible at: https://www.mdpi.com/topics/modelling_metallic.

For citation purposes, cite each article independently as indicated on the article page online and as indicated below:

Lastname, A.A.; Lastname, B.B. Article Title. *Journal Name* **Year**, *Volume Number*, Page Range.

ISBN 978-3-7258-3881-3 (Hbk)
ISBN 978-3-7258-3882-0 (PDF)
https://doi.org/10.3390/books978-3-7258-3882-0

Contents

metals

MDPI

Editorial

Numerical Modelling on Metallic Materials

Shuwen Wen [1,*], Yongle Sun [2] and Xin Chen [2]

1 AMRC Castings, University of Sheffield, Advanced Manufacturing Park, Brunel Way, Rotherham S60 5WG, UK
2 Welding and Additive Manufacturing Centre, Faculty of Engineering and Applied Sciences, Cranfield University, Cranfield MK43 0AL, UK; yongle.sun@cranfield.ac.uk (Y.S.); x.chen1@cranfield.ac.uk (X.C.)
* Correspondence: shuwen.wen@amrc.co.uk

1. Introduction

Numerical modelling of metallic materials has emerged as a pivotal research area in modern materials science and engineering [1]. With the advancement of computational power and the development of sophisticated modelling techniques, numerical simulation has become an indispensable tool for studying the microstructural evolution, enhancing process understanding and optimisation, and predicting the mechanical properties of metallic materials [2–9]. By employing accurate mathematical models and efficient computational methods, researchers can simulate the complex behaviours of real-world materials in a virtual environment. This approach not only reduces experimental costs and boosts research efficiency, but also uncovers physical phenomena that are challenging to observe directly through experiments [10–15].

This Topic is dedicated to the "Numerical Modelling on Metallic Materials" and presents a curated collection of twenty cutting-edge research papers in the field. These studies span multiple scales, from microstructural evolution to macroscopic mechanical behaviour, and explore the response of various metallic materials under diverse processing and service conditions. At the microscopic level, numerical modelling is applied extensively to explore phase transformations, diffusion behaviours, and microstructural evolution in metallic materials. At the mesoscopic and macroscopic levels, numerical simulations are employed to predict the mechanical properties and failure mechanisms of metallic materials, offering insights into performance and durability. The studies featured in this collection highlight the vital role of advanced modelling techniques, including finite element analysis (FEA), molecular dynamics simulations (MDS), computational fluid dynamics (CFD), discrete element method (DEM), and phase-field (PF) and other modelling techniques, in optimising engineering designs, predicting material performance, and improving manufacturing processes.

Additionally, this Topic also showcases emerging numerical modelling approaches, such as topology optimisation and the applications of various other optimisation algorithms, in metallic material simulations. These innovative approaches, when combined with traditional physical modelling techniques, extend the capabilities of numerical modelling to encompass intelligent computation and multi-scale analysis. Such advancements pave the way for more efficient and effective strategies for the design and optimisation of future materials.

2. Outline of This Topic

This Topic includes twenty technical papers contributed by the authors working in the field and is organised into three main sections, each offering a comprehensive exploration

check for updates

Received: 26 March 2025
Accepted: 7 April 2025
Published: 9 April 2025

Citation: Wen, S.; Sun, Y.; Chen, X. Numerical Modelling on Metallic Materials. *Metals* **2025**, *15*, 423. https://doi.org/10.3390/met15040423

of numerical modelling applications in metallic materials. The three sections comprise atomic-scale and microstructural evolution modelling, meso- and macro-scale mechanical behaviour simulations, and emerging computational techniques and advanced applications in metallic materials research. These contributions are summarised and listed below.

2.1. Atomic-Scale and Microstructural Evolution Modelling

Baidak, S.T. et al. (contribution 1) analysed the electronic band structures of intermetallic compounds and their topological characteristics using an ab initio method. Xiang, G., et al. (contribution 2) investigated the diffusion of atoms and crystal structure evolution at the Fe-Ti interface through MDS. Zheng, H., et al. (contribution 3) employed MDS to examine how different orientations, temperatures, and strain rates influence the diffusion behaviour and mechanical properties of the Fe/Cu solid–liquid interface. Lv, B., et al. (contribution 4) developed computational interatomic potentials for accurately modelling the behaviours of multi-component metallic systems. Fashu, S., et al. (contribution 5) explored the impact of intragranular nanoparticles on precipitate-coarsening behaviours in metallic alloys based on PF modelling. Kumnorkaew, T., et al. (contribution 6) presented a kinetic model for predicting bainitic transformation kinetics in low-carbon steels under different thermomechanical treatments. Chen, X., et al. (contribution 7) investigated how multi-directional forging (MDF) at 1150 °C affects the microstructure of SDP1 steel and compared it to traditional forging methods.

2.2. Meso- and Macro-Scale Mechanical Behaviour Simulations

Ciepielewski, R., et al. (contribution 8) investigated the energy-absorbing properties of aluminium honeycomb structures, particularly focusing on the impact of entrapped air within the cells during dynamic loading conditions. Zhao, Y., et al. (contribution 9) developed a model combining CFD with DEM and investigated the mechanism of laser polishing and its influence on the surface finish of additively manufactured nickel alloys. Alshoaibi, A.M., et al. (contribution 10) applied FEA to predict fatigue crack growth behaviours under cyclic loading conditions. Ma, W., et al. (contribution 11) modelled the buckling and post-buckling behaviours of thin-walled structures under axial loads. Lopez-Garcia, R.D., et al. (contribution 12) studied how quenching parameters influence distortion and residual stresses in high-strength steel components. Peng, Y., et al. (contribution 13) examined the effectiveness of different stiffener geometries in enhancing the stability of steel girders. Huang, Z., et al. (contribution 14) simulated fracture expansion behaviours in steel cylindrical shells subjected to explosive loads. In addition, He, T., et al. (contribution 15) analysed the effects of mould design on shrinkage porosity in titanium alloy casting.

2.3. Emerging Computational Techniques and Advanced Applications

Kim, M.S., et al. (contribution 16) introduced a novel constitutive model to predict ductile fracture initiation and progression. Wallat, L., et al. (contribution 17) explored computational methods for designing gyroid-based porous structures with tuneable mechanical properties. Liu, M., et al. (contribution 18) enhanced the accuracy of temperature prediction for steel slabs in industrial heating processes using an optimised particle swarm algorithm. Zhao, T., et al. (contribution 19) modelled and analysed the optical properties of nanochain aggregates on silicon surfaces. Chen, Y., et al. (contribution 20) used smooth particle hydrodynamic simulation to examine how steel fibre reinforcement enhances the tensile and bonding properties of explosion-welded aluminium–steel composites.

3. Concluding Remarks

This Topic is intended for researchers, engineers, and graduate students engaged in metallic materials research. Through these research papers, readers will gain valuable insights into the latest advancements in numerical modelling, explore cutting-edge computational techniques, and discover diverse applications of numerical modelling and simulation in materials science and engineering. This Topic serves as a comprehensive reference for both academic research and industrial practice, advancing the field of metallic materials.

Funding: This research received no external funding.

Conflicts of Interest: The authors declare no conflicts of interest.

List of Contributions

1. Baidak, S.T.; Lukoyanov, A.V. Common topological features in band structure of RNiSb and RSb compounds for R= Tb, Dy, Ho. *Materials* **2023**, *16*, 242. https://doi.org/10.3390/ma16010242.
2. Xiang, G.; Luo, X.; Cao, T.; Zhang, A.; Yu, H. Atomic diffusion and crystal structure evolution at the Fe-Ti interface: molecular dynamics simulations. *Materials* **2022**, *15*, 6302. https://doi.org/10.3390/ma15186302.
3. Zheng, H.; Sun, J.; Guo, N.; Wang, M. Atomic Research on the Diffusion Behavior, Mechanical Properties and Fracture Mechanism of Fe/Cu Solid–Liquid Interface. *Coatings* **2022**, *12*, 1299. https://doi.org/10.3390/coatings12091299.
4. Lv, B.; Chen, C.; Zhang, F.; Poletaev, G.M.; Rakitin, R.Y. Potentials for describing interatomic interactions in γFe-Mn-CN system. *Metals* **2022**, *12*, 982. https://doi.org/10.3390/met12060982.
5. Fashu, S.; Huang, B.; Wang, N. Modification of Precipitate Coarsening Kinetics by Intragranular Nanoparticles—A Phase Field Study. *Metals* **2022**, *12*, 892. https://doi.org/10.3390/met12060892.
6. Kumnorkaew, T.; Lian, J.; Uthaisangsuk, V.; Bleck, W. Kinetic model of isothermal bainitic transformation of low carbon steels under ausforming conditions. *Alloys* **2022**, *1*, 93–115. https://doi.org/10.3390/alloys1010007.
7. Chen, X.; Wu, B.; Li, J.; Zhang, X.; Zuo, P.; Wu, X.; Li, J. Microstructural Evolution in Large-Section Plastic Mould Steel during Multi-Directional Forging. *Metals* **2022**, *12*, 1175. https://doi.org/10.3390/met12071175.
8. Ciepielewski, R.; Miedzińska, D. A Study of Aluminum Honeycomb Structures under Dynamic Loading, with Consideration Given to the Effects of Air Leakage. *Materials* **2023**, *16*, 2211. https://doi.org/10.3390/ma16062211.
9. Zhao, Y.; Du, C.; Wang, P.; Meng, W.; Li, C. The mechanism of in-situ laser polishing and its effect on the surface quality of nickel-based alloy fabricated by selective laser melting. *Metals* **2022**, *12*, 778. https://doi.org/10.3390/met12050778.
10. Alshoaibi, A.M. Fatigue crack growth analysis under constant amplitude loading using finite element method. *Materials* **2022**, *15*, 2937. https://doi.org/10.3390/ma15082937.
11. Ma, W.; Sun, Z.; Wu, H.; Xu, L.; Zeng, Y.; Wang, Y.; Huang, G. Buckling Analysis of Thin-Walled Circular Shells under Local Axial Compression using Vector Form Intrinsic Finite Element Method. *Metals* **2023**, *13*, 564. https://doi.org/10.3390/met13030564.
12. Lopez-Garcia, R.D.; Medina-Juárez, I.; Maldonado-Reyes, A. Effect of quenching parameters on distortion phenomena in AISI 4340 steel. *Metals* **2022**, *12*, 759. https://doi.org/10.3390/met12050759.
13. Peng, Y.; Kong, Z.; Dinh, B.H.; Nguyen, H.H.; Cao, T.S.; Papazafeiropoulos, G.; Vu, Q.V. Web Bend-Buckling of Steel Plate Girders Reinforced by Two Longitudinal Stiffeners with Various Cross-Section Shapes. *Metals* **2023**, *13*, 323. https://doi.org/10.3390/met13020323.
14. Huang, Z.; Yu, X. Numerical simulation study of expanding fracture of 45 steel cylindrical shell under different detonation pressure. *Materials* **2022**, *15*, 3980. https://doi.org/10.3390/ma15113980.
15. He, T.; Chen, Y. Influence of Mold Design on Shrinkage Porosity of Ti-6Al-4V Alloy Ingots. *Metals* **2022**, *12*, 2122. https://doi.org/10.3390/met12122122.
16. Kim, M.S.; Kim, H.T.; Choi, Y.H.; Kim, J.H.; Kim, S.K.; Lee, J.M. A new computational method for predicting ductile failure of 304L stainless steel. *Metals* **2022**, *12*, 1309. https://doi.org/10.3390/met12081309.

17. Wallat, L.; Altschuh, P.; Reder, M.; Nestler, B.; Poehler, F. Computational design and characterisation of gyroid structures with different gradient functions for porosity adjustment. *Materials* **2022**, *15*, 3730. https://doi.org/10.3390/ma15103730.
18. Liu, M.; Yan, P.; Liu, P.; Qiao, J.; Yang, Z. An improved particle-swarm-optimization algorithm for a prediction model of steel slab temperature. *Appl. Sci.* **2022**, *12*, 11550. https://doi.org/10.3390/app122211550.
19. Zhao, T.; Gao, Y.; Shi, R.; Li, Z.; Shi, Q. Structure of Randomly Distributed Nanochain Aggregates on Silicon Substrates: Modeling and Optical Absorption Characteristics. *Materials* **2022**, *15*, 4778. https://doi.org/10.3390/ma15144778.
20. Chen, Y.; Gao, Y.; Guo, C.; Guo, Y.; Guo, Z.; Liu, Y.; Liu, T. Effect of the Addition of Steel Fibers on the Bonding Interface and Tensile Properties of Explosion-Welded 2A12 Aluminum Alloy and SS-304 Steel. *Materials* **2023**, *16*, 116. https://doi.org/10.3390/ma16010116.

References

1. Rappaz, M.; Bellet, M.; Deville, M.O.; Snyder, R. *Numerical Modeling in Materials Science and Engineering*; Springer: Berlin, Germany, 2003; Volume 20.
2. Liu, H.; Xu, X.; Zhang, J.; Liu, Z.; He, Y.; Zhao, W.; Liu, Z. The state of the art for numerical simulations of the effect of the microstructure and its evolution in the metal-cutting processes. *Int. J. Mach. Tools Manuf.* **2022**, *177*, 103890. [CrossRef]
3. DebRoy, T.; Mukherjee, T.; Wei, H.L.; Elmer, J.W.; Milewski, J.O. Metallurgy, mechanistic models and machine learning in metal printing. *Nat. Rev. Mater.* **2021**, *6*, 48–68. [CrossRef]
4. Bandyopadhyay, A.; Traxel, K.D. Invited review article: Metal-additive manufacturing—Modeling strategies for application-optimized designs. *Addit. Manuf.* **2018**, *22*, 758–774. [CrossRef] [PubMed]
5. Klingenberg, W.; Singh, U.P. Further observations and review of numerical simulations of sheet metal punching. *Int. J. Adv. Manuf. Technol.* **2006**, *30*, 638–644. [CrossRef]
6. Dal, M.; Fabbro, R. An overview of the state of art in laser welding simulation. *Opt. Laser Technol.* **2016**, *78*, 2–14. [CrossRef]
7. Thomas, B.G. Review on modeling and simulation of continuous casting. *Steel Res. Int.* **2018**, *89*, 1700312. [CrossRef]
8. Sun, Y.; Li, Q.M. Dynamic compressive behaviour of cellular materials: A review of phenomenon, mechanism and modelling. *Int. J. Impact Eng.* **2018**, *112*, 74–115. [CrossRef]
9. Zhang, Z.; Wang, Y.; Ge, P.; Wu, T. A review on modelling and simulation of laser additive manufacturing: Heat transfer, microstructure evolutions and mechanical properties. *Coatings* **2022**, *12*, 1277. [CrossRef]
10. Kiran, A.; Hodek, J.; Vavřík, J.; Urbánek, M.; Džugan, J. Numerical simulation development and computational optimization for directed energy deposition additive manufacturing process. *Materials* **2020**, *13*, 2666. [CrossRef]
11. Chen, X.; Wang, C.; Ding, J.; Bridgeman, P.; Williams, S. A three-dimensional wire-feeding model for heat and metal transfer, fluid flow, and bead shape in wire plasma arc additive manufacturing. *J. Manuf. Process.* **2022**, *83*, 300–312. [CrossRef]
12. Gu, H.; Li, L. Computational fluid dynamic simulation of gravity and pressure effects in laser metal deposition for potential additive manufacturing in space. *Int. J. Heat Mass Transf.* **2019**, *140*, 51–65. [CrossRef]
13. Taghizadeh, M.; Zhu, Z.H. A comprehensive review on metal laser additive manufacturing in space: Modeling and perspectives. *Acta Astronaut.* **2024**, *222*, 403–421. [CrossRef]
14. Zhou, N.; Lv, D.C.; Zhang, H.L.; McAllister, D.; Zhang, F.; Mills, M.J.; Wang, Y. Computer simulation of phase transformation and plastic deformation in IN718 superalloy: Microstructural evolution during precipitation. *Acta Mater.* **2014**, *65*, 270–286. [CrossRef]
15. Zepeda-Ruiz, L.A.; Stukowski, A.; Oppelstrup, T.; Bulatov, V.V. Probing the limits of metal plasticity with molecular dynamics simulations. *Nature* **2017**, *550*, 492–495. [CrossRef] [PubMed]

materials

MDPI

Article

Common Topological Features in Band Structure of RNiSb and RSb Compounds for R = Tb, Dy, Ho

Semyon T. Baidak [1,2] and Alexey V. Lukoyanov [1,2,*]

1 Institute of Physics and Technology, Ural Federal University Named after the First President of Russia B.N. Yeltsin, 620002 Ekaterinburg, Russia

2 M.N. Mikheev Institute of Metal Physics of Ural Branch of Russian Academy of Sciences, 620108 Ekaterinburg, Russia

* Correspondence: lukoyanov@imp.uran.ru; Tel.: +7-3433783886

Abstract: The electronic and band structures of ternary RNiSb and binary RSb compounds for R = Tb, Dy, Ho, have been investigated using an ab initio method accounting for strong electron correlations in the 4f shell of the rare-earth metals. These ternary compounds are found to be semiconductors with the indirect gap of 0.21, 0.21, and 0.26 eV for Tb, Dy, and Ho(NiSb), respectively. In contrast, in all binary RSb compounds, bands near the Fermi energy at the Γ and X points are shifted relatively to RNiSb and form hole and electron pockets, so the energy gap is closed in RSb. The band structure typical for semimetals is formed in all RSb compounds for R = Tb, Dy, Ho. For the first time, we identify similar features near the Fermi level in the considered binary semimetals, namely, the presence of the hole and electron pockets in the vicinity of the Γ and X points, the nonsymmetric electron pocket along Γ–X–W direction and hole pockets along the L–Γ–X direction, which were previously found experimentally in the other compound of this series GdSb. The magnetic moment of all considered compounds is fully determined by magnetic moments of the rare earth elements, the calculated effective magnetic moments of these ions have values close to the experimental values for all ternary compounds.

Keywords: electronic structure; topologic structure; alloys; intermetallics; first principles calculations

Citation: Baidak, S.T.; Lukoyanov, A.V. Common Topological Features in Band Structure of RNiSb and RSb Compounds for R = Tb, Dy, Ho. *Materials* **2023**, *16*, 242. https://doi.org/10.3390/ma16010242

Academic Editors: Shuwen Wen, Yongle Sun, Xin Chen and Luca Spiridigliozzi

Received: 16 November 2022
Revised: 7 December 2022
Accepted: 23 December 2022
Published: 27 December 2022

1. Introduction

The RTX is a family of compounds, where R is a rare earth element, T designates a transition metal, and X is a s/p element. These types of compounds have been known for their unusual electronic and magnetic properties, such as heavy fermion behavior [1,2], magnetic superconductivity [3,4], the presence of Weyl fermions [5], magnetocaloric effect [6–8], large magnetoresistance and negative temperature coefficient of resistivity [9,10], and more. Another outstanding characteristic of many RNiSb compounds is giant magnetoresistance [11,12] found in half-Heusler structure [13]. Such ternary materials with remarkable thermoelectric properties can be used for thermoelectric power generation and for conversion of waste heat in electricity with high efficiency [14]. High values of ZT, which is thermoelectric figure of merit, were observed in TmNiSb and Sn-doped ErNiSb with the value of 0.25 for TmNiSb at 700 K [15] and even higher value of 0.29 at 669 K for ErNiSb [16] and other compounds of this series [15,17–19].

Magnetic susceptibility measurements for RNiSb compounds have also been conducted and experiments show that most of the compounds exhibit Curie-Weiss behavior where magnetism is dominated by the magnetic moments of rare earth elements [20]. Theoretical electronic structure calculations for GdNiSb [21], as well as for the close GdNiGe compound [22], indeed show that the dominant contribution to the total magnetic moment is due to R while the contribution of Ni and Sb (Ge) is negligible. There is also a close binary compound GdSb that is found to have a pair of Weyl fermions, the presence of which can lead to chiral anomaly-induced negative longitudinal magnetoresistance under external magnetic field [23]. High-resolution angle-resolved photoemission spectroscopy (ARPES) measurements for DySb and HoSb were conducted and revealed at least two concentric hole pockets at the Γ point and two intersecting electron pockets at the X point [24], similar features were found in the band structure of LuSb calculated within GGA not accounting for electron correlations reported in [24]. Binary compound DySb was found to have extremely large positive magnetoresistance from experimental results with the suggested non-trivial band topology from DFT calculations with an inversion point seated about 0.34 eV below the Fermi level [25]. A recent experimental study of the binary compound HoSb also found extremely large magnetoresistance, the close interplay between conduction electrons and magnetism was suggested [26].

The ternary Tb/Ho/DyNiSb intermetallics, which are studied in this work, are half-Heusler compounds and crystallize in the cubic MgAgAs-type structure. This structure may be regarded as NaCl-type structure, where the rare earth and Sb atoms take the positions of the Na and Cl atoms, while the Ni atoms occupy one half of the tetrahedral voids formed by the Sb atoms [13]. It has also been shown that compounds in question exhibit large negative magnetoresistance at low temperatures which is caused by the reduction of spin disorder scattering due to the alignment of the moments under a magnetic field [11,12]. The ternary DyNiSb compound is found to be a narrow-gap intrinsic p-type semiconductor with the experimental value of the energy gap of 0.130–0.171 eV [9], 0.089–0.130 eV [15], for HoNiSb it is estimated as 0.08–0.11 eV [9].

In this study, we consider in detail the band (and electronic) structure, as well as magnetic properties, of the RSb and RNiSb compounds for R = Tb, Dy, Ho, in order to identify common topological features of the band structure, electronic structure and magnetic properties.

2. Materials and Methods

The ternary Tb/Dy/HoNiSb compounds have MgAgAs-type half-Heusler structure (space group F-43m, number 216) [13] with the lattice parameters: a = b = c = 6.304 Å for TbNiSb, a = b = c = 6.298 Å for DyNiSb and a = b = c = 6.262 Å for the HoNiSb compound [20] with following atomic positions Tb/Dy/Ho in 4*a* (0, 0, 0), Ni in 4*c* (1/4, 1/4, 1/4) and Sb in 4*b* (1/2, 1/2, 1/2). Binary RSb compounds have similar cubic crystal structure with the same atomic positions of rare earth element and Sb atoms with the following lattice parameters: a = b = c = 6.170 Å for TbSb, a = b = c = 6.150 Å for DySb and a = b = c = 6.130 Å for the HoSb compound [27].

The crystal structure of Tb/Dy/HoNiSb is plotted in Vesta [28] in Figure 1. The unit cell of RNiSb contains 1 rare-earth atom, 1 nickel atom and 1 antimony atom. The Sb atom has an environment of four Ni atoms in the form of a tetrahedron.

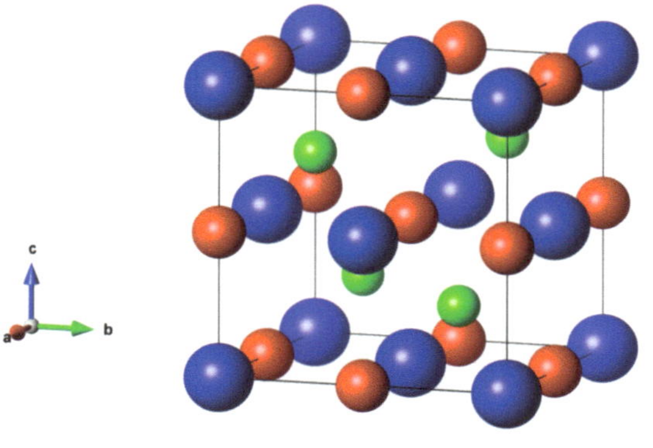

Figure 1. Crystal structure of RNiSb compounds. R atoms are shown in blue, Ni—in green, Sb—in red.

Electronic structure calculations were conducted in the Quantum Espresso package [29,30] using GGA+U version of LSDA+U method. Such method is widely used to take into account strong electron correlations between electrons of 4f shells in ions of rare earth elements. Parameters in GGA+U method have following values: Hund's exchange parameter J = 0.7 eV for all three considered elements and direct Coulomb interaction U is equal to 5.4 eV for Tb, 5.8 eV for Dy and 5.9 eV for Ho [2,31]. In this work we assume that magnetic moments of rare earth elements have ferromagnetic ordering. The exchange correlation potential was employed in generalized gradient approximation (GGA) of Perdew-Burke-Ernzerhof (PBE) [32]. The calculations used the standard ultrasoft potentials from the pseudopotential library of Quantum ESPRESSO for Ni and Sb [33], projected augmented wave method (PAW) scalar-relativistic potentials for rare-earth elements from work [34]. Wave functions were expanded in plane waves, Bloechl's tetrahedron method was employed for Brillouin-zone integration on a $12 \times 12 \times 12$ k-point mesh, interactions between ions and valence electrons were taken into account within the framework of the method of plane augmented waves.

3. Results

3.1. TbNiSb and TbSb Intermetallic Compounds

In Figure 2, the total and partial densities of electronic states of the TbNiSb and TbSb intermetallic compounds are given for two opposite spin directions. Two intense peaks in the total density of states for the majority spin direction of the TbNiSb (TbSb) compound in Figure 2a are formed by the 4f states of Tb at following energies: −7.2 (−7.8) eV and −6.1 (−6.6) eV below the valence band. Another two noticeable intense peaks in Figure 2a for the minority spin projection are found in the conduction band at energies 2.9 (2.3) eV and 3.3 (2.7) eV. One can see that valence band in TbNiSb compound is mostly formed by non-spin-polarized Ni 3d states Figure 2b with some contribution from Tb-5d and Sb-5p states. In binary TbSb compound the biggest contribution to the valence band is due to Sb-5p states Figure 2c. For both compounds and both spin projections Tb-5d states lay in the conduction band mostly unoccupied Figure 2b. The other electronic states are not shown in this figure due to their negligible contribution.

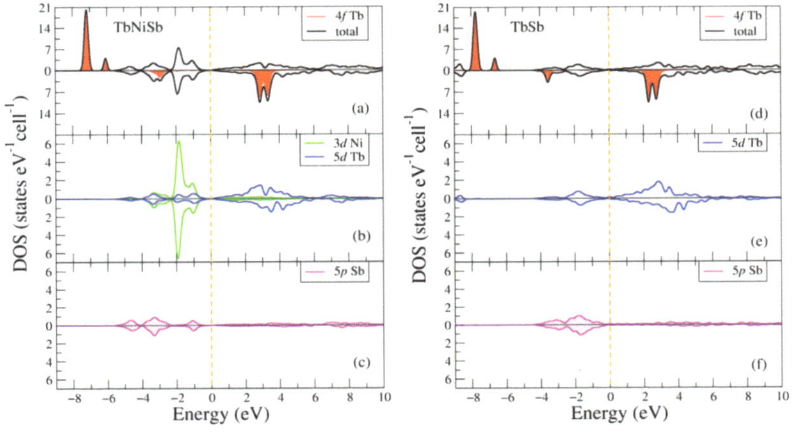

Figure 2. Densities of electronic states from DFT+U for (**a–c**) TbNiSb; (**d–f**) TbSb. (**a**,**d**) Total and partial Tb-4f densities of states; (**b**,**e**) Partial density of states for Tb-5d (Ni-3d); (**c**,**f**) Partial density of states for Sb-5p. The plot is shifted relatively to the Fermi energy shown at zero as a vertical line.

The band structure for the majority and minority spin projections of TbNiSb is shown in Figure 3. One can see the energy gap of 0.43 eV in the majority spin projection in Figure 3a and energy gap of 0.21 eV for the other spin projection in Figure 3b. Point Γ is the highest point in the valence band and the lowest point in the conduction band is X, so the compound is a semiconductor with an indirect gap. There are localized bands at the energies of eV −7.2 eV and −6.1 eV in Figure 3a and at the energies of (2.9; 3.3) eV in Figure 3b which correspond to intense peaks at the same energies in the density of states Figure 2a. There is also the presence of the occupied states near the Γ point at the Fermi level.

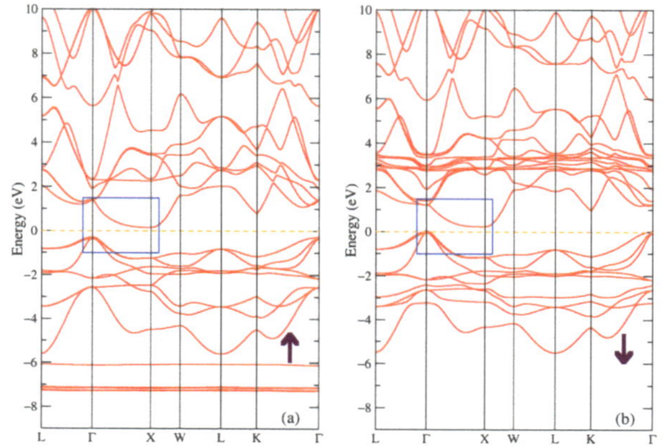

Figure 3. Band structure of TbNiSb: (**a**) majority and (**b**) minority spin projections. The blue rectangle points out the bands involved in the band gap formation.

The band structure for the majority and minority spin projections of TbSb is shown in Figure 4. The bands near the Fermi energy at the Γ and X points are shifted relatively to the band structure of TbNiSb shown in Figure 3 and here they form hole and electron pockets (see the blue rectangle in Figure 4), so there is no energy gap for both spin projections. Such a band structure is typical for a semimetal. There is also a presence of similar to Figure 3 localized bands from 4f states of Tb at pretty much the same energies, only a few tenths of eV lower, one can find narrow intense peaks at these energies in the total density of states in Figure 2d.

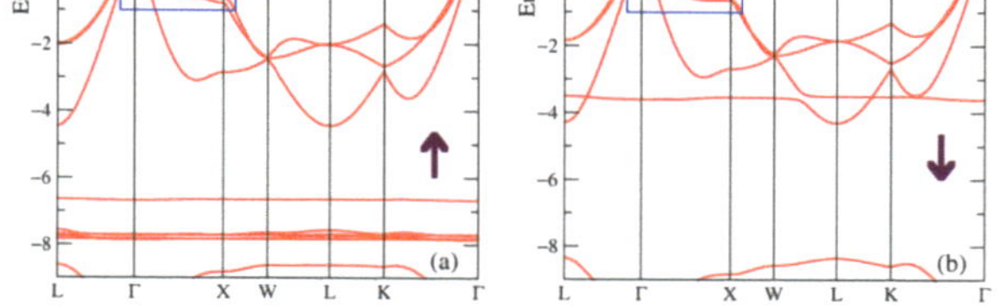

Figure 4. Band structure of TbSb: (**a**) majority and (**b**) minority spin projections. The blue rectangle points out the bands involved in the formation of pockets, see in the text.

3.2. *DyNiSb and DySb Intermetallic Compounds*

In Figure 5, the total and partial densities of electronic states of the DyNiSb and DySb intermetallic compounds are given for two opposite spin directions. In Figure 5a,d one can find similar to Figure 2a,d narrow intense peaks which manifest 4f states of rare earth metals. For the DyNiSb (DySb) compound and the majority spin projection such peaks are found at following energies: −7.8 (−8.3) eV and −6.6 (−7.3, −6.9) eV below the valence band. For the minority spin projection intense peaks are found at 2.4 (1.7) eV and 3.0 (2.2) eV. There is also a noticeable peak at −4.1 (4.6) eV contrary to Figure 2a,d for the minority spin projection for both DyNiSb and DySb in the bottom part of the valence band. The densities of Ni-3d, Dy-5d, and Sb-5p states in Figure 5 exhibit the behavior identical to those plotted in Figure 2.

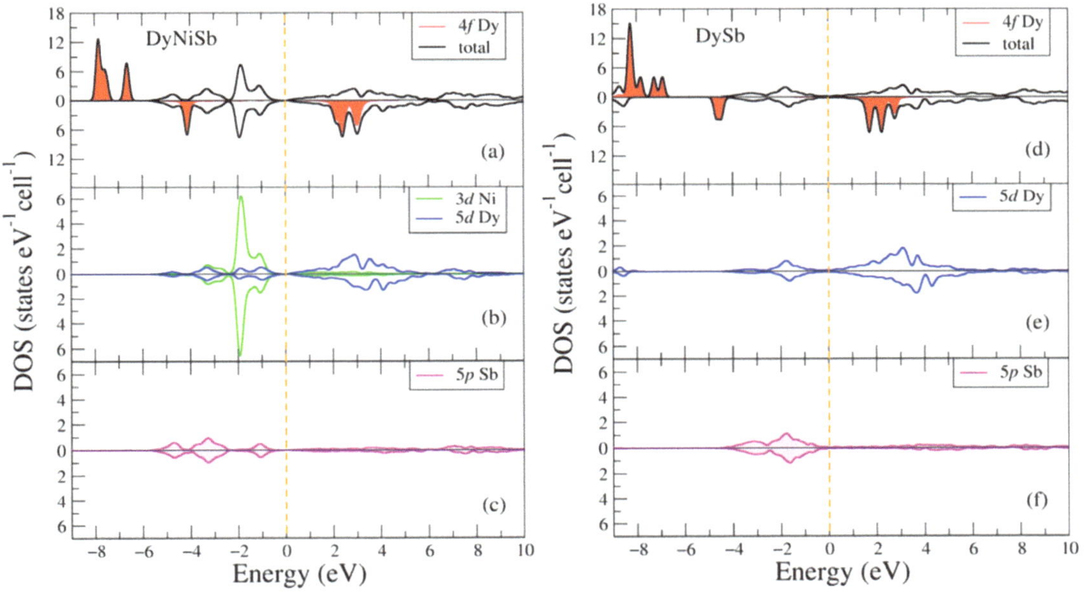

Figure 5. Densities of electronic states from DFT+U for (**a–c**) DyNiSb; (**d–f**) DySb. (**a**,**d**) Total and partial Dy-4f densities of states; (**b**,**e**) partial density of states for Dy-5d (Ni-3d); (**c**,**f**) partial density of states for Sb-5p. The plot is shifted relatively to the Fermi energy shown at zero as a vertical dashed line.

The band structure for the majority and minority spin projections of DyNiSb compound is shown in Figure 6. The electronic states near the Fermi energy are mostly dominated by the Ni-3d and Dy-5d states, so the band structure near this level looks similar to the one in Figure 3 with the main difference being the value of the energy gap (see the blue rectangle in Figure 6). Thus, we conclude that the ternary DyNiSb compound is a semiconductor with an indirect gap of 0.39 eV for the majority spin projection and of 0.21 eV for the minority spin projection. It is in a good agreement with the experimental value of the energy gap of 0.130–0.171 eV [9], 0.089–0.130 eV [15] for DyNiSb. The localized bands from the 4f states of dysprosium similarly to previous Tb-compounds in Figures 3 and 4 produce intense peaks at the same energies in the density of states in Figure 5a.

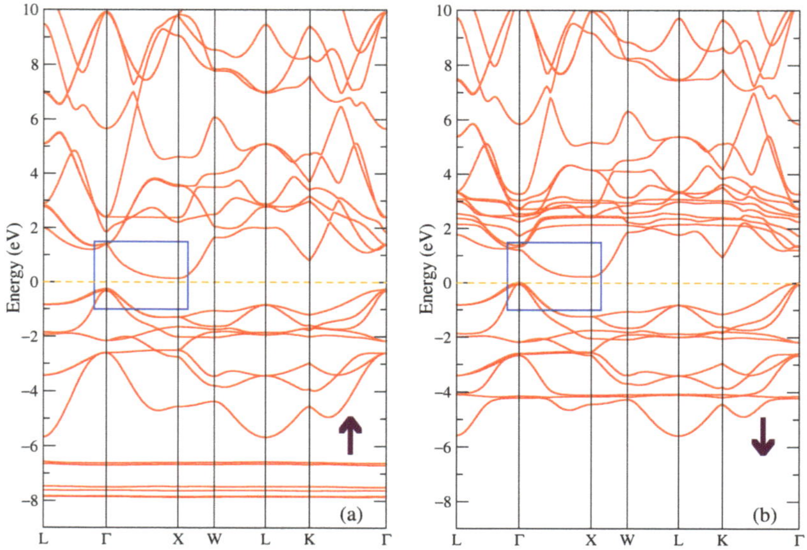

Figure 6. Band structure of DyNiSb: (**a**) majority and (**b**) minority spin projections. The blue rectangle points out the bands involved in the band gap formation.

The band structure for DySb is shown in Figure 7. One can see that this picture is resembling Figure 4 where the band structure for the TbSb compound is shown, since the largest contribution to the states near the Fermi level is due to the Sb 5p and Dy 5d electronic states. The calculated band structure is similar to the one proposed for DySb in [25]. From Figure 7 and the blue rectangle pointing out the bands near the Fermi level, one can conclude that DySb is a semimetal with the hole and electron pockets in the band structure around Γ and X points in the Brillouin zone.

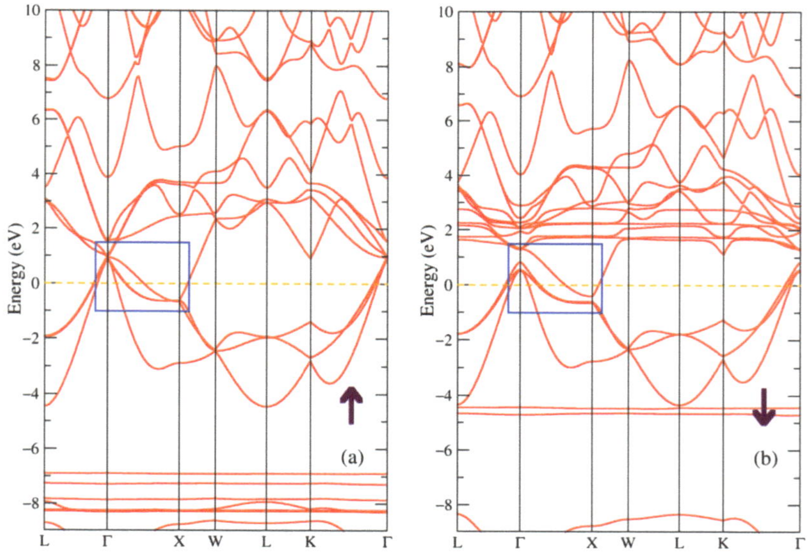

Figure 7. Band structure of DySb: (**a**) majority and (**b**) minority spin projections.

3.3. HoNiSb and HoSb Intermetallic Compounds

In Figure 8, the total and partial densities of electronic states of the HoNiSb and HoSb intermetallic compounds are given for two opposite spin directions. Densities of states for both compounds look similar to ones in Figures 2 and 5 with the main difference being due to the holmium 4f shell. The positions of the intense peaks for this shell in the energy spectrum Figure 8a,d are following: −7.9 (−8.4) eV and −6.7 (−7.4) eV below the valence band for the majority spin projection and 2.0, 2.8 (1.7) eV for the minority spin projection. There is also another Ho-4f intense peak for the minority spin direction at −4.8 eV at the lower part of the valence band for the HoNiSb compound in Figure 8a, such as a peak which seems to be isolated from the Ho-5d and Sb-5p states in the binary HoSb compound at −5.5 eV (Figure 8d), it is not similar to that in the previous compounds.

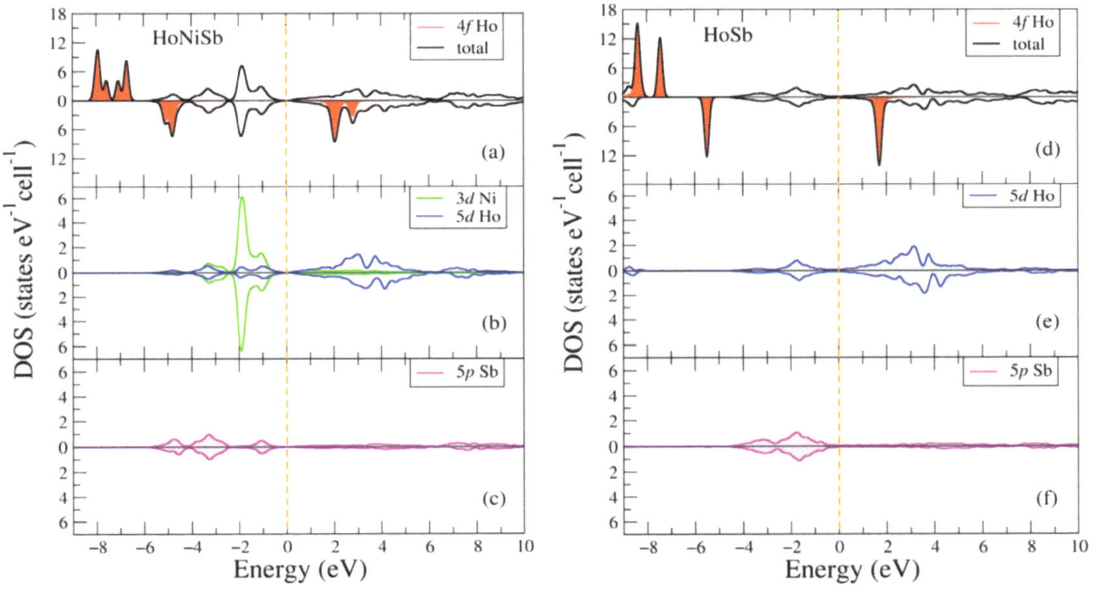

Figure 8. Densities of electronic states from DFT+U for (**a–c**) HoNiSb; (**d–f**) HoSb per cell. (**a**,**d**) Total and partial Ho-4f densities of states; (**b**,**e**) partial density of states for Ho-5d (Ni-3d); (**c**,**f**) partial density of states for Sb-5p. The plot is shifted relatively to the Fermi energy shown at zero as a vertical dashed line.

The band structure for the majority and minority spin projections of HoNiSb compound is shown in Figure 9. Once again, we can see that ternary HoNiSb compound is a semiconductor with an indirect gap in the band structure resembling that of TbNiSb and DyNiSb in Figures 3 and 6, respectively. The value for the energy gap is 0.37 eV for the majority spin projection and 0.26 eV for the minority spin projection. It is in a good agreement with the experimental value of the energy gap of 0.08–0.11 eV [9] for HoNiSb. Worth mentioning that the localized bands above the Fermi level from the 4f states of holmium are closest to the Fermi energy for HoNiSb in Figure 9b among all three ternary compounds.

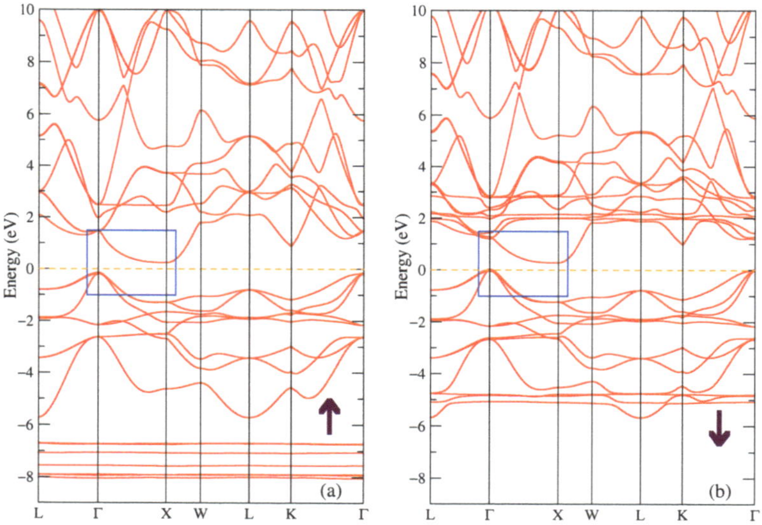

Figure 9. Band structure of HoNiSb: (**a**) majority and (**b**) minority spin projections. The blue rectangle points out the bands involved in the band gap formation.

The band structure for the majority and minority spin projections of the HoSb compound is shown in Figure 10. There is a noticeable flat band from the Ho-4f states in the minority spin projection Figure 10b only 1.8 eV above the Fermi level which is closest of all compounds. We can also see another localized band below the valence band in the minority spin projection which was discussed earlier. In the band structure near the Fermi level several bands are touching at X, see the blue rectangle in Figure 10. It is clearly seen that the binary HoSb compound is a semimetal with the band structure near the Fermi level very similar to that of the TbSb and DySb compounds shown in Figures 4 and 7.

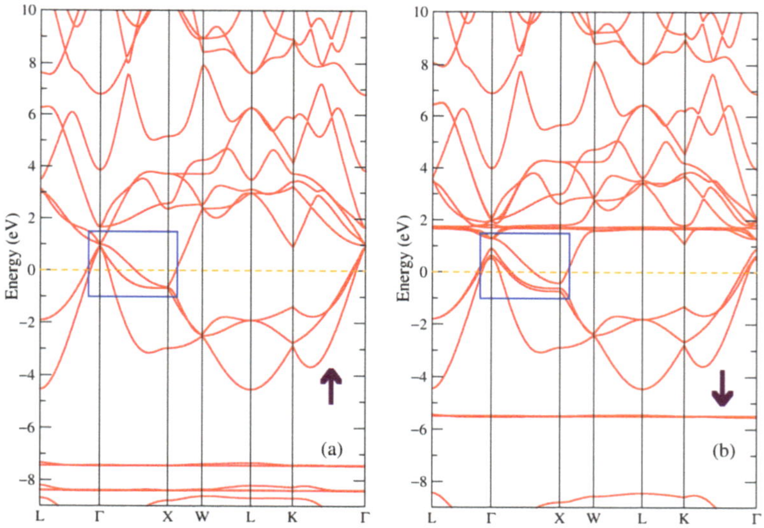

Figure 10. Band structure of HoSb: (**a**) majority and (**b**) minority spin projections.

3.4. Magnetic Moments

In addition to the densities of states and band structures, the values of the magnetic moments of compounds were calculated within the framework of GGA+U. Magnetic moments of the nickel and stibium ions in the compounds are found to be negligible so we can consider it zero. Then all of the magnetic properties of considered RNiSb and RSb compounds are determined by those of rare earth elements [20]. The spin moments of rare earth ions, which are mostly defined by the 4f electronic states [22], as calculated in the present work, have following values: 5.92 μ_B for the Tb ions in both TbSb and TbNiSb compounds, 4.92 μ_B for Dy in DySb and DyNiSb compounds and 3.94 μ_B for the Ho ions in HoSb and HoNiSb. These values are close to the ones for the corresponding R^{3+} ions, but do not exactly coincide, see [20]. We can get the values of orbital momentum as for R^{3+} ions, L = 3, 5, and 6 for Tb, Dy, and Ho, respectively. Now, we are able to calculate the effective magnetic moments, and the results are 9.64 μ_B for TbSb and TbNiSb, 10.56 μ_B for DySb and DyNiSb, 10.55 μ_B for HoSb and HoNiSb vs. the experimental values reported as 10.2(4) μ_B for TbNiSb, 10.6(4) μ_B for DyNiSb, 10.7(4) μ_B for HoNiSb [20], 10.8 μ_B for HoSb [26].

4. Conclusions

In this work, we investigated the electronic and band structures of three ternary compounds Tb,Dy,Ho(NiSb) and three binary compounds Tb,Dy,Ho(Sb). Calculations were carried out in the framework of the GGA+U method and revealed that these ternary compounds are semiconductors with an indirect gap and the binary compounds are semimetals. The Tb,Dy,Ho(NiSb) semiconductors have the following values of the energy gap: 0.43, 0.39, and 0.37 eV for the majority spin projection and 0.21, 0.21, and 0.26 eV for the minority spin projection, respectively, which are in a good agreement with the published experimental values. From the band structure, we identify similar to each other topological features near the Fermi level in the Tb,Dy,Ho(Sb) binary semimetals, these are the hole and electron pockets in the vicinity of Γ and X points, the non-symmetric electron pocket along Γ-X-W direction and hole pockets along the L-Γ-X direction. It is emphasized that the corresponding band structures can be found experimentally in the other compound of this series GdSb. It was also shown that the magnetic moment of all considered compounds is fully determined by magnetic moments of rare earth elements, the calculated effective magnetic moments of such ions have values close to experimental values.

Author Contributions: Conceptualization, methodology, software, investigation, writing—review and editing, S.T.B. and A.V.L.; project administration, supervision, A.V.L. All authors have read and agreed to the published version of the manuscript.

Funding: This research was supported by Russian Science Foundation (project No. 22-42-02021) for the electronic structure calculations in Sections 3.1 and 3.2, the magnetic values (Sections 3.3 and 3.4) were calculated within the state assignment of Ministry of Science and Higher Education of the Russian Federation (theme "Electron" No. 122021000039-4).

Institutional Review Board Statement: Not applicable.

Informed Consent Statement: Not applicable.

Data Availability Statement: The data presented in this study are available on request from the corresponding author.

Conflicts of Interest: The authors declare no conflict of interest. The funders had no role in the design of the study; in the collection, analyses, or interpretation of data; in the writing of the manuscript, or in the decision to publish the results.

References

1. Hundley, M.F.; Thompson, J.D.; Canfield, P.C.; Fisk, Z. Electronic transport properties of the semimetallic heavy fermion YbBiPt. *Phys. Rev. B* **1997**, *56*, 8098–8102. [CrossRef]
2. Gupta, S.; Suresh, K.G.; Nigam, A.K.; Lukoyanov, A.V. Magnetism in RRhGe (R = Tb, Dy, Er, Tm): An experimental and theoretical study. *J. Alloys Compd.* **2015**, *640*, 56–63. [CrossRef]

3. Nakajima, Y.; Hu, R.; Kirshenbaum, K.; Hughes, A.; Syers, P.; Wang, X.; Wang, K.; Wang, R.; Saha, S.R.; Pratt, D.; et al. Topological RPdBi half-Heusler semimetals: A new family of noncentrosymmetric magnetic superconductors. *Sci. Adv.* **2015**, *1*, e1500242. [CrossRef] [PubMed]
4. Pavlosiuk, O.; Kaczorowski, D.; Fabreges, X.; Gukasov, A.; Wisniewski, P. Antiferromagnetism and superconductivity in the half-Heusler semimetal HoPdBi. *Sci. Rep.* **2016**, *6*, 18797. [CrossRef] [PubMed]
5. Hirschberger, M.; Kushwaha, S.; Wang, Z.; Gibson, Q.; Liang, S.; Belvin, C.A.; Bernevig, B.A.; Cava, R.J.; Ong, N.P. The chiral anomaly and thermopower of Weyl fermions in the half-Heusler GdPtBi. *Nat. Mater.* **2016**, *15*, 1161–1165. [CrossRef]
6. Gupta, S.; Suresh, K.G.; Nigam, A.K.; Knyazev, Y.V.; Kuz'min, Y.I.; Lukoyanov, A.V. The magnetic, electronic and optical properties of HoRhGe. *J. Phys. D Appl. Phys.* **2014**, *47*, 365002. [CrossRef]
7. Singh, N.K.; Suresh, K.G.; Nirmala, R.; Nigam, A.K.; Malik, S.K. Correlation between magnetism and magnetocaloric effect in the intermetallic compound DyNiAl. *J. Appl. Phys.* **2006**, *99*, 08K904. [CrossRef]
8. Singh, N.K.; Suresh, K.G.; Nirmala, R.; Nigam, A.K.; Malik, S.K. Effect of magnetic polarons on the magnetic, magnetocaloric, and magnetoresistance properties of the intermetallic compound HoNiAl. *J. Appl. Phys.* **2007**, *101*, 093904. [CrossRef]
9. Ciesielski, K.; Synoradzki, K.; Wolanska, I.; Stachowiak, P.; Kepinski, L.; Jezowski, A.; Tolinski, T.; Kaczorowski, D. High-temperature power factor of half-Heusler phases RENiSb (RE = Sc, Dy, Ho, Er, Tm, Lu). *J. Alloys Compd.* **2020**, *816*, 152596. [CrossRef]
10. Gnida, D.; Ciesielski, K.; Kaczorowski, D. Origin of the negative temperature coefficient of resistivity in the half-Heusler antimonides LuNiSb and YPdSb. *Phys. Rev. B* **2021**, *103*, 174206. [CrossRef]
11. Pierre, J.; Karla, I. Giant magnetoresistance in RENiSb semiconductors (RE = Tb, Dy, Ho). *J. Magn. Magn. Mater.* **2000**, *217*, 74–82. [CrossRef]
12. Pierre, J.; Karla, I.; Kaczmarska, K. Giant magnetoresistance in Heusler-type rare earth and 3d semiconductors. *Physica B Condens. Matter.* **1999**, *259–261*, 845–846. [CrossRef]
13. Pecharskii, V.K.; Pankevich, Y.V.; Bodak, O.I. Crystal structures of the compounds RNiSb with various rare earth elements. *Sov. Phys. Crystallogr.* **1983**, *28*, 97–98.
14. Zhu, H.; He, R.; Mao, J.; Zhu, Q.; Li, C.; Sun, J.; Ren, W.; Wang, Y.; Liu, Z.; Tang, Z.; et al. Discovery of ZrCoBi based half Heuslers with high thermoelectric conversion efficiency. *Nat. Commun.* **2018**, *9*, 2497. [CrossRef]
15. Ciesielski, K.; Synoradzki, K.; Veremchuk, I.; Skokowski, P.; Szymański, D.; Grin, Y.; Kaczorowski, D. Thermoelectric Performance of the Half-Heusler Phases RNiSb (R=Sc,Dy,Er,Tm,Lu): High Mobility Ratio between Majority and Minority Charge Carriers. *Phys. Rev. Appl.* **2020**, *14*, 054046. [CrossRef]
16. Kawano, K.; Kurosaki, K. Effect of Sn doping on the thermoelectric properties of ErNiSb-based *p*-type half-Heusler compound. *Appl. Phys. Lett.* **2007**, *91*, 062115. [CrossRef]
17. Ciesielski, K.; Wolańska, I.; Synoradzki, K.; Szymański, D.; Kaczorowski, D. Mobility Ratio as a Probe for Guiding Discovery of Thermoelectric Materials: The Case of Half-Heusler Phase $ScNiSb_{1-x}Te_x$. *Phys. Rev. Appl.* **2021**, *15*, 044046. [CrossRef]
18. Jia, X.; Deng, Y.; Bao, X.; Yao, H.; Li, S.; Li, Z.; Chen, C.; Wang, X.; Mao, J.; Cao, F.; et al. Unsupervised machine learning for discovery of promising half-Heusler thermoelectric materials. *Npj Comput. Mater.* **2022**, *8*, 34. [CrossRef]
19. Satyam, J.K.; Saini, S.M. Role of R-f states on electronic structure and thermoelectric performance of RNiSb (R = Gd, Er and Lu) half Heusler compounds: Narrow gap thermoelectric materials. *Appl. Phys. A* **2021**, *127*, 828. [CrossRef]
20. Hartjes, K.; Jeitschko, W. Crystal structures and magnetic properties of the lanthanoid nickel antimonides LnNiSb (Ln = LaNd, Sm, GdTm, Lu). *J. Alloys Compd.* **1995**, *226*, 81–86. [CrossRef]
21. Sahariya, J.; Kumar, P.; Bhamu, K.C.; Soni, A. Electronic structure of Gd based transition metal antimonides GdTSb (T = Ni, Pt). *AIP Conf. Proc.* **2018**, *1953*, 110010. [CrossRef]
22. Baglasov, E.D.; Lukoyanov, A.V. Electronic structure of intermetallic antiferromagnet GdNiGe. *Symmetry* **2019**, *11*, 737. [CrossRef]
23. Li, Z.; Xu, D.-D.; Ning, S.-Y.; Su, H.; Iitaka, T.; Tohyama, T.; Zhang, J.-X. Predicted Weyl fermions in magnetic GdBi and GdSb. *Int. J. Mod. Phys. B* **2017**, *31*, 1750217. [CrossRef]
24. Wu, Y.; Lee, Y.; Kong, T.; Mou, D.; Jiang, R.; Huang, L.; Bud'ko, S.L.; Canfield, P.C.; Kaminski, A. Electronic structure of RSb (R = Y, Ce, Gd, Dy, Ho, Tm, Lu) studied by angle-resolved photoemission spectroscopy. *Phys. Rev. B* **2017**, *96*, 035134. [CrossRef]
25. Liang, D.D.; Wang, Y.J.; Xi, C.Y.; Zhen, W.L.; Yang, J.; Pi, L.; Zhu, W.K.; Zhang, C.J. Extreme magnetoresistance and Shubnikov-de Haas oscillations in ferromagnetic DySb. *APL Mater.* **2018**, *6*, 086105. [CrossRef]
26. Xia, Z.-L.; Tang, F.; Xu, C.-Q.; Cong, S.; Zhao, W.; Zhang, L.; Han, Z.-D.; Qian, B.; Jiang, X.-F.; Ke, X.; et al. Influence of magnetization anisotropy on angular magnetoresistance in the antiferromagnetic topological semimetal HoSb. *Phys. Rev. B* **2022**, *106*, 115137. [CrossRef]
27. Abdusalyamova, M.N.; Shokirov, H.S.; Rakhmatov, O.I. Investigation of the rare earth monoantimonides. *J. Less Common Metals* **1990**, *166*, 221–227. [CrossRef]
28. Momma, K.; Izumi, F. VESTA 3 for three-dimensional visualization of crystal, volumetric and morphology data. *J. Appl. Crystallogr.* **2011**, *44*, 1272–1276. [CrossRef]
29. Giannozzi, P.; Baroni, S.; Bonini, N.; Calandra, M.; Car, R.; Cavazzoni, C.; Ceresoli, D.; Chiarotti, G.L.; Cococcioni, M.; Dabo, I.; et al. Quantum ESPRESSO: A modular and open-source software project for Quantum simulations of materials. *J. Phys. Condens. Matter* **2009**, *21*, 395502. [CrossRef]

30. Giannozzi, P.; Andreussi, O.; Brumme, T.; Bunau, O.; Buongiorno Nardelli, M.; Calandra, M.; Car, R.; Cavazzoni, C.; Ceresoli, D.; Cococcioni, M.; et al. Advanced capabilities for materials modelling with Quantum ESPRESSO. *J. Phys. Condens. Matter* **2017**, *29*, 465901. [CrossRef]
31. Knyazev, Y.V.; Lukoyanov, A.V.; Kuz'min, Y.I.; Kuchin, A.G. Electronic structure and optical spectroscopy studies of $HoNi_5$ and $ErNi_5$ compounds doped with Cu. *Phys. Status Solidi B* **2012**, *249*, 824–828. [CrossRef]
32. Perdew, J.P.; Burke, J.P.; Ernzerhof, M. Generalized gradient approximation made simple. *Phys. Rev. Lett.* **1996**, *77*, 3865–3868. [CrossRef]
33. Quantum ESPRESSO. Available online: https://www.quantum-espresso.org/pseudopotentials (accessed on 7 December 2022).
34. Topsakal, M.; Wentzcovitch, R. Accurate projected augmented wave (PAW) datasets for rare-earth elements (RE = La–Lu). *Comput. Mater. Sci.* **2014**, *95*, 263–270. [CrossRef]

materials

MDPI

Article

Atomic Diffusion and Crystal Structure Evolution at the Fe-Ti Interface: Molecular Dynamics Simulations

Guojin Xiang [1,2], Xu Luo [2], Tianxu Cao [1], Ankang Zhang [1] and Hui Yu [1,*]

1 School of Mechanical Engineering, Yanshan University, Qinhuangdao 066000, China
2 Pangang Group Research Institute Co., Ltd., Panzhihua 617000, China
* Correspondence: yuhui@ysu.edu.cn; Tel.: +86-137-8035-0181

Abstract: The diffusion bonding method is one of the most essential manufacturing technologies for Ti-steel composite plates. In this paper, the atomic diffusion behavior at the Fe-Ti interface during the bonding process of Ti-steel composite plates is studied using classical diffusion theory and molecular dynamics (MD) simulation. Henceforth, the diffusion mechanism of Fe and Ti atoms at the bonding interface is obtained at the atomic scale. The results show that Fe and Ti atoms diffused deeply into each other during the diffusion process. This behavior consequently increased the thickness of the diffusion layer. Moreover, the diffusion quantity of Fe atoms to the Ti side was much greater than that of Ti atoms to the Fe side. Large plastic deformation and shear strain occurred at the diffusion interface during diffusion. The crystal structure of the diffusion zone was damaged and defects were generated, which was beneficial to the diffusion behavior of the interface atoms. As the diffusion time and temperature increased, the shear strain of the atoms at the interface also increased. Furthermore, there is a relationship between the mutual diffusion coefficient and the temperature. Subsequently, after the diffusion temperature was raised, the mutual diffusion coefficient and atomic disorder (Fe atom and Ti atom) increased accordingly.

Keywords: diffusion bonding; molecular dynamics; diffusion mechanism; shear strain; diffusion coefficient

Citation: Xiang, G.; Luo, X.; Cao, T.; Zhang, A.; Yu, H. Atomic Diffusion and Crystal Structure Evolution at the Fe-Ti Interface: Molecular Dynamics Simulations. *Materials* **2022**, *15*, 6302. https://doi.org/10.3390/ma15186302

Academic Editors: Shuwen Wen, Yongle Sun and Xin Chen

Received: 9 August 2022
Accepted: 7 September 2022
Published: 11 September 2022

1. Introduction

The Ti-steel composite plate is a kind of composite metal plate that applies heating, pressure, or two types of combination procedures to create strong metallurgical bonding at the interface between the titanium plate and steel plate. In some circumstances, the combination of titanium and steel produces components with both strong corrosion resistance and the price advantages of titanium and steel, thus, effectively utilizing the complementary advantages of titanium and steel in performance and economy. Researchers have conducted a lot of research work to solve the problem of insufficient bonding strength of the composite plates, which is caused by bonding interface defects in Ti-steel composite plates. According to a large number of studies, the intermediate compounds formed on the composite interface are the primary motivation for the interface bonding limitation of the composite plate, which leads to the formation of interface defects. The more kinds of interfacial compounds, the greater the influence on interfacial bonding strength [1,2].

Jiang et al. [3] performed heat treatment on the Ti-steel composite plate produced by the explosive composite method. It was observed that the major compound generated at the composite interface at 850 °C was TiC. At 950 °C, the main products were Fe_2Ti, FeTi, and a trace of TiC. Momono et al. [4] investigated the influence of C element mass fraction on Ti-steel diffusion welding performance. When the steel with 0.01 % C element was heated at temperatures over 900 °C, FeTi and Fe_2Ti compounds were formed at the composite interface, reducing composite strength. TiC, FeTi, and Fe_2Ti compounds were produced at the composite interface in the steel with a 0.19 % mass fraction of C when the temperature exceeded 900 °C and the composite strength dropped. The grain orientation of titanium has a great influence on the interfacial reaction behavior of titanium and steel. When the basal plane of a titanium cell is parallel to the Fe/Ti interface, it has a strong inhibitory impact on the diffusion of C atoms and the TiC layer is thin, according to Li et al. [5] The TiC layer is thicker when the titanium cell base surface is perpendicular to the titanium/steel interface. This is owing to the anisotropy of the gap carbon atom transport in titanium cells.

Molecular dynamics (MD) is a powerful approach for simulating molecular systems using Newtonian mechanics. It can explain some macroscopic properties of matter and conduct an experiment at the micro level, which is significant for revealing the development of atomic-scale structures. Nowadays, MD simulation has become an effective tool for studying the interface diffusion bonding process.

Firstly, Leo Miglio et al. [6] investigated the mathematical simulation of Si diffusion in TiS_2 compounds and discovered that the diffusion of Si in this compound was mostly owing to its low formation energy. Since then, numerous researchers have investigated the diffusion phenomena by MD simulation. Chen et al. [7] simulated the formation process of the Cu-Ag diffusion bond using the MD method and discovered that the thickness of the interface area is mainly dictated by stress. In the process of diffusion bonding, the interface region is transformed into an amorphous form. However, it often transitions from an amorphous to a crystal structure when it cooled to room temperature. Chen et al. [8] used MD simulation to investigate atomic diffusion during explosive welding of $Ni_{50}Ti_{50}$-Cu (at. %). The geometric similarity of concentration distribution curves at different points during diffusion was applied to compute the distribution of atomic concentration at any time throughout the explosive welding process and the simulation results are virtually identical to the experimental results. Similarly, Xiu et al. [9] applied MD to simulate W/Cu diffusion bonding. The atomic diffusion behavior of W and Cu atoms was investigated in their study and the diffusion coefficient and radial distribution function (RDF) were also calculated. The diffusion mechanism of the W/Cu diffusion bond is as follows: when the temperature and diffusion time rise, the thickness of the diffusion layer and atomic disorder increase as W and Cu atoms propagate along the crystal defect surface. Solid-state wetness (SSW) happens not only at the micron scale but also at the nanoscale, according to research by M. Samsonov et al. [10], who simulated the spread of solid copper (Cu) and gold (Au) nanoparticles over the same metal (100) surface. The temperature effect is stronger than the pressure effect during diffusion, according to research by Zhang et al. [11], which looked at the effects of temperature, pressure, and surface roughness on the diffusion welding of stainless steel and pure Ni. By simulating self-diffusion along the screw dislocation core in the metals aluminum, nickel, copper, and silver, Soltani et al. [12] simulated the effect of screw dislocation on boosting self-diffusion for each metal mentioned above. In a nanostructured Cu/Ag system, Béjaud et al. [13] investigated the interaction between interface and deformation twins and precisely analyzed the effects of misfit dislocation on twin nucleation and thickening in a face-centered cubic structure. Yang et al. [14] investigated the diffusion behavior of Al and Cu atoms in ultrasonic welding using MD modeling. Their findings revealed that asymmetric diffusion occurred at the Al/Cu interface during the ultrasonic welding (UW)process. Simultaneously, the recovery of disordered Al blocks at low temperatures was observed and the thickness of the diffusion layer increased as welding time increased. Because of the significant trapping effect of the grain boundaries in nanocrystalline Fe, Zhou et al.'s study [15] on the dependency of the hydrogen diffusion coefficient on grain size revealed that smaller grain sizes correspond to reduced hydrogen diffusion coefficients. The effects of temperature and pressure on atomic interdiffusion along the direction perpendicular to the Cu/Al solid–liquid interface were revealed by Mao et al. [16] and found that while the thickness of the diffusion layer exhibits a parabolic connection with diffusion time, the Cu atoms' depth of diffusion exhibits a linear relationship with system temperatures. Using the MD approach, Chen et al. [17] investigated the influence of temperature on the diffusion rate and mechanical characteristics of the nano-scale TiAl/Ti_3Al interface. The elastic modulus, yield strength, and flow stress all reduced as the temperature increased from 1273 K to 1473 K. Wang et al. [18] investigated the atomic diffusion behavior of the Mo/Au interface using classical diffusion theory and MD simulation and the mutual diffusion coefficient and radial distribution function of Mo and Au atoms were obtained. In a recycled asphalt mixture, Zhan et al. [19] described the behavior of new asphalt and aged asphalt diffusion, which revealed that the diffusion direction was mostly from virgin asphalt to aged asphalt and that the diffusion efficiency improved with temperature. Zurhelle et al. [20] studied the oxygen atoms' diffusion characteristics in platinum atoms with extended defects and their results demonstrated that platinum vacancies prevent oxygen atom diffusion across the platinum lattice by trapping oxygen atoms. Amorphization of Cu atoms during diffusion was noted in this study by Zhang et al. [21], who simulated the diffusion process and atomic structure of the interface between metallic glass $Cu_{50}Zr_{50}$ and crystalline Cu.

In this paper, MD simulation was used to study the atomic diffusion behavior and crystal structure evolution at the the Fe-Ti interface in the diffusion bonding process of the Ti-steel composite plates and classical diffusion theory. The diffusion process of Fe-Ti interface was characterized through the calculation of profiles for atomic trajectory, atomic concentration distribution, diffusion layer thickness, atomic shear strain, radial distribution function (RDF), mean square displacement (MSD), and diffusion coefficient. The relationship between the calculation profiles and diffusion temperature was also presented, which will benefit our understanding of the atomic diffusion mechanism in the Fe-Ti interface.

2. Simulation Method

2.1. Potential Function

Potential function describes the interaction of atoms or molecules, also known as a force field. The correctness of the potential function has a significant impact on the dependability of MD simulation results. Daw and Baskes [22] developed the embedded atom method (EAM) based on density functional theory and effective medium theory in 1983. It could solve the inadequacies in the two-body potential model that are incompatible with particle interaction in metal systems. Furthermore, in order to implement the EAM potential to covalent bond materials, the non-spherical symmetric distribution of electrons should be considered. Baskes et al. [23] developed a modified embedded atom method (MEAM). The potential function of MEAM is represented as follows:

$$E = \sum_i \{F_i(\vec{\rho_i}) + \frac{1}{2}\sum_{i \neq j} S_{ij}\phi_{ij}(R_{ij})\} \tag{1}$$

$$F_i(\vec{\rho}) = A_i E_i^0 \vec{\rho_i} \ln \vec{\rho_i} \tag{2}$$

where E is the total energy in the system; F_i is the embedding function for an atom i embedded in a background electron density ρ_i; and S_{ij} and ϕ_{ij} (R_{ij}) are the screening function and the pair interaction between atoms i and j separated by a distance R_{ij}; E_i^0 is the binding energy of atom I; and A_i is the structure parameter.

The embedded atomic method (MEAM) interatomic potential of the Fe-Ti-C binary alloy system was developed by Kim et al. [24]. The MEAM potentials were validated in the literature by Prasanthi et al. [25]. At an equilibrium distance of 2.879 Å, they calculated that the cohesive energy for the Fe–Ti system was 4.360 eV, which was quite compatible with MEAM potentials. As well, the calculated bulk moduli of Fe (153 GPa) and Ti (110 GPa) were in good agreement with the experimentally measured values of 166 GPa for "Armco Iron" and 108 GPa for Ti [26,27].

In addition, by examining the overall structural consistency of atoms in the system before and after the relaxation process, it is demonstrated that utilizing this potential function for relaxation yields a relatively stable starting model, which is adequate for the simulation system.

2.2. Simulation Model

The lattice type of Fe and Ti at home temperature is a body-centered cubic and hexagonal close-packed structure, respectively. The lattice constants of Fe and Ti are 0.286 nm and 0.295 nm, respectively. Where the size of Fe is 90 Å × 90 Å × 50 Å and the size of Ti is 90 Å × 90 Å × 50 Å, as depicted in Figure 1. Among them, the blue atoms are Fe atoms and the yellow atoms are Ti atoms. The total numbers of the Fe and Ti atoms in the model were 33,597 and 22,971, respectively. There is an initial gap of 2 Å between the Fe and Ti samples, which aims to reduce the strong interaction force between the two atoms at the interface, causing the samples to fit better.

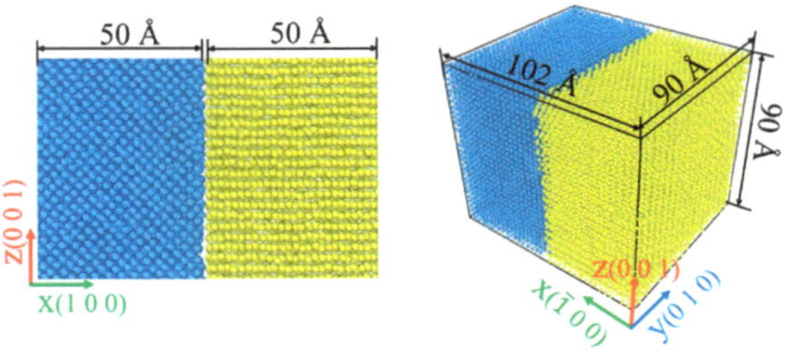

Figure 1. Simulation model.

The large-scale atomic molecular massively parallel simulator (LAMMPS) program is used in this paper [28]. To begin with, the two models are completely relaxed by the normal pressure and temperature (NPT) ensemble with a goal temperature at the time of 500 ps to reach a rather stable state. Then, they are merged in the y-z surface and the periodic boundary conditions in the y and z directions are applied, while the shrinking boundary conditions are applied in the x-direction. The ensemble is set to NVT (N: number of particles; V: volume; T: temperature) and the Nosé–Hoover method is used to maintain a constant system temperature of 1123 K and no external pressure is applied to the whole system. The timestep is 0.001 ps and the initial velocity of atoms follows the Maxwell distribution.

3. Result and Discussion

3.1. Interface Diffusion Behavior

Figure 2 shows the atomic diffusion behavior of the Fe/Ti diffusion interface at the temperature T = 1123 K, (a) 0.001 ns; (b) 0.5 ns; (c) 1.5 ns; (d) 2 ns. The diffusion area between Fe and Ti is defined as the region between the farthest diffused Fe atom (i.e., the Fe atom with the largest x coordinate) and the farthest diffused Ti atom (i.e., the Ti atom with the smallest x coordinate). A distinct interface was produced at the time of t = 0.001 ns and t = 0.5 ns, as shown in Figure 2a,b, in which a tiny quantity of Fe atoms diffused into the Ti layer and a few Ti atoms diffused into the Fe layers. As the diffusion time increased to t = 1.5 ns and t = 2 ns, more Fe atoms diffused into the other side and the diffusion area became thicker, as shown in Figure 2c,d. Additionally, the interface between the Fe layer and Ti layer began to diminish as the diffusion time increased. This phenomenon can be explained as follows: binding energy was produced as the Fe layer contacted with the Ti layer, which enhanced the interaction and movement between Fe and Ti atoms. Figure 3 illustrates the atomic concentration distribution at different diffusion times. As the time increased from t = 0.001 ns to t = 1.5 ns, the thickness of the diffusion area increased from 4.47 Å to 11.22 Å. While the diffusion area just increased to 1.32 Å as the time rose from t = 1.5 ns to t = 2 ns, as shown in Figure 3a,d, whereas the number of Fe atoms (or Ti atoms) entering the titanium lattice (or Fe lattice) is still increasing as the time increases from t = 1.5 ns to t = 2 ns, as shown in Figure 3c,d. This result indicates that when the diffusion time is sufficient, the thickness of the diffusion area will no longer increase, but atoms will always enter the diffusion region.

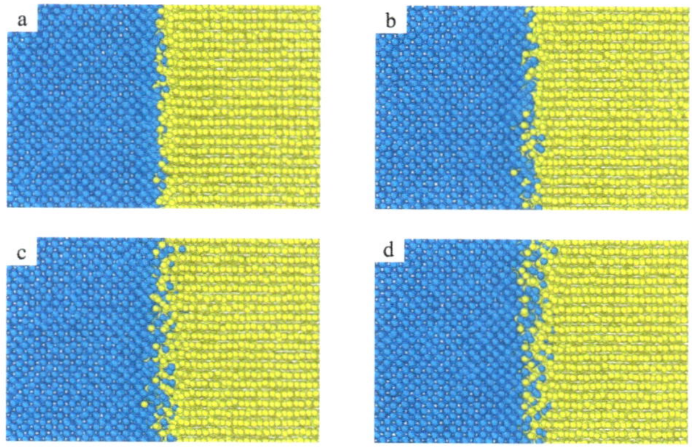

Figure 2. Atomic diffusion behavior of Fe/Ti diffusion interface at the temperature T = 1123 K ((**a**) 0.001 ns; (**b**) 0.5 ns; (**c**) 1.5 ns; (**d**) 2ns).

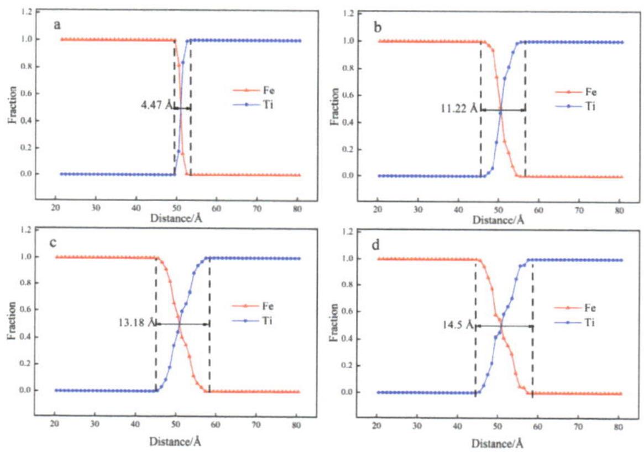

Figure 3. Atomic concentration distribution at the temperature T = 1123 K ((**a**) 0.001 ns; (**b**) 0.5 ns; (**c**) 1.5 ns; (**d**) 2 ns).

Figure 4 shows the atomic diffusion behavior of the Fe/Ti diffusion interface at different temperatures and the atom diffusion behavior is temperature dependent. The number of diffusion atoms rises with the increase in diffusion temperature. At a temperature of 973 K, a few Fe atoms diffuse into the Ti layer and nearly no Ti atoms diffuse into the Fe layer, as shown in Figure 3a. A growing number of atoms diffuse into the other side as the diffusion temperature increases and the diffusion area becomes thicker, as shown in Figure 3b,d. Furthermore, the overall number of Ti atoms that diffused and the diffusion depth are significantly smaller than those of Fe atoms in the titanium lattice, indicating asymmetric diffusion.

The atomic concentration distribution curves at different temperatures are depicted in Figure 5. The thickness of the diffusion and the number of atoms in the diffusion region increase with the rise in the diffusion temperature. The thickness of the diffusion zone increases from 9.5 Å to 10 Å as the diffusion temperature increases from 973 K to 1023 K. Similarly, when the diffusion temperature is 1073 K and 1123 K, the thickness of the diffusion zone increases to 11.41 Å and 14.5 Å, respectively. Further, the higher the temperature, the more pronounced the effect. All in all, diffusion temperature plays an effective role in promoting interface diffusion between Fe and Ti layers.

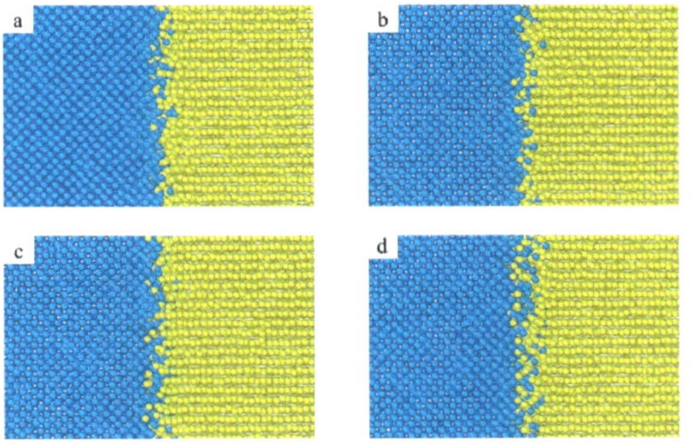

Figure 4. Atomic diffusion behavior of Fe/Ti diffusion interface at the time t = 2 ns ((**a**) 973 K; (**b**) 1023 K; (**c**) 1073 K; (**d**) 1123 K).

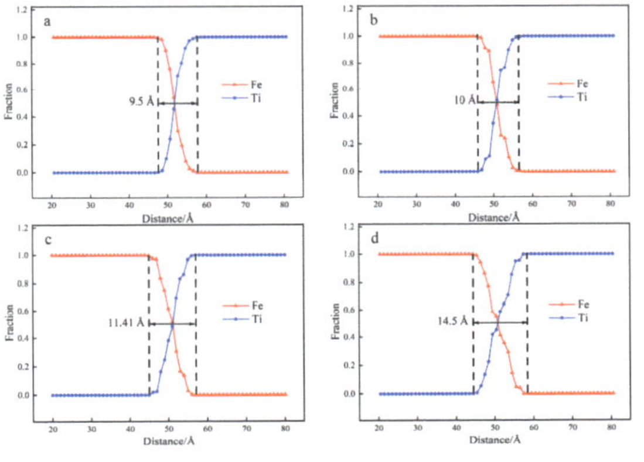

Figure 5. Atomic concentration distribution at the time t = 2 ns ((**a**) 973 K; (**b**) 1023 K; (**c**) 1073 K; (**d**) 1123 K).

The number of diffused atoms was counted by the means of the "selection" function in the open visualization tool (OVITO) [29]. Figure 6 shows the number of diffused atoms of Fe and Ti after each specific time and temperature and the number of diffusion atoms increases with the increase in temperature and diffusion time and the curve gradually becomes gentle over time, which showed a linear relationship with the diffusion time after t = 0.5 ns. In addition, the number of diffused Fe atoms is always greater than that of Ti atoms, which is consistent with the previous analysis results.

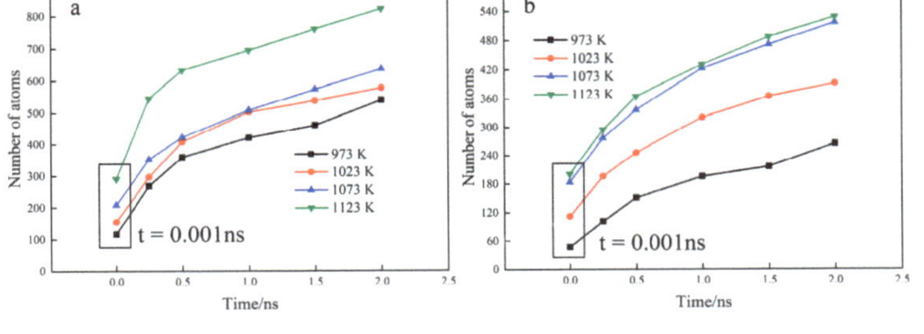

Figure 6. Number of diffused atoms with specific time and temperature ((**a**) Fe atoms; (**b**) Ti atoms).

3.2. Crystal Structure Analysis

Figure 7 depicts the shear strain distribution at different diffusion times and a narrow shear plastic deformation zone was observed at the bonding interface. At the time of t = 0.001 ns, a limited number of atoms are influenced by the plastic deformation and the value of the shear strain is negligible, as depicted in Figure 7a. Furthermore, from t = 0.001 ns to t = 2 ns, the shear strain gradually increases and the plastic deformation band spreads to both sides, as shown in Figure 7a–d, indicating the deformation of the interface area during interface bonding. As a result, a large number of Fe and Ti atoms migrate from the original equilibrium position to a new position on both sides of the adjacent interface, so that the vacancies appear near the binding interface and the vacancy concentration increases continuously. Since the activation energy of high-density diffusion is much lower than that of lattice diffusion, the vacancy can be considered as the best place for large-scale diffusion of dissimilar atoms, so mutual diffusion between Fe and Ti atoms occurs when the vacancy on the iron side (or Ti side) reaches a new equilibrium. It can also be found in Figure 8 that temperature has a significant effect on the atomic shear strain in the diffusion area. The atomic shear strain in the diffusion zone rises with the increase in temperature and the effect range expands with the increase in diffusion zone thickness.

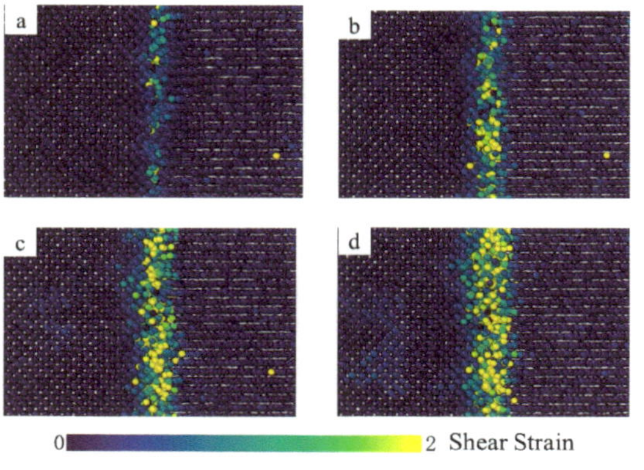

Figure 7. Shear strain distribution at the temperature T = 1123 K ((**a**) 0.001 ns; (**b**) 0.5 ns; (**c**) 1.5 ns; (**d**) 2 ns).

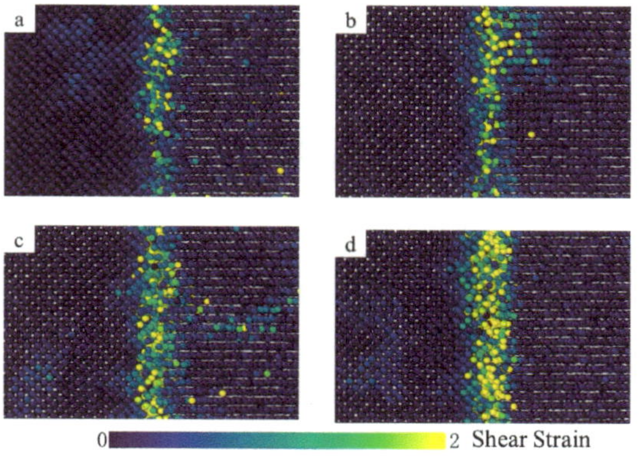

Figure 8. Shear strain at the time t = 2 ns ((**a**) 973 K; (**b**) 1023 K; (**c**) 1073 K; (**d**) 1123 K).

Radial distribution function (RDF) is an effective method for describing system structure, which can characterize the disorder degree of the structure [30]. The RDF refers to the probability of finding another atom at a distance r from one atom, which can be used to represent information, such as the relationship between atoms and the interaction intensity, so as to study the atomic structure.

Figures 9 and 10 show the RDF curves (g(r)) of Fe and Ti atoms at different diffusion times and diffusion temperatures, respectively. At different diffusion times and temperatures, the RDF curves of Fe and Ti atoms are essentially the same. There is only one main peak, representing the binding strength between the first nearest neighbor atoms of the central atom. Its sharp shape indicates that the number density of atoms in this radius range is much higher than the average density and the binding strength between the central atom and the nearest neighbor atom is also relatively large. As presented in Figure 9a, the main peak of g(rFe-Fe) (the RDF curve of the Fe-Fe atom) appears at r = 2.55 Å, suggesting that the nearest atoms of Fe atoms appear at r = 2.55 Å and the bond length of Fe-Fe can be obtained at 2.55 Å. Furthermore, Figure 9b depicts that the nearest atom of Ti atom could be found at r = 2.93 Å and the Ti-Ti bond length can be gained at 2.93 Å. Similarly, the bond length of Fe-Ti can be calculated at 2.65 Å, as shown in Figure 9c, which is consistent with the previous simulation results [31]. Furthermore, it could be found that the value of r(Fe-Ti) (the bond length of Fe-Ti) is between r(Fe-Ti) and r(Fe-Ti) and this problem is caused by the electronic interaction between metals [32]. In addition, the RDF curve in Figure 9a,b has another peak at r > 2.55 Å and 2.93 Å, except for the first main peak, which indicates that the Fe and Ti crystal structures are ordered.

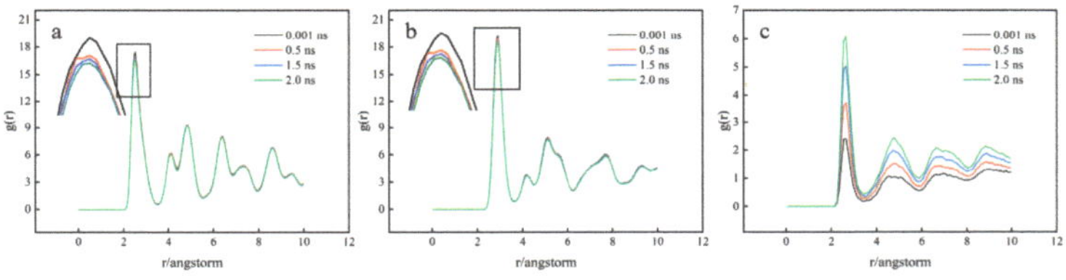

Figure 9. RDF curves of Fe and Ti atoms at T = 1123 K ((**a**) Fe-Fe; (**b**) Ti-Ti; (**c**) Fe-Ti).

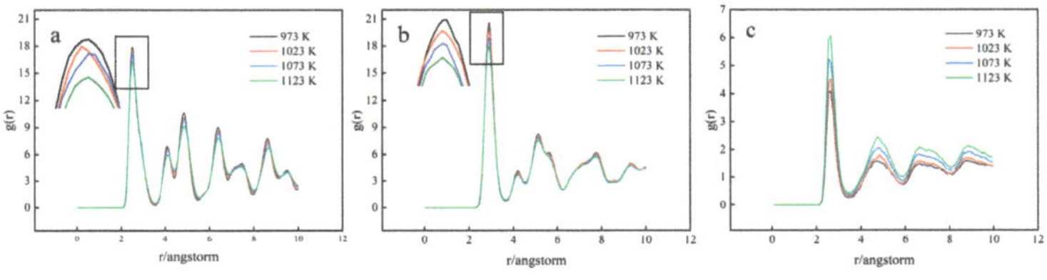

Figure 10. RDF curves of Fe and Ti atoms at t = 2 ns ((**a**) Fe-Fe; (**b**) Ti-Ti; (**c**) Fe-Ti).

The peak values of g(rFe-Fe) and g(rTi-Ti) are much larger than that of g(rFe-Ti), because the bonding strength of Fe-Fe and Ti-Ti is greater than that of Fe-Ti. With the increase in diffusion time, the peak strength of g(rFe-Fe) and g(rTi-Ti) decreases, whereas the peak strength of g(rFe-Ti) increases. It can be explained as follows: during the diffusion process, the atoms in the crystal migrate, resulting in the crystal structure becoming disordered and the bonding strength between the same atoms decreasing. While the bonding reaction between different atoms occurs, the Fe atoms and Ti atoms combine with each other, resulting in an increase in bonding strength. In addition, raising the temperature will also result in (rFe-Fe) and g(rTi-Ti) peak decreases and g(rFe-Ti) peak increases. It indicates that the increase in temperature will accelerate the disorder of crystal structure at the interface and the mutual diffusion of atoms at the interface.

3.3. Diffusion Coefficient Analysis

The Einstein method has been used to investigate the atom diffusion behavior at the interface. Firstly, the mean square displacement (MSD) of the atom was calculated.

Through the relationship between the MSD curves and the diffusion coefficient, the diffusion coefficient D of the particle can be obtained, in which the MSD can be calculated as [33]:

$$MSD = \frac{1}{N}\sum_{i=1}^{N}\left\langle |r_i(t) - r_0(0)|^2 \right\rangle \tag{3}$$

where N is the number of atoms; $r_i(t)$ and $r_i(0)$ are displacement vectors of i atom at zero and t moment, respectively; the calculated symbol $\langle\rangle$ is temporal correlation.

When the running time tends to be infinite:

$$\lim_{t\to\infty} MSD = c + 2dDt \tag{4}$$

where d is the dimension of the system; D is the diffusion coefficient; c is constant. The mean square displacement is listed as a function of time and the slope of the curve is $2dDt$. The system discussed in this paper is three-dimensional, so the diffusion coefficient can be expressed as [9]:

$$D = \frac{1}{6}\lim_{t\to\infty}\left(\frac{d}{dt}\left\langle |r_i(t) - r_0(0)|^2 \right\rangle\right) \tag{5}$$

The temperature was discovered to be a key factor influencing the diffusion rate in the preceding investigation. The general expression of diffusion coefficient can be established by Arrhenius relation [33]:

$$D = D_0 \exp\left(-\frac{Q}{RT}\right), \tag{6}$$

where D is the diffusion coefficient; D_0 is diffusion factor; Q is the diffusion activation energy; R is the gas constant, R = 8.314 J/(mol K).

Taking the logarithm of Equation (6), the relationship between D and T could be expressed as:

$$\ln D = \ln D_0 - \frac{Q}{RT}, \tag{7}$$

The MSD curves of the Fe and Ti atoms were obtained by the MD simulations, illustrated in Figure 11. The diffusion of Fe atoms is relatively stable, while the movement of Ti atoms is relatively intense at the beginning of diffusion and then gradually tends to be stable. Furthermore, the MSD of Fe and Ti atoms gradually increases as temperature rises.

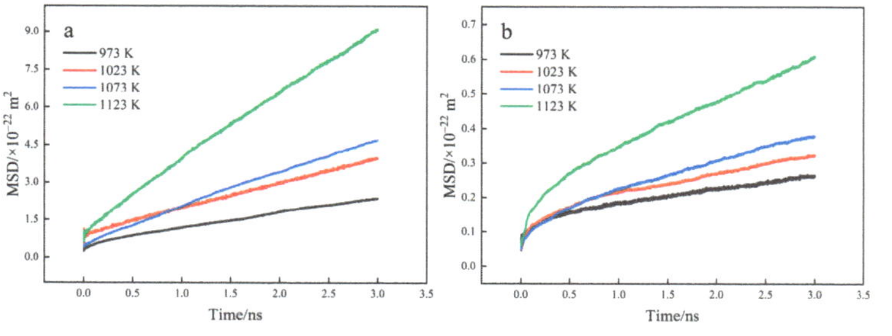

Figure 11. MSD curves of Fe and Ti atoms at different temperature ((**a**) Fe atoms; (**b**) Ti atoms).

The diffusion coefficients of Fe and Ti atoms at the interface with different temperatures are calculated according to the MSD curves and Equation (5). It can be found from Figure 12 that the diffusion coefficient is exponentially related to the diffusion temperature, which is consistent with the results of Equation (6) and Equation (7). The D-value of Fe atoms is higher than that of Ti atoms and the D-value of Fe atoms and Ti atoms increases with the rise in temperature. When the diffusion temperature is 973 K, the diffusion coefficients of Fe and Ti atoms are 9.57×10^{-15} m^2/s and 8.12×10^{-16} m^2/s, respectively. In addition, when the diffusion temperature is less than 1073 K, the diffusion coefficients of Fe and Ti atoms are small and the atoms are in the low-speed diffusion stage. When the diffusion temperature is greater than 1073 K, the diffusion coefficient of Fe atoms and Ti atoms increases obviously and the atoms are in the rapid diffusion stage. As the diffusion temperature reaches 1123 K, the diffusion coefficients of iron and titanium atoms reach 4.45×10^{-14} m^2/s and 2.43×10^{-15} m^2/s, respectively. As a result, atomic diffusion behavior occurs mostly at high temperatures, whereas there are no discernible diffusion phenomena at low temperatures. This is because atoms at low temperatures cannot get sufficient energy to break the potential barrier, yet atoms at high temperatures can obtain sufficient kinetic energy.

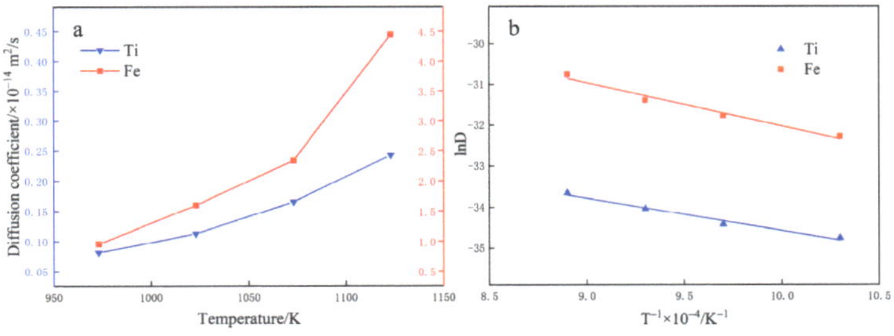

Figure 12. Relationship between diffusion coefficient and temperature.

4. Conclusions

The atomic diffusion behavior of the Fe-Ti composite interface at different times and temperature was simulated by using MD simulation. The following conclusions can be obtained:

1. As diffusion time and temperature rise, the boundary between the Fe and Ti layer dissolves, the thickness of the diffusion layer grows, and the diffusion amount of Fe atoms to the Ti layer exceeds that of Ti atoms to the Fe layers.

2. During the diffusion process, the diffusion boundary experienced significant plastic deformation and shear strain. The structure of the diffusion area was disrupted and flaws were formed, which enhanced atom diffusion at the interface. The shear strain of atoms at the contact rose as diffusion time and temperature increased.

3. There is only one primary peak in the RDF curves of Ti and Fe atoms. Both short- and long-range order may be seen in the crystal structure. As the diffusion time and temperature increase, the peak intensity decreases and the order of the crystal structure is destroyed during diffusion.

4. The MSD curves were extracted and utilized to calculate Ti and Fe atom's diffusion coefficients and the diffusion rate of Fe atoms is greater than that of Ti atoms.

5. As the diffusion temperature rises, the diffusion coefficients and the thickness of the diffusion area in the interaction area increase and there is a linear relationship between the logarithm of the diffusion coefficient

Author Contributions: Conceptualization, X.L. and H.Y.; methodology, G.X. and T.C.; software, G.X. and X.L.; validation, A.Z.; formal analysis, G.X.; investigation, G.X.; resources, T.C.; data curation, G.X.; writing—original draft preparation, G.X.; writing—review and editing, T.C.; visualization, A.Z.; supervision, H.Y.; project administration, H.Y.; funding acquisition, H.Y. All authors have read and agreed to the published version of the manuscript.

Funding: This research was funded by the Natural Science Foundation of Hebei Province, grant number E2021203237 and the Central Government Guides Local Science and Technology Development Fund Projects, grant number 216Z1002G.

Institutional Review Board Statement: Not applicable.

Informed Consent Statement: Not applicable.

Data Availability Statement: Not applicable.

Conflicts of Interest: The authors declare no conflict of interest.

References

1. Luo, Z.; Wang, G.; Xie, G.; Wang, L.; Zhao, K. Effect of niobium interlayer on microstructure and properties of vacuum rolling composite Ti-stainless steel plate. *J. Nonferrous Met. China* **2013**, *23*, 3335–3340.
2. Yadegari, M.; Ebrahimi, A.R.; Karami, A. Effect of Heat Treatment on Interface Microstructure and Bond Strength in Explosively Welded Ti/304L Stainless Steel Clad. *Mater. Sci. Technol.* **2013**, *29*, 69–75. [CrossRef]
3. Jiang, H.; Yan, X.; Liu, J.; Zeng, S.; Duan, X. Diffusion behavior and mathematical model of titanium-steel explosive composite plate during heat treatment. *Rare Met. Mater. Eng.* **2015**, *44*, 972–976.
4. Momono, T.; Enjo, T.; Ikeuchi, K. Effect of Carbon Content on the Diffusion Bonding of Iron and Steel to Titanium. *ISIJ Int.* **1990**, *30*, 978–984. [CrossRef]
5. Li, B.; Chen, Z.; He, W.; Zhou, T.; Wang, Y.; Peng, L.; Li, J.; Liu, Q. Effect of titanium grain orientation on the growth of compounds at diffusion bonded titanium/steel interfaces. *Mater. Charact.* **2019**, *148*, 243–251. [CrossRef]
6. Miglio, L.; Iannuzzi, M.; Raiteri, P.; Celino, M. Silicon diffusion in competitive $TiSi_2$ phases by molecular dynamics simulations. *Microelectron. Eng.* **2001**, *55*, 83–92. [CrossRef]
7. Chen, S.D.; Soh, A.K.; Ke, F.J. Molecular dynamics modeling of diffusion bonding. *Scr. Mater.* **2005**, *52*, 1135–1140. [CrossRef]
8. Chen, S.Y.; Wu, Z.W.; Liu, K.X. Atomic diffusion across Ni50Ti50—Cu explosive welding interface: Diffusion layer thickness and atomic concentration distribution. *Chin. Phys. B* **2014**, *23*, 066802. [CrossRef]
9. Xiu, L.; Wu, J.F. Atomic diffusion behavior in W/Cu diffusion bonding process. *J. Fusion Energy* **2015**, *34*, 769–773. [CrossRef]
10. Samsonov, V.M.; Bembel, A.G.; Popov, I.V.; Vasilyev, S.A.; Talyzin, I.V. Solid-state wetting at the nanoscale: Molecular dynamics and surface diffusion approach. *Surf. Innov.* **2017**, *5*, 161–169. [CrossRef]
11. Zhang, Y.; Jiang, S. Atomistic investigation on diffusion welding between stainless steel and pure Ni based on molecular dynamics simulation. *Mater.* **2018**, *11*, 1957. [CrossRef] [PubMed]
12. Soltani, S.; Abdolrahim, N.; Sepehrband, P. Molecular dynamics study of self-diffusion in the core of a screw dislocation in face centered cubic crystals. *Scr. Mater.* **2017**, *133*, 101–104. [CrossRef]
13. Béjaud, R.; Durinck, J.; Brochard, S. Twin-interface interactions in nanostructured Cu/Ag: Molecular dynamics study. *Acta Mater.* **2018**, *144*, 314–324. [CrossRef]
14. Yang, J.; Zhang, J.; Qiao, J. Molecular dynamics simulations of atomic diffusion during the Al–Cu ultrasonic welding process. *Mater.* **2019**, *12*, 2306. [CrossRef]
15. Zhou, X.; Zhu, J.; Wu, H. Molecular dynamics studies of the grain-size dependent hydrogen diffusion coefficient of nanograined Fe. *Int. J. Hydrogen Energy* **2021**, *46*, 5842–5851. [CrossRef]
16. Mao, A.; Zhang, J.; Yao, S.; Wang, A.; Wang, W.; Li, Y.; Qiao, C.; Xie, J.; Jia, Y. The diffusion behaviors at the Cu-Al solid-liquid interface: A molecular dynamics study. *Results Phys.* **2020**, *16*, 102998. [CrossRef]
17. Chen, J.; Chen, W.; Wang, C. Modeling and investigation for atomic diffusion and mechanical properties of TiAl/Ti_3Al interface: Temperature effect. *Appl. Phys. A* **2020**, *126*, 1–9. [CrossRef]
18. Wang, K.; Bai, H.; Hu, W.; Wu, S.; Wang, H.; Fan, H. Investigation of atomic diffusion behavior of Mo/Au interface. *Mater. Chem. Phys.* **2021**, *257*, 123839. [CrossRef]
19. Zhan, Y.; Wu, H.; Song, W.; Zhu, L. Molecular Dynamics Study of the Diffusion between Virgin and Aged Asphalt Binder. *Coat* **2022**, *12*, 403. [CrossRef]
20. Zurhelle, A.F.; Stehling, W.; Waser, R.; De Souza, R.A.; Menzel, S. Oxygen Diffusion in Platinum Electrodes: A Molecular Dynamics Study of the Role of Extended Defects. *Adv. Mater. Interfaces* **2022**, *9*, 2101257. [CrossRef]
21. Zhang, W.; Huang, J.; Yu, X.; Liu, G.; Fan, D. Atomic-level diffusion at the amorphous Zr50Cu50/crystalline Cu interface: A molecular dynamics study. *J. Adv. Join. Processes* **2022**, *6*, 100120. [CrossRef]
22. Daw, M.S.; Baskes, M.I. Embedded-atom method: Derivation and application to impurities, surfaces, and other defects in metals. *Phys. Rev. B* **1984**, *29*, 6443. [CrossRef]
23. Baskes, M.I. Modified embedded-atom potentials for cubic materials and impurities. *Phys. Rev. B* **1992**, *46*, 2727. [CrossRef] [PubMed]
24. Kim, H.K.; Jung, W.S.; Lee, B.J. Modified embedded-atom method interatomic potentials for the Fe–Ti–C and Fe–Ti–N ternary systems. *Acta Mater.* **2009**, *57*, 3140–3147. [CrossRef]
25. Prasanthi, T.N.; Sudha, C.; Saroja, S. Molecular Dynamics Simulation of Diffusion of Fe in HCP Ti Lattice. *Trans. Indian Inst. Met.* **2018**, *71*, 1951–1955. [CrossRef]
26. Ledbetter, H.M.; Reed, R.P. Elastic properties of metals and alloys, I. Iron, nickel, and iron-nickel alloys. *J. Phys. Chem. Ref. Data* **1973**, *2*, 531–618. [CrossRef]
27. Song, Y.; Yang, R.; Li, D.; Guo, Z. A first-principles study of the theoretical strength and bulk modulus of hcp metals. *Philos. Mag. A* **2001**, *81*, 321–330. [CrossRef]
28. Plimpton, S. Fast parallel algorithms for short-range molecular dynamics. *J. Comput. Phys.* **1995**, *117*, 1–19. [CrossRef]

29. Stukowski, A. Visualization and analysis of atomistic simulation data with OVITO–the Open Visualization Tool. *Model. Simul. Mater. Sci. Eng.* **2009**, *18*, 015012. [CrossRef]
30. Terban, M.W.; Billinge, S.J.L. Structural analysis of molecular materials using the pair distribution function. *Chem. Rev.* **2021**, *122*, 1208–1272. [CrossRef]
31. Sa, I.; Lee, B.J. Modified embedded-atom method interatomic potentials for the Fe-Nb and Fe-Ti binary systems. *Scr. Mater.* **2008**, *59*, 595–598. [CrossRef]
32. Trong, D.N.; Long, V.C.; Ţălu, Ş. Effects of Number of Atoms and Doping Concentration on the Structure, Phase Transition, and Crystallization Process of Fe1-x-yNixCoy Alloy: A Molecular Dynamic Study. *Appl. Sci.* **2022**, *12*, 8473. [CrossRef]
33. Yuan, J.; Yu, A.; Zou, P.; Zhao, S.; Li, Y.; Wu, Q. Molecular dynamics simulation of hot pressing diffusion behavior between diamond and titanium. *Rare Met. Mater. Eng.* **2022**, *51*, 159–165.

coatings

MDPI

Article

Atomic Research on the Diffusion Behavior, Mechanical Properties and Fracture Mechanism of Fe/Cu Solid–Liquid Interface

Hongyu Zheng *, Jingwen Sun, Na Guo and Mingjie Wang *

School of Intelligent Manufacturing, Huanghuai University, Zhumadian 463000, China
* Correspondence: 20191887@huanghuai.edu.cn (H.Z.); 15513882577@163.com (M.W.)

Abstract: A molecular dynamics simulation was applied to investigate the diffusion behavior and mechanical properties of a Fe/Cu solid–liquid interface with different orientations, temperatures, and strain rates. The results show that the displacement distance of Fe atoms' diffusion into the Cu matrix was obviously larger than that of Cu atoms' diffusion into the Fe matrix at any diffusion temperature and diffusion time. Moreover, the diffusion coefficient and diffusion distance both increase with temperature and time, and reach the highest value when the temperature and diffusion time are 1523 K and 3 ns, respectively. Additionally, the diffusion coefficients of the Fe atoms are arranged in the following order: Fe (100) < Fe (110) < Fe (111). The diffusion coefficients of the Cu atoms are arranged in the following order: Cu (110) > Cu (111) > Cu (100), when temperature and time are 1523 K and 3 ns, respectively. The yield strength and fracture strain of the bimetallic interface is positively correlated with the strain rate, but negatively correlated with the tensile temperature. Moreover, the yield strength of the three orientations can be arranged as follows: Fe (110)/Cu (110) > Fe (100)/Cu (100) > Fe (111)/Cu (111), and the yield strength and fracture strain of Fe (110)/Cu (110) diffusion interface are 12.1 GPa and 21% when the strain rate was 1×10^9/s and the tensile temperature was 300 K. The number of stacking faults and dislocations of the diffused Fe/Cu interface decreased significantly in comparison to the undiffused Fe/Cu interface, even in the length of Stair-rod dislocation and Shockley dislocation. All these results lead to a decrease in the tensile yield strength after interface diffusion.

Keywords: diffusion behavior and mechanical properties; Fe/Cu solid–liquid interface; molecular dynamics; orientation; temperature and strain rate

Citation: Zheng, H.; Sun, J.; Guo, N.; Wang, M. Atomic Research on the Diffusion Behavior, Mechanical Properties and Fracture Mechanism of Fe/Cu Solid–Liquid Interface. *Coatings* **2022**, *12*, 1299. https://doi.org/10.3390/coatings12091299

Academic Editor: Gianni Barucca

Received: 13 August 2022
Accepted: 1 September 2022
Published: 4 September 2022

Publisher's Note: MDPI stays neutral with regard to jurisdictional claims in published maps and institutional affiliations.

1. Introduction

Over the last few years, high-strength and highly wear resistant bimetallic materials have been attracted worldwide attention and demonstrated important application prospects in aerospace, ships, automobiles, and other fields due to their excellent dual performance structure and the comprehensive high-strength and high wear resistance properties [1–5]. The steel–copper bimetallic materials are one of the typical representatives, combining the advantages of both steel (high strength and stiffness) and copper (high wear resistance), and have been used to fabricate military aerospace rotors [6]. Additionally, for steel–copper bimetallic materials with a high bonding strength, the relationship between the diffusion mechanism, the diffusion distance, and the tensile stress-strain of the interface atoms play a more important role in determining the bonding strength and the diffusion coefficient of the bimetallic interface [7,8]. However, the pivotal technique of steel–copper bimetallic materials, the interfacial diffusion and interface enhancement mechanism of steel–copper bimetallic materials, has hardly been published and investigated.

Up to now, atomic calculations have been demonstrated to be an effective method to comprehensibly investigate the solid–liquid interface, as it is difficult to carry out an experimental investigation at high temperature [9–11]. According to the relevant reports, several theoretical investigations of solid–liquid interfaces and multi-layer interfaces have been proposed, such as the Cu/Pb interface [12,13], Al/Cu interface [14,15], Ni/Al multi-layer [16], Al/Al_2O_3 interface [17], and Fe/Li interface [18]. Especially, Mao et al. [19] studied the diffusion behavior of the Cu/Al solid–liquid interface using the molecular dynamic (MD) method, and found that the influence of temperature and pressure on the diffusion behavior of Cu/Al solid–liquid interface is greater than that of time, and all of these results depend on the formation of $\theta-Al_2Cu$. S. Raman et al. [20] investigated the thermodynamic and kinetic properties of Fe/Mn solid–liquid interface utilizing MD simulations, and successfully revealed that the average value of solid–liquid interfacial free energy and anisotropy parameters remain constant with fluctuating temperature and the variation in diffusion speed is higher than that of the bulk liquid diffusion coefficient with increasing temperature. Moreover, a number of researchers have explored some incurable factors associated with the mechanical properties of solid–liquid interface, such as stress–strain curves and microstructure evolution in the course of the tensile process [21–23]. Su et al. [24] investigated the tensile behavior of Ti/Ni multilayered films and predicted the tensile process, stress–strain curve, and other tensile behaviors through MD simulations. Liu et al. [25] simulated the diffusion process of the Al/Cu interface, analyzed the tensile behavior, and improved the mechanical properties of the Al/Cu solid–liquid interface after diffusion solidification and cooling. All these mechanical property investigations of the solid–liquid interface have been studied by quite a few scholars, but there has been no systematic analysis of the Fe/Cu solid–liquid interface.

In this paper, we aimed to propose a theoretical framework based on the diffusion behavior and mechanical properties of Fe/Cu solid–liquid interface using MD simulation. In detail, we analyzed the diffusion coefficient, the diffusion distance, the atomic concentration distribution, and the diffusion mechanism of Fe/Cu solid–liquid interfaces at different temperatures and diffusion times. Meanwhile, the mechanical properties, i.e., tensile strain–stress curve, influence of tensile orientation, strain rate, and tensile temperature of the Fe/Cu solid–liquid interface, are also investigated. Although the diffusion and bonding characteristics of steel–copper bimetallic interfaces cannot be directly characterized in time scale and space scale in this study, our calculations provide a theoretical guidance for the production and manufacture of high-performance Fe/Cu bimetallic materials.

2. Simulation Methodology

All MD simulations are calculated using the Large-scale Atomic/Molecular Massively Parallel Simulator (LAMMPS) [26] with a time step of 1.0 fs. Additionally, an embedded atomic method (EAM) potential developed by Byeong-Joo Lee et al. [27] was used for modeling the atomic interaction between FCC Fe atoms and FCC Cu atoms near the interface, which has been employed in exploring diffusion properties, mechanical properties [28], and magnetic properties [29–31] of the Fe–Cu alloys. An open visualization tool (OVITO) [32] was used to observe diffusion behavior and the deformation process. Moreover, periodic boundary conditions were applied in three transverse (x, y, and z) directions. The initial velocity of atoms was assumed to accord with Maxwell–Boltzmann random distribution. Additionally, the Verlet integration algorithm was introduced to solve the Newtonian equation of motion integration.

In this study, three surface orientations ((100), (110), and (111)) of the γ-Fe/Cu interface were used to calculate the diffusion behavior and mechanical properties of Fe/Cu bimetallic materials. An additional advantage of using a case study approach is that our simulation models are based on our steel–copper bimetallic casting experiment at high temperature [33,34]. The modeling consists of FCC Fe bulk and FCC Cu bulk, and the initialization configurations of a cross-section are shown in Figure 1. As seen in Figure 1, the initial Fe/Cu interface model consisted of two parts: an upper Cu block containing 6400 atoms with the size of $10a_{Cu} \times 5a_{Cu} \times 30a_{Cu}$ and a lower Fe block with the same number of atoms and the size of $10a_{Fe} \times 5a_{Fe} \times 30a_{Fe}$ (the lattice constant of the a_{Cu} and a_{Fe} atoms are 0.36147 and 0.36457 nm respectively). To perform the solid–liquid interface calculations, five different temperatures—1323, 1373, 1423, 1473, and 1523 K—were selected according to the melting point of Cu and Fe. Additionally, the overall diffusion process was relaxed at 300 K under a constant pressure-temperature (NPT) ensemble for 100 ps, and then relaxed at the target diffusion temperature under a constant pressure-temperature (NPT) ensemble for 1000 ps. Finally, the system was quickly cooled to 300 K, shown in Figure 1.

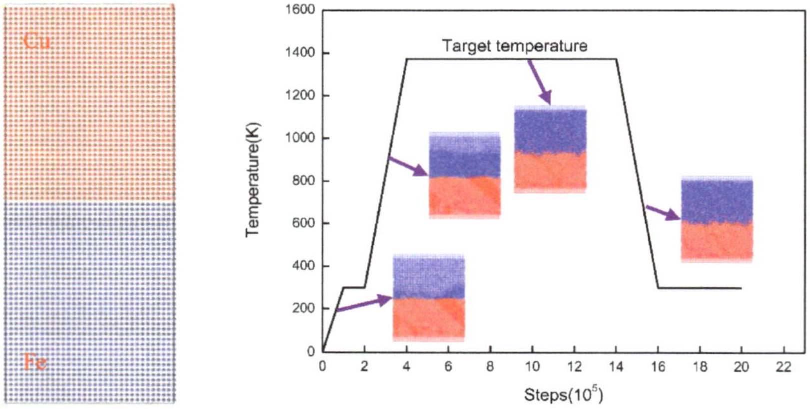

Figure 1. Atomic configuration of Fe/Cu interface model and the temperature–time relationship of the diffusion process.

According to [35], the mean square displacement (MSD) of the solid–liquid interface can be approximately calculated in such a complicated system. Thus, the MSD of Cu and Fe was calculated first to forecast the diffusion coefficient. The MSD can be defined by the following equation [36]:

$$\mathrm{MSD} = \left\langle r^2(t) \right\rangle = \left\langle |r_i(t) - r_i(0)|^2 \right\rangle$$

where $r_i(t)$ is the position vector of atom i at time t, which represents the average ensemble of the atoms in the simulated time. Accordingly, the diffusion coefficient is defined by the following equation [37]:

$$\mathrm{D} = \lim_{t\to\infty} \frac{1}{2Nt} \left\langle |r(t) - r(0)|^2 \right\rangle$$

where D is the diffusion coefficient, N denotes the dimension of the simulated system, and $N = 3$ for the simulation of blocks.

3. Result and Discussion

3.1. *Diffusion Behavior*

Temperature and diffusion time, as the two main factors determining the Fe/Cu interface, were considered first in our simulation. Thus, the snapshots of the cross-section diffusion process of Fe bulk and Cu bulk at different temperatures and times are distinctively illustrated in Figure 2. By comparing the four snapshots under the same temperature with the diffusion time elevated, the portion of atoms diffusing across the initial Fe/Cu interface can be seen to increase gradually and the diffusion depth of the Fe atoms diffusing into the Cu bulk is deeper, while that of Cu atoms diffusing into the Fe bulk is shorter. From these snapshots, it can be seen that only a small amount of Fe atoms cross through the initial interface, while much less Cu atoms cross the interface at 0.5 ns. After 3.0 ns, the diffusion distance of Cu atoms and Fe atoms both increase, the percentage of Fe atoms that diffuse across the interface increases to 1.5% from 0.3% at 0.5 ns, and the percentage of Cu atoms increases to 0.3% from 0.05% at 0.5 ns. These asymmetrical diffusion phenomena have similarly appeared in Al/Cu [38], Mo/Ti [39], and Fe/W [40] interfaces. Likewise, by comparing five different temperatures at the same diffusion time, one can see that only a few Fe atoms diffuse into the Cu bulk at 1323 K, the interfacial structure is relatively smooth, and the degree of atomic confusion is lower. However, when temperature is increased, the portion of Fe and Cu atoms that cross initial interface increases and the diffusion distance of Fe atoms diffusing into the Cu bulk is also deeper than that of the Cu atoms diffusing into the Fe bulk at every temperature. Meanwhile, when the temperature is increased to 1523 K from 1323 K with a 50 K temperature interval, the percentages of the number of Fe atoms diffusing into the Cu bulk are 0.1%, 0.13%, 0.15%, 0.17%, and 0.2%, while that of the Cu atoms diffusing into Fe bulk are 0.03%, 0.035%, 0.05%, 0.08%, and 0.1%. Therefore, the influence of time on the diffusion distance is greater than that of temperature.

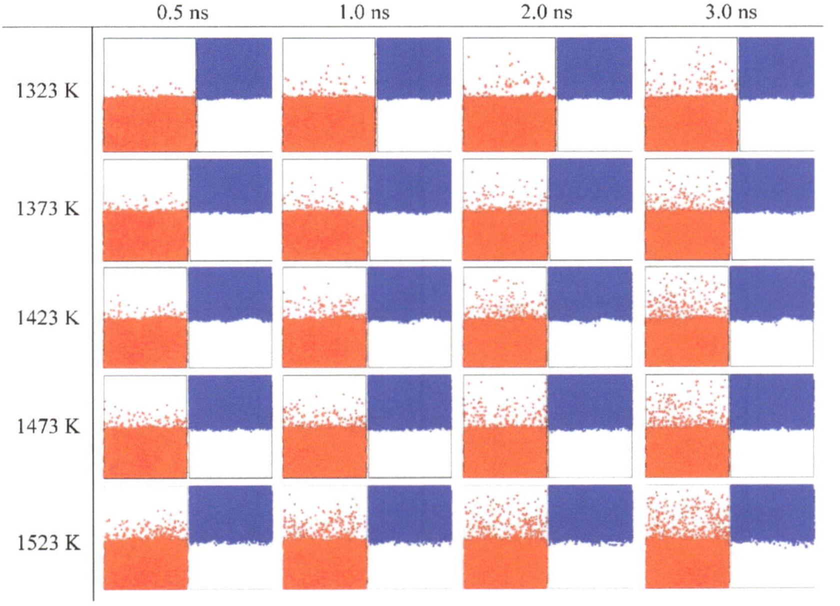

Figure 2. Snapshots of Fe atoms and Cu atoms under different diffusion times and temperatures. Red is Fe, blue is Cu.

Figure 3 illustrates the snapshots of the planar views of the equilibrium diffusion Fe/Cu interface at different temperature after 3 ns, respectively. As shown in Figure 3, when the temperature is increased, the Cu bulk presents an obvious liquid amorphous structure due to the disordered arrangement of atoms, while the Fe bulk maintains a solid crystalline structure. Furthermore, after the Cu atoms diffuse into the Fe lattice, the Cu atoms show an ordered state and occupy the lattice position of the Fe bulk, while the majority of the Fe atoms are located in the vacancies of the Cu lattice after diffusing into the inner Cu lattice. Thus, the Cu atoms gradually form agglomerations due to the increasing number of Fe atoms occupying more and more vacant positions in Cu bulk, explaining why the diffusion distance of Fe atoms diffusing into the Cu bulk is deeper than that of the Cu atoms diffusing into the Fe bulk.

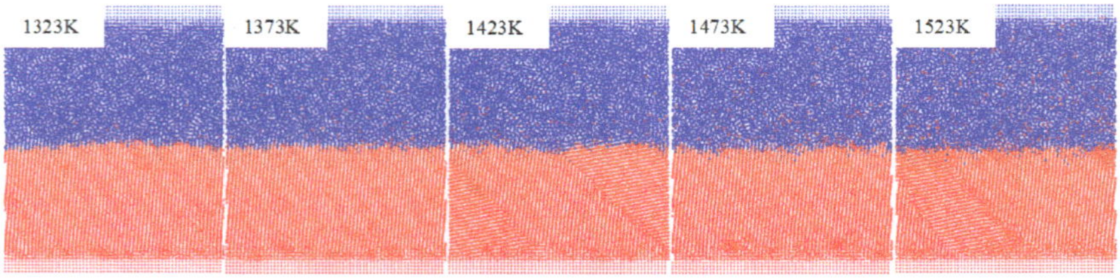

Figure 3. Atomic structure of Fe/Cu interface at different temperatures.

To deeply understand the non-uniform diffusion phenomena of the three Fe/Cu interface models, the MSDs of the Fe surface and the Cu surface in different directions at 1523 K were determined. As shown in Figure 4, the MSD curves of the Cu atoms in all orientations increase linearly, while the MSD curves of Fe atoms in different orientations areas fluctuate upward, which is consistent with the liquid and solid diffusion characteristics of the Cu and Fe matrixes. For the MSD curves of Fe atoms and Cu atoms in different directions, the MSD value of the diffusion of Cu atoms along the y-axis is higher than that along the x-axis and z-axis, indicating that Cu atoms mainly diffuse along the y-axis. The diffusion of Fe atoms on Fe (100) and Fe (111) surfaces is mainly along the x-axis, while on Fe (110) surfaces, the diffusion of Fe atoms is mainly along the y-axis. It can be seen that the degree of diffusion of the Fe/Cu interface in the diffusion process along the perpendicular direction is lower than that along the parallel direction of the interface.

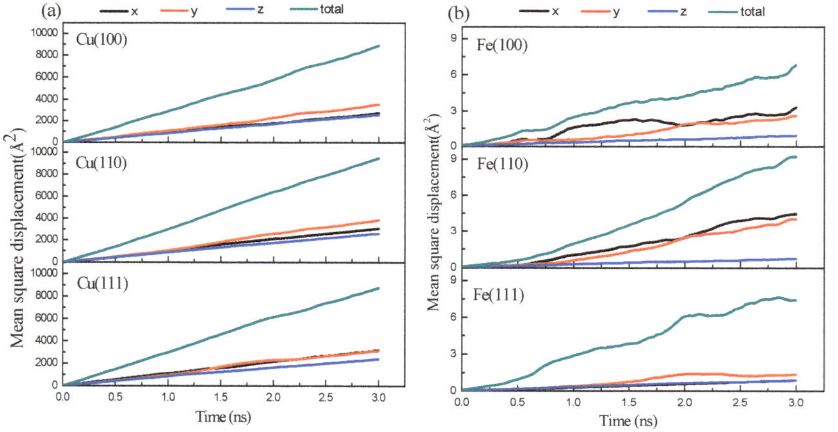

Figure 4. The mean square displacement curves of Fe/Cu interface at 1523 K with different orientation: (**a**) Cu (100), Cu (110), and Cu (111); (**b**) Fe (100), Fe (110), and Fe (100).

Table 1 shows the atomic diffusion coefficients of different surfaces and directions, and it can be seen that the diffusion coefficients of the Cu slabs in the z-axis direction can be ordered as (110) < (111) < (100), and the diffusion coefficients of Fe atoms in the z-axis direction can be ordered as (111) < (100) < (110). For all the Cu slabs, the diffusion coefficient of Cu atoms along y-axis is greater than that along the x-axis and z-axis, which can be ranked as follows: z < x < y, indicating that the blocking effect of Fe matrix on Cu atoms reduces the movement of Cu atoms along z-axis. However, for all the Fe slabs, the diffusion coefficient of Fe atoms along the z-axis is also smaller than other directions, which indicates that the transverse movement is greater than the axial movement during the diffusion process of Fe/Cu interface, and the Fe (110)/Cu (110) interface displays the best diffusion effect for the Fe atoms diffusing into the Cu matrix.

Table 1. Diffusion coefficient of atoms under different surface and orientation.

Diffusion Direction	Cu (100)	Cu (110)	Cu (111)	Fe (100)	Fe (110)	Fe (111)
D_x (10^{-11} m^2/s)	409.5	523.6	528.6	0.803	0.296	0.745
D_y (10^{-11} m^2/s)	605.7	735.2	626.6	0.456	0.575	0.813
D_z (10^{-11} m^2/s)	431.1	421.8	389.3	0.126	0.159	0.117
D_{total} (10^{-11} m^2/s)	491.7	549.6	512.6	0.353	0.464	0.554

In the in-depth investigation of the specific diffusion information of the Fe (110)/Cu (110) interface, the mean square displacement (MSD) of Cu atoms and Fe atoms along z-direction were calculated as shown in Figure 5. Additionally, the diffusion coefficient and the concentration distributions along the z-direction of Fe and Cu atoms was calculated to characterize the atomic diffusion at the Fe (100)/Cu (100) interface with temperature increase, as shown in Figure 5. One can see that all the MSD curves of the Cu atoms and Fe atoms are not only increased with increasing diffusion time at different temperatures, but also increased with increasing temperature due to the higher kinetic energy of atomic diffusion. Additionally, as shown in Figure 5a, the MSD curves of the Cu atoms present as straightly increasing trends as the diffusion time is incremented, but that of the Fe atoms does not maintain a straight line, as shown in Figure 5b. This is possibly due to the fact that the solid Fe matrix exhibits obvious volume instability when the ambient temperature approaches the melting point. From Figure 5c, it can be seen that the diffusion coefficient of Fe and Cu atoms exhibits a significantly increasing trend with increasing temperature, this is possibly due to an increase in the internal energy of the system. The diffusion coefficient of Cu atoms is far smaller than that of Fe atoms, seen in Figure 5c, but the diffusion depth of Fe atoms the diffuse into the Cu area is much extensive than that of Cu atoms, as seen in Figure 5d. This phenomenon concurs with the above results (see the snapshots in Figure 1), and is possibly due to the lower melting point of Cu than that of Fe. In contrast, the interface migrates to the Fe sides during the diffusion process. As shown in Figure 5d, the moving distance of the red dotted line is greater than that of the black dotted line. Thus, the interface migration may appear in these Fe/Cu interface models.

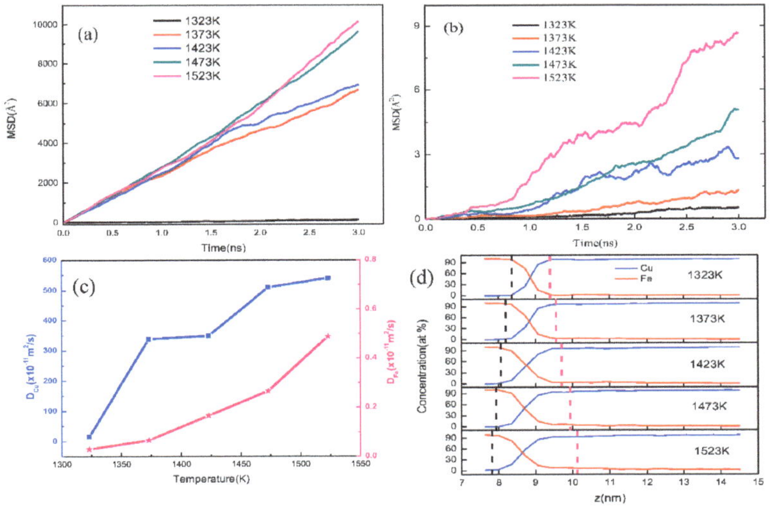

Figure 5. The diffusion behavior of atoms and the diffusion distance of Fe (110)/Cu (110) interface under different temperatures. (**a**) MSD curves of Cu atoms under different temperatures; (**b**) MSD curves of Fe atoms under different temperatures; (**c**) diffusion coefficient of Fe and Cu atoms; (**d**) atomic concentration of Fe and Cu atoms.

To clearly observe the phenomenon of interface migration, we compared the snapshots of the diffusion Fe/Cu interface at the initial simulation time and the final simulation time, as shown in Figure 6. Compared with Figure 6a,b, the Cu atoms become disordered in comparison to the initial model. The Fe atoms keep the original crystal structure, yet only a few Cu atoms replaced the position of Fe atoms, which elucidates the results of the concentration distributions. Furthermore, the red dashed line in Figure 6 migrates to the Fe side after the diffusion of Fe/Cu bimetallic material, indicating that Cu atoms are difficult to diffuse into the Fe matrix, regardless of displacement diffusion and gap diffusion. Hence, it has been demonstrated that the interface migration is authentically consistent in these Fe/Cu diffusion interface systems.

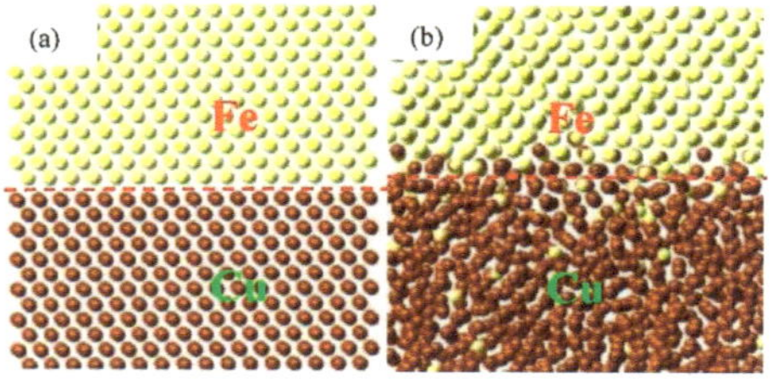

Figure 6. Atomic snapshots concerning Kirkendall effect of the Fe/Cu bimetallic interface. (**a**) Initial atomic structures; (**b**) the final atomic structures after the diffusion at 1523 K.

To deeply understand the inner diffusion mechanism of the Fe/Cu interface, the transition state search (TSS) tool was calculated to find the minimum energy and saddle points of the diffusion process. Figure 7 shows the diffusion energy barrier of five diffusion paths and the schematic diagram of the possible atomic diffusion paths near the interface boundary of the Fe and Cu atoms. Comparing five diffusion energy barriers in Figure 7, we can find that the diffusion of Cu and Fe atoms is mainly dominated by first-nearest-neighbor diffusion, and the diffusion activation energy of Fe atoms in the Cu matrix is greater than that of Cu atoms in the Fe matrix, indicating that the diffusion of Fe atoms in Cu matrix is easier than that of Cu atoms in Fe matrix.

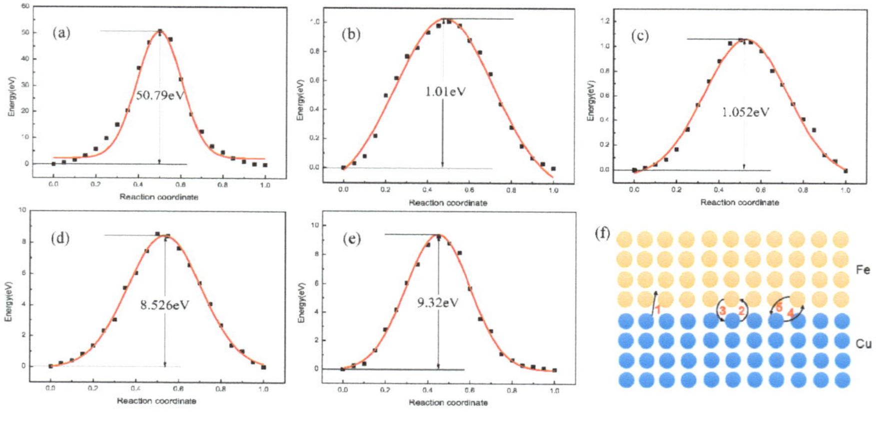

Figure 7. Atomic snapshots concerning Kirkendall effect of the Fe/Cu bimetallic interface. (**a**) the diffusion energy of the interstitial diffusion path of Cu atom; (**b**) the diffusion energy of the first nearest neighbor diffusion path of Cu atom; (**c**) the diffusion energy of the first nearest neighbor diffusion path of Fe atom; (**d**) the diffusion energy of the second nearest neighbor diffusion path of Cu atom; (**e**) the diffusion energy of the second nearest neighbor diffusion path of Cu atom; (**f**) schematic diagram of diffusion mechanism of Fe/Cu diffusion interface.

3.2. Mechanical Properties

3.2.1. Strain Rate and Orientation Effect

The strain rate's dependence of the stress–strain relationships of the Fe (100)/Cu (100) interface, the Fe (110)/Cu (110) interface, and the Fe (111)/Cu (111) interface under uniaxial tension are shown in Figure 8. As shown in Figure 8a, the strain–stress curves of the Fe (100)/Cu (100) interface are almost linear at the initial stage of uniaxial tension, which indicates that the Fe (100)/Cu (100) interface undergoes elastic deformation. After that, the stress decreases slowly with the increase in strain when the tension reaches ultimate strength, indicating that the Fe (100)/Cu (100) interface presented a plastic deformation ability and ductile fracture characteristics. From Figure 8b,c, one can see that the strain–stress curves are almost increasing linearly at the early stage in all the strain rates, which indicates that the Young's modulus is less sensitive to the strain rate. However, with the application of increasing strain, the curves present apparent ups and downs at the end of elastic deformation, which indicates that fracture occurred. Additionally, all the strain–stress curves decrease suddenly at the stress peak point at different strain rates, and no obvious plastic deformation can be observed. These results imply that the brittle fracture of the Fe (100)/Cu (100) interface and the Fe (111)/Cu (111) interface occurred at all simulated strain rates. Meanwhile, the yield strength and fracture strain both increase with the increasing strain rates, and the yield strength of three orientations can be arranged in the following order: Fe (111)/Cu (111) > Fe (110)/Cu (110) > Fe (100)/Cu (100), which reaches a remarkable agreement with the diffusion behavior of the three models.

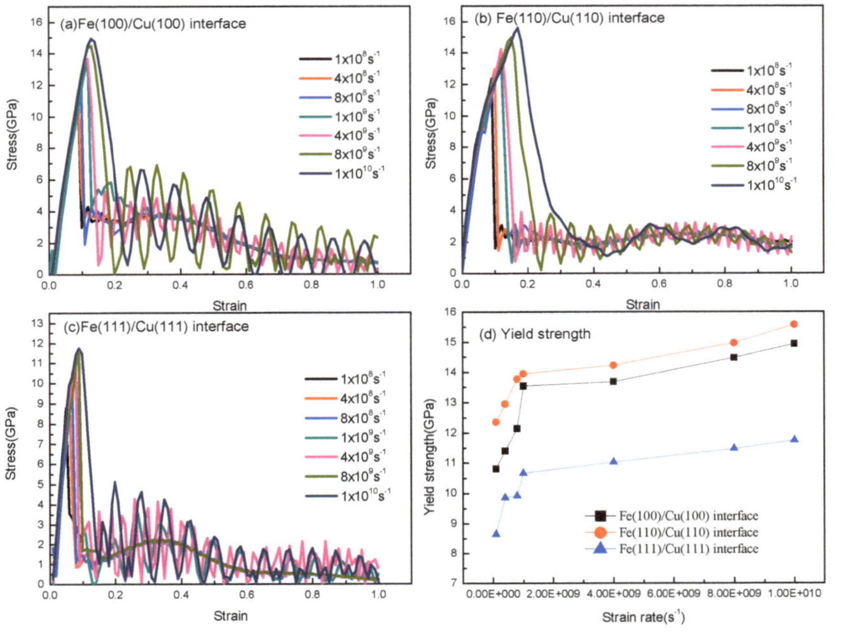

Figure 8. Stress–strain curves of (**a**) Fe (100)/Cu (100) interface at different strain rates from 1×10^8/s to 1×10^{10}/s; (**b**) Fe (110)/Cu (110) interface at different strain rates from 1×10^8/s to 1×10^{10}/s; (**c**) Fe (111)/Cu (111) interface at different strain rates from 1×10^8/s to 1×10^{10}/s; (**d**) the yield strength of three different interface models under different strain rates.

Referring to Figure 8a–c, the yield strength and fracture strain curves with different strain rates are described in Figure 8d. As shown in Figure 8d, the yield strength and fracture strain are particularly susceptible to strain rate when the strain stress is less than

$1 \times 10^9\ s^{-1}$. Nevertheless, when the strain rate exceeds $1 \times 10^9\ s^{-1}$, the tensile strength still increases slowly with the growth of strain rate, this is possibly due to the fact that the reinforcement effect is not statistically significant. Furthermore, the brittle fracture behavior of the three Fe/Cu interface models at a high strain rate is similar to that of single-crystal titanium [41].

All the atomic visualizations of the Fe/Cu interface are described using the Open Visualization Tool (OVITO), and all the inner structure of Fe/Cu bimetallic materials is characterized to observe the defect's behavior from HCP environments with common neighbor analysis (CNA) [42]. Additionally, the centrosymmetric parameter (CSP) and dislocation extraction algorithm (DXA) [43] analysis are powerful modules that were used to analyze the dislocation motion and defects of the Fe/Cu interface. To clearly describe the fracture mechanism of the Fe (100)/Cu (100) interface, the Fe (110)/Cu (110) interface and the Fe (111)/Cu (111) interface, the atomic snapshots (rendered by CNA, CSP, and DXA results) of the three interface models under uniaxial loading at different strain are described and shown in Figures 9–11.

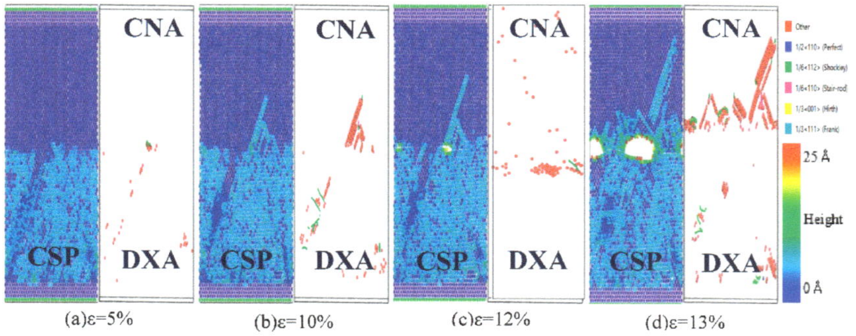

Figure 9. Atomic CSP, CNA, and DXA analysis of the Fe (100)/Cu (100) undiffused interface during z-axis tensile process.

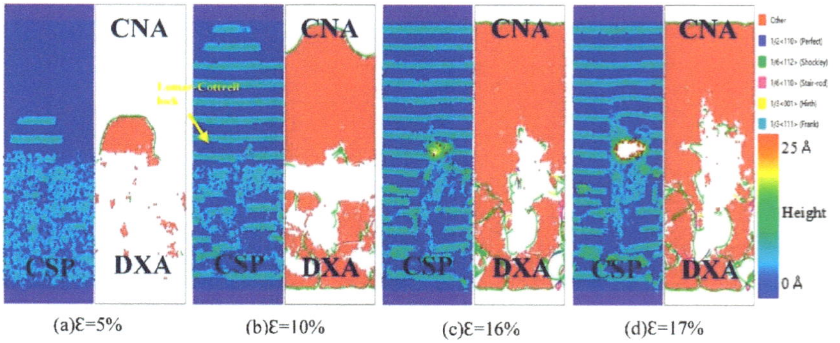

Figure 10. Atomic CSP, can, and DXA analysis of the Fe (110)/Cu (110) undiffused interface during z-axis tensile process.

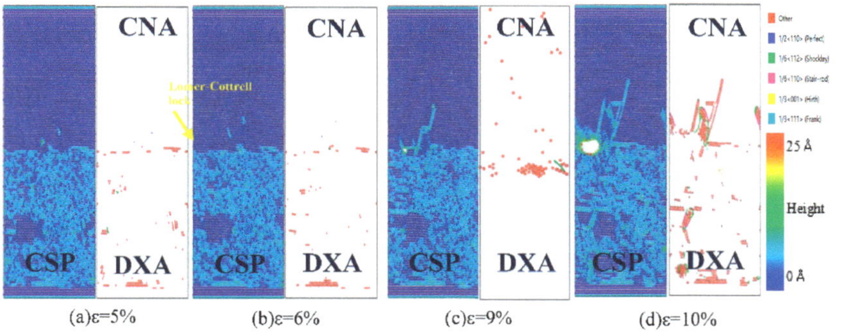

Figure 11. Atomic CSP, can, and DXA analysis of the Fe (111)/Cu (111) undiffused interface during z-axis tensile process.

From the CSP analysis in Figure 9, the atoms of the Fe (100)/Cu (100) interface are arranged neatly and represent an elastic deformation process on both sides of the interface at the beginning of the tension, and the interface structure hardly changes when the strain is lower than 5%. With gradually increasing stress, the Fe (100)/Cu (100) interface emits surface angular dislocations to the Cu side when the strain reaches 10%. Then, micropores are formed at the surface's angular dislocations when the strain reaches 12%. With continuous increase in strain, the interface gradually expands into a microcrack along the micropore, which results in a sudden drop of the interface stress value, and brittle fracture occurs. In addition, according to the CNA and DXA analysis, the stacking faults on the Cu side increase gradually with the progression of the Fe (100)/Cu (100) interface tension, and the evolution of the 1/6 <112> Shockley dislocations occupied a dominant position during the dislocation movement in the tensile process. Additionally, the number of Shockley dislocations near the interface increases sharply when a large number of cracks appear at the interface, while the number of Shockley dislocations is less before the fracture, indicating that brittle fracture begins at the strain of 13% in the Fe (100)/Cu (100) interface.

From Figure 10, it can be seen that a large number of stacking faults appear on both sides of the interface during the tensile fracture of Fe (110)/Cu (110) interface. Firstly, according to the CSP analysis, the steady stacking fault forms on the Cu side when the strain is 5%, but a large number of stacking faults appear on both sides of the Fe/Cu interface and surface angle dislocations are emitted to the Cu side when the strain increases to 10%. These results cause the development of micropores and cracks, resulting in interface fracture. Similarly, through integrated DXA analysis and CNA analysis, we can observe that a large number of stacking faults appear in the tensile process of Fe (110)/Cu (110) interface, and more 1/6 <112> Shockley dislocations appear near the interface. These phenomena hinder the progress of z-axis loading tensile, but improve the maximum stress value of the Fe (110)/Cu (110) interface.

In Figure 11, it can be seen that the Fe (111)/Cu (111) interface starts to emit Lomer–Cottrell lock to the Cu side when the strain reaches 6%, expands into micropores when the strain is 9%, and then expands into a microcrack when the strain is 10%. Comparing the tensile process with the two above interface models, one can see that there are fewer interlayer faults and Shockley dislocations at the Fe (111)/Cu (111) interface in the tensile process, resulting in lower yield strength and lower interface bonding strength of the Fe (111)/Cu (111) interface. To summarize, the yield strengths of the three orientation's interface models can be ranked as follows: Fe (110)/Cu (110) > Fe (100)/Cu (100) > Fe (111)/Cu (111). In addition, the Fe/Cu bimetallic interface shows brittle fracture characteristics in the tensile process, and the Fe (110)/Cu (110) interface shows a certain fracture toughness compared with the other two interface models.

3.2.2. Temperature Effect

To systematically study the influence of z-axis tensile temperature on the mechanical properties of Fe/Cu interface, the diffused Fe (110)/Cu (110) interface (the diffusion temperature is 1523 K, and the time is 3 ns) and undiffused Fe (110)/Cu (110) interface models were selected to carry out the z-axis tensile test at different temperatures (50, 100, 200, 300, 400, and 500 K). The stress–strain tensile results are shown in Figure 12; all these strain–stress curves show a linear and sharp increasing trend when the strain value is less than 10%, consistent with the characteristics of elastic deformation. Additionally, the slope of the stress–strain curve decreases with the increase in temperature, which indicates that the elastic modulus of the Fe/Cu interface decreases with the increase in temperature. Moreover, the Fe/Cu interface tension enters the plastic deformation stage after the yield strength, and the dislocation and the slip phenomenon are more prominent, and the tensile stress drops sharply and then changes slowly, indicating that brittle fracture occurs in the Fe (110)/Cu (110) interface before and after diffusion at all temperatures. Additionally, the maximum yield stress of the undiffused Fe (110)/Cu (110) interface at 50 and 500 K are 15.8 and 13.9 GPA, respectively, and the corresponding strain values are 18.9% and 16.1%, respectively. However, for the diffused Fe (110)/Cu (110) interface, the maximum yield stress values are 13.4 and 10.4 GPa, respectively, and the corresponding strain values are 11.9% and 10.7%, respectively. Moreover, the yield strength of the Fe (110)/Cu (110) interface is lower after interfacial diffusion, which is caused by the weakening of the interatomic bonding energy caused by the bimetal interface reconstruction after diffusion. When temperature increases from 50 to 100 K, the yield strength of the Fe (110)/Cu (110) interface decreases from 13.5 to 13.3 GPa respectively. However, when the temperature is raised from 400 to 500 K, the yield strength decreases from 11.2 to 10.3 GPa. Thus, the decreasing rates were 1.5% and 8.0%, respectively, indicating that the tensile Fe (110)/Cu (110) interface is more sensitive to temperature changes.

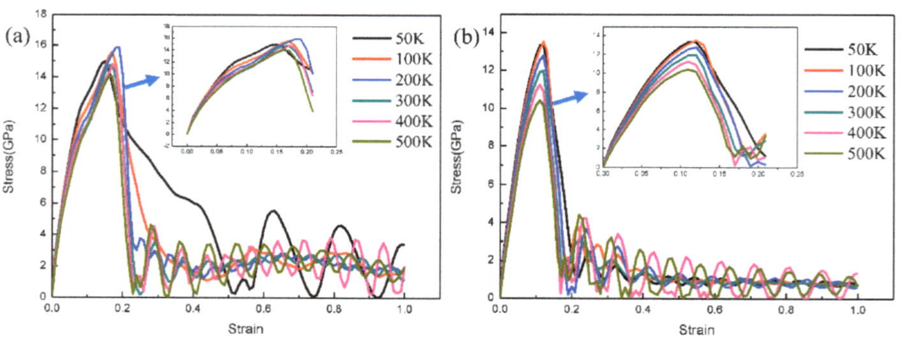

Figure 12. Strain–stress curves of z-axis tensile of Fe (111)/Cu (111) interface before and after diffusion under different temperature. (**a**) z-axis tensile stress–strain curve before diffusion; (**b**) z-axis tensile stress–strain curve after diffusion.

Thus, the atomic oscillation amplitude of Fe/Cu bimetallic interface is small when the tensile temperature is 50 K, resulting in a stable structure and a higher bonding strength. However, due to the increase in movement speed of atoms at the Fe/Cu bimetallic interface and the interaction force between atoms decreases when the tensile temperature is 500 K, a reduction in the energy required for dislocation, nucleation and emission, and lattice deformation occurred easily under the same load.

To reveal the fracture mechanism of the Fe (110)/Cu (110) interface at different working temperatures, the undiffused Fe (110)/Cu (110) interface at 300 and 500 K were selected to analyze the changes and atomic defects of the interface structure, and the results are shown in Figures 10 and 13 respectively. By comparing Figures 10 and 13, it can be found that a large number of stacking faults also appear at the undiffused Fe (110)/Cu (110) interface at 500 K, which indicates that the main factor affecting stacking faults is the orientation relationship instead of tensile temperature. Additionally, the undiffused Fe (110)/Cu (110) interface emits a plane angular dislocation at the Cu side, and then evolves into a micropore when the strain is 10% and the stacking faults in the interface tensile process gradually decrease with the increase in temperature. This is possibly due to the higher tensile temperature, resulting in disordered structure arrangement and interface amorphization [44], which hinder the extended motion of dislocation and reduce the yield strength of the interface model. Although the number and length of dislocations increase or decrease, the types of the dislocations are all Shockley dislocations. Moreover, the fracture strain of the undiffused Fe (110)/Cu (110) interface decreases gradually with the gradual increase in temperature, possibly due to the early occurrence of interface fracture with the gradual increase in stress and the obstruction of dislocation propagation.

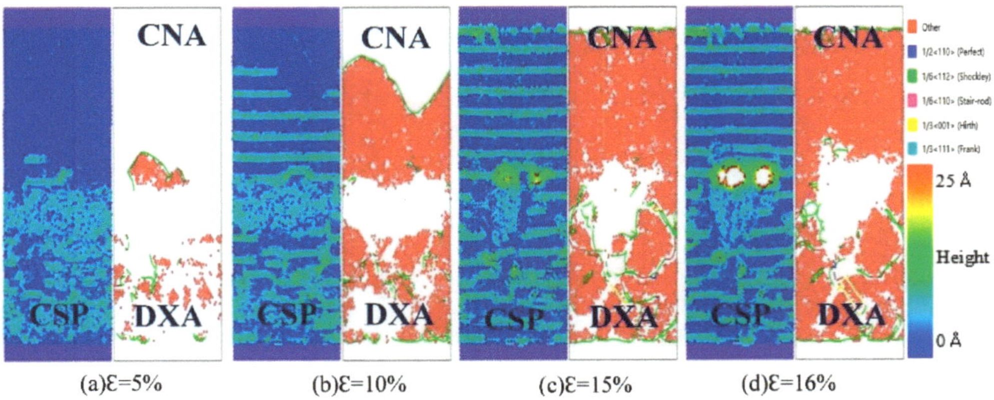

Figure 13. Atomic CSP, CNA, and DXA analysis of the Fe (110)/Cu (110) undiffused interface under z-axis tensile process at 500 K.

Figures 14 and 15 show the interface structure and atomic defect analysis during the tensile process of the diffused Fe (110)/Cu (110) interface at 300 and 500 K, respectively. From Figure 14, one can see that the micropores and microcrack at the diffused Fe (110)/Cu (110) interface appeared far from the interface on the Cu side, indicating that the bonding strength of the diffused Fe (110)/Cu (110) interface is greater than that of the Cu matrix. Similarly, the amorphization near the interface increase with increase in temperature, resulting in the reduction of interface stacking faults and the yield strength. From Figure 15, compared with the tension at 300 K, the fracture strain of the diffused Fe (110)/Cu (110) interface decreases from 20% to 19%, but the dislocation types are mainly 1/6 <112> Shockley dislocation at both 300 and 500 K. In summary, temperature will not change the dislocation and defect types of the Fe (110)/Cu (110) interface model under a certain temperature range, but will reduce or improve the yield strength and fracture strain of bimetallic interface by hindering or promoting the growth of dislocation.

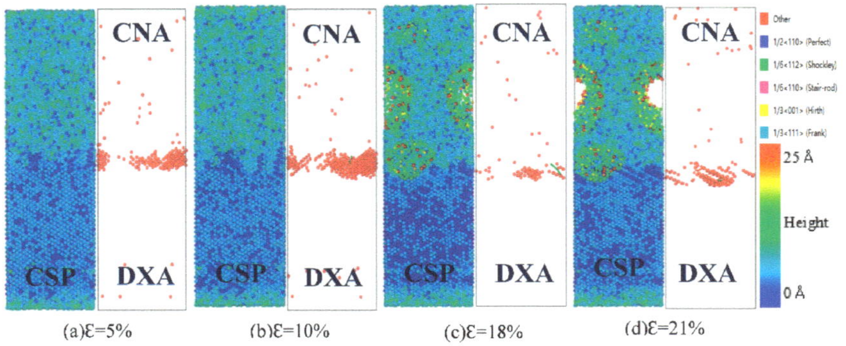

Figure 14. Atomic CSP, CNA, and DXA analysis of the Fe (110)/Cu (110) diffused interface under z-axis tensile process at 300 K.

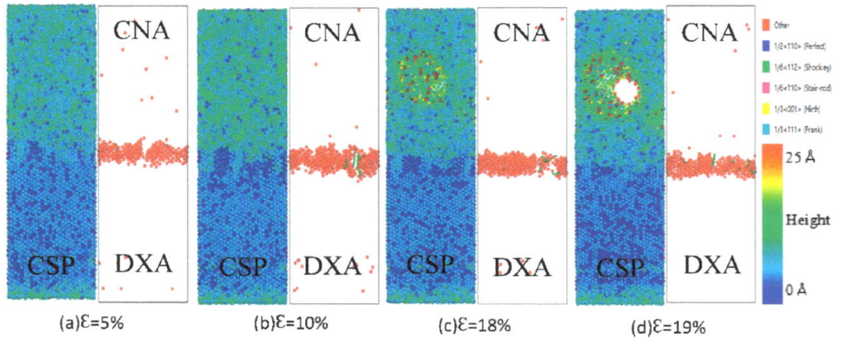

Figure 15. Atomic CSP, CNA, and DXA analysis of the Fe (110)/Cu (110) diffused interface under z-axis tensile process at 500 K.

4. Conclusions

In this research, the diffusion behavior, mechanical properties, and fracture mechanism of Fe/Cu solid–liquid interfaces were calculated by using a molecular dynamics simulation. The main conclusions obtained are as follows:

(1) Regarding the diffusion phenomenon of the Fe/Cu interface, it was observed that the diffusion distance increases with the increase in diffusion temperature and diffusion time. In addition, the diffusion distance of the Cu atoms diffusing into the Fe matrix is obviously less than that of the Fe atoms diffusing into the Cu matrix. The diffusion coefficient and the diffusion distance reach their maximums when the solid–liquid temperature is 1523 K, and the diffusion time is 3 ns.

(2) The diffusion coefficients of the Fe atoms when the temperature and time is 1523 K and 3 ns, respectively, are arranged in the following order: Fe (100) < Fe (110) < Fe (111). The diffusion coefficients of the Cu atoms are arranged in the following order: Cu (110) > Cu (111) > Cu (100).

(3) The yield strength and fracture strain of Fe/Cu bimetallic interface increase with the increase in strain rate and gradually decrease with the increase in tensile temperature. The yield strength of the three orientations can be ranked in the following order: Fe (110)/Cu (110) > Fe (100)/Cu (100) > Fe (111)/Cu (111). The Fe/Cu bimetallic interface shows brittle fracture characteristics during the tensile process, and the Fe (110)/Cu (110) interface shows a certain fracture toughness compared with the other two interface models.

(4) The yield strength of undiffused Fe/Cu bimetallic interface is higher than that of the diffused Fe/Cu interface. The number and length of 1/6<112> Shockley dislocations in the tensile process of the Fe/Cu bimetallic interface decreased after diffusion and the bonding performance of the Fe/Cu bimetallic interface is also reduced. In the undiffused Fe (110)/Cu (110) interface, bimetallic interface fractures occur along the interface, while in the diffused Fe (110)/Cu (110) interface, these occur on the Cu side, away from the interface.

Author Contributions: H.Z.: conceptualization, methodology, software; J.S.: visualization, investigation and supervision; N.G.: investigation; M.W.: investigation, writing, reviewing, and editing. All authors have read and agreed to the published version of the manuscript.

Funding: This research was funded by the Natural Science Foundation of Henan Province, China, under Grant No.: 212300410205.

Institutional Review Board Statement: Not applicable.

Informed Consent Statement: Not applicable.

Data Availability Statement: Some or all data, models, or code generated or used during the study are proprietary or confidential in nature and may only be provided with restrictions.

Conflicts of Interest: The authors declare no conflict of interest. The founders had no role in the design of the study; in the collection, analyses, or interpretation of the data; in the writing of the manuscript, or in the decision to publish the results.

References

1. Lozhkomoev, A.S.; Lerner, M.I.; Pervikov, A.V.; Kazantsev, S.O.; Fomenko, A.N. Development of Fe/Cu and Fe/Ag Bimetallic nanoparticles for promising biodegradable materials with antimicrobial effect. *Nanotechnol. Russ.* **2018**, *13*, 18–25. [CrossRef]
2. Kaczmar, J.W.; Pietrzak, K.; Wlosinski, W. The production and application of metal matrix composite materials. *J. Mater. Process. Technol.* **2000**, *106*, 58–67. [CrossRef]
3. Ghosh, M.; Chatterjee, S. Effect of interface microstructure on the bond strength of the diffusion welded joints between titanium and stainless steel. *Mater. Charact.* **2006**, *54*, 327–337. [CrossRef]
4. Hai, C.B.; Zhang, H.P.; Du, L.Y.; Wang, D.X.; Xing, D.J.; Zhang, M.Z. Nickel/iron based bimetallic MOF-derived nickel ferrite materials for triethylamine sensing. *Cryst. Eng. Comm.* **2020**, *22*, 1286–1293. [CrossRef]
5. Cheng, J.; Zhao, J.H.; Zhang, J.Y.; Guo, Y.; He, K.; Wen, F.L. Microstructure and mechanical properties of galvanized-45 steel/AZ91D bimetallic material by liquid-solid compound casting. *Materials* **2019**, *12*, 1651. [CrossRef]
6. Chen, Y.; Wang, Y. Calculating method for the leakage between slipper and swashplate in spherical swashplate type axial piston pump with conical cylinder. *Adv. Mater. Res.* **2013**, *753–755*, 2736–2741. [CrossRef]
7. Li, Y.; Gao, J.; Xu, N.; Li, P.; Gong, M.; Tong, W. Fabrication of a high chromium cast iron/low carbon steel bimetal: Diffusion behavior and bonding strength. *J. Mater. Eng. Perform.* **2019**, *28*, 6904–6911. [CrossRef]
8. Tan, C.; Zhou, K.; Ma, W.; Min, L. Interfacial characteristic and mechanical performance of maraging steel-copper functional bimetal produced by selective laser melting based hybrid manufacture. *Mater. Des.* **2018**, *155*, 77–85. [CrossRef]
9. Guo, Y.; Donatas, S.; Yoshiaki, K.; Hiroki, M.; Liu, X. A molecular dynamics study on the effect of surfactant adsorption on heat transfer at a solid-liquid interface. *Int. J. Heat Mass Transf.* **2019**, *135*, 115–123. [CrossRef]
10. Debenedetti, P.G.; Chen, M.; Vella, J.R.; Furstenberg, S.; Carter, E.A.; Stillinger, F.H.; Panagiotopoulos, A.Z. Characterization of the liquid-solid Mo (110) interface from classical molecular dynamics for plasma-facing applications. *Nucl. Fusion* **2017**, *57*, 11. [CrossRef]
11. Jesse, L.; Kern, P.R.; Barry, B.B.L. Characterization of the Al-Ga solid-liquid interface using classical and ab initio molecular dynamics simulation. *Phys. Rev. Mater.* **2020**, *4*, 043604. [CrossRef]
12. Hoyt, J.J. Molecular dynamics study of equilibrium concentration profiles and the gradient energy coefficient in Cu-Pb nanodroplets. *Phys. Rev. B. Condens. Matter Mater. Phys.* **2007**, *76*, 094102. [CrossRef]
13. Palafox-Hernandez, J.P.; Laird, B.B. Orientation dependence of heterogeneous nucleation at the Cu-Pb solid-liquid interface. *J. Chem. Phys.* **2016**, *145*, 211914. [CrossRef] [PubMed]
14. Tanaka, Y.; Kajihara, M.; Watanabe, Y. Growth behavior of compound layers during reactive diffusion between solid Cu and liquid Al. *Mater. Sci. Eng. A* **2007**, *445–446*, 355–363. [CrossRef]
15. Tanaka, Y.; Kajihara, M. Numerical analysis for migration of interface between liquid and solid phases during reactive diffusion in the binary Cu-Al system. *Mater. Sci. Eng. A* **2007**, *459*, 101–110. [CrossRef]
16. Scott, W.N.; Rice, B.M. A molecular dynamics study of the role of relative melting temperatures in reactive Ni/Al nanolaminates. *J. Phys. Condens. Matter* **2011**, *23*, 5701. [CrossRef]

17. Yan, R.; Sun, W.; Ma, S.; Jing, T.; Dong, H. The orientation dependence of liquid ordering at a-Al_2O_3/Al solid-liquid interfaces: A molecular dynamics study. *Comput. Mater. Sci.* **2020**, *174*, 109489. [CrossRef]
18. Gan, X.; Xiao, S.; Deng, H.; Sun, X.; Li, X.; Hu, W. Atomistic simulations of the Fe(001)-Li solid-liquid interface. *Fusion Eng. Des.* **2014**, *89*, 2894–2901. [CrossRef]
19. Mao, A.; Zhang, J.; Yao, S.; Wang, A.; Wang, W.; Li, Y.; Qiao, C.; Xie, J.; Jia, Y. The diffusion behaviors at the Cu-Al solid-liquid interface: A molecular dynamics study. *Results Phys.* **2020**, *16*, 102998. [CrossRef]
20. Raman, S.; Hoyt, J.J.; Saidi, P.; Asta, M. Molecular dynamics study of the thermodynamic and kinetic properties of the solid-liquid interface in FeMn. *Comput. Mater. Sci.* **2020**, *182*, 109773. [CrossRef]
21. Liu, L.; Deng, D.; Su, M.; An, M.; Wang, R. Strain rate and temperature effects on tensile behavior of Ti/Al multilayered nanowire: A molecular dynamics study. *Superlattices Microstruct.* **2019**, *135*, 106272. [CrossRef]
22. Nishida, S.; Surblys, D.; Yamaguchi, Y.; Kuroda, K.; Kagawa, M.; Nakajima, T.; Fujimura, H. Molecular dynamics analysis of multiphase interfaces based on in situ extraction of the pressure distribution of a liquid droplet on a solid surface. *J. Chem. Phys.* **2014**, *140*, 074707. [CrossRef] [PubMed]
23. Yan, R.; Sun, W.; Ma, S.; Davidchack, R.L.; Di Pasquale, N.; Hai, Q.; Jing, T.; Dong, H. Structural and mechanical properties of homogeneous solid-liquid interface of Al modelled with COMB3 potential. *Comput. Mater. Sci.* **2018**, *155*, 136–143. [CrossRef]
24. Su, M.; Deng, Q.; An, M.; Liu, L.; Ma, C. Molecular dynamics study of the tensile behaviors of Ti(0 0 0 1) / Ni(1 1 1) multilayered nanowires. *Comput. Mater. Sci.* **2019**, *158*, 149–158. [CrossRef]
25. Liu, H.; Ke, F.; Pan, H.; Zhou, M. Molecular dynamics simulation of the diffusion bonding and tensile behavior of a Cu-Al interface. *Acta Phys. Sin.* **2008**, *56*, 407–412. [CrossRef]
26. Plimpton, S. Fast parallel algorithms for short-range molecular dynamics. *J. Comput. Phys.* **1995**, *117*, 1–19. [CrossRef]
27. Lee, B.J.; Wirth, B.D.; Shim, J.H.; Kwon, J.; Kwon, S.C.; Hong, J.H. Modified embedded-atom method interatomic potential for the Fe-Cu alloy system and cascade simulations on pure Fe and Fe-Cu alloys. *Phys. Rev. Ser. B* **2005**, *71*, 4205. [CrossRef]
28. Guo, Q.; Greer, J.R. Compressive properties of interface-containing Cu-Fe nano-pillars. *Scr. Mater.* **2017**, *66*, 272–275. [CrossRef]
29. Lee, S.G.; Chung, Y.C. Atomic investigation of Fe-Cu magnetic thin films by molecular dynamics simulation. *Jpn. J. Appl. Phys.* **2007**, *46*, 6309–6311. [CrossRef]
30. Lee, S.G.; Chung, Y.C. The early stage of deposition process for Fe-Cu magnetic multilayer systems: Molecular dynamics simulation. *J. Phys. D Appl. Phys.* **2009**, *42*, 135305–135309. [CrossRef]
31. Lee, S.G.; Chung, Y.C. Molecular-dynamics investigation of the surface characteristics of Fe-Cu magnetic thin-film layers. *J. Vac. Sci. Technol. A Vac. Surf. Film.* **2008**, *26*, 1392–1396. [CrossRef]
32. Stukowski, A. Visualization and analysis of atomistic simulation data with OVITO-the open visualization tool. *Model. Simul. Mater. Sci. Eng.* **2010**, *18*, 015012. [CrossRef]
33. Zhang, G.; Kang, Y.; Wang, M.; Xu, H.; Jia, H. Atomic diffusion behavior and diffusion mechanism in Fe–Cu bimetal casting process studied by molecular dynamics simulation and experiment. *Mater. Res. Express* **2020**, *7*, 096519. [CrossRef]
34. Wang, M.; Zhang, G.; Xu, H. Atomistic simulations and experimental investigations of the diffusion behavior of steel/ZCuPb20Sn5 bimetals. *Coatings* **2020**, *10*, 549. [CrossRef]
35. Meis, C.; Fleche, J.L. Study of the solubility limit of oxygen vacancies in TiO_{2-x} using molecular dynamics. *Solid State Ion.* **1997**, *1*, 333–335. [CrossRef]
36. Trong, D.N.; Long, V.C.; Tălu, S. Molecular dynamics simulation of bulk Cu material under various factors. *Appl. Sci.* **2022**, *12*, 4437. [CrossRef]
37. Quoc, T.T.; Long, V.C.; Tălu, S.; Trong, D.N. Molecular dynamics study on the crystallization process of cubic Cu–Au alloy. *Appl. Sci.* **2022**, *12*, 946. [CrossRef]
38. Li, C.; Li, D.; Tao, X.; Chen, H.; Ouyang, Y. Molecular dynamics simulation of diffusion bonding of Al-Cu interface. *Model. Simul. Mater. Sci. Eng.* **2014**, *22*, 1. [CrossRef]
39. Luo, M.; Liang, L.; Lang, L.; Xiao, S.; Hu, W.; Deng, H. Molecular dynamics simulations of the characteristics of Mo/Ti interfaces. *Comput. Mater. Sci.* **2018**, *141*, 293–301. [CrossRef]
40. Wei, J.; Xu, C.; Li, P.; Deng, H.; Xiao, S.; Hu, W. Molecular dynamics simulations of the diffusion characteristics on the Fe-W interfaces system. *Fusion Eng. Des.* **2020**, *159*, 111850. [CrossRef]
41. Chang, L.; Zhou, C.; Liu, H.; Li, J.; He, X. Orientation and strain rate dependent tensile behavior of single crystal titanium nanowires by molecular dynamics simulations. *J. Mater. Sci. Technol.* **2018**, *34*, 864–877. [CrossRef]
42. Honeycutt, J.D.; Andersen, H.C. Molecular dynamics study of melting and freezing of small Lennard-Jones clusters. *J. Phys. Chem.* **1987**, *91*, 4950–4963. [CrossRef]
43. Tukowski, A.; Albe, K. Extracting dislocations and non-dislocation crystal defects from atomistic simulation data. *Model. Simul. Mater. Sci. Eng.* **2010**, *18*, 025016. [CrossRef]
44. Wang, J.; Shi, J.; Lu, Y.; Jin, G.; Wang, J.; Jiang, Y.; Zhou, Q. Deformation evolution of Cu/Ta nanoscale multilayer during nanoindentation by a molecular dynamics study. *Surf. Coat. Technol.* **2022**, *441*, 128562. [CrossRef]

metals

MDPI

Communication

Potentials for Describing Interatomic Interactions in γFe-Mn-C-N System

Bo Lv [1], Chen Chen [2,3,*], Fucheng Zhang [2,3], Gennady M. Poletaev [4] and Roman Y. Rakitin [5]

1 College of Environmental and Chemical Engineering, Yanshan University, Qinhuangdao 066004, China; lvbo@ysu.edu.cn
2 State Key Laboratory of Metastable Materials Science and Technology, Yanshan University, Qinhuangdao 066004, China; zfc@ysu.edu.cn
3 National Engineering Research Center for Equipment and Technology of Cold Strip Rolling, Yanshan University, Qinhuangdao 066004, China
4 Information Technologies Faculty, Altai State Technical University, Lenin Str. 46, 656038 Barnaul, Russia; gmpoletaev@mail.ru
5 College of ASU, Altai State University, Lenin Str. 61, 656049 Barnaul, Russia; movehell@gmail.com
* Correspondence: chenchen@ysu.edu.cn

Abstract: Potentials for describing interatomic interactions in a γFe-Mn-C-N multicomponent system, modified Hadfield steel, where face-centered cubic (f.c.c.) iron is the main component, are proposed. To describe the Fe-Fe interactions in austenite, it is proposed to use Lau EAM potential. For all other interactions, Morse potentials are proposed, the parameters of which were found from various experimental characteristics: in particular, the energy of dissolution and migration of an impurity in an f.c.c. iron crystal, the radius of atoms, their electronegativity, mutual binding energy, etc. The found potentials are intended for modeling the atomic structures and processes occurring at the atomic level in Hadfield steel using relatively large computational cells by the molecular dynamics method.

Keywords: molecular dynamics; interatomic potential; Hadfield steel; potential parameters; impurity

Citation: Lv, B.; Chen, C.; Zhang, F.; Poletaev, G.M.; Rakitin, R.Y. Potentials for Describing Interatomic Interactions in γFe-Mn-C-N System. *Metals* **2022**, *12*, 982. https://doi.org/10.3390/met12060982

Academic Editors: Shuwen Wen, Yongle Sun and Xin Chen

Received: 28 April 2022
Accepted: 2 June 2022
Published: 7 June 2022

1. Introduction

Despite the interest in Hadfield steel and a long history of research on its unique properties, mainly associated with its excellent work hardening ability [1–3], there are very few works devoted to the modeling of its atomic structures and the processes occurring in it under deformation conditions at the atomic level. At the same time, computer modeling can help answer a number of questions regarding the mechanisms and regularities of the deformation behavior of this steel, which relates to the peculiarities of the interaction of dislocations with twins and grain boundaries.

The present work is devoted to the search for potentials for describing interatomic interactions in Hadfield steel for their subsequent application in molecular dynamics simulation. Hadfield steel is a multicomponent system and, in addition to classical iron, manganese, and carbon, it may contain some other alloying elements [1,3]. To describe interatomic interactions even in the three-component system, it is necessary, at least, to specify six potentials (or pair components for many-body potentials) to describe the interactions of different pairs of atoms in a given system: Fe-Fe, Fe-Mn, Mn-Mn, Fe-C, Mn-C, C-C. Unfortunately, there are very few works, within the framework of one approach, where potentials are proposed simultaneously for several of these bonds.

Lee et al., developed modified EAM potentials that take into account two coordination spheres and anisotropy of electron clouds (2N MEAM potentials). Despite the physical validity, the Lee potentials [4–7] have two disadvantages. Firstly, these potentials are relatively very cumbersome. They include at least 11 parameters and additional calculations of the mutual orientation of atoms, which is used to calculate the screening functions of electron clouds. This significantly slows down the calculation of a computer experiment and is critical for models that include a large number of atoms. Secondly, while strictly taking into account many factors that affect interatomic interactions, the 2N MEAM Lee potentials take into account the interaction of atoms no further than only the 2nd coordination sphere. This is not enough according to our previous work [8]. There are only six atoms in the second coordination sphere of an f.c.c. lattice, at a distance of $\sqrt{3/2}a$, where a is the lattice parameter. There are 24 atoms in the third sphere at a distance of $\sqrt{2}a$, i.e., only 0.19a beyond the 2nd coordination sphere. According to the calculations previously made [8], the third coordination sphere contributes about 20% to the calculation of atomic energy depending on the metals, and the second even less, about 15%. Xie et al. [9] proposed a set of Morse pair potentials for describing all the desired interactions, except for the Fe-Mn bond. However, they are not suitable for modeling interactions in Hadfield steel, since they were selected not for steels, but for Mn_7C_3 and Fe_7C_3 carbides. For example, when using one potential for the Fe-Fe bond [9], it was not possible to obtain satisfactory results for the basic characteristics of f.c.c. iron. In the present paper, we propose a set of Morse potentials to describe all interactions in the four-component γFe-Mn-C-N system.

2. Choice of Potential for Describing Interatomic Interactions in F.C.C. Iron

In view of the foregoing, we decided to independently select the potentials for the γFe-Mn-C system, which must satisfy two important criteria: (1) not be mathematically cumbersome, so that relatively large computational cells can be simulated (including 10^5–10^6 atoms); (2) satisfactorily describe the known experimental characteristics of the γFe-Mn-C system. The potential for describing the Fe-Fe bond is a main potential in the γFe-Mn-C system. As a basis, it was decided to take one of the known and tested EAM potentials that well describe the properties of austenite. Most of the known potentials for iron were created to describe its body-centered cubic (b.c.c.) modification, and in the majority of cases, the f.c.c. phase is considered as secondary. As a result, not all known potentials describe austenite well. For this reason, we abandoned the well-known EAM potential of Ackland and Mendelev [10–12].

Of all the potentials that we have examined, the potentials for Fe-Fe and Fe-C bonds in austenite proposed by Lau et al. [13] are best suited for our purposes. These are classic EAM potentials, where the energy of the i-th atom is calculated as the sum of the pair and multiparticle components:

$$\begin{aligned} E_{\alpha,i} &= -A_\alpha \sqrt{\sum_{j\neq i} \rho_{\beta\alpha}(r_{ij})} + \frac{1}{2}\sum_{j\neq i} \phi_{\beta\alpha}(r_{ij}) \\ \rho_{\beta\alpha}(r_{ij}) &= t_1(r - r_{c,\rho})^2 + t_2(r - r_{c,\rho})^3, \quad r \leq r_{c,\rho} \\ \phi_{\beta\alpha}(r_{ij}) &= (r - r_{c,\phi})^2(k_1 + k_2 r + k_3 r^2), \quad r \leq r_{c,\phi} \end{aligned} \quad (1)$$

The potential for the Fe-Fe bond in the article of Lau et al. [13] is a modified potential from the work of Rosato [14]. We drew attention to this potential primarily because it was tested to describe the structural, energy, and elastic characteristics of austenite [15]. The obtained values are in satisfactory agreement with the experiment: the lattice parameter at 0 K is 3.573 Å, the sublimation energy (per atom) is 4.228 eV, and the bulk modulus is 156.7 GPa. In addition, good agreement with the reference value of thermal expansion coefficient of $16{\cdot}10^{-6}$ K^{-1} was obtained.

3. Fe-C, C-C, Fe-Mn, Mn-Mn, and Mn-C Bonds in the Crystal Lattice of F.C.C. Iron

An adjustment was necessary for the Fe-C bond, since such values as the energy of carbon dissolution in f.c.c. iron; the difference between energies of carbon impurity in octahedral and tetrahedral voids of the lattice; the migration energy of a carbon atom in f.c.c. iron; and the binding energy with a vacancy were unsatisfactorily described by potentials proposed by Lau et al. [13]. It was decided for all the other five bonds to build pair Morse potentials that would satisfactorily describe the known characteristics. The Morse potential is a pair potential, but it is often used in molecular dynamics calculations, including the description of interatomic interactions in metals. Pair potentials are often used by various researchers to describe interatomic interactions in metal-impurity systems [9,16–23]. The procedure for calculating forces using many-body and pair potentials in a computer program is fundamentally different, and therefore, the latter (with the same methods of optimizing the computer code) always wins in the speed of calculating forces in the model. The Morse potential determines the interaction energy of a pair of atoms located at a distance r from each other:

$$\varphi(r) = D\beta e^{-\alpha r}\left(\beta e^{-\alpha r} - 2\right) \tag{2}$$

where α, β, D are the potential parameters.

When determining the parameters of the Morse potential for the Fe-C bond, we relied on the already known potentials from [13,24], only slightly correcting them to obtain good agreement with the experimental data and first-principle calculations on the energy of carbon dissolution in f.c.c. iron (E_{sol}); the difference between the energies of carbon impurity in octahedral and tetrahedral voids of the lattice (ΔE_{OT}); the migration energy of a carbon atom in f.c.c. iron (E_m); and the binding energy with a vacancy (E_{bv}). These energy characteristics are listed in Table 1. In addition, the radius of the Fe and C atoms was also taken into account. As can be seen, the potentials proposed here simultaneously well describe the considered energy characteristics of the carbon impurity in austenite, even better than the known EAM potentials for this system [13,24].

Table 1. Energy characteristics of carbon impurity in f.c.c. iron.

Energy Characteristics	Our Model	Experiments or FP-Calc.	Potentials from [24]	Potentials from [13]
E_{sol} (eV/atom)	0.38	0.25–0.48 [25,26] (experiment, FP-calc.)	1.01	0.78
ΔE_{OT} (eV)	1.36	1.48 [27] (FP-calc.)	1.03	1.12
E_m (eV)	1.20	1.40–1.53 [4,28,29] (experiment)	0.33	0.86
E_{bv} (eV)	0.41	0.37–0.41 [30] (experiment)	0.50	0.54

To describe the interaction between carbon atoms in an austenite lattice, we transformed the potentials [16,24] into Morse potentials. The justification given in these works seemed quite reasonable. When searching for parameters for the Mn-Mn bond, we focused on the radius of Mn atoms, the sublimation energy, and the bulk modulus of metallic Mn [31]. The equilibrium distance for the Fe-Mn bond was calculated as the sum of the radii of Fe and Mn atoms [31], the elastic modulus, as the mean of the modules for Fe and Mn. The depth of the potential, i.e., the parameter D in Formula (2), was determined by the value of the mixing energy of Mn in f.c.c. Fe, where it was calculated by first principles (ab initio) method [32]. For a concentration of 13 wt.% Mn, the mixing energy of Mn in Fe was −0.0182 eV/atom [32]. A minus sign indicates that mixing is energetically favorable, and a relatively small value refers that the tendency to order the atomic structure of the alloy is very weak. The parameter D was selected so that this value was exactly obtained.

The last of the six necessary potentials, for the Mn-C bond, was chosen, taking into account the potentials proposed in [5,7,9], the radii of the Mn and C atoms [31], as well as the binding energy between Mn and C atoms in f.c.c. Fe. The potential parameters were selected in such a way as to accurately reproduce the binding energy of Mn and C atoms experimentally obtained in a f.c.c. Fe lattice, 0.35 eV [33]. This energy seems very large and comparable with the binding energy of a carbon atom with a vacancy in iron. However, if we pay attention to the difference in electronegativities of Mn and C atoms compared to Fe and C, 1.00 and 0.72, respectively [31], such a strong bond becomes clear, which is comparable in strength, probably, to an Al-C bond, for which the difference in electronegativities is close to the Mn-C bond and amounts to 0.94 [31].

When calculating the binding energy of a C atom with an Mn atom in a f.c.c. lattice of Fe, the carbon atom was introduced into the octahedral interstitial closest to the Mn atom, as shown in Figure 1. It is known that impurity atoms of light elements (such as C, N, O, etc.) occupy octahedral voids in f.c.c., h.c.p., and b.c.c. lattices of metals [20,34].

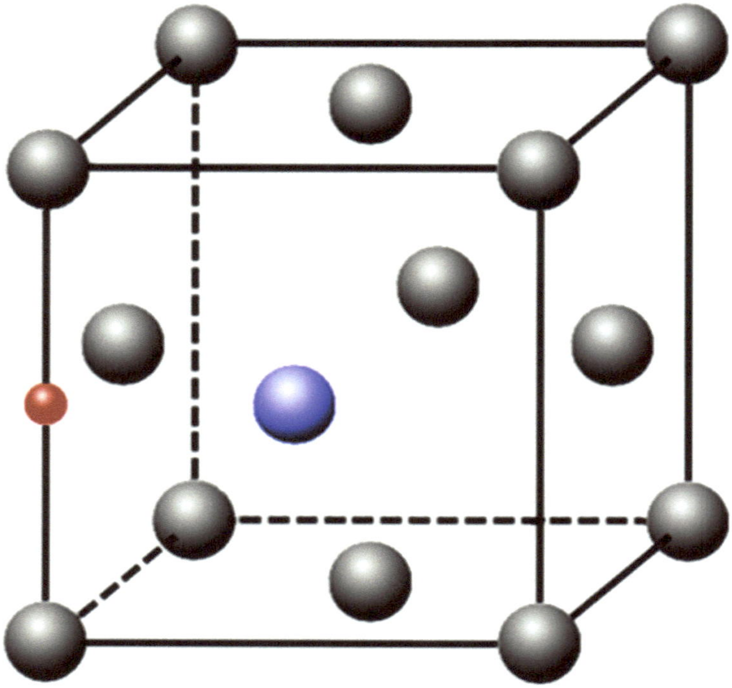

Figure 1. The positions of Mn (blue) and C (red) atoms in a f.c.c. Fe lattice (gray atoms). (The positions of the atoms are shown without taking into account structural relaxation).

4. Interatomic Potentials for Describing Nitrogen Impurity in Hadfield Steel

Nitrogen impurity is not a classic component of Hadfield steel. At the same time, it is sometimes considered as an additional alloying element that stabilizes austenite and increases strength [1]. The radius of N atom is slightly less than that of C atom: 0.71 and 0.77 Å, respectively [34]. The difference between the electronegativities of iron and nitrogen is 1.21, while for iron and carbon, it is 0.72 [31]. The high difference between the electronegativity means a strong interatomic bond, and its character is close to the covalent type.

The dissolution energy of nitrogen in f.c.c. iron is an important characteristic by which the depth of the interatomic potential can be determined. Unfortunately, there is a very large run-off of this value in the literature: −0.13 eV [35], −0.27 eV [36], and −0.53 eV [37]. Another important energy characteristic is the activation energy of migration of an impurity atom in the metal lattice. Based on the analysis of data on the migration energy of C and N impurity atoms in α-Fe and γ-Fe [6,28,29,38,39], we took a benchmark value of migration energy of nitrogen impurity in the f.c.c. iron lattice of 1.29 eV. Therefore, we were guided by the radius of iron and nitrogen atoms, the energy of dissolution, the energy of migration, and the assumed depth of the potential found from the empirical dependencies [20]. Table 2 shows the data by which we were guided when selecting the potentials and the values obtained in our work.

Table 2. Some energy characteristics of nitrogen impurity in f.c.c. iron.

Energy Characteristics	Our Model	Experiment or FP-Calc.
E_{sol} (eV/atom)	−0.13	−0.13 [35] (experiment) −0.27 [36] (ab initio calculations) −0.53 [37] (thermodynamic evaluation)
E_m (eV)	1.29	1.29 [6,28,29,38,39]
E_{bv} (eV)	0.45	–

Mn atoms have a high binding energy with N atoms and a large difference in electronegativities compared to iron atoms, which causes a high effect of Mn impurity on the solubility of nitrogen in iron [40]. When searching for the potential parameters for the Mn-N bond in f.c.c. iron, we took into account the radius of Mn and N atoms, empirical regularities [20], and the data on the solubility of Mn and N in iron [40]. To describe the interactions of nitrogen atoms with each other in metals, the considerations and pair potentials proposed by Vashishta et al. [17] and San et al. [19] were taken as a basis. According to these potentials, impurity atoms, such as nitrogen and oxygen, repel each other in the crystal lattice of metals. To describe the N-N and O-O bonds in metals, the parameters of the Morse potentials were calculated [20]. The radius of atoms and their electronegativity were also taken into account to determine the parameters of the C-N potential. The binding energy of C and N atoms in the metal lattice was calculated using Lorentz–Berthelot mixing rules [41], taking into account the ratio of the C-C bond energies in the metal and graphene, as well as the C-N bond in the organic molecule.

5. Potential Parameters

Table 3 shows the obtained parameters of Morse potentials. Figure 2 illustrates graphs of the found potentials. Table 3 also contains the parameters for the Fe-Fe bond, found from the values of the γ-Fe lattice parameter, sublimation energy, and bulk modulus obtained using the potential from [13]. The Morse potential for the Fe-Fe bond is a simpler analogue of the EAM potential [13] and similarly well describes the above characteristics, including thermal expansion. It can be used in models with a large number of atoms and in lengthy computer experiments requiring large resources. All Morse potentials were found for a cutoff radius of 4.7 Å, i.e., taking into account the three coordination spheres in f.c.c. Fe.

Table 3. Parameters of Morse potentials for the considered interactions.

Bond	α (Å^{-1})	β	D (эB)
Fe-Fe	1.285	35.878	0.433
Fe-C	1.82	41	0.41
C-C	1.97	50	0.65
Mn-Mn	1.321	39.792	0.373
Mn-Fe	1.306	38.030	0.413
Mn-C	1.87	43	0.777
Fe-N	1.788	34.046	0.579
Mn-N	1.812	36.482	0.940
C-N	2.140	58.323	0.230
N-N	1.556	700	0.001

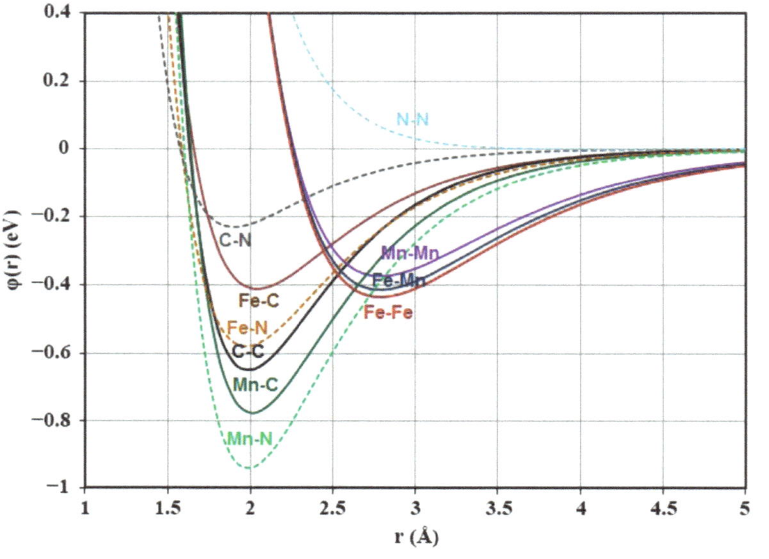

Figure 2. Morse potentials for bonds in the γFe-Mn-C-N system (the potentials describing interactions with a nitrogen impurity are shown by a dotted line).

For the Fe-Fe and Mn-Mn bonds, the classical method by Girifalco and Weiser [42] was used to determine the potential parameters from the crystal lattice parameter, bulk modulus, and sublimation energy, so these characteristics are reproduced quite accurately. For other bonds describing the interaction of C, Mn, and N impurities with atoms of the γ-Fe lattice and with each other, the potential parameters were chosen in such a way as to reproduce, as best as possible, above-mentioned experimental characteristics: in particular, the energy of dissolution and the energy of migration of an impurity in an f.c.c. iron crystal, the radius of atoms, their electronegativity, mutual binding energy, etc.

It should be borne in mind that in complex multicomponent systems such as γFe-Mn-C-N, the bonds of impurity atoms depend on the mutual arrangement of neighboring atoms of different types in the austenite lattice space. However, strict consideration of this in our case, when the system consists of four components, will lead to a significant complication in calculations. It will be not allowed to perform molecular dynamics simulations using relatively large computational cells.

6. Conclusions

Potentials for describing interatomic interactions in a γFe-Mn-C-N multicomponent system, modified Hadfield steel, where f.c.c. iron is the main component, are proposed. To describe the Fe-Fe interactions in austenite, it is proposed to use Lau EAM potential, which reproduces well the structural, energy, and elastic characteristics of austenite. For all other nine interactions, Morse potentials are proposed, the parameters of which were found from various experimental characteristics: in particular, the energy of dissolution and the energy of migration of an impurity in a f.c.c. iron crystal, the radius of atoms, their electronegativity, mutual binding energy, etc. The found potentials are intended for modeling the atomic structure and processes occurring at the atomic level in Hadfield steel using relatively large computational cells by the molecular dynamics method.

Author Contributions: Resources, B.L. and F.Z.; data curation, G.M.P. and R.Y.R.; writing—original draft preparation, B.L.; writing—review and editing, C.C. and G.M.P.; supervision, F.Z.; funding acquisition, B.L. All authors have read and agreed to the published version of the manuscript.

Funding: The authors gratefully acknowledge the National Natural Science Foundation of China (Nos. 52171049, 51831008 and U1810207), the Key Research & Development Program of Hebei Province (No. 19211018D).

Institutional Review Board Statement: Not applicable.

Informed Consent Statement: Not applicable.

Data Availability Statement: No new data were created or analyzed in this study. Data sharing is not applicable to this article.

Conflicts of Interest: The authors declare no conflict of interest.

References

1. Chen, C.; Zhang, F.; Wang, F.; Liu, H.; Yu, B. Effect of N+Cr alloying on the microstructures and tensile properties of Hadfield steel. *Mater. Sci. Eng. A* **2016**, *679*, 95–103. [CrossRef]
2. Zhang, F.C.; Lv, B.; Wang, T.S.; Zheng, C.L.; Zhang, M.; Luo, H.H.; Liu, H.; Xu, A.Y. Explosion hardening of Hadfield steel crossing. *Mater. Sci. Technol.* **2010**, *26*, 223–229. [CrossRef]
3. Chen, C.; Lv, B.; Ma, H.; Sun, D.; Zhang, F. Wear behavior and the corresponding work hardening characteristics of Hadfield steel. *Tribol. Int.* **2018**, *121*, 389–399. [CrossRef]
4. Lee, B.J.; Baskes, M.I.; Kim, H.; Cho, Y.K. Second nearest-neighbor modified embedded atom method potentials for B.C.C. transition metals. *Phys. Rev. B.* **2001**, *64*, 184102. [CrossRef]
5. Lee, B.-J.; Lee, J.W. A modified embedded atom method interatomic potential for carbon. *Calphad* **2005**, *29*, 7–16. [CrossRef]
6. Lee, B.-J. A modified embedded-atom method interatomic potential for the Fe–C system. *Acta Mater.* **2006**, *54*, 701–711. [CrossRef]
7. Kim, Y.-M.; Shin, Y.-H.; Lee, B.-J. Modified embedded-atom method interatomic potentials for pure Mn and the Fe–Mn system. *Acta Mater.* **2009**, *57*, 474–482. [CrossRef]
8. Poletaev, G.M.; Starostenkov, M.D.; Zorya, I.V. Morse potentials for F.C.C. metals taking into account the interaction of the five coordination spheres. *Fundam. Probl. Sovrem. Materialoved.* **2017**, *14*, 70–75.
9. Xie, J.-Y.; Chen, N.-X.; Shen, J.; Teng, L.; Seetharaman, S. Atomistic study on the structure and thermodynamic properties of Cr7C3, Mn7C3, Fe7C3. *Acta Mater.* **2005**, *53*, 2727–2732. [CrossRef]
10. Numakura, H.; Yotsui, G.; Koiwa, M. Calculation of the strength of Snoek relaxation in dilute ternary B.C.C. alloys. *Acta Met. Mater.* **1995**, *43*, 705–714. [CrossRef]
11. Mendelev, M.I.; Han, S.; Srolovitz, D.J.; Ackland, G.J.; Sun, D.Y.; Asta, M. Development of new interatomic potentials appropriate for crystalline and liquid iron. *Philos. Mag.* **2003**, *83*, 3977–3994. [CrossRef]
12. Ackland, G.J.; Mendelev, M.I.; Srolovitz, D.J.; Han, S.; Barashev, A.V. Development of an interatomic potential for phosphorus impurities in α-iron. *J. Phys. Condens. Matter.* **2004**, *16*, S2629. [CrossRef]
13. Lau, T.T.; Först, C.J.; Lin, X.; Gale, J.D.; Yip, S.; van Vliet, K.J. Many-Body Potential for Point Defect Clusters in Fe-C Alloys. *Phys. Rev. Lett.* **2007**, *98*, 215501. [CrossRef] [PubMed]
14. Rosato, V. Comparative behavior of carbon in B.C.C. and F.C.C. iron. *Acta Met.* **1989**, *37*, 2759–2763. [CrossRef]
15. Oila, A.; Bull, S.J. Atomistic simulation of Fe-C austenite. *Comp. Mater. Sci.* **2009**, *45*, 235–239. [CrossRef]
16. Ruda, M.; Farkas, D.; Abriata, J. Interatomic potentials for carbon interstitials in metals and intermetallics. *Scr. Mater.* **2002**, *46*, 349–355. [CrossRef]
17. Vashishta, P.; Kalia, R.K.; Nakano, A.; Rino, J.P. Interaction potentials for alumina and molecular dynamics simulations of amorphous and liquid alumina. *J. Appl. Phys.* **2008**, *103*, 083504. [CrossRef]
18. Liu, S.; Shi, S.; Huang, H.; Woo, C. Interatomic potentials and atomistic calculations of some metal hydride systems. *J. Alloy. Compd.* **2001**, *330–332*, 64–69. [CrossRef]
19. San Miguel, M.A.; Sanz, J.F. Molecular-dynamics simulations of liquid aluminum oxide. *Physical Review B.* **1998**, *58*, 2369–2371. [CrossRef]
20. Poletaev, G.M.; Zorya, I.V.; Rakitin, R.Y.; Iliina, M.A. Interatomic potentials for describing impurity atoms of light elements in F.C.C. metals. *Mater. Phys. Mech.* **2019**, *42*, 380–388.
21. Poletaev, G.M.; Starostenkov, M.D.; Dmitriev, S.V. Interatomic potentials in the systems Pd-H and Ni-H. *Mater. Phys. Mech.* **2016**, *27*, 53–59.
22. Poletaev, G.M.; Zorya, I.V. Effect of Light Element Impurities on the Edge Dislocation Glide in Nickel and Silver: Molecular Dynamics Simulation. *J. Exp. Theor. Phys.* **2020**, *131*, 432–436. [CrossRef]
23. Poletaev, G.M.; Zorya, I.V. Influence of light impurities on the crystal-melt interface velocity in Ni and Ag. Molecular dy-namics simulation. *Tech. Phys. Lett.* **2020**, *46*, 575–578. [CrossRef]
24. Ruda, M.; Farkas, D.; Garcia, G. Atomistic simulations in the Fe–C system. *Comput. Mater. Sci.* **2009**, *45*, 550–560. [CrossRef]
25. Gustafson, P. A thermodynamic evaluation of the Fe-C System. Scand. *J. Metall.* **1985**, *14*, 259–267.
26. Ridnyi, Y.M.; Mirzoev, A.A.; Mirzaev, D.A. Ab initio simulation of dissolution energy and carbon activity in F.C.C. Fe. *Phys. Solid State* **2017**, *59*, 1279–1284. [CrossRef]

27. Jiang, D.E.; Carter, E.A. Carbon dissolution and diffusion in ferrite and austenite from first principles. *Phys. Rev. B* **2003**, *67*, 214103. [CrossRef]
28. Askill, J. *Tracer Diffusion Data for Metals, Alloys and Simple Oxides*; Springer: Berlin/Heidelberg, Germany, 1970.
29. Brandes, E.A.; Brook, G.B. *Smithells Metals Reference Book*; Butterworth-Heinemann: Oxford, UK, 1992.
30. Slane, J.; Wolverton, C.; Gibala, R. Carbon–vacancy interactions in austenitic alloys. *Mater. Sci. Eng. A* **2003**, *370*, 67–72. [CrossRef]
31. Lide, D.R. (Ed.) *CRC Handbook of Chemistry and Physics (Internet Version 2005)*; CRC Press: Boca Raton, FL, USA, 2005.
32. Mirzoev, A.A.; Yalalov, M.M.; Mirzaev, D.A. Energy of mixing and magnetic state of components of Fe-Mn alloys: A first-principles calculation for the ground state. *Phys. Met. Met.* **2006**, *101*, 341–348. [CrossRef]
33. Massardier, V.; Merlin, J.; Le Patézour, E.; Soler, M. Mn-C interaction in Fe-C-Mn steels: Study by thermoelectric power and internal friction. *Met. Mater. Trans. A* **2005**, *36*, 1745–1755. [CrossRef]
34. Goldschmidt, H.J. *Interstitial Alloys*; Butterworths: London, UK, 1967.
35. Mogutnov, B.M.; Tomilin, I.A.; Shvartsman, L.A. *Thermodynamics of Iron Alloys*; Metallurgiya: Moscow, Russia, 1984. (In Russian)
36. Ridniy, Y.M.; Mirzoev, A.A.; Mirzaev, D.A. Ab initio modeling the energy of dissolution of nitrogen in F.C.C. lattice of iron. *Vestn. SUSU. Seriya Metall.* **2014**, *14*, 59–63. (In Russian)
37. Frisk, K. A Thermodynamic evaluation of the Cr-N, Fe-N, Mo-N and Cr-Mo-N systems. *CALPHAD* **1991**, *15*, 79–106. [CrossRef]
38. Rochana, P.; Lee, K.; Wilcox, J. Nitrogen adsorption, dissociation, and subsurface diffusion on the vanadium (110) surface: A DFT study for the nitrogen-selective catalytic membrane application. *J. Phys. Chem. C.* **2014**, *118*, 4238–4249. [CrossRef]
39. Domain, C.; Becquart, C.S.; Foct, J. Ab initio study of foreign interstitial atom (C, N) interactions with intrinsic point defects in α-Fe. *Phys. Rev. B.* **2004**, *69*, 144112. [CrossRef]
40. You, Z.; Paek, M.-K.; Jung, I.-H. Critical Evaluation and Optimization of the Fe-N, Mn-N and Fe-Mn-N Systems. *J. Phase Equilibria Diffus.* **2018**, *39*, 650–677. [CrossRef]
41. Yin, X.; Opara, A.; Du, H.; Miller, J.D. Molecular dynamics simulations of metal-cyanide complexes: Fundamental considera-tions in gold hydrometallurgy. *Hydrometallurgy* **2011**, *106*, 64–70. [CrossRef]
42. Girifalco, L.A.; Weiser, V.G. Application of the Morse potential function to cubic metals. *Phys. Rev.* **1959**, *114*, 687–790. [CrossRef]

metals

MDPI

Article

Modification of Precipitate Coarsening Kinetics by Intragranular Nanoparticles—A Phase Field Study

Simbarashe Fashu [1], Binting Huang [1,2] and Nan Wang [1,2,3,*]

[1] Guangdong Technion-Israel Institute of Technology, #241 Daxue Rd, Shantou 515063, China; sfashu04@gmail.com (S.F.); huang.binting@gtiit.edu.cn (B.H.)
[2] Technion-Israel Institute of Technology, Technion City, Haifa 32000, Israel
[3] Guangdong Provincial Key Laboratory of Materials and Technologies for Energy Conversion, Guangdong Technion-Israel Institute of Technology, Shantou 515063, China
* Correspondence: nan.wang@gtiit.edu.cn; Tel.: +86-0752-88077177

Abstract: Precipitate coarsening is a major mechanism responsible for the degradation in mechanical properties of many precipitation-hardened alloys at high temperatures. With recent developments in processing of nanocomposite materials, a substantial volume fraction of inert second phase ceramic nanoparticles can be introduced into the grain interiors of polycrystalline materials. These intragranular nanoparticles can have synergistic effects of impeding dislocation motion and interacting with coarsening precipitates to modify the coarsening rate. In this work, the precipitate coarsening behavior of an alloy in the presence of intragranular inert nanoparticles was studied using the phase field method. Two key measurements of coarsening kinetics, precipitate size distribution and coarsening rate, were found to be affected by the volume fraction and the size of nanoparticles. Two novel mechanisms related to geometric constraints imposed by inter-nanoparticle distance and the blockage of solute diffusion path by nanoparticle–matrix interfaces were proposed to explain the observed changes in precipitate coarsening kinetics. The simulation results in general suggest that the use of small nanoparticles with large number density is effective in slowing down the coarsening kinetics.

Keywords: precipitates; coarsening; nanoparticles; phase field modeling; high temperature strength

Citation: Fashu, S.; Huang, B.; Wang, N. Modification of Precipitate Coarsening Kinetics by Intragranular Nanoparticles—A Phase Field Study. *Metals* **2022**, *12*, 892. https://doi.org/10.3390/met12060892

Academic Editors: Shuwen Wen, Yongle Sun, Xin Chen and John D. Clayton

Received: 12 April 2022
Accepted: 12 May 2022
Published: 24 May 2022

1. Introduction

High performance alloys, including Al, Ni, and Co–based superalloys are strengthened by the heat treatment process to produce nano-sized, hard, and coherent precipitates through hindering of both dislocation movement and grain growth. However, at elevated temperatures, rapid increase in diffusion rate accelerates precipitate coarsening rate and results in the loss of mechanical strength. Precipitate coarsening refers to growth of larger particles at the expense of smaller shrinking particles in the matrix. Coarsening occurs when finely dispersed nanoprecipitates change their sizes, shapes, and distribution due to microstructural evolution driven by reduction in interfacial energy [1]. The kinetics of precipitate coarsening influence the strength of alloys such that this area continues to attract the interests of many researchers. Techniques like solute segregation at precipitate interfaces in Al alloys [2–6], formation of stacking fault ribbons in Ni alloys [7–9], introduction of slow diffusing elements in superalloys [10–14], and application of external compressive stresses [15,16] were proposed as measures to retard the precipitate coarsening process. The mechanisms responsible for precipitate coarsening resistance are thermodynamic and kinetic in nature and involve interfacial energy reduction, solute drag against precipitate growth, development of strain fields around growing precipitates, slowing down of solute diffusion [17], etc. Through proper incorporation of these effects during processing, the precipitate coarsening behavior can be forced to deviate significantly from the normal coarsening, described by the classical Lifshitz–Slyozov–Wagner (LSW) theory [18].

The practice of dispersing fine ceramic nanoparticles like titanium carbon nitride (TiCN), alumina (Al_2O_3), zirconia (ZrO_2), and yttria (Y_2O_3) for the synergistic purposes of retarding solidification coarsening [19–21] and pinning grain growth [22–27] at elevated service temperatures is now established. During solidification, the introduced mobile second–phase particles migrate and form assemblies around the established nuclei and retard dendritic growth through combined pushing and drag effects. Retarded grain growth causes many nuclei to form in the melt, thereby refining the microstructure. Due to slow solidification velocities found in conventional casting processes, the majority of the nanoparticles are incorporated along grain boundaries (due to particle pushing) and become intergranular particles in the final microstructure, and only a few are engulfed by the moving solidification front to become intragranular in the final microstructure. However, with small sized nanoparticles and high cooling rates such as those encountered in laser additive manufacturing, it is possible for a large volume fraction of nanoparticles to be engulfed into the grains owing to fast moving solidification fronts which are larger than the critical velocity required for engulfment [28]. This technique was recently applied to produce an aluminum metal containing a high content of uniformly dispersed TiCN nanoparticles of a volume fraction of 35% using selective laser melting method [28]. Ceramic nanoparticles can also be introduced in grain interiors during grain growth and recrystallization [29], and the presence of second–phase inert particles in grain interiors during recrystallization was also predicted using 3D simulations [27]. Several studies have suggested that the engulfed nanoparticles inside the grains are more effective in strengthening the alloy through interaction with dislocations [19,20] when compared with those segregated on the grain boundaries. It should be noted that in the presence of nano-ceramic particles of lower thermal conductivity like Al_2O_3, SiO_2, SiC, and ZrO_2, it was observed [19,20] that during solidification cooling process the temperature of the particles is higher than that of the melt, and nucleation does not occur on the particles surface.

Many researchers to date have focused on the influence of nanoparticles on grain growth control, however at high temperatures, both grain growth and precipitate coarsening control are of considerable importance. The strengthening mechanism in conventionally solidified aluminum alloy containing 0.2–0.3 vol% of small sized coherent Al_3Sc precipitates and 30 vol% incoherent grain boundary Al_2O_3 ceramic particles (300 nm) was comprehensively studied by Karnesky et al. [26]. The results demonstrated synergistic strengthening due to dislocation pinning at the departure side of Al_2O_3 particles and back strain effects imposed on dislocations by the Al_3Sc precipitates due to lattice mismatch, but the coarsening kinetics of precipitates was not affected by the presence of intergranular ceramic particles. Although the presence of nanoparticles in grain interiors has been experimentally confirmed, according to our knowledge, their influence on precipitate coarsening kinetics has not yet been comprehensively studied. Since detailed experimental study of the influence of grain interior nanoparticles is not available, our present work begins by attempting to provide some insights, and this will guide future work.

The phase field method is a powerful continuum modelling method now extensively employed to simulate microstructural evolutions through characterizing phases using a set of conserved and non-conserved order parameters [30]. It has been used to model complex interface evolutions in a wide variety of materials by describing interfaces using smooth order parameters [30–33]. The coarsening kinetics of precipitates has been studied using the phase field method in many previous works [17,34–36], and the validity of the method has been established through experimental comparisons [17,37]. In this work, the effect of inert nanoparticles on precipitate coarsening kinetics is studied by extending the coarsening model [34] to include inert second–phase particles.

The article is arranged as follows: after the introduction section, the phase field model used in our simulations is formulated in the next section, then simulation results for the influence of various nanoparticle volume fractions and sizes on the coarsening kinetics are presented, together with discussions on potential mechanisms related to these results. This is then followed by conclusions.

2. Phase Field Model

In comparison with macroscopic models relying on thermodynamic equilibrium calculations, the phase field model consists of kinetic equations capable of predicting realistic mesoscale morphological and microstructural evolution rather than merely relying on final states, which is the case in macroscopic models. The phase field method was initially developed for simulating solidification microstructures [33] and solid-state phase transformation processes [30,31].

Coarsening of precipitates is satisfactorily explained by the Lifshitz–Slyozov–Wagner (LSW) theory in which large particles grow at the expense of small particles driven by interfacial energy reduction. In the classic LSW theory, the precipitate coarsening process occurs as the result of solute diffusion and the Gibbs–Thomson effect. The phase field method captures these two fundamental mechanisms by incorporating solute diffusions using the Cahn–Hilliard type generalized diffusion equation and the Gibbs–Thomson effect using an order parameter equation for precipitate–matrix interface evolution. Validity of the original phase field model from which the current work was developed has been previously verified by comparing experimental results with simulation results [35,38].

The phase field model used in this work is an extension of the previous coarsening model for binary alloys [34]. The free energy functional of the system is expressed using the following volume integral:

$$F = \int \left[f_0(C, \eta_i) + \frac{k_c}{2}(\nabla C)^2 + \sum_{i=1}^{4} \frac{k_\eta}{2}(\nabla \eta_i)^2 \right] dv, \tag{1}$$

where $C(r)$ is the concentration field which is equal to the equilibrium concentration C_α, in the host matrix, and, in the precipitate phase, it is set to the equilibrium concentration of the precipitate phase C_β, v is the system volume, η_i is the phase field parameter for tracking precipitate surface evolution of the grain with ith lattice orientation, and k_c and k_η are constant coefficients. The local term $f_0(C, \eta_i)$ is

$$f_0 = H\left[f_1(C) + \sum_{i=1}^{4} f_2(C, \eta_i) + \sum_{i=1}^{4}\sum_{j\neq i} f_3(\eta_i, \eta_j) + \phi \sum_{i=1}^{4} (\eta_{nano})^2 (\eta_i)^2 \right] \tag{2}$$

where,

$$f_1(C) = -\frac{A}{2}(C - C_m)^2 + \frac{B}{4}(C - C_m)^4 + \frac{D_\alpha}{4}(C - C_\alpha)^4 + \frac{D_\beta}{4}(C - C_\beta)^4 \tag{3}$$

$$f_2(C, \eta_i) = -\frac{\gamma}{2}(C - C_\alpha)^2(\eta_i)^2 + \frac{\delta}{4}(\eta_i)^4 \tag{4}$$

$$f_3(\eta_i, \eta_j) = \frac{\varepsilon_{ij}}{2}(\eta_i)^2(\eta_j)^2 \tag{5}$$

The terms $C_m = \left(\frac{C_\alpha + C_\beta}{2}\right)$, A, B, D_α, D_β describe the concentration dependence of the thermodynamic free energy, γ and δ are used to describe the interfacial energy of precipitate matrix interfaces, η_{nano} is the nanoparticle phase field parameter, η_j is the phase field parameter for tracking precipitate surface evolution of the grain with jth lattice orientation, and ε_{ij} describes the interfacial energy between precipitates with different lattice orientations [34]. H is a dimensional constant with the unit of energy per volume which was introduced to quantify the total free energy of the system. The inert nanoparticles were introduced into the model through the phase field parameter, η_{nano}, which is equal to 1 inside the nanoparticle and 0 outside. The interfacial energy between nanoparticles and precipitates was incorporated in the model through the term $\phi \sum_{i=1}^{4} (\eta_{nano})^2 (\eta_i)^2$, with ϕ being the interaction parameter term. This nanoparticle–precipitate interfacial term has the same form as the precipitate interfacial term $f_3(\eta_i, \eta_j)$ and the inert nanoparticle was treated as a non-evolving precipitate phase, as will be seen in the evolution equations below. The free energy functional above does not include the elastic energy associated with coherent interfaces and transformation induced volume change. Evolutions of the concentration and precipitate surfaces were studied using the standard variational dynamics for conserved (Equation (6)) and non-conserved (Equation (7)) fields:

$$\frac{\partial C(r,t)}{\partial t} = \nabla \cdot \left[M_C \nabla \left(\frac{\delta F}{\delta C} \right) \right], \tag{6}$$

$$\frac{\partial \eta_i(r,t)}{\partial t} = -L_i \frac{\delta F}{\delta \eta_i}, \; i = 1,2,3,4, \tag{7}$$

where, L_i and M_C are kinetic coefficients, t is the time, and F is the free energy functional which is given in Equation (1). Since there is no solute diffusion inside inert nanoparticles, the concentration kinetic coefficient was chosen to be $M_C = Mg(\eta_{nano})$ where $g(\eta_{nano}) = \left(1 - \left(\eta_{nano}^3\right) * \left(6\eta_{nano}^2 - 15\eta_{nano} + 10\right)\right)$, M is the solute mobility in the matrix (the solute mobility in the precipitate is assumed to be the same as in the matrix) such that the mobility M_C vanishes inside the nanoparticle and varies smoothly between either the nanoparticle–precipitate or nanoparticle–matrix interface. For simplicity, the kinetic coefficients of all the phase field parameters were set to be the same, i.e., $L_i = L$. One may notice that there are only four phase field parameters for precipitate surfaces appearing in the formulation. This choice is solely made for improving computational efficiency. It is acceptable when the precipitate volume fraction is low, but, at high precipitate volume fraction, unphysical precipitate coalescence may occur, since, by using only four phase field parameters, all the precipitates are assumed to have only 8 possible different lattice orientations ($\eta = \pm 1$). It was found from simulation tests that using 4 phase field parameters maintained the simulation efficiency while avoiding significant coalescence problems at the moderate precipitate volume fraction in our simulations.

Numerical implementation of the model was based on non-dimensionalized forms of Equations (8) and (9) as follows:

$$\frac{\partial C(r^*,t^*)}{\partial t^*} = \nabla^* \cdot \left[M^* g(\eta_{nano}) \nabla^* \left(\frac{\partial f_0^*}{\partial C} - k_C^* (\nabla^*)^2 C \right) \right] \tag{8}$$

$$\frac{\partial \eta_i(r^*,t^*)}{\partial t^*} = -\left(\frac{\partial f_0^*}{\partial \eta_i} - k_\eta^* (\nabla^*)^2 \eta_i \right) \tag{9}$$

where, dimensionless time $t^* = LtH$, particle radius $r^* = \frac{r}{l}$, $k_\eta^* = \frac{k_\eta}{(l^2 H)}$, $k_c^* = \frac{k_c}{(l^2 H)}$, $f^* = \frac{f}{H}$, $\nabla^* = l\nabla$ and the relative kinetic coefficient $M^* = \frac{M}{Ll^2}$. Here, $l = \sqrt{k_\eta / H}$ is the interface width. The mobility M is related to solute diffusivity D [39] and the choice of L does not significantly affect the coarsening result [31].

A single grain in the illustrated microstructure in Figure 1 was set to be the study domain where the coarsening of existing precipitates occurs. Since the purpose of the current study was to explore the general precipitate coarsening behavior in the presence of inert nanoparticles, only 2-dimensional (2D) simulations were performed, but it can be easily extended to 3 dimensions (3D). The nanoparticle volume fraction in this 2D case was equivalent to the area fraction such that in this article these two terms are used interchangeably. A square simulation box with a grid size of 1000 × 1000 and spatial discretization of $\frac{h}{l} = 1$ was used. The simulation time step was set to be $dt^* = 0.002$. The following model parameters $C_\alpha = 0.05$, $C_\beta = 0.95$, $C_m = 0.5$, A = 2.0, B = 9.8, $\gamma = 2.5$, $D_\alpha = D_\beta = 2.5$, $\delta = 1.0$, $\varepsilon_{ij} = 3.0$, $k_i^* = k_c^* = 3.0$ were adapted from ref. [34]. For simplicity, the kinetic coefficient M^* was set to have the same value equal to 1 in both the matrix and the precipitate. It has been shown that selecting M^* within the range of 1 to 100 does not significantly affect the numerical result [39]. Initially, a large number of small nuclei with concentration of $C = 0.95$ were randomly seeded into the system following an approach introduced by Simmons et al. [40]. The initial matrix concentration was set to be 0.22

which was enough for the precipitate volume fraction to reach about 20%. This precipitate volume fraction was small such that it was possible to vary the volume fraction of the inert nanoparticles within a large range. No further nucleation events were considered except those given as the initial nuclei. Equations (6) and (7) were solved numerically using center difference spatial discretization and forward Euler time integration. Periodic boundary conditions were applied to all the simulations. To properly capture the coarsening statistics, a simulation setup was repeated three times and coarsening statistics were averaged over the three simulations. For convenience, the volume fraction, equivalent average radii and particle size distributions of precipitates were post-processed from raw data with the aid of Image J software.

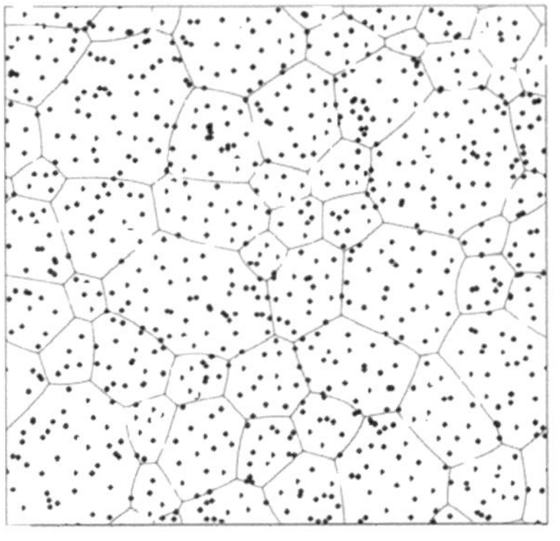

Figure 1. A schematic microstructure showing grains with randomly dispersed nanoparticles in both grain interiors and grain boundaries. Solid lines are grain boundaries, and dark dots are nanoparticles.

3. Results and Discussion

3.1. Coarsening Behavior in Absence of Nanoparticles

To evaluate the model, the precipitate coarsening process without inert nanoparticles was simulated first. Figure 2 shows the coarsening behavior of precipitates in the absence of nanoparticles at aging times of t^* = 2000, 12,000 and 24,000. The red particles are precipitates and the blue regions are the matrix phase. The spherical precipitate shape is the result of the simplified model formulation where surface energy anisotropy and elastic energy are both ignored [34]. The precipitate number declines from more than 300 at a time step of t^* = 2000 to less than 100 at a time step of t^* = 24,000, whilst the average precipitate size increases. Here, coarsening is mainly controlled by inter-precipitate diffusion through the matrix phase driven by free energy differences between small and large precipitates due to the curvature effect, and this results in simultaneous shrinkage and growth of precipitates. The coarsening kinetics was studied by plotting the cube of the average precipitate radius versus aging time according to the power law (i.e., $\overline{R}^n \propto t$) whilst fixing the value of the exponent n to 3. The precipitate radius is obtained from the relation $Radius = \sqrt{A/\pi}$ where A is the precipitate area. We adopted the cubic growth law, since it is the standard exponent from the LSW theory and was reported by many researchers in literature. Such an approach makes it easy to compare our results with previous simulation and experimental results. Besides a brief introduction in Section 2, detailed background of the LSW theory can be found elsewhere [1,18,41]. The plot of the cube of average precipitate size ($\overline{R}^3$) with coarsening time step (t) is given in Figure 3a. In the absence of nanoparticles, the cube of the particle size evolves linearly with time following the power growth law with the coarsening exponent of n = 3. The coarsening behavior satisfies the classical LSW theory [18] and this verifies that the process is controlled by bulk matrix diffusion. The variation of the precipitate area fraction with aging time step in absence of nanoparticles is shown in Figure 3b; it is evident that with time step, the area fraction rapidly decreases before reaching a steady state. The sluggish behavior observed in the transient stage before the steady state condition is due to the seeding method used to introduce precipitates in the system. Figure 3c show the changes in a number of precipitates during coarsening and reflects that the numbers are continuously decreasing due to simultaneous growth and shrinkage of precipitates.

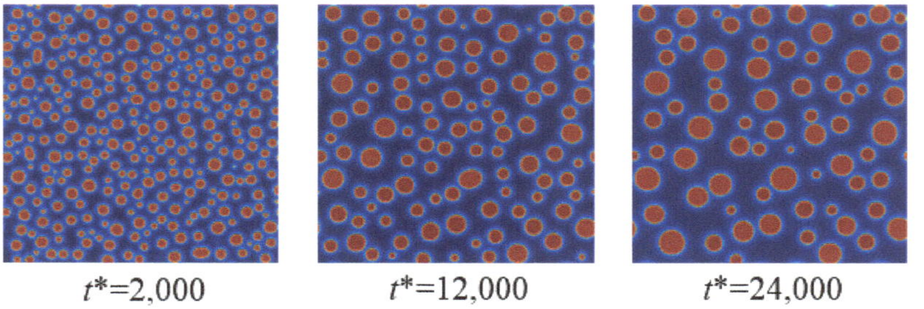

Figure 2. Simulation snapshots of the precipitate evolution in absence of nanoparticles at different times. The precipitates are colored red, and the matrix is in dark blue.

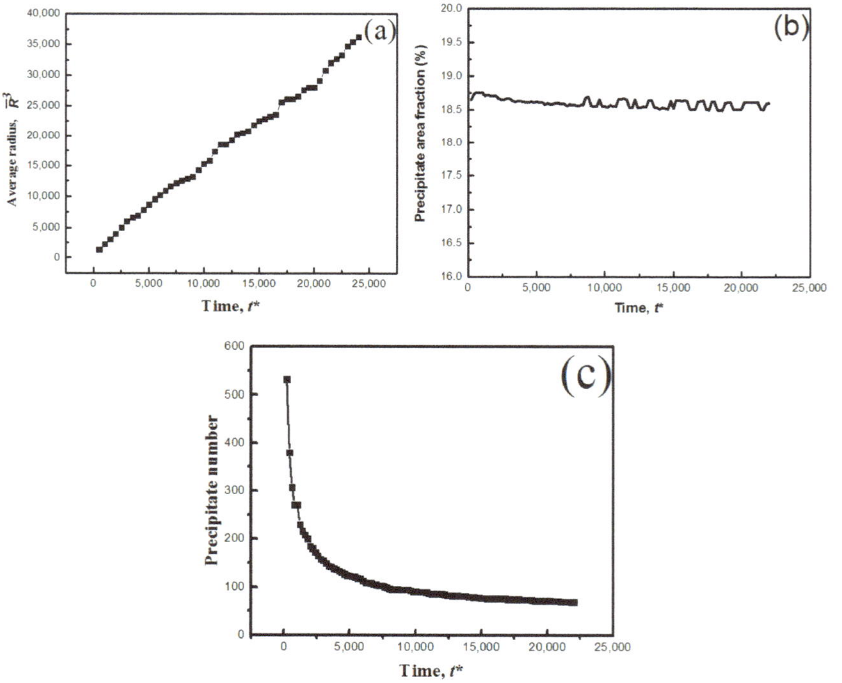

Figure 3. (**a**) Plots of the cube of the average precipitate radius with time in absence of nanoparticles, (**b**) the change in precipitate area fraction with time in absence of nanoparticles, and (**c**) the change of precipitate numbers with time in absence of nanoparticles in the system.

The LSW theory shows that regardless of the system initial particle size distribution, the particle size distribution (PSD) will always converge towards self-similarity with time [42]. The self-similarity can be examined by either calculating the structural function to verify dynamical scaling [36] or through analyzing the PSD shape that changes with time step. In this work, the latter was used to verify the self-similarity of our system. The normalized PSD shapes with fitted curves for precipitate coarsening at selected time steps in absence of nanoparticles are shown in Figure 4a–d. Data for plotting histograms was extracted from coarsening results using Image J software before plotting histograms Microsoft Excel. The peak frequency of the PSD gradually decreases and shifts to the right with time due to coarsening and this is consistent with normal coarsening behavior [36]. The figure also shows that the size distribution range becomes broader with an increase in time, as expected in a normal coarsening process. Observations show that the PSD did not significantly change in shape with time step and this indicates that, for the time steps considered, a scaling behavior during coarsening has already been reached.

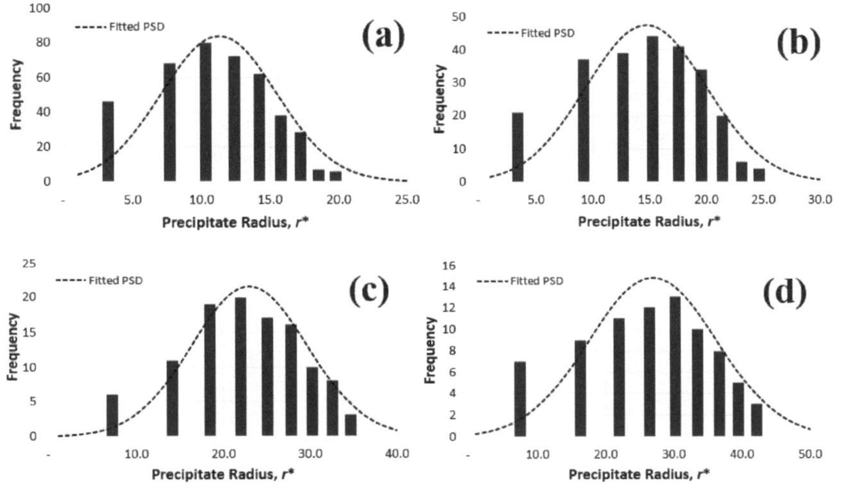

Figure 4. Histograms of the PSD in absence of nanoparticles at different times t^* of: (**a**) 1000, (**b**) 2500, (**c**) 10,000, and (**d**) 20,000.

To validate the model, our simulation results for precipitate coarsening in absence of nanoparticles were compared with results by Wu et al. [43], Wang et al. [44], and those from Zhu et al. [36] for coarsening of binary Ni Al alloys. The PSD shapes and curves of $\overline{R}^3$ *vs t* from our simulation results (Figures 3a and 4a–c) are similar with those from these previous results, i.e., they both obey the linear cube growth law and have similar unimodal skewed PSD shapes. Quantitative comparisons are necessary in the future to further clarify this aspect.

3.2. Coarsening Behavior in Presence of Nanoparticles

To understand the effect of nanoparticles on the precipitate coarsening process, inert nanoparticle phases with volume fractions of 10%, 20%, and 30% were introduced into the domain. A previous study on grain boundary pinning [24] showed that the size distribution of nanoparticles is not important in controlling grain growth. Thus, for a given nanoparticle volume fraction, mono-sized spherical nanoparticles with three radii sizes, r^*, of 10, 20 and 30 were used. All the nanoparticles were randomly distributed in the system. It should be noted that introduction of nanoparticles reduced the total solute content within the matrix, and, therefore, the volume fraction of the precipitate phase in the coarsening stage is also reduced.

Figure 5a–c shows the evolution of the precipitates with different nanoparticle volume fractions at a fixed nanoparticle size of $r^* = 20$. The relationships of the cube of the average precipitate size with the coarsening time under different volume fractions of nanoparticles, including zero volume fraction, are merged for comparison and plotted in Figure 6a. At a given time, the average precipitate size decreases with increasing nanoparticle volume fractions. Similar simulations with different nanoparticle sizes are also shown in Figure 6b,c.

In general, the average precipitate size curves deviate from the linear cubic growth law and become more nonlinear at higher volume fractions of nanoparticles. This behavior is expected, since low nanoparticle volume fraction systems are closer to the classic LSW coarsening theory, and one should expect a larger deviation of the average precipitate size from the LSW theory as more nanoparticles are added into the system. One may also notice that, at the same nanoparticle volume fraction, the deviation from linear LSW coarsening rate is more significant for smaller nanoparticles. This particle size effect will be discussed in a later section.

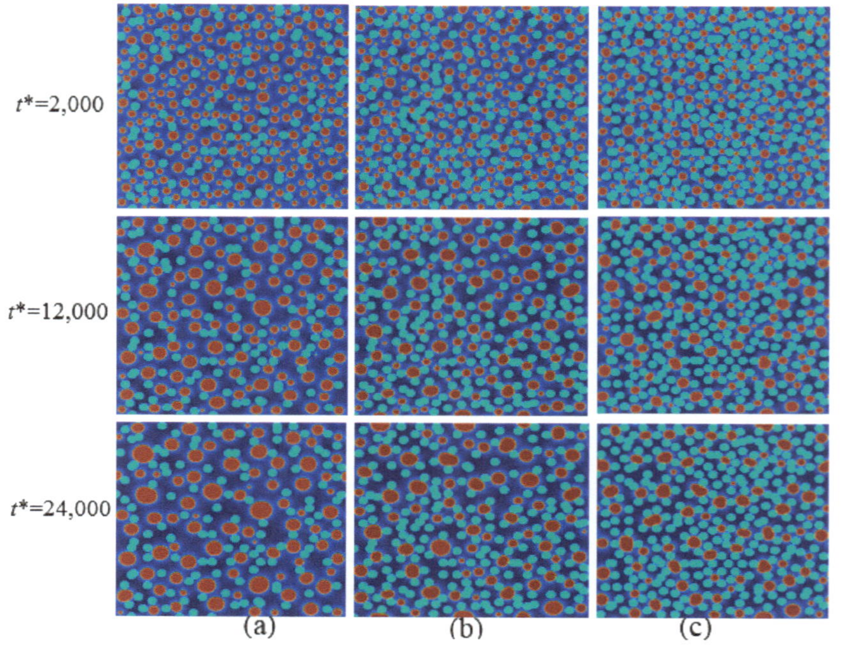

Figure 5. Simulation snapshots of the system at different times are shown in vertical columns. Particle size, r^*, is fixed to 20. Volume fractions of nanoparticles are: (**a**) 10%, (**b**) 20%, and (**c**) 30%. The nanoparticles are shown in light blue, precipitates are colored red, and the matrix is in dark blue.

Figure 6 also shows that, at a given time, the gradients of the curves are non-constant and gradually decrease with an increase in nanoparticle volume fraction, and this signifies the slowing down of the coarsening rate. The local slopes of the curves are the coarsening rate constants which indicate the kinetic and thermodynamic changes in terms of the diffusion coefficient, interfacial energy, and molar volume according to the coarsening rate constant relationship given below [37].

$$k = \frac{8}{9} D \frac{C_e \gamma_s V_m^2 f(\phi)}{RT} \tag{10}$$

where, V_m is the molar volume of the precipitates, D is the bulk diffusion of solutes in the matrix, C_e is the equilibrium solute concentration of the matrix, γ_s is the interfacial energy of the precipitate–matrix interface, R is the particle radius, T is the temperature, and $f(\phi)$ is the ratio of the coarsening rate with finite precipitate volume fraction to the coarsening rate with zero volume fraction. After examining the precipitate shrinkage and growth behavior in detail, it was found that the decelerated coarsening behavior experienced in this study is not attributed to inverse coarsening like the one experienced in the previous study [45], but the behavior is in agreement with results obtained by Ryu et al. [46] for regular coarsening deceleration.

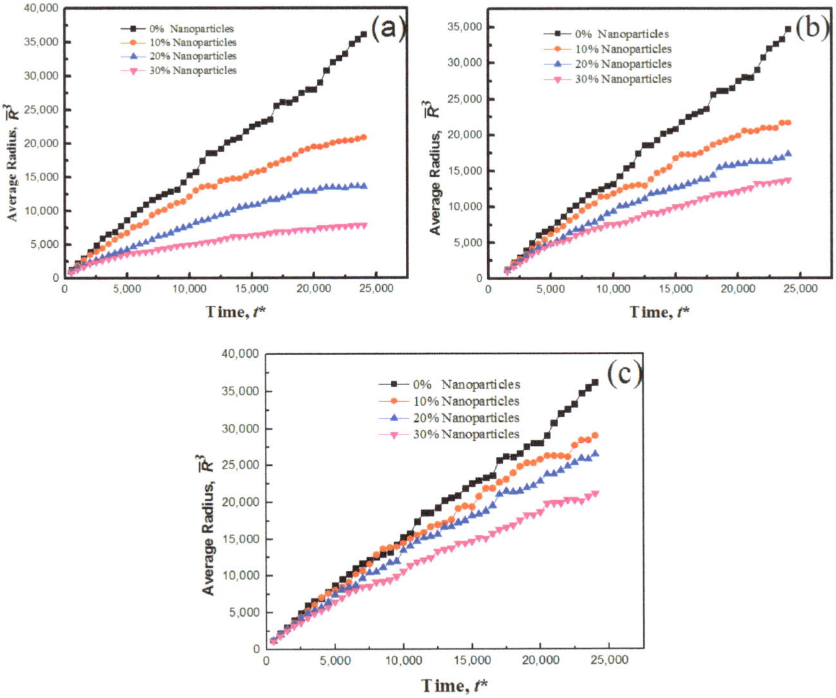

Figure 6. Plots of the cube of the average precipitate radius versus time. Nanoparticle area fractions are labeled in the figures. Nanoparticle sizes are: (**a**) 10, (**b**) 20 and (**c**) 30.

The PSD at different times in presence of nanoparticles of radius size $r^* = 10$ with different volume fractions in the system are given in Figures 7–9. The graphs show that with an increase in the system nanoparticle area fraction, the particle size distributions change and significant differences from the non-nanoparticles PSD in Figure 4 can be identified. Firstly, with an increase in volume fraction of nanoparticles, the PSD shapes become more disordered and peak positions do not significantly move to the right. This signifies much departure from the regular PSD observed in the absence of nanoparticles. Secondly, the rate of increase of the broadness of the precipitate size ranges becomes slower at higher volume fractions of precipitates (Figure 9) when compared to that in the absence of nanoparticles (Figure 4). After examining the coarsening simulation results carefully, it is found that, at high nanoparticle volume fraction, the shapes of the precipitates become irregular due to contact with nanoparticles. As seen in the high volume fraction PSD in Figure 9, the precipitate size range becomes narrow, and such PSD shapes are quite similar to those obtained for decelerated coarsening in the presence of stresses [15]. Figures 7–9 show that in the presence of nanoparticles in the system, the PSD shapes are constantly changing with time and are not self-similar according to the classic LSW theory. This behavior can be understood by identifying another precipitate size–limiting mechanism, which is the distance between randomly distributed inert nanoparticles. As shown in Figure 5, when the precipitates are in contact with nanoparticles, further growth of precipitates is constrained by the surrounding nanoparticles. Since the nanoparticles are randomly distributed in the matrix, the size distribution of the inter-particle space where the precipitates can grow is also random, with the average inter-particle space being inversely proportional to the number of nanoparticles. The particle size distribution, particularly in the late stages of coarsening, is the combined result of the classic coarsening kinetics that pushes the distribution towards the standard LSW PSD shapes and the geometric constraint imposed by inter-particle, distance which leads to random size distribution. Thus, with less size constraints at low nanoparticle volume fraction, the system will relax towards the standard PSD given by the LSW theory, and with strong size constraints at high nanoparticle volume fractions, precipitates will form a PSD that centers around the average inter-particle distance.

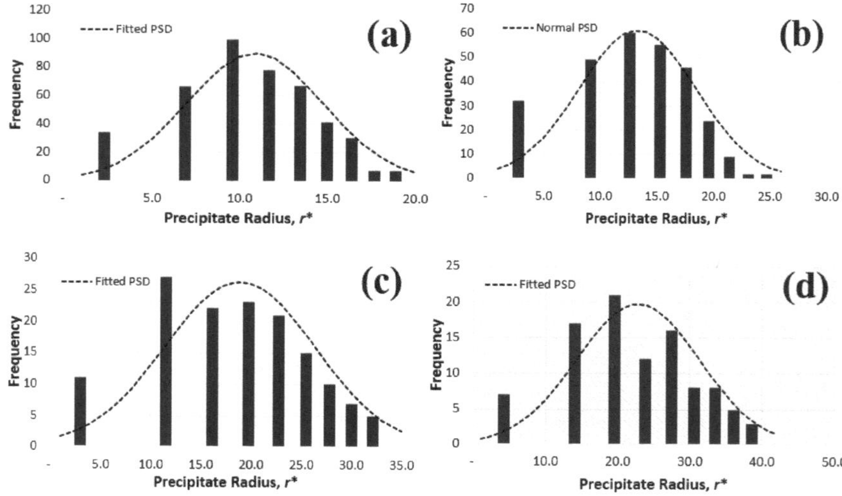

Figure 7. Histograms of the PSD in presence of 10% nanoparticles of size 10 at different times t^* of: (**a**) 1000, (**b**) 2500, (**c**) 10,000, and (**d**) 20,000.

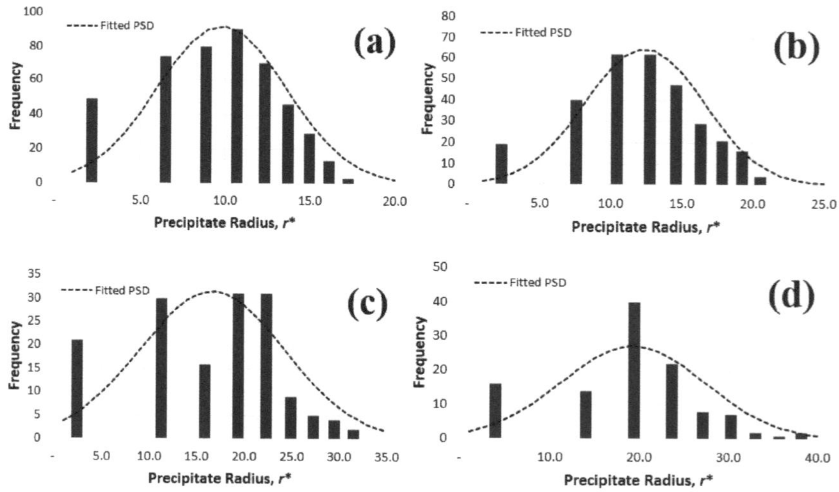

Figure 8. Histograms of the PSD in presence of 20% nanoparticles of size 10 at different times, t^*, of: (**a**) 1000, (**b**) 2500, (**c**) 10,000, and (**d**) 20,000.

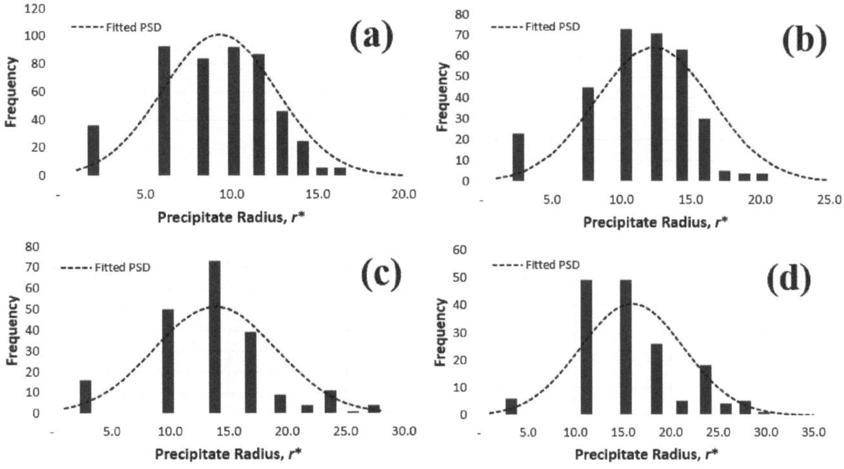

Figure 9. Histograms of the PSD in presence of 30% nanoparticles of size 10 at different times t^* of: (**a**) 1000, (**b**) 2500, (**c**) 10,000, and (**d**) 20,000.

To investigate the effect of nanoparticle sizes on the precipitate coarsening rate, the nanoparticle volume fractions were fixed and the nanoparticle sizes were varied. Figure 10a–c shows the evolution of precipitates with time steps for nanoparticles of a fixed volume fraction of 30% and different sizes, r^*, of 10, 20, and 30. In the presence of small sized nanoparticles, the late stage precipitate shape changes from circular to irregular, whilst for large nanoparticles the changes are insignificant. Figure 11a–c shows the plots of the cube of average precipitate radius with time for nanoparticles with fixed volume fractions and different particle sizes. It is shown that small-sized nanoparticles are more effective in suppressing precipitate coarsening when compared to large ones. The results show that, for the same volume fraction, the precipitate sizes and gradients of the curves decreases with a decrease in the nanoparticle size. The smallest precipitate sizes are obtained in the presence of a high volume fraction of nanoparticles of the smallest size, as shown in Figure 11c. On the contrary, a small area fraction of large nanoparticles, like those shown in Figure 11a, only slightly affects the precipitate' coarsening rate. At the same volume fraction, small nanoparticles possess larger overall surface area. Since the coarsening process involves the transportation of solute atoms from small precipitates to large precipitates through diffusion, the large amount of particle–matrix interfaces produced by small nanoparticles block more solute transportation path and, therefore, effectively slows down the coarsening process.

The effect of nanoparticle sizes on PSD are shown in Figures 9, 12 and 13 for nanoparticles with 30% volume fraction and size radii, R^*, of 10, 20, and 30. The figures show that with a decrease in the nanoparticle size, the PSD shapes become more random due to the inter-particle distance constraints from the increasing nanoparticle number density. With larger nanoparticle size, like in Figure 13, the PSD shape is close to the one in absence of nanoparticles since the inter-particle distance is also large. A similar behavior was observed in NiAl and Fe-Cr alloys, where it was shown that the coarsening rate can be decelerated by solute segregation, stresses and strains [10,47].

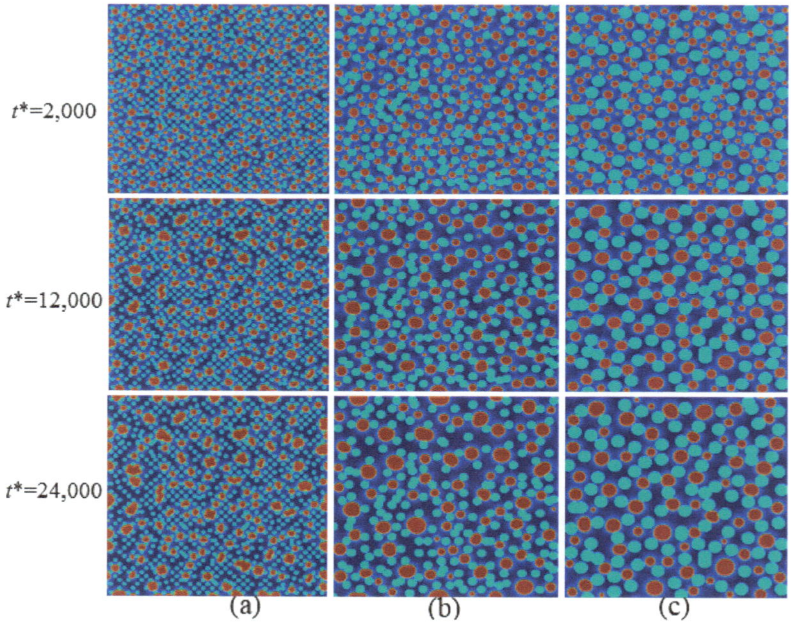

Figure 10. Simulation snapshots of the system at different times are shown in vertical columns. Particle volume fraction is set to 30%. Particle sizes are: (**a**) 10, (**b**) 20, and (**c**) 30. The same coloring scheme in Figure 5 is used.

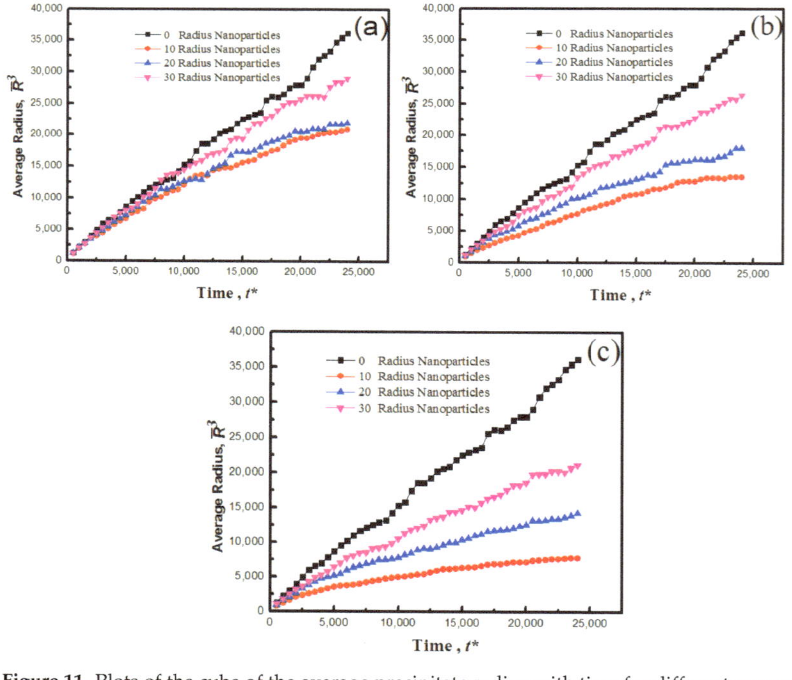

Figure 11. Plots of the cube of the average precipitate radius with time for different nanoparticles area fractions of: (**a**) 10%, (**b**) 20%, and (**c**) 30%. Particle sizes are labeled in the figures.

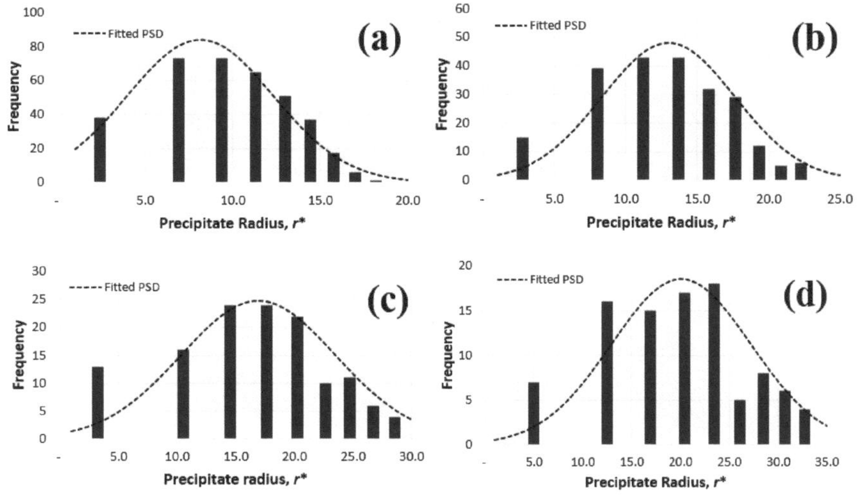

Figure 12. Histograms of the PSD in presence of 30% nanoparticles of size 20 at different times, t^*, of: (**a**) 1000, (**b**) 2500, (**c**) 10,000 and (**d**) 20,000.

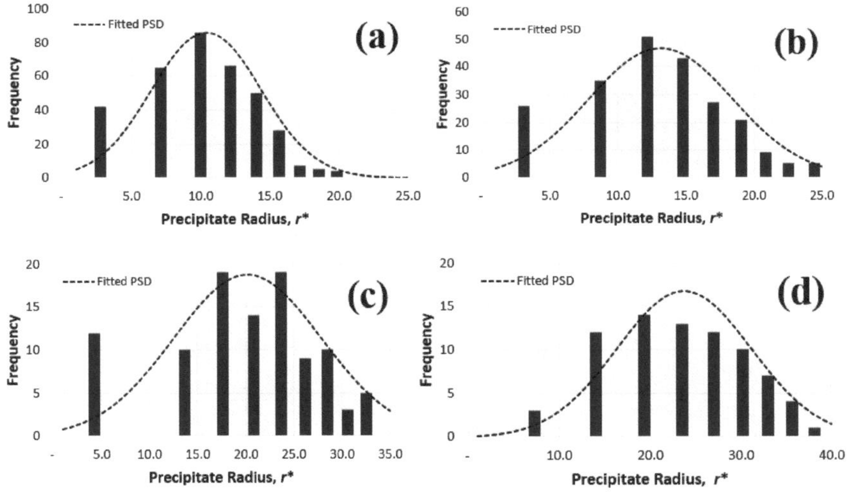

Figure 13. Histograms of the PSD in presence of 30% nanoparticles of size 30 at different times, t^*, of: (**a**) 1000, (**b**) 2500, (**c**) 10,000 and (**d**) 20,000.

4. Conclusions

The effects of intra-granular nanoparticles on early-stage coarsening behavior of precipitates were studied computationally by generalizing a previously proved phase field model for precipitate coarsening. Inert nanoparticles of various sizes and volume fractions were incorporated in the matrix phase using a non-evolving order parameter. It is found that both the precipitate size distribution and the coarsening rate may be affected by the presence and characteristics of nanoparticles. Two distinct mechanisms for these nanoparticle effects are proposed. The new length scale related to the inter-nanoparticle distance introduced by incorporating random nanoparticles may change the standard LSW precipitate size distribution by geometrically constraining the precipitate size, and the reduction of available diffusion paths due to the blockage of nanoparticle–matrix interfaces can significantly slow down the coarsening process. As the first systematic computational study of the effects of nanoparticles on the precipitate coarsening process, this work sheds light on potential approaches to control the coarsening process and further improve the mechanical properties of nanoparticle–reinforced metal matrix composites.

The simulation results in general suggests that the use of smaller nanoparticles with large number density is effective in slowing down of the coarsening kinetics. However, since the current work was performed in 2D only, the diffusion path blockage mechanism that leads to this conclusion may become less significant in 3D due to an increasing transport path. Computationally intensive 3D simulations need to be carried out in the future in order to confirm the result. In addition, other potential contributing factors like the precipitate–matrix misfit stress and surface diffusion along the particle–matrix interface should be included when generalizing the results to real materials.

Author Contributions: Conceptualization and study design, N.W.; Model development, simulations, manuscript writing, S.F.; Data interpretation, data collection and manuscript revision, B.H. All authors have read and agreed to the published version of the manuscript.

Funding: This research received no external funding.

Data Availability Statement: Not applicable.

Acknowledgments: The authors acknowledge the computer time support from the High Performance Computing Center of Guangdong Technion—Israel Institute of Technology.

Conflicts of Interest: The authors of this manuscript declare that there is no any conflict of interest in this paper.

References

1. Baldan, A. Review Progress in Ostwald ripening theories and their applications to the γ'-precipitates in nickel-base superalloys Part II Nickel-base superalloys. *J. Mater. Sci.* **2002**, *37*, 2379–2405. [CrossRef]
2. Yang, C.; Cao, L.; Gao, Y.; Cheng, P.; Zhang, P.; Kuang, J.; Zhang, J.; Liu, G.; Sun, J. Nanostructural Sc-based hierarchy to improve the creep resistance of Al–Cu alloys. *Mater. Des.* **2020**, *186*, 108309. [CrossRef]
3. Shower, P.; Morris, J.; DShin Radhakrishnan, B.; Poplawsky, J.; Shyam, A. Mechanisms for stabilizing $\theta'(Al_2Cu)$ precipitates at elevated temperatures investigated with Phase Field modeling. *Materialia* **2019**, *6*, 100335. [CrossRef]
4. Devaraj, A.; Wang, W.; Vemuri, R.; Kovarik, L.; Jiang, X.; Bowden, M.; Trelewicz, J.R.; Mathaudhu, S.; Rohatgi, A. Grain boundary segregation and intermetallic precipitation in coarsening resistant nanocrystalline aluminum alloys. *Acta Mater.* **2019**, *165*, 698–708. [CrossRef]
5. Gao, Y.H.; Guan, P.F.; Su, R.; Chen, H.W.; Yang, C.; He, C.; Cao, L.F.; Song, H.; Zhang, J.Y.; Zhang, X.F.; et al. Segregation-sandwiched stable interface suffocates nanoprecipitate coarsening to elevate creep resistance. *Mater. Res. Lett.* **2020**, *8*, 446–453. [CrossRef]
6. Gao, Y.H.; Yang, C.; Zhang, J.Y.; Cao, L.F.; Liu, G.; Sun, J.; Ma, E. Stabilizing nanoprecipitates in Al-Cu alloys for creep resistance at 300 °C. *Mater. Res. Lett.* **2019**, *7*, 18–25. [CrossRef]
7. Smith, T.M.; Good, B.S.; Gabb, T.P.; Esser, B.D.; Egan, A.J.; Evans, L.J.; McComb, D.W.; Mills, M.J. Effect of stacking fault segregation and local phase transformations on creep strength in Ni-base superalloys. *Acta Mater.* **2019**, *172*, 55–65. [CrossRef]
8. Bian, H.; Xu, X.; Li, Y.; Koizumi, Y.; Wang, Z.; Chen, M.; Yamanaka, K.; Chiba, A. Regulating the coarsening of the γ' phase in superalloys. *NPG Asia Mater.* **2015**, *7*, e212. [CrossRef]
9. Feng, L.; Rao, Y.; Ghazisaeidi, M.; Mills, M.J.; Wang, Y. Quantitative prediction of Suzuki segregation at stacking faults of the γ'phase in Ni-base superalloys. *Acta Mater.* **2020**, *200*, 223–235. [CrossRef]
10. Mushongera, L.T.; Fleck, M.; Kundin, J.; Wang, Y.; Emmerich, H. Effect of Re on directional γ'-coarsening in commercial single crystal Ni-base superalloys: A phase field study. *Acta Mater.* **2015**, *93*, 60–72. [CrossRef]
11. Wang, W.Z.; Jin, T.; Liu, J.L.; Sun, X.F.; Guan, H.R.; Hu, Z.Q. Role of Re and Co on microstructures and γ' coarsening in single crystal superalloys. *Mater. Sci. Eng. A* **2008**, *479*, 148–156. [CrossRef]
12. Lu, F.; Antonov, S.; Lu, S.; Zhang, J.; Li, L.; Wang, D.; Zhang, J.; Feng, Q. Unveiling the Re effect on long-term coarsening behaviors of γ' precipitates in Ni-based single crystal superalloys. *Acta Mater.* **2022**, *233*, 117979. [CrossRef]
13. Pyczak, F.; Devrient, B.; Neuner, F.C.; Mughrabi, H. The influence of different alloying elements on the development of the γ/γ' microstructure of nickel-base superalloys during high-temperature annealing and deformation. *Acta Mater.* **2005**, *53*, 3879–3891. [CrossRef]
14. Chen, Y.; Slater, T.J.; Bai, M.; Mitchell, R.; Ciuca, O.; Preuss, M.; Haigh, S.J. An investigation of diffusion-mediated cyclic coarsening and reversal coarsening in an advanced Ni-based superalloy. *Acta Mater.* **2016**, *110*, 295–305. [CrossRef]

15. Prikhodko, S.V.; Ardell, A.J. Coarsening of γ' in Ni–Al alloys aged under uniaxial compression: I. Early-stage kinetics. *Acta Mater.* **2003**, *51*, 5001–5012. [CrossRef]
16. Ardell, A.J.; Prikhodko, S.V. Coarsening of γ' in Ni–Al alloys aged under uniaxial compression: II. Diffusion under stress and retardation of coarsening kinetics. *Acta Mater.* **2003**, *51*, 5013–5019.
17. Kundin, J.; Mushongera, L.; Goehler, T.; Emmerich, H. Phase-field modeling of the γ'-coarsening behavior in Ni-based superalloys. *Acta Mater.* **2012**, *60*, 3758–3772. [CrossRef]
18. Voorhees, P.W. The theory of Ostwald ripening. *J. Stat. Phys.* **1985**, *38*, 231–252. [CrossRef]
19. Chen, L.Y.; Xu, J.Q.; Li, X.C. Controlling phase growth during solidification by nanoparticles. *Mater. Res. Lett.* **2015**, *3*, 43–49. [CrossRef]
20. Wang, K.; Jiang, H.Y.; Jia, Y.W.; Zhou, H.; Wang, Q.D.; Ye, B.; Ding, W.J. Nanoparticle-inhibited growth of primary aluminum in Al–10Si alloys. *Acta Mater.* **2016**, *103*, 252–263. [CrossRef]
21. Chen, L.Y.; Xu, J.Q.; Choi, H.; Konishi, H.; Jin, S.; Li, X.C. Rapid control of phase growth by nanoparticles. *Nat. Commun.* **2014**, *5*, 3879. [CrossRef] [PubMed]
22. Vanherpe, L.; Moelans, N.; Blanpain, B.; Vandewalle, S. Pinning effect of spheroid second-phase particles on grain growth studied by three-dimensional phase-field simulations. *Comput. Mater. Sci.* **2010**, *49*, 340–350. [CrossRef]
23. Chakrabarti, T.; Manna, S. Zener pinning through coherent precipitate: A phase-field study. *Comput. Mater. Sci.* **2018**, *154*, 84–90. [CrossRef]
24. Chang, K.; Feng, W.; Chen, L.Q. Effect of second-phase particle morphology on grain growth kinetics. *Acta Mater.* **2009**, *57*, 5229–5236. [CrossRef]
25. Moelans, N.; Blanpain, B.; Wollants, P. Phase field simulations of grain growth in two-dimensional systems containing finely dispersed second-phase particles. *Acta Mater.* **2006**, *54*, 1175–1184. [CrossRef]
26. Karnesky, R.A.; Meng, L.; Dunand, D.C. Strengthening mechanisms in aluminum containing coherent Al3Sc precipitates and incoherent Al_2O_3 dispersoids. *Acta Mater.* **2007**, *55*, 1299–1308. [CrossRef]
27. Moelans, N.; Blanpain, B.; Wollants, P. Pinning effect of second-phase particles on grain growth in polycrystalline films studied by 3-D phase field simulations. *Acta Mater.* **2007**, *55*, 2173–2182. [CrossRef]
28. Lin, T.C.; Cao, C.; Sokoluk, M.; Jiang, L.; Wang, X.; Schoenung, J.M.; Lavernia, E.J.; Li, X. Aluminum with dispersed nanoparticles by laser additive manufacturing. *Nat. Commun.* **2019**, *10*, 4124. [CrossRef]
29. Suwa, Y.; Saito, Y.; Onodera, H. Phase field simulation of grain growth in three dimensional system containing finely dispersed second-phase particles. *Scr. Mater.* **2006**, *55*, 407–410. [CrossRef]
30. DeWitt, S.; Thornton, K. Phase field modeling of microstructural evolution. *Comput. Mater. Syst. Des.* 2018, pp. 67–87. Available online: https://link.springer.com/chapter/10.1007/978-3-319-68280-8_4 (accessed on 30 March 2022).
31. Moelans, N.; Blanpain, B.; Wollants, P. An introduction to phase-field modeling of microstructure evolution. *Calphad* **2008**, *32*, 268–294. [CrossRef]
32. Bellemans, I.; Moelans, N.; Verbeken, K. Phase-field modelling in extractive metallurgy. *Crit. Rev. Solid State Mater. Sci.* **2018**, *43*, 417–454. [CrossRef]
33. Boettinger, W.J.; Warren, J.A.; Beckermann, C.; Karma, A. Phase-field simulation of solidification. *Annu. Rev. Mater. Res.* **2002**, *32*, 163–194. [CrossRef]
34. Fan, D.; Chen, S.P.; Chen, L.Q.; Voorhees, P.W. Phase-field simulation of 2-D Ostwald ripening in the high volume fraction regime. *Acta Mater.* **2002**, *50*, 1895–1907. [CrossRef]
35. Vaithyanathan, V. Phase-Field Simulations of Coherent Precipitate Morphologies and Coarsening Kinetics. Ph.D. Thesis, The Pennsylvania State University, Centre County, PA, USA, 2002.
36. Zhu, J.Z.; Wang, T.; Ardell, A.J.; Zhou, S.H.; Liu, Z.K.; Chen, L.Q. Three-dimensional phase-field simulations of coarsening kinetics of γ' particles in binary Ni–Al alloys. *Acta Mater.* **2004**, *52*, 2837–2845. [CrossRef]
37. Li, X.; Saunders, N.; Miodownik, A.P. The coarsening kinetics of γ' particles in nickel-based alloys. *Metall. Mater. Trans. A* **2002**, *33*, 3367–3373. [CrossRef]
38. Ardell, A.J.; Nicholson, R.B. The coarsening of γ' in Ni-Al alloys. *J. Phys. Chem. Solids* **1966**, *27*, 1793–1794. [CrossRef]
39. Vaithyanathan, V.; Wolverton, C.; Chen, L.Q. Multiscale modeling of precipitate microstructure evolution. *Phys. Rev. Lett.* **2002**, *88*, 125503. [CrossRef]
40. Simmons, J.P.; Shen, C.; Wang, Y. Phase field modeling of simultaneous nucleation and growth by explicitly incorporating nucleation events. *Scr. Mater.* **2000**, *43*, 935–942. [CrossRef]
41. Marqusee, J.A.; Ross, J. Theory of Ostwald ripening: Competitive growth and its dependence on volume fraction. *J. Chem. Phys.* **1984**, *80*, 536–543. [CrossRef]
42. Sun, Y.; Andrews, W.B.; Thornton, K.; Voorhees, P.W. Self-similarity and the dynamics of coarsening in materials. *Sci. Rep.* **2018**, *8*, 17940. [CrossRef]
43. Wu, X.; Li, Y.; Huang, M.; Liu, W.; Hou, Z. Precipitation kinetics of ordered γ' phase and microstructure evolution in a NiAl alloy. *Mater. Chem. Phys.* **2016**, *182*, 125–132. [CrossRef]
44. Wang, T.; Sheng, G.; Liu, Z.K.; Chen, L.Q. Coarsening kinetics of γ' precipitates in the Ni–Al–Mo system. *Acta Mater.* **2008**, *56*, 5544–5551. [CrossRef]

45. Miyazaki, T.; Koyama, T. Stability bifurcation against coarsening in elastically constrained many-particle systems. *Mater. Sci. Eng. A* **1993**, *169*, 159–165. [CrossRef]
46. Ryu, H.J.; Hong, S.H.; Weber, J.; Tundermann, J.H. Effect of elastic interaction energy on coarsening of γ' precipitates in a mechanically alloyed ODS Ni-base superalloy. *J. Mater. Sci.* **1999**, *34*, 329–336. [CrossRef]
47. Zhu, L.; Li, Y.; Shi, S.; Yan, Z.; Chen, J.; Maqbool, S. Morphology and kinetics evolution of nanoscale phase in Fe–Cr alloys under external strain. *Nanomaterials* **2019**, *9*, 294. [CrossRef] [PubMed]

alloys

Article

Kinetic Model of Isothermal Bainitic Transformation of Low Carbon Steels under Ausforming Conditions

Theerawat Kumnorkaew [1,*], Junhe Lian [2], Vitoon Uthaisangsuk [3] and Wolfgang Bleck [1]

1 Steel Institute, RWTH Aachen University, 52072 Aachen, Germany; bleck@iehk.rwth-aachen.de
2 Department of Mechanical Engineering, Aalto University, 02150 Espoo, Finland; junhe.lian@aalto.fi
3 Center for Lightweight Materials, Design and Manufacturing, Department of Mechanical Engineering, King Mongkut's University of Technology Thonburi, Bangkok 10140, Thailand; vitoon.uth@kmutt.ac.th
* Correspondence: theerawat.kumnorkaew@iehk.rwth-aachen.de

Abstract: Carbide-free bainitic steels show attractive mechanical properties but are difficult to process because of the sluggish phase transformation kinetics. A macroscopic model based on the classical nucleation theory in conjunction with the modified Koistinen–Marburger relationship is proposed in this study to simulate the kinetics of incomplete bainitic and martensitic phase transformations with and without austenite deformation. A 0.26C-1Si-1.5Mn-1Cr-1Ni-0.003B-0.03Ti steel and a 0.18C-1Si-2.5Mn-0.2Cr-0.2Ni-0.02B-0.03Ti steel were investigated with different levels of ausforming. The concept of ausforming is expected to accelerate the onset of the bainitic transformation and to enhance the thermodynamic stability of austenite by increased dislocation density. The phase transformation kinetics of both steels is quantitatively analyzed in the study by dilatometry and X-ray diffraction so that the carbon concentration in the retained austenite and bainitic ferrite, as well as their volume fractions, is determined. A critical comparison of the numerical and experimental data demonstrates that the isothermal kinetics of bainite formation and the variation of driving energy can be satisfactorily described by the developed model. This model captures the incompleteness of the bainite phase transformation and the carbon enrichment in the austenite well. A fitting parameter can be used to elucidate the initial energy barrier caused by the ausforming. An increase in austenite stability can be described by the nucleation reaction and the thermodynamic energies associated with the change of dislocation density. The proposed model provides an in-depth understanding of the effect of ausforming on the transformation kinetics under different low-carbon steels and is a potential tool for the future design of heat treatment processes and alloys.

Keywords: bainitic steels; phase transformation; ausforming; carbon enrichment; activation energy; dislocation density

Citation: Kumnorkaew, T.; Lian, J.; Uthaisangsuk, V.; Bleck, W. Kinetic Model of Isothermal Bainitic Transformation of Low Carbon Steels under Ausforming Conditions. *Alloys* **2022**, *1*, 93–115. https://doi.org/10.3390/alloys1010007

Academic Editors: Damien Fabrègue, Shuwen Wen, Yongle Sun and Xin Chen

Received: 15 May 2022
Accepted: 8 June 2022
Published: 13 June 2022

Publisher's Note: MDPI stays neutral with regard to jurisdictional claims in published maps and institutional affiliations.

1. Introduction

Controlling the thermodynamic stability of austenite has been a challenging issue in the development of carbide-free bainitic (CFB) steels when considering carbon as an essential element for a bainitic transformation at low temperatures. With an addition of 1.5%wt.%Si, the decomposition of austenite occurs when supersaturated carbon from bainitic ferrite is rejected into adjacent austenitic regions and becomes robustly available for stabilization during the bainitic transformation instead of forming cementite (Fe3C) [1–4]. The increased stability of the neighboring austenite via the enrichment of carbon thus leads to an incompleteness of the transformation that leaves two different features of the retained austenite (RA) as the secondary phase, namely film-like RA and blocky-type RA. These features of RA can be characterized by their stabilities in accordance with chemical contributions that can be statistically estimated by atom probe tomography (APT) [5,6]. The carbon-rich, thin-film RA is always more stable compared to the other type [7–10]. In light of the thermodynamic stability, Garcia Mateo et al. [11] reported that the blocky

RA could also be present in various granular morphologies, depending on the carbon concentration. The morphologies with low carbon content, particularly in the central zone, are prone to decompose further into fresh martensite (FM) during ambient cooling. Even though the formation of FM is beneficial in some applications that require material with high hardness, strength, and wear resistance (e.g., railway material and agricultural parts, etc.), the existence of such a hard phase is usually an impairment when the transformation-induced plasticity (TRIP) effect is desired [3]. Therefore, the recent progress in bainitic forging steel development is aimed at refining the microstructures by means of replacing the large granular blocky RA with nano-structured film-like RA. For instance, Caballero et al. and Garcia-Mateo et al. [12,13] exploited the T_0 concept as a thermodynamic limit of CFB transformation to design advanced carbon CFB steels with an ultra-fine structure (RA thickness < 100 nm). They achieved prominent steel properties, including strength and elongation of about 2.2–2.5 GPa and 20–30%, respectively. However, the concept is successful only in high- and medium-carbon steels (0.4–1.0 wt.%C), whereby transformation at very low temperatures above the martensite start temperature is necessary. Despite the attempt to take low-carbon steels (<0.2% C) into account, higher Gibbs free energy, associated with the insufficient C enrichment in austenite affected by the lower bulk density of C, has promoted the thermal instability of the austenite during the cooling process after isothermal holding.

Yao et al. [14] proposed an alloying modification strategy and considered a chemical Gibbs free energy change of the FCC to BCC transformation. An addition of 1.5%Ni encourages the austenite stability by means of enlarging the austenitic phase region so that an increase in the free energy difference retards the kinetics of the bainitic transformation and induces a reduction in the bainitic ferrite (BF) plate thickness. However, although the energy reduction due to another supplement of 1.5%Cr favors the bainitic transformation, a formation of iron carbide (Fe3C) consequently deteriorates the thermal stability of the austenite. Herein, the stability of austenite likely depends on the free energy difference between the parent and the child phases. As a consequence, both Cr and Ni alloyed in nearly equivalent compositions, which causes another reduction in the driving energy by ~400 J/mol; thus, this is expected to enhance the thermal stability of austenite and stimulate bainitic transformation simultaneously. Hence, the enhanced stabilization of austenite is feasible by adjusting an appropriate amount of Cr and Ni addition in the concerned steel. Changle et al. [15] stated that steels alloyed with Mn content over 2.2 wt.% provide an excellent hardenability and a lower bainitic start temperature (Bs), which in turn yields an 18% volume fraction of retained austenite with a reduction in BF laths. Mn as a solid solution element is evidently effective in raising the free energy of ferrite and reducing that of austenite, causing a delay of austenite decomposition to bainite at low temperatures. Nevertheless, if the Mn content in bainitic steel exceeds its limitation and segregation occurs at the prior austenite grain boundaries, it may lead to an increase in transformation temperature and undesired brittleness.

Ausforming is one of the thermomechanical treatment processes, in which the structure of austenite is refined at its metastable temperature prior to phase transformation. The deformation plays an important part in accelerating the bainitic transformation due to increased bainitic nucleation sites, whereas raised dislocation density encourages the thermal stability of austenite. The process parameters of ausforming, such as deformation temperature, strain, and strain rate, are also reported as being the essential controlling parameters of the kinetics of isothermal bainitic transformation [16]. A sophisticated result showed that the stability of austenite is not improved when a severe ausforming strain of 50% is applied, because of a high BF volume fraction. It is noticed that such a result is confirmed in a few works concerning low-carbon steels [17,18]. Although these qualitative conclusions can be applied to optimize the microstructure of CFB steels, a reliable physical-based model for correlating the alloying composition with processing parameters is still required because the variations of both the alloy design and the process parameters are very high. Several models have been developed for describing phase transformation of alloys, but one proposed by Bhadeshia and his co-workers has been widely employed [19,20]. Their model is based on a displacive approach, by which bainite growth without diffusion of any alloying elements is considered. In this approach, bainite formation is assumed to begin at prior nucleation sites on austenite grain boundaries and to successively form by autocatalytic nucleation at the newly formed sheaves. Consequently, the transformation of bainite is a nucleation-controlled process. The prior austenite grain size and the maximum driving energy are essential factors controlling the nucleation kinetics of bainitic transformation. Later, Magee [21] revealed that the number of nucleation sites as a function of prior austenite grain size should also be considered and can be determined by the driving energy difference. It is increased with a rise in undercooling. Van Bohemann formulated this concept to estimate the number of potential embryos for martensitic nucleation [22]. However, these displacive models use empirical constants by which the activation energy is only represented a net energy used to activate the transformation. In the model of isothermal martensitic transformation proposed by Ghosh and Olsen [23], an energy barrier consistent with the critical driving energy needed for the nucleation should be incorporated by the sum of the strain energy, the defect-size-dependent interfacial energy, and the composition-dependent interfacial work [24]. They also introduced a model that takes autocatalytic factors into account as material constants. The effectiveness of the γ/α interface boundaries and the thickness of the bainitic plate play a significant role in the overall transformation kinetics. Meng et al. [25] reported that morphological features also affect internal stresses by the autocatalytic nucleation of martensite. A stress field, which disperses outside a transformed martensite plate, is potentially capable of triggering unstable martensite embryos to become stable nuclei and then grow up. Zou et al. [26] studied the effects of prior deformation and undercooling on the isothermal bainitic transformation. They found that the pre-deformation provides a mechanical driving force, which further enhances the nucleation rate of bainite transformation, in addition to stresses internally generated by the dislocation density when new bainite plates are formed. Moreover, the difference in the activation energies between grain boundary nucleation and autocatalytic nucleation is proposed by Ravi et al. [27]. The autocatalytic nucleation, which is considered to have a dynamic nature, has been controlled by the degree of carbon enrichment in austenite during the transformation. Nevertheless, an empirical constant, which involves the autocatalytic term, remains undefined with any physical significance.

From the previous studies, it can be concluded that a more precise model of the bainitic transformation, especially for an ausforming process of low-carbon steels, is still to be developed. Thus, in this work, a unified physically based model is derived to thoroughly elucidate the isothermal bainitic transformation, while taking into consideration varying ausforming strain and alloy modifications. The model is derived from the theoretical basis of the displacive transformation concerning a modified T_0 concept and the empirical Koistinen–Marburger (KM) equation. The activation energy, nucleation activity, and carbon enrichment variations caused by the process contributions are also correlated with the kinetics of isothermal bainitic transformation. By means of the model, the thermal stability of austenite can be appropriately adjusted with an optimal design of the processing parameters of the ausforming and alloy modification of low-carbon CFB steels.

2. Materials and Methods

2.1. As-Received Materials

Two as-received low-carbon steels of different compositions were chosen. The steels are identified as MC1.5Mn1NiCr and LC2.5Mn0.2NiCr steels. The chemical compositions of both steels are represented in Table 1. Manganese as the former austenite element was added to increase the stability of the retained austenite and properly elevate the hardenability of the steels. Silicon was alloyed in the amount of 1 wt.% for retarding and suppressing the formation of cementite in the bainitic structure so that retained austenite as a secondary phase was promoted. The addition of boron was to provide a shift of the diffusion-controlled ferrite/pearlite transformation to a longer time. A precipitation of boron nitride in solid solution was suppressed by alloying titanium of about 0.03 wt.%, whereas the interaction of nitrogen and titanium in the form of titanium nitride (TiN) could be formed. The steels were individually melted in a laboratory-scale vacuum arc furnace and cast into a square ingot with a dimension of 140 × 140 × 525 mm. Afterwards, the ingots were homogenized at 1250 °C for 2 h, then hot-forged into square billets with a dimension of 60 × 60 × 1000 mm at a finishing temperature of 950 °C. The billets were slowly cooled down to room temperature. Finally, the homogenized billets were cut parallel to the forged direction to manufacture cylindrical specimens of 10 ± 0.1 mm height and 5 ± 0.1 mm diameter.

Table 1. Chemical composition of the investigated steels in wt.%.

Steel	Fe	C	Si	Mn	Cr	Ni	B	Ti
MC1.5Mn1NiCr	Bal.	0.26	1.07	1.46	0.99	0.98	0.0031	0.027
LC2.5Mn0.2NiCr	Bal.	0.18	0.97	2.50	0.20	0.21	0.0018	0.033

2.2. Experiment

The cylindrical specimens were subjected to three different heat treatment conditions. These heat treatments are distinguished between quenching and isothermal tempering, with and without austenite deformation, as follows: direct quenching (DQ), pure isothermal tempering (PIT), and ausforming followed by isothermal tempering (AIT). All heat-treating experiments were conducted on a Bähr DIL805A/D dilatometer. The machine was additionally equipped with an optical module for cross-sectional measurement, which enabled a precise investigation of the phase transformations by monitoring changes in the length of the specimen in the longitudinal and radial directions. A Pt/Pt/−10 Rh thermocouple (type S) was spot-welded at the central edge of the specimens so that the temperature signal was accurately gathered from the machine. The experiments were conducted within a vacuum chamber where the specimens were located in the middle between two quartz rods inside an induction coil. A cooling gas, e.g., helium, was directly fed into the chamber through a pressure control valve. The desired temperature could be achieved by balancing the heating coil and the cooling gas. A deformation module was also installed for uniaxial compression tests. As a result, the relative volume strains were determined by considering the profile variations of the specimens in the radial and longitudinal directions, as follows [28,29].

$$\Delta V/V_0 = (1+\varepsilon_L)(1+\varepsilon_R)^2 - 1 \tag{1}$$

where ΔV represents the volume change of specimen, V_0 is the initial volume, ε_L is the longitudinal strain, and ε_R is the radial strain of the specimen.

For all the experiments, the specimens were first heated to the austenitizing temperature at 950 °C, at the rate of 18 °C/s, and soaked for 5 min for homogenization. The PIT specimens were subsequently cooled to 400 °C at a cooling rate of 50 °C/s and isothermally treated for 1 h before cooling to room temperatures at 20 °C/s. The AIT specimens were cooled after homogenization to a deformation temperature of 650 °C at the same rate and held for 10 s. The specimens were subsequently deformed with an ausforming strain of either of 0.15 (AIT0.15) or 0.35 (AIT0.35), at a rate of 1 s^{-1}, and cooled to the isothermal temperature within the same period, as was conducted for the PIT samples. To examine changes in the martensite start (M_s) temperature, a specimen of each material was directly quenched from the austenitizing temperature. It was defined as the DQ specimen, and the volumetric expansion result was set as a reference. The M_s locus of the DQ specimens was captured from the first deviation of the dilation curve during cooling, whereas that of the AIT specimens was traced in the same manner specifically during the secondary stage of cooling after isothermal tempering. In the case of the DQ specimens, the volumetric transformations of the martensite were calculated by using a total volumetric expansion with respect to the relative tangent of the dilatation curve as a reference, bearing in mind that in this research the M_s temperature was also empirically estimated, using the following equation [30,31].

$$M_s = 539 - 423x_C - 30.4x_{Mn} - 7.5x_{Si} + 30x_{Al} \quad (2)$$

where x_C, x_{Mn}, x_{Si}, and x_{Al} are the carbon, manganese, silicon, and aluminum contents in wt.%, respectively. The specimens for evaluating the prior austenite grain (PAG) were directly quenched from the ausforming stage to room temperature.

2.3. Characterization

In the beginning, the specimens were ground and polished with abrasive papers (grids no. 600, 1200, 2400, and 4000) and 0.1 μm diamond paste, respectively. For the PAG observation, they were chemically etched using 4 g of sodium dodecylbenzene sulfonic acid in 100 mL of aqueous saturated picric acid diluted in 100 mL of distilled water in a water bath. The optimized temperature and time for etching were 60 °C and 30 s, respectively. After etching, the average grain size of PAG (d_γ) was measured on a light optical microscope (LOM) and calculated by using the linear intercept method based on the ASTM112-12 standard [32]. The same preparation techniques were applied to the specimens for phase quantification by X-ray diffraction (XRD) measurement, except for the solution etching. Hence, after the polishing step, all the specimens were further electro-polished by means of a TenuPol-5 single-jet electro-polishing device. An A2 electrolyte was employed at room temperature with a voltage of 32 V and a flow rate of 15 mm/s. The XRD machine was operated using a filtered CrKα radiator at 40 kV and 30 mA, under the collection range between 60° and 165°, at a step width of 0.08° and a counting time of 2 s. The phase fractions of face-centered cubic (FCC) RA and body-centered cubic (BCC) bainitic ferrite or/and martensite were analyzed by the Rietveld's refinement method, using MAUD software. The carbon concentration of the retained austenite (x_C in wt.%) was estimated from its lattice parameter (a_γ), as given in Equation (3) [33]. The parameters x_{Ni}, x_{Cr}, x_N, x_{Co}, x_{Cu}, x_{Nb}, x_{Mo}, x_{Ti}, x_V, and x_W are the nickel, chromium, nitrogen, cobalt, copper, niobium, titanium, vanadium, and tungsten contents in wt.%, respectively.

$$\begin{aligned} a_\gamma(\dot{A}) = 3.5780 \; & + 0.0330x_C + 0.00095x_{Mn} - 0.0002x_{Ni} + 0.0006x_{Cr} \\ & + 0.0056x_{Al} + 0.0220x_N + 0.0004x_{Co} + 0.0015x_{Cu} \\ & + 0.0051x_{Nb} + 0.0031x_{Mo} + 0.0039x_{Ti} + 0.0018x_V \\ & + 0.0018x_W \end{aligned} \quad (3)$$

Apart from phase quantification, the Rietveld method, which involves the Fourier analysis of the broadened peaks, was used to evaluate the microstructure parameters, such as the effective crystallite size and the average microstrain within the crystal [34]. Considering the plastically deformed materials, the Popa model was used to deconvolute the size and strain effect in the deformed crystals in accordance with the anisotropic size–strain broadening [35]. The peak shifting, broadening, and asymmetry experienced by the line profiles because of the accumulation of irradiation defects were analyzed using the Warren model [36]. The dislocation density due to the crystallite size contribution (ρ_C) was defined as the length of dislocation line per unit volume of crystal and could then be estimated using the Williamson and Smallman relation [37].

$$\rho_C = 3/D^2 \quad (4)$$

where D is the average crystal size. On the other hand, the dislocation density due to the contribution of the microstrain (ρ_S) was evaluated by the following relation.

$$\rho_S = k\frac{\langle \varepsilon_1^2 \rangle}{\bar{b}^2} \quad (5)$$

where k is a material constant (k = 0.9 for cubic crystal), ε_l is the microstrain within the crystal domain, and b is the Burger vector. Likewise, the average dislocation density (ρ) could be estimated from the relationship given as follows.

$$\rho = \sqrt{\rho_C \rho_S} \tag{6}$$

In order to determine the dislocation density for the bainitic transformation range, only the PIT, AIT0.15, and AIT0.35 specimens were taken into consideration. The dislocation density value of the PIT specimen was used as a reference and further compared with that of each AIT specimen so that the dislocation density influenced by the ausforming strain was calculated.

3. Transformation Models

3.1. Transformation Models

According to the thermodynamic principle of bainitic transformation [3,38,39], it has been suggested that if bainite forms under the displacive approach, the transformation can occur when the criteria of $\Delta G_m < \Delta G_N$ and $\Delta G^{\gamma \to \alpha} < -G_{SB}$ are satisfied. The first criterion is defined for the nucleation process, and the latter is for displacive growth. ΔG_m represents the maximum driving force available for the para-equilibrium nucleation. ΔG_N is the universal nucleation function, which was experimentally determined by Ali and Bhadeshia [40]. $\Delta G^{\gamma \to \alpha}$ is associated with a stored energy difference between austenite and bainite, and G_{SB} is a stored energy of bainite, which is usually considered to be 400 J/mol. A further suggestion is that ΔG_m and $\Delta G^{\gamma \to \alpha}$ are dependent on the chemical compositions of steel and undercooling [3]. The temperatures at which $\Delta G_m = \Delta G_N$ and $\Delta G^{\gamma \to \alpha} = -G_{SB}$ are called the T_h and T_0' temperatures, respectively. Hence, it is understood that bainitic transformation can only occur when the transformation temperature is below both temperatures. In any silicon-rich steels in which the formation of cementite is most likely impossible, the concept of T_0' can be used to indicate the incompleteness phenomena of bainitic transformation. This is with respect to carbon, which is only enriched into the adjacent austenite, instead of forming cementite, during the decomposition of austenite into bainite. The temperature is thus defined as a locus where the enrichment process of carbon ceases [10,41]. For steels subjected to ausforming, a prior austenite grain is subdivided into several sub-grains by plastic deformation such that it gives a greater defects per unit volume in an austenite grain, as shown in Figure 1a. The sub-grain boundaries acting as additional defects give rise to more nucleation sites which will then facilitate the bainitic transformation. Therefore, the deformation leads to an increase in austenite free energy and enables more carbon enrichment in austenite at the beginning stage of the transformation. However, on the other hand, the dislocations generated by the growth process will arrest the transformation as it reduces the austenite free energy, as presented in Figure 1b.

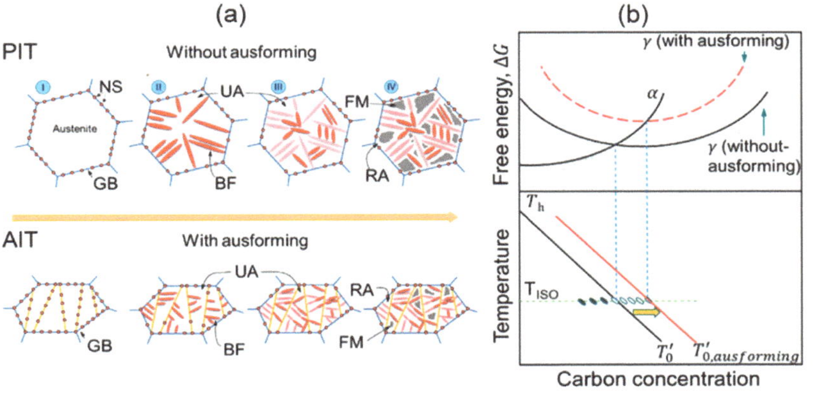

Figure 1. (**a**) Graphical illustrations of the isothermal decomposition of austenite into bainite for PIT and AIT specimens, and (**b**) schematic diagram showing relationships of Gibbs free energy, temperature, and composition of steels under various heat treatments. NS, GB, UA, BF, and FM stand for nucleation site, grain boundary, untransformed austenite, bainitic ferrite, and fresh martensite, respectively.

3.1.1. Nucleation Rate Model

By the displacive approach, the kinetics of the bainitic transformation is controlled by a nucleation rate reaction. Firstly, sub-units of bainitic ferrite (BF) are formed by activated nucleation at the interfaces of the austenite grain boundary (γ/γ interface) and the subsequent autocatalytic nucleation on the pre-existing platelets surrounding the UA (γ/α interface) [27]. The former process incorporates prior austenite grain, while the latter is essential for the growth of BF sheaves, which might arise spontaneously at the tip of previously formed sub-units due to elastic and plastic strain generations within the surrounding austenite [42]. According to this hypothesis, the processes have successively continued to form bainitic sheaves until their evolutions were arrested by the mechanical stabilization of austenite. Therefore, the development of these nucleation mechanisms can be expressed as follows.

$$\left(\frac{dN}{dt}\right)_{\mathrm{t}} = \left(\frac{dN}{dt}\right)_{\mathrm{GB}} + \left(\frac{dN}{dt}\right)_{\mathrm{AN}} \tag{7}$$

where (dN/dt) is the total nucleation rate per unit volume, $(dN/dt)_{\mathrm{GB}}$ and $(dN/dt)_{\mathrm{AN}}$ are the nucleation rate per unit volume arising at the austenite grain boundaries and that caused by the autocatalytic nucleation, respectively. Considering the individual nucleation rate reaction, the density of the potential nucleation sites and the activation energy play a vital role in describing the principle of the nucleation kinetics, which is expressed by Equation (8).

$$\left(\frac{dN}{dt}\right)_{\mathrm{i}} = \frac{k_{\mathrm{B}}T}{h} \mathrm{N_i} exp\left(\frac{Q_{\mathrm{i}}^*}{RT}\right) \tag{8}$$

where the subscript i represents either a nucleation process at the grain boundary (GB) or the successive nucleation by autocatalytic reaction (AN). k_{B} is the Boltzmann's constant, T is the isothermal transformation temperature, h is the Plank's constant, N_{i} is the number of potential nucleation sites at the very beginning stage, R is the gas constant, and Q_{i}^* is the activation energy required for the individual nucleation process. With respect to the original Koistinen–Marburger (KM) equation [43], Van Bohemann and Seitma indicated that the nucleation density is associated with the net driving energy and undercooling [22]. The driving energy is associated with thermodynamic activation, and it thus requires two atomic processes to trigger the reaction. On the one hand, the dissociation of certain dislocation defects may partially facilitate sites for BF embryos. On the other hand, the enrichment of carbon into the surrounding austenite matrix allows a necessarily available driving force for the nucleation.

3.1.2. Activation Energy

As the activation energy accounts for the thermodynamic requirements of the nucleation processes, any energies stored within the parent phase can be presumed to be the parts of the energy barrier necessary for the isothermal transformation. As reported by Olsen and Cohen [44], the activation energy can be determined independently with the free energy sum, as expressed by Equation (9).

$$Q_i^* = Q_0^* + K_1 \Delta G_i \tag{9}$$

where Q_0^* is considered to be the activation energy when the nucleation process is absent, given here as a fitting parameter. However, Q_0^* can be also used to estimate the existing energy required for the phase transformation when no chemical energy is involved and only the effects of undercooling and ausforming are occupied; K_1 is a fitting parameter in the model; and ΔG_i is the total molar driving force of the associated event of nucleation.

In the present study, the extra stored energy due to the deformation of austenite is formulated based on the assumption that no dynamic recrystallization occurred. Thereby, the deformation likely introduces only two energy portions to the austenite, namely the grain boundary and the change of dislocation energy. The variation of these energies depends on two factors. Firstly, the deformation elongates the austenite grains and enhances the grain boundary areas. Secondly, the deformation increases the disorder of the grain structure, which leads to an increase in the grain boundary energy per unit area ($\sigma_{\gamma/\gamma}$), roughly 10% [45,46]. Note that it is not significantly increased if the deformation stress is relatively small. However, the total driving energy required for the grain boundary nucleation is not included the effect of an increase in the number of nucleation sites because it is already involved in the initial activation energy term. The total molar driving force can thus be represented by Equation (10).

$$\Delta G_{GB} = \Delta G_{ch} + \Delta G_{\gamma/\gamma} \tag{10}$$

where ΔG_{ch} is the difference in thermodynamic Gibbs free energy between the austenite and the bainite, determined by using the MUCG83 database [47], and $\Delta G_{\gamma/\gamma}$ is the semi-coherent interfacial energy of the γ/γ interfaces required to overcome its barrier, which can be expressed as the following relationship.

$$\Delta G_{\gamma/\gamma} = 2V_{mol} \frac{\sigma_{\gamma/\gamma}}{nd} \tag{11}$$

V_{mol} is the molar volume of austenite, $\sigma_{\gamma/\gamma}$ is the semi-coherent interfacial energy at the γ/γ interfaces, n is the number of FCC cubic planes along the nucleus thickness, and d is the spacing between the cubic planes, which equals 2.5×10^{-10} m for the FCC planes [44].

Considering the spontaneous association of the nucleation events, the driving force is related to the elastic and plastic stresses generated in the adjacent austenite. In addition to the chemical driving energy change owing to the variations of carbon in BF, the stored energy associated with the stress field can also provide a particular strain accompanying the stress-energy in the surrounding austenite, due to the dislocation barrier that significantly affects the bainitic transformation. Consequently, the semi-coherent interfacial energy between austenite and bainite (γ/α interfaces), established by Dong et al. [45], is a straightforward representation of the subsequent stage of the transformation. Therefore, the total driving energy required for the iterative dynamic nature of the autocatalytic nucleation is represented by Equation (12).

$$\Delta G_{\mathrm{AN}} = \Delta G_{\mathrm{chA}} + \Delta G_{\gamma/\alpha} + G_{\mathrm{dislo}} \tag{12}$$

Parameter ΔG_{chA} is the chemical free energy change depending on the degree of carbon enrichment in the austenite. $\Delta G_{\gamma/\alpha}$ is the semi-coherent interfacial energy between the austenite and bainite (γ/α interfaces). G_{dislo} is the dislocation interaction energy because of the bainitic stress field. The carbon rejected from BF during the transformation is thus specified by the chemical driving energy reduction, while the interfacial energy at the γ/α interfaces is neglected afterwards because its influence is negligible compared to the other factors [44]. Rees and Bhadeshia [20] suggested that the decrease in total energy change is caused by a mechanically induced stabilization surrounding the enrichment zone. Thus, the chemical energy change is then determined as shown in Equation (13).

$$\Delta G_{\mathrm{chA}} = \Delta G_{\mathrm{chAi}}\left(1 - d_{\mathrm{s}} \times \left(\overline{w} + f_{\alpha}\frac{(\overline{w} - w_{\alpha})}{(1 - f_{\alpha})}\right)\right) \tag{13}$$

ΔG_{chAi} is the initial free energy change at the beginning of the transformation or grain boundary nucleation. d_{s} is a constant considered as a fitting parameter in the model. $\overline{w}$ is the average carbon concentration of steel in wt.%. f_{α} is the fraction of bainite formed during the transformation. w_{α} denotes the fraction of carbon entrapped in the bainite, either in solid solution or in a form of iron carbide, depending on the condition of the applied heat treatment and the ability of the carbon enrichment in the austenite due to the alloy addition. For the interaction energy of the bainitic stress field mentioned above, it can be given as follows.

$$G_{\mathrm{dislo}} \approx -\frac{3E^{\gamma}}{2(1+\nu)}\left(\varepsilon_{22}^{\mathrm{tr}}\right)^{2} \tag{14}$$

where $\varepsilon_{22}^{\mathrm{tr}}$ represents the internal elastic strain state resulting from the local strain incompatibility between the bainite and austenite, E^{γ} is the elastic modulus of austenite, and ν is the Poisson's ratio of austenitic steel (0.27). From Equation (9)–(14), the formation of the total energy difference required for activating the transformation at each iteration can be determined as follows.

$$\Delta Q^{*} = Q_{\mathrm{GB}}^{*} - Q_{\mathrm{AN}}^{*} \tag{15}$$

It is noted that the physical basis of such autocatalytic factors has been not adequately given in the current reported literature. Hence, the proposed model provides more details describing the particular net free energy changes caused by the elemental partitioning, particularly when the effect of ausforming is considered.

3.1.3. Potential Nucleation Site Density

In this model, the number density of the potential nucleation sites is estimated based on the original concept of Magee for athermal martensitic transformation [21]. It is reported to be linearly dependent on the undercooling temperature. In this regard, Van Bohemann and Seitsma [48] validated the increase in undercooling with a reduction in the M_{s} temperature due to carbon content. The higher density of nucleation sites results from the driving energy, which is increased when decreasing the transformation temperature of martensite. This concept is later reformulated to be applicable to an isothermal transformation and is given by the following expression.

$$N_{0} = \varnothing(G_{\mathrm{N}}(T_{\mathrm{h}}) - G_{\mathrm{N}}(T_{\mathrm{iso}})) = \xi\Gamma(T_{\mathrm{h}} - T_{\mathrm{iso}}) \tag{16}$$

By analogy with the work of Magee on martensite nucleation, ξ is equal to $\alpha/(V_{\alpha}\Gamma)$. α is a material constant, and Γ is determined by the slope of $d(\Delta G_{\mathrm{m}})/dT$, which is approximately constant in the temperature range of martensitic/bainitic transformation. V_{α} represents the average volume of the bainite sub-unit [20,22]. T_{h} is the highest temperature at which the displacive transformation occurs. Because of the experimental investigations, the number density of the pre-existing defects can be calculated by Equation (17).

$$N_i = \frac{m}{V_\alpha}(T_h - T_{iso}) \tag{17}$$

m is the proportionality constant between the number density of the bainite nucleation sites and the degree of undercooling. In the case of martensite formation, m and α are assumed to be identical and are used as material parameters. Their values are between 0.01 and 0.07 K^{-1} and depend slightly on the chemical composition [21,43]. A fundamental difference between the nucleation of martensite and bainite is that the density of the pre-existing defects for martensite nucleation is governed by the prior austenitic grain size, whereas the bainitic nucleation is also controlled by the structural interfaces, namely the γ/γ and γ/α interfaces. According to Van Bohemann and Seitma's report [22], m can be replaced by b_{GB} and b_{AN} with consideration of the effects of the γ/γ and γ/α interfaces, respectively. The density of the available γ/γ interfaces is dependent on the volume fraction of the remaining available austenite and the austenite grain size. The b_{GB} parameter is thus given as follows.

$$b_{GB} = \frac{Z\delta}{d_\gamma} m f_\gamma \tag{18}$$

The proportional relationship of $Z\delta/d_\gamma$ is defined as the austenite grain boundary area per unit volume, in which δ and d_γ denote the effective thickness of the prior austenitic grain boundary and the austenitic grain size, respectively. f_γ represents the fraction of the UA. In addition, when the effective thickness of the austenitic grain boundary is considered as the atomic layers of a grain in the grain boundary region, a few of the outermost atomic layers in each grain can be presumed to participate in the nucleation process. Hereby, it is suggested that an austenitic grain size is composed of two atomic layers in each grain. With regard to the autocatalytic parameter represented by b_{AN}, it is controlled by the remaining available austenitic phase fraction, at which the α/γ interfaces are formed while the bainitic transformation progresses. The parameter can therefore be given by the following equation.

$$b_{AN} = \frac{Z\delta}{d_\gamma} m f_\gamma f_\alpha \tag{19}$$

where f_α represents the volume fraction of BF. In summary, the size of the BF sub-units has an inverse relationship with the density of the nucleation sites and the remaining available austenite grains.

3.1.4. Austenitic Phase Fraction as a Function of Carbon Enrichment

As shown in Equations (17) and (18), the variations of f_γ and f_α are the critical factors when estimating the density of the nucleation sites. It is kinetically changed during the formation of BF due to the carbon enrichment in the austenite. If the stored energy of bainite caused by shear transformation is involved, the process should be completely terminated at the T_0' locus where the free energies of austenite and bainite are identical, as presented in Section 3.1. By this concept, the conservation of mass balance in the bainite and austenite can be applied, and the variation of carbon concentration in the austenite associated with the locus can be given as follows.

$$w_\gamma = \overline{w} + f_\alpha \frac{(\overline{w} - w_\alpha)}{(1 - f_\alpha)} \tag{20}$$

$\overline{w}$ represents the bulk carbon concentration of steel, and w_α is the composition of carbon in a BF sub-unit. Ravi et al. [27] used the same concept to determine the bainitic transformation temperature and found that the temperature linearly reduces with the increased carbon content in austenite, as provided in Equation (21).

$$T_h = T_{h\overline{X}} - C_1 \frac{(\overline{w} - f_\alpha w_\alpha)}{(1 - f_\alpha)} \tag{21}$$

where $T_{h\overline{X}}$ is the T_h temperature at the beginning of the transformation and is determined using thermodynamic calculation software called MUCG83 [47], and C_1 is a proportional constant relating T_h and the carbon concentration.

Moreover, the total fraction of UA, f_γ, is estimated by considering the T_0' locus as being determined for the carbon enrichment. In this regard, a certain amount of austenite may not participate in the bainitic reaction and yields the incomplete phenomenon. It is assumed that the unavailable austenite certainly contains no potential nucleation site due to its stability. This fraction is thus subtracted while calculating the overall nucleation rate. Using the principles of the incomplete reaction phenomenon, f_γ and T_0' can be expressed as follows.

$$f_\gamma = (1 - f_\alpha) \frac{(T_0' - T_{iso})}{\left(T_{0\overline{X}}' - T_{iso}\right)} \tag{22}$$

$$T_0' = T_{0\overline{X}}' - C_2 f_\alpha \frac{(\overline{w} - w_\alpha)}{(1 - f_\alpha)} \tag{23}$$

where $T_{0\overline{X}}'$ is the T_0' temperature at the beginning of the transformation, and C_2 is a proportional constant relating to T_0' and the carbon concentration. Note that the influences of ausforming on the free energy change of austenite are not incorporated here due to the complexity of the implementation.

3.1.5. Bainitic Transformation Model

From Equations (7) to (23), the overall nucleation kinetics of the isothermal bainitic transformation can be summarized by the following equation.

$$\left(\frac{dN}{dt}\right)_t = \left(\frac{k_B T}{h} \frac{Z\delta}{d_\gamma} \frac{m}{V_\alpha} (1 - f_\alpha) \frac{(T_0' - T_{iso})}{\left(T_{0\overline{X}}' - T_{iso}\right)} (T_h - T_{iso}) \left\{exp\left(-\frac{Q_{GB}^*}{RT}\right)\right\} \{1 - \lambda f_\alpha\}\right) \tag{24}$$

The variable λ in this equation is represented by $exp(\Delta Q^*/RT)$ as an autocatalytic factor. It plays an important role in the kinetics of bainitic transformation and mainly relies on the activation energy difference of the nucleation process. Thus, the autocatalytic function termed by $\{1 - \lambda f_\alpha\}$ can be given as the β parameter, which is similar to that defined in the literature [20,49]. As discussed earlier, the parameters w_α, d_s, ΔQ^*, and K_1 are given as the fitting parameters in this model, from which their physical significances can be defined.

In order to estimate the volume fraction of BF, it can be presumed that BF sheaves are formed by stacking sub-unit layers on the iterative sites of the nucleation and are therefore dependent on the number of nucleated sites. Nonetheless, the volume of a BF sub-unit examined by Matsuda and Bhadeshia [19] is not a constant value, it changes depending on the transformation temperature, which is given as follows.

$$V_\alpha = \left(2 \times 10^{-17}\right) \times \left(\frac{T_{iso} - 528}{150}\right)^3 \tag{25}$$

where V_α represents the average volume of a BF sub-unit in m^3. Consequently, the kinetics of the isothermal bainitic transformation can be calculated by means of a numerical integration of the associated nucleation rates. The product of the integration is given by the following equation.

$$f_\alpha = \int \frac{dN}{dt} V_\alpha dt \tag{26}$$

3.2. Martensitic Transformation

As reported in a previous study [16], the UA after isothermally formed BF is not thermodynamically stable. It can be transformed to FM during ambient cooling because of the low carbon content. Therefore, only the overall kinetics equation for predicting all existing phases in the steel may not be satisfactory. To incorporate the course of martensitic transformation, the conventional KM equation [21,43,50] was modified based on Lee's equation [51] and further developed according to the fact that martensite does not directly transform from a fully austenite. The certain area of the untransformed product of austenite can be expressed as follows.

$$f_{\mathrm{FM}} = (f_{\mathrm{UA}}) - e^{-\alpha_{\mathrm{FM}}(M_{\mathrm{s,UA}} - T_{\mathrm{RM}})^n} \tag{27}$$

f_{UA} is the phase fraction of UA and is equivalent to $1 - f_\alpha$. α_{FM} is a material parameter, which is calculated from the data reported for low-carbon steels with a high concentration of silicon and chromium [51], as given in Equation (28); $M_{\mathrm{s,UA}}$ is a martensite start temperature after the isothermal tempering, indicated as being equal to M_S'; and T_{RM} is room temperature. Moreover, n is also a material constant that depends on the chemical composition, as shown in Equation (29). Finally, the empirical equation of martensitic transformation can be written by Equation (30), where f_{RA} and f_{FM} are the volume fraction of the retained austenite and martensite after cooling, respectively.

$$\alpha_{\mathrm{FM}}\left(K^{-1}\right) = 0.0231 - 0.0105x_{\mathrm{C}} + 0.0074x_{\mathrm{Cr}} - 0.0017x_{\mathrm{Ni}} - 0.0193x_{\mathrm{Mo}} \tag{28}$$

$$n = 1.4304 - 1.1836x_{\mathrm{C}} + 0.7527x_{\mathrm{Cr}}^2 - 0.739x_{\mathrm{Si}} - 0.0258x_{\mathrm{Ni}} + 0.3108x_{\mathrm{Mo}} \tag{29}$$

$$f_{\mathrm{UA}} = f_{\mathrm{RA}} + f_{\mathrm{FM}} \tag{30}$$

4. Results and Discussion

4.1. Experimentally Determined Phase Fractions

In our previous studies [16,52], no carbide precipitation was found in the examined steels with minimized alloys of around 1 wt.%Si. Therefore, the final microstructures of steel should comprise BF and RA ($\alpha + \gamma$), unless the enrichment of carbon in the UA after the bainitic transformation is thermodynamically unstable. In this circumstance, the UA can be partially decomposed to FM (α'), and the heterogeneous microstructure composed of $\alpha + \gamma + \alpha'$ consequently appears. Under thermodynamic equilibrium, the average carbon concentration in the microstructure components of steel could be expressed by $w_\alpha f_\alpha + w_{\gamma,\mathrm{RA}} f_{\gamma,\mathrm{RA}} + w_{\alpha'} f_{\alpha'} = \overline{w}$, where the quantity of w_α, $w_{\gamma,\mathrm{RA}}$, and $w_{\alpha'}$ represents the carbon distribution in BF, RA, and FM, respectively, and $\overline{w}$ is the bulk carbon content of the steels. Figure 2a illustrates the results of the XRD measurements of MC1.5Mn1NiCr steel with and without ausforming. Only the phases belonging to the body-centered and face-centered structures were quantified. The bainitic, ferritic, and martensitic phases that possessed the BCC structure were displayed all together on the same crystallographic planes. The RA whose peaks appeared at the (111), (200), (220) planes was apparently observed, especially in the case of the PIT and AIT specimens, whereas the identical peaks of BF and FM exposed at the (110), (200), and (211) planes were not distinguishable. Thus, the volumetric strain response in the DQ and PIT specimens were exploited together with the volumetric fraction of the BCC obtained by the XRD measurement so that the volumetric fractions of all the constituents could be quantified rationally. Figure 2b shows that the M_s temperature could be determined at the point where the cooling curve started to deviate from its tangent. A vertically measured distance between the tangent and the cooling curve at the final cooling temperature of the DQ results was given as the reference and accounted for the maximum volumetric expansion of the martensitic transformation. By comparing the reference value with the volumetric expansions for the AIT specimen using the lever rule, the martensitic phase fraction of the AIT specimen could be estimated by Equation (31).

$$f_{\alpha'}(\mathrm{AIT}) = \frac{f_{\alpha'}(\mathrm{DQ, XRD}) \times \alpha'(\mathrm{AIT, DIL})}{\alpha'(\mathrm{DQ, DIL})} \tag{31}$$

where $f_{\alpha'}(\mathrm{DQ, XRD})$ represents the volume fraction of martensite acquired from the XRD measurement, and $\alpha'(\mathrm{AIT, DIL})$ and $\alpha'(\mathrm{DQ, DIL})$ are the volumetric strains obtained from the dilatation curves of the AIT and DQ specimens during cooling, respectively. As aforementioned, the total phase fraction of BF was further evaluated from the curve for the change in volume strain at the time where the BF transformation was terminated (distance AB, shown in Figure 3). With help of Equation (30), the amount of UA in the AIT specimen was calculated from the summation of the RA fraction quantified by the XRD analysis and the FM fraction from Equation (31). Subsequently, the distance BC, with regard to the UA volume fraction, was obtained by referring the distance AC to the volume fraction of all the phases. Finally, the proportionality of the dilatation locus was applied to determine the kinetics evolution of the bainitic phase transformation.

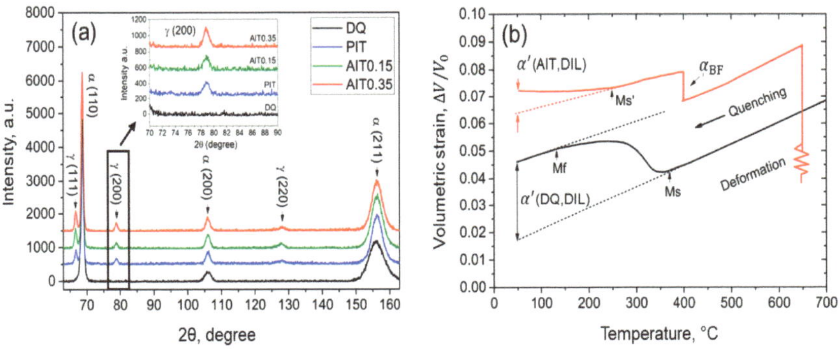

Figure 2. (**a**) Phase quantifications of examined steels using XRD measurement and (**b**) volumetric strain of DQ and AIT specimens after cooling from austenitizing and deformation stage.

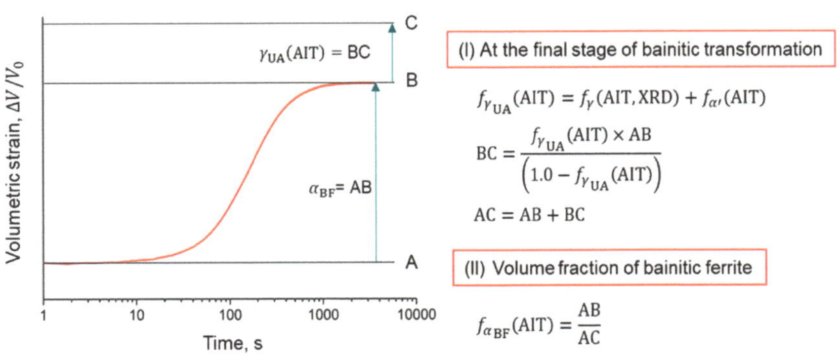

Figure 3. Determination of the volume fraction of bainitic ferrite during tempering at 400 °C.

From the calculations, as shown, the percentage of the developed phases, the amount of carbon enrichment in the austenite in wt.% ($w_{\gamma,\mathrm{RA}}$), the martensite start temperature ($M_{\mathrm{s,exp}}$) of the steels subjected to the DQ, PIT, and the AIT heat treatments were gathered, as listed in Table 2. It is noteworthy that the prior deformation and its degree have played the primary role in enhancing the stability of austenite and have resulted in a decrease in the BF fraction. However, the stabilized austenite in MC1.5Mn1NiCr steel, which had a higher carbon content and low alloyed Cr and Ni contents, was not always sustained along the AIT process, especially when some UA decomposed into FM during cooling. The ausforming strains caused extreme deterioration of the thermodynamic stability of austenite, leading to a greater formation of FM with a slight reduction in the RA fraction. It seems that ausforming only influenced the FM formation, whereas the variation of RA was negligible.

Table 2. Volume phase fraction determined by means of quantitative analysis of XRD measurement combined with the dilatometry results.

Material/Condition	Volume Fraction, %			$w_{\gamma,RA}$(=x_C) Equation (3)
	RA	BF	FM	
MC1.5Mn1NiCr/DQ	-	-	100	-
MC1.5Mn1NiCr/PIT	15.6 ± 3.3	77.4 ± 4.2	7.0 ± 2.8	0.78
MC1.5Mn1NiCr/AIT0.15	14.6 ± 5.1	62.0 ± 6.7	23.4 ± 6.3	0.68
MC1.5Mn1NiCr/AIT0.35	13.5 ± 5.8	47.9 ± 4.5	38.6 ± 3.4	0.56
LC2.5Mn0.2NiCr/DQ	1.2 ± 0.8	-	98.5 ± 2.8	0.04
LC2.5Mn0.2NiCr/PIT	8.5 ± 2.5	83.6 ± 4.6	7.9 ± 3.9	0.45
LC2.5Mn0.2NiCr/AIT0.15	11.3 ± 3.7	77.6 ± 3.4	11.1 ± 4.5	0.51
LC2.5Mn0.2NiCr/AIT0.35	16.9 ± 3.5	74.3 ± 4.9	8.8 ± 4.1	0.89

Meanwhile, it is somewhat surprising that increasing the ausforming strain in the LC2.5Mn0.2NiCr specimens undergoing AIT treatment was quite effective for suppressing the phase transformation. The enrichment of the dislocation density within the plastically deformed austenite had not only accumulated while ausforming, but remained propagating during the BF formation, resulting in the suppression of the formation of FM, with the leaving of more UA as a residual product. It is also noted that the degree of carbon enrichment in austenite during bainitic transformation was directly subjected to UA stability. It was resisted once the structural stability of UA was improved. The carbon concentrations listed in the table represent the overall carbon contents in RA, which take both the products of carbon enrichment and the partitioning processes into account. Therefore, the $w_{\gamma,RA}$ values should be evaluated further to quantify the exact value at individual stages of the transformations.

4.2. Modelling Results

4.2.1. M_s Temperature

The kinetics parameters and the theoretical start temperatures of martensite determined using the modified KM equation for both MC1.5Mn1NiCr and LC2.5Mn0.2NiCr steels under PIT and AIT conditions are presented in Table 3. Note that the parameters, α_{FM} and n, were dependent on the alloying composition and were not affected by the heat treatment conditions. The calculated values of α_{FM} and n for the MC1.5Mn1NiCr steel were 0.0205 and 0.96, respectively, whereas those for the LC2.5Mn0.2NiCr steel were 0.0243 and 1.06, respectively. The discrepancies between the experimental and the calculated values were highly acceptable. Clearly, the martensitic start temperature significantly depended on the heat treatment conditions and the material. For both steels, the M_s temperatures were decreased markedly with the application of ausforming. In the case of MC1.5Mn1NiCr steel, increasing the ausforming strain caused a slightly higher M_s value. The increased strain deteriorated the stability of the austenite and subsequently led to a higher fraction of FM. In contrast, the stability of the austenite in the LC2.5Mn0.2NiCr steel seemed to be effectively promoted by the enhancing of the ausforming strain in accordance with the lowered transformation temperature of the martensite. It implies that the chemical composition was of importance for adjusting the austenite stability of steel, particularly when carbon as an austenite former element was only the most important factor in controlling the bainitic transformation at low temperature.

Table 3. Determined kinetics parameters using the modified KM equation for the examined MC1.5Mn1NiCr and LC2.5Mn0.2NiCr steels under PIT and AIT conditions.

Material/Condition	α_{FM}, K^{-1}	n,−	$M_{s,exp}$, C	$M_{s,cal}$,C
MC1.5Mn1NiCr/DQ	0.0205	0.96	354 ± 5.1	344
MC1.5Mn1NiCr/PIT	0.0205	0.96	347 ± 4.7	332
MC1.5Mn1NiCr/AIT0.15	0.0205	0.96	260 ± 7.1	255
MC1.5Mn1NiCr/AIT0.35	0.0205	0.96	265 ± 6.3	260
LC2.5Mn0.2NiCr/DQ	0.0243	1.06	388 ± 2.4	380
LC2.5Mn0.2NiCr/PIT	0.0243	1.06	351 ± 4.2	345
LC2.5Mn0.2NiCr/AIT0.15	0.0243	1.06	270 ± 3.6	263
LC2.5Mn0.2NiCr/AIT0.35	0.0243	1.06	192 ± 5.8	184

4.2.2. Model Parameters

The input parameters and the critical temperatures calculated by Bhadeshia's program (see Section 3.1.4) are represented in Table 4. It has been seen that the PAGs are decreased by the ausforming treatments and by adding either carbon or other austenite stabilizer elements due to the reduction in the driving energy for grain growth. It is suggested in [45] that the condition of the parent phase controls the driving energy for lower-temperature phase transformation. In general, the energy stored for the nucleation reaction increases with rising the number of nucleation sites and can be considerably dependent on the chemical compositions, undercooling, and prior deformation. If austenite is plastically deformed and more defects are generated as nucleation sites, the primary driving energy of the phase transformation is most probably controlled by nucleation sites increased rather than the other factors. Furthermore, the locus of $T_{h,\overline{X}}$ is hereby broadened as $T'_{0,\overline{X}}$ is suppressed. However, changes in the critical temperature were observed only as a result of the chemical contributions and the undercooling, regardless of the ausforming effect.

Table 4. Parameters used for the transformation models.

Parameter	MC1.5Mn1NiCr			LC2.5Mn0.2NiCr		
	PIT	AIT0.15	AIT0.35	PIT	AIT0.15	AIT0.35
d_γ, μm	48 ± 1.5	43 ± 3.3	35 ± 2.1	56 ± 0.9	49 ± 1.4	44 ± 2.2
T_{iso}, K		673			673	
$T_{h\overline{X}}$, K		753			983	
C_1		2304			2205	
$T'_{0\,\overline{X}}$, K		763			778	
C_2		8911			8537	

4.2.3. Kinetics of Bainitic Phase Transformation

Figure 4 displays the kinetics of the bainitic transformation of MC1.5Mn1NiCr and LC2.5Mn0.2NiCr steels under PIT and AIT conditions at different ausforming strains. It shows that the calculation results fitted very well with the experimental data. The transformation kinetics presented in Figure 4a,b could be divided into three stages, namely onset, growth, and cessation. The onset was indicated at a locus, at which the first BF sub-units were formed. After that, the growth driven by the shear was kinetically captured by the progress of the BF sheaves. Finally, the transformation stopped when the decomposition of austenite was supersaturated as it reached its plateau. Ausforming had a strong influence on an acceleration of the transformation onset and a reduction in the BF volume fraction. However, for MC1.5Mn1NiCr/AIT0.35 steel, the transformation was somewhat sluggish compared to the other conditions for the same material. This circumstance could be precisely explained by the nucleation rate activity and the driving energy, as given in the next section. According to the fact that ausforming enhances the austenite stability, even though the deformation simultaneously accelerated the transformation, the drastic increase in defect density in UA during the initial progress of bainitic transformation resulted in a reduction in the BF fraction. The results were consistent with those reported in [18,53,54], particularly in the MC1.5Mn1NiCr steel. It can be seen that increasing the ausforming strain in LC2.5Mn0.2NiCr steel gave a slight difference in the lowering of the fraction of BF. In light of the chemical variations, LC2.5Mn0.2NiCr steel with a lower carbon content with adjusted Cr and Ni alloying elements possessed higher BF fractions than MC1.5Mn1NiCr steel, regardless of the applied ausforming. It could be verified from their bulk carbon contents and the effective enrichment of carbon in austenite. Hence, it was not surprising that the lower carbon steel exhibited smaller fractions of BF, which is similar to [55].

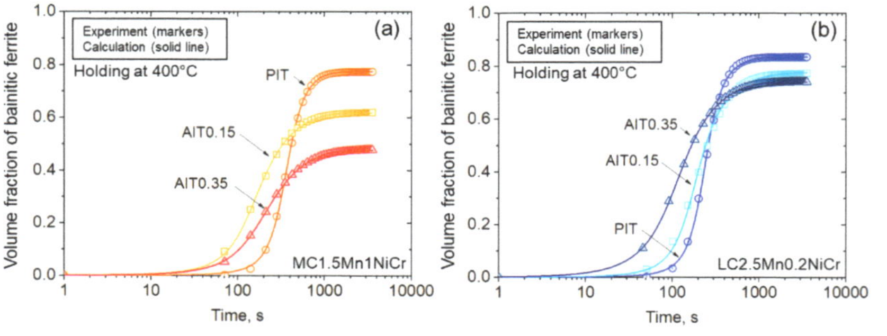

Figure 4. The kinetics of bainitic transformation of (**a**) MC1.5Mn1NiCr and (**b**) LC2.5Mn0.2NiCr steels under PIT and AIT with the strain of 0.15 and 0.35.

The fitted values of w_α, d_s, Q_0^*, and K_1 obtained from the proposed model are shown in Table 5. Only w_α and Q_0^* were well explicable with their physical meaning. The reduction of Q_0^* values signified a decreased energy barrier of the nucleation reaction and seemed remarkably diminished by the ausforming. This parameter was associated with the kinetics acceleration of the BF transformation due to the increased number of existing nucleation sites per unit volume in UA. Considering the lower value of Q_0^* presented in the MC1.5Mn1NiCr/AIT0.15 compared with that in the MC1.5Mn1NiCr/AIT0.35, it could be described by the higher number of nucleation sites or defect densities, which essentially contributed in the later stage of the bainitic transformation. In this respect, it was not surprising that the average amount of carbon concentration in the BF supersaturation (w_α) became constricted with regard to the higher kinetics acceleration at the beginning and was most likely associated with more carbon enrichment. However, the w_α values gathered for the present model were invariably smaller than the mean carbon concentration in the mole fraction of both of the examined MC1.5Mn1NiCr and LC2.5Mn0.2NiCr steels, namely 0.01185 and 0.00824, respectively, which is similar to Ravi's work [27]. The silicon and aluminum additions allowed the specific energy for carbon enrichment and enabled austenite to consume with a certain amount of carbon during the enrichment process, as illustrated in Figure 5. By this model, the evolution of carbon enrichment could be evaluated by the rejection of carbon from the supersaturated BF sub-unit during the austenite decomposition. The presence of RA evidenced the effectiveness of the incomplete reaction that occurred when the austenite was stabilized by the carbon enrichment. Nonetheless, the net carbon value calculated from this model could not be directly compared with the corresponding experimental values in Table 2, because the remaining available carbon may be further partitioned during the subsequent transformation of FM.

Table 5. Fitting parameters obtained for the used model.

Parameter	MC1.5Mn1NiCr			LC2.5Mn0.2NiCr		
	PIT	AIT0.15	AIT0.35	PIT	AIT0.15	AIT0.35
w_{α}, mole fraction	0.0091	0.0056	0.00081	0.0060	0.0048	0.0031
Q_0^*, kJ/mole	172.98	166.48	170.9	167.53	164.83	145.74
d_s	1.47	19.26	12.03	1.81	4.36	9.73
K_1	0.46	4.86	7.32	0.36	0.69	1.29

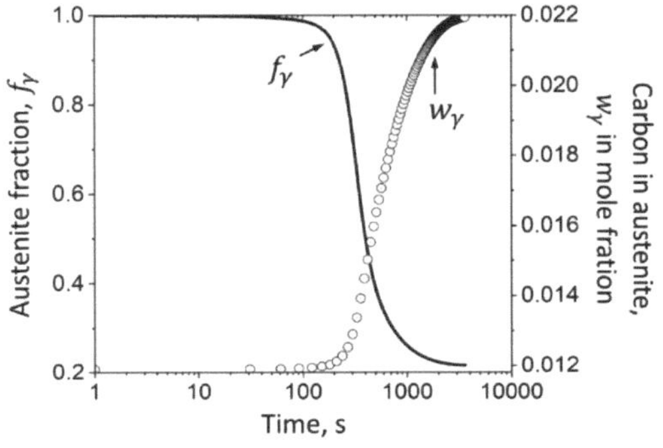

Figure 5. Relationship between the austenite decomposition and the degree of carbon enrichment in austenite of MC1.5Mn1NiCr steel with PIT condition.

As previously mentioned in Section 3.1.1, the kinetics of bainitic transformation was governed by the nucleation rate mechanism, i.e., the grain boundary nucleation and the autocatalytic nucleation. The initial nucleation rate was activated by the grain boundary nucleation, which depended only on the density of the nucleation sites. Subsequent nucleation rates were then controlled by the autocatalytic nucleation with regard to the spontaneous dissociation of the specific dislocation defects and the accumulated stored energy of the corresponding dislocations [56]. Moreover, a generation of additional defects by ausforming presumably provided more available sites while simultaneously deteriorating the activation energy of the nucleation. Enhancing the defect densities thus increased the nucleation rate by triggering the transformation onset. The nucleation kinetics of examined steels subjected to different conditions are presented in Figure 6. It was hereby agreed that at the earliest stage of BF formation the driving energy for nucleation was higher than the stabilization effect, owing to the potential nucleation sites, which were generated during either the ausforming or the cooling to the transformation temperature. The formation of BF continued until the driving energy for nucleation and stored energy due to the mechanically stabilized austenite, which became identical at the maximum nucleation rate. Hereafter, the deceleration kinetics occurred, implying that the stability of the neighboring austenite was increased by the successive formations of the BF sub-units, which resulted in a reduction in stored energy for the transformation. This occurrence could be verified by the relationship between the driving energy for the autocatalytic nucleation and the corresponding number of nucleation sites, as depicted in Figure 7. It is noteworthy that the nucleation rates and the driving energy values of LC2.5Mn0.2NiCr steel were all lower than those of the MC1.5Mn1NiCr steels, regardless of the heat-treated condition. This result was likely caused by the elemental distribution, which was similar to that observed in an experimental validation proposed by Wang et al. [4].

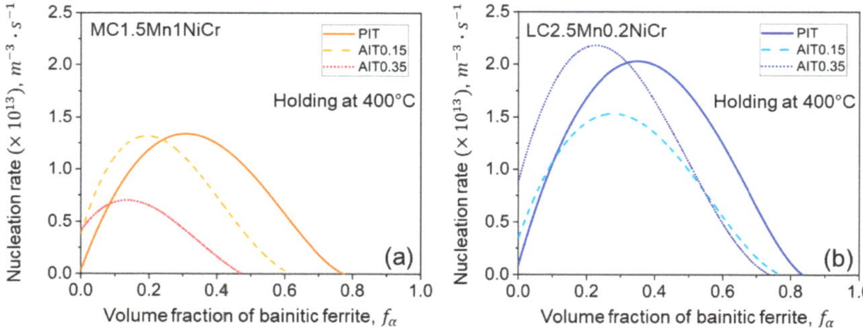

Figure 6. Nucleation rate as a function of BF volume fraction for (**a**) MC1.5Mn1NiCr and (**b**) LC2.5Mn0.2NiCr steels under various conditions.

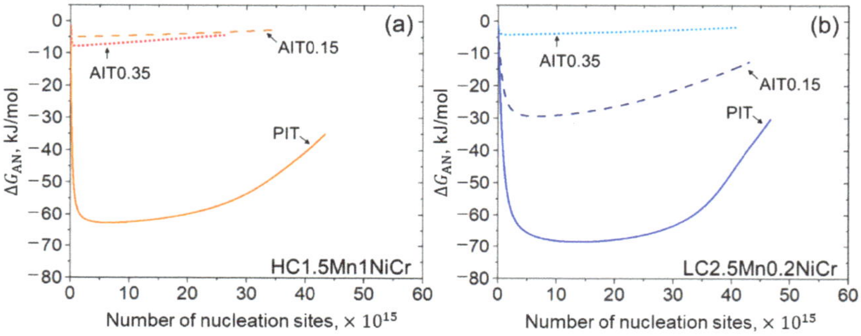

Figure 7. Driving force for autocatalytic nucleation as a function of density of nucleation sites of (**a**) MC1.5Mn1NiCr and (**b**) LC2.5Mn0.2NiCr steels under various conditions.

The autocatalytic driving energy could be also considered as an important factor for controlling the nucleation event. When the austenite stability became more dominant, the total driving energy stored for the phase transformation decreased. It can be observed that the driving energy was drastically increased at the initial stage of nucleus formation and then decreased gradually during the developing of the nucleation sites. The effect of ausforming almost caused degenerations of the nucleus development, accordingly, except that they occurred in MC1.5Mn1NiCr/AIT0.35 steel. The number of nucleation sites developed in the MC1.5Mn1NiCr/AIT0.15 steel was more extensive than in the MC1.5Mn1NiCr/AIT0.35 steel, which corresponded with lower driving energy, as presented in Figure 7a. Even though the MC1.5Mn1NiCr steel consumed less driving energy than the LC2.5Mn0.2NiCr steel, it had a small number of nucleation sites due to the existence of more dislocations. Interestingly, the driving energy of the LC2.5Mn0.2NiCr steels shown in Figure 7b was significantly sensitive to ausforming, and was well described, especially at the high degree of ausforming. From the correlation between the activation energy and the driving force defined in Equation (9), it was inevitable that the initial activation energy required to overcome any obstacles played an essential role in all the nucleation events. Nevertheless, in order to extract and consider only the activation energy influenced by the other mechanisms, regardless of the energy contributed by ausforming, it was reasonable to take the activation energy difference in Equation (15) into account so that a discussion of only the effect of chemical contribution on the variation of the activation energy difference, ΔQ^* was conceivable.

In Figure 8, the variations of the total activation energy difference with the evolution of the BF formation of MC1.5Mn1NiCr and LC2.5Mn0.2NiCr steels under PIT and AIT conditions are illustrated. The carbon enrichment was presumed to be the main contributor for BF formation. It was found that the ΔQ^* energies required in all the conditions of the MC1.5Mn1NiCr steel were somewhat higher than those of the LC2.5Mn0.2NiCr steel. Such a tendency corresponded well with those shown in [27,56], in which it was reported that the activation energy was dependent, on one hand, on the undercooling temperature. It was decreased with the increasing of the temperature as a higher defect density was generated. On the other hand, ΔQ^* was partially governed by the chemical element addition. In particular, those containing more carbon content acting as an austenite stabilizing element could directly lead to suppression of the bainitic transformation, accompanied by the raising of the energy required for the nucleation event. The maximum ΔQ^* energies obtained for both the MC1.5Mn1NiCr and the LC2.5Mn0.2NiCr steels with PIT treatment were consistently higher than those with other heat-treating conditions. It was because of the small number of potential nucleation sites in PIT steel. In this sense, the reduction of the ΔQ^* energy with respect to the evolution of the BF formation should be associated with the increased number of nucleation sites, while the termination of BF development could be associated with an overconsumption of nucleation sites in UA. Hence, the result gave rise to more stable austenite.

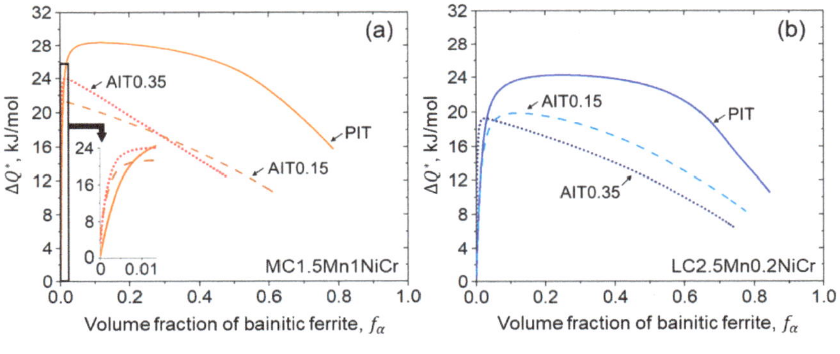

Figure 8. Variation of total activation energy difference with bainite formation evolution of (**a**) MC1.5Mn1NiCr and (**b**) LC2.5Mn0.2NiCr steels under various conditions.

4.2.4. Dislocation Density Estimation

Figure 9a,b shows the estimated dislocation density and the PAGs of the steels under PIT, AIT0.15, and AIT0.35 conditions, respectively. As can be seen, the decreases in the PAGs were not the only factors affecting the density development of the dislocations in this study. The contribution of hard phases such as BF and FM should also be included as the formation of such structures was in fact accompanied by the accumulation of stress [57], particularly when the average dislocation density values estimated from the PIT specimens of individual steels were compared. However, the effect of carbon content on such PAG reduction could be described by the carbon-controlled nucleation mechanism during the solid solution treatment [58]. The influence of the PAGs was more pronounced in the ausformed specimens than in the PIT specimen, and thus it would be reasonable to describe it by the Hall–Petch relation [59].

Under the ausforming conditions, besides the PAGs reduction that came along with the pre-existing density of dislocations, the enhancement of the dislocation density was also associated with the fractions of the hard phase, as mentioned previously. The density of the pre-existing dislocation in the austenite was directly dependent on the ausforming strain, while the formations of BF and FM were governed by the thermodynamic stability of austenite and the degree of carbon enrichment, respectively. In addition, the pre-existing dislocation density affected by ausforming could be evaluated by subtracting the average density of the dislocations of any AIT specimen from the value estimated from the PIT specimen. For the LC2.5Mn0.2NiCr steel, the additional dislocations introduced by the 0.15 and 0.35 strains were 2.4×10^{14} and 3.6×10^{14} m^{-2}, respectively. Likewise, the estimated values of the MC1.5Mn1NiCr steel in accordance with the ausforming strains were 1.3×10^{15} and 1.5×10^{15} m^{-2}, accordingly. Such extra dislocations can be inherited further to the tempering region and change the nucleation activity during the bainitic transformation. Therefore, it could be concluded that the presence of higher dislocations in the MC1.5Mn1NiCr steel was likely due to a strong hindrance of the nucleation reaction, in which the effect of the dislocation-induced mechanical stabilization of austenite was more pronounced. However, the conclusion with respect to ausforming the enhanced thermal stability of austenite may not be applicable for the MC1.5Mn1NiCr steel as more FM was formed. In contrast, the deformation encouraged a higher nucleation activity and was mostly available for a greater formation of nucleation sites during bainitic transformation for the LC2.5Mn0.2NiCr steel. The results hereby enabled UA to be better stabilized after the transformation, although the ausformed LC2.5Mn0.2NiCr steels had a much lower density of dislocations. As a consequence, it seems that the relationship between nucleation sites and dislocation density evolutions could not be concluded as two different trends were seen in two different materials. Nevertheless, the LC2.5Mn0.2NiCr steel, with a substantial BF fraction due to the large degree of ausforming, effectively resisted the formation of FM by improving the thermal stability of austenite.

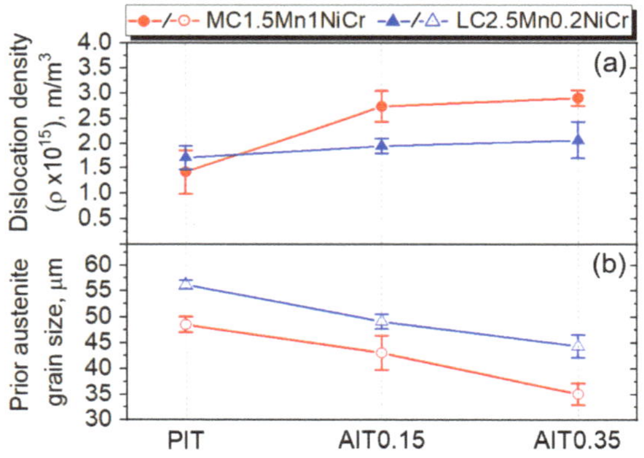

Figure 9. (**a**) Dislocation density estimation and (**b**) prior austenite grain size of MC1.5Mn1NiCr and LC2.5Mn0.2NiCr steels under PIT, AIT0.15, and AIT0.35 treatments.

5. Conclusions/Summary

A model was formulated in this study to describe the effect of ausforming on the kinetics of isothermal bainitic transformation. The behavior of two low-carbon steels can be described based on nucleation control and the Koistinen–Marburger relationship. Different levels of ausforming were systematically characterized in experiments and further used to develop, calibrate, and validate the model. The model distinguishes grain boundary and autocatalytic nucleation. The associated driving energies are controlled by the chemical energy, interfacial energy, and stress-field energy caused by the formation of bainitic ferrite sub-units. The major conclusions are drawn in the following:

- The formation of bainitic ferrite is mainly governed by two factors: carbon enrichment in austenite and the activation energy as an energy barrier required for nucleation.
- Ausforming accelerates the onset of the bainitic phase transformation but results in sluggish transformation due to the mechanical stabilization of austenite. A higher degree of ausforming is more applicable in the steel with lower carbon content. With the substantial development of nucleation sites, even though they provide a slightly lower fraction of bainitic ferrite, the result effectively resists the formation of fresh martensite by improving the thermal stability of austenite.

- A fitting parameter representing the initial energy barrier can be used to examine the activation energy change caused by ausforming. A decrease in the energy barrier allows the acceleration of the transformation. While the transformation progresses, the driving energy for autocatalytic nucleation becomes smaller due to the enhancement of the dislocation density.
- The impact of carbon content plays a slight role in the onset period, but it is more pronounced during the progress of bainitic transformation. Minimizing carbon concentration in steel gives rise to a decrease in the net activation energy difference with the increasing of the nucleation rate. The result allocates a higher density of nucleation sites with more bainitic ferrite fractions.

Author Contributions: Conceptualization, T.K.; methodology, T.K.; validation, T.K., W.B., J.L. and V.U.; formal analysis, T.K.; data curation, T.K. and J.L.; writing—original draft preparation, T.K.; writing—review and editing, T.K., W.B., J.L. and V.U. All authors have read and agreed to the published version of the manuscript.

Funding: This research received no external funding.

Institutional Review Board Statement: Not applicable.

Informed Consent Statement: Not applicable.

Data Availability Statement: Not applicable.

Acknowledgments: The authors wish to thank Jiali Zhang from the Institute for Materials Applications in Mechanical Engineering, RWTH-Aachen University, for her assistance in using XRD measurement.

Conflicts of Interest: The authors declare no conflict of interest.

References

1. Bhadeshia, H.; Edmonds, D.V. Bainite in silicon steels: New composition–property approach Part 1. *Met. Sci. J.* **1983**, *17*, 411–419. [CrossRef]
2. Caballero, F.G. Carbide-free bainite in steels. In *Phase Transformations in Steels*; Elsevier: Amsterdam, The Netherlands, 2012; pp. 436–467, ISBN 9781845699703.
3. Bhadeshia, H. *Bainite in Steels: Transformations, Microstructure and Properties*, 2nd ed.; IOM Communications: London, UK, 2001; ISSN 9781861251121.
4. Wang, S.C.; Yang, J.R. Effects of chemical composition, rolling and cooling conditions on the amount of martensite/austenite (M/A) constituent formation in low carbon bainitic steels. *Mater. Sci. Eng. A* **1992**, *154*, 43–49. [CrossRef]
5. Hofer, C.; Bliznuk, V.; Verdiere, A.; Petrov, R.; Winkelhofer, F.; Clemens, H.; Primig, S. Correlative microscopy of a carbide-free bainitic steel. *Micron* **2016**, *81*, 1–7. [CrossRef] [PubMed]
6. Hofer, C.; Leitner, H.; Winkelhofer, F.; Clemens, H.; Primig, S. Structural characterization of "carbide-free" bainite in a Fe–0.2C–1.5Si–2.5Mn steel. *Mater. Charact.* **2015**, *102*, 85–91. [CrossRef]
7. Hu, F.; Wu, K.M. Isothermal transformation of low temperature super bainite. *Adv. Mater. Res.* **2010**, *146–147*, 1843–1848. [CrossRef]
8. Li, C.W.; Han, L.Z.; Luo, X.M.; Liu, Q.D.; Gu, J.F. Fine structure characterization of martensite/austenite constituent in low-carbon low-alloy steel by transmission electron forward scatter diffraction. *J. Microsc.* **2016**, *264*, 252–258. [CrossRef]
9. Shimanov, M.; Korpala, G.; Terzic, A.; Kawalla, R. Bainitic steels: Their characteristics and applications. *Key Eng. Mater.* **2016**, *684*, 104–110. [CrossRef]
10. Takahashi, M.; Bhadeshia, H. A model for the microstructure of some advanced bainitic steels. *Mater. Trans.* **1991**, *32*, 689–696. [CrossRef]
11. Garcia-Mateo, C.; Sourmail, T.; Caballero, F.G. Bainitic Steel: Nanostructured. In *Encyclopedia of Iron, Steel, and Their Alloys*; Colás, R., Totten, G.E., Eds.; CRC Press: Boca Raton, FL, USA, 2016; pp. 271–290, ISBN 978-1-4665-1104-0.
12. Caballero, F.G.; Santofimia, M.J.; García-Mateo, C.; Chao, J.; De Andrés, C.G. Theoretical design and advanced microstructure in super high strength steels. *Mater. Des.* **2009**, *30*, 2077–2083. [CrossRef]
13. Garcia-Mateo, C.; Caballero, F.G. Advanced high strength bainitic steels. In *Comprehensive Materials Processing*; Elsevier: Amsterdam, The Netherlands, 2014; pp. 165–190, ISBN 9780080965338.
14. Yao, Z.; Xu, G.; Hu, H.; Yuan, Q.; Tian, J.; Zhou, M. Effect of Ni and Cr addition on transformation and properties of low-carbon bainitic steels. *Trans. Indian Inst. Met.* **2019**, *72*, 1167–1174. [CrossRef]
15. Changle, Z.; Hanguang, F.; Shengqiang, M.; Dawei, Y.; Jian, L.; Zhenguo, X.; Yongping, L. Effect of Mn content on microstructure and properties of wear-resistant bainitic steel. *Mater. Res. Express* **2019**, *6*, 86581. [CrossRef]

16. Kumnorkaew, T.; Lian, J.; Uthaisangsuk, V.; Bleck, W. Effect of ausforming on microstructure and hardness characteristics of bainitic steel. *J. Mater. Res. Technol.* **2020**, *9*, 13365–13374. [CrossRef]
17. Lee, C.H.; Bhadeshia, H.; Lee, H.-C. Effect of plastic deformation on the formation of acicular ferrite. *Mater. Sci. Eng. A* **2003**, *360*, 249–257. [CrossRef]
18. Hase, K.; Garcia-Mateo, C.; Bhadeshia, H. Bainite formation influenced by large stress. *Mater. Sci. Technol.* **2004**, *20*, 1499–1505. [CrossRef]
19. Matsuda, H.; Bhadeshia, H. Kinetics of the bainite transformation. *Proc. R. Soc. Lond. A* **2004**, *460*, 1707–1722. [CrossRef]
20. Rees, G.I.; Bhadeshia, H. Bainitic transformation kinetics: Part 1. Modified model. *Mater. Sci. Technol.* **1992**, *8*, 985–993. [CrossRef]
21. Magee, C.L. The kinetics of martensite formation in small particles. *Metall. Mater. Trans. B* **1971**, *2*, 2419–2430. [CrossRef]
22. van Bohemen, S.M.C.; Sietsma, J. Modeling of isothermal bainite formation based on the nucleation kinetics. *Int. J. Mater. Res.* **2008**, *99*, 739–747. [CrossRef]
23. Ghosh, G.; Olson, G.B. Kinetics of F.C.C. → B.C.C. heterogeneous martensitic nucleation-I. The critical driving force for athermal nucleation. *Acta Metall. Mater.* **1994**, *42*, 3361–3370. [CrossRef]
24. Ghosh, G.; Olson, G.B. Kinetics of F.C.C. → B.C.C. heterogeneous martensitic nucleation-II. Thermal activation. *Acta Metall. Mater.* **1994**, *12*, 3371–3379. [CrossRef]
25. Meng, Q.P.; Rong, Y.H.; Hsu, T.Y. Effect of internal stress on autocatalytic nucleation of martensitic transformation. *Metall. Mater. Trans. A* **2006**, *37*, 1405–1411. [CrossRef]
26. Zou, H.; Hu, H.; Xu, G.; Xiong, Z.; Dai, F. Combined effects of deformation and undercooling on isothermal bainitic transformation in an Fe-C-Mn-Si alloy. *Metals* **2019**, *9*, 138. [CrossRef]
27. Ravi, A.M.; Sietsma, J.; Santofimia, M.J. Exploring bainite formation kinetics distinguishing grain-boundary and autocatalytic nucleation in high and low-Si steels. *Acta Mater.* **2016**, *105*, 155–164. [CrossRef]
28. Rees, G.I.; Shipway, P.H. Modelling transformation plasticity during the growth of bainite under stress. *Mater. Sci. Eng. A* **1997**, *223*, 168–178. [CrossRef]
29. Zhou, M.; Xu, G.; Wang, L.; Xue, Z.; Hu, H. Comprehensive analysis of the dilatation during bainitic transformation under stress. *Met. Mater. Int.* **2015**, *21*, 985–990. [CrossRef]
30. Lee, S.; Lee, S.J.; de Cooman, B.C. Austenite stability of ultrafine-grained transformation-induced plasticity steel with Mn partitioning. *Scr. Mater.* **2011**, *65*, 225–228. [CrossRef]
31. Mahieu, J.; de Cooman, B.C.; Maki, J. Phase transformation and mechanical properties of si-free CMnAl transformation-induced plasticity-aided steel. *Metall. Mater. Trans. A* **2002**, *33*, 2573–2580. [CrossRef]
32. Abrams, H. Grain size measurement by the intercept method. *Metallography* **1971**, *4*, 59–78. [CrossRef]
33. Dyson, D.J.; Holmes, B. Effect of Alloying Additions on the Lattice Parameter of Austenite. *J. Iron Steel Inst.* **1970**, *208*, 469–474.
34. Lutterotti, L.; Scardi, P. Simultaneous structure and size–strain refinement by the Rietveld method. *J. Appl. Crystallogr.* **1990**, *23*, 246–252. [CrossRef]
35. Popa, N.C. The (hkl) dependence of diffraction-line broadening caused by strain and size for all Laue groups in Rietveld refinement. *J. Appl. Crystallogr.* **1998**, *31*, 176–180. [CrossRef]
36. Warren, B.E. *X-ray Diffraction*; Dover Publications Inc.: New York, NY, USA, 1969; ISBN 9780201085242.
37. Williamson, G.K.; Smallman, R.E., III. Dislocation densities in some annealed and cold-worked metals from measurements on the X-ray debye-scherrer spectrum. *Philos. Mag.* **1956**, *1*, 34–46. [CrossRef]
38. Fielding, L.C.D. The Bainite Controversy. *Mater. Sci. Technol.* **2013**, *29*, 383–399. [CrossRef]
39. Caballero, F.G. Bainitic Steel: Transformation Mechanisms and Properties. In *Encyclopedia of Iron, Steel, and Their Alloys*; Colás, R., Totten, G.E., Eds.; CRC Press: Boca Raton, FL, USA, 2016; pp. 291–305, ISBN 978-1-4665-1104-0.
40. Ali, A.; Bhadeshia, H. Nucleation of Widmanstätten ferrite. *Mater. Sci. Technol.* **1990**, *6*, 781–784. [CrossRef]
41. Bhadeshia, H.; Edmonds, D.V. The mechanism of bainite formation in steels. *Acta Mater.* **1980**, *28*, 1265–1273. [CrossRef]
42. Tszeng, T.C. Autocatalysis in bainite transformation. *Mater. Sci. Eng. A* **2000**, *293*, 185–190. [CrossRef]
43. Koistinen, D.; Marburger, R. A general equation prescribing the extent of the Austenite-Martensite transfor-mation in pure iron-carbon alloys and plain carbon steels. *Acta Metall.* **1959**, *7*, 59–60. [CrossRef]
44. Olsen, G.B.; Cohen, M. Dislocation theory of martensitic transformations. In *Dislocations in Solids*; Nabarro, F.R.N., Ed.; Elsevier Science Publishers B.V.: Amsterdam, The Netherlands, 1986; pp. 293–408.
45. Dong, H.; Sun, X. Deformation Induced Ferrite Transformation. In *Ultra-Fine Grained Steels*; Weng, Y., Ed.; Springer: Berlin/Heidelberg, Germany, 2009; pp. 86–136, ISBN 978-3-540-77230-9.
46. Ghosh, G.; Raghavan, V. The kinetics of isothermal martensitic transformation in an Fe-23.2wt.%Ni2.8wt.%Mn alloy. *Mater. Sci. Eng. A* **1986**, *80*, 65–74. [CrossRef]
47. Bhadeshia, H. MAP Program MAP_STEEL_MUCG83. Available online: http://www.phase-trans.msm.cam.ac.-uk/map/steel/programs/mucg83.html (accessed on 12 July 2021).
48. van Bohemen, S.M.C.; Sietsma, J. Kinetics of martensite formation in plain carbon steels: Critical assessment of possible influence of austenite grain boundaries and autocatalysis. *Mater. Sci. Technol.* **2014**, *30*, 1024–1033. [CrossRef]
49. Santofimia, M.J.; van Bohemen, S.; Sietsma, J. Combining bainite and martensite in steel microstructures for light weight application. *J. S. Afr. Inst. Min. Metall.* **2013**, *113*, 143–148.

50. Capdevila, C.; Caballero, F.G.; De Andrés, C.G. Determination of Ms temperature in steels: A bayesian neural network model. *ISIJ Int.* **2002**, *42*, 894–902. [CrossRef]
51. Lee, J.S.; van Tyne, C.J. A kinetics model for martensite transformation in plain carbon and low-alloyed steels. *Metall. Mater. Trans. A* **2012**, *43*, 422–427. [CrossRef]
52. Wirths, V. Prozessführung und Zyklisches Werkstoffverhalten von Karbidfreien Bainitischen Stählen. Ph.D. Thesis, Rheinisch-Westfällischen Technischen Hochschule Aachen, Aachen, Germany, 2016.
53. Shipway, P.H.; Bhadeshia, H. Mechanical stabilisation of bainite. *Mater. Sci. Technol.* **1995**, *11*, 1116–1128. [CrossRef]
54. Hu, H.; Xu, G.; Wang, L.; Zhou, M.; Xue, Z. Effect of ausforming on the stability of retained austenite in a C-Mn-Si bainitic steel. *Met. Mater. Int.* **2015**, *21*, 929–935. [CrossRef]
55. Soliman, M.; Palkowski, H. Development of the low temperature bainite. *Arch. Civ. Mech. Eng.* **2016**, *16*, 403–412. [CrossRef]
56. Shah, M.; Kumar, D.S.; Ankita, P. Phenomenological kinetic model of the nano-bainitic steels to characterize the dynamics of the autocatalytic nucleation process. *SN Appl. Sci.* **2020**, *2*, 635. [CrossRef]
57. PEET, M.J.; Bhadeshia, H. Surface relief due to bainite transformation at 473 K (200 °C). *Metall. Mater. Trans. A* **2011**, *42*, 3344–3348. [CrossRef]
58. Castro Cerda, F.M.; Sabirov, I.; Goulas, C.; Sietsma, J.; Monsalve, A.; Petrov, R.H. Austenite formation in 0.2% C and 0.45% C steels under conventional and ultrafast heating. *Mater. Des.* **2017**, *116*, 448–460. [CrossRef]
59. Narutani, T.; Takamura, J. Grain-size strengthening in terms of dislocation density measured by resistivity. *Acta Metall. Mater.* **1991**, *39*, 2037–2049. [CrossRef]

metals

MDPI

Article

Microstructural Evolution in Large-Section Plastic Mould Steel during Multi-Directional Forging

Xuan Chen, Boya Wu, Jiayuan Li, Xiaoxiao Zhang, Pengpeng Zuo, Xiaochun Wu and Junwan Li *

School of Materials Science and Engineering, Shanghai University, Shanghai 200444, China; chenxuan9311@163.com (X.C.); boya03@163.com (B.W.); lijiayuan_0129@163.com (J.L.); zhangxx1020@163.com (X.Z.); zpp0513@163.com (P.Z.); wuxiaochun@shu.edu.cn (X.W.)

* Correspondence: lijunwan@shu.edu.cn

Abstract: To obtain excellent mechanical properties from large cross-sections of plastic mould steel (SDP1), we conducted multi-directional forging (MDF) to control the microstructure of ingots. To investigate the microstructural evolution of SDP1 steel during MDF, we performed hot forging at 1150 °C using a THP01–500A hydraulic press. The dimensions of the specimens were Φ38 mm × 80 mm. The microstructure of the specimens after forging was observed under a metallographic microscope. Furthermore, the results of the finite element method (FEM) simulations were employed to improve the quality of the forgings. The predicted results agreed well with the experimental ones, indicating that FEM is effective for analysing microstructural evolution during MDF. Thus, MDF for large cross-sections of SDP1 steel (Φ1000 mm × 2200 mm) was simulated. The results showed that the average grain size of SDP1 steel at the core of an ingot after MDF ranged from 40.6 to 43.3 μm. Although this was slightly higher than the grain size of the sample after traditional upsetting and stretching forging (TUSF) (35.7–46.0 μm), the microstructure of the SDP1 steel sample after MDF was more uniform than that after TUSF. Compared with TUSF, MDF not only refines the grain size but also improves the microstructure uniformity of the sample.

Keywords: multi-directional forging; large cross-section plastic mould steel; microstructure evolution; FEM simulation

Citation: Chen, X.; Wu, B.; Li, J.; Zhang, X.; Zuo, P.; Wu, X.; Li, J. Microstructural Evolution in Large-Section Plastic Mould Steel during Multi-Directional Forging. *Metals* **2022**, *12*, 1175. https://doi.org/10.3390/met12071175

Academic Editors: Shuwen Wen, Yongle Sun and Xin Chen

Received: 12 June 2022
Accepted: 8 July 2022
Published: 10 July 2022

1. Introduction

During the fabrication of large plastic mould steel blooms, the low impact toughness of the product is a major quality issue. Through microstructural analysis, this phenomenon is attributed to the coarse grain and nonuniform grain size [1–3]. As the most important manufacturing process for plastic mould steel, microstructural control in free forging processes determines product quality. Discontinuous dynamic recrystallisation (DRX), which occurs during hot deformation processes, is an important microstructural evolution behaviour. Recently, severe plastic deformation (SPD) methods have been widely studied owing to their ability to provide fairly good grain refinement in metallic alloys [4]. These techniques improve mechanical fatigue, corrosion, wear and creep properties of various alloys [5,6]. Several SPD methods, such as accumulative roll bounding, equal-channel angular pressing, friction stir processing and constrained groove pressing and multi-directional forging (MDF) [7–17], have found industrial applications. MDF has various advantages, such as repeatability, cost effectiveness, simplicity and the possibility of fabricating large parts [18–21].

MDF is a strong plastic deformation process improved by hammer forging compared with traditional unidirectional forming processes [22–25]. During MDF, the ingot is continuously elongated or compressed along the *x*-, *y*- and *z*-axes with changing loading direction, which has a great influence on the flow stress behaviour and the microstructure [26–28]. Although MDF of a large number of ferrous and nonferrous alloys have already been studied, there are few investigations on plastic mould steel available in the literature. For example, Moghanaki et al. [24] studied the effect of solution treatment on the mechanical properties of Al-Cu-Mg alloy during multi-directional forging. The hardness of solution treated samples after MDF was strongly increased from 84 to 163 HV, which is attributed to both dynamic formation of Guinier-Preston-Bagaryatsky(GPB) zones/Cu-Mg co-clusters and dislocation storage. Nakao et al. [27], based on experiments on microstructure and mechanical properties, proved that the MDF can improve the level of grain refinement of stainless steel. In the process of MDF, the grain fragmentation degree increased due to the coordination of mechanical twins and martensitic transformation. With the increase in strain, the tensile strength significantly improved. Soleymani et al. [29] conducted multi-directional forging experiments on low-carbon steel modules and found that the grain refinement mechanism was static recrystallisation when the forging temperature was 500 °C. The new grains nucleated within the deformed grains, grew and consumed the deformed tissues until the material was completely recrystallised.

MDF has been studied and verified by many scholars, but in the manufacturing process of plastic mould steel, MDF technology has not been studied and applied. In this paper, both MDF and the traditional upsetting and stretching forging (TUSF) were carried out on the forging of large cross-sections of plastic mould steel. Based on the numerical simulation of the forging processes, the strain state and the evolutionary law of the microstructure during MDF and TUSF were investigated and discussed to provide effective guidance for the production of forgings.

2. Materials and Methods

2.1. Material

The material used in this study was as-forged micro-alloyed plastic mould steel (SDP1®), and its normal chemical composition is listed in Table 1. The steel sheet was manufactured by air and spray cooling after forging, and its thickness was 700 mm. The microstructure was observed and captured using an optical microscope (OM, Nikon, LV 150, Tokyo, Japan) and a field-emission scanning electron microscope (Carl Zeiss SUPRA 40, Oberkochen, Germany). For OM observation, the samples were mechanically ground using sandpaper, polished, and etched in a 4% nital solution.

Table 1. Normal chemical composition of SDP1 plastic mould steel (wt.%).

Element	C	Cr	Mn	Mo	Si	N	Nb	S	P	Fe
Wt.%	0.30	1.40	2.00	0.30	0.20	0.008	0.035	≤0.003	≤0.015	Bal.

2.2. MDF Procedure

Cylindrical samples (38 mm in diameter and 80 mm in length) were prepared by wire-electrode cutting, followed by homogenisation at 1150 °C for 2 h. MDF was performed to refine the grains of the SDP1 steel, and it was performed based on three-step forging (Figure 1a) using a THP01-500A hydraulic press(Tianduan, Tianjin, China). MDF was performed at an initial forging temperature of 1150 °C and a speed of 5 mm·s^{-1} with a controllable deformation speed. During the process, the reduction was controlled at 50%. The ingots were reheated to 1150 °C for 0.5 h between every pass. In Figure 1a, P1 is located in the centre of the upper surface of the cylinder, P2 in the centre of the cylinder body, and P3 in the middle of the cylinder side surface. The samples were repeatedly deformed up to three passes by changing the loading direction in each pass by 90°. Prior to the second forging step, the pancake produced in the first step was rotated by 90° around the x-axis to press the y-plane of the ingot. The third forging process step was similar to the second step. Thus, the z-, y- and x-planes of the sample were compressed, in turn, to obtain fine grains. After the MDF process, to prevent the forged samples from cracking due to the rapid cooling, the ingots were precooled until the temperature of the centre surface was below 850 °C, after which they were quenched in water. Samples 1–3 (Figure 1b) corresponded to the first, second and third steps of the forged ingots. To investigate the microstructure of the MDF samples at different regions, three positions on the samples were selected for microstructural observation (Figure 1a). The samples were cut from the middle section of the forged billets. The middle area was representative of the entire sample and convenient for comparison. Austenite grain boundaries were shown by the oxidation method. The average grain size of the samples under different hot deformation positions was calculated based on the volume average of the grain using the three-circle truncated-point method. Several metallographic photographs were recorded at each sampling point to create grain size data statistics, and Image-pro Plus (version 6.0, Media Cybernetics, Rockville, MD, USA) was used to revise the data to obtain accurate and reliable average grain-size data.

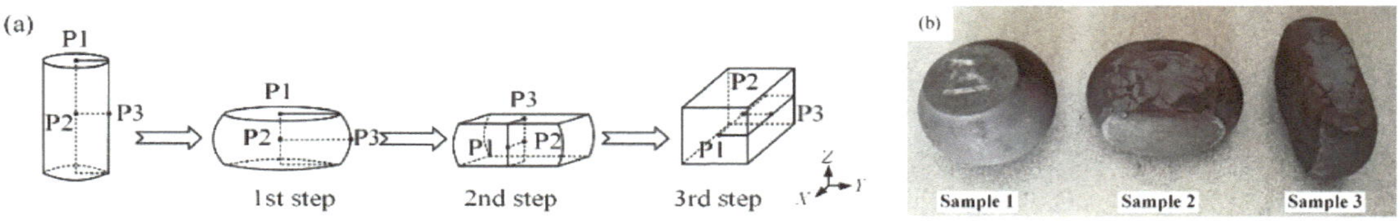

Figure 1. MDF experiment: (**a**) sequence of MDF steps; (**b**) state of SDP1 steel after MDF.

3. Experimental Results and Analysis

Figure 2 shows the grain size and morphology of the ingot before forging. The microstructure revealed fairly equiaxed grain morphology delineated with definite grain boundaries. The microstructure was characterised dominantly by coarse grains; the average grain size of the ingots was 120 μm, indicating coarse grains before the forging process. The ingots were collected from the hydraulic press after plastic deformation, but no crack was observed at the surface of the billets, suggesting that the alloy exhibited good workability at 1150 °C.

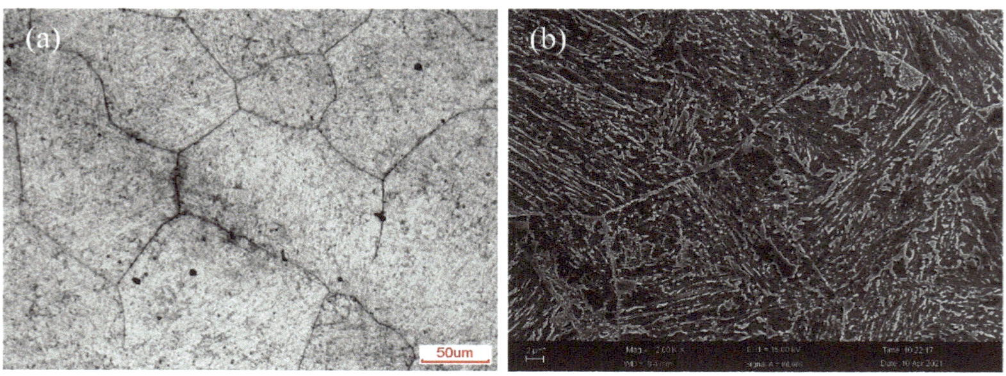

Figure 2. (**a**) initial grain size and (**b**) morphology of the SDP1 steel.

Figure 3 shows deformation microstructures of the samples after MDF, and Figure 4 shows the grain size distribution statistics of the SDP1 steel samples after MDF. The black area in the metallographic photographs shows an unpolished oxide layer, which does not affect the evaluation of the grain size. The grain size varied significantly with location during the MDF process (Figure 4a). Many deformed fine grains were observed in Sample 1, indicating recrystallisation during the MDF process. After the first step in the MDF process, many coarse grains were observed at point P1, and it was difficult to obtain more grain refinement. However, the grain size of points P2 and P3 decreased sharply. The average grain sizes of points P1, P2 and P3 of Sample 1 decreased from 120 to 119.0, 28.2 and 29.7 μm, respectively. The MDF first step slightly affected the average grain size of point P1, and more grain refinement could only be achieved by increasing the number of MDF passes. After the second step in MDF, the size distribution of point P1 showed significant refinement, and the distribution was dominated by 30–40 μm grains (approximately 42.86%); in addition, grains in the range of <10 μm and ≥40 μm constituted approximately 5.71% and 22.86%, respectively. After the third MDF step, the size distribution of point P1 was dominated by 20–40 μm grains (approximately 70.73%), and grains ≥ 40 μm constituted only 2.44%.

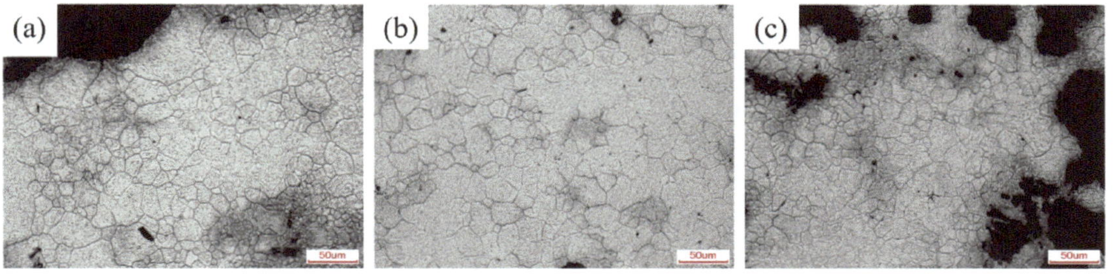

Figure 3. Microstructures of point P2 in (**a**) Sample 1; (**b**) Sample 2; (**c**) Sample 3.

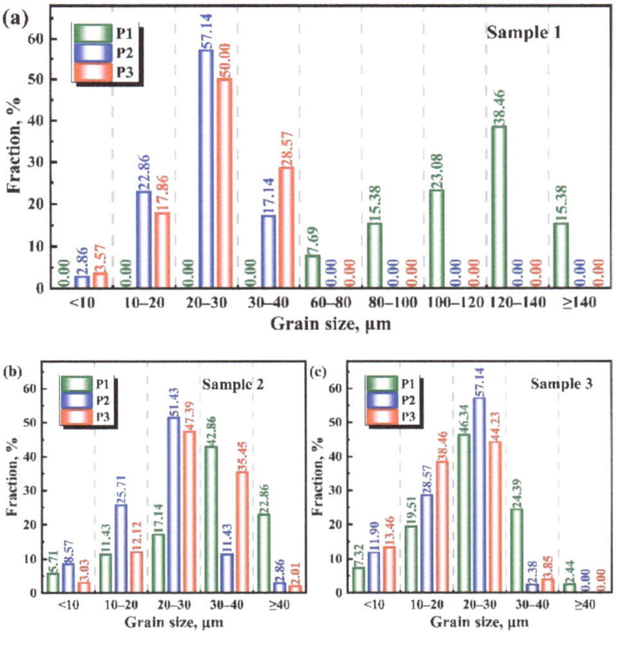

Figure 4. Grain size distribution and statistics of (**a**) Sample 1; (**b**) Sample 2; (**c**) Sample 3 at different tracking points.

All MDF steps resulted in remarkable grain refinement. With an increase in the MDF passes, the number fraction of the fine recrystallised grains increased (Figure 4b), and the recrystallisation grains became much finer than those obtained after one MDF pass. The grain size distribution of point P2 (Figure 4b) showed a broad distribution between 20 and 30 μm, and the average grain size was 26.5 μm. The number fraction of fine grains at point P2 increased compared with that of Sample 1. With a further increase in MDF passes, the number fraction of fine recrystallised grains further increased (Figure 4c). A uniform fine grain structure evolved in Sample 3. The deformation microstructure was almost composed of mainly equiaxed fine grains considered to be recrystallisation grains. Homogeneous microstructures with average grain sizes of 28.7, 25.0 and 22.9 μm were obtained at points P1, P2 and P3, respectively. Few grains larger than 40 μm were obtained only at P1. The corresponding grain size distribution was characterised by a relative peak against the grain sizes ranging from 20 to 30 μm. The number fractions of the grains ranging from 20 to 30 μm in Sample 3 processed at points P1, P2 and P3 were 46.34%, 57.14% and 44.23%, respectively. The results show that MDF improved grain refinement. In addition to local coarse grains, the grains of the ingot were fine and uniform.

4. Numerical Procedure

4.1. Microstructural Evolution Model

Recrystallisation is a process by which deformed grains are replaced by new sets of grains that nucleate and grow until the original grains have been entirely consumed. Recrystallisation is important in developing microstructures in deformation-processed metals. Recrystallisation may occur during or after deformation; the former is termed dynamic, and the latter is termed static. The microstructure evolution model of SDP1 steel used in this simulation was derived from our previous calculations (Table 2) [30]. Combining the recrystallisation kinetic model and FE simulation of SDP1 steel, including the DRX, static recrystallisation (SRX) and meta recrystallisation (MRX), the effects of MDF and TUSF on the microstructural evolution of the ingots were investigated.

Table 2. Microstructural evolution model of SDP1 steel.

Quantity	Equation
Zener–Holloman parameter	$Z = 1.06 \times 10^{12}[\sinh(0.0094\sigma_P)]^{5.35}$
DRX grain size evolution	$D_{\mathrm{DRX}} = 21400.5\left[\dot{\varepsilon}\exp\left(\frac{257325}{RT}\right)\right]^{-0.26}$
DRX rate	$\varepsilon_{0.5} = 0.0059\dot{\varepsilon}^{0.23}\exp\left(\frac{46187}{RT}\right)$
Peak strain	$\varepsilon_P = 0.0011\left[\dot{\varepsilon}\exp\left(\frac{257325}{RT}\right)\right]^{0.22}$
DRX fraction	$X = 1 - \exp\left[-0.693\left(\frac{\varepsilon-\varepsilon_c}{\varepsilon_{0.5}-\varepsilon_c}\right)^2\right]$
Time for 50% SRX	$t_{0.5} = 4.5 \times 10^{-8}\varepsilon^{-1.27}\dot{\varepsilon}^{-0.25}\exp\left(\frac{151231}{RT}\right)$
SRX volume fraction	$X_{\mathrm{SRX}} = 1 - \exp\left[-0.693\left(\frac{t}{t_{0.5}}\right)^{1.1}\right]$
Time for 50% MRX	$t_{0.5} = 1.22 \times 10^{-6}\dot{\varepsilon}^{-0.41}\exp\left(\frac{126605}{RT}\right)$
MRX volume fraction	$X_{\mathrm{MRX}} = 1 - \exp\left[-0.693\left(\frac{t}{t_{0.5}}\right)^{0.6}\right]$
Grain growth	$d_g^{8.18} = d_0^{8.18} + 6.03 \times 10^{54} t \exp\left(\frac{-800866}{RT}\right)$

σ_P—peak stress, MPa; ε—strain; $\dot{\varepsilon}$—strain rate; ε_c and $\varepsilon_{0.5}$—critical strain rate and the DRX rate; R—gas constant, 8.314 J mol^{-1} K^{-1}; T—deformation temperature, K; t—time, s; d_0—initial grain size, μm.

4.2. FEM Model

A comparative experiment was conducted to verify the accuracy of the recrystallisation model. To compare with the hot compression test, a cylindrical ingot (Φ38 mm × 80 mm) was used for MDF. The speed of the press was approximately 5 mm s^{-1}. During MDF, the height of the ingot after each pass was reduced to 40, 27 and 35 mm along the z, y and x central axes of the ingot, respectively.

As shown in Figure 5, a hexahedron element was used to discrete the ingot during FEM simulations. The suitable mesh ensured simulation precision and saved computing resources. The as-homogenised SDP1 steel had an average grain size of 120 μm. Based on the experimental data, the average initial grain size of the ingot was set to 120 μm, and other simulation parameters were also consistent with the experimental values. Combining the recrystallisation kinetic model and the FEM of SDP1 steel, the effects of MDF on the microstructural evolution in forgings were simulated.

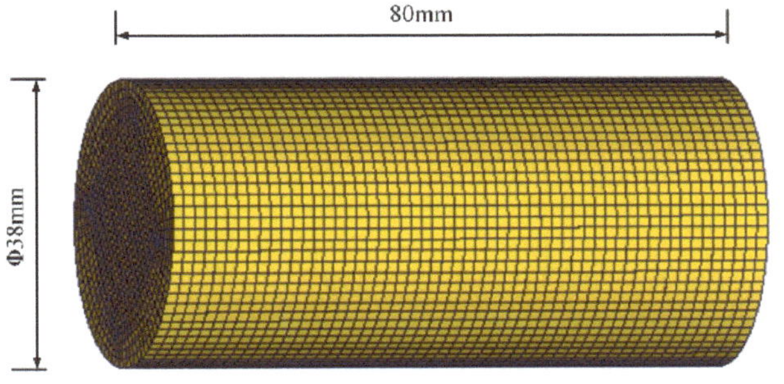

Figure 5. Three-dimensional finite element (FE) model of the specimen.

Microstructural evolution is not a major concern in the FE simulation of hot forging; thus, hot forging numerical theory only considers the coupling effect of the temperature and stress/strain fields. The coupling effect between the temperature and stress/strain evolution is described as follows. On the one hand, the plastic-deformation work accumulated in forging is converted into an internal heat source at a certain proportion. On the other hand, because of the existence of a temperature gradient, the expansion is different. Finally, a thermal strain is induced.

The instantaneous temperature field of the forging blank in a forging process can be expressed by the Fourier heat-transfer equation in a rectangular coordinate system.

$$\frac{\partial}{\partial X}(\lambda\frac{\partial T}{\partial X}) + \frac{\partial}{\partial Y}(\lambda\frac{\partial T}{\partial Y}) + \frac{\partial}{\partial Z}(\lambda\frac{\partial T}{\partial Z}) + \frac{\partial Q_2}{\partial t} = \rho c\frac{\partial T}{\partial t} \quad (1)$$

where X, Y and Z are the three directions of the rectangular coordinate system; ρ, c and λ are the density, specific heat capacity, and thermal conductivity of the materials, respectively; and Q_2 and t are the internal heat source and time, respectively. The change rate of the heat source can be expressed as follows:

$$\frac{\partial Q_2}{\partial t} = \dot{q} = kq \cdot \overline{\sigma} \cdot \dot{\overline{\varepsilon}} \quad (2)$$

where kq is the conversion ratio of the heat-to-plastic deformation work; $\overline{\sigma}$ is the effective stress; $\dot{\overline{\varepsilon}}$ is the effective strain rate. This equation realizes unilateral coupling of the stress/strain field to the temperature field.

Initial and boundary conditions must be determined to solve the transient-heat-conduction problem. The initial temperature of the forging blank was uniform and not less than 1150 °C. It can be expressed as:

$$T(X,Y,Z)|_{t=0} = T_0(X,Y,Z) \quad (3)$$

The boundary condition of the forging blank can be divided into two categories depending on whether the forging blank is in contact with the die. The boundary condition of a free surface that is not in contact with the model is expressed as follows:

$$q\mathrm{f} = H\mathrm{f}(T_\mathrm{f} - T_\mathrm{E})(S \in S\mathrm{f}) \quad (4)$$

where S is the module surface; Sf is a free surface; q_f and T_f are the heat flow and temperature of the free surface, respectively; T_E is the environment temperature; Hf is the total heat-transfer coefficient including convection and radiation heat transfer. Thus, the boundary condition of the free surface belongs to the second boundary condition.

For the surface in contact with the die, the heat-transfer mechanism was very complex, which can be roughly divided into the following two heat-transfer mechanisms. On the one hand, heat conduction was achieved through the contact points between the blank and anvil and solid interstitials (oxide skin, etc.). In addition, the friction heat generation between the die and forging blank should be considered. Thus, the boundary condition of the contact surface can be written as:

$$qc = Hc(T_c - T_d) + q\mu(S \in Sc) \tag{5}$$

where Sc represents the contact surface; q_c and q_μ are the total and friction heat flows of the contact surface, respectively; H_c denotes the total heat-transfer coefficient of the contact surface; T_c and T_d are the temperatures of the contact and die surfaces, respectively.

The stress/strain field of hot forging satisfies the assumption and basic equation of a rigid-plastic material. In addition, we need to specify the coupling effect of the temperature field on the stress/strain field.

$$\Delta\varepsilon^T = \alpha_2(\Delta T) \tag{6}$$

where ε^T and ΔT are the thermal strain and temperature increments, respectively; α_2 is the thermal expansion coefficient of the forging blank.

The forging blank is cylindrical with dimensions of Φ1000 mm × 2200 mm (Figure 6). Because the forging blank is formed in different directions, instead of simplifying the model using a symmetrical relationship, we used the entire model. The geometry of the forging blank was meshed using a hexahedral element, and more than 10,000 hexahedral elements were obtained. To simulate the heat transfer between the forging blank and the mould, meshing the forging blank is insufficient; thus, a discrete mould geometry was employed.

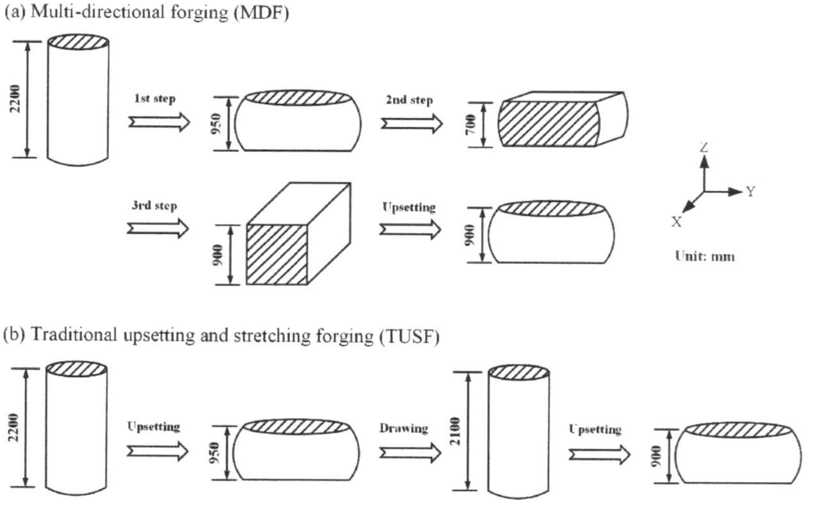

Figure 6. Schematic of (**a**) MDF and (**b**) traditional upsetting and stretching forging (TUSF) processes.

Furthermore, TUSF was conducted to compare with MDF. In these two forging processes, the first step was to upset the z-direction. Thus, the forging blank height could be forged from 2200 to 950 mm. In the second TUSF step, the blank was forged in the y-direction using a small smith anvil, and the height was reversed to 2200 mm. In the drawing process, the feed and reduction rates were maintained at 50% and 25%, respectively. In the second and third MDF steps, the blank was forged to 700 and 900 mm in the y- and x-direction, respectively, by successively using a large Smith anvil. A TUSF pass consisted of upsetting and elongation in the z- and y-directions, respectively, whereas one MDF pass involved upsetting in the z-direction and compression in the y- and x-directions. Finally, the MDF and TUSF samples were upset to a height of 900 mm. Thus, TUSF and MDF differ in terms of the steps after z-direction upset. Upset in the z-direction has attracted attention because it is the first step and results in maximum deformation.

For both TUSF and MDF, a compressive speed of 10 mm/s was adopted. The friction and heat transfer coefficients of the anvil and blank were 0.3 and 11, respectively. The initial temperature of the forging blank was 1150 °C, and the temperature of the anvil and the ambient temperature were 150 °C and 20 °C, respectively. Before each step, the temperature of the blank was reset to 1150 °C. Based on the austenite grain size in the high-temperature homogenisation state, the average initial grain size of the ingot was set to 200 μm. Combining these two forging procedures and the FE model of hot forging, we simulated two-pass TUSF and MDF to analyse the distribution of temperature, effective stress and effective strain in the forging blank.

5. Results and Discussion

When the sample was compressed with a flat anvil, the metal flew around as the height decreased because of the friction between the specimen and the anvils, making it difficult for the metal to deform near the contact surface. Thus, the compressed specimen formed a drum shape, resulting in different grain sizes at different positions on the specimen. During the first MDF step (i.e., upsetting), the deformed specimen is divided into three regions (Figure 7). Region I is located near the interface between the specimen and the anvils, which is greatly affected by friction, reducing plastic deformation [25]. Region II is far from the end surface and is subject to less friction. It is in the most favourable deformation area at an angle of 45° to the vertical force and prone to plastic deformation. Region III is a small deformation zone, which is close to the surface of the specimen and around region I, and the extent of deformation is between that in regions I and II.

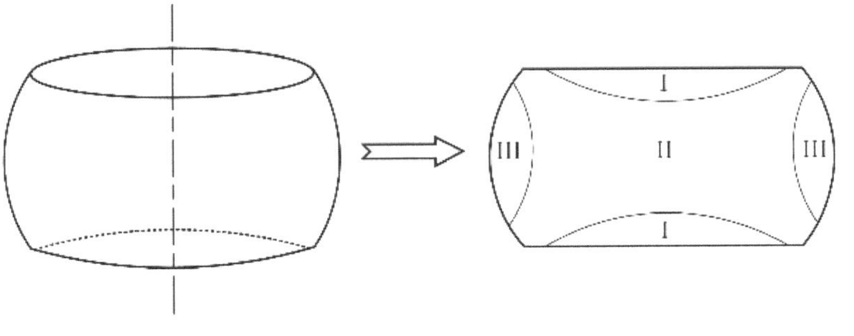

Figure 7. Schematic of difficult (I), large (II) and small (III) deformation areas of the specimen after upsetting.

5.1. Comparative Analysis of Experimental and Simulation Results

Figure 8 shows the experimental and simulated specimens after MDF. The arrows in Figure 8a–c indicate the flow state of the material during plastic deformation. Comparing the two sets of data, we find that FEM could well predict the shape and size of deformation forgings, which can guide the forging process. Figure 9 shows the average grain size distribution of the specimen after different forging steps. The samples show regular shapes after different forging steps, and a drum shape was formed on the side of each sample, which was similar to the shape of the ingot used in the experiment. Figure 9a shows the grain size distribution after the first step of MDF. Compared with Figure 4, the average grain size near the interface was quite large. The grain size near the centre of the interface (i.e., point P1) was the largest (119 μm), which is consistent with the initial grain size without refinement. The core of Sample 1 had the smallest grain size of 24.2 μm. The grain size of point P3 was close to but not as uniform as that of point P2. After MDF, the grain of Sample 3 was tiny and distributed uniformly. For some areas with uneven grain distribution, secondary forging or other heat-treatment processes can be carried out. The simulation results were in good agreement with the experimental results. The FEM simulation results showed that the evolutionary law of the average grain size can well reflect the microstructural evolution of recrystallisation. Meanwhile, FEM is an effective method for analysing the thermal compression process, and the obtained recrystallisation model of SDP1 is reliable and effective.

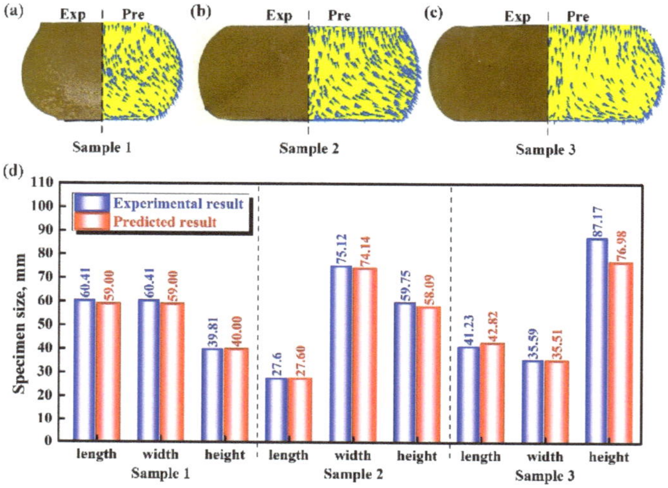

Figure 8. Experimental and simulated specimen sizes after MDF. (**a**) sample 1, (**b**) sample 2, (**c**) sample 3, (**d**) three-dimensional comparison.

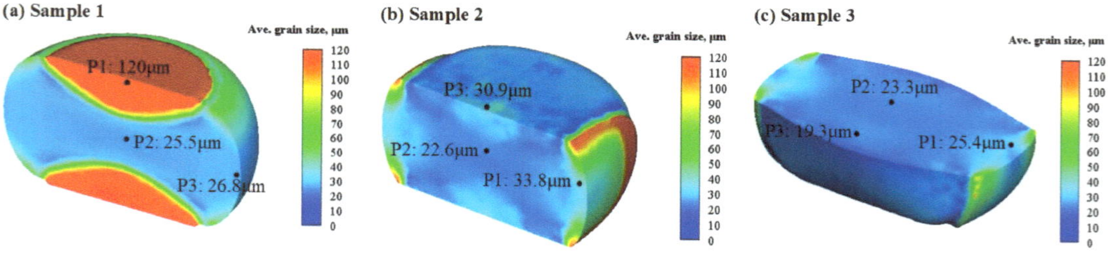

Figure 9. Average grain size of (**a**) Sample 1; (**b**) Sample 2; (**c**) Sample 3 at different tracking points obtained from FEM results.

5.2. Microstructural Evolution during the First MDF Step

Figure 10 shows the effective strain and stress contour plots of the deformed specimen after MDF at 1150 °C. The results show an inhomogeneous distribution of effective strain in the samples during MDF. The nonuniform strain distribution after upsetting agreed with the characteristics of the three deformation zones in the specimen. The accumulated strain in region I was higher than that in the other two regions. During upsetting, the strain distribution was extremely uneven, and the strain value decreased gradually from the centre along the radius to the side surface. In region I, where the specimen was in contact with the anvils, due to the friction resistance and heat transfer, the deformation was very small, because the magnitude of friction varied inversely with the distance, and region I was a cone. The outer drum part of region III was a free surface, which was less affected by friction resistance and depended on the expansion of region I; thus, longitudinal bulging was observed. With the same reduction, the deformation decreased from the centre to the end face. The maximum effective strain inside the sample reached 1.78, and the minimum effective strain was 0.03 after upsetting.

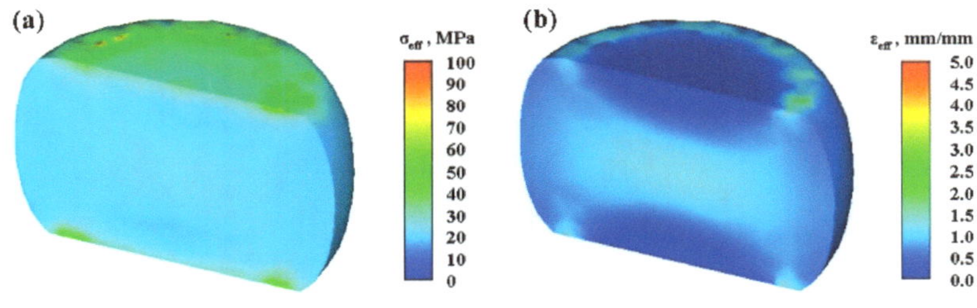

Figure 10. (**a**) Effective stress and (**b**) strain distributions of specimens after upsetting.

Grain refinement due to the fact of recrystallisation during the forging process determines the final grain size of an alloy. As shown in Figure 11a, the uneven distribution of deformation resulted in a difference in grain size in different regions. The as-homogenised coarse grain structure underwent significant grain refinement during the upsetting process. The recrystallisation volume fraction of region I was small because of the small strain, and there was no recrystallisation at the centre of the region. The grain size of the region was the same as the initial grain size, and recrystallisation occurred only in some regions. Due to the severe deformation in region II, there was adequate recrystallisation in the region, the grain size was uniform, and the centre of the region was completely recrystallised. The grain size of region III was not uniform, unlike in regions I and II. Compared with other regions, region II showed the largest deformation and the most sufficient recrystallisation. Compared to the edge region, the accumulated strain at the centre was larger, and recrystallisation occurred at the initial stage of the MDF process, providing more potential sites for recrystallisation nuclei. A higher accumulated strain indicates higher stored energy and an increased driving force for recrystallisation [31]. Consequently, a finer and more homogeneous grain structure was formed at the centre region after upsetting. In addition, due to the heat transfer between the specimen and the dies and that between the specimen and the environment, the final temperature around the specimen was much lower than that of the specimen core after upsetting. Consequently, finer and more homogeneous grain with an average grain size range of 45.7–49.9 μm developed at the centre region after upsetting.

Recrystallisation is a process that can induce stress relaxation to varying extents in a deformed metal by releasing the stored energy arising from the deformation process when heat-treated at an appropriate temperature [32]. Recrystallisation is important in microstructural development in deformation-processed metals [33]. During upset simulation, three types of recrystallisation behaviour, including DRX, SRX and MRX, may occur within the sample. DRX occurs during high-temperature deformation, and it aids grain refinement and homogenisation. Figure 11b shows that the DRX volume fraction in region II was sufficient for grain refinement. The DRX fraction at the core of the sample reached 100%, and the grains were uniformly distributed. The microstructure agreed with the stress–strain field. SRX occurs in high-temperature multi-pass deformation gaps of materials. MRX differs from SRX. It occurs with DRX at the core and continues to grow within the gap time. As shown in Figure 11c,d, neither SRX nor MRX occurred during upsetting. This was because upsetting was a one-off finish, and there was no time for deformation gaps. Thus, it cannot provide sufficient conditions for SRX or MRX [34–40].

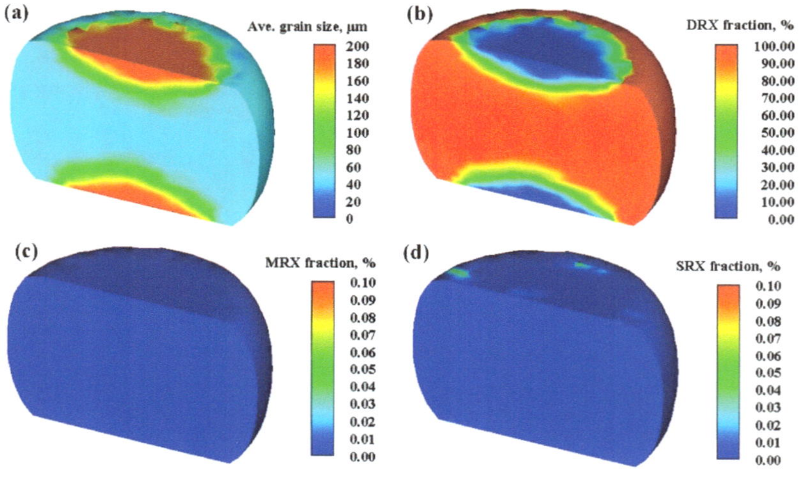

Figure 11. (**a**) Average grain size; (**b**) dynamic recrystallisation (DRX); (**c**) meta-recrystallisation (MRX); (**b**,**d**) static recrystallisation (SRX) fractions in the specimen after upsetting.

To further investigate the grain evolution, different points on the specimen were observed as shown in Figure 12a. With an increase in reduction during the MDF process, both the effective strain at the centre (P2) and side (point P3) regions increased, and the difference in the accumulated strain between the centre and side regions increased, but there was a little change in the effective strain at P1. After upsetting, the effective strain at points P1, P2 and P3 were approximately 0.03, 1.24 and 0.72, respectively. The average grain sizes at points P2 and P3 decreased as the reduction exceeded a certain value (Figure 12c). Point P2 recrystallised in 17 s, which was earlier than that at P3 (18 s), and the rate of grain refinement at P2 was greater than that at P3. Compared with P3, P2 showed a better grain refinement effect, and the final average grain size was obtained first. In contrast, P1 was restricted by friction resistance, the strain was always lower than the critical strain, and no recrystallisation occurred. The final average grain size at P1, P2 and P3 was 200, 45.4 and 45.6 μm, respectively. The DRX volume fractions at P2 and P3 (Figure 12d) increased when the reduction exceeded a certain value, which is consistent with Figure 12b,c. At 17 s, that is, when the strain reached 0.15, the recrystallisation fraction at P2 increased first. Complete recrystallisation occurred at P2 and P3 after upsetting, but no DRX occurred at P1, which is in a difficult deformation zone. The final DRX fractions at P1, P2 and P3 were 0%, 100% and 100%, respectively.

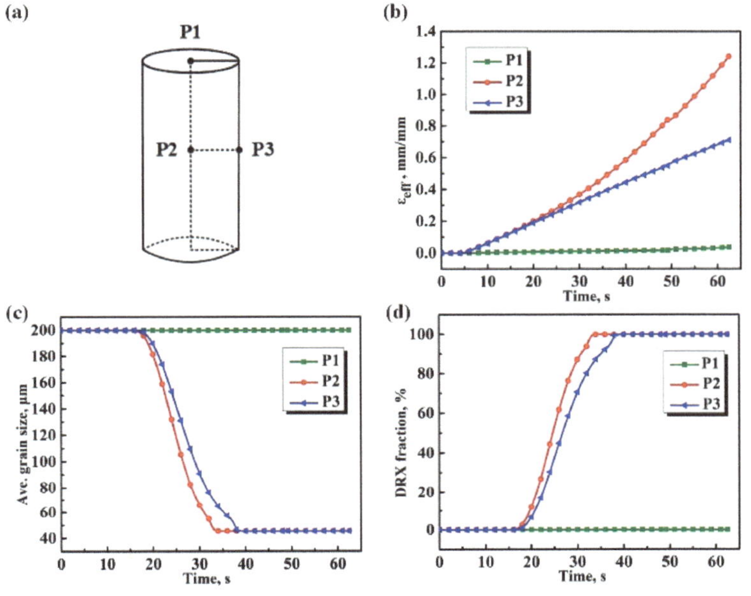

Figure 12. (**a**) Different tracking points; (**b**) effective strain; (**c**) average grain size; (**d**) DRX fraction curves.

5.3. Comparison of Microstructural Evolution between MDF and TUSF

Drawing is a forging process in which the cross-sectional area of a blank is reduced, and the length is increased by deformation. According to the strain during each compression deformation, drawing increases the length through continuous accumulation of axial strain. The end surface of the specimen was in contact with an anvil; thus, it was affected by friction resistance, resulting in limited deformation and smaller strain at the region. On the other hand, friction resistance decreased towards the specimen centre, resulting in more deformation at the centre (Figure 13). Herein, the maximum and minimum strains after MDF were 3.86 and 0.6, respectively. Compared with MDF, the feeding amount of each drawing step was small in TUSF, and it was necessary to rotate the specimen to achieve deformation. Therefore, the TUSF process was more complex and inefficient. After TUSF, the maximum and minimum strains were 3.43 and 0.32, respectively.

The average grain size varied from 40.6 to 43.3 μm at the centre of the specimen after MDF (Figure 14a) and from 35.7 to 46.0 μm after TUSF (Figure 14b). The average grain size of the specimen core after TUSF was slightly smaller, but the range was higher than that of the sample after MDF. The DRX of MDF was more adequate, and the grain at the centre was more uniform; meanwhile, the surface of the sample was neater after MDF than after TUSF. The minimum and maximum grain sizes were 1 and 60 μm, respectively after MDF. However, the surface of the sample after TUSF was not smooth. The minimum and maximum grain sizes were 12 and 132 μm, respectively. The large span of the grain size and incomplete recrystallisation in the local region after upsetting and drawing affected the comprehensive performance of forging and, subsequent, processing. The SRX and MRX volume fractions of the MDF ingot were much lower than those of traditional forging, because MDF is a continuous compression deformation process. The ingot showed DRX during deformation, and the DRX fraction in most regions of the ingot reached 100%. In contrast, the conditions were provided for SRX and MRX during the TUSF process. In general, grain refinement was mainly achieved by DRX; grain refinement and dynamic softening caused by the SRX and MRX were relatively low.

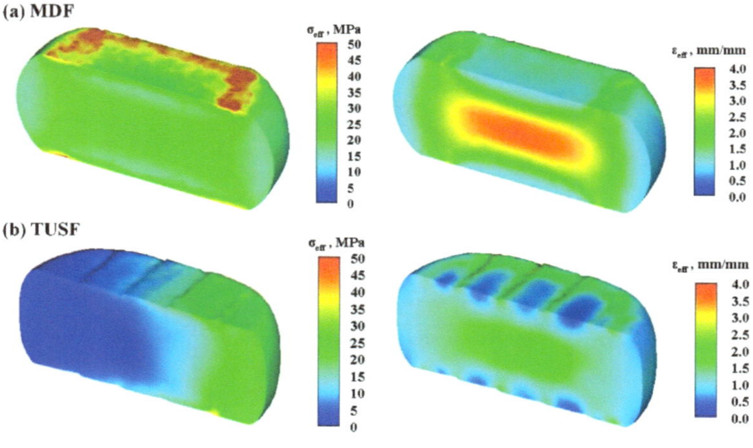

Figure 13. Effective stress and strain distributions of specimens after (**a**) MDF and (**b**) TUSF.

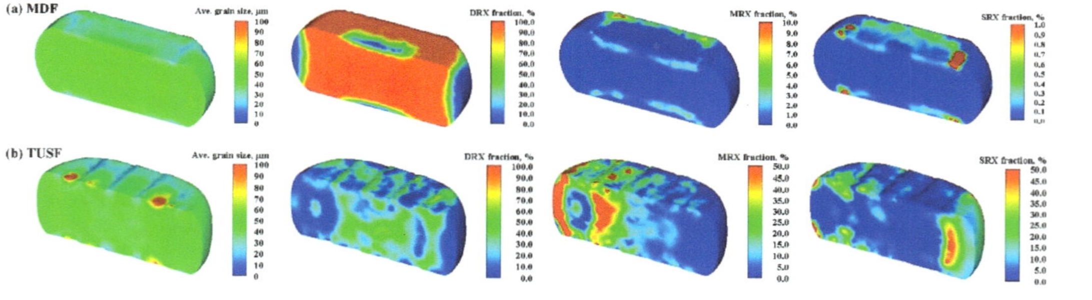

Figure 14. Schematic diagram of the microstructural distribution and recrystallisation behaviour after (**a**) MDF and (**b**) TUSF.

Second upsetting deformation was carried out after a complete MDF and TUSF to compare and analyse the effects of MDF and TUSF on subsequent deformation of ingots. Due to the influence of friction resistance on the end surface of the specimen, the nonuniform deformation of the specimen showed a drum shape. The distribution of effective stress and strain was consistent with the characteristics of the three deformation regions in the upsetting process (Figure 15). As shown in Figure 16, with an increase in reduction, strain at the three points increased gradually and reached a maximum. The initial effective strains at P1, P2 and P3 in the MDF sample were 0.86, 3.89 and 1.88, respectively, and the final strains were 1.15, 5.02 and 2.10, respectively. The initial values at P1, P2 and P3 in the TUSF sample were 0.91, 2.07, 1.18, and the final strain values were 1.07, 3.18 and 1.63, respectively.

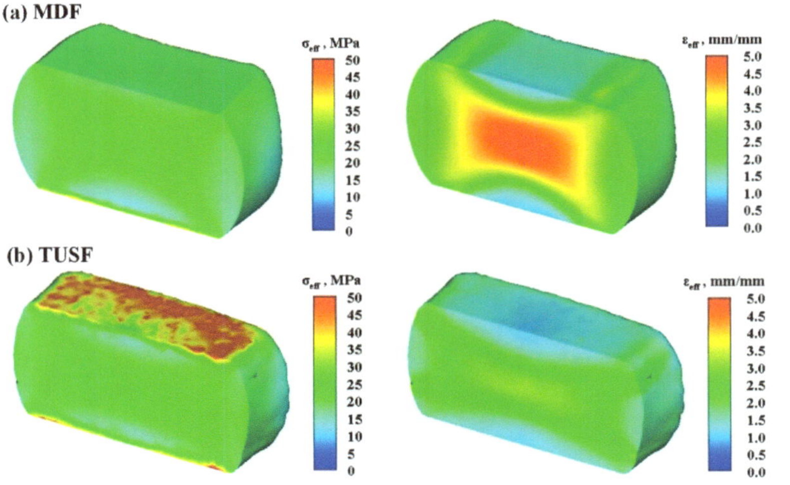

Figure 15. Effective stress and strain distributions in (**a**) MDF and (**b**) TUSF specimens.

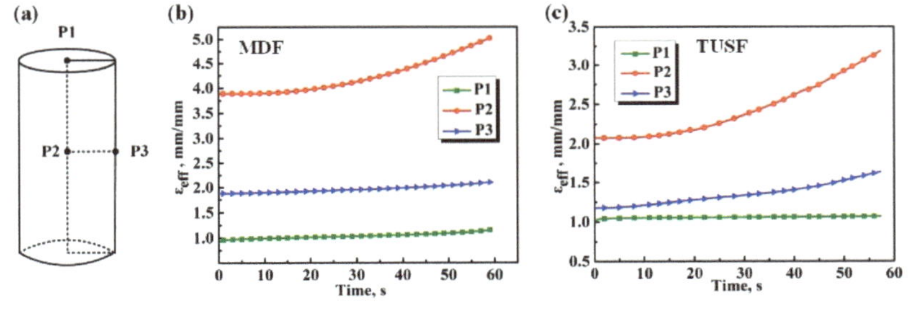

Figure 16. Time-dependent effective strain curves for (**b**) MDF and (**c**) TUSF specimens at (**a**) different tracking points.

The average grain size ranged from 40.2 to 41.8 μm in the core of the MDF specimen and from 38.7 to 42.2 μm in that of the TUSF specimen. Their average values were similar. The grain size in the core of the TUSF specimen was slightly smaller, but the range was slightly higher. Analysing and comparing Figure 17a,b, we can find that DRX in the MDF sample was more sufficient. The recrystallisation volume fraction in the core of the MDF specimen reached 100%, and the grain size was more uniform than that of the TUSF specimen. The minimum and maximum grain sizes of the entire sample were 12 and 52 μm, respectively. However, the surface of the TUSF specimen was not smooth, and the minimum and maximum grain sizes were 13 and 63 μm, respectively. The grain size of a traditional forging specimen was large, the grain of the sample centre was not uniform and there were mixed crystals, which can affect the overall performance of forged samples and subsequent processing.

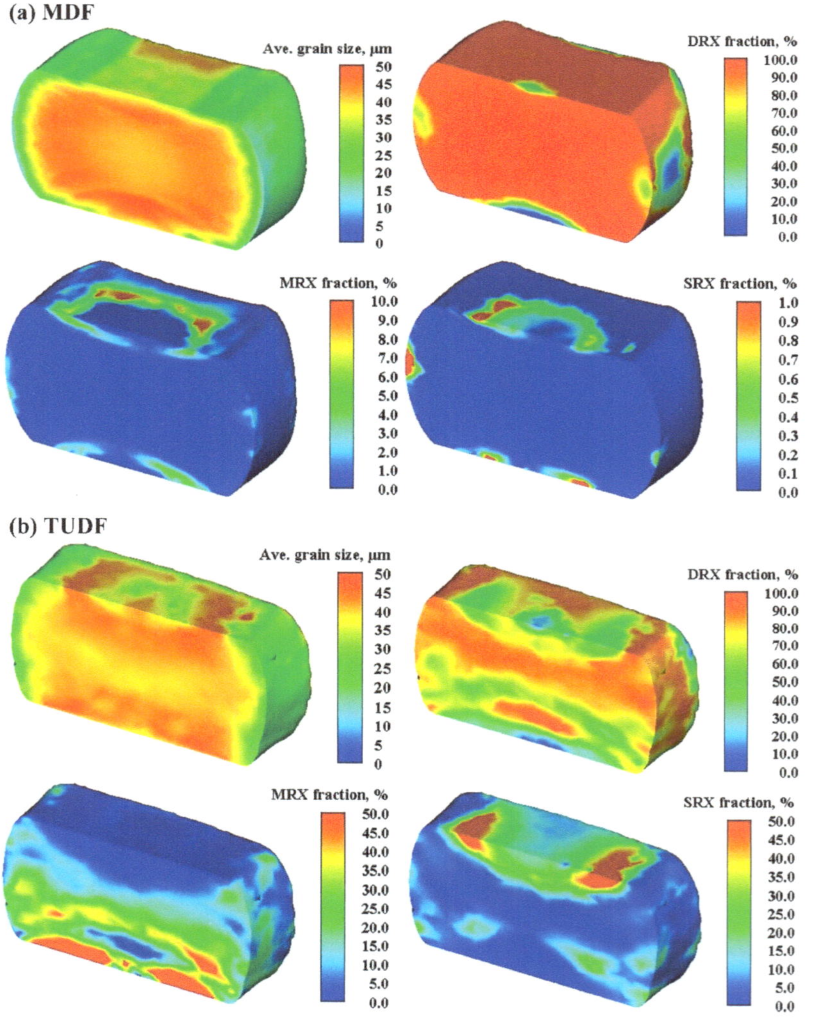

Figure 17. Microstructural distribution contours for the (**a**) MDF and (**b**) TUSF specimens.

The initial grain sizes at P1, P2 and P3 in the MDF specimen were 39, 40 and 17 μm, and the final grain sizes were 36, 40 and 14 μm, respectively (Figure 18). The grain size at P2 was almost unchanged, and that at P1 and P3 decreased significantly, possibly because the deformation on the side surface was relatively large, and the grains on the surface were easily broken. However, the temperature of the core remained at the initial temperature of 1150 °C, the grain plasticity was good, and the grains would not fracture during plastic deformation. On the other hand, the initial grain sizes at P1, P2 and P3 on the TUSF specimen were 46, 39 and 41 μm, and the final grain sizes were 42, 39 and 24 μm, respectively. After further treatment, the grain structure at P2 remained unchanged, and the average grain size was 39 μm. There are more variations in the average grain size and DRX volume fraction in the TUSF specimen (Figure 18b), attributed to the uneven distribution of grain size and DRX during drawing.

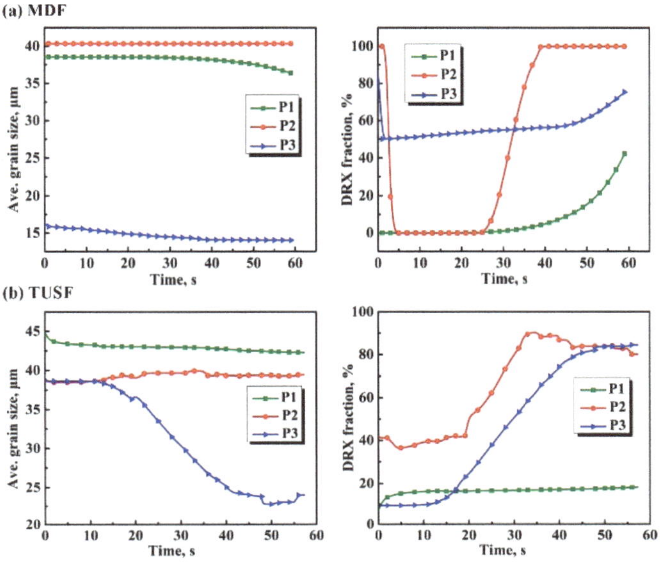

Figure 18. Time-dependent average grain size and DRX fraction curves for (**a**) MDF and (**b**) TUSF specimens at different tracking points.

6. Conclusions

(1) The grain size distribution in the MDF specimen was inhomogeneous. During the first step of MDF, the microstructure undergoes significant grain refinement, but an inhomogeneous grain structure was obtained due to the inhomogeneous strain distribution at the centre and edge regions. Finer grains developed at the centre region during forging. Based on experimental results, the microstructural evolutionary model used in this study is reliable;

(2) The forged ingot had an inhomogeneous grain distribution, and the average grain size at the core of the SDP1 steel sample after MDF was approximately 40.6–43.3 μm. Although this was slightly greater than the post-TUSF grain size (35.7–46.0 μm), the microstructure of the SDP1 steel sample after MDF was more uniform than that after TUSF;

(3) After MDF, the grains at the core of the sample were completely recrystallised (a recrystallisation volume fraction of 100%). In contrast, after TUSF, the grains at the centre of the sample were completely recrystallised, resulting in a nonuniform grain structure. Insufficient recrystallisation can result in mixed crystals;

(4) The average grain size of the SDP1 steel sample after MDF was 11.0–60.0 μm, which was smaller than that of the post-TUSF sample (12.0–132.0 μm). In addition, the surface of the sample after TUSF was not smooth, which can affect the overall performance of the sample and subsequent processing.

Author Contributions: Conceptualisation, X.C. and J.L. (Junwan Li); methodology, X.Z.; software, X.Z.; validation, X.C., B.W. and J.L. (Jiayuan Li); formal analysis, X.C.; investigation, X.C.; resources, X.W.; data curation, B.W.; writing—original draft preparation, X.C. and X.Z.; writing—review and editing, J.L. (Junwan Li) and P.Z.; visualisation, X.C.; supervision, J.L. (Junwan Li); project administration, X.C.; funding acquisition, X.W. All authors have read and agreed to the published version of the manuscript.

Funding: This research was funded by National Key R&D Program of China (grant numbers: 2016YFB0300400 and 2016YFB0300404).

Data Availability Statement: The data presented in this study are available upon request from the corresponding author. The data are not publicly available due to the fact of technical or time limitations.

Acknowledgments: The authors are grateful to Xiaocheng Li., Xijuan He. and Jie Ji. for their contribution to the discussion.

Conflicts of Interest: The authors declare no conflict of interest.

References

1. Zheng, Y.-F.; Wu, R.-M.; Li, X.-C.; Wu, X.-C. Continuous cooling transformation behaviour and bainite formation kinetics of new bainitic steel. *Mater. Sci. Technol.* **2016**, *33*, 454–463. [CrossRef]
2. Wu, R.M.; Zheng, Y.; Wu, X.C.; Li, X.C. Effect of titanium on the microstructure and hardness uniformity of non-quenched and tempered prehardened steel for large-section plastic mould. *Ironmak. Steelmak.* **2016**, *44*, 17–22. [CrossRef]
3. Xiaoxiao, Z.; Xuan, C.; Xiaocheng, L.; Yushu, H.; Junwan, L.; Xiaochun, W. Numerical Study on Microstructure Evolution of Large Cross-section SDP1 Plastic Die Steel during Multi-directional Forging. *Shanghai Metals* **2020**, *42*, 57–62. [CrossRef]
4. Xiao, Z.; Yang, X.; Wang, J.; Fang, Z.; Guo, C.; Zhang, D.; Yang, Y.; Zhang, X. Influence of Fe addition on annealing behav-iors of a phosphorus containing brass. *J. Alloys. Compd.* **2017**, *712*, 268–276. [CrossRef]
5. Tang, L.; Liu, C.; Chen, Z.; Ji, D.; Xiao, H. Microstructures and tensile properties of Mg–Gd–Y–Zr alloy during multidirec-tional forging at 773K. *Mater Des.* **2013**, *50*, 587–596. [CrossRef]
6. Bahmani, A.; Arthanari, S.; Shin, K.S. Achieving a high corrosion resistant and high strength magnesium alloy using multi directional forging. *J. Alloys Compd.* **2020**, *856*, 158077. [CrossRef]
7. Heidarzadeh, A.; Saeid, T.; Klemm, V. Microstructure, texture, and mechanical properties of friction stir welded commer-cial brass alloy. *Mater. Charact.* **2016**, *119*, 84–91. [CrossRef]
8. Miura, H.; Yu, G.; Yang, X. Multi-directional forging of AZ61Mg alloy under decreasing temperature conditions and im-provement of its mechanical properties. *Mat. Sci. Eng. A-Struct.* **2011**, *528*, 6981–6992. [CrossRef]
9. Zherebtsov, S.V.; Salishchev, G.A.; Galeyev, R.M.; Valiakhmetov, O.R.; Mironov, S.Y.; Semiatin, S.L. Production of submicrocrystalline structure in large-scale Ti–6Al–4V billet by warm severe deformation processing. *Scripta Mater.* **2004**, *51*, 1147–1151. [CrossRef]
10. Miura, H.; Maruoka, T.; Yang, X.; Jonas, J.J. Microstructure and mechanical properties of multi-directionally forged Mg–Al–Zn alloy. *Scripta Mater.* **2012**, *66*, 49–51. [CrossRef]
11. Gubicza, J.; Dobatkin, S.; Khosravi, E.; Kuznetsov, A.; Lábár, J.L. Microstructural stability of Cu processed by different routes of severe plastic deformation. *Mater. Sci. Eng. A* **2011**, *528*, 1828–1832. [CrossRef]
12. Sitdikov, O.; Sakai, T.; Miura, H.; Hama, C. Temperature effect on fine-grained structure formation in high-strength Al alloy 7475 during hot severe deformation. *Mater. Sci. Eng. A* **2009**, *516*, 180–188. [CrossRef]
13. Nunes, L.G.d.S.; Machado, M.L.P. Simulation of the Accumulative Roll-Bonding process through warm torsion test. *J. Mater. Res. Technol.* **2018**, *7*, 326–330. [CrossRef]
14. Wang, H.; Su, L.; Yu, H.; Lu, C.; Tieu, A.K.; Liu, Y.; Zhang, J. A new finite element model for multi-cycle accumulative roll-bonding process and experiment verification. *Mater. Sci. Eng. A-Struct.* **2018**, *726*, 93–101. [CrossRef]
15. Kulagin, R.; Beygelzimer, Y.; Ivanisenko, Y.; Mazilkin, A.; Straumal, B.; Hahn, H. Instabilities of interfaces between dissim-ilar metals induced by high pressure torsion. *Mater. Lett.* **2018**, *222*, 172–175. [CrossRef]
16. Li, P.; Lin, Q.; Wang, X.; Tian, Y.; Xue, K.-M. Recrystallization behavior of pure molybdenum powder processed by high-pressure torsion. *Int. J. Refract. Met. Hard Mater.* **2018**, *72*, 367–372. [CrossRef]
17. Martynenko, N.; Lukyanova, E.; Serebryany, V.; Gorshenkov, M.; Shchetinin, I.; Raab, G.; Dobatkin, S.; Estrin, Y. Increasing strength and ductility of magnesium alloy WE43 by equal-channel angular pressing. *Mater. Sci. Eng. A-Struct.* **2018**, *712*, 625–629. [CrossRef]
18. Wei, J.; You, J.; Zhang, D.; Jiang, S.; Chen, Z.; Liu, C. Reducing yield asymmetry in a wrought Mg–9Al alloy by random-ized texture achieved via multi-directional forging. *Mat. Sci. Eng. A-Struct.* **2020**, *796*, 140003. [CrossRef]
19. Manjunath, G.A.; Shivakumar, S.; Fernandez, R.; Nikhil, R.; Sharath, P.C. A review on effect of multi-directional forg-ing/multi-axial forging on mechanical and microstructural properties of aluminum alloy. *Mater. Today. Proc.* **2021**, *47*, 2565–2569. [CrossRef]
20. Bagherpour, E.; Pardis, N.; Reihanian, M.; Ebrahimi, R. An overview on severe plastic deformation: Research status, techniques classification, microstructure evolution, and applications. *Int. J. Adv. Manuf. Technol.* **2018**, *100*, 1647–1694. [CrossRef]
21. Xing, J.; Yang, X.; Miura, H.; Sakai, T. Mechanical Properties of Magnesium Alloy AZ31 after Severe Plastic Deformation. *Mater. Trans.* **2008**, *49*, 69–75. [CrossRef]
22. Zhang, R.; Wang, D.; Yuan, S. Effect of multi-directional forging on the microstructure and mechanical properties of TiBw/TA15 composite with network architecture. *Mater. Des.* **2017**, *134*, 250–258. [CrossRef]
23. Zhang, Z.; Qu, S.; Feng, A.; Shen, J. Achieving grain refinement and enhanced mechanical properties in Ti–6Al–4V alloy produced by multidirectional isothermal forging. *Mater. Sci. Eng. A-Struct.* **2017**, *692*, 127–138. [CrossRef]
24. Khani Moghanaki, S.; Kazeminezhad, M.; Logé, R. Mechanical behavior and texture development of over-aged and solu-tion treated Al-Cu-Mg alloy during multi-directional forging. *Mater. Charact.* **2018**, *135*, 221–227. [CrossRef]

25. Huang, H.; Zhang, J. Microstructure and mechanical properties of AZ31 magnesium alloy processed by multi-directional forging at different temperatures. *Mater. Sci. Eng. A-Struct.* **2016**, *674*, 52–58. [CrossRef]
26. Seshacharyulu, T.; Medeiros, S.; Morgan, J.; Malas, J.; Frazier, W.; Prasad, Y. Hot deformation mechanisms in ELI Grade Ti-6a1-4V. *Scr. Mater.* **1999**, *41*, 283–288. [CrossRef]
27. Nakao, Y.; Miura, H. Nano-grain evolution in austenitic stainless steel during multi-directional forging. *Mater. Sci. Eng. A-Struct.* **2011**, *528*, 1310–1317. [CrossRef]
28. Zherebtsov, S.; Murzinova, M.; Salishchev, G.; Semiatin, S. Spheroidization of the lamellar microstructure in Ti–6Al–4V alloy during warm deformation and annealing. *Acta Mater.* **2011**, *59*, 4138–4150. [CrossRef]
29. Soleymani, V.; Eghbali, B. Grain Refinement in a Low Carbon Steel Through Multidirectional Forging. *J. Iron Steel Res. Int.* **2012**, *19*, 74–78. [CrossRef]
30. Li, X.; Duan, L.; Li, J.; Wu, X. Experimental study and numerical simulation of dynamic recrystallization behavior of a mi-cro-alloyed plastic mold steel. *Mater. Des.* **2015**, *66*, 309–320. [CrossRef]
31. Kim, W.J.; Hong, S.I.; Kim, Y.S.; Min, S.H.; Jeong, H.T.; Lee, J.D. Texture development and its effect on mechanical proper-ties of an AZ61 Mg alloy fabricated by equal channel angular pressing. *Acta Mater.* **2003**, *51*, 3293–3307. [CrossRef]
32. Sakai, T.; Belyakov, A.; Kaibyshev, R.; Miura, H.; Jonas, J.J. Dynamic and post-dynamic recrystallization under hot, cold and severe plastic deformation conditions. *Prog. Mater. Sci.* **2014**, *60*, 130–207. [CrossRef]
33. Lu, Y.; Ma, R.; Wang, Y.-n. Texture evolution and recrystallization behaviors of Cu–Ag alloys subjected to cryogenic roll-ing. *T. Nonferr. Metal. Soc.* **2015**, *25*, 2948–2957. [CrossRef]
34. Dehghan-Manshadi, A.; Barnett, M.R.; Hodgson, P.D. Recrystallization in AISI 304 austenitic stainless steel during and af-ter hot deformation. *Mat. Sci. Eng. A-Struct.* **2008**, *485*, 664–672. [CrossRef]
35. Montheillet, F.; Lurdos, O.; Damamme, G. A grain scale approach for modeling steady-state discontinuous dynamic re-crystallization. *Acta Mater.* **2009**, *57*, 1602–1612. [CrossRef]
36. Eghbali, B. Effect of strain rate on the microstructural development through continuous dynamic recrystallization in a microalloyed steel. *Mater. Sci. Eng. A-Struct.* **2010**, *527*, 3402–3406. [CrossRef]
37. Li, D.; Guo, Q.; Guo, S.; Peng, H.; Wu, Z. The microstructure evolution and nucleation mechanisms of dynamic recrystalli-zation in hot-deformed Inconel 625 superalloy. *Mater. Des.* **2011**, *32*, 696–705. [CrossRef]
38. Vervynckt, S.; Verbeken, K.; Thibaux, P.; Houbaert, Y. Recrystallization–precipitation interaction during austenite hot de-formation of a Nb microalloyed steel. *Mat. Sci. Eng. A-Struct.* **2011**, *528*, 5519–5528. [CrossRef]
39. BadJena, S.; Park, J. Effect of particles on the dynamic recrystallization behavior of Al–V–N micro-alloyed medium carbon steel. *Mater. Sci. Eng. A-Struct.* **2012**, *548*, 126–133. [CrossRef]
40. Mehtonen, S.; Karjalainen, L.; Porter, D. Hot deformation behavior and microstructure evolution of a stabilized high-Cr ferritic stainless steel. *Mater. Sci. Eng. A-Struct.* **2013**, *571*, 1–12. [CrossRef]

materials

MDPI

Article

A Study of Aluminum Honeycomb Structures under Dynamic Loading, with Consideration Given to the Effects of Air Leakage

Radosław Ciepielewski and Danuta Miedzińska *

Faculty of Mechanical Engineering, Military University of Technology, Kaliskiego 2 Street, 00-908 Warsaw, Poland; radoslaw.ciepielewski@wat.edu.pl
* Correspondence: danuta.miedzinska@wat.edu.pl

Abstract: Aluminum honeycomb structures are used in the construction of protective materials due to the positive relationship between their mass and their energy-absorbing properties. Applying such materials in the construction of large machinery, such as military vehicles, requires the development of a new method of finite element modeling, one that considers conditions with high strain rates, because a material model is currently lacking in the available simulation software, including LS-DYNA. In the present study, we proposed and verified a method of numerically modeling honeycomb materials using a simplified Y element. Results with a good level of agreement between the full core model and the Y element were achieved. The obtained description of the material properties was used in the subsequent creation of a homogeneous model. In addition, we considered the influence of increases in pressure and the leakage of the air entrapped in the honeycomb cells. As a result, we were able to attain a high level of accuracy regarding the stress values across the entire range of progressive failure, from the loss of stability to full core densification, and across a wide range of strain rates.

Keywords: honeycomb; numerical modelling; dynamic compression; structure; air leakage

Citation: Ciepielewski, R.; Miedzińska, D. A Study of Aluminum Honeycomb Structures under Dynamic Loading, with Consideration Given to the Effects of Air Leakage. *Materials* **2023**, *16*, 2211. https://doi.org/10.3390/ma16062211

Academic Editors: Shuwen Wen, Yongle Sun and Xin Chen

Received: 27 January 2023
Revised: 6 March 2023
Accepted: 7 March 2023
Published: 9 March 2023

1. Introduction

Functional, cell-oriented structures, including honeycombs, have a long history of successful application. For example, they were used in the initial period of space exploration, specifically in the Apollo 11 [1] project, to build elements of the crew and technical compartments. NASA and others [2–4] concluded that the honeycomb structure has the best energy-absorbing properties in relation to its specific mass. Due to this feature, honeycomb structures are used across a wide range of industries, including the automotive industry [5,6] and the military [7,8].

Sandwich structures with honeycomb cores are commonly used because of their high stiffness-to-weight ratio. Moreover, they have predictable progressive folding (stress–strain) characteristics in the case of static loading. This feature has made it possible to derive a mathematical model that enables the calculation of the so-called average value of crushing stress (plateau) to the core (occurring from the moment of stability loss to the beginning of the increase in force due to the structure's densification) on the basis of the known, basic values characterizing honeycomb structures, such as the yield point of the core material, the wall thickness, and the cell size [9]. However, in [10,11], the authors show that the strength of the structure can increase by up to 50% with dynamic loading. In addition, there are a number of aspects that, if simplified, may make it impossible to effectively predict a honeycomb's structural behavior, such as the air trapped inside the cells [12].

Modern tools, such as the finite element method (FEM) used in industry and science, allow for the effective prediction of the effects of dynamic loading, including fast-changing ones such as those that a shock wave from the detonation of an explosive may have on protective structures. In order to ensure a high-quality representation of the real conditions and the properties of the materials used, the data used in the preparation of the model should be as reliable as possible.

Scientific publications dealing with the numerical modeling of honeycomb structures can be divided into several groups, according to the subject covered by the authors. The most popular subjects include modeling the phenomenon of the loss of stability, modeling the local phenomena that occur during structural deformation, the determination of strength characteristics based on full geometric models and submodels, and observations of the phenomenon of energy absorption by structures modeled as homogeneous, using special constitutive models.

In [13], the authors focused on modeling the buckling process and progressive folding by comparing the effects obtained by using geometry to represent the samples, containing 16 single cells, used in their experimental tests, with results obtained from a four-cell compression analysis. The authors indicate that, by applying the appropriate boundary conditions—that is, by ensuring that the free edges of the walls maintained symmetry with respect to the plane perpendicular to them—the authors were able to obtain the same characteristics regardless of the number of individual cells that were modeled.

Similarly, in [14], the authors focused on determining the characteristics of dynamically loaded aluminum cores. The structures were modeled by retaining the full geometry corresponding to that which was used in the experimental research. Their test results show that successive walls inside the core deformed symmetrically in relation to their longer axes of symmetry.

In scientific nomenclature, the behavior of the core as a single-cell sector is referred to as the Y element. Among others, this modeling method was used in [15]. The authors presented the results of numerical analyses, the subject of which was the static compression of honeycomb cores as modeled in two different ways—by fully mapping the honeycomb geometry, and by solely modeling one single-unit sector. The main advantage of the method, in addition to the simple construction of the model and the significant reduction in the time needed to perform the calculations, is the ability to determine the full characteristics of the core, including the elastic range and the compaction range. The oscillations observed in the area of progressive folding represent a noticeable disadvantage.

A different approach to reducing the so-called numerical cost of the analysis of the strength of a cell core in an aluminum honeycomb is presented by the authors of [16], which is devoted to modeling the structure by using a structure consisting of joint-connected short beams. The authors used a Discrete Beam Method (DBM) and proved that the approach is correct and that it provides results that demonstrate a high level of agreement with those obtained experimentally.

Another approach often found in the literature [17,18]—which allows for a significant simplification of the metallic honeycomb geometric modeling process and a reduction of the time needed to perform the calculations—is the use of one of the constitutive models designed to reproduce the behavior of the loaded structure, which is modeled as homogeneous, usually using eight-node cubic elements. This method allows for a significant simplification of the core-modeling process, especially when it is applied to structures with large dimensions in relation to the characteristic dimensions of the structure. Modeling fine walls using elements with a small characteristic dimension, thus ensuring a reliable representation of the deformation of the core, reduces the time step in the case of dynamic analyses, leading to a significant increase in the time needed to perform the calculations.

In such cases, the MAT_HONEYCOMB constitutive material model, which is applied in FEM calculation systems (including LS-DYNA), can be used [19]. Its application allows for a full representation of the global stiffness of a structure, both axial and shear, in terms of elasticity, progressive folding, and complete compression, accompanied by its linear elastic characteristics. This model has been used to successfully predict the behavior of materials in which thin-walled, anisotropic-oriented structures were used, e.g., honeycombs, as well as in foamed metals and plastics. However, a significant simplification of the structural modeling process using MAT_HONEYCOMB requires the knowledge of a number of the parameters and characteristics of a particular core.

On the basis of the work presented in [20] produced by researchers at Toyota Motors®, the LS-DYNA system has been equipped with an additional constitutive model called MAT_MODIFIED_HONEYCOMB, which is designed to describe transversely isotropic bodies, the mechanical properties of which are symmetrical with respect to the axis normal to the isotropic plane. Honeycomb-type geometric structures have such a feature. However, industrially manufactured structures of this type are double-walled in one direction; therefore, treating them as transversely isotropic may be considered an oversimplification. Nevertheless, due to trends in the design of energy-consuming elements, including ones with regular spatial structures that have been created using additive manufacturing techniques, it is necessary to mention this material model as well. Because none of the material models described above have the ability to account for increases in stress caused by the pressure of compressed air trapped inside the cells, there is a need for a solution to that problem.

The aim of the present study was to develop and verify a method of modeling a homogenous honeycomb structure—as subjected to dynamic loads—using the characteristics obtained from a fast Y-element model. Such an approach has allowed us to obtain valuable results from FEM analysis, based on material models of the MAT_HONEYCOMB type, across a wide range of strain rates, thus minimizing the need for experimental research. The authors of [21] present experimental studies on the static and dynamic compression of honeycomb structures. The conclusion we drew from their work was that the strengthening of the structure is related to the volume of air trapped inside the closed cells.

2. Materials

Seven types of structures, all available from the catalogs of the manufacturers of these materials, were selected. The geometrical parameters of the individual structures are presented in Table 1. The name given to each structural type includes the following information: cell size–aluminum type–wall thickness. All dimensions used in the names are expressed in inches. An example of the studied material, along with the characteristic dimensions of its core, is shown in Figure 1.

Table 1. Geometrical parameters of the studied honeycomb structures.

No.	Type	Declared * Cell Size *S* [mm]	Declared * Wall Thickness *t* [mm]	Height *T* [mm]
1	1/8-5052-0.0007	3.1750	0.01778	10
2	1/8-5052-0.001	3.1750	0.02540	10
3	1/8-5052-0.0015	3.1750	0.03810	10
4	3/16-5052-0.0007	4.7625	0.01778	10
5	3/16-5052-0.001	4.7625	0.02540	10
6	3/16-5052-0.0015	4.7625	0.03810	10
7	1/4-5052-0.001	6.3500	0.02540	10

* data provided by the producer's catalogue [21].

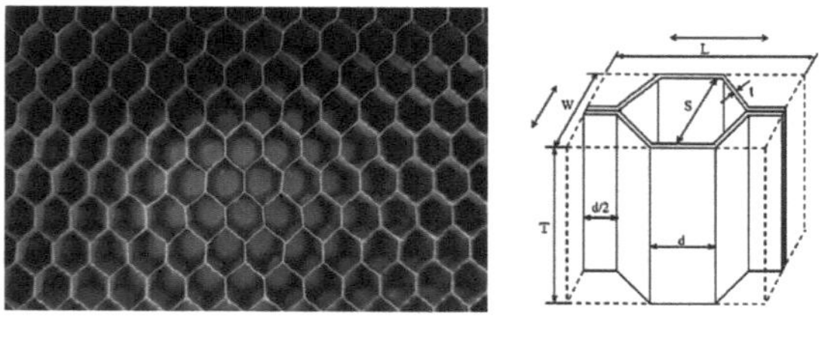

Figure 1. An example of the analyzed honeycomb structure (**left**), and the basic designations of the honeycomb's structural elements (**right**) [9].

3. Research Methodology

The research was conducted in the following stages:

First, a numerical analysis of the static compression of a simplified model of the honeycomb structure (the Y element) was completed with the introduction of imperfections related to the geometry of the real structure (based on computed tomography (CT) studies); validation of the model using experimental research was achieved.

Second, numerical analyses of the honeycomb structures were performed, using a homogeneous model and characteristics obtained from the calculations of the Y element, including the influence of the air trapped within the cells of the structure.

The scheme of the research stages is presented in Figure 2.

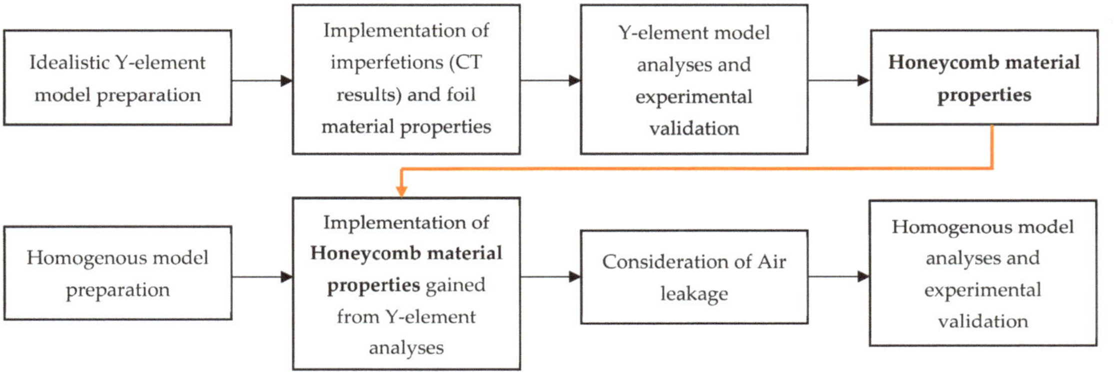

Figure 2. Research methodology scheme.

4. Models

In this section, the preparation of the model and its validation are presented. The conditions for the numerical analyses of static compression using the simplified Y element and those using the homogenous model are presented in Sections 4.1 and 4.2, respectively.

4.1. Numerical Analyses of Static Compression Using the Simplified Y Element

The smallest, repetitive subarea that can be separated from a structure is a cell with the base of an equilateral triangle, the sides of which are perpendicular to the walls of three adjacent honeycomb cells. A separated single cell (the Y element) is shown in Figure 3.

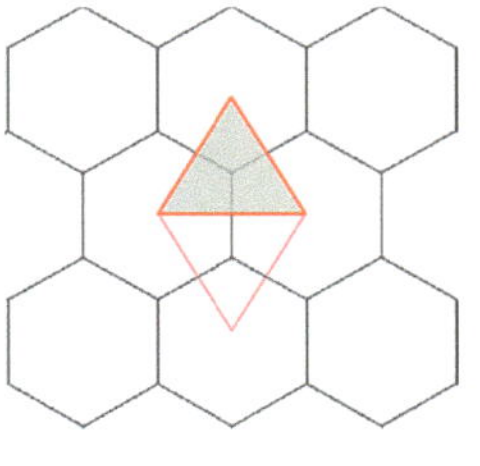

Figure 3. A single cell (Y element), as contained within a honeycomb structure.

As indicated in [13–15], structures consisting of repeating elements can be analyzed by solely focusing on a model that has been simplified down to the level of a segment, using appropriate boundary conditions. In the case of the analyzed structures, these conditions were limited to the treatment of the sides of the subarea as planes of symmetry.

The geometric model shown in Figure 4a, when prepared for discretization, consisted of three walls connected to each other in line with one of the longer edges. The walls were connected to each other at an opening angle of 120°.

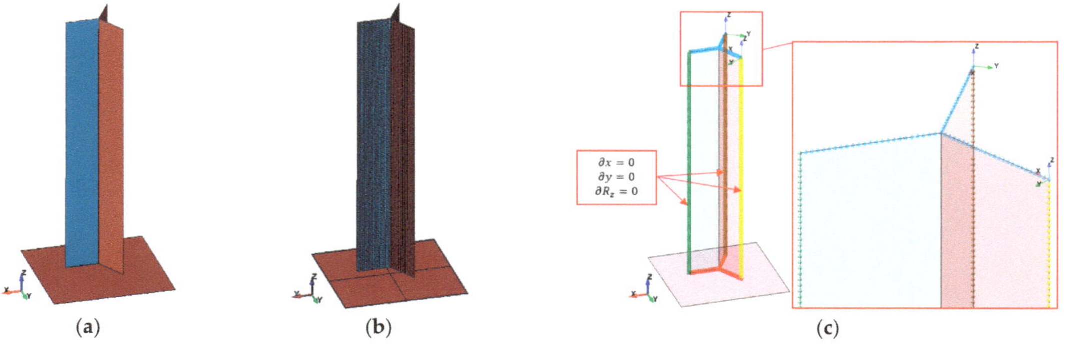

Figure 4. Y element: (**a**) geometric model, (**b**) finite element model, (**c**) boundary conditions.

Four-node, fully integral shell elements with five integration points across the element thickness were used to build the finite element mesh, as presented in Figure 4b. The characteristic dimension of the element was 0.23 mm, and there were 100 of them at the wall height. The selection of the element size was preceded by an analysis of the impacts that changing this parameter would have. At levels below the dimensions described above, no improvements were observed. The Belytschko–Tsay elements were used. To carry out the static compression analysis, all degrees of freedom were fixed in the top nodes and in the nodes next to the plane that forces motion (except that which was translational in the Z-axis). The boundary conditions, along with the symmetry conditions and the local coordinate systems, are shown in Figure 4c.

In order to accurately reproduce the actual structure and identify imperfections, the tested samples were subjected to computed tomography (CT) studies.

A SkyScan 1174 tomograph was used for this task. The obtained point clouds were processed to reduce noise, and then they were used to generate a polygonised structure; discrete, triangular surfaces were created between the closest points. In this way, spatial models of the cores were obtained, which was useful for making measurements. The axonometric views of the CT models for the 1/8-0.0015 and 3/16-0.0015 samples are shown in Figure 5.

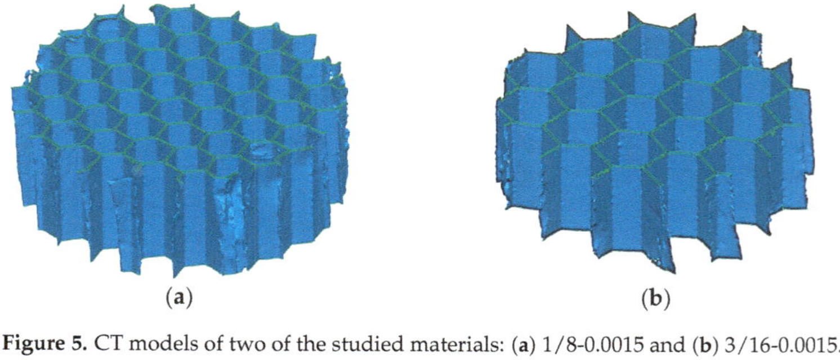

Figure 5. CT models of two of the studied materials: (**a**) 1/8-0.0015 and (**b**) 3/16-0.0015.

Analyzing the actual structures of the cores via the creation of three-dimensional models using computed tomography allowed for the examination of the basic geometric parameters, e.g., the length of individual walls, the size of the cell, and the bending radii of the walls near the joints. Exemplary measurement results are shown in Figure 6.

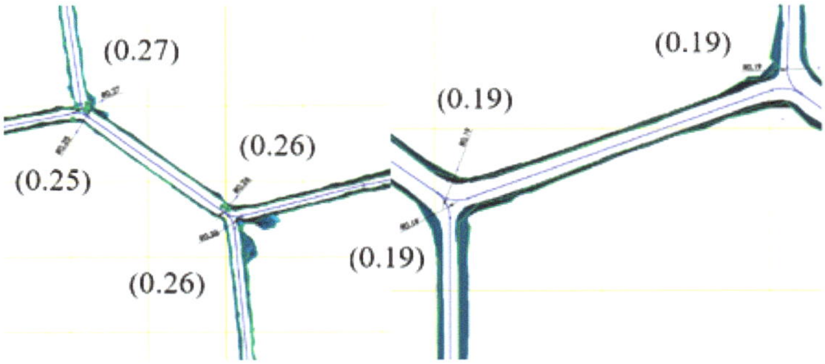

Figure 6. Examples of the measurement results of the geometrical parameters of the actual cores.

The results of the geometric measurements, averaged for individual samples, are presented in Table 2.

Table 2. List of characteristic dimensions obtained by measuring actual structures.

Measured Value	1/4-5052-0.001	3/16-5052-0.0007	3/16-5052-0.001	3/16-5052-0.0015	1/8-5052-0.0007	1/8-5052-0.001	1/8-5052-0.0015
R [mm]-bending radius of a single wall	0.27	0.22	0.28	0.39	0.19	0.26	0.37
d_{theor} [mm] theoretical wall width *	3.67	2.75	2.75	2.75	1.83	1.83	1.83
d_1 [mm] single-wall width	3.57	3.02	3.06	3.14	2.19	2.11	2.14
d_2 [mm] double-wall width	3.84	2.72	2.62	2.53	1.63	1.75	1.71

* provided in the manufacturer's catalog card [21].

On the basis of the data presented above, it can be seen that the double walls have a smaller width and the single walls have a higher width than those declared in the catalog cards. Moreover, the portion of the wall that is arched is a distinct part of it. The obtained measurement results allowed us to conclude that the bend radii of the walls depend mainly on their thickness, and that the dimensions are linked by a ratio of approximately 10:1.

In order to check the impact of the above-described geometric parameters, models of the structure of the sample designated as 1/8-0.0015 were built, one using a simple, idealized version, and the other constructed using the mapped shape of the walls according to the parameters described in Table 2. Both versions are shown in Figure 7, which demonstrates that, in addition to the visible rounding of the edges of the single wall (marked in red), the width of the double wall (marked in blue) has been reduced in favor of the single wall.

The introduced change in the geometry resulted in differences in the form of structural deformation that were apparent as early as the initial loss-of-stability phases. As can be seen in Figure 8, there are seven ridges on the flat core walls, and there are six smaller ridges on the rounded sides.

Minor differences in the geometry of the compared models were also reflected in the stress–strain characteristics obtained via the compression test. The value of the critical stress decreased by 11.6% (from 8.65 MPa to 7.64 MPa), and the value of the mean breaking stress decreased by 15.5% (from 5.98 MPa to 5.05 MPa). The value of the longitudinal modulus of elasticity before loss of stability also changed by 28.1% (from 3.11 GPa to 3.24 GPa). These are significant differences, thus the change in the model's geometry should be taken into account in simulations. A comparison of the discussed characteristics, with reference to those obtained experimentally [21], is presented in Figure 9.

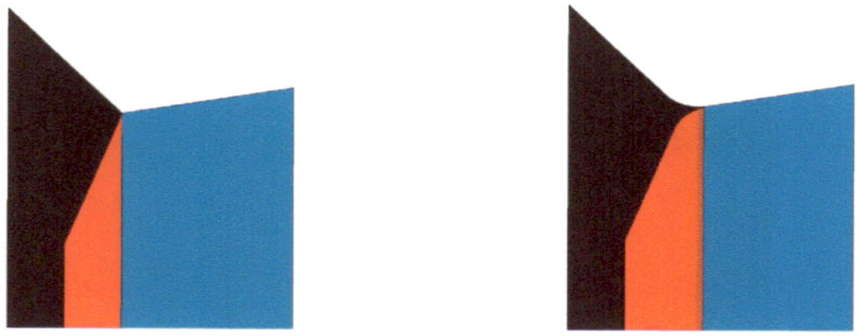

(**a**) (**b**)

Figure 7. Geometric models of the core for the following versions: (**a**) simple/idealized; (**b**) showing the mapped shape of the rounded walls.

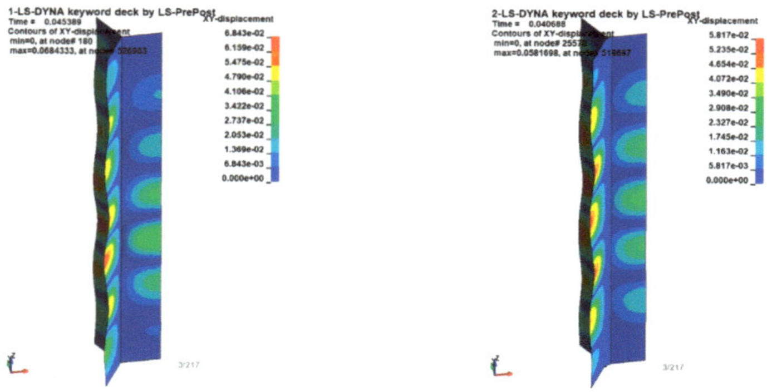

(**a**) (**b**)

Figure 8. Maps of displacements [in mm] in the X–Y plane of the model in (**a**) the straight/idealized version; and (**b**) with the mapped shape of the rounded walls.

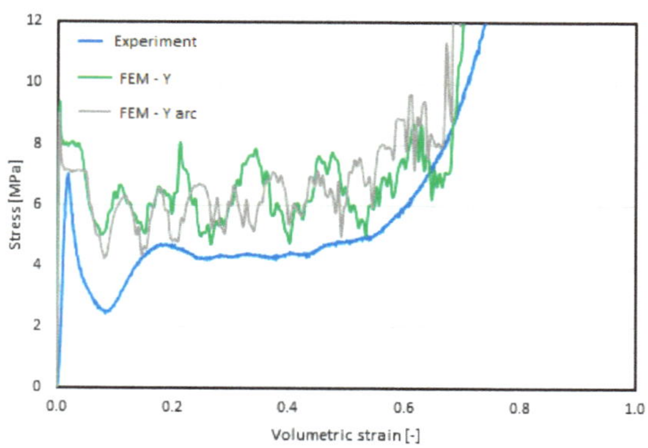

Figure 9. Comparison of the stress–volumetric strain characteristics of material 1/8-5052-0.0015 obtained for the following versions of the models: straight/idealized (FEM–Y, green line); with the mapped shape of the rounded walls (FEM–Y arc, gray line); and experimentally obtained characteristics, blue line [21].

The constitutive model MAT_003_PLASTIC_KINEMATIC [19] was used to reflect the behavior of the material from which the tested Y element was made: aluminum alloy Al 5052-H39. It is a bilinear model, the stress–strain relationship of which is described by defining the tangent of the angle of the elastic part and the part of the kinematic strengthening. The parameters and mechanical properties of the material were taken from a report published by NASA [22] because the parameters of the thin, multi-rolled foil significantly differ from the parameters of the material from which it was made. In [22], the results of tests of the foil composed of the Al 5052-H39 alloy, as well as those of the foil fragments cut from the finished structure of the cellular core, are presented.

The basic parameters of aluminum Al 5052-H39, as used to make cores with a honeycomb topology in the untreated form [23], in the form of foil [22], and in the form of a fragment of the core [24], are presented in Table 3. After plastic processing, the material is characterized by significantly lower strength parameters. In the form of a film, it has a 19% lower Young's modulus, a 30% lower yield point, and a 25% lower tensile strength. The sample cut from the core fragment has a Young's modulus that is up to 46% lower, a 32% lower yield point, and a 35% lower tensile strength.

Table 3. Comparison of the mechanical properties of aluminum Al 5052-H39, a foil composed of Al 5052-H39, and a fragment of a honeycomb core composed of that foil.

Parameter	Symbol	Unit	5052-H39 [23]	Foil 5052-H39 [24]	Core Fragment [24]
Density	ρ	kg/m^3	2.7×10^3	2.7×10^3	2.7×10^3
Elastic modulus	E	GPa	70.00	56.53	37.92
Yield stress	R_e	MPa	325.0	227.5	220.6
Tensile strength	R_m	MPa	330.0	248.2	234.4
Elongation at break	ε_u	%	4.0	1.6	4.7

Due to the fact that the developed Y-element models were used to determine the mechanical properties of the statically compressed aluminum honeycombs without the need for experimental testing, the properties of the Al 5052-H39 foil were considered in the subsequent numerical calculations.

4.2. Numerical Analysis of the Honeycomb Structures Using the Homogenous Model

The aim of the tests described below was to assess the effectiveness of using the results of the simulation of the Y element in models that describe the global response of this type of material without a detailed analysis of the behavior of their internal structure.

Geometric models, as shown in Figure 5, were analyzed in order to identify fully closed cells and the possible damage resulting in the formation of open cells. Then, the volumes of the solids bounded by the closed walls of the cores were measured. Measurement results for the same types of samples were averaged. A visualization of an example of the discussed process is presented in Table 4. The samples with smaller core sizes had a noticeably higher proportion of closed-cell volume, considering that they had the same global dimensions.

The previous experimental research of honeycomb structures under various levels of strain rate loading carried out by the authors [21] was used to develop the assessments for this paper. By presenting the experimentally obtained, absolute increases in plateau stress [21] in the domain of the air volume locked in by the cell walls, an almost-linear characteristic was obtained, which can be observed in Figure 10. This observation may indicate that the main factor influencing the final shape of the stress–volumetric strain characteristic is the increase in the pressure of the air trapped inside the cells, especially for dynamic testing (strain rate of 3.8×10^2 and 3.3×10^3 1/s).

Another argument for the correctness of this assertion is the conclusion made on the basis of the observation of images recorded with a high-speed camera in the dynamic test using a split Hopkinson pressure bar (3.3×10^3 1/s). As shown in Figure 11, the gas began to leak from the inside of the sample during the final compression phase (t = 333.30 μs). It became visible due to a dust cloud—a mixture of air escaping from the inside of the sample and fine particles of the damaged core and resin. This phenomenon took place in each case after reaching approximately 50% of the deformation; it continued after the maximum displacement of the initiating bar face, and this continued even after its plane lost contact with the sample plane. The described phenomenon may prove that, during the test, strong air compression takes place inside the core, which is released only after complete compression.

Table 4. Measurement results of the volume of air limited by walls in the real models of the samples.

	Graphical Representation	Wall Thickness t [mm]	Cell Size S [mm]	Closed Air Volume V_a [mm^3]
1/4-5052-0.001	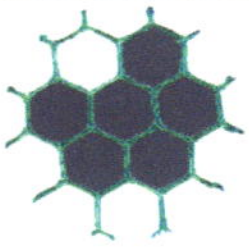	0.02540	6.3500	2264.5
3/16-5052-0.0007	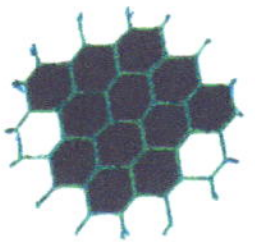	0.01778	4.7625	2556.8
3/16-5052-0.001	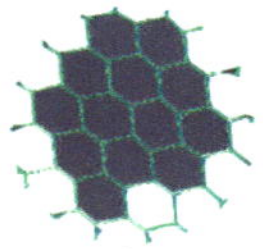	0.02540	4.7625	2702.0
1/8-5052-0.0015	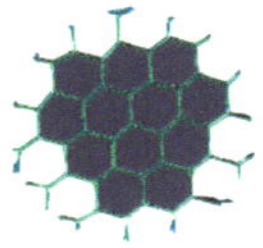	0.03810	4.7625	2744.9
1/8-5052-0.0007	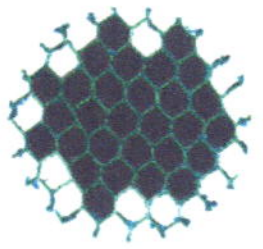	0.01778	3.1750	3426.4
1/8-5052-0.001	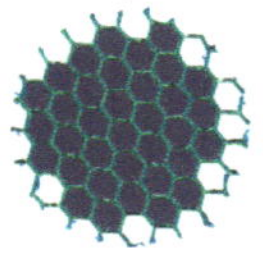	0.02540	3.1750	3548.8
1/8-5052-0.0015	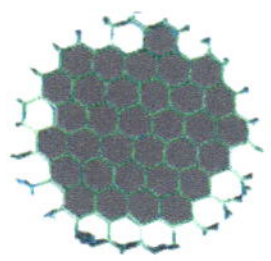	0.0381	3.175	3662.1

The above conclusion is important because none of the constitutive models available in commercially used computing environments take into account this kind of phenomenon's influence on the change in the global mechanical properties of the structure.

Therefore, in order to describe the behavior of aluminum honeycomb structures in terms of strain rates higher than quasi-static, it is necessary to take an approach that will account for both the reaction of the deformed, thin-walled core and that of the air pressure inside the cells.

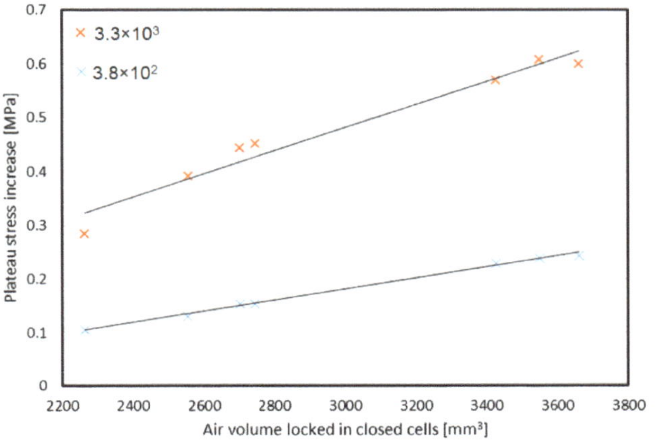

Figure 10. Increase in plateau stress depending on the volume of air trapped inside cells (based on results obtained in [21]).

(**a**) (**b**)

Figure 11. Deformation of sample 1/8-5052-0.0015 in the following time steps: (**a**) t = 0, and (**b**) t = 333.30 μs (based on results obtained in [21]).

Assuming, as a simplification, that the air closed inside the cells is compressed without heat exchange, the increase in pressure ΔP can be described by a simple equation [25],

$$\Delta P = P_0\left(\frac{V_0}{V} - 1\right) \quad (1)$$

where P_0 is the initial pressure (atmospheric).

Then, the indirectly measured stress value of the real structure would be the sum of the stress value in the structure (treating it as homogeneous according to the assumption made) and the pressure at a given time step. Due to the fact that the deformation of the core takes place only in the direction of displacement, the change in volume can be treated as a change in the height of the sample; as a result, the expected value of the measured stress is described by the formula [25]

$$\sigma = \sigma_r + P_0\left(\frac{T_0}{T_0 - u} - 1\right) \quad (2)$$

where σ_r is the stress in the core of the sample treated as a continuous medium, and T_0 is the initial core height.

The assumptions formulated above lead to the conclusion that the air trapped inside the cells causes an increase in the noted stress value, regardless of the strain rate. There are studies [24] which prove that the key aspect linking the structure's response with the duration of its destruction are air leakages resulting from imperfections in the structure and the material discontinuities formed in the process of deformation. The assumption can be modified in accordance with research carried out by Xu et al. [24] and Hu et al. [25], in which it was shown that the cross-sectional area of a honeycomb block also changes during axial compression. The air leakage δ is determined by the relationship [21]

$$\delta = 1 - \frac{PV}{P_0 V_0} \tag{3}$$

Therefore, the pressure value can be determined by the formula

$$P = P_0 \frac{1-\delta}{1-\varepsilon_v} \tag{4}$$

where ε_v is the volumetric strain.

It is assumed that δ is the core failure time of the t' function and differentiating Equation (4) over time, assuming that one obtains

$$\dot{P} = \frac{P_0}{1-\varepsilon_v}\left(\frac{1-\delta}{1-\varepsilon_v}\cdot\dot{\varepsilon_v} - \dot{\delta}\right) \tag{5}$$

and the leakage rate is

$$\dot{\delta} = \frac{1}{P_0}\left[P\dot{\varepsilon_v} - \dot{P}(1-\dot{\varepsilon_v})\right] \tag{6}$$

Based on the relations presented above, it can be noticed that the pressure value and the rate of the leakage are related to the strain rate $\dot{\varepsilon_v}$. Researchers developing analytical models for this type of issue [12], using the assumption that the strain rate is constant during the test, derived a relationship that links the pressure change inside the core with the strain and the leakage rate

$$\Delta P = P_0\left(\frac{1}{1-\varepsilon_v} - 1\right)\cdot\left(1 - \frac{\dot{\delta}}{\dot{\varepsilon_v}}\right) \tag{7}$$

Equation (16) does not contain any unknown parameters other than the leakage rate. The authors of [25] state that this rate should be determined empirically by comparing the stress–volumetric strain results obtained during the compression of the samples with the cores tightly closed between covers, and those with openings releasing air during the test. The leakage intensity is very similar when the results of testing honeycomb structures with the same t/d ratio are compared, and its value depends on the strain rate. The obtained characteristics of the leakage intensity as a function of the strain rate for one of the cases analyzed in [25] (1/8-5052-0.001) is presented in Figure 12. In this case, the leakage rate increased almost in direct proportion to the value of the strain rate. This means that a tenfold increase in strain rate resulted in at least a tenfold increase in the leakage rate.

To perform simulation tests of the uniaxial compression of the homogenous honeycomb materials, a model with a cylinder geometry with a diameter of 25 mm and a height of 10 mm was used, reflecting the global geometry of the sample core. It consisted of 525 elements with a hexagonal topology.

The material model MAT_26_HONEYCOMB, available in the LS-DYNA system, was used to map the core, which was treated as a homogeneous material [19].

In MAT_26, in the uncompressed state (initial state, loss of stability, and progressive folding), the material retains its orthotropic properties, and the stress tensor components remain unconnected from each other so that the strain component in one local direction does not cause reaction forces in the others. Modules of the longitudinal and shear stiffness in particular directions depend on the given modules of the initial stiffness and the stiffness of a fully compressed (compacted) structure. These dependencies are as follow [19]:

$$E_{aa} = E_{aau} + \beta(E - E_{aau}) \tag{8}$$

$$E_{bb} = E_{bbu} + \beta(E - E_{bbu}) \tag{9}$$

$$E_{cc} = E_{ccu} + \beta(E - E_{ccu}) \tag{10}$$

$$G_{ab} = G_{abu} + \beta(G - G_{abu}) \tag{11}$$

$$G_{bc} = G_{bcu} + \beta(G - G_{bcu}) \tag{12}$$

$$G_{ca} = G_{cau} + \beta(G - G_{cau}) \tag{13}$$

where

$$\beta = \max\left[\min\left(\frac{1-V}{1-V_f}, 1\right), 0\right] \tag{14}$$

and E is the modulus of elasticity of the core material; G is the shear modulus of the core material; E_{aau}, E_{bbu}, and E_{ccu} are the modules of elasticity of the uncompressed cores; G_{abu}, G_{bcu}, and G_{cau} are the shear modules of the uncompressed cores; V is the relative volume (the ratio of the current volume to the initial volume), and V_f is the relative volume at which the core is considered fully compressed and it transforms into a linear elastic characteristic (the relative volume of total compaction).

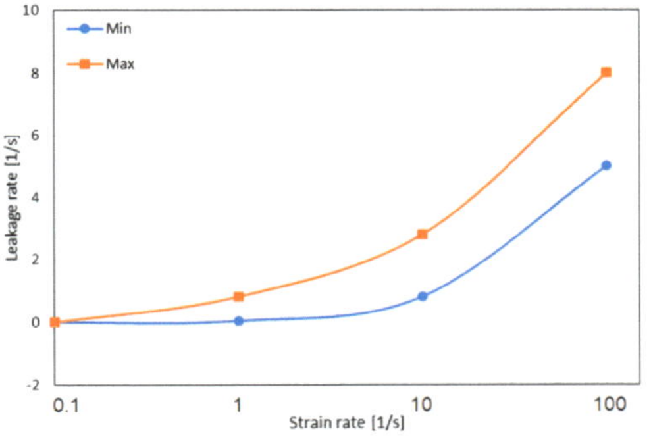

Figure 12. Leakage rate vs. strain rate for 1/8-5052-0.001 (based on [25]).

In addition, the material model requires the definition of a set of curves that present the material characteristics obtained in the experimental tests: compression in each of the basic directions and shear in each of the base planes. There are two ways to define these characteristics. The first one is to determine the magnitude of stresses as a function of the relative volume (V). It is also possible to determine the magnitude of stresses as a function of volumetric strains, defined as

$$\varepsilon_V = 1 - V \tag{15}$$

Finally, the components of the stress tensor are calculated according to the following relationship:

$$\sigma_{ij}^{n+1} = s_{ij}^{n+1} - p^{n+1}\delta_{ij} \tag{16}$$

After the process of updating the stress values is completed, they are converted into a global form.

For the modeling of individual structure types, the parameters presented in Table 4 were adopted. In each case, the $\sigma_{ij} - \varepsilon_{ij}$ characteristics were developed by selecting characteristic points from the curves obtained by the FEM simulation using the Y-element model.

The development of a model capable of taking into account the change in the nature of the response of the change of the initial conditions causing the strain-rate increase was based on one of the available methods describing the behavior of air-filled elements.

The model should allow us to describe the change in gas pressure, along with the change in the volume inside where it was located. It should also take into account the leakage in the calculation of the pressure changes and the application to the surface and the spatial boundary of the element. All of the possibilities mentioned above are offered by one of the simplest models: AIRBAG_SIMPLE_AIRBAG_MODEL [19]. The current value of the pressure acting on the boundaries of the vessel domain is calculated using the equation of state [19]:

$$P = (\gamma - 1)\rho e \tag{17}$$

where P is the pressure, ρ is the density, and e is the internal energy of gas.

The γ coefficient is the adiabatic exponent: the ratio of specific heat at a constant pressure c_p to specific heat at a constant volume c_v [25]

$$\gamma = \frac{c_p}{c_v} \tag{18}$$

The rate of changes in the air mass m inside the volume, with time, is described by the relationship [26]

$$\frac{dm}{dt} = \frac{dm_i}{dt} - \frac{dm_o}{dt} \tag{19}$$

The value of the air mass, which flows in via subsequent time steps m_i, is defined by the appropriate characteristic. There are two options for determining the mass of the outflowing air m_o: by defining the area of the holes through which the air leaks and their shape coefficient, and by defining the mass characteristics over time. There is also a possibility of making the size of the surface of the holes and the shape factor dependent on the value of the pressure inside.

In the discussed approach, the components of the energy balance [26]

$$\dot{e} = \dot{e}_i - \dot{e}_o - P\dot{V} \tag{20}$$

are as follows: $\dot{e}_i$ is energy change caused by the mass of inflowing gas, $\dot{e}_o$ is the change in energy caused by the mass of the outflowing gas, $P\dot{V}$ is work achieved by the pressure per volume change.

The described boundary condition was applied to the surfaces of the elements on the outer walls of the cylinder, as shown in Figure 13. This allowed for a direct transfer of the forces resulting from the increase in pressure to the sample boundaries. The parameters used to describe the air enclosed inside the sample are presented in Table 5. All parameters related to the influence of the air mass were omitted.

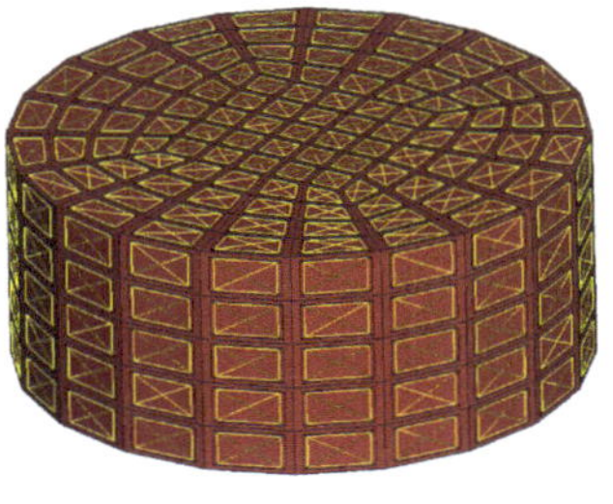

Figure 13. Numerical model of the homogeneous honeycomb material, with surfaces used as air-leakage boundaries.

Table 5. Parameters of the air enclosed inside the cores, according to the assumptions of the AIRBAG_SIMPLE_AIRBAG_MODEL [27].

Parameter	Symbol	Unit	Value
Density	ρ	kg/m^3	1.2
Specific heat at constant volume	c_v	J/(kg·K)	713.00
Specific heat at constant pressure	c_p	J/(kg·K)	1000.00
Ambient pressure	P_e	MPa	0.1
Temperature	T	K	293
Area of holes	A_o	m^2	0.004

The initial boundary conditions are presented in Figure 14. An explicit scheme of the integration of the equations of motion was used for the calculations.

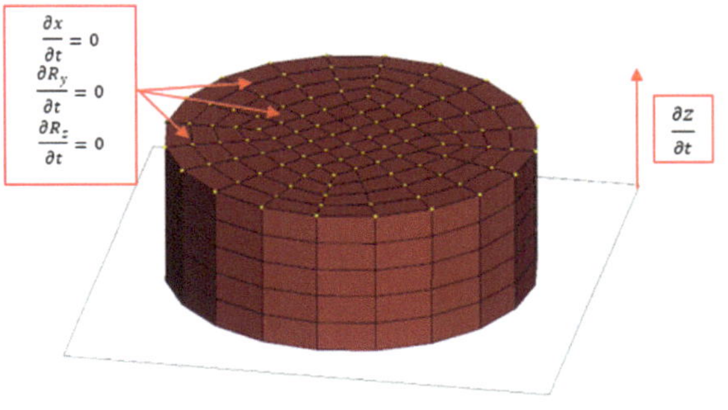

Figure 14. Scheme of the initial boundary conditions applied in the model.

5. Results

The results are presented in two subsections in accordance with the descriptions of models presented in Section 3. The results of the numerical analyses for the quasi-static compression tests are described, first for the simplified Y element, and then for the homogenous model, in which the material characteristics gained via analyses of the Y element were implemented.

5.1. Results of the Numerical Analyses of the Static Compression of the Y-Element

The basic result of the performed analyses was the characteristic of the change in the stress value as a function of the volumetric strain and the list of resulting parameters. The characteristics obtained in the static and experimental FEM analyses [21], presented in Figure 15, reveal a consistent course of changes in the stress value, with an increase in the strain value. The similarity is especially visible in the elastic phase until it reaches critical stress, progressive folding, and compaction. The most pronounced difference in the stress values occurred just after the loss of stability. In each of the experimentally tested cases, a subsequent deep decrease occurred, up to about one-half of the subsequent progressive folding stress. The characteristics obtained by the FEM analysis did not include this feature. After the loss of stability, the force value decreased and stabilized until the compaction phase. Another effect that was recorded only during the simulation was the oscillation of the stress value in the progressive folding and compaction phases. The first of the causes mentioned above should be viewed as being involved with the local nature of such phenomena: as a loss of stability, the formation of a single fold, its closing, the mutual contact of walls, the subsequent loss of stability, etc. In the real structure, the local effects compensated for each other and there were no oscillations. Another noteworthy feature of the recorded oscillations of the stress value in the progressive folding phase was the increase in the amplitude with an increase in the wall thickness and a decrease in frequency with an increased cell size. In contrast, the oscillations in the densification phase were related to the occurrence of contact forces and friction due to the contact of the progressively larger surfaces of the walls with each other and their sliding over one another.

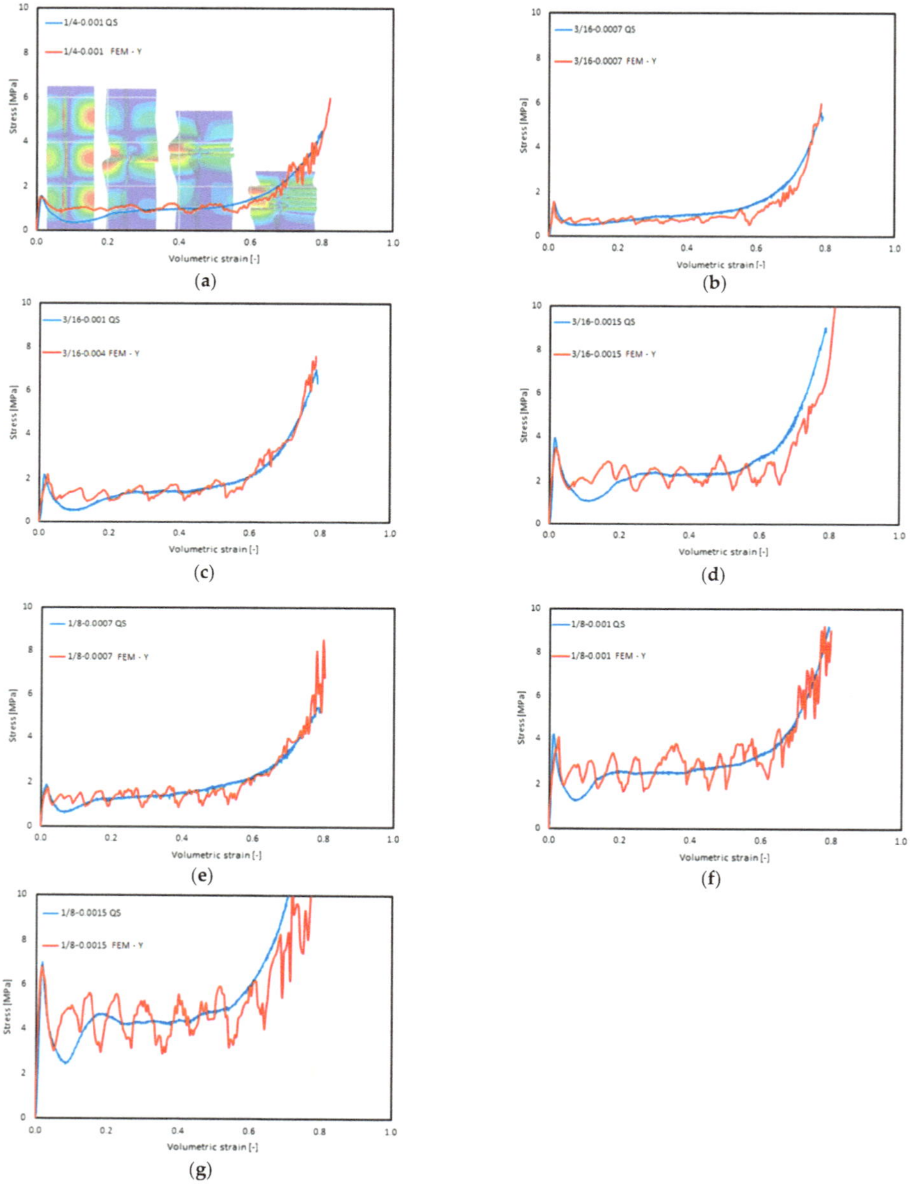

Figure 15. Summary of the stress–volumetric strain characteristics obtained using the FEM–Y model (FEM–Y) and experimentally in the quasi-static compression (QS) test [21] of the following material types: (**a**) 1/4-5052-0.001; (**b**) 16-5052-0.0007; (**c**) 3/16-5052-0.001; (**d**) 3/16-5052-0.0015; (**e**) 1/8-5052-0.0007; (**f**) 1/8-5052-0.001; and (**g**) 1/8-5052-0.0015.

The obtained characteristics of the full range of deformations allowed for the identification and determination of the basic parameters of the individual types of honeycomb cores. Table 6 presents a summary and comparison between our results and those obtained through empirical research. Values included within parentheses denote the percentage of difference between those obtained through experimentation and those obtained through FEM analyses.

Table 6. List of the basic parameters of the examined structures.

Parameter	Plateau Stress σ_{pl} [MPa]	Elastic Modulus E_u [MPa]	Critical Stress σ_{cr} [MPa]	Relative Volume of Total Compaction V_F [-]
1/4-5052-0.001	0.99 (−8.8%)	261.16 (−6.5%)	1.42 (9.6%)	0.28 (9.7%)
3/16-5052-0.0007	0.95 (3.1%)	168.22 (−6.7%)	1.60 (−6.0%)	0.28 (−7.7%)
3/16-5052-0.001	1.42 (3.4%)	247.73 (5.8%)	2.22 (−1.4%)	0.29 (−7.4%)
1/8-5052-0.0015	2.28 (2.6%)	397.34 (4.7%)	3.54 (10.6%)	0.21 (27.6%)
1/8-5052-0.0007	1.49 (2.6%)	251.34 (3.8%)	1.72 (9.5%)	0.27 (−8.0%)
1/8-5052-0.001	2.70 (−2.3%)	474.06 (8.8%)	4.11 (3.1%)	0.24 (7.7%)
1/8-5052-0.0015	4.41 (1.3%)	591.11 (−1.9%)	6.82 (3.0%)	0.21 (30.0%)

5.2. Results of the Numerical Analyses of the Honeycomb Structures Using the Homogenous Model

The stress–volumetric strain characteristics obtained through experimental tests [21] and those obtain via numerical analyses using the solid (homogenous) model are presented in Figure 16. Each of the graphs contain the results obtained during the tests carried out at three strain rates: 8.3×10^{-2}, marked as QS; 3.8×10^{2}, marked as DH; and 3.3×10^{3}, marked as SHPB. The summaries are grouped according to the type of structure under study. In order to emphasize the characteristics obtained in the FEM calculations, the previously shown characteristics from empirical studies [21] are marked with dashed lines.

Characteristics obtained by means of the FEM analyses were consistent with their equivalents obtained experimentally, and the tendency to strengthen with an increase in the strain rate was maintained. Additionally, the nature of this tendency is similar; that is, in the plateau stress range, no hardening of a constant, proportional value occurred. It can therefore be assumed that the use of the material characteristics obtained via the static tests of the model—and the inclusion of the air enclosed inside the core under the test conditions in the model—brought about the intended results.

No vibrations occurred in any of the analyzed cases, and no other symptoms of wave phenomena were observed, which were recorded during the experiments using the split Hopkinson pressure bar. However, it should be mentioned that the method of determining the stress value was indirect and consisted of the measurement of the deformation of a long, slender bar that is also susceptible to oscillation in directions other than longitudinal. Therefore, the oscillations did not result from the response of the material being tested, but from the method of measurement and reading. The characteristics obtained in the FEM tests were smooth, devoid of sinusoidal components characterized by high amplitudes or low frequencies.

It should be pointed out that the characteristics obtained experimentally and numerically differ significantly in the initial range. While the elastic range had a similar course in all cases, as well as in terms of the nature of the stress build-up in the initial range, after reaching the value defined by the $\sigma_{ij} - \varepsilon_{ij}$ curve characteristic for the MAT_HONEYCOMB constitutive model used, the stress value stabilized; in the remaining further range of strain, it followed the values defined by it. Importantly, we observed no phenomenon where a clear plateau value was achieved that was much higher than that of the critical stress, followed (in the case of a real structure) by a loss of stability and the kind of breakdown of the structure's stability, as manifested by a decrease in the stress value to a level below the plateau value that was achieved and maintained at a later stage in most of the compression tests. A more precise observation of this issue is shown in Figure 17, which presents the stress–volumetric strain characteristics of the 1/8-5052-0.0015 sample with the displayed range of volumetric strain limited to 20%.

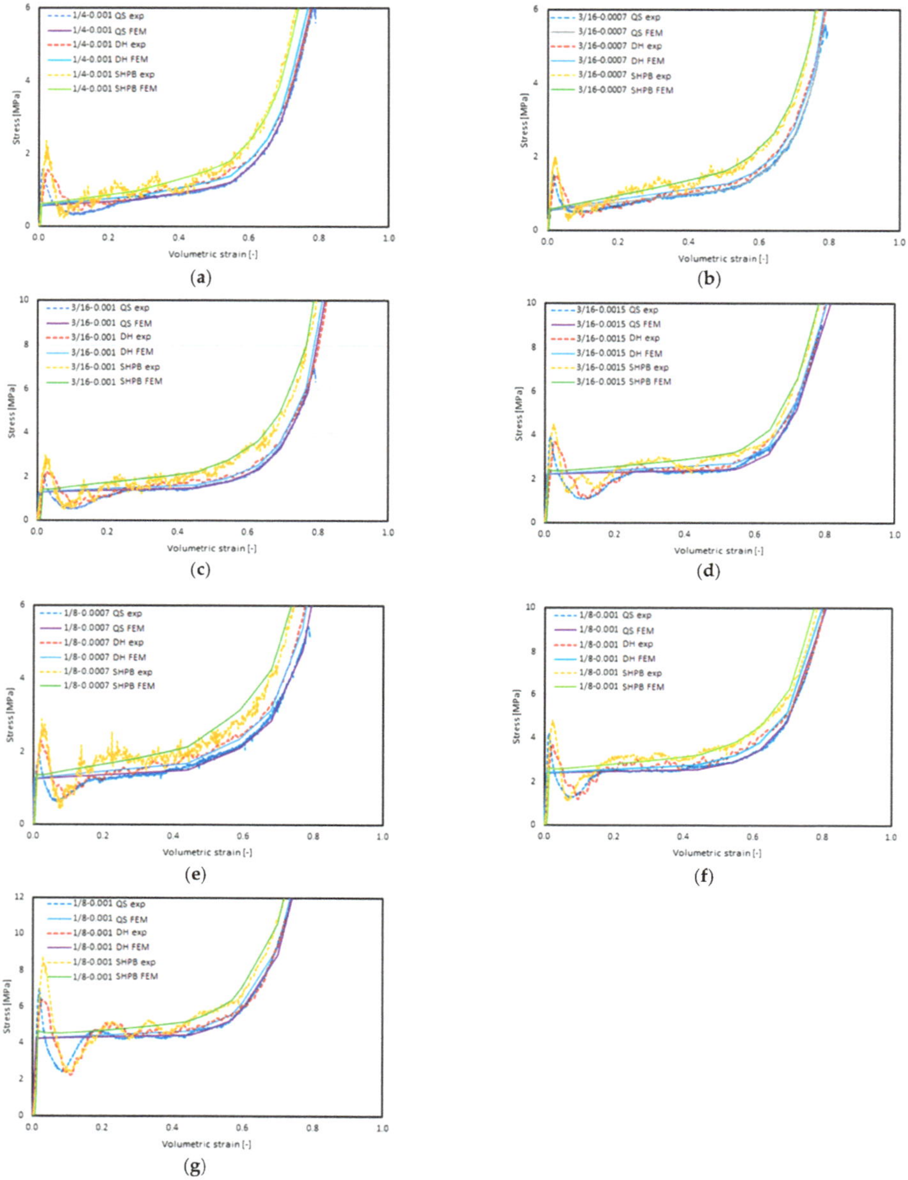

Figure 16. Summary of the stress–volumetric strain characteristics obtained using the FEM–homogeneous model (FEM) and experimentally (exp) [21] at a quasi-static load (QS) strain rate of 8.3×10^{-2} 1/s, at a dynamic load (DH) strain rate of 3.8×10^{2} 1/s, and using a Split Hopkinson Pressure Bar (SHPB) with a strain rate of 3.3×10^{3} 1/s) of the materials, including (**a**) 1/4-5052-0.001; (**b**) 3/16-5052-0.0007; (**c**) 3/16-5052-0.001; (**d**) 3/16-5052-0.0015; (**e**) 1/8-5052-0.0007; (**f**) 1/8-5052-0.001; and (**g**) 1/8-5052-0.0015.

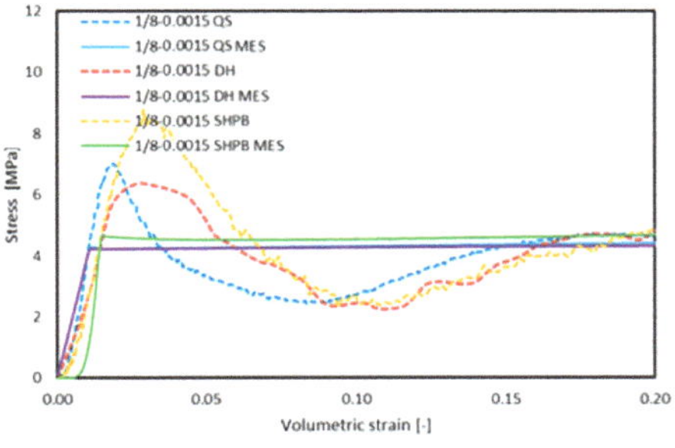

Figure 17. Summary of stress–volumetric strain characteristics obtained using the FEM–homogeneous model (FEM) and experimentally (exp) [21] at a quasi-static load (QS) strain rate of 8.3×10^{-2} 1/s, at a dynamic load (DH) strain rate of 3.8×10^{2} 1/s), and using a Split Hopkinson Pressure Bar (SHPB) at a strain rate of 3.3×10^{3} 1/s of the sample 1/8-5052-0.0015 for a strain limited to <0;0.2> scope.

6. Conclusions

The MAT_HONEYCOMB model does not include an option that allows for the consideration of the phenomenon by indicating a scalar stress value, up to which the material response would be dependent on the modulus of longitudinal elasticity, followed by assuming a value consistent with the indicated characteristic.

In order to obtain the full stress–strain characteristics of a metallic honeycomb, a numerical model can be used, the geometry of which will be limited to a single Y element. In order to obtain descriptions of the stress–volumetric strain characteristics and the scalar parameters, one must make sure that the following conditions are met. Regardless of the assumed strain rate, it is appropriate to apply the characteristics from static tests and achieve correct simulation results under a wide range of strain rates, assuming that the model takes into account the pressure change due to the potential compression of the air enclosed in the core volume.

In this study, the effectiveness of one of the simpler methods was tested: the definition of the boundary condition of uniform pressure distribution over a limited area. This enables the development of a model that faithfully reproduces the responses of a honeycomb structure at different strain rates.

Author Contributions: Conceptualization, R.C. and D.M.; methodology, R.C. and D.M.; validation, R.C.; formal analysis, R.C. and D.M.; investigation, R.C.; resources, R.C. and D.M.; data curation, R.C.; writing—original draft preparation, R.C. and D.M.; writing—review and editing, R.C. and D.M.; visualization, R.C.; supervision, D.M. All authors have read and agreed to the published version of the manuscript.

Funding: The paper was written as part of the implementation of the university research grant No. 22-838.

Institutional Review Board Statement: Not applicable.

Informed Consent Statement: Not applicable.

Data Availability Statement: The data presented in this study are available on request from the corresponding author.

Conflicts of Interest: The authors declare no conflict of interest.

References

1. Dotts, R. *Apollo Experience Report—Spacecraft Heating Environment and Thermal Protection for Launch Through the Atmosphere of the Earth, NASA TN D-7085*; National Aeronautics and Space Administration: Washington, DC, USA, 1973.
2. McFarland, R. *The Development of Metal Honeycomb Energy-Absorbing Elements*; Technical Report no. 32-639; California Institute of Technology: Pasadena, CA, USA, 1964.

3. Xiang, Y.; Yu, T.; Yang, L. Comparative analysis of energy absorption capacity of polygonal tubes, multi-cell tubes and honeycombs by utilizing key performance indicators. *Mater. Des.* **2016**, *89*, 689–696. [CrossRef]
4. Zhang, D.; Lu, G.; Ruan, D.; Fei, Q.; Duan, W. Quasi-static combined compression-shear crushing of honeycombs: An experimental study. *Mater. Des.* **2019**, *167*, 107632. [CrossRef]
5. Quoc, P.M.; Krzikalla, D.; Mesicek, J.; Petru, J.; Smiraus, J.; Sliva, A.; Poruba, Z. On Aluminum Honeycomb Impact Attenuator Designs for Formula Student Competitions. *Symmetry* **2020**, *12*, 1647. [CrossRef]
6. Zhang, S.; Chen, W.; Gao, D.; Xiao, L.; Han, L. Experimental study on dynamic compression mechanical properties of aluminum honeycomb structures. *Appl. Sci.* **2020**, *10*, 1188. [CrossRef]
7. Reyno, T.; Underhill, P.R.; Krause, T.W.; Marsden, C.; Wowk, D. Surface Profiling and Core Evaluation of Aluminum Honeycomb Sandwich Aircraft Panels Using Multi-Frequency Eddy Current Testing. *Sensors* **2017**, *17*, 2114. [CrossRef] [PubMed]
8. Jedral, A. Review of testing methods dedicated for sandwich structures with honeycomb core. *Trans. Aerosp. Res.* **2019**, *2*, 7–20. [CrossRef]
9. Wierzbicki, T. Crushing analysis of metal honeycombs. *Int. J. Impact Eng.* **1983**, *1*, 157–174. [CrossRef]
10. Goldsmith, W.; Sackman, L. An experimental study of energy absorption in impact on sandwich plates. *Int. J. Impact Eng.* **1992**, *12*, 241–262. [CrossRef]
11. Baker, W.E.; Togami, T.C.; Weydert, J.C. Static and dynamic properties of high-density metal honeycombs. *Int. J. Impact Eng.* **1998**, *21*, 149–163. [CrossRef]
12. Xu, S.; Beynona, J.H.; Ruan, D.; Yu, T.X. Strength enhancement of aluminium honeycombs caused by entrapped air under dynamic out-of-plane compression. *Int. J. Impact Eng.* **2012**, *47*, 1–13. [CrossRef]
13. Wilbert, A.; Jang, W.Y.; Kyriakides, S.; Floccari, J.F. Buckling and progressive crushing of laterally loaded honeycomb. *Int. J. Solids Struct.* **2011**, *48*, 803–816. [CrossRef]
14. Deqiang, S.; Weihong, Z.; Yanbin, W. Mean out-of-plane dynamic plateau stresses of hexagonal honeycomb cores under impact loadings. *Compos. Struct.* **2010**, *92*, 2609–2621. [CrossRef]
15. Khoshravan, M.R.; Najafi Pour, M. Numerical and experimental analyses of the effect of different geometrical modelings on predicting compressive strength of honeycomb core. *Thin-Walled Struct.* **2014**, *84*, 423–431. [CrossRef]
16. Jost, T.; Heubrandtner, T.; Ruff, C.; Fellner, B. A new method to model Aluminium honeycomb based crash barriers in lateral and frontal crash load cases. *LS-DYNA Anwend.* **2008**, *B-III*, 13–24.
17. Shkolnikov, M.B. Honeycomb modeling for side impact moving deformable barrier (MDB). In Proceedings of the 7th International LS-DYNA®Users Conference, Detroit, MI, USA, 1–14 July 2002.
18. Jackson, K.E. Predicting the Dynamic Crushing Response of a Composite Honeycomb Energy Absorber Using Solid-Elem Based Models in LS-DYNA®. In Proceedings of the 11th International LS-DYNA®Users Conference, Detroit, MI, USA, 1–10 January 2020.
19. Hallquist, J. *LS-DYNA®Theory Manual*; Livermore Software Technology Corporation: Livermore, CA, USA, 2006.
20. Kojima, S.; Yasuki, R.; Mikutsu, S.; Takasudo, T. A study on yielding function of aluminum honeycomb. In Proceedings of the 5th European LS_DYNA Users Conference, Birmingham, UK, 25–26 May 2005.
21. Ciepielewski, R.; Gieleta, R.; Miedzińska, D. Experimental Study on Static and Dynamic Response of Aluminum Honeycomb Sandwich Structures. *Materials* **2022**, *15*, 1793. [CrossRef]
22. Yamashita, M.; Gotoh, M. Impact behavior of honeycomb structures with various cell specifications—Numerical simulation and experiment. *Int. J. Impact Eng.* **2005**, *32*, 618–630. [CrossRef]
23. Matweb—Your Source for Materials Information. 2015. Available online: https://www.matweb.com/ (accessed on 10 November 2022).
24. *HexWeb® Rigicell™, Corrosion Resistant Aluminum Corrugated Honeycomb*; Product Data; Hexcel Corporation: Stamford, CT, USA, 2020; p. 24.
25. Shanqing, X.; Beynona, J.; Ruana, D.; Lu, G. Experimental study of the out-of-plane dynamic compression of hexagonal honeycombs. *Compos. Struct.* **2012**, *94*, 2326–2336.
26. Mattern, S. *Overview on Airbag-Modelling Possibilities in LS-DYNA*; DYNAmore Express, DYNAmore GmbH: Stuttgart-Vaihingen, Germany, 2020.
27. Available online: engineeringtoolbox.com/individual-universal-gas-constant-d_588v (accessed on 20 November 2021).

metals

MDPI

Article

The Mechanism of In-Situ Laser Polishing and Its Effect on the Surface Quality of Nickel-Based Alloy Fabricated by Selective Laser Melting

Yanhua Zhao [1,*], Chuanbin Du [1], Peifu Wang [2], Wei Meng [1] and Changming Li [3]

1 School of Mechanical and Electronic Engineering, Shandong Jianzhu University, Fengming Road, Jinan 250101, China; 2020070101@stu.sdjzu.edu.cn (C.D.); 2020075115@stu.sdjzu.edu.cn (W.M.)
2 Jinan Special Equipment Inspection Institute, Jinan 250100, China; wangpeifu2001@sohu.com
3 Hangzhou Optimax Tech Co., Ltd., Hangzhou 310000, China; h2378649304@163.com
* Correspondence: zyh@sdjzu.edu.cn; Tel.: +86-189-5416-9696

Abstract: Laser polishing (LP) is an effective method to improve the surface quality of an additively manufactured nickel-based alloy. In this paper, the in-situ laser polishing (ILP) experiment is performed on the selective laser melting (SLM) IN718 samples. The white light interferometer is used to test the three-dimensional surface profile and surface roughness of samples. The results show that the surface quality of as-SLMed samples by ILP is improved. In particular, the surface roughness is decreased by 33.5%. To reveal the mechanism of ILP, a three-dimensional numerical model is established based on the finite volume method (FVM). The model can accurately simulate the mesoscopic scale physical phenomena when the laser interacts with the metal. The temperature field, the melt pool flow, and the evolution of the surface morphology during the ILP process are predicted using this model. The mechanism of ILP is revealed based on the dynamics of the molten pool. The contribution of capillary and thermocapillary forces to the reduction of bulge curvature at different stages is studied. Furthermore, the effect of ILP power on the surface quality is investigated, and the mechanism of bulges and depressions on the track surface during high-power ILP is revealed.

Keywords: selective laser melting; in-situ laser polishing; mesoscopic simulation; temperature change; surface morphology evolution

Citation: Zhao, Y.; Du, C.; Wang, P.; Meng, W.; Li, C. The Mechanism of In-Situ Laser Polishing and Its Effect on the Surface Quality of Nickel-Based Alloy Fabricated by Selective Laser Melting. *Metals* **2022**, *12*, 778. https://doi.org/10.3390/met12050778

Academic Editors: Shuwen Wen, Yongle Sun, Xin Chen and Alexandre Emelyanenko

Received: 17 March 2022
Accepted: 29 April 2022
Published: 30 April 2022

1. Introduction

Nickel-based alloys have been widely used in the nuclear industry [1], aerospace [2], and other industries due to their excellent yield strength, structural stability, and resistance to oxidation and corrosion [3] at elevated temperatures. Nickel-based alloy parts often have complex structures that are difficult to manufacture by traditional methods. Additive manufacturing (AM) technology is one of the most efficient means of manufacturing complex parts. Among them, SLM technology is particularly suitable for the manufacture of complex components [4], such as centrifugal compressor blades [5]. However, the surface quality of SLM alloys is still poorer compared to subtractive manufacturing. Subsequent subtractive processing is required to meet the needs of assembly [6]. It also seriously affects the fatigue performance of parts [7]. These factors limit the popularization and application of SLM to a certain extent [8].

1.1. Polishing and LP

Polishing can effectively improve the surface quality of parts. The traditional polishing methods for IN718 materials mainly include electrochemical polishing and abrasive flow polishing. Electrochemical polishing can reduce the surface roughness of AM IN718 parts to 0.25 microns [9]. Abrasive flow polishing is good at grinding the inner surface of complex parts [10]. However, traditional polishing methods also have their disadvantages.

Traditional mechanical polishing is not suitable for the complex parts manufactured by SLM [8]. Chemical polishing is dangerous and has environmental concerns [11]. In addition, abrasive flow polishing has material loss [9].

As an eco-friendly manufacturing method, LP does not pollute the environment, does not harm the health of workers [12], does not affect mechanical properties, and has a high degree of automation [13]. LP has begun to be used to improve the surface quality of SLM parts [14]. Li et al. [3] studied the thermodynamic behavior and porosity change of SLM IN718 parts subjected to LP. The results showed that the polishing zone temperature was above 1683 K, the cooling rate was about 2.46×10^6 K/s, and the porosity decreased by 65.7% after polishing. Fang et al. [15] carried out LP on the laser additive manufacturing of IN718 parts. In addition, it was reported that the surface roughness can be reduced from 7 μm to less than 1 μm by the LP process. The microhardness increased from 345 HV to 440 HV. Cwikła et al. [16] used LP to reduce the surface roughness of AM IN718 parts by 87.9% and proved that the argon environment is more conducive to the improvement of surface quality than the air environment. Ma et al. [17] performed LP on AM TC4, and the surface roughness was decreased from 5 μm to 1 μm after polishing. Due to the formation of the α' martensitic phase in the polishing zone, the surface microhardness was increased by 32%, and the wear resistance was also greatly improved. Chen et al. [18] used LP as a post-processing technique to improve the mechanical properties of 316 L stainless steel parts fabricated by SLM. The research showed that the average grain size of the polished zone was 17.3 μm. A hardened layer of 50–70 μm was formed. In addition, the tensile strength and ductility were also significantly improved. Liang et al. [19] studied the effect of LP on the surface quality of SLM aluminum alloy. The results showed that LP can repair the cracks on the surface of the alloy, fuse the pores on the surface, melt the powders on the surface, and greatly reduce the surface roughness. It can be seen that LP can effectively improve the surface quality of SLM parts. Immediately before the work environment is cooled and ventilated, the surface of parts made with SLM is subjected to an LP process, namely, ILP. ILP technology can not only improve the surface quality of parts but also improve processing efficiency and avoid oxidation during LP [20].

1.2. Mechanism of LP

In order to improve the surface quality of parts more effectively, many scholars have devoted themselves to the study of the LP mechanism. Because it is difficult to observe the interaction between laser and material during LP-only by experimental means, it is even more difficult to reveal the mechanism of LP. Therefore, some scholars have carried out LP numerical simulation research to try to explain the mechanism of LP. Marimuthu et al. [8] used experiments and numerical simulation to prove the importance of energy input on LP surface quality. The surface roughness of the LP region strongly depends on the flow velocity of the molten pool, which is also directly related to the energy input. Xu et al. [21] established a two-dimensional numerical model to study the mechanism of bulge formation during the LP process. The main reason for this was that the molten material on both sides of the track flows toward the center of the track. Zhang et al. [22] established a model to describe the flow behavior of the molten pool during LP and mainly studied the effects of capillary force and thermocapillary force in the LP process. The results showed that capillary forces play a leading role and were responsible for eliminating surface defects with large curvatures. Li et al. [23] used numerical simulation to explain why the defects cannot be completely eliminated during LP. Due to the rapid cooling rate of the molten pool, the flow time of the molten pool was too short to completely eliminate the surface defects. Li et al. [24] established a two-dimensional axisymmetric numerical model for LP. The process of cooling was simulated. The model observed that the top of the peak starts to melt first, and the molten liquid flows from the peak to the trough. At the same time, the flow velocity would suddenly increase, and the molten pool was in an unstable state. As the study revealed, an unstable melt pool might prevent the flow of liquid from the bulge to the depression, which is detrimental to the improvement of surface quality.

1.3. Scope of This Research

The surface morphology of SLM-fabricated parts has a significant impact on the process of LP and the final surface quality [25]. Few current LP numerical simulation studies take this into account. The purpose of this study is to fill in the knowledge gap so as to reveal the mechanism of the effect of LP on the surface quality of SLM parts more realistically and effectively.

To this end, the fabrication of SLM samples and ILP experiments were carried out first. The macro- and micro-surface morphologies of the samples were observed. Based on FVM, a three-dimensional numerical model, including the SLM and LP stages, was established. The physical phenomena, such as capillary force, thermocapillary force, vaporization, recoil pressure, convective heat transfer, thermal radiation, and latent heat of phase change, were fully considered. Based on this model, the temperature field, molten pool dynamics, and surface morphology evolution during the ILP were simulated. The mechanism of the reduction of bulge curvature and the generation of new bulges and depressions after ILP on the SLM surface was revealed.

2. In-Situ Laser Polishing (ILP) Experiment

2.1. Experiment Methods

The material used in this study was the IN718 powder obtained by gas atomization. The size of the powder was 15–53 μm, and the base material was 304 stainless steel. The SLM and ILP equipment was a LATEC LAM-150V 3D printer, with a maximum power of 1000 W, and the oxygen content could be controlled within 100 ppm.

The purpose of the experimental work was to study the influence of LP on the surface quality of parts fabricated using SLM. Similar process parameters (i.e., power, speed, etc.) were used for both the SLM and ILP processes. According to our previous experimental exploration, the SLM processing parameters were determined to be 185 W of laser power and a 0.6 m/s scanning speed, and the ILP processing parameters were 180 W of laser power and a 0.6 m/s scanning speed. The spot diameter was 0.05 mm. The laser energy distribution was a Gaussian distribution. The layer thickness was 0.04 mm. In addition, the hatch spacing was 0.01 mm. The time-lapse between SLM and ILP was about 0.023 s. The IPL was carried out immediately after the SLM. The size of the samples was 10 mm × 10 mm × 5 mm. The samples were printed in 125 layers, each with 500 tracks. The ILP scanning was consistent with the SLM. In this experiment, a single laser was used, and the ILP was performed after the SLM was completed. The as-SLMed and ILPed samples are shown in Figure 1a,b, respectively.

Figure 1. Macroscopic morphology of sample. (**a**) As-SLMed; (**b**) ILPed. The parameters of SLM were 185 W, 0.6 m/s and 0.05 mm and 0.01mm (power, speed, spot diameter and hatch spacing). The parameters of ILP were 180 W, 0.6 m/s and 0.05 mm and 0.01mm (power, speed, spot diameter and hatch spacing).

A total of 8 samples were prepared, as-SLMed samples 1—4, ILPed samples 1—4, and the surface roughness was tested, respectively. Sample No. 1 was measured four times each.

All measurement results are shown in Figure 2. The errors of the measurement data are all within the allowable range, which proves the repeatability of processing and measurement is good. All as-SLMed and ILPed samples' test results are averaged together and obtained surface roughness with uncertainty.

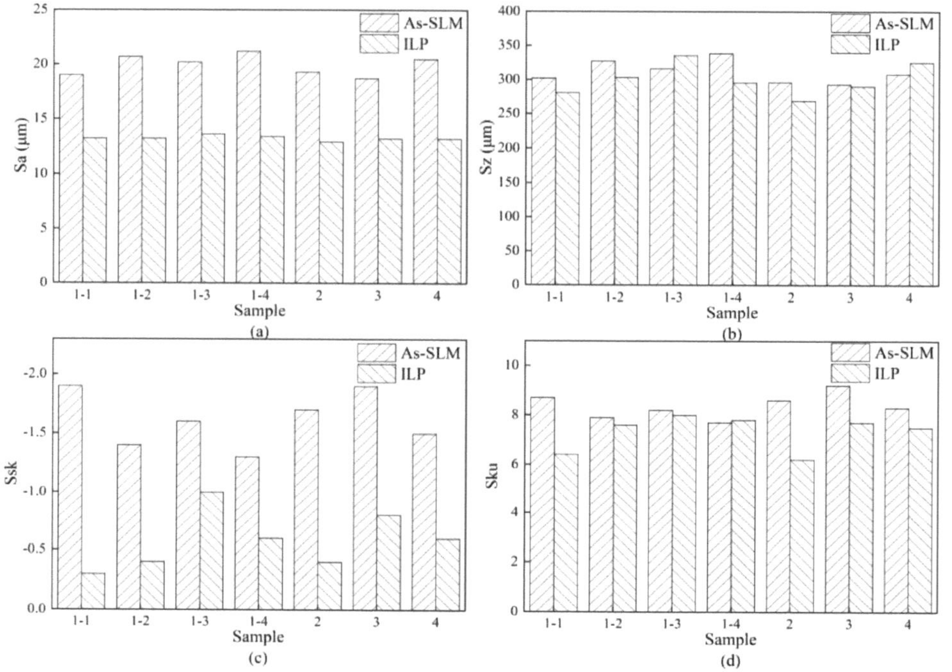

Figure 2. Surface roughness data. (**a**) Sa; (**b**) Sz; (**c**) Ssk; (**d**) Sku. The parameters of SLM were 185 W, 0.6 m/s and 0.05 mm and 0.01mm (power, speed, spot diameter and hatch spacing). The parameters of ILP were 180 W, 0.6 m/s and 0.05 mm and 0.01mm (power, speed, spot diameter and hatch spacing).

A Morphology Elite K white light interferometer was used to test the surface morphology and surface roughness of the samples. The surface area of the samples is 10 mm × 10 mm. Due to the limitation of the measuring range of the equipment, and in order to facilitate the positioning of the sample during measurement, the measurement was carried out in an area of 1.302 mm × 1.735 mm in the center of the sample. There were 305,000 data points. According to the research results of scholar Przemysław [26], high-frequency noise will seriously affect the measurement results of the surface topography. When using the white light interferometer for 3D surface measurement, the main sources of noise are shot noise, positioning noise, and discretization noise [27]. In this study, the Fourier filtering technology was used to deal with high-frequency noise in the raw data.

2.2. Results of Experiment

Figure 3 shows the microscopic three-dimensional contour of the samples. Figure 3a,b are as-SLMed and ILPed, respectively. Compared with an as-SLMed sample, the height difference in the ILPed sample is reduced, and the number of bulges is also reduced. The surface roughness results are shown in Figure 2. Sa represents the difference between the height of each point and the arithmetic mean of the entire surface. The calculation formula is the following:

$$Sa = \frac{1}{A}\iint_{A} |Z(x, y)|dxdy \quad (1)$$

where A is the area of the sample, and Z(x,y) is the height of the point.

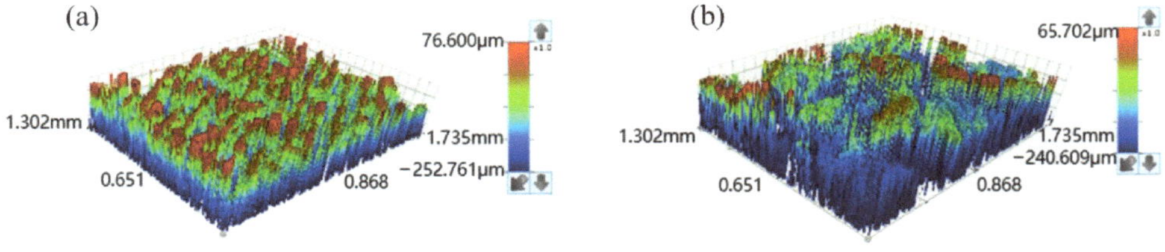

Figure 3. Three−dimensional surface morphology measured by white light interferometer. (**a**) As−SLMed; (**b**) ILPed. The parameters of SLM were 185 W, 0.6 m/s and 0.05 mm and 0.01mm (power, speed, spot diameter and hatch spacing). The parameters of ILP were 180 W, 0.6 m/s and 0.05 mm and 0.01mm (power, speed, spot diameter and hatch spacing).

Sa is the extension of Ra from a straight line to a curved surface, which is more suitable for evaluating the surface quality of a curved surface. In addition, compared with other surface quality parameters, Sa is more comprehensive and can better represent the smoothness of the entire surface. Sz represents the height difference of the surface. As shown in Table 1, compared with the as-SLMed sample, the ILPed samples have a great decrease in Sa, but not much in Sz. Even some ILP samples have a larger Sz. The Sa of the additive part is 20.0 ± 1.25 μm, and the Sz is 315.9 ± 22.75 μm. The Sa after ILP is 13.3 ± 0.35 μm, and the Sz is 302.2 ± 33.35 μm. The skewness (Ssk) of the two samples is less than 0, and the kurtosis (Sku) is greater than 3, indicating that the surface morphology of SLMed sample is mainly characterized by 'peak'. The shape of the peak is narrow and sharp, and LP does not change this surface morphology feature. The Ssk increased by 62.5%, closer to 0, indicating a more uniform height distribution of the ILP sample. The Sku decreased by 13.1%, indicating that the ILP sample was smoother. LP improves the surface quality that is, the metal surface is remelted under the irradiation of the laser. The molten metal liquid is redistributed on the surface by the combined driving of capillary force and thermocapillary force, and the molten material at the bulge will fill the depression, making the surface becomes smooth. This ultimately reduces surface roughness. The experimental results show that ILP can improve the surface quality of the SLMed samples.

Table 1. Average value of surface roughness.

Label	As-SLMed	ILP	Change Percentage
Sa (μm)	20.0 ± 1.25	13.3 ± 0.35	33.5%
Sz (μm)	315.9 ± 22.75	302.2 ± 33.35	4.3%
Ssk	−1.6 ± 0.3	−0.6 ± 0.3	62.5%
Sku	8.4 ± 0.75	7.3 ± 0.90	13.1%

3. Numerical Model and Validation

3.1. Model Establishment

The processes of SLM and LP include heat conduction between laser and metal, phase transformation of metal, molten pool flow, heat conduction and convection, vaporization, thermal radiation, latent heat of phase change, and other mechanisms [28]. These factors constitute the numerical model.

To simplify the physical model, the following assumptions are made: (1) The fluid is assumed to be an incompressible Newtonian fluid. (2) The heat flux is assumed to be Gaussian distribution.

The three-dimensional model of the computational domain is shown in Figure 4. The size of the computational domain was 1 mm × 0.4 mm × 0.24 mm. The size of the powder bed was 1 mm × 0.4 mm × 0.04 mm. The thickness of the powder bed was the same as in the experiment. The powder bed was established by the discrete element software EDEM 2018. Its thickness was controlled by the distance between the blade and the base. The powder bed was placed on the base. The model was placed in an argon environment and contained 868,602 grids with a uniform grid size of 5 μm. A single-track simulation and three-track simulation were performed in this study. The three-track simulation only appeared in Section 4.4. The simulation used a single laser, and the ILP scan would wait until the SLM scan of the whole layer was completed.

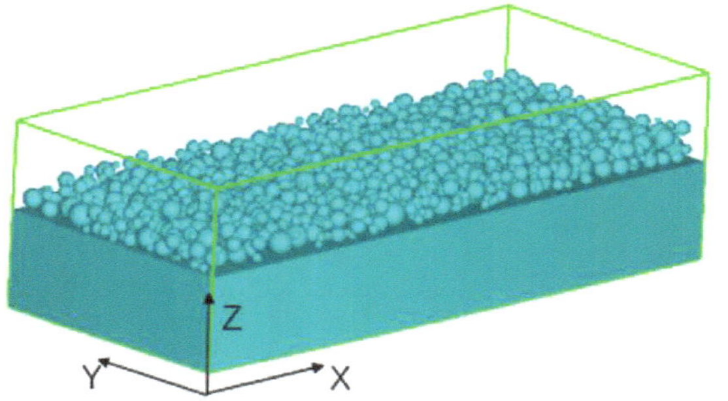

Figure 4. Computational domain.

The modeling was carried out using FLOW-3D V11.1 [29], a commercial computational fluid dynamics (CFD) code, on a GTS2-RIS212Q workstation with 24 cores and 24 threads. Using the FVM method to simulate the SLM and ILP, information such as surface morphology evolution, molten pool flow, and temperature distribution can be obtained. However, the information on stress evolution during processing cannot be obtained. This is a limitation of the FVM approach.

3.2. Governing Equations

The metal solid is melted under the role of laser. In addition, the fluid flow in the molten pool satisfies continuity, momentum conservation, and energy conservation.

Continuity equation [30]:

$$\nabla\cdot(\rho_{\vec{v}}) = 0 \tag{2}$$

Momentum conservation equation [30]:

$$\frac{\partial}{\partial t}(\rho_{\vec{v}}) + \nabla(\rho_{\vec{v}} \oplus_{\vec{v}}) = \nabla\cdot(\mu\nabla_{\vec{V}}) - \nabla P + \rho_{\vec{g}} \tag{3}$$

Energy conservation [31]:

$$\frac{\partial h}{\partial t} + (\vec{v}\cdot\nabla)h = \frac{1}{\rho}(\nabla\cdot k\nabla T) \tag{4}$$

In Equations (2)–(4), ρ is the density of the material, μ is the viscosity, v is the fluid velocity, g is the acceleration of gravity, h is enthalpy, t is the time, k is the thermal conductivity, and T is temperature of the material.

The volume of fluid (VOF) method is used to capture the free surface. The free surface is viewed as a material surface, and the fluid volume fraction is used to represent the proportion of fluid occupying space in the grid, so as to realize the capturing of the free surface. The volume fraction function satisfies the following transportation equation [32]:

$$\frac{\partial F}{\partial t} + \nabla(F\vec{v}) = 0 \tag{5}$$

where F is the volume fraction of the fluid, F = 1, F = 0, 0 < F < 1 represent fluid-filled, fluid-free, and partially fluid-filled with free surface, respectively [32].

As the laser moves, the temperature in the computational domain changes sharply. The heat conduction obeys the energy conservation law and the Fourier law. It is a typical three-dimensional transient heat conduction process, which satisfies the following equation [29]:

$$\rho c\frac{\partial T}{\partial t} = \frac{\partial}{\partial x}\left(k\frac{\partial T}{\partial x}\right) + \frac{\partial}{\partial y}\left(k\frac{\partial T}{\partial y}\right) + \frac{\partial}{\partial z}\left(k\frac{\partial T}{\partial z}\right) + Q \tag{6}$$

In the formula, c is the specific heat capacity of the material and Q is the source term.

3.3. Gaussian Heat Source and Boundary Conditions

Since the energy of free photoelectrons in the laser is relatively low, it is difficult to reach a deeper position directly through the metal surface, so the surface heat source model is adopted. The following is the heat source formula [33]:

$$q(r) = \frac{2\alpha P}{\pi\omega^2}\exp\left(-2\frac{r^2}{\omega^2}\right) \tag{7}$$

where q (r) is the heat flux, P is the laser power, α is absorption, ω is the heat source radius, and r is the distance from any point in the plane to the laser center point. In this study, α was 0.38 and ω was 0.025 mm. α is cited from the paper by Li et al. [31], ω is consistent with the experiment. They are used for all simulations. (These parameters are used for all performed simulations).

Thermal boundary conditions at the free surface of the molten pool are as follows [34]:

$$q_{in} = q(r) - h_c(T - T_0) - \varepsilon_r\sigma(T^4 - T_0^4) - q_{ev} \tag{8}$$

In the formula, q_{in} is the heat flux variation at the free interface of the molten pool, h_c is the natural convection heat transfer coefficient, ε_r is the emissivity, σ_s is the Stefan-Boltzmann constant, T is the temperature, T_0 is the ambient temperature, and q_{ev} is the evaporation heat.

In the process of SLM and ILP, the metal phase is melted rapidly and even reach the boiling temperature, evaporation occurs, and heat dissipation to the environment. The following is the evaporative heat equation [31,35]:

$$q_{ev} = 0.82\frac{\Delta H^*}{\sqrt{2\pi MRT}}p_0\exp\left(\Delta H^*\frac{T - T_{lv}}{RTT_{lv}}\right) \tag{9}$$

In the formula, ΔH^* is the escape metal vapor, M is the molar mass of the material, R is the ideal gas constant, p_0 is the environmental pressure, and T_{lv} is the boiling temperature.

Metal vapor will exert recoil pressure on free interface as follows [35]:

$$p_r = 0.54_{p_0}\exp\left(L_{lv}\frac{T - T_{lv}}{RTT_{lv}}\right) \tag{10}$$

L_{lv} is the latent heat of vaporization of metallic materials.

3.4. Process Parameters and Material Properties

The SLM process and four ILP process parameters are shown in Table 2. The purpose of the modeling study is to reveal the mechanism of ILP. Modeling should try to ensure the quality of the track. And there must be no excessive computing time. The typical computing time of a single track SLM is about 40 hours (h) [30]. To choose suitable parameters, the parameters of Yuan et al. [36], using 0.8 m/s and 200 W, were referred to. A series of simulations were performed by varying the power. Finally, the scanning speed was 0.8m/s and the power used was 150W. Using these parameters, good track quality is obtained with suitable computing time.

Table 2. Process parameters used in the modelling computation of SLM followed by ILP with different power.

Sample	Power (W)	Scanning Speed (m/s)
SLM	150	0.8
ILP	50	0.8
ILP	90	0.8
ILP	130	0.8
ILP	170	0.8

The thermophysical properties of IN718 are shown in Figure 5. The other material properties are shown in Table 3.

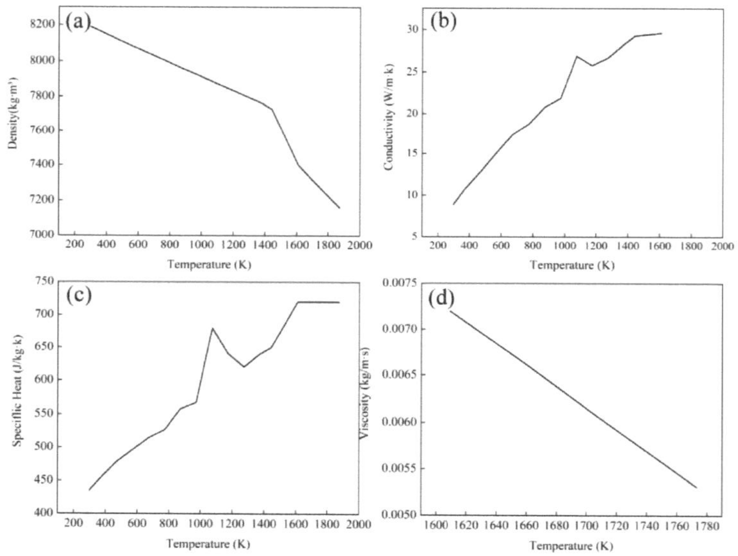

Figure 5. Temperature-dependent thermophysical properties of IN718. (**a**) Density; (**b**) Conductivity; (**c**) Specific Heat; (**d**) Viscosity. Reprinted with permission from [31]. Copyright 2016 Elsevier.

Table 3. IN718 material properties. Data from [30].

Property	Value	Units	Property	Value	Units
Reflection coefficient	0.7	K	Marangoni coefficient	1.1×10^{-4}	$N{\cdot}m^{-1}{\cdot}K^{-1}$
Solidus temperature	1523	K	Latent heat of fusion	270	$kJ{\cdot}kg^{-1}$
Liquidus temperature	1608	K	Latent heat of vaporisation	6300	$kJ{\cdot}kg^{-1}$
Boiling temperature	3188	K	Surface tension coefficient	1.89	$N{\cdot}m^{-1}$

3.5. Validation

To demonstrate the viability of the modelling techniques adopted in the current study, a single-layer, single-track SLM process was simulated, using the same processing parameters as for the SLM experiment in Section 2.1. The simulated surface topography was compared with that of the experiment that was characterized using the white light interferometer. The section location, line profile and line roughness of the experimental surface are shown in Figure 6a. The section location, line profile and line roughness of the simulated surface are shown in Figure 6b. The roughness of the experimental surface was compared with the roughness of the simulated surface. The error of Ra is 6.06%; the error of Rz is 1.33%, and the variation trend of the line profile of the simulated and experimental surfaces is consistent. That is, the simulated surface is in good agreement with the experimental surface.

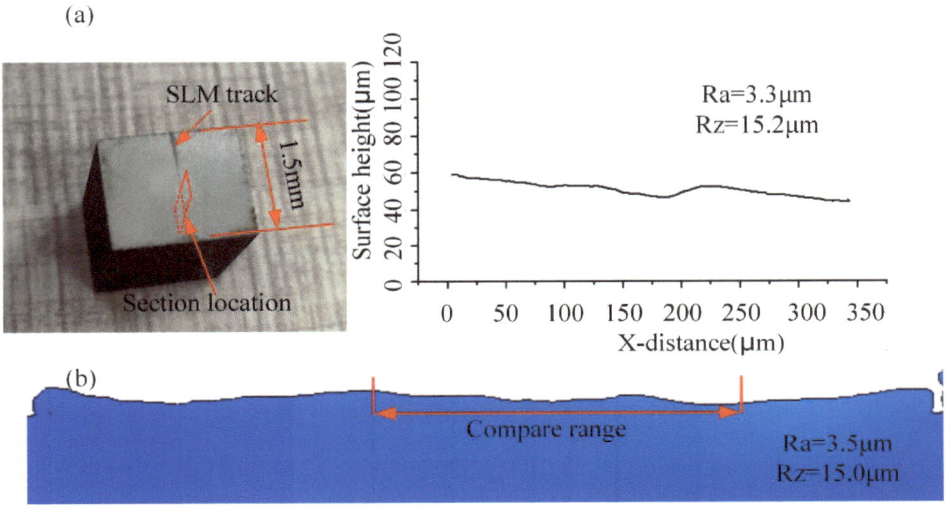

Figure 6. Simulation and experimental surface comparison. (**a**) experiment; (**b**) simulation. Obtained at 185 W, 0.6 m/s and spot diameter were 0.05 mm (power, speed and spot diameter).

In order to further verify the validity of the simulation model. The surface topography of a 1 mm-long SLM single track obtained by the simulated was compared with the surface topography of the track obtained by the SLM test [37]. The simulation and experiment use the same parameters, as shown in Figure 7. The contour of Figure 7a represents the boundary of the track. It can be seen that the width of the track obtained by experiment and simulation is gradually reduced along the laser scanning direction. The shape of the width reduction area is approximately conical. The experimental track and the simulated track show good agreement. The variation trend of the track width is related to the inherent heat and mass transfer characteristics during the SLM process [37], and the LP also depends on the heat and mass transfer of the molten pool [36,38]. Therefore, the numerical model established in this paper can well predict the changing trend of the SLM track and can be used to simulate the ILP process.

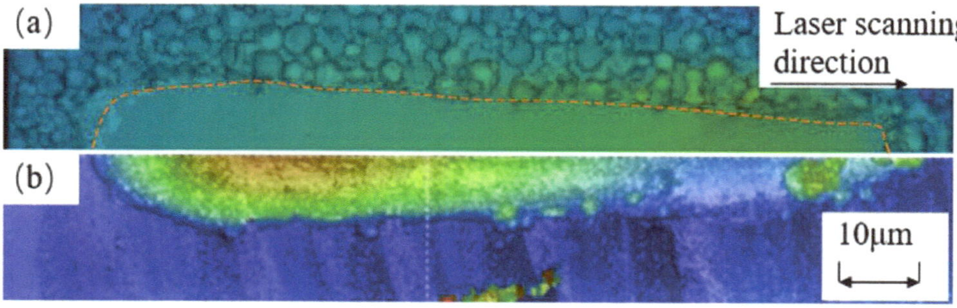

Figure 7. Track morphology comparison. (**a**) The simulated track morphology, obtained at 195 W, 0.4 m/s and 0.05 mm; (**b**) Experimental track morphology, obtained at 195 W, 0.4 m/s and 0.05 mm (power, speed and spot diameter). Reprinted with permission from [37]. Copyright 2021 The Society of Manufacturing Engineers.

4. Results and Discussion of Numerical Simulation

4.1. Temperature Change in ILP Process

The temperature change curve of the monitoring point under different ILP powers is shown in Figure 8. The monitoring point is located below the center point of the top surface of the track. The simulation was performed in the computational domain shown in Figure 4. The temperature in the powder bed and the solidified layers changed rapidly during SLM. The peak temperature of the SLM stage is 4309 K and reaches its peak at 0.55 ms. The peak temperature is above the evaporation temperature, indicating that the monitoring point is in the direct-action area of the laser at this time. The liquid at the monitoring point is evaporated. At 1.65 ms, because the ILP laser moves to this point, the temperature of the monitoring point reaches a peak again. With the increase in ILP power, the peak temperature of the monitoring point of the ILP stage gradually increased.

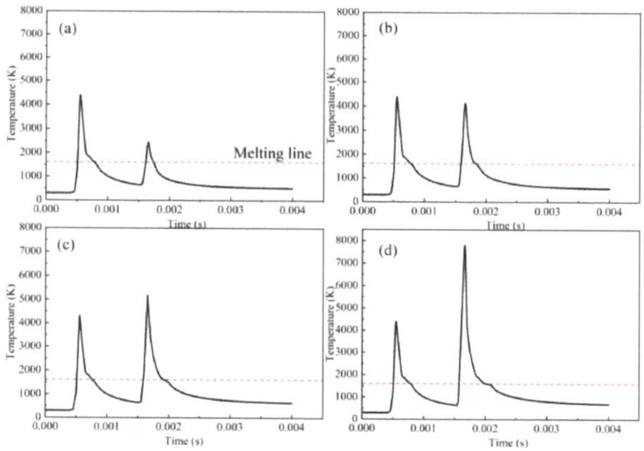

Figure 8. The temperature change of ILP with different ILP power. (**a**) 50 W; (**b**) 90 W; (**c**) 130 W; (**d**) 170 W.

When the ILP power is 50 W, although there is heat residue in the SLM stage, the second temperature peak is only 2448 K, which is far lower than the first peak of 4309 K (as shown in Figure 8a). Mainly due to the energy density of the ILP laser being much lower than the SLM laser. When the ILP power is 90 W, the second peak temperature is increased greatly to 4052 K, which is still below the first peak temperature. When the molten pool of the monitoring point cools down, the cooling rate tends to decrease gradually. From

Figure 8b, it can be seen that the slope of the curve above the liquid line in the ILP stage is gradually decreasing. This is because as the laser moves, the monitoring point is positioned further back in the molten pool.

The temperature contours on the vertical plane through the center of the ILP track predicted by the numerical simulation under different powers are shown in Figure 9. The black horizontal lines are the paths along which the temperature gradient curves are derived and plotted in Figure 10. As shown in Figure 10, in the molten pool, the more backward (the opposite direction of the X direction), the smaller the temperature gradient, and the slower the temperature change rate of the monitoring point. When the ILP power is 50 W, the cooling rate does not change significantly, because, under this processing condition, the temperature gradient in the molten pool changes very little. After the molten pool at the monitoring point is solidified, the cooling rate is significantly accelerated. From Figure 8, it can be seen that the sudden change in the slope of the curve at the liquid line in the ILP stage. This is because the heat transfer rate from liquid to solid is lower than the heat transfer rate between solids, so the metal phase dissipates heat to the base faster after solidification. When the ILP power is 130 W, the first peak temperature is exceeded by the second, and the second peak temperature is 5184 K. Although the ILP power is less than the SLM power, the working temperature is higher than the latter. This phenomenon occurs because the thermal conductivity of a continuous solid is greater than that of powder [39], and the heat of the laser is more easily transferred to the monitoring point. In addition, there is residual heat on the track after SLM, and the initial temperature of the monitoring point during ILP is 660 °C. When the ILP power is 170 W, the second peak temperature is 7710 K due to the increase in input energy density.

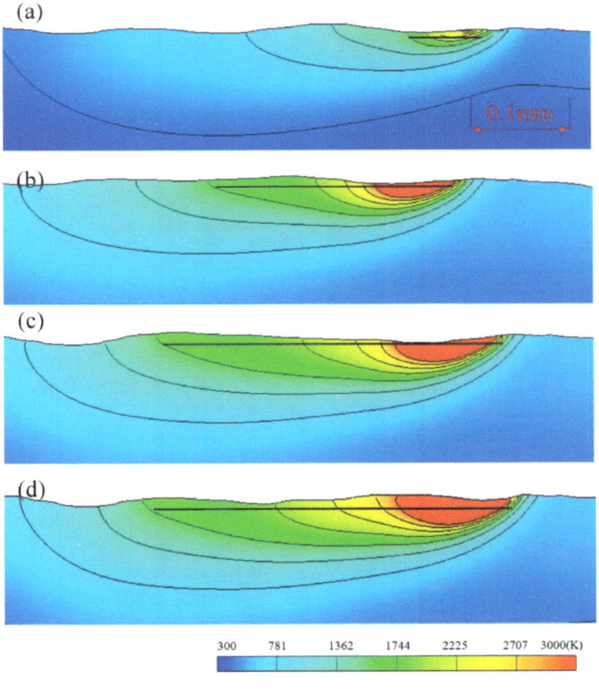

Figure 9. The temperature contours of ILP with different ILP power. (**a**) 50 W; (**b**) 90 W; (**c**) 130 W; (**d**) 170 W. The scanning speed was 0.8 m/s, and the spot diameter was 0.05 mm. The entire domain is 0.6 mm long and 0.13 mm high.

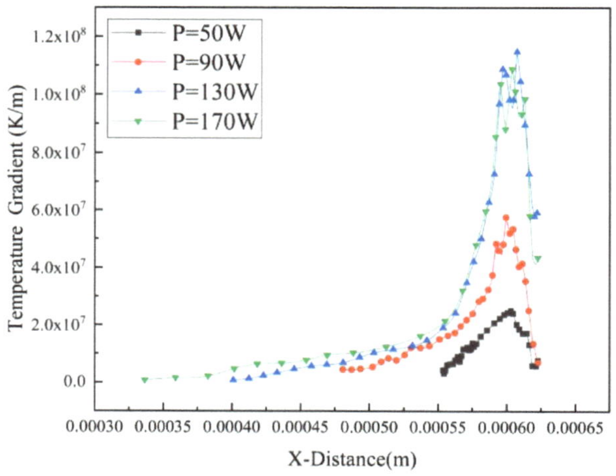

Figure 10. The temperature gradient in the X direction of the molten pool during the ILP process with four processing parameters.

In the ILP process, the working temperature of the molten pool is increased with increasing power. The larger working temperature can increase the fluidity of the molten pool so that the liquid at the bulge can fully flow in the molten pool and redistribute evenly. However, too high of a working temperature also greatly increases the instability of the molten pool [40], which may impair the surface quality.

4.2. Surface Morphology Evolution and Molten Pool Dynamics during ILP

To reveal the mechanism of ILP, the evolution of the surface morphology and the molten pool dynamics in the ILP process were studied. Taking the large bulge in the middle of Figure 11c as an example, it is described how the surface topography of the SLMed track is altered by ILP. Figure 11 is a x-z view of the center of the track, the contour is the temperature contour, and the arrow is the molten pool velocity vector. Figure 11a shows the initial state of the powder bed. Figure 11b shows the surface profile after SLM. When t = 1.35 ms, the track of the SLM has been completely solidified, but the temperature of the original molten pool area is still about 1000 K, and the temperature of other areas is also higher than the ambient temperature. The temperature accumulation in the SLM stage is beneficial for the ILP process. At this time, the ILP molten pool has not yet touched the bulge. The laser is just acting on the track, and the energy input is low, and the temperature gradient of the molten pool is low.

Among the internal forces in the molten pool, surface tension (capillary force) plays a leading role. As shown in Figure 11, the velocity vectors of the molten pool surface mostly point to the center of the molten pool. The surface area of the molten pool tends to be reduced by surface tension, which may lead to balling [35]. When t = 1.40 ms, as shown in Figure 11d, the molten pool is elongated at this time, and the Marangoni flow effect appears behind it. These indicate that there is a surface tension gradient at the gas-liquid interface. A vortex appears in the middle of the molten pool. The circulation of the liquid in the molten pool can be accelerated by the vortex, thereby the cooling and solidification of the molten metal are also accelerated [28]. At the front of the molten pool, the velocity vector points to the bulge. The molten pool begins to wet the bulge and the heat transfers to the bulge. This is the accumulation of heat for the molten pool to swallow the bulge. When t = 1.45 ms, as shown in Figure 11e, with the forward movement of the laser and the influence of the wetting behavior of the molten pool, the left half of the bulge has been swallowed by the molten pool [41]. The molten pool does not diffuse downward to the SLM track at the bulge area but flows into the valley under the action of the capillary force [21,22,24], thereby filling the valley. This flow trend can be seen in Figure 11, which is in the surface shallow melting (SSM) state of LP [42]. When t = 1.50 ms, as shown in Figure 11f, most of the bulge area has been melted. Since the liquid in the bulge area flows to the lower left of the bulge under the action of capillary force, the bulge in Figure 11e has been flattened at this moment.

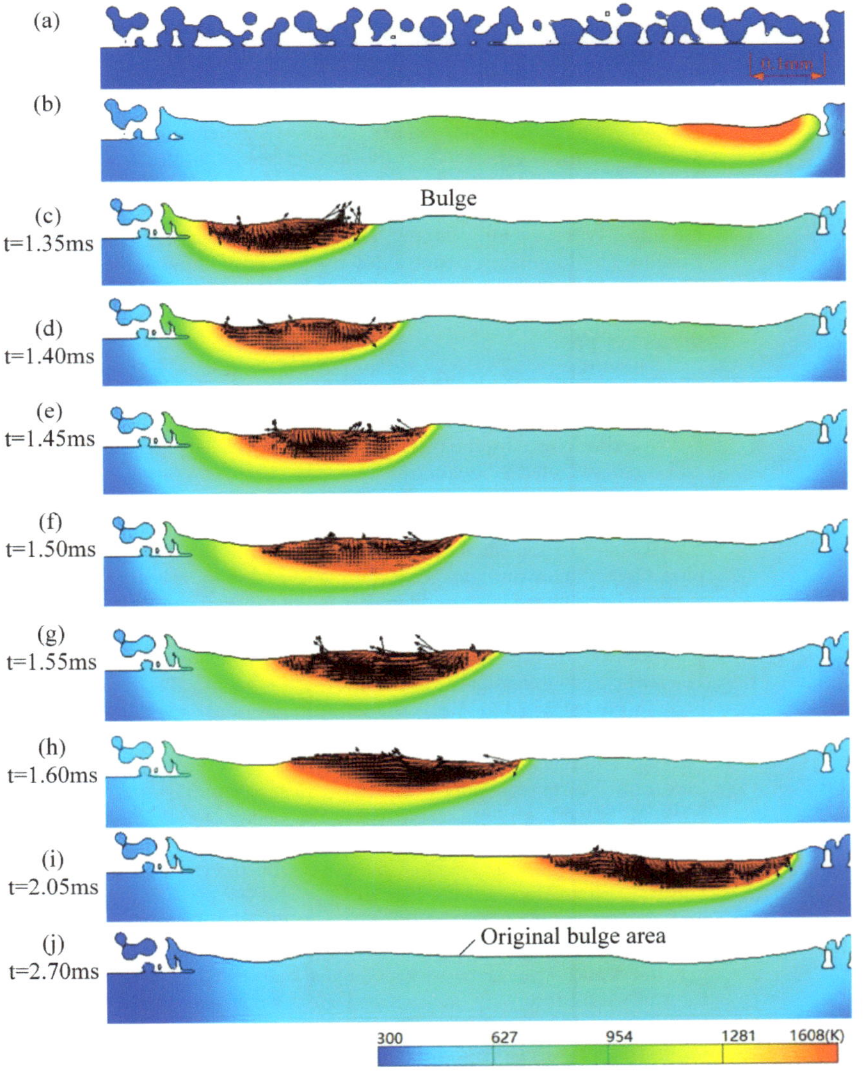

Figure 11. Surface topography evolution. (**a**) Powder bed; (**b**) surface profile after SLM; (**c**–**j**) are the ILP process. The parameters of SLM were 150 W, 0.8 m/s and 0.05 mm (power, speed and spot diameter). The parameters of ILP were 130 W, 0.8 m/s and 0.05 mm (power, speed and spot diameter). The entire domain is 1 mm long and 0.15 mm high.

At this moment, the remaining bulge area wetted by the molten pool also has this flow to the lower left. When t = 1.55 ms, as shown in Figure 11g, the bulge has been completely swallowed by the molten pool, which is in the surface over melting (SOM) state of LP [42]. The capillary force and thermocapillary force work together [22,24,41] in the molten pool. There is a downward flow driven by capillary force and a tangential flow driven by thermocapillary force. The capillary force plays a major role, which gradually reduces the curvature of the bulge. When t = 1.60 ms, the original bulge was eliminated under the driving of capillary force. At this time, the thermocapillary force plays a major role, as shown in Figure 11h. When t = 2.05 ms, as shown in Figure 11i, the molten pool at the original bulge area is completely solidified and becomes flat under the action of ILP. When t = 2.7 ms, as shown in Figure 11j, the laser no longer acts on the track, and the molten pool is completely solidified. Several bulges in the as-SLMed track disappear, and the track becomes smooth. However, there are depressions at both ends of the track, indicating that new surface defects will also be generated during the ILP process.

The main driving force in the molten pool changes with the development of the molten pool. When the bulge begins to be swallowed by the molten pool, capillary force dominates, causing the liquid in the molten pool to flow to the lower left of the bulge along the slope. When the bulge is completely swallowed by the molten pool, the capillary force continues to dominate, causing a downward normal flow of the molten pool to reduce the curvature of the bulge. As the curvature of the bulge decreases, the effect of capillary force is gradually weakened, and the thermocapillary force becomes the main driving force in the molten pool.

The surface roughness of SLMed parts is reduced by ILP. That is, as the laser advances and under the action of the wettability of the molten pool, the bulges on the surface of the SLMed part are constantly swallowed by the molten pool. The molten material at the bulge is moved backward by the combined action of capillary force and thermal capillary force and is redistributed in the molten pool. The surface becomes flat.

4.3. Effect of ILP Power on the Surface Quality of as-SLMed Track and Its Mechanism

ILP simulations at 50 W, 90 W, 130 W, and 170 W were carried out, and the surface morphology characteristics of all samples and the underlying mechanism were analyzed. Figure 12 shows the cross-sectional of the samples. Figure 12a shows the morphology of the track before ILP, and the track shows an uneven wavy shape, which is caused by the bulge caused by the laser scanning and the lack of fusion and spatter of the powder [18,42]. The surface roughness of the as-SLMed track is 4.9 μm. The coordinate data of the surface points were extracted in the simulation software, and Sa was calculated by the formula (1). The track morphology at an ILP power of 50 W is shown in Figure 12b. It can be seen that the length and height of each bulge are reduced, but the overall surface roughness changes little, only 0.1 μm less than the as-SLMed surface roughness. When the power of the ILP is 90 W, as shown in Figure 12c, the length and height of each bulge are further reduced. Due to the short life of the molten pool, the bulges do not completely disappear [23], and the surface roughness is reduced to 4.0 μm. From the three-dimensional surface morphology of Figure 12, the surface morphology of 50 W and 90 W has no significant change compared with the as-SLMed. When the ILP power is 130 W, the life of the molten pool is increased. As shown in Figure 12d, the original bulges are eliminated after ILP, but new bulges are generated, and depressions are generated at both ends of the track surface. From Figure 12d, the surface morphology has changed greatly compared with the as-SLMed, and a new surface morphology has been formed. The surface morphology is closely related to the molten pool flow velocity, and too fast flow velocity is not conducive to the formation of good surface morphology [8]. The molten pool flow velocity during ILP was investigated. The location of the monitoring of the molten pool flow velocity point is shown in Figure 12a. This point is melted under all processing conditions. In addition, in the fluctuating melt pool, this point is always in a liquid state. Figure 13 shows the velocity fluctuation during the ILP. As the laser moves, the position of the monitoring point in the molten pool is constantly changing. The flow velocity of the fluid at the monitoring point is also constantly changing, and the flow velocity curve fluctuates accordingly. When the flow velocity is 0, the monitoring point is solidified. The ILP powers are 50 W, 90 W, 130 W, and 170 W, and they solidify at 0.0018 s, 0.0019 s, 0.0020 s, and 0.0022 s, respectively. When the ILP power is at 170 W, the melt pool flow velocity is increased greatly, and the growth range is also greater compared to the first two groups. The large increase in the molten pool velocity makes the molten pool extremely unstable. During the ILP process, the surface morphology of the molten pool changes rapidly, and some bulges will be generated on the surface of the molten pool. Some of the molten materials at the bulges will be brought to other areas during the flow of the molten pool, and some will have solidified before they are redistributed in the molten pool. Periodic waves are generated at the track surface due to the high energy input, as shown in Figure 12e. These waves severely affect the surface roughness. At this time, the surface roughness is 13.4 μm.

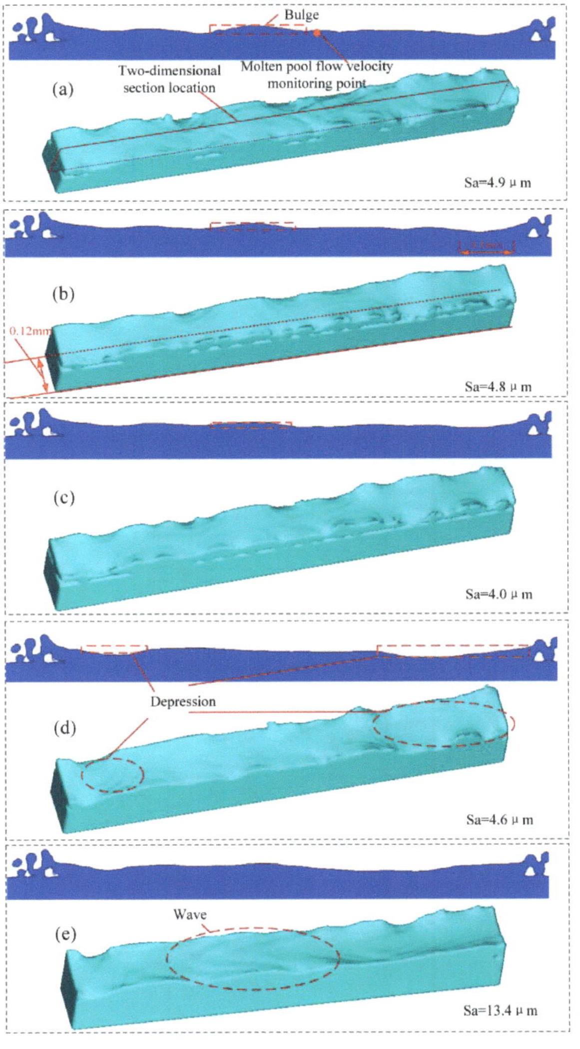

Figure 12. Surface morphology. (**a**) As received; (**b**) 50 W; (**c**) 90 W; (**d**) 130 W; (**e**) 170 W. The parameters of SLM were 150 W, 0.8 m/s and 0.05 mm. The other parameters of ILP were 0.8 m/s and 0.05 mm. The length of the plane domain is 1 mm, and the height is 0.15 mm. The size of the 3D domain is 1 mm × 0.12 mm × 0.2 mm.

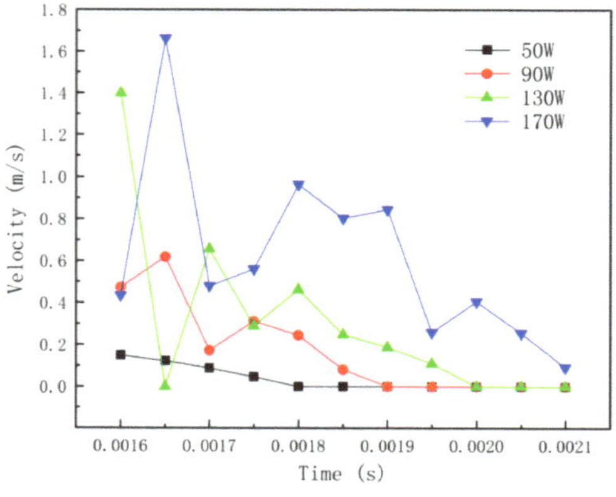

Figure 13. The flow velocity of the molten pool of the monitoring point fluctuates with time under different ILP power.

The mechanism of depression generation during ILP can be revealed by analyzing the melt pool fluid dynamics results from modelling with 130W laser power. As shown in Figure 14, when t = 1.30 ms, the molten pool at the left end of the track generates a depression under the action of capillary force. When t = 1.40 ms, there is a liquid flow from the bulge to the depression. When t = 1.60 ms, the depression has been solidified but not flattened. This is because when t = 1.40 ms, due to the insufficient heat accumulation, the molten pool size is small and the life is short, which causes the depression to not have enough time to be filled [23]. At t = 1.60 ms, the depression is further deepened due to the accumulation of material on the right side of the depression by the Marangoni flow. The depression on the right side of the track is generated because the material is continuously transported to the left by the Marangoni flow during the laser movement. The depression area transports material to the left area, but the depression area does not receive the material, resulting in a depression, is generated at the right end of the track.

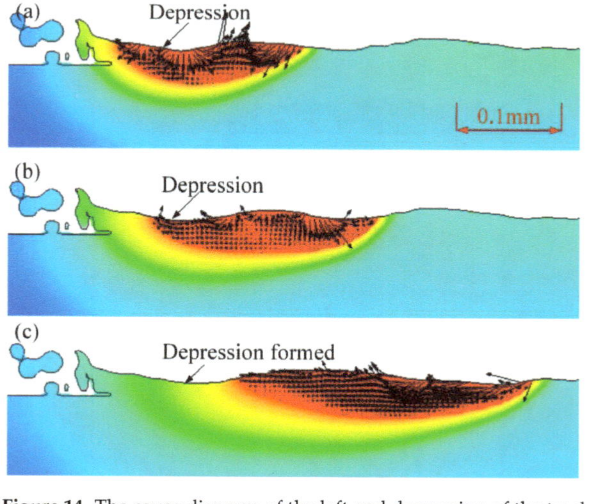

Figure 14. The cause diagram of the left end depression of the track when the ILP power is 130 W. (**a**) t = 1.30 ms; (**b**) t = 1.40 ms; (**c**) t = 1.60 ms. The parameters of SLM were 150 W, 0.8 m/s and 0.05 mm. The parameters of ILP were 130 W, 0.8 m/s and 0.05 mm. The entire domain is 0.54 mm long and 0.15 mm high.

The molten pool flow is used to reveal the mechanism of the formation of the new bulges in the ILP process. Figure 15 shows the velocity fields of the y-z section when the ILP power is 130 W. As shown in Figure 15a, when t = 1.7 ms, the surface tension gradient of the molten pool is greater than 0. The velocity vector points from both sides of the laser scanning track to the center of the track. The molten material on both sides of the track is delivered to the center of the track [21]. At this time, it is in the heating stage of molten pool formation. As shown in Figure 15b, when t = 1.85 ms, the size of the molten pool is the largest, and the molten pool begins to cool at this time. The surface tension gradient is still greater than 0. Both sides of the track transport material to the center of the track, and a bulge has been generated in the center at this time. When t = 1.90 ms, as shown in Figure 15c, the surface tension gradient of the molten pool becomes negative. The bulge in the center of the track begins to transport material to both sides of the track. When t = 1.95 ms, as shown in Figure 15d, the molten pool surface becomes flat under the action of the previous negative surface tension gradient. However, at this time, the surface tension gradient of the molten pool becomes positive, and both sides of the track transport materials to the center of the track. At t = 2.0 ms, a bulge is generated in the center of the track in the melt pool, as in Figure 15e, and the bulge is dragged up by the velocity vector. When the molten pool is completely solidified, as shown in Figure 15f, the bulge still exists. The formation of the bulge has a great relationship with the surface tension gradient, and the final formation of the surface bulge depends on the surface tension gradient during the cooling stage of the molten pool. The thermal capillary force affected by surface-active elements is the driving force for the tangential flow of the molten pool [24], and it plays a major role in the formation of bulges [21,37]. If the bulge created in the transverse direction is hemispherical, it will also be seen in the longitudinal direction. If it is semi-cylindrical, it will not be seen in a longitudinal direction. However, both will increase the surface roughness. The bulge in Figure 15 resembled a semi-cylindrical shape. When the ILP power is 130 W, there is no further reduction in surface roughness due to the presence of new depressions and new bulges. However, the surface roughness is still reduced to 4.6 μm.

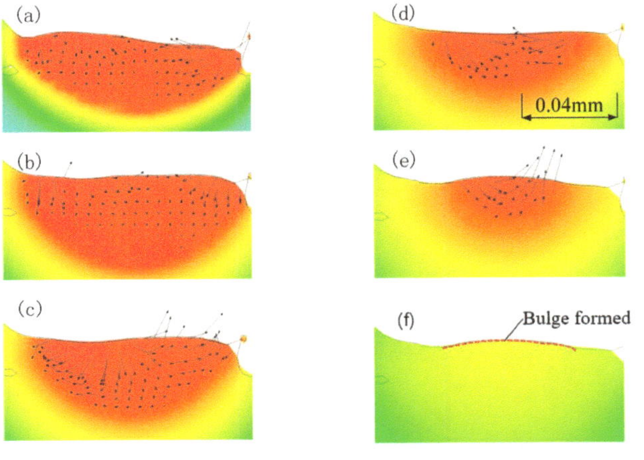

Figure 15. The formation of bulge in LP process. (**a**) t = 1.70 ms; (**b**) t = 1.85 ms; (**c**) t = 1.90 ms; (**d**) t = 1.95 ms; (**e**) t = 2.00 ms; (**f**) 2.15 ms. The parameters of SLM were 150 W, 0.8 m/s and 0.05 mm. The parameters of ILP were 130 W, 0.8 m/s and 0.05 mm. The entire domain is 0.11 mm long and 0.06 mm high.

The ILP power is a key factor in improving surface quality. At the appropriate power, the molten material will be redistributed on the metal surface under the combined action of capillary and thermocapillary forces. The curvature of the bulges is reduced, and the depressions are filled, thereby the surface roughness is reduced, as shown in Figure 12c. As the laser power increases, the life of the molten pool is increased and the molten material has more time to be redistributed on the surface, resulting in the surface roughness being further reduced. However, after the laser power exceeds a certain value, the surface roughness is not reduced but increased. Increasing the input energy increases the life and instability of the molten pool. During the flow of the molten pool, although the original defects are eliminated, new defects are also created, as shown in Figure 12d,e. Eventually, this leads to the surface roughness being increased.

4.4. Influence of the Interaction between the Tracks on Surface Quality

In order to investigate the effect of the interaction between the tracks on the final surface quality, two three-track simulations were performed. One was for SLM, and the other was for ILP after the SLM with an elapsed time of 0 ms. The processing parameters were the same as those of SLM and Sample-2 in Table 1. The results are shown in Figure 16. The multiple-track has a great improvement in the surface quality compared with single-track. In the single-track simulation, the Sa of the as-SLM sample is 4.9 μm, and the Sa of the ILPed sample is 4.0 μm. While in the multi-track simulation, the two numbers are 3.9 μm and 3.0 μm. In the multiple-track simulation, the surface roughness after ILP is reduced by 23%, which is greater than the 18.4% of the single track. This is because in the multiple-track simulation, the laser heats up and melts the tracks on both sides, which results in the surface having been remelted several times and the surface quality being improved. Therefore, the interaction between the tracks is beneficial to the improvement of the surface quality.

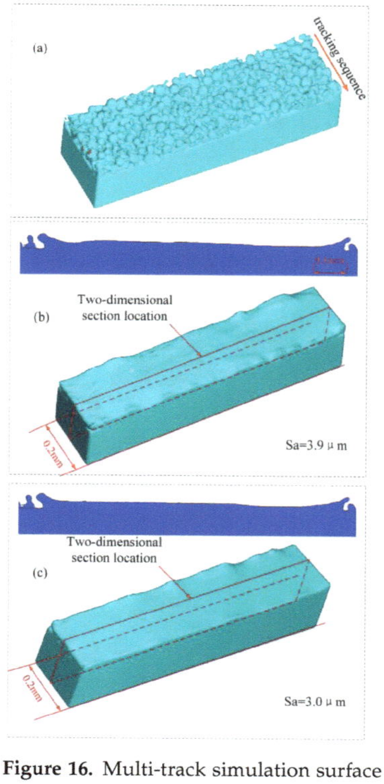

Figure 16. Multi-track simulation surface profile. (**a**) Powder bed; (**b**) As-SLMed; (**c**) ILPed. The parameters of SLM were 150 W, 0.8 m/s and 0.05 mm. The parameters of ILP were 90 W, 0.8 m/s and 0.05 mm. The hatch spacing was 0.01 mm. The length of the plane domain is 1 mm, and the height is 0.15 mm. The size of the 3D domain is 1 mm × 0.2 mm × 0.2 mm.

In addition, during SLM and ILP, the interaction between tracks is affected by hatch spacing and spot diameter. If the hatch spacing is too large or the spot diameter is too small, the tracks will not overlap, and they will not affect each other. So, the surface quality of the samples is affected by the hatch spacing and spot diameter, which deserves further study.

5. Conclusions

This paper demonstrates the effectiveness of ILP to improve the surface quality of nickel-based alloys fabricated by SLM. The temperature field, molten pool flow, and surface morphology evolution of the ILP process were studied by numerical simulation. The mechanisms of ILP for surface quality improvement and new defect formation were revealed. The following results were obtained:

(1) SLM and ILP experiments were performed, respectively. The surface roughness of the SLM samples after ILP is reduced from 20.0 ± 1.25 μm to 13.3 ± 0.35 μm. The quality of the surface is significantly improved, which proves the effectiveness of ILP in improving the surface quality of nickel-based alloys fabricated by SLM;

(2) During the ILP process, the peak temperature in the ILP stage increases with increasing power, and the life of the molten pool also increases. When the ILP power is 130 W, the peak temperature in the ILP stage exceeds the peak temperature in the SLM stage. The high temperature of the molten pool will make the molten pool unstable, which is not conducive to the improvement of surface quality;

(3) The mechanism of ILP improving the surface quality of SLM parts is that capillary force and thermal capillary force drive the molten pool flow to reduce the curvature of the bulge. During the initial stage of the molten pool swallowing the bulge, capillary force drives the liquid to the left of the bulge. When the bulge is completely swallowed by the molten pool, capillary force drives the molten pool to flow downward. As the curvature of the bulge decreases, the thermocapillary force becomes the main driving force, and the liquid is driven to the back of the molten pool;

(4) The ILP power has a great influence on the surface quality. If the ILP power is too small, the molten pool life will be short, which is not enough to make the surface bulge disappear, and the effect on the surface quality improvement is limited. If the ILP power is too large, new bulges and depressions will be produced on the track surface, which will affect the surface quality;

(5) Capillary force during the ILP process leads to the formation of the depression in the molten pool. If the mass transfer caused by the thermocapillary force cannot fill the depression in time, it will cause the formation of a recessed defect on the track surface. The material exchange between the center and both sides of the track caused by the surface tension gradient leads to the formation of the bulges. The final surface bulge depends on the surface tension gradient in the cooling stage of the molten pool.

Author Contributions: Y.Z. designed the experimental plan and supervised the entire work. C.D. established the numerical model and carried out a numerical simulation analysis. P.W. completed the sample preparation, W.M. completed the model verification part, and C.L. completed the sample test. All authors have read and agreed to the published version of the manuscript.

Funding: This research was funded by National Natural Science Foundation of China (Grant No. 51975339,51605262); China Postdoctoral Science Foundation (Grant No. 2019T120602,2017M610439); Youth Innovation and Technology Support Program for University in Shandong Province (Grant No. 2019KJB003).

Informed Consent Statement: Not applicable.

Data Availability Statement: Not applicable.

Acknowledgments: The authors would like to thank Shandong Leishi Intelligent Manufacturing Co., Ltd. for providing the technical support.

Conflicts of Interest: The authors declare no conflict of interest.

References

1. Wang, Z.; Guan, K.; Gao, M.; Li, X.; Chen, X.; Zeng, X. The microstructure and mechanical properties of deposited-IN718 by selective laser melting. *J. Alloys Compd.* **2011**, *513*, 518–523. [CrossRef]
2. Qiu, C.; Chen, H.; Liu, Q.; Yue, S.; Wang, H. On the solidification behaviour and cracking origin of a nickel-based superalloy during selective laser melting. *Mater. Charact.* **2018**, *148*, 330–344. [CrossRef]
3. Li, Y.; Zhang, Z.; Guan, Y. Thermodynamics analysis and rapid solidification of laser polished Inconel 718 by selective laser melting. *Appl. Surf. Sci.* **2020**, *511*, 145–152. [CrossRef]
4. Li, N.; Huang, S.; Zhang, G.; Qin, R.; Liu, W.; Xiong, H.; Shi, G.; Blackburn, J. Progress in additive manufacturing on new materials: A review. *J. Mater. Sci. Technol.* **2018**, *35*, 242–269. [CrossRef]
5. Zhao, Y.; Sun, J.; Li, J.; Yan, Y.; Wang, P. A comparative study on Johnson-Cook and modified Johnson-Cook constitutive material model to predict the dynamic behavior laser additive manufacturing FeCr alloy. *J. Alloys Compd.* **2017**, *723*, 179–187. [CrossRef]
6. Zhao, Y.; Sun, J.; Li, J.; Wang, P.; Zheng, Z.; Chen, J.; Yan, Y. The stress coupling mechanism of laser additive and milling subtractive for FeCr alloy made by additive–subtractive composite manufacturing. *J. Alloys Compd.* **2018**, *769*, 898–905. [CrossRef]
7. Cao, L.; Zhou, Q.; Han, Y.; Song, B.; Nie, Z.; Xiong, Y.; Xia, L. Review on intelligent monitoring of defects and process control of selective laser melting additive manufacturing. *Acta Aeronaut. Astronaut. Sin.* **2020**, *65*, 199–233. [CrossRef]
8. Jain, S.; Corliss, M.; Tai, B.; Hung, W. Electrochemical polishing of selective laser melted Inconel 718. *Procedia CIRP* **2019**, *34*, 239–246. [CrossRef]
9. Marimuthu, S.; Triantaphyllou, A.; Antar, M.; Wimpenny, D.; Morton, H.; Beard, M. Laser polishing of selective laser melted components. *Int. J. Mach. Tools. Manuf.* **2015**, *95*, 97–104. [CrossRef]
10. Tsai, M.; Yang, W. Combined ultrasonic vibration and chemical mechanical polishing of copper substrates. *Int. J. Mach. Tools Manuf.* **2012**, *53*, 69–76. [CrossRef]
11. Guo, J.; Au, K.; Sun, C.; Goh, M.; Kum, C.; Liu, K.; Wei, J.; Suzuki, H.; Kang, R. Novel rotating-vibrating magnetic abrasive polishing method for double-layered internal surface finishing. *J. Mater. Process. Technol.* **2019**, *264*, 422–437. [CrossRef]
12. Pereira, O.; Rodríguez, A.; Fernández-Abia, A.; Barreiro, J.; López de Lacalle, L.N. Cryogenic and minimum quantity lubrication for an eco-efficiency turning of AISI 304. *J. Clean. Prod.* **2016**, *139*, 440–449. [CrossRef]
13. Habibzadeh, S.; Li, L.; Shum-Tim, D.; Davis, E.C.; Omanovic, S. Electrochemical polishing as a 316L stainless steel surface treatment method: Towards the improvement of biocompatibility. *Corros. Sci.* **2014**, *87*, 89–100. [CrossRef]
14. Yao, Y.; Zhou, R.; Zhang, C.; Mei, T.; Wu, M. Surface polishing technology for additive manufacturing complex metal components. *Acta Aeronaut. Astronaut. Sin.* **2020**, *78*, 1–14. [CrossRef]
15. Fang, Z.; Lu, L.; Chen, L.; Guan, F. Laser Polishing of Additive Manufactured Superalloy. *Procedia CIRP* **2018**, *71*, 150–154. [CrossRef]
16. Cwikła, M.; Dziedzic, R.; Reiner, J. Influence of Overlap on Surface Quality in the Laser Polishing of 3D Printed Inconel 718 under the Effect of Air and Argon. *Materials* **2021**, *14*, 1479. [CrossRef] [PubMed]
17. Ma, C.; Guan, Y.; Zhou, W. Laser polishing of additive manufactured Ti alloys. *Opt. Laser. Eng.* **2017**, *93*, 171–177. [CrossRef]
18. Chen, L.; Richter, B.; Zhang, X.; Bertsch, K.; Thoma, D.; Pfefferkorn, F. Effect of laser polishing on the microstructure and mechanical properties of stainless steel 316L fabricated by laser powder bed fusion. *Mater. Sci. Eng. A* **2021**, *802*, 140–154. [CrossRef]
19. Liang, C.; Hu, Y.; Liu, N.; Zou, X.; Wang, H.; Zhang, X.; Fu, Y.; Hu, J. Laser Polishing of Ti6Al4V Fabricated by Selective Laser Melting. *Metals* **2020**, *10*, 191. [CrossRef]
20. Zhou, J.; Han, X.; Li, H.; Liu, S.; Shen, S.; Zhou, X.; Zhang, D. In-Situ Laser Polishing Additive Manufactured AlSi10Mg: Effect of Laser Polishing Strategy on Surface Morphology, Roughness and Microhardness. *Materials* **2021**, *14*, 393. [CrossRef]
21. Xu, J.; Zou, P.; Kang, D.; Wang, W. Theoretical and experimental study of bulge formation in laser polishing of 304 stainless steel. *J. Manuf. Process.* **2021**, *66*, 39–52. [CrossRef]
22. Zhang, C.; Zhou, J.; Shen, H. Role of Capillary and Thermocapillary Forces in Laser Polishing of Metals. *J. Manuf. Sci. Eng. Trans. ASME* **2017**, *139*, 41–56. [CrossRef]
23. Li, K.; Zhao, Z.; Zhou, H.; Zhou, H.; Yin, J.; Zhang, W.; Zhou, G. Numerical Simulation of Effect of Different Initial Morphologies on Melt Hydrodynamics in Laser Polishing of Ti6Al4V. *Micromachines* **2021**, *12*, 581. [CrossRef] [PubMed]
24. Li, K.; Zhao, Z.; Zhou, H.; Zhou, H.; Jin, J. Numerical analyses of molten pool evolution in laser polishing Ti6Al4V. *J. Manuf. Process.* **2020**, *58*, 574–584. [CrossRef]
25. Song, Y. *Research on Laser Polishing Mechanism and Process of SLM Additive Manufacturing Surface*; Shandong University: Jinan, China, 2021. [CrossRef]
26. Przemysław, P. Reduction of Influence of the High-Frequency Noise on the Results of Surface Topography Measurements *Materials* **2021**, *14*, 333. [CrossRef]
27. Pavel, P.; Ondřej, H. White-light interferometry on rough surfaces—measurement uncertainty caused by noise. *Appl. Opt.* **2012**, *51*, 465–473.
28. Lv, F.; Liang, H.; Xie, D.; Mao, Y.; Wang, C.; Shen, L.; Tian, Z. On the role of laser in situ re-melting into pore elimination of Ti–6Al–4V components fabricated by selective laser melting. *J. Alloys Compd.* **2021**, *854*, 156–168. [CrossRef]
29. Khairallah, S.; Anderson, A.; Rubenchik, A.; King, W. Laser powder-bed fusion additive manufacturing: Physics of complex melt flow and formation mechanisms of pores, spatter, and denudation zones. *Acta Mater.* **2016**, *108*, 36–45. [CrossRef]

30. Ye, W. *Numerical Simulation of Temperature Field and Morphology Evolution of Molten Pool during Laser Selective Melting Process*; Xi'an University of Technology: Xi'an, China, 2019.
31. Lee, Y.; Zhang, W. Modeling of heat transfer, fluid flow and solidification microstructure of nickel-base superalloy fabricated by laser powder bed fusion. *Addit. Manuf.* **2016**, *12*, 178–188. [CrossRef]
32. Hu, Y.; Zhu, K.; Xi, B. A 3-dimensional volume of fluid method-PLIC3D. *Chin. Q. Mech.* **2003**, *18*, 238–243. [CrossRef]
33. Dai, D.; Gu, D. Influence of thermodynamics within molten pool on migration and distribution state of reinforcement during selective laser melting of AlN/AlSi10Mg composites. *Int. J. Mach. Tools Manuf.* **2016**, *100*, 14–24. [CrossRef]
34. Xiang, Y.; Zhang, S.; Li, J.; Wei, Z.; Yang, L.; Jiang, L. Numerical simulation and experimental verification for selective laser single track melting forming of Ti6Al4V. *J. Zhejiang Univ. (Eng. Sci.)* **2019**, *53*, 2102–2109. [CrossRef]
35. Gürtler, F.; Karg, M.; Leitz, K.H.; Schmidt, M. Simulation of Laser Beam Melting of Steel Powders using the Three-Dimensional Volume of Fluid Method. *Phys. Procedia* **2013**, *41*, 881–886. [CrossRef]
36. Yuan, W.; Chen, H.; Cheng, T.; Wei, Q. Effects of laser scanning speeds on different states of the molten pool during selective laser melting: Simulation and experiment. *Mater. Des.* **2020**, *189*, 108–129. [CrossRef]
37. Shrestha, S.; Chou, K. A study of transient and steady-state regions from single-track deposition in laser powder bed fusion. *J. Manuf. Process.* **2021**, *61*, 226–235. [CrossRef]
38. Xu, J.; Zou, P.; Wang, W.; Kang, D. Study on the mechanism of surface topography evolution in melting and transition regimes of laser polishing. *Opt. Laser. Technol.* **2021**, *139*, 106–118. [CrossRef]
39. Roberts, I.; Wang, C.J.; Esterlein, R.; Stanford, M.; Mynors, D. A three-dimensional finite element analysis of the temperature field during laser melting of metal powders in additive layer manufacturing. *Int. J. Mach. Tools Manuf.* **2009**, *49*, 916–923. [CrossRef]
40. Bayat, M.; Thanki, A.; Mohanty, S.; Yang, A.S.; Tiedje, J.N.; Hattel, J. Keyhole-induced porosities in Laser-based Powder Bed Fusion (L-PBF) of Ti6Al4V: High-fidelity modelling and experimental validation. *Addit. Manuf.* **2019**, *30*, 100–112. [CrossRef]
41. Leung, C.; Marussi, S.; Towrie, M.; Atwood, R.; Withers, P.; Lee, P. The effect of powder oxidation on defect formation in laser additive manufacturing. *Acta Mater.* **2019**, *166*, 294–305. [CrossRef]
42. Basha, S.; Bhuyan, M.; Basha, M.; Venkaiah, N.; Sankar, M. Laser polishing of 3D printed metallic components: A review on surface integrity. *Mater. Today Proc.* **2020**, *26*, 2047–2054. [CrossRef]

materials

MDPI

Article

Fatigue Crack Growth Analysis under Constant Amplitude Loading Using Finite Element Method

Abdulnaser M. Alshoaibi

Mechanical Engineering Department, Faculty of Engineering, Jazan University, P.O. Box 114, Jazan 45142, Saudi Arabia; alshoaibi@jazanu.edu.sa

Abstract: Damage tolerant design relies on accurately predicting the growth rate and path of fatigue cracks under constant and variable amplitude loading. ANSYS Mechanical R19.2 was used to perform a numerical analysis of fatigue crack growth assuming a linear elastic and isotropic material subjected to constant amplitude loading. A novel feature termed Separating Morphing and Adaptive Remeshing Technology (SMART) was used in conjunction with the Unstructured Mesh Method (UMM) to accomplish this goal. For the modified compact tension specimen with a varied pre-crack location, the crack propagation path, stress intensity factors, and fatigue life cycles were predicted for various stress ratio values. The influence of stress ratio on fatigue life cycles and equivalent stress intensity factor was investigated for stress ratios ranging from 0 to 0.8. It was found that fatigue life and von Mises stress distribution are substantially influenced by the stress ratio. The von Mises stress decreased as the stress ratio increased, and the number of fatigue life cycles increased rapidly with the increasing stress ratio. Depending on the pre-crack position, the hole is the primary attraction for the propagation of fatigue cracks, and the crack may either curve its direction and grow towards it, or it might bypass the hole and propagate elsewhere. Experimental and numerical crack growth studies reported in the literature have validated the findings of this simulation in terms of crack propagation paths.

Keywords: fatigue analysis; equivalent stress intensity factor; linear elastic fracture mechanics; ANSYS; constant amplitude loading

Citation: Alshoaibi, A.M. Fatigue Crack Growth Analysis under Constant Amplitude Loading Using Finite Element Method. *Materials* **2022**, *15*, 2937. https://doi.org/10.3390/ma15082937

Academic Editors: Shuwen Wen, Yongle Sun and Xin Chen

Received: 23 March 2022
Accepted: 15 April 2022
Published: 18 April 2022

1. Introduction

One of the most common catastrophic failures in mechanical structures is fatigue. Over the last few decades, researchers have strived to comprehend the mechanism of fatigue loading in materials that were exposed to dynamic loading, starting with the stress and strain life methodologies proposed by many researchers [1–6], which were curve-fitting-based approaches that used nominal and local stress–strain values. Another approach is the energy-based approach proposed by [7], which has since been used as the starting point for several experimental studies. However, such techniques are mostly limited to calculating the permissible number of load cycles before material failure instead of presenting characteristics of fatigue crack nucleation and propagation mechanisms. To properly study fatigue failure, several parameters, such as stress level, loading frequency, stress ratio (R = min/max), and material type, must be considered. In several studies, it has been demonstrated that the level of stress applied has a major impact on the fatigue failure of materials [8–11]. The linear elastic fracture mechanics (LEFM) theory was developed to identify the issue of fatigue crack growth [12–14]. The LEFM methods are commonly adopted for use on long cracks within small-scale yielding. The LEFM techniques are commonly used on long fractures with small-scale yielding behaviors near the crack tip, i.e., the Paris regime near the crack tip, i.e., the Paris regime [12,15]. Meanwhile, the Boundary Element Method [16,17], Meshless Method [18], Finite Difference Method [19], Finite Element Method (FEM), and Extended Finite Element Method (XFEM) [20–22] are the most used methodologies for modeling crack propagation. The most common computational approach for simulating damage and failure under both static and dynamic loadings is the FEM, which obtained stress, strain, displacement, and stress intensity factor (SIF) solutions for a wide range of engineering problems. The FEM, commonly known as adaptive remeshing procedures, has proved to be highly effective and reliable. The adaptive remeshing procedures consist of four main steps: (1) existence of a demonstrative 3D finite element framework; (2) prediction of the equivalent SIFs along the crack front; (3) prediction of crack front progressions using appropriate fatigue crack growth law; and (4) specification of a new 3D finite element model considering the new crack front. These procedures are continued until a predetermined crack length or ultimate fracture is reached. Using the 3D FEM to compute the stress intensity factor at a set of points on the crack front, the fatigue crack growth analysis can be accomplished precisely. Nowadays, there is a variety of software to deal with the problem of fatigue crack growth, e.g., FRANC3D [23], ABAQUS [24], ANSYS [25–30], ZENCRACK [31], COMSOL [32], BEASY [33], and NASTRAN [34]. Three approaches have been commonly used to describe material fatigue analysis: the method of fracture mechanics proposed by Paris and Erdogan [35], the method of strain–life introduced by Coffin [36], and the method of stress–life introduced by Wöhler [37]. In this work, the first technique was used to estimate fatigue life, in which the crack tip was entirely described by the stress intensity factors. Various experimental procedures have been reported; however, the procedures are generally time-consuming and costly to implement. A numerical analysis approach such as the ANSYS Mechanical R19.2 is an effective process to save both time and money in the laboratory by reducing the amount of work, time, and expenses. Alternatively, there was also an analytical-based technique that was efficient in simulating fatigue growth [38,39]. The main motivation for this study was to make a significant contribution to the use of ANSYS as an effective tool for simulating crack growth under mixed-mode loading situations and monitoring the influence of the holes and crack location on the crack growth trajectory.

2. SMART Crack Growth Procedure

SMART is an efficient fracture mechanics simulation approach based on an adaptive meshing strategy in the surrounding area of the crack propagation path. At a certain loading level, the crack begins to grow as soon as a critical value is reached. The crack propagates either to a certain limit specified by the user or to the point where generating a new mesh is impossible, which generally corresponds to the total split of the body into sections. The Unstructured Mesh Method (UMM) was employed in ANSYS to reduce the consumption time in the pre-processing using the tetrahedral mesh generated automatically for the crack front instead of using the ideal hex mesh configuration, reducing the computational time from a few days to a few minutes. The UMM approach is described in detail in [40]. Tetrahedron meshes were used for the crack fronts in the SMART analysis, which were automatically updated as the crack front changed due to the crack growth. The crack propagation path is defined by an angle θ, which is estimated by the ratio of modes of SIF at the crack tip [41–43]. A mixed-mode loading condition is considered by ANSYS, and the maximum circumferential stress is used as a crack growth criterion in the present study. Based on this criterion, the following formula is used for the crack growth path in ANSYS [25,44]:

$$\theta = \cos^{-1}\frac{3(K_{II}^{\max})^2 + (K_I^{\max})\sqrt{(K_I^{\max})^2 + 8(K_{II}^{\max})^2}}{(K_I^{\max})^2 + 9(K_{II}^{\max})^2} \quad (1)$$

where:

$K_I^{\max}$ = maximum values of the first mode of SIF under cyclic loading, and
$K_{II}^{\max}$= maximum values of the second mode of SIF under cyclic loading.

The SIFs were calculated via interaction-integral evaluation at the solution phase of the analysis, and then the values were stored in the results file. The crack propagation simulation in this ANSYS simulation is confined to region II of the typical crack propagation under fatigue loading, which may be expressed as:

$$\frac{da}{dN} = C(\Delta K_{eq})^m \tag{2}$$

where a = crack length, n = the number of fatigue life cycles, C = Paris constant, m = Paris exponent, and ΔK_{eq}= the equivalent range of stress intensity factor, which may be represented as [44,45]:

$$\Delta K_{eq} = \frac{1}{2}\cos\left(\frac{\theta}{2}\right)[\Delta K_I(1+\cos\theta) - 3\Delta K_{II}\sin\theta] \tag{3}$$

where:

$$\begin{array}{c} \Delta K_I = K_I^{\max} - K_I^{\min} = (1-\mathrm{R})K_I^{\max} \\ \Delta K_{II} = K_{II}^{\max} - K_{II}^{\min} = (1-\mathrm{R})K_{II}^{\max} \end{array} \tag{4}$$

as R represents the load ratio.

According to Equation (2), with a crack growth increment Δa, the fatigue life cycles can be expressed as:

$$\int_0^{\Delta a} \frac{da}{C(\Delta K_{eq})^m} = \int_0^{\Delta N} dN = \Delta N \tag{5}$$

Figure 1 illustrates a simplified flow chart for the ANSYS SMART procedures for fatigue crack propagation.

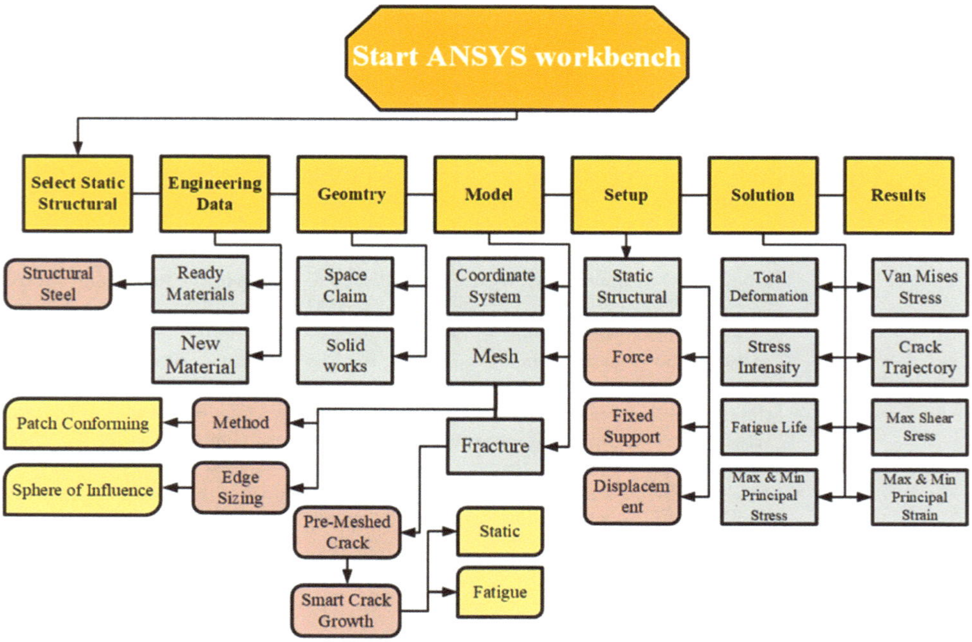

Figure 1. Flowchart for the ANSYS SMART procedures.

3. Results of Numerical Simulations

Modified Compact Tension with Different Pre-Crack Location

The modified compact tension specimen was studied in three distinct configurations in this study. The modified specimens vary from standard specimens in that they have three extra holes, as shown in Figure 2, which violate the standard specimens' symmetry and result in curvilinear fatigue crack pathways. The actual crack initiation locations are compared to the nominal position of the notch tip in the geometries, as shown in Table 1. The considered material was a nickel-based superalloy with the following material properties shown in Table 1. The amount of the applied load was p = 3.6 kN with a stress ratio of R = 0 and cyclic frequency of 20 Hz. Changing the vertical location of the original notch (H) up or down its normal midline position, as illustrated in Table 2, leads to altering the path and ultimate destination of the crack growth. As shown in Figure 2, the vertical notch location (H) is defined relative to the geometry's top edge. The initial mesh generated by ANSYS, which had a 1 mm element size and generated 292,160 nodes and 192,860 elements, is shown in Figure 3, which employed the sphere of influence at the crack tip area. There are three different scenarios for the crack growth trajectory based on the nominal notch positions.

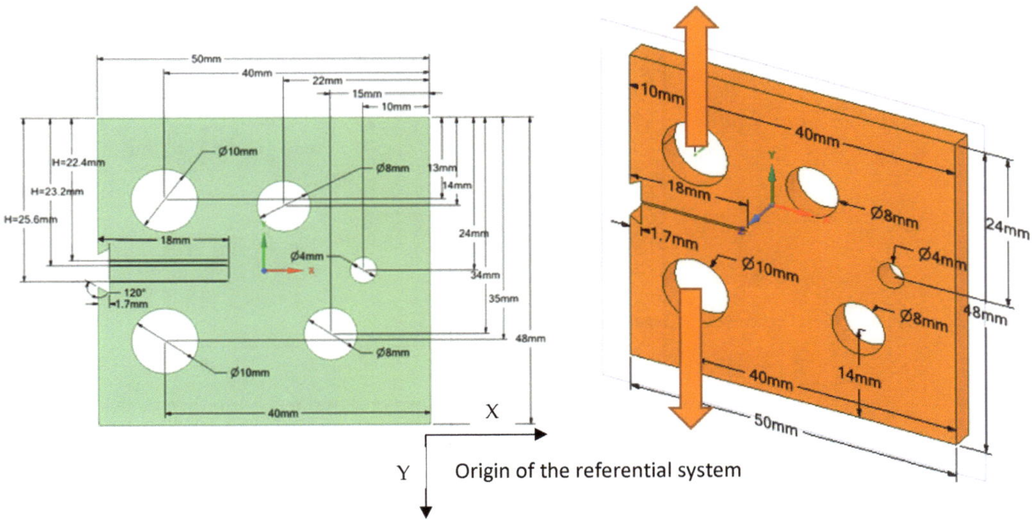

Figure 2. Modified compact tension geometrical dimensions.

Table 1. Mechanical properties of the nickel-based superalloy material.

Properties	Metric Units Value
Elasticity modulus, E	211 GPa
Poisson's ratio, v	0.3
Yield strength, σ_y	422 MPa
Ultimate strength, σ_u	838 MPa
Fracture toughness, K_{IC}	130 MPa $\sqrt{m}$
Paris' law coefficient, C	1.02×10^{-11}
Paris' law exponent, m	2.5

Table 2. Pre-crack position for the modified compact tension.

Specimen Number	Crack Tip Position (mm)		
	(H)	(x)	(y)
1	22.4	−32	25.6
2	25.6	−32	22.4
3	23.2	−32	24.8

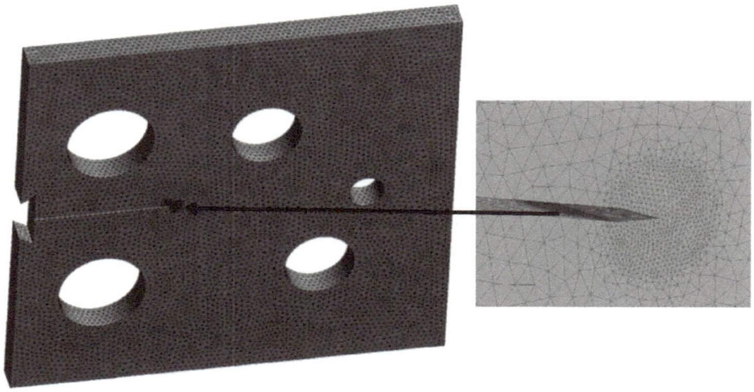

Figure 3. Initial mesh for the modified compact tension.

Specimen 1

The initial crack in this specimen was located at 22.4 mm from the specimen's top edge. Comparisons of the simulated crack propagation trajectory using ANSYS to the reference experimental [46] and numerical [47] paths are shown in Figure 4a–c, respectively. Crack propagation trajectories in the numerical findings provided by [47] were predicted in three steps: the first step is to use the hyper-complex FEM trial energy response function (ZFEM-TERF) technique for crack trajectory estimation; at each step of crack growth, the model is updated with curvilinear crack path segments that are generated by the trial energy response function (TERF) approach. A finite element model was generated using the FRANC3D program in the second step before being solved using the Abaqus software in the final step. In comparison to the numerical crack growth paths presented in Figure 4c applying the ZFEM-TERF approach and FRANC3D [31], Figure 4a–b indicate that the estimated crack propagation trajectory in this study is very consistent with the experimental trajectory [46].

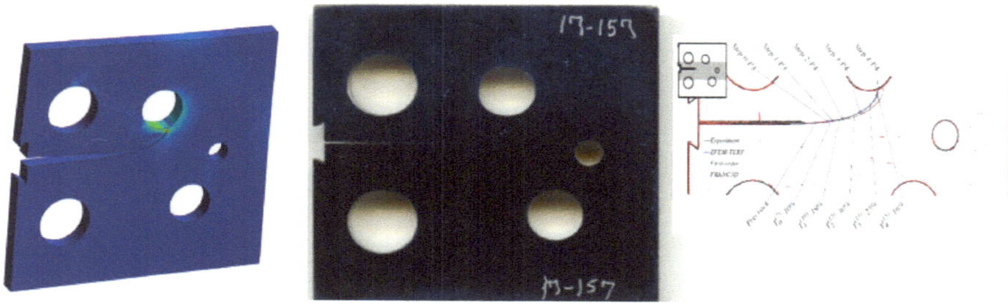

(a) (b) (c)

Figure 4. Specimen 1, crack propagation path (**a**) ANSYS results, (**b**) experimental results [46], and (**c**) numerical results [47].

Specimen 2

The initial crack in the second specimen was located at 25.6 mm from the specimen's top edge. The predicted crack propagation trajectory using ANSYS has matched the experimental trajectory reported by [46] more closely than the predicted trajectories estimated by [47], which had tighter curvature trajectories, as illustrated in Figure 5.

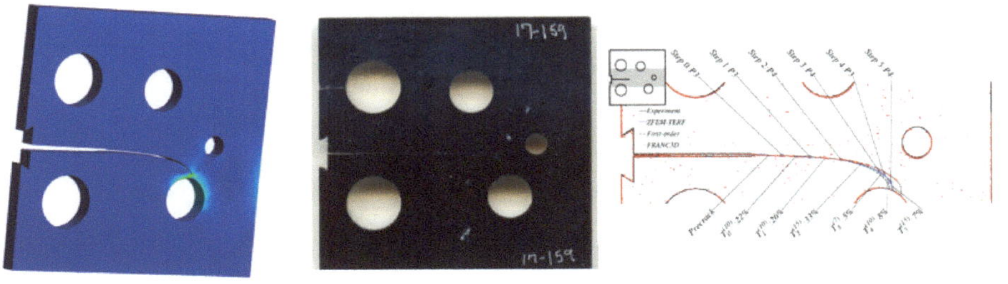

(a) (b) (c)

Figure 5. Specimen 2, crack propagation path (**a**) ANSYS results, (**b**) experimental results [46], and (**c**) numerical results [47].

Specimen 3

The initial crack in the second specimen was located at 23.2 mm from the specimen's top edge. As can be seen in Figure 6, the estimated crack propagation trajectory tightly matches the experimental crack growth trajectory reported by [46] compared to the predicted paths from the numerical results using ZFEM-TERF and FRANC3D conducted by [47], which deviated from the experimental path [46].

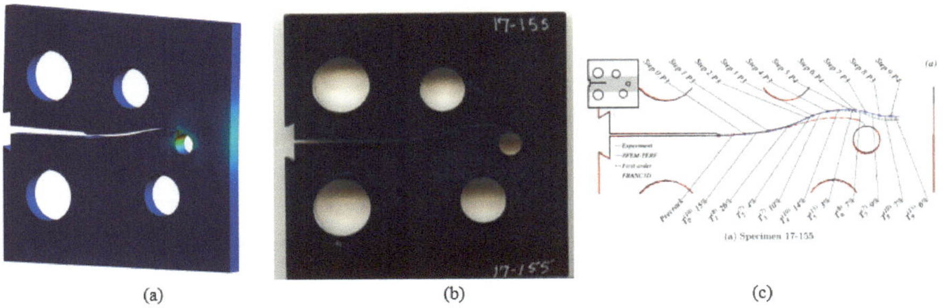

(a) (b) (c)

Figure 6. Specimen 3, crack propagation path (**a**) ANSYS results, (**b**) experimental results [46], and (**c**) numerical results [47].

Considering that, the von Mises stress as well as the maximum principal stress are essential parameters for crack propagation assessment, which indicate the regions of maximum and minimum stresses on the geometry. Figures 7 and 8 show the von Mises stress distribution stress contour as well as the maximum principal stress for each of the three specimens. The von Mises stresses and the maximum principal stress were higher in specimen one, where the top hole was located closer to the crack based on the original crack location. As the crack also sinks on the smallest hole near the right edge of the specimen, specimen two had the lowest values of the von Mises stresses and the maximum principal stress, whereas specimen three had the intermediate values of both stresses, as the crack also sinks on the second lower hole near the right edge of the specimen.

The results of the opening mode of SIF (K_I) for the three samples are shown in Figure 9. The maximum values of K_I are 1205 MPa mm$^{0.5}$, 4136 MPa mm$^{0.5}$, and 5800 MPa mm$^{0.5}$ for a crack length of 12.97, 18.33 mm, and 21.161 mm for specimens one, two, and three, respectively. Similarly, Figure 10 also displays the estimated values for the second mode of stress intensity factor (K_{II}). As the crack follows a curving trajectory toward the top hole, the K_{II} values for the first specimen increase to a maximum of 96.133 MPa mm$^{0.5}$ at the boundary of the hole. However, in specimens two and three, the values of K_{II} decreased with negative values as the crack propagated on a curved path in the opposite direction of specimen one, with minimum values of −243 MPa mm$^{0.5}$ and −230 MPa mm$^{0.5}$ for specimens two and three, respectively. In the mixed-mode situations, the direction of the tangential component of the applied load is attributed to the negative mode II stress intensity factor. The signs of SIFs depend on the orientation of the crack with the loading.

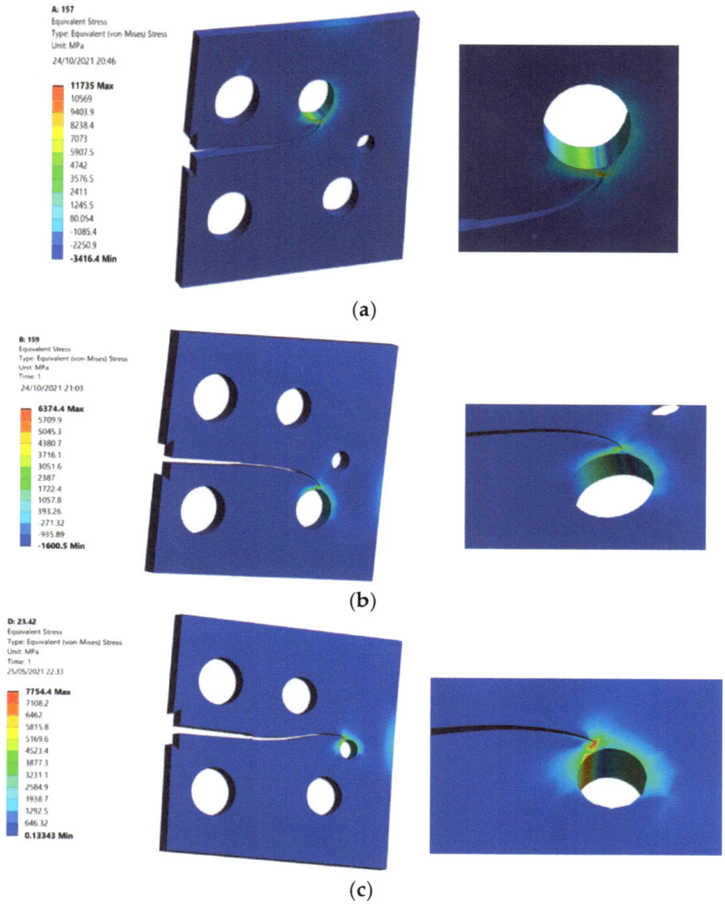

Figure 7. Von Mises stress distribution of (**a**) specimen 1, (**b**) specimen 2, and (**c**) specimen 3.

To determine fatigue life under constant amplitude loading circumstances with a stress ratio of R = 0, a step-by-step simulation of crack propagation was performed according to the associated SIFs. Figure 11 displays the predicted fatigue life cycles for each specimen; as seen in this figure, the fatigue life cycles were gradually increased from specimens one to three, since the stress intensity factors were also increased to the same extent for all of the specimens.

The third specimen was simulated at various stress ratios ranging from R = 0.1 to 0.8 to correlate the stress ratio effects on the equivalent stress intensity factor as well as fatigue crack growth rates. Almost at a given applied cyclic equivalent stress intensity, an increase in load ratio leads to an increase in fatigue crack growth rate. Equivalently, the observed equivalent stress intensity factor for fatigue crack growth decreases as the load ratio is increased, as shown in Figure 12. In other words, at high-stress ratios, less accumulated fatigue energy is necessary to support crack growth than at lower stress ratios. In contrast, the number of load cycles with respect to the crack growth extension increases as the stress ratio increases, as shown in Figures 13 and 14 for the stress ratios ranging from 0.1 to 0.8. This effect is proportional to the maximum concentration of von Mises stress and hence to the driving force of mode I cracking. According to the results shown in Figures 13 and 14, the percentages of increase in the equivalent stress intensity factors for different stress ratios ranging from 0.1 to 0.8 are not equal to the percentages of increase in the fatigue life cycles. Damage distributions differed depending on the stress ratios. Damage was equally distributed along with the specimens with larger stress ratios, but it was severe and concentrated at lower stress ratios (0.1–0.4), resulting in higher self-generated temperatures and specimen failure at shorter lifetimes. The von Mises stress distribution for specimen three under different stress ratios R = 0.1–0.8 is shown in Figure 15. It is found that von Mises stress decreases as the stress ratio increases, which was also related to the increase in the fatigue life cycles.

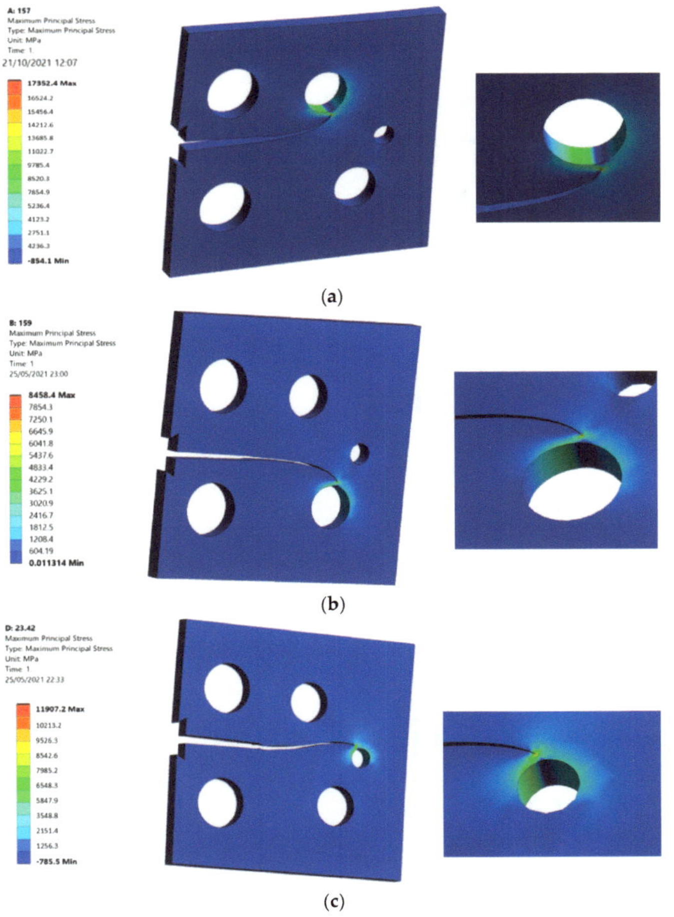

Figure 8. Maximum principal stress of (**a**) specimen 1, (**b**) specimen 2, and (**c**) specimen 3.

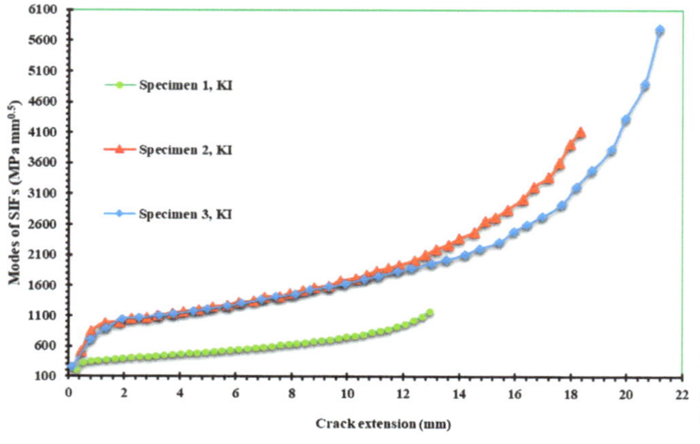

Figure 9. The first mode of SIFs versus crack length for the three specimens.

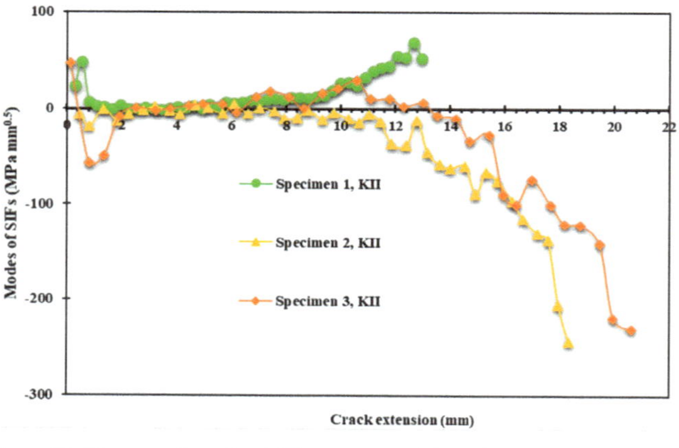

Figure 10. The second mode of SIFs versus crack length for the three specimens.

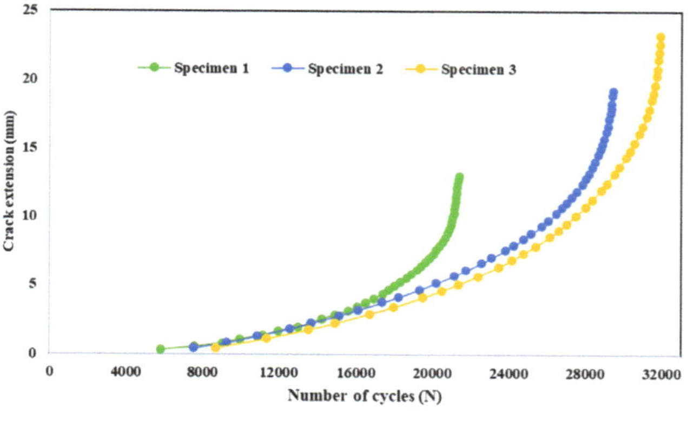

Figure 11. Predicted fatigue life cycles for the three specimens under stress ratio R = 0.

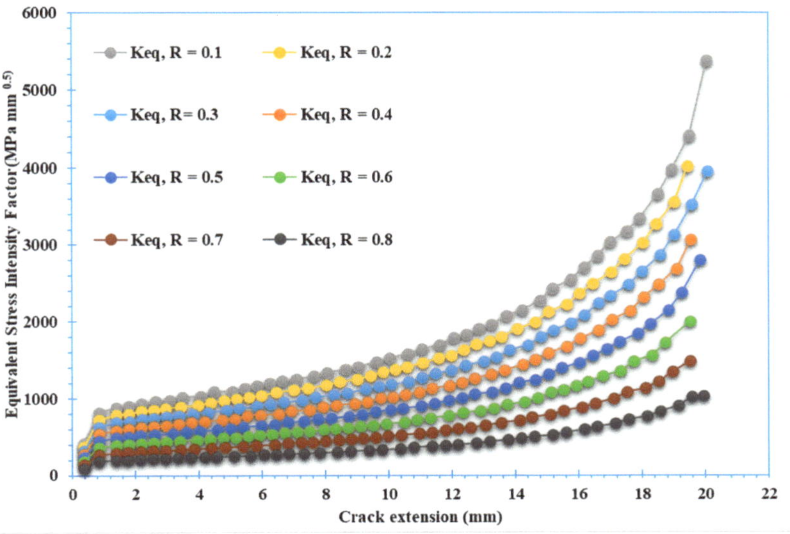

Figure 12. Equivalent stress intensity factor for specimen 3 with different stress ratios.

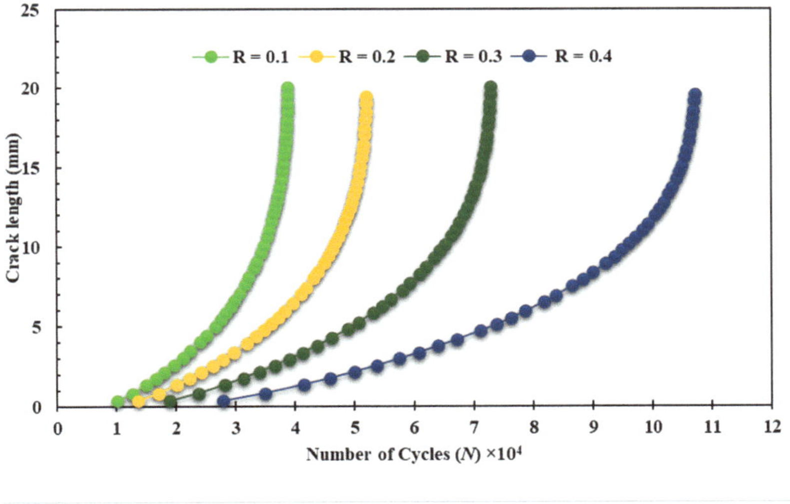

Figure 13. Predicted fatigue life cycles for specimen 3 with stress ratios, R = 0.1–0.4.

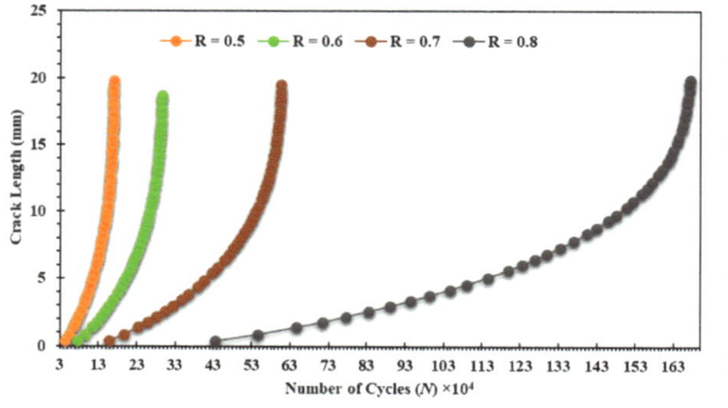

Figure 14. Predicted fatigue life cycles for specimen 3 with stress ratios, R = 0.5–0.8.

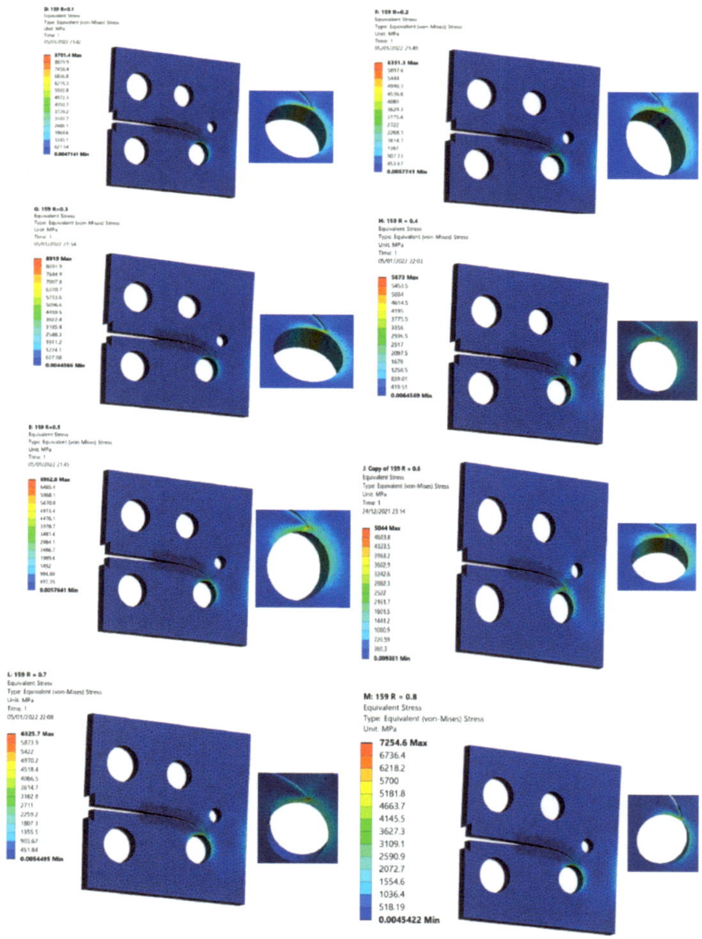

Figure 15. Von Mises stresses distribution for specimen 3 with various stress ratios, R = 0.1–0.8.

4. Conclusions

This study investigated the fatigue crack growth in isotropic linear elastic materials under constant amplitude loading where some defects such as holes are intentionally introduced in the material and different load ratios are applied. For this purpose, the finite element software ANSYS was used. The topic of study is certainly very important and can result in a deeper understanding of crack propagation and material design. The fatigue crack propagation of a modified compact tension specimen with various pre-crack locations was simulated using the ANSYS SMART methodology. Based on the Paris law, the crack growth simulation in SMART used tetrahedral meshes for the crack fronts that were updated automatically when the crack front was modified as a consequence of crack propagation. Based on the position of the hole and the starting position of the crack tip, the growth of the crack was either attracted to the hole and changed its trajectory to reach the hole "sink in the hole behavior" or deviated away from the hole and grew when the hole was missing "missed hole behavior". The influence of a wide range of load ratios (R = 0, 0.1, 0.2, 0.3, 0.4, 0.5, 0.6, 0.7, 0.8) on fatigue crack growth, fatigue life, and equivalent range of SIF was investigated. According to the predicted results, it was found that as the stress ratio increased and the fatigue life cycles rapidly increased, whereas von Mises stress decreased.

Funding: This research received no external funding.

Institutional Review Board Statement: Not applicable.

Informed Consent Statement: Not applicable.

Data Availability Statement: The data presented in this study are available upon request from the corresponding author.

Conflicts of Interest: The authors declare no conflict of interest.

References

1. Gerber, W. Bestimmung der zulossigenin eisen construcionen. *Z. Bayer Arch. Ing. Ver.* **1874**, *6*, 101–110.
2. Smith, K.N. A stress-strain function for the fatigue of metals. *J. Mater.* **1970**, *5*, 767–778.
3. Socie, D.; Morrow, J. Review of contemporary approaches to fatigue damage analysis. In *Risk and Failure Analysis for Improved Performance and Reliability*; Springer: Berlin/Heidelberg, Germany, 1980; pp. 141–194.
4. Manson, S.; Halford, G.R. Practical implementation of the double linear damage rule and damage curve approach for treating cumulative fatigue damage. *Int. J. Fract.* **1981**, *17*, 169–192. [CrossRef]
5. Ince, A.; Glinka, G. A modification of Morrow and Smith–Watson–Topper mean stress correction models. *Fatigue Fract. Eng. Mater. Struct.* **2011**, *34*, 854–867. [CrossRef]
6. Ince, A.; Glinka, G. A generalized fatigue damage parameter for multiaxial fatigue life prediction under proportional and non-proportional loadings. *Int. J. Fatigue* **2014**, *62*, 34–41. [CrossRef]
7. Weertman, J. Theory of fatigue crack growth based on a BCS crack theory with work hardening. *Int. J. Fract.* **1973**, *9*, 125–131. [CrossRef]
8. Mansor, N.; Abdullah, S.; Ariffin, A. Effect of loading sequences on fatigue crack growth and crack closure in API X65 steel. *Mar. Struct.* **2019**, *65*, 181–196. [CrossRef]
9. Stewart, A.T. The influence of environment and stress ratio on fatigue crack growth at near threshold stress intensities in low-alloy steels. *Eng. Fract. Mech.* **1980**, *13*, 463–478. [CrossRef]
10. Hudson, C.M. *Effect of Stress Ratio on Fatigue-Crack Growth in 7075-T6 and 2024-T3 Aluminum-Alloy Specimens*; National Aeronautics and Space Administration: Washington, DC, USA, 1969.
11. Noroozi, A.; Glinka, G.; Lambert, S. A study of the stress ratio effects on fatigue crack growth using the unified two-parameter fatigue crack growth driving force. *Int. J. Fatigue* **2007**, *29*, 1616–1633. [CrossRef]
12. Paris, P.C.; Tada, H.; Donald, J.K. Service load fatigue damage—A historical perspective. *Int. J. Fatigue* **1999**, *21*, S35–S46. [CrossRef]
13. Pook, L.P. *Linear Elastic Fracture Mechanics for Engineers: Theory and Applications*; WIT Press: London, UK, 2000.
14. Irwin, G.R. Analysis of stresses and strains near the end of a crack transversing a plate. *Trans. ASME Ser. E J. Appl. Mech.* **1957**, *24*, 361–364. [CrossRef]
15. Bang, D.; Ince, A.; Noban, M. Modeling approach for a unified crack growth model in short and long fatigue crack regimes. *Int. J. Fatigue* **2019**, *128*, 105182. [CrossRef]

16. Giannella, V.; Dhondt, G.; Kontermann, C.; Citarella, R. Combined static-cyclic multi-axial crack propagation in cruciform specimens. *Int. J. Fatigue* **2019**, *123*, 296–307. [CrossRef]
17. Giannella, V.; Fellinger, J.; Perrella, M.; Citarella, R. Fatigue life assessment in lateral support element of a magnet for nuclear fusion experiment "Wendelstein 7-X". *Eng. Fract. Mech.* **2017**, *178*, 243–257. [CrossRef]
18. Anaei, M.M.; Khosravifard, A.; Bui, T. Analysis of fracture mechanics and fatigue crack growth in moderately thick plates using an efficient meshfree approach. *Theor. Appl. Fract. Mech.* **2021**, *113*, 102943. [CrossRef]
19. Nejad, R.M.; Liu, Z.; Ma, W.; Berto, F. Reliability analysis of fatigue crack growth for rail steel under variable amplitude service loading conditions and wear. *Int. J. Fatigue* **2021**, *152*, 106450. [CrossRef]
20. Huynh, H.D.; Nguyen, M.N.; Cusatis, G.; Tanaka, S.; Bui, T.Q. A polygonal XFEM with new numerical integration for linear elastic fracture mechanics. *Eng. Fract. Mech.* **2019**, *213*, 241–263. [CrossRef]
21. Surendran, M.; Natarajan, S.; Palani, G.; Bordas, S.P. Linear smoothed extended finite element method for fatigue crack growth simulations. *Eng. Fract. Mech.* **2019**, *206*, 551–564. [CrossRef]
22. Rozumek, D.; Marciniak, Z.; Lesiuk, G.; Correia, J. Mixed mode I/II/III fatigue crack growth in S355 steel. *Procedia Struct. Integr.* **2017**, *5*, 896–903. [CrossRef]
23. Carter, B.; Wawrzynek, P.; Ingraffea, A. Automated 3-D crack growth simulation. *Int. J. Numer. Methods Eng.* **2000**, *47*, 229–253. [CrossRef]
24. *Abaqus User Manual*; Version 6.14; Abacus: Providence, RI, USA, 2020.
25. ANSYS. Academic Research Mechanical, Release 19.2, Help System. In *Coupled Field Anal. Guide*; ANSYS, Inc.: Canonsburg, PA, USA, 2020.
26. Lee, H.-H. *Finite Element Simulations with ANSYS Workbench 2021: Theory, Applications, Case Studies*; SDC publications: Mission, KS, USA, 2021.
27. Alshoaibi, A.M. Computational Simulation of 3D Fatigue Crack Growth under Mixed-Mode Loading. *Appl. Sci.* **2021**, *11*, 5953. [CrossRef]
28. Alshoaibi, A.M.; Fageehi, Y.A. 3D modelling of fatigue crack growth and life predictions using ANSYS. *Ain Shams Eng. J.* **2022**, *13*, 101636. [CrossRef]
29. Alshoaibi, A.M. Numerical Modeling of Crack Growth under Mixed-Mode Loading. *Appl. Sci.* **2021**, *11*, 2975. [CrossRef]
30. Fageehi, Y.A.; Alshoaibi, A.M. Numerical Simulation of Mixed-Mode Fatigue Crack Growth for Compact Tension Shear Specimen. *Adv. Mater. Sci. Eng.* **2020**, *2020*, 1–14. [CrossRef]
31. Hou, J.; Wescott, R.; Attia, M. Prediction of fatigue crack propagation lives of turbine discs with forging-induced initial cracks. *Eng. Fract. Mech.* **2014**, *131*, 406–418. [CrossRef]
32. Jafari, A.; Broumand, P.; Vahab, M.; Khalili, N. An eXtended Finite Element Method Implementation in COMSOL Multiphysics: Solid Mechanics. *Finite Elem. Anal. Des.* **2022**, *202*, 103707. [CrossRef]
33. Giannella, V.; Sepe, R.; De Michele, G.; Esposito, R. Deterministic fatigue crack-growth simulations for a railway axle by Dual Boundary Element Method. In *IOP Conference Series: Materials Science and Engineering*; IOP Publishing: Bristol, UK, 2021; p. 012080.
34. Fajdiga, G.; Sraml, M. Fatigue crack initiation and propagation under cyclic contact loading. *Eng. Fract. Mech.* **2009**, *76*, 1320–1335. [CrossRef]
35. Paris, P.; Erdogan, F. A critical analysis of crack propagation laws. *J. Fluid. Eng.* **1963**, *85*, 528–534. [CrossRef]
36. Coffin, L. *Cyclic deformation and fatigue of metals*; Elsevier: Amsterdam, The Netherlands, 1963; pp. 257–272.
37. Wöhler, A. Versuche zur Ermittlung der auf die Eisenbahnwagenachsen einwirkenden Kräfte und die Widerstandsfähigkeit der Wagen-Achsen. *Z. Bauwes.* **1860**, *10*, 583–614.
38. Zakavi, B.; Kotousov, A.; Branco, R. An analytical-based approach for simulating fatigue crack growth in round bars. *Int. J. Fract.* **2021**, *7*, 1–12. [CrossRef]
39. Zakavi, B.; Kotousov, A.; Khanna, A.; Branco, R. A new method for analysis of part-elliptical surface cracks in structures subjected to fatigue loading. *Theor. Appl. Fract. Mech.* **2019**, *103*, 102258. [CrossRef]
40. Alshoaibi, A.M.; Fageehi, Y.A. Finite Element Simulation of a Crack Growth in the Presence of a Hole in the Vicinity of the Crack Trajectory. *Materials* **2022**, *15*, 363. [CrossRef]
41. Wawrzynek, P.; Carter, B.; Banks-Sills, L. *The M-Integral for Computing Stress Intensity Factors in Generally Anisotropic Materials*; National Aeronautics and Space Administration, Marshall Space Flight Center: Huntsville, AL, USA, 2005.
42. Citarella, R.; Giannella, V.; Lepore, M.; Dhondt, G. Dual boundary element method and finite element method for mixed-mode crack propagation simulations in a cracked hollow shaft. *Fatigue Fract. Eng. Mater. Struct.* **2018**, *41*, 84–98. [CrossRef]
43. Dhondt, G.; Hackenberg, H.-P. Use of a rotation-invariant linear strain measure for linear elastic crack propagation calculations. *Eng. Fract. Mech.* **2021**, *247*, 107634. [CrossRef]
44. Bjørheim, F. Practical comparison of crack meshing in ANSYS mechanical APDL 19.2. Master's Thesis, University of Stavanger, Stavanger, Norway, 2019.
45. Xiangqiao, Y.; Shanyi, D.; Zehua, Z. Mixed-mode fatigue crack growth prediction in biaxially stretched sheets. *Eng. Fract. Mech.* **1992**, *43*, 471–475. [CrossRef]

46. Wagner, D. A finite element-based adaptive energy response function method for curvilinear progressive fracture. Ph.D. Thesis, The University of Texas at San Antonio, San Antonio, TX, USA, 2018.
47. Wagner, D.; Garcia, M.J.; Montoya, A.; Millwater, H. A Finite Element-based Adaptive Energy Response Function Method for 2D Curvilinear Progressive Fracture. *Int. J. Fatigue* **2019**, *127*, 229–245. [CrossRef]

metals

MDPI

Article

Buckling Analysis of Thin-Walled Circular Shells under Local Axial Compression using Vector Form Intrinsic Finite Element Method

Wenliang Ma [1,2], Zihan Sun [2,3,*], Han Wu [3], Leige Xu [2], Yong Zeng [4], Yanxing Wang [5] and Guangyin Huang [5]

1 School of Water Conservancy, North China University of Water Resources and Electric Power, Zhengzhou 450046, China
2 School of Civil Engineering and Communication, North China University of Water Resources and Electric Power, Zhengzhou 450046, China
3 Key Laboratory for Mechanics in Fluid Solid Coupling Systems, Institute of Mechanics, Chinese Academy of Sciences, Beijing 100190, China
4 Henan Provincial Communications Planning, Survey & Design Institute Co., Ltd., Zhengzhou 450046, China
5 Department of Capital Construction, North China University of Water Resources and Electric Power, Zhengzhou 450046, China
* Correspondence: sunzihan621@163.com

Abstract: The buckling failure of thin-walled circular shells under local axial compression is common in engineering. This study uses the vector form intrinsic finite element (VFIFE) method to investigate the buckling behavior of thin-walled circular shells under local axial compression by introducing a multilinear hardening model, taking into account geometric and material nonlinearity. A buckling analysis program based on the VFIFE method was developed and verified by comparison with experimental results. The buckling mode and postbuckling behavior of thin-walled circular shells were studied by using the verified program. The results show that the VFIFE method with a multilinear hardening model can accurately calculate the buckling load of local axially compressed thin-walled circular shells, and effectively simulate the buckling development process, which offers great advantages in predicting the postbuckling of structures.

Keywords: vector form intrinsic finite element; thin-walled circular shells; local axial compression buckling; postbuckling; multilinear hardening model

Citation: Ma, W.; Sun, Z.; Wu, H.; Xu, L.; Zeng, Y.; Wang, Y.; Huang, G. Buckling Analysis of Thin-Walled Circular Shells under Local Axial Compression using Vector Form Intrinsic Finite Element Method. *Metals* **2023**, *13*, 564. https://doi.org/10.3390/met13030564

Academic Editors: Shuwen Wen, Yongle Sun and Xin Chen

Received: 15 February 2023
Revised: 6 March 2023
Accepted: 8 March 2023
Published: 10 March 2023

1. Introduction

As a basic structural unit, a thin-walled circular shell has strong functionality and is simple to process which is widely used in aerospace, marine engineering, civil engineering, and other fields [1–4]. Thin-walled circular shells are prone to buckling failure under axial loading, and this kind of buckling is essentially the buckling of thin-shell structures. The whole buckling process involves geometric nonlinearity, material nonlinearity, and contact nonlinearity, and its buckling behavior is difficult to predict. To accurately predict the buckling behavior of thin-shell structures, Horrigmoe et al. [5] deduced the buckling analysis of thin-shell structures under the condition of linear elasticity in the early stage, Combescure [6] studied the buckling behavior of thin-shell structures with elastoplastic constitutive equations while considering large deformations. Spagnoli [7] deduced the postbuckling behavior of linear elastic conical shells. Abambres [8] used generalized beam theory(GBT) to predict steel pipe buckling and compared the results with the finite element calculation. Kadkhodayan [9] predicted the buckling of a thin-walled shell under uniform and linearly varying inplane loading by incremental theory. In practical engineering, the stress of thin-walled circular shells is mainly subject to axial pressure and is very prone to uneven force phenomenon [10,11]. Jiao et al. [12,13] studied the influence of local axial compression of different degrees on the buckling load and buckling mode of thin-walled circular shells through experiments and the finite element method (FEM). Hossein Nassiraei et al. [14], based on the FEM, conducted and analyzed a total of 138 collar plates with strengthened and unstrengthened tubular X-joints, and a theoretical equation was proposed based on the yield volume model to predict the ultimate capacity of the X-connections strengthened with collar plates under brace compression, Then, the formula was confirmed by the acceptance criteria of the UK Department of Energy [15]. Wu et al. [16] used modified couple stress theory to study the buckling and postbuckling of symmetric functionally graded microplate lying on a nonlinear elastic foundation. Liu et al. [17] studied the buckling of hydroformed toroidal pressure hulls with octagonal cross sections. Peng et al. [18] studied the numerical and experimental buckling and postbuckling analyses of sphere-segmented toroidal shell subject to external pressure.

In the process of buckling, the materials of thin-walled circular shells will undergo a transition from the elastic stage to the plastic stage. The stress–strain relationship in the plastic stage is complex, and it is difficult to accurately simulate the complete stress–strain relationship of materials. Generally, the hardening stage during plastic is commonly simplified as ideal elastoplasticity, the bilinear hardening model [19–21], the power hardening model [22,23], and the Ramberg–Osgood model [24–26]. Different hardening models have different characteristics, among which is ideal elastoplasticity, which will not harden after the material enters the plastic stage, and the stress–strain curve is parallel to the strain axis. The bilinear hardening model simplifies the constitutive curve into two broken lines, and the stress–strain curve in the hardening stage is a broken line with a fixed slope. In the power hardening model, the stress–strain curve in the hardening stage of the material is a power curve. The Ramberg–Osgood model simulates the stress–strain curve as a power curve for the material as a whole. The above four models all simplify the full elastoplastic curve into a finite polyline or curve, which is not accurate enough to describe the stress–strain relationship of the entire elastoplastic stage. To describe the stress–strain relationship in the entire elastoplastic stage more accurately, this paper uses a multilinear hardening model, that is, a model based on the incremental elastic-plastic flow law, where the constitutive curve after the material enters the plastic stage is discretized into multiple points, and then each point is connected with a straight line.

FEM is often used to predict the structural buckling behavior of elastic-plastic materials [27–29]. Structural buckling is a large deformation problem, and the FEM needs to constantly update the element stiffness matrix and boundary conditions to solve this kind of problem, which makes the calculation complicated. For the elastic-plasticity of materials, only the above four simplified hardening models can be used in most cases, so compatibility with the multilinear hardening method of constantly updating the tangent modulus of the constitutive is difficult. Additionally, the thin-shell structure will collapse after local buckling, resulting in a large degree of displacement of the structure. In this case, the simple statics problem-solving error of the FEM is so large that the small deformation assumption is no longer applicable. Statics and dynamics are often combined to deal with such geometric nonlinear problems, although Shih et al. pointed out that if the FEM calculation of rigid body displacement is much larger than the pure deformation, this may lead to unstable results [30]. Due to the deficiency of the FEM in analyzing large deformation and displacement problems such as the local axial buckling of thin-shell structures, the VFIFE is used for simulation analysis.

VFIFE is an innovative method developed based on vector mechanics and the finite particle method, which was first proposed by Ting et al. [30–32]. The VFIFE discretizes the structure into particles, calculates the displacement within the path element through Newton's second law, and then calculates the pure deformation and internal force of the structure through the inverse motion. The VFIFE does not need to solve the complex stiffness matrix and has its geometric nonlinearity, so it has a great advantage in dealing with large deformation and large displacement. Currently, the VFIFE has been developed into a beam element [33], membrane element [34], plate element [35], solid element [36], etc. Wu et al. [37] introduced the Cowper–Symonds (C-S) viscoplastic constitutive model into the VFIFE, and Wang formed the shell element by linear superposition of the plate element and membrane element [38].

Due to the characteristics of dynamic analysis, compared with the FEM which can only calculate the structural modes through the singular matrix method, the VFIFE can directly predict the development of postbuckling of the structure and has a great advantage over the FEM in predicting the postbuckling behavior of the structure. The method of predicting structural buckling by VFIFE was originally derived by Wang [39,40], who calculated the buckling of a thin-shell structure under linear elasticity by force-control and displacement-control methods, respectively, then compared the calculated results with the classical linear elastic buckling derivation and ABAQUS software, which verified that the calculation accuracy of the buckling of a thin-shell structure by VFIFE with the displacement-control method was higher. Xu et al. introduced the von Mises yield criterion to correctly predict the local buckling of buried pipelines subjected to reverse fault motion [41]. Yu et al. [42] discussed the buckling of subsea pipelines with integral buckle arrestors and then introduced soil springs to study the buckling failure of buried subsea pipelines under reverse fault displacement [43]. The above research enriches the content of VFIFE and demonstrates its advantages.

This paper, based on the vector form intrinsic finite thin-shell element, simulates the buckling of a thin-walled circular shell under local axial compression, calculates the tensile internal force of the thin-shell element using the constant strain triangle membrane element (CST) and the bending internal force of the thin-shell element using the discrete Kirchhoff triangle thin-plate element (DKT), and the two elements are linearly superimposed to form a complete shell element. Geometric nonlinearity and material nonlinearity are considered in the buckling process of the structure. The multilinear hardening model was applied to the triangular shell element, and the corresponding tangent modulus is found according to the equivalent von Mises stress calculated in each calculation step to carry out the elastoplastic calculation. The Fortran calculation program for the vector form intrinsic finite thin-shell element and circular tube buckling was compiled and OpenMP [44] technology was used to improve the calculation efficiency. The simulation results were compared with existing experiments to verify the reliability of the analysis method and program. Then, the postbuckling process of the thin-walled circular shell was analyzed, and the reliability of thin-shell structure buckling was verified by the introduction of a multilinear hardening model, which provides a new and reliable idea for the buckling analysis of a thin-shell structure.

2. Analysis Model

2.1. Circular Tube under Localized Axial Compression Loads

The research object of this paper is a thin-walled circular shell under axial compression loads. Figure 1 is a schematic diagram of a thin-walled circular shell under local axial compression loads. The left figure is a three-dimensional diagram of the circular tube, with red arrows representing the local axial compression loads exerted on the end face of the circular tube in a certain angle range. On the right are a top-down view of the circular tube, the black areas where it is not stressed, and the red areas representing the axial compression loads at an axial pressure angle of $2 \times \theta$, which simulates the experiment in reference [12], by adding a convex flange with a certain angle to the top of the cylindrical shell, the specimen was subjected to the axial compression buckling test system.

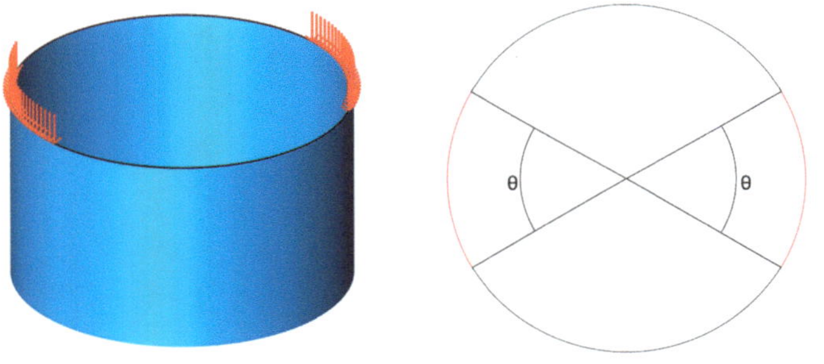

Figure 1. Tube under localized axial compression loads.

The tube has a thickness of 1.2 mm, a diameter of 1000 mm, and a height of 600 mm, and is made of DC01 cold-rolled steel with an elastic modulus of 206 GPa and a yield strength of 158.7 MPa.

2.2. Governing Equations

Different from the FEM, the VFIFE discretizes the structure into finite particles to describe the deformation and motion of the structure under loads, and the mass (moment of inertia) of each particle is the point at which the structure is concentrated after equivalence. The motion of a particle is described by Newton's second law:

$$\mathbf{M}\ddot{x} + c\mathbf{M}\dot{x} = \mathbf{F} \tag{1}$$

$$\mathbf{I}\ddot{\boldsymbol{\theta}} + c_\theta \mathbf{I}\dot{\boldsymbol{\theta}} = \mathbf{F}_\theta \tag{2}$$

where **M** and **I** are the mass and moment of the inertia matrix of the particle, $\ddot{x}$ and $\ddot{\boldsymbol{\theta}}$ are the acceleration and angular acceleration of the particle, $\dot{x}$ and $\dot{\boldsymbol{\theta}}$ are the velocity and angular velocity of the particle, and **F** and $\mathbf{F}_\theta$ are the resultant force and the external force on the particle. The damping of the particle on translational and rotational degrees of freedom are c and c_θ, and generally, the particle is damped the same in all directions.

Refer to the derivation of existing scholars, substituting Equations (1) and (2) into the central difference method, the particle motion formula can be obtained as follows:

$$\boldsymbol{x}_{n+1} = c_1(\frac{\Delta t^2}{\boldsymbol{m}})\mathbf{F}_n + 2c_1\boldsymbol{x}_n - c_2\boldsymbol{x}_{n-1} \tag{3}$$

$$\boldsymbol{\theta}_{n+1} = c_1\Delta t^2\mathbf{I}^{-1}\mathbf{F}_{\theta n} + 2c_1\boldsymbol{\theta}_n - c_2\boldsymbol{\theta}_{n-1} \tag{4}$$

where $\boldsymbol{x}_n$ and $\boldsymbol{\theta}_n$ are the position vector and the corner vector of the particle at step n, $\mathbf{F}_n$ and $\mathbf{F}_{\theta n}$ are the resultant force and resultant moment of the particle at step n, Δt is the time step of a single loading step, and c_1 and c_2 are the constants associated with damping.

2.3. Internal Force

Unlike the FEM, which solves the internal force by solving linear equations, the VFIFE solves the internal force by decomposing the element displacement into the rigid body displacement and solving the internal force by using pure deformation through the virtual work principle.

2.3.1. Inverse Motion

VFIFE solves the pure deformation of the structure by inverse motion. The inverse motion of a shell element is represented in Figure 2:

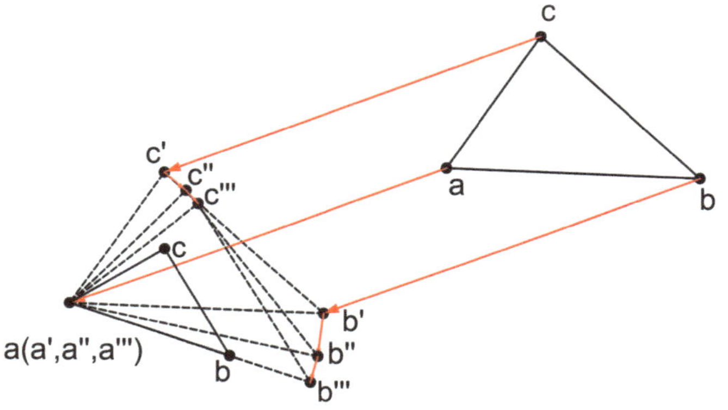

Figure 2. Inverse motion of a shell element.

The inverse motion in space undergoes rigid body translation (abc→a′b′c′), out of the plane to inplane (a′b′c′→a″b″c″), and rotation inplane (a″b″c″→a‴b‴c‴), and finally, the pure deformation of each particle on the element is obtained:

$$\begin{cases} \Delta\boldsymbol{\eta}_{\mathrm{a}}^{\mathrm{d}} = 0, \Delta\boldsymbol{\eta}_{i}^{\mathrm{d}} = (u_i - u_a) + \Delta\boldsymbol{\eta}_{i-op}^{r} + \Delta\boldsymbol{\eta}_{i-ip}^{r}, i = b, c \\ \Delta\boldsymbol{\eta}_{\theta i}^{\mathrm{d}} = u_{\theta i} + \Delta\boldsymbol{\eta}_{\theta i-op}^{r} + \Delta\boldsymbol{\eta}_{\theta i-ip}^{r}, i = a, b, c \end{cases} \tag{5}$$

where $\Delta\boldsymbol{\eta}_{i}^{\mathrm{d}}$ and $\Delta\boldsymbol{\eta}_{\theta i}^{\mathrm{d}}$ represent the pure deformation line displacement and angular displacement of the particle, $\Delta\boldsymbol{\eta}_{i-op}^{r}$ and $\Delta\boldsymbol{\eta}_{\theta i-op}^{r}$ represent the pure deformation line and angular displacements out of the plane, and $\Delta\boldsymbol{\eta}_{i-ip}^{r}$ and $\Delta\boldsymbol{\eta}_{\theta i-ip}^{r}$ represent the pure deformation line and angular displacements in the plane.

2.3.2. Solution of Internal Force

The internal force calculation on VFIFE is based on pure deformation by the virtual work principle. Through the virtual work principle, the internal forces at each node of the shell element can be obtained by linear superposition of the bending internal forces obtained by the DKT thin plate element and the tensile and compressive internal forces obtained by the CST membrane element. The internal forces are solved by the following formula:

$$\sum_{i} \delta(\hat{\boldsymbol{u}}_i)^{\mathrm{T}} \hat{f}_i + \sum_{i} \delta(\hat{\boldsymbol{u}}_{\theta i})^{\mathrm{T}} \hat{f}_{\theta i} = \int_{\mathrm{V}} \delta(\Delta\hat{\boldsymbol{\varepsilon}}_m)^{\mathrm{T}} \hat{\boldsymbol{\sigma}}_m dV + \int_{\mathrm{V}} \delta(\Delta\hat{\boldsymbol{\varepsilon}}_p)^{\mathrm{T}} \hat{\boldsymbol{\sigma}}_p dV \tag{6}$$

where $\Delta\hat{\varepsilon}$ and $\hat{\sigma}$ represent the strain increment and stress of the node, respectively, and m stands for the membrane element and p for the plate element.

2.4. Determination of Buckling

This paper determines whether the structure is buckling according to the axial–force-displacement relationship. Before the structure buckling, the axial force gradually increases with the increase of displacement, with the two having a linear relationship. When the structure buckles, the axial force at the buckling point will decrease with the increase of displacement, and the axial force at the buckling point is considered as the buckling load of the structure. Due to the dynamic characteristics of the process of VFIFE analysis, the method of gradually increasing the axial force needs to wait until the result converges, which is not effective for the prediction of buckling. Therefore, the method of imposing forced displacement on the structure is adopted to push back the axial force of the structure, to draw the axial-force–displacement relationship curve of the buckling process. Formula (7) is the calculation formula of the axial force in reverse derivation:

$$\begin{cases} f_n^{ext} = \frac{1}{c_1 \Delta t^2}\mathbf{M}(\mathbf{x}_{n+1} + c_2\mathbf{x}_{n-1} - 2c_1\mathbf{x}_n) - f_n^{int} \\ f_{\theta n}^{ext} = \frac{1}{c_1 \Delta t^2}\mathbf{M}(\boldsymbol{\theta}_{n+1} + c_2\boldsymbol{\theta}_{n-1} - 2c_1\boldsymbol{\theta}_n) - f_{\theta n}^{int} \end{cases} \quad (7)$$

where f_n^{ext} and $f_{\theta n}^{ext}$ represent the external force and moment of the particle at loading step n, and f_n^{int} and $f_{\theta n}^{int}$ represent the internal force and internal moment of the particle at loading step n.

2.5. *Nonlinearity*

2.5.1. Geometric Nonlinearity

Based on Newton's second law, VFIFE discretizes the structure into a finite number of particles to describe the deformation and motion state of the structure. The state of the structure is related to the velocity and acceleration direction of the particles under the force. The motion state of the points in the path element can be clearly described, so it is not necessary to correct the geometric nonlinearity of a structure with a large deformation and a large displacement.

2.5.2. Material Nonlinearity

When the deformation is small, the material will recover its original state after the load is removed, so this stage belongs to the elastic stage of the material. When the deformation exceeds the range of elastic deformation, it will produce permanent deformation from which the material cannot recover. However, the stress–strain relationship of the material in the plastic stage is relatively complex, and the plasticity of the material in the numerical simulation is a material nonlinear problem, which requires a separate method to describe.

In the existing VFIFE study, there is only the bilinear hardening model and the power hardening model, which cannot accurately simulate the elastoplasticity of the materials. In order to accurately describe the strain state of the element under different stress states, a multilinear hardening model was used to discretize the material constitutive curve into multiple points, and the points were connected into broken lines. The elastic-plastic matrix under the state was updated according to the stress state of different load steps to accurately predict the elastic-plastic stress–strain of the material.

The von Mises yield criterion was used to judge whether it was plastic or not, and each integration layer is considered as a plane stress state while integrating according to thickness. When the stress satisfies the following formula,

$$F = \overline{\sigma} - \sigma_s^2 = 0 \quad (8)$$

The material is considered to have reached the yield point, if $F > 0$, and the material enters the plastic stage. Equation (8), $\overline{\sigma}$ is equivalent to von Mises stress, and σ_s is the yield limit;

At time t, the strain increment $\Delta\varepsilon$ can be calculated according to the existing node displacement. Assuming that the increment step is elastic, the stress increment can be calculated according to the existing strain increment, and the elastic predicted stress at a time $t + \Delta t$ can be calculated by superimposing the stress at time t, as shown in Equation (9),

$$^{t+\Delta t}\sigma^{pr} = {}^{t}\sigma + \mathbf{D}_e\Delta\varepsilon \quad (9)$$

where $^{t+\Delta t}\sigma^{pr}$ represents elastic predictive stress at the time $t + \Delta t$, $^{t}\sigma$ represents the stress at time t, and $\mathbf{D}_e$ represents the elastic matrix;

The equivalent treatment of elastic predicted stress solved by the elastic prediction method is introduced into the von Mises yield formula. If the result is greater than zero, the node is in an elastic state and continues to be solved according to linear elasticity. If the result is less than zero, it means that it has entered a plastic state. Here, the elastic factor m, which is the proportion of elasticity at the integration point, is introduced. The solution of m is:

$$m = (-a_1 + \sqrt{a_1{}^2 + 4a_0a_2})/2a_2 \tag{10}$$

where, for isotropic hardening materials, a_0 is the von Mises yield function at time t. Therefore, a_1 and a_2 are obtained by the following formula:

$$\begin{cases} a_1 = \mathbf{S}^{\mathrm{T}}\Delta\mathbf{S} \\ a_2 = \frac{1}{2}\Delta\mathbf{S}^{\mathrm{T}}\Delta\mathbf{S} \end{cases} \tag{11}$$

where $\mathbf{S}$ presents the tensor of deviation stress at time t and $\Delta\mathbf{S}$ represents the increment of deviation stress at time t, and the superscript T stands for matrix transpose.

It should be noted that the m value calculated here should be greater than zero and less than one. If it is not in this interval, all the values are set as plastic, that is, m = 0;

The update of and material hardening modulus of the elastic-plastic matrix are closely related, and the mathematic expression of the hardening modulus is:

$$\boldsymbol{E}_p = \frac{\mathrm{EE_t}}{\mathrm{E} - \mathrm{E_t}} \tag{12}$$

where $\mathrm{E_t}$ represents the tangent modulus, which is the slope of the tangent at the point corresponding to the constitutive curve of the material under this stress state. The multilinear hardening model in this paper simplifies the constitutive curve of materials into multiple broken lines, and the tangent modulus is taken as the slope of the corresponding interval broken line on the simplified constitutive curve where the stress is at time $t - \Delta t$. At the same time, according to the flow rule of isotropic hardening, the hardening of materials can only go outward along the circular tangent line of von Mises stress, not inward, and the slope of the actual constitutive curve will be less than zero when it reaches tensile strength, and therefore, in this paper, the tangent modulus of the part with a slope less than zero in the constitutive curve is taken as 1000, which can not only ensure the continuous hardening of the material but also keep a small error with the actual material.

After solving the hardening modulus, the elastic-plastic matrix can be updated. The plastic matrix is calculated as follows:

$$\mathbf{D}_\mathrm{p} = \frac{\mathrm{E}}{\mathrm{B}(1-v^2)}\begin{bmatrix} (\mathrm{s_x} + v\mathrm{s_y})^2 & (\mathrm{s_x} + v\mathrm{s_y})(\mathrm{s_y} + v\mathrm{s_x}) & (1-v)(\mathrm{s_x} + v\mathrm{s_y})\tau_{\mathrm{xy}} \\ (\mathrm{s_x} + v\mathrm{s_y})(\mathrm{s_y} + v\mathrm{s_x}) & (\mathrm{s_y} + v\mathrm{s_x})^2 & (1-v)(\mathrm{s_y} + v\mathrm{s_x})\tau_{\mathrm{xy}} \\ (1-v)(\mathrm{s_x} + v\mathrm{s_y})\tau_{\mathrm{xy}} & (1-v)(\mathrm{s_y} + v\mathrm{s_x})\tau_{\mathrm{xy}} & (1-v)^2\tau_{\mathrm{xy}}{}^2 \end{bmatrix} \tag{13}$$

where,

$$\mathrm{B} = \mathrm{s_x}^2 + \mathrm{s_y}^2 + 2\nu\mathrm{s_x}\mathrm{s_y} + 2(1-\nu)\tau_{\mathrm{xy}}{}^2 + \frac{2(1-\nu)\mathrm{E}_p\sigma_\mathrm{s}}{9\mathrm{G}} \tag{14}$$

In the formula, G is the shear modulus and $\mathrm{s_x}$ and $\mathrm{s_y}$ is the deviation stress.

After the plastic matrix is calculated, the elastic-plastic matrix can be calculated:

$$\mathbf{D}_{ep} = \mathbf{D}_e - \mathbf{D}_p \tag{15}$$

First, calculate the plastic strain increment according to the strain increment

$$\Delta\boldsymbol{\varepsilon}_p = (1 - \mathrm{m})\Delta\boldsymbol{\varepsilon} \tag{16}$$

Finally, the plastic stress is calculated:

$${}^{t+\Delta t}\sigma = {}^{t}\sigma + \mathrm{m}^{t+\Delta t}\sigma^{pr} + \mathbf{D}_{ep}\Delta\boldsymbol{\varepsilon}_p \tag{17}$$

Based on the VFIFE principles, the buckling program of the thin-shell structure is compiled, and the model data, boundary conditions, and loads are input into the calculation program. In each analysis step, the element displacement and internal force are calculated first, and then the equivalent stress and strain are calculated and introduced into the subprogram of the multilinear hardening model. First, the elastic prediction is carried out to determine whether the plastic stage has been entered, and if it has entered into the plastic stage, plastic correction is required. Finally, the next analysis step is carried out. The program flow of each elastic-plastic incremental step is shown in Figure 3.

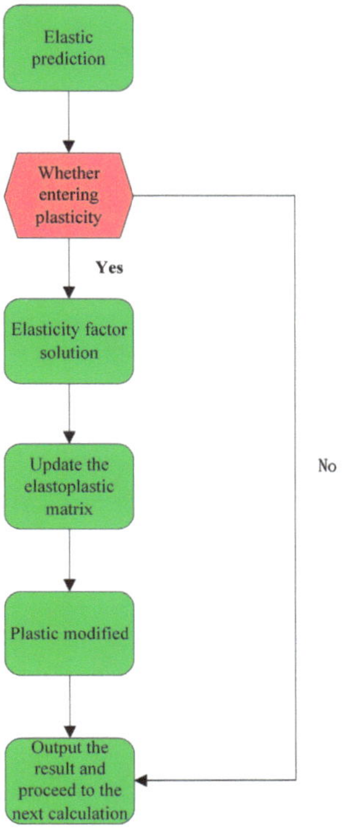

Figure 3. Multilinear hardening model program. Green represents process flow and red represents logical judgment.

3. Model Verification and Parameter Setting

3.1. Multilinear Hardening Model Verification

To verify the feasibility of the multilinear hardening model, the axial tensile process of cold-rolled steel was simulated by an analysis program, and the simulated stress–strain curve was compared with the experimental stress–strain curve. The simulation material selected was DC01 cold rolled steel, with an elastic modulus of E = 206 GPa, a yield strength of σ_s = 158.7 MPa, and an established specimen model, as shown in Figure 4. The simulation results were compared with the tensile test results [12]. To simulate the situation close to the test, the bottom of the specimen was fixed, and the top of the specimen was restricted except for the degree of freedom in the tensile direction. A forced upward displacement was applied to the top of the specimen. The stress–strain curve drawn by taking the equivalent stress–strain of the central node of the specimen was compared with the test results, as shown in Figure 5.

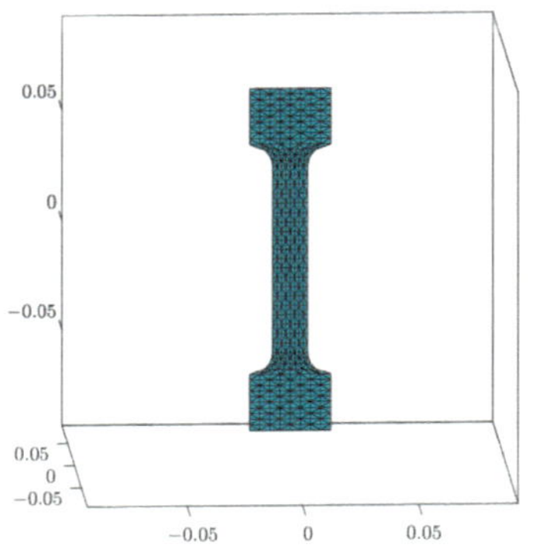

Figure 4. Tensile test model.

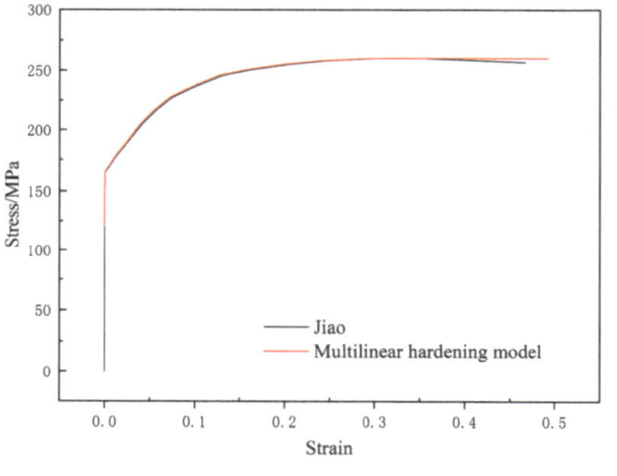

Figure 5. Comparison of stress–strain curves.

It can be seen from the above comparison results that the simulation results are in good agreement with the experimental results, which proves that the multilinear hardening model introduced in this paper can accurately predict the stress value of the material under a given strain in the VFIFE.

3.2. Calculation Parameters

Based on the above description, the buckling behavior of a circular tube under local axial compression is simulated by the VFIFE method. The axial compression of the structure is derived by applying the forced displacement, that is, a fixed constraint is applied to the bottom node of the tube, and a forced axial displacement is applied to the top node, limiting the other degrees of freedom except the forced axial displacement. The time step is $\Delta t = 1 \times 10^{-6}$s. The multilinear hardening model is adopted. The material of the tube is DC01 cold-rolled steel, and the material parameters are set according to the test in Section 3.1.

A triangular shell element is adopted for the circular tube model. As shown in Figure 6, the element is discretized into three mass points distributed at the element nodes. Each triangular element is an isosceles right triangle.

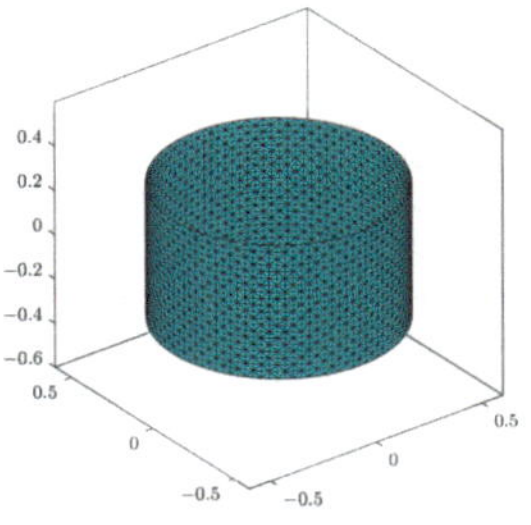

Figure 6. Cylindrical shell model under VFIFE.

Based on the analysis of the buckling load of a thin-walled circular shell under a $2 \times 90°$ local axial compression load under different mesh sizes, the appropriate mesh is determined. The simulation calculation refers to the experimental setting of Jiao [12], with triangular meshes with side lengths of 20 mm, 30 mm, 40 mm, and 50 mm were analyzed respectively, loaded at 2 mm/s. The simulation and experimental errors obtained are shown in Table 1.

Table 1. Buckling loads and errors were calculated from the mesh of different sizes.

Meshing Size/mm	Buckling Load/kN	Error/%
50	272.7	5.50
40	263.6	2.24
30	258.6	0.35
20	258.1	0.15

In the experiment, the first buckling load is 257.7 kN, as can be seen in Table 1, when the mesh size is 30 mm, the calculation error becomes small and the margin of error decreases slightly as the mesh size decreases, so the mesh size in this paper is 30 mm.

The forced displacement of the top node needs to be loaded at a certain speed, and the loading speed will affect the calculation results. The axial force–displacement curves at different speeds are analyzed, the optimal loading speed is selected by comparison., and the loading angle is $2 \times 90°$, with a triangular shell element, and the length of the right angle side of the element is 30 mm, loaded the top displacement at the speeds of 30 mm/s, 10 mm/s, 2 mm/s, and 1 mm/s, respectively, and the calculated displacement axial force curve is shown in Figure 7.

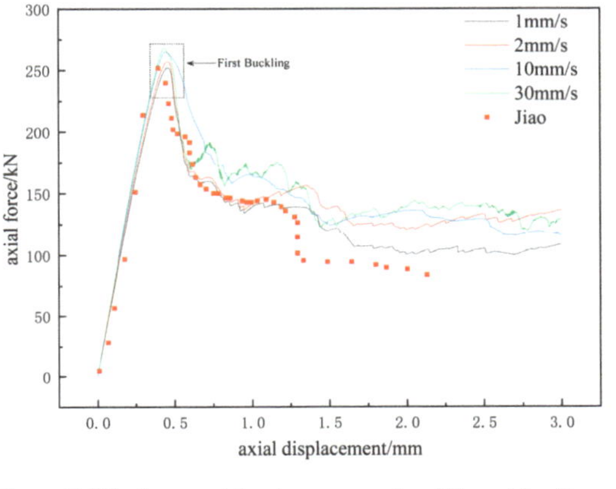

Figure 7. Displacement-load curves under different loading rates.

In the experiment, the first buckling load is 257.7 kN, and the first buckling load loaded at 30 mm/s is 265.6 kN, with an error of 3.07%, the first buckling load loaded at 10 mm/s is 263.6 kN, with an error of 2.28%, the first buckling load loaded at 2 mm/s is 258.6 kN, with an error of 0.34%, and the first buckling load loaded at 1 mm/s is 256.9 kN, with an error of 0.31%. It can be seen from the analysis results that before the buckling of the structure, each loading rate can be very close to the experimental results. However, due to the influence of inertial force, a faster loading rate means a greater buckling load. When the loading rate is reduced to 2 mm/s, the agreement between the first buckling load and the experiment reaches a relatively ideal level, however, there is still a large gap between the subsequent curve and the experiment. After loading at the rate of 1 mm/s, the result is improved, so this paper adopted the rate of 1 mm/s for loading.

It should be pointed out that there are often many small gaps between the experimental process and the expected design, such as the initial defects of the specimen and the uneven control of the load. These conditions are in an ideal state in the numerical simulation, and the small gap between the two will cause non-negligible errors in the buckling structure. For the subsequent buckling behavior after the first buckling, it is very difficult to simulate results that are very consistent with the experiment. However, if subsequent buckling similar to the experiment can be simulated, it has a very important value for guiding the structural design.

4. Results and Discussion

In this paper, based on the simulation program, buckling loads under different degrees of local axial compression are calculated, and the buckling modes are predicted. Finally, postbuckling is visualized by drawing a displacement contour of the tube at different stages in the deformation process, and the development of buckling under local axial compression load is explored.

4.1. First Buckling Load Buckling Modes

Simulations were performed for circular tubes with axial compression angles of $2 \times 30°, 2 \times 60°, 2 \times 90°, 2 \times 120°, 2 \times 150°$, and $2 \times 180°$ (overall axial compression), and the calculated results were plotted as displacement–axial force curves to compare with the experimental results, as shown in Figure 8.

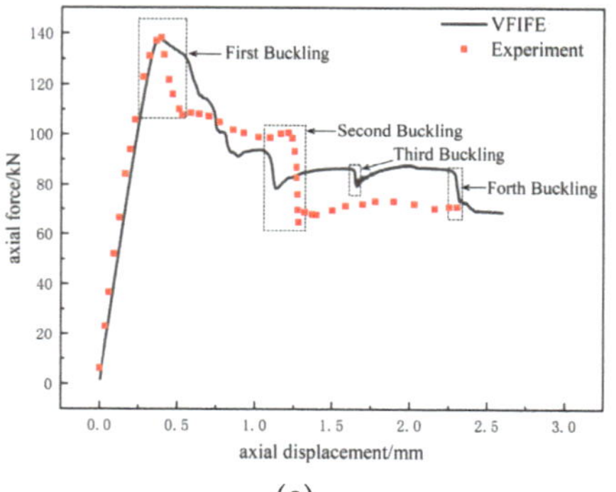

(a)

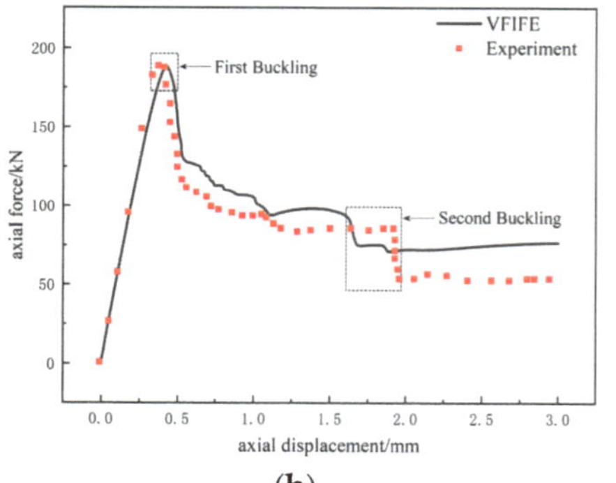

(b)

Figure 8. *Cont.*

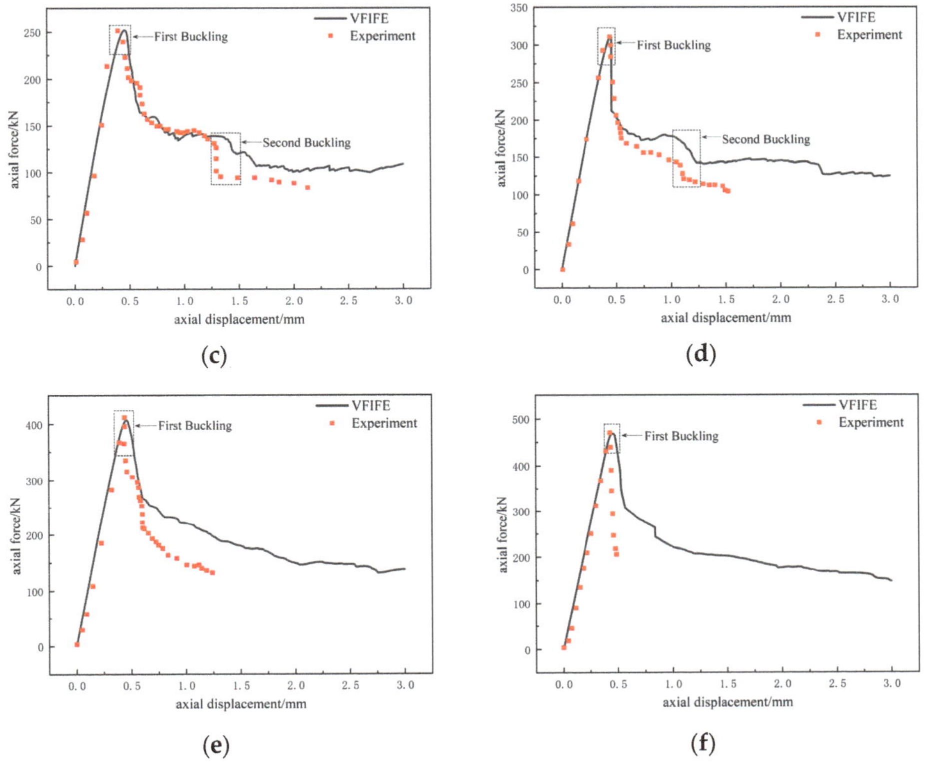

Figure 8. Axial displacement-force curve: (**a**) 2 × 30°, (**b**) 2 × 60°, (**c**) 2 × 90°, (**d**) 2 × 120°, (**e**) 2 × 150°, (**f**) 2 × 180°.

Table 2 shows the comparison between the first buckling load calculated by VFIFE and the experiments in the literature. The error between the first buckling load predicted by VFIFE and the experimental results is less than 1%.

Table 2. Comparison of results.

Axial Compression Angle	Experiment/kN	VFIFE/kN	Error/%
2 × 30°	138.9	137.9	0.72
2 × 60°	191.0	192.3	0.68
2 × 90°	257.7	256.9	0.31
2 × 120°	310.3	313.1	0.99
2 × 150°	409.1	406.3	0.68
2 × 180°	466.3	468.4	0.45

When the angle of local axial comparison is 2 × 30°, 2 × 60°, 2 × 90°, and 2 × 120°, secondary buckling occurs in both the experimental and the simulation results, as seen in the abrupt change of the displacement-axial force curve in Figure 9. VFIFE predicts four buckling of the structure for a local axial comparison angle of 2 × 30°, two buckling for local axial comparison angles of 2 × 60°, 2 × 90°, and 2 × 120°, and only one buckling for axial comparison angles of 2 × 150° and 2 × 180°. It is inferred that the smaller the local axial comparison angle, the more likely the structure is to buckle several times, so every engineering project should try to avoid a local axial pressure load on the circular tube.

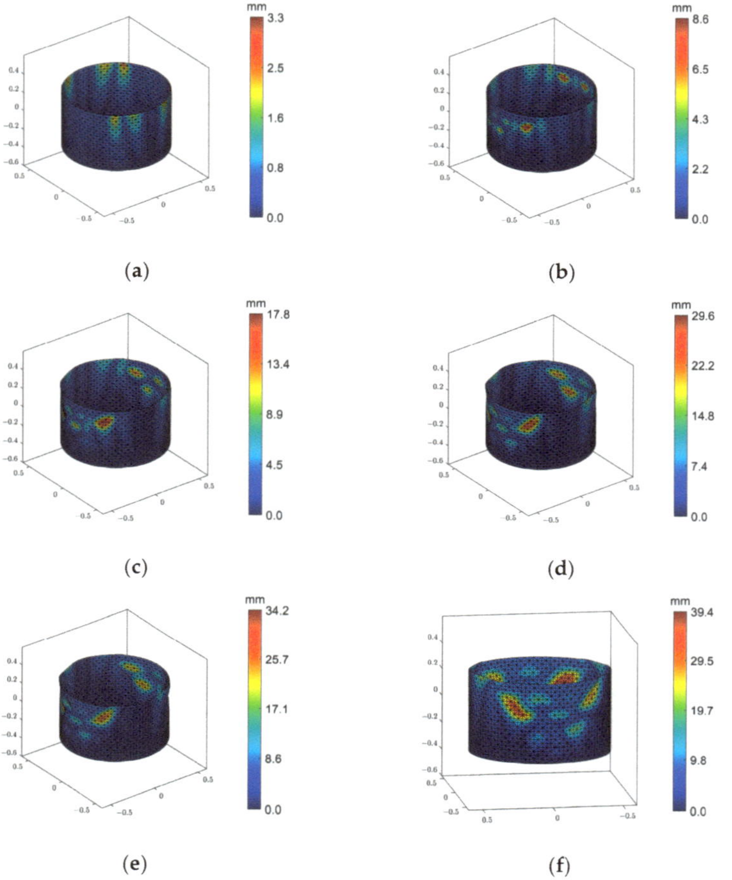

Figure 9. Postbuckling of 2 × 90° local axial compression: (**a**) before buckling, (**b**,**c**) first buckling, (**d**–**f**) second buckling.

The prediction of modal and postbuckling behavior of thin-shell structure buckling has far-reaching significance for practical engineering. The FEM prediction of postbuckling buckling is done by first solving a linear equation system, then calculating different modal orders according to different eigenvalues, and then adding different modalities to the structure according to geometric defects, all in order to predict the postbuckling behavior of the structure, which is a complicated process to solve. VFIFE with its dynamic solution predicts the buckling behavior of thin-shell structures after buckling without additional modal analysis operations and predicts the structural modalities at different stages of buckling and accurately predicts the loads in this state.

Table 3 shows the comparison between the buckling modes calculated by VFIFE near the first buckling of the circular tube under 2 × 90° and 2 × 120° local axial compression loads, and 2 × 180° overall axial compression load, and the experimental results [12]. In the table, the buckling modes calculated by VFIFE are represented by a displacement contour, in which the color bar from blue to red represents the change of the overall displacement from large to small, and the buckling deformation is generally larger, and it is easy to distinguish the deformation direction, so deformation size is taken as an absolute value.

Table 3. Comparison of buckling mode. Pictures of the experiment are reproduced with the permission of [12], copyright@Thin-Walled Structures, 2021.

Angle	Experiment	VFIFE
2 × 90°	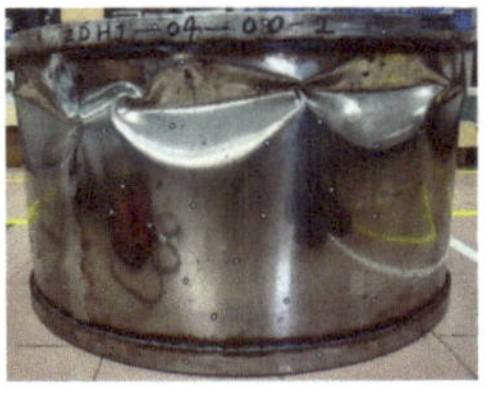	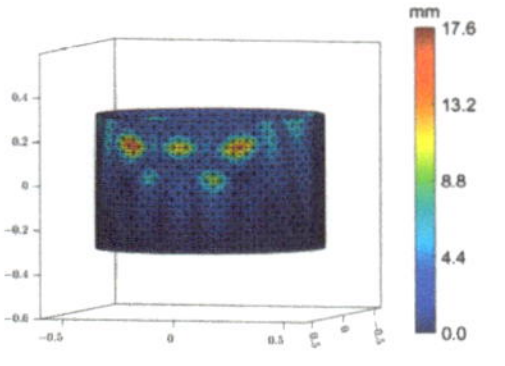
2 × 120°		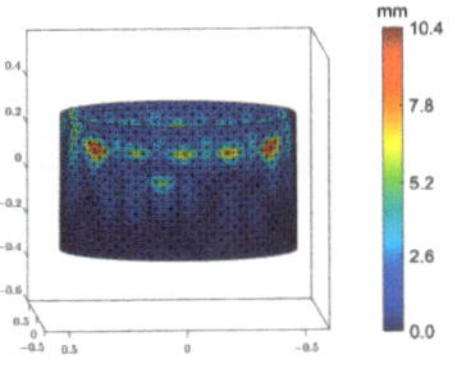
2 × 180°		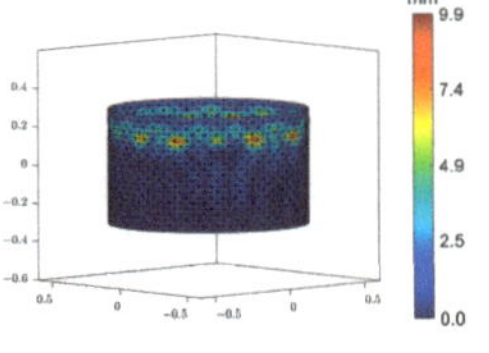

Based on the buckling modes near the first buckling, it can be seen that the location and deformation of the buckling deformation calculated by VFIFE are more similar to the experiments, and the predictions of the buckling deformation of the thin-shell structure by VFIFE are more appropriate to the actual situation.

4.2. Buckling Modes and Postbuckling

The calculation of the first buckling load is of great value for guiding the structural design, however, it is still difficult to avoid buckling of the structure in actual engineering. The prediction of the postbuckling behavior of the structure is also of high value for guiding engineering. Most commercial software based on FEM can solve the structural buckling load by solving linear equations with the eigenvalue method and solve the postbuckling modes by the explicit dynamics method, though the process is cumbersome, and VFIFE can do these calculations directly. According to previous studies, the validity and accuracy of the VFIFE method proposed in this paper are verified in the buckling analysis of thin-walled circular shells. Based on developed analytical procedures, this section investigates the postbuckling behavior of thin-walled circular shells.

The postbuckling behavior under local axial compression load was analyzed, and the postbuckling of 2 × 90° local axial compression and 2 × 120° local axial compression were analyzed, respectively. The forced displacement of 8 mm downward was applied to the end of the pipe through different local axial compression angles, and the postbuckling behavior was studied according to the displacement contour after deformation, with the color in the displacement contour the same as in Section 3, the blue to red representing the change of the total displacement from large to small, and the deformation size is taken as an absolute value.

Figure 9 depicts the postbuckling of the circular tube under the local axial compression load of 2 × 90°. Figure 9a–e is the unified view angle, and in Figure 9f, the view angle is adjusted to the side of the axial compression applied more conveniently. Figure 9a shows the state of the structure when buckling is about to occur, under a local axial compression load of 2 × 90°, the figure shows that the locations where obvious folds appear were not under compression, and the larger deformations occur in the regions outside the axial compression locations where folds appear. Figure 9b,c are after the first buckling occurs, as the second buckling of the structure appears and develops, and the third buckling occurs in Figure 9d,e, where the depression deformation only becomes larger inward, no new depressions appear, and there is no load position where the fold of the structure gradually becomes larger. Similar to the overall axial compression, depressions still occur in the structure from above and downwards, however, the depressions only occur near the axial compression load. Figure 9f is the state of the structure after the 8 mm forced displacement is applied, the depression finally closes and occurs at the location of the axial compression load, and the rest is folded deformation. It can be predicted that if the load continues to be applied to it, the depression deformation at the position where the forced displacement is applied and the fold at the position where the load is not applied will continue to increase, and eventually, all the deformation will coincide, and then new deformation or even structural fracture may occur.

Figure 10 depicts the postbuckling of the circular tube under a local axial compression load of 2 × 120°. Figure 10a shows the state of the structure when buckling is about to occur, and the location where a clear fold appears can be seen in the figure as being without pressure, Figure 10b shows the depressed deformation of the structure after the first buckling, Figure 10b–d show the second buckling of the structure, and the deformations are developing and tend to merge, Figure 10e shows the third buckling of the structure, however, the buckling deformation is smaller than the first two bucklings, and Figure 10f shows the state of the structure after 8 mm forced displacement is applied, as the depressions finally merge and occur at the locations of the axial compression load, while at the locations without axial compression load, the structure has a large inward fold, and the deformation of the structure is already very serious at this time. As can be seen from the postbuckling of 2 × 90° local axial compression, the development of buckling is very similar to the postbuckling of 2 × 90° local axial compression and folds appear in areas where no axial compression is applied before buckling occurs. When the fold continues to expand after buckling, depression deformation occurs at the location where the axial compression is applied.

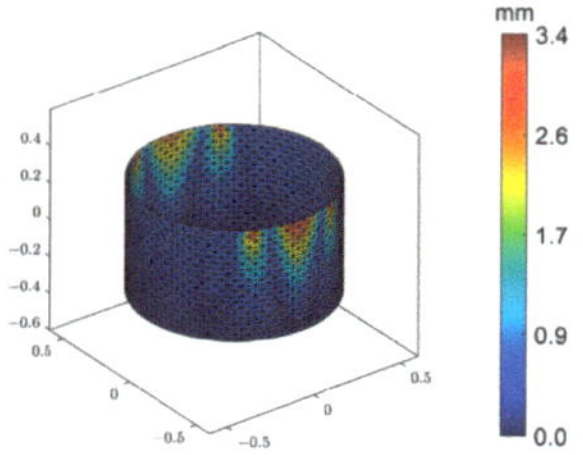

(a)

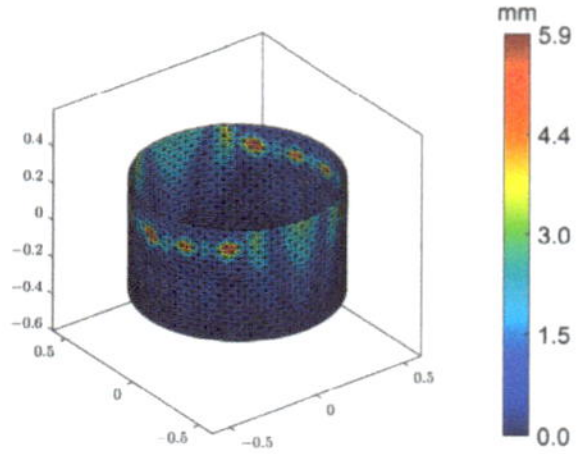

(b)

Figure 10. *Cont.*

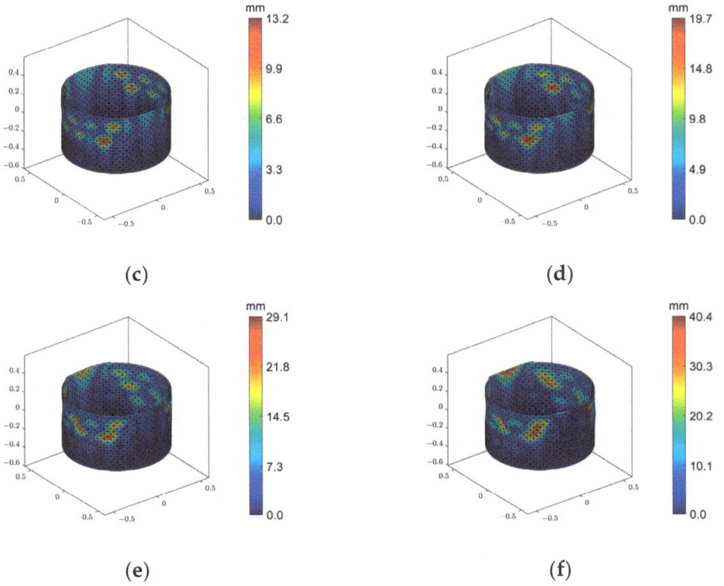

Figure 10. Postbuckling of 2 × 120° local axial compression: (**a**) before buckling, (**b**) first buckling, (**c**–**f**) second buckling.

Finally, the postbuckling behavior of overall axial compression is simulated as a control, and the postbuckling of the circular tube is achieved by applying a forced axial displacement of 8 mm downward from the end of the tube to the circular tube. Figure 11 is the displacement contour of the structure in the whole loading process after buckling, Figure 11a shows that the first buckling occurs just at the beginning, the element extrusion deformation first appears at the location where the depression is about to occur, Figure 11b depicts the gradual formation of the depression at the extrusion location to the inside of the tube, the deformation of the depression gradually expands and then the second buckling appears at Figure 11c and the third buckling at Figure 11d. At this time, the depression formed by the third buckling has covered the entire round tube. The deformation of the depression in Figure 11e gradually increases and eventually merges into a larger depression in Figure 11f, and as the buckling deformation continues to develop, contact between elements will appear. At this time, boundary nonlinearity needs to be introduced. The shell element adopted in this paper does not include boundary nonlinearity and fracture determination, so subsequent developments are not within the scope of this paper.

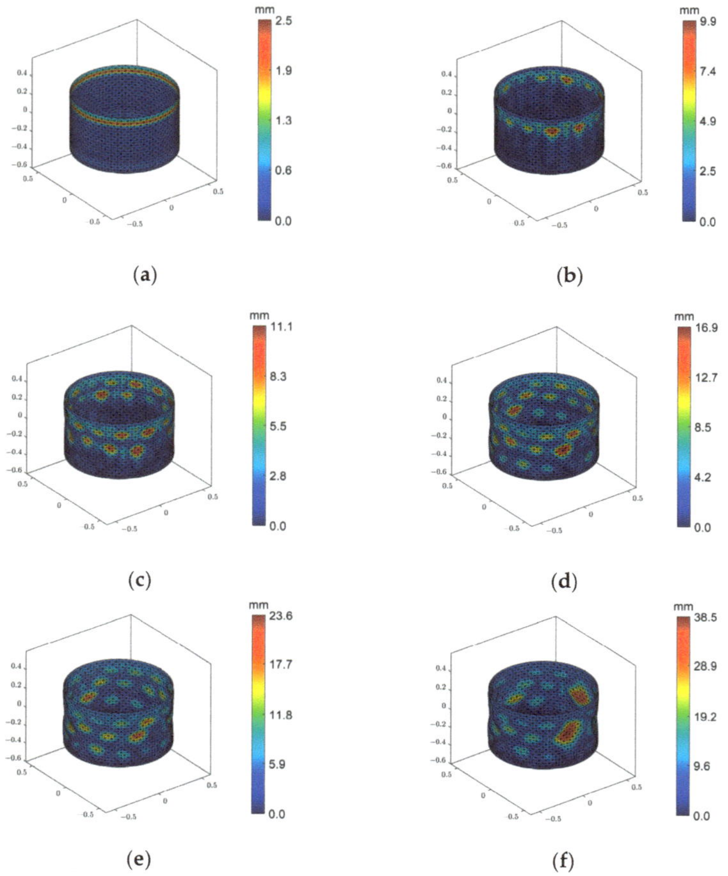

Figure 11. Postbuckling of the overall axial compression: (**a**) before buckling, (**b**) first buckling, (**c**) second buckling, (**d–f**) third buckling.

The three examples of postbuckling above show that a circular tube subjected to axial compression loading will deform in a top-down depression after buckling occurs. Compared with the overall axial compression, the deformation of the local axial compression circular tube is more irregular, the deformation of the region under load is mostly an inward sag, the region without axial compression will fold, and the fold will gradually increase with the axial displacement, resulting in the shape of the top of the circular tube becoming irregular, and as the mechanical characteristics of such irregular shapes are difficult to predict, such phenomenon should be avoided as much as possible in practical engineering.

5. Conclusions

Based on the vector form intrinsic finite element theory, this paper developed a program for calculating and analyzing the buckling of the structure, considering the multilinear hardening model, verified the multilinear hardening model according to the experimental data, and then simulated the buckling of a thin-walled circular shell under a local axial compression load. Compared with the experimental results, the postbuckling behavior of the thin-walled circular shell was as predicted. The following conclusions can be drawn:

(1) The multilinear hardening model based on the von Mises yield criterion can divide the elastic-plastic constitutive curve of materials into multiple broken lines and can approach the elastic-plastic constitutive curve of some materials infinitely closely. The stress–strain relationship of materials can be simulated accurately when the model is applied to VFIFE. Compared with the ideal elastoplasticity, bilinear hardening model, power hardening model, and Ramberg–Osgood model used in most numerical simulations, the structural strain under a given stress can be simulated more accurately.
(2) The VFIFE introduces the multilinear hardening model, which predicts the first buckling load of the local axial compression tube with high accuracy, and the error between the simulation and experiment in this paper is less than 1%. For the buckling mode, after the first buckling occurs, the results of the VFIFE simulation are more consistent with the experiment, which verifies the effectiveness and accuracy of VFIFE in predicting structural buckling.
(3) For the prediction of the postbuckling of structures, VFIFE has great advantages. Different from the complex settings of most commercial software based on FEM, VFIFE can directly predict the postbuckling behavior of structures under the action of force.
(4) When a circular tube is subjected to local axial compression loads, the smaller the local axial compression angle, the smaller the buckling load will be. If the load is in the form of overall axial compression, the depression deformation will appear uniformly along the circular tube ring, while if the load is in the form of local axial compression, the depression will occur at the location of the load, and the rest of the location of the fold, so that the geometry of the force side of the circular tube becomes irregular, affecting the mechanical properties of the structure. Such situations should be avoided in engineering wherever possible.

Author Contributions: Conceptualization, W.M. and Z.S.; Data curation, Z.S.; Funding acquisition, W.M. and L.X.; Methodology, W.M., Z.S., H.W., L.X. and Y.Z.; Project administration, W.M.; Software, Z.S. and L.X.; Supervision, W.M.; Validation, H.W., Y.Z. and G.H.; Visualization, Z.S. and H.W.; Writing—original draft, Z.S.; Writing—review & editing, W.M., Z.S., H.W. and Y.W. All authors have read and agreed to the published version of the manuscript.

Funding: This research was supported by the Science and Technology Project of Henan Province (Grant No. 192102310210); The Science and Technology Project of Henan Province (Grant No. 212102310951).

Data Availability Statement: Data will be available upon reasonable request.

Conflicts of Interest: The authors declare no conflict of interest.

References

1. Wagner, H.N.R.; Hühne, C. Robust knockdown factors for the design of cylindrical shells under axial compression: Potentials, practical application and reliability analysis. *Int. J. Mech. Sci.* **2018**, *135*, 410–430. [CrossRef]
2. Trivedi, N.; Singh, R.K. Fracture characterization studies of concrete structures through experiments on reinforced concrete cylindrical shell specimens. *Ann. Nucl. Energy* **2020**, *140*, 107338. [CrossRef]
3. Su, R.; Li, X.; Xu, S.-Y. Axial behavior of circular CFST encased seawater sea-sand concrete filled PVC/GFRP tube columns. *Construct. Build. Mater.* **2022**, *353*, 129159. [CrossRef]
4. Chung, C.-C.; Lee, K.-L.; Pan, W.-F. Collapse of Sharp-Notched 6061-T6 Aluminum Alloy Tubes under Cyclic Bending. *Int. J. Struct. Stab. Dyn.* **2016**, *16*, 1550035. [CrossRef]
5. Horrigmoe, G.; Bergan, P.G. Nonlinear analysis of free-form shells by flat finite elements. *Comput. Methods Appl. Mech. Eng.* **1978**, *16*, 11–35. [CrossRef]
6. Combescure, A.; Galletly, G.D. Plastic buckling of complete toroidal shells of elliptical cross-section subjected to internal pressure. *Thin-Walled Struct.* **1999**, *34*, 135–146. [CrossRef]
7. Spagnoli, A.; Chryssanthopoulos, M.K. Elastic buckling and postbuckling behaviour of widely-stiffened conical shells under axial compression. *Eng. Struct.* **1999**, *21*, 845–855. [CrossRef]
8. Abambres, M.; Camotim, D.; Silvestre, N. GBT-based elastic–plastic post-buckling analysis of stainless steel thin-walled members. *Thin-Walled Struct.* **2014**, *83*, 85–102. [CrossRef]

9. Kadkhodayan, M.; Maarefdoust, M. Elastic/plastic buckling of isotropic thin plates subjected to uniform and linearly varying in-plane loading using incremental and deformation theories. *Aerosp. Sci. Technol.* **2014**, *32*, 66–83. [CrossRef]
10. Song, C.Y.; Teng, J.G.; Rotter, J.M. Imperfection sensitivity of thin elastic cylindrical shells subject to partial axial compression. *Int. J. Solids Struct.* **2004**, *41*, 7155–7180. [CrossRef]
11. Changyong, S. Buckling of un—Stiffened cylindrical shell under non—Uniform axial compressive stress. *J. Zhejiang Univ. Sci.* **2002**, *3*, 520–531. [CrossRef]
12. Jiao, P.; Chen, Z.; Ma, H.; Ge, P.; Gu, Y.; Miao, H. Buckling behaviors of thin-walled cylindrical shells under localized axial compression loads, Part 1: Experimental study. *Thin-Walled Struct.* **2021**, *166*, 108118. [CrossRef]
13. Jiao, P.; Chen, Z.; Ma, H.; Ge, P.; Gu, Y.; Miao, H. Buckling behaviors of thin-walled cylindrical shells under localized axial compression loads, Part 2: Numerical study. *Thin-Walled Struct.* **2021**, *169*, 108330. [CrossRef]
14. Nassiraei, H.; Zhu, L.; Gu, C. Static capacity of collar plate reinforced tubular X-connections subjected to compressive loading: Study of geometrical effects and parametric formulation. *Ships Offshore Struct.* **2019**, *16*, 54–69. [CrossRef]
15. Nassiraei, H.; Rezadoost, P. Static capacity of tubular X-joints reinforced with fiber reinforced polymer subjected to compressive load. *Eng. Struct.* **2021**, *236*, 112041. [CrossRef]
16. Wu, C.; Lou, J.; He, L.; Du, J.; Wu, H. Buckling and Post-Buckling of Symmetric Functionally Graded Microplate Lying on Nonlinear Elastic Foundation Based on Modified Couple Stress Theory. *Int. J. Struct. Stab. Dyn.* **2018**, *18*, 1850110. [CrossRef]
17. Liu, X.; Zhang, J.; Di, C.; Zhan, M.; Wang, F. Buckling of Hydroformed Toroidal Pressure Hulls with Octagonal Cross-Sections. *Metals* **2022**, *12*, 1475. [CrossRef]
18. Peng, Y.; Kong, Z.; Dinh, B.H.; Nguyen, H.-H.; Cao, T.-S.; Papazafeiropoulos, G.; Vu, Q.-V. Web Bend-Buckling of Steel Plate Girders Reinforced by Two Longitudinal Stiffeners with Various Cross-Section Shapes. *Metals* **2023**, *13*, 323. [CrossRef]
19. Barham, W.S.; Idris, A.A. Flexibility-based large increment method for nonlinear analysis of Timoshenko beam structures controlled by a bilinear material model. *Structures* **2021**, *30*, 678–691. [CrossRef]
20. Zhang, C.; Yang, X. Bilinear elastoplastic constitutive model with polyvinyl alcohol content for strain-hardening cementitious composite. *Constr. Build. Mater.* **2019**, *209*, 388–394. [CrossRef]
21. Turkalj, G.; Lanc, D.; Brnic, J. Large displacement beam model for creep buckling analysis of framed structures. *Int. J. Struct. Stab. Dyn.* **2009**, *9*, 61–83. [CrossRef]
22. Xiao, G.; Yang, X.; Qiu, J.; Chang, C.; Liu, E.; Duan, Q.; Shu, X.; Wang, Z. Determination of power hardening elastoplastic constitutive relation of metals through indentation tests with plural indenters. *Mech. Mater.* **2019**, *138*, 103173. [CrossRef]
23. Lu, X.; Huang, F.; Zhao, B.; Keer, L.M. Contact Behaviors of Coated Asperity with Power-Law Hardening Elastic–Plastic Substrate During Loading and Unloading Process. *Int. J. Appl. Mech.* **2018**, *10*, 1850034. [CrossRef]
24. Chen, L.; Yu, Y.; Song, W.; Wang, T.; Sun, W. Stability of geometrically imperfect struts with Ramberg-Osgood constitutive law. *Thin-Walled Struct.* **2022**, *177*, 109438. [CrossRef]
25. Huang, Z.; Chen, Y.; Bai, S.-L. An Elastoplastic Constitutive Model for Porous Materials. *Int. J. Appl. Mech.* **2013**, *5*, 50035. [CrossRef]
26. Mourlas, C.; Khabele, N.; Bark, H.A.; Karamitros, D.; Taddei, F.; Markou, G.; Papadrakakis, M. Effect of Soil–Structure Interaction on Nonlinear Dynamic Response of Reinforced Concrete Structures. *Int. J. Struct. Stab. Dyn.* **2020**, *20*, 2041013. [CrossRef]
27. Ruocco, E.; Reddy, J.N. Buckling analysis of elastic–plastic nanoplates resting on a Winkler–Pasternak foundation based on nonlocal third-order plate theory. *Int. J. Non-Linear Mech.* **2020**, *121*, 103453. [CrossRef]
28. Zhou, W.; Shi, Z.; Li, Y.; Rong, Q.; Zeng, Y.; Lin, J. Elastic-plastic buckling analysis of stiffened panel subjected to global bending in forming process. *Aerosp. Sci. Technol.* **2021**, *115*, 106781. [CrossRef]
29. Zou, Z.; Liu, D.; Song, C.; Jin, M.; Guo, B.; Zhang, H. Elastic-Plastic Finite Element New Method for Lower Bound Shakedown Analysis. *Int. J. Struct. Stab. Dyn.* **2022**, *22*, 2250171. [CrossRef]
30. Shih, C.; Wang, Y.-K.; Ting, E.C. Fundamentals of a Vector Form Intrinsic Finite Element: Part III. Convected Material Frame and Examples. *J. Mech.* **2011**, *20*, 133–143. [CrossRef]
31. Ting, E.C.; Shih, C.; Wang, Y.-K. Fundamentals of a Vector Form Intrinsic Finite Element: Part I. Basic Procedure and A Plane Frame Element. *J. Mech.* **2011**, *20*, 113–122. [CrossRef]
32. Ting, E.C.; Shih, C.; Wang, Y.-K. Fundamentals of a Vector Form Intrinsic Finite Element: Part II. Plane Solid Elements. *J. Mech.* **2011**, *20*, 123–132. [CrossRef]
33. Wu, H.; Zeng, X.; Xiao, J.; Yu, Y.; Dai, X.; Yu, J. Vector form intrinsic finite-element analysis of static and dynamic behavior of deep-sea flexible pipe. *Int. J. Nav. Archit. Ocean Eng.* **2020**, *12*, 376–386. [CrossRef]
34. Wu, T.-Y.; Ting, E.C. Large deflection analysis of 3D membrane structures by a 4-node quadrilateral intrinsic element. *Thin-Walled Struct.* **2008**, *46*, 261–275. [CrossRef]
35. Wu, T.-Y. Dynamic nonlinear analysis of shell structures using a vector form intrinsic finite element. *Eng. Struct.* **2013**, *56*, 2028–2040. [CrossRef]
36. Zhen, W.; Yang, Z.; Xuelin, Y. Nonlinear behavior analysis of entity structure based on vector form intrinsic finite element. *J. Build. Struct.* **2015**, *36*, 133–140. [CrossRef]
37. Wu, T.-Y.; Tsai, W.-C.; Lee, J.-J. Dynamic elastic–plastic and large deflection analyses of frame structures using motion analysis of structures. *Thin-Walled Struct.* **2009**, *47*, 1177–1190. [CrossRef]

38. Zhen, W.; Yang, Z.; Xuelin, Y. Collision-contact, crack-fracture and penetration behavior analysis of thin-shell structures based on vector form intrinsic finite element. *J. Build. Struct.* **2016**, *37*, 53–59. [CrossRef]
39. Zhen, W.; Yang, z.; Xuelin, Y. Vector form intrinsic finite element method for buckling analysis of thin-shell structures. *J. Cent. South Univ. (Sci. Technol.)* **2016**, *47*, 2058–2064.
40. Zhen, W.; Yang, Z.; Xuelin, Y. Analisis of buckling behavior of planar membrane structures based on vector form intrinsic finite element. *J. Zhejiang Univ. (Eng. Sci.)* **2015**, *49*, 1116–1122.
41. Xu, L.; Lin, M. Analysis of buried pipelines subjected to reverse fault motion using the vector form intrinsic finite element method. *Soil Dyn. Earthq. Eng.* **2017**, *93*, 61–83. [CrossRef]
42. Yu, Y.; Li, Z.; Yu, J.; Xu, L.; Zhao, M.; Cui, Y.; Wu, H.; Duan, Q. Buckling analysis of subsea pipeline with integral buckle arrestor using vector form intrinsic finite thin shell element. *Thin-Walled Struct.* **2021**, *164*, 107533. [CrossRef]
43. Yu, Y.; Li, Z.; Yu, J.; Xu, L.; Cheng, S.; Wu, J.; Wang, H.; Xu, W. Buckling failure analysis for buried subsea pipeline under reverse fault displacement. *Thin-Walled Struct.* **2021**, *169*, 108350. [CrossRef]
44. The OpenMP API Specification for Parallel Programming. Available online: https://www.openmp.org (accessed on 1 June 2022).

metals

MDPI

Article

Effect of Quenching Parameters on Distortion Phenomena in AISI 4340 Steel

Ricardo Daniel Lopez-Garcia [1,*], Israel Medina-Juárez [2,*] and Araceli Maldonado-Reyes [1]

[1] Department of Mechanical Engineering, Tecnológico Nacional de México/Campus Cd. Victoria, Blvd. Emilio Portes Gil No. 1301 Pte., C.P. 87010 Cd. Victoria, Tamaulipas, Mexico; araceli.mr@cdvictoria.tecnm.mx

[2] Postgraduate and Research Department, Tecnológico Nacional de México/Campus Cd. Victoria, Blvd. Emilio Portes Gil No. 1301 Pte., C.P. 87010 Cd. Victoria, Tamaulipas, Mexico

* Correspondence: ricardo.lg@cdvictoria.tecnm.mx (R.D.L.-G.); israel.mj@cdvictoria.tecnm.mx (I.M.-J.)

Abstract: During quenching heat treatment, the formation of high residual stress values and the presence of distortion are phenomena which are difficult to control and accurately predict, their effects being extremely important to the components or pieces of complex and robust geometry that are commonly used in the industry. The latter is mainly due to the mixture of the high temperature levels formed between the surface and the cores of the components and the martensitic transformation during quenching. In this research, an experimental and simulated analysis of the process of the quenching heat treatment of AISI 4340 steel, using geometrically complex components, was undertaken with the objective of studying and understanding the effect of quenching process parameters on distortion, stress generation, and mechanical properties. A model that applied the finite elements method (FEM), in which entry data such as thermo-physical and mechanical properties were obtained through experimental techniques that were reported in the literature, made it possible to simulate the cooling process under different conditions, which helped to explain the origins of the distortion in the quenched parts. The results show a close relationship between various quenching parameters such as heat extraction speed, the immersion orientation in the liquid, and the component's geometry. The data obtained could contribute to accelerating the design process of the heat processing routes for quenching components by taking into consideration both the classic process variables and, due to the increased precision resulting from mathematical modeling, additional factors such as the geometry of real applications.

Keywords: distortion; quenching; FEM; residual stresses

Citation: Lopez-Garcia, R.D.; Medina-Juárez, I.; Maldonado-Reyes, A. Effect of Quenching Parameters on Distortion Phenomena in AISI 4340 Steel. *Metals* **2022**, *12*, 759. https://doi.org/10.3390/met12050759

Academic Editors: Shuwen Wen, Yongle Sun and Xin Chen

Received: 23 March 2022
Accepted: 22 April 2022
Published: 28 April 2022

1. Introduction

The purpose of the quenching heat treatment is to modify the initial microstructure of steel by means of cooling the heated parts through the austenitizing temperatures of different liquid quenchants, such as brine, water, or oil. The high cooling rate achieved during the quenching process suppresses diffusion-controlled phase transformations (ferrite, perlite, and bainite) and favors non-diffusional transformations such as martensite, a desirable phase in quenched steel which is responsible for reaching optimum levels of mechanical properties [1,2], followed by a complementary tempering heat treatment. Quenching heat treatment on steel adequately meets the requirements that the modern industry demands. This is mainly in the automotive and aerospace industries, where the strict control of the specifications of mechanical properties and residual stresses are required in order to reduce the possibility of failures and to increase the service life of various components with differing complex geometries. For this reason, distortion control plays a determining role in the dimensional precision of massive production parts when it comes to reducing waste.

Most of the rejection problems of hardened parts are related to the quenching process and caused by poor heat treatment design; because of this, in the metal–mechanical industry, optimal control of the quenching process becomes essential in reducing economic losses [3]. Quenching process design must consider the strict control of the involved process variables, such as the cooling rate, the orientation of the component during immersion, the agitation of the quenching media, the quenching bath temperature, and the component's geometry, all of which influence the dimensional accuracy of hardened parts [4]. The AISI 4340 alloy is a widely used heat-treatable steel because of its high fatigue strength and toughness. Therefore, this type of steel is widely used for the manufacturing of components with high demands for mechanical properties; such components include gears, bolts, torsion bars, and crankshafts, among others [5,6].

On an industrial level, the manufacturing of robust, long or geometrically complex parts is traditionally sensitive to the distortion effects present in the manufacturing process. Therefore, the control of distortion is a research topic widely studied by academics and researchers. The development of distortion not only involves the effect of volumetric expansion induced by the phase transformation (austenite–martensite) but also the complex interaction between different variables such as the austenitizing temperature, immersion velocity and direction, the temperature and agitation of the quenching media, and the geometry of the treated components, among other factors, which results in a complex phenomenon that is difficult to predict and control [7–10]. Due to the complex geometry and the non-homogeneous temperature distribution, the uncontrolled variables of the quenching process could cause the appearance of various undesirable phenomena such as high residual stresses, variations in dimensional precision, cracking and fractures that compromise the component's integrity during its service life [11], and the many other types of waste that can occur during the manufacturing process. Although a significant number of studies have been conducted on this topic, and their results have helped to improve processes in manufacturing industries, failures still occur in the quenching process when the parts or components are heavy or have complex geometries, as is the case with gears, molds, crankshafts, and springs, among many others [12].

In recent years, an important number of numerical tools, such as the finite element method (FEM), have been developed and can be used for the analysis and understanding of the behavior of materials under different processing conditions, and which currently allow a first approach in the design of new heat treatments, the evaluation of residual stress, and the understanding of distortion [13–18]. These investigations have focused on studying variables such as chemical composition, the heat transfer coefficient (HTC), austenitic grain size, component geometry, the temperature of the quenching bath, the agitation of the liquid quenchants, and the immersion direction and immersion velocity. However, despite numerous studies' attempts to understand the relationship between such variables, there are still questions related to their influence on the distortion phenomenon.

Currently, there is little information in the literature on the experimental validation of distortion formation during the quenching of geometrically complex steel parts [19], and these studies have focused on small parts with simple geometries that can be simplified to model and measure the generated distortion [20–25]. Some of these studies that used FEM to predict distortion did not perform experimental measurements [10,26,27].

In the present investigation, a study of the quenching process of AISI 4340 steel samples of complex geometry and also uses an FEM model is presented in order to predict the dimensional behavior, the evolution of internal stresses, and the mechanical properties obtained during the quenching heat treatment, in which the magnitude of the distortion was experimentally measured. The study focuses on the relationships between the immersion speed, the direction of immersion into the quenching media, the mechanical properties, the microstructure, the thermal history, and the final stress profile and distortion.

2. Materials and Methods

2.1. Materials and Experimentral Procedure

The material used in this research was an AISI 4340 steel of medium carbon content, of which its main alloyings were Cr and Ni. The chemical composition of the steel is presented in Table 1. For the quenching tests, probes were fabricated with a geometry similar to the one shown in Figure 1. There was a dimensional control of the specimens prior to and after the heat treatment with an interior gauge, a Starrett 700MA with a precision of 0.01 mm, for the purpose of quantifying the dimensional changes in the A, B, and C gaps of the specimens that were caused by the distortion during quenching; see Figure 1.

Table 1. Chemical composition of AISI 4340 steel.

Element	C	Cr	Mo	Ni	Mn	Si	S	P	Fe
% wt.	0.4	0.89	0.25	1.80	0.7	0.3	0.021	0.005	Balance

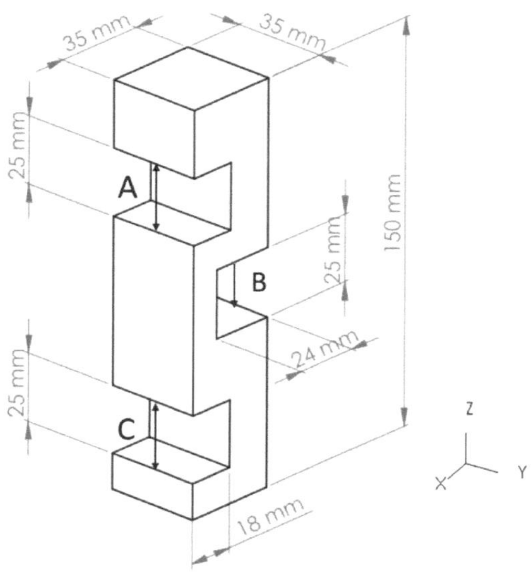

Figure 1. Geometry and dimensions of the part studied.

2.2. Heat Treatment Process

For all the quenching tests, the samples used were instrumented with type K thermocouples attached to their surfaces and connected to a data acquisition card with 10 channels: OMB-DAQ 54. The pieces were heated at a rate of 5 °C/min in a Thermolyne 3500 muffle-type furnace (Thermo Scientific, Waltham, MA, USA) until reaching an austenitization temperature of 860 °C; once reached, it was kept for 45 min to achieve a homogeneous austenitic microstructure. The quenching was performed in two liquid quenching media, oil and water. The commercial quenching oil Equiquench 770 of Equimsa brand was used at a temperature of 60 °C, while the aqueous media was tap water at a temperature of 25 °C. In both cases, the quenching media were kept in constant agitation by the action of a peripheral pump of $\frac{1}{2}$ HP, and the immersion speed used was 40 mm/s, controlled through a robotic arm with stepper motor controls; the conditions of the process are summarized in Table 2. The direction of the immersions occurred in the direction "-z" in the case of the quenching probes in a vertical mode and in the "x" direction for the quenching probes in a horizontal mode; see Figure 1.

Table 2. Quenching process conditions for AISI 4340 steel.

Sample	Temp. (°C)	Soaking Time	Quenching Media	Immersion Rate (mm/s)	Orientation
V-OQ	860	45	Oil	40	Vertical
H-OQ	860	45	Oil	40	Horizontal
V-WQ	860	45	Water	40	Vertical
H-WQ	860	45	Water	40	Horizontal

2.3. *Microstructural Analysis*

The microstructure of the heat-treated samples was analyzed using an optical microscope (Velab, Ecatepec, Mexico) with a prior standard metallographic preparation according to the norm ASTM E3 [28], while the microstructure was revealed using a 2% Nital etchant according to ASTM E407 [29].

2.4. *Hardness Evaluation*

The hardness of the treated specimens was measured with a TIME TH-500 model hardness tester (Time Group Inc., Beijing, China) in Rockwell C scale along an axial cut in the "z" axis.

3. Modeling of the Quenching Process

The simulation of the finite elements in the quenching process involved three main aspects: the heat transfer coefficient, the phase transformations, and the residual stresses plus the deformations that had to be properly taken into account [14,16,30,31].

The mathematical model presented in this research considered the relationships between the thermal phenomena, the microstructural changes, mechanical properties, and the residual stresses generated both by thermal origin stress and those due to the phase transformations. The research considered a component of complex geometry through FEM simulation; therefore, thermal gradients existed at the same instant in time. This temperature distribution in the component depends on factors such as the quenching severity, thermal conductivity, heat capacity and latent heat.

Modeling the phase transformations was considered through the evolution of the phase volume fraction during its solid-state transformation as a function of the cooling time and temperature; because of this, diagrams for time–temperature–transformation (TTT) were necessary in modeling. On the other hand, internal stresses produced in the material were calculated by elasto-plastic analysis that assumed small deformations in the part.

3.1. *Heat Transfer*

The quenching heat treatment can be defined as a transient heat conduction problem that involves all possible means of heat extraction (conduction, radiation, and convection). However, the effect of thermal radiation was not considered in the simulation model used because the treated piece was exposed to the environment for an extremely short time, and the loss of heat by radiation was not considerable; therefore, only the heat conduction and convection equations are described below, using the law of Fourier [32] in Equation (1):

$$q = -k\nabla T \quad (1)$$

where q is the heat flux, k is the thermal conductivity, and ∇T is the temperature gradient field inside the part. According to Fourier´s law, the heat conduction equation of the transient problem that contains the phase transformation can be defined using the conservation of energy balance in the rectangular coordinate system presented in Equation (2):

$$\frac{\partial}{\partial x}\left(k\frac{\partial T}{\partial x}\right) + \frac{\partial}{\partial y}\left(k\frac{\partial T}{\partial y}\right) + \frac{\partial}{\partial z}\left(k\frac{\partial T}{\partial z}\right) + q_v = \left(\frac{\partial T}{\partial t}\right)\rho C_p \quad (2)$$

where k is the thermal conductivity, T is the quenching part temperature, q_v is the heat generation rate from the steel phase transformations, is the material density, C_p is the specific heat capacity at a constant pressure, t is the time, and (x, y, z) are the rectangular coordinates.

To calculate the heat transmission by convection between the part's surface and the cooling media, the cooling equation of Newton is used, as described in Equation (3):

$$Q = hA\Delta T \tag{3}$$

where Q is the heat flux density, h is the heat transfer coefficient, A is the surface area of the part, and ΔT is the temperature difference between the part's surface and the quenching media.

3.2. Phase Transformation

Because phase transformations have a strong relationship with the thermal and mechanical behavior of the material, they should be considered when finite element simulations are used. In the first place, the temperature ranges where the phase transformations occur and which are limited by the critical temperatures must be defined. Commonly these temperatures can be calculated using time–temperature–transformation (TTT) diagrams or through analytical expressions.

TTT diagrams describe the relationship between the beginning and end of a transformation and indicate a transformed volume fraction during the isothermal process at different temperatures. The isothermal kinetic equation, known as the Johnson–Mehl equation [33], is a fundamental variable in the numerical simulation of thermal processes, although it cannot be directly applied to calculate the volume fraction during non-isothermal processes. Due to this restriction, the Avrami equation was proposed, which has been widely used in these type of processes [34], Equation (4):

$$\xi = 1 - exp(-bt^n) \tag{4}$$

where ξ is the volume fraction of the new phase, t is the isothermal time duration, b is a temperature, chemical composition of parent phase, and grain size dependent constant, and n is a constant dependent on the type of phase transformation, varying from 1 to 4.

In the case of displacive transformation (martensite), there is a stage of nucleation and growth; however, the growth rate is so high that the volume of transformation of the phase is almost entirely controlled by nucleation, and as a result, its transformation kinetics are not influenced by the cooling speed. Because of this, it cannot be explained by Avrami's equation. Therefore, the amount of martensite formed is calculated using the equation established by Koistinen and Marburger [35], Equation (5):

$$\xi = 1 - exp[-\alpha(M_s - T)] \tag{5}$$

where ξ is the martensite transformed volume fraction, T is the temperature, M_s is the martensite transformations beginning temperature, and α is a constant that indicates the transformation rate and depends on the chemical composition of steel. It is important to mention that through the TTT and CCT diagrams it is possible to obtain the martensitic transformation temperature at a critical cooling rate.

3.3. Mechanical Interactions

The formation of residual stresses during quenching can occur in different manners—high temperature gradients, martensitic transformations, or the combination of both. In the first, the differences in temperature between the surface and the core of the part cause the surface to cool faster than the core, and therefore a volume contraction of the part begins on the surface with the presence of tension from the residual stresses, while in the core, to balance the entire part state of the residual stresses, there should be compression. In the second case, when a martensitic transformation is involved, stresses appear immediately after the martensitic transformation occurs on the surface of the piece, causing compressive residual stresses and tension-types in the core. The end of the residual stress behavior finishes as soon as the martensitic transformation occurs in the core of the part; at this point, the surface of the part is completely transformed and has reached room temperature [36]. Assuming that steel behaves like a thermo-elasto-plastic material, the total strain rate on the steel during quenching can be expressed in terms of the five deformation sources in Equation (6).

$$\dot{\varepsilon}_{ij} = \dot{\varepsilon}_{ij}^{e} + \dot{\varepsilon}_{ij}^{p} + \dot{\varepsilon}_{ij}^{th} + \dot{\varepsilon}_{ij}^{pt} + \dot{\varepsilon}_{ij}^{tr} \tag{6}$$

where $\dot{\varepsilon}_{ij}^{e}$, $\dot{\varepsilon}_{ij}^{p}$, $\dot{\varepsilon}_{ij}^{th}$, $\dot{\varepsilon}_{ij}^{pt}$, $\dot{\varepsilon}_{ij}^{tr}$ terms are the elastic, plastic, thermal, phase transformation, and plasticity transformation strain rates, respectively. Equations (7)–(11) are used individually in the simulation model to calculate the addition of deformations due to the different physical origins considered in this study.

$$\dot{\varepsilon}_{ij}^{e} = \frac{1}{E}\left[(1+v)\sigma_{ij} - \delta_{ij}v\sigma_{ij}\right] \tag{7}$$

$$\dot{\varepsilon}_{ij}^{p} = d\lambda\frac{\partial\varphi}{\partial\sigma_{ij}} \tag{8}$$

where E, v, σ_{ij}, $d\lambda$, δ_{ij}, and φ are the elastic modulus, Poisson´s ratio, Cauchy stress tensor, the plastic multiplier, Kronecker delta, and the yield functional using temperature, respectively.

$$\dot{\varepsilon}_{ij}^{th} = \sum_{\kappa=1}^{p} \zeta_{\kappa} \int_{0}^{T} \alpha_{\kappa} dT \tag{9}$$

where α_{κ} is the termal expansión coefficient of the phase κ, and ζ_{κ} is the volume fraction of phase κ.

$$\dot{\varepsilon}_{ij}^{pt} = \sum_{\kappa=1}^{p} \frac{1}{3}\delta_{ij}\Delta_{\kappa}\zeta_{\kappa} \tag{10}$$

where Δ_{κ} is the structural dilation due to the phase transformation.

$$\dot{\varepsilon}_{ij}^{tr} = \frac{3}{2}K_{\kappa}\dot{\zeta}_{k}(1-\zeta_{\kappa})S_{ij} \tag{11}$$

where K_{κ} is a constant due to the transformation-induced plasticity (TRIP), $\dot{\zeta}_{k}$ is the transformation rate of the phase κ, and S_{ij} is the stress deviator tensor.

Once the previous sources of deformation have been established, the internal stresses can be reconstructed using Hooke's law for isotropic materials.

Hardness values can be calculated using a simple rule of mixtures, assuming a constant hardness value for each phase.

$$\overline{H} = f_1(H_1) + f_2(H_2) + \cdots f_n(H_n) \tag{12}$$

where $\overline{H}$ is the weighted average of hardness in any element of the simulated geometry, $H_{(1-n)}$ are typical values of hardness for each phase, and $f_{(1-n)}$ values are the calculated volume fraction for each present phase.

3.4. Finite Element Simulation Conditions

The quenching process was simulated using the finite elements method (FEM). Table 3 summarizes the simulation parameters used. The compute domain of geometry is shown in Figure 2, representing a tridimensional model with 165,000 tetrahedral elements and 36,500 nodes, which was chosen so the results did not depend on the refinement degree of the mesh. The input data of the thermal and transformation properties of the material can be appreciated in Figure 3; these thermophysical properties of the material were the ones available in the literature for AISI 4340 steel. The model considered the thermal interactions, the formation of internal stress, the elasto-plastic deformation, and the microstructural evolution.

Table 3. Simulation parameters.

Simulation Parameters	Value (s)
Number of simulation steps	800
Number of elements	165,000
Number of nodes	36,500
Initial temperature of the nodes (°C)	860
Environment temperature (°C)	25
Quenching oil temperature (°C)	55
Cooling water temperature (°C)	25
Immersion rate (mm/s)	40
Immersion direction	−Z and X
Iteration method	Newton–Raphson
Heat transfer coefficient	$f(T)$, shown in Figure 3e,f

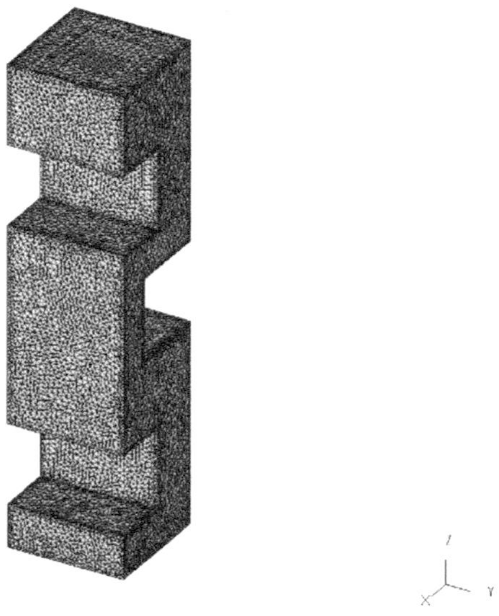

Figure 2. Mesh used in FEM simulation.

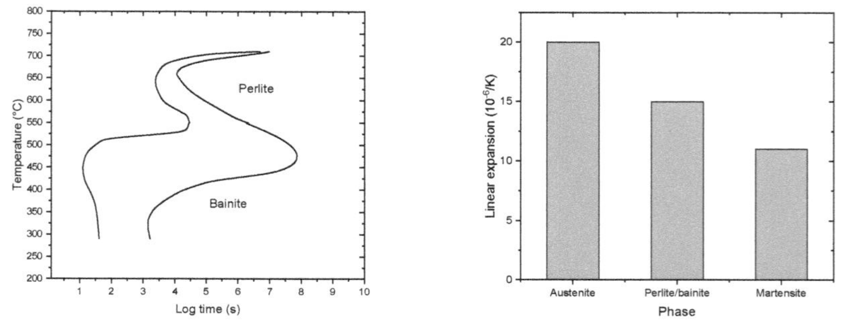

(**a**) (**b**)

Figure 3. *Cont.*

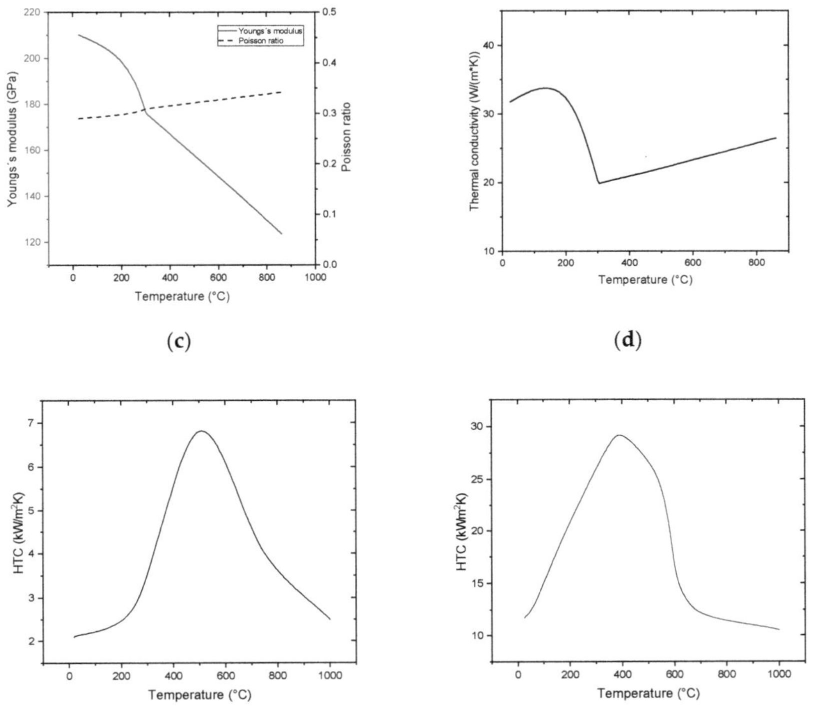

(c) (d)

(e) (f)

Figure 3. Thermophysical properties for AISI 4340 steel used in FEM simulation: (**a**) TTT diagram [37]; (**b**) thermal expansion [38]; (**c**) Young´s modulus and Poisson's ratio [38]; (**d**) thermal conductivity [38]; (**e**) HTC (oil) adapted with permission from [39] 2022, Elsevier; and (**f**) HTC (water) [40].

4. Results and Discussion

4.1. Thermal Analysis

In Figure 4, the experimental results of the temperature evolution during the cooling on the surface of the quenched parts from 860 °C until reaching the quenching media temperature are shown; oil and water with temperatures of 55° and 25°, respectively, in the two directions of immersion, vertical and horizontal.

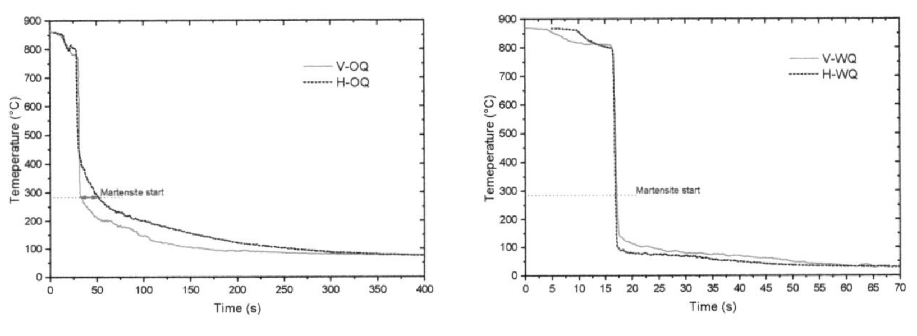

(a) (b)

Figure 4. Thermal history of studied pieces in tow immersion modes: (**a**) oil quenching and (**b**) water quenching.

At the beginning of the cooling for all the studied cases, it was observed that there was a downfall in the cooling curve temperature of 860 °C until 770 °C, and this downfall was related to the stage of convection cooling followed by a severe cooling caused by the immersion into the used liquid. Just as expected, the water-quenched samples showed an extremely rapid decrease in temperature, quickly reaching the martensitic transformation, as shown in Figure 4b, while the time required for the samples to reach thermal equilibrium with the quenching media was 70 s in the case of the quenching in oil, with the time extending asymptotically until 400 s, as shown in Figure 4a. Additionally, it could be observed that the relative cooling times of the oil immersion mode were reduced when the specimen was submerged in a vertical position (V-OQ), which is explained by the austenite–martensite transformation beginning before expected and by the transformation being completed in a shorter span of time for all the specimens. However, for the horizontal immersion (H-OQ), this timing was delayed for an approximated period of 30 s to begin the martensite transformation (see Figure 4a). Figure 5 shows the thermal history and the maximum cooling rates reached in each stage until the completion of cooling for all the studied conditions. Figure 5a,b represents the oil immersion conditions of the tested specimens, the maximum cooling rates of 55° and 45 °C/s for the vertical direction immersion (V-OQ) and horizontal (H-OQ), respectively. On the other hand, the maximum cooling rate was observed in the vertical immersion position, with water as the cooling medium. The maximum cooling rates were 150 and 250 °C/s for the vertical direction immersions (V-WQ) and horizontal (H-WQ), respectively; see Figure 5c,d. Cooling rates in that order of magnitude typically cause problems such as cracking or fracture of the samples because of the high temperature gradient formed at different points in the component.

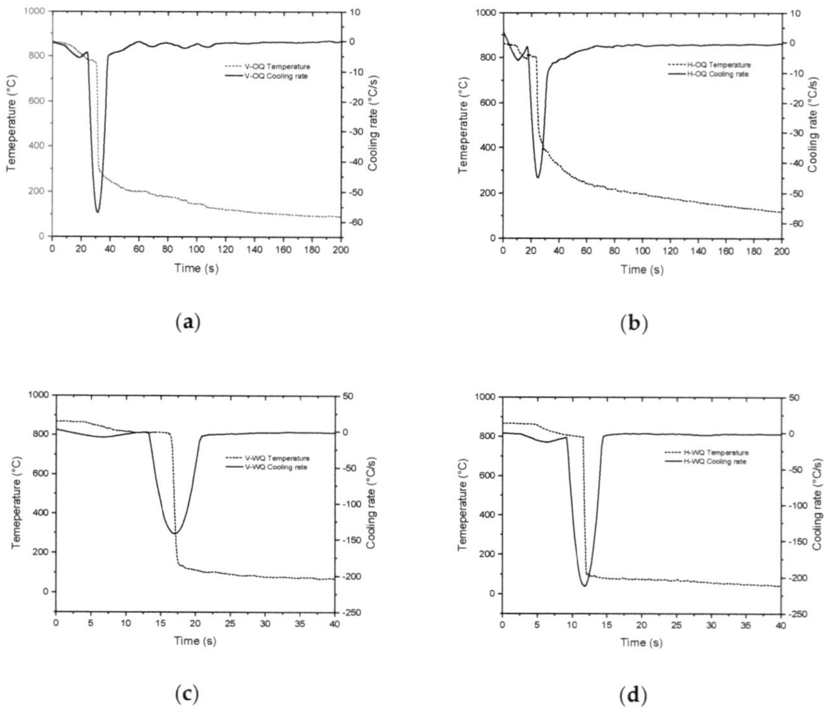

Figure 5. Maximum cooling rates in quenched AISI 4340 steel samples: (**a**) V-OQ; (**b**) H-OQ; (**c**) V-WQ; and (**d**) H-WQ.

During the cooling of the quenched samples, it was desirable that the temperature changes along the samples occured in a uniform way between the surface and the core in order to avoid the formation of high thermal gradients, which can be the origin of high stress of thermal origin. Even though the formation of residual stresses depends on a complex interaction among the diverse thermal–physical–mechanical phenomena, for the specific case of the quenching treatment, the homogeneous distribution of temperature along the component is a critical condition. Because of this, it becomes an issue of great interest to evaluate the temperature gradient intensity formed between the surface and core zones during the cooling, which can partially inform the stress profiles found through the mathematical model at the end of the heat treatment and in the presence of the diverse unwanted phenomena of excessive distortion and the occurrence of fractures. These thermal gradients were evaluated in six positions (surface and core) during the whole cooling stage for all the simulated cases, and the results are presented in Figure 6. The results show that the highest thermal gradients were present when water was used as the quenching medium; see Figure 6c,d. Values of $\Delta T \approx 650$ °C were found between the surface and the core of the piece, showing non-homogeneous temperature profile behavior due to the fast heat extraction in its surface and a heat extraction slower in the core, increasing the possibility of the presence of cracks and subsequent fractures. Figure 6a,b shows the thermal gradients in the samples quenched in oil as the quenching medium. It can be observed that the lowest thermal gradients, $\Delta T \approx 350$ °C, are found on the points (P3 vs. P6) which corresponded to the thinnest zone of the piece, while in the other studied points, the maximum values of ΔT were ≈ 440 °C, increasing the probability of a major concentration of thermal origin stresses and the presence of distortions that compromised the dimensional precision of the pieces and caused the generation of cracks and fractures in real quenched components.

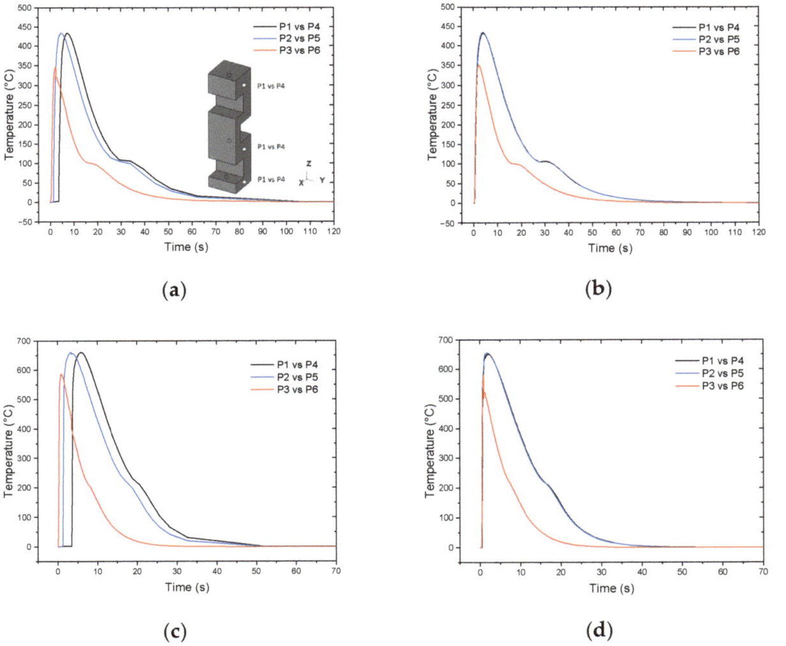

Figure 6. Temperature gradient (ΔT) between surface and core points during cooling of AISI 4340: (**a**) VOQ; (**b**) H-OQ; (**c**) V-WQ; and (**d**)H-WQ.

The cooling profiles obtained experimentally were compared with the simulation results and are presented in Figure 7. It is evident that the model could predict the necessary time for the piece to reach the thermal balance with the quenching media; however, the model needs to be perfected to obtain the same cooling rates in the inferior parts of the cooling curves corresponding to the convection cooling stage. In a general way, the implemented model overestimated the cooling rates in the oil cases and underestimated the cooling rates in the water quenched case estimations.

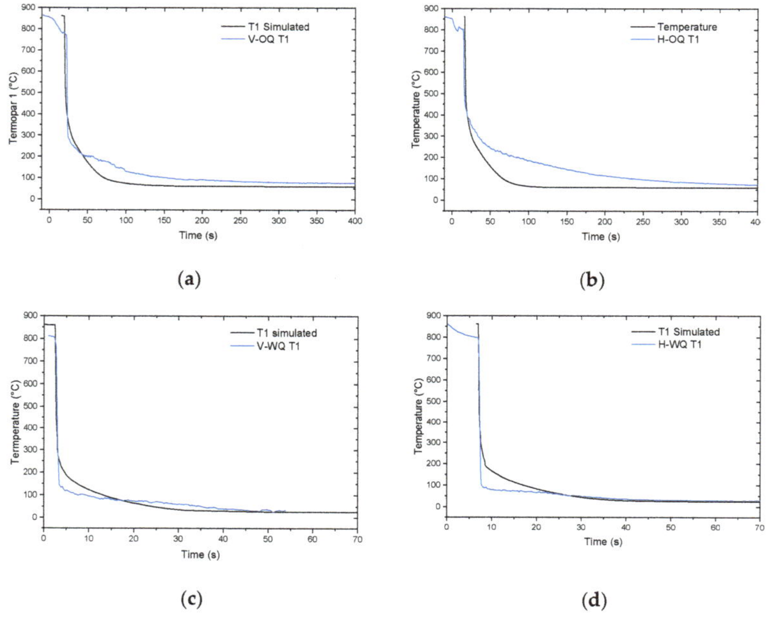

Figure 7. Comparison between simulated cooling curves and those measured with thermocouples in different quenching media and immersion modes: (**a**) V-OQ; (**b**) H-OQ; (**c**) V-WQ; and (**d**) H-WQ.

4.2. Microstructure and Hardness

Figure 8 shows images through optical microscopy of the obtained microstructure from the quenched AISI 4340 steel in oil and water and two immersion directions. In all studied cases, an almost complete martensite transformation could be observed, which was expected according to the steel chemistry and the used quenching conditions.

After the microstructural characterization, a simulation analysis through FEM modeling was made to compare the results of the austenite–martensite transformation during the cooling stage. According to the results shown in the histograms in Figure 9a,b, which represents the martensite distribution in all the 36,500 nodes of the model, 93% transformed martensite was obtained for the quenching condition in oil; these results are similar for the vertical immersion condition (V-OQ) and horizontal condition (H-OQ). Moreover, with the quenching treatment using water as the quenching medium, the martensite transformation reached 97% in both vertical immersion (V-WQ) and horizontal immersion modes (H-WQ). In addition to the histograms in Figure 9, for the distribution of martensite in the pieces studied at the end of the modeled cooling for the different studied conditions of immersion, see Figure 10. It can be seen that despite the similarities in the amount of final martensite transformation for both immersion cases, the most homogeneous distribution was found in the vertical immersion cases without consideration of the quenching media, indicating a more uniform cooling under this condition when compared to the cooling in a horizontal position that generates less symmetrical martensite distributions.

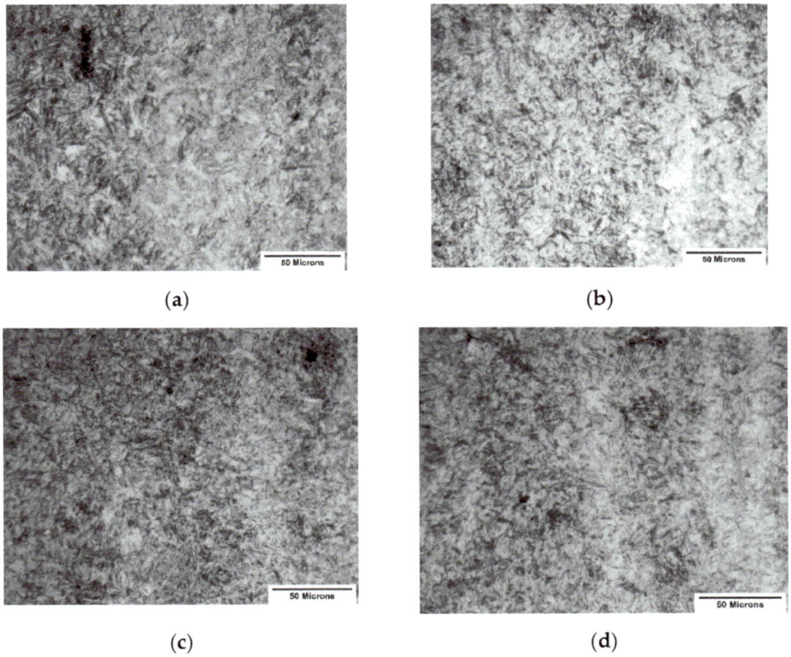

Figure 8. Obtained microstructure after quenching treatment of AISI-4340 steel at 500×: (**a**) V-OQ; (**b**) H-OQ; (**c**) V-WQ; and (**d**) H-WQ.

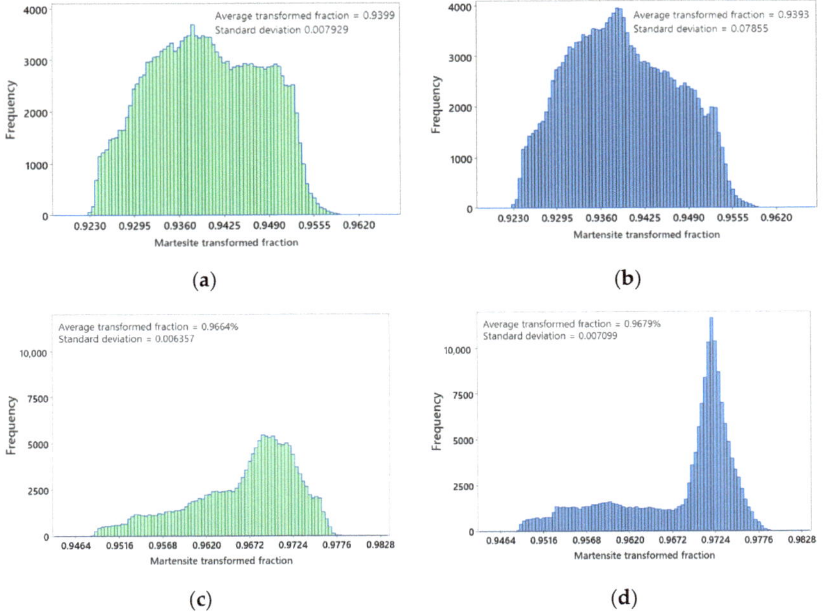

Figure 9. Nodal transformed fraction in quenched samples: (**a**) V-OQ; (**b**) H-OQ; (**c**) V-WQ; and (**d**) H-WQ.

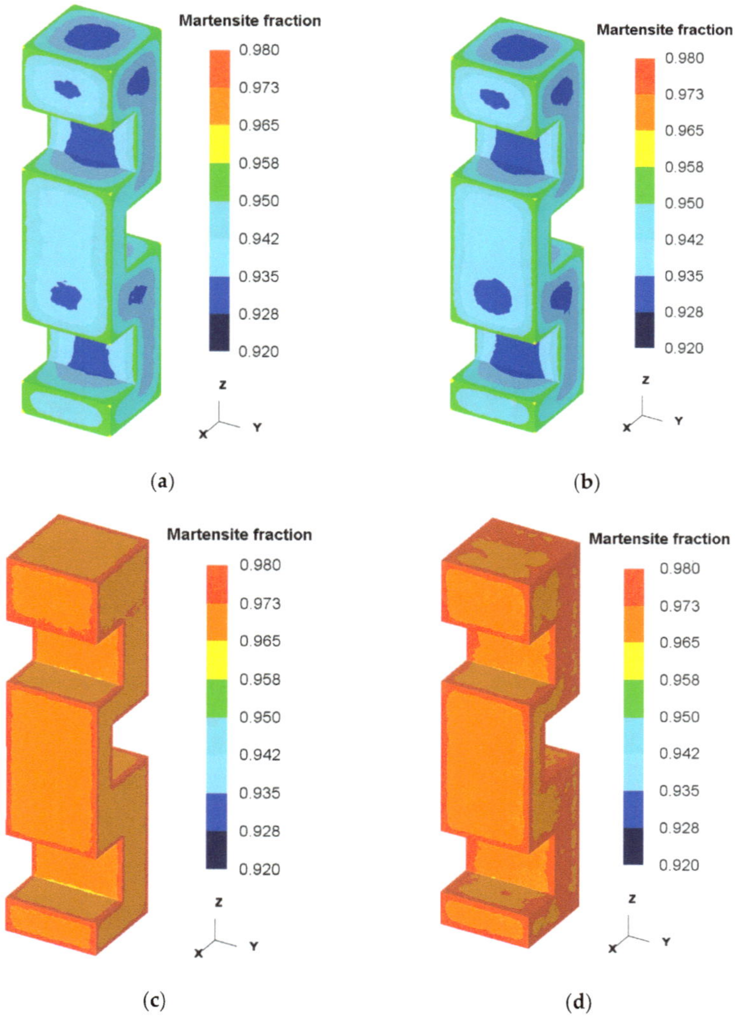

Figure 10. Martensite distribution in samples: (**a**) vertical oil quench; (**b**) horizontal oil quench; (**c**) vertical water quench; and (**d**) horizontal water quench.

The difference between the results lies mainly in the quenching media; when water was used, the quenching severity was more aggressive and a major cooling rate was obtained, promoting the austenite–martensite phase transformation to occur early in most nodes in the simulation model. By contrast, for the oil quenched cases, the cooling rate was minor, and the martensite complete transformation was affected by variables such as the oil temperature, agitation of the liquid, and the rate and the immersion direction of the pieces.

The hardness of the quenched pieces was measured in a longitudinal section along the axis "z" in the coordinate system. Figure 11 shows the average hardness in three zones in the studied samples of AISI 4340 steel under different immersion modes in oil and water. The water-cooled pieces showed more hardness in both immersion modes in the three analyzed zones, and the hardness levels obtained were 57.3–59.1 HRC. By contrast, for the oil cooling and the vertical immersion conditions, a major hardness was obtained with hardness magnitudes between 51.9 and 54 HRC. On the other hand, the quenching condition in oil and horizontal immersion showed the lowest hardness magnitude for the experiments performed. The hardness results gobtained through FEM showed a more uniform distribution compared with the experimental results for all the types of immersion and quenching media, and the latter can be observed in Figure 12, which shows the hardness distribution of the 36,500 nodes that formed the mesh through histograms. These hardness fluctuations could be explained by the variations in the transformation kinetics used in the model in addition to the HTC used, which could be optimized using the experimental data collected in this research for future works.

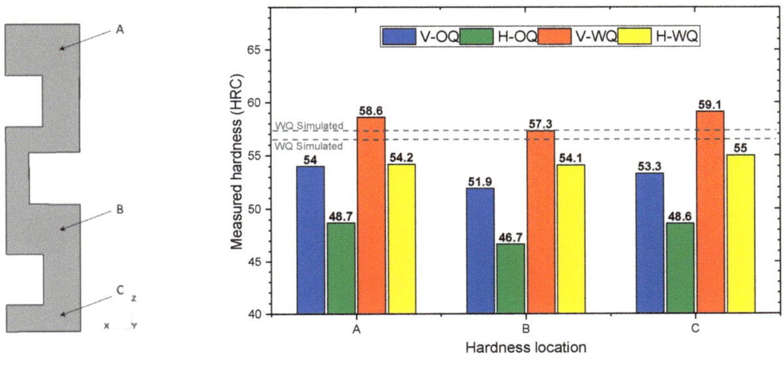

(a) (b)

Figure 11. Measured and calculated hardness in quenched samples: (**a**) measured zones and (**b**) experimental hardness results.

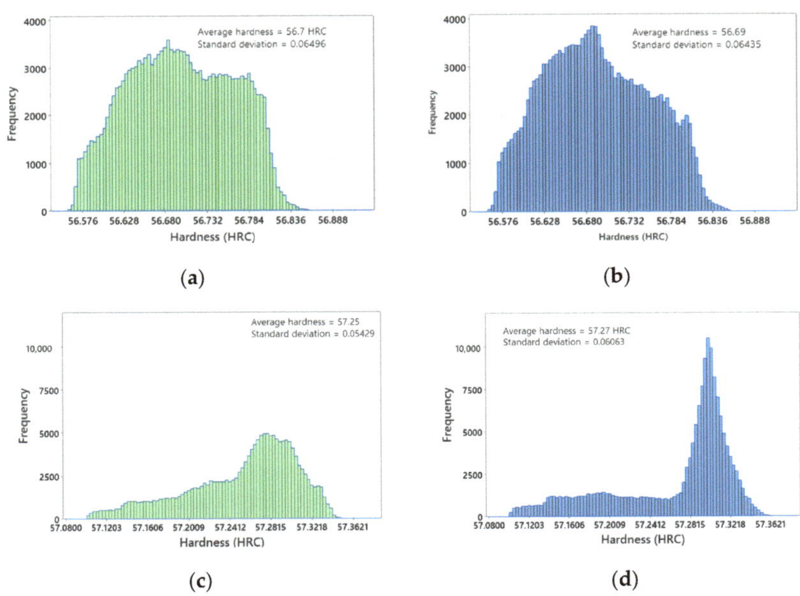

(a) (b) (c) (d)

Figure 12. Nodal hardness distribution in entire FEM model: (**a**) V-OQ; (**b**) H-OQ; (**c**) V-WQ; and (**d**) H-WQ.

4.3. Distortion

Figure 13 shows the distortion generated the studied samples after the quenching treatment measured in a quantitative way in three zones identified as: A, B, and C, which correspond to the nominal 25 mm gaps prior to the heat treatment (see Figure 1).

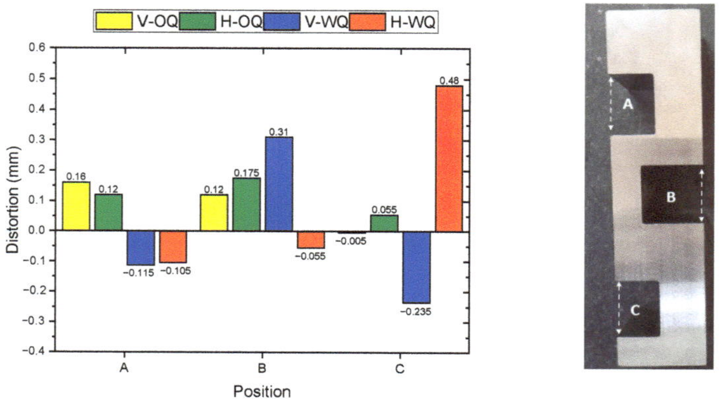

(**a**) (**b**)

Figure 13. (**a**) Measured distortion after quench and (**b**) measured gaps in geometry.

It can be seen that the major distortion was presented in the water-quenched samples in the direction of horizontal immersion (H-WQ), and the magnitude of the measured distortion was ≈ 0.48 mm, while the minor distortion was found for the case of the oil-quenchied samples and when the piece enters the quenching media in a vertical direction (V-OQ), with a magnitude of just |0.005| mm. In both cases, these values occurred in the thinnest region of the piece, identified as zone C. Furthermore, it was observed that for the quench in water and in the direction of the vertical immersion (V-WQ), a distortion was obtained in the range of 0.23–0.31 mm for the gaps B and C, while for A it corresponded to the thickest region of the piece and presented the lowest distortion levels, regardless the cooling media used. In a general way, the tendency to increase the magnitude of the distortion was presented when using the direction of horizontal immersion and using water as the quenching medium, while when using oil, which is the less severe quenching medium, and a vertical direction immersion, the decrease of the distortion effect is favored.

In contrast, the distortion values obtained in the simulation presented in Figure 14 are in good agreement with the experimental results in the V-OQ and H-OQ specimens. Less consistent values are found with specimen V-WQ due to the difference found in the gap "A" that shows an inverted distortion with respect to the corresponding experimental result. Finally, the H-WQ specimen was the one with the worst concordance with the experimental results. This difference between results shows the difficulty in modeling processes with many variables, and in this case, even the slightest deviation during immersion in the quenching medium creates an inadequate value for the heat transfer coefficient due to the different convection conditions generating such resultant discrepancies.

The austenite to martensite phase transformation is accompanied by a volumetric expansion and promotes the formation of internal stresses, and similarly, the thermal gradients during the cooling process produce the correspondent thermal origin stresses. In such a way, the dimensional change in the geometry of the studied pieces was related to the combined effect of these phenomena. Both are difficult to avoid; however, these are possible to predict and control through the design of a quenching system that allows the manipulation of each one of the critical variables along with numeric techniques, such as FEM models, which make the design of heat treatments as efficient and controllable as possible.

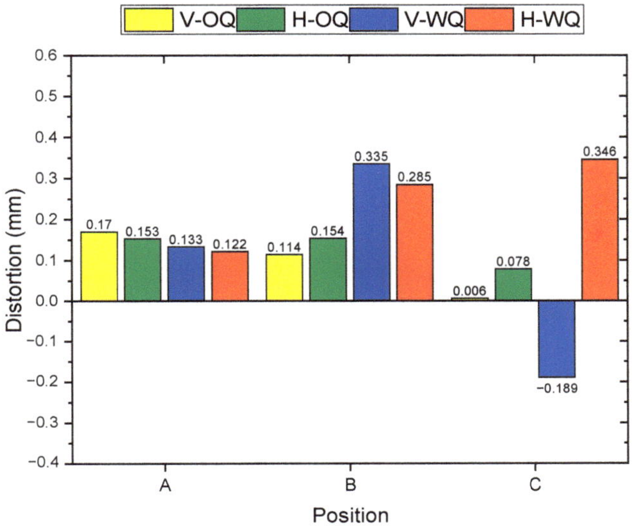

Figure 14. Simulated distortion after quench.

4.4. Simulated Residual Stresses and Precense of Cracks

The accumulation of internal stresses of thermal origins due to the high temperature gradients is added to the stresses generated by the volumetric expansion during the martensitic transformation, generating not only a decrease in the dimensional precision due to distortion but also, if these stresses overcome the yield stress, cracks or fractures in the material may occur. That is the reason why it becomes important to study the evolution of these stresses to know their magnitude and distribution related to cooling time. Figure 15 shows the stress distribution in the pieces at the end of the simulation, pointing out the residual stress concentrators obtained in each case studied.

Figure 15 shows the stress concentration along the geometry of the piece for all the studied cases. It can be appreciated that the vertical quenching modes produced more minor stress values than the corresponding horizontal quenching cases. Showing the mode V-OQ, Figure 15a shows minor stresses at the end of the quenching. The mode H-WQ, in Figure 15d, is the sample that reached stress values on the order of 530 MPa. This last value was far from overcoming the elastic limit of AISI 4340 steel; however, analyzing the stress history along the cooling time, peak values can be found very close to the yield limit of the material, at least in the water-quenched simulated samples.

In Figure 16 are shown the maximum values of residual stresses related to the simulation time in different zones of the studied pieces in both directions of immersion for water-quenched samples. Figure 16a represents the vertical immersion condition when the maximum effective stresses were reached at 10 s after beginning the quenching process, while, for the horizontal direction, the average time after the immersion started was 7 s; see Figure 16c. This difference in times is caused by the fact that in the horizontal immersion mode, the time for the total immersion of the sample is less than that for the the horizontal way, exposing a greater heat transfer area that accelerates the cooling process of the entire piece in relation to the vertical immersion, and reaching major cooling rates; see Figure 16c,d. In Figure 16a,b (V-WQ), it can be observed that there are a few critical zones that can compromise the integrity of the entire piece, and the magnitudes of the stresses in these zones are found between 800 and 1100 MPa; in contrast, observe zone P4, where even though it underwent a quick cooling process, it was not exposed to a sustained thermal gradient such as that in zones P1–P3, which developed high values for stress. On the contrary, the horizontal mode immersion in Figure 16c,d, shows a major number of zones with elevated stress values along the whole piece due to the immersion condition. The magnitudes of such stresses are in the range of 750–1100 MPa and are found in all the geometrical concentrators (sharp edges corners), unlike the vertical immersion case.

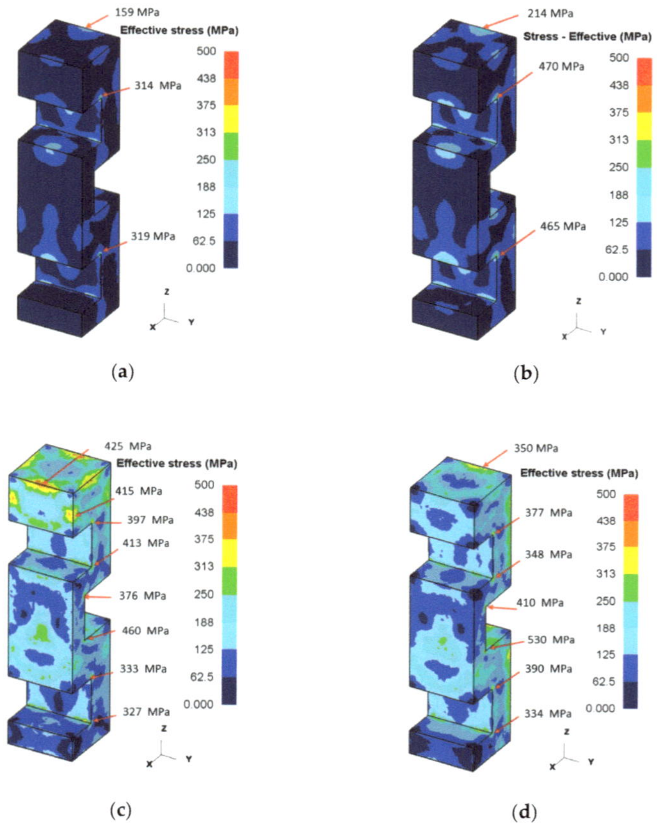

Figure 15. Final stress profiles after quench: (**a**) vertical oil quench; (**b**) horizontal oil quench; (**c**) vertical water quench; and (**d**) horizontal water quench.

Most of the steels are susceptible to fragilization when they are exposed to severe temperature changes, such as those in the quenching process. In this research, the formation of cracks in the water-quenched samples was detected for both modes of immersion, and this effect was not observed in the oil-quenched samples. In Figure 17 are shown some images of fractures found in a full section cut along the "z" axis during the metallographic preparation of the pieces; in both cases, they were of the intergranular type, finding their origins in the surface with a propagation direction towards the core of the samples.

The appearance and location of the cracks could be explained from the previously exposed results in which the evolution of the residual stresses are related to the occurrence time, therefore reaching values close to the yield limit of AISI 4340 steel in the quenched condition, as reported by Li [41]. It is possible that the real stresses in the pieces are bigger than the calculated values by the FEM model; however, such simulated stresses predict the zones of the piece where the material is prone to fail and correspond to the zones of high thermal gradients and geometric concentrators. Because of this, the accuracy of the model is able to be improved in future works, including the experimental measurement data of residual stresses, the mechanical properties, and heat transference coefficient optimization.

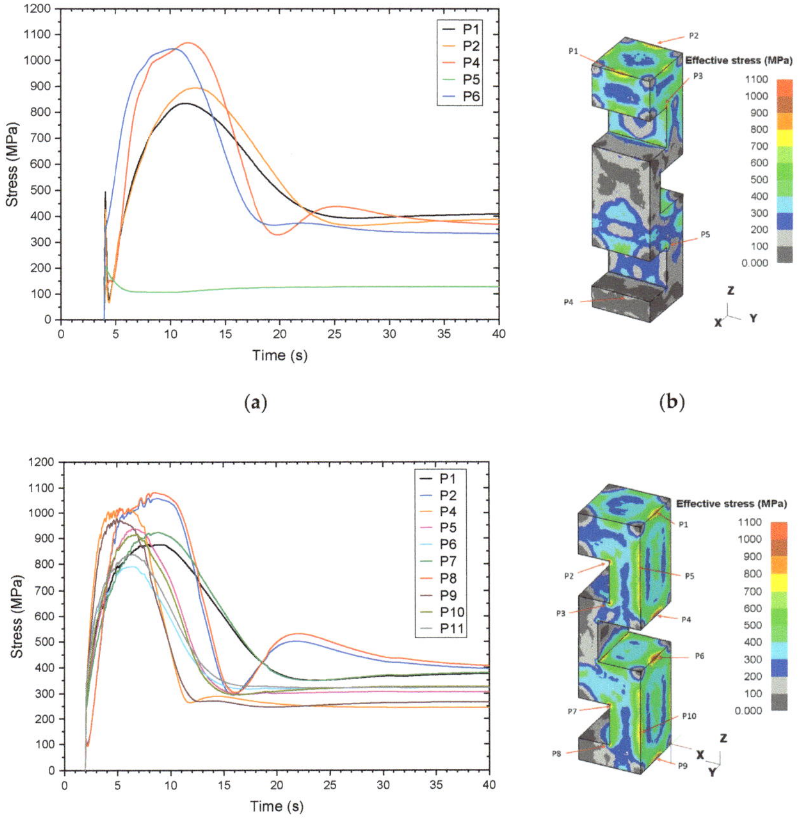

(c) (d)

Figure 16. Stress evolution during quenching: (**a**) internal stress vs quenching time in V-WQ sample; (**b**) stress profile at 10 s after quenching starts in V-WQ sample; (**c**) internal stress vs. quenching time in H-WQ sample; (**d**) stress profile at 7 s after quenching starts in H-WQ sample.

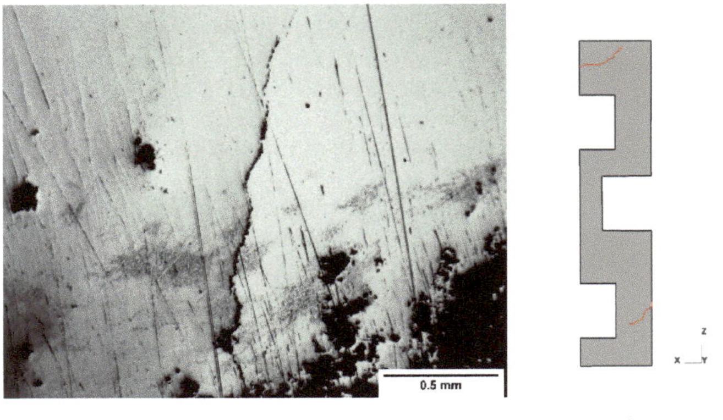

(a) (b)

Figure 17. *Cont.*

(c) (d)

Figure 17. Developed cracks during water quenching: (**a**) V-WQ sample crack; (**b**) location of V-WQ crack; (**c**) H-WQ sample crack; and (**d**) location of H-WQ crack.

5. Conclusions

Quenching heat treatment was characterized under controlled conditions relying on an FEM model which is able to reproduce the thermal history, phase transformations, hardness, and the formation of residual stresses during the simulated quenching process of a geometrically complex piece of AISI 4340 steel. The direction of immersion and the cooling media used were the studied variables. Considering the obtained results, it can be concluded:

- The highest magnitude of thermal gradients was present when using water as the cooling medium and the direction of horizontal immersion, increasing the presence of high values of stress with thermal origin and decreasing the dimensional precision and its mechanical properties;
- There are favorable conditions during the quenching process to reduce those phenomena that affect the quality of the quenched pieces. The condition of vertical immersion and the employment of oil at 60 °C presented a lower magnitude of effective stresses according to the FEM model used, as well as more minor distortion and higher hardness values when compared with the horizontal immersion in both quenching media;
- The present research allowed us to analyze the distortion behavior in a geometrically non-conventional piece under two different quenching media using FEM modeling, becoming a potent tool in the design of heat treatments for real engineering elements with complex geometries;
- Once certain variables have been defined, such as the immersion direction and the quenching media, in which a low level of residual stresses and distortion can obtained, it is advisable to continue researching other critical conditions in the process that contribute to reducing and predicting the distortion and cracking in quenched steels.

Author Contributions: Conceptualization, R.D.L.-G. and I.M.-J.; methodology, R.D.L.-G. and I.M.-J.; formal analysis, R.D.L.-G., I.M.-J. and A.M.-R.; investigation, I.M.-J.; resources, R.D.L.-G.; writing—original draft preparation, R.D.L.-G., I.M.-J. and A.M.-R.; writing—review and editing, R.D.L.-G., I.M.-J. and A.M.-R.; supervision, I.M.-J.; project administration, R.D.L.-G. and I.M.-J.; funding acquisition, R.D.L.-G. All authors have read and agreed to the published version of the manuscript.

Funding: This research was funded by Tecnológico Nacional de México under the grant number 11358.21-P.

Institutional Review Board Statement: Not applicable.

Informed Consent Statement: Not applicable.

Data Availability Statement: Not applicable.

Acknowledgments: The authors are very grateful to the Tecnológico Nacional de México campus Ciudad Victoria for the use of its facilities and laboratories.

Conflicts of Interest: The authors declare no conflict of interest.

References

1. Denis, S.; Sjostrom, S.; Simon, A. Coupled Temperature, Stress, Phase Transformation Calculation; Model Numerical Illustration of the Internal Stresses Evolution During Cooling of a Eutectoid Carbon Steel Cylinder. *Metall. Trans. A Phys. Metall. Mater. Sci.* **1987**, *18*, 1203–1212. [CrossRef]
2. Li, Z.; Freborg, A.M.; Hansen, B.D.; Srivatsan, T.S. Modeling the Effect of Carburization and Quenching on the Development of Residual Stresses and Bending Fatigue Resistance of Steel Gears. *J. Mater. Eng. Perform.* **2012**, *22*, 664–672. [CrossRef]
3. Šolić, S.; Podgornik, B.; Leskovšek, V. The occurrence of quenching cracks in high-carbon tool steel depending on the austenitizing temperature. *Eng. Fail. Anal.* **2018**, *92*, 140–148. [CrossRef]
4. Da Silva, A.; Pedrosa, T.; Gonzalez-Mendez, J.; Jiang, X.; Cetlin, P.; Altan, T. Distortion in quenching an AISI 4140 C-Ring—Predictions and experiments. *Mater. Des.* **2012**, *42*, 55–61. [CrossRef]
5. Jami, M.; Gil, E.; Ushiña, E.; Cabrera, V.H.; Cartagena, A. Determine hardness and torsional resistance of AISI/SAE 4340 steel, treated by quenching at 860 °C and tempering at 300 °C, 350 °C and 400°C. *Mater. Today Proc.* **2021**, *49*, 35–42. [CrossRef]
6. Fakir, R.; Barka, N.; Brousseau, J.; Caron-Guillemette, G. Analysis of the Mechanical Behavior of AISI 4340 Steel Cylindrical Specimens Heat Treated with Fiber Laser. *J. Manuf. Process.* **2020**, *55*, 41–56. [CrossRef]
7. Huang, D.; Arimoto, K.; Lee, K.; Lambert, D.; Narazaki, M. Prediction of Quench Distortion on Steel Shaft with Keyway by Computer Simulation. In Proceedings of the ASTM Proceedings: Heat Treating, St. Louis, MO, USA, 9 –12 October 2000; pp. 708–712.
8. Watanabe, K.; Yamada, M.; Nakasaki, M.; Matsumoto, R.; Utsunomiya, H. Determination of Transformation Transformation Plasticity Plasticity Coefficient Coefficient of Steel Steel by Horizontal Quenching of Shaft. *Procedia Manuf.* **2020**, *50*, 498–502. [CrossRef]
9. De Oliveira, W.P.; Savi, M.A.; Pacheco, P.M.C.L. Finite element method applied to the quenching of steel cylinders using a multi-phase constitutive model. *Arch. Appl. Mech.* **2013**, *83*, 1013–1037. [CrossRef]
10. Kang, S.-H.; Im, Y.-T. Three-dimensional thermo-elastic–plastic finite element modeling of quenching process of plain-carbon steel in couple with phase transformation. *Int. J. Mech. Sci.* **2007**, *49*, 423–439. [CrossRef]
11. He, L.; Li, H. FEM simulation of quenching residual stress for the plane strain problems. In Proceedings of the 2010 International Conference on Computer Design and Applications, Qinhuangdao, China, 25–27 June 2010; Volume 3, pp. 119–123. [CrossRef]
12. Decroos, M.; Seefeldt, M. The Effect of Size on the Distortion Behavior After Carburisation and Quenching Processes of Gears. *Int. J. Met. Mater. Eng.* **2017**, *139*, 1–10. [CrossRef]
13. Jung, M.; Kang, M.; Lee, Y.-K. Finite-element simulation of quenching incorporating improved transformation kinetics in a plain medium-carbon steel. *Acta Mater.* **2012**, *60*, 525–536. [CrossRef]
14. Basak, A.; Levitas, V.I. Finite element procedure and simulations for a multiphase phase field approach to martensitic phase transformations at large strains and with interfacial stresses. *Comput. Methods Appl. Mech. Eng.* **2019**, *343*, 368–406. [CrossRef]
15. Lopez-Garcia, R.D.; Garcia-Pastor, F.A.; Castro-Roman, M.J.; Alfaro-Lopez, E.; Acosta-Gonzalez, F.A. Effect of Immersion Routes on the Quenching Distortion of a Long Steel Component Using a Finite Element Model. *Trans. Indian Inst. Met.* **2016**, *69*, 1645–1656. [CrossRef]
16. Şimşir, C.; Gür, C.H. 3D FEM simulation of steel quenching and investigation of the effect of asymmetric geometry on residual stress distribution. *J. Mater. Process. Technol.* **2008**, *207*, 211–221. [CrossRef]
17. Lee, S.-J.; Lee, Y.-K. Finite element simulation of quench distortion in a low-alloy steel incorporating transformation kinetics. *Acta Mater.* **2008**, *56*, 1482–1490. [CrossRef]
18. Medina-Juárez, I.; Araujo de Oliveira, J.; Moat, R.J.; García-Pastor, F.A. On the Accuracy of Finite Element Models Predicting Residual Stresses in Quenched Stainless Steel. *Metals* **2019**, *9*, 1308. [CrossRef]
19. Şimşir, C. Modeling and Simulation of Steel Heat Treatment-Prediction of Microstructure, Distortion, Residual Stress, and Cracking. In *Steel Heat Treating Technologies*; Dossett, J.L., Totten, G.E., Eds.; ASM International: Metals Park, OH, USA, 2014; pp. 409–466. ISBN 978-1-62708-166-5.
20. Nie, Z.; Wang, G.; Lin, Y.; Rong, Y. (Kevin) Precision Measurement and Modeling of Quenching-Tempering Distortion in Low-Alloy Steel Components with Internal Threads. *J. Mater. Eng. Perform.* **2015**, *24*, 4878–4889. [CrossRef]
21. Sonar, T.; Lomte, S.; Gogte, C.; Balasubramanian, V. Minimization of Distortion in Heat Treated AISI D2 Tool Steel: Mechanism and Distortion Analysis. *Procedia Manuf.* **2018**, *20*, 113–118. [CrossRef]
22. Li, Z.; Grandhi, R.V.; Srinivasan, R. Distortion minimization during gas quenching process. *J. Mater. Process. Technol.* **2006**, *172*, 249–257. [CrossRef]
23. Li, Z.; Zhan, M.; Fan, X.; Wang, X.; Ma, F.; Li, R. Multi-mode distortion behavior of aluminum alloy thin sheets in immersion quenching. *J. Mater. Process. Technol.* **2019**, *279*, 116576. [CrossRef]
24. Farivar, H.; Prahl, U.; Hans, M.; Bleck, W. Microstructural adjustment of carburized steel components towards reducing the quenching-induced distortion. *J. Mater. Process. Technol.* **2019**, *264*, 313–327. [CrossRef]

25. Cho, J.; Kang, W.; Kim, M.; Lee, J.; Lee, Y.; Bae, W. Distortions induced by heat treatment of automotive bevel gears. *J. Mater. Process. Technol.* **2004**, *153*, 476–481. [CrossRef]
26. Bouissa, Y.; Bohlooli, N.; Shahriari, D.; Champliaud, H.; Morin, J.-B.; Jahazi, M. FEM modeling and experimental validation of quench-induced distortions of large size steel forgings. *J. Manuf. Process.* **2020**, *58*, 592–605. [CrossRef]
27. Nallathambi, A.K.; Kaymak, Y.; Specht, E.; Bertram, A. Sensitivity of material properties on distortion and residual stresses during metal quenching processes. *J. Mater. Process. Technol.* **2010**, *210*, 204–211. [CrossRef]
28. ASTM Standard E3. In *Standard Guide for Preparation of Metallographic Specimens*; ASTM International: West Conshohocken, PA, USA, 2017.
29. ASTM Standard E407. In *Standard Practice for Microetching Metals and Alloys*; ASTM International: West Conshohocken, PA, USA, 2017.
30. Mirzakhani, S.; Javanbakht, M. Phase Field-Elasticity Analysis of Austenite-Martensite Phase Transformation at the Nanoscale: Finite Element Modeling. *Comput. Mater. Sci.* **2018**, *154*, 41–52. [CrossRef]
31. Feulvarch, E.; Fontaine, M.; Bergheau, J.-M. XFEM investigation of a crack path in residual stresses resulting from quenching. *Finite Elements Anal. Des.* **2013**, *75*, 62–70. [CrossRef]
32. Fourier, J.B.J. *Théorie Analytique de La Chaleur*; Cambridge University Press: Cambridge, UK, 2009; ISBN 9780511693229.
33. Johnson, W.A. Reaction Kinetics in Processes of Nucleation and Growth. *Am. Inst. Min. Met. Petro. Eng.* **1939**, *135*, 416–458.
34. Avrami, M. Kinetics of Phase Change. I General Theory. *J. Chem. Phys.* **1939**, *7*, 1103–1112. [CrossRef]
35. Koistinen, D.P.; Marburger, R.E. A general equation prescribing the extent of the austenite-martensite transformation in pure iron-carbon alloys and plain carbon steels. *Acta Metall.* **1959**, *7*, 59–60. [CrossRef]
36. Todinov, M.T. Mechanism for formation of the residual stresses from quenching. *Model. Simul. Mater. Sci. Eng.* **1998**, *6*, 273–291. [CrossRef]
37. Vander Voort, G.F. *Atlas of Time-Temperature Diagrams for Irons and Steels*; ASM International: Metals Park, OH, USA, 1991; ISBN 978-0871704153.
38. JMatPro 13.0 Demo Version. Available online: https://www.sentesoftware.co.uk/demo (accessed on 20 March 2022).
39. Buczek, A.; Telejko, T. Investigation of heat transfer coefficient during quenching in various cooling agents. *Int. J. Heat Fluid Flow* **2013**, *44*, 358–364. [CrossRef]
40. Sugianto, A.; Narazaki, M.; Kogawara, M.; Shirayori, A. Validity of Heat Transfer Coefficient Based on Cooling Time, Cooling Rate, and Heat Flux on Jominy End Quench Test. In Proceedings of the MS&T 2007, Materials Science and Technology, Detroit, MI, USA, 16–20 September 2007; pp. 171–180.
41. Li, H.F.; Duan, Q.Q.; Zhang, P.; Zhang, Z.F. The Relationship between Strength and Toughness in Tempered Steel: Trade-off or Invariable? *Adv. Eng. Mater.* **2019**, *21*, 1801116. [CrossRef]

metals

Article

Web Bend-Buckling of Steel Plate Girders Reinforced by Two Longitudinal Stiffeners with Various Cross-Section Shapes

Yongli Peng [1], Zhengyi Kong [2], Ba Huu Dinh [3], Huu-Hue Nguyen [4], Truong-Son Cao [4], George Papazafeiropoulos [5,*] and Quang-Viet Vu [6,*]

1 Department of Civil Engineering, Ma'anshan University, Ma'anshan 243100, China
2 Department of Civil Engineering, Anhui University of Technology, Ma'anshan 243032, China
3 Department of Civil Engineering, Chonnam National University, Yongbong-ro 77, Buk-gu, Gwangju 61186, Korea
4 Faculty of Civil Engineering, Thuyloi University, 175 Tay Son, Dong Da, Hanoi 100000, Vietnam
5 Department of Structural Engineering, National Technical University of Athens, Zografou, 15780 Athens, Greece
6 Faculty of Civil Engineering, Vietnam Maritime University, 484 Lach Tray Street, Le Chan District, Hai Phong 180000, Vietnam
* Correspondence: gpapazafeiropoulos@yahoo.gr (G.P.); vietvq@vimaru.edu.vn (Q.-V.V.); Tel.: +30-6983507762 (G.P.); +84-974-282222 (Q.-V.V.)

Abstract: This work performs an investigation into the optimal position of two longitudinal stiffeners with different cross-section shapes such as open section (L-shaped and T-shaped) and closed section (rectangular and triangular shapes) shapes of stiffened plate girders under bending loading through an optimization procedure using a gradient-based interior point (IP) optimization algorithm. The stiffener optimum locations are found by maximizing the bend-buckling coefficient, k_b, generated from eigenvalue buckling analyses in Abaqus. The optimization procedure efficiently combines the finite element method and the IP optimization algorithm and is implemented using the Abaqus2Matlab toolbox which allows for the transfer of data between Matlab and Abaqus and vice versa. It is found that the proposed methodology can lead to the optimum design of the steel plate girder for all stiffener cross-section types with an acceptable accuracy and a reduced computational effort. Based on the optimization results, the optimum positions of two longitudinal stiffeners with various cross-section shapes are presented for the first time. It is reported that the optimum locations of two longitudinal stiffeners with open cross-section shapes (T- and L-shaped) are similar to that of flat cross-section, while the optimum positions of two longitudinal stiffeners with closed cross-section types (rectangular and triangular sections) are slightly different. One of the main findings of this study is that the bend-buckling coefficient of the stiffened girder having stiffeners with triangular cross-section shape is highest while that with flat cross-section shape is lowest among all considered stiffener types and this latter case has minimum requirement regarding the web thickness.

Keywords: Abaqus2Matlab; longitudinal stiffeners; optimization procedure; steel plate girders; web bend-buckling

Citation: Peng, Y.; Kong, Z.; Dinh, B.H.; Nguyen, H.-H.; Cao, T.-S.; Papazafeiropoulos, G.; Vu, Q.-V. Web Bend-Buckling of Steel Plate Girders Reinforced by Two Longitudinal Stiffeners with Various Cross-Section Shapes. *Metals* **2023**, *13*, 323. https://doi.org/10.3390/met13020323

Academic Editor: Xin Chen

Received: 11 January 2023
Revised: 31 January 2023
Accepted: 2 February 2023
Published: 5 February 2023

1. Introduction

Longitudinal stiffeners have been extensively used to improve the buckling strength of steel plates or steel plate girders subject to different loading conditions such as compression, patch loading, combined bending and shear, pure bending, etc. As a result of the significant increase in strength that stiffeners offer when placed at steel plates or steel plate girders, research related to members of this type has been widely conducted. Regarding steel plates reinforced by one or more longitudinal stiffeners under compression, Haffar et al. [1] proposed two new mathematical models for buckling resistance prediction of a steel plate with a closed longitudinal stiffener. Both proposed methods gave similar results, leading to load resistance values of satisfactory precision. Kovesdi et al. [2] investigated the buckling resistance of longitudinally stiffened plates subjected to compression using the shell finite element (FE) method. The author suggested an alternative design procedure to improve the economy of the practical design. Regarding plate girders under patch loading, Loaiza et al. [3] investigated buckling and post buckling behavior of longitudinally stiffened I-girders using an FE simulation. Various hypotheses regarding the effect of vertical and out-of-plane displacements of the web panel on the determination of the critical buckling load of the stiffened plate girder were taken into consideration. The analysis results showed that a full restriction of the vertical and out-of-plane displacements at the stiffener location led to improved patch load resistance at the ultimate load level. Demari et al. [4] performed a numerical study of slender I-girders strengthened with one longitudinal stiffener under patch loading. They reported that the optimum stiffener location for patch loading resistance is closer to the loaded flange when compared to girders under pure bending. Recently, based on an experimental database, Truong et al. [5] proposed an efficient machine learning method, namely the XGBoost algorithm, for predicting the patch load resistance of longitudinally stiffened plate girders. The efficiency and accuracy of the proposed method were demonstrated by comparing its performance with other machine learning methods as well as design equations from the existing standards. Regarding stiffened girders under combined bending and shear, by analyzing various FE models, Truong et al. [6] investigated the influence of multiple longitudinal stiffeners on the ultimate strength of plate girders. It was reported in this research that the variation in the ultimate strength of the girder was almost constant against the various dimensionless geometric parameters. Chen and Yuan [7] conducted a comprehensive experimental and numerical investigation into the local buckling behavior of longitudinally stiffened stainless steel plate girders subjected to combined bending and shear loading. It was observed that the existing M–V interaction curves recommended in EN 1993-1-4 for determining the bending and shear endpoints provide safe-sided estimations with a good level of consistency and accuracy for such structures.

Regarding the case of stiffened girders subjected to pure bending loads, in recent decades, longitudinal stiffeners have been widely utilized in girder webs to enhance the bending strength of the stiffened girder with slender sections. It has been reported that longitudinal stiffeners with various cross-section types, consisting of open cross-sections (flat, T, and L sections) and closed sections (rectangular, triangular, and trapezoidal sections) have been used for this purpose. Research related to the bending response of plate girders strengthened by longitudinal stiffeners has been extensively conducted all over the world, especially for flat stiffeners. Regarding the optimization problem of a single longitudinal stiffener with a flat cross-section, many researchers have proposed that the optimum position of a single longitudinal stiffener is placed at $0.2D$ from the girder compression flange (D is the depth of girder web), assuming the longitudinal edges of the girder web are simply supported [8–11]. Recently, through the finite element method, several researchers have found that the optimum location of a single stiffener for the stiffened girder under bending loading is at about $0.42D_c$ from the compression flange (D_c is the depth of girder web in compression), regardless of any asymmetric cross-section [12–15]. By investigating the elastic bend-buckling response of symmetric and asymmetric I-section girders with a single longitudinal stiffener using FE modeling, Cho and Shin [16] suggested the optimum stiffener position to be $0.425D_c$ from the compression flange. These optimal values are slightly different with those mentioned in AASHTO LRFD [17], in which the optimum stiffener position is at $0.4D_c$. On the other hand, research related to the optimization problems of multiple longitudinal stiffeners are still limited in the literature. Based on theoretical solutions, Rockey and Cook [18,19] proposed an optimum placement of multiple longitudinal stiffeners with flat cross-section plate girders. It was reported in these studies that the boundary conditions of longitudinal edges of the girder web were presumed to be either simply supported or clamped, whilst its vertical edges were assumed to be simply supported. By using numerical simulations, Kim et al. [20] proposed the optimum placement of two stiffeners with a flat cross-section girder under bending. An empirical formula to calculate the buckling coefficient of the stiffened girder under bending was recommended as well. Kim et al. [15] conducted a comprehensive work related to the optimum location of a single and two longitudinal stiffener(s) of a stiffened girder subjected to pure bending. Finally, the optimum stiffener locations and the minimum flexural rigidity for both the single and two stiffener(s) were suggested and presented a good comparison with the previous works. All the research mentioned above adequately provided the optimum position of a stiffener with a flat shape.

Furthermore, research related to the optimal stiffener position considering various stiffener cross-section types is still limited. Through finite element analysis, Maiorana et al. [21] investigated the buckling behavior of stiffened plate subjected to bending loading. Based on analysis results, the authors suggested the optimum position for all considered cross-section types of the stiffeners (flat, T, L, rectangular, triangular, and trapezoidal sections) to be at $0.2D$. However, the presence of flanges of the girder affecting the bend-buckling response and optimal stiffener placement was not taken into consideration. Recently, George et al. [22] suggested the optimal location of a single stiffener with open and closed cross-section types for stiffened girders subjected to bending loading. The presence of girder flanges was taken into consideration. However, the optimum location of two longitudinal stiffeners with various cross-section shapes has not been considered elsewhere.

Although the issue of the optimum stiffener position on steel plate girders has been addressed as mentioned in the previous paragraphs, consideration of multiple stiffeners has been limited only to flat stiffener shapes, whereas the studies that have investigated the effect of the stiffener shape on the buckling response of the girder have not taken into account the case of two or more stiffeners. This work tries to bridge this research gap, i.e., explore the case of multiple stiffeners with various cross-section shapes and their effect on the buckling load capacity and design efficiency of the steel plate girder. In this study, the optimal positions of two longitudinal stiffeners with open and closed section types along the web height of the stiffened girder subjected to bending loading are investigated by maximizing the critical buckling load of the latter. We develop various optimum designs depending on the cross-section type. The efficiency of the latter among the various aforementioned optimum designs is explored in terms of the maximum buckling coefficient and the minimum web thickness of the stiffened girder. The gradient-based interior point (IP) optimization algorithm, coupled with an appropriate FE model, is used for calculating the aforementioned optimum designs. The proposed numerical procedure proves to have low requirements in terms of implementation and computational effort, given that the Abaqus2Matlab [23] toolbox which automatically combines Abaqus [24] and Matlab [25] in a loop is employed. Based on the analysis results, the optimum stiffener locations and minimum web thickness for various stiffener types are suggested.

2. Existing Design Standards

2.1. AASHTO LRFD Standards

The AASHTO LRFD standards [17] for optimum stiffener position and the bend-buckling coefficient of a stiffened girder web were based on the research reported by Frank and Helwig [26], in which the boundary conditions of the longitudinal edges of the girder web were assumed to be simply supported at flanges, and its vertical edges were presumed to be simply supported by vertical stiffeners as well. The critical buckling load recommended by the AASHTO LRFD standards is presented in the following equation:

$$F_{crw} = \frac{0.9kE}{\left(\frac{D}{t_w}\right)^2} \tag{1}$$

where k is the bend-buckling coefficient, E represents the steel elastic modulus, D is the web depth, and t_w is the web thickness.

In the AASHTO LRFD standards, the bend-buckling coefficient was recommended as follows:

$$k = \begin{cases} \frac{11.64}{\left(\frac{D_c - d_s}{D}\right)} & if\, \frac{d_s}{D_c} < 0.4 \\ \frac{5.17}{\left(\frac{d_s}{D}\right)^2} & if\, \frac{d_s}{D_c} \geq 0.4 \end{cases} \tag{2}$$

where D_c and d_s are the web depth in compression in the elastic range and the distance of the stiffener from the compression flange, respectively. The optimum position of a single flat stiffener is at $0.4D_c$ from the compression flange, regardless of the asymmetry of the girder section. It is noted that in the AASHTO LRFD standards, Equation (2) can be conservatively utilized for girder webs with multiple longitudinal stiffeners. No specific equations were provided for girder webs with two or more stiffeners.

2.2. Eurocode 3 Standard

In the Eurocode 3 standard [27], the bending strength of the stiffened girder can be calculated by taking into consideration the combination of the effective widths of the stiffened girder web and the compression flange. In this standard, the buckling coefficients are also determined based on similar assumptions as the AASHTO LRFD standards. The bend-buckling coefficients were defined as a function of $\psi = \sigma/\sigma_c$, in which σ is the maximum stress at other web edges and σ_c represents the maximum compressive stress. The bend-buckling coefficient is given as follows:

$$k = \begin{cases} \frac{8.2}{(1.05+\psi)} & for \quad 0 < \psi < 1 \\ 7.81 - 6.29\psi + 9.78\psi^2 & for \quad -1 < \psi < 1 \\ 5.98(1-\psi)^2 & for \quad -3 < \psi < -1 \end{cases} \tag{3}$$

The optimum position of a single stiffener is consistent with that recommended by the AASHTO LRFD standards (at $0.4D_c$ from the compression flange). However, the design philosophy for bend-buckling resistance mentioned in the Eurocode 3 standard is intrinsically different from that considered in the AASHTO LRFD standards.

3. Methodology

3.1. Elastic Buckling Analysis

In this work, a linear elastic buckling analysis is implemented for the evaluation of the critical load, F_{cr}, of the longitudinally stiffened plate girders with open (T and L sections) and closed (rectangular and triangular sections) cross-sections of the stiffeners subject to bending. The lowest positive value of λ, which is the buckling eigenvalue, called λ_{cr}, can be obtained by solving Equation (1) as follows:

$$(K + \lambda K_G)u = 0 \tag{4}$$

where K presents the model stiffness matrix, K_G is the geometric stiffness matrix, λ stands for the multiplier of the reference load pattern F, and u is the buckling mode shape.

The buckling load is computed by Equation (5) as follows:

$$F_{cr} = \lambda_{cr} F \tag{5}$$

On the other hand, based on the classical buckling theory of plates under pure compression, the critical buckling load of a plate girder under bending loading can be computed as follows:

$$F_{cr} = k_b \frac{\pi^2 E t_w^3}{12(1-\nu^2)D} \tag{6}$$

where E represents the elastic modulus, t_w stands for the thickness of girder web, ν represents the Poisson's ratio, and D is the depth of girder web.

Based on Equations (5) and (6), the buckling coefficient, k_b, can be determined.

3.2. Problem Statement for Optimization of Longitudinally Stiffened Plate Girders

It is known that the optimum stiffener location can be obtained when the critical buckling load of a stiffened plate girder is maximized. Hence, the optimal stiffener position is determined by maximizing the bend-buckling coefficient, k_b. The optimization problem can be presented in the following form:

Find:

$$x = [d_{s1}, d_{s2}] \tag{7}$$

so that

$$k_b(x) = \frac{12(1-\nu^2)D}{\pi^2 E t_w^3} F_{cr}(x) \tag{8}$$

is maximized, subject to:

$$0.1D \le d_{s1} \le 0.5D \tag{9}$$

$$0.1D \le d_{s2} \le 0.5D \tag{10}$$

$$d_{s2} - d_{s1} \ge 18t_w \tag{11}$$

with the following values assigned to parameters:

$$b_{S,1} = b_{S,2} = 0.08D \tag{12}$$

$$\frac{b_{S,1}}{t_{S,1}} = \frac{b_{S,2}}{t_{S,2}} = 8 \tag{13}$$

In Equations (9) and (10), the limits are selected based on structural constraints, i.e., the stiffener location cannot exceed the half-depth of the web and it must be at least 10% of the height depth apart from the compression flange. In the latter case, for lower distances from the web, it is generally preferable to increase the stiffness of the plate girder through increasing the cross-section of the compression flange rather than placing a stiffener, which will require additional material and workmanship while additionally not contributing much to the increase in the girder plate stiffness. Moreover, Equation (11) takes into account the fact that each stiffener cross-section integrates with part of the web section to which it is attached, equal to $9t_w$, as designated in AASHTO LRFD standard part 6.10.11.3.3 [17]. Equations (12) and (13) specify the dimensions of the stiffeners in relation to the web depth. The dimensions of the stiffener cross-section remain constant during the optimization procedure and are selected so that the stiffeners are assumed to form a nodal line at the stiffener–plate junction to provide the highest buckling coefficient. To ensure the condition of nodal line formation, the out of plane displacements along the nodal line are restrained. The out of plane displacements do not exceed the following nonzero positive tolerances:

$$r \le r_{tol} \tag{14}$$

where r_{tol} represents the tolerance and the normalized parameter, r, is provided by the relation:

$$r = \frac{\max(|w_S|)}{\max(|w_w|)} \tag{15}$$

In Equation (15), w_s represents the out-of-plane displacement of the stiffener and w_w stands for the out-of-plane displacement of the girder web.

3.3. Optimization Procedure

In order to provide the solution of the optimization problem presented in Equations (4)–(7), an interior point algorithm (IPA), adopted in some references [28–30], was utilized. The optimization procedure is conducted in Abaqus [24] and Matlab [25] through Abaqus2Matlab toolbox [23,31] that integrates these software within the optimization loop. Detailed steps are presented as follows:

1. Establish a Matlab function so that an Abaqus input file (*inp) is automatically created when it runs inside Matlab;
2. Define input variables (longitudinal stiffener positions) in the Matlab function above;
3. Define the objective function mentioned in Section 3.2 for the optimization problem;
4. Build a main Matlab function consisting of the starting point value for the solution, lower and upper bound values of the stiffener position, and an optimization algorithm (using the fmincon function available in Matlab). The starting point is an initial guess and can be any arbitrary selection which satisfies the lower and upper bounds as well as any other constraints that may apply;
5. Compute the objective function defined in step 3;
6. Perform the optimization procedure;
7. Check the stopping criterion. This criterion is a maximization of k_b. If the criterion is satisfied, the optimization procedure will complete. Otherwise, it will go to the next step;
8. Change the design variable value to create a new Abaqus input file;
9. Run the analysis in Abaqus again;
10. Repeat steps 6–9 until convergence is attained, satisfying the specific tolerance;
11. The optimization result obtained is the final solution.

The optimization procedure for finding the optimal position of two longitudinal stiffeners is described in Figure 1. The optimization problem is essentially convex, without any local optima. For the case of more than one stiffener, this has been suggested by simplified analytical methods published in the literature (see, e.g., [19]). It is assumed that there are no large deviations in the optimization space considered in this study from the optimization space of such simplified approaches, since their results can approximate the actual result very well. On the other hand, it has been shown in many studies (see, e.g., [32,33]) that the variation in the buckling coefficient of a plate girder with a single stiffener with varying stiffener locations does not yield local optima, and the optimum position of the single stiffener is unique. Therefore, from the aforementioned points it can be deduced that the optimum configuration in the case of two stiffeners is unique and independent of the starting guess of the solution.

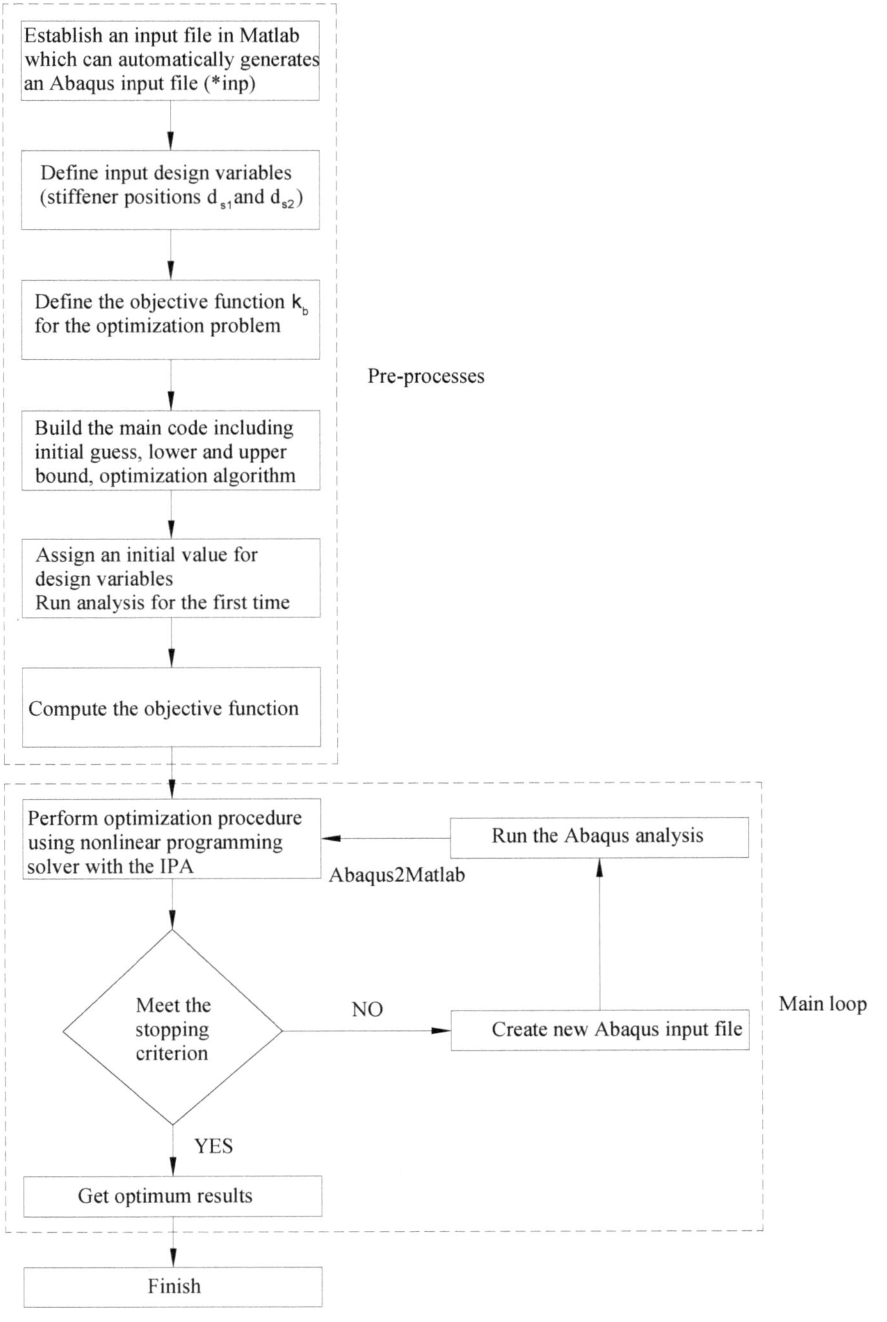

Figure 1. The optimization procedure flowchart [10,11].

4. Finite Element Modeling

The bend-buckling behavior of a stiffened plate girder presented in Figure 2 was computed based on finite element (FE) analysis of the structure using ABAQUS commercial software [24]. In this work, FE models of the girder with two longitudinal stiffeners with open (T-shaped and L-shaped) and closed (triangular and rectangular) cross-sections are based on the FE model mentioned in [14,15,22,34]. For instance, all descriptions of geometric dimensions of the girder (except the dimensions of longitudinal stiffeners), material properties, and FE modeling procedure are consistent with those of model 2 reported in [14,15]. Particularly, the web depth was selected as 3.0 m, while the web thickness was 9.0 mm. The flange width and flange thickness were 600 mm and 54 mm, respectively. The length-to-depth ratio (panel aspect ratio) of the girder was chosen to be 1.0. All materials were considered to be in the elastic range with an elastic modulus of 210 GPa and a Poisson's ratio of 0.3. The vertical edges of the girder web were assumed to be simply supported. All elements were simulated using 4-node shell elements S4R with a mesh size of 40 mm [14,15]. Figures 3 and 4 display the loading and boundary conditions for all stiffener cross-section types. Based on these FE models developed and the procedure mentioned in Section 3.3, the optimum stiffener position of two longitudinal stiffeners with open and closed cross-sections will be investigated in Sections 5 and 6 of this study. It is noted that, although the load distribution, which is specified in the Abaqus model, follows the linear shape which appears in Figure 3, in Figure 4, due to the notation followed in Abaqus/CAE interface, the length of the force vectors appears as fixed for visualization purposes.

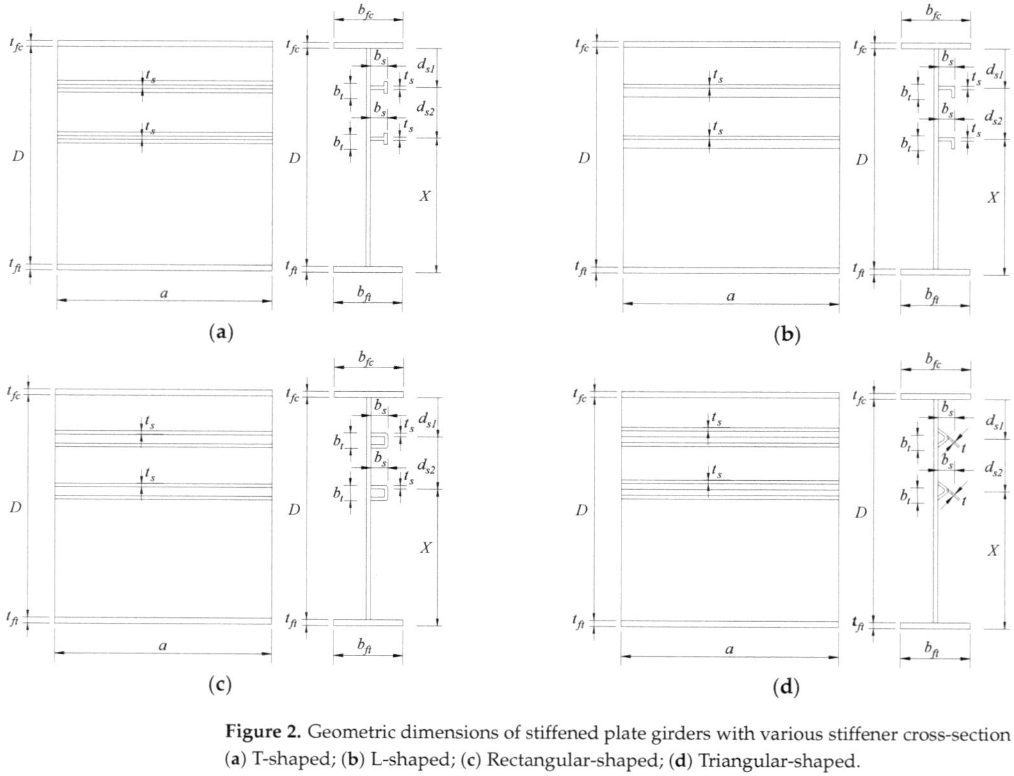

Figure 2. Geometric dimensions of stiffened plate girders with various stiffener cross-section shapes. (**a**) T-shaped; (**b**) L-shaped; (**c**) Rectangular-shaped; (**d**) Triangular-shaped.

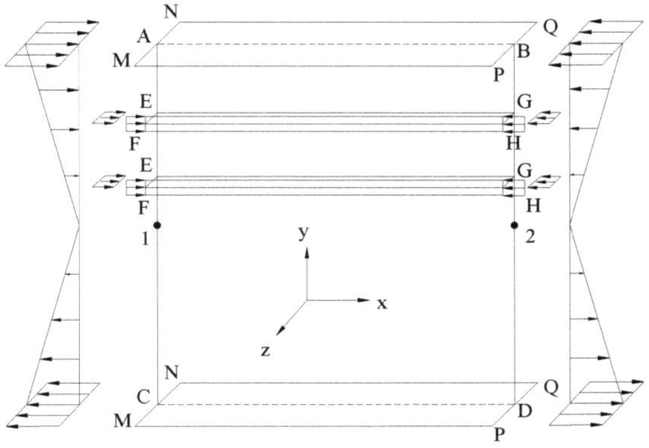

Edge	U_x	U_y	U_z	θ_x	θ_y	θ_z
AB & CD	–	–	–	–	–	–
AC & BD	–	–	R	–	–	–
MN & PQ	–	–	R	R	R	–
EF & GH	–	–	R	R	R	–
Point 1	R	R	–	–	–	–
Point 2	–	R	–	–	–	–
- Free			R Restrained			

(**a**)

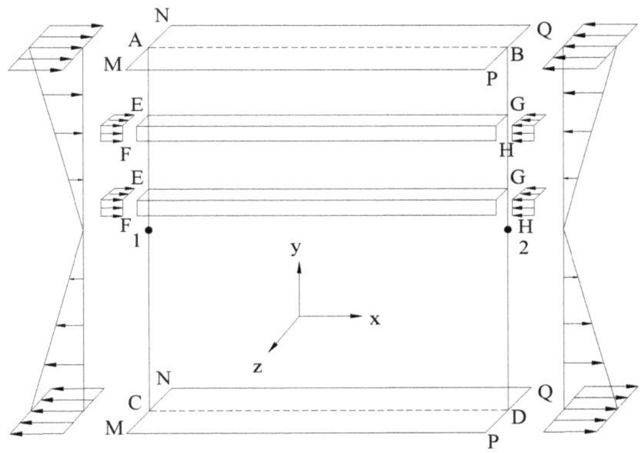

Edge	U_x	U_y	U_z	θ_x	θ_y	θ_z
AB & CD	–	–	–	–	–	–
AC & BD	–	–	R	–	–	–
MN & PQ	–	–	R	R	R	–
EF & GH	–	–	R	R	R	–
Point 1	R	R	–	–	–	–
Point 2	–	R	–	–	–	–
- Free			R Restrained			

(**b**)

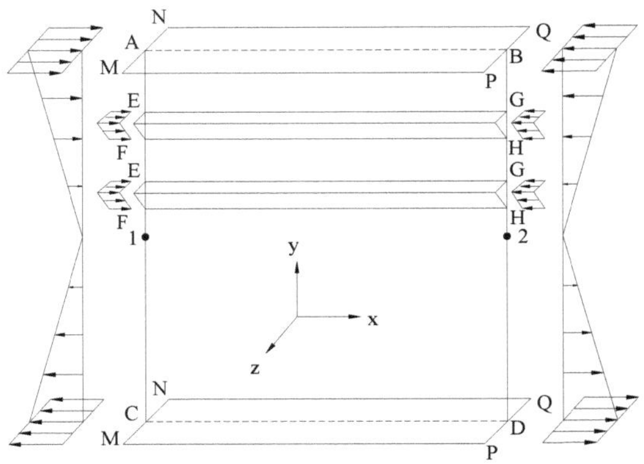

Edge	U_x	U_y	U_z	θ_x	θ_y	θ_z
AB & CD	–	–	–	–	–	–
AC & BD	–	–	R	–	–	–
MN & PQ	–	–	R	R	R	–
EF & GH	–	–	R	R	R	–
Point 1	R	R	–	–	–	–
Point 2	–	R	–	–	–	–
- Free			R Restrained			

(**c**)

Figure 3. *Cont.*

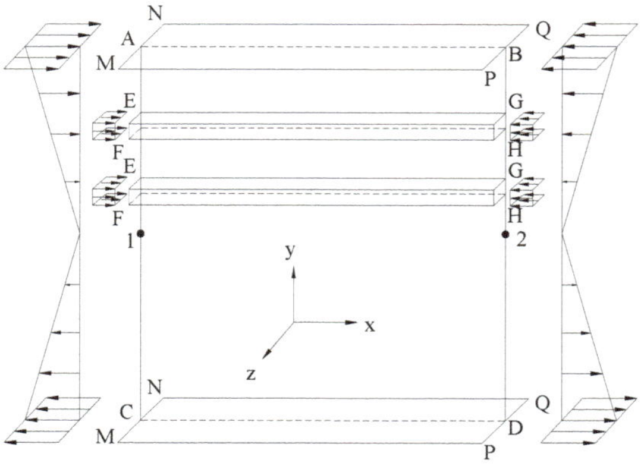

Edge	U_x	U_y	U_z	θ_x	θ_y	θ_z
AB & CD	-	-	-	-	-	-
AC & BD	-	-	R	-	-	-
MN & PQ	-	-	R	R	R	-
EF & GH	-	-	R	R	R	-
Point 1	R	R	-	-	-	-
Point 2	-	R	-	-	-	-
- Free			R Restrained			

(**d**)

Figure 3. Loading and boundary conditions for various plate girder configurations: (**a**) T section, (**b**) L section, (**c**) triangular section, (**d**) rectangular section.

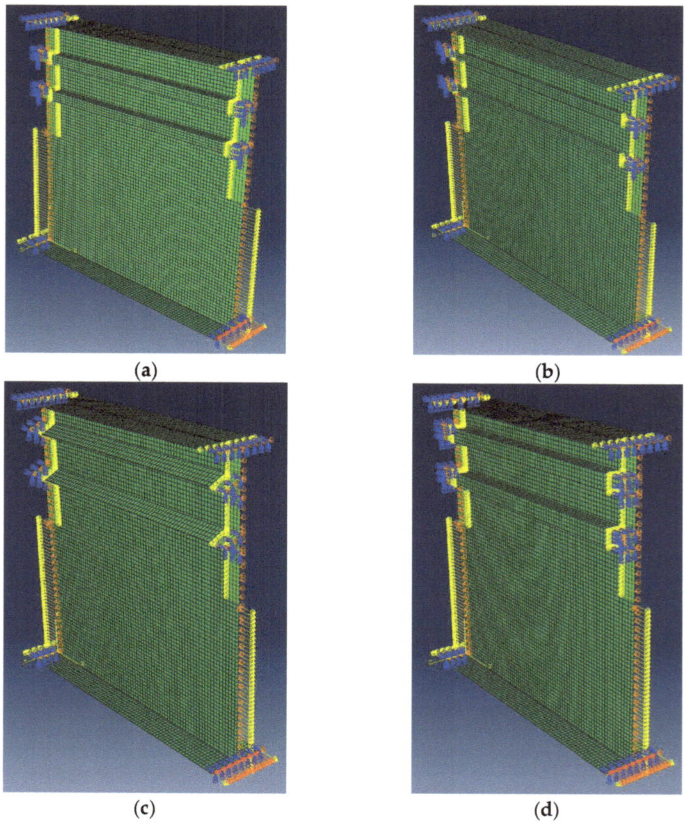

(**a**) (**b**) (**c**) (**d**)

Figure 4. Loading and boundary conditions for various plate girder configurations in Abaqus: (**a**) T section, (**b**) L section, (**c**) triangular section, (**d**) rectangular section.

5. Optimum Location of 2 Stiffeners with Open Cross-Section Types

In this section, the optimization procedure presented in Section 3.3 is employed for finding the optimal positions of two longitudinal stiffeners with open cross-section types (T and L sections) along the web depth of the stiffened girder subjected to pure bending loading.

The optimum placement of two stiffeners having flat, T-shaped and L-shaped cross-sections is presented in Table 1, in which the results for the flat shape were taken from Kim et al. [15] for comparison. The aspect ratio of the panel ($\varphi = a/D$) was fixed as 1.0 and the slenderness ratio of girder web (D/t_w) was fixed as 333 for all stiffener cross-section types. From Table 1, it can be observed that the optimal positions of longitudinal stiffeners 1 and 2 for both T-shaped and L-shaped cross-sections are at around $0.25D_c$ and 0.55 D_c, respectively. It is apparent that the optimum values are similar with those obtained for the flat stiffener cross-section type reported by Kim et al. [15]. In addition, it can be observed that the buckling coefficient, k_b, of the stiffener with a T-shaped cross-section is slightly higher than that of the stiffener with an L-shaped cross-section. However, both stiffeners with T-shaped and L-shaped cross-sections yield much higher buckling coefficients compared with the flat cross-section. Therefore, using stiffeners with T-shaped or L-shaped cross-sections significantly enhances the bend-buckling strength of the stiffened plate girder compared to stiffeners with a flat cross-section. Figure 5 shows the convergence history for stiffeners with open cross-section shapes, while Figure 6 shows the mode shapes of the girders with stiffeners of T-shaped and L-shaped cross-section types obtained from the optimization procedure.

Table 1. Optimal values for two longitudinal stiffeners for both cross-section types.

Stiffener Type	φ	d_{s1}/D_c	d_{s2}/D_c	k_b	Note
Flat-shaped	1.0	0.244	0.566	437.02	[15]
T-shaped	1.0	0.25	0.55	501.62	Present study
L-shaped	1.0	0.25	0.55	500.25	Present study

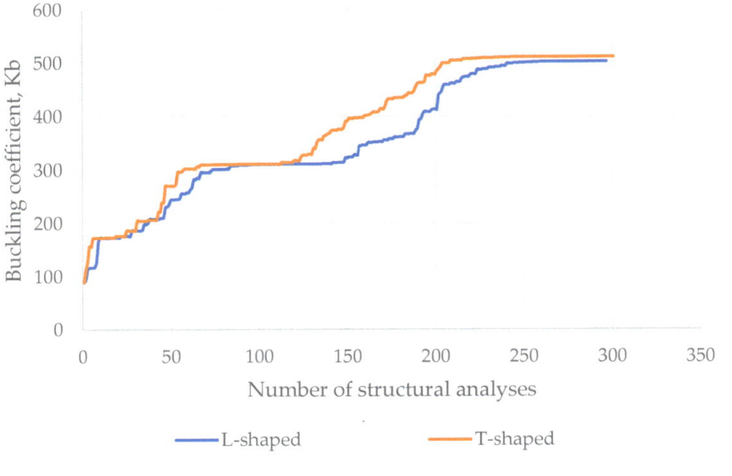

Figure 5. Convergence history for stiffeners with open cross-section shapes.

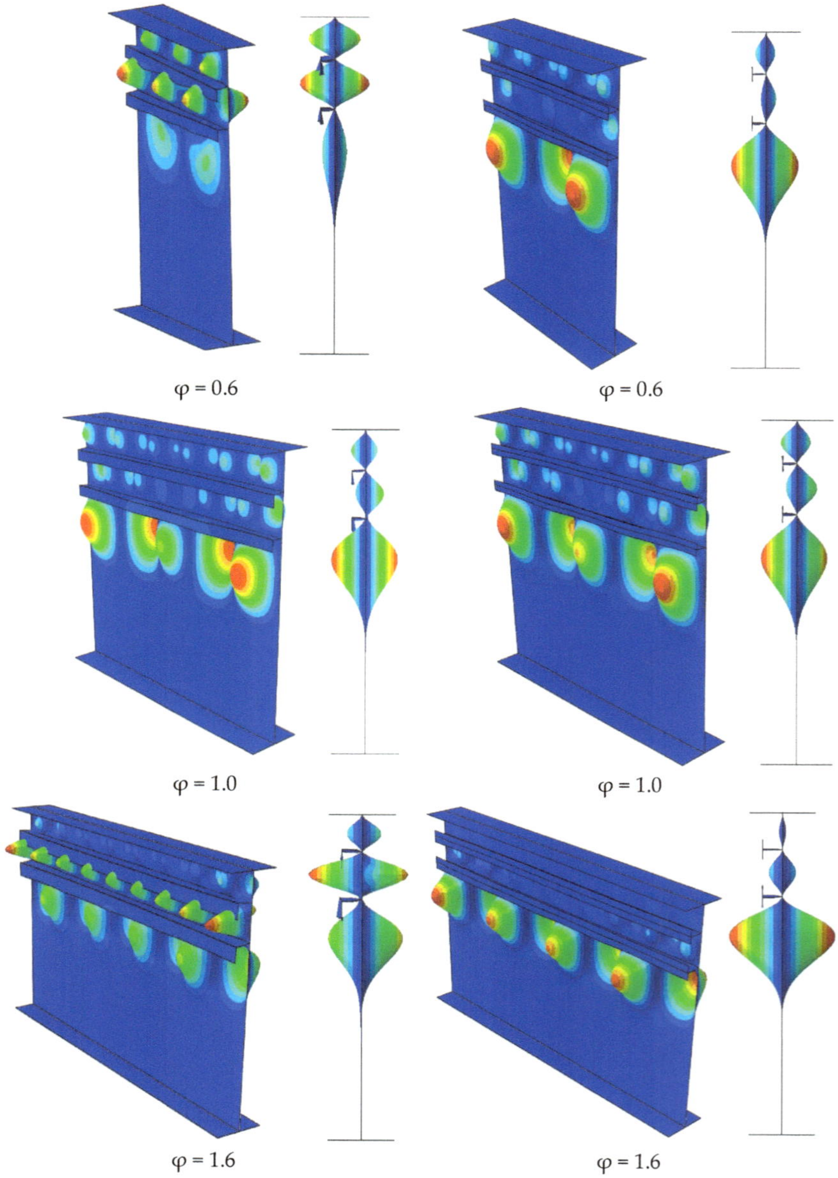

Figure 6. *Cont.*

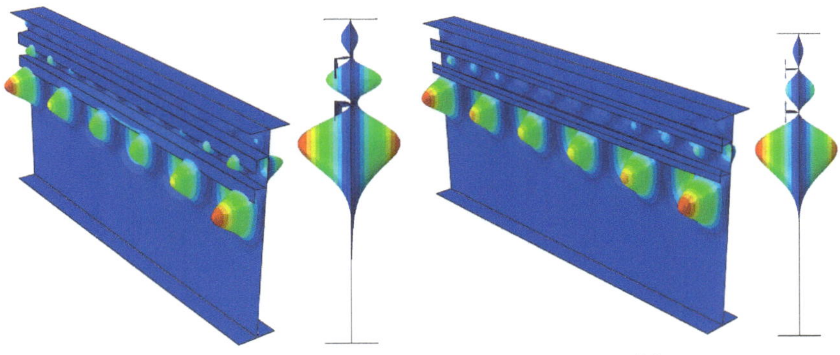

Figure 6. Buckling mode shapes of the stiffened girder with T-shaped and L-shaped stiffeners.

Table 2 discloses the effect of the panel aspect ratio on the optimal stiffener locations with the aspect ratios of 0.6, 1.0, 1.6, and 2.0. It can be seen that there is only a small effect of the panel aspect ratio on the optimal stiffener locations for both stiffener types. It was also observed that the buckling coefficients of the stiffeners corresponding to $\varphi = 0.6$ and $\varphi = 1.0$ were almost identical, while the buckling coefficients of the stiffeners corresponding to $\varphi = 1.6$ and $\varphi = 2.0$ were almost identical as well. The reason is because the flexural rigidify used for the stiffeners with $\varphi = 1.6$ and 2.0 is higher than that for the stiffeners with $\varphi = 0.6$ and 1.0.

Table 2. Effect of panel aspect ratio on the optimum stiffener locations.

φ	Stiffener Type	d_{s1}/D_c	d_{s2}/D_c	k_b
0.6	T-shaped	0.25	0.55	501.48
	L-shaped	0.25	0.55	500.1
1.0	T-shaped	0.25	0.55	501.62
	L-shaped	0.25	0.55	500.25
1.6	T-shaped	0.24	0.53	582.76
	L-shaped	0.24	0.53	578.59
2.0	T-shaped	0.24	0.53	582.96
	L-shaped	0.24	0.53	579.32

6. Optimum Location of Two Stiffeners with Closed Cross-Section Types

This section examines the optimum placements of two longitudinal stiffeners with closed cross-section types consisting of triangular and rectangular shapes for a stiffened girder subjected to bending by performing the procedure presented in Section 3.3.

The optimal positions of two longitudinal stiffeners with rectangular and triangular cross-section types are indicated in Table 3 for different aspect ratios. It is seen from this table that the optimal stiffener positions for these cross-section types are similar regardless of the aspect ratio. The optimum placements of stiffeners 1 and 2 are at around $0.23D_c$ and $0.55D_c$ from the compression flange of the girder, respectively. These optimum values are slightly different from the optimum locations of stiffeners with open cross-section types. Therefore, it can be concluded that the optimum positions of two longitudinal stiffeners with open and closed cross-section configurations are quite similar. In addition, it can be observed from Table 3 that the bend-buckling coefficients of the stiffeners with triangular shape are about 5% higher than those with rectangular shape. The convergence histories obtained from the optimization procedure for stiffeners with triangular and rectangular cross-section shapes are presented in Figure 7, while the mode shapes of these stiffeners are illustrated in Figure 8. It is noted that the mode shapes obtained from these stiffener types are similar.

Table 3. Optimum results for the longitudinal stiffeners with closed cross-section types.

φ	Stiffener Types	d_{s1}/D_c	d_{s2}/D_c	k_b
0.6	Triangular	0.23	0.55	1109.33
	Rectangular	0.23	0.54	1049.47
1.0	Triangular	0.23	0.55	1112.28
	Rectangular	0.23	0.54	1050.66

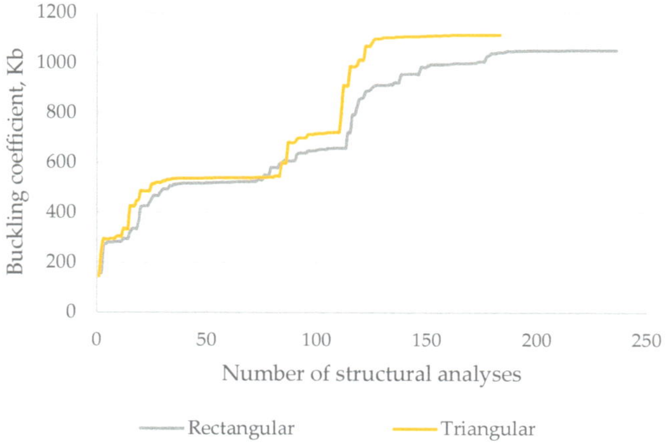

Figure 7. Convergence history for stiffeners with closed cross-section shapes.

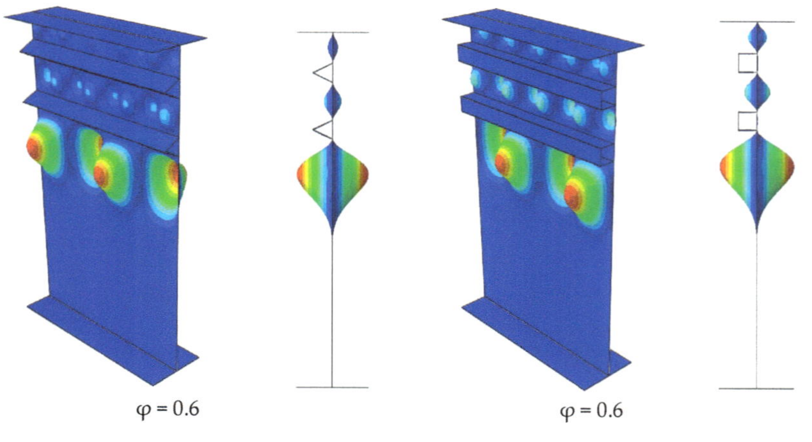

Figure 8. *Cont.*

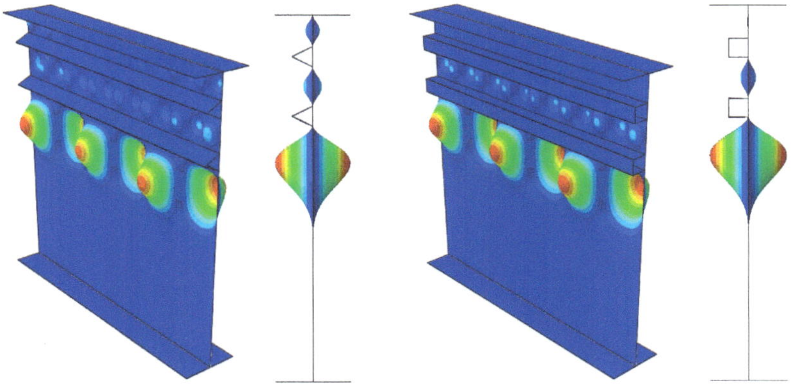

Figure 8. Bend-buckling mode shapes of a stiffened girder with triangular and rectangular stiffener cross-sections.

7. Comparison of the Efficiency of Longitudinal Stiffener Types

This section compares the efficiency of two longitudinal stiffeners with various cross-section shapes in terms of buckling coefficient and minimum web thickness of the stiffened girder. Regarding the buckling coefficient, Figure 9 presents a comparison of the buckling coefficient for two longitudinal stiffeners with flat, T, L, rectangular, and triangular cross-section shapes with respect to a panel aspect ratio of 1. It can be observed that the buckling coefficients of stiffeners with closed section shapes are significantly higher than those with open cross-section shapes. In particular, the triangular shape provides the highest buckling coefficient, while the flat shape gives the lowest buckling coefficient.

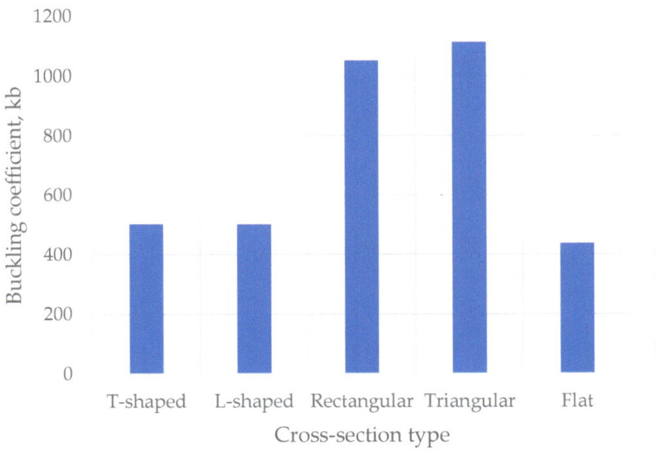

Figure 9. Comparison of buckling coefficient for two longitudinal stiffeners with various cross-section shapes (φ = 1.0).

Regarding the minimum web thickness of the stiffened girder, the limit of the slenderness ratio of stiffened webs should satisfy the requirements mentioned in AASHTO LRFD [17] as follows:

$$\frac{D}{t_w} \leq 0.95\sqrt{\frac{Ek_b}{F_y}} \tag{16}$$

where

k_b represents the bend-buckling coefficient of the stiffened girder;
F_y stands for the steel yield strength;
F_y is assumed to be 315 MPa;
E is elastic modulus, where E = 210 GPa.

From Equation (13), the minimum thickness of the stiffened web is computed as follows:

$$t_w \geq \frac{D}{0.95\sqrt{\frac{Ek_b}{F_y}}} \tag{17}$$

From Equation (17), a comparison of the minimum thicknesses of the girder web computed for various stiffener shapes is given in Table 4. It is observed in Table 4 that the flat stiffener needs the highest minimum web thickness, while the triangular stiffener requires the lowest minimum web thickness among all stiffener types considered. It is noteworthy that the required web thickness of the girder web reinforced by two stiffeners with closed cross-section shapes is significantly reduced compared to those reinforced by two stiffeners with open cross-section shapes. In particular, when the web girder is reinforced by two stiffeners with a triangular section shape, the required web thickness decreases by at least 37.26% compared with the case in which the web is reinforced by two stiffeners with a flat cross-section shape.

Table 4. Comparison of minimum stiffened web thickness.

Stiffener Type	D (mm)	k_b	t_{min} (mm)
Flat-shaped	3000	437.02	5.85
T-shaped	3000	501.62	5.46
L-shaped	3000	500.25	5.47
Triangular	3000	1112.28	3.67
Rectangular	3000	1050.66	3.77

8. Conclusions

In this work, the optimum positions of two longitudinal stiffeners with different cross-section shapes placed at the web of stiffened girders under bending are examined through an optimization procedure performed by coupling Abaqus and Matlab through the Abaqus2Matlab toolbox. Based on the optimization results, the optimum locations of two longitudinal stiffeners with open and closed cross-section types are obtained. An advantage of the proposed methodology is that it simulates the structural optimization problem with a robust numerical procedure which combines FEA and optimization, and it proves to be able to yield meaningful results for all structural configuration cases with an acceptable accuracy and a reduced computational effort. Some conclusions can be drawn as follows:

- The optimum positions of the stiffeners with open cross-section shapes (T- and L-shaped) are around 0.25*Dc* and 0.55*Dc*, which are similar to the optimum location of the flat stiffener.
- The optimum positions of the stiffeners with closed cross-section shapes (triangular and rectangular shapes) are around 0.23*Dc* and 0.54*Dc*, which are slightly different to the stiffeners with open cross-sections.

- The bend-buckling coefficient of the stiffened girder with stiffeners with a triangular cross-section shape is highest, while that with a flat cross-section shape is lowest in all considered stiffener types.
- The required web thickness of the girder web reinforced by two stiffeners with closed section shapes is remarkably reduced compared with those reinforced by two stiffeners with open cross-section shapes.

The main objective of this study is to investigate the effect of the stiffener location and shape on the buckling load and configuration of steel plate girders reinforced by two longitudinal stiffeners due to bending loading. Maximizing the buckling coefficient leads to the optimum design, since this maximizes the load capacity in each structural configuration of the steel plate girder. The optimization procedure that is implemented in this study has led to the discovery of optimum configurations which maximize the buckling load capacity. Therefore, the optimum locations of the two stiffeners proposed in this study should be taken into account for maximizing the safety of the structure, as should other constraints in construction. A major observation is that stiffeners with triangular cross-sections lead to the highest buckling coefficient compared with other cross-section shapes. Apart from this, it is proven in this study that a suitable selection of stiffener cross-section type and location can lead to a substantial construction cost reduction compared to the usual state of practice designs, since the web thickness can be reduced by as much as 37.26%.

Future work could address issues such as investigating the effect on the buckling load capacity of the plate girder of various loading types (shear loading, patch loading, biaxial bending, etc.), structural constraints (e.g., presence of bolts at the web or flanges), stiffener orientations (vertical or oblique), and cutouts (circular or rectangular) at the web body.

Author Contributions: Conceptualization, Y.P.; methodology, Z.K.; software, G.P.; validation, Z.K.; formal analysis, Y.P.; investigation, Q.-V.V.; resources, B.H.D.; data curation, T.-S.C.; writing—original draft preparation, G.P.; writing—review and editing, H.-H.N.; visualization, B.H.D.; supervision, Q.-V.V.; project administration, Q.-V.V.; funding acquisition, Q.-V.V. All authors have read and agreed to the published version of the manuscript.

Funding: This research was funded by Vietnam National Foundation for Science and Technology Development (NAFOSTED) under grant number 107.01-2019.322.

Data Availability Statement: Data will be available upon reasonable request.

Conflicts of Interest: The authors declare no conflict of interest.

References

1. Haffar, M.Z.; Kövesdi, B.; Adany, S. Buckling of compressed plates with closed-section longitudinal stiffeners: Two new mathematical models for resistance prediction. *Structures* **2021**, *33*, 3526–3539. [CrossRef]
2. Kovesdi, B.; Haffar, M.; Adany, S. Buckling resistance of longitudinally stiffened plates: Eurocode-based design for col-umn-like and interactive behavior of plates with closed-section stiffeners. *Thin-Walled Struct.* **2021**, *159*, 107266. [CrossRef]
3. Loaiza, N.; Graciano, C.; Casanova, E. Web slenderness for longitudinally stiffened I-girders subjected to patch loading. *J. Constr. Steel Res.* **2019**, *162*, 105737. [CrossRef]
4. Demari, F.E.; Mezzomo, G.P.; Pravia, Z.M.C. Numerical study of slender I-girders with one longitudinal stiffener under patch loading. *J. Constr. Steel Res.* **2020**, *167*, 105964. [CrossRef]
5. Truong, V.-H.; Papazafeiropoulos, G.; Vu, Q.-V.; Pham, V.-T.; Kong, Z. Predicting the patch load resistance of stiffened plate girders using machine learning algorithms. *Ocean Eng.* **2021**, *240*, 109886. [CrossRef]
6. Truong, V.-H.; Papazafeiropoulos, G.; Pham, V.-T.; Vu, Q.-V. Effect of multiple longitudinal stiffeners on ultimate strength of steel plate girders. *Structures* **2019**, *22*, 366–382. [CrossRef]
7. Chen, Z.; Yuan, H. Local buckling behaviour of longitudinally stiffened stainless steel plate girders under combined bending and shear. *Thin-Walled Struct.* **2023**, *184*, 110541. [CrossRef]
8. Azhari, M.; Bradford, M. Local buckling of I-section beams with longitudinal web stiffeners. *Thin-Walled Struct.* **1993**, *15*, 1–13. [CrossRef]
9. Alinia, M.; Moosavi, S. A parametric study on the longitudinal stiffeners of web panels. *Thin-Walled Struct.* **2008**, *46*, 1213–1223. [CrossRef]
10. Vu, Q.-V.; Papazafeiropoulos, G.; Graciano, C.; Kim, S.-E. Optimum linear buckling analysis of longitudinally multi-stiffened steel plates subjected to combined bending and shear. *Thin-Walled Struct.* **2018**, *136*, 235–245. [CrossRef]

11. Vu, Q.-V.; Truong, V.-H.; Papazafeiropoulos, G.; Graciano, C.; Kim, S.-E. Bend-buckling strength of steel plates with multiple longitudinal stiffeners. *J. Constr. Steel Res.* **2019**, *158*, 41–52. [CrossRef]
12. Elbanna, A.; Ramadan, H.; Mourad, S. Buckling enhancement of longitudinally and vertically stiffened plate girders. *J. Eng. Appl. Sci.* **2014**, *61*, 351–370.
13. HKim, S.; Park, Y.; Kim, B.; Kim, K. Numerical investigation of buckling strength of longitudinally stiffened web of plate girders subjected to bending. *Struct. Eng. Mech.* **2018**, *65*, 141–154.
14. Hoàn, P.; Trung, P.; Vi, V. Nghiên cứu xác định vị trí tối ưu của sườn tăng cường dọc của dầm cầu thép chịu uốn. Tạp chí Khoa học Công nghệ Xây dựng. *NUCE* **2020**, *14*, 29–38.
15. Kim, S.-E.; Papazafeiropoulos, G.; Graciano, C.; Truong, V.-H.; Do, Q.T.; Kong, Z.; Vu, Q.-V. Optimal design of longitudinal stiffeners of unsymmetric plate girders subjected to pure bending. *Ocean Eng.* **2021**, *221*, 108374. [CrossRef]
16. Cho, E.-Y.; Shin, D.-K. Elastic web bend-buckling analysis of longitudinally stiffened I-section girders. *Int. J. Steel Struct.* **2011**, *11*, 297–313. [CrossRef]
17. AASHTO. *AASHTO LRFD Bridge Design Specifications*, 7th ed.; American Association of State Highway and Transportation Officials: Washington, DC, USA, 2014.
18. Rockey, K.; Cook, I. Optimum reinforcement by two longitudinal stiffeners of a plate subjected to pure bending. *Int. J. Solids Struct.* **1965**, *1*, 79–92. [CrossRef]
19. Rockey, K.; Cook, I. The buckling under pure bending of a plate girder reinforced by multiple longitudinal stiffeners. *Int. J. Solids Struct.* **1965**, *1*, 147–156. [CrossRef]
20. Kim, B.J.; Park, Y.; Kim, K.; Choi, B. Web bend-buckling strength of plate girders with two longitudinal web stiffeners. *Struct. Eng. Mech.* **2019**, *69*, 383–397.
21. Maiorana, E.; Pellegrino, C.; Modena, C. Influence of longitudinal stiffeners on elastic stability of girder webs. *J. Constr. Steel Res.* **2010**, *67*, 51–64. [CrossRef]
22. Papazafeiropoulos, G.; Vu, Q.-V.; Nguyen, V.-S.; Truong, V.-H. Optimum location of a single longitudinal stiffener with various cross-section shapes of steel plate girders under bending loading. *J. Sci. Technol. Civ. Eng. (STCE)—NUCE* **2022**, *16*, 65–75. [CrossRef]
23. Papazafeiropoulos, G.; Muñiz-Calvente, M.; Martínez-Pañeda, E. Abaqus2Matlab: A suitable tool for finite element post-processing. *Adv. Eng. Softw.* **2017**, *105*, 9–16. [CrossRef]
24. ABAQUS. *Analysis User's Manual*, Version 6.14; Dassault Systems: Los Angeles, CA, USA, 2014.
25. MathWorks, Inc. *MATLAB R2017b*; MathWorks, Inc.: Natick, MA, USA, 2017.
26. Frank, K.H.; Helwig, T.A. Buckling of webs in unsymmetric plate girders. *Eng. J. Second Quart.* **1995**, *32*, 43–53.
27. CEN. *EN 1993-1-5*; Eurocode 3: Design of Steel Structures-Part 1-5: Plated Structural Elements. European Committee for Standardization: Brussels, Belgium, 2006.
28. Byrd, R.H.; Gilbert, J.C.; Nocedal, J. A trust region method based on interior point techniques for nonlinear programming. *Math. Program.* **2000**, *89*, 149–185. [CrossRef]
29. Byrd, R.H.; Hribar, M.E.; Nocedal, J. An Interior Point Algorithm for Large-Scale Nonlinear Programming. *SIAM J. Optim.* **1999**, *9*, 877–900. [CrossRef]
30. Waltz, R.; Morales, J.; Nocedal, J.; Orban, D. An interior algorithm for nonlinear optimization that combines line search and trust region steps. *Math. Prograhm.* **2006**, *107*, 391–408. [CrossRef]
31. Pham, V.T.; Vu, Q.V.; Papazafeiropoulos, G.; Ngo, V.T. Efficiency of Abaqus2Matlab toolbox for structural optimization problems. *IOP Conf. Ser. Mater. Sci. Eng.* **2020**, *869*, 022025. [CrossRef]
32. Ghorashi, M.; Askarian, A.; Gashtasby, M. Optimal design of stiffened plates for buckling under in-plane forces and bending moments. In Proceedings of the Eighth International Conference on the Application of Artificial Intelligence to Civil and Structural Engineering Computing, Stirling, UK, 19–21 September 2001; pp. 83–84.
33. Silva, D.A.B.; Filho, J.O.F.; Barreto, R. Numerical study for optimization of the buckling behavior of longitudinally stiffened plates under pure bending. In Proceedings of the 1st International Congress on Structural Integrity and Maintenance—SIM 2021, Online, 8–9 April 2021.
34. Papazafeiropoulos, G.; Vu, Q.-V.; Truong, V.-H.; Luong, M.-C.; Pham, V.-T. Prediction of buckling coefficient of stiffened plate girders using deep learning algorithm. *Lect. Notes Civ. Eng.* **2019**, *54*, 1143–1148. [CrossRef]

materials

MDPI

Article

Numerical Simulation Study of Expanding Fracture of 45 Steel Cylindrical Shell under Different Detonation Pressure

Zhenwei Huang [1] and Xinlu Yu [1,2,*]

1 Key Laboratory of Impact and Safety Engineering, Ningbo University, Ministry of Education, 818 Fenghua Road, Jiangbei District, Ningbo 315211, China; hzw961020@163.com
2 College of Science & Technology, Ningbo University, Wenwei Road 521, Ningbo 315300, China
* Correspondence: yuxinlu@nbu.edu.cn; Tel.: +86-152-574-92195

Abstract: Detonation and fragmentation of ductile cylindrical metal shells is a complicated physical phenomenon of material and structural fracture under a high strain rate and high-speed impact. In this article, the smoothed particle hydrodynamics (SPH) numerical model is adopted to study this problem. The model's reliability is initially tested by comparing the simulation findings with experimental data, and it shows that different fracture modes of cylindrical shells can be obtained by using the same model with a unified constitutive model and failure parameters. By using this model to analyze the explosive fracture process of the cylindrical shells at various detonation pressures, it shows that when the detonation pressure decreases, the cylindrical metal shell fracture changes from a pure shear to tensile–shear mixed fracture. When the detonation pressure is above 31 GPA, a pure shear fracture appears in the shell during the loading stage of shell expansion, and the crack has an angle of 45° or 135° from the radial direction. When the pressure is reduced to 23 GPA, the fracture mode changes to tension–shear mixing, and the proportion of tensile cracks is about one-sixth of the shell fracture. With the explosion pressure reduced to 13 GPA, the proportion of tensile cracks is increased to about one-half of the shell fracture. Finally, the failure mechanism of the different fracture modes was analyzed under different detonation pressures by studying the stress and strain curves in the shells.

Keywords: cylindrical shell fracture; 45 steel; SPH numerical simulation

Citation: Huang, Z.; Yu, X. Numerical Simulation Study of Expanding Fracture of 45 Steel Cylindrical Shell under Different Detonation Pressure. *Materials* **2022**, *15*, 3980. https://doi.org/10.3390/ma15113980

Academic Editors: Shuwen Wen, Yongle Sun and Xin Chen

Received: 25 April 2022
Accepted: 31 May 2022
Published: 3 June 2022

1. Introduction

In modern explosion research, the study of expanding fracture mode of cylindrical shells under high-speed impact loading has always been a hot topic [1–5]. A large number of researchers have observed the fragmentation process by changing variables in detonation experiments and used the recovered fragments to study the influencing factors of the fracture mode of the cylindrical shell. The detonation of a cylindrical shell is a process under a high-impact, high-strain rate. Cylindrical shell fracture is a difficult subject that involves both material and structural damage. Based on experiments and simulations, several researchers have proposed related assumptions and findings.

As early as 1943, Gurney [6] proposed an empirical formula for the velocity of cylindrical shell fragments based on the law of energy balance. The velocity relates to the mass of explosives, the mass of cylindrical shells, and the Gurney energy. In 1944, Taylor [7] observed the fracture process of the metal cylindrical shell under a certain explosion pressure through a high-speed camera, believed that the fracture of the cylindrical shell was a tensile fracture process and he proposed the Taylor criterion. However, the criterion did not consider other influencing factors, such as explosion pressure, shell dimensions, and loading method. Besides, the control mechanism is relatively simple. Based on the Taylor criterion, Hoggatt et al. [8] found another mode of expansion fracture under different explosion pressure. They thought that the fracture process of the cylindrical shell was a competitive fracture between tensile fracture and shear fracture, under high explosive pressure, the shell is prone to pure shear fracture. Through experiments and numerical simulation studies, Martineau [9] found that the cylindrical shell had an adiabatic shear phenomenon in the expansion process when the strain rate was 104 s^{-1} and the strain was about 150%, the thickness of the cylindrical shell also affected adiabatic shear. Studies on microscopic characterization of metal materials [10], metal thermodynamic properties [11], and the influence of impurities in steel materials on the properties of the material [12] have shown the metal failure process is affected by many factors. In recent years, Liang et al. [13] studied the expansion ring's fracture by numerical simulation, and found that with low notch depth conditions, the expansion ring is prone to pure shear fracture. Tensile fracture dominated when the notch depth was higher, this result indicated that fracture mode was affected by defect factors. Under the guidance of the dynamic fragmentation theory of solids and the fragmentation theory of continuum energy, Grady [5] analyzed that the distribution of shell fragments was related to the ductility of the material itself. Hu et al. [14] and Hu et al. [15] carried out experimental research on the damage to metal cylindrical shells of 45 steel, TC4, and WTG05 tungsten alloys with different geometric dimensions driven by explosive detonation. They found that factors such as the thickness of the shell and loading detonation pressure would affect the fracture process of cylindrical shells. Generally speaking, thin-walled shells are prone to pure shear fracture under high detonation pressure loading, and fracture strain and strain rate would be relatively higher. When the explosion pressure is decreased or the shell thickness is increased, the fracture mode of the shell would change to shear–tensile mixed fracture, and the fracture strain and strain rate would be lower. Through numerical simulation research, Liu et al. [16] also found that different initiation methods of the explosion would affect the fracture of the cylindrical shell because the incident angle of the detonation wave to the inner wall of the cylindrical shell varied under different initiation methods of explosion; the pressure experienced by the inner surface of the cylindrical shell differs greatly. The results show that the peak pressure of the centerline detonation is the maximum, the single-point detonation is the second, and the plane detonation is the minimum. Arnold [4] studied the experimental results of metal cylindrical shells with different materials, wall thicknesses, and processing techniques, and proposed a calculation formula for predicting the distribution of shell fragments. Arnold found that medium-strength steel casings produced the largest fragments. In the axial direction, medium-strength steel casings produced the longest fragments. In terms of numerical simulation, Yu et al. [17] studied the pure shear fracture mechanism of TA2 cylindrical metal shells loaded with different explosives. Liu et al. [18] considered heterogeneity of the material and introduced a material shear failure constitutive model with probability, and discussed the initiation of cracks and propagation of multiple adiabatic shears in metal cylindrical shells.

However, the traditional numerical simulation method, especially the finite element method (FEM), can only analyze the deformation process, stress, and strain state of a cylindrical shell under an explosive shock wave, but by using the FEM method, various fracture modes such as pure shear fracture and the tensile–shear mixed fracture cannot be reproduced at present [17–19]. When using the traditional method (FEM) to analyze the fracture of a cylindrical shell, because the element produces extreme deformation and sudden change in the interface, it always appears that the failed element is deleted, which affects the fracture evolution process of a cylindrical metal shell in a subsequent calculation [20,21]. In addition, a large number of deleted elements lead to excessive loss of mass, momentum, and energy in the system. The method of smoothed particle hydrodynamics (SPH) adopted in this paper is a meshless method with Lagrangian particle configuration [22]; it avoids the mesh distortion of the Lagrangian method and the difficulty of the Euler method to capture the boundary. In the SPH method, the state of the system is described by discrete particles, and the smooth particles are used to bear their unique material properties so that they can work according to the law of the governing equation. By solving the integral expression of kernel function interpolation in the form of discrete particle weighted summation, a stable and smooth approximation is obtained. In addition, when the distance between two particles exceeds a failure threshold, the particles are not deleted when the material fractures. The conservation of mass, momentum, and energy of the system is guaranteed; thus, the SPH simulation method has great applications in the fields of ultra-high-speed impact, explosion, crack propagation, and metal forming [23–25]. Therefore, this article selected the smoothed particle hydrodynamics method to establish the cylindrical shell explosion model.

2. Experimental Phenomenon of Fracture of Metal Cylindrical Shell

Many experimental studies have been carried out in China and worldwide on the fracture phenomenon and fragmentation characteristics of metal cylindrical shells under explosive loads [26–28]. Research shows that, for different materials and different detonation pressures, the fracture of the cylindrical shell demonstrates the following characteristics (in Figure 1): (1) Under high explosive pressure, the cylindrical shell undergoes shear fracture. When the detonation pressure decreases, it becomes a tensile–shear mixed fracture. The proportions of tensile cracks in the fracture are different; the lower the detonation pressure, the higher the proportion of tensile cracks in the fracture. (2) The shear cracks are formed by the unstable slip of the adiabatic shear belt. The direction is 45° or 135° to the radial direction. The cross-section is relatively flat (S area); the direction of the tensile cracks is along the radial direction, and the cross-section is uneven and rough (R area). (3) Cylindrical shell fragments can be divided into two types: large fragments of type A and small fragments of type B, as shown in Figure 1. Large fragments of type A include two surfaces of the cylindrical shell: the inner surface and the outer surface. Small fragments of type B only contain one surface of the cylindrical shell, according to whether the contained surface is the inner surface or the outer surface; it can be divided into B′ and B′′.

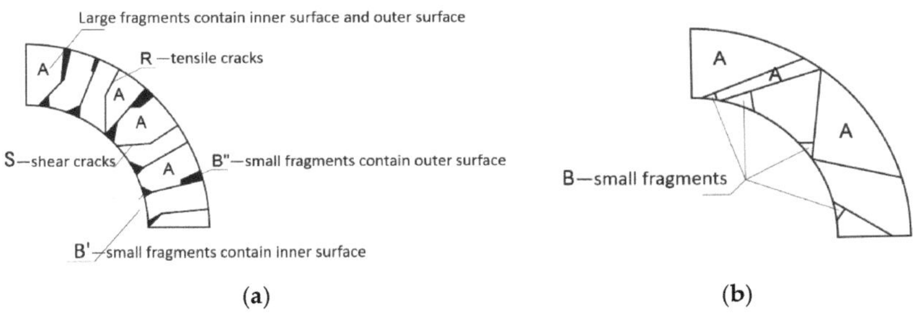

Figure 1. Main fragments: "A"—large fragments contain inner surface and outer surface and "B"—small fragments only contain one surface of the cylindrical shell; (**a**) tensile–shear mixed fracture mode; (**b**) pure shear fracture mode.

However, scholars have different understandings of the fracture process and the formation mechanism of fragments [3,29,30]. It is generally acknowledged that under low detonation pressure, the radial cracks initially occur on the outer surface under circumferential stress, and the thermoplastic shear band appears close to the inner wall, forming shear–tensile mixed fracture phenomenon. Under high detonation pressure, the inner wall undergoes thermoplastic adiabatic shear instability under compressive stress. The inward development of the radial cracks in the outer surface is inhibited and it develops along the shear direction; it shows pure shear fracture at last [6], However, Zhang et al. [31] used numerical simulation to analyze the equivalent plastic strain evolution of the cylindrical metal shell in its explosion. They believed that the fracture of the ideal cylindrical shell cannot start from the outer surface. The tensile fracture starting from the outer surface may have a geometrical relationship with the defects in the surface of the shell. Zhen et al. [32] analyzed the metallographic structure of the exploding fracture fragments of the thick-walled cylinder, and believed that under low explosive detonation pressure, the microcrack damage zone was first generated inside the wall thickness of the cylindrical shell, with the explosion pressure loading, the shell fractures. Since the zone near the inner wall of the cylindrical shell is under a state of compressive stress and the outer wall is under a state of tensile stress, the crack expands along radially outwards and expands along the shear concentration zone toward the inner surface, forming a tensile–shear mixed fracture.

Since most experimental research does not systematically design the experimental loading conditions, it is difficult to carry out a univariate comparison by verification analysis of the literature results. In China, Hu et al. [15], Tang et al. [33], and Hu et al. [34] carried out a series of comparative experimental studies on cylindrical shells and found different fracture modes, such as shear fracture and tensile–shear mixed fracture; there is also a single-rotation adiabatic shear fracture phenomenon under high detonation pressure. It can be seen that there is broad understanding of the fracture process of cylindrical shells under different explosion pressures, but the fracture mechanism is not yet clear.

In Tang's [34] experiment, JOB-9003 and hollow RHT-901 explosives were used to carry out a series of experimental studies on the explosion and fragmentation of a 45 steel cylindrical shell. Under different detonation pressures and charging conditions, the cylindrical shells of the same material, explosion, and fragmentation had pure shear fractures and tensile–shear mixed fractures.

The 45 steel cylindrical shell shows pure shear fracture under JOB-9003 explosive loading. A cylindrical shell with 4 mm thickness shows tensile–shear mixed fracture under the hollow RHT-901 explosive loading; the proportion of tensile cracks is extremely low, so it is almost pure shear fracture; when the wall thickness is 5 mm, the proportion of tensile cracks accounts for one-third of the shell thickness [33]. The cylindrical shell size, loading conditions, and fracture phenomenon are listed in Table 1. (Relative sized δ_d = shell thickness h/inner radius R of cylindrical shell). The average strain and strain rate of the cylindrical shell at time t were calculated using the formula: $\varepsilon_t = (R(t) - R_0)/R_0$, $\dot{\varepsilon} = d\varepsilon_t/dt \approx V_t/R_0$. The statistical experimental data are shown in Table 1 [15,33,34].

Table 1. Experimental cylindrical shell, loading conditions, and explosion–expansion fracture phenomenon, Refs. [15,33,34].

Types of Explosives	Wall Thickness/Inner Radius/d/R	δ_d	ε_c	t_c/μs	ε_f	t_f/μs	$\dot{\varepsilon}/10^4\ s^{-1}$	Fracture Mode
RHT-901	5/30	0.17	0.18	7.8	0.37	15.8	2.5	tensile–shear mixed fracture
	4/30	0.13	0.24	8.8	0.43	15.4	2.9	tensile–shear mixed fracture
	3/30	0.10	0.30	10.1	0.45	14.1	3.5	pure shear
JOB9003	6/20	0.30	0.44	11.5	0.89	23.5	4.5	pure shear
	5/20	0.25	0.42	9.5	1.24	22.5	6.0	pure shear
	4/20	0.20	0.40	7.5	1.31	19.5	7.1	pure shear
	3/20	0.15	0.38	6.5	0.88	13.5	8.4	pure shear

Relative sized δ_d = shell thickness h/inner radius R of cylindrical shell ε_c—initial fracture strain, the radial strain at the moment t_c when cracks appear on the outer wall of the cylindrical shell. ε_f—product leakage strain, the radial strain at the moment t_f when the explosion products leak $\dot{\varepsilon}$—strain rate.

3. SPH Numerical Simulation Model and Verification

3.1. Model Construction

When building the SPH model, the arrangement method of the particles is very important, which affects the accuracy of the result. LS-DYNA software provides automatic polar coordinate point distribution mode and the rectangular coordinate point distribution mode, but these two methods cause local uniform particle dispersion on the contact surface of the explosive metal that are not in one-to-one correspondence. Therefore, while building the cylindrical metal shell model, it is necessary to simultaneously satisfy the uniform dispersion of particles in the explosive propagation process and the symmetry of the interface contact between the annular explosive and the metal cylinder shell. It is required that the particle arrangement and spacing of the SPH model should be kept as uniform as possible, and the interparticle symmetrical arrangement of the interface should be maintained. For this reason, the initial geometrical arrangement of particles is controlled by geometric grid cell division during modeling. Then, the method of generating particles at the grid cell center is used to ensure that the final arrangement and distribution of SPH particles are uniform and the particle pairs at the interface are symmetrically arranged, as shown in Figure 2.

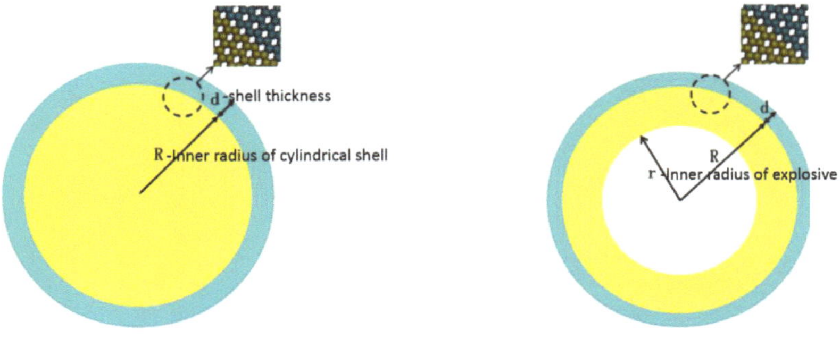

(a) (b)

Figure 2. SPH numerical model: (**a**) JOB9003 packing (R = 30 mm) and (**b**) RHT-901 hollow packing (R/r = 30 mm/20 mm).

According to the literature [35,36], SPH particles have the same performance as atoms, that is, they diverge when they are close to each other and attract each other when they are far apart, thus resulting in particle aggregation during the process of calculation. Therefore, the artificial stress represented by Monaghan can effectively eliminate the tensile instability caused by SPH, so that the SPH method can maintain continuity in higher-order calculations. Monaghan artificial viscosity is controlled by parameters α and β in practical application. The β coefficient controls the penetration of interfacial particles (as shown in Figure 3). When the simulated detonation pressure is very high, it is necessary to increase the value of β to prevent nonphysical penetration, but the increase in β decreases the peak detonation pressure, increases the calculation time, and also changes the shear band development.

To improve the accuracy and tensile stability, moving least-squares-based formulation is used. This method is suitable for large deformation models, but it requires a large computational cost. MPP simulation needs to set the maximum and minimum smooth length scale factor at 1.00 to ensure that the calculation is correct, then by adjusting a quasi-linear approximation term QL in the algorithm. To the combination of precision and stability in an extremely large deformation simulation, if QL is set too small, the model will have higher accuracy, but the stability problem is outstanding; if QL is too large, the model will inhibit instability but have lower precision. By comparison, QL = 0.01 was chosen to achieve a good compromise between stability and accuracy.

The setting of β changes the detonation pressure. With the increase in β, the value of interface load decreases (as shown in Figure 4), changing the distribution, distance, and development of the shear band in the shell thickness. When the stability of contact between particles is satisfied, no particle penetration occurs at the interface; β is required to be as small as possible to obtain a clear shear band. Therefore, $\beta = 1$ was selected as the value of its viscosity coefficient in this paper. In this way, the stability of the particles can be controlled and the failure of the shell can be accurately reflected.

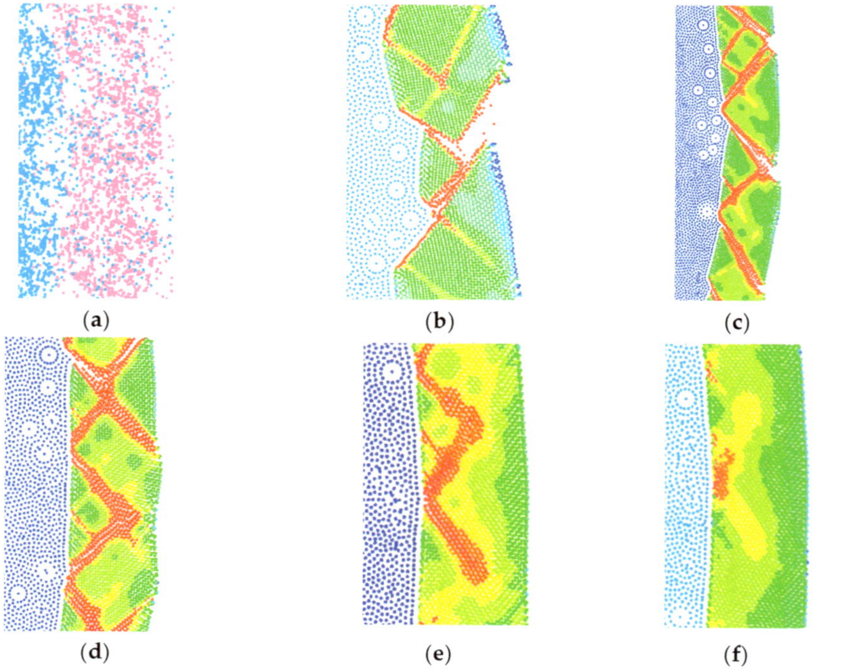

Figure 3. Comparison of the effects of different β values: (**a**) 14.2 μs, β = 0.8; (**b**) 14.2 μs, β = 1; (**c**) 14.2 μs, β = 2; (**d**) 14.2 μs, β = 3; (**e**) 14.2 μs, β = 7; (**f**) 14.2 μs, β = 10.

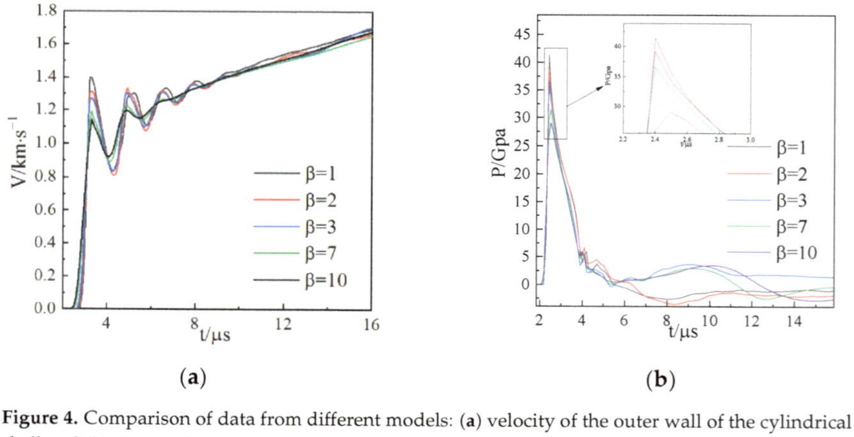

Figure 4. Comparison of data from different models: (**a**) velocity of the outer wall of the cylindrical shell and (**b**) internal hydrostatic pressure.

The SPH distribution distance has an impact on the model accuracy and internal stress propagation. If the particle spacing distribution distance decreases, the model accuracy is improved slightly, but the calculation amount is increased greatly. Finally, the model SPH distribution distance of 50 μm is chosen; it is close to the experimental result and requires less calculation.

3.2. Material Parameters of the Model

The 45 steel's rate of change under high pressure is described by the Gruneison equation of state [37]. The plasticity is described by the Johnson–Cook constitutive equation considering strain hardening, strain rate hardening, and thermodynamic softening effects.

$$\widetilde{\sigma} = \left(A + B\overline{\sigma}^{pn}\right)\left(1 + Cln\frac{\dot{\varepsilon}}{\dot{\varepsilon}_0}\right)\left(1 - T^{*m}\right) \tag{1}$$

Adapting the experimental results of Hu [38], from the same institute as Tang, he conducted an experimental study on the strain hardening, strain rate hardening, and thermodynamic softening characteristics of 45 steel, and obtained the parameters of the Johnson–Cook constitutive equation by fitting. Since the quasi-static yield strength $\sigma_{0.2}$ = 350 MPa, the fracture strength σ_b = 600 MPa, and the elongation of 45 steel used for the experimental cylindrical shell are 17%, the yield strength is lower than that for materials reported in the literature [38]. The strain hardening term in the Johnson–Cook constitutive equation is $\left(A + B\overline{\sigma}^{pn}\right)$; the quasi-static yield strength and fracture strength properties of 45 steel used in the experiment were used for fitting and correction. The following were obtained: parameter A = 350 MPa, B = 600 MPa, n = 0.307. Other parameters of strain rate hardening and thermal softening are adopted directly from those reported in the literature [38], as shown in Table 2.

Table 2. Constitutive parameter values of 45 steel cylindrical shell [38].

Johnson–Cook						Grüneison State Equation		
A/MPa	B/MPa	n	m	C	$\dot{\varepsilon_0}$	C/m·s^{-1}	s	γ_0
350	600	0.307	0.804	0.07	2×10^{-4}	4600	1.49	2.17

To facilitate the analysis, the failure model of 45 steel adopts the simplified plastic damage accumulation softening failure criterion and does not consider the influence of stress state, strain rate, or temperature on the damage. The damage degree D is:

$$D = \begin{cases} 0, & \overline{\varepsilon}^P < \varepsilon_{cr} \\ \frac{\sum \Delta\overline{\varepsilon}^P - \varepsilon_{cr}}{\varepsilon_f - \varepsilon_{cr}}, & \overline{\varepsilon}^P > \varepsilon_{cr} \end{cases} \tag{2}$$

where D is the damage parameter, $\Delta\overline{\varepsilon}^P$ is the plastic variable increment of a time step, ε_{cr} is the material plastic initial damage strain, and ε_f is the material failure strain. We take ε_{cr} = 0.48 and material failure strain ε_f = 1.0. Once damage occurs, the material softens:

$$\sigma = \widetilde{\sigma}(1 - D) \tag{3}$$

The explosive uses the JWL equation of state. The specific parameters are shown in Table 3 [39].

Table 3. JWL EOS parameters of the explosives [39].

Types of Explosives	A/GPA	B/GPA	ω	R_1	R_2	E_0/GJm^{-3}	P_{cj}/GPA	ρ/Kg·m^{-3}	D/m·s^{-1}
JOB9003	842.0	21.81	0.28	4.6	1.35	1	35	1884	8740
RHT-901	503.0	9.065	0.35	4.3	1.1	7.6	27	1658	7800

3.3. Validation of Numerical Models

The experimental results of detonation and expansion of 45 steel cylindrical shell given in Table 1 are taken as the object for the SPH numerical simulation. Figure 5 is the plastic strain diagram of the fracture process for 45 steel under the loading of JOB9003.

The beginning of time in the figure is defined uniformly as the moment when the detonation wave reaches the inner wall of the cylindrical shell. The SPH numerical results show that the plastic damage and fracture of the material first start in the middle of the shell thickness. The crack expands to the inner and outer surfaces along the shear direction of 45° or 135° to the radial direction. When t = 7.5 μs, the shear crack penetrates the shell; at that time, the shear crack surface is loaded by the internal detonation products, so the detonation products cannot leak immediately under the action of compressive stress. When t = 17.9 μs, the shell fragments on both sides of the crack disengaged and the detonation products leaked from the opening. The numerical simulation fracture mode is consistent with the trend of the experimental phenomenon.

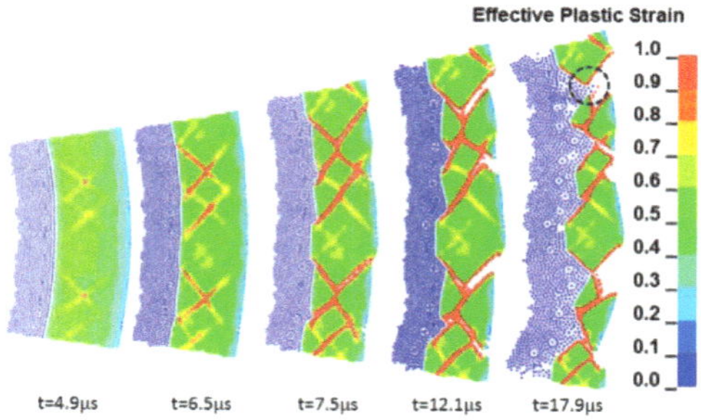

Figure 5. The evolution process of pure shear fracture under JOB9003 loading obtained by numerical simulation.

Figure 6 shows the comparison between the velocity–time history curve of the outer wall of the cylindrical shell obtained by SPH numerical simulation and the experimental DPS test results. They fit well; the time t_c of visible cracks on the outer surface of the cylindrical shell and the time of detonation particle leakage t_f are marked.

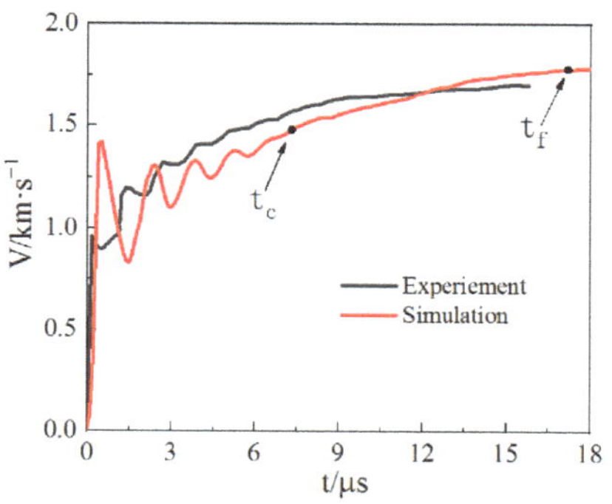

Figure 6. The radial expansion velocity for the outer surface of the cylindrical shell.

A numerical simulation of the fracture of 45 steel cylindrical shell with an inner diameter of 30 mm and wall thickness of 4 and 5 mm under the loading of hollow RHT-901 was carried out. The fracture model is a tensile–shear mixed fracture. Figure 7 shows the equivalent plastic strain and fracture evolution process of cylindrical shells with 4 and 5 mm thickness. It can be seen that the fracture of the cylindrical shell still starts from the middle of the wall thickness of the cylindrical shell and expands in the shear direction. In contrast to the shear evolution fracture under the action of the JOB9003 explosive, a similar "necking" phenomenon occurs at the joint of the shell when the shear crack develops close to the outer surface of the cylindrical shell. The crack direction turns to radial development. The final fracture presents a tensile–shear mixed fracture.

Among them, the shear fracture mainly occurs in the cylindrical shells with a thickness of 4 mm; tensile features can be seen locally, and the proportion of radial tensile cracks is very small, as shown in Figure 7a. The proportion of the cylindrical shell with a thickness of 5 mm has a significant increase in radial cracks, accounting for nearly one-third of the shell thickness, as shown in Figure 7b. This result is consistent with the experimental phenomenon and fracture characteristics described in the experiment [29].

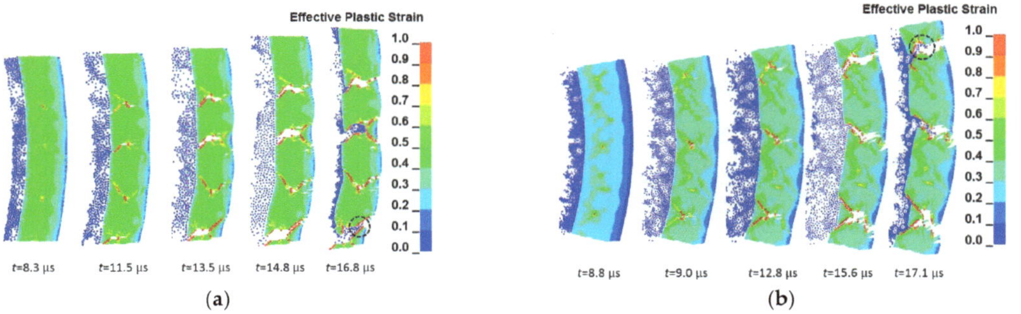

Figure 7. Simulation results of cylindrical shell expansion and fracture process with different R/h under a load of hollow RHT-901 explosive: (**a**) inner diameter 30 mm, shell thickness 4 mm; (**b**) inner diameter 30 mm, shell thickness 5 mm.

Taking the detonation wave reaching the outer surface of the cylindrical shell as 0 μs, the numerical simulation results above are used in the same method as the experiment to obtain the apparent expansion strain of the cylindrical shell at the characteristic time: $\varepsilon_t = (R(t) - R_0)/R_0$. R_0 is the initial radius of the cylindrical shell. $R(t)$ is the expansion radius of the cylindrical shell at the characteristic time t of the detonation loading process. Finally, the initial fracture strain ε_c at time t_c is obtained when visible cracks are on the outer surface, together with the shell fracture strain ε_r at time t_r and the product leakage strain ε_f at the explosion product leakage time t_f. The expansion strain rate $\dot{\varepsilon}$ of the cylindrical shell is approximated by the formula: $\dot{\varepsilon} = d\varepsilon_t/dt \approx V_t/R_0$. V_t is the expansion velocity when the crack penetrates the shell thickness. Under the loading of JOB9003 and hollow RHT-901 explosives, the comparison between the SPH numerical experimental results and the experiment results [29] shows that the apparent initial fracture strain ε_c and the product leakage strain ε_f are in good agreement, as shown in Table 4.

Table 4. Comparison of numerical simulation and experiment results.

Types of Explosives	R/h		ε_c	t_c/μs	ε_r	t_r/μs	ε_f	t_f/μs	$\dot{\varepsilon}/10^4$ s^{-1}	Fracture Mode
RHT-901	30/3	experiment *	0.30	10.1	/	/	0.45	14.1	3.5	pure shear
		simulation	0.39	13.2	0.39	13.3	0.45	15.2	3.1	pure shear
	30/4	experiment *	0.24	8.8	/	/	0.43	15.4	2.9	tensile–shear mixed fracture
		simulation	0.30	13.1	0.31	13.5	0.41	16.9	2.7	tensile–shear mixed fracture
	30/5	experiment *	0.18	7.8	/	/	0.37	15.8	2.5	tensile–shear mixed fracture
		simulation	0.21	11.9	0.23	12.8	0.30	15.6	2.1	tensile–shear mixed fracture
JOB9003	20/3	experiment *	0.38	6.5	/	/	0.88	13.5	8.4	pure shear
		simulation	0.36	6.0	0.37	6.1	0.89	12.9	8.2	pure shear
	20/4	experiment *	0.40	7.5	/	/	1.31	19.5	7.1	pure shear
		simulation	0.37	7.3	0.37	7.3	1.07	16.7	7.1	pure shear

Table 4. *Cont.*

Types of Explosives	R/h		ε_c	t_c/μs	ε_r	t_r/μs	ε_f	t_f/μs	$\dot{\varepsilon}$/10^4 s^{-1}	Fracture Mode
	20/5	experiment *	0.42	9.5	/	/	1.24	22.5	6.0	pure shear
		simulation	0.41	9.4	0.42	9.6	1.21	22.3	6.4	pure shear
	20/6	experiment *	0.44	11.5	/	/	0.89	23.5	4.5	pure shear
		simulation	0.43	11.4	0.45	11.7	0.90	20.0	5.4	pure shear

ε_c—initial fracture strain, the radial strain at the moment t_c when cracks appear on the outer wall of the cylindrical shell; ε_f—products leakage strain, the radial strain at the moment t_f when the explosion products leak $\dot{\varepsilon}$—strain rate; experiment *—data from the literature [30–32].

The results show that SPH can simulate the fracture mode of 45 steel cylindrical shell under different ways of filling and detonation pressure of JOB-9003 explosive and hollow RHT-901 explosive, which is consistent with the shear fracture, tensile–shear mixed fracture mode in experimental results. The changing trend of the fracture strain with the loading strain rate is also consistent with the experiment. The results show that the experimental end-state fragmentation phenomenon of shear and the tensile–shear mixed fracture mode under different explosive conditions can be well simulated by the SPH numerical model with the unified material constitutive and fracture model.

4. Fracture Results and Analysis of 45 Steel Cylindrical Shell under Different Detonation Pressure

4.1. Numerical Simulation Models

The abovementioned SPH simulation model (Figure 2a) was used to simulate the expansion fracture process of 45 steel cylindrical shell under four detonation pressures; the models of explosives are all fully filled SPH models with an inner diameter of 40 mm and shell thickness of h = 4 mm.

The waveform is a triangular wave obtained by fitting the JWL high-pressure state equation (Figure 8). The explosive parameters are shown in Table 5.

Table 5. Parameters related to the JWL constitutive equation of explosives.

Types of Explosives	A/MPA	B/MPA	ω	R_1	R_2	E_0/GJm^{-3}	P_{cj}/GPA	ρ/Kg·m^{-3}	D/m·s^{-1}
JOB9003	842.0	21.81	0.28	4.6	1.35	10	35	1884	8740
RHT-901	503.0	9.065	0.35	4.3	1.1	7.6	27	1658	7800
TNT	371.2	3.231	0.30	4.15	0.95	3	21	1730	6930
HNS	162.7	10.82	0.25	5.4	1.8	4.1	7.9	1000	5100

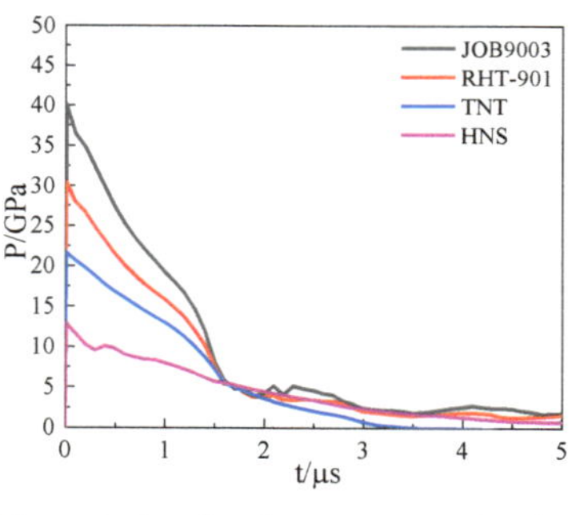

Figure 8. Loading history curves for the inner wall of the cylindrical shell under different explosion pressures.

It can be seen from the loading curve that the pulse width of the loading wave is the same, but the peaks of the loading pressures are different; the pressure peaks of the explosive detonation wave are about 40, 31, 21, and 13 GPA, and time to reach the shell inner wall was 2.2, 2.5, 2.8, and 3.8 μs. The following results take the time when the detonation wave reaches the inner wall as the starting time. Through the following results, the expansion and fracture evolution process of 45 steel under different explosive loads are analyzed.

4.2. Numerical Results and Analysis

Under the loading of JOB9003 explosive with high detonation pressure, the pressure propagation characteristics of the inner surface, the outer surface, and the mid-shell thickness of the 45 steel cylindrical shell with an inner diameter of 40 mm and shell thickness of 4 mm can be seen in Figure 9. Note that the inner surface is loaded with a peak value of about 40 GPA; t_i represents the initial damage time, t_c represents the cylinder shell penetration fracture time, and t_f represents the detonation products leakage time. (1) The inner surface of the shell is loaded by detonation products and is in a state of hydrostatic stress, the outer surface is in a state of tension under the tensile stress, and the middle of the shell thickness is in a tension–compression fluctuation state. The penetration fracture of the cylindrical shell occurs in the loading stage (the inner surface of the cylindrical shell is subjected to compressive stress and regarded as the loading stage; when the pressure drops to 0, the loading stage ends), as shown in Figure 9a. (2) Before the detonation shock wave is transmitted to the outer surface of the cylindrical shell, the effective plastic strain on the inner surface accumulates the most; the effective plastic strain has a decreasing distribution from the inner wall to the outer wall. When the shock wave reaches the outer surface and is reflected, the effective plastic strain in the middle of the metal shell is the largest, and the overall distribution is convex for the secondary plastic accumulation; the strain in the middle is always the position with the largest plastic strain. It is stipulated in the previous model establishment that, when the strain accumulation reaches $D = 0.48$, the cylindrical shell material becomes damaged and softened; when the strain accumulates rapidly to $D = 1.00$, the initial failure occurs, as shown in Figure 9b; when $t = 2.5$ μs, the material damage first forms in the middle of the wall thickness. When $t = 4.9$ μs, the failure particles appear in the middle of the shell thickness and expand to the inner and outer surfaces along the 45° and 135° directions; they extend to the inner and outer surfaces of the cylindrical shell when $t_c = 7.5$ μs, forming shear cracks and penetrating the cylindrical shell. When $t_f = 17.0$ μs, the detonation products leak from the opening fracture of the cylindrical shell, and the fragment fracture finally shows a pure shear shape, as indicated in Figure 9c.

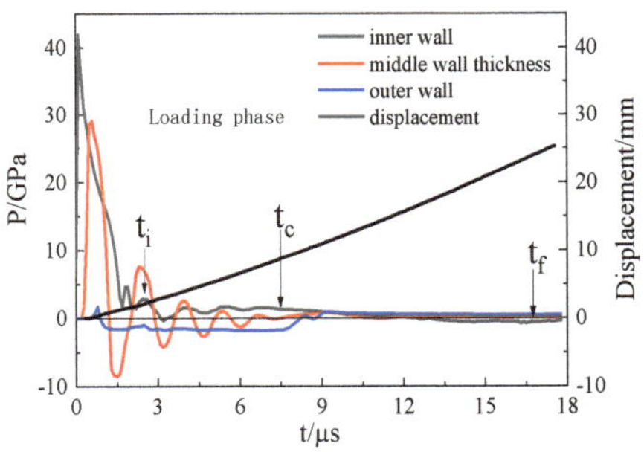

(a)

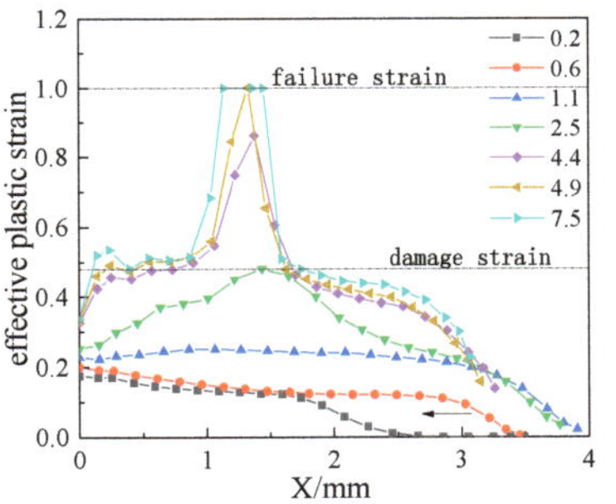

(b)

Figure 9. *Cont.*

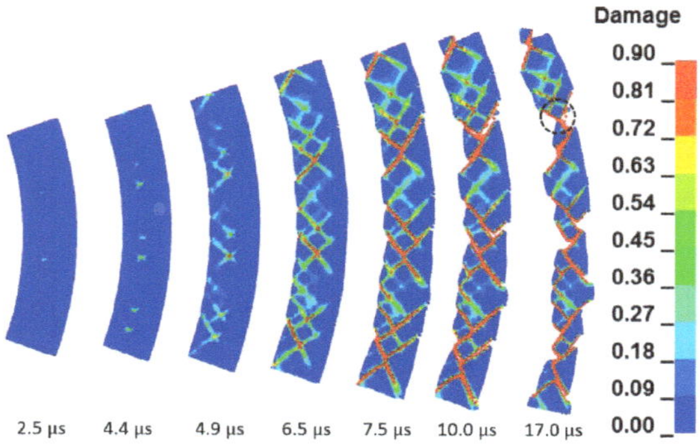

(c)

Figure 9. The explosive pressure, expanding deformation, and fracture for 45 steel cylinder shell under JOB9003 loading: (**a**) detonation wave propagation curves; (**b**) development of effective plastic strain in the shell thickness; (**c**) damage evolution for pure shear fracture.

The fracture process of 45 steel shell under RHT-901 loading is similar to JOB9003 loading. It shows pure shear fracture and fracture occurs in the loading stage. Under RHT−901 loading, the fracture and detonation product leakage time is later, as shown in Figure 10.

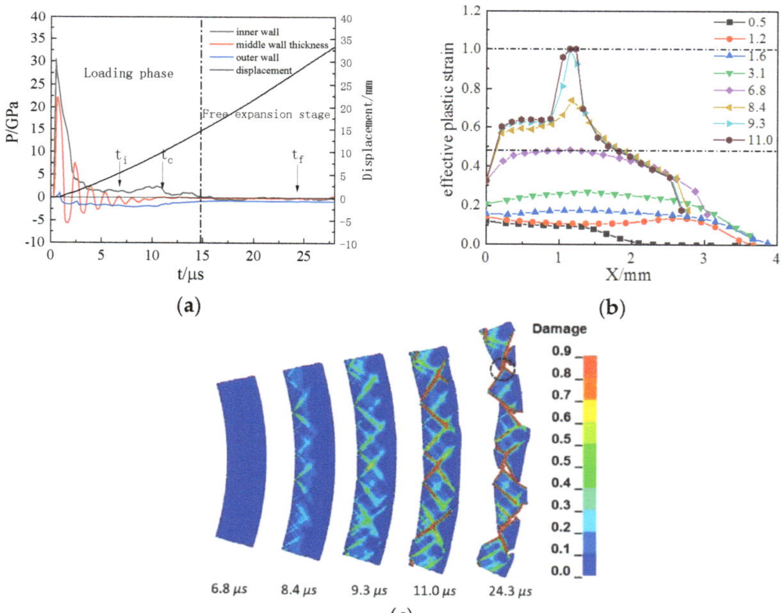

(c)

Figure 10. The explosive pressure, expanding deformation, and fracture for 45 steel cylinder shell under RHT−901 loading: (**a**) detonation wave propagation curves; (**b**) development of effective plastic strain in the shell thickness; (**c**) damage evolution for fracture.

Under the loading of TNT with a lower detonation pressure, the fracture mode of the cylindrical shell shows tensile–shear mixed fracture mode. The inner surface loaded by compressive stress ends at time t = 11.2 μs, and then the shell enters the free expansion stage, as shown in Figure 11a. In the development of effective plastic strain in shell thickness, as shown in Figure 11b, the initial crack first forms near the inner wall; at this time, the cylindrical shell is still in the loading stage, and the crack along the 45° or 135° shear direction expands to the inner and outer surfaces. When the cylindrical shell enters the free expansion stage, t = 11.2 μs, the time–history curve of the triaxiality stress law at any position in the unbroken area is similar to position "A", as shown in Figure 11d; the unbroken area of the cylindrical shell appears similar to the phenomenon of tensile "necking"—the crack development turns to the radial direction, finally becoming a tensile–shear mixed fracture. The proportion of tensile cracks is relatively small, about one-sixth, at time t = 12.6 μs. The cylindrical shell becomes a penetrative fracture when t = 21.8 μs, and detonation products leak.

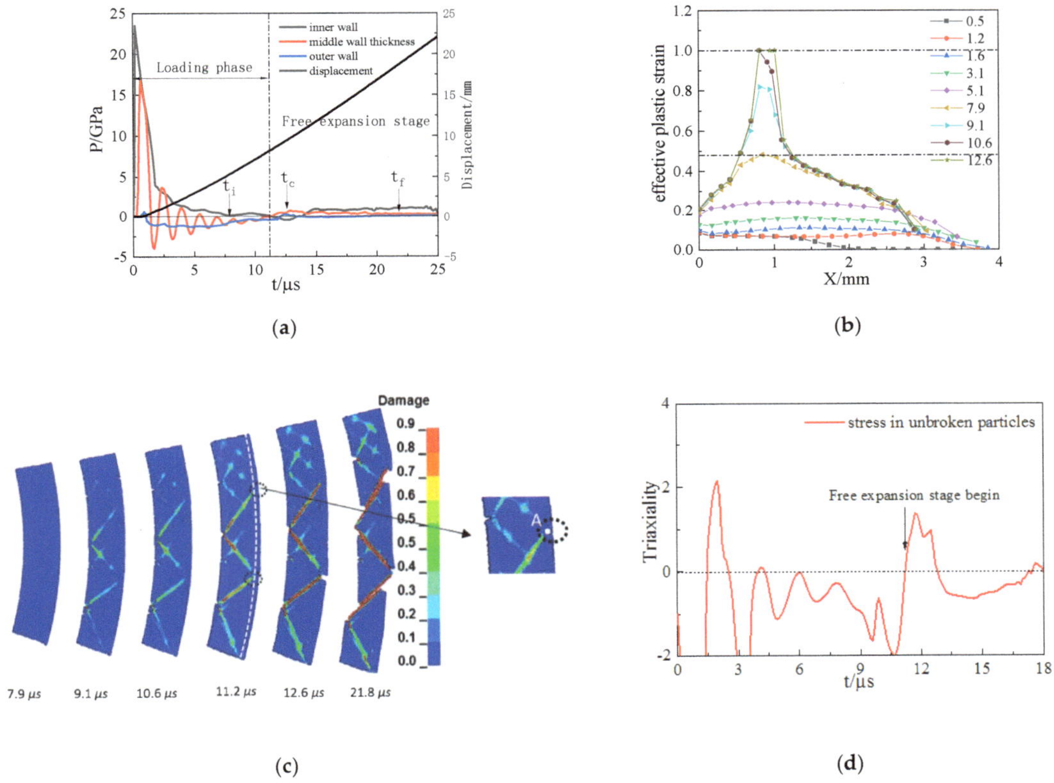

Figure 11. The explosive pressure, expanding deformation, and fracture for 45 steel cylinder shell under TNT loading: (**a**) detonation wave propagation curves; (**b**) development of effective plastic strain in the shell thickness; (**c**) damage evolution for fracture; (**d**) triaxial stress.

Under the loading of HNS with low detonation pressure, the 45 steel shell also undergoes tensile–shear mixed fracture, but tensile cracks accounted for a higher proportion in the fracture, about one-half. Shear cracks are less developed because fracture development takes less time during the loading phase, as shown in Figure 12.

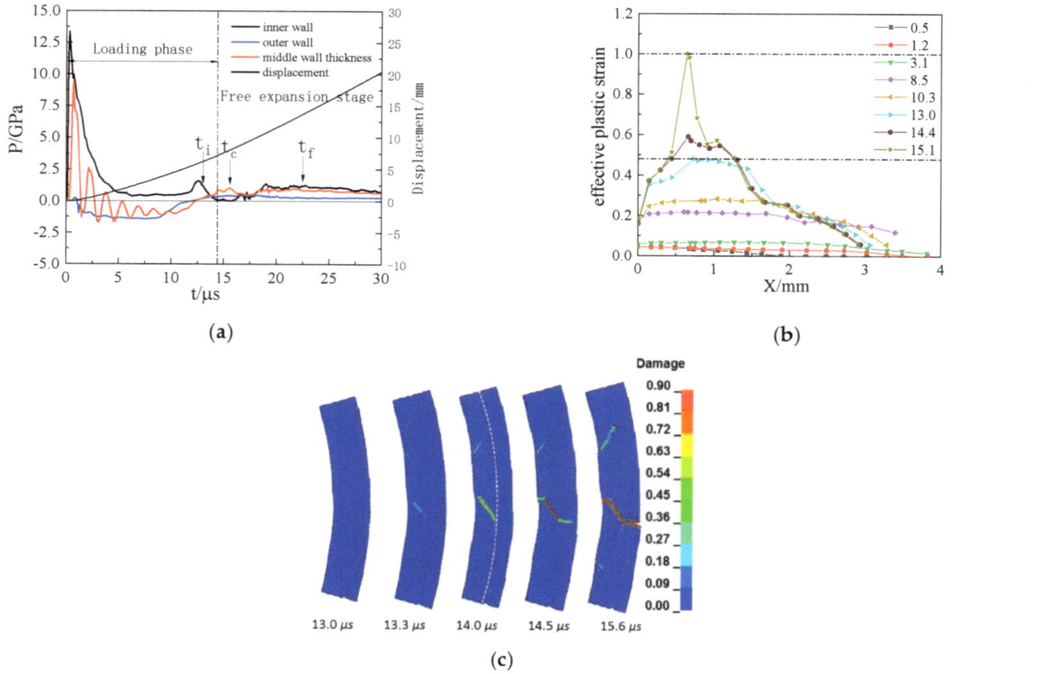

Figure 12. The explosive pressure, expanding deformation, and fracture for 45 steel cylinder shell under HNS loading: (**a**) detonation wave propagation curves; (**b**) development of effective plastic strain in shell thickness; (**c**) fracture damage evolution.

5. Conclusions

The following conclusions were drawn from the SPH numerical simulation analysis of the expansion fracture process of 45 steel cylindrical shell under different detonation pressure loading:

1. The experimental end-state fragmentation phenomenon of shear and tensile–shear mixed fracture under different explosive conditions can be simulated well by the SPH numerical model with a unified material constitutive and fracture model.
2. As the detonation pressure decreases, the fracture mode of 45 steel cylindrical shell changes from pure shear to tensile–shear mixed fracture. Cracks always form in the middle of shell thickness.
3. Under higher detonation pressure loading (JOB9003, RHT-901), the crack initiation and fracture of the 45 steel cylindrical shell occur during the loading stage; the failure of the cylindrical shell begins at mid-thickness in which the plastic strain accumulation is at its maximum. It spreads in the shear direction to the inner and outer walls, resulting in pure shear fracture mode.
4. As the detonation loading pressure decreases, loaded by explosives of TNT and HNS, on account of the initial fracture occurring during the loading stage, the fracture still begins from the middle of the shell thickness and spreads along the shear direction to the inner and outer walls. When the cylindrical shell expands to the free expansion stage, the unbroken area is under circumferential tensile stress, resulting in a phenomenon known as "necking". The growing crack turns to the radial direction along the "necking" area, resulting in a tensile–shear mixed fracture mode. Furthermore, the lower the explosion pressure, the earlier the cylindrical shells enter the free expansion stage and the higher proportion of final tensile cracks.

Author Contributions: Conceptualization, Z.H., X.Y.; methodology, Z.H.; software, Z.H.; validation, Z.H.; formal analysis, Z.H.; investigation, Z.H.; resources, Z.H.; data curation, Z.H.; writing—original draft preparation, Z.H.; writing—review and editing, X.Y.; visualization, Z.H.; supervision, X.Y.; project administration, X.Y.; funding acquisition, Z.H. All authors have read and agreed to the published version of the manuscript.

Funding: This research received no external funding.

Institutional Review Board Statement: Not applicable.

Informed Consent Statement: Not applicable.

Data Availability Statement: Not applicable.

Conflicts of Interest: The authors declare no conflict of interest.

References

1. Mott, N.F. Fragmentation of shell cases. *Proc. R. Soc. Ser. A. Math. Phys. Sci.* **1947**, *189*, 300–308.
2. Beetle, J.C. SEM/1971(Part I). In Proceedings of the 4th Annual SEM Symposium, Chicago, IL, USA, 27–29 April 1971; IIT Research Institute: Chicago, IL, USA; pp. 137–144.
3. Singh, M.; Suneja, H.R.; Bola, M.S.; Prakash, S. Dynamic tensile deformation and fracture of metal cylinders at high strain rates. *Int. J. Impact Eng.* **2002**, *27*, 939–954. [CrossRef]
4. Arnold, W.; Rottenkolber, E. Fragment mass distribution of metal cased explosive charges. *Int. J. Impact Eng.* **2008**, *35*, 1393–1398. [CrossRef]
5. Grady, D.E. Length scales and size distributions in dynamic fragmentation. *Int. J. Fract.* **2010**, *163*, 85–99. [CrossRef]
6. Gurney, R.W. The initial velocities of fragments from bombs, shell, grenades. *Ballist. Res. Lab.* **1943**, *405*, 1–12.
7. Taylor, G.I. Analysis of the Explosion of a Long Cylindrical Bomb Detonated at One End. In *Mechanics of Fluids, Scientific Papers of GI Taylor*; Technical Report HO 195/10/193; Cambridge University Press: Cambridge, UK, 1941; Volume 2, pp. 277–286.
8. Hoggatt, C.R.; Recht, R.F. Fracture behavior of tubular bombs. *J. Appl. Phys.* **1968**, *39*, 1856–1862. [CrossRef]
9. Martineau, R.L.A. Viscoplastic Model of Expanding Cylindrical Shells Subjected to Internal Explosive Detonations. Ph.D. Thesis, Colorado State University, Fort Collins, CO, USA, 1998.
10. Jovičević-Klug, M.; Verbovšek, T.; Jovičević-Klug, P.; Šetina Batič, B.; Ambrožič, B.; Dražić, G.; Podgornik, B. Revealing the Pb whisker growth mechanism from Al-AlloySurface and morphological dependency on material stress and growth environment. *Materials* **2022**, *15*, 2574. [CrossRef]
11. Hofmeister, A.M.; Criss, E.M.; Criss, R.E. Thermodynamic relationships for perfectly elastic solids undergoing steady-state heat flow. *Materials* **2022**, *15*, 2638. [CrossRef]
12. Sidorova, E.; Karasev, A.; Kuznetsov, D.; Jönsson, P.G. Investigation of the initial corrosion destruction of a metal matrix around different non-metallic inclusions on surfaces of pipeline steels. *Materials* **2022**, *15*, 2530. [CrossRef]
13. Liang, M.Z.; Li, X.; Lu, F. Effects of U-notches on the dynamic fracture and fragmentation of explosively driven cylinders. *Theor. Appl. Fract. Mech.* **2015**, *77*, 50–58. [CrossRef]
14. Hu, B.Y.; Dong, Q.D.; Han, C.S.; Hu, H.B.; Wang, D.S.; Gao, X.H. Macroscopic study on adiabatic shear fracture of explosive metal tubes. *Explos. Shock.* **1992**, *4*, 319–325.
15. Hu, H.B.; Tang, T.G.; Hu, B.Y.; Wang, D.S.; Han, C.S. The single-spin phenomenon of metallic cylindrical shells under blast loading and fracture. *Explos. Shock.* **2004**, *24*, 97–107.
16. Liu, M.T.; Tang, T.G.; Hu, H.B.; Li, Q.Z.; Hu, X.Z.; Li, Y.C. Numerical Simulation of explosive-driven cylindrical shell expansion fracture under different detonation modes. *Explos. Shock.* **2014**, *34*, 415–420.
17. Yu, X.L.; Dong, X.L.; Pan, S.J. Fracture behaviors of explosively driven TA2 alloy cylinders under different loadings. *Explos. Shock. Waves* **2018**, *38*, 148–154. [CrossRef]
18. Liu, M.T.; Ren, G.W.; Fan, C.; Tang, T.G.; Wang, X.Y.; Hu, H.B. Experimental and numerical studies on the expanding fracture behavior of an explosively driven 1045 steel cylinder. *Int. J. Impact Eng.* **2017**, *109*, 240–252. [CrossRef]
19. Zhang, S.W.; Jin, S.; Liu, C.L. Numerical simulation of detonation expansion fracture of a thick-walled circular tube with defects. *Chin. J. Appl. Mech.* **2010**, *27*, 622–625, 653.
20. Zhai, M.H.; Wang, Z.K.; Jiao, H.W. Comparative analysis of SPH method and finite element method in meshless method. *Mach. Des. Manuf.* **2008**, *7*, 32–34.
21. Wu, S.S.; Dong, X.L.; Yu, X.L. SPH simulation analysis of explosive expansion fracture of 45 steel cylindrical shell. *Explos. Shock.* **2021**, *41*, 67–77.
22. Liu, G.R.; Liu, M.B.; Li, S. *Smoothed Particle Hydrodynamics—A Meshfree Method*; World Science: Hackensack, NJ, USA, 2004.
23. Remington, T.P.; Owen, J.M.; Nakamura, A.M.; Miller, P.L.; Bruck Syal, M. Numerical simulations of laboratory-scale, hypervelocity-impact experiments for asteroid-deflection code validation. *Earth Space Sci.* **2020**, *7*, e2018EA000474. [CrossRef]
24. Jankowiak, T.; Łodygowski, T. Smoothed particle hydrodynamics versus finite element method for blast impact. *Bull. Pol. Acad. Sciences. Tech. Sci.* **2013**, *61*, 111–121. [CrossRef]

25. Kong, X.; Wu, W.; Li, J.; Liu, F.; Chen, P.; Li, Y. A numerical investigation on explosive fragmentation of metal casing using Smoothed Particle Hydrodynamic method. *Mater. Eng.* **2013**, *51*, 729–741. [CrossRef]
26. Zhou, F.; Molinari, J.F.; Ramesh, K.T. A cohesive model-based fragmentation analysis: Effects of strain rate and initial defects distribution. *Int. J. Solids Struct.* **2005**, *42*, 5181–5207. [CrossRef]
27. Wright, T.W.; Ockendon, H. A scaling law for the effect of inertia on the formation of adiabatic shear bands. *Int. J. Plast.* **1996**, *12*, 927–934. [CrossRef]
28. Grady, D.E.; Kipp, M.E. The growth of unstable thermoplastic shear with application to steady-wave shock compression in solids. *J. Mech. Phys. Solids* **1987**, *35*, 95–119. [CrossRef]
29. Xiao, Y.H. Normalized SPH without boundary deficiency and its application to transient solid mechanics problems. *Meccanica* **2020**, *55*, 2263–2283. [CrossRef]
30. Goto, D.M.; Becker, R.; Orzechowski, T.J.; Springer, H.K.; Sunwoo, A.J.; Syn, C.K. Investigation of the fracture and fragmentation of explosively driven rings and cylinders. *Int. J. Impact Eng.* **2008**, *35*, 1547–1556. [CrossRef]
31. Zhang, S.W.; Liu, C.L.; Yu, J.Q. Influence of microdefects on expansion fracture of round tube. *Explos. Shock.* **2008**, *28*, 316–323.
32. Zhen, K.; Dong, X.L. Tensile-shear failure of 20 steel cylindrical shells subjected to explosive loading. *Ordnance Mater. Sci. Eng.* **2018**, *41*, 61–64. [CrossRef]
33. Tang, T.G.; Li, Q.Z.; Sun, X.L.; Sun, Z.F.; Jin, S.; Gu, Y. Strain rate effect on expansion fracture of 45 steel cylindrical shell. *Explos. Shock.* **2006**, *26*, 129–133.
34. Hu, B.Y.; Dong, Q.D.; Han, C.S.; Hu, H.B.; Wang, D.S. A mesoscopic study on adiabatic shear fracture of explosive metal pipes. *Explos. Shock.* **1993**, *13*, 305–312.
35. Gray, J.P.; Monaghan, J.J.; Swift, R.P. SPH elastic dynamics. *Comput. Methods Appl. Mech. Eng.* **2001**, *190*, 49–50. [CrossRef]
36. Belytschko, T.; Yong, G.; Shao, P.X. A unified stability analysis of meshless particle methods. *Int. J. Numer. Methods Eng.* **2000**, *48*, 1359–1400. [CrossRef]
37. Wang, L.L. *Basics of Stress Waves*; National Defense Industry Press: Beijing, China, 2005.
38. Hu, C.M.; He, H.L.; Hu, S.S. A study on dynamic mechanical behaviors of 45 steel. *Explos. Shock. Waves* **2003**, *23*, 188–192. [CrossRef]
39. Sun, C.W. *Applied Detonation Physics*; National Defense Industry Press: Beijing, China, 2000.

metals MDPI

Article

Influence of Mold Design on Shrinkage Porosity of Ti-6Al-4V Alloy Ingots

Tongzheng He [1,2] and Yuyong Chen [1,3,*]

1 School of Materials Science and Engineering, Harbin Institute of Technology, Harbin 150001, China
2 Pangang Group Panzhihua Research Institute of Iron & Steel Co., Ltd., Panzhihua 617000, China
3 College Vanadium and Titanium, Panzhihua University, Panzhihua 617000, China
* Correspondence: yychen@hit.edu.cn; Tel.: +86-0451-86418802

Abstract: Mold design is one of the important ways to control shrinkage porosity. In this study, four mold forms with different tapers were first designed, the corresponding three-dimensional finite element models were built using the ProCAST software, and the influence of mold design on the filling and solidification processes of Ti-6Al-4V alloy was investigated. The results showed that the titanium alloy ingots exhibit typical characteristics of layer-by-layer solidification, and that the removal of the riser results in: (a) shortening the time it takes for molten metal to reach the bottom of the mold and the time needed to complete mold filling; (b) decreasing the maximum flow velocity and improving the filling stability; and (c) moving the shrinkage cavities up along the central axis of the ingot and decreasing the cavity volume. Meanwhile, it was also found that the shrinkage cavity volume decreases significantly with increasing mold taper, meaning a significant increase in ingot utilization rate. The shrinkage cavity formation mechanism was revealed through macrostructure analysis. During solidification, a grain frame is formed as a large number of equiaxed crystals intersect, thus creating an isolated liquid phase zone. When the liquid in this zone solidifies, the last zone to do so, its volume shrinkage cannot be compensated, thus leading to the formation of a shrinkage cavity.

Keywords: mold design; Ti-6Al-4V alloy ingot; mold taper; numerical simulation; shrinkage porosity

Citation: He, T.; Chen, Y. Influence of Mold Design on Shrinkage Porosity of Ti-6Al-4V Alloy Ingots. *Metals* **2022**, *12*, 2122. https://doi.org/10.3390/met12122122

Academic Editors: Shuwen Wen, Yongle Sun and Xin Chen

Received: 11 November 2022
Accepted: 6 December 2022
Published: 9 December 2022

1. Introduction

Ti-6Al-4V (wt.%) alloy has been widely applied in aerospace, ocean engineering, biomedicine, and other fields owing to its good comprehensive mechanical properties and low production costs [1,2]. However, the defect of shrinkage porosity in Ti-6Al-4V alloy castings seriously reduces the serviceability of the alloy products and reduces the material utilization rate [3]. At present, researchers use various methods to control the formation of shrinkage cavities, including the hot isostatic pressing technique [3–5], optimizing the smelting method [6–11] and improving the casting process [12–16], and certain positive results have been achieved.

Research shows that the use of the hot isostatic pressing technique can effectively reduce shrinkage porosity in titanium alloy castings. Atkinson et al. [4] demonstrated clearly that reducing shrinkage porosity in as-cast microstructure is one of the main benefits of the hot isostatic pressing technique. Feng et al. [3] investigated the influence of the hot isostatic pressing technique on the shrinkage porosity of large, thin-walled Ti-6Al-4V alloy castings, finding that the method can not only effectively suppress the formation of shrinkage cavities, but can also significantly improve the bonding force of β grain boundaries, thus improving the toughness of the castings. However, the hot isostatic pressing process causes dimensional deformation of castings, and using the method it is difficult to eliminate shrinkage porosity completely [8,17–19]. With the continuous progress in casting technology in recent years, advanced processes such as centrifugal casting [6–8], anti-gravity casting [9,10], and vacuum suction casting [11] have been widely used in the preparation of titanium alloy castings. Yang et al. [6] compared the effects of centrifugal and gravity casting on the shrinkage porosity of TiAl alloy castings by means of numerical simulation and experimental verification. The results show that centrifugal casting is more effective at suppressing the formation of isolated liquid phase zones and at reducing the volume of the shrinkage cavity. Suzuki et al. [7] investigated the effect of centrifugal speed on the shrinkage porosity of Ti-6Al-4V alloy castings, revealing that increasing centrifugal speed can effectively suppress the formation of shrinkage cavities. However, when the centrifugal speed is too high, the flow of molten metal may stop, resulting in the formation of shrinkage cavities [8]. In their study on the anti-gravity casting process of Ti-6Al-4V alloy, Ma et al. [9] clearly illustrated that the formation of shrinkage cavities can be effectively controlled by improving mold filling and adopting pressurized solidification. By examining the effect of filling pressure on the shrinkage porosity of TiAl alloy during anti-gravity casting, Yang et al. [10] clarified that increasing filling pressure can significantly increase the fluidity of molten metal and reduce heat loss, helping to suppress the formation of pre-solidification zones and effectively improve the feeding capacity of the molten metal. However, excessively high filling pressure is not conducive to the control of shrinkage porosity. It is worth noting that proper design of the casting process, such as setting a suitable pouring temperature and mold preheating temperature, is always important in reducing shrinkage porosity. Research shows [12–16] that these factors can effectively improve the fluidity of the molten metal and reduce heat loss, ensuring the high feeding capacity of the molten metal and suppressing the formation of shrinkage cavities. In addition, Gao et al. [20] significantly reduced the shrinkage cavity volume in a TiAl alloy ingot by adding a temperature gradient to the mold, pioneering a new path for controlling shrinkage porosity in titanium alloy ingots.

Unfortunately, relatively little effort has been made to study the influence of mold design on the shrinkage porosity of titanium alloy castings. In this study, the influence of mold design on the shrinkage porosity of Ti-6Al-4V alloy ingots was systematically investigated using four casting mold forms with slightly different shapes. Adopting an approach combining numerical simulation and experimental verification, the influence of a riser on alloy filling and solidification processes was studied, the quantitative relationship between the mold taper and shrinkage cavity (including shrinkage cavity volume and material utilization rate) was analyzed in depth, and the formation mechanism of shrinkage cavities in titanium alloy ingots was elucidated, thus providing an effective way to control shrinkage porosity in titanium alloy ingots.

2. Materials and Methods

2.1. Finite Element Model

In this study, the casting process of Ti-6Al-4V (wt.%) alloy was systematically explored using the ProCAST 2018software (ESI Group, Paris, France). First, four common mold forms I–IV were designed (Figure 1) based on the casting equipment available, and the sample ingots prepared with the four mold forms were named alloys I–IV, respectively. Mold I features a riser with a height of 50 mm, and molds II–IV have no risers. The four ingots and molds have relative mass ratios of 0.56, 0.51, 0.55, and 0.47. Then, the sample ingots and molds were modeled in three dimensions and partitioned into meshes. The mesh of each ingot has an average cell size of 5 mm to ensure that there are at least two meshes at the thinnest part of the ingots, the mesh of each mold has an average cell size of 10 mm, and the number of volume mesh cells is about 334,737.

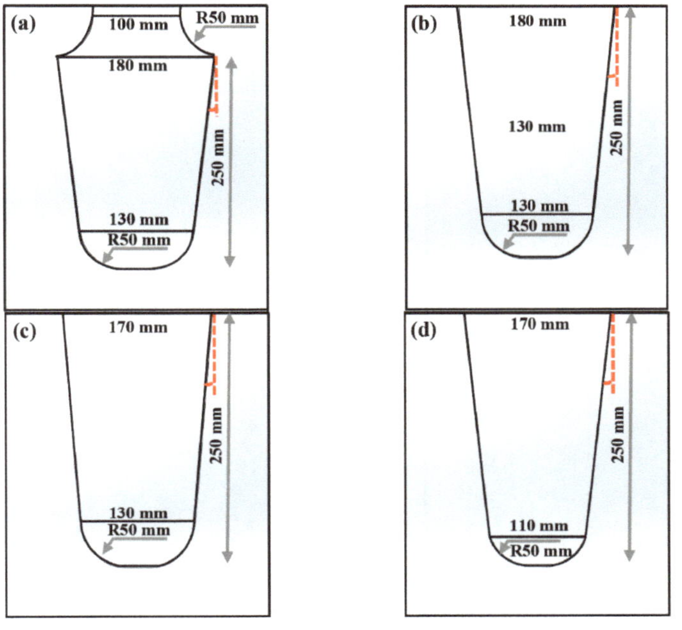

Figure 1. Mold design: (**a**) mold I; (**b**) mold II; (**c**) mold III; (**d**) mold IV.

2.2. Thermophysical Parameters and Boundary Conditions

Ti-6Al-4V alloy was chosen as the material for preparing the ingots and graphite was used as the mold material. The liquidus and solidus temperatures of Ti-6Al-4V alloy are 1660 °C and 1604 °C, respectively [5], and other thermophysical parameters were calculated using the ProCAST software, as shown in Table 1. Graphite has a density of 1690–1720 kg·m^{-3}, a specific heat capacity of 0.72–2.89 kJ·kg^{-1}·K^{-1}, and a thermal conductivity of 12–108 W·m^{-1}·K^{-1}. In the calculation process, the pouring temperature was set to 1700 °C, the mold preheating temperature was 300 °C, and the pouring speed was 4.7 kg·s^{-1}. Meanwhile, the stress type of the Ti-6Al-4V alloy was set to "linear-elastic", the stress type of graphite was "rigid", and the cooling method was "air cooling", with a heat transfer coefficient between the casting and the mold shell of 800 W·m^{-2}·K^{-1}.

Table 1. Physical parameters of Ti-6Al-4V based on ProCAST calculation.

Temperature (°C)	Density (kg·m^{-3})	Enthalpy (kJ·kg^{-1})	Viscosity (Pa·s)	Thermal Conductivity (W·m^{-1}·K^{-1})
25	4430	11.8	-	7.87
200	4322	153.6	-	10.92
400	4297	214.4	-	12.54
600	4262	369.8	-	14.47
800	4231	464.3	-	16.49
1000	4189	585.9	-	18.68
1200	4138	741.3	-	20.95
1400	4094	856.1	-	24.56
1600	4039	1052.0	2.14	28.09
1700	3938	1416.7	1.76	33.56

In this study, the flowing molten metal was regarded as an incompressible Newtonian fluid, and the governing Equations (1)–(5) in the process of mold filling and solidification are as follows [6,10,21]:

The Navier-Stokes equations for viscous fluids are [6,21]:

$$\rho\left(\frac{\partial u}{\partial t}+u\frac{\partial u}{\partial x}+v\frac{\partial u}{\partial y}+w\frac{\partial u}{\partial z}\right)=-\frac{\partial P}{\partial x}+\rho g_x+\mu\left(\frac{\partial^2 u}{\partial x^2}+\frac{\partial^2 u}{\partial y^2}+\frac{\partial^2 u}{\partial z^2}\right) \tag{1}$$

$$\rho\left(\frac{\partial v}{\partial t}+u\frac{\partial v}{\partial x}+v\frac{\partial v}{\partial y}+w\frac{\partial v}{\partial z}\right)=-\frac{\partial P}{\partial y}+\rho g_y+\mu\left(\frac{\partial^2 v}{\partial x^2}+\frac{\partial^2 v}{\partial y^2}+\frac{\partial^2 v}{\partial z^2}\right) \tag{2}$$

$$\rho\left(\frac{\partial w}{\partial t}+u\frac{\partial w}{\partial x}+v\frac{\partial w}{\partial y}+w\frac{\partial w}{\partial z}\right)=-\frac{\partial P}{\partial z}+\rho g_z+\mu\left(\frac{\partial^2 w}{\partial x^2}+\frac{\partial^2 w}{\partial y^2}+\frac{\partial^2 w}{\partial z^2}\right) \tag{3}$$

Continuity equation [10,21]:

$$\frac{\partial(\rho u_x)}{\partial x}+\frac{\partial(\rho u_y)}{\partial y}+\frac{\partial(\rho u_z)}{\partial z}=0 \tag{4}$$

Heat conduction equation [10,21]:

$$\rho Cp\left(\frac{\partial T}{\partial t}+u\frac{\partial T}{\partial x}+v\frac{\partial T}{\partial y}+w\frac{\partial T}{\partial z}\right)=\lambda\left(\frac{\partial^2 T}{\partial x^2}+\frac{\partial^2 T}{\partial y^2}+\frac{\partial^2 T}{\partial z^2}\right)+L\frac{\partial f_s}{\partial t} \tag{5}$$

where, ρ is the density of the molten metal; u, v, and w are the velocity vectors in the directions of x, y, and z respectively; P is pressure; g is the gravitational acceleration; μ is the viscosity; t is time; Cp is the specific heat capacity; T is the temperature; λ is the thermal conductivity coefficient; L is the latent heat of crystallization, and f_s is the solid fraction in the solidification stage.

2.3. Ingot Preparation and Macrostructure Characterization

Titanium alloy ingots weighing 20 kg each were prepared using a vacuum induction melting furnace. The raw materials include sponge titanium, pure Al, and Al-V alloy (V: 58.18wt.%). The iCAP 7400 ICP-OES analyzer (Thermo, Waltham, MA, USA) and the TCH600 N, H, and O analyzer (LECO, St. Joseph, MI, USA) were used to analyze the main elements (Al and V) and impurity elements (N, H, and O) of the ingots, respectively. The above test samples were all $\varphi 4 \times 10$ mm discs weighing less than 30 mg, which were mechanically ground and ultrasonically cleaned in alcohol for 10 min. For each ingot, at least three positions were chosen for analysis, and the results are displayed in Table 2 to guarantee the correctness of the test results. Then, the ingots were dissected along the center line to show the distribution of shrinkage cavities. Meanwhile, the as-cast macrostructure of each ingot was observed. The metallographic etching solution used in the observation was Kroll reagent (5% HF + 10% HNO_3 + 85% H_2O).

Table 2. Measured chemical compositions (wt.%) of the Ti-6Al-4V.

Alloys	Al	V	H	N	O	Ti
I	6.2	4.3	0.002	0.011	0.096	Bal.
II	5.9	4.1	0.002	0.015	0.104	Bal.
III	6.1	4.2	0.003	0.011	0.101	Bal.
IV	6.2	4.2	0.002	0.010	0.098	Bal.

3. Results

3.1. Numerical Simulation of Mold I

Figure 2 shows the filling process and the flow velocity of molten metal when mold I is used. At 0.24 s, the molten metal quickly reached the bottom of the mold under the action of gravity (Figure 2a), and the mold was completely filled at 4.09 s, with a filling rate of 98vol.% (Figure 2e). At the initial stage of mold filling, the molten metal has a high flow velocity (2.394 m·s^{-1}) and an obvious velocity gradient in the height direction (Figure 2g,h), but the mold filling stability is poor, which can easily lead to gas entrapment and molten metal splashing at the bottom of the mold. Xiong et al. [11] studied the influence of filling speed on the filling process of a TiAl alloy exhaust valve, reaching the following two conclusions. (1) A high filling speed causes a violent collision between the molten metal and the sprue, leading to the splashing of the molten metal. (2) A high filling speed causes serious turbulence at the surface of the molten metal, resulting in the entrapment of a large amount of gas. Liu et al. [22] also found in the investment casting process of a ZG310-570 steel bracket that an increase in filling speed significantly increases the probability of gas entrapment at the bottom of the casting. As the mold filling process progresses, the filling speed decreases obviously, the speed gradient weakens significantly, and the mold filling process stabilizes gradually (Figure 2d,i). At the late filling stage, the speed gradient disappears completely, and the filling proceeds stably (Figure 2j).

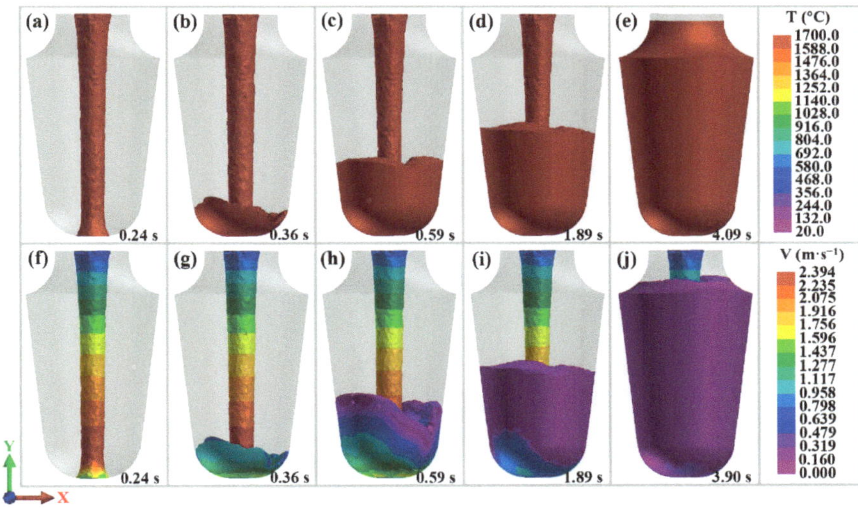

Figure 2. Filling process (**a–e**) and fluid velocity (**f–j**) of Ti-6Al-4V alloy in mold I.

Figure 3 shows the solidification process of molten metal and the distribution of shrinkage cavities when mold I is used. It can be seen from Figure 3a that at 1.13 s the part of the molten metal in contact with the mold wall solidifies first. As the mold-filling process progresses, the solidification zone gradually expands along the mold wall (the expansion direction is indicated by the gray arrow in Figure 3b). At 9.12 s, the part close to the mold wall approaches complete solidification (Figure 3c). This phenomenon further proves that the heat radiation of molten metal along the vertical wall is faster than in other directions. As the solidification process progresses, the part of the molten metal at the riser area reaches complete solidification first (Figure 3d), which means that the riser designed in this study cannot effectively fulfill the purpose of feeding. Meanwhile, the bottom part of the alloy ingot (where the cross-sectional area is smaller), under conditions of relatively good heat dissipation, reaches complete solidification earlier than the top part of the alloy ingot (where the cross-sectional area is larger). This leads to the formation of an isolated liquid phase zone at the central region of the top part of the alloy ingot, as shown in Figure 3e. Generally speaking, titanium alloy ingots exhibit typical characteristics of layer-by-layer solidification. It can be seen from Figure 3f that the shrinkage cavities are concentrated at the top part of the ingot, with a volume of 3.29 cm^3, and the ingot utilization rate is

63.6% (where ingot utilization rate = the distance between the shrinkage cavity and the bottom of the ingot/mold height × 100%, and mold height = ingot height + riser height) It is worth noting that the zone of the shrinkage cavity almost completely coincides with the isolated liquid phase zone. Therefore, it can be concluded that the formation of the shrinkage cavity is closely related to the formation of the isolated liquid phase zone. The formation mechanism of the isolated liquid phase zone will be analyzed in detail later in this article.

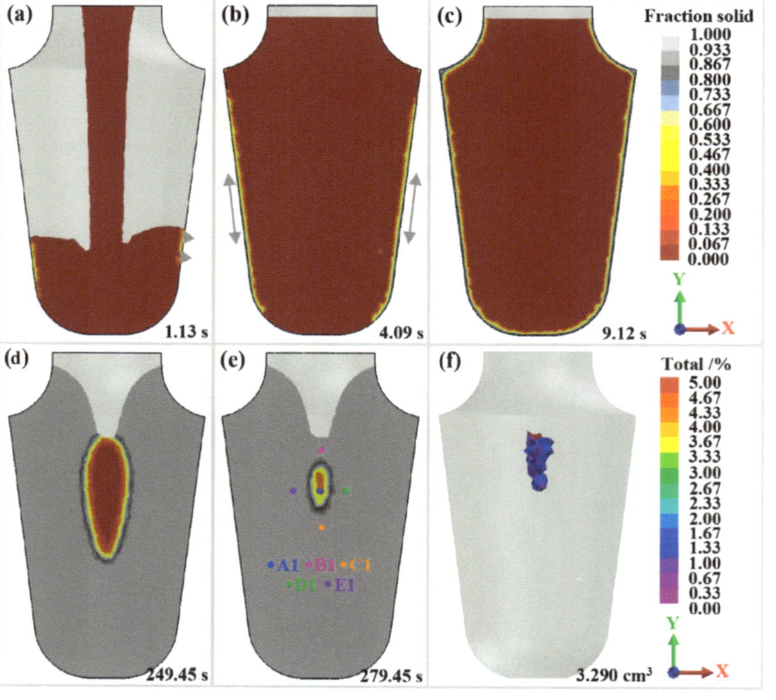

Figure 3. Solidification process (**a**–**e**) and shrinkage porosity (**f**) of Ti-6Al-4V alloy in mold I.

3.2. Numerical Simulation of Molds II, III, and IV

Mold I was optimized based on the above simulation results. Usually, to achieve proper feeding, the riser design can be optimized to ensure that it solidifies later than the casting. However, the use of a riser will significantly reduce the ingot utilization rate, especially when the mass of the ingot is small. Therefore, the riser in mold I was removed, resulting in mold II (Figure 1b). Figure 4a–e shows the mold filling process when mold II is used. It can be seen that when mold II is used, the molding filling has characteristics similar to those of mold I, but the bottom-reaching time of molten metal and filling time are shortened to 0.23 s and 3.57 s, respectively. As can be seen from the simulation results of molten metal flow velocity, Figure 5a–e, the maximum flow velocity of the molten metal decreases to 2.228 m·s^{-1} and the speed gradient difference at the initial stage of mold filling decreases, indicating that the mold II filling process is more stable than that of mold I. As demonstrated by the solidification process simulation results in Figure 6a–e, switching from mold I to mold II results in the formation position of the isolated liquid phase zone moving towards the top of the ingot along the central line. This is a desirable result. Figure 7a further proves that the use of mold II can not only effectively suppress the formation of shrinkage cavities (alloy II: 0.061 cm^3) but can also move the shrinkage cavity formation position towards the top of the ingot along the central line. In alloy I, the shrinkage cavity is 46.8 mm from the top of the ingot, while this distance is shortened to 34.1 mm in alloy II, increasing the utilization rate of ingot II to 86.4%.

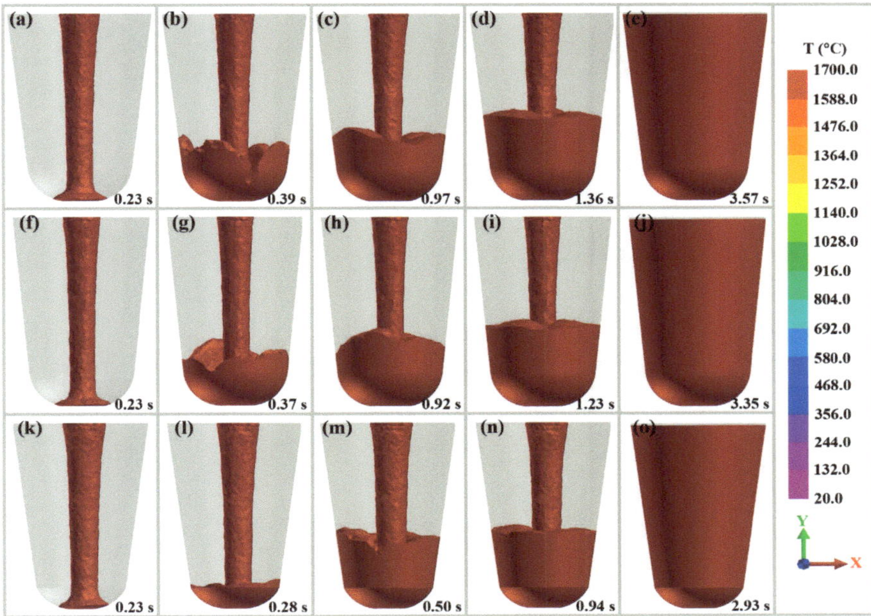

Figure 4. Filling process of Ti-6Al-4V alloy in mold II (**a–e**), III (**f–j**), and IV (**k–o**).

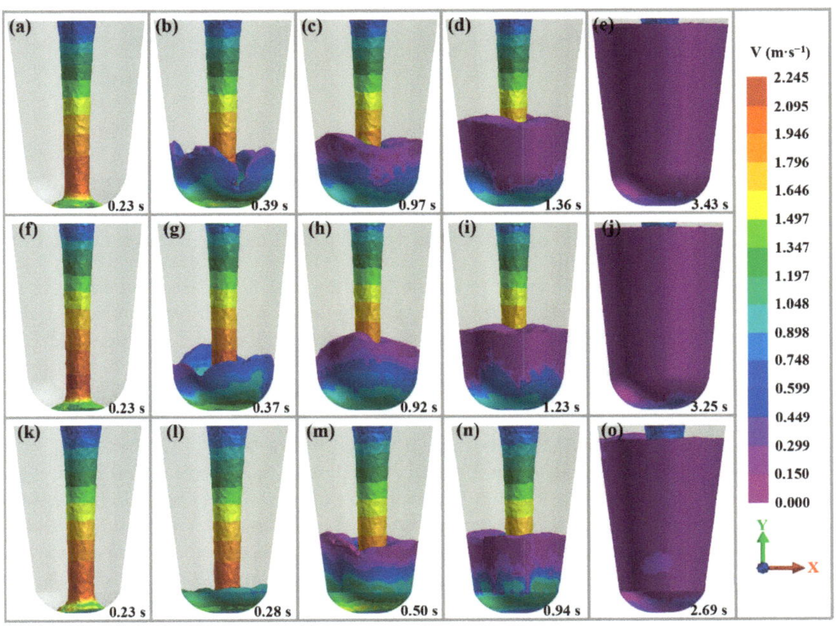

Figure 5. The fluid velocity of Ti-6Al-4V alloy in mold II (**a–e**), III (**f–j**), and IV (**k–o**).

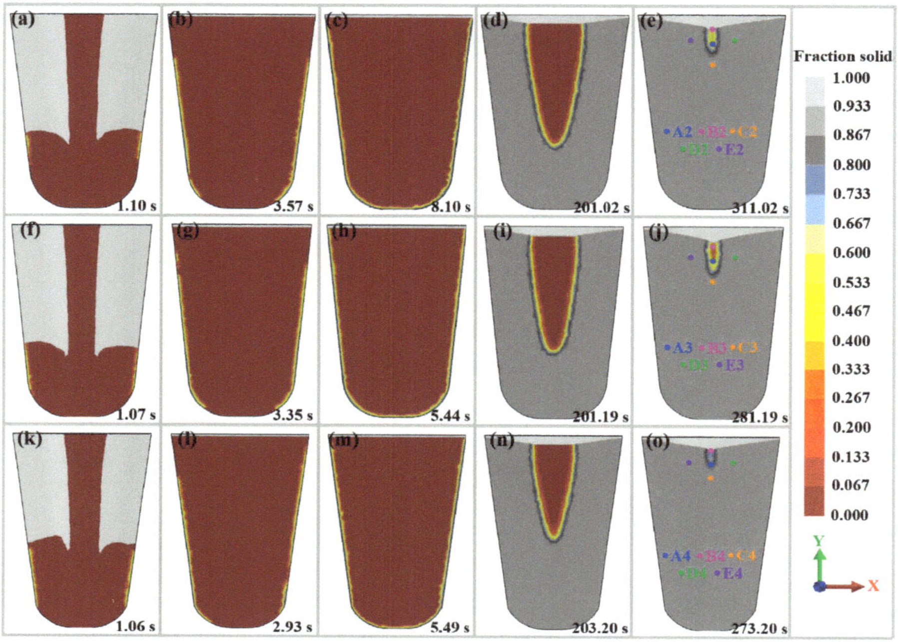

Figure 6. Solidification process of Ti-6Al-4V alloy in mold II (**a–e**), III (**f–j**), and IV (**k–o**).

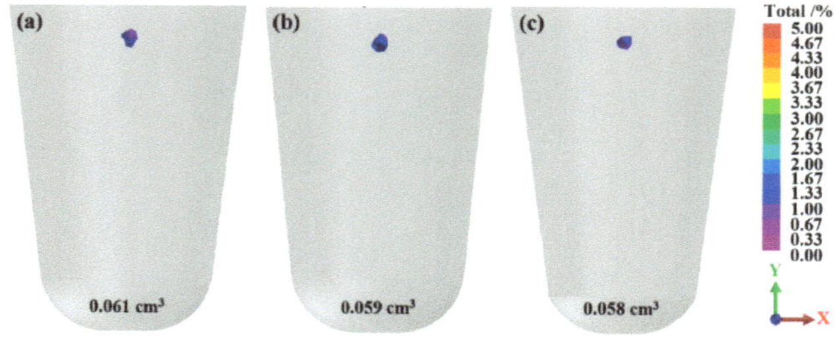

Figure 7. Distribution of shrinkage porosity for Ti-6Al-4V alloy in mold II–IV: (**a**) II; (**b**) III; (**c**) IV.

To further improve the mold filling and alloy casting solidification processes, an in-depth investigation on the key parameters of mold II was carried out, designing molds III and IV as shown in Figure 1c,d. The design of mold III reduces the top size of the ingot to 170 mm, and the design of mold IV reduces the bottom size of the ingot to 110 mm (smaller than mold III). It can be seen from Figures 4f–j and 5f–j that, when mold III is used, the molding filling has characteristics similar to those of mold II, but the maximum flow velocity of molten metal is slightly increased (2.245 m·s^{-1}). It can be seen from Figure 6f–j that, when mold III is used, the ingot solidification process exhibits no unique characteristics and that the shrinkage cavity volume is basically unchanged (0.059 cm^3), but that the ingot utilization rate is reduced to 85.8% (Figure 7b). It is worth noting that reducing the bottom size of the ingot lowers the maximum flow velocity of molten metal to 2.198 m·s^{-1}, thus promoting smooth mold filling, as shown in Figures 4k–o and 5k–o. At the same time, this change has no obvious impact on shrinkage cavity volume (0.058 cm^3) and ingot utilization rate (86.7%), as shown in Figure 7c.

3.3. Experimental Verification

Figure 8 shows the Ti-6Al-4V alloy ingots prepared using molds I–IV. It can be seen that the actual distribution and volume variation pattern of shrinkage cavities in ingots I–IV are basically consistent with the simulation results. However, ingot I contains a large number of small shrinkage cavities and the riser can play a few roles in feeding. Ingot IV contains a small number of small shrinkage cavities. In order to visualize the morphology and distribution of the small shrinkage cavities, the blue rectangular boxed area in Figure 8a is enlarged appropriately and its important features are drawn at a certain scale, as shown in Figure 8b,c. The large shrinkage cavity is represented by the blue region, while the small shrinkage cavity is represented by the orange area. The statistical results show that ingot I has four large shrinkage cavities with an aspect ratio of 1.2–2.0 and nearly 100 small shrinkage cavities with a length of 0.37–2.94 mm that are concentrated around the large shrinkage cavities; in contrast, ingot IV has two large shrinkage cavities with an aspect ratio of 1.1 or 1.4 and a significantly smaller number (about 30) of small shrinkage cavities with a length of 0.5–1.94 mm. Noticeably, the utilization rates of ingots I–IV are 60.5%, 73.6%, 74.5%, and 80.5%, respectively, slightly different from the simulation results. Nonetheless, these figures prove that a rational mold design can effectively improve the material utilization rate. On the whole, numerical simulation results can provide some guidance for controlling the shrinkage porosity of titanium alloy ingots.

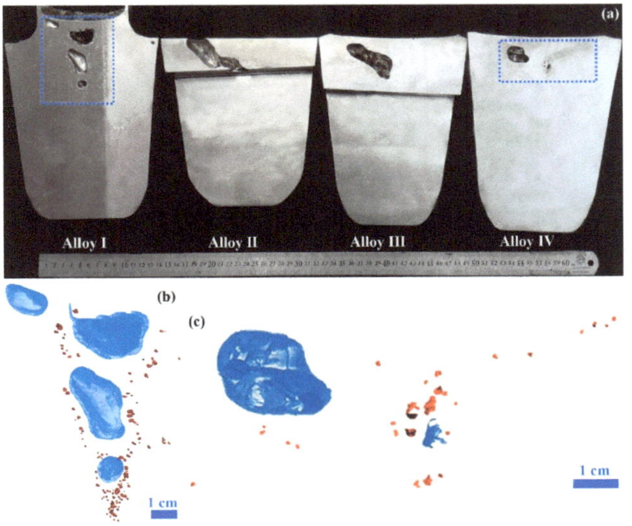

Figure 8. Shrinkage porosity of actual ingots (**a**) and local enlargement of ingot I (**b**) and IV (**c**).

4. Discussion

4.1. Formation Mechanism of Shrinkage Porosity

It can be seen from Figure 3, Figure 6, and Figure 7 that, at the initial stage of solidification, the molten metal solidifies layer by layer from the mold wall to the center of the ingot. The areas with good heat dissipation, such as the bottom of the ingot (where the cross-sectional area is relatively small), solidify first to form a solid surface layer. As the solidification process progresses, the solid shrinkage in the solidified area, the solidification shrinkage caused by solidification, and the liquid shrinkage in the core of the ingot caused by the temperature decrease occur simultaneously. At the late stage of solidification, an isolated liquid phase zone gradually forms in the top central part of the ingot (the area with a larger cross-sectional area). When the sum of the liquid shrinkage and solidification shrinkage of the alloy is greater than the solid shrinkage, a shrinkage cavity will be formed in this area.

Zhang et al. [9] studied the anti-gravity casting process of Ti-6Al-4V alloy and discovered, through numerical simulation and experiment, that shrinkage cavities mostly occur in areas with relatively large cross sections, and that their locations are close to the sprue and riser. This is mainly because the area with a large cross-sectional area can better preserve heat. When other areas are completely solidified, cutting off the feeding channel, an isolated liquid phase zone will be formed in this area [11]. By studying the influence of TiB_2 on the fluidity of TiAl alloy, Han et al. [23] revealed that shrinkage cavities usually occur in the area that is the last part to solidify. In other words, at the late stage of solidification, the temperature of the isolated liquid phase zone is higher than its surrounding area. For each ingot, five points (A, B, C, D, and E) in the isolated liquid phase zone and the surrounding areas (Figures 3e and 6e,j,o) were picked, and the temperature variation of each point with time was analyzed. The results are shown in Figure 9. As the solidification process progresses, the temperature of point A gradually becomes higher than those of points B, C, D, and E, indicating that these solidify earlier than point A. This means that an isolated liquid phase region will be formed at point A and a shrinkage cavity will be formed at this location. In the same way, Yang et al. [24] explored the formation mechanism of shrinkage cavities in the casting process of TiAl alloy blades, finding that the formation of a shrinkage cavity is related to the presence of an isolated liquid phase zone (in the central region of the blade with larger thickness), and that the temperature of this zone was significantly higher than its surrounding area. Similarly, this formation mechanism was also discovered by Zheng et al. [25] in a study on the vacuum casting process of a brake drum made of Al-based composite materials.

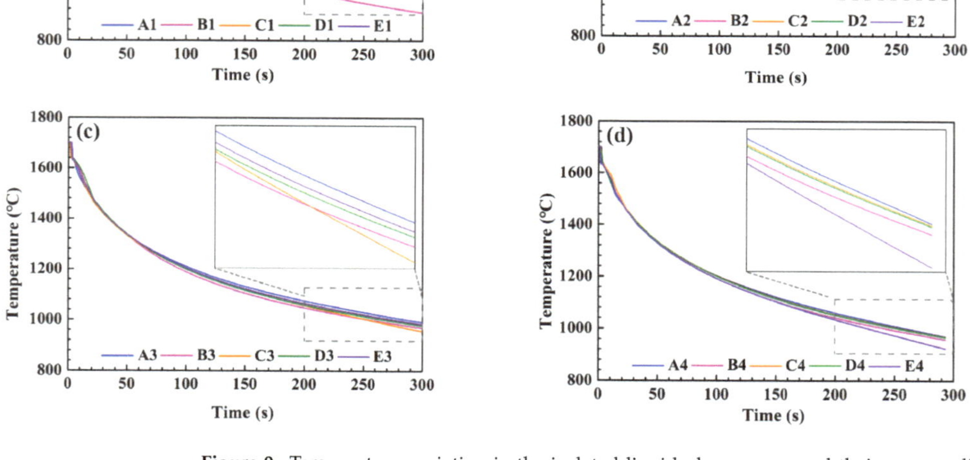

Figure 9. Temperature variation in the isolated liquid phase zones and their surrounding areas: (**a**) alloy I; (**b**) alloy II; (**c**) alloy III; (**d**) alloy IV.

Research shows [26] that the termination of melt metal flow of Ti-6Al-4V alloy exhibits the characteristics of alloys with a narrow crystallization temperature range. The formation of a shrinkage cavity in casting can be largely attributed to an isolated liquid phase zone resulting from the formation of a crystal grain frame comprising a large number of columnar crystal grains and a small number of equiaxed crystal grains. It was found through macrostructure observation (Figure 10) that the basic structure of the ingots prepared in this study is almost completely composed of equiaxed crystal grains. Therefore, it can be concluded that Ti-6Al-4V alloy ingots exhibit the characteristics of alloys with a wide crystallization temperature range. Analysis reveals that the difference in macrostructure characteristics is mainly attributed to the difference in the casting cooling rate which, in turn, can be traced to differences in the specifications of the castings. In the study [26], a spiral-shaped casting with a relatively small cross-sectional area was used, so the cooling speed was relatively high, providing favorable conditions for the formation of columnar crystal grains. By comparison, the near-cylindrical ingot in this study has a large thickness and relatively slow cooling speed (especially at the core), which seriously restricts the nucleation of columnar crystal grains. According to the characteristics of alloys with wide crystallization temperature ranges introduced by Dahle et al. [27], the formation of shrinkage cavities in this study is shown in Figure 11. In summary, the formation of shrinkage cavities in Ti-6Al-4V alloy ingots in this study is mainly due to (a) the formation of isolated liquid phase zones resulting from a grain frame comprising the intersection of a large number of equiaxed crystals, and (b) the inability of the last crystallization part to compensate for the volume shrinkage.

Figure 10. Macrostructure of Ti-6Al-4V alloy ingots.

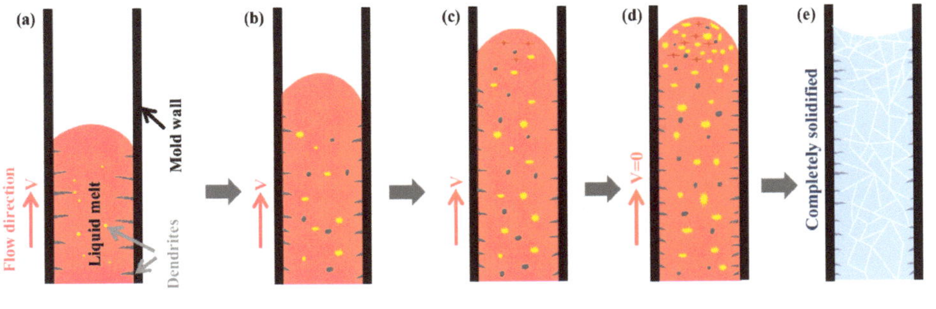

Figure 11. The cessation mechanism of flow for Ti-6Al-4V ingot: (**a–c**) filling step in which the liquid metal is in motion; (**d**) stagnant step in which the velocity of liquid metal is zero; (**e**) completely solidified step.

4.2. Influence of Mold Design on Shrinkage Porosity

In this study, the influence of riser removal on mold filling and the solidification process of Ti-6Al-4V alloy mainly causes the following three effects. (1) The bottom-reaching time of the molten metal and the mold filling completion time are shortened (Figures 2e and 4e). (2) The maximum flow velocity of the molten metal is decreased and the mold filling stability is improved (Figures 2f–j and 5 a–e). (3) The shrinkage cavities move up along the central axis of the ingot and the cavity volume decreases significantly (Figures 3f and 7a). On the one hand, the removal of the riser shortens the distance that the molten metal covers before reaching the bottom of the mold and reduces the total mass of the ingot, factors conducive to shortening the bottom-reaching time and the mold filling completion time. On the other hand, reducing the mold filling distance has the effect of decreasing the initial speed at which the molten metal reaches the bottom, slowing down the flow speed of the molten metal and improving the mold filling stability. Research shows that shortening the mold-filling time can effectively reduce heat loss in the mold-filling process and ensure good feeding during the solidification process [22,28]. These factors play an important role in reducing shrinkage cavity volume and changing the position of the shrinkage cavity. Based on this finding, Liu et al. [14] suggested that the filling time should be shortened as much as possible to ensure filling stability. In addition, after riser removal, the top area of the ingot can serve as a pool for feeding which is the fundamental reason why the shrinkage cavity moves up along the center line and its volume decreases. For mold I, the riser part solidifies earlier than any other part of the ingot in conformance with the layer-by-layer solidification pattern of titanium alloy ingots, cutting off the feeding channel earlier and promoting the formation of a shrinkage cavity. Huang et al. [29] explored the influence of the casting system on shrinkage porosity, and found that feeding becomes difficult when the sprue part solidifies earlier than the casting, resulting in a significantly increased probability of forming shrinkage cavities.

It can be seen from Figure 7 that when the top size of the ingot is reduced from 180 mm to 170 mm, while the volume of the shrinkage cavity is reduced from 0.061 cm^3 to 0.059 cm^3, the utilization rate of the ingot is reduced from 86.4% to 85.8%. When the bottom size of the ingot is reduced from 130 mm to 110 mm, the volume of the shrinkage cavity barely changes but the utilization rate of the ingot is increased to 86.7%. It is worth noting that changing the top or bottom size of the ingot means changing the mold taper. To further optimize the mold design, the influence of mold taper on the volume and distribution of shrinkage cavities was studied (Figure 12). It can be seen from Figure 12a,b that as the top size increases or the bottom size decreases (i.e., the mold taper increases), the shrinkage cavity volume decreases, and the ingot utilization rate increases, both significantly. Analysis reveals that increasing the top size is equivalent to increasing the volume of the pool for high-temperature molten metal, and decreasing the bottom size has the effect of increasing the solidification rate, both of which play an important role in promoting effective feeding, suppressing the formation of shrinkage cavities and making these move up along the central axis. Figure 12c,d show the quantitative relationship between mold taper and the shrinkage cavity, respectively. It is also found that the shrinkage cavity volume decreases significantly with increasing the mold taper, significantly increasing the ingot utilization rate.

In summary, it is advisable to remove the riser in the design of casting molds of titanium alloy ingots of small mass. This not only ensures that shrinkage cavities are concentrated at the top part of the ingot, thereby improving the utilization rate of the ingot, but also reduces the production cost. However, the removal of the riser increases the design requirements of the mold in order to suppress the formation of shrinkage cavities and to move them to the top part of the ingot.

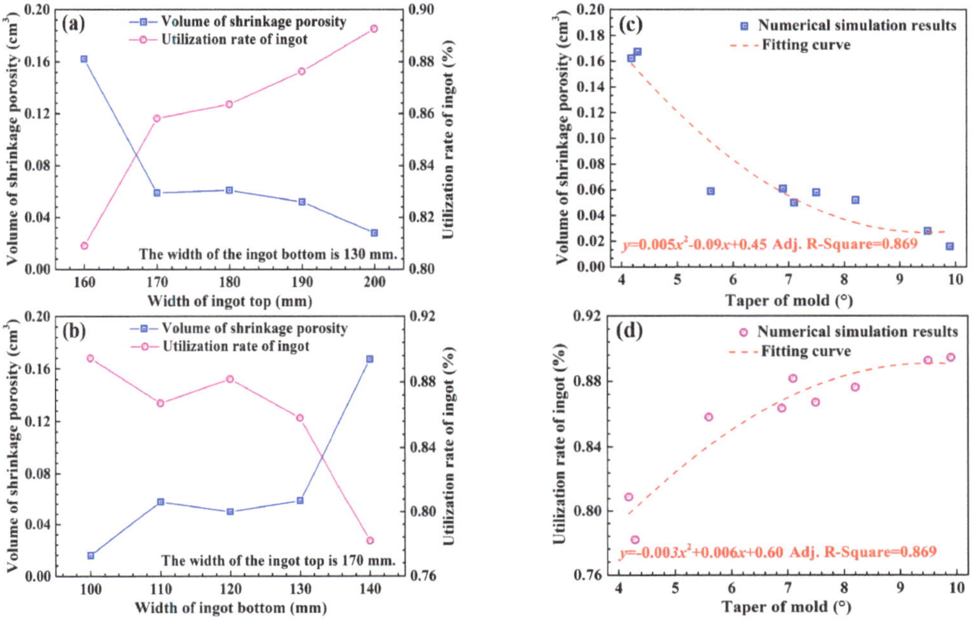

Figure 12. Effect of top width (**a**), bottom width (**b**), and mold taper (**c**,**d**) for ingots on shrinkage porosity and utilization rate of ingots.

5. Conclusions

The effect of mold design on the shrinkage porosity of Ti-6Al-4V alloy ingots has been thoroughly investigated using numerical simulation and experimental verification. The main conclusions are the following.

1. The presence of a riser in mold I results in a relatively high flow rate (2.394 m·s^{-1}) of molten metal, significantly reducing the stability of the mold-filling process. In addition, the riser part solidifies earlier than the main body of the ingot in conformance with the layer-by-layer solidification pattern of titanium alloy. The riser, therefore, cannot serve the purpose of feeding effectively, resulting in a significantly larger shrinkage cavity volume (3.29 cm^3) and a relatively low ingot utilization rate (63.6%).
2. The removal of the riser shortens the bottom-reaching time of the molten metal and the mold-filling completion time. Therefore, the maximum flow velocity decreases (alloy II–IV: 2.228 m·s^{-1}, 2.245 m·s^{-1}, 2.198 m·s^{-1}) and the filling stability increases, effectively suppressing the formation of shrinkage cavities (alloy II–IV: 0.061 cm^3, 0.059 cm^3, 0.058 cm^3) and improving the ingot utilization rate significantly (alloy II–IV: 86.4%, 85.8%, 86.7%).
3. The formation of shrinkage cavities in Ti-6Al-4V alloy ingots in this study is mainly due to the formation of an isolated liquid phase zone. This zone results from the formation of a grain frame caused by the intersection of a large number of equiaxed crystals. The inability of this last crystallization part to compensate for volume shrinkage ultimately leads to the formation of shrinkage cavities.
4. With increasing top size or decreasing bottom size of the ingot (i.e., increasing mold taper), the volume of the shrinkage cavities decreases and the ingot utilization rate increases, both significantly. For instance, the shrinkage cavity volume of the ingot decreases from 0.162 cm^3 to 0.016 cm^3, and the utilization rate increases from 80.9% to 89.5% when the mold taper is increased from 4.2° to 9.9°. Analysis reveals that increasing the top size is equivalent to increasing the volume of the pool of high-

temperature molten metal, and decreasing the bottom size has the effect of increasing the solidification rate, both of which play an important role in promoting effective feeding and suppressing the formation of shrinkage cavities.

Author Contributions: Conceptualization, T.H. and Y.C.; methodology, T.H.; software, T.H. and Y.C.; validation, T.H. and Y.C.; formal analysis, T.H.; investigation, T.H.; resources, Y.C.; data curation, T.H.; writing—original draft preparation, T.H.; writing—review and editing, T.H. and Y.C.; visualization, T.H.; supervision, Y.C.; project administration, Y.C.; funding acquisition, Y.C. All authors have read and agreed to the published version of the manuscript.

Funding: This work was financially supported by the Sichuan Science and Technology Program (No. 2019YFG0082).

Institutional Review Board Statement: Not applicable.

Informed Consent Statement: Not applicable.

Data Availability Statement: The data that support the findings of this study are available from the corresponding author, Yuyong Chen, upon reasonable request.

Conflicts of Interest: The authors declare that they have no conflict of interest.

References

1. Williams, J.C.; Boyer, R.R. Opportunities and issues in the application of titanium alloys for aerospace components. *Metals* **2020**, *10*, 705. [CrossRef]
2. Zhang, S.Y.; Li, J.S.; Kou, H.C.; Yang, J.R.; Yang, G.; Wang, J. Effects of thermal history on the microstructure evolution of Ti–6Al–4V during solidification. *J. Mater. Process. Technol.* **2016**, *227*, 281–287. [CrossRef]
3. Feng, X.; Qiu, J.K.; Ma, Y.J.; Lei, J.F.; Cui, Y.Y.; Wu, X.H.; Yang, R. Influence of processing conditions on microstructure and mechanical properties of large thin-wall centrifugal Ti–6Al–4V casting. *J. Mater. Sci. Technol.* **2015**, *32*, 362–371. [CrossRef]
4. Atkinson, H.V.; Davies, S. Fundamental aspects of hot isostatic pressing: An overview. *Metall. Mater. Trans. A* **2000**, *31*, 2981–3000. [CrossRef]
5. Zhang, S.Y.; Li, J.S.; Kou, H.C.; Yang, J.R.; Yang, G.; Wang, J. Effect of mold temperature and casting dimension on microstructure and tensile properties of counter-gravity casting Ti–6Al–4V alloys. *China Foundry* **2016**, *13*, 9–14. [CrossRef]
6. Yang, L.; Chai, L.H.; Liang, Y.F.; Zhang, Y.W.; Bao, C.L.; Liu, S.B.; Lin, J.P. Numerical simulation and experimental verification of gravity and centrifugal investment casting low pressure turbine blades for high Nb-TiAl alloy. *Intermetallics* **2015**, *66*, 149–155. [CrossRef]
7. Suzuki, K.; Yao, M. Simulation of mold filling and solidification during centrifugal precision casting of Ti–6Al–4V alloy. *Met. Mater. Int.* **2004**, *10*, 33–38. [CrossRef]
8. Jia, Y.; Xiao, S.L.; Tian, J.; Xu, L.J.; Chen, Y.Y. Modeling of TiAl alloy grating by investment casting. *Metals* **2015**, *5*, 2328–2339. [CrossRef]
9. Zhang, S.Y.; Li, J.S.; Kou, H.C.; Yang, J.R.; Yang, G.; Wang, J. Numerical modeling and experiment of counter-gravity casting for titanium alloys. *Int. J. Adv. Manuf. Technol.* **2016**, *85*, 1877–1885. [CrossRef]
10. Yang, J.R.; Wang, H.; Wu, Y.L.; Zhang, K.R.; Wang, X.Y.; Hu, R. Numerical calculation and experimental evaluation of counter-gravity investment casting of Ti–48Al–2Cr–2Nb alloy. *Int. J. Adv. Manuf. Technol.* **2018**, *96*, 3295–3309. [CrossRef]
11. Xiong, C.; Ma, Y.C.; Chen, B.; Liu, K.; Li, Y.Y. Modeling of filling and solidification process for TiAl exhaust valves during suction casting. *Acta. Metall. Sin. Engl. Lett.* **2013**, *26*, 33–48. [CrossRef]
12. Qi, Y.S.; Chen, L.L.; Chen, G.; Li, K.F.; Li, C.; Du, Z.M. Preparation and properties of special vehicle cover via a novel squeeze casting quantitative feeding system of molten metal. *Metals* **2020**, *10*, 266. [CrossRef]
13. Dai, H.L.; Zhang, C.L.; Zhao, Q.; Zhu, H. The study of titanium alloy precision casting turbine blades based on Procast. *IOP Conf. Ser. Mater. Sci. Eng.* **2019**, *677*, 022090. [CrossRef]
14. Liu, X.J.; Hao, Z.J.; Huang, M. Optimization of vacuum counter-pressure casting process for an aluminum alloy casing using numerical simulation and defect recognition techniques. *Int. J. Adv. Manuf. Technol.* **2020**, *107*, 2783–2795. [CrossRef]
15. Shangguan, H.L.; Kang, J.W.; Yi, J.H.; Deng, C.Y.; Hu, Y.Y.; Huang, T. Controlled cooling of an aluminum alloy casting based on 3D printed rib reinforced shell mold. *China Foundry* **2018**, *15*, 210–215. [CrossRef]
16. Zhao, J.; Zhang, Z.Y.; Liu, S.B.; Shi, K.; Bao, C.L.; Ning, Z.S.; Yan, P.; Wang, L.; Lou, Y.C. Elimination of misrun and gas hole defects of investment casting TiAl alloy turbocharger based on numerical simulation and experimental study. *China Foundry* **2020**, *17*, 29–34. [CrossRef]
17. Eylon, D.; Keller, M.M.; Jones, P.E. Development of permanent-mold cast Ti–Al automotive valves. *Intermetallics* **1998**, *6*, 703–708. [CrossRef]
18. Rishel, L.L.; Biery, N.E.; Raban, R.; Gandelsman, V.Z.; Pollock, T.M.; Cramb, A.W. Cast structure and property variability in gamma titanium aluminides. *Intermetallics* **1998**, *6*, 629–636. [CrossRef]

19. Lasalmonie, A. Why is it so difficult to introduce them in gas turbine engines? *Intermetallics* **2006**, *14*, 1123–1129. [CrossRef]
20. Gao, Y.; Zhang, L.J.; Gao, W.L.; Zhang, H. Prediction and improvement of shrinkage porosity in TiAl based alloy. *China Foundry* **2011**, *8*, 19–24.
21. Mi, G.F.; Liu, X.Y.; Wang, K.F.; Fu, H.Z. Application of numerical simulation technique to casting process of valve block. *J. Iron Steel Res. Int.* **2009**, *16*, 12–17. [CrossRef]
22. Liu, J.G.; Yang, L.; Fang, X.G.; Li, B.; Yang, Y.W.; Fang, L.Z.; Hu, Z.B. Numerical simulation and optimization of shell mould casting process for leaf spring bracket. *China Foundry* **2020**, *17*, 35–41. [CrossRef]
23. Han, J.C.; Dong, J.; Zhang, S.Z.; Zhang, C.J.; Xiao, S.L.; Chen, Y.Y. Effect of TiB2 addition on microstructure and fluidity of cast TiAl alloy. *Vacuum* **2020**, *174*, 109210. [CrossRef]
24. Yang, L.; Chai, L.H.; Zhang, L.Q.; Lin, J.P. Numerical simulation and process optimization of investment casting of the blades for high Nb containing TiAl alloy. *Mater. Sci. Forum.* **2013**, *747*, 105–110. [CrossRef]
25. Zheng, H.S.; Zhang, Z.F.; Bai, Y.L. Numerical simulation and experimental study on compound casting of layered aluminum matrix composite brake drum. *Materials* **2021**, *14*, 1412. [CrossRef]
26. Wang, J.H.; Guo, X.L.; Wang, L.Q.; Lu, W.J. The influence of B_4C on the fluidity of Ti-6Al-4V-$x$$B_4C$ composites. *Mater. Trans.* **2014**, *55*, 1367–1371. [CrossRef]
27. Dahle, A.K.; Karlsen, S.; Arnberg, L. Effect of grain refinement on the fluidity of some binary Al-Cu and Al-Mg alloys. *Int. J. Cast Metal. Res.* **1996**, *9*, 103–112. [CrossRef]
28. Xuan, D.P.; Zhou, C.; Zhou, Y.; Jiang, T.L.; Zhu, B.J.; Fan, W.H.; Jia, Y.G. Numerical simulation of the top side-pouring twin-roll casting of 6.5wt.% Si steel process. *Int. J. Adv. Manuf. Technol.* **2022**, *119*, 2355–2368. [CrossRef]
29. Huang, P.H.; Cheng, C.Y.; Huang, W.J.; Chou, C.S. Optimal design of investment casting system for toothed chain joint: Computer simulations and experimental verification. *Int. J. Adv. Manuf. Technol.* **2020**, *106*, 1931–1943. [CrossRef]

metals

Article

A New Computational Method for Predicting Ductile Failure of 304L Stainless Steel

Myung-Sung Kim [1], Hee-Tae Kim [2], Young-Hwan Choi [2], Jeong-Hyeon Kim [3], Seul-Kee Kim [3,*] and Jae-Myung Lee [2,3,*]

[1] Department of Reliability Assessment, Korea Institute of Machinery and Materials, Daejeon 34103, Korea
[2] Department of Naval Architecture and Ocean Engineering, Pusan National University, Busan 46241, Korea
[3] Hydrogen Ship Technology Center, Pusan National University, Busan 46241, Korea
* Correspondence: skkim@pusan.ac.kr (S.-K.K.); jaemlee@pusan.ac.kr (J.-M.L.); Tel.: +82-51-510-7953 (S.-K.K.); +82-51-510-2342 (J.-M.L.); Fax: +82-51-512-8836 (S.-K.K. & J.-M.L.)

Abstract: Austenitic stainless steel is useful for storing and transporting liquefied natural gas (LNG) at temperatures below −163 °C due to its superior low-temperature applications. This study develops a computational method for the failure prediction of 304L stainless steel sheet to utilize its usability as a design code for industrial purposes. To consider material degradation in a phenomenological way during the numerical calculation, the combined Swift–Voce equation was adopted to describe the nonlinear constitutive behavior beyond ultimate tensile strength. Due to the stress state-dependent fracture characteristics of ductile metal, a modified Mohr–Coulomb fracture criterion was adopted using stress triaxiality and Lode angle parameter. The numerical formulation of the elastoplastic-damage coupled constitutive model with fracture locus was implemented in the ABAQUS user-defined subroutine UMAT. To identify the material and damage parameters of constitutive models, a series of material tests were conducted considering various stress states. It has been verified that the numerical simulation results obtained by the proposed failure prediction methodology show good agreement with the experimental results for plastic behavior and fractured configuration.

Keywords: ductile fracture; elastoplastic constitutive model; modified Mohr–Coulomb model; numerical implementation; austenitic stainless steel

Citation: Kim, M.-S.; Kim, H.-T.; Choi, Y.-H.; Kim, J.-H.; Kim, S.-K.; Lee, J.-M. A New Computational Method for Predicting Ductile Failure of 304L Stainless Steel. *Metals* **2022**, *12*, 1309. https://doi.org/10.3390/met12081309

Academic Editors: Shuwen Wen, Yongle Sun and Xin Chen

Received: 29 June 2022
Accepted: 27 July 2022
Published: 4 August 2022

1. Introduction

Austenitic stainless steel is recognized as a functional material in various industries due to its excellent strength, toughness, and superior corrosion resistance even in low-temperature environments [1–3]. In particular, 304L austenitic stainless steel, which is generally known as 18/8 steel, is useful for storing and transporting liquefied natural gas (LNG) at temperatures below −163 °C due to its superior low-temperature applications. Zheng et al. (2018) reported that the mechanical strength of low-temperature treated 304 stainless steel was increased up to 2.7 times compared to conventional as-received samples [4]. Mallick et al. (2017) reported that 10–20% low-temperature (−196 °C) deformation leads to a higher level of strength (1306–1589 MPa) owing to the formation of a higher volume fraction of strain-induced martensite [5]. Singh et al. (2018) reported that the low-temperature mechanical strength was increased to 1200 MPa, which is much more than test results under ambient conditions. In addition, in low-temperature treatment, the micro-hardness was increased from 208 VHN to 520 VHN, which is more than double that of the as-received sample [6]. Oh et al. (2018) reported the low-temperature fatigue strength of 304 stainless steel was significantly improved compared to the ambient fatigue strength [7]. Thanks to these valuable studies, it has been established that 304(L) austenitic stainless steel is the optimal material in low-temperature applications because an enhanced mechanical performance was observed in terms such as strength, hardness, and fatigue strength at low temperatures.

Due to its superior low-temperature mechanical performance, 304L stainless steel is adopted as the main material for LNG carrier cargo holds. The primary barrier is corrugated because it undergoes repeated thermal shrinkage and contraction during operation in the LNG loading and unloading processes [8–11]. Therefore, some researchers have focused on the deformation and pressure resistance of 304L stainless steel-made corrugated membranes in experimental approaches. Kim et al. (2011) reported the pressure levels that induced the collapse of the corrugated walls via plastic buckling were six and twelve bar for the large and small corrugated containers, respectively [12]. Lee et al. (2015) reported the pressure-resisting capability of 304L-made stainless steel plate. The maximum deformation of the weakest corrugation was measured with respect to an applied exterior pressure and the pressure at the start of buckling failure was chosen as the pressure-resisting capability of the corrugations [13]. Jeong et al. (2021) reported the pressure-resisting capabilities of a 304L stainless steel corrugated membrane under hydrodynamic load [14]. Those research outcomes have helped improve the shape of the corrugated membrane. However, most evaluations of the pressure resistance performance and the deformation of the primary barrier of the LNG CCS were all conducted at room temperature. This is because it is very difficult to build a pressure test environment for large structures in low-temperature environments. Even so, it is important to evaluate the primary barrier in low-temperature environments, which is the main operating condition of 304L stainless steel primary barriers.

If difficulties are encountered in the experimental evaluation of low-temperature performance, failure evaluation techniques based on computational methods can be an excellent alternative. Evaluating the failure of ductile materials requires predicting the elastoplastic behavior and occurrence of failure. Among failure prediction approaches, the phenomenological failure model is defined as failure occurrence when the stress or strain of an element reaches a specific value. The strain-based failure model is more suitable than the stress-based failure model for dealing with structures undergoing severe plastic deformation and has proven quite useful in collision and failure problems [15–22].

Early studies to predict ductile failure explained the relationship between stress triaxiality and equivalent plastic strain [23–26]. In addition, it was found that the Lode angle, defined as the third invariant of the deviation stress tensor, also affects ductile failure [27–30]. The Mohr–Coulomb criterion is based on the maximum shear stress and is mainly used to determine the failure of rock, soil, and concrete. To eliminate the shortcoming of the absence of pressure dependency, Bai and Wierzbicki (2010) developed a modified Mohr–Coulomb criterion that is transformed from a local representation in terms of shear stress and normal stress to a mixed strain–stress representation of stress triaxiality, Lode angle parameter, and equivalent plastic strain for monotonic loading conditions [28]. The MMC criterion can predict the crack initiation point and the direction of crack propagation and its usefulness has already been verified in many previous studies [31–37].

Several studies have been conducted to predict the ductile fracturing of 304L stainless steel. Othmen et al. (2020) carried the prediction of the onset of rupture of austenitic stainless steel during its forming process [38]. Various fracture criteria, implemented in the finite element code Abaqus/Implicit via a user subroutine USDFLD, have been investigated. Pham and Iwamoto (2018) proposed the numerical fracture prediction of 304 stainless steel with the modified Johnson–Cook damage model [39]. Kim et al. (2013) proposed a viscoplastic model for 304L stainless steel considering the pre-strain and temperature effects [40]. These studies accurately predicted the occurrence of fractures and crack propagation but provided limited failure predictions due to their lack of consideration for various ranges of stress states.

Thus, the present study proposes the failure prediction methodology of 1.2 t 304L stainless steel sheet. An elastoplastic-damage coupled constitutive equation was developed to establish the failure criterion for the primary barrier and to propose a failure analysis technique. To establish the phenomenological ductile failure criterion for 304L stainless steel sheet, a series of material tests were performed considering various stress states. A modified

Mohr–Coulomb model was adopted to formulate ductile fracture criteria in accordance with the stress triaxiality and Lode angle of ductile materials. Numerical analysis with the completed ductile fracture criterion shows good agreement with experimental results.

2. Phenomenological Ductile Fracture Criteria

2.1. Characterization of Stress State

For a certain stress state $\{\sigma_1, \sigma_2, \sigma_3\}$ of an isotropic material, the stress tensor can be expressed as hydrostatic and deviatoric parts. The three main invariants of the stress tensor can be expressed as follow.

$$\boldsymbol{\sigma} = \boldsymbol{s} + p\boldsymbol{I} \tag{1}$$

$$I_1 = tr[\boldsymbol{\sigma}] \tag{2}$$

$$J_2 = \left(\frac{1}{2}\boldsymbol{s} : \boldsymbol{s}\right) \tag{3}$$

$$J_3 = det[\boldsymbol{s}] \tag{4}$$

where $\boldsymbol{s}$, p is the deviatoric stress tensor and hydrostatic stress, respectively. $\boldsymbol{I}$ is the second-order identity tensor. From principal stress space, von-Mises yield surface circumscribes three-dimensional cylinder orthogonal to deviatoric plane (π-plane). To indicate a certain stress state on the deviatoric plane, a cylindrical coordinate system can be used to define the Lode angle from the hydrostatic stress and principal stress directions. The Lode angle can be defined as the angle of the principal stress axis on the deviatoric plane. The Lode angle is related to the normalized third invariant as follows [28].

$$\xi = cos(3\theta) = \left(\frac{r}{q}\right)^3 \quad (0 \le \theta \le \pi/3) \tag{5}$$

$$r = \left[\frac{27}{2}(\sigma_1 - p)(\sigma_2 - p)(\sigma_3 - p)\right]^{1/3} = \left[\frac{27}{2}\det[s]\right]^{1/3} = \left[\frac{27}{2}J_3\right]^{1/3} \tag{6}$$

$$\theta = \frac{1}{3}arccos\left(\frac{27}{2}\frac{J_3}{[3J_2]^{3/2}}\right) \tag{7}$$

where ξ, r is the normalized third invariant and the third invariant, respectively. θ is the Lode angle expressed as stress invariants. Stress triaxiality is expressed as the ratio of hydrostatic stress ($\overline{OO'}$) and equivalent stress ($\overline{O'P}$) as follows.

$$\eta = \frac{p}{q} = \frac{\sqrt{2}}{3}cot\left(\arctan\frac{\overline{O'P}}{\overline{OO'}}\right) \tag{8}$$

where q is the equivalent stress. From a viewpoint of principal stress coordinates, stress triaxiality represents the dominance of the hydrostatic stress in a certain stress state. The normalized Lode angle can be expressed as follows through the modified Haigh–Westergaard coordinate system [41].

$$\overline{\theta} = 1 - \frac{6\theta}{\pi} \quad (-1 \le \overline{\theta} \le 1) \tag{9}$$

In the planar stress condition, it is possible to convert a three-dimensional stress space into two dimensions.

2.2. Modified Mohr–Coulomb Model

Bai and Wierzbicki (2010) modified and extended Mohr–Coulomb (MC) fracture criterion to describe ductile fracture of isotropic crack-free solids in terms of equivalent plastic strain [28]. The Mohr–Coulomb fracture criterion has been widely used in rock and soil mechanics [42,43]. This criterion states that fracture occurs at a certain plane when the linear combination of shear and normal stress reaches a critical value [44]. Bai and

Wierzbicki (2010) assumed that the behavior of ductile materials can be described by the von-Mises yield condition and hardening power law [28]. The modified Mohr–Coulomb (MMC) fracture model can be expressed as follows in terms of the stress triaxialiy and the normalized Lode angle.

$$\bar{\varepsilon}_i^p(\eta, \bar{\theta}) = \left\{ \frac{A}{c_2}\left[c_3 + \frac{\sqrt{3}}{2-\sqrt{3}}(1-c_3)\left(sec\left(\frac{\bar{\theta}\pi}{6}\right)-1\right)\right] \times \left[\sqrt{\frac{1+c_1^2}{3}}cos\left(\frac{\bar{\theta}\pi}{6}\right) + c_1\left(\eta + \frac{1}{3}sin\left(\frac{\bar{\theta}\pi}{6}\right)\right)\right]\right\}^{-\frac{1}{n}}. \quad (10)$$

where A, n are material parameters of swift equation, and $c_i(i = 1, 2, 3)$ is fracture parameters. Three fracture parameters need to be calibrated from experimental results. In this study, with the following condition satisfied, the damage accumulation of the element is initiated.

$$\int_0^{\bar{\varepsilon}_i^p} \frac{d\bar{\varepsilon}^p}{\bar{\varepsilon}_i^p(\eta, \bar{\theta})} = 1 \quad (11)$$

where $\bar{\varepsilon}_i^p$ is equivalent plastic strain at damage initiation, and $\bar{\varepsilon}^p$ is equivalent plastic strain.

3. Elastoplastic-Damage Coupled Constitutive Model

3.1. Hardening Function

In this study, the Swift-Voce equation was adopted as a hardening function that expresses the isotropic hardening behavior according to the equivalent plastic strain for general ductile metal materials [41,45,46]. Swift-Voce equation can be described as follow.

$$\sigma_y(\bar{\varepsilon}^p) = \alpha k_s + (1-\alpha)k_v k \quad (12)$$

$$k_s = A(\varepsilon_0 + \bar{\varepsilon}^p)^n \quad (13)$$

$$k_v = \sigma_{y0} + Q(1 - exp(-\beta\bar{\varepsilon}^p)) \quad (14)$$

where k_s, k_v, and α is Swift equation, Voce equation, and weight parameter, respectively. A, ε_0, n, σ_{y0}, Q, β is material parameters for Swift-Voce equation. Some ductile materials show yield plateau after yielding. Considering yield plateau strain (ε_{plat}), Swift-Voce equation can be described as follow.

$$k_s = \begin{cases} A(\varepsilon_0)^n & \bar{\varepsilon}^p \le \varepsilon_{plat} \\ A\left(\varepsilon_0 + \bar{\varepsilon}^p - \varepsilon_{plat}\right)^n & \bar{\varepsilon}^p > \varepsilon_{plat} \end{cases} \quad (15)$$

$$k_v = \begin{cases} \sigma_{y0} & \bar{\varepsilon}^p \le \varepsilon_{plat} \\ Q\left\{1 - exp\left[-\beta\left(\bar{\varepsilon}^p - \varepsilon_{plat}\right)\right]\right\} & \bar{\varepsilon}^p > \varepsilon_{plat} \end{cases} \quad (16)$$

3.2. Damage Evolution Rules

In the present study, ductile materials undergo damage after damage initiation. For the isotropic hardening ductile material, damage manifests itself in two forms; softening of yield stress and degradation of elastic modulus as shown in Figure 1. Lemaitre (1985) explained that damage to the material affects the cross-sectional area due to the growth of pores and micro-cracks inside the material, leading to a decrease in the modulus of elasticity [47]. In this study, in order to define the phenomenological fracture, the damage variable is simply expressed in terms of the equivalent plastic strain rate as follows.

$$D = \begin{cases} 0 & (\bar{\varepsilon}^p < \bar{\varepsilon}_i^p) \\ D_s \times \dot{\bar{\varepsilon}}^p & (\bar{\varepsilon}_i^p \le \bar{\varepsilon}^p < \bar{\varepsilon}_f^p) \\ D_c & \left(\bar{\varepsilon}_f^p \le \bar{\varepsilon}^p\right) \end{cases} \quad (17)$$

where D_s, D_c is damage accumulation control parameter and critical damage, respectively. Damage accumulation control parameter adjusts the degree of damage accumulation. Critical damage defines the thresholds for damage. $\bar{\varepsilon}_i^p$ and $\bar{\varepsilon}_f^p$ represents equivalent plastic strain at damage initiation and fracture, respectively.

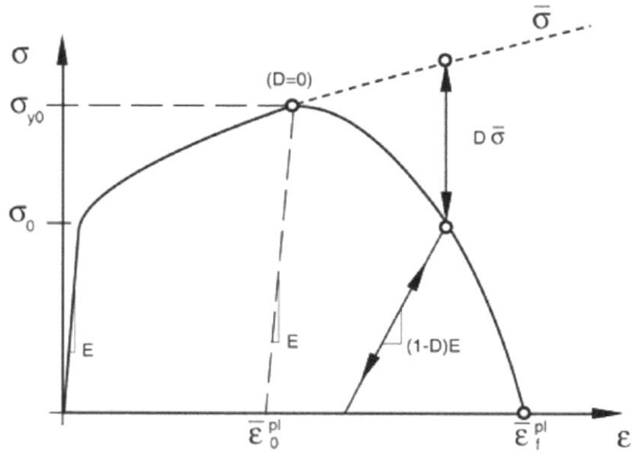

Figure 1. Stress-strain relationship with damage accumulation. With damage accumulation, critical equivalent plastic strain provokes damage initiation. After the onset of damage initiation, elastic modulus and strength are dramatically decreased.

3.3. Constitutive Model

In this study, an elastoplastic-damage coupled constitutive model is proposed to predict the damage of ductile material. The total strain tensor and strain rate tensor can be decomposed into elastic part and plastic part as follows.

$$\varepsilon = \varepsilon^e + \varepsilon^p \tag{18}$$

$$\dot{\varepsilon} = \dot{\varepsilon}^e + \dot{\varepsilon}^p \tag{19}$$

Using the concept of effective area and stress of damaged material proposed by Lemaitre (1985) [47], the general Hooke's law can be expressed as follows.

$$\sigma = (1 - \mathrm{D})\boldsymbol{D} : \varepsilon^e \tag{20}$$

where σ, $\boldsymbol{D}$ represents the stress tensor and the stiffness tensor for isotropic materials, respectively. D is damage variable. Deviatoric stress and hydrostatic stress are as follows, respectively.

$$\boldsymbol{s} = (1 - \mathrm{D})2G\varepsilon_d^e \tag{21}$$

$$p = (1 - \mathrm{D})K\varepsilon_v^e \tag{22}$$

where ε_d^e, ε_v^e is deviatoric strain tensor and volumetric strain, respectively. G and K represent shear modulus and volume modulus, respectively. Under the constant loading direction, the effect of kinematic hardening can be ignored. The yield function according to the von-Mises yield criterion is as follows.

$$\Phi = q - (1 - \mathrm{D})\sigma_y(\bar{\varepsilon}^p) \tag{23}$$

$$q = \sqrt{3J_2} = \sqrt{\frac{3}{2}\boldsymbol{s} : \boldsymbol{s}} = \sqrt{\frac{3}{2}}||\boldsymbol{s}|| \tag{24}$$

where q is the von-Mises equivalent stress expressed as the deviatoric stress, and $\sigma_y(\bar{\varepsilon}^p)$ represents the isotropic hardening function expressing according to the equivalent plastic strain. According to the Prandtl-Reuss plastic law, flow rule is defined as follows [48].

$$\dot{\varepsilon}^p = \dot{\gamma}\frac{\partial\Phi}{\partial\sigma} = \dot{\gamma}\sqrt{\frac{3}{2}}\frac{s}{||s||} \tag{25}$$

where $\dot{\varepsilon}^p$, $\dot{\gamma}$ is the plastic strain rate and the plastic multiplier, respectively. Plastic strain rate is expressed as Prandtl-Reuss flow vector and plastic multiplier. Equivalent plastic strain rate is defined as follows.

$$\dot{\bar{\varepsilon}}^p = \sqrt{\frac{2}{3}\dot{\varepsilon}^p : \dot{\varepsilon}^p} = \sqrt{\frac{2}{3}}||\dot{\varepsilon}^p|| = \dot{\gamma} \tag{26}$$

According to Equation (25), the equivalent plastic strain rate is the same as the plastic multiplier. The loading/unloading conditions of the constitutive model is as follows.

$$\Phi \leq 0,\ \ \dot{\gamma} \geq 0,\ \ \dot{\gamma}\Phi = 0 \tag{27}$$

3.4. Numerical Implementation Algorithm

In order to formulate the proposed elastoplastic-damage coupled constitutive model with fracture locus, ABAQUS/STANDARD, a commercial finite element analysis software, was adopted. ABAQUS with subroutine UMAT (user subroutine to define a material's mechanical behavior) provides the user to define material properties and provides values calculated at the integration point of each element.

In this study, the elastoplastic-damage coupled constitutive model was formulated with a fully implicit backward Euler integration scheme. A return mapping scheme with elasticity prediction and plastic correction was adopted. When the total strain increment ($\Delta\varepsilon$) according to the time interval $[t_n, t_{n+1}]$ is determined at each integration point, the subroutine UMAT calculates unknown variables $\sigma_{n+1}, \varepsilon^p_{n+1}, \mathrm{D}_{n+1}, t_{n+1}$ using the known variables $\sigma_n, \varepsilon^p_n, \mathrm{D}_n, t_n$ [48,49]. When the total strain increment is determined, the trial stress and strain components can be expressed as follows.

$$\varepsilon^{e\ trial}_{n+1} = \varepsilon^e_n + \Delta\varepsilon \tag{28}$$

$$\bar{\varepsilon}^{p\ trial}_{n+1} = \bar{\varepsilon}^p_n \tag{29}$$

$$\mathrm{D}^{trial}_{n+1} = \mathrm{D}_n \tag{30}$$

$$\mathbf{s}^{trial}_{n+1} = (1-\mathrm{D}_n)2G\varepsilon^{e\ trial}_{v\ n+1} \tag{31}$$

$$p^{trial}_{n+1} = (1-\mathrm{D}_n)K\varepsilon^{e\ trial}_{v\ n+1} \tag{32}$$

The corresponding trial yield function and trial equivalent stress are as follows.

$$\Phi^{trial} = q^{trial}_{n+1} - (1-\mathrm{D}_n)\sigma_y\left(\bar{\varepsilon}^p_n\right) \tag{33}$$

$$q^{trial}_{n+1} = \sqrt{3J_2\left(\mathbf{s}^{trial}_{n+1}\right)} = \sqrt{\frac{3}{2}\mathbf{s}^{trial}_{n+1} : \mathbf{s}^{trial}_{n+1}} = \sqrt{\frac{3}{2}}||\mathbf{s}^{trial}_{n+1}|| \tag{34}$$

When the trial yield function is $\Phi^{trial} \leq 0$, the trial stress exists within the yield function, so it is regarded as an elastic region in which plastic increment does not occur. The state variables are updated as Equations (28)–(32) and the stress is updated as follows.

$$\sigma_{n+1} = \mathbf{s}^{trial}_{n+1} + p^{trial}_{n+1}I \tag{35}$$

If the trial yield function is resulted in $\Phi^{trial} > 0$, plastic correction is required according to the incremental calculation of the equivalent plastic strain. First, strain and damage parameters can be defined by the Backward Euler method as follows.

$$\varepsilon^{e}_{n+1} = \varepsilon^{e\ trial}_{n+1} - \varepsilon^{p}_{n+1} = \varepsilon^{e\ trial}_{n+1} - \Delta\gamma\sqrt{\frac{3}{2}}\frac{\boldsymbol{s}_{n+1}}{\|\boldsymbol{s}_{n+1}\|} \tag{36}$$

$$\bar{\varepsilon}^{p}_{n+1} = \bar{\varepsilon}^{p}_{n} + \Delta\gamma \tag{37}$$

$$\mathrm{D}_{n+1} = \mathrm{D}_{n} + \mathrm{D}_{s}\Delta\gamma \tag{38}$$

$$\varepsilon^{e}_{d\ n+1} = \varepsilon^{e\ trial}_{d\ n+1} - \Delta\gamma\sqrt{\frac{3}{2}}\frac{\boldsymbol{s}_{n+1}}{\|\boldsymbol{s}_{n+1}\|} \tag{39}$$

$$\varepsilon^{e}_{v\ n+1} = \varepsilon^{e\ trial}_{v\ n+1} \tag{40}$$

According to the definition of the strain tensor in the next step shown in Equation (36), deviatoric strain tensor and the volumetric strain of the next step are defined in Equations (39)–(40). The deviatoric stress and the hydrostatic stress are defined as follows.

$$\boldsymbol{s}_{n+1} = (1-\mathrm{D}_{n+1})2G\varepsilon^{e\ trial}_{d\ n+1} - (1-\mathrm{D}_{n+1})2G\Delta\gamma\sqrt{\frac{3}{2}}\frac{\boldsymbol{s}_{n+1}}{\|\boldsymbol{s}_{n+1}\|} \tag{41}$$

$$p_{n+1} = (1-\mathrm{D}_{n+1})K\varepsilon^{e}_{v\ n+1} \tag{42}$$

In order to represent the plastically corrected yield function at t_{n+1}, the deviatoric stress of the next step shown in Equation (43) must be calculated. This can be expressed as Equation (43) using the trial deviatoric stress shown in Equation (33), and because the trial deviation stress is proportional to the deviatoric stress of the next step, it can be summarized as Equation (44).

$$\boldsymbol{s}_{n+1} = \frac{1-\mathrm{D}_{n+1}}{1-\mathrm{D}_{n}}\boldsymbol{s}^{trial}_{n+1} - (1-\mathrm{D}_{n+1})2G\Delta\gamma\sqrt{\frac{3}{2}}\frac{\boldsymbol{s}_{n+1}}{\|\boldsymbol{s}_{n+1}\|} \tag{43}$$

$$\boldsymbol{s}_{n+1} = (1-\mathrm{D}_{n+1})\left(\frac{1}{1-\mathrm{D}_{n}} - \frac{3G\Delta\gamma}{q^{trial}_{n+1}}\right)\boldsymbol{s}^{trial}_{n+1} \tag{44}$$

The yield function in the next step is defined as follows because it must satisfy the consistency condition.

$$\Phi_{n+1} = q_{n+1} - (1-\mathrm{D}_{n+1})\sigma_{y}\left(\bar{\varepsilon}^{p}_{n} + \Delta\gamma\right) \tag{45}$$

$$q_{n+1} = (1-\mathrm{D}_{n+1})\left(\frac{1}{1-\mathrm{D}_{n}}q^{trial}_{n+1} - 3G\Delta\gamma\right) \tag{46}$$

The yield function can be expressed as a function of the plastic multiplier and the damage of the next step, and the return mapping method must be performed to calculate the plastic multiplier and the damage of the next step.

With Equation (38), Equation (45) can be simplified as Equation (47). The plastic multiplier is calculated through the Newton-Raphson method. The simplified yield function does not require the process of estimating the initial value of the plastic multiplier [48] in calculating the damage energy release rate like Lemaitre's damage composition equation.

$$\Phi_{n+1} = \frac{1}{1-\mathrm{D}_{n}}q^{trial}_{n+1} - 3G\Delta\gamma - \sigma_{y}\left(\bar{\varepsilon}^{p}_{n} + \Delta\gamma\right) \tag{47}$$

$$\sigma_{n+1} = \boldsymbol{s}_{n+1} + p_{n+1}\boldsymbol{I} \tag{48}$$

When all stress and state variables are updated, the yield function of the next step reaches an elastic region as a value close to zero. In finite element analysis, in order to calculate the tangent stiffness matrix of each element, a consistent tangent modulus of the material reaching the last updated plastic region is required. Through the relationship between stress and strain tensor, the consistent tangent modulus ($\boldsymbol{D}^{ep}$) of the elastoplastic region is defined as follows [48].

$$\boldsymbol{D}^{ep} \equiv \frac{\partial \sigma_{n+1}}{\partial \varepsilon^{e\ trial}_{n+1}} \tag{49}$$

$$\sigma_{n+1} = \left[\boldsymbol{D}^e - \frac{\Delta\gamma 6G^2}{q^{trial}_{n+1}} \boldsymbol{I}_d \right] : \varepsilon^{e\ trial}_{n+1} \tag{50}$$

$$\boldsymbol{D}^e = 2G\boldsymbol{I}_d + \left(K - \frac{2}{3}G \right) \boldsymbol{I} \otimes \boldsymbol{I} \tag{51}$$

$$\boldsymbol{I}_d = \boldsymbol{I}_s - \frac{1}{3}\boldsymbol{I} \otimes \boldsymbol{I} \tag{52}$$

$$\boldsymbol{I}_s = I_{ijkl} = \frac{1}{2}\left(\delta_{ik}\delta_{jl} + \delta_{il}\delta_{jk} \right) \tag{53}$$

where $\boldsymbol{I}_s$ and $\boldsymbol{I}_d$ represent the fourth symmetric identity tensor and the deviatoric projection tensor, respectively, and δ_{ij} is Krönecker delta. $\boldsymbol{D}^e$ is the consistent tangent modulus derived through Hooke's law in the elastic region where plastic correction was not performed. $\boldsymbol{D}^{ep}$ can be expressed as follows by partial unification of Equation (50).

$$\boldsymbol{D}^{ep} = \boldsymbol{D}^e - \frac{\Delta\gamma 6G^2}{q^{trial}_{n+1}} \boldsymbol{I}_d - \frac{6G^2}{q^{trial}_{n+1}} \varepsilon^{e\ trial}_{d\ n+1} \otimes \frac{\partial \Delta\gamma}{\partial \varepsilon^{e\ trial}_{n+1}} + \frac{\Delta\gamma 6G^2}{\left(q^{trial}_{n+1}\right)^2} \varepsilon^{e\ trial}_{d\ n+1} \otimes \frac{\partial q^{trial}_{n+1}}{\partial \varepsilon^{e\ trial}_{n+1}}. \tag{54}$$

$$\boldsymbol{D}^{ep} = \boldsymbol{D}^e - \frac{\Delta\gamma 6G^2}{q^{trial}_{n+1}} \boldsymbol{I}_d + 6G^2 \left(\frac{\Delta\gamma}{q^{trial}_{n+1}} - \frac{1}{3G + \left.\frac{d\sigma_y}{d\bar{\varepsilon}^p}\right|_{\bar{\varepsilon}^p_n + \Delta\gamma}} \right) \frac{\boldsymbol{s}^{trial}_{n+1}}{\|\boldsymbol{s}^{trial}_{n+1}\|} \otimes \frac{\boldsymbol{s}^{trial}_{n+1}}{\|\boldsymbol{s}^{trial}_{n+1}\|}. \tag{55}$$

$$\boldsymbol{D}^{ep} = 2G\left(1 - \frac{\Delta\gamma 6G^2}{q^{trial}_{n+1}} \right) \boldsymbol{I}_d + 6G^2 \left(\frac{\Delta\gamma}{q^{trial}_{n+1}} - \frac{1}{3G + \left.\frac{d\sigma_y}{d\bar{\varepsilon}^p}\right|_{\bar{\varepsilon}^p_n + \Delta\gamma}} \right) \frac{\boldsymbol{s}^{trial}_{n+1}}{\|\boldsymbol{s}^{trial}_{n+1}\|} \otimes \frac{\boldsymbol{s}^{trial}_{n+1}}{\|\boldsymbol{s}^{trial}_{n+1}\|} + K\boldsymbol{I} \otimes \boldsymbol{I}. \tag{56}$$

4. Comparison with Experimental Results

4.1. Specimen

In order to perform the failure analysis of the primary barrier of Mark-III type LNG CCS, a series of material tests were performed on the same material applied to the primary barrier. 304L stainless steel was collected from 3500 mm × 1271 mm × 1.2 t of STS304L, and the chemical composition is presented in Table 1. As shown in Figure 2, five types of tensile specimen were prepared. The DB specimen, which means dogbone type tensile specimen, was fabricated to obtain the flow stress of 304L stainless steel. The width is 6 mm and the length of the reduced area is 30 mm. For NT05, NT10, and NT15 specimens, the radius of curvature of the notch in the middle was 5 mm, 10 mm, and 15 mm to obtain high-stress triaxiality. The minimum width at the center of the NT specimen is 6 mm. The CH03 specimen has a radius of curvature of 3 mm in the center hole. The central width of the NT and CH specimens is equal to 16 mm. The total length of all specimens is 110 mm, and the length and rolling direction of the specimens are the same.

Table 1. Chemical composition of 304L stainless steel.

C	Si	Mn	P	S	Cr	Ni
0.0152	0.379	1.130	0.0227	0.0017	18.653	10.178

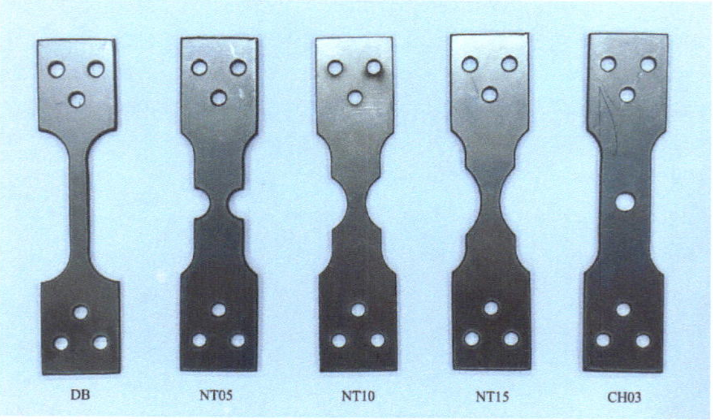

Figure 2. Material test specimens of 304L stainless steel sheet. DB specimen is used to acquire flow stress of 304L stainless steel, which has 6 mm width and 30 mm the length of reduced area. Other specimens are to identify the fracture strain. NT specimens have the radius of curvature of the notch to obtain high-stress triaxiality. The CH specimen has a radius of curvature of 3 mm in the center hole.

4.2. Experimental Set-Up

In this study, a universal testing machine was adopted to perform material testing. The maximum load capacity of UTM is 50 kN, and the speed of the crosshead can be controlled from 0.001 to 400.0 mm/min. In this study, the speed of the crosshead was controlled at 1.5 mm/min so that the initial strain rate for the DB specimen was 0.001/s. The displacement of all specimens was measured using an extensometer, and the gauge length was set to 25 mm. All material tests were performed at room temperature (13 °C). To verify repeatability, all tests were repeated three times and the results were shown as an average value.

4.3. Experimental Results

Figure 3a shows the results of the tensile test of the DB specimen at room temperature in terms of engineering stress and strain relationship. The elongation of the DB specimen was 0.7464, the 0.2% offset yield strength was 282 MPa, and the tensile strength was 679 MPa. The experimental result of the DB specimen was adopted only as a flow stress calculation. Figure 3b shows the force–displacement relationships of all specimens performed at room temperature. The smaller notch radius of the NT specimen increased the stress triaxiality and resulted in rapid failure. The CH03 specimen had a higher load because of the larger cross-sectional area at the center of the specimen.

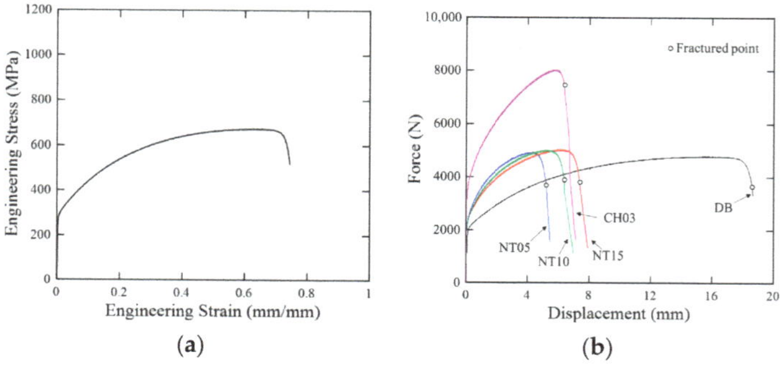

Figure 3. (**a**) Engineering stress–strain relationship at room temperature for DB specimen and (**b**) force–displacement relation. Given that the DB and NT specimens have the same minimum width, the maximum strength is almost the same. The smaller notch radius of the NT specimen caused increased stress triaxiality and rapid failure.

In order to observe the failure pattern of each specimen, the picture just before failure in the material test is shown in Figure 4. In the NT specimen, localized necking was observed in the center of the specimen, after which fracture progression was observed in the outer direction of the specimen. The CH03 specimen also begins to crack on both sides of the center hole and propagates outward. The fracture angle of all STS304L specimens was observed irregularly and it was judged that there was no tendency. As the purpose of the material test is to establish the ductile fracture criteria, no examination of the fracture surface of the specimen was conducted, nor was the effect of the notch radius on the material behavior analyzed through experimental results.

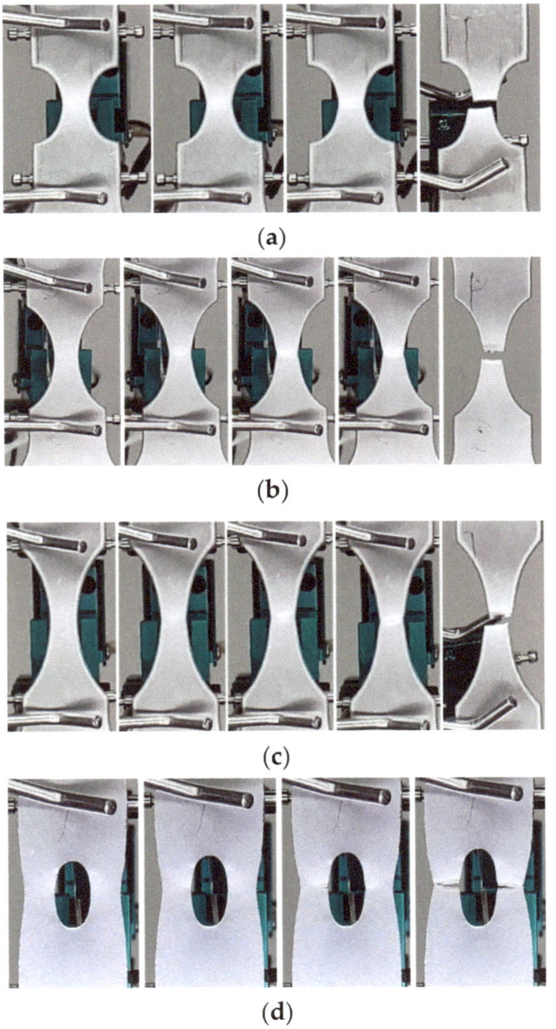

(a) (b) (c) (d)

Figure 4. Ductile fracture configuration of 304L stainless steel: (**a**) NT05 specimen, (**b**) NT10 specimen, (**c**) NT15 specimen, and (**d**) CH03 specimen. Localized necking was clearly observed. The CH specimen showed a localized neck followed by crack propagation.

5. Calibration of Ductile Fracture Model

5.1. FE Model

In order to establish a ductile fracture model, the equivalent plastic strain at the fracture location is required during material testing. Bao-Wierzbicki (2004) proposed the procedure of comparing experimental results with detailed numerical simulations because it is difficult to obtain experimentally [50]. This procedure is very useful and easy to predict in evaluating the failure of a structure through a commercial finite element analysis program. With recent technological advances, many researchers are adopting a method of obtaining the strain contour of a material using a digital image correction technique without comparing numerical simulation and experimental results [31,51,52]. In this study, in order to evaluate the effectiveness of the proposed elastoplastic-damage coupled constitutive model, the DIC method is not adopted, and the equivalent plastic strain is obtained by comparing the experimental results and the numerical analysis results, and the ductile fracture model is formulated.

Parallel numerical simulations of all material tests were carried out using commercial finite element code ABAQUS/Standard. Material specimen modeling was performed as shown in Figure 5. Although all specimens had symmetry conditions in the thickness direction, width direction, and length direction, no symmetric model (1:1 modeling) was considered for the failure prediction. The modeling range was to be included from the center point of the specimen to 25 mm. Since the length of the reduced section of the DB specimen exceeded 25 mm, the entire specimen was modeled only for the DB specimen. Displacement control was performed by applying coupling constraints as reference points to the upper and lower surfaces of the finite element analysis model.

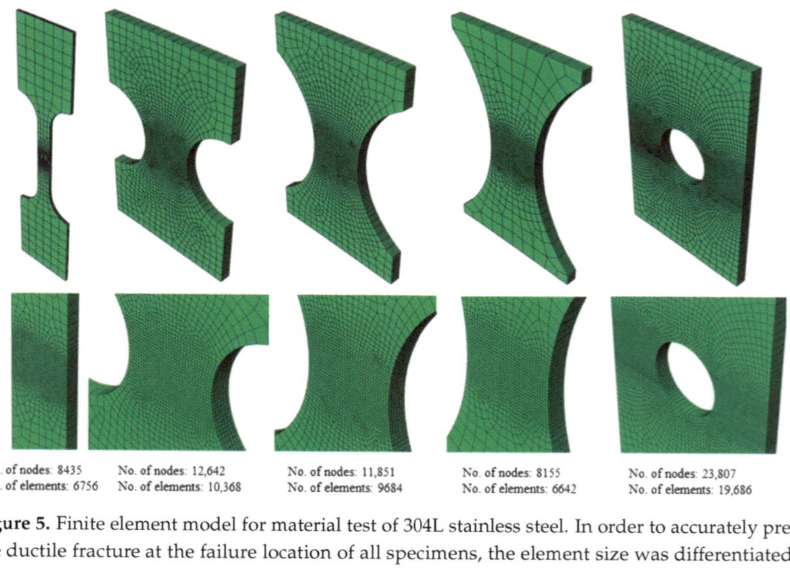

Figure 5. Finite element model for material test of 304L stainless steel. In order to accurately predict the ductile fracture at the failure location of all specimens, the element size was differentiated. In the region of interest, the element size was selected as 200 μm, and the number of elements in the thickness direction of the specimen was 6. To reduce computational cost, a coarse mesh was made outside the region of interest.

Eight-node brick element with reduced integration (C3D8R) was adopted for the finite element analysis model. When the ductile material undergoes plastic deformation and enters the necking range, the cross-sectional area decreases and the stress in the thickness direction cannot be ignored. Therefore, for precise prediction, the FE model was established using solid elements, not shell elements. The element size of the region of interest of each element was selected as 200 μm, and the number of elements in the thickness direction of the specimen was 6. To simulate the specimen's fracture pattern, it is necessary to select a smaller element size [53–55]. However, this requires considerable computation cost, and it was difficult to observe a regular fracture pattern in the material test of 304L stainless steel.

5.2. Calculation of Flow Stress

Flow stress is calculated through the experimental results of 304L stainless steel. The engineering stress-strain relation obtained through the tensile test was converted into a true stress-strain relation, which was expressed by the Swift-Voce equation, a hardening function of the constitutive model proposed in this study. The experimental results and hardening equation fitting results are shown in Figure 6. Material parameters of 304L stainless steel are listed in Table 2.

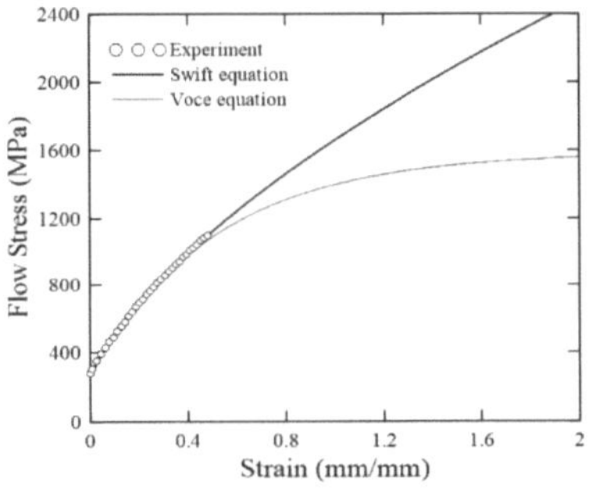

Figure 6. Flow stress of 304L stainless steel. Experimental results with engineering stress-strain relation are valid until the onset of the diffuse neck. In the domain of diffuse neck, the true stress is estimated through Swift equation and Voce equation.

Table 2. Material parameters of 304L stainless steel. In this study, 304L stainless steel did not show a yield plateau, so the yield plateau was set to zero.

Swift equation part	A	ε_0	n	ε_{plat}
	1610.0	0.0496	0.6	0.0
Voce equation part	α	σ_{y0}	Q	β
	1.0	282.0	1300.0	1.95

The Swift equation and Voce equation just estimate the post-necking behavior. The necessity of stress correction was reviewed by comparing the experimental results and the finite element analysis results applying flow stress. Figure 7 shows the comparison between the analysis result and the experimental result applying hardening equations. Through the flow stress calculated by the Swift equation, it was shown that the mechanical behavior of all specimens was well simulated. No yield plateau was observed in the tensile test at room temperature for 304L stainless steel, and failure occurred immediately after reaching the tensile strength.

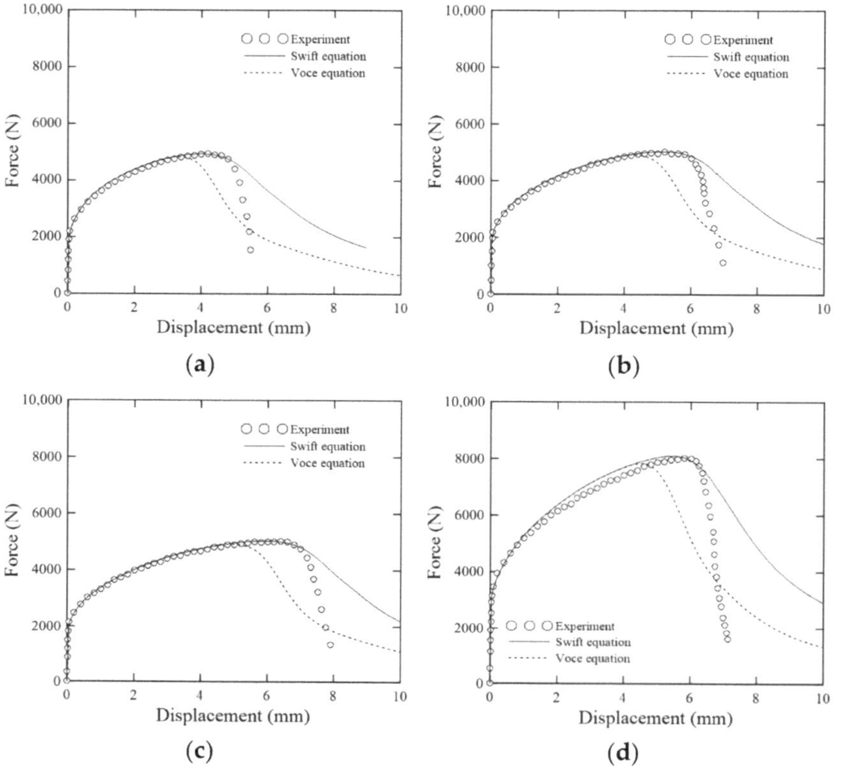

Figure 7. Comparison between experimental and simulation results of 304L stainless steel: (**a**) NT05 specimen, (**b**) NT10 specimen, (**c**) NT15 specimen, and (**d**) CH03 specimen. The flow stress using the Voce equation showed a strength drop before reaching the maximum strength. The flow stress calculation using the Swift equation is most appropriate.

5.3. Loading Path to Failure

To calibrate the ductile fracture model, the loading history at the predicted point of failure was investigated. In this study, when damage was initiated in the material test, the position of the largest equivalent plastic strain of the FE model was regarded as the predicted point of failure. The position of the highest equivalent plastic strain for the FE model is shown in Figure 8. This phenomenon can be observed at the same location in the experimental results shown in Figure 4.

As the equivalent plastic strain increases, the stress state is shown in Figure 9. If the growth of the equivalent plastic strain grows with a uniform stress triaxiality and Lode angle parameter, the fracture model can be easily calibrated. However, the stress triaxiality and Lode angle parameter of most ductile materials constantly fluctuate. Therefore, to consider the history of stress triaxiality and Lode angle parameter that appears as the specimen is deformed, and to reduce the sensitivity to fluctuations, the average value was introduced as follows.

$$\eta_{av} = \frac{1}{\bar{\varepsilon}_i^p} \int_0^{\bar{\varepsilon}_i^p} \eta(\bar{\varepsilon}^p) d\bar{\varepsilon}^p \tag{57}$$

$$\bar{\theta}_{av} = \frac{1}{\bar{\varepsilon}_i^p} \int_0^{\bar{\varepsilon}_i^p} \bar{\theta}(\bar{\varepsilon}^p) d\bar{\varepsilon}^p \tag{58}$$

where η_{av}, $\overline{\theta}_{av}$ are average stress triaxiality and average normalized Lode angle, respectively. $\overline{\varepsilon}_i^p$ is the equivalent plastic strain at damage initiation and $\overline{\varepsilon}^p$ is the equivalent plastic strain of the element. The average stress triaxiality, average normalized Lode angle, and equivalent plastic strain at damage initiation according to the specimen are listed in Table 3.

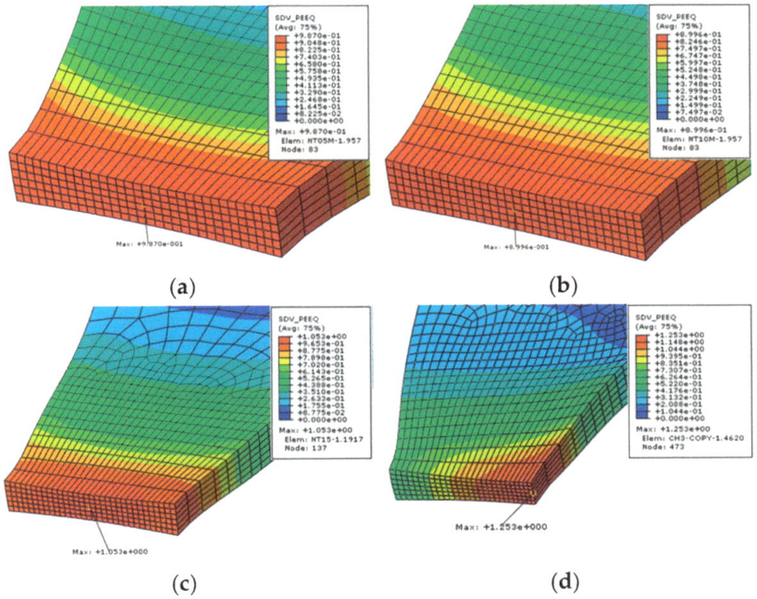

Figure 8. Equivalent plastic strain contour at damage initiation; (**a**) NT05 specimen, (**b**) NT10 specimen, (**c**) NT 15 specimen, and (**d**) CH03 specimen. In the NT specimen, when the stiffness drop occurred rapidly, the maximum equivalent plastic strain appeared at the center of the specimen. In the CH specimen, the maximum equivalent plastic strain was observed in the direction of the hole diameter due to symmetric structure of specimen. A decrease in thickness was observed in the numerical analysis model of all specimens.

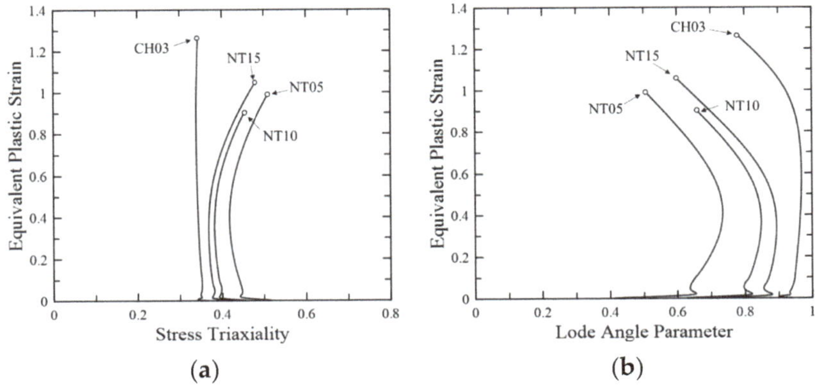

Figure 9. Equivalent plastic strain in accordance with (**a**) stress triaxiality and (**b**) Lode angle parameter. Black line white dot indicates equivalent plastic strain at damage initiation. As plastic strain accumulates, the stress triaxiality and Lode angle parameters were continuously changed.

Table 3. Average stress triaxiality, average normalized Lode angle, and equivalent plastic strain at damage initiation of 304L stainless steel. The average values were derived considering the loading history.

Specimen	Average Stress Triaxiality, η_{av}	Average Normalized Lode Angle, $\overline{\theta}_{av}$	Equivalent Plastic Strain at Damage Initiation, $\overline{\varepsilon}_i^p$
NT05	0.442	0.665	0.987
NT10	0.399	0.799	0.900
NT15	0.395	0.817	1.053
CH03	0.343	0.936	1.262

5.4. Determination of Fracture Parameters

The fracture parameters of the modified Mohr–Coulomb fracture model were determined using the average stress triaxiality, average normalized Lode angle, and equivalent plastic strain at damage initiation for each specimen shown in Table 3. Fracture parameters were selected as the value with the least error from experimental data among fracture loci. The finally determined fracture parameters are summarized in Table 4. Figure 10 shows the 3D modified Mohr–Coulomb fracture locus of 304L stainless steel projected on the $\eta - \overline{\varepsilon}^p$ plane and $\overline{\theta} - \overline{\varepsilon}^p$ plane.

Table 4. Fracture parameters of the modified Mohr–Coulomb fracture model for 304L stainless steel sheet.

Fracture parameter	c_1	c_2	c_3
	0.016	961	1.05

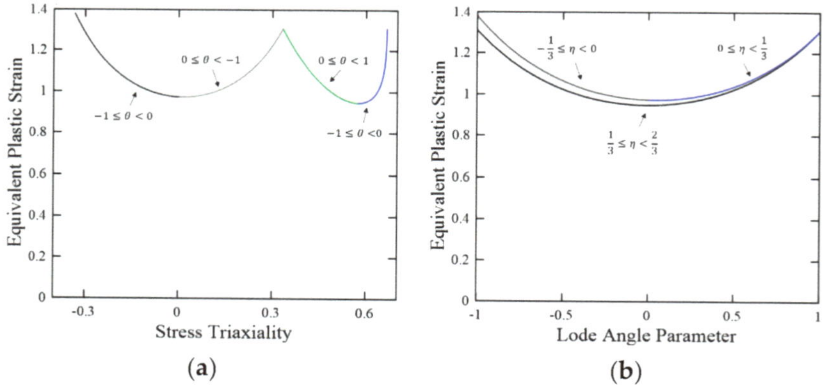

Figure 10. Fracture loci of the modified Mohr–Coulomb fracture model in accordance with (**a**) stress triaxiality and (**b**) Lode angle parameter of 304L stainless steel.

6. Prediction of Ductile Fracture

A ductile fracture simulation was performed to verify that the completed fracture locus was valid. The analysis model and boundary conditions are the same as those shown in chapter 5.1 FE model. Damage accumulation control parameter (D_s) and critical damage (D_c), which are damage parameters related to damage evolution, were set to 2.0 and 0.9, respectively. Figure 11a–d shows the comparison of the experimental results and analysis results for each test piece. In the deformed configuration shown in Figure 11e–h, the simulation results are shown after removing the element whose critical damage reached 0.9. All of the simulations show good agreement with experimental results.

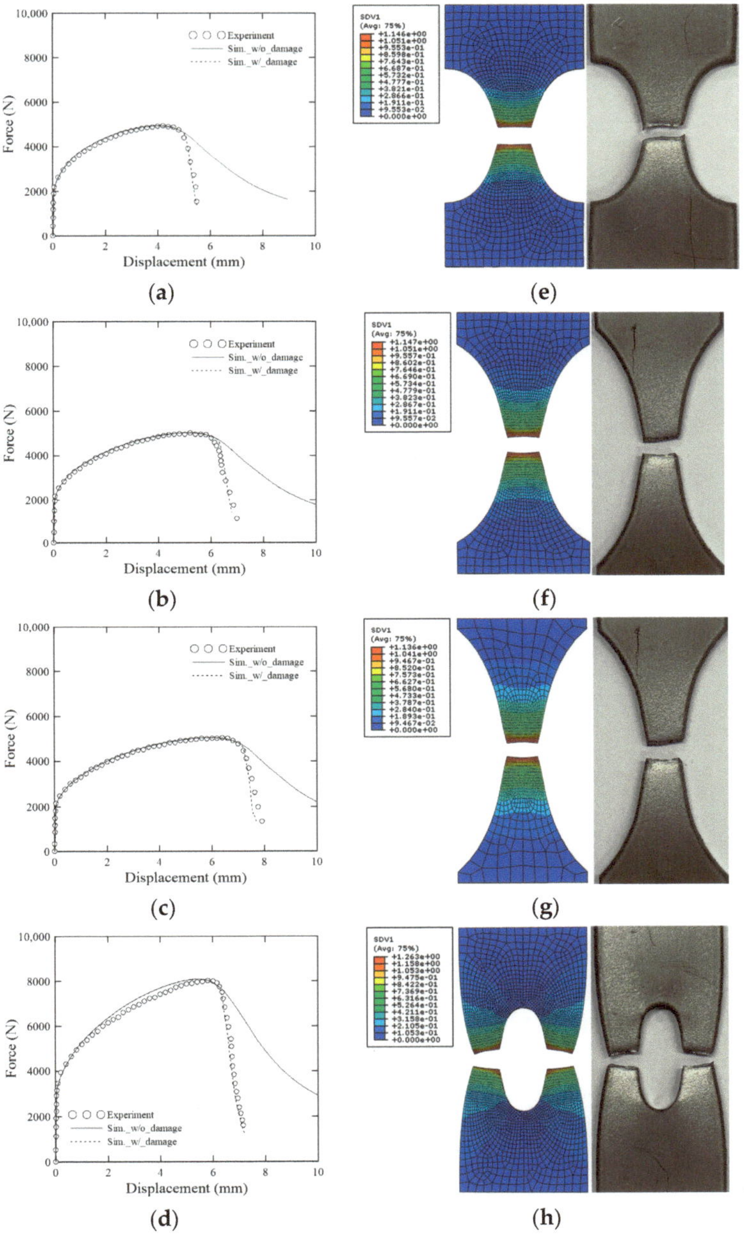

Figure 11. Comparison of force–displacement curve of (**a**) NT05, (**b**) NT10, (**c**) NT15, and (**d**) CH03 specimens and deformed configuration of (**e**) NT05, (**f**) NT10, (**g**) NT15, and (**h**) CH03 between experimental and simulation results (SDV1: equivalent plastic strain). With the Modified Mohr–Coulomb fracture criterion satisfied, crack propagation of the specimen occurs and complete failure follows. The fractured configurations between the numerical analysis and experiments are almost consistent.

7. Conclusions

In the present study, the failure prediction methodology was numerically developed to predict the ductile fracture of 304L stainless steel sheet. First of all, an elastoplastic constitutive model for a 304L stainless steel sheet was developed. To describe the constitutive behavior after diffuse necking, the combined Swift-Voce equation was adopted as a hardening function. The modified Mohr–Coulomb criterion based on the equivalent plastic strain was also adopted to describe the sudden fracture of the 304L stainless steel sheet. Numerical formulation of the elastoplastic-damage coupled constitutive model with fracture locus was implemented into ABAQUS user-defined subroutine UMAT.

To identify elastoplastic behavior and establish ductile fracture criterion, a series of material tests considering various stress states was performed. Five types of specimens were processed with 1.2 t 304L stainless steel, the raw material of the primary barrier of Mark-III type LNG CCS. Due to the ductility of 304L stainless steel, a considerable deformation occurred, and then a fracture was reached. After localized necking, fracture propagation was observed from the center point of the specimen. Since the modified Mohr–Coulomb model defines ductile fracture based on equivalent plastic strain, the equivalent plastic strain was obtained by comparing numerical analysis and experimental results in parallel. As plastic deformation accumulates, the stress triaxiality and Lode angle parameters fluctuate. To correct this problem, the average stress triaxiality and average Lode angle parameters were introduced. The modified Mohr–Coulomb fracture locus of 304L stainless steel was determined based on the material test results. Numerical analysis with ductile fracture criterion shows good agreement with experimental results.

The stainless steel exhibits outstanding mechanical performance in terms of the yield and tensile strength under cryogenic temperatures rather than at room temperature. However, ductility at a cryogenic temperature significantly decreases compared to room temperature. A fairly optimistic fracture analysis was performed based on fracture criteria based on equivalent plastic strain to the actual conditions in a LNG cargo tank. Therefore, the fracture criterion based on the experimental results at room temperature considered in this study is a fairly optimistic failure assessment result. Further studies will include failure prediction of the primary barrier of Mark-III type LNG CCS using the proposed numerical methodology in this study.

Author Contributions: Conceptualization, M.-S.K. and J.-M.L.; Data curation, M.-S.K., H.-T.K., Y.-H.C., J.-H.K. and S.-K.K.; Formal analysis, H.-T.K. and Y.-H.C.; Investigation, M.-S.K., H.-T.K., Y.-H.C., J.-H.K. and S.-K.K.; Methodology, M.-S.K.; Project administration, J.-M.L.; Supervision, J.-M.L.; Validation, H.-T.K., Y.-H.C., J.-H.K. and S.-K.K.; Writing—original draft, M.-S.K.; Writing—review & editing, J.-H.K., S.-K.K. and J.-M.L. All authors have read and agreed to the published version of the manuscript.

Funding: This work was supported by the Materials/Parts Technology Development Program (20017575, Development of Applicability Evaluation Technology for Cryogenic Insulation Material and Storage Vessel considering Operating Condition of Hydrogen Commercial Vehicle) funded By the Ministry of Trade, Industry & Energy (MOTIE, Korea). This work was supported by the R&D Platform Establishment of Eco-Friendly Hydrogen Propulsion Ship Program (No. 20006644) funded by the Ministry of Trade, Industry & Energy (MOTIE, Korea). This research was supported by the Development and demonstration of the on-board marine debris disposal modules program of the Korea Institute of Marine Science & Technology Promotion (KIMST) funded by the Ministry of Oceans and Fisheries (KIMST-20220494).

Institutional Review Board Statement: Not applicable.

Informed Consent Statement: Not applicable.

Data Availability Statement: The data presented in this study are available on request from corresponding author.

Conflicts of Interest: The authors declare no conflict of interest.

References

1. Jia, S.; Tan, Q.; Ye, J.; Zhu, Z.; Jiang, Z. Experiments on dynamic mechanical properties of austenitic stainless steel S30408 and S31608. *J. Constr. Steel Res.* **2021**, *179*, 106556. [CrossRef]
2. Kim, J.-H.; Park, W.-S.; Chun, M.-S.; Kim, J.-J.; Bae, J.-H.; Kim, M.-H.; Lee, J.-M. Effect of pre-straining on low-temperature mechanical behavior of AISI 304L. *Mater. Sci. Eng. A* **2012**, *543*, 50–57. [CrossRef]
3. Kim, S.-K.; Lee, C.-S.; Kim, J.-H.; Kim, M.-H.; Lee, J.-M. Computational evaluation of resistance of fracture capacity for SUS304L of liquefied natural gas insulation system under cryogenic temperatures using ABAQUS user-defined material subroutine. *Mater. Des.* **2013**, *50*, 522–532. [CrossRef]
4. Zheng, C.; Liu, C.; Ren, M.; Jiang, H.; Li, L. Microstructure and mechanical behavior of an AISI 304 austenitic stainless steel prepared by cold- or cryogenic-rolling and annealing. *Mater. Sci. Eng. A* **2018**, *724*, 260–268. [CrossRef]
5. Mallick, P.; Tewary, N.; Ghosh, S.; Chattopadhyay, P. Effect of cryogenic deformation on microstructure and mechanical properties of 304 austenitic stainless steel. *Mater. Charact.* **2017**, *133*, 77–86. [CrossRef]
6. Singh, R.; Sachan, D.; Verma, R.; Goel, S.; Jayaganthan, R.; Kumar, A. Mechanical behavior of 304 austenitic stainless steel processed by cryogenic rolling. *Mater. Today Proc.* **2018**, *5*, 16880–16886. [CrossRef]
7. Oh, D.-J.; Kim, N.-K.; Song, S.-W.; Kim, Y.-D.; Kim, M.-H. Investigation of fatigue performance for new membrane-type LNG CCS at cryogenic temperature. *Mar. Struct.* **2018**, *62*, 90–105. [CrossRef]
8. Park, S.Y.; Kim, W.T.; Choi, J.H.; On, S.Y.; Kim, S.S. Effects of process-induced residual stress and geometric characteristics on pressure-resisting capability of corrugation in primary barriers of liquefied natural gas carriers. *Ocean Eng.* **2021**, *237*, 109613. [CrossRef]
9. Kim, J.-H.; Kim, S.-K.; Kim, M.-H.; Lee, J.-M. Numerical model to predict deformation of corrugated austenitic stainless steel sheet under cryogenic temperatures for design of liquefied natural gas insulation system. *Mater. Des.* **2014**, *57*, 26–39. [CrossRef]
10. Kim, M.-S.; Kwon, S.-B.; Kim, S.-K.; Kim, J.-H.; Lee, J.-S.M. Impact failure analysis of corrugated steel plate in LNG containment cargo system. *J. Constr. Steel Res.* **2019**, *156*, 287–301. [CrossRef]
11. Kim, M.-S.; Kim, J.-H.; Kim, S.-K.; Lee, J.-M. Experimental Investigation of Structural Response of Corrugated Steel Sheet Subjected to Repeated Impact Loading: Performance of LNG Cargo Containment System. *Appl. Sci.* **2019**, *9*, 1558. [CrossRef]
12. Kim, B.C.; Yoon, S.H.; Gil Lee, D. Pressure resistance of the corrugated stainless steel membranes of LNG carriers. *Ocean Eng.* **2011**, *38*, 592–608. [CrossRef]
13. Lee, D.; Kim, K.H.; Choi, I.; Gil Lee, D. Pressure-resisting capability of the knot area of the primary barrier for a LNG containment system. *Ocean Eng.* **2015**, *95*, 128–133. [CrossRef]
14. Jeong, Y.-J.; Kim, H.-T.; Kim, S.-K.; Kim, J.-H.; Kim, Y.-T.; Heo, W.-S.; Lee, J.-M. Evaluation of the pressure-resisting capability of membrane-type corrugated sheet under hydrodynamic load. *Thin-Walled Struct.* **2021**, *162*, 107388. [CrossRef]
15. Deng, X.; Wang, B. Peridynamic modeling of dynamic damage of polymer bonded explosive. *Comput. Mater. Sci.* **2020**, *173*, 109405. [CrossRef]
16. Depraetere, R.; De Waele, W.; Hertelé, S. Fully-coupled continuum damage model for simulation of plasticity dominated hydrogen embrittlement mechanisms. *Comput. Mater. Sci.* **2021**, *200*, 110857. [CrossRef]
17. Varandas, L.F.; Catalanotti, G.; Melro, A.R.; Falzon, B.G. On the importance of nesting considerations for accurate computational damage modelling in 2D woven composite materials. *Comput. Mater. Sci.* **2020**, *172*, 109323. [CrossRef]
18. Paredes, M.; Wierzbicki, T.; Zelenak, P. Prediction of crack initiation and propagation in X70 pipeline steels. *Eng. Fract. Mech.* **2016**, *168*, 92–111. [CrossRef]
19. Björklund, O.; Larsson, R.; Nilsson, L. Failure of high strength steel sheets: Experiments and modelling. *J. Mater. Process. Technol.* **2013**, *213*, 1103–1117. [CrossRef]
20. Kim, S.-K.; Kim, J.-H.; Kim, J.-H.; Lee, J.-M. Numerical Model for Mechanical Nonlinearities of High Manganese Steel Based on the Elastoplastic Damage Model. *Metals* **2018**, *8*, 680. [CrossRef]
21. Baltic, S.; Magnien, J.; Gänser, H.-P.; Antretter, T.; Hammer, R. Coupled damage variable based on fracture locus: Modelling and calibration. *Int. J. Plast.* **2019**, *126*, 102623. [CrossRef]
22. Charoensuk, K.; Panich, S.; Uthaisangsuk, V. Damage initiation and fracture loci for advanced high strength steel sheets taking into account anisotropic behaviour. *J. Mater. Process. Technol.* **2017**, *248*, 218–235. [CrossRef]
23. Hancock, J.W.; Mackenzie, A.C. On the mechanisms of ductile failure in high-strength steels subjected to multi-axial stress-states. *J. Mech. Phys. Solids* **1976**, *24*, 147–160. [CrossRef]
24. Theocaris, P.S. Failure criteria for isotropic bodies revisited. *Eng. Fract. Mech.* **1995**, *51*, 239–255, 257–264. [CrossRef]
25. Schiffmann, R.; Bleck, W.; Dahl, W. The influence of strain history on ductile failure of steel. *Comput. Mater. Sci.* **1998**, *13*, 142–147. [CrossRef]
26. Schlüter, N.; Grimpe, F.; Bleck, W.; Dahl, W. Modelling of the damage in ductile steels. *Comput. Mater. Sci.* **1996**, *7*, 27–33. [CrossRef]
27. Bai, Y.; Wierzbicki, T. A new model of metal plasticity and fracture with pressure and Lode dependence. *Int. J. Plast.* **2008**, *24*, 1071–1096. [CrossRef]
28. Bai, Y.; Wierzbicki, T. Application of extended Mohr–Coulomb criterion to ductile fracture. *Int. J. Fract.* **2010**, *161*, 1–20. [CrossRef]
29. Kim, S.-K.; Lee, C.-S.; Kim, J.-H.; Kim, M.-H.; Noh, B.-J.; Matsumoto, T.; Lee, J.-M. Estimation of Fatigue Crack Growth Rate for 7% Nickel Steel under Room and Cryogenic Temperatures Using Damage-Coupled Finite Element Analysis. *Metals* **2015**, *5*, 603–627. [CrossRef]

30. Kim, S.-K.; Kang, K.-Y.; Kim, M.-S.; Lee, J.-M. Low-Temperature Mechanical Behavior of Super Duplex Stainless Steel with Sigma Precipitation. *Metals* **2015**, *5*, 1732–1745. [CrossRef]
31. Paredes, M.; Grolleau, V.; Wierzbicki, T. On ductile fracture of 316L stainless steels at room and cryogenic temperature level: An engineering approach to determine material parameters. *Materialia* **2020**, *10*, 100624. [CrossRef]
32. Dunand, M.; Mohr, D. On the predictive capabilities of the shear modified Gurson and the modified Mohr–Coulomb fracture models over a wide range of stress triaxialities and Lode angles. *J. Mech. Phys. Solids* **2011**, *59*, 1374–1394. [CrossRef]
33. Luo, M.; Dunand, M.; Mohr, D. Experiments and modeling of anisotropic aluminum extrusions under multi-axial loading—Part II: Ductile fracture. *Int. J. Plast.* **2012**, *32–33*, 36–58. [CrossRef]
34. Chen, X.; Shi, M.F.; Shih, H.-C.; Luo, M.; Wierzbicki, T. AHSS Shear Fracture Predictions Based on a Recently Developed Fracture Criterion. *SAE Int. J. Mater. Manuf.* **2010**, *3*, 723–731. [CrossRef]
35. Granum, H.; Morin, D.; Børvik, T.; Hopperstad, O.S. Calibration of the modified Mohr-Coulomb fracture model by use of localization analyses for three tempers of an AA6016 aluminium alloy. *Int. J. Mech. Sci.* **2021**, *192*, 106122. [CrossRef]
36. Quach, H.; Kim, J.-J.; Nguyen, D.-T.; Kim, Y.-S. Uncoupled ductile fracture criterion considering secondary void band behaviors for failure prediction in sheet metal forming. *Int. J. Mech. Sci.* **2020**, *169*, 105297. [CrossRef]
37. Qian, L.-Y.; Fang, G.; Zeng, P. Modeling of the ductile fracture during the sheet forming of aluminum alloy considering non-associated constitutive characteristic. *Int. J. Mech. Sci.* **2017**, *126*, 55–66. [CrossRef]
38. Ben Othmen, K.; Haddar, N.; Jegat, A.; Manach, P.-Y.; Elleuch, K. Ductile fracture of AISI 304L stainless steel sheet in stretching. *Int. J. Mech. Sci.* **2020**, *172*, 105404. [CrossRef]
39. Pham, H.; Iwamoto, T. An evaluation of fracture properties of type-304 austenitic stainless steel at high deformation rate using the small punch test. *Int. J. Mech. Sci.* **2018**, *144*, 249–261. [CrossRef]
40. Kim, J.-H.; Lee, C.-S.; Kim, M.-H.; Lee, J.-M. Prestrain-dependent viscoplastic damage model for austenitic stainless steel and implementation to ABAQUS user-defined material subroutine. *Comput. Mater. Sci.* **2013**, *67*, 273–281. [CrossRef]
41. Mohr, D.; Marcadet, S.J. Micromechanically-motivated phenomenological Hosford–Coulomb model for predicting ductile fracture initiation at low stress triaxialities. *Int. J. Solids Struct.* **2015**, *67–68*, 40–55. [CrossRef]
42. Palchik, V. Application of Mohr–Coulomb failure theory to very porous sandy shales. *Int. J. Rock Mech. Min. Sci.* **2006**, *43*, 1153–1162. [CrossRef]
43. Zhao, J. Applicability of Mohr–Coulomb and Hoek–Brown strength criteria to the dynamic strength of brittle rock. *Int. J. Rock Mech. Min. Sci.* **2000**, *37*, 1115–1121. [CrossRef]
44. Kofiani, K.; Nonn, A.; Wierzbicki, T. New calibration method for high and low triaxiality and validation on SENT specimens of API X70. *Int. J. Press. Vessel. Pip.* **2013**, *111–112*, 187–201. [CrossRef]
45. Swift, H.W. Plastic Instability under Plane Stress. *J. Mech. Phys. Solids* **1952**, *1*, 1–18. [CrossRef]
46. Voce, E. The relationship between stress and strain for homogeneous deformation. *J. Inst. Met.* **1948**, *74*, 537–562.
47. Lemaitre, J. A continuous damage mechanics model for ductile fracture. *J. Eng. Mater. Technol.* **1985**, *107*, 83–89. [CrossRef]
48. Neto, E.A.; de Eduardo, S.; Perić, D.; Owen, D.R.J. *Computational Methods for Plasticity: Theory and Applications*; Wiley: Chichester, UK, 2008; ISBN 9780470694527.
49. Lian, J.; Wu, J.; Münstermann, S. Evaluation of the cold formability of high-strength low-alloy steel plates with the modified Bai–Wierzbicki damage model. *Int. J. Damage Mech.* **2015**, *24*, 383–417. [CrossRef]
50. Bao, Y.; Wierzbicki, T. On fracture locus in the equivalent strain and stress triaxiality space. *Int. J. Mech. Sci.* **2004**, *46*, 81–98. [CrossRef]
51. Sancho, A.; Cox, M.; Cartwright, T.; Davies, C.; Hooper, P.; Dear, J. An experimental methodology to characterise post-necking behaviour and quantify ductile damage accumulation in isotropic materials. *Int. J. Solids Struct.* **2019**, *176–177*, 191–206. [CrossRef]
52. Lou, Y.; Huh, H. Extension of a shear-controlled ductile fracture model considering the stress triaxiality and the Lode parameter. *Int. J. Solids Struct.* **2013**, *50*, 447–455. [CrossRef]
53. Xue, Z.; Pontin, M.; Zok, F.; Hutchinson, J. Calibration procedures for a computational model of ductile fracture. *Eng. Fract. Mech.* **2010**, *77*, 492–509. [CrossRef]
54. Erice, B.; Gálvez, F. A coupled elastoplastic-damage constitutive model with Lode angle dependent failure criterion. *Int. J. Solids Struct.* **2014**, *51*, 93–110. [CrossRef]
55. Kubík, P.; Šebek, F.; Hůlka, J.; Petruška, J. Calibration of ductile fracture criteria at negative stress triaxiality. *Int. J. Mech. Sci.* **2016**, *108–109*, 90–103. [CrossRef]

materials

MDPI

Article

Computational Design and Characterisation of Gyroid Structures with Different Gradient Functions for Porosity Adjustment

Leonie Wallat [1,*], Patrick Altschuh [2,3], Martin Reder [2,3], Britta Nestler [2,3] and Frank Poehler [1,*]

1 Institute of Materials and Processes, Karlsruhe University of Applied Sciences, Molkestr. 30, 76133 Karlsruhe, Germany
2 Institute for Digital Materials Research, Karlsruhe University of Applied Sciences, Molkestr. 30, 76133 Karlsruhe, Germany; patrick.altschuh@h-ka.de (P.A.); martin_dominik.reder@h-ka.de (M.R.); britta.nestler@kit.edu (B.N.)
3 Institute for Applied Materials-Microstructure Modelling and Simulation, Karlsruhe Institute of Technology, Kaiserstraße 12, 76131 Karlsruhe, Germany
* Correspondence: leonie.wallat@h-ka.de (L.W.); frank.Poehler@h-ka.de (F.P.)

Abstract: Triply periodic minimal surface (TPMS) structures have a very good lightweight potential, due to their surface-to-volume ratio, and thus are contents of various applications and research areas, such as tissue engineering, crash structures, or heat exchangers. While TPMS structures with a uniform porosity or a linear gradient have been considered in the literature, this paper focuses on the investigation of the mechanical properties of gyroid structures with non-linear porosity gradients. For the realisation of the different porosity gradients, an algorithm is introduced that allows the porosity to be adjusted by definable functions. A parametric study is performed on the resulting gyroid structures by performing mechanical simulations in the linear deformation regime. The transformation into dimensionless parameters enables material-independent statements, which is possible due to linearity. Thus, the effective elastic behaviour depends only on the structure geometry. As a result, by introducing non-linear gradient functions and varying the density of the structure over the entire volume, specific strengths can be generated in certain areas of interest. A computational design of porosity enables an accelerated application-specific structure development in the field of engineering.

Keywords: TPMS structures; sheet-based gyroid; mechanical simulation; modelling; PACE3D

Citation: Wallat, L.; Altschuh, P.; Reder, M.; Nestler, B.; Poehler, F. Computational Design and Characterisation of Gyroid Structures with Different Gradient Functions for Porosity Adjustment. *Materials* **2022**, *15*, 3730. https://doi.org/10.3390/ma15103730

Academic Editors: Shuwen Wen, Yongle Sun and Xin Chen

Received: 20 April 2022
Accepted: 17 May 2022
Published: 23 May 2022

1. Introduction

Triply periodic minimal surfaces (TPMS) are three-dimensional cell structures that occur in nature in many forms: for example, in butterfly wings [1] or on the skeletal plate of a sea urchin [2]. There are a variety of different structures: for example, gyroid, Schwarz diamond, and Schwarz primitive structures [3], which are defined by a mathematical periodic function and whose surfaces have a mean curvature of zero. This results in a smoothly curved surface, while the periodic cells are divided into two disjointed continuous channels that are intertwined. In addition to their lightweight potential, these cell structures are characterised by unique properties and shapes that make them attractive for a wide range of engineering applications. For example, the high surface-to-volume ratio and the two-phase system are very preferable properties for the development of heat exchangers [4–6]. In particular, the work by Weihong Li et al. [5] has shown that a comparison between a printed circuit heat exchanger (PCHE) and a heat exchanger with TPMS structures (Schwarz diamond and gyroid) shows both a higher thermal performance and a higher Nusselt number [5]. Furthermore, TPMS structures are of great interest in the field of tissue engineering, as their topological structure is similar to that of trabecular bone [7]. The introduction of a porosity gradient on the TPMS structures opens up new engineering possibilities. In Dawei Li's research [8], for example, it was shown that sheet-based linear graded gyroid structures have a high energy absorption potential, which is a crucial property in applications with regard to crash safety. Moreover, the introduction of the gradient offers a new freedom of design. As such, the work of [9] aims to use linear graded cell gradients to replicate the natural environment of bones. Since structures with gradients have so far mostly been investigated with linear gradients and are interesting for a variety of applications, it is desirable to put more emphasis on non-linear porosity adjustments. In the following, the group of double gyroid structures is considered, which is exemplified in Figure 1, with the characteristic two-tunnel system. For example, in [8,10], this structure is referred to as 'sheet-based gyroid' (in the remainder of this article, it will only be referred to as 'gyroid').

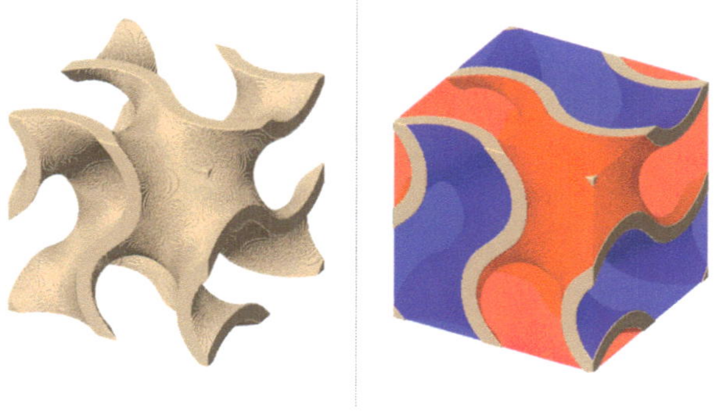

a b

Figure 1. (**a**) Sheet-based gyroid structure; (**b**) gyroid structure with labelled two-tunnel system.

The following equation is used to approximate the surface of the gyroid structure by trigonometrical functions [4].

$$0 = \left[\sin\frac{2\pi x}{L_x} \cdot \cos\frac{2\pi y}{L_y} + \sin\frac{2\pi y}{L_y} \cdot \cos\frac{2\pi z}{L_z} + \sin\frac{2\pi z}{L_z} \cdot \cos\frac{2\pi x}{L_x} \right]^2 - t^2 \quad (1)$$

The number of cell repetitions in the x-, y-, and z-directions and the absolute sizes of the unit cells L_x, L_y, and L_z define the cell space [11]. The thickness of the cell wall is controlled by the variable t. Thus, t has an effect on the volume fraction (v^*) of the lattice structure [12]. According to [13,14], the volume fraction (v^*) and the closely related parameter porosity Φ are defined as follows:

$$\Phi = (1 - v^*) \cdot 100[\%] \quad (2)$$

with

$$v^* = \frac{v}{v_s}, \quad (3)$$

where v and v_s denote the volume of the pore structure and the volume of the solid structure, respectively [13,14]. In the literature, v^* is also referred to as 'relative density' [14,15]. As can be seen from equation (2), the higher the porosity, the thinner the cell walls. According to Gibson and Ashby [16], the mechanical properties of porous structures of the same topology are directly influenced by their porosity. They propose a correlation between the effective Young modulus and the relative density, which is known as Gibson–Ashby correlation. In addition to the porosity, the mechanical properties are also strongly influenced by the topology of the structure [11,17]. With respect to gyroid structures with imposed porosity gradients, the question therefore arises as to how different geometries with non-linear porosity gradients influence the resulting effective behaviour of the structures. Since the complex manufacture of gyroid structures is cost and time intensive, it is desirable to answer these questions through digital modelling and simulations.

In this work, an algorithm for generating gyroid structures with imposed porosity gradients is introduced, and the resulting mechanical properties of the gyroids are investigated by performing a simulation study. For the structure generation, a constant porosity and two different functions are considered: a linear and a quadratic function. The numerical simulations are performed in the linear elastic regime, which is a common approach in the field of open cell foams. Kaoua et al. [18] use finite element (FE) simulations on Kelvin unit cells to investigate different ligament cross section geometries. In the work by Gan et al. [19] and Zhu et al. [17], elastic FE simulations are also applied to Voronoi-based foams, whereby in the latter work, the influence of geometry irregularities is investigated. The aim of the work is to enable the digital design of gyroid structures with tailored porosity gradients for specific applications.

2. Computational Design

Before mechanical simulations of the structures can be performed, the digital structures are created on the basis of a spatial algorithm. For the structure creation, a MatLab [20] source was programmed, which enables the creation of TPMS structures with and without gradients. The aim of the MatLab program is to create gyroid structures with adjustable porosities and definable porosity gradients, using mathematical functions. The TPMS structures are stored in vtk files, while the further preprocessing of the structures as well as the simulations are realised with the simulation framework PACE3D [21]. The simulation software "Parallel Algorithms for Crystal Evolution in 3D" (PACE3D) is a massive parallel in-house software package [21] which is developed at the Institute for Digital Materials Science (IDM) of the Karlsruhe University of Applied Sciences, Germany. The objective of PACE3D is to provide a package for large-scale multiphysics simulations, so as to solve coupled problems such as solidification, grain growth, mass and heat transport, fluid flow and mechanical forces (elasticity, plasticity), etc. The use of dimensionless quantities enables a scale-independent representation of the results, so that the simulations are performed with a non-dimensionalisation. With the help of a conversion table, physical quantities can be obtained from the results. The corresponding flowchart from the creation of the structures in the MatLab program to the mechanical simulation is summarised in Figure 2.

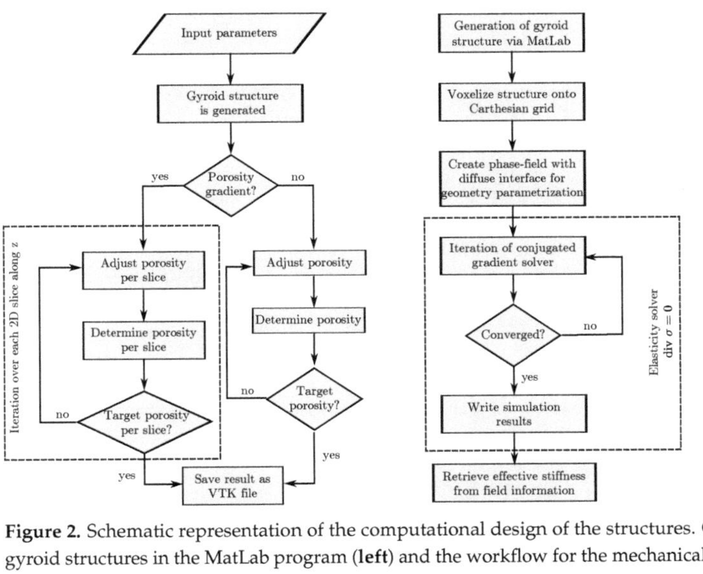

Figure 2. Schematic representation of the computational design of the structures. Generation of the gyroid structures in the MatLab program (**left**) and the workflow for the mechanical simulations with PACE3D (**right**).

2.1. Structure Generation

2.1.1. Input Parameters

The MatLab program offers the possibility of creating structures with constant and graded porosity. Table 1 lists and briefly describes the required input parameters for the the structure generation, while the individual parameters and their influence are discussed more specifically in the following.

Table 1. Input parameters of the MatLab program and their function for the creation of a gyroid structure.

Input Parameter	Function
num_x num_y num_z	Number of unit cells to be repeated in the x-, y-, and z-direction
$unitCellSize$	Size of the unit cells (in mm)
n_{steps}	Resolution of the unit cell
Φ_{max} Φ_{min}	Maximum and minimum porosity of the cell
$func$	Gradient function
$grad$	With/without gradient function (1, 0)
$delta$	Tolerance range

The input parameters num_x, num_y, and num_z specify the number of unit cells to be repeated in the corresponding direction, while the parameter $unitCellSize$ defines the physical dimension of the created gyroid unit cell. By multiplying $unitCellSize$ with the input parameters num_x, num_y, and num_z, the quantities L_x, L_y, and L_z of Equation (1) are obtained, which also represent the total domain size of the resulting structure in physical dimensions. The parameter n_{steps} defines the number of voxels that are used to discretise the gyroid in all directions of one spatial unit cell.

With Φ_{min} and Φ_{max}, the range of the minimum and maximum porosity is defined. In the MatLab program, the porosity is specified in the range between [0; 1], which represents the more commonly used expression of 0% and 100% for the porosity. When creating a gyroid structure with a constant overall porosity, the value from Φ_{max} is used. The porosity function is defined with the parameter $func$. The polynomial degree of the gradient function can be selected between 0 and 2, corresponding to a constant (0), a linear (1), and a quadratic (2) representation. The input parameter $grad$ determines whether the gyroid structure is generated with or without gradient by the integer values 1 and 0, respectively. Per default, the gradient occurs in the z-direction.

The parameter $delta$ is mainly responsible for the iterative adjustment of the actual porosity to the target porosity. A tolerance range is defined that describes the maximum permissible deviation between the actual porosity of the current layer and the target porosity. For instance, if the parameter is set to a value of 0.02, this corresponds to a deviation of the actual data of 2%, compared to the target porosity function. The smaller the number of this parameter, the more accurate and longer the program takes to calculate. The target porosity per cell layer is calculated by the gradient function.

Figure 3 shows four different gyroid unit cell structures with a cell size of 2.5 mm, which is generated with 200 steps and a $delta$ parameter of 0.02. Figure 3a,d illustrate a gyroid unit cell with a constant porosity of 0.8 (a) and 0.4 (d). By looking at the two structures, the influence of the porosity on the cell thickness becomes evident. The higher the porosity, the thinner the wall thickness. The structures Figure 3b,c refer to different gradient functions: linear (b) and quadratic (c) functions. The structures are in a porosity interval between 0.4 and 0.8.

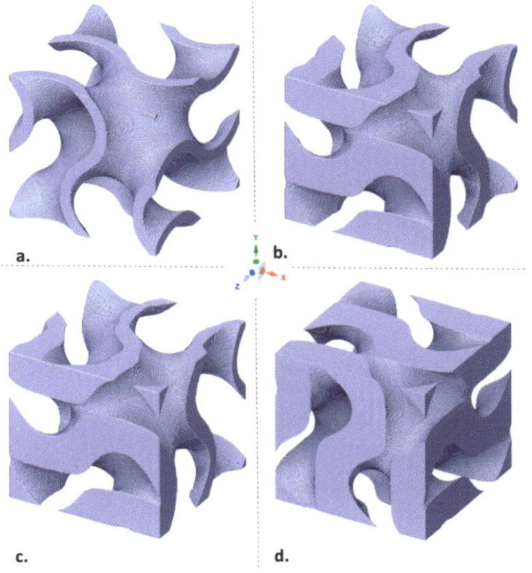

Figure 3. Unit cell of the gyroid structure; gradient with (**a**) constant function, with a porosity of 0.8; (**b**) linear function, with a porosity from 0.4 to 0.8; (**c**) quadratic function, with a porosity from 0.4 to 0.8; (**d**) constant function, with a porosity of 0.4.

2.1.2. Algorithm

In the algorithm, the porosity is adjusted step by step. First, the total volume of an initially generated cell structure is calculated as a starting point, while the discrete voxel values in the domain vary between −1 and 1, according to Equation (1). Once the gradient function is selected, the superimposition begins in an iterative process. Three encapsulated for-loops are used to iterate over the spatial domain and a while-loop is responsible for adjusting the porosity to the target porosity. The adjustment is made by applying a threshold with values between −1 and 1, which divides the domain into structure space and tunnels. The specified tolerance limits are used as the termination criterion of the while loop. In this way, the target porosity for each 2D layer of the 3D structure can be adjusted according to the gradient function. If no gradient is selected, the porosity adjustment is not applied per layer but to the entire cell structure. For the definition of the gradient function (Φ_{target}), a choice between the following three functions is possible so far.

Constant function:

$$\Phi_{\text{target}} = \Phi_{\text{max}} \tag{4}$$

Linear function:

$$\Phi_{\text{target}} = -\frac{\Phi_{\text{max}} - \Phi_{\text{min}}}{n_{\text{steps}_z}} \cdot q + \Phi_{\text{max}} \tag{5}$$

Quadratic function:

$$\Phi_{\text{target}} = -\frac{\Phi_{\text{max}} - \Phi_{\text{min}}}{(n_{\text{steps}_z} - 3)^2} \cdot (q-2)^2 + \Phi_{\text{max}} \tag{6}$$

The constant function calculates a structure with a constant porosity along each spatial direction. For the linear cell gradient, a linear function with the usual linear structure $y = a * x + b$ is used. The first part of the equation calculates the stepwise increase in the cell volume in each cell level (or the decrease in the porosity). Here, the calculation depends on the number of discrete points (n_{steps}) in whose direction the gradient is imposed. In this case, the gradient extends into the z-direction. The second part of the formula is used to determine the initial porosity $\Phi_{\max}$. In the first layer, the structure has the porosity of $\Phi_{\max}$, which is gradually decreased until the final porosity $\Phi_{\min}$ is reached. q is the index parameter for the outermost for-loop and at the same time the spatial position of the current 2D layer.

For the quadratic cell gradient, a quadratic function of the structure $y = a * x^2 + b * x + c$ is used with $(b = 0)$. As with the linear function, a stepwise decrease in the porosity (increase in the cell volume) is determined at each cell level, except that the decrease should be quadratic. Since the quadratic gradient is defined as an inverted parabola with an offset on the y-axis, at the level of maximum porosity, the gyroid structure with a quadratic gradient is thickened more slowly than those with a thickening linear structure, as can be seen in Figure 4.

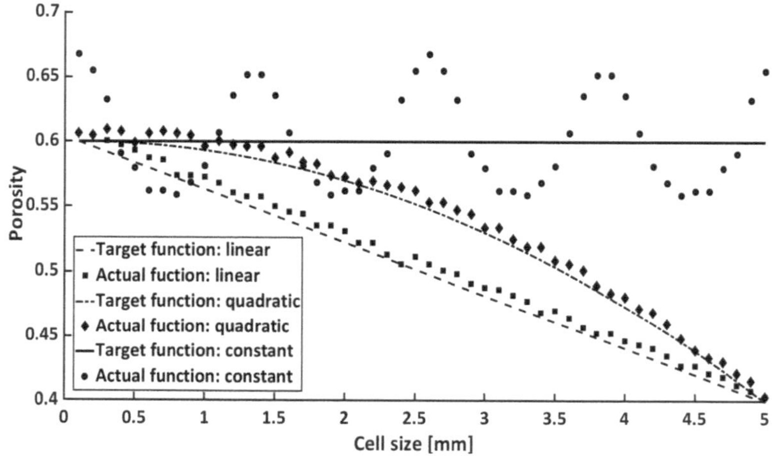

Figure 4. Porosity function of the constant, linear, and quadratic function.

In Figure 4, the three different gradients (constant, linear, and quadratic) are compared with their target and actual values of a single-cell gyroid structure. In each case, 50 actual and 50 target values per porosity function (input parameter: $n_{\text{steps}} = 50$) are mapped over 'Cell size' [5 mm], in the z-direction [x-axis], and 'Porosity' [y-axis]. The maximum porosity is 0.6, and the minimum porosity is 0.4. The actual values are an approximation of the target values. As already mentioned, the fit of the objective function mainly depends on the $delta$ parameter.

2.2. Model and Setup for Mechanical Simulations

For the mechanical simulations, the static momentum balance in the small deformation regime is solved with a finite element discretisation. This is done using the PACE3D framework, which employs a phase-field method for the geometry parametrization. Therefore, the structures generated in the MatLab program (see Section 2.1) are discretised on a Cartesian grid, and a diffuse interface is employed between the metal and the surrounding air.

From the micromechanics-microstrcture simulations, the stress tensor σ and the strain tensor ϵ are obtained as full field information. This gives rise to the normalised von Mises stresses σ_{VM}, whose maximum value determines the start of local plastification if it reaches the yield strength of the materials. Through homogenisation, an effective Young modulus can additionally be obtained from the stress and strain field [22]. This is done using the volume-averaged stress and strain over the whole computational domain and relating them via the effective Young modulus.

Simulations of compression tests are performed with the specified stress σ_{BC}, which is applied in the z-direction, as the boundary condition on both sides of the simulation domain. All other boundaries are set to be stress-free. The domain is discretised using a Cartesian grid with $200 \times 200 \times 200$ elements. The air phase between the structure is modelled with a stiffness of zero, while the solid phase is considered to exhibit an isotropic elastic material behaviour.

3. Results and Discussion

3.1. Structure Consideration

Table 2 lists all 13 gyroid unit cells created for the subsequent investigation by mechanical simulations. Overall, the structures differ only in their porosity and gradient function. As can be seen from Table 2, the structures range in porosity from 0.4 to 0.8. For the structures with graded porosity (linear and quadratic), the thickening of the structures always ends at a porosity of 0.4. It should be mentioned that when the volume decreases, the influence on the mechanical stability should be considered, so as to optimise the lightweight potential. Since the porosity has a significant influence on the surface-to-volume ratio, and thus on the mechanical stability, it is also taken into account. The other input parameters that do not change are listed in Table 3. Assuming that the gyroid structures are characterised by the periodicity of their unit cell in all directions, it should be possible to apply the results of the mechanical simulation to multicell structures. In all three directions (x, y, z), the size of the analysed cell is set to 2.5 mm. A unit cell is divided into 200 voxels per spatial direction ($n_{\text{steps}} = 200$). The possible deviation from the target function is 2% ($delta = 0.02$).

Table 2. Created gyroid structures: constant gradient, linear gradient, quadratic gradient.

Constant Gradient	Linear Gradient	Quadratic Gradient
0.4	-	-
0.5	0.4 to 0.5	0.4 to 0.5
0.6	0.4 to 0.6	0.4 to 0.6
0.7	0.4 to 0.7	0.4 to 0.7
0.8	0.4 to 0.8	0.4 to 0.8

Table 3. Non-varying input parameters across all structures.

Input Parameter	Value
num_x	1
num_y	1
num_z	1
$unitCellSize$ [mm]	2.5
n_{steps}	200
$delta$	0.02

3.2. Surface Area-to-Volume Ratio

The surface area-to-volume (SA/V) ratio is an important technical aspect. A high SA/V ratio, for example, favours more efficient heat exchange [23] but usually has negative effects on the mechanical properties, which is why the SA/V ratio of the gyroid cells is investigated. To calculate the surface area-to-volume ratio, the stl files created in the MatLab program were imported into the Ansys workbench [24], where the volume and surface area of each structure were output.

The bar chart Figure 5 lists the SA/V ratio from the gyroid structures in ascending order, with and without gradients. There, it can be seen that a high porosity favours the ratio. For this reason, the gyroid cell with a constant porosity of 0.8 has the highest SA/V ratio of all structures. In contrast, the gyroid with a porosity of 0.4 has the lowest ratio. The structures with a quadratic gradient have a higher SA/V ratio than those with a linear gradient. This is due to the fact that the structure with a quadratic gradient thickens more slowly, as can be seen in Figure 4. Between the gyroid structures with a constant porosity of 0.5 and 0.6, there are four structures with gradients. In the course of this work, a possible correlation between the SA/V ratio and the mechanical properties will be considered.

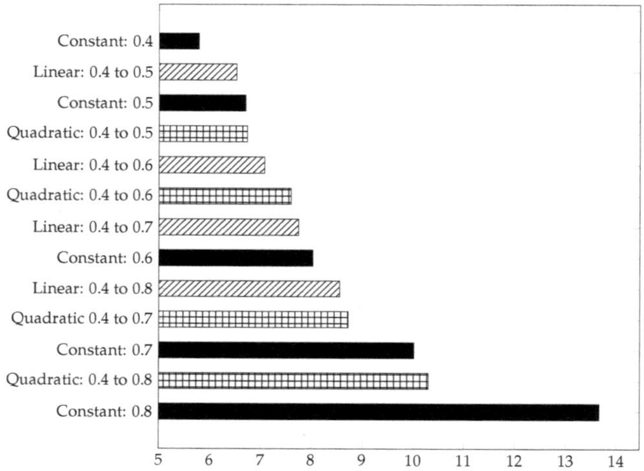

Figure 5. Surface [mm^2]/volume [mm^3] of gyroid structures in ascending order.

3.3. Mechanical Simulation

Material-independent and relative statements depending on the porosity type can be made about the structures with the same load scenario. For this reason, scaled data are used. For the analysis of the structures, a steady-state case with an applied load of $\sigma_{BC} = 400\,\text{MPa}$ is considered. Since the simulations are performed within the linear-elastic regime, the results are independent of the structure material and the characteristic length, due to the linear scalability. The effective Young modulo as well as the maximum and mean values of the von Mises stress are evaluated. The latter values correspond respectively to the volume average $\bar{\sigma}_{VM}$ of the von Mises equivalent stress field and its maximum value $\sigma_{VM,max}$ within the domain. Note that both quantities are given normalised with the load σ_{BC} and can thus be interpreted as mean and maximum values of a stress amplification factor. Dividing the yield strength of the material under consideration (e.g., AlMg7Si0.6) by this amplification factor gives the actual limit for local plastification and thus an effective yield strength. The effective Young modulus is given normalised with the one of the structure materials. In order to obtain physical quantities, a multiplication can be carried out with the material value under consideration. For example, $\bar{E}_{AlMg7Si0.6} = E_{AlMg7Si0.6}\bar{E}$, with $E_{AlMg7Si0.6} = 59\,\text{GPa}$, if the structure is made of the alloy AlMg7Si0.6. The use of these normalised quantities allows a comparison between the structures without specifying the material or length scale.

Tables 4–6 represent the material-independent and scaled values of the respective structures. For better clarity and comprehensibility of the results, they are also shown in the bar charts Figures 6–8. The structures in the charts are all labelled according to the following pattern: 'Type of gradient function: Porosity interval'. 'Quadratic: 0.4 to 0.8', for example, means that a gyroid structure with a quadratic gradient function and a porosity interval between 0.4 and 0.8 is considered.

Table 4. Scaled results of the gyroid structure with a constant gradient [dimensionless] of the normalised effective Young modules $\bar{E}$, a mean von Mises stress $\bar{\sigma}_{VM}$, and a maximum von Mises stress $\sigma_{VM,max}$, which is given for different porosities.

Porosity	$\bar{E}$	$\bar{\sigma}_{VM}$	$\sigma_{VM,max}$
0.4	0.10	0.88	10.40
0.5	0.05	0.99	14.92
0.6	0.03	1.09	19.50
0.7	0.02	1.18	23.94
0.8	0.01	1.24	28.74

Table 5. Scaled results of the gyroid structure with a linear gradient [dimensionless] of the normalised effective Young modules $\bar{E}$, a mean von Mises stress $\bar{\sigma}_{VM}$, and a maximum von Mises stress $\sigma_{VM,max}$, which is given for different porosities.

Porosity	$\bar{E}$	$\bar{\sigma}_{VM}$	$\sigma_{VM,max}$
from 0.4			
to			
0.5	0.06	0.97	14.99
0.6	0.04	1.02	18.94
0.7	0.03	1.07	22.82
0.8	0.02	1.11	28.01

Table 6. Scaled results of the gyroid structure with a quadratic gradient [dimensionless] of the normalised effective Young modules $\bar{E}$, a mean von Mises stress $\bar{\sigma}_{VM}$, and a maximum von Mises stress $\sigma_{VM,max}$, which is given for different porosities.

Porosity	$\bar{E}$	$\bar{\sigma}_{VM}$	$\sigma_{VM,max}$
from 0.4			
to			
0.5	0.06	0.99	15.07
0.6	0.04	1.05	18.75
0.7	0.02	1.11	23.86
0.8	0.02	1.16	29.64

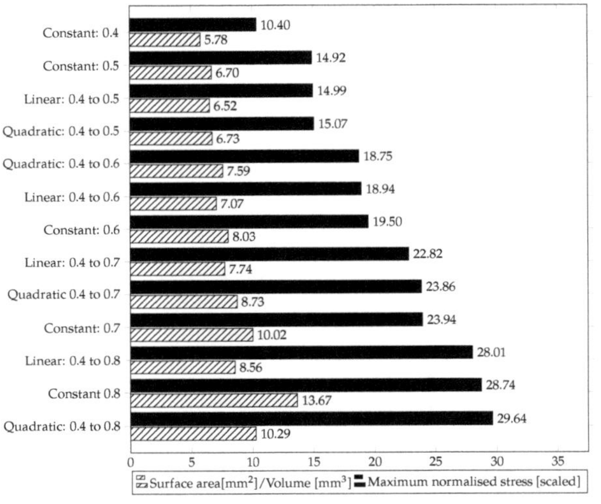

Figure 6. Dimensionless, scaled maximum normalised stress $\sigma_{VM,max}$ [scaled] of gyroid structures in ascending order, in comparison to the SA/V ratio of the same structure.

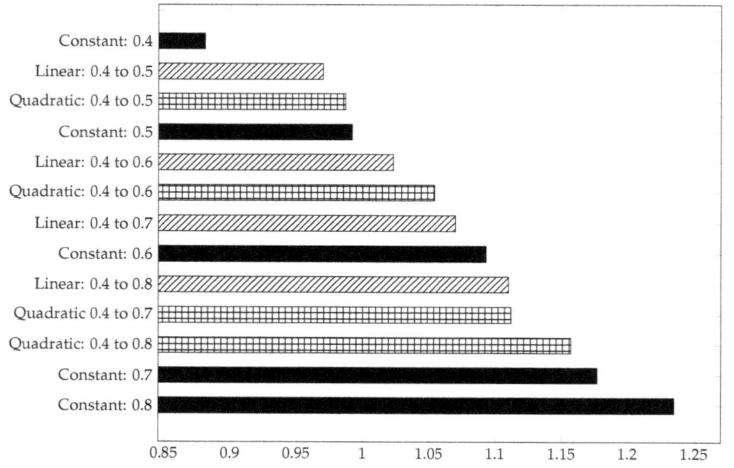

Figure 7. Dimensionless, scaled mean normalised stress $\bar{\sigma}_{VM}$ of gyroid structures in ascending order.

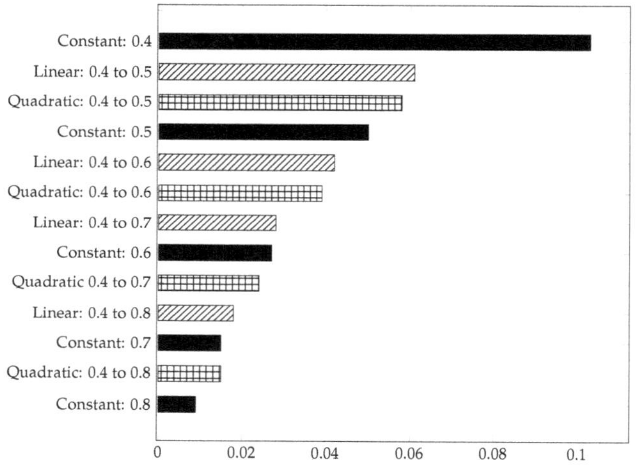

Figure 8. Dimensionless, scaled effective Young modules $\bar{E}$ of gyroid structures in decreasing order.

In Figure 6, the black bars represent the maximum normalised stresses ($\sigma_{VM,max}$) of the gyroid structures, sorted in ascending order. In addition, the corresponding SA/V ratio is shown in striped bars.

The gyroid structures with a quadratic porosity have the highest (quadratic: 0.4 to 0.8) and the gyroid structure with constant porosity has the lowest (constant: 0.4) maximum stresses, respectively. In addition, the bar graph illustrates that the linear gradient structures have lower scaled normalized maximum stresses than the quadratic gradient structures, but higher than the constant gradient structures. A structure with a linear porosity of 0.4 to 0.6 corresponds to an average of a constant porosity of 0.5. Here, it becomes clear that by adjusting the gradient, a higher SA/V ratio is achieved, but also higher maximum stresses. This observation also applies to the structure with a linear porosity from 0.4 to 0.8, which corresponds to an average porosity of 0.6.

As in Figure 6, the dimensionless, scaled, and normalised stresses [$\bar{\sigma}_{VM}$] of the structures are also sorted in ascending order in Figure 7. This allows for a faster comparison of the diagrams. The structural arrangements of the two diagrams are not in the same order, since there is a deviation between the highest and the lowest values of the maximum normalised stress and the mean normalised stress.

Compared to the structures with quadratic or linear porosity, the structures with constant porosity have higher normalised stresses. The higher the porosity level of the structures, the higher the normalised stresses. This may indicate a more uniform stress distribution in the structures with gradients or that the structure has more unstressed regions. The gyroid structure 'quadratic: 0.4 to 0.8', for example, has a higher SA/V ratio than 'constant: 0.7', but a lower mean normalised stress.

The structures with a linear gradient generally exhibit the lowest mean stresses, compared to the other structures, which is also reflected in the scaled, dimensionless elastic modulus (see Figure 8).

The structures with a linear gradient have a higher effective Young modulus [$\bar{E}$] than the other structures in the same porosity range. This can be partly explained by the fact that the linear gradient structures are thickened more quickly and more evenly, which means that the initiating force can be better distributed. The Figure 8 lists the scaled and dimensionless effective Young modulus of the considered structures in decreasing order. It becomes clear that the porosity has a high influence on the effective Young modulus. Between the structures 'constant: 0.4' and 'constant: 0.8', for example, the effective Young modulus is reduced by a factor of about 11. In contrast, the effective Young modulus for the structures 'quadratic: 0.4 to 0.5' ('linear: 0.4 to 0.5') and 'quadratic: 0.4 to 0.8' ('linear: 0.4 to 0.8') decreases by a factor of 3.9 (quadratic) or 3.4 (linear). A clear relationship between the SV/A ratio and the effective Young modulus can be seen when comparing Figures 5 and 8. The sorted effective Young modulus is almost in the same order as the sorted SV/A ratio.

The evaluation of the von Mises stress field on the gyroid structures revealed that it is generally located at the rounded edges of the structure. It was noticed that the stress peaks for the gyroid structures with gradients are on the side with the highest porosity, due to the difference in porosity, while the stress peaks for the structures with constant porosity are on both sides, which can be seen in Figure 9. The stressed areas are marked in red and are located at the rounded edges, as described previously. Accordingly, the loaded edges are likely to fail first in compression tests. In addition, the one-sided loading of the structures with gradients would explain the higher stress distributions given in Figure 6.

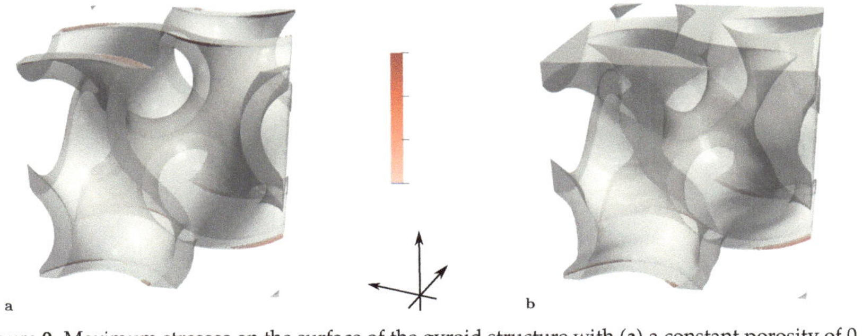

Figure 9. Maximum stresses on the surface of the gyroid structure with (**a**) a constant porosity of 0.8; (**b**) a quadratic porosity function between 0.4 and 0.8.

The mechanical simulation has shown that the structures with gradients enable new design and lightweight construction possibilities. The structures with gradients can be better adapted to the required properties. Depending on the choice of the new and the original structure, one property can be specifically improved, while another property can be reduced. In general, it can be said that the SA/V ratio increases with increasing porosity, while the effective Young modulus decreases. By choosing the structure 'linear: 0.4 to 0.5', for example, instead of 'constant: 0.5', the effective Young modulus increases by about 22% [quadratic gradient 16%], while the surface-to-SA/V ratio and the mean normalised stress would decrease by about 2% [quadratic about 0%].

If only the mechanical stability is of interest and not the lightweight potential or the surface area-to-volume ratio, the gyroid with the lowest porosity—in our case 0.4—is still unbeatable.

Likewise, the gyroid with a constant porosity of 0.8 would be interesting for applications where only the surface-to-volume ratio is of central importance, but not the mechanical properties.

4. Conclusions

In this study, gyroid structures, which are associated with the TPMS family, were investigated. In addition to the used structures with constant porosity, graded structures were produced. For the graded structures, a distinction was made between structures with linear and quadratic gradients. For the mechanical simulation, the created structures were imported into PACE3D. The results were converted into dimensionless, material-independent indices, so that a general statement could be made. In addition to the mechanical simulation, the SV/A ratio was also analysed.

The mechanical simulation shows that the introduction of the gradient multiplies the range of engineering design possibilities. Depending on the desired property and application, it is worthwhile to integrate a gradient into the structure. In general, it can be said that the structures with gradients usually have higher stress peaks, but lower mean normalised stresses.

For future applications, it would be interesting to create more gradient functions. One possibility, for example, would be to create a structure with a gradient which is thickened hourglass-like in terms of volume fraction, since the stress peaks all occur in the outer edge region. In addition, an experimental validation of the results would also be important, which is part of the ongoing work. The combination of experimental data and dimensionless indices will allow a tailor-made design for individual parts in the future.

Author Contributions: Conceptualization, B.N. and F.P.; Data Curation, P.A. and M.R.; Investigation, L.W.; Methodology, L.W.; Software, P.A. and M.R.; Supervision, B.N. and F.P.; Writing – original draft, L.W. All authors have read and agreed to the published version of the manuscript.

Funding: The paper was written as part of the project "InSeL—Innovative Schaumstrukturen für den effizienten Leichtbau" (Innovative foam structures for efficient lightweight construction), which is funded by the European Regional Development Fund (EFRE) and the state of Baden-Württemberg, as part of the Centre for Applied Research ZAFH. Furthermore, the authors would like to thank the Ministry of Science, Research, and Art Baden-Württemberg (MWK-BW) in the project MoMaF–Science Data Center, with funds from the state digitization strategy digital@bw (project number 57).

Institutional Review Board Statement: Not applicable.

Informed Consent Statement: Not applicable.

Data Availability Statement: Not applicable.

Acknowledgments: The authors thank Leon Geisen for his editorial support.

Conflicts of Interest: The authors declare no conflict of interest.

References

1. Michielsen, K.; Stavenga, D. Gyroid cuticular structures in butterfly wing scales: Biological photonic crystals. *J. R. Soc. Interface* **2008**, *5*, 85–94. [CrossRef] [PubMed]
2. Lai, M.; Kulak, A.N.; Law, D.; Zhang, Z.; Meldrum, F.C.; Riley, D.J. Profiting from nature: Macroporous copper with superior mechanical properties. *Chem. Commun.* **2007**, *34*, 3547–3549. [CrossRef] [PubMed]
3. Kladovasilakis, N.; Tsongas, K.; Tzetzis, D. Mechanical and FEA-Assisted Characterization of Fused Filament Fabricated Triply Periodic Minimal Surface Structures. *J. Compos. Sci.* **2021**, *5*, 58. [CrossRef]
4. Alketan, O.; Abu Al-Rub, R. Multifunctional mechanical-metamaterials based on triply periodic minimal surface lattices: A review. *Adv. Eng. Mater.* **2019**, *21*, 1900524. [CrossRef]
5. Li, W.; Yu, G.; Yu, Z. Bioinspired heat exchangers based on triply periodic minimal surfaces for supercritical CO_2 cycles. *Appl. Therm. Eng.* **2020**, *179*, 115686. [CrossRef]

6. Torquato, S.; Donev, A. Minimal surfaces and multifunctionality. *Proc. R. Soc. A Math. Phys. Eng. Sci.* **2004**, *460*, 1849–1856. [CrossRef]
7. Dong, Z.; Zhao, X. Application of TPMS structure in bone regeneration. *Eng. Regen.* **2021**, *2*, 154–162. engreg.2021.09.004. [CrossRef]
8. Li, D.; Liao, W.; Dai, N.; Xie, Y.M. Comparison of Mechanical Properties and Energy Absorption of Sheet-Based and Strut-Based Gyroid Cellular Structures with Graded Densities. *Materials* **2019**, *12*, 2183. [CrossRef]
9. Liu, F.; Mao, Z.; Zhang, P.; Zhang, D.Z.; Jiang, J.; Ma, Z. Functionally graded porous scaffolds in multiple patterns: New design method, physical and mechanical properties. *Mater. Des.* **2018**, *160*, 849–860. [CrossRef]
10. Jin, Y.; Kong, H.; Zhou, X.; Li, G.; Du, J. Design and Characterization of Sheet-Based Gyroid Porous Structures with Bioinspired Functional Gradients. *Materials* **2020**, *13*, 3844. [CrossRef]
11. Maskery, I.; Sturm, L.; Aremu, A.; Panesar, A.; Williams, C.; Tuck, C.; Wildman, R.; Ashcroft, I.; Hague, R. Insights into the mechanical properties of several triply periodic minimal surface lattice structures made by polymer additive manufacturing. *Polymer* **2018**, *152*, 62–71. SI: Advanced Polymers for 3DPrinting/Additive Manufacturing,
12. Maskery, I.; Aboulkhair, N.; Aremu, A.; Tuck, C.; Ashcroft, I. Compressive failure modes and energy absorption in additively manufactured double gyroid lattices. *Addit. Manuf.* **2017**, *16*, 24–29. [CrossRef]
13. Chen, Z.; Xie, Y.; Wu, X.; Wang, Z.; Li, Q.; Zhou, S. On hybrid cellular materials based on triply periodic minimal surfaces with extreme mechanical properties. *Mater. Des.* **2019**, *183*, 108109. [CrossRef]
14. Feng, J.; Liu, B.; Lin, Z.; Fu, J. Isotropic porous structure design methods based on triply periodic minimal surfaces. *Mater. Des.* **2021**, *210*, 110050. [CrossRef]
15. Zaharin, H.; Abdul-Rani, A.M.; Azam, F.; Ginta, T.; Sallih, N.; Ahmad, A.; Yunus, N.A.; Zulkifli, T.Z.A. Effect of Unit Cell Type and Pore Size on Porosity and Mechanical Behavior of Additively Manufactured Ti6Al4V Scaffolds. *Materials* **2018**, *11*, 2402. [CrossRef]
16. Gibson, L.J.; Ashby, M.F. *Cellular Solids: Structure and Properties*, 2nd ed.; Cambridge Solid State Science Series; Cambridge Univ. Press: Cambridge, UK, 1997. [CrossRef]
17. Zhu, H.; Hobdell, J.; Windle, A. Effects of cell irregularity on the elastic properties of open-cell foams. *Acta Mater.* **2000**, *48*, 4893–4900. [CrossRef]
18. Kaoua, S.A.; Boutaleb, S.; Dahmoun, D.; Azzaz, M. Numerical modelling of open-cell metal foam with Kelvin cell. *Comput. Appl. Math.* **2016**, *35*, 977–985. [CrossRef]
19. Gan, Y.; Chen, C.; Shen, Y. Three-dimensional modeling of the mechanical property of linearly elastic open cell foams. *Int. J. Solids Struct.* **2005**, *42*, 6628–6642. [CrossRef]
20. MATLAB. *Version 9.6.0.1072779 (R2019a)*; The MathWorks Inc.: Natick, MA, USA, 2019.
21. Hötzer, J.; Reiter, A.; Hierl, H.; Steinmetz, P.; Selzer, M.; Nestler, B. The parallel multi-physics phase-field framework Pace3D. *J. Comput. Sci.* **2018**, *26*, 1–12. [CrossRef]
22. John, A.; John, M. Foam metal and honeycomb structures in numerical simulation. *Ann. Fac. Eng. Hunedoara* **2016**, *14*, 27–32.
23. Planinsic, G.; Vollmer, M. The surface-to-volume ratio in thermal physics: From cheese cube physics to animal metabolism. *Eur. J. Phys.* **2008**, *29*, 369. [CrossRef]
24. Ansys. *Version 2021 R2*; Ansys Inc.: Canonsburg, PA, USA, 2021.

applied sciences

MDPI

Article

An Improved Particle-Swarm-Optimization Algorithm for a Prediction Model of Steel Slab Temperature

Ming Liu [1,2], Peng Yan [3], Pengbo Liu [1,2], Jinwei Qiao [1,2] and Zhi Yang [1,2,*]

1 School of Mechanical & Automotive Engineering, Qilu University of Technology (Shandong Academy of Sciences), Jinan 250353, China
2 Shandong Institute of Mechanical Design and Research, Jinan 250353, China
3 Key Laboratory of High-Efficiency and Clean Mechanical Manufacture, Ministry of Education, School of Mechanical Engineering, Shandong University, Jinan 250061, China
* Correspondence:yangzhi@qlu.edu.cn; Tel.: +86-13066769976

Abstract: Aiming at the problem of the low accuracy of temperature prediction, a mathematical model for predicting the temperature of a steel billet is developed. For the process of temperature prediction, an improved particle-swarm-optimization algorithm (called XPSO) is developed. XPSO was designed based on a multiple swarm scheme to improve the global search capability and robustness; thus, it can improve the low accuracy of prediction and overcome the problem of easy entrapment into local optima. In the XPSO, the multiple swarm scheme comprises four modified components: (1) the strategy of improving the positional initialization; (2) the mutation strategy for particle swarms; (3) the adjustment strategy of inertia weights; (4) the strategy of jumping out local optima. Based on widely used unimodal, multimodal and composite benchmark functions, the effectiveness of the XPSO algorithm was verified by comparing it with some popular variant PSO algorithms (PSO, IPSO, IPSO2, HPSO, CPSO). Then, the XPSO was applied to predict the temperatures of steel billets based on simulation data sets and measured data sets. Finally, the obtained results show that the XPSO is more accurate than other PSO algorithms and other optimization approaches (WOA, IA, GWO, DE, ABC) for temperature prediction of steel billets.

Keywords: optimization; particle swarm optimization algorithm; reheating furnace; temperature prediction

Citation: Liu, M.; Yan, P.; Liu, P.; Qiao, J.; Yang, Z. An Improved Particle-Swarm-Optimization Algorithm for a Prediction Model of Steel Slab Temperature. *Appl. Sci.* **2022**, *12*, 11550. https://doi.org/10.3390/app122211550

Academic Editors: Shuwen Wen, Yongle Sun and Xin Chen

Received: 10 September 2022
Accepted: 6 November 2022
Published: 14 November 2022

1. Introduction

In the steel industry, the reheating furnace must reheat the material (slabs) to the desired uniformity temperature profiles at the exit. However, the slab reheating furnace's operation is a complex physical and chemical process [1]. To better control and optimize the furnace's operations, there should be a suitable temperature prediction model to predict the accurate temperatures for the slabs inside the furnace. Given the continuous development of artificial intelligence techniques, the demand for a suitable temperature prediction model is increasing [2]. Therefore, a suitable mathematical model which can predict the discharge temperatures of billets accurately and quickly should be proposed for the control and optimization of the reheating furnaces. In general, the prediction models can be divided into two categories [3]: the mechanism models based on first principles [4] and the empirical models based on the production data and black-box approaches [5]. The first kind of model needs to fully understand the physical and chemical processes inside the reheating furnace—e.g., [1,6–8]. The computational requirements of these models vary widely depending on the level of complexity [3]. Finally, the heat-transfer process is often summarized by partial differential equations (PDEs). Usually, mechanism models are complicated and nonlinear (such as those in ([9–14]). Hong et al. [13] investigated the sequential function specification coupled with the Broyden combined method (BC-SFSM) to obtain the temperature field of a steel billet based on the inverse heat-conduction problem. The results illustrate that the majority of relative errors during the whole reheating time are less than 5%. Chen et al. [14] presented a novel method to obtain these parameters by combining a "black box" test with a billet heat transfer model. The surface temperature and center temperature's relative error were 2.34% and 3.51%, respectively. The numerical computation of PDEs will require a lot of computational time, which will exceed a practical time criterion. Therefore, this kind of model cannot satisfy the requirements for an online system.

The empirical models are often determined by identification methods involving the genetic algorithm (GA), support vector machines (SVMs), neural networks (NNs), particle swarm optimization (PSO) and other intelligent techniques ([15–18]). For instance, Thanawat et al. [19] investigated a prediction model using the production data from Ratchasima Steel Products Company. The GA method was used to identify the model's unknown factors. Tang et al. [20] presented the SVM predictive model for the temperature control problem. The PSO was applied to determine proper parameters for the SVM model and finally obtained good performance. Wang et al. [21] constructed a prediction model of slab discharging temperature by combining GA with a BP neural network. The mean-square error of the network was 72.3477, and the error was lower than 20 °C. Yang et al. [22] used the relevance vector machine (RVM) method to predict the slab temperature. The maximum prediction error of slab temperature was 10.46 °C. In general, the empirical model is often simplified to a simple formula, so the calculation time is small enough to satisfy online production [23]. The production data and the performance of the intelligent algorithm used will determine the advantages and disadvantages of an empirical model. As the industrial software and database techniques continue developing, more production data are being obtained [24]. Thus, an intelligent algorithm with excellent performance is indispensable, and it behooves us to study the intelligent algorithms carefully.

The PSO algorithm has the characteristics of high solution accuracy and a fast approach to the optimal solution. However, the basic PSO varies in its ability to solve problems in different application contexts. It also easily falls into local optima. Researchers have studied various strategies for the improvement of PSO. For instance, Alsaidy et al. [25] proposed the longest job to fastest processor PSO (LJFP-PSO) algorithm and the minimum completion time PSO (MCT-PSO) algorithm for the task-scheduling problem. The effective performance of the two algorithms was proved by simulation results. Yue et al. [26] presented a novel multi-objective PSO that has ring topology for solving multimodal multi-objective optimization problems. The ring topology is used to form stable niches and locate multiple optima. Peng et al. [27] proposed a symbiotic PSO (SPSO) algorithm to control a water-bath-temperature system. A multiple swarm scheme was proposed for the SPSO algorithm. Three major components (create initial swarms, evaluating fitness values and updating each particle) are used to escape from a locally optimal solution.

Inspired by these algorithms, we propose an improved PSO (named XPSO) algorithm that uses the multiple swarm scheme in this paper. The scheme comprises four modified components: (1) improving the positional initialization of the particle swarms: one randomly generated and the other uniformly generated; (2) adding the mutation strategy for the particle swarm to increase its population diversity; (3) adjusting the inertia weight through a "stepped" adaptive model; (4) adding the strategy of escaping from the local minimized point.

This paper is presented as follows: the prediction model of the slab temperature is established in Section 2; the detailed strategies for improvement of XPSO are described in Section 3; the simulation and discussion are given in Section 4; Section 5 summarizes the conclusions.

2. The Prediction Model of the Slab Temperature

2.1. The Structure of a Reheating Furnace

The heat transfer processes in the furnace mainly consist of radiation heat transfer and convection heat exchange. The main function of reheating furnaces is to help the slabs reach the desired discharging temperatures for the next rolling process. The slab passes through the preheating zone, multistage heating zone and soaking zone from the furnace's inlet to the outlet, as shown in Figure 1. The multi-stage heating zone can be divided into four subzones (upper zones 1 and 2; bottom zones 4 and 5), and the soaking zone is divided into two subzones (upper zone 3, bottom zone 6). The key components of a subzone are a series of regenerative burners, the corresponding furnace nozzle temperatures TN_j and the furnace zone temperature TP_k. Here, TN_j is easily obtained by the thermocouples nearby the location of burner's nozzle j and $TP_k = \frac{1}{n}\sum_{j=1}^{n} TN_j$, which stands for the average value of the nozzle temperatures TN_j in zone k.

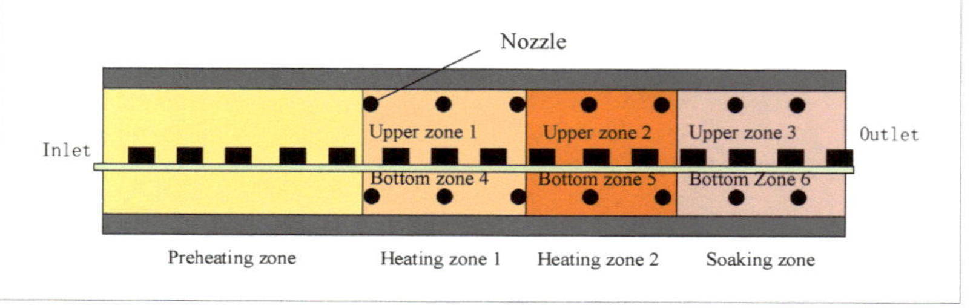

Figure 1. Schematic diagram of a heating furnace's structure.

2.2. The Prediction Model and Optimization Problem

The most important quality criterion of reheating furnace is the average outlet temperature of the steel slab. There are many factors affecting the slab's discharging temperature. By analyzing the field data and some related research ([22,23]), we can confirm some key factors: the initial temperature of the billet, T_0; the furnace nozzle temperature, T_j; the mean temperature of each zone, T_p; the reheating time of the billet in the furnace, θ; the material of the slab, α; the thickness of the slab, d. Thus, the inputs of the prediction model are $X = [X_1, X_2, X_3]^T$, which consist of the slab's physical parameters, $X_1 = [\alpha, d, \theta, T_0]^T$; the vector of the furnace nozzle temperatures, $X_2 = [TN_1, TN_2, \cdots, TN_m]^T$; and the vector of furnace zone temperatures, $X_3 = [TP_1, TP_2, \cdots, TP_6]^T$. Notice, m is the total number of burners in the reheating furnace. The outputs are the predicted temperature $Y = TE$, which is the vector of the predicted temperatures of the slabs when discharged out from the furnace. Finally, the formula of predicting model is constructed as:

$$Y = f(X, W) = w_0 d + w_1 \alpha + w_2 T_0^1 + w_3 T_0^2 + \sum_{i=1}^{m} TN_i w_{(i+3)} TN_i + \sum_{j=1}^{6} w_{(j+m+3)} cos(TP_j \theta) + \sum_{j=1}^{6} w_{(j+m+9)} sin(TP_j \theta) \quad (1)$$

where $W = [w_0, w_1, \cdots, w_{m+15}]$ represents the vector of unknown weight parameters, which will be determined by the proposed XPSO algorithm. The formula f is composed of a polynomial and a Fourier function, which was designed by the author based on experience. To make the analytical response Y equal to the measured value Y^*, an optimization problem should be established with the objective of minimizing the error between the theoretically calculated values and the measured data. To alleviate overfitting and improve generalization performance, the strategy of regularization is employed for the optimization problem. Finally, the formula of the optimization problem is shown as follows.

$$\begin{aligned} Minimize \quad & J = \|Y - Y^*\|^2 + \lambda_1 \|W\| + \lambda_2 \|W\|^2 \\ st. \quad & Y = f(X, W) \end{aligned} \quad (2)$$

where Y^* represents the measured temperature of the slab, and λ_1, λ_2 are the regularization parameters.

3. Improved Particle-Swarm-Optimization Algorithm

3.1. The Basic PSO Algorithm

The PSO algorithm originates from a researcher observing the social behavior of a flock of birds or fish [28]. Each particle in the population is a potential solution to the objective function. The particle has two attributes: velocity $V_i = [v_{i1}, v_{i2}, \cdots, v_{ij}, \cdots, v_{ND}]$ and position $X_i = [x_{i1}, x_{i2}, \cdots, x_{ij}, \cdots, x_{ND}]$, where $i = 1, 2, \cdots, N$—N is the size of particle swarm; and $j = 1, 2, \cdots, D$—D is the dimension of the solution space.

In each iteration, the information of individual particle is filtered to find the information about the historically optimal position p of the individual and the historically optimal position g of the population. The known information is brought into the velocity updating equation, Equation (3), and the position updating equation, Equation (4), to adjust the search direction of the population, so that the particle swarm approaches the global optimal solution, which is the optimal position of the population.

$$v_{ij}(k+1) = wv_{ij}(k) + r_1c_1(p_i - x_{ij}(k)) + r_2c_2(g - x_{ij}(k)) \tag{3}$$

$$\dot{x_{ij}}(k+1) = x_{ij}(k) + v_{ij}(k+1) \tag{4}$$

where w represents the inertia weight; $r_1, r_2 \in [0,1]$ are two uniformly distributed random values; c_1, c_2 are the acceleration parameters, which are non-negative constants; k represents the current iteration; and $k = 1, 2, 3, \ldots, G$, where G is the maximum number of iterations. The velocity updating of the particle is influenced by three factors: the current moment particle velocity $V_{ij}(k)$, the particle's self-experience ΔV_P and the experience of particle swarm ΔV_g. ΔV_P is the part of the particle that learns from its historical information, and ΔV_g represents the part of the particle that learns from the historical information of other particles within the population.

3.2. Improvement Strategies of the XPSO Algorithm

The basic PSO algorithm is a non-globally-convergent optimization algorithm [29]. To reduce the premature probability of falling into a local optimal solution and improve the convergence speed of the basic PSO, an improved PSO (named XPSO) algorithm is proposed based on the multi-strategy co-evolutionary approach. Four specific improvements are described as follows.

1. Improving the positional initialization of the particle swarm: one randomly generated and the other uniformly generated.
   In a basic PSO algorithm, all of the particles are randomly initialized. The expression is given as follows:

$$X_{ND} = \begin{bmatrix} lx_{11} & x_{12} & \cdots & x_{1D} \\ x_{21} & x_{22} & \cdots & x_{2D} \\ \vdots & \vdots & \ddots & \vdots \\ x_{N1} & x_{N2} & \cdots & x_{ND} \end{bmatrix} \tag{5}$$

   An increase in the positional diversity of particle swarms can facilitate the exploration of global range. However, increasing the diversity of particle swarms also makes it difficult to converge to the global optimum every time. Hence, an improved approach, based on the "double-edged sword" nature of particle swarm's diversity, is proposed to improve the algorithm's stability. During the initialization of the particle swarm, a dimension called X-Dim in the position matrix is randomly selected. The positional information of the particle in X-Dim is generated according to a uniform distribution, as shown in Equation (6).

$$x_{ij} = \frac{O_{\max} - O_{\min}}{N} * i \tag{6}$$

   where j is the randomly selected dimension; and O_{max} and O_{min} are the upper and lower limits of the value range of independent variables in different dimensions, respectively.

2. The mutation strategy is introduced into the position updating of particle swarm to compensate for the decline in the overall diversity of particle swarm after improved initialization.
   Unlike some other meta-heuristic algorithms, standard PSO has no evolution operators such as crossover or mutation. The mutation strategy will be implemented by screening the particles in each iteration. If the corresponding fitness function value of someone particle is lower than the average fitness function value, the mutation strategy is performed in the current iteration. The formula of positional mutation is:

$$x_{ij}^{*}(k) = x_{ij}(k) - wv_{ij}(k) - w(g - p_i) \tag{7}$$

   where x^* denotes the position of the particle after mutation.
3. The adjustment of inertia weight is given to improve the flexibility of particle flight speed change, and the idea of "stepped" adaptive change is injected into the updating of inertia weight.
   Inertia weight w is directly related to the convergence speed. Most researches use the subtraction function as its updating formula for inertia weight [30]. Some others use the "stepped" improvement method to update the inertia weight [31]. In our method, the inertia weight is adjusted by combining the strategy of decreasing function and the "stepped" improvement. The specific change is given as follows:
   - A "three-step" strategy is proposed to switch the range $[w_s, w_e]$ of inertia weight by determining the fitness function value of the best position so far. The switching formula is given as follows:

$$[w_s, w_e] = \begin{cases} [w_{s1}, w_{e1}] & f(g) \gg Fitness1 \\ [w_{s2}, w_{e2}] & Fitness2 < f(g) < Fitness1 \\ [w_{s3}, w_{e3}] & f(g) \ll Fitness2 \end{cases} \tag{8}$$

     where $[w_s, w_e]$ is the range of inertia weight; $f(g)$ is the fitness function value corresponding to the global optimal solution; $Fitness1$ and $Fitness2$ are the autonomous set values. The values of $([w_{s1}, w_{e2}], [w_{s2}, w_{e2}]$ and $[w_{s3}, w_{e3}])$ need to be adjusted according to the conditions of the objective function in different application contexts.
   - After the ranges of the inertia weight $[w_s, w_e]$ have been determined, a decreasing function is introduced to adjust the w. The switching condition is related to the fitness function value of the best position so far. The update formula is given as follows:

$$w = \begin{cases} \frac{r\sin(w_s \pi)}{4}, & f(g) \gg Fitness3 \\ w_s - (w_s - w_e)\sqrt{\frac{K}{G}}, & f(g) < Fitness3 \end{cases} \tag{9}$$

     where $r \in [0, 4)$ is a uniformly distributed random number; and w_s and w_e are the initial and final values of the range $[w_s, w_e]$ of the inertia weight, respectively. $Fitness3$ is the autonomous set value, k is the current iteration and G is the maximum iterations.

4. The strategy of jumping out local optimum is proposed.
   A slope parameter tr is given to judge whether the particle swarm has fallen in the local optimum. Here, tr is the count of the condition when $Slope$ is less than the value ε in five iterations. The slope is calculated as follows:

$$Slope = \frac{\left(f_k(g) - f_{(k-5)}(g)\right)}{5} \tag{10}$$

   If the value of tr equals the value of s, which is an autonomous set value, the particle swarm is trapped in a local optimum. Then, the particle swarm will perform a "jumping out local optimum" operation, which is done to change its position. The specific formula of a particle jumping out of a local optimum is given as follows:

$$x_{ij}(k) = x_{ij}(k) - r_1c_1(g - x_{ij}(k)) + r_2c_2(bad - x_{ij}(k)) \tag{11}$$

where *bad* represents the information of the global worst position. The core of this strategy is that the particle swarm should be nearest to the global worst position while staying away from the local optimum.

Finally, the pseudo-code of the XPSO algorithm is demonstrated in Algorithm 1.

Algorithm 1: The pseudo-code of the XPSO algorithm

```
Initialize the parameters: (N, G, D, O_max, O_min, V_max, V_min, t, s, ε)
Combine uniform and random distribution to initialize position matrix X_{N×D}
Generate the initial velocity V_i of each particle randomly
Evaluate the fitness value of each particle
Set p_i with a copy of X_i
Initialize g and bad with the best and worst fitness value among population
While k < G
  If k ≥ 6
    Update the slope of the fitness function curve
    Slope = (f(g)_k − f(g)_{k−5})/5
    If Slope ≤ ε
      t = t + 1
    End If
  End If
  Update inertia weight ω by Equations (8) and (9)
  For i = 1 : N
    Update the velocity V_i
    v_ij(k+1) = wv_ij(k) + r1c1(p_i − x_ij(k)) + r2c2(g − x_ij(k))
    Update the velocity X_i
    If t = s
        For m = 1 : 50
          x_ij(k) = x_ij(k) − r1c1(g − x_ij(k)) + r2c2(bad − x_ij(k))
        End For
    Else
      x_ij(k+1) = x_ij(k) + v_ij(k+1)
    End If
    Calculate the fitness values of the new particle X_i
    Execute position mutation
      x*_ij(k) = x_ij(k) − wv_ij(k) − w(g − p_i)
    Calculate the fitness values of the new particle X*
    If X_i or X* is better than p_i
      Update p_i
    End If
    If X_i or X* is better than g
      Update g
    End If
    If X_i or X* is worse than bad
      Update bad
    End If
  End For
  k = k + 1
End While
```

4. Simulations and Discussion

To verify the performance of the XPSO algorithm, some simulations are designed to involve both the performance and application of the algorithm. To obtain an unbiased CPU time comparison, all simulations were programmed by MATLAB R2017b and implemented on a computer with an Intel i5-11400F GPU, 2.60 GHz, 16 GB RAM.

4.1. Validation of XPSO by Benchmark Test Functions

The XPSO algorithm is compared with some popular variant PSO algorithms (PSO [32] IPSO [33], IPSO2 [30], HPSO [34], CPSO [35]) on a series of widely used optimization benchmark functions. A set of benchmark functions were selected from papers ([36,37]).

The benchmark set consisted of three main groups of benchmark functions: 4 unimodal (UM) functions, 2 multimodal (MM) functions and 3 composition (CM) functions. The UM functions (f1–f4) with a unique global best can expose the intensification capacities of different algorithms. The MM functions (f5–f6) can expose the diversification of algorithms The CM functions (f7–f9) were selected from the IEEE CEC 2005 competition [37], which are also utilized in many papers to test the performances (balancing the exploration and exploitation inclinations and escaping from local optima) of algorithms.

The mathematical formulation and characteristics of UM and MM functions are shown in Table 1. Details of the CM functions are shown in Table 2. The parameters of both the PSO algorithms and the optimization problem (2) were as follows. The specific parameter combinations for the inertia weight were $[w_{s1}, w_{e1}] = [0.9, 0.4]$, $[w_{s2}, w_{e2}] = [0.65, 0]$ and $[w_{s3}, w_{e3}] = [0.55, 0.05]$. The maximum and minimum velocities of the particle were $V_{max} = 0.1$ and $V_{min} = -0.1$, respectively. In addition, the acceleration coefficients c_1 and c_2 were selected to be 2.5 and 1.5, respectively. The inertia weight of the HPSO varied randomly in the range (0, 1). The parameters related to jumping out of local optimal were $s = 270$ and $\varepsilon = 0.001$. The values of the regularization parameters in the optimization problem were $\lambda_1 = 1.2, \lambda_2 = 1.0$.

Table 1. Descriptions of unimodal and multimodal benchmark functions.

Function	Name	Function's Expressions	Search Range	Global opt. [1]
f1	Sphere	$f_1 = \sum_{i=1}^{n} x_i^2$	$[-100, 100]^n$	0
f2	Schwefel's 1.2	$f_2 = \sum_{i=1}^{n} \sum_{j=1}^{i} x_j^2$	$[-100, 100]^n$	0
f3	Schwefel's 2.21	$f_3 = \max\{\lvert x_i \rvert, 1 \le i \le n\}$	$[-100, 100]^n$	0
f4	Quartic Noise	$f_4 = \sum_{i=1}^{n} i x_i^4 + random[0.1)$	$[-1.28, 1.28]^n$	0
f5	Generalized Rastrigin	$f_5 = \sum_{i=1}^{n} \left[x_i^2 - 10\cos(2\pi x_i) + 10\right]$	$[-5.12, 5.12]^n$	0
f6	Generalized Penalized Function 2	$f_6 = 0.1\sin^2(3\pi x_1) + 0.1\sum_{i=1}^{n} (x_i - 1)^2\left[1 + \sin^2(3\pi x_i + 1)\right] + 0.1(x_n - 1)^2\left[1 + \sin^2(2\pi x_n)\right] + \sum_{i=1}^{n} u(x_i, 5, 100, 4)$	$[-50, 50]^n$	0

[1] Global opt.: global optimal solution.

The maximum number of iterations for all benchmark functions (f1–f9) was selected as 8000. The dimensions of these benchmark functions (f1–f9) were selected as 10, 30 and 50. Thus, the performances of the six variant PSO algorithms can be compared in different dimensions.

Each algorithm was run individually 10 times, and the average statistical error was calculated. The mean of objective values (Mean) and standard deviation of its solving error (S.D.) were chosen as the performance measures for each algorithm. The simulation results are shown in Table 3 and Figures 2–10. Table 3 shows best values in bold.

Table 2. Details of hybrid composition functions.

Function (CEC2005-ID)	Description	Properties	Range	Global opt.
f7(C16)	Rotated Hybrid Composition Function	MM [1], R [2], NS [3], S [4]	$[-5,5]^n$	120
f8(C18)	Rotated Hybrid Composition Function	MM, R, NS, S	$[-5,5]^n$	10
f9(C21)	Rotated Hybrid Composition Function	MM, R, NS, S	$[-5,5]^n$	360

[1] MM: Multi-modal, [2] R: Rotated, [3] NS: Non-Separable, [4] S: Scalable.

Table 3. Results of benchmark functions (Dim = 10, 30, 50).

F [1]	D [2]	XPSO		CPSO		IPSO		IPSO2		PSO		HPSO	
		Mean [3]	S.D. [4]	Mean	S.D.	Mean	S.D.	Mean	S.D.	Mean	S.D.	Mean	S.D.
	10	**1.29×10^{-102}**	**2.89×10^{-102}**	3.42×10^{2}	2.07×10^{2}	5.81×10^{-8}	2.39×10^{-8}	1.07×10^{-3}	2.38×10^{-4}	7.95×10^{-9}	1.63×10^{-8}	2.33×10^{-8}	6.15×10^{-8}
f1	30	**6.41×10^{-57}**	**1.92×10^{-56}**	4.40×10^{2}	2.25×10^{2}	4.15×10^{-6}	1.22×10^{-6}	0.31	0.41	7.51×10^{-6}	4.04×10^{-6}	4.22×10^{-7}	2.57×10^{-7}
	50	**1.59×10^{-8}**	**3.89×10^{-8}**	4.57×10^{2}	2.80×10^{2}	1.93×10^{-5}	5.87×10^{-6}	6.73	13.42	2.00×10^{-5}	1.30×10^{-5}	1.02×10^{-6}	3.08×10^{-7}
	10	**1.06×10^{-77}**	**1.50×10^{-77}**	1.06×10^{3}	8.87×10^{2}	5.93×10^{-7}	3.33×10^{-7}	2.95×10^{2}	3.70×10^{2}	4.08×10^{-7}	4.74×10^{-7}	1.35×10^{-8}	1.71×10^{-8}
f2	30	**8.67×10^{-12}**	**1.23×10^{-11}**	4.36×10^{3}	1.95×10^{3}	1.44×10^{-3}	4.36×10^{-4}	9.07×10^{2}	8.11×10^{2}	5.74×10^{-3}	2.33×10^{-3}	3.11×10^{-5}	1.98×10^{-5}
	50	**1.36×10^{-5}**	1.35×10^{-2}	7.74×10^{3}	6.45×10^{3}	1.88×10^{-2}	3.33×10^{-3}	1.25×10^{3}	2.74×10^{3}	2.49×10^{-2}	9.20×10^{-3}	1.91×10^{-4}	**8.34×10^{-5}**
	10	**1.11×10^{-21}**	**1.91×10^{-21}**	4.59	1.13	2.70×10^{-4}	1.47×10^{-4}	6.41×10^{-2}	7.51×10^{-2}	4.45×10^{-4}	2.61×10^{-4}	8.41×10^{-4}	7.47×10^{-4}
f3	30	**1.03×10^{-19}**	**1.79×10^{-19}**	6.15	1.65	9.01×10^{-3}	2.77×10^{-3}	0.71	0.16	3.03×10^{-2}	1.62×10^{-2}	1.96×10^{-2}	1.99×10^{-2}
	50	**6.84×10^{-5}**	**1.37×10^{-4}**	4.87	0.65	0.23	0.22	5.76	0.41	0.36	0.33	0.13	0.11
	10	**0.24**	**0.16**	0.43	0.26	0.44	0.26	0.46	0.32	0.61	0.24	0.5	0.29
f4	30	**0.27**	0.2	0.58	0.28	0.53	0.32	0.35	0.3	0.64	0.31	0.52	**0.18**
	50	**0.35**	0.24	0.65	0.3	0.54	0.31	0.38	**0.23**	0.53	**0.23**	0.58	0.3
	10	12.93	3.98	12.08	4.46	10.45	4.43	**9.3**	**3.4**	10.94	4.09	9.83	4.16
f5	30	**18.05**	**4.16**	48.43	17.81	26.96	6.91	24.81	6.5	27.16	9.67	28.56	6.13
	50	**20.9**	**2.54**	67.23	18.56	38.51	11.94	34.18	4.93	39.99	14.36	33.73	10.42
	10	**1.45**	**0.94**	9.41	4.09	2.13	1.37	2.62	2.47	2.81	3.26	1.48	1.46
f6	30	**5.08**	**3.56**	25.39	5.56	14.51	8.42	22.09	5.69	9.49	5.42	6.59	6.24
	50	9.83	**6.98**	37.21	7.54	24.36	9.83	31.42	8.14	14.71	8.78	**9.29**	8.89
	10	**196.35**	**26.50**	502.96	179.38	381.96	94.03	731.48	80.86	362.11	36.18	428.34	59.63
f7	30	**542.93**	**46.56**	867.86	204.51	584.27	122.68	1223.15	110.51	505.28	80.91	623.86	90.26
	50	**563.92**	**97.82**	937.18	241.44	593.34	132.26	1378.35	129.68	599.16	106.17	651.98	126.37
	10	**1090.72**	41.53	1364.68	23.43	1135.18	**22.51**	1492.86	39.30	1136.88	65.80	1098.15	59.30
f8	30	**1147.03**	123.61	1408.69	**50.61**	1182.64	70.86	1561.77	56.23	1194.76	120.68	1287.80	84.09
	50	**1167.73**	134.83	1486.24	**56.47**	1202.59	132.75	1620.53	70.99	1249.79	131.05	1464.34	151.77
	10	**1066.48**	28.84	1387.39	19.11	1120.63	38.33	1538.21	**9.21**	1286.09	23.52	1287.69	39.38
f9	30	**1291.78**	38.31	1440.31	31.75	1319.86	43.93	1604.27	41.32	1303.97	**30.27**	1309.14	52.25
	50	**1316.15**	54.37	1485.75	**41.98**	1338.22	68.71	1633.61	82.42	1326.48	43.43	1524.77	68.02

[1] F: Function, [2] D: dimensional, [3] Mean: mean of objective values, [4] S.D.: standard deviation.

According to Table 3, the XPSO algorithm is superior to other PSO algorithms in terms of solution accuracy. Considering the UM benchmark functions, the results of (f1–f3) solved by XPSO are better than those of the other algorithms. For f4, the best value of Mean was obtained by the XPSO algorithm for the (10, 30, 50)-dimensional cases. The best values of S.D for the (10, 30, 50)-dimensional cases were obtained by XPSO, HPSO and IPSO2. Except f5 in the 10-dimensional case and f6 in the 50-dimensional case, the results for the MM benchmark functions by XPSO are superior to those of the other algorithms. Based on values of the Mean of hybrid CM functions (f7–f9) in Table 3, high-quality solutions can be obtained by the XPSO algorithm.

The convergence curves of the mean of best function were plotted in Figures 2–10 to enable clearer visualization of each algorithm's performance. In Figures 2–10, the XPSO algorithm does not show outstanding performance in the early stage of the solution, but its search capability is better in the late stage of the solution, and it can significantly improve the convergence speed in the late evolution. The reason is that the updating strategy of inertia weight includes the formula based on chaotic mapping; consequently, the global search has a high convergence speed for particles. In addition, the updating formula for inertia weight switched in local search, which resulted a slow flying speed of particles but improved the accuracy of the solution. Finally, it can be concluded that the XPSO algorithm has excellent performance in numerical optimization problems and can be used to solve problems in various application contexts.

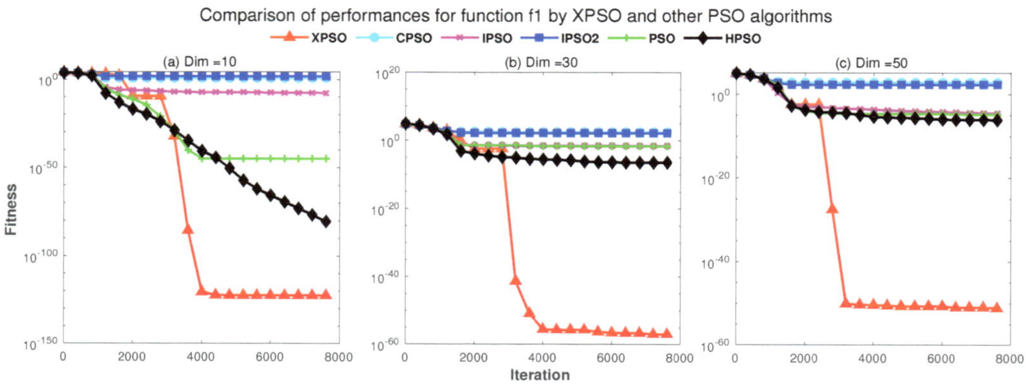

Figure 2. Comparison of performances on f1 by XPSO and other PSO algorithms.

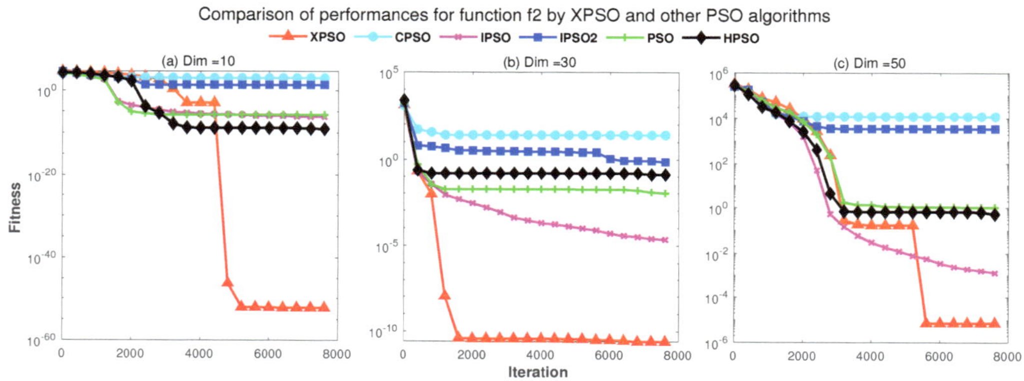

Figure 3. Comparison of performances on f2 by XPSO and other PSO algorithms.

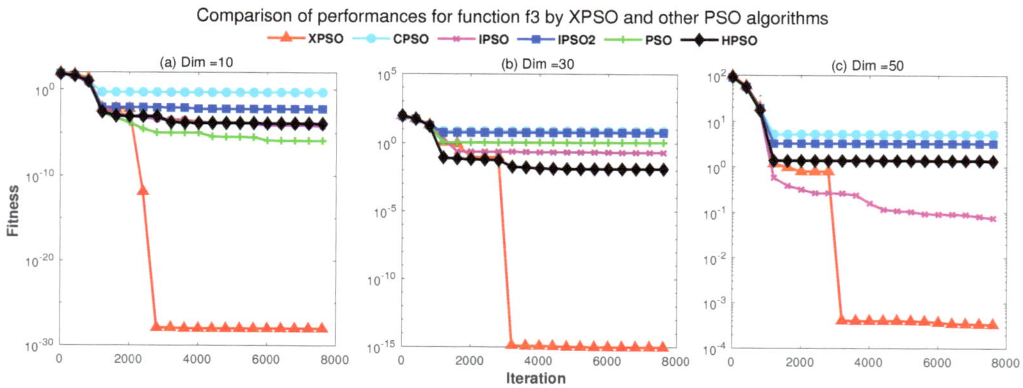

Figure 4. Comparison of performances on f3 by XPSO and other PSO algorithms.

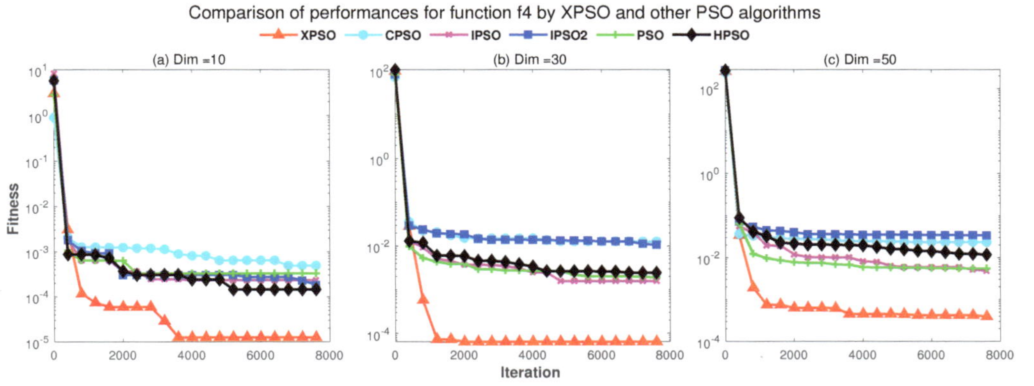

Figure 5. Comparison of performances on f4 by XPSO and other PSO algorithms.

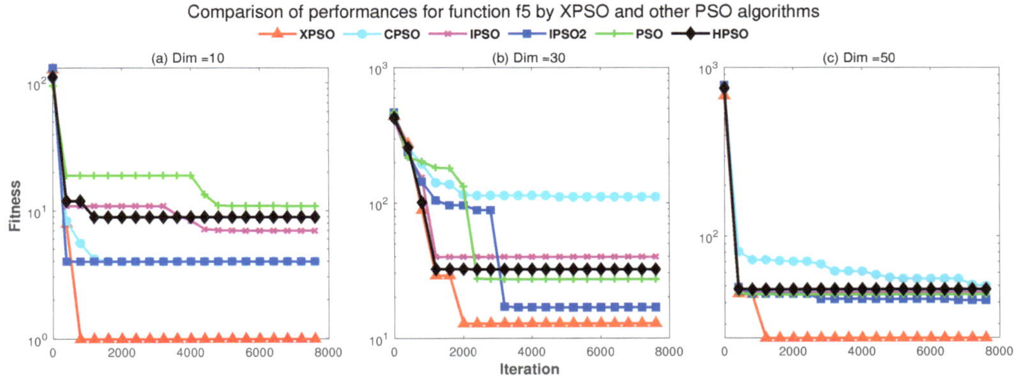

Figure 6. Comparison of performances on f5 by XPSO and other PSO algorithms.

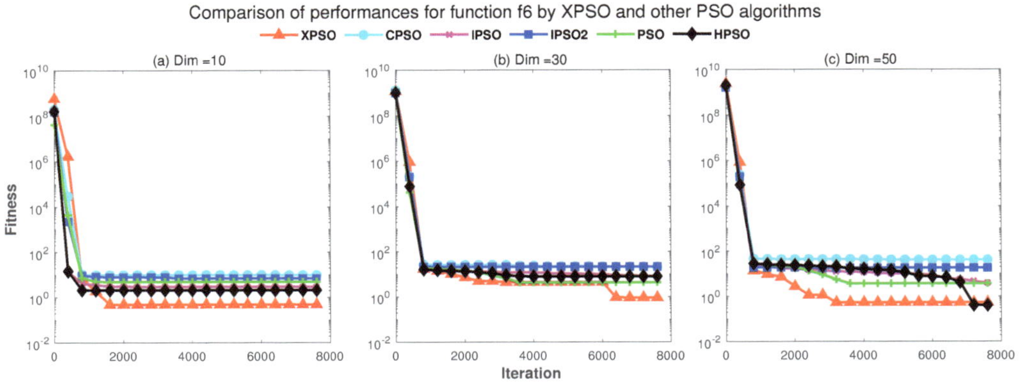

Figure 7. Comparison of performances on f6 by XPSO and other PSO algorithms.

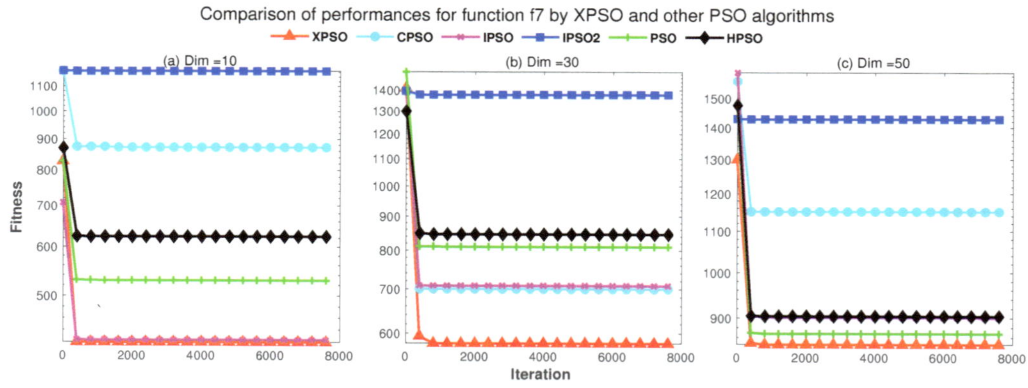

Figure 8. Comparison of performances on f7 by XPSO and other PSO algorithms.

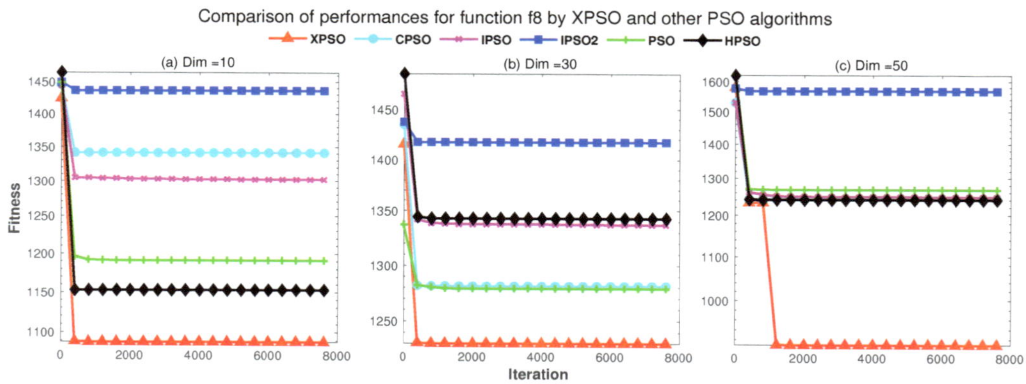

Figure 9. Comparison of performances on f8 by XPSO and other PSO algorithms.

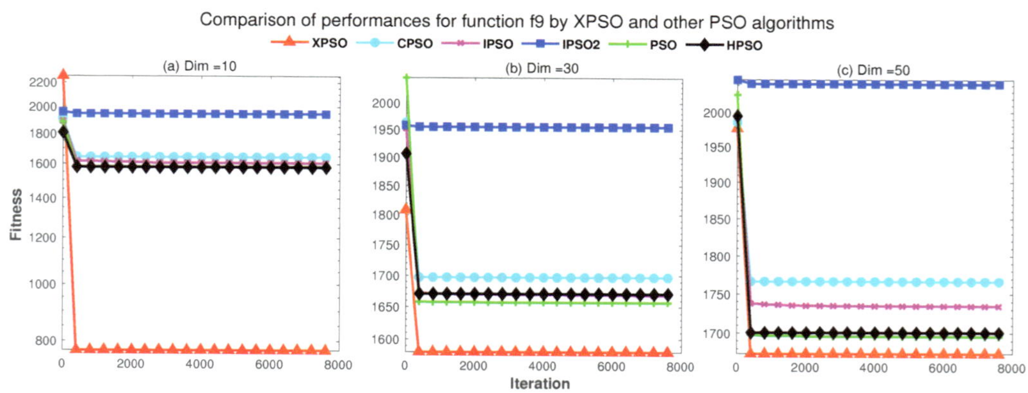

Figure 10. Comparison of performances on f9 by XPSO and other PSO algorithms.

4.2. Validation Of XPSO by Benchmark Test Functions

Firstly, 1280 simulation data sets were generated based on the existing mechanism model to verify the validity of the XPSO algorithm. In total, 1200 data sets were randomly selected as training sets and the remaining 80 as test sets. Here, the XPSO algorithm is compared with not only other PSO algorithms (PSO [32], IPSO [33], IPSO2 [30], HPSO [34], CPSO [35]), but also the other optimization algorithms (WOA [38], IA [39], GWO [40], DE [41], ABC [42]). Secondly, the actual data sets were collected from a reheating furnace in Angang's building. Fourty-three sets were randomly selected as the training sets. The remaining 10 sets were used as the test sets.

4.2.1. Comparison of XPSO and Other PSO Algorithms

Due to the random initialization of the PSO algorithms, each PSO algorithm independently ran 10 times. The relevant parameters for the PSO algorithms are shown in Table 4. Each algorithm was evaluated by the mean, maximum, median, variance and standard deviation of the errors. The simulation results are shown in Table 5 and Figures 11 and 12.

Table 4. Classical benchmark functions.

Symbol	Name	Size
N	Particle swarm size	125
D	Particle Swarm Dimension	35
G	Maximum number of iterations	8000
w_s	Initial value of inertia weights	0.8
w_e	Final value of inertia weights	0.05
c_1	Acceleration coefficient 1	2.5
c_2	Acceleration coefficient 2	1.5
V_{max}	Value of maximum particle's velocity	0.1
V_{min}	Value of minimum particle's velocity	−0.1

Table 5. Classical benchmark functions.

Algorithm	Mean	Maximum	Median	Variance	S.D.
XPSO	**0.55**	**2.29**	**0.46**	**0.216**	**0.465**
CPSO	3.9	13.99	3.41	8.098	2.846
IPSO	7	23.75	5.85	26.582	5.156
IPSO2	3.76	13.58	3	8.388	2.896
PSO	3.58	11.65	2.77	6.818	2.611
HPSO	0.78	2.94	0.61	0.446	0.668

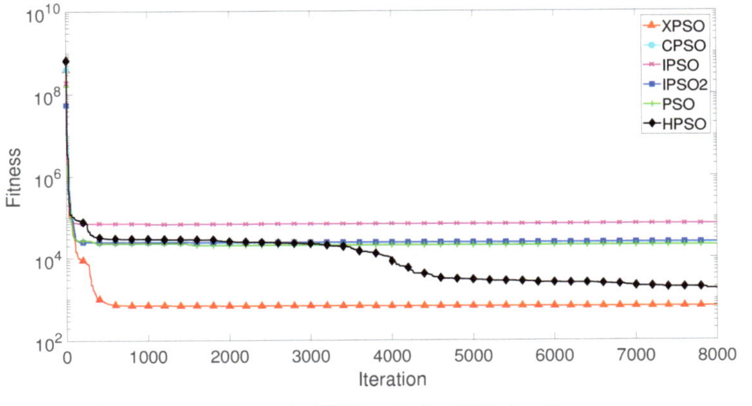

Figure 11. Comparison of fitness for XPSO and other PSO algorithms.

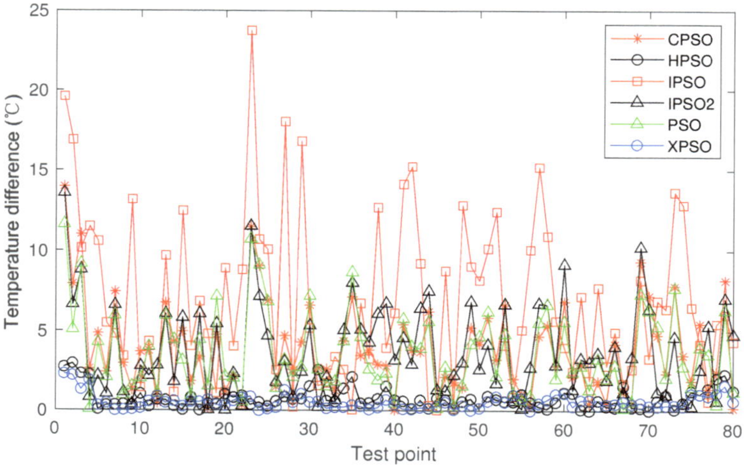

Figure 12. Prediction error of XPSO and other PSO algorithms.

In Figure 11, the XPSO algorithm clearly has a faster search speed than other algorithms in the early iterations and can quickly move to convergence. Table 5 shows that XPSO is also more accurate in terms of computational accuracy at the later stages of the iterations. Figure 12 ensures the accuracy of the proposed XPSO method for temperature prediction. The IPSO algorithm gave the worst results, for which the maximum error value was almost 24 °C. The mean and median prediction errors by the XPSO algorithm were 0.55 and 0.46 °C, and 99% of the prediction errors by the XPSO algorithm were within 2 °C.

4.2.2. Comparison of XPSO and Other Optimization Algorithms

In this section, the XPSO algorithm is compared with the other optimization algorithms (WOA [38], IA [39], GWO [40], DE [41], ABC [42]) that have been proposed in recent years. The parameters of these algorithms are listed in Table 6. Each algorithm was tested 10 times independently to reduce statistical errors. The mean, maximum, median, variance and standard difference of simulation results were recorded and are shown in Table 7. The best results are shown in bold type. The convergence graph of each algorithm is shown in Figure 13. The slab temperature prediction errors of the XPSO algorithm and other optimization algorithms are shown in Figure 14.

Table 6. Parameters of other optimization algorithms.

Algorithms	Population	Maximum Iteration	Dim	Other
WOA	125	8000	35	$r_1, r_2 \in [0,1]$ are random numbers
IA	125	8000	35	$p_m = 0.7, \alpha = \beta = 1, \delta = 0.2, ncl = 10$
GWO	125	8000	35	$r_1, r_2 \in [0,1]$are random numbers
DE	125	8000	35	$F0 = 0.4, CR = 0.1$
ABC	125	8000	35	$\alpha = 1$

Table 7. Results of XPSO and other optimization algorithms.

Algorithm	Mean	Maximum	Median	Variance	S.D.
XPSO	**0.55**	**2.29**	**0.46**	**0.216**	**0.465**
WOA	4.53	28.27	3.45	18.94	4.3519
IA	6.92	27.52	6.43	28.0032	5.2918
GWO	2.28	13.39	1.2	6.6249	2.5739
DE	2.3	10.53	2.1	2.8971	1.7021
ABC	6.59	27.23	9.79	24.2055	4.9199

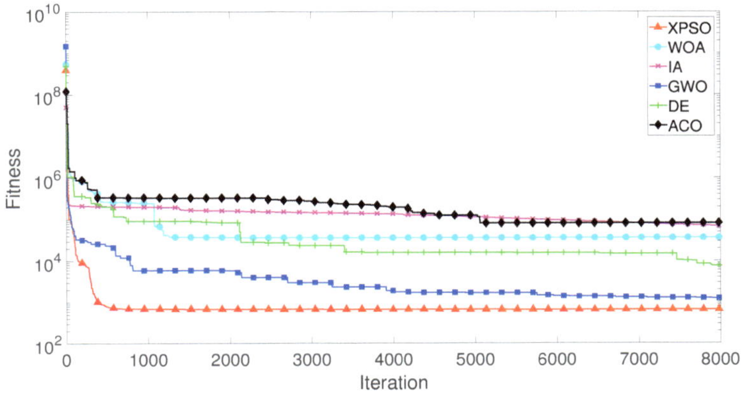

Figure 13. Comparison of XPSO with other optimization algorithms.

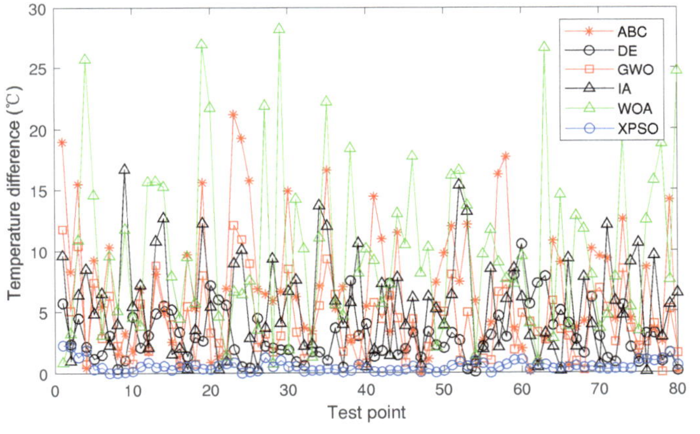

Figure 14. Prediction error of XPSO and other optimization algorithms.

Table 7 proves that the solution of the XPSO algorithm gives the best value. In Figure 13, the XPSO algorithm is more successful than all of the other optimization approaches, and the algorithm determined the global optimal solution after approximately 500 generations. As shown in Figure 14, the temperature prediction errors by the XPSO algorithm were much lower than the errors by the other optimization algorithms. The WOA algorithm gave the worst results; its maximum error value was almost 29 °C. In summary, the proposed XPSO algorithm exhibited fast search performance and accuracy when predicting the billet temperature based on simulation data sets. The results and figures show that this prediction model of billet temperature is credible and reliable. Its accurate prediction is expected to satisfy the future control needs of industrial reheating furnace systems, which will let operators adjust production plans in time to ensure efficiency and reliability.

4.2.3. Validation of the Temperature Prediction Model With Measured Data

The proposed XPSO algorithm was applied to predict slabs' discharging temperatures based on actual data sets from Angang (company). The algorithm was independently run 20 times, and then the average prediction error was calculated. The predicted temperatures are compared with the actual temperatures in Figure 15. The errors between prediction results and measured data $|Y - Y^*|$ are shown in Table 8.

Table 8. Prediction errors between calculation results and measured data.

Points	1 (°C)	2 (°C)	3 (°C)	4 (°C)	5 (°C)	6 (°C)	7 (°C)	8 (°C)	9 (°C)	10 (°C)	Mean (°C)
XPSO	5.43	2.91	8.93	7.32	3.47	5.17	1.17	5.09	4.09	6.43	4.99

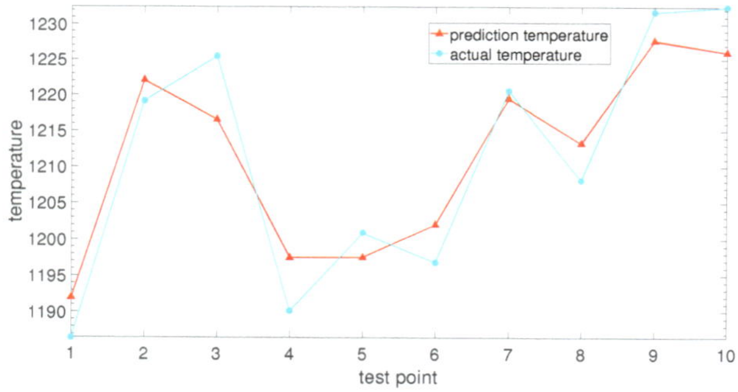

Figure 15. The slab temperature prediction by the XPSO algorithm.

From Figure 15 and Table 8, the minimum error of the XPSO algorithm can be seen to be 1.17 °C, and the average error was 4.99 °C. For this actual reheating furnace company, the error between the calculation results and measured data should be lower than 10 °C. Thus, the accuracy of the proposed method is high enough. As the actual discharging temperature was between 1150–1250 °C, the relative error was only 0.4%. Finally, the slab temperature prediction by the XPSO algorithm has achieved the desired effect.

5. Conclusions

In this paper, the XPSO algorithm was proposed to establish a prediction model of slab temperature in a reheating furnace. A novel weight-updating strategy that combines a decreasing function and the adaptive "stepped" strategy was introduced into the XPSO algorithm, so it can greatly improve the search capabilities at a later stage. The validity and feasibility of the XPSO were verified by nine classical benchmark functions, simulation data sets generated by the existing mechanism model and actual data sets from Angang. The following conclusions are given.

1. The benchmark results indicate that the XPSO algorithm has a superior performance to other PSO algorithms (PSO, IPSO, IPSO2, HPSO, CPSO).
2. The XPSO algorithm, which can accurately predict the billet temperature (99% of the prediction errors were less than 2 °C) while ensuring faster convergence, was more successful than all of the other optimization approaches (WOA, IA, GWO, DE, ABC).
3. The prediction model based on the XPSO algorithm can predict more accurate discharging temperatures for the operators. Consequently, the paper verifies the feasibility of the XPSO algorithm and the success of the establishment of the prediction model of slab temperature, and provides a theoretical basis for subsequent research.

Author Contributions: Conceptualization, M.L., P.Y. and. Z.Y.; methodology, M.L.; software, M.L. and P.L.; validation, M.L. and J.Q.; formal analysis, J.Q.; investigation, P.L. and Z.Y.; resources, M.L. and P.L.; data curation, M.L.; writing—original draft preparation, M.L.; writing—review and editing, M.L. and Z.Y.; visualization, Z.Y.; supervision, P.Y. and Z.Y.; project administration, P.Y. and Z.Y.; funding acquisition, Z.Y. All authors have read and agreed to the published version of the manuscript.

Funding: This research was funded by the PhD research startup foundation of Qilu University of Technology, grant number (81110535), and the Industry-University-Research Collaborative Innovation Fund (grant number 2020-CXY46)—Development of an automated system for casting post-processing processes.

Institutional Review Board Statement: Not applicable.

Informed Consent Statement: Not applicable.

Data Availability Statement: Not applicable.

Conflicts of Interest: The authors declare no conflict of interest.

References

1. Gu, M.; Chen, G.; Liu, X.; Wu, C.; Chu, H. Numerical simulation of slab heating process in a regenerative walking beam reheating furnace. *Int. J. Heat Mass Transf.* **2014**, *76*, 405–410. [CrossRef]
2. Gao, Q.; Pang, Y.; Sun, Q.; Liu, D.; Zhang, Z. Modeling approach and numerical analysis of a roller-hearth reheating furnace with radiant tubes and heating process optimization. *Case Stud. Therm. Eng.* **2021**, *28*, 101618. [CrossRef]
3. Hu, Y.; Tan, C.; Broughton, J.; Roach, P.A.; Varga, L. Model-based multi-objective optimisation of reheating furnace operations using genetic algorithm. *Energy Procedia* **2017**, *142*, 2143–2151. [CrossRef]
4. Pantelides, C.C.; Renfro, J.G. The online use of first-principles models in process operations: Review, current status and future needs. *Comput. Chem. Eng.* **2013**, *51*, 136–148. [CrossRef]
5. Staalman, D.F.; Kusters, A. On-line slab temperature calculation and control. *Manuf. Sci. Eng.* **1996**, *4*, 307–314.
6. Ji, W.; Li, G.; Wei, L.; Yi, Z. Modeling and determination of total heat exchange factor of regenerative reheating furnace based on instrumented slab trials. *Case Stud. Therm. Eng.* **2021**, *24*, 100838. [CrossRef]
7. Emadi, A.; Saboonchi, A.; Taheri, M.; Hassanpour, S. Heating characteristics of billet in a walking hearth type reheating furnace. *Appl. Therm. Eng.* **2014**, *63*, 396–405. [CrossRef]
8. Tang, G.; Wu, B.; Bai, D.; Wang, Y.; Bodnar, R.; Zhou, C.Q. Modeling of the slab heating process in a walking beam reheating furnace for process optimization. *Int. J. Heat Mass Transf.* **2017**, *113*, 1142–1151. [CrossRef]
9. Kim, M.Y. A heat transfer model for the analysis of transient heating of the slab in a direct-fired walking beam type reheating furnace. *Int. J. Heat Mass Transf.* **2007**, *50*, 3740–3748. [CrossRef]
10. Singh, V.K.; Talukdar, P. Comparisons of different heat transfer models of a walking beam type reheat furnace. *Int. Commun. Heat Mass Transf.* **2013**, *47*, 20–26. [CrossRef]
11. Morgado, T.; Coelho, P.J.; Talukdar, P. Assessment of uniform temperature assumption in zoning on the numerical simulation of a walking beam reheating furnace. *Appl. Therm. Eng.* **2015**, *76*, 496–508. [CrossRef]
12. Casal, J.M.; Porteiro, J.; Míguez, J.L.; Vázquez, A. New methodology for CFD three-dimensional simulation of a walking beam type reheating furnace in steady state. *Appl. Therm. Eng.* **2015**, *86*, 69–80. [CrossRef]
13. Hong, D.; Li, G.; Wei, L.; Yi, Z. An improved sequential function specification coupled with Broyden combined method for determination of transient temperature field of the steel billet. *Int. J. Heat Mass Transf.* **2022**, *186*, 122489. [CrossRef]
14. Chen, D.; Xu, H.; Lu, B.; Chen, G.; Zhang, L. Solving the heat transfer boundary condition of billet in reheating furnace by combining "black box" test with mathematic model. *Case Stud. Therm. Eng.* **2022**, *40*, 102486. [CrossRef]
15. Kim, Y.I.; Moon, K.C.; Kang, B.S.; Han, C.; Chang, K.S. Application of neural network to the supervisory control of a reheating furnace in the steel industry. *Control. Eng. Pract.* **1998**, *6*, 1009–1014. [CrossRef]
16. Laurinen, P.; Röning, J. An adaptive neural network model for predicting the post roughing mill temperature of steel slabs in the reheating furnace. *J. Mater. Process. Technol.* **2005**, *168*, 423–430. [CrossRef]
17. Liao, Y.; Wu, M.; She, J.H. Modeling of reheating-furnace dynamics using neural network based on improved sequential-learning algorithm. In Proceedings of the 2006 IEEE Conference on Computer Aided Control System Design, 2006 IEEE International Conference on Control Applications, 2006 IEEE International Symposium on Intelligent Control, Munich, Germany, 4–6 October 2006; pp. 3175–3181.
18. Tan, C.; Wilcox, S.; Ward, J. Use of artificial intelligence techniques for optimisation of co-combustion of coal with biomass. *J. Energy Inst.* **2006**, *79*, 19–25. [CrossRef]
19. Pongam, T.; Khomphis, V.; Srisertpol, J. System modeling and temperature control of reheating furnace walking hearth type in the setting up process. *J. Mech. Sci. Technol.* **2014**, *28*, 3377–3385. [CrossRef]
20. Tang, Z.; Yang, Y. Two-stage particle swarm optimization-based nonlinear model predictive control method for reheating furnace process. *ISIJ Int.* **2014**, *54*, 1836–1842. [CrossRef]
21. Aoxiang, W.; Xiaohua, L.; Xiaolin, W. Temperature optimization setting model of the reheating furnace on 1700 line in tangsteel. In Proceedings of the 2018 Chinese Control Additionally, Decision Conference (CCDC), Shenyang, China, 9–11 June 2018; pp. 4099–4103.
22. Yang, Y.; Liu, Y.; Liu, X.; Qin, S. Billet temperature soft sensor model of reheating furnace based on RVM method. In Proceedings of the 2011 Chinese Control and Decision Conference (CCDC), Mianyang, China, 23–25 May 2011; pp. 4003–4006.
23. Yi, Z.; Su, Z.; Li, G.; Yang, Q.; Zhang, W. Development of a double model slab tracking control system for the continuous reheating furnace. *Int. J. Heat Mass Transf.* **2017**, *113*, 861–874. [CrossRef]
24. Chen, Y.W.; Chai, T.Y. Modelling and prediction for steel billet temperature of heating furnace. *Int. J. Adv. Mechatron. Syst.* **2010**, *2*, 342–349. [CrossRef]

25. Alsaidy, S.A.; Abbood, A.D.; Sahib, M.A. Heuristic initialization of PSO task scheduling algorithm in cloud computing. *J. King Saud -Univ.-Comput. Inf. Sci.* **2020**, *34*, 2370–2382. [CrossRef]
26. Yue, C.; Qu, B.; Liang, J. A multiobjective particle swarm optimizer using ring topology for solving multimodal multiobjective problems. *IEEE Trans. Evol. Comput.* **2017**, *22*, 805–817. [CrossRef]
27. Peng, C.C.; Chen, C.H. Compensatory neural fuzzy network with symbiotic particle swarm optimization for temperature control. *Appl. Math. Model.* **2015**, *39*, 383–395. [CrossRef]
28. Eberhart, R.; Kennedy, J. Particle swarm optimization. In Proceedings of the IEEE International Conference on Neural Networks, Perth, WA, Australia, 27 November–1 December 1995; Volume 4, pp. 1942–1948.
29. Kennedy, J. The particle swarm: Social adaptation of knowledge. In Proceedings of the 1997 IEEE International Conference on Evolutionary Computation (ICEC'97), Indianapolis, IN, USA, 13–16 April 1997; pp. 303–308.
30. Ravi, K.; Rajaram, M. Optimal location of FACTS devices using improved particle swarm optimization. *Int. J. Electr. Power Energy Syst.* **2013**, *49*, 333–338. [CrossRef]
31. Zhang, L.; Zhao, L. High-quality face image generation using particle swarm optimization-based generative adversarial networks. *Future Gener. Comput. Syst.* **2021**, *122*, 98–104. [CrossRef]
32. Ouyang, A.; Tang, Z.; Zhou, X.; Xu, Y.; Pan, G.; Li, K. Parallel hybrid pso with cuda for ld heat conduction equation. *Comput. Fluids* **2015**, *110*, 198–210. [CrossRef]
33. Gao, Z.; Lu, H. Logistics Route Optimization Based on Improved Particle Swarm Optimization. In Proceedings of the Journal of Physics: Conference Series, Diwaniyah, Iraq, 21–22 April 2021; Volume 1995, p. 012044.
34. Wu, J.; Long, J.; Liu, M. Evolving RBF neural networks for rainfall prediction using hybrid particle swarm optimization and genetic algorithm. *Neurocomputing* **2015**, *148*, 136–142. [CrossRef]
35. Liu, B.; Wang, L.; Jin, Y.H.; Tang, F.; Huang, D.X. Improved particle swarm optimization combined with chaos. *Chaos, Solitons Fractals* **2005**, *25*, 1261–1271. [CrossRef]
36. Yao, X.; Liu, Y.; Lin, G. Evolutionary programming made faster. *IEEE Trans. Evol. Comput.* **1999**, *3*, 82–102.
37. Suganthan, P.N.; Hansen, N.; Liang, J.J.; Deb, K.; Chen, Y.P.; Auger, A.; Tiwari, S. *Problem Definitions and Evaluation Criteria for the CEC 2005 Special Session on Real-Parameter Optimization*; KanGAL Report Number 2005005; Nanyang Technological University: Singapore, 2005.
38. Mirjalili, S.; Lewis, A. The whale optimization algorithm. *Adv. Eng. Softw.* **2016**, *95*, 51–67. [CrossRef]
39. Hong, G.; Zong-Yuan, M. Immune algorithm. In Proceedings of the 4th World Congress on Intelligent Control and Automation (Cat. No. 02EX527), Shanghai, China, 10–14 June 2002; Volume 3, pp. 1784–1788.
40. Mirjalili, S.; Mirjalili, S.M.; Lewis, A. Grey wolf optimizer. *Adv. Eng. Softw.* **2014**, *69*, 46–61. [CrossRef]
41. Arslan, M.; Çunkaş, M.; Sağ, T. Determination of induction motor parameters with differential evolution algorithm. *Neural Comput. Appl.* **2012**, *21*, 1995–2004. [CrossRef]
42. Karaboga, D.; Basturk, B. A powerful and efficient algorithm for numerical function optimization: Artificial bee colony (ABC) algorithm. *J. Glob. Optim.* **2007**, *39*, 459–471. [CrossRef]

materials

Article

Structure of Randomly Distributed Nanochain Aggregates on Silicon Substrates: Modeling and Optical Absorption Characteristics

Tianze Zhao [1,2,3], Yanze Gao [1,2,3,*], Rui Shi [1,2,3], Zhuo Li [1,2,3] and Qingfeng Shi [1,2,3]

1 School of Optics and Photonics, Beijing Institute of Technology, Zhongguancun South Street 5, Beijing 100081, China; zhaotz12@163.com (T.Z.); bitsr@126.com (R.S.); lizhuo@bit.edu.cn (Z.L.); 81908027@bit.edu.cn (Q.S.)
2 Beijing Key Laboratory for Precision Optoelectronic Measurement Instrument and Technology, Zhongguancun South Street 5, Beijing 100081, China
3 Analysis & Testing Center, Beijing Institute of Technology, Zhongguancun South Street 5, Beijing 100081, China
* Correspondence: gao_yanze@bit.edu.cn

Abstract: Nanoparticle aggregate structures allow for efficient photon capture, and thus exhibit excellent optical absorption properties. In this study, a model of randomly distributed nanochain aggregates on silicon substrates is developed and analyzed. The Gaussian, uniform, and Cauchy spatial distribution functions are used to characterize the aggregate forms of the nanochains and their morphologies are realistically reconstructed. The relationships between the structural parameters (thickness and filling factor), equivalent physical parameters (density, heat capacity, and thermal conductivity), and visible absorptivity of the structures are established and analyzed. All the above-mentioned parameters exhibit extreme values, which maximize the visible-range absorption; these values are determined by the material properties and nanochain aggregate structure. Finally, Al nanochain aggregate samples are fabricated on Si substrates by reducing the kinetic energy of the metal vapor during deposition. The spectral reflection characteristics of the samples are studied experimentally. The Spearman correlation coefficients for the calculated spectral absorption curves and those measured experimentally are higher than 0.82, thus confirming that the model is accurate. The relative errors between the calculated visible-range absorptivities and the measured data are less than 0.3%, further confirming the accuracy of the model.

Keywords: optical absorber; nanoparticle clusters; morphological reconstruction; equivalent physical characteristics; visible-range absorption

Citation: Zhao, T.; Gao, Y.; Shi, R.; Li, Z.; Shi, Q. Structure of Randomly Distributed Nanochain Aggregates on Silicon Substrates: Modeling and Optical Absorption Characteristics. *Materials* **2022**, *15*, 4778. https://doi.org/10.3390/ma15144778

Academic Editors: Shuwen Wen, Yongle Sun and Xin Chen

Received: 30 May 2022
Accepted: 6 July 2022
Published: 7 July 2022

1. Introduction

Nanostructures consisting of metal nanoparticle clusters can be described as aggregates of nanocells. These sparse and porous nanostructures can efficiently capture photons, and thus exhibit excellent optical properties [1–3]. Hence, optical absorbers based on nanoaggregate structures are widely used in microbolometers [4], photothermal converters [5,6], solar cells [7,8], and photocatalysis [9]. In addition, nanoaggregate structures show several desirable physical properties, such as low heat capacity, high specific surface area, and low density. As a result, these structures have shown significant potential in the fields of tunable excitation radiation and photothermoacoustics [10].

Several quasianalytical models have been developed for characterizing the nanoaggregate structures, including the effective medium model [11], cluster-cluster aggregation (CCA) model [12], and fractal lossy antennas (FLA) model [13]. The effective medium model simplifies complex nanoclusters into hypothetical ellipsoids. However, the permittivity elucidated by the effective medium model is based on statistical parameters, and the model lacks a description of the nanostructures. The CCA model represents the aggregate morphology of spherical units and the growth process of clusters based on the theory of diffusion-limited aggregation [14–16]. The FLA model describes the nanoaggregate structure as a forest-like structure composed of chains. The length of the chains in the FLA model is the order of microns.

In this paper, a new model of randomly distributed nanochain aggregate structures is proposed to describe their micromorphology. The relationships between the structural parameters (thickness and filling factor), equivalent physical parameters (density, heat capacity, and thermal conductivity), and visible-range absorptivity of the model were established and analyzed. These relationships can serve as a theoretical reference for further research on areas such as the study of the photothermoacoustic effect of nanoaggregate structures. To evaluate the model, Al nanochain aggregate structures were fabricated on Si substrates, and the spectral reflection characteristics of the samples were determined experimentally. The experimental results showed that the model can realistically reconstruct the morphology of the structures and allows for accurate calculations of their spectral absorptivity.

2. Materials and Methods

2.1. Fabrication of Nanoaggregate Structures

To fabricate nanoaggregate samples, we chose two types of substrates, Si wafers and polyimide (PI) films. The Si wafers were cleaned using an oxygen plasma. The PI films were fabricated on Si wafers by spin coating. A 6 mL polyamic acid (PAA) solution was spin-coated on the Si wafers at a low speed (800 r min^{-1}) for 1 min and then at a high speed (7000 r min^{-1}) for 3 min. The PAA coating was then air-dried at 25 °C for 1 h and cured in an oven filled with N_2 gas for 3 h at 300 °C to chemically transform PAA into PI. The process for fabricating a layer of the nanoaggregate structure is shown in Figure 1. Metal Al was evaporated in a He atmosphere at a pressure of 980 Pa, and the Al metal vapor was made to pass through the He atmosphere to reduce its kinetic energy before it was deposited on the substrate.

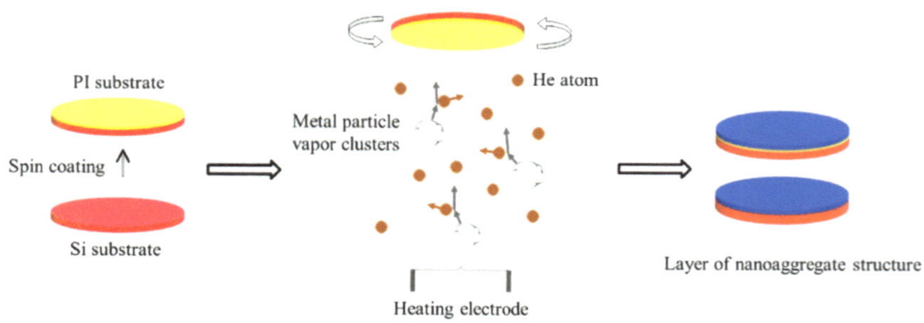

Figure 1. Process for fabricating layer of nanoaggregate structure.

2.2. Model of Nanochain Aggregates

2.2.1. Basic Structural Unit of Model

We obtained scanning electron microscopy (SEM, Supra55, Zeiss, Oberkochen, Germany) images of the fluffy, "smoke-like" metal deposition structure, as shown in Figure 2. The fluffy structure is porous, micron sized, and composed of a large number of clusters, which, in turn, consist of randomly distributed nanochains. These nanochains are composed of metal nanoparticles, which can be considered as nanospheres.

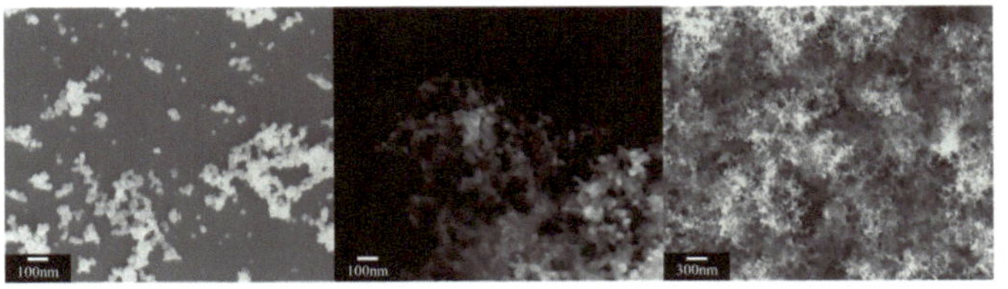

Figure 2. SEM images of nanoaggregate structure.

2.2.2. Spatial Distribution of Nanochains

A large number of nanochains were generated in a three-dimensional space to form clusters. The distribution of nanochains is disordered, but this disordered distribution shows statistical regularity. Therefore, we try to use classical probability distribution functions, Gaussian, uniform and Cauchy distributions, to characterize such statistical regularity. The spatial distribution function represents the aggregate form of the nanochains and reflects the probability density distribution of the nanochains in the clusters, as shown in Figure 3. In the spatial coordinate system with the cluster centroid as the origin, the probability density functions corresponding to the three spatial distributions are as follows:

$$\mathrm{G}(r) = \frac{1}{\sqrt{2\pi}\,\sigma} \exp\left(-\frac{r^2}{2\sigma^2}\right), \tag{1}$$

$$\mathrm{U}(r) = \frac{1}{2\sigma\sqrt{3}}, \tag{2}$$

$$\mathrm{C}(r) = \frac{1}{\pi}\left(\frac{\sigma}{r^2+\sigma^2}\right), \tag{3}$$

where G(r), U(r), and C(r) represent the probability density functions of the Gaussian, uniform, and Cauchy distributions, respectively; r represents the distance between the nanochain and the cluster centroid; σ is the size parameter of the spatial distribution, which determines the radius of the cluster.

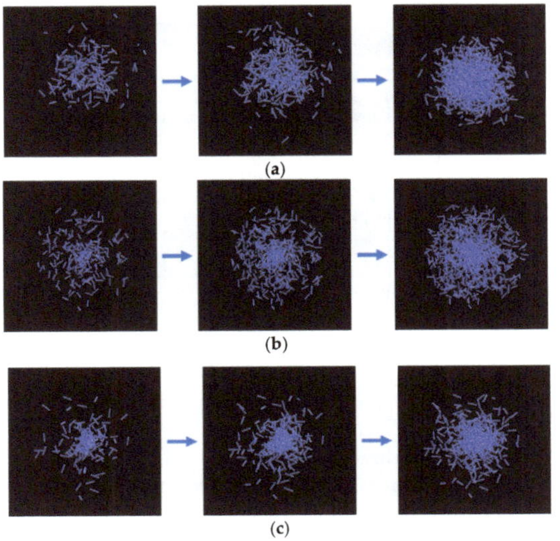

Figure 3. Nanochain cluster models generated using different distribution functions. (**a**) Clusters whose form is represented by Gaussian distribution. (**b**) Clusters whose form is represented by uniform distribution. (**c**) Clusters whose form is represented by Cauchy distribution.

As a structural parameter, the distribution function used affects the propagation path of photons within the structure. Thus, the selection of the appropriate distribution function is essential for accurately calculating the optical absorption. Given that the actual spatial distribution of nanochains is highly random, we add a weight to each distribution function to improve the accuracy of the model, demonstrated by the following equation:

$$W_G + W_U + W_C = 1, \quad (4)$$

where W_G is the weight of the nanochain clusters based on the Gaussian distribution function; W_U is the weight of the nanochain clusters based on the uniform distribution function, and W_C is the weight of the nanochain clusters based on the Cauchy distribution function. The weight of the distribution function reflects the number (percentage) of clusters aggregated by a specific distribution function to all clusters in the model.

2.2.3. Aggregation Model of Clusters

In addition to the spatial distribution of nanochains in a cluster, we use the uniform distribution function to describe the spatial distribution of the clusters in the aggregation model as well. Similar to the spatial distribution of nanochains, the spatial distribution of the clusters in the aggregation model can be described as $U(x, y, z) = \frac{1}{2\sigma\sqrt{3}}$. This completes the physical modeling of the nanochain aggregate structure, as shown in Figure 4.

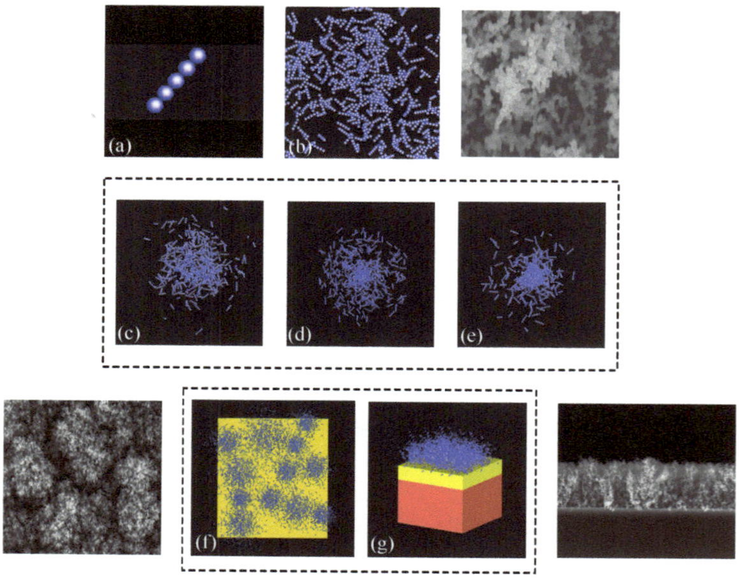

Figure 4. Modeling of nanochain aggregate structure. (**a**) Basic nanochain structure. (**b**) Nanochains with random three-dimensional orientations. (**c**) Cluster based on Gaussian distribution. (**d**) Cluster based on uniform distribution. (**e**) Cluster based on Cauchy distribution. (**f**) Top view of physical model. (**g**) Sectional view of physical model.

We define the dimension of the model in the direction perpendicular to the substrate as the model thickness, *d*. In addition to the thickness, we define a structural parameter, namely, the filling factor (γ), to describe the relative density of the nanochain aggregate structure. The filling factor was calculated as the ratio of the structural density to the material density. It also reflects the porosity of the nanochain aggregate structure. In the proposed model, the filling factor was varied by changing the number of nanospheres contained in each cluster.

2.2.4. Finite-Difference Time-Domain (FDTD) Method

The FDTD method was used to calculate the spectral absorptivity of the model [17–20]. The FDTD method discretized the time-domain Maxwell's equations by central difference approximation of the spatial and temporal partial derivatives. The finite difference equations with the following form were obtained:

$$E_x^{n+1}(i+1,j,k) = \frac{1-\frac{\sigma(i+1/2,j,k)\Delta t}{2\varepsilon(i+1/2,j,k)}}{1+\frac{\sigma(i+1/2,j,k)\Delta t}{2\varepsilon(i+1/2,j,k)}} \cdot E_x^n(i+1/2,j,k) + \frac{\Delta t}{\varepsilon(i+1/2,j,k)} \cdot \frac{1}{1+\frac{\sigma(i+1/2,j,k)\Delta t}{2\varepsilon(i+1/2,j,k)}} \cdot \left[\frac{H_z^{n+1/2}(i+1/2,j,k)-H_z^{n+1/2}(i+1/2,j-1/2,k)}{\Delta y} + \frac{H_y^{n+1/2}(i+1/2,j,k-1/2)-H_y^{n+1/2}(i+1/2,j,k+1/2)}{\Delta z}\right] \quad (5)$$

where i, j, and k are the grid numbers in x, y, and z directions, respectively; n is the number of time domain iterations; Δt is the time step; σ is the conductivity; and ε is the dielectric constant. The dielectric constant and conductivity were wavelength-dependent parameters taken from the literature [21]. The reflectivity and transmittivity of the model were calculated using frequency domain power monitors with the following formula:

$$T(f) = \frac{\text{Re}(P(f)) \cdot d\vec{S}}{\text{Re}(P_{source}(f)) \cdot d\vec{S}} \quad (6)$$

where $T(f)$ is the normalized transmittivity of the monitor; $P(f)$ and $P_{source}(f)$ are the Poynting vectors on the surfaces of the monitor and light source, respectively; and $d\vec{S}$ is the differential element of normal direction. The incident light source was set as a plane wave with a wavelength step of 2 nm. To simplify the model and allow for faster computations, the periodic boundary condition was used for the FDTD calculations. The spectral reflectivity (R_λ) and transmittance (T_λ) of the model were obtained using power monitors. The spectral absorptivity (A_λ) of the model was calculated using Equation (7), and the visible absorptivity (A) of the model was calculated by integrating the spectral absorptivity.

$$A_\lambda = 1 - R_\lambda - T_\lambda, \quad (7)$$

We designed the Gaussian, uniform, and Cauchy models based on the spatial distribution of the nanochains. The spatial distribution of the clusters within the computational region was uniform for all three models, as described in the previous subsection.

3. Results and Discussion

3.1. Optical Absorption Properties of Model

3.1.1. Relationship between Filling Factor and Optical Absorption

We first analyzed the relationship between the filling factor and optical absorption. We set the nanochain materials as Al, Au, and Cr and used a model thickness of 1 μm and filling factor of 0.5–8%. The spectral absorption curves in the wavelength range of 400–800 nm, as well as the visible-range absorptivity of the model for different filling factors, are shown in Figure 4. The filling factor affects the motion path of the incident photons, and thus the efficiency of the model in capturing photons. An extremely sparse or dense distribution of nanochains leads to a reduction in the model absorptivity. For all three models, the filling factor and visible-range absorptivity exhibited a quadratic relationship. The extreme values of the filling factor, which enhanced the visible-range absorption, are listed in Table 1. We propose the following equation to describe the relationship between the filling factor and visible-range absorptivity:

$$A = c + c_1 \cdot \gamma + c_2 \cdot \gamma^2, \quad (8)$$

where c_1, c_2, and c are constant coefficients, whose values are determined from the FDTD results. The aggregate form of the nanochains has a significant effect on the visible-range absorptivity, as shown in Figure 5. The Gaussian model possesses the highest visible-range absorptivity, indicating that the nanostructure corresponding to this model is more efficient at capturing photons.

Table 1. Relationships between model parameters and visible-range absorptivity.

Parameter	Type of Relationship Curve	Material of the Models	Extreme Values for Gaussian, Uniform and Cauchy Models
Filling factor (%)	Quadratic	Al	$\gamma_G = 5.4$, $\gamma_U = 4.4$, $\gamma_C = 4.6$
		Au	$\gamma_G = 7.4$, $\gamma_U = 5.0$, $\gamma_C = 5.6$
		Cr	$\gamma_G = 6.0$, $\gamma_U = 4.4$, $\gamma_C = 5.2$
Thickness (μm)	Exponential	Al	$d_G = 21$, $d_U = 29$, $d_C = 36$
		Au	$d_G = 47$, $d_U = 49$, $d_C = 59$
		Cr	$d_G = 37$, $d_U = 42$, $d_C = 43$
Density ($\times 10^3$ kg m^{-3})	Quadratic	Al	$\rho_G = 0.146$, $\rho_U = 0.113$, $\rho_C = 0.124$
		Au	$\rho_G = 1.430$, $\rho_U = 0.996$, $\rho_C = 1.121$
		Cr	$\rho_G = 0.431$, $\rho_U = 0.316$, $\rho_C = 0.374$
Thermal conductivity (W m^{-1} K^{-1}))	Cubic	Al	$k_G = 33.86$, $k_U = 28.64$, $k_C = 30.43$
		Au	$k_G = 55.87$, $k_U = 43.02$, $k_C = 47.50$
		Cr	$k_G = 14.36$, $k_U = 11.68$, $k_C = 13.05$
Volumetric heat capacity ($\times 10^3$ J m^{-3} K^{-1})	Quadratic	Al	$s_G = 128.3$, $s_U = 99.79$, $s_C = 109.3$
		Au	$s_G = 183.15$, $s_U = 123.75$, $s_C = 143.55$
		Cr	$s_G = 194.13$, $s_U = 123.75$, $s_C = 143.55$

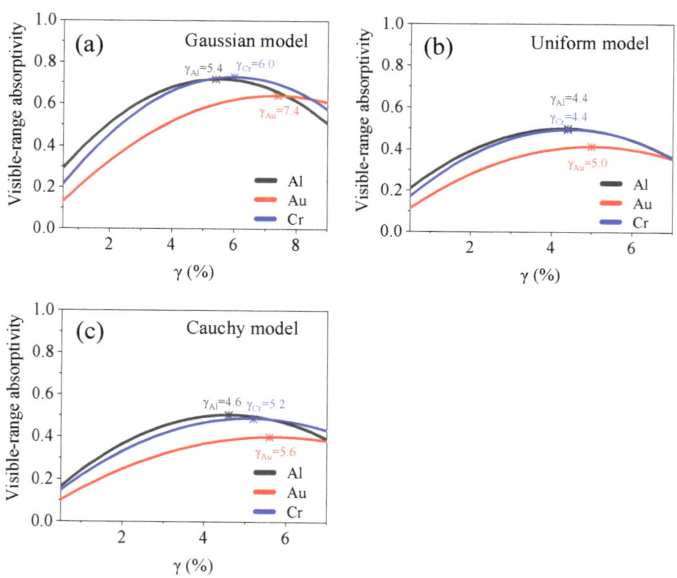

Figure 5. Relationship between optical absorption and filling factor for (**a**) Gaussian, (**b**) uniform, and (**c**) Cauchy models. Filling factor of each model and maximum absorptivity are shown in the figures.

3.1.2. Relationship between Thickness and Optical Absorption

The nanoaggregate structure demonstrates ultralow surface reflection in the visible band [1–3]. For nanochain aggregate structures with the same filling factor, a higher value of thickness would mean that the expected motion path of the incident photons would be longer. Once the thickness of the model exceeds a threshold value, the visible-range transmittance of the model decreases to zero. When the model thickness is more than this threshold value, it no longer affects the visible-range absorptivity of the model. Therefore, the relationship between the thickness and absorptivity takes the form of an exponential function, as shown in Figure 6. We set the nanochain material as Al, Au, and Cr and used the model filling factor of 0.6% and thickness of 1–60 μm. The extreme values of the thickness, which enhanced the visible-range absorption, are listed in Table 1. The critical thickness for the Gaussian model is much smaller than those for the uniform and Cauchy models because of the higher photon-capturing efficiency of the former. The maximum visible-range absorptivities of the three models are similar, indicating that the aggregate form of the nanochains has a negligible effect on the surface reflection of the model. Based on Equation (6), we obtained the following fitting equation:

$$A = c + C_1 \cdot (\gamma + C_2 \cdot \gamma^2) \cdot e^{-d\ /\ C_3}, \tag{9}$$

where C_1, C_2, and C_3 are constant coefficients, whose values can be obtained from the FDTD results. Equation (7) represents the relationship between the visible-range absorptivity, filling factor, and model thickness. It can be observed that $c_1 = C_1 \cdot e^{-d\ /\ C_3}$ and $c_2 = C_1 \cdot C_2 \cdot e^{-d\ /\ C_3}$.

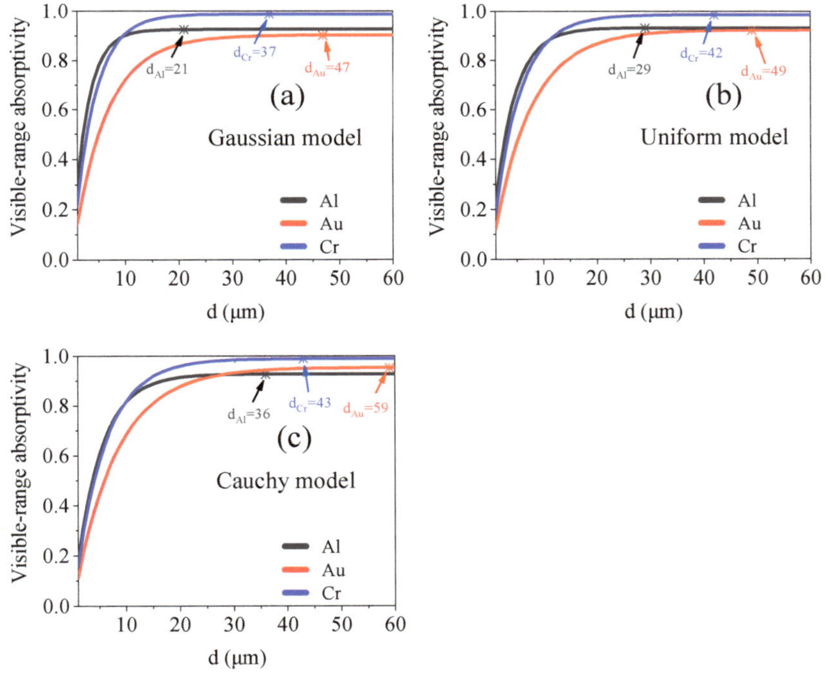

Figure 6. Relationship between optical absorption and thickness for (**a**) Gaussian, (**b**) uniform, and (**c**) Cauchy models. Critical thicknesses of models are shown in the figures.

3.1.3. Relationship between Equivalent Density and Optical Absorption

High optical absorption can be achieved by fabricating nanochain aggregate structures from various materials. Specific application scenarios would result in limitations in terms of the density of the optical absorber. Therefore, it is essential to study the relationship between the equivalent density and optical absorption of nanochain aggregate structures Based on the above-stated definition of the filling factor, the relationship between the filling factor and equivalent density of the structure can be described as follows:

$$\rho_{eff} = \rho_0 \cdot \gamma + \rho_m \cdot (1 - \gamma), \tag{10}$$

where ρ_{eff} is the equivalent density of the structure; ρ_0 is the density of the material used, and ρ_m is the density of the medium. The structural equivalent densities are affected by the filling factor and material density, as shown in Figure 6a. For vacuum, $\rho_m = 0$. Therefore, the relationship between ρ_{eff} and A is as follows:

$$A = c + C_1 \cdot [\frac{\rho_{eff}}{\rho_0} + C_2 \cdot (\frac{\rho_{eff}}{\rho_0})^2] \cdot e^{-d\ /\ C_3}, \tag{11}$$

We set the model thickness as 1 μm and the filling factor as 0.6–8%. The nanochain materials included Al, Au, and Cr. The relationship between the equivalent density and visible-range absorptivity of the nanochain aggregate models is shown in Figure 7. The models use the same structural parameters. Thus, the material density of the model has a distinct effect on its equivalent density. The extreme values of the equivalent density, which enhanced the visible-range absorption, are listed in Table 1.

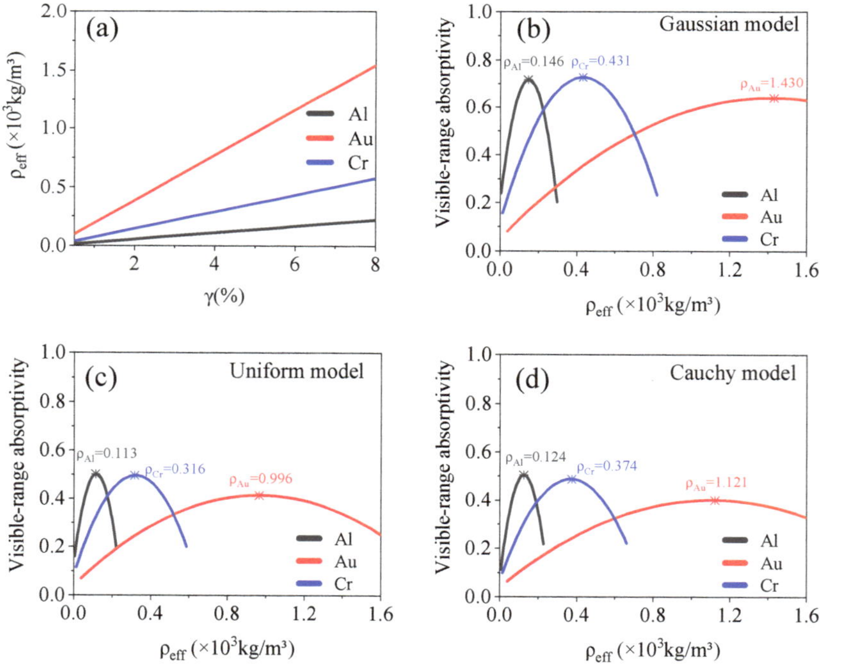

Figure 7. (**a**) Relationship between filling factor and equivalent density of model used. (**b–d**) Relationship between equivalent density and visible absorptivity for Gaussian, uniform, and Cauchy models. Equivalent density of model with maximum absorptivity is marked in the figures.

3.1.4. Relationship between Equivalent Thermal Conductivity and Optical Absorption

Thermal conductivity is an important physical parameter for optical absorbers, with respect to infrared scene generation and photoacoustic effects. Optical absorbers with high thermal conductivities exhibit fast heat dissipation and high-sensitivity time-domain temperature responses. Optical absorbers with low thermal conductivities can generate high-temperature radiation signals at relatively low incident laser powers. The relationship between the equivalent thermal conductivity of the model and its filling factor can be estimated from the classical cheese model [22], which is as follows:

$$k_{eff} = k_0 \cdot \varepsilon + k_m \cdot (1-\varepsilon), \tag{12}$$

where k_{eff} is the equivalent thermal conductivity of the model; k_0 is the thermal conductivity of the material used; k_m is the thermal conductivity of the medium, and $\varepsilon = \gamma^{2/3}$ represents the two-dimensional (2D) porosity in the direction of heat conduction. After combining Equations (7) and (10), we obtain the following equation:

$$A = c + C_1 \cdot [(\frac{k_{eff}}{k_0})^{\frac{3}{2}} + C_2 \cdot (\frac{k_{eff}}{k_0})^{3}] \cdot e^{-d\ /\ C_3}, \tag{13}$$

Figure 8 shows the relationship between the filling factor, equivalent thermal conductivity, and visible-range absorptivity for the Gaussian, uniform, and Cauchy models. The extreme values of the equivalent thermal conductivity, which enhanced the visible-range absorption, are listed in Table 1. Thus, Au nanoaggregate structures are more suitable as optical absorbers that must exhibit efficient absorption and fast heat dissipation. However, Cr nanoaggregate structures allow for efficient energy capture and accumulation.

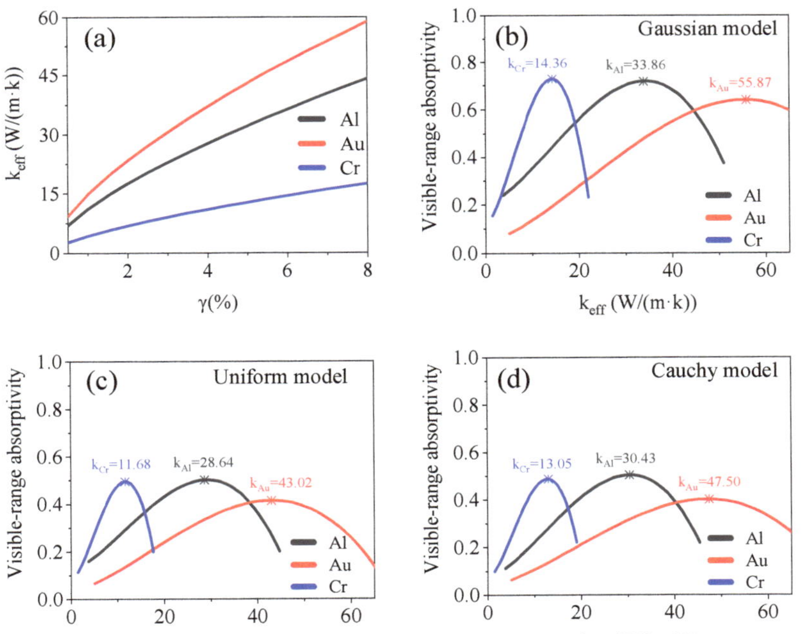

Figure 8. (**a**) Relationship between filling factor and equivalent thermal conductivity of model used. (**b–d**) Relationship between equivalent thermal conductivity and visible-range absorptivity for Gaussian, uniform, and Cauchy models. Equivalent thermal conductivity of model with maximum absorptivity is marked in the figures.

3.1.5. Relationship between Equivalent Volumetric Heat Capacity and Optical Absorption

The specific heat capacity affects the time-domain temperature response rate of optical absorbers. High optical absorption and low specific heat capacity are essential material properties for ensuring a strong photoacoustic effect [10,23]. Similar to the case for the equivalent density, the relationship between the filling factor and equivalent volume heat capacity of the model can be described as follows:

$$s_{eff} = s_0 \cdot \gamma + s_m \cdot (1-\gamma), \tag{14}$$

$$A = c + C_1 \cdot [\frac{s_{eff}}{s_0} + C_2 \cdot (\frac{s_{eff}}{s_0})^2] \cdot e^{-d\,/\,C_3}, \tag{15}$$

where s_{eff} is the equivalent volumetric heat capacity of the model; s_0 is the volumetric heat capacity of the material used, and s_m is the volumetric heat capacity of the medium. The equivalent volumetric heat capacities of the Al and Au models were almost equal, while that of the Cr model was slightly higher, as shown in Figure 9. The extreme values of the equivalent volumetric heat capacity, which enhanced the visible-range absorption, are listed in Table 1. Because its absorptivity is higher, the Al nanoaggregate structure is more suitable as an optical absorber that shows a low volumetric heat capacity and high-sensitivity time-domain temperature response.

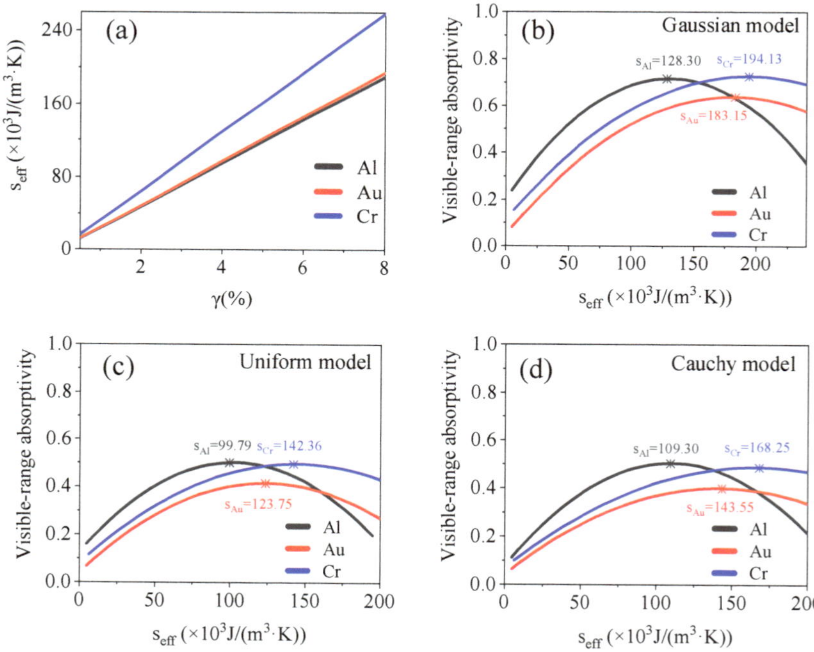

Figure 9. (**a**) Relationship between filling factor and equivalent volumetric heat capacity of model. (**b**–**d**) Relationship between equivalent volumetric heat capacity and visible-range absorptivity of Gaussian, uniform, and Cauchy models. Equivalent volumetric heat capacity of model with maximum absorptivity is marked in the figures.

3.2. Model Validation

3.2.1. Sample Characterization

We fabricated six different samples with nanoaggregate-structured layers. Photographs and SEM images of the samples are shown in Figure 10. The thickness and

2D porosity of the layers of the nanochain aggregates were measured. The filling factor of the samples was calculated using the expression $\varepsilon = \gamma^{2/3}$. The structural parameters of the samples are listed in Table 2.

Figure 10. Photographs and SEM images of nanoaggregate samples. Samples #1–3 were formed on 500 μm Si substrates. Samples #4–6 were formed on 300 nm PI layer; PI layer was fabricated by spin-coating.

Table 2. Structural parameters of nanoaggregate samples.

Number	1#	2#	3#	4#	5#	6#
Thickness of the absorbed layer (μm)	4.78	7.34	8.37	4.83	5.07	5.56
Filling factor (%)	6.1	5.1	4.1	5.9	6.3	6.8
Substrate thickness and material		500 μm Si			300 nm PI + 500 μm Si	

Energy dispersive spectroscopy (EDS, Ultim Extreme, Oxford, UK) detection is performed on the sample and Si substrate to quantitatively analyze the oxide content in Al nano aggregates. Figure 11a shows the types (O, Al and Si) and the relative weight ratio (0.175:1:0.416) of the elements contained in the sample. According to the relative atomic mass of each element, the atomic ratio of O, Al and Si is calculated as 0.0109:0.037:0.0149. Since EDS detection is carried out in vacuum, the O element is derived from aluminum oxide and silicon oxide. The Al element is derived from aluminum and oxide in the nano aggregates, and the Si element is derived from silicon and oxide in the substrate. Figure 11b shows the types (O and Si) and the relative weight ratio (0.011:0.416) of the elements contained in the substrate. Results of the substrate are normalized based on the Si weight ratio in the sample (0.416). The atomic ratio of O and Si is calculated as 0.0007:0.0149. The O element in the substrate is derived only from silicon oxide. Therefore, it can be calculated that 93.6% of the O element in the sample is derived from aluminum oxide and 6.4% from silicon oxide. In addition, 81.5% of the Al element is derived from aluminum; 18.5% from aluminum oxide. The molecular ratio of aluminum to its oxide is calculated as 8.81:1.

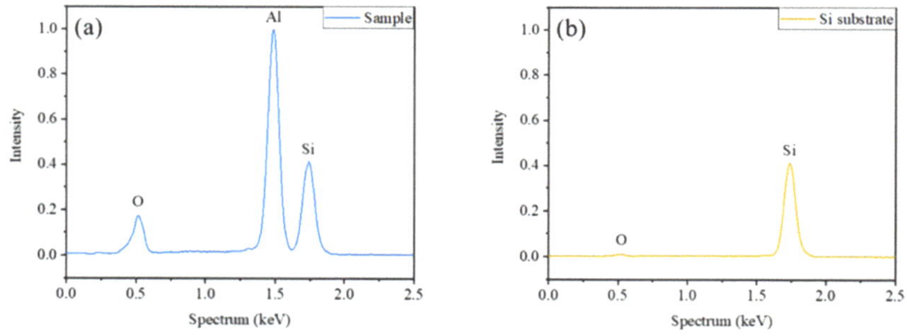

Figure 11. EDS results of the sample and Si substrate. (**a**) Results of the sample are normalized based on Al peak intensity, and the relative weight ratio of O, Al and Si is 0.175:1:0.416. (**b**) Results of Si substrate are normalized based on the Si weight ratio in the sample (0.416), and the relative weight ratio of O and Si is 0.011:0.416.

The spectral absorption curves of the samples in the 400–800 nm band were measured using a dual optical path ultraviolet–visible (UV–vis) spectrophotometer (TU-1901, Persee, Beijing, China). Two standard reflectance plates were used for correction. All the samples were supported by a 500 µm Si wafer, which was completely opaque in the visible range. The spectral absorptivities of the samples were calculated using Equation (7). The spectral absorption curves of the Si substrates, PI substrates, and samples are shown in Figure 12. The root mean square error (RMSE) between the measured and fitted curves was less than 0.7%, as shown in Table 3. The absorptivities of the samples were higher than 0.97 in the 400–800 nm band.

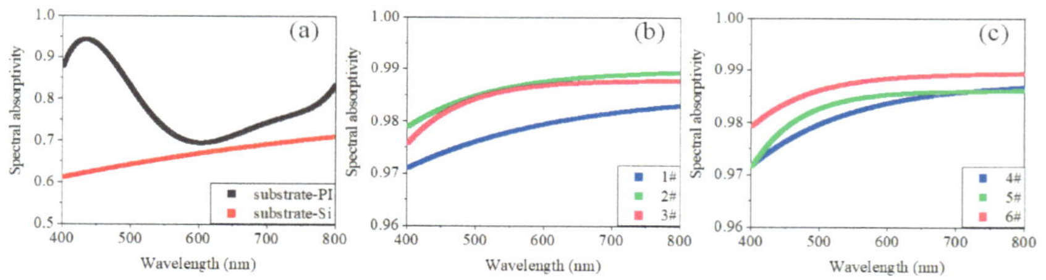

Figure 12. Measured spectral absorption curves of various substrates and nanoaggregate samples. (**a**) Measured spectral absorption curves of Si and PI substrates. (**b**) Measured spectral absorption curves of samples #1–3. (**c**) Measured spectral absorption curves of samples #4–6.

Table 3. Statistics of measured error.

Object	RMSE between Measured and Fitted Curves
Si	0.0044
1#	0.0008
2#	0.0001
3#	0.0007
PI	0.0071
4#	0.0002
5#	0.0007
6#	0.0001

3.2.2. Analysis of Calculation and Experimental Results

We constructed a physical model of the nanoaggregate samples based on the structural parameters listed in Table 2. The spectral reflectivity and absorptivity of the substrate used were included in the model as the boundary conditions. The weights of the three distribution functions in the aggregation model were adjusted to optimize the calculation results. The values of W_G, W_U, and W_C were set at 0.6, 0.2, and 0.2, respectively. A comparison of the calculation and experimental results is shown in Figure 13. The Spearman correlation coefficient was used to evaluate the correlation between the calculation and measurement curves; the values obtained are listed in Table 4.

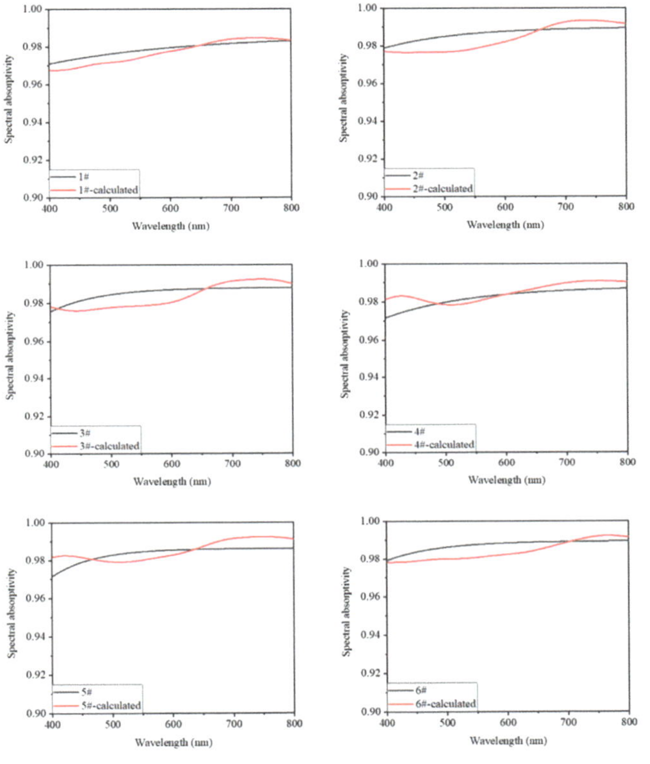

Figure 13. Spectral absorption curves determined based on weighted Gaussian, uniform, and Cauchy distribution models and those obtained experimentally.

Table 4. Comparison of calculation and measurement results.

Object	Spearman Correlation Coefficient of Spectral Absorption Curves	Relative Error of Visible Absorptivity
1#	0.976	0.0018
2#	0.954	0.0031
3#	0.962	0.0017
4#	0.846	0.0028
5#	0.822	0.0013
6#	0.996	0.0029

To evaluate the accuracy of the model described by Equation (9), we compared the calculated values of the visible-range absorptivity with the measured ones, as shown in Figure 14. The relative errors are presented in Table 4.

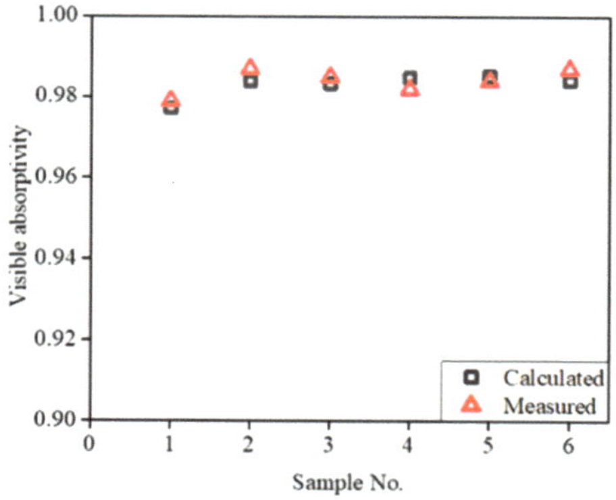

Figure 14. Calculated and measured values of visible-range absorptivity of various samples.

The calculated spectral absorption curves of the samples are in good agreement with the experimental curves, with the Spearman correlation coefficients being higher than 0.82. Within the range of values used for the structural parameters of the samples, the relative errors between the calculated visible-range absorptivities and those determined experimentally were less than 0.3%. This confirmed that the model was accurate.

3.3. Discussion

The model of nanoaggregate structure presented here is only a preliminary model, which is mainly established and verified for Al. Due to the limitation of the experimental conditions, the samples have a relatively small range of fill factor and thickness compared to the range of the presented model predictions. Therefore, the accuracy of the model is only verified in a relatively small range. We will study and improve the fabrication method to expand the range of sample thickness and filling factor in subsequent research. In addition, the model should be further improved to make it universal. The general form of nanoaggregate structure can be applied to metal absorbers obtained by various processes. This model is suitable for the study of surface absorption of various optical sensors, photothermal effect and photovoltaic, as well as terahertz generation and detection.

4. Conclusions

In this study, the structures of randomly distributed nanochain aggregates on silicon substrates were modeled, and the model was evaluated. The relationship between the structural parameters (thickness and filling factor), equivalent physical parameters (density, heat capacity, and thermal conductivity), and visible-range absorptivity of the model were established and analyzed. All the above-mentioned parameters exhibited extreme values, which enhanced the visible-range absorptivity. The accuracy of the model was verified experimentally. The following conclusions were drawn based on the results obtained.

(1) The visible-range absorptivity of the structure was quadratically related to the filling factor. The filling factor affects the motion path of incident photons and, thus, the efficiency of capturing photons. An extremely sparse or dense distribution of nanochains leads to a reduction in the model absorptivity.

(2) The visible-range absorptivity of the modeled structure is exponentially related to its thickness. The critical thickness of the Gaussian model is much smaller than those of the uniform and Cauchy models because of the higher photon-capturing efficiency of the former. The maximum visible-range absorptivities of the three models are similar,

indicating that the aggregate form of the nanochains has a negligible effect on the surface reflection of the model.

(3) The visible-range absorptivity of the modeled structure is quadratically related to its equivalent density. The Al nanochain aggregate structure is more suitable as an optical absorber that exhibits a low density and high visible-range absorption.

(4) The visible-range absorptivity of the modeled structure is also related to its equivalent thermal conductivity. The Au nanochain aggregate structure allows for efficient optical absorption and fast heat dissipation. Meanwhile, the Cr nanoaggregate structure allows for efficient energy capture and accumulation.

(5) Finally, the visible-range absorptivity of the modeled structure is quadratically related to its equivalent volumetric heat capacity. The Al nanoaggregate structure is more suitable as an optical absorber with a low volumetric heat capacity and high-sensitivity time-domain temperature response.

Actual nanochain aggregate samples were fabricated by reducing the kinetic energy of the deposited Al nanoparticle clusters. The visible-range spectral absorption curves of the fabricated samples were measured using a Fourier spectrometer. The Spearman correlation coefficients for the calculated spectral absorption curves and those measured experimentally were higher than 0.82; this confirmed the accuracy of the model. In addition, the relative errors between the calculated visible-range absorptivities and the measured values were less than 0.3%; this confirmed that the model is suitable for calculating the absorptivity.

Author Contributions: Conceptualization, Z.L.; Funding acquisition, Y.G.; Investigation, T.Z. and Q.S.; Methodology, T.Z. and Y.G.; Project administration, Y.G. and R.S.; Resources, Y.G. and R.S.; Software, T.Z.; Validation, T.Z. and Q.S.; Writing—original draft, T.Z.; Writing—review and editing, Y.G. and Z.L. All authors have read and agreed to the published version of the manuscript.

Funding: This research was funded by China Postdoctoral Science Foundation, grant number 2020TQ0036.

Institutional Review Board Statement: Not applicable.

Informed Consent Statement: Not applicable.

Data Availability Statement: Not applicable.

Acknowledgments: The authors thank Yao Zhaozhao for her help with this study.

Conflicts of Interest: The authors declare no conflict of interest.

References

1. More-Chevalier, J.; Yudin, P.V.; Cibert, C.; Bednyakov, P.; Lančok, J. Black aluminum-coated Pt/Pb (Zr0.56Ti0.44) O3/Pt thin film structures for pyroelectric energy harvesting from a light source. *J. Appl. Phys.* **2019**, *126*, 214501. [CrossRef]
2. Li, N.; Yang, D.J.; Shao, Y.; Liu, Y.; Tang, J.; Yang, L.; Sun, T.; Zhou, W.; Liu, H.; Xue, G. Nanostructured Black Aluminum Prepared by Laser Direct Writing as a High-Performance Plasmonic Absorber for Photothermal/ Electric Conversion. *ACS Appl. Mater. Interfaces* **2021**, *13*, 4305–4315. [CrossRef]
3. Kwon, N.; Oh, H.; Kim, R.; Sinha, A.; Kim, J.; Shin, J.; Chon, J.W.M.; Lim, B. Direct Chemical Synthesis of Plasmonic Black Colloidal Gold Superparticles with Broadband Absorption Properties. *Nano. Lett.* **2018**, *18*, 5927–5932. [CrossRef] [PubMed]
4. Georgieva, E.; Priestley, K.J.; Dunn, B.; Cageao, R.; Barki, A.; Osmundsen, J.; Turczynski, C.; Abedin, N. Radiation budget instrument (RBI) for JPSS-2. In Proceedings of the Conference on Characterization and Radiometric Calibration for Remote Sensing, Logan, UT, USA, 24 August 2015.
5. Xu, C.; Liu, D.; Zhou, L.; Shi, Q.; Gao, Y.; Wang, X.; Li, Z. Thermodynamics characteristics of MEMS infrared thin film. *Opt. Express.* **2019**, *27*, 32779–32788. [CrossRef] [PubMed]
6. Wang, X.; Li, D.F.; Li, Z.; Yang, S.H.; Zhang, J.Y.; Zhao, Q. Performance improvement of an infrared scene generation chip by in-plane microstructures. *Opt. Express.* **2020**, *28*, 26807–26822. [CrossRef] [PubMed]
7. Arinze, E.; Qiu, B.; Nyirjesy, G.; Thon, S.M. Plasmonic Nanoparticle Enhancement of Solution-Processed Solar Cells: Practical Limits and Opportunities. *ACS. Photonics* **2016**, *3*, 158–173. [CrossRef]
8. Liu, C.; Zhang, D.; Liu, Y.; Wu, D.; Chen, L.; Ma, R.; Yu, Z.; Yu, L.; Ye, H. Numerical Study of an Efficient Solar Absorber Consisting of Metal Nanoparticles. *Nanoscale. Res. Lett.* **2017**, *12*, 601. [CrossRef]
9. Mascaretti, L.; Schirato, A.; Fornasiero, P.; Boltasseva, A.; Shalaev, V.M.; Alabastri, A.; Naldoni, A. Challenges and prospects of plasmonic metasurfaces for photothermal catalysis. *Nanophotonics* **2022**, *11*, 3035–3056. [CrossRef]

10. Giorgianni, F.; Vicario, C.; Shalaby, M.; Tenuzzo, L.D.; Marcelli, A.; Zhang, T.; Zhao, K.; Chen, Y.; Hauri, C.; Lupi, S. High-Efficiency and Low Distortion Photoacoustic Effect in 3D Graphene Sponge. *Adv. Funct. Mater.* **2017**, *28*, 1702652. [CrossRef]
11. Morales-Luna, G.; Morales-Luna, M. Effective medium theory to the description of plasmonic resonances: Role of Au and Ti nanoparticles embedded in MoO3 thin films. *Sci. Rep.* **2020**, *10*, 1–12. [CrossRef]
12. Schmitt, A.; Odriozola, G.; Moncho-Jordá, A.; Callejas-Fernández, J.; Martínez-García, R.; Hidalgo-Álvarez, R. Multiple contact kernel for diffusionlike aggregation. *Phys. Rev. E* **2000**, *62*, 8335–8343. [CrossRef]
13. Munir, N.B.; Mahan, J.R.; Priestley, K.J. First-principle model for the directional spectral absorptivity of gold-black in the near infrared. *J. Opt. Soc. Am. A* **2019**, *36*, 1675–1689. [CrossRef] [PubMed]
14. Witten, J.T.A.; Sander, L.M. Diffusion-Limited Aggregation, a Kinetic Critical Phenomenon. *Phys. Rev. Lett.* **1981**, *47*, 1400–1403. [CrossRef]
15. Jensen, P.; Barabási, A.L.; Larralde, H.; Havlin, S.; Stanley, H.E. Connectivity of diffusing particles continually deposited on a surface: Relation to LECBD experiments. *Physica A* **1994**, *207*, 219–277. [CrossRef]
16. Jensen, P.; Barabási, A.L.; Larralde, H.; Havlin, S.; Stanley, H.E. Controlling nanostructures. *Nature* **1994**, *368*, 22. [CrossRef]
17. Atteia, F.; Rouzo, J.; Denaix, L.; Duché, D.; Berginc, G.; Simon, J.J.; Escoubas, L. Morphologies and optical properties of black silicon by room temperature reactive ion etching. *Mater. Res. Bull.* **2020**, *131*, 110973. [CrossRef]
18. Chen, C.; Youngblood, N.; Peng, R.; Yoo, D.; Mohr, A.D.; Johnson, W.T.; Oh, S.H.; Li, M. Three-Dimensional Integration of Black Phosphorus Photodetector with Silicon Photonics and Nanoplasmonics. *Nano. Lett.* **2017**, *17*, 985–991. [CrossRef]
19. Ryu, Y.; Kim, C.; Ahn, J.; Urbas, A.M.; Park, W.; Kim, K. Material-Versatile Ultrabroadband Light Absorber with Self-Aggregated Multiscale Funnel Structures. *ACS. Appl. Mater. Interfaces* **2018**, *10*, 29884–29892. [CrossRef]
20. Deng, G.; Dereshgi, S.A.; Song, X.; Aydin, K. Polarization dependent, plasmon-enhanced infrared transmission through gold nanoslits on monolayer black phosphorus. *J. Opt. Soc. Am. B. Opt. Phys* **2019**, *36*, F109–F116. [CrossRef]
21. Palik, E.D. *Handbook of Optical Constants of Solids*; Academic Press: New York, NY, USA, 1998.
22. Boomsma, K.; Poulikakos, D. On the effective thermal conductivity of a three-dimensionally structured fluid-saturated metal foam. *Int. J. Heat. Mass. Transf.* **2001**, *44*, 827–836. [CrossRef]
23. Xin, G.Z.; Chen, D.Q.; Cai, Y.; Huang, Y.; Wang, L.; Bai, T.Z.; Wang, L.X. Direct observation of widely tunable mid-infrared emission of graphene foam induced by modulated laser diode light. *Carbon* **2021**, *179*, 486–492. [CrossRef]

materials

MDPI

Article

Effect of the Addition of Steel Fibers on the Bonding Interface and Tensile Properties of Explosion-Welded 2A12 Aluminum Alloy and SS-304 Steel

Yao Chen [1], Yonghong Gao [1,*], Chuanxiang Guo [1], Yanping Guo [2], Zhijun Guo [3], Yingbin Liu [1] and Tiansheng Liu [1]

1 School of Environment and Safety Engineering, North University of China, Taiyuan 030051, China
2 Department of Environmental and Safety Engineering, Taiyuan Institute of Technology, Taiyuan 030008, China
3 Shanxi Fenxi Heavy Industry Co., Ltd., Taiyuan 030027, China
* Correspondence: gyh54gyh@163.com

Abstract: First of all, the explosion-welding method was adopted to prepare steel fiber-reinforced steel-aluminum composite plates. Secondly, the smooth particle hydrodynamic (SPH) method was used to investigate the effect of introducing steel fibers to a vortex region created at the bonding interface of the steel-aluminum composite plate. Thirdly, the following conclusions were drawn through an analysis of the vortex region with the assistance of scanning electron microscopy and energy-dispersive X-ray spectroscopy. A brittle intermetallic compound FeAl was produced in the vortex region in an environment characterized by high temperature, high pressure, and high strain rate, resulting in cracks, holes and pores. In addition, the hardness of the vortex area was less than the estimated value, which is mainly because the main element in the vortex area was 2A12 aluminum with low hardness, and there were cracks, holes, pores and other defects that caused hardness reduction. Although the addition of steel fibers caused defects at the bond interface, the addition of steel fibers was effective in improving the tensile resistance performance of steel-aluminum composite panels to a certain extent. In addition, the larger the fiber diameter, the more significant the increase in tensile resistance.

Keywords: steel fiber; explosive welding; numerical simulation; bonding interface

Citation: Chen, Y.; Gao, Y.; Guo, C.; Guo, Y.; Guo, Z.; Liu, Y.; Liu, T. Effect of the Addition of Steel Fibers on the Bonding Interface and Tensile Properties of Explosion-Welded 2A12 Aluminum Alloy and SS-304 Steel. *Materials* **2023**, *16*, 116. https://doi.org/10.3390/ma16010116

Academic Editor: Matthias Bönisch

Received: 12 November 2022
Revised: 14 December 2022
Accepted: 20 December 2022
Published: 22 December 2022

1. Introduction

With the rapid development of industry, more and more regions need to produce special equipment by machining high-quality metals into sheets with excellent anticorrosion, antioxidation, and mechanical properties. If all these devices are made of high-quality metals, their costs will be inevitably high, resulting in unnecessary waste. Therefore, replacing high-quality metals with bimetallic materials is the most ideal solution. SS-304 stainless steel is characterized by high strength, impact resistance, reliable performance, and good welding and riveting performance, but it has a lot of shortcomings, such as high maintenance cost, heaviness, and large production energy consumption. However, 2A12 aluminum material has a lot of advantages, including lightness, good electrical conductivity, ease of extension, corrosion resistance, and few toxic and side effects. Bimetallic materials have the advantages of two metal components, so the application of steel-aluminum composite plates can reduce the mass of SS-304 stainless steel and improve its corrosion resistance. Explosive welding is a solid-phase welding method, which can be used for the welding of double-layer or multilayer composite plates of the same or different metals [1–6]. Compared with traditional diffusion welding [7,8], light welding [9,10], magnetic pulse welding [11,12], and hot rolling welding [13,14], the explosive welding method still has very good welding quality in the case of large differences in the points or mechanical properties of the welded metal, so it is extensively used in metal welding. Because of its high tensile strength, low price, and outstanding corrosion resistance, steel fiber is widely used to reinforce concrete and improve its protective performance. In this paper, explosive welding was used to produce a new fiber-reinforced steel aluminum composite plate by combining the advantages of SS-304 stainless steel, 2A12 aluminum, and steel fiber.

Extensive studies have been conducted on steel-aluminum composite plates, with outstanding achievements. In 1983, Kotov V A studied [15] unidirectional fiber-reinforced composites composed of steel wires based on Amr6 aluminum alloy, and the results revealed that under uniaxial or biaxial loads, the strength of the fiber-reinforced composites was mostly dependent on the bonding strength of the wire and the base plate, as well as the internal geometric installation, and when the load was along the wire, the effect was relatively obvious. Therefore, fiber-reinforced composites and tubular products prepared by explosive welding have sufficient strength under various loads. Zhou et al. [16] successfully prepared steel fiber-reinforced composite plates and used numerical simulation and experimental methods to study the impact of the addition of steel fiber on the antipenetration performance of the composite plate. It was concluded that the addition of the fiber-reinforced phase was improved the antipenetration performance of the composite target plate, and the reduced fiber distribution spacing and orthogonal arrangement distribution were helpful in improving the antipenetration performance of the target plate Wang et al. [17] applied numerical simulation and experimental methods to study the impact of carbon fiber-reinforced polymer laminates under high-speed impact. The results showed that within a certain impact velocity range, carbon fiber composite laminates had the advantage of replacing metal plates to resist high-speed impact. Mahfuz et al. [18] comprehensively studied the antipenetration performance of multilayer ceramic–rubber–glass fiber composite target plates against high-velocity projectiles by numerical simulation and hydrogen gun experiments, and discussed the failure mode of the target plate at ballistic limit velocity. The addition of fiber can significantly increase the difficulty of explosive welding, so the experiment needed to calculate the dynamic parameters of explosive welding in advance to guarantee the rationality of the parameters.

The effects of steel fiber with different diameters on the interface and mechanical properties of steel-aluminum composite sheets were studied. In the early stage, the SPH method in Autodyn software was used to study the influence of high-temperature and high-pressure environments on steel fiber and composite interface to simulate the explosive welding experiment. Later, scanning electron microscopy and energy-dispersive spectroscopy were adopted to systematically study the changes in the bonding interface and the formation of intermetallic compounds after adding steel fibers. The effect of the addition of steel fiber on the hardness of the base plate was further studied by microhardness tests. Finally, a universal testing machine was used to verify the impact of the addition of steel fibers on the tensile properties of steel-aluminum composite plates.

2. Materials and Methods

Ansys/Autodyn software was used to carry out simulation experiments on SS-304 and 2A12 aluminum composite plates prepared by explosive welding. The SPH method can extensively simulate large deformation problems such as disintegration, fragmentation, solid spalling, and brittle fracture of continuum structures and avoid algorithm coupling, so it is very suitable for numerical simulation of explosive welding of multilayer metal plates [19,20]. Therefore, the SPH method can simulate the interaction of jet particles in the preparation of steel fiber-reinforced steel-aluminum composite plates by explosive welding and provide reasonable and effective experimental parameters for the following explosive welding experiment. As the contrast between the plate thickness and length width was large, a two-dimensional plane model was constructed. To save the calculation time of the model, the length of the base plate, aluminum composite plate, and explosive was all set to 40 mm. The particle size was set to 10 μm, and a total of 1.1815 million particles were generated to obtain accurate numerical results. The specific model is shown in Figure 1, where the density of the ANFO explosive is 0.6821 g/cm^3, the explosive height is 20 mm, and the detonation velocity is 2430 m/s. The specific parameters of ANFO are shown in Table 1.

Figure 1. Explosive welding model.

Table 1. Performance parameters of ANFO.

Material	Density ρ/(g cm^{-3})	Detonation Velocity D/(m s^{-1})	Specific Internal Energy E0/(kg cm^{-3})	Detonation Pressure /MPa	Heat Capacity Ratio
ANFO	0.681	2430	2.484	1160	2.5

Experimental Procedure

Explosive welding mainly includes: directly in the atmosphere, in water, and in rough vacuum. Compared with explosive welding experiments in low vacuum and water, explosive welding experiments in the air have a simpler assembly process, use lower detonation velocity of explosives, and are safer and more efficient. Therefore, in this paper, explosive welding experiments in the air [21–23] were selected. As shown in Figure 2, taking the SS-304 stainless steel with a size of 200 mm × 300 mm × 1 mm as the base plate, and the 2A12 aluminum plate with a size of 200 mm × 300 mm × 1 mm as the flyer plate, a steel fiber was evenly wound on the base plate every 5 mm to ensure that the length of the steel fiber was the same as that of the base plate, and the standoff distance between the base plate and cladding plate was 4 mm. Then, the base plate, flyer plate, and steel fiber were welded together by explosive welding. The performance parameters of steel and aluminum are shown in Table 2. S20910 is the material parameter of steel fiber and SS 304 is the material parameter of the base plate. In this experiment, the diameter of the steel fibers needs to be strictly controlled and the fibers are exposed to extrusion and extreme thermomechanical conditions in the welding process. The mechanical properties of the fibers therefore need to be good and the hot drawing process produces steel fibers of a more uniform size and high precision. Because the working temperature is above the recrystallization temperature, the resulting process hardening phenomenon is eliminated by recrystallization, the internal stress of the steel fibers is also eliminated, the toughness and plasticity will be better, so that the welded composite plate can better reflect the impact of the fibers on the composite plate. While cold rolled out of the steel plate as the base plate of the explosion welding, it ensures the strength and hardness of the base plate, in line with the experimental design. SS 304 steel as the base plate ensures the strength and hardness of the base plate, in line with the experimental design. The abovementioned metal materials were purchased by the teachers of the research group in Shenzhen Hongwang Mold Co., Ltd., Guangzhou, China.

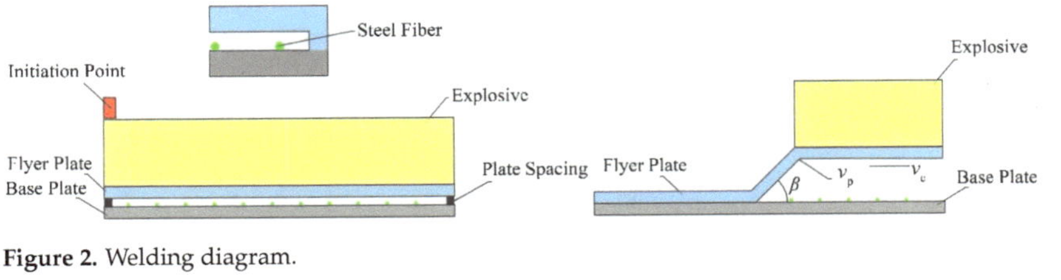

Figure 2. Welding diagram.

Table 2. Performance parameters of metal materials [24,25].

Material	Density ρ/(g·cm^{-3})	Yield Strength σ_b/MPa	Vickers Hardness /Hv	Wave Velocity C_0/(m·s^{-1})	Melting Point T/K	Specific Heat C_p /(J·kg^{-1}·K^{-1})	Thermal Conductivity K/(W·m^{-1}·K^{-1})
2A12 aluminum	2.79	354	110	5328	933	940	237
S20910 steel	7.88	380	201	4569	1450	476	14
SS-304 steel	7.93	515	200	5790	1400	500	21.5

Explosive welding aims to connect weldments by using the impact force generated by the explosion to cause the rapid collision of the weldments. The three most critical parameters are the impact velocity of the composite plate v_p, the collision angle β, and the moving speed of the collision point v_c. The moving speed of the collision point is equal to the detonation velocity of the explosive. Using 2# rock ammonium nitrate explosive, the parallel explosive welding experiment was used, satisfying the following relationship [26]:

$$R = \frac{\rho_e h_e}{\rho_f h_f} \tag{1}$$

$$R = \frac{C}{m} \tag{2}$$

$$M = C \cdot S_e \tag{3}$$

$$v_p = \sqrt{2E}\left[\frac{(1+2/R)^3+1}{6(1+1/R)} + \frac{1}{R}\right]^{-\frac{1}{2}} \tag{4}$$

$$\sqrt{2E} = \frac{v_d}{3.08} \tag{5}$$

$$v_p = 2v_d \sin\frac{\beta}{2} \tag{6}$$

where v_p is the impact velocity; $2E$ is the Gurney energy; m is the mass of the flyer plate; R is the explosion ratio; ρ_e is the density of the explosive; ρ_f is the density of the flyer plate; h_e is the height of the explosive; h_f is the thickness of the flyer plate; C is the explosive mass per unit area; S_e is the explosive area, which is the same as the surface area of the flyer plate; M is the explosive payload; v_d = v_c equals the detonation velocity of the explosive. Therefore, the detonation velocity is 2400 m/s [27]. v_p = 668 m/s, β = 17.5°.

3. Results and Discussion

3.1. Interface Evolution Mechanism

The bonding of explosive welding is divided into three categories: (1) direct bonding of metals; (2) forming a uniform and continuous melting layer; and (3) undulate bonding, which is the most common form. In undulate bonding, because the molten material at the bonding interface is retained in the vortex and is periodically and discontinuously separated, when there is an external load, the microcrack source generated by the interface defects in the melting tank is not easily propagated. Therefore, the most ideal method is tiny undulate bonding. Many experiments and studies have shown that to realize high-quality explosive welding, the following three requirements should be met: (1) undulate bonding can be obtained under certain collision conditions, that is, the impact velocity v_p and the impact angle β meet the explosive welding window; (2) there must be jet formation when welding, so that the interface can be self-cleaning, exposing the fresh surface; and (3) a fine and uniform corrugated interface or a flat interface with sufficient strength is formed. Finally, to better understand this problem, SPH simulation was introduced to study the detailed evolution process of the steel-aluminum interface.

Figure 3a–c shows the evolution mechanism of the influence of steel fibers on jet particles. As shown in Figure 3a, during the welding process, the incident jet consisted of two layers on the surface. These particles ejected from the flyer plate moved obliquely downwards and acted on the base plate surface, thereby compressing the base plate, causing it to form a depression, forming a corresponding bulge on the base plate surface, and generating a forward jet. The incident jet could also remove oxides and other impurities on the surface, and help to establish ideal welding conditions under the circumstance of original clean contact [28]. However, after the detonation of the explosive, the explosive product formed a high-voltage pulse load, which directly acted on the flyer plate. Then, the flyer plate accelerated and reached a speed of several hundred meters/second in a few microseconds, starting from the starting end, collided with the base plate in turn, and formed a certain angle. At this time, the fixed included angle formed between the flyer plate, the steel fiber, and the base plate prevented the jet from continuing to act on the surface of the base plate, so that a large number of particles converged on the side of the steel fiber, as shown in Figure 3b. As the steel fiber spacing was 5 mm, when the jet was blocked, the new jet continued to generate. According to Figure 3c, the process of the jet from being blocked to being regenerated was repeated. Figure 3d is a schematic diagram for simulating jet particles without steel fibers. By comparing with Figure 3a,c, both the number of jet particles and the distance of propagation in Figure 3d were much larger than those in Figure 3a,c. It should be emphasized that Yang et al. [29] studied the evolution mechanism of the Ag–Fe welding interface by SPH simulation. The results showed that the movement direction of these jet particles was the main reason for the formation of the undulating interface, which explained why the undulating structure formed at the interface of the steel-aluminum composite plate with a steel fiber diameter of 0.5 mm in Figure 4a was more uniform than that without steel fibers in Figure 4b, and the wavelength and wave height in Figure 4b were greater than those in Figure 4a.

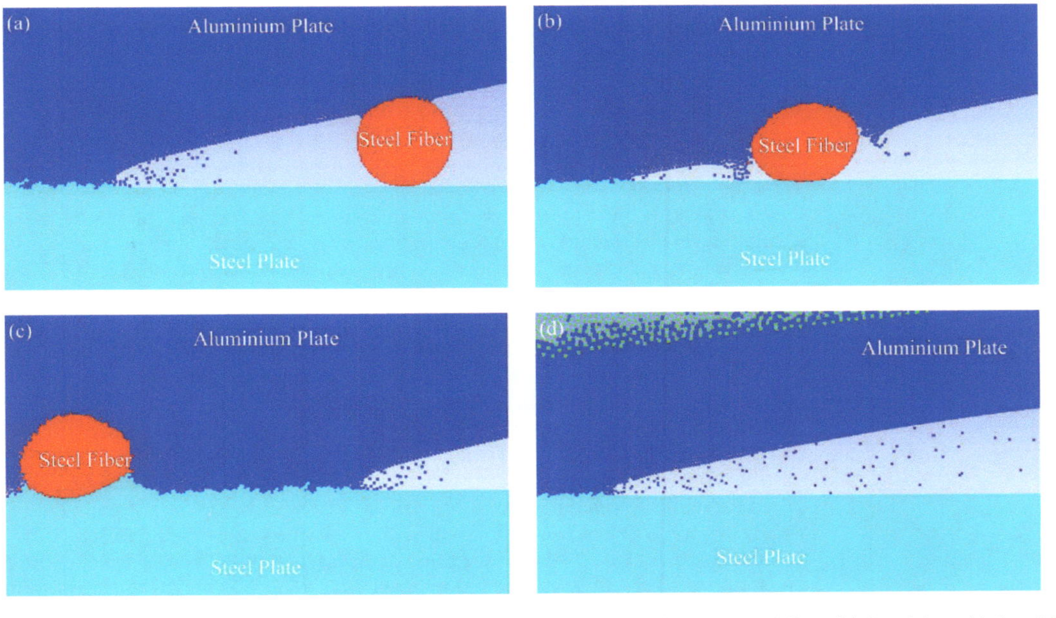

Figure 3. Evolution mechanism of welding interface: (**a**) time 3.5 μs; (**b**) time 4.0 μs; (**c**) time 4.5 μs; (**d**) time 4.0 μs.

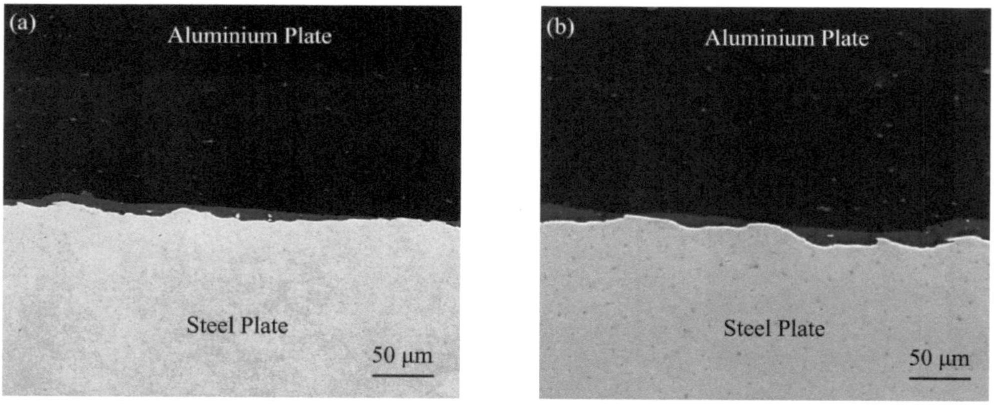

Figure 4. SEM images of explosion welded interfaces: (**a**) for steel fiber diameter 0.5 mm; (**b**) without steel fiber.

The thermodynamic state of the explosive welding process was simulated by the SPH method to better understand the interface evolution. Figure 5 shows the cross-sections of steel fibers with diameters of 0.5 mm, 0.35 mm, and 0.25 mm under an optical microscope In the figure, three groups of steel fibers all showed a deformation to varying degrees, but all maintained a complete fiber structure. Among them, the addition of steel fibers with a diameter of 0.5 mm produced the most obvious vortex area and cracks, so it was taken as an example to analyze the impact of the addition of steel fibers on the bonding interface of the composite plate by combining with numerical simulation. Figure 6 is a schematic diagram of the thermodynamic state during the explosive welding process. As shown in Figure 6a, when the jet particles filled the unilateral side of the steel fiber, a high-temperature and high-pressure environment was formed, with a pressure of up to 10 GPa. For the three materials, the pressure was much greater than the yield strength of 235 MPa, so the steel fiber directly contacting the jet particles would behave as a fluid and undergo a strong plastic deformation. Figure 6b shows the temperature field. A high-temperature region was constituted by the molten jet particles on the steel fiber side, and combined with the uneven thin melt formed by the molten jet particles at the interface, a lot of heat accumulated in these two areas because time was not enough to spread heat. Figure 6c shows the maximum plastic deformation on one side of the steel fiber, where the maximum deformation occurred on the steel fiber side blocking the jet advance, with a maximum strain rate of 5. Figure 6d shows that the ultrahigh strain rate occurred at the steel fiber surface and the welding interface was up to 900/s. To sum up, the explosive welding process occurred in an extreme environment with high temperature, high pressure, large strain, and superhigh strain rate, and this extreme thermodynamic state was closely related to the bonding interface. These high-temperature molten particles driven by the kinetic energy of the composite plate were correlated with the combined effects of the two materials, including plastic deformation, friction, shear, and stirring [30,31]. Since the time was not enough to reduce the large-scale heat, these molten particles could not be rapidly solidified. Then, the residual velocity forced the molten particles on one side of the steel fiber to stir and mix strongly, and finally a vortex region with cracks was formed, as shown in the yellow region in Figure 7b. Yang, Zhang, Bazarnik et al. [32–34] proved that the formation of the vortex structure might be due to the good ductility and high density of steel. Figure 7a shows the cross-section of the steel fiber-reinforced steel-aluminum composite plate with a steel fiber diameter of 0.5 mm. For the shape and cracks in the vortex region in the figure, by comparing it to Figure 7b, it was found that the numerical simulation results were in good agreement with the experimental observations.

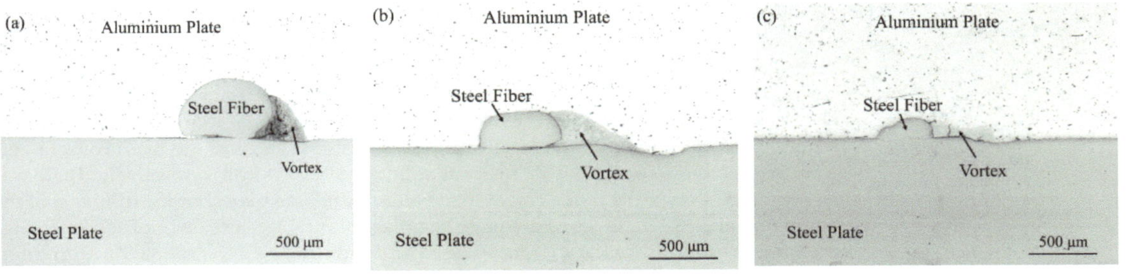

Figure 5. Schematic cross-section of steel fibers with different diameters: (**a**) steel fiber diameter: 0.5 mm; (**b**) steel fiber diameter: 0.35 mm; (**c**) steel fiber diameter: 0.25 mm.

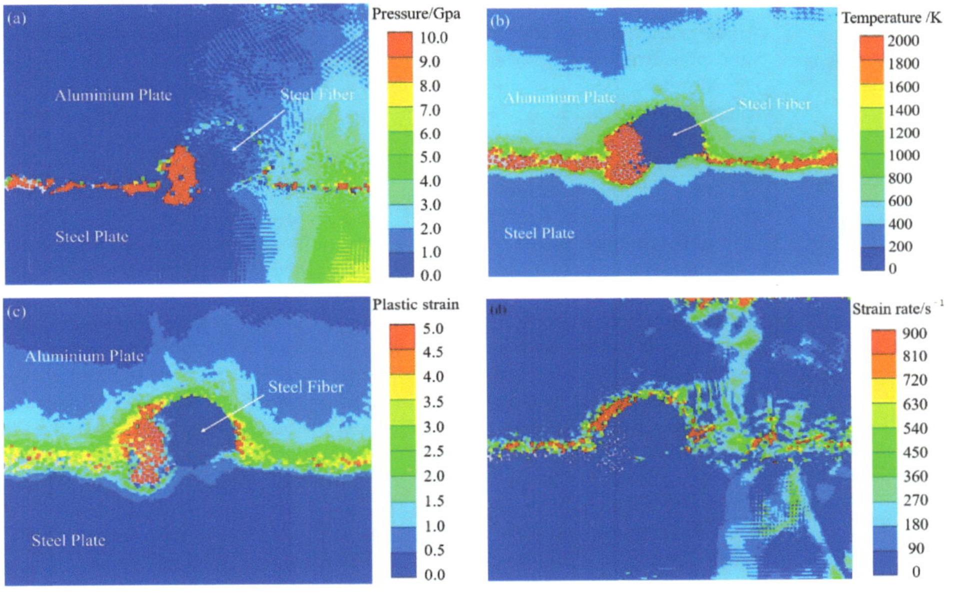

Figure 6. Thermodynamic state of bonding interface during explosive welding: (**a**) pressure; (**b**) temperature; (**c**) plastic strain; (**d**) strain rate.

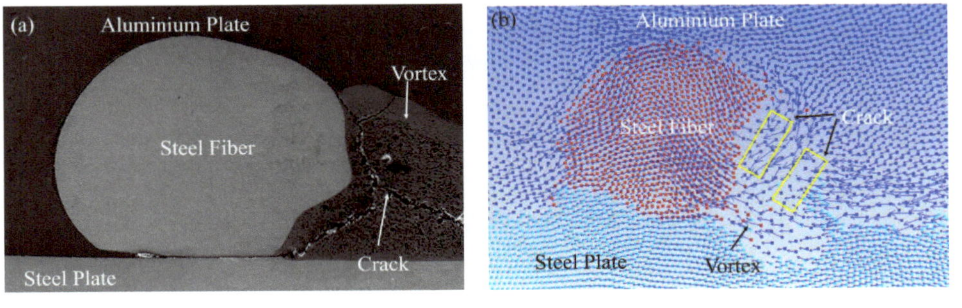

Figure 7. SEM and simulation results of vortex region: (**a**) SEM of steel fiber with a diameter of 0.5 mm; (**b**) numerical simulation of steel fiber diameter of 0.5 mm.

3.2. Energy Spectrum Analysis

To further analyze the distribution of elements in the unilateral vortex area of steel fiber, 8 points were selected for EDS point scanning in this area, as shown in Figure 8. The percentage of iron and aluminum is shown in Table 3. The first point and the second points were in the iron base plate, and the main components were 66.2% iron and 64.8% iron. The seventh and eighth points were in the aluminum base plate, and the main components were 88.6% aluminum, 1.3% iron, 86.9% aluminum, and 0.9% iron, respectively. There was element diffusion of the iron element in the aluminum base plate, but the diffusion of the aluminum element in the iron base plate near the interface was not obvious. The elements measured near the iron base plate were 55% aluminum, 14.4% iron, 32.3% aluminum, and 10% iron, while the elements near points 5 and 6 were 71.8% aluminum, 9.7% iron, 73.9% aluminum, and 9.3% iron. The results revealed that the elements in the vortex region were mainly aluminum and iron, and the closer to the middle region of the vortex, the higher the iron content. The main reason is that under the circumstance of high temperature and high pressure, the molten iron and molten aluminum would occur in a mutual melting phenomenon to form an iron aluminum alloy, and with the gradual decrease in temperature and pressure, the proportion of iron and aluminum in the alloy gradually changed until the aluminum and iron separated. However, because the cooling rate was too fast, the iron element in the middle of the molten region had been solidified before it was separated, so the content of iron element near the bonding interface was less than that in the middle of the vortex, which also indirectly confirmed the research results obtained by Zeng [35] and Zhang et al. [36]: intermetallic compounds were mainly formed in the vortex region on the base plate side and rarely formed on other sides of the interface. Wu Tong et al. [37,38] found that a large number of brittle intermetallic compounds produced during explosive welding were the main causes of cracking at the bonding interface. These intermetallic compounds formed cracks, pores, and voids in the vortex region, resulting in a decrease in the mechanical properties of metal composites [39–41]. To explore the types of metal compounds in the vortex region, EDS scanning was performed on cracks, holes, and air pores in the vortex region, and the results are shown in Figure 9. The four groups of dotted lines in Figure 9 are the line scanning results corresponding to the SEM images of pores and air holes in the vortex region. By comparing the SEM and EDS results in the figure, it was found that the content of iron and aluminum in the EDS scanning results corresponding to cracks, holes, and air pores decreased at the same time, while other elements increased. Therefore, the metal compound formed in the vortex region might be FeAl.

To further explore the impact of the addition of steel fibers on the interface of the steel-aluminum composite plate prepared by explosive welding, EDS scanning was carried out on the interface of the steel-aluminum composite plate without steel fibers and the steel-aluminum composite plate with a steel fiber diameter of 0.5 mm, and the results are shown in Figure 10a,d, respectively, forming an obvious contrast. There was a gradual element change in Figure 10a, transiting from iron element to aluminum element. The mixing amount of elements in the transition zone reached about 500, and the content of aluminum and iron elements was 41.6% and 34.3%, respectively, as shown in Figure 10c. In Figure 10d, the transition from iron to aluminum was very fast. The mixing amount of elements in the transition zone was about 300, and the aluminum and iron contents were 64.2% and 17.5%, respectively, as shown in Figure 10f. The element transition in the group without steel fibers was more obvious and the mixing amount of the two elements in the transition zone was larger than in the steel fiber group. Studies have shown that the formation of the transition zone of chemical elements was caused by the strong stirring of jet particles, and the diffusion rate increased significantly with the increase in the defect density caused by plastic deformation [42–44]. Therefore, the simulation results and experimental results also simultaneously verified that the addition of steel fiber blocked the continuous action of jet particles on the base plate surface and affected the element transition at the bonding interface of the composite plate.

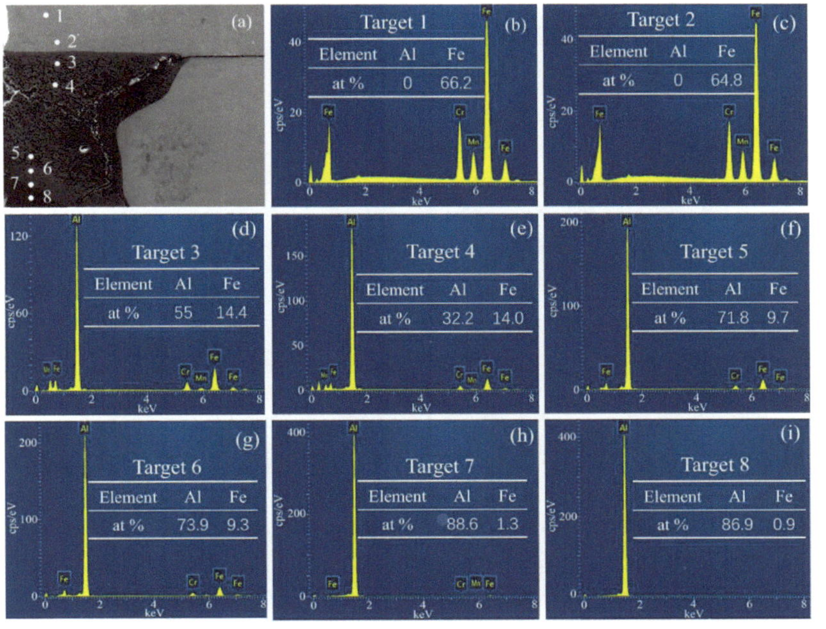

Figure 8. Position points and numerical results of welding interface analyzed by EDS in SEM image: (**a**) Schematic diagram of SEM positions of 8 points; (**b**) Scan result of point 1; (**c**) Scan result of point 2; (**d**) Scan result of point 3; (**e**) Scan result of point 4; (**f**) Scan result of point 5; (**g**) Scan result of point 6; (**h**) Scan result of point 7; (**i**) Scan result of point 8.

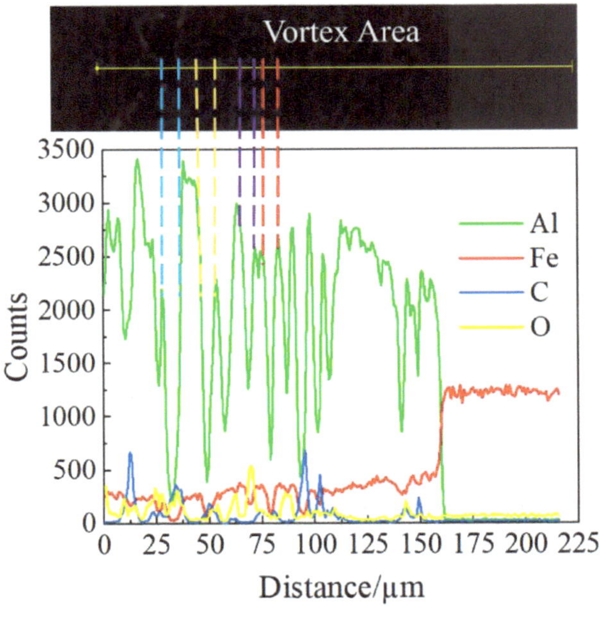

Figure 9. SEM scanning and EDS analysis of vortex, crack and hole.

Table 3. Iron and aluminum contents at 8 EDS points.

Element	1	2	3	4	5	6	7	8
iron	66.2%	64.8%	14.4%	10%	9.7%	9.3%	1.3%	0.9%
aluminum	0	0	55%	32.2%	71.8%	73.9%	88.6%	86.9%

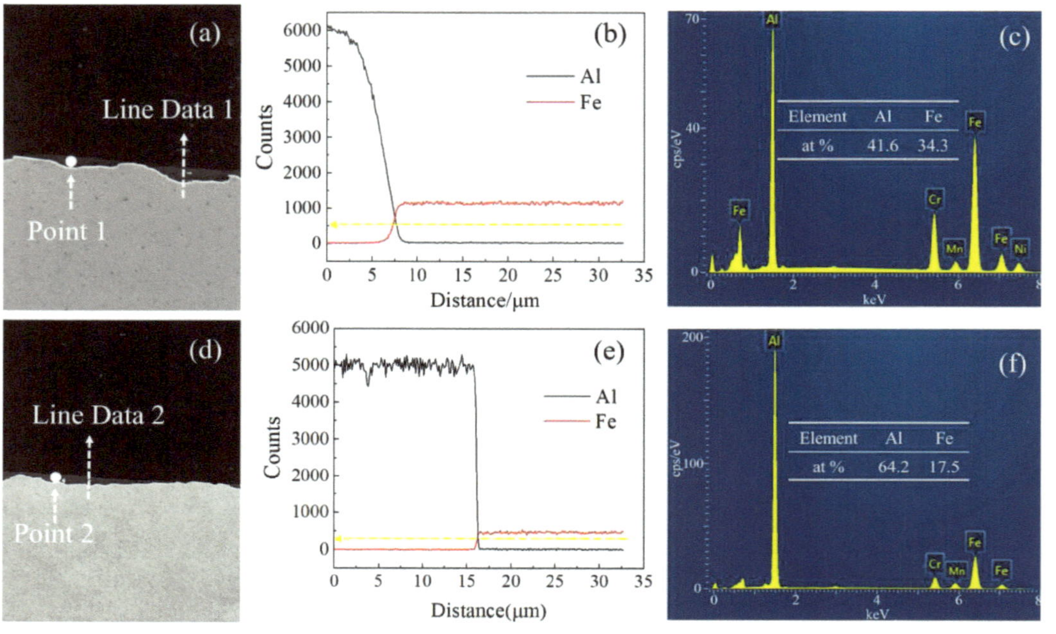

Figure 10. SEM and EDS results of the bonding interface of steel-aluminum composite plate: (**a**) no bonding interface of steel fiber; (**b**) figure a line scan results; (**c**) figure a point scan results; (**d**) bonding interface with steel fiber diameter of 0.5 mm; (**e**) figure d line scan results; (**f**) figure d point scan results.

3.3. Microhardness Analysis

To explore the micromechanical properties of the steel fiber-reinforced steel-aluminum composite plate, a microhardness test of the cross-section of the steel-aluminum composite plate with a steel fiber diameter of 0.5 mm was carried out. The corresponding test area was divided into five parts, as shown in Figure 11a. Among them, Zone 1 is the microhardness test area of the cross-section of the 2A12 aluminum, Zone 2 is the microhardness test area of the interface between the vortex region and 2A12 aluminum, Zone 3 is the microhardness test area of the vortex region, Zone 4 is the microhardness test area of the interface between vortex region and SS-304 stainless steel, and Zone 5 is the microhardness test area of SS-304 stainless steel. Figure 11b shows the numerical results of the microhardness values of the five zones in Figure 11a. The highest hardness was 419 Hv for the SS-304 stainless steel base plate, the lowest hardness was 126 Hv for 2A12 aluminum, the interface between the vortex region and 2A12 aluminum was 190 Hv, the interface between the vortex region and SS-304 stainless steel was 381 Hv, and the vortex region was 276 Hv. According to previous studies, the highest hardness of the composite plate prepared by explosive welding should be in the intermetallic compound layer [45,46], that is, the vortex region. However, the experimental data obtained in this paper deviated from the previous conclusions, and there might be two reasons for the decrease in the hardness in the vortex region. First of all, during the experimental process of explosive welding, a large number of high-temperature molten jet particles were strongly stirred and mixed on one side of the steel fiber to form the intermetallic compound FeAl. As the time was not enough to reduce the large-scale heat, these molten particles could not rapidly solidify, and the continuous stirring and mixing led to a large number of cracks, holes, and air pores in the vortex region, as shown in Figure 8, so the hardness of the vortex region decreased. Secondly, it can be seen from the EDS point scanning results of points 3 and 7 in Figure 8 that the interface between the vortex zone and the SS-304 stainless steel was composed of 14.4% SS-304 stainless steel and 55% aluminum, and the interface between the vortex zone and the 2A12 aluminum was composed of 1.3% SS-304 stainless steel and 88.6% 2A12 aluminum. According to the point scanning results of points 5 and 6, it can be seen that the vortex region was mainly composed of about 32% 2A12 aluminum and about 10% SS-304 stainless steel, while the microhardness of 2A12 aluminum was 126 Hv, so the microhardness of the vortex region decreased. The microhardness of SS-304 stainless steel in this test reached 419 Hv. The main reason for the significant increase in the microhardness of SS-304 stainless steel was that austenitic steel is one of the most obvious metal materials subjected to explosive strengthening. Therefore, explosive strengthening was more widely used in the strengthening process of austenitic steel than mechanical forging.

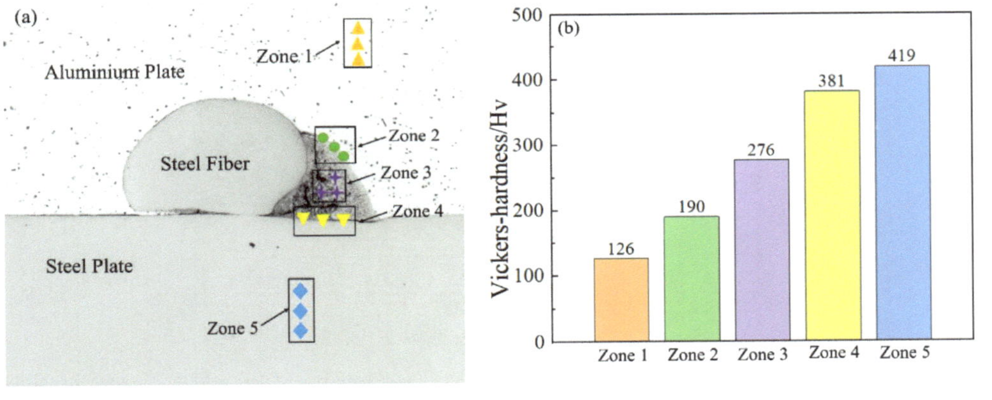

Figure 11. Microhardness at the interface of steel fiber-reinforced steel-aluminum composite plate: (**a**) microhardness test area; (**b**) microhardness test value.

3.4. Mechanical Property Testing

A 0.5 mm notch was opened on one side of the sample to measure the tensile strength of the sample, then the impact of the addition of steel fibers on the tensile properties of steel-aluminum composite plates prepared by explosive welding was explored. Three specimens were taken in the steel fiber parallel to the tensile load direction and the steel fiber perpendicular to the tensile load direction. The specific location and size parameters of the sample are shown in Figure 12. Then, the tensile test of the specimen was performed at room temperature by the UTM5000 series microcomputer-controlled electronic multipurpose testing machine at a loading rate of 2 mm/min. The accuracy of this series of equipment is class 0.5, and the error of test force indication of load parameters is within ±0.5% of the indicated value. Finally, the tensile strength of the specimen was calculated by the following formula [47].

$$\sigma = \frac{F}{S} \tag{7}$$

where σ is the tensile strength, F is the loading load, and S is the stressed area of the sample.

The samples in each group were from the same position in the center of four steel-aluminum composite plates prepared by explosive welding, aiming to avoid the influence of boundary effect on tensile samples. In Figure 13 there were no steel fibers in 1^- and 5^+

samples, and the fiber diameters in 2^- and 6^+, 3^- and 7^+, 4^- and 8^+ samples were 0.25 mm, 0.35 mm, and 0.5 mm, respectively. To ensure the accuracy of the experimental data, each sample of the steel fiber parallel to the tensile load direction contained four steel fibers.

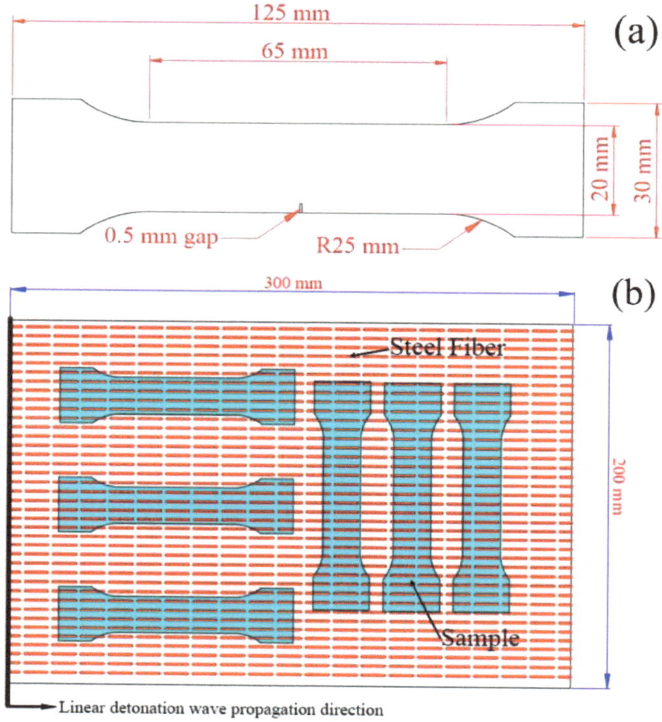

Figure 12. Tensile specimen parameters: (**a**) dimensional drawing of tensile specimen; (**b**) composite plate size, sampling position and linear detonation wave propagation direction diagram.

Table 4 shows the specific parameters of the tensile resistance of the steel fiber-reinforced steel-aluminum composite plate. "−" and "+" indicate that the steel fiber is parallel and perpendicular to the tensile load direction, respectively. In the direction of the steel fiber parallel to the tensile load, the average tensile strength of sample 1^- was 420.53 MPa. The average tensile strength of sample 2^- increased by 8.5% compared with sample 1^-, with the highest increase of 9.8% and the lowest increase of 7.0%; the average tensile strength of sample 3^- increased by 15.4% compared with sample 1^-, with the highest increase of 18.3% and the lowest increase of 13.8%. The average tensile strength of sample 4^- increased by 33.7% compared with sample 1^-, with the highest increase of 38.6% and the lowest increase of 29.9%. In the direction of the steel fiber perpendicular to the tensile load, the average tensile resistance of sample 5^+ was 435.48 MPa, and the average tensile resistance strength of sample 6^+ was 2.9% higher than that of sample 5^+, with the highest increase of 3.5% and the lowest increase of 2.2%. Compared with sample 5^+, the average tensile resistance strength of sample 7^+ increased by 6.5%, with the highest increase of 8.9% and the lowest increase of 2.2%. Compared with sample 5^+, the average tensile resistance strength of sample 8^+ increased by 11.2%, with the highest increase of 14.8% and the lowest increase of 8.4%. The main difference between samples 1^-, 2^-, 3^-, 4^- and samples 5^+, 6^+, 7^+, 8^+ was that the first four groups of samples were parallel to the direction of the tensile load and along the direction of the detonation wave propagation, while the last four groups were perpendicular to the direction of the tensile load and the detonation wave propagation. Among them, the sample groups 1^- and 5^+ were from the steel-aluminum composite plates without steel fibers. When the highest and lowest tensile resistance properties were removed from the two groups of data, they were 435.35 MPa and 438.93 MPa, respectively. The difference in the tensile resistance between the two groups could be neglected, so the influence of the propagation direction of the detonation wave on the tensile resistance was excluded. In summary, the addition of steel fibers could improve the tensile resistance of steel-aluminum composite plates prepared by explosive welding. With the increase in steel fiber diameter, the tensile resistance strength increased gradually. Agraw Ryuichi [15] used explosive welding to produce a fiber composite consisting of a high-strength plastic steel wire and a pure aluminum or titanium matrix, and then verified the strength of the composite by tensile tests. The results revealed that the optimum tensile resistance load was determined by the volume of the fiber part, that is, the larger the volume of fibers added, the stronger the tensile resistance. Taking AMr6 aluminum alloy as the base plate and unidirectional fiber composited by steel wire as the reinforced composite material, Kotov V A [15] pointed out that the strength of this material depended on the internal geometric installation under uniaxial and biaxial loads in many cases. When the load was parallel to the fiber, the fiber had a particularly large effect on the composite. The results revealed that the fiber-reinforced composites and tubular products prepared by explosive welding had high enough strength under various loads. Explosion strengthening is using the force generated by explosives to replace mechanical forging so that the strength of steel is improved to a certain extent, and the effect of austenitic steel is more obvious when subjected to explosion strengthening process. Therefore, steel fiber was selected as the steel fiber-reinforced phase. The strengthening of metal composite materials prepared by explosive welding mainly includes two aspects: (1) after explosive welding, the base plate, composite plate, and steel fiber were all subjected to explosion strengthening, so their tensile strength was improved; (2) the overall tensile resistance strength of the composite material after explosive welding was better than that of the base plate metal [48]. With the addition of steel fibers, the tensile resistance of steel-aluminum composite plates was greatly increased. The tensile resistance of the steel fiber parallel to the tensile load direction was much larger than that perpendicular to the tensile load direction. Therefore, the tensile resistance of steel fiber-reinforced steel-aluminum composite plates was greatly improved. The main reasons are as follows. (1) During the explosive welding process, the tensile resistance of the steel fiber was improved due to the explosion strengthening. (2) The addition of steel fiber improved the overall tensile resistance of steel fiber-reinforced steel-aluminum composite plates.

Table 4. Tensile performance parameters of steel-aluminum composite plates.

Serial No.	First Group of Data/MPa	Second Group of Data/MPa	Third Group of Data/MPa	Mean Value/MPa
1^-	449.53	435.35	406.72	420.53
2^-	462.11	456.41	450.11	456.21
3^-	497.36	479.85	478.65	485.29
4^-	582.79	557.06	546.59	562.15
5^+	442.36	438.93	425.15	435.48
6^+	450.77	448.27	444.95	447.99
7^+	474.48	468.88	448.09	463.80
8^+	500.05	480.38	472.14	484.19

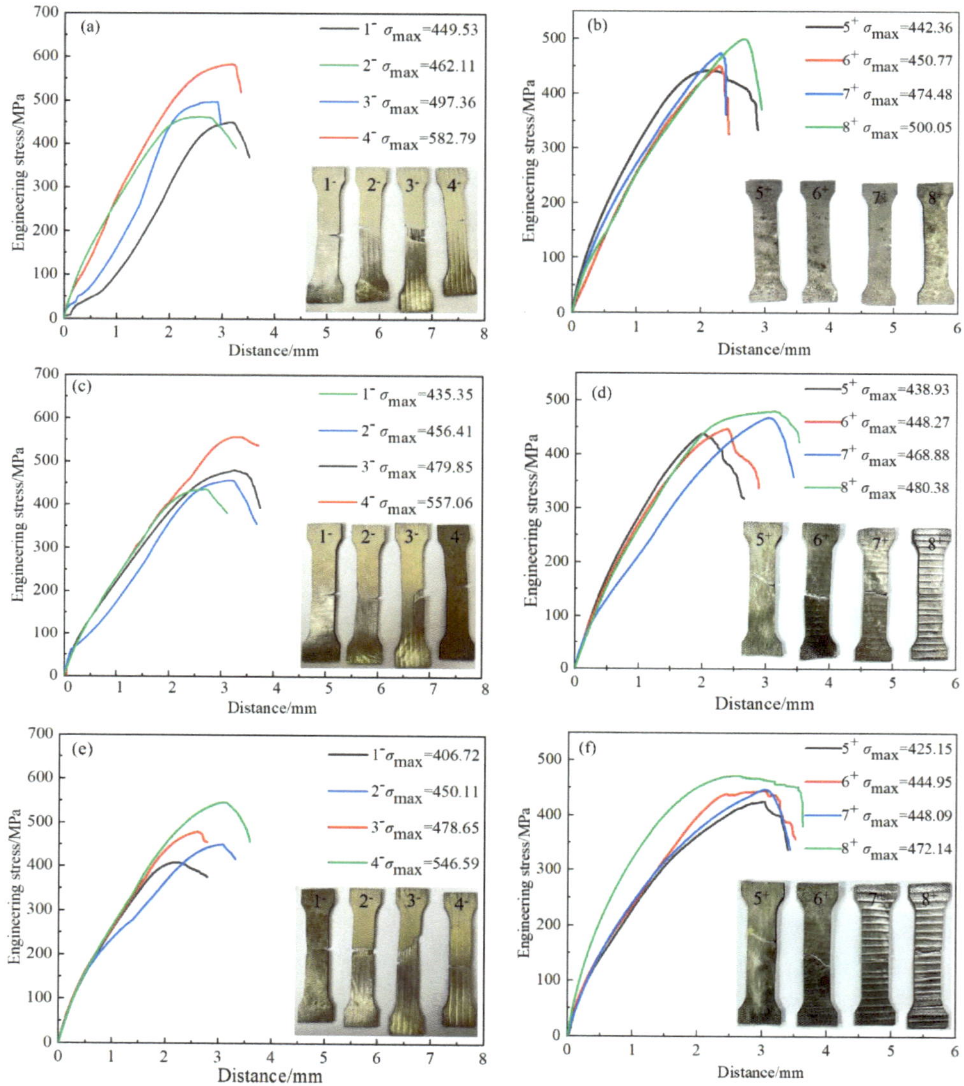

Figure 13. Tensile test and results: (**a**) the first group of tensile test samples and tensile data parallel to the fiber direction; (**b**) the first group of tensile test samples and tensile data orthogonal to the fiber direction; (**c**) the second group of tensile test samples and tensile data parallel to the fiber direction; (**d**) the second group of tensile test samples and tensile data orthogonal to the fiber direction; (**e**) the third group of tensile test samples and tensile data parallel to the fiber direction; (**f**) the third group of tensile test samples and tensile data orthogonal to the fiber direction.

3.5. Fracture Appearance Analysis

SEM was used to analyze the fracture of the tensile specimen and study the failure mechanism of steel fiber-reinforced steel-aluminum composite plates. Figure 14a shows the overall morphology of the fracture of the tensile specimen. From top to bottom, the layers are 2A12 aluminum, S20910 steel fibers and SS 304 stainless steel. The overall fracture morphology is relatively flat, but small tough nests and quasi-dissociative fracture openings can still be observed, which proves that different fractures have occurred in the different materials. Meanwhile, it can be observed that the steel fibers have separated from the aluminum layer and remain bonded to the steel layer. Figure 14b shows the fracture morphology of the 2A12 aluminum layer, from which it can be observed that the fracture surface is hole-like, the hole is shallow and the edge of the hole is parabolic, showing an obvious dimple shape. Therefore, the fracture of the 2A12 aluminum layer is a ductile fracture, so it is worth noting that near the edge of the welding line, the dimple morphology of the 2A12 aluminum layer gradually changed, and the depth of the hole gradually became shallow. The main reason is that the composite plate was subjected to explosion strengthening during the explosive welding process, and as the deformation increased, the strength of the explosive strengthening also gradually increased, which also verified that the maximum deformation in the numerical simulation results occurred at the bonding interface, as shown in Figure 6c. The microhardness at the bonding interface between the 2A12 aluminum and steel fiber was also greater than that of the 2A12 aluminum, which further confirmed that the bonding interface was subjected to explosion strengthening, as shown in Figure 11. Figure 14c shows the fracture morphology of the steel layer under high multiples. Obvious tearing edges and holes can be observed, and there is no flat cleavage section, so the steel layer has a quasi-cleavage fracture. The fracture morphology of the steel fiber is shown in Figure 14d, from which it can be seen that the fracture of steel fiber is relatively flat, and there are small dimples and obvious tearing edges on the surface, so the steel fiber also has a quasi-cleavage fracture. Meanwhile, it can also be observed that not all the steel fiber is extracted, and the local section is reduced while maintaining the connection with the base plate and the composite plate, and it was judged that the steel fiber had a necking phenomenon. This is because the steel fiber has not completely broken after the fracture behavior of the plate during the stretching process, and the necking phenomenon was finally broken by stretching.

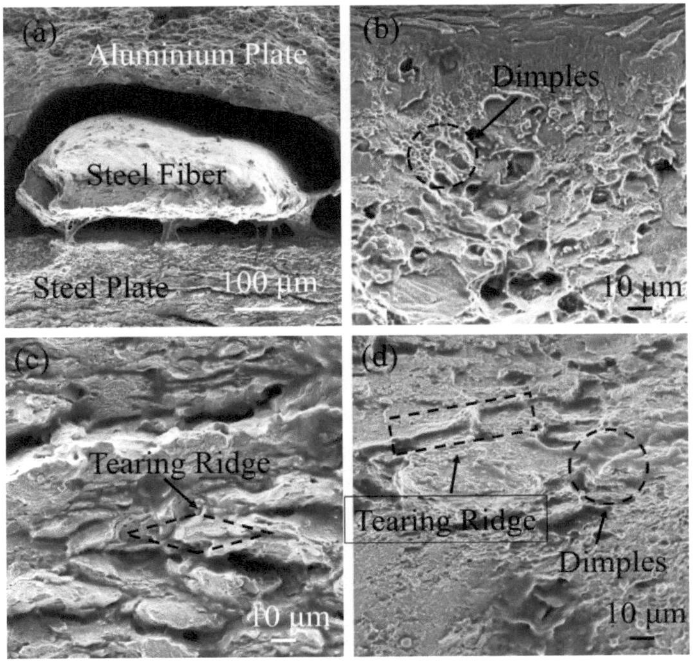

Figure 14. Fracture SEM of fiber-reinforced composite plate: (**a**) fracture SEM; (**b**) 2A12 aluminum fracture SEM; (**c**) SS-304 stainless steel fracture SEM; (**d**) fiber fracture SEM.

4. Conclusions

The steel fiber-reinforced steel-aluminum composite plate was successfully prepared by explosive welding, and the interface evolution mechanism of the steel-aluminum composite interface after adding steel fiber was simulated by the SPH method. Then, steel fiber-reinforced steel-aluminum composite plates were further studied using SEM, EDS and tensile tests, and the following conclusions were drawn.

(1) The SPH simulation method proved that there was a certain angle formed by the steel fiber, the base plate, and the composite plate, which blocked the direction of the jet so that the jet accumulated on one side of the steel fiber to form an environment with high temperature, high pressure, and high strain rate, thereby forming a vortex area. The residual stress of the jet particles in the vortex region forced them to undergo strong stirring, producing cracks, holes, and air pores in the vortex region.

(2) SEM and EDS analysis proved the following two points. 1. The main reason for cracks, holes, and air pores in the vortex area was the formation of a large number of brittle intermetallic compounds, and the brittle intermetallic compound was mainly FeAl 2. due to the blocking effect of steel fiber on the jet, the transition of elements at the interface of the composite plate with steel fiber was relatively fast.

(3) Microhardness test results revealed that the cracks, holes, and air pores in the vortex region influenced the microhardness of the bonding interface to a certain extent. As the microhardness of 2A12 aluminum was relatively small, the microhardness increased with the decrease in 2A12 aluminum content in the vortex region.

(4) The tensile test on the universal specimen machine proved that adding steel fibers could improve the antitensile properties of the steel-aluminum composite plates. The larger the diameter of steel fiber, the more obvious the tensile properties are. The main reason is that the steel fiber material is S20910 stainless steel, which was significantly strengthened during the explosion process.

(5) Fracture morphology analysis indicated that 2A12 aluminum had a typical ductile fracture. The ductile fracture far from the bonding interface was more obvious because of explosion strengthening. The SS-304 stainless steel layer and steel fiber were quasi-cleavage fractures, and the steel fibers showed a necking phenomenon during fracture.

Author Contributions: Conceptualization, Y.G. (Yonghong Gao) and Y.C.; methodology, Y.G. (Yonghong Gao), Y.C. and C.G.; software, Y.C. and C.G.; validation, Y.G. (Yonghong Gao), Y.C., C.G. and Y.G. (Yanping Guo); formal analysis, Y.C., C.G. and Z.G.; investigation, Y.G. (Yonghong Gao), Y.C. and C.G.; resources, Y.G. (Yonghong Gao); data curation, Y.C., C.G. and Y.L.; writing—original draft preparation, Y.C.; writing—review and editing, Y.G. (Yonghong Gao), Y.C. and C.G.; visualization, T.L.; supervision, Y.G. (Yonghong Gao), Y.C., C.G. and T.L.; project administration, Y.G. (Yonghong Gao), Y.C. and C.G.; funding acquisition, Y.G. (Yonghong Gao). All authors have read and agreed to the published version of the manuscript.

Funding: This research received no external funding.

Data Availability Statement: Data are contained within this article.

Acknowledgments: The authors thank the North University of China for fabrication and characterizations, and specially Chuanxiang Guo for the co-supervision of Yao Chen.

Conflicts of Interest: The authors declare no conflict of interest.

References

1. Zeng, X.Y.; Li, X.J.; Chen, X.; Wang, X.-H.; Yan, H.-H. Numerical and experimental studies on the explosive welding of plates with different initial strength. *Weld. World* **2019**, *63*, 967–974. [CrossRef]
2. Wu, X.; Shi, C.; Fang, Z.; Lin, S.; Sun, Z. Comparative study on welding energy and Interface characteristics of titanium-aluminium explosive composites with and without interlayer. *Mater. Des.* **2021**, *197*, 109279. [CrossRef]
3. Zhang, Z.L.; Liu, M.B. Numerical studies on explosive welding with ANFO by using a density adaptive SPH method. *J. Manuf. Process.* **2019**, *41*, 208–220. [CrossRef]

4. Zhang, H.; Jiao, K.X.; Zhang, J.L.; Liu, J. Microstructure and mechanical properties investigations of copper-steel composite fabricated by explosive welding. *Mater. Sci. Eng. A* **2018**, *731*, 278–287. [CrossRef]
5. Gulenc, B. Investigation of interface properties and weldability of aluminum and copper plates by explosive welding method. *Mater. Des.* **2008**, *29*, 275–278. [CrossRef]
6. Chu, Q.; Tong, X.; Xu, S.; Zhang, M.; Li, J.; Yan, F.; Yan, C. Interfacial investigation of explosion-welded titanium/steel bimetallic plates. *J. Mater. Eng. Perform.* **2020**, *29*, 78–86. [CrossRef]
7. Kundu, S.; Ghosh, M.; Laik, A.; Bhanumurthy, K.; Kale, G.; Chatterjee, S. Diffusion bonding of commercially pure titanium to 304 stainless steel using copper interlayer. *Mater. Sci. Eng. A* **2005**, *407*, 154–160. [CrossRef]
8. Ghosh, M.; Bhanumurthy, M.; Kale, G.B.; Krishnan, J.; Chatterjee, S. Diffusion bonding of titanium to 304 stainless steel. *J. Nucl. Mater.* **2003**, *322*, 235–241. [CrossRef]
9. Gao, X.L.; Zhang, L.J.; Liu, J.; Zhang, J.-X. Porosity and microstructure in pulsed Nd:YAG laser welded Ti6Al4V sheet. *J. Mater. Process. Technol.* **2014**, *214*, 1316–1325. [CrossRef]
10. Akbari, M.; Saedodin, S.; Toghraie, D.; Shoja-Razavi, R.; Kowsari, F. Experimental and numerical investigation of temperature distribution and melt pool geometry during pulsed laser welding of Ti6Al4V alloy. *Opt. Laser Technol.* **2014**, *59*, 52–59. [CrossRef]
11. Yu, H.P.; Xu, Z.D.; Fan, Z.S.; Zhao, Z.; Li, C. Mechanical property and microstructure of aluminum alloy-steel tubes joint by magnetic pulse welding. *Mater. Sci. Eng. A* **2013**, *561*, 259–265. [CrossRef]
12. Geng, H.; Mao, J.; Zhang, X.; Li, G.; Cui, J. Strain rate sensitivity of Al-Fe magnetic pulse welds. *J. Mater. Process. Technol.* **2018**, *262*, 1–10. [CrossRef]
13. Liu, J.G.; Cai, W.C.; Liu, L.; Han, J. Investigation of interfacial structure and mechanical properties of titanium clad steel sheets prepared by a brazing-rolling process. *Mater. Sci. Eng. A* **2017**, *703*, 386–398. [CrossRef]
14. Xie, G.M.; Yang, D.H.; Luo, Z.A.; Li, M.; Wang, M.-K.; Misra, R.D.K. The determining role of Nb interlayer on interfacial microstructure and mechanical properties of Ti/steel clad plate by vacuum rolling cladding. *Materials* **2018**, *11*, 1983. [CrossRef]
15. Zheng, Y.M. *Explosive Welding and Explosive Composite Material*; National Defense Industry Press: Beijing, China, 2017.
16. Zhou, N. Study on Penetration Resistance and Damage Mechanism of Explosive Composite Functional Gradient Target. Ph.D. Thesis, Nanjing University of Science and Technology, Nanjing, China, 2014.
17. Wang, B.; Xiong, J.; Wang, X.J.; Ma, L.; Zhang, G.-Q.; Wu, L.-Z.; Feng, J.-C. Energy absorption efficiency of carbon fiber reinforced polymer laminates under high velocity impact. *Mater. Des.* **2013**, *50*, 140–148. [CrossRef]
18. Mahfuz, H.; Zhu, Y.; Haque, A.; Abutalib, A.; Vaidya, U.; Jeelani, S.; Gama, B.; Gillespie, J.; Fink, B. Investigation of high-velocity impact on integral armor using finite element method. *Int. J. Impact Eng.* **2000**, *24*, 203–217. [CrossRef]
19. Yang, M.; Xu, J.F.; Chen, D.G.; Ma, H.; Shen, Z.; Zhang, B.; Tian, J. Understanding interface evolution during explosive welding of silver foil and Q235 substrate through experimental observation coupled with simulation. *Appl. Surf. Sci.* **2021**, *566*, 150703.1–150703.10. [CrossRef]
20. Sun, Z.R.; Shi, C.G.; Shi, H.; Li, F.; Gao, L.; Wang, G. Comparative study of energy distribution and interface morphology in parallel and double vertical explosive welding by numerical simulations and experiments. *Mater. Des.* **2020**, *195*, 109027. [CrossRef]
21. Sun, W.; Li, X.J.; Yan, H.H.; Hokamoto, K. Effect of Initial Hardness on Interfacial Features in Underwater Explosive Welding of Tool Steel SKS3. *J. Mater. Eng. Perform.* **2014**, *23*, 421–428. [CrossRef]
22. Chen, X.Q.; Zhang, K.Y.; Zhan, F.M. Experiment and analysis of underwater explosive welding repair. *Blasting* **2004**, *21*, 3.
23. Liang, H.; Luo, N.; Li, X.J.; Sun, X.; Shen, T.; Ma, Z. Joining of Zr 60 Ti 17 Cu 12 Ni 11 bulk metallic glass and aluminium 1060 by underwater explosive welding method. *J. Manuf. Process.* **2019**, *45*, 115–122.
24. Zhang, Z.H.; Xue, J.X. Influence of droplet impact direction on weld quality. *J. South China Univ. Technol.* **2019**, *47*, 138–144.
25. Wu, Y.K.; Wu, R.; Zhang, M.Y.; Tan, S.; Lu, J.; Jiang, F. Enhancing explosive weldability of thick precipitate-hardened alloys. *J. Manuf. Process.* **2022**, *77*, 339–347. [CrossRef]
26. Zhou, Q.; Liu, R.; Chen, P.W.; Zhu, L. Microstructure characterization and tensile shear failure mechanism of the bonding interface of explosively welded titanium-steel composite. *Mater. Sci. Eng. A* **2021**, *820*, 141559. [CrossRef]
27. Zheng, Z.M.; Yang, Z.S. *Explosive Processing*; National Defense Industry Press: Beijing, China, 1981.
28. Zhou, Q.; Liu, R.; Zhou, Q.; Ran, C.; Fan, K.S.; Xie, J.; Chen, P.W. Effect of microstructure on mechanical properties of titanium-steel explosive welding interface. *Mater. Sci. Eng. A* **2022**, *830*, 142260. [CrossRef]
29. Yang, M.; Ma, H.; Shen, Z.; Chen, D.-G.; Deng, Y.-X. Microstructure and mechanical properties of Al-Fe meshing bonding interfaces manufactured by explosive welding. *Trans. Nonferrous Met. Soc. China* **2019**, *29*, 680–691. [CrossRef]
30. Chu, Q.; Zhang, M.; Li, J.; Yan, C. Experimental and numerical investigation of microstructure and mechanical behavior of titanium/steel interfaces prepared by explosive welding. *Mater. Sci. Eng. A* **2017**, *689*, 323–331. [CrossRef]
31. Gloc, M.; Wachowski, M.; Plocinki, T.; Kurzydlowski, K.J. Microstructural and microanalysis investigations of bond titanium grade1/low alloy steel st52-3N obtained by explosive welding. *J. Alloys Compd.* **2016**, *671*, 446–451. [CrossRef]
32. Yang, M.; Ma, H.; Shen, Z.; Huang, Z.; Tian, Q.; Tian, J. Dissimilar material welding of tantalum foil and Q235 steel plate using improved explosive welding technique. *Mater. Des.* **2020**, *186*, 108348. [CrossRef]
33. Zhang, H.; Jiao, K.X.; Zhang, J.L.; Liu, J. Experimental and numerical investigations of interface characteristics of copper/steel composite prepared by explosive welding. *Mater. Des.* **2018**, *154*, 140–152. [CrossRef]

34. Bazarnik, P.; Adamczyk, C.B.; Gałka, A.; Płonka, B.; Sniezek, L.; Cantoni, M.; Lewandowska, M. Mechanical and microstructural characteristics of Ti6Al4V/AA2519 and Ti6Al4V/AA1050/AA2519 laminates manufactured by explosive welding. *Mater. Des.* **2016**, *111*, 146–157. [CrossRef]
35. Zeng, X.; Wang, Y.; Li, X.; Li, X.-J.; Zhao, T.-J. Effects of gaseous media on interfacial microstructure and mechanical properties of titanium/steel explosive welded composite plate. *Fusion Eng. Des.* **2019**, *148*, 111292.1–111292.9. [CrossRef]
36. Ning, J.; Zhang, L.J.; Xie, M.X.; Yang, H.-X.; Yin, X.-Q.; Zhang, J.-X. Microstructure and property inhomogeneity investigations of bonded Zr/Ti/steel trimetallic sheet fabricated by explosive welding. *J. Alloys Compd.* **2017**, *698*, 835–851. [CrossRef]
37. Wu, T.; Yang, C.L. Influence of pulse TIG welding thermal cycling on the microstructure and mechanical properties of explosively weld titanium/steel joint. *Vacuum* **2022**, *197*, 110817. [CrossRef]
38. Hoseini, M.M.; Tolaminejad, B. Weldability window and the effect of interface morphology on the properties of Al/Cu/Al laminated composites fabricated by explosive welding. *Mater. Des.* **2015**, *86*, 516–525. [CrossRef]
39. Jiang, S.; Shen, J.; Nagasaka, T.; Muroga, T.; Sagara, A.; Ohnuki, S.; Hokamoto, K.; Tanaka, S.; Inao, D.; Morizono, Y.; et al. Interfacial characterization of dissimilar-metals bonding between vanadium alloy and Hastelloy X alloy by explosive welding. *J. Nucl. Mater.* **2020**, *539*, 152322. [CrossRef]
40. Akbari, M.S.A.A.; Farhadi, S.P. Experimental investigation of explosive welding of cp-titanium/AISI 304 stainless steel. *Mater. Des.* **2009**, *30*, 459–468. [CrossRef]
41. Aceves, S.M.; Espinosa, L.F.; Elmer, J.W.; Huber, R. Comparison of Cu, Ti and Ta interlayer explosively fabricated aluminum to stainless steel transition joints for cryogenic pressurized hydrogen storage. *Int. J. Hydrogen Energy* **2015**, *40*, 1490–1503. [CrossRef]
42. Zhang, Z.L.; Feng, D.; Liu, M.B. Investigation of explosive welding through whole process modeling using a density adaptive SPH method. *J. Manuf. Process.* **2018**, *35*, 169–189. [CrossRef]
43. Nassiri, A.; Abke, T.; Daehn, G. Investigation of melting phenomena in solid-state welding processes. *Scr. Mater.* **2019**, *168*, 61–66. [CrossRef]
44. Song, J.; Kostka, A.; Veehmayer, M.; Raabe, D. Hierarchical microstructure of explosive joints: Example of titanium to steel cladding. *Mater. Sci. Eng. A* **2011**, *528*, 2641–2647. [CrossRef]
45. Carvalho, G.; Galvo, I.; Mendes, R.; Leal, R.; Loureiro, A. Explosive welding of aluminium to stainless steel using carbon steel and niobium interlayers. *J. Mater. Process. Technol.* **2020**, *283*, 116707. [CrossRef]
46. Venkateswaran, P.; Xu, Z.H.; Li, X.; Reynolds, A.P. Determination of mechanical properties of Al–Mg alloys dissimilar friction stir welded interface by indentation methods. *J. Mater. Sci.* **2009**, *44*, 4140–4147. [CrossRef]
47. Wang, Y.X.; Beom, H.G.; Sun, M.; Lin, S. Numerical simulation of explosive welding using the material point method. *Int. J. Impact Eng.* **2011**, *38*, 51–60. [CrossRef]
48. Zheng, Y.M.; Huang, G.R.; Chen, S.H. Strengthening of metals in explosive composites. *Eng. Blasting* **2000**, *6*, 7–25.

MDPI AG
Grosspeteranlage 5
4052 Basel
Switzerland
Tel.: +41 61 683 77 34

MDPI Books Editorial Office
E-mail: books@mdpi.com
www.mdpi.com/books

www.ingramcontent.com/pod-product-compliance
Lightning Source LLC
Chambersburg PA
CBHW041424210926
51850CB00013B/3353

9783725838813

Special Edition Using Corel® WordPerfect® 8 for Linux

Roderick W. Smith

QUE®

A Division of Macmillan Computer Publishing, USA
201 W. 103rd Street
Indianapolis, Indiana 46290

Contents at a Glance

Special Edition Using Corel® WordPerfect® 8 for Linux®

International Standard Book Number: 0-7897-20329

Library of Congress Catalog Card Number: 99-60644

Printed in the United States of America

First Printing: August 1999

01 00 99 4 3 2 1

Trademarks

Warning and Disclaimer

Executive Editor
Dean Miller

Executive Editor
John Pierce

Acquisitions Editors
Jamie Milazzo
Jenny Watson

Development Editors
Jill Hayden
Susan Hobbs

Managing Editor
Tom Hayes

Project Editor
Rebecca Mounts

Copy Editor
June Waldman

Indexer
Eric Schroeder

Proofreaders
Wendy Ott
Mary Ellen Stephenson

Technical Editor
Gary Krakow

Software Development Specialist
Aaron Price

Interior Design
Ruth Harvey

Cover Design
Dan Armstrong
Ruth Harvey

Copy Writer
Eric Borgert

Layout Technicians
Stacey DeRome
Ayanna Lacey
Heather Hiatt Miller
Staci Somers
Mark Walchle

CONTENTS

IV Complex Formatting with Boxes for Text, Graphics, and Equations

VII Web Publishing

VIII Appendixes

About the Author

Roderick W. Smith is a long-time user of WordPerfect, dating back to version 4.1 in 1988. He's used WordPerfect on Atari ST, DOS, Windows, Macintosh, OS/2, and Linux systems. His experience with Linux, although not as temporally extended as his acquaintance with WordPerfect, is still extensive, dating back to 1994 when the 1.0.x kernel series was current. Rod has publications in several computer magazines, most notably *Linux Journal*, and maintains several Web pages devoted to Linux, including one on configuring fonts and printers for WordPerfect for Linux. He holds a Ph.D. in cognitive psychology from Tufts University and has done postdoctoral research in that field at various institutions. He currently resides in Malden, Massachusetts, just outside Boston.

Dedication

To Jennie, the best sister I could ask for.

Acknowledgments

I'd like to thank acquisitions editors Jenny Watson and Jamie Milazzo, who shepherded this book through its formative stages and dealt with a few bumps along the way; development editors Jill Hayden and Susan Hobbs, who tirelessly read and corrected my prose; and project coordinators Vicki Harding and Laurie Morgan, who did much of the work of holding it all together. Without their efforts this book would not be possible.

A book like this doesn't just slide through the works without being checked for accuracy. Gary Krakow deserves thanks for doing this generally thankless job. Any errors that remain are, of course, my own.

In order to write this book, I used the GIMP image manipulation software by Spencer Kimball and Peter Mattis to capture screenshots. I used Marko Macek's icewm window manager for controlling onscreen windows (icewm is what you want if you really like the look of the windows in this book). Of course, I used the Linux kernel by Linus Torvalds and others, along with far too many additional programs and utilities to mention by name.

Tell Us What You Think!

As the reader of this book, *you* are our most important critic and commentator. We value your opinion and want to know what we're doing right, what we could do better, what areas you'd like to see us publish in, and any other words of wisdom you're willing to pass our way.

As a Publisher for Que, I welcome your comments. You can fax, email, or write me directly to let me know what you did or didn't like about this book—as well as what we can do to make our books stronger.

Please note that I cannot help you with technical problems related to the topic of this book, and that due to the high volume of mail I receive, I might not be able to reply to every message.

When you write, please be sure to include this book's title and author as well as your name and phone or fax number. I will carefully review your comments and share them with the author and editors who worked on the book.

Fax: 317.581.4666

Email: `que_desktop@mcp.com`

Mail: John Pierce
Publisher
Que
201 West 103rd Street
Indianapolis, IN 46290 USA

INTRODUCTION

WordPerfect has long been a major force in the word processing world. Most people probably think of it as having a DOS heritage. It *was*, after all, the dominant word processor on DOS for many years, and that's where many users first learned of WordPerfect's strengths at handling large documents, including what were at the time advanced features such as automatic placement of footnotes and support for laser printers. WordPerfect also has a UNIX background, however; at the time WordPerfect was dominant on DOS, UNIX versions of the program were also available. As time passed and DOS gave way to Windows, UNIX development of WordPerfect continued, for the most part quietly and unnoticed by most desktop computer users.

In the past few years, another little-noticed project has been underway. Started at the University of Helsinki in Finland by Linus Torvalds, the Linux kernel was initially a project to reimplement parts of the UNIX operating system on inexpensive PC hardware. It quickly gathered around it the fruit of years of software development for UNIX, especially what has come to be known as *open-source software*, much of it from Richard Stallman's Free Software Foundation (FSF). The result of this accretion of software, ranging from the Linux kernel itself to the XFree86 windowing system, the Emacs editor, the gcc C language compiler, and others, is frequently referred to as *Linux*, though some prefer the term *GNU/Linux* to spread the credit for the nonkernel utilities, many of which were developed by the FSF's GNU project. By whatever name (I use *Linux* in this book), the result is a very complete UNIX-like operating system that is freely available, stable, and flexible.

With the rise in popularity of Linux, WordPerfect's Linux version has been gaining attention. WordPerfect is perhaps the most widely recognized commercial application available for Linux, though it is by no means the only one. Corel's announcement that it would make a version of WordPerfect 8 for Linux freely available for download sparked considerable enthusiasm in the Linux community.

WordPerfect has actually been available for Linux for some time, in version 6.1 and 7.0 incarnations, which were implementations based strongly upon the UNIX versions of the program. Version 8.0 of WordPerfect for Linux is also based upon the UNIX version of the program. Version 8.0 of WordPerfect for Linux shares many characteristics with its Windows counterpart, but also differs in some important ways. To get the most out of the program, it's useful to understand the interaction between WordPerfect and Linux.

WHO SHOULD BUY THIS BOOK

If you want to use WordPerfect for Linux and need a good reference or learning tool, this book is for you. WordPerfect 8 for Linux comes in three versions: a free downloadable version (which also appears on the CD accompanying this book), a personal retail version, and a retail server edition. The downloadable version comes with limited documentation, and the frequently rather cryptic user's guide that comes with the retail versions doesn't clearly show how WordPerfect's features integrate with one another. This book can therefore fill the gap between what you have in documentation and what you need.

This book covers the personal retail version of WordPerfect, but whenever there's a difference between it and the downloadable version, I point this out. In general, the downloadable version is the same as the retail version except that the former is missing some features, such as the equation editor and font installer. Chapter 1, "Introduction to WordPerfect for Linux," details these differences.

Because WordPerfect for Linux is essentially one specific variant of WordPerfect for UNIX, this book also applies to WordPerfect for UNIX. Mentally replace any reference to "Linux" in the text with "UNIX," and you won't go far wrong. Your particular implementation of UNIX, however, may differ from Linux in some way that is important for certain functions, but for the most part they work the same.

I've written this book with the goal of explaining WordPerfect for Linux to somebody with prior experience with other word processors, probably on operating systems other than Linux. You might have experience with Microsoft Word in Windows or DeScribe in OS/2 or Nisus Writer in MacOS. You might even be familiar with WordPerfect on one of the many platforms on which it has existed throughout the years. You should be familiar with basic word processing features such as fonts, text attributes, cut-and-paste operations, and so on. You might not know the details of how to do these things in WordPerfect, however.

You also might not know much about Linux or about how a program like WordPerfect interacts with Linux's unique features. You may have managed printer queues in Windows, but how do you do it in Linux? (See Chapter 6, "Producing Output.") You may have accessed files on floppy in MacOS, but how do you do that in Linux? (See Chapter 3, "Managing Files and Folders.") Some of the most Linux-specific material is in the first few chapters of this book, with another helping in Appendixes A, "Installation and System Configuration," and B, "Fonts and Printing." Linux-relevant tips are scattered throughout the entire book.

If you're completely new to Linux, you'd do well to get a good book on the topic, such as Que's *Special Edition Using Linux*. This background will help you through rough spots with Linux, especially if you're installing Linux from scratch for the first time and therefore need to configure everything from X to networking. Although I briefly discuss some Linux configuration issues, particularly in Appendixes A and B, a separate book on the topic is invaluable if you're managing your own Linux system for the first time.

I haven't forgotten the Linux pros, though. If you're familiar with Linux or other UNIX-like operating systems, you'll get advanced information on configuring printers and fonts for use with WordPerfect, and you'll learn enough about WordPerfect to be able to use it for just about any text-editing need. If you're into programming, you might want to check out Linux's macro facility (Chapter 19, "Letting the Computer Do the Work: Macros"), which could easily be the subject of an entire book.

No matter what your computer experience, the one assumption I make is that you want to accomplish some real-world task with WordPerfect running on Linux. Perhaps you need to write a report, prepare a letter for mass mailing, or create a newsletter. Whatever your word processing task, WordPerfect for Linux provides tools that can help you accomplish your goals, and this book can help you learn those tools. You can use this book as a reference if you're already at least somewhat familiar with WordPerfect, or you can read chapters that cover features you want to learn better.

Of course, no program of WordPerfect's size and complexity is entirely without blemishes, and WordPerfect for Linux is no exception. It has bugs and limitations, and I point these out. Whenever possible, I provide a workaround to these problems or at least suggest some alternative way to accomplish an affected task.

HOW THIS BOOK IS ORGANIZED

This book is divided into eight parts of from two to six chapters. Each part covers a class of features or procedures, and each chapter covers one particular feature or procedure. I've tried to make each chapter as self-contained as possible, but of course some features interact strongly with others, so you may need to read one chapter before you can get the most out of another. If a chapter has a "prerequisite," I mention that at the start. Most of the chapters do assume a basic familiarity with Linux and with WordPerfect basics. If you're new to either, then, you should read the appropriate sections of Chapter 1 before doing anything else, and you may find the rest of Part I, "Program Overview," helpful, as well. I'm getting ahead of myself, though. Here's a summary of the sections of this book:

- Part I, "Program Overview," introduces WordPerfect for Linux, including basic tasks. It's the largest section of the book, and if you're new to word processing, you may want to read it all. If you're familiar with word processing in general, you may want to skim most of this material to get a feel for WordPerfect's way of doing things. If you're new to Linux, be sure to read the "If You Are New to Linux" section of Chapter 1, and read Chapter 6, which is about printing and other ways of producing output.

- Part II, "Enhancing Text by Formatting," tells you how to format your document—how to use fonts, set margins, create headers and footers, use columns and tables, and so on. Knowing how to use these core features of WordPerfect (or of any word processor), will allow you to make effective use of more advanced features.
- Part III, "Creating Consistent Documents with Styles and Templates," tells you how to use two of WordPerfect's most powerful labor-saving tools. You place formatting commands in styles and templates in order to apply these commands quickly and easily, either to sections of text or to entire documents. A small investment in creating a style or a template can pay off massively over months or years of use.
- Part IV, "Complex Formatting with Boxes for Text, Graphics, and Equations," introduces WordPerfect's *box* feature. You use elements called boxes to place text, graphics, or equations on the page, embedded within ordinary text but manipulable separately. These features are particularly helpful for designing complex layouts like newsletters.
- Part V, "Automating Repetitive Tasks," tells you how to use two of WordPerfect's most powerful features, macros and mail merge. With these, you can do in a few minutes what might otherwise take hours or days, by telling the computer *how* to manipulate information in a document. That way, you don't need to insert every character and select every feature yourself; you simply tell WordPerfect what to do and then let it do all the work.
- Part VI, "Managing Large Documents," covers features that you'll need when creating large documents. In this context, "large" can mean anything from a five-page school paper to an encyclopedia. You'll learn how to create an outline with WordPerfect, track document revisions, create footnotes and endnotes, and organize lists and references.
- Part VII, "Web Publishing," covers using WordPerfect to create and manage the latest form of written document, Web pages. You can design your own Web page with WordPerfect or convert an existing document to Web page format. Some of WordPerfect's features behave differently when you are working with Web pages than they do for traditional documents, so be sure to check this section if you want to use the program for this purpose.
- Part VIII, "Appendixes," covers program installation and configuration, including integrating WordPerfect into some common Linux desktop environments. WordPerfect for Linux uses a complex printer and font model, and you'll learn how to make the best use of your printers and fonts here. Finally, you'll learn how to customize many of WordPerfect's defaults, including its color scheme, where it looks for files, and how it interprets keystrokes.

What's on the CD

The CD that accompanies this book contains a copy of the downloadable version of WordPerfect 8 for Linux, so if you want to try out WordPerfect without buying the retail version, you can do so easily, without doing a lengthy download of the software. (You will need to register the program with Corel to use it for more than 90 days, but registration is free.)

In addition, the CD includes an assortment of programs, utilities, and sample files, located in directories named after their respective chapters. Each chapter describes the files relevant to it.

To access a CD in Linux, you need to use the `mount` command, either directly at a command prompt or via some tool in a GUI environment such as the K Desktop Environment (KDE; `http://www.kde.org`). This topic, and related issues of file management and Linux's directory structure, are discussed at length in Chapter 3.

CONVENTIONS USED IN THIS BOOK

This book uses certain conventions to help you get the most from the text and from WordPerfect 8 for Linux.

TEXT CONVENTIONS

Various typefaces in this book identify terms and other special objects. These special typefaces include the following:

Type	Meaning
Italic	New terms or phrases when initially defined. An italic term followed by a page number indicates the page where that term is first defined.
Underline	Menu and dialog box options that appear underlined onscreen, indicating shortcut keys (hotkeys).
`Mono`	Information that you type.
Capitalization	The capitalization of menu names, dialog box names, dialog box elements and commands in the text matches the way they appear onscreen.

Key combinations are represented with a plus sign. For example, if the text calls for you to enter Ctrl+S, you would press the Ctrl key and the S key at the same time.

WordPerfect menu hotkeys require you to use both the Alt key and the Shift key; use the Alt key for the first menu item, and the Shift key for subsequent items. For instance, if the text refers to File, Open, you would press Alt+F followed by Shift+O to perform this sequence by keyboard.

SPECIAL ELEMENTS

Throughout this book, you'll find Tips, Notes, Cautions, Sidebars, Cross-References, and Troubleshooting Tips. These elements provide a variety of information, ranging from warnings you shouldn't miss to ancillary information that will enrich your WordPerfect experience, but isn't required reading.

"TUX" TIPS

Tip #1001 from
Tux Tips

Tips are designed to point out features, annoyances, and tricks of the trade that you might otherwise miss. Tips generally help you to make better use of WordPerfect by providing a quicker or more effective way of doing something than might otherwise be immediately obvious.

NOTES

Note

Notes point out items that you should be aware of, though you can skip these if you're in a hurry. Generally, I've added notes as a way to give you some extra information on a topic without weighting you down.

CAUTIONS

Caution

Pay attention to Cautions! These could save you precious hours in lost work. Don't say I didn't warn you.

TROUBLESHOOTING TIPS

These elements call attention to common pitfalls that you're likely to encounter. When you see a Troubleshooting Tip, you can flip to the end of the chapter to learn how to solve or avoid a problem.

CROSS-REFERENCES

Cross-references are designed to point you to other locations in this book (or other books in the Que family) that provide supplemental or supporting information. Cross-references appear as shown below:

→ **See** "Organizing a Document with Outlining," **p. xxx**

PART I

Program Overview

CHAPTER 1

INTRODUCTION TO WORDPERFECT FOR LINUX

In this chapter

OVERVIEW

WordPerfect for Linux provides Linux users with a powerful traditional word processor. Unlike formatting systems such as TeX or LaTeX, which are popular among certain groups of technically oriented Linux users, WordPerfect and other traditional word processors provide an easy-to-use, what-you-see-is-what-you-get (*WYSIWYG*) approach to document creation. This approach brings the advantages of Linux, such as its robust multitasking, multiuser, and networking features, to one of the more popular word processors from the DOS and Windows worlds.

In this chapter, you learn

- How to use the Linux operating system for running WordPerfect
- Which hardware and software you need to effectively run WordPerfect 8 for Linux
- How to navigate around the WordPerfect environment, and how to create, edit, and print simple documents

IF YOU ARE NEW TO LINUX

Certain aspects of Linux may be baffling to people who have used operating systems such as Windows or MacOS. Compared to Windows, Linux can be a more complex operating system to install and configure, but once set up, Linux can be quite easy to use. Linux applications such as WordPerfect can be as easy to use as their Windows counterparts. This book assumes that Linux is already up and running on your system.

Note

If you need help getting Linux installed, check out "Installing Linux" in *Special Edition Using Linux*, published by Que.

Linux is most often installed using one of a number of prepackaged *distributions*. Common Linux distributions include Red Hat, Debian, Slackware, and SuSE. Each distribution is a collection of a stable Linux kernel and specific versions of literally hundreds of different programs. Although distributions can vary somewhat in configuration and the programs they include, all distributions are capable of running the same set of Linux applications, including WordPerfect. Nothing in this book depends on your having a specific Linux distribution installed.

LOGGING IN TO LINUX

When you start a Windows computer, you may or may not be asked to log in to the computer, depending on whether you're hooked up to a network and what your configuration settings are. With Linux, you are *always* required to log in. The login requirement adds security and flexibility to the system.

Caution

With Windows 98, if you forget or mistype your password, you can still use the computer. With Linux, that's not the case; if you forget the password, you cannot log in, so remembering your password is vital!

Depending on your configuration, you may see either a text-mode or graphical login prompt. The text-mode prompt is often simply a line that reads "login." Type your username here and type your password at the prompt that follows (your password does not echo to the screen, for security reasons). You then see a command prompt from which you start the X Window System.

A graphical login, which is handled by a system called *XDM (X Display Manager)*, also presents login and password prompts, but these are generally centered in the screen and are displayed in a proportional font. After you enter a valid username and password, you see whatever X Window System environment you have configured.

Users and the Superuser

Most of the time when you use Linux, you are logged in as an ordinary *user*. This designation refers to an account that's used by one person for normal, day-to-day operation. An ordinary user can log in, run a wide variety of programs, including WordPerfect, alter files in his or her own directory, and log off. On Linux systems, users generally have *home directories* in the /home directory tree, and they're named after the users' login names. For instance, my login name on my personal Linux system is rodsmith, and my home directory is /home/rodsmith. I'm using a separate account for writing this book, however, and its directory is /home/book.

➔ If you want to learn more about the Linux directory structure, **see** "Understanding the Linux Filesystem Model," **p. 73**

Ordinary users cannot perform certain key functions, such as installing software in anything but their own home directories. The reason is that, in a UNIX system with tens, hundreds, or even thousands of users, most users typically don't know enough to perform these tasks. Therefore, UNIX systems and, by derivation, Linux, have a special account, known as the *root account* or simply *root*. The root account, or the person who uses it, is also referred to as the *superuser* or the *system administrator*. The superuser installs software and performs other important system maintenance activities. If you installed Linux on your own system, you're the superuser, but you should also have an ordinary account for day-to-day work because one of the few safeguards that Linux provides against accidentally destroying the entire system with a stray command is that ordinary users don't have the permissions do so.

Most of the things you'll be doing with WordPerfect require only an ordinary user account. On rare occasion, I'll mention that you need to be root to perform some task. If somebody else is the system administrator for the Linux computer you're using, you'll have to ask that person for help with these matters.

Tip #1 from *Tux Tips*

You can become the superuser in one of two ways. First, you can log in with the username *root* and whatever the appropriate password is. Second, you can use the `su` command at an xterm—just type `su` and then press the Enter key. When Linux asks for the root password, type it in; anything you perform in that xterm is then done with root privileges. Type `exit` to divest yourself of superuser privileges in that xterm.

➔ If you want to learn more about the xterm, **see** "Running the xterm: Your Key to Linux from X," **p. 18**

STARTING X

If you logged in to Linux via a text-mode prompt, you must start the X Window System, often referred to simply as *X*, before proceeding. Linux is built in layers, from the kernel at the base to user applications such as WordPerfect at the apex. Somewhere in between comes X, which itself has multiple layers. X is Linux's graphical user interface (GUI) environment and is what displays windows and graphics on the screen, providing support for complex GUI programs such as WordPerfect. Linux can run without X, much as MS-DOS could run without Windows 3.1. WordPerfect for Linux, however, is an X-based application, so you need to have X running to use WordPerfect.

To start X from a Linux text-mode prompt, simply type `startx` and then press the Enter key. If your system is configured correctly, this step starts the X Window System and any associated programs. If you type `startx` and get an error message but no GUI environment, you need to configure X for your hardware.

Note

Configuring X for your hardware is beyond the scope of this book. However, Que's *Special Edition Using Linux* can help you (see Chapter 2, "System Administration").

If you used an XDM graphical login, X was already running on your system. At this point you are where you would have been if you'd logged in using text mode and typed `startx`.

Unlike Windows, the X environment has few fixed features; instead, X relies on a number of ancillary programs to provide much of the "look and feel" of the environment. The most important of these are the window manager and the file manager.

UNDERSTANDING WINDOW MANAGERS

By itself, an X window is boring and difficult to use, consisting only of a box on the screen. By supplying a border, drag bar, and other optional elements, an X window manager adds visual interest and utility. There are literally dozens, if not hundreds, of window managers available for Linux, and they vary substantially in how they look and are used. A few examples of window managers available for Linux are presented in Figures 1.1 through 1.3. Many more are available, both on Linux CDs and from the Web. The Web site `http://www.plig.org/xwinman/` contains summaries and links to many window managers. Most are open source, like Linux itself, so you can download them from the Web and use them at no cost. A few are commercial, however. Most of the screen shots in this book were taken using the icewm window manager running its Wigren theme. The most important thing to remember is that although your displays may look different from the ones shown here, the behavior of WordPerfect itself remains the same.

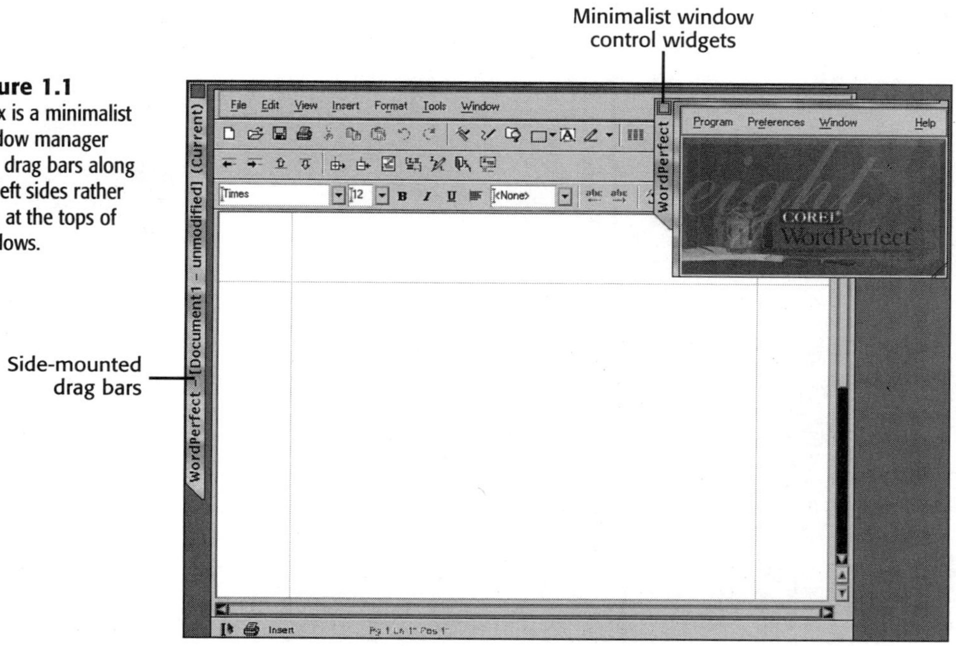

Figure 1.1
Wmx is a minimalist window manager with drag bars along the left sides rather than at the tops of windows.

In addition to providing adornment for your windows, window managers make available features for window manipulation, program launching, and other functions. The window manager controls are often referred to as *widgets*. As in Microsoft Windows, the area above a window, known as the *drag bar* or *title bar*, is typically used to drag a window around on the screen (but a few window managers, such as wmx, place the drag bar on the window's side). There may be a window-closing widget with an X-like symbol in it, minimize and maximize widgets, and so on. The details of how these widgets work vary considerably from window manager to window manager, so you should consult the documentation that came with your window manager to see what is available.

Many window managers provide functions or program-launch capabilities in some sort of status bar along the side or bottom of the screen.

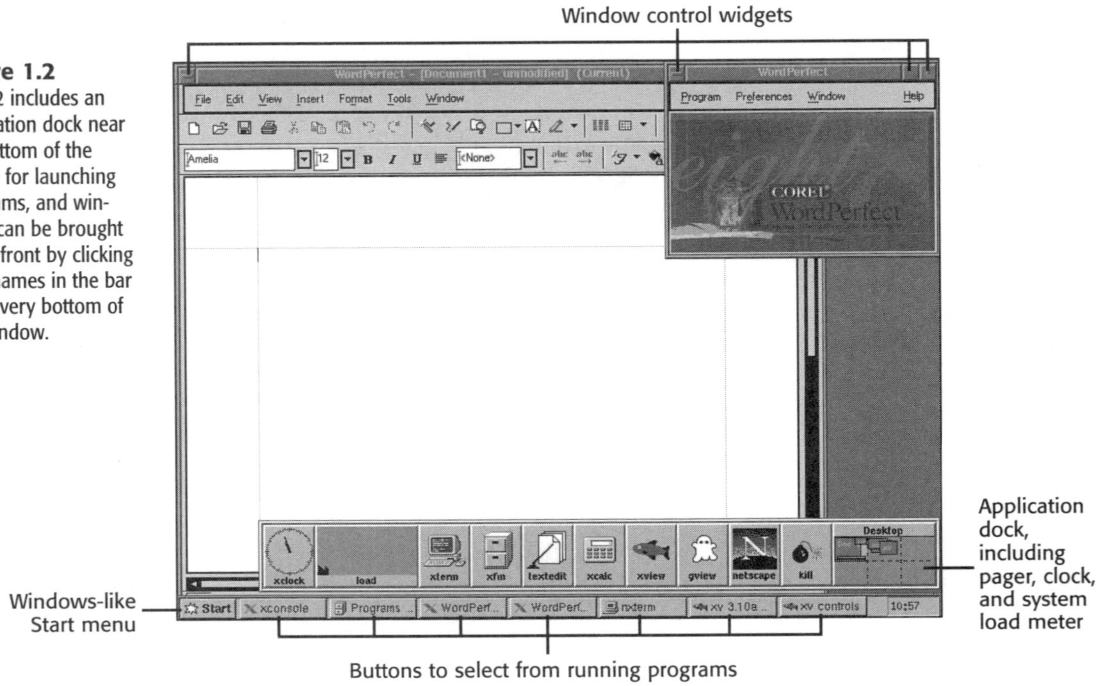

Figure 1.2
Fvwm2 includes an application dock near the bottom of the screen for launching programs, and windows can be brought to the front by clicking their names in the bar at the very bottom of the window.

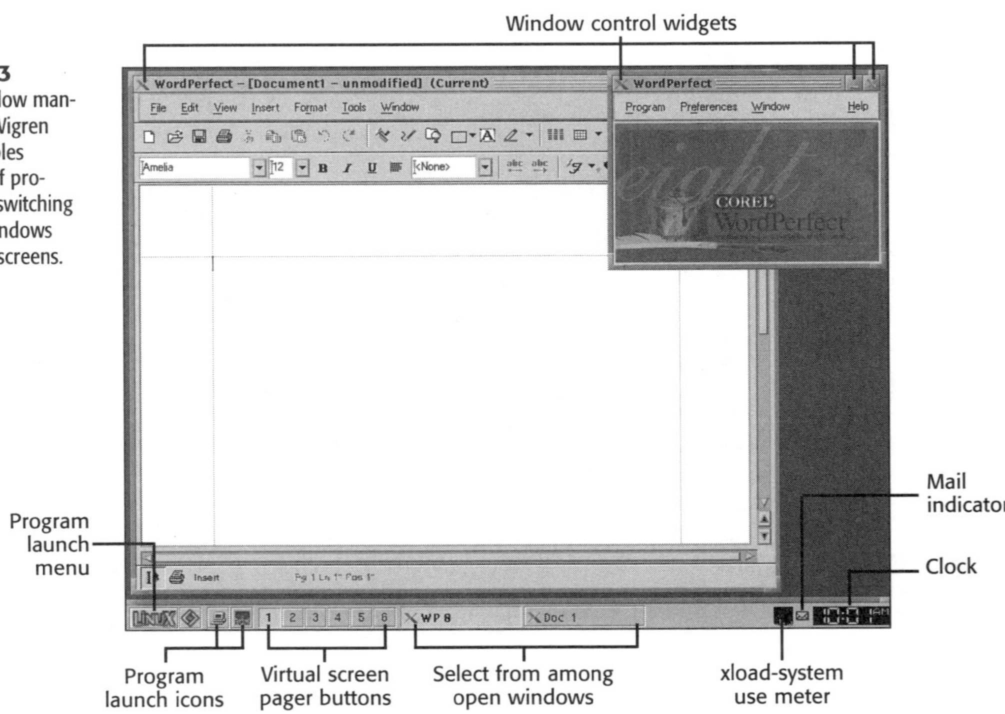

Figure 1.3
Icewm window manager in its Wigren theme enables launching of programs and switching between windows and virtual screens.

Note

Program-launch tools will almost certainly not be configured to launch WordPerfect, so if you want to make WordPerfect appear in such a menu, you have to do that yourself. Consult your window manager's documentation for instructions.

You can click on various icons and on the bare desktop to see which functions may be available in your window manager. Be sure to try clicking with the left, right, and middle buttons in turn. If your mouse has only two buttons, using those two buttons together will usually produce the same effect as the middle button on a three-button mouse.

Tip #2 from *Tux Tips*

A three-button mouse can be a very worthwhile investment for Linux users because many X programs make use of all three buttons. Although you can simulate the effects of the middle button by clicking the two buttons of a two-button mouse at the same time, this movement is awkward and more error prone than having a true middle button. (The only time you use the middle button in WordPerfect is to paste text selected from other applications.)

Window managers provide a number of ways of selecting (or *focusing*, in X parlance) a window. If you're familiar with Windows, the method you probably use there is called *click to focus*. When a window manager uses click to focus, you click on a window to make it the center of attention. After clicking on a window, anything you type on the keyboard is directed at that window. Typically, the window's drag bar changes color to signify that it has focus. Many Linux window managers also offer another focus model, called *focus follows mouse*. In this mode, whatever window is under the mouse pointer gets the focus automatically, without the need to click on it. This feature can be handy if you want to quickly switch between two or more windows, but it can be annoying if you like to fidget with the mouse.

In addition to focus methods, window managers differ in how and when they bring a window to the front so that it obscures any other windows occupying the same space on the screen. Microsoft Windows uses a method in which clicking on any part of the window brings it to the front, and this technique is available in many Linux window managers, too. Another model available in many window managers is to bring a window to the front only when the drag bar is clicked. The second model can be more restrictive than the first in that you may not be able to make a window fully visible if its drag bar is obscured, but the drag-bar approach can be useful if you want to perform some mouse operations on a partially obscured window without bringing it to the front. A few window managers also automatically bring a window to the front whenever it obtains focus, even when using the focus-follows-mouse model. Automatically bringing the window to the front can make for a very dynamic display as you move the mouse across a crowded screen!

You may want to familiarize yourself with the focus and "window fronting" methods used by your window manager. If you don't like what you've been given, you can either learn to live with it or you can read the window manager's documentation to learn how to change it (or change to a different window manager). In practice, most people can get used to just about anything in window manager focus and fronting methods within a few days.

With the exception of possibly adding a way of exiting from WordPerfect (by clicking on a close widget), your choice of window manager will not affect WordPerfect's functionality; the application program controls everything inside the window. Using one window manager or another may indirectly affect how you use WordPerfect, however. For instance, many window managers include a "pager" feature, which allows you to maintain several virtual desktops. You can have WordPerfect running in one desktop, a Web browser in another, and a spreadsheet in a third. It is possible to switch from one desktop to another by clicking on a button or selecting a menu item from a pop-up menu. If you use a window manager that supports this feature, you may find that your work environment is less cluttered and more useable overall.

Finally, window managers do sometimes intercept certain keystrokes. If WordPerfect uses one of these keystrokes, you'll get the window manager's function for that key instead of WordPerfect's. To correct this problem, you can check your window manager's documentation to see if you can adjust which keystrokes it uses, or you can adjust WordPerfect's keyboard mapping to reassign the intercepted keystroke.

→ If you want to learn how to adjust WordPerfect's keyboard mapping, **see** "Keyboard," **p. 788**

USING FILE MANAGERS

Another part of Windows that a number of different programs in Linux can handle is the file manager. Like window managers, file managers provide helpful additional functionality for the X environment. Unlike window managers, file managers are not a practical necessity, and your Linux system may or may not have come configured with a file manager. Fortunately, WordPerfect for Linux includes some file management features of its own, but you may also want to familiarize yourself with any file manager you have on your system.

→ If you want to learn more about file management in Linux, **see** "Managing Files and Folders," **p. 71**

Some file managers, such as the one built into the K Desktop Environment (KDE) package, emulate the Windows look and feel to a greater or lesser extent (see Figure 1.4). When properly configured, they may even be able to launch WordPerfect when you select a WordPerfect document in the file manager's window. These file managers typically work much like Windows, allowing you to select files, copy them by dragging, and so on.

One thing to keep in mind is that, unlike in Windows, in Linux the file and window managers do not automatically learn about an application when you install it. Thus you are likely to need to edit some configuration files to get WordPerfect to show up on your window manager's menus or to have your file manager launch WordPerfect when you select a WordPerfect document in a file browser.

→ If you want to learn more about setting up WordPerfect on your system, **see** "Installation and System Configuration," **p. 711**

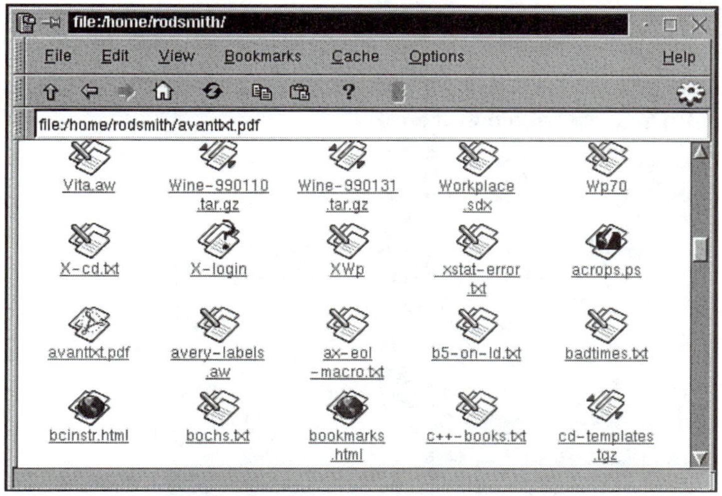

Figure 1.4
The KDE file manager uses a Web browser–style interface similar to that in Windows 98, including icons for various document types and the capability to launch a program when an associated document file is selected.

Sampling WordPerfect's Look and Feel

No GUI is complete without menus, dialog boxes, and other elements. Following the modular tradition of UNIX, X itself does not provide these key program components. Instead, individual programs use GUI libraries, often called *toolkits*, to provide these program staples. The choice of toolkit is up to the programmer, not the user, so the toolkit used by WordPerfect has already been chosen for you. WordPerfect uses a toolkit known as *Motif*, which is popular among commercial programs for Linux. Other programs may use different toolkits, and this variety can result in user-interface elements, such as menus and dialog boxes, that have a distinctly different "look and feel" from one program to another. Those unfamiliar with Linux may find this feature disconcerting at first, but it's not too difficult to get used to. Fewer than half a dozen toolkits are in wide use in the Linux world, although maverick programs that use rarer toolkits add to the variability.

For the most part, the Motif toolkit, and therefore WordPerfect for Linux, should not seem very alien to somebody familiar with Windows, MacOS, or OS/2. A few elements of the Motif interface deserve description, though. Figure 1.5 illustrates a Motif dialog box. Specifically, it is the Print dialog box from WordPerfect.

The Print dialog box contains a number of buttons similar to those in Windows. These also function much as they do in Windows, bringing up additional dialog boxes or terminating the current one.

- Spin boxes—Spin boxes can be edited directly or their values can be adjusted by clicking on the adjacent small buttons with up and down arrowheads.
- Option buttons—These buttons present menus that allow you to select a single item from a fixed set of options.

- Radio buttons—Only one button from a set may be active at any given time, and clicking one automatically deactivates any others that may be active.
- Check buttons—These buttons can be selected or deselected independently of one another, unlike radio buttons.
- Ordinary buttons—You click one of these buttons to start or cancel processing or to open another dialog box.

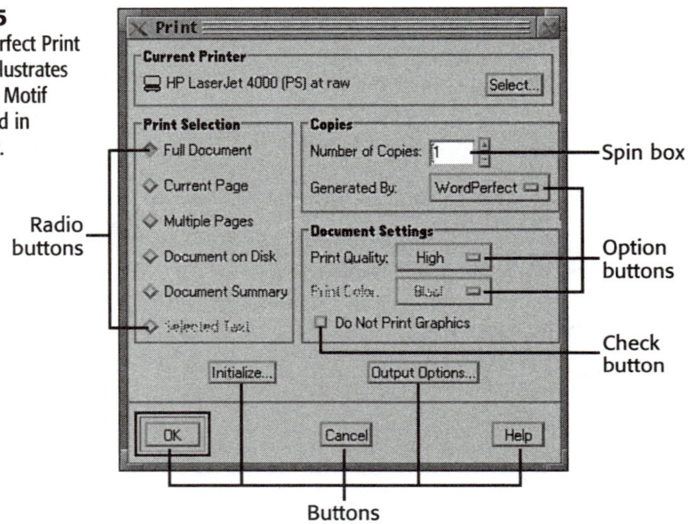

Figure 1.5 The WordPerfect Print dialog box illustrates many of the Motif controls used in WordPerfect.

Both radio buttons and check buttons use a 3D look; when they appear to be depressed, they are active. This effect is relatively subtle, however, so if you're unfamiliar with Motif applications, you may find it difficult to tell when an item is active and when it's not. Any type of element may be deactivated by context. For instance, the final radio button in the Print Selection area, Selected Text, is not active in Figure 1.5 because no text was selected when I told WordPerfect to print. The Print Color item in the Document Settings area is inactive because the selected printer, the HP LaserJet 4000, is a black-and-white laser printer incapable of producing color output. Some of these elements, such as the diamond-shaped radio button and the square check buttons, may appear in menus as well as in dialog boxes.

RUNNING THE XTERM: YOUR KEY TO LINUX FROM X

When using Linux from within X, one standard program is absolutely indispensable: the *xterm*. Several variants on this program are available, but they all do much the same thing. Their task is to provide a command-line interface to Linux from within X. From an xterm, you can run any text-mode program you can run from a purely textual login to Linux. You can also launch X-based programs from an xterm, and unless your window manager or file manager happens to be configured to recognize WordPerfect, doing so is the quickest way to get WordPerfect running.

The xterm actually runs within it a *shell* program, that is, a program that takes your typed input and interprets it as valid commands. As with window and file managers, Linux offers a variety of shells, such as `bash` and `csh`. Fortunately, for simple operations most shells operate in much the same way, so the examples given here work on most systems.

Most Linux systems are configured at install time to start up one or more xterms when X is launched, so you will probably see one on your desktop already. If not, look for a menu item or desktop icon called xterm or some variant on that, such as nxterm. It might also be called X terminal, terminal, or something similar. When running, the xterm will display some variety of prompt, which you can configure by editing a file for your shell. By default, this prompt may contain your username, machine name, current directory, all of these, or none of these. Try typing `ls` and pressing the Enter key after you locate your xterm and give it focus. This step produces a listing of files in the current directory, which is probably your home directory. If you are using this machine for the first time, the directory is likely to be empty, so you simply get another prompt. Type `ls /` to show the contents of the *root directory*, which is guaranteed to contain files. Although the number of commands available in a typical Linux shell is huge, a few such commands (summarized in Table 1.1) are extremely useful. Note that Linux is case sensitive in its shell commands, so `Ls` is not the same as `ls`.

TABLE 1.1 THE MOST USEFUL COMMANDS FOR AN XTERM

Command	Description
`cd`	Changes from one directory to another.
`cp`	Copies a file. Give the original file first after the `cp` command and the destination file second.
`diff`	Displays differences between two files. If the files are the same, returns nothing.
`exit`	Closes the current xterm.
`less`	Displays a file in the xterm window one full window at a time. Press the spacebar to see the next section of the file; press Q to quit and return to the prompt.
`ln`	Creates a *link* from one location to another so that you can easily access a file in one location from another without typing the complete pathname.
`ls`	Displays the contents of the current directory or of the directory given as an argument.
`mv`	Moves or renames a file or directory. Give the original filename first and the destination or new filename second.
`pwd`	Displays the current working directory.
`rm`	Removes (deletes) a file or files.

In DOS and Windows, the backslash (\)separates directories. For instance, C:\Documents\Paper.wpd might be the complete path to a WordPerfect file. Linux uses the standard slash character (/) to separate directories and does not use drive letters, so you might use /home/fred/Paper.wpd instead. A single dot (.) indicates the current directory, and a double dot (..) refers to the current directory's parent. Thus, if your xterm's current working directory is /home/fred, typing `cd ..` changes that to /home.

➔ If you want to learn more about how Linux handles directories, **see** "Understanding the Linux Filesystem Model," **p. 72**

If you type the name of a program in an xterm window, that program executes and takes over the window, even if it doesn't produce any textual output in the window. The result is that the xterm can't be used for anything else until you quit from the program you launched. You can get around this problem in two ways. First, you may append an ampersand (&) to the end of the line you use to launch the program. This method is the easier alternative if you know ahead of time that you want full access to both the xterm and the program when you launch the program, but you may not always have that information or you may forget to include the ampersand. Therefore, if you've already launched a program, the second way to keep both the program and the xterm active is to give the xterm window focus and press Ctrl+Z to reactivate the xterm and deactivate the program you launched. To use the program and the xterm simultaneously, you can then type `bg` and press the Enter key in the xterm to reactivate the launched program.

Remote Access to WordPerfect for Linux

Because Linux's heritage is in multiuser UNIX, most things that you can do when sitting at the *console* (the keyboard and monitor that are attached directly to the computer) you can also do remotely. Remote access can be extremely useful if you are, say, a college student in need of working from assorted locations around campus or if you run a small home network and want to work in a room other than the one in which your Linux box is located. Two or more people can even use the same Linux computer simultaneously by working at separate remote locations. A great way to get more life out of aging computers is to use them as terminals for a single more modern Linux machine!

To use Linux effectively from a remote location, you must have some way to access Linux remotely. In this context, "remotely" can mean from a computer physically next to the Linux box or halfway around the planet. A necessary prerequisite for this sort of access is that both the Linux computer running WordPerfect and whatever you use for remote access be connected via some type of network, which could be a simple serial cable or the whole of the Internet.

One common and inexpensive means of remote access is through a character-mode terminal, be it a standalone terminal or a terminal program on another computer. This method provides text-only access, which is not enough to run the full GUI version of WordPerfect, though it is enough to run the text-mode version that comes with the server variety, described shortly. To run the full GUI version of WordPerfect from a remote location, you need an *X terminal* or an *X windowing environment* for another computer.

X terminals, not to be confused with the xterm window, are essentially limited computers that can run X but no programs; for that, they rely on a host UNIX or Linux system. X terminals are common on college campuses and in certain businesses, but you aren't likely to find one in your local computer superstore. X windowing environments for Windows or MacOS are essentially a way to give these computers the functionality of X terminals. Compared to X terminal hardware, X windowing environments are more commonly available and can be obtained from a variety of sources. With any of these environments, you can run WordPerfect for Linux remotely.

Note Learn everything you need to know about networking essentials in *Upgrading and Repairing Networks*, published by Que.

As with many Linux features, however, you may need to tweak the default installed configuration, so be prepared to dig into some Linux documentation for this project. Also, keep in mind that the WordPerfect license forbids the use of the software by more than one person at a time unless you purchase the server license or several individual licenses. Therefore, using this capability to give everybody in your home or office *simultaneous* access to WordPerfect may not be legal.

Security Considerations

If your computer is connected to the Internet, you need to take some extra security precautions. Here are two things you can do to make your system more secure:

- Keep your security-related programs up-to-date. The company or organization that produced your distribution probably has an errata page on its Web site. Check occasionally and look for anything that says it's a security improvement.
- Choose your password carefully. The best passwords are complete gibberish, but you may have trouble remembering them. One helpful strategy is to take two unrelated short words, distort them in some minor way, and throw in digits or punctuation. For instance, ca9TpE3n is a combination of *cat* and *pen* with the digits 9 and 3. Linux passwords are case sensitive, so the uppercase T and E can help add to the security of this password. Using words from two or more languages or rare technical words can also help. The key is to make your password something that another person or a computer using a dictionary to search for potential passwords cannot easily guess.

More Useful Linux Programs

Chances are, you don't want to use Linux just to use WordPerfect for Linux. Many useful Linux programs are available, and an entire encyclopedia could be written about them. The Web site `http://www.linux.org` includes pointers to many of these programs. The CD from which you installed Linux also has many useful programs, though what they are may not be obvious from their filenames. A few that may be particularly appropriate for use with WordPerfect include

- The GIMP—The GIMP (or GNU Image Manipulation Program) is Linux's most powerful image-editing program. It includes basic image-manipulation features, plus plug-in and scripting capabilities to allow advanced users to do very sophisticated image manipulations. Its Web site is `http://www.gimp.org`.
- GNU Plot—You can use this scientific and technical graphing package to display numeric data in graphics form. One Web site dedicated to GNU Plot is `http://www.cs.dartmouth.edu/gnuplot_info.html`.
- NExS—NExS is a commercial X-based spreadsheet program. Corel has announced that it will produce a Linux version of Quattro Pro, but until then NExS may be useful to Linux users. The NExS Web site is `http://www.xess.com`.
- Netscape—The popular multiplatform Web browser is available for Linux. The personal retail version of WordPerfect includes an older version of Netscape, but you can obtain the latest version from Netscape's Web site at `http://www.netscape.com`.
- Adobe Acrobat Reader—This program is a multiplatform viewer for so-called Acrobat (or PDF) files. Product documentation and other files are increasingly being distributed in Acrobat form. As with Netscape, the personal retail version of WordPerfect includes Adobe Acrobat Reader, but you can check for a newer version at `http://www.adobe.com`.
- Ghostscript—This utility comes with all major Linux distributions, but because of licensing requirements, the distributions don't include the most up-to-date version. Ghostscript converts PostScript, which is a computer language intended to describe a printed page for printers, into a format that can be printed by non-PostScript printers. Ghostscript can also produce Acrobat files from WordPerfect's PostScript output. The main Ghostscript Web site is at `http://www.cs.wisc.edu/~ghost`.

Some of these programs partially duplicate functionality that's present in WordPerfect; for instance, WordPerfect includes plotting and graphics capabilities similar to those in GNU Plot and the GIMP. These other programs are dedicated to their tasks, however, and so include more functionality in their specialized areas.

In some of these cases, the preceding programs interact with WordPerfect in some way; for example, Ghostscript takes WordPerfect's PostScript printer output, and Netscape or Acrobat are called for certain WordPerfect help functions. In other cases, you may want to include the output of one of these programs in a WordPerfect document, as in loading a GIMP-created graphic into a WordPerfect flyer or newsletter.

→ For more information on Ghostscript and printer configuration, **see** "Fonts and Printing," **p. 739**

→ For more information on WordPerfect's help systems, **see** "When You're Stuck: Online Help," **p. 35**

→ For more information on importing graphics into WordPerfect, **see** "Inserting Graphics from External Sources," **p. 379**

If You Are New to WordPerfect

WordPerfect supports a plethora of features, ranging from common word processing functions such as text justification and columns through complex features such as basic spreadsheet functionality, macros, and on-the-fly grammar and spell checking. WordPerfect for Linux is a modern *WYSIWYG (p. 740)* word processor which should be immediately useable (though perhaps not optimally) to anybody familiar with other word processors in Linux or on other platforms. If you're looking for a specific feature, try skimming through WordPerfect's menus or browsing the contents or index to this book to find it.

WYSIWYG Editing

As a WYSIWYG program, WordPerfect displays text on the screen as it will appear on paper, within limitations imposed by the technology (for example, printers generally produce much higher resolution output than do computer monitors). Linux experts may be familiar with programs such as TeX, LaTeX, or LyX, which produce printed output that may be formatted completely differently from what's on the screen. By default, WordPerfect's display is what might be called "very WYSIWYG," showing not just fonts and line breaks as they will appear, but also margins, headers, footers, and other page design elements. Some other word processing programs may show the bulk of the text in a WYSIWYG fashion but omit headers, footers, page breaks, and so on by default. WordPerfect does have options to reduce the level of its WYSIWYG display to eliminate headers, footers, and other elements that may sometimes clutter the display.

Better Editing with Codes

WordPerfect is a *code-based* word processor—that is, it works by embedding codes to perform various operations within a stream of text. Unlike most modern word processors, WordPerfect gives you access to these codes through its *Reveal Codes* feature, described in detail in Chapter 4, "Making Changes to a Document: Basic Editing." You can use the power of Reveal Codes to determine why a document isn't displaying or printing in the way you expect. You can click on and edit an individual code, delete a code, or add one precisely where you want it. Simple features such as commands to change a font or center a line of text are standard codes, whereas more complex features such as styles may appear as a compound code that contains separate codes. To the casual user, WordPerfect's code-based nature need not obtrude itself, but to an advanced user codes are an extremely flexible and valuable tool.

In fact, the combination of WYSIWYG editing and WordPerfect's codes gives the program much of the power of both WYSIWYG word processors and page layout languages such as TeX or LaTeX. TeX and LaTeX advocates often claim that WYSIWYG word processors rob the user of precise control over the formatting commands because the word processor must necessarily make assumptions about the placement and extent of formatting commands. With WordPerfect's Reveal Codes, you can regain some of the precision lost to WYSIWYG convenience.

EXTENSIVE PRINTER AND FONT SUPPORT

WordPerfect for Linux includes one of the largest selections of printer drivers available for Linux. It is possible to print to many printers using a printer driver designed specifically for that printer, using that printer's built-in fonts and other features. This approach may speed up printing relative to using a PostScript driver and printing via Ghostscript, which is a common printing strategy in Linux. Alternatively, you can use a WordPerfect PostScript driver and print to a regular Linux printer queue that directs its output through Ghostscript. This dual-driver strategy gives you flexibility and options you may not have with other Linux word processing software.

WordPerfect for Linux uses PostScript Type 1 fonts (even when printing with its own drivers to non-PostScript printers). Because X and Ghostscript both use Type 1 fonts, most Linux computers already come with a number of Type 1 fonts, and more are available on commercial font CDs or the Internet. If you print using a WordPerfect driver for the printer, you have access to that printer's built-in fonts, as well as any Type 1 fonts you install.

Tip #3 from
Tux Tips

The version of WordPerfect 8 for Linux that can be easily downloaded from the Web contains no font installer and thus can handle only a few fonts. For greater font choices, purchase the retail version of the program. This version comes with a selection of fonts and has a font installer so that you can install more.

→ For more information on installing fonts, **see** "Installing and Managing Fonts," **p. 754**

EXCHANGING FILES WITH OTHER PROGRAMS AND COMPUTERS

WordPerfect includes an extensive collection of file import/export filters, which allows the program to read files from foreign applications and to write files readable by them. If you've previously used another word processor, there's a good chance that WordPerfect can read your files. The Linux version even includes filters for Applix Word, a competing Linux word processor. Other filters include Microsoft Word, rich text format (RTF), ASCII text, and FrameMaker. If you need to convert files from some other word processor, some "neutral" format, such as RTF, may work well; or because WordPerfect has been a major word processor for years, the other program may support export to WordPerfect format.

WordPerfect for Linux 8 and WordPerfect for Windows versions 6.0 through 8.0 use the same file format, so file exchange between Windows and Linux boxes is a breeze. Keep this feature in mind if you need to export to WordPerfect format, too; if an application can save to WordPerfect 6 format, WordPerfect 8 for Linux should also be able to read that format. WordPerfect for the Mac uses a completely different file format, though, so if you've used WordPerfect on the Mac, be sure to export your files to WordPerfect 6/7/8 format before trying to use them in Linux.

→ For more information on exchanging document files with other word processors, **see** "Document Exchange," **p. 88**

SYSTEM REQUIREMENTS

The official requirements for running WordPerfect 8 for Linux are

- 43MB to 87MB hard disk space
- 9MB memory for the first user
- 2.5MB memory for each additional user

As a practical matter, of course, WordPerfect benefits from having even more resources. Keep in mind that these are the requirements for WordPerfect only; Linux, X, your window manager, and other programs all require RAM and disk space. As a practical matter, a Linux system being used primarily for word processing with WordPerfect could probably get by with a 500MB hard disk and 16MB of RAM. One GB of disk space and 32MB of RAM make for a comfortable working environment, even including some additional tools.

WordPerfect 8 for Linux currently runs only on x86-based computers and those using the StrongARM CPU used in the NetWinder computers, so if you're running Linux on a Mac, Alpha, or other computer, you may be out of luck. Corel may eventually port WordPerfect to other Linux platforms, however.

In principle, WordPerfect for Linux should run on any x86 or compatible CPU that runs Linux, meaning anything from a 386 on up. For best performance, I recommend a fast 486 or better. CPUs from Intel, AMD, NexGen, Cyrix, IBM, and IDT should all work fine.

In practice, you want to run WordPerfect for Linux with a display size of at least 800×600. The program can run in 640×480, but some dialog boxes, including the file load and save dialog boxes, are over 640 pixels high, so using the program at such a low resolution is annoying at best.

Tip #4 from *Tux Tips*

If you're restricted to 640×480, consider using an X *virtual screen* of larger dimensions. This feature of X allows you to maintain a desktop that's larger than your monitor can physically display. You then scroll through this larger desktop with the mouse, displaying only part of it at a time. The scrolling process can be awkward but is better than not being able to access parts of a dialog box because it's too large for the screen.

On some systems, running WordPerfect in 24-bit color depth distorts the display of graphics and icons. However, 8-, 16-, and 32-bit modes generally work fine, and sometimes changing to another X server can fix the 24-bit depth problem. With 8-bit displays, X often runs out of colors when you run color-intensive programs, a condition that can make WordPerfect or other programs look bland and colorless.

CHANGES FROM VERSION 7

If you have used WordPerfect 7 for Linux, you may be wondering what is new in version 8. Features in version 8 that were not present in version 7 include

- Grammar-As-You-Go grammar checking for checking your grammar as you type

→ For more information on Grammar-As-You-Go, **see** "Proofreading with Spell-As-You-Go and Grammar-As-You-Go," **p. 118**

- The shadow cursor to place your cursor anywhere on a page, even where there is no text

→ For more information on the shadow cursor, **see** "Entering Text," **p. 44**

- Enhanced file management features

→ If you want to learn more about WordPerfect's file management features, **see** "Linux File Manipulation," **p. 80**

- Enhancements to table management features

→ If you want to learn more about creating and manipulating tables, **see** "Tables," **p. 275**

- Version Control to manage multiple revisions of a document, allowing you to go back to a previous version if you decide to undo a large number of changes

→ To learn more about managing multiple versions of a document, **see** "Document Revision Tracking," **p. 591**

- Expanded and improved support for Web page creation and editing

→ To learn more about creating and editing Web pages, **see** "Web Publishing," **p. 659**

- WYSIWYG selection of recently used fonts

→ If you want to learn more about selecting fonts, **see** "Entering Text," **p. 44**

In addition to these major changes, WordPerfect 8 includes changes to its default menu and icon bar layouts, as well as minor cosmetic changes.

VARIETIES OF WORDPERFECT 8 FOR LINUX

One of the changes that has garnered much discussion in the Linux community does not relate to the program's functionality but to its method of distribution. WordPerfect 8 is now available in three versions: a free version available for download from Corel's Web site, a personal retail version available from computer dealers, and a retail server version for use at multiuser sites.

This book primarily covers the downloadable and personal retail versions of the program. Compared to the downloadable version, the personal retail version adds

- More uses permitted by the license
- Drawing, charting, and equation editing features

→ To learn more about WordPerfect's Draw module, **see** "Using WP Draw," **p. 387**
→ For more information on WordPerfect's charting capabilities, **see** "Creating a Chart," **p. 411**
→ For more information on creating equations with WordPerfect, **see** "Equations," **p. 433**

- 130 included fonts and a font installer utility

→ For more information on installing fonts in WordPerfect, **see** "Installing and Managing Fonts," **p. 754**

- Sample clip art, textures, and photos included in the box
- Sample WordPerfect documents
- Print queue manager
- Custom hyphen and supplemental dictionary creation utilities

→ For more information on creating a custom dictionary file, **see** "Creating a Custom Dictionary," **p. 143**

- Online and printed reference manuals

→ For more information on WordPerfect's online help systems, **see** "When You're Stuck: Online Help," **p. 35**

- Adobe Acrobat Reader and Netscape Web browser

Many of these features are covered in this book. The relevant chapters mention when a feature is not present in the downloadable version of the program.

The server edition of WordPerfect for Linux adds the following features:

- Site license management software
- Character-based binary for non-GUI operation
- System and application customization utilities
- Support for the Oracle database engine

These topics are beyond the scope of this book.

If you download WordPerfect, you are asked for a registration number. This request is not an error. You can register the software at Corel's Web site and receive a registration number at no cost. Registration allows you to use WordPerfect for Linux indefinitely, albeit without the extra features of the personal retail version. If you choose not to register with Corel, the program stops working in 90 days. The license to the downloadable version of WordPerfect allows you to use it for personal noncommercial uses or to evaluate its suitability to commercial uses. You should read the license if you have any doubts about whether you can legally use the downloadable version in the way that you intend.

LAUNCHING WORDPERFECT FOR LINUX

If WordPerfect for Linux is not already installed on your computer, you need to install it.

→ To learn how to install WordPerfect for Linux, **see** "Installation and System Configuration," **p. 711**

Because of the differences in window and file managers described earlier, the way you start WordPerfect for Linux usually depends on the environment you're using. The one method that works in all environments, however, is to use an *xterm (p. 44)* to start WordPerfect by typing its name. Follow these steps:

- If necessary, start an xterm window.
- In that window, type the complete path to the WordPerfect for Linux executable. This file is called xwp, and it is stored in the wpbin subdirectory in the WordPerfect installation directory. Thus, if WordPerfect is installed in the /opt/wp8 directory, you would type `/opt/wp8/wpbin/xwp` to start WordPerfect.

Tip #5 from *Tux Tips*

> Your system may be configured so that you can launch WordPerfect simply by typing `xwp` in an xterm, rather than by typing the entire path; or your window manager may be configured so that you can launch WordPerfect by clicking on an icon or selecting the program from a menu. If your system is not configured in either of these ways, see Appendix A, "Installation and System Configuration," for configuration instructions.

However you do it, after you launch WordPerfect you should see two windows on your screen. The larger one is for entering and editing text, and the smaller one is for managing documents. The windows will look something like the windows shown in Figures 1.1 through 1.3, though, of course, the details depend on your window manager.

A Window and Menu Quick Tour

WordPerfect for Linux displays two windows at startup. One has the title WordPerfect and the other, if no document is loaded, is titled WordPerfect - [Document1 - Unmodified (Current)]. The first window is the WordPerfect Control window. It provides functions relating to overall program and document management, such as exiting from WordPerfect and creating a new document. The second (and larger) window is the window you use for actual text editing, including all the functions you apply to text.

Note

> If you're familiar with WordPerfect for Windows, you may be puzzled by the presence of the Control window because WordPerfect for Windows folds those functions into the editing window. This separation in WordPerfect for Linux has few practical consequences, though; it's just a minor difference in the user interface.

The Program Control Window

The WordPerfect Control window contains four menu items, as shown in Figure 1.6: Program, Preferences, Window, and Help. The WordPerfect logo in the window is there for show and increases the size of the window so that it's less easily lost in your desktop.

- Program—The Program menu item contains options for creating new or loading old files (the menu uses *window* as a synonym for "file" or "document"), for manipulating files and printers, and for exiting from the program.

- Preferences—Clicking on the Preferences menu item immediately brings up a dialog box used for setting various program defaults.

➔ To learn about the preferences you can set, **see** "Customizing Preferences," **p. 767**

- Window—The Window item provides options for automatically positioning Document windows on the screen and for switching among multiple open documents.
- Help—The Help item provides options for getting help from the online help systems.

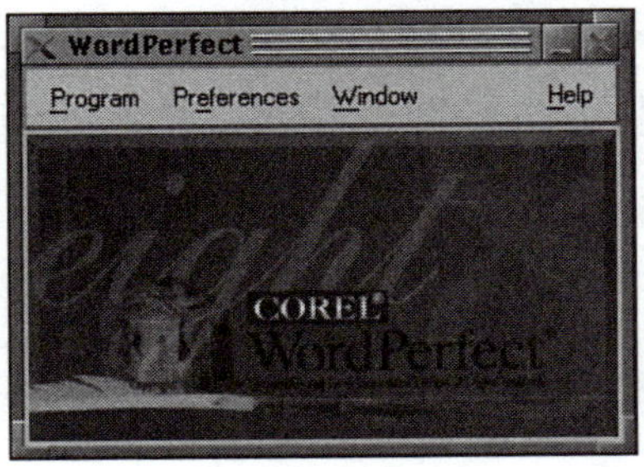

Figure 1.6
The WordPerfect Control window provides functions for managing files, documents, and windows whether they're open or not, as well as for setting programwide preferences.

Note

In Linux, it's common, though not universal, for Help menu items to be located at the far right side of menu bars. WordPerfect follows this convention.

THE DOCUMENT WINDOW

The Document window is the one you work in most when you use WordPerfect for Linux. The dominant feature of the Document window is its editing area, which is large and white, with horizontal and vertical guidelines marking out the margins, as shown in Figure 1.7.

THE INSERTION POINT AND SHADOW CURSOR

The editing area sports a blinking vertical cursor (also called an *insertion point*) that marks where text you type appears on the screen. As you move your mouse pointer around the window, a blue cursor moves with the pointer. This cursor is the *shadow cursor*, which allows you to place the insertion point anywhere on the page, even where there's no existing text. You can use this feature, for instance, to offset a block of text without using either the Tab key or the spacebar or adjusting the margins.

THE STATUS BAR

At the bottom of the window is the *status bar*, which provides useful one-click controls and status information. The shadow cursor toggle disables or enables the shadow cursor feature, which is on by default. Thus, if you don't normally like to use this feature, you can easily disable it from the status bar. The printer selector icon allows you to select a printer. The

Insert mode toggle icon switches WordPerfect between Insert and Typeover modes—the former inserts text at the insertion point, whereas the latter replaces text. The status bar also contains information on the current page, line number, and horizontal position of the insertion point.

Figure 1.7
The WordPerfect Document window provides a plethora of controls and features.

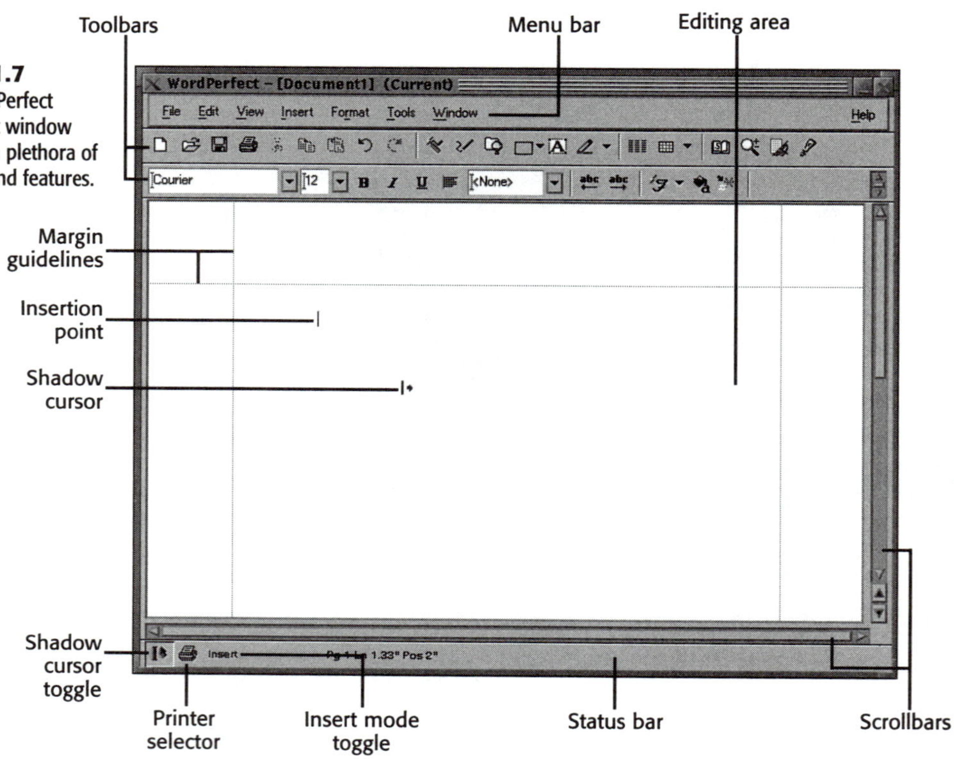

THE SCROLLBARS

Along the right and bottom edges of the Document window are the *scrollbars*. These allow you to scroll through a document top to bottom or pan the window's view left to right. You generally don't need to use the horizontal scrollbar to view documents created for 8.5×11-inch paper, but the scrollbar is available if you need to expand the view for some reason or if you need to work on wider documents. At the bottom of the vertical scrollbar is a pair of arrowhead icons. Clicking on these scrolls the document up or down by one page, placing the insertion point and document view at the top of the page.

TOOLBARS

One *toolbar* and one *property bar* appear immediately above the editing area. For simplicity's sake, I refer to both items as *toolbars* because they provide similar types of functions and interfaces. These toolbars provide one-click access to commonly used features, represented by icons.

The lower of the two default toolbars provides settings related to text formatting. The items on this toolbar (see Figure 1.8) are

- Font selector—The font selector lets you select a font. The default varies with your printer and printer settings but is usually Courier or something similar. Click on the triangular button to the right of the font name to change the font. A drop-down menu with a selection of fonts appears, as in Figure 1.8. If you have many fonts installed, this list contains its own scrollbar, which you can manipulate to see fonts earlier and later in the list.

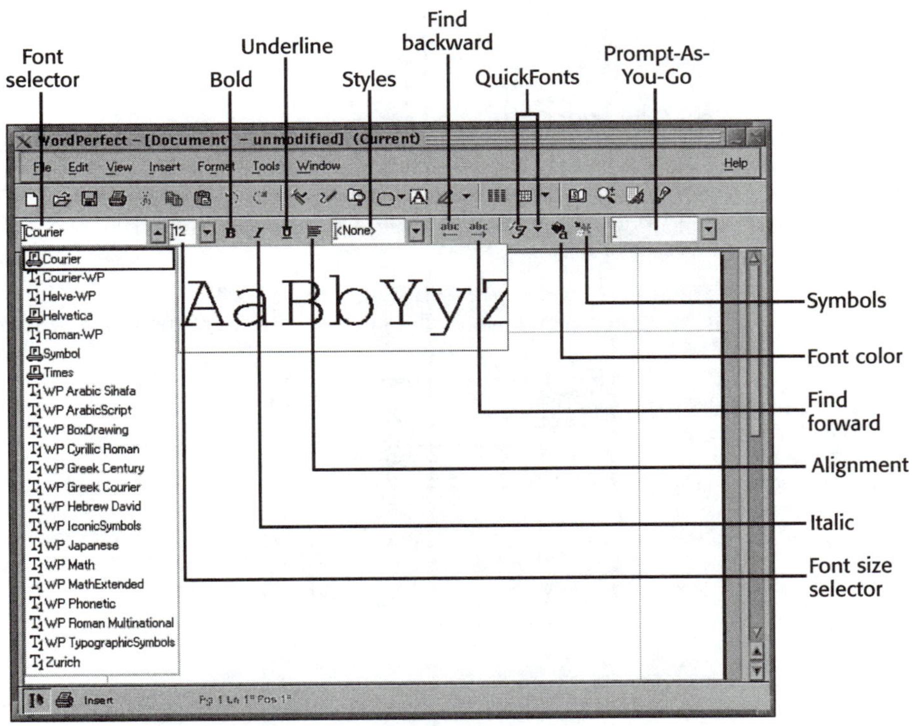

Figure 1.8
Fonts with a printer icon next to their names are built into the printer or downloaded to a PostScript printer in raw form, whereas WordPerfect renders those with a T1 next to their names.

- Font size selector—The font size selector, not surprisingly, enables you to set the font's size in points. Point sizes of between 10 and 12 are common in most documents.
- Bold, Italic, and Underline—The Bold, Italic, and Underline icons toggle boldface, italic, and underlined attributes, respectively.
- Alignment—If you click on the Alignment icon, you'll see different options for *justification*—whether text is aligned to the left, center, right, or both edges of the paper.
- Styles—The Styles toolbar item, which by default is `<None>`, allows you to select a style. *Styles* are collections of formatting options used to help create consistent documents.

→ To learn about styles, **see** "Ensuring Consistency Within Your Document with Styles," **p. 305**

- Find—The Find forward and Find backward icons are quick search tools. Try typing a few words, being sure to repeat a word several times. Then double-click on one of the repeated words so that it's highlighted. Click on the Find tools and watch the highlight jump from one word to another instance of the same word.
- QuickFonts —The QuickFonts icon allows you to quickly select a font, complete with attributes, from among some that you've recently used, as shown in Figure 1.9.

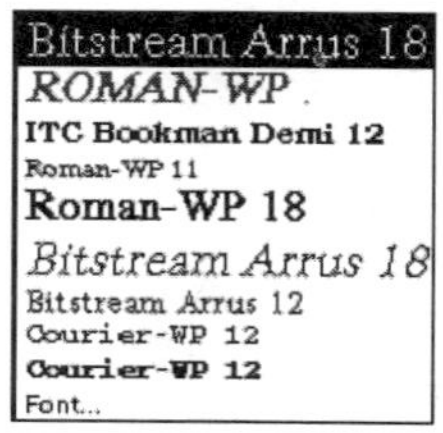

Figure 1.9
The QuickFonts list displays the fonts as they will appear in your document.

- Font color—The Font color icon pops up a rectangular color selection chart for setting the color of text.
- Symbols—The Symbols icon brings up a font insertion dialog box to allow you to insert special characters, such as mathematical or typesetting symbols, as in Figure 1.10.

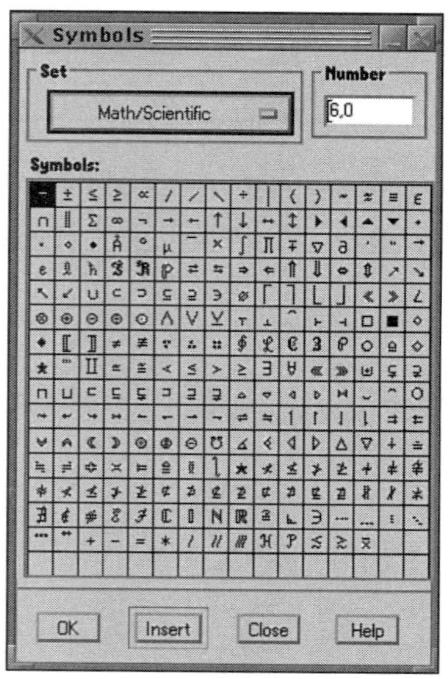

Figure 1.10
To insert a symbol in your document, select the set (such as Math/Scientific in this example), click on a symbol, and click OK.

- Prompt-As-You-Go—The Prompt-As-You-Go element displays suggested spelling and corrections or synonyms for the word on which the insertion point is placed. You may need to widen your WordPerfect Document window to see the Prompt-As-You-Go element.

The upper default toolbar (Figure 1.11) provides shortcuts to less textually oriented functions.

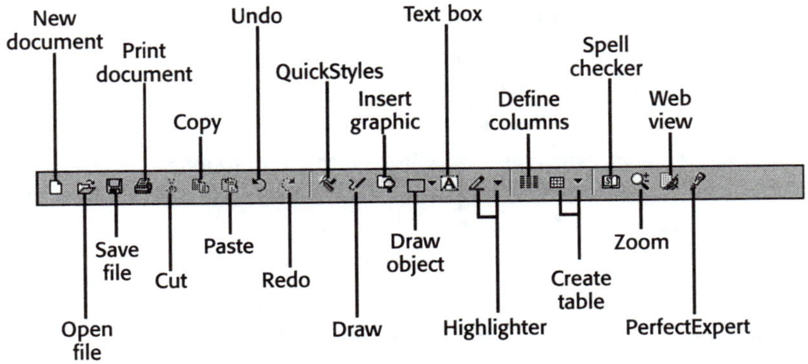

Figure 1.11
You may need to widen your window to see all the icons on a toolbar.

- New document—The New document item opens a new empty Document window.
- Open file—The Open file icon loads a file into the existing window.
- Save file—The Save file icon saves the current file to disk.
- Print document—The Print document icon prints the current file or a selected subset of it.
- Cut and Copy—The Cut and Copy items are active only if some text is selected. They remove the text and copy it to the Clipboard, respectively.
- Paste—The Paste icon is active only if text is currently in the Clipboard. Paste retrieves that text and places it at the insertion point in the document.
- Undo and Redo—The Undo and Redo icons undo and redo actions, respectively. If you make a mistake, say by deleting something you did not intend to delete, you can undo that action with the counterclockwise arrow. If you subsequently decide that you do want to delete it, you can select the clockwise arrow to redo that deletion.
- QuickStyles—The QuickStyles tool creates a style based on the current formatting.

➔ To learn about styles, **see** "Ensuring Consistency Within Your Document with Styles," **p. 305**

- Draw—The Draw tool starts the drawing module, which is not present in the downloadable version of WordPerfect.
- Insert graphic—The Insert graphic icon loads a graphic from a disk file.
- Draw object—If you're careful to select the triangle in the Draw object icon, you get a small menu of shapes. Select one, and you can insert it directly into the current document, without loading the regular graphics module.
- Text box—The Text box icon creates a text box. Text boxes enable you to offset a piece of text from the main body of a document, as in a sidebar in a magazine article.

➔ If you want to learn more about text boxes, **see** "Using Text Boxes to Add Variety to Text Layout," **p. 357**

- Highlighter—The highlighter allows you to *highlight* a section of text. Click on the triangle to select a color and then select some text. When you click on the marker icon, the text's background turns into whatever color you selected.
- Define columns—The Define columns icon enables or disables column formatting for the selected text or the text after the insertion point only.
- Create table—The Create table item, unlike other icons with triangles, is one item, and it inserts a table into the text. Click on the item to display a grid. Keep your left mouse button depressed and drag it over the grid to select the size of the table; when you release the mouse button, the table appears.

➜ If you want to learn more about tables, **see** "Tables," **p. 275**

- Spell checker—The Spell checker icon launches the WordPerfect spell checker.

➜ To learn about the WordPerfect spell checker, **see** "Checking an Entire Document at Once," **p. 131**

- Zoom—The Zoom icon alters the magnification of the window view to allow you to see more or less of the document at once.
- Web view—The Web view icon converts the document to a Web page.

➜ If you want to learn more about converting a conventional WordPerfect document into a Web page, **see** "Exporting Existing Documents as Web Pages," **p. 673**

- PerfectExpert—The PerfectExpert icon starts the PerfectExpert, which is a tool designed to help guide you to an appropriate WordPerfect function to accomplish a given task.

WordPerfect's toolbars can be customized. Try right-clicking on a blank spot to the right of either toolbar. A new menu allows you to change the toolbar to a different one or otherwise edit it. Some WordPerfect features put up new toolbars for you automatically. If you regularly use certain features, you may want to change the default toolbars to ones that are more suited to your style of working. If you remove all your toolbars but want one or more back, choose View, Toolbars to bring up a list of toolbars to display.

THE MENU BAR

The WordPerfect Document window's menu bar provides eight selections. Many of the most commonly used functions can be accessed via toolbars, as well.

- The File menu provides access to file-related functions, such as reading and saving documents, printing, and obtaining information about files.
- The Edit menu allows you to manipulate text and other document elements by cutting, pasting, selecting, and so on. The Find and Replace function (discussed in Chapter 4) is also located on this menu.
- The View menu contains options that influence your view of the document, such as whether a ruler is present, whether graphics are displayed, and whether the document is magnified by a "zoom" factor.

- The Insert menu lets you insert various elements into your document, such as headers or footers, graphics, or a new page command.
- The Format menu contains operations that influence the appearance and layout of text on the page, such as margins, line spacing, or font selection.
- The Tools menu provides tools that you can use to check your documents for errors and to create sections such as tables of contents or indexes.
- The Window menu allows you to manipulate the window as a whole, in particular as it relates to other open WordPerfect Document windows. You can have WordPerfect place windows side by side, top to bottom, or cascaded. You can also switch among windows so that if several documents are open at once, you can easily find the one you want.
- The Help menu provides access to WordPerfect's various help systems.

Pop-up Menus

If you right-click your mouse in the main body of the Document window, a pop-up menu appears, as in Figure 1.12. These menus are *context sensitive*, meaning that their contents depend on their context—the type of text under the mouse pointer, in this case, along with the objects that are selected. Pop-up menus are handy for quickly accessing relevant WordPerfect features without having to find them in the main menu structure.

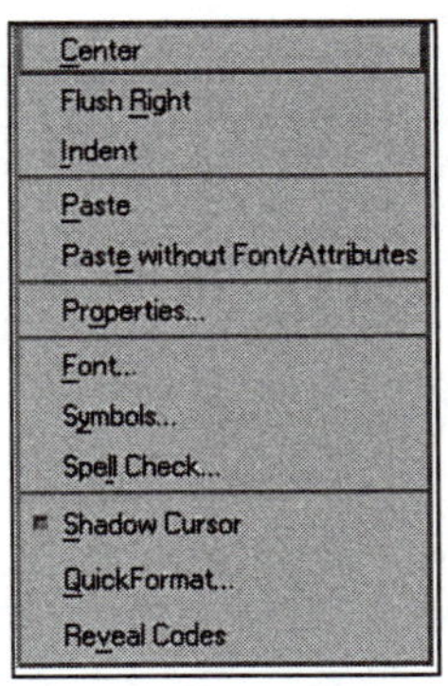

Figure 1.12
Pop-up menus allow you to access key features without moving your mouse from the text you're editing.

When You're Stuck: Online Help

WordPerfect's online help systems can provide useful guidance if you can't remember how to access some feature or perform some task or if you never knew how to do these things in the first place. The hypertext help system is useful for getting a quick clue on how to use some feature but is often lacking in the sort of detail that this book provides. The online manual provides more detail, but where to find things is not always obvious.

NAVIGATING THE HYPERTEXT HELP SYSTEM

From either the Control or the Document window, choose Help, Contents... or choose Help, Search for Help on... to bring up the main hypertext help system, as shown in Figure 1.13. This system is similar to the help system in Windows programs and allows you to click on highlighted terms to go to a page that describes them. The two methods of starting this help system differ in that the Help, Search for Help on... method brings up the system's Search dialog box immediately, whereas the Help, Contents... sequence brings up the introductory page shown in Figure 1.13.

Figure 1.13
The hypertext help system allows you to search for information in a number of ways.

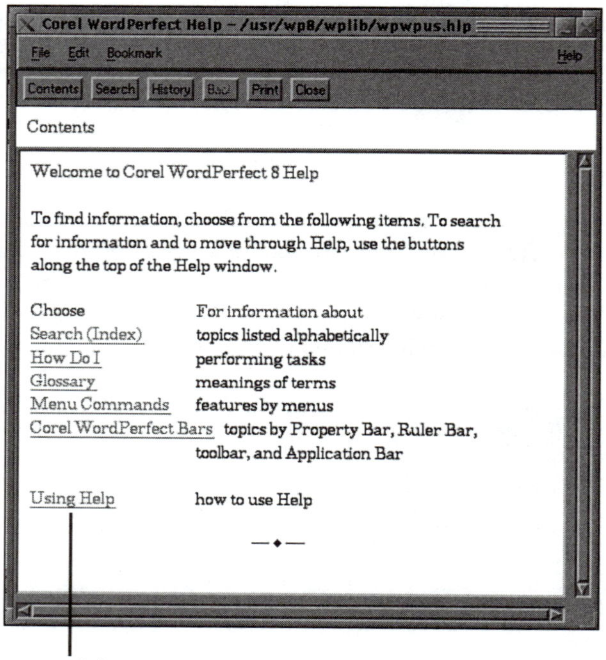

When you load the hypertext help system, you can immediately click one of the six links listed in the Choose column to obtain information. For instance, if you click on Glossary, you can look up the meanings of terms ranging from *active* to *zoom*. One of the best aspects of the help system is the search feature, which allows you to search for a term anywhere in the help database. For instance, you could search on `font` to find information on fonts. When you type `font`, Search displays a list of keywords that match your query (if that word is in the keyword list, of course), as shown in Figure 1.14.

You can select one of the keywords by clicking Find Topics or by pressing Enter. The help system then shows a list of specific topics in which your term appears (Figure 1.15). Select a topic and then click on Go To to see the entry.

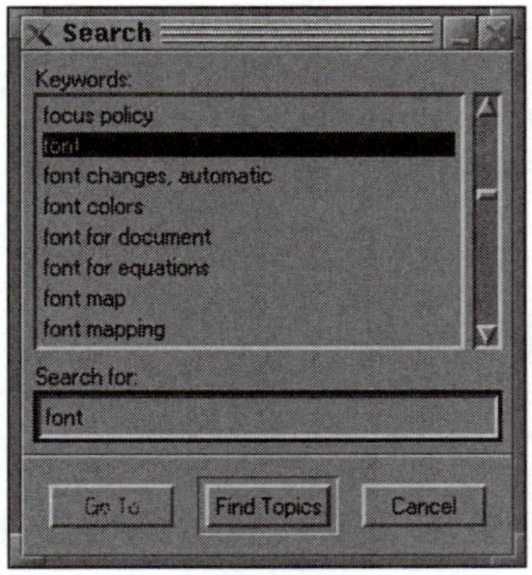

Figure 1.14
Typing one word may produce multiple matches to a query.

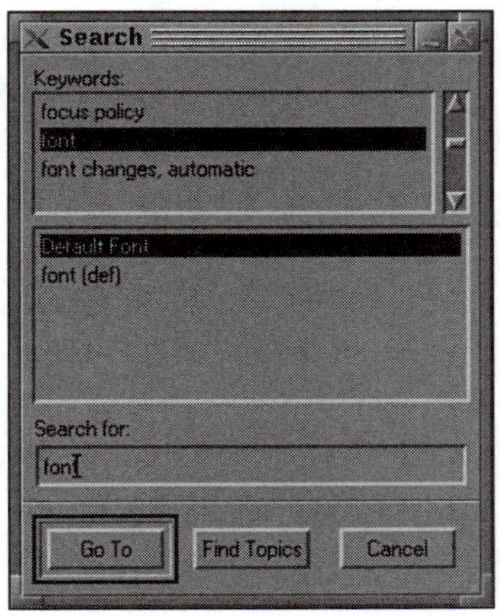

Figure 1.15
One keyword may appear in multiple pages of the help system.

Finding Help on Macros

A *macro* is a collection of operations that you can perform at a single time. For instance, you might want to routinely do a search and replace on certain words and then spell check a document. Macros allow you to combine these two operations into one step. The macro system has its own help system, accessed from the Help, Macros menu. The macro help system is the same as the general hypertext help system except that the former is specific to WordPerfect's macro system.

➔ If you'd like to learn more about macros, **see** "Letting the Computer Do the Work: Macros," **p. 483**

The PerfectExpert

The PerfectExpert is a task-oriented help system. When you start the PerfectExpert, it displays a series of tasks to be accomplished in a writing endeavor. Then click on one task to see more choices, and so on until you reach a point where the PerfectExpert displays the relevant WordPerfect dialog box to set some option. For instance, if from the main PerfectExpert menu (Figure 1.16) you select Typing, Change Tab Settings, the PerfectExpert displays a dialog box for setting tabs. You can set your document's tabs from here and then return to the PerfectExpert dialog box to continue. If you build a document in the way the system's designers anticipated, the PerfectExpert can be helpful, especially if you're unfamiliar with WordPerfect.

The PerfectExpert is a good way to build a document if you've no idea what to do with WordPerfect, but you may find the structure limiting or confusing, especially if you're already familiar with WordPerfect—or any other word processor. Although the PerfectExpert menu structure guides you through the steps of writing a document, where you'd go in the PerfectExpert to accomplish a task is often no more obvious than it is in WordPerfect's main menu system. Also, because whenever you select an option from among the PerfectExpert's buttons those buttons disappear and are replaced by the new options, you can lose track of where you are in the PerfectExpert menu hierarchy.

Figure 1.16
You can go through the PerfectExpert document creation steps top to bottom, or you can omit steps that aren't relevant to your document or which you've handled in other ways.

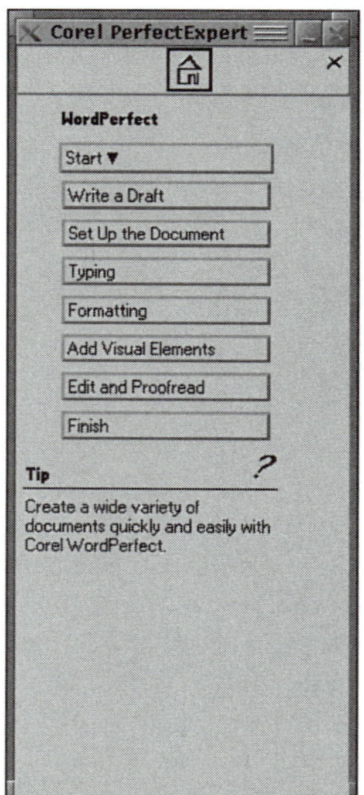

Tip #6 from *Tux Tips*

The ? symbol brings up WordPerfect's hypertext help system, the icon of a home returns the PerfectExpert to its starting point, and the X symbol closes the PerfectExpert. These symbols remain in view as long as the PerfectExpert is running, though the other buttons in the dialog box change as you perform a task.

Online Help

If you purchased the retail version of WordPerfect for Linux and did a full installation, the WordPerfect online manual is at your disposal. This online version is a copy of the printed manual that also came with your package but is accessible from your computer via Adobe Acrobat Reader, as shown in Figure 1.17. Choose Help, Online Help, WP Online Manual... to access the manual. You can use Acrobat's search function to search the online manual.

Figure 1.17
The online manual allows you to look up terms in the contents or index or to browse through the manual's contents.

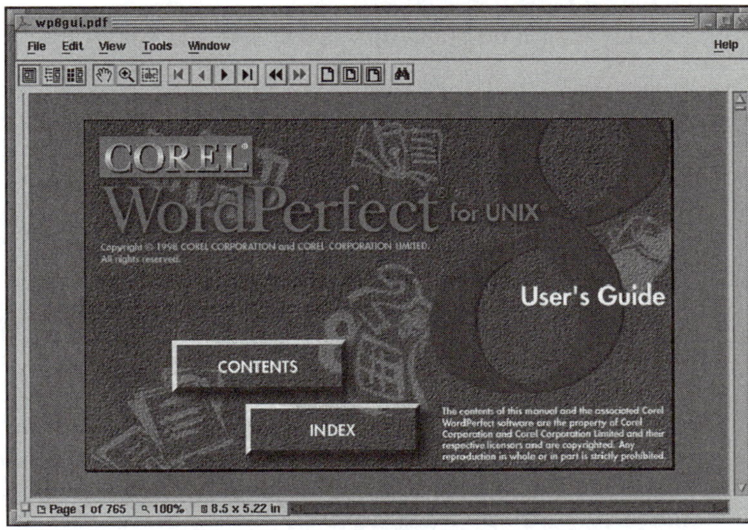

Tip #7 from *Tux Tips*

The personal retail edition of WordPerfect for Linux includes a copy of the Adobe Acrobat Reader. If you did not install it because you already had Acrobat or another PDF file viewer installed, the online manual may not be accessible from WordPerfect's menu. You can fix this problem by creating a symbolic link from the executable you already have to the acroread/bin/acroread file within the WordPerfect installation directory. For instance, if WordPerfect is installed in /opt/wp8 and your existing Acrobat binary is in /usr/local/acrobat/bin, as root you could type `ln -s /usr/local/acrobat/bin/acroread/ opt/wp8/acroread/bin/acroread`.

The three WWW items in the Help, Online Help menu are intended to provide help from online sources on Corel's Web site. Unfortunately, the Web links don't work. Try using your browser to check Corel's Web site directly instead.

CLOSING DOCUMENTS AND QUITTING WORDPERFECT

When you're done using WordPerfect, you may want to exit from it. This step isn't strictly required, however. Linux is very good about managing its memory, and with the pager features in many Linux window managers, you can easily move WordPerfect out of the way. On the other hand, if you work in an environment where others have access to your computer, you should close all applications and log out for security reasons whenever you're not using the computer.

If you want to close the current document only, you can do so from the Document window. Choose File, Close, or press Ctrl+F4 to close the current Document window. If you've made changes to your document since saving it, WordPerfect asks whether you want to save them. Clicking Yes saves the changes and closes the document, clicking No closes the document without saving the changes, and clicking Cancel stops the close operation, allowing you to continue editing.

Whether a Document window is open or not, selecting Program, Exit from the Control window's menu or pressing Shift+Ctrl+F4 quits from WordPerfect entirely. As with closing a single Document window, if you have unsaved changes, WordPerfect asks you what to do about them.

Another way to close a document or exit from WordPerfect is to use the features for window closing in your window manager. These can be a close box (typically marked with an X or similar symbol near the upper-right corner of the window), double-clicking in another control widget, or selecting a Close or Exit option from a menu obtained by clicking on a window manager widget. Some window managers also have keyboard shortcuts to their window-closing features. Using the window manager window-closing features sometimes activates prompts to save unsaved work, but not always. Fvwm, for instance, includes a "close" option that produces the check for unsaved work and a much less polite "destroy" option, which shuts down WordPerfect most unceremoniously. If in doubt, therefore, it's best to use the menu options or WordPerfect's keyboard shortcuts to close documents or quit from the program.

TROUBLESHOOTING

X WON'T START

When you type `startx` at a Linux prompt, text scrolls across the screen and an error message indicates that X won't start.

Your X server isn't properly configured for your video hardware. Configuring X is beyond the scope of this book, but fortunately, other resources are available.

Note

Que's *Special Edition Using Linux* can help you configure X (see Chapter 2, "System Administration").

SYMBOLIC LINKS TO WORDPERFECT

You know how to install and configure Linux programs; have created a symbolic link to the WordPerfect executable file, xwp; and have placed the link on your path. When you try to launch WordPerfect in this way, though, it doesn't work.

WordPerfect is quirky that way. Instead of a symbolic link, try creating a script file that calls the WordPerfect executable using its complete path.

→ To learn about more ways to configure your system to launch WordPerfect, **see** "Installation and System Configuration," **p. 711**

VARIABLE POP-UP MENUS

You can't find something in a pop-up menu that you know you've seen before.

The pop-up menus are *context sensitive*, so they show different menus depending on what sort of text or other element is under your mouse when you call them up. Think about what type of item you saw and to what types of objects it would apply. Position your mouse over that type of object and try it there. Also, be sure you don't have any objects selected that might use different pop-up menus. If necessary, click in the main text area to deselect any text boxes, graphics, or other elements.

CHAPTER 2

Working Within a Document

In this chapter

OVERVIEW

No matter what types of documents you create with WordPerfect for Linux, certain text manipulation techniques and principles are always helpful. After all, WordPerfect is a *word processor*, and it's used to create primarily textual documents. Therefore, you use the same text entry principles with all the documents you create with it. This chapter covers these principles, including how to get the most out of WordPerfect's particular text entry and editing features.

CREATING A DOCUMENT

WordPerfect makes creating a new document exceptionally easy—if you start WordPerfect without specifying any parameters (by typing a filename after `xwp` in an *xterm* *(p. 28)* or by using your file manager to load a specific WordPerfect document), WordPerfect starts with a blank Document window already open. You can begin typing in this window immediately.

If you already have a Document window open and want to start a second one, or if you've created one document, closed it, and want to start a new one, you can do so by choosing Program, New Window from the program Control window (Figure 2.1).

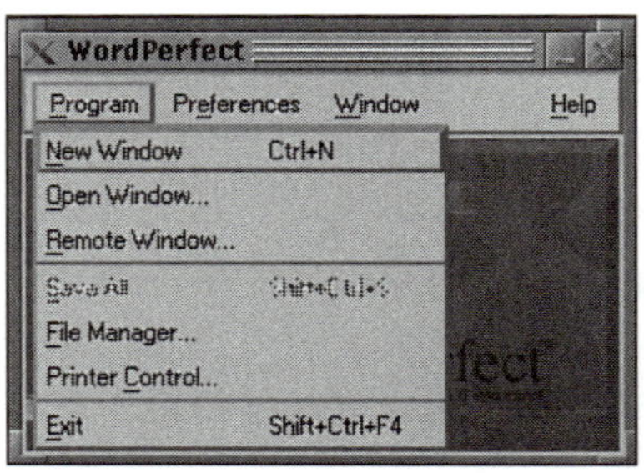

Figure 2.1
The options in the Program menu use "Window" to refer to a document.

Tip #8 from *Tux Tips*

Although Ctrl+N is listed as a shortcut for the Program, New Window option in the program Control window, this shortcut works only from a Document window. If you have no Document window open, you *must* use the menus, not the shortcut.

ENTERING TEXT

BASIC PRINCIPLES OF TEXT ENTRY

If you're familiar with other word processors, either in Linux or in another operating system (OS), text entry with WordPerfect for Linux will be familiar. If you're used to typing on a typewriter, or if you've only used a word processor for small tasks, you should keep some things in mind when using WordPerfect:

- You don't normally press Enter at the end of each line. WordPerfect uses word wrapping and automatically "wraps" the line around to the beginning of the next one. WordPerfect readjusts this wrapping as required when you insert or remove text, change the margins, or make other alterations that necessitate a paragraph reformat.
- Unlike most typewriters, WordPerfect for Linux gives you a choice of several fonts. A font is a set of characters displayed in a particular style, and your choice of font can greatly influence the feel of your document. WordPerfect's default font is usually Courier or something similar (the default varies from one printer to another). Courier is a boring monospaced font similar to that used by many typewriters. You can easily select other fonts, however, many of which are *proportional*—that is, their characters vary in width so that a *W* takes up more space than an *i*.

→ If you want to learn more about using fonts, **see** "Fonts," **p. 178**

→ If you want to learn how to install fonts, as well as how they interact with your printer, **see** "Fonts and Printing," **p. 739**

- You *can* create layouts such as tables, hanging indents, centering, and so on by using tabs and spaces, the way you would with a typewriter; but in most cases, WordPerfect provides features to do this sort of formatting automatically. Of course, if you need to accomplish something right away, you may initially find it easier to perform tasks as you have with other word processors or with a typewriter. In the long run, though, learning to use WordPerfect's formatting tools is helpful.

CURSORS SHOW WHERE ACTIONS OCCUR

WordPerfect for Linux uses three types of *cursors*, as shown in Figure 2.2. These onscreen indicators show where you can perform various operations. The cursor types are the insertion point, the mouse pointer, and the shadow cursor.

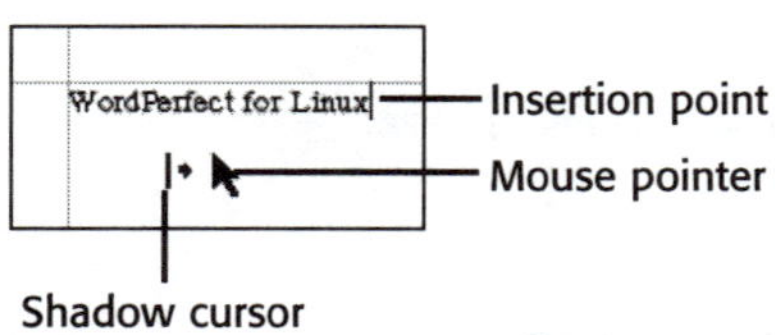

Figure 2.2
The three types of cursors show you where various operations occur.

- The *insertion point*—The insertion point is a blinking text-entry cursor; it shows WordPerfect where to add the text you type into the document. Your text either appears at the insertion point or overwrites what comes after it, depending on whether Insert or Typeover appears in the status bar. Click on that word or press the Ins key on your keyboard to change this mode.
- The *mouse pointer*—The mouse pointer is the pointer with which you're presumably already familiar. It is usually arrow shaped, though in WordPerfect it changes shape depending on where it is and what you're doing with it. Use it to select menu items, click buttons, select text, move the insertion point, and otherwise manipulate the

onscreen display. If you click the left mouse button when the mouse pointer is in the text-editing area, the insertion point moves to the mouse pointer's location.

- The *shadow cursor*—The shadow cursor is a WordPerfect feature that allows you to automatically insert *whitespace* (tabs and carriage returns) as needed to position the insertion point in a given location. The shadow cursor is blue on the screen and follows the mouse pointer around within the confines of the text-entry area—but in a jerky sort of way because the current tab settings, line height, and so on restrict the possible locations of the insertion point. If you click the left mouse button when the shadow cursor is active, WordPerfect adds enough whitespace to position the insertion point at that location, and the insertion point moves there. When you move your mouse pointer over existing text, the shadow cursor changes to a blue I-beam and shows exactly where the insertion point would appear if you were to left-click. This feature can be handy for placing the insertion point in an exact position. Unlike the other cursors, you can deactivate the shadow cursor by selecting Display from the Preferences dialog box on the Control window.

SELECTING FONTS AND FONT ATTRIBUTES

As described earlier, WordPerfect for Linux gives you a variety of fonts from which to choose. Although the downloadable version comes with a relatively small number of fonts, the retail version comes with more fonts and allows you to install an even wider variety, as discussed in Appendix B, "Fonts and Printing." Before you get very far in creating a document, you may want to select an appropriate font. The easiest way to do so is to pick one from one of the font selection widgets in the toolbar. As described in Chapter 1, "Introduction to WordPerfect for Linux," you can select either from a listing of all available fonts in the font selector on the left side of the toolbar or from a WYSIWYG display of recently used fonts in the QuickFonts selector widget near the right of the toolbar (Figure 2.3).

→ To learn more about fonts in WordPerfect, **see** "Fonts," **p. 178**

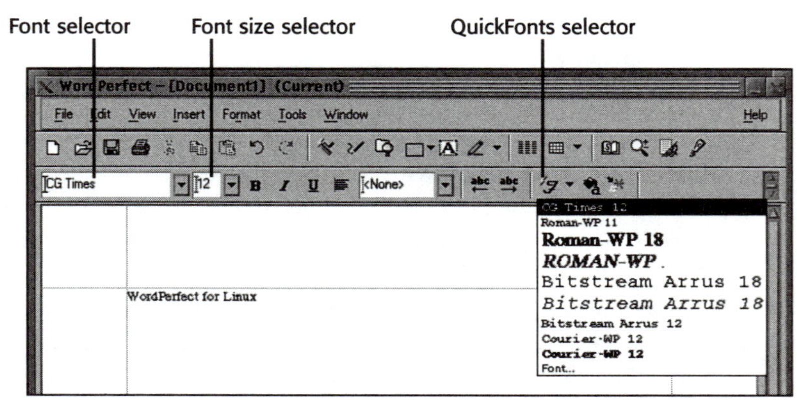

Figure 2.3
Select a font and size that are appropriate for your document. Font sizes of between 10 and 12 points are appropriate for standard body text, whereas larger sizes are useful for headlines and flyers.

Another way of selecting a font is through the Font dialog box, accessed by choosing Format, Font or by pressing the F9 key (Figure 2.4). This dialog box provides many options to fine-tune the appearance of your text.

Figure 2.4
The Font dialog box offers more than you need most times you select a font, but the options can sometimes be useful.

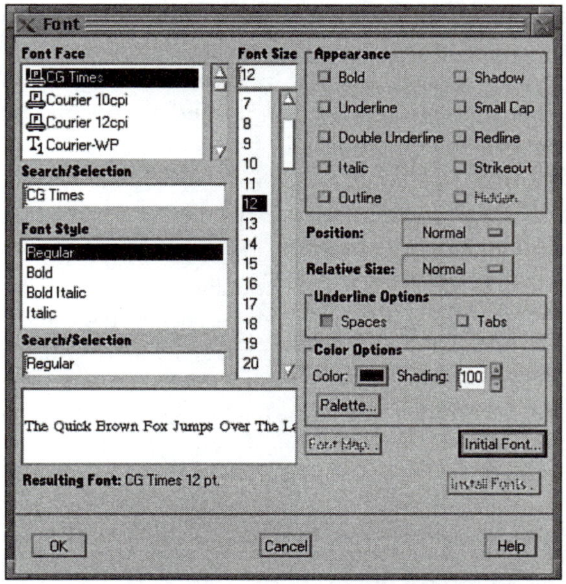

- The Font Face box lists all available fonts. The printer icon next to some fonts indicates either that these fonts are present in the printer or that they're Type 1 fonts that are downloaded as such to a PostScript printer. These printer fonts may differ from one printer to another, so if you use them, your document's appearance may change if you change printers or try to read the document on another computer. The T1 icon next to other fonts indicates that they are WordPerfect graphics fonts. These fonts can be used with any printer, but in some cases they result in slower or lower-quality printing than printer fonts.
- The Search/Selection box allows you to type in the name of a specific font so WordPerfect can find it quickly, rather than making you scroll through the list in the Font Face box.
- The Font Size section lets you select the size of the font. Normally, you'd select the font size from one of the options in the scrollable list. Certain printer fonts may be available in only a few sizes, so your choices are limited to those in the list with such fonts. With graphics fonts and certain other printer fonts, however, you can select any size you like, even including fractional sizes such as 11.3. This capability can be handy if you need to fine-tune the font size, but if you need to do such fine-tuning to control the size of your document, check out WordPerfect's Make It Fit feature.

➔ To learn more about Make It Fit, **see** "Meeting Page-Length Goals," **p. 233**

- The Font Style section of the dialog box lets you select variants of a font. Many fonts come in a "standard" form called Regular, but have variants with assorted names, such as Italic, Oblique, or Bold. Occasionally, these variants may be combined, as in Bold Italic. Normally, these variants are used for emphasis, though occasionally they may be used for other purposes. You can select these variants here. Note that these variant font styles are technically separate fonts; they reside on your disk or in the printer in separate files, and they're rendered separately by WordPerfect or your printer. Like the Font Face, the Font Style has an associated Search/Selection field so you can type in the style name you want to find quickly.
- Below the Font Style area is a font preview pane. The preview shows you what the font looks like. This feature is helpful if you're not sure just what appearance goes with which font or font attribute.
- The Appearance section of the dialog box contains a number of check buttons to enable or disable various font attributes. Some of these, such as Italic and Bold, may duplicate variant font styles; however, if a font does not contain an appropriate variant style, the appearance check buttons allow WordPerfect to "fake it." The WordPerfect-generated attribute generally doesn't look as good as one generated by a variant font style but is better than nothing. If an appropriate variant font style exists, WordPerfect uses it when creating the font with an appearance attribute set.
- The Position option lets you select Normal, Superscript, or Subscript. These can be useful for creating mathematical or chemical formulas, among other things (though complex equations are better handled with WordPerfect's equation editor). If you want to create footnotes, WordPerfect has other methods of doing this.

➔ If you want to learn about creating equations, **see** "Equations," **p. 433**

➔ If you want to learn how to create footnotes or endnotes, **see** "Footnotes and Endnotes," **p. 607**

- The Relative Size option button lets you alter the size of characters relative to the text around them. This option can be more flexible than simply altering the point size directly: You can change the point size of the entire document, and the size of the text created with the Relative Size option changes automatically.
- The Underline Options area lets you tell WordPerfect whether or not to underline the spaces and tabs in an underlined section of text.
- The Color Options section lets you specify a color for the text. Click on the square next to the word Color to open a rectangular color array. Select a color, and any text you've selected becomes that color, or any new text you type appears in that color to begin with. You can lighten the color by reducing the percentage in the Shading spin box. If the color pop-up menu doesn't present enough options, try clicking the Palette button for a menu that lets you define a color with even more precision, as shown in Figure 2.5.
- The Initial Font button lets you specify the font that's used from the start of a document. This setting may be particularly helpful if you've already entered text and want to change the font for all of it.

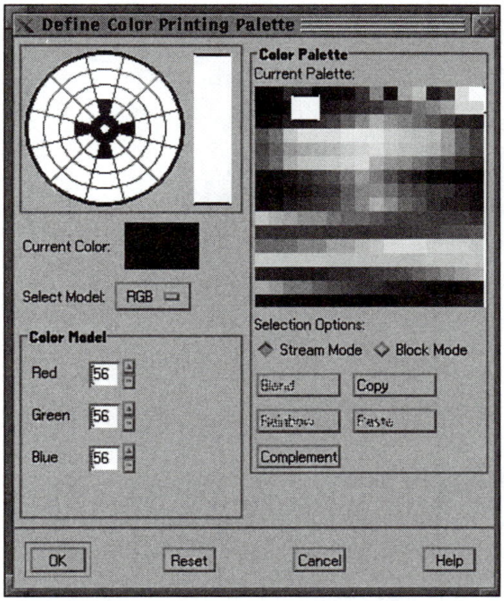

Figure 2.5
You can select a color by designating its exact value in any of several color-specification models.

Note

The Install Fonts and Font Map buttons are grayed out unless you're running WordPerfect with the `adm` switch. These functions relate to program administration—namely, installing and configuring fonts.

→ To learn how to install fonts, **see** "Installing and Managing Fonts," **p. 754**

You can handle most of your font-changing needs without using the Font dialog box. The default toolbars contain not only items for selecting fonts, as I've just described, but also options for selecting the bold, italic, and underline font attributes. When possible, bold and italic attributes actually select their appropriate alternative font styles; when not, WordPerfect does its best to simulate this style. Another way to select the bold, italic, and underline font attributes is to press Ctrl+B, Ctrl+I, and Ctrl+U, respectively. Pressing these keys a second time deactivates them.

Tip #9 from *Tux Tips*

The B, I, and U toolbar items are highlighted when bold, italic, and underline styles are selected. You can use this feature to tell whether your text will appear in one of these styles.

WordPerfect applies your font or font attribute change in various ways, depending on which text is selected, where the insertion point is positioned in the text, and what type of change you're requesting:

- If you haven't selected any text and if your insertion point is at the end of the document, your font change affects only new text that you type after the current position of the insertion point. If you reposition your insertion point to a location before the font change, any new text you type takes on the old font style.
- If you haven't selected any text and if your insertion point is in the middle of text between words, a change in font or font size affects all text after that point up to any other preexisting font change. Under these conditions, applying a font attribute such as bold has no visible effect.
- If you've selected text, any font-related changes apply only to the selected text.

➔ To learn more about selecting text, **see** "Selecting Text," **p. 96**

- If the insertion point is in the middle of a word, font changes occur as if that word were selected.

CENTERING AND RIGHT-JUSTIFYING TEXT

You may occasionally want to center a line or two of text on the page or position the text flush against the right edge of the page, also called right justification. Chapter 7, "Character and Line Formatting," deals with these topics in more detail, but for now you can use the Format, Line, Center and Format, Line, Flush Right options, respectively. (Alternatively, use the keyboard shortcuts of Shift+F7 and Shift+Ctrl+F7, respectively.) Either enter these commands before you type the text for the line or place your insertion point at the start of the line before issuing them. These commands affect only one line of text, unless you first select more than one line.

➔ If you would like to learn more about centering and right-justifying text, **see** "Justification," **p. 191**

Tip #10 from *Tux Tips*

If you use the center or flush right commands in the middle of a line of text, only the part of the line following the insertion point is affected. You can also use both commands on a single line. This technique can be useful if you want to spread something across the entire width of the page, as shown in Figure 2.6.

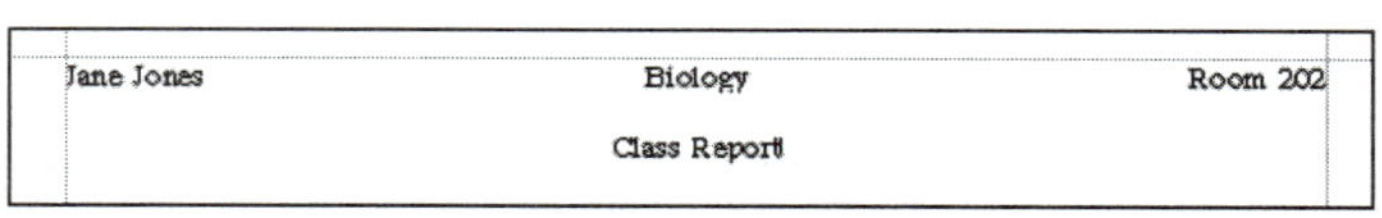

Figure 2.6
The center and right-justification features are useful for fitting information into a small vertical space.

Tip #11 from *Tux Tips*

Pressing Shift+F7 or Shift+Ctrl+F7 a second and subsequent times toggles the presence of leaders, or a series of periods leading from the text on the left to the text on the right, as shown in Figure 2.7.

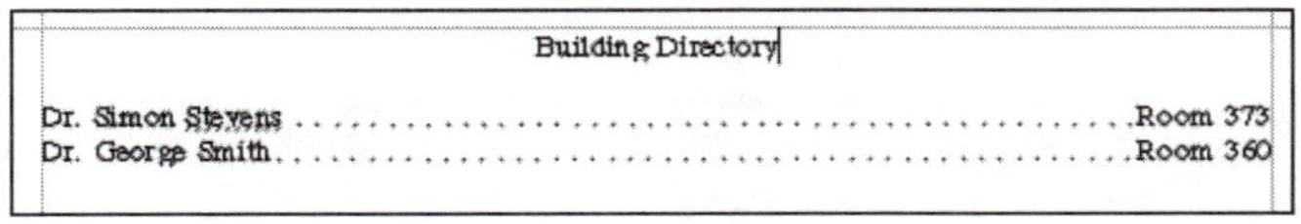

Figure 2.7
Use leaders for directories and other lists.

Moving Through Text

When your document is short, you can easily see the whole document or position your insertion point where you want it by clicking the left mouse button after positioning the mouse pointer. As your document grows in size, however, you need to adjust your view of the document to see more than the first page in order to read it or position your insertion point.

As you read this section, you may want to try out these methods. It's helpful when you do so to have a document of several pages loaded into WordPerfect. You can easily import an ASCII text file if you have one that you know is of suitable size on your system. If not, the CD accompanying this book includes the file gwfarewell.wpd, which is the text of George Washington's farewell address to the nation at the end of his second term as president. To read this file, you first need to mount the CD. To do so, log in as root and type `mount /dev/cdrom /mnt` at an xterm. You may need to change the device file from /dev/cdrom to /dev/scd0, /dev/hdb, or something else, and you may need to change the mount point from /mnt to /mnt/cdrom or some other location. Consult your Linux distribution's documentation for details. When the CD is mounted, find the file in the chapter02 directory.

➔ For more information on mounting CDs or other removable media, **see** "Accessing Removable Devices: Using the `mount` Command," **p. 76**

WordPerfect for Linux offers both mouse- and keyboard-driven methods of moving through a document. You'll probably find yourself using a combination of both as you use WordPerfect.

On the smallest scale, the keyboard's arrow keys are useful in moving the insertion point a character or line at a time. Depending on your X configuration, you may not be able to use the arrow key duplicate functions on your numeric keypad; however, the dedicated arrow keys found on most PC keyboards should work fine. If you move the insertion point to a tab, it jumps over the intervening space. Similarly, if you move to the start of a line that has been centered with Shift+F7, the insertion point jumps to the end of the preceding line. If you move to the last or first line on the screen and press the down- or up-arrow key, respectively, the screen scrolls one line at a time. You can also scroll the window one line at a time by clicking the left mouse button on the scroll arrows at the top and bottom of the vertical scrollbar (Figure 2.8). You can continue scrolling by holding the arrow key or pressing the scroll arrow. When a document is wider than can be displayed in one window, you can use the left- and right-arrow keys and the scroll arrows on the horizontal scrollbar to scroll the window in small increments. Using the Ctrl key in conjunction with the left- and right-arrow keys moves the insertion point by one word, rather than by one character.

Figure 2.8
Scrolling controls for WordPerfect are found near the right and bottom edges of the Document window.

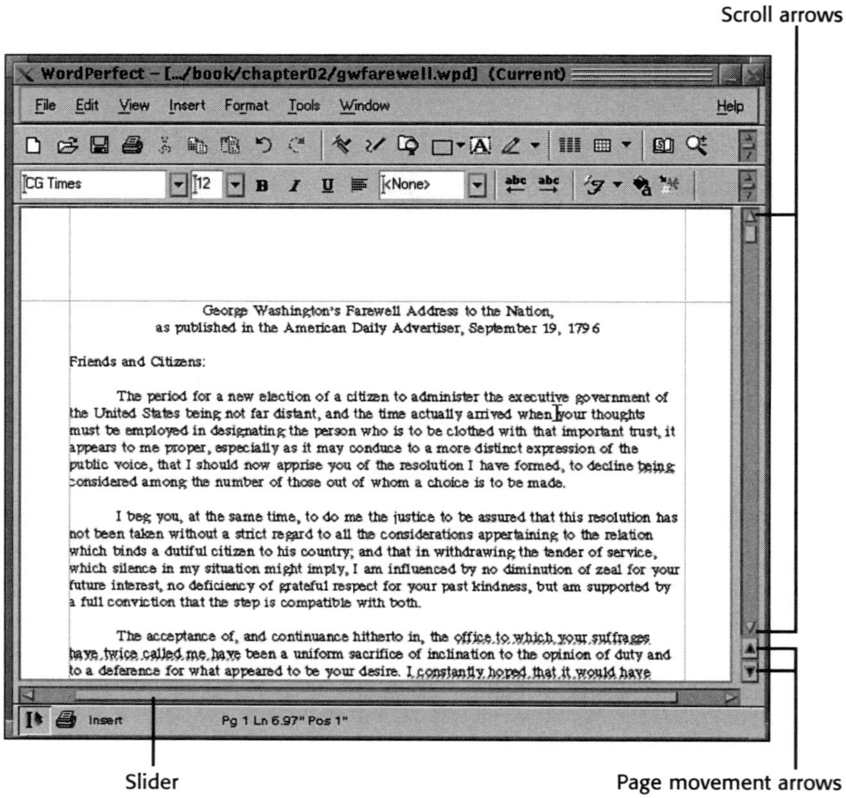

You can move the view up or down by one screen by using the Page Up or Page Down keys on your keyboard or by clicking in the area of the scrollbar above or below the scrollbar slider. If you use the keyboard, note what happens with the insertion point: When the insertion point appears in the middle of the window, pressing Page Up or Page Down once moves the point to the top or bottom line of the window, respectively. Only when you press the key again does the page actually scroll.

You can move the insertion point to the beginning or end of a line by pressing the Home or End keys, respectively. Unlike scrolling up or down with Page Up and Page Down, Home and End move all the way; if the window is too narrow for the current document, Home and End scroll it completely right or left the first time they're pressed.

If you want to move the insertion point up or down by a paragraph, try Ctrl+up arrow or Ctrl+down arrow. This step positions the insertion point at the beginning of the preceding or following paragraph, respectively. It scrolls the screen to keep the insertion point in view, as well. Aside from clicking, there is no equivalent mouse operation.

You may want to move up or down by one printed page at a time. If so, you'll find the page movement arrows at the bottom of the vertical scrollbar handy. These move the insertion point and scroll the display to the top of the following or preceding page. The keystrokes Alt+Page Up and Alt+Page Down have the same effect.

You can move to the start or end of the document by pressing Ctrl+Home or Ctrl+End, respectively. You can drag the slider in the vertical scrollbar all the way to the top or bottom to move the view in a similar way.

Note

The keyboard operations always operate on the insertion point and affect the view of the document by shifting the view to include the insertion point. If you use the scrollbars, however, the document view is affected but the insertion point *is not*. Thus, if you scroll with the scrollbar and then type something, what you type appears where the insertion point is, and the screen display scrolls back to that point. Keyboard scrolling commands behave inconsistently after a scrollbar-based movement, however; some, such as Page Up and Page Down, shift the insertion point to the presently viewable area and then operate from there. Others, such as the arrow keys, operate from the insertion point's location. The page up and page down widgets at the bottom of the scrollbar operate on the insertion point itself, unlike the other scrollbar operations.

If you have word fragments that disappear when you scroll, see "Word Fragments at the Ends of Lines" in the "Troubleshooting" section at the end of this chapter.

Tip #12 from *Tux Tips*

Most keyboard-based movement operations can be combined with the Shift key to select a region of text. For instance, Shift+Ctrl+Home selects the entire region from the insertion point to the top of the document. Not all of the keyboard movement functions work with Shift, however.

Tip #13 from *Tux Tips*

When you select text by clicking with the mouse and moving up or down, if you move the mouse beyond the confines of the editing area, the screen automatically scrolls. You can therefore select large amounts of text in this way. The scrolling is rapid, though, so be careful not to select too much text!

Choosing Edit, Go To; pressing Ctrl+G; or clicking the position display in the status bar brings up the Go To dialog box, shown in Figure 2.9. This dialog box allows you to go to a specific point in a document in a number of ways. The scrollable list at the top gives several often-used options, such as the last position, preceding or next table, and so on. The most-used option may be the default of a specific page number, which you can type in or enter with the spin controls. Figure 2.9 also shows three grayed-out fields, relating to specific bookmarks, tables, or cells within a table. Those options are active only when the relevant document elements are present.

Figure 2.9
The Go To dialog box is a convenient way to find specific pages or elements in a large document.

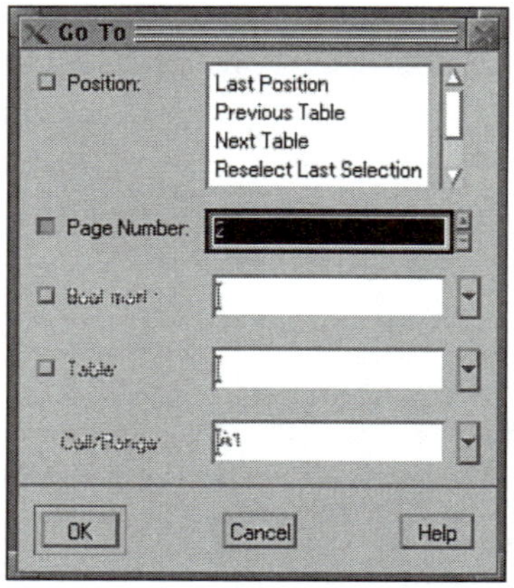

Document Views

WordPerfect allows you to alter the way you view a document in various ways that affect how you interact with the document without influencing how it prints. Scrolling through a document is one such operation. Others affect the amount of detail shown, the magnification applied to the text, and which other helpful tools occupy the screen (or clutter it, depending on your point of view).

Draft Versus Page Views

By default, WordPerfect for Linux starts up in Page view mode. In this mode, you see everything that will appear on a printed page, subject to limitations of the screen resolution and window size. For instance, you see the page's margin areas, headers and footers, and graphics elements such as equations and pictures.

When you choose View, Draft—or press Ctrl+F5—WordPerfect switches from Page view to Draft view. Figure 2.10 shows the same file in two WordPerfect Document windows. The window on the left is in Draft view, and the one on the right is in Page view. Draft view removes the page number and the top and bottom margins from the display, replacing them by a single thin line. Consequently, Draft view supports an efficient use of screen space, which may be particularly important if your screen is small.

The effects of Draft versus Page view go further, however. In Page view, you can edit directly in the main Document window many of the elements that are invisible in Draft view. For instance, Figure 2.11 shows direct editing of a header in Page view. Because the header is not shown in Draft view, this job is impossible; you must click on menus to get to the header, and then you edit it in a separate window, as in Figure 2.12. You must edit certain other WordPerfect elements, such as text boxes, in separate windows when you use Draft view, as well. The window opening and closing can make Draft view awkward for certain types of editing, but the simplified screen view can be beneficial for other types of editing.

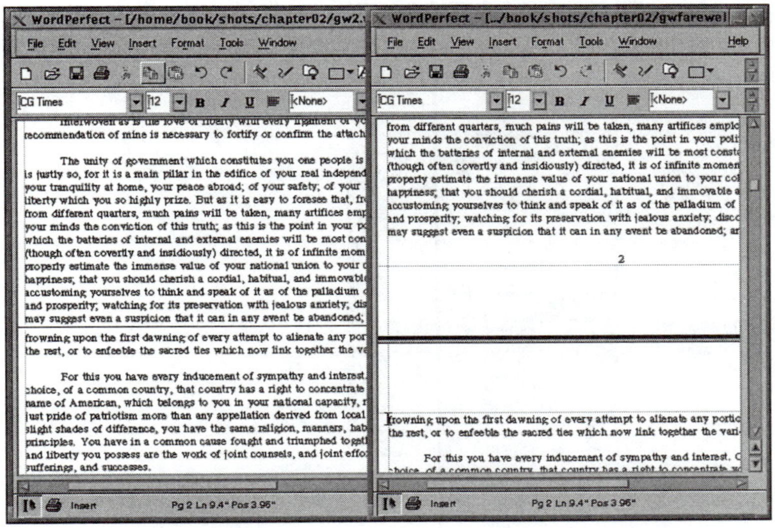

Figure 2.10
Page view (right) shows the page as it will print, whereas Draft view (left) shows more text per square inch of screen space.

➔ For more information on headers and footers, **see** "Headers, Footers, and Page Numbers," **p. 235**

➔ If you want more information on text boxes, **see** "Using Text Boxes to Add Variety to Text Layout," **p. 357**

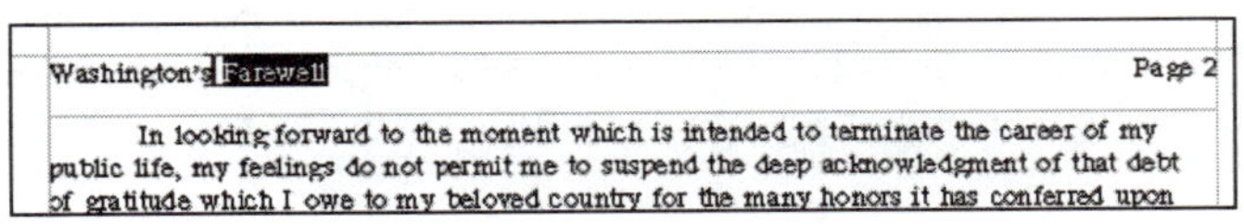

Figure 2.11
Page view provides easy access for editing a header.

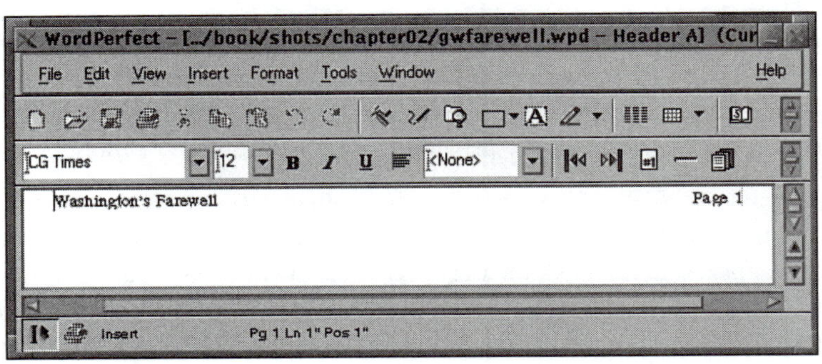

Figure 2.12
The title in the drag bar indicates that this window contains a header.

Zoom

If you want to increase or decrease the size of WordPerfect's display, you can use the zoom feature: View, Zoom. Figure 2.13 shows the Set Zoom dialog box. It has a number of preset zoom factors, ranging from 50% to 200%, plus an option to configure the zoom factor to any value in 1% increments. You can also set the zoom factor so that the window displays the contents of the page from margin to margin, from one edge of the page to another, or an entire page. Click OK to see the zoom factor you selected.

Figure 2.13
In theory, 100% zoom matches the size of a printed page, but in practice this representation may be off, depending on the resolution of your screen and the size of your monitor.

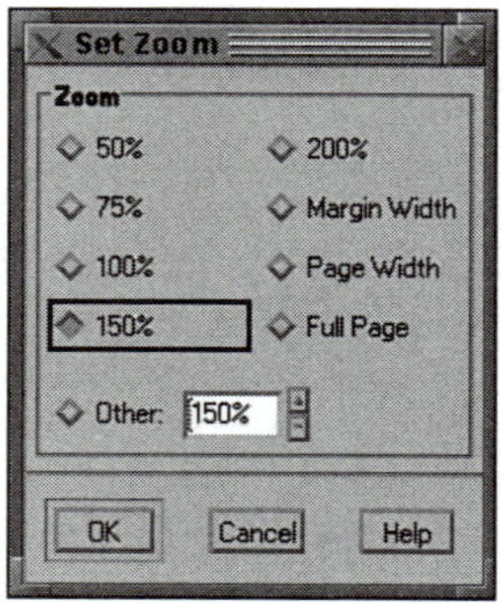

Tip #14 from *Tux Tips*

WordPerfect's fonts look better when they're bigger. You may find that setting the zoom factor to Margin Width makes text more legible.

Tip #15 from *Tux Tips*

You can change the default zoom factor from the Control window's Preferences menu. Click Display and then select the View/Zoom radio button. From here you can set the zoom, as well as the default view (Draft versus Page).

Toolbars

Some people like to work in a very sparse environment, with few distractions or redundancies. Others prefer to work with lots of embellishments and minor tools at their fingertips. WordPerfect caters to both work styles. It allows you to change your working style at will for different types of tasks by providing options for different toolbars and rulers.

By default, WordPerfect starts with two toolbars (one of which is technically a "property bar"). If you position your mouse pointer in a toolbar and click the right button, a pop-up menu appears. If you click on a property bar, the options available will be Hide Property Bar, Edit, and Settings. If you click a toolbar instead, you see the last two property bar items, plus a series of predefined toolbars. Figure 2.14 shows a WordPerfect Editing window with many of the available toolbars active, plus the pop-up menu for selecting them. WordPerfect remembers any changes you make to the toolbar and restores them when you restart the program. In addition to displaying the toolbar overkill situation shown in Figure 2.14, you can remove both default toolbars to increase the available text editing space.

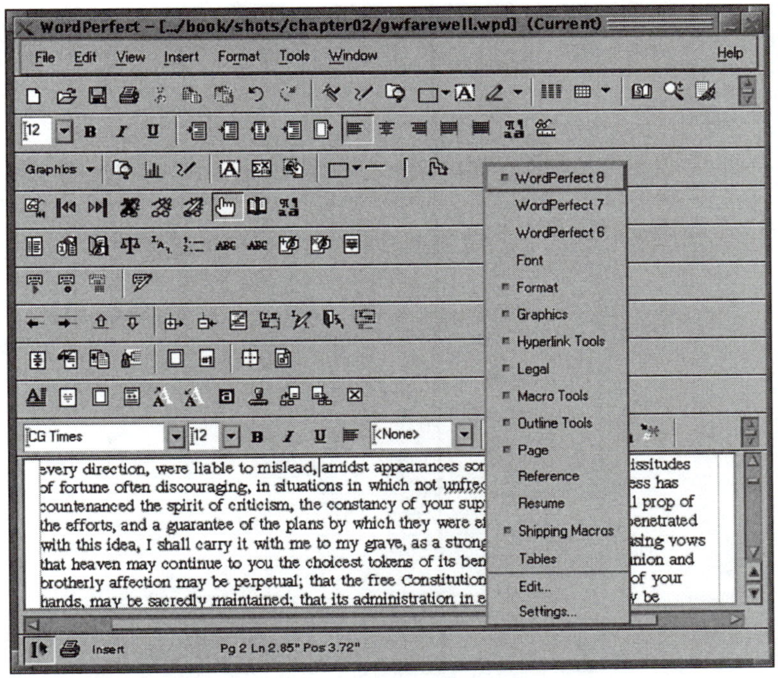

Figure 2.14
Between the various toolbars, most of WordPerfect's features can be accessed by a single mouse click—but at the cost of reduced editing space.

If you're not satisfied with the available toolbars, you can edit one to your liking. From the pop-up menu, choose Edit, and WordPerfect displays the Toolbar Editor dialog box as shown in Figure 2.15. The precise features available depend on which of the four radio buttons at the top is active. When Activate a Feature is selected, as in Figure 2.15, you can select from a wide variety of predefined WordPerfect functions in any of several categories similar to the main menu items. When you select Play a Keyboard Script, you're creating a button that inserts whatever text you type into the text entry field. With Launch a Program selected, you're given a button to press to bring up a file selection dialog box with which to select an external program. Whenever you press the button bar icon that results, the program you selected earlier runs. Finally, when you select the Play a Macro radio button, you have the opportunity to select a WordPerfect macro. You select a macro much as you do an external program.

→ If you'd like to learn more about macros, **see** "Letting the Computer Do the Work: Macros," **p. 483**

When you click OK, WordPerfect adds the feature you selected to the end of the toolbar. If you want to add a vertical line to separate logically related groups of functions, you can do so with the Separator element. Click and drag the Separator to the desired toolbar location and then release.

Figure 2.15
The features in the Toolbar Editor dialog box change with the radio button selections.

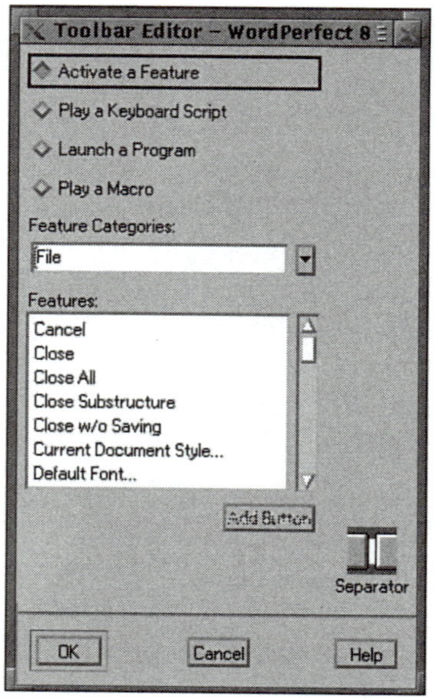

Tip #16 from *Tux Tips*

When the Toolbar Editor dialog box is open, you can double-click on a toolbar icon to customize the button's text and pop-up help prompt.

You can further alter toolbar settings by selecting the Settings item from the pop-up menu, which brings up the Customize Settings dialog box (Figure 2.16). From this dialog box, you can reset a toolbar to its defaults; bring up the Toolbar Editor dialog box; or change options related to whether the toolbar items appear as icons, text, or both and which font is used for textual display in the toolbar (Figure 2.17). For toolbars proper, but not for property bars, you can also create a new toolbar from this menu.

Figure 2.16
You can use the Customize Settings dialog box to select which toolbars are visible, as well as to modify them and their options.

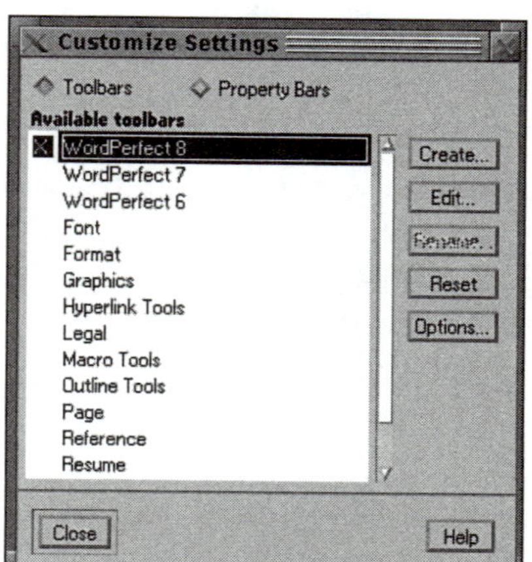

Figure 2.17
If you install additional fonts in WordPerfect for Linux, more fonts become available for use in toolbar text.

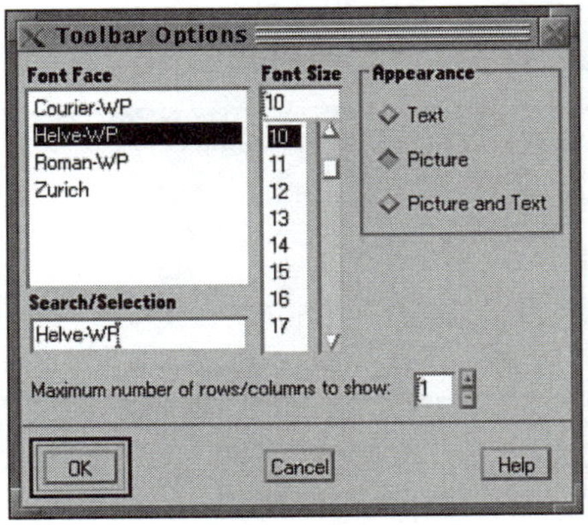

The Ruler

You use the WordPerfect ruler to see and set tabs and the left and right margins (Figure 2.18). The ruler itself contains three lines. The center line shows measurements in inches, centimeters, millimeters, or points. You can set your preference from the Display Preferences dialog box, as described in Appendix C, "Customizing Preferences." The lower line displays your tab stops, and the upper line shows margins. To illustrate how these work, consider the text in Figure 2.18. The first four lines of text are a single paragraph, typed without any tabs or other formatting features aside from the margin settings. The last line of text uses three tabs to position the text.

→ To learn how to set display options, including those related to the Ruler, **see** "Display," **p. 768**

The margin widgets control the margins that correspond to the margin guidelines in the text entry area. The large triangle just to the left of the right margin widget controls the right margin *adjustment*. It can be set independently of the *main* right margin to allow for a brief section of text that should be formatted differently from the rest. The left page margin widget is comparable to the right one, but the left margin adjustment widget is split into two parts. In Figure 2.18, these parts are separated. The upper part controls the indent for the first line of text in a paragraph, and the lower part controls the margin for all subsequent lines of text. You can see the result in Figure 2.18.

Note

Unlike some word processors, WordPerfect does not allow you to create a hanging indent, such as is often used in bibliographies, by dragging the upper portion of the left margin adjustment widget to the left of the lower portion of that widget. WordPerfect has other ways to accomplish this effect.

→ For more information on creating and using hanging indents, **see** "Hanging Indents," **p. 210**

Figure 2.18
The WordPerfect for Linux ruler displays tabs and margins.

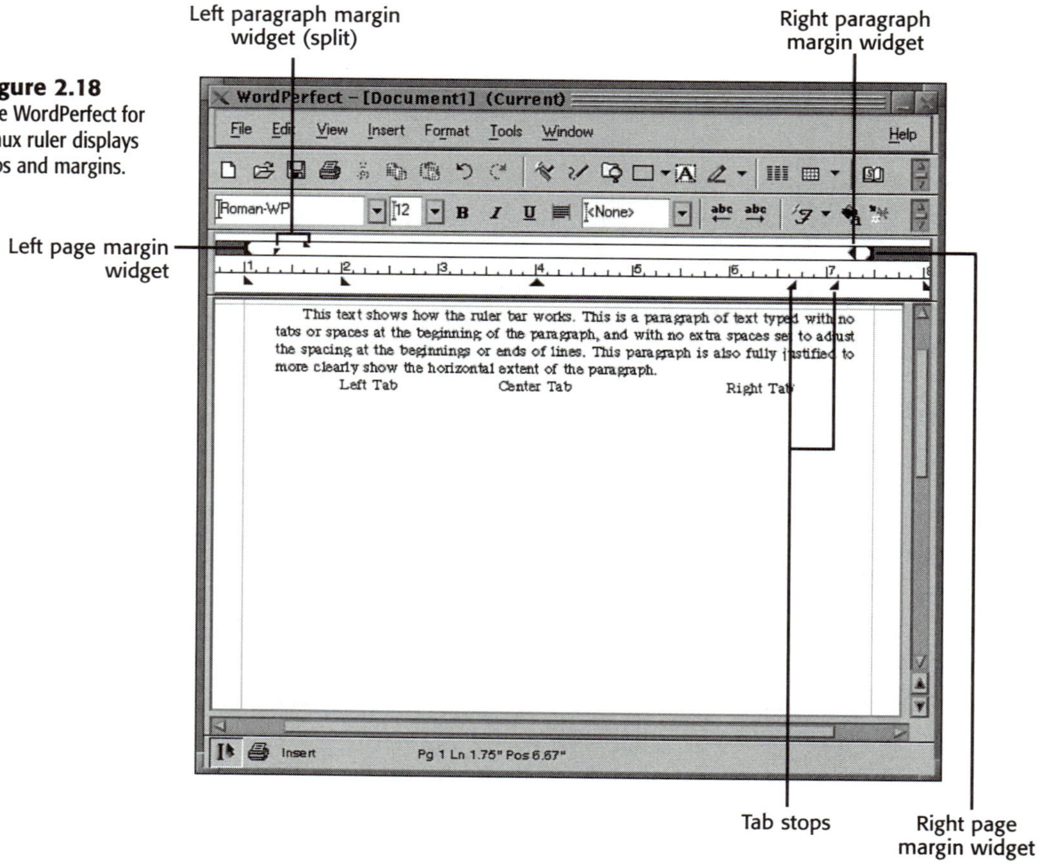

To set a tab, you first set the type of tab by right-clicking in the tab area of the ruler; then, from the tab options context menu (Figure 2.19), select the type of tab you need and left-click to set it.

→ If you'd like more information on tabs, **see** "Tabs," **p. 194**

When you set a tab or margin, the setting affects the current paragraph and all subsequent paragraphs unless you've highlighted some text. In that case, the setting affects only the paragraph(s) in which you've highlighted text. (If you've highlighted *part* of a paragraph, the *entire* paragraph is affected by your tab or margin change.)

ADDITIONAL VIEW ELEMENTS

In addition to the major viewing options discussed so far, WordPerfect provides a number of other, relatively minor, options that influence the way a document appears on screen.

Figure 2.19
You can set the tab type and leader options in the tab options context menu.

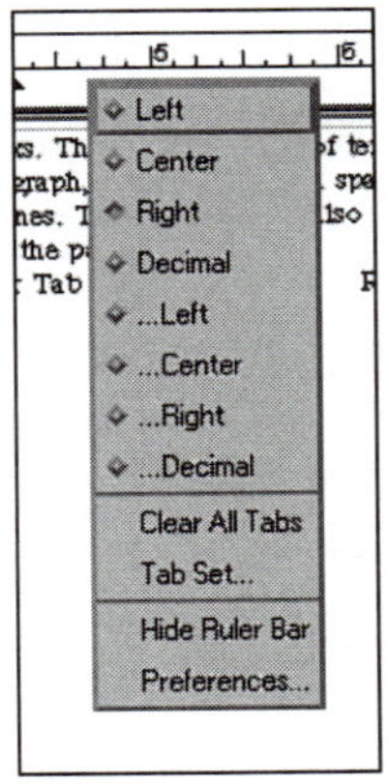

Application Bar

One additional display option is the presence of the application bar (also sometimes referred to as the status bar) at the bottom of the window. By default, this bar displays the insertion point location, whether WordPerfect is in Insert or Typeover mode, and buttons to let you select a printer and toggle the shadow cursor on or off. You can turn the application bar on or off by choosing View, Toolbars and checking or unchecking the Application Bar check box in the Toolbars dialog box, as shown in Figure 2.20. As you might guess by looking at the Toolbars dialog box, you can also control which toolbars are displayed from this menu.

Figure 2.20
The application bar is grouped in with toolbars in the Toolbars dialog box.

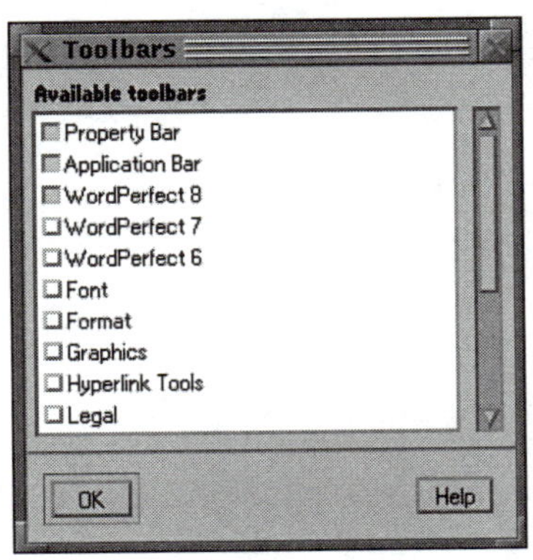

Another way to remove the application bar is to right-click on it and select Hide Application Bar from the resulting context menu. This menu also contains a Settings option, which brings up the Application Bar Preferences dialog box, that allows you to add or remove elements from the application bar (Figure 2.21).

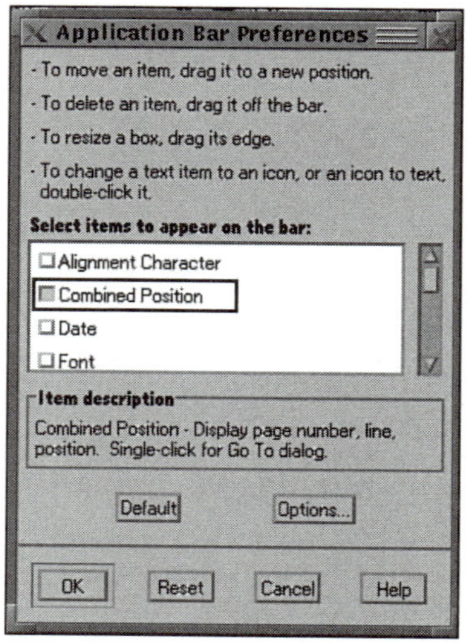

Figure 2.21
You can change the font used in the application bar by clicking the Options button from the Applications Bar Preferences dialog box.

GUIDELINES

As mentioned earlier, guidelines are the faint lines that outline margins or other document elements on the screen. The View, Guidelines menu command opens the Guidelines dialog box, which allows you to determine which of four types of guidelines WordPerfect displays (Figure 2.22). By default, all guideline types are displayed. A simple document might contain only margin guidelines. Table guidelines outline tables, column guidelines add extra lines to outline columns, and header/footer guidelines show where headers and footers are placed. If you check the Use as Default check button, the settings you create become the default for any future documents.

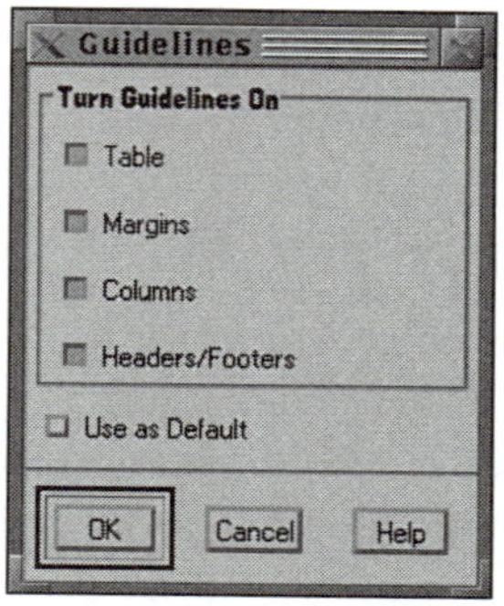

Figure 2.22
The Guidelines dialog box lets you select any combination of the four types of guidelines to include or not include.

GRAPHICS DISPLAY

As you might guess, the View, Graphics option in the Document window's menu toggles the display of graphics on the screen. It simultaneously toggles the display of other graphics-type objects, such as equations and text boxes. I discuss these document elements in Part IV, "Complex Formatting with Boxes for Text, Graphics, and Equations." If your document has box elements, setting them to not display can speed up WordPerfect's operation but at the obvious cost of your not being able to see how they fit into the overall design.

→ For more information on boxed elements, **see** "Complex Formatting with Boxes for Text, Graphics, Tables, and Equations," **p. 355**

Show Paragraph

The View, Show Paragraph (¶) option in the Document window's menu toggles the display of whitespace codes such as paragraph, space, and tab. Figure 2.23 shows the results of enabling this feature. This type of display is most useful when you need to determine which methods a document's author used to create a formatting effect or when you need to maintain precise control over the placement of text in a series of spaces or other nonprinting characters.

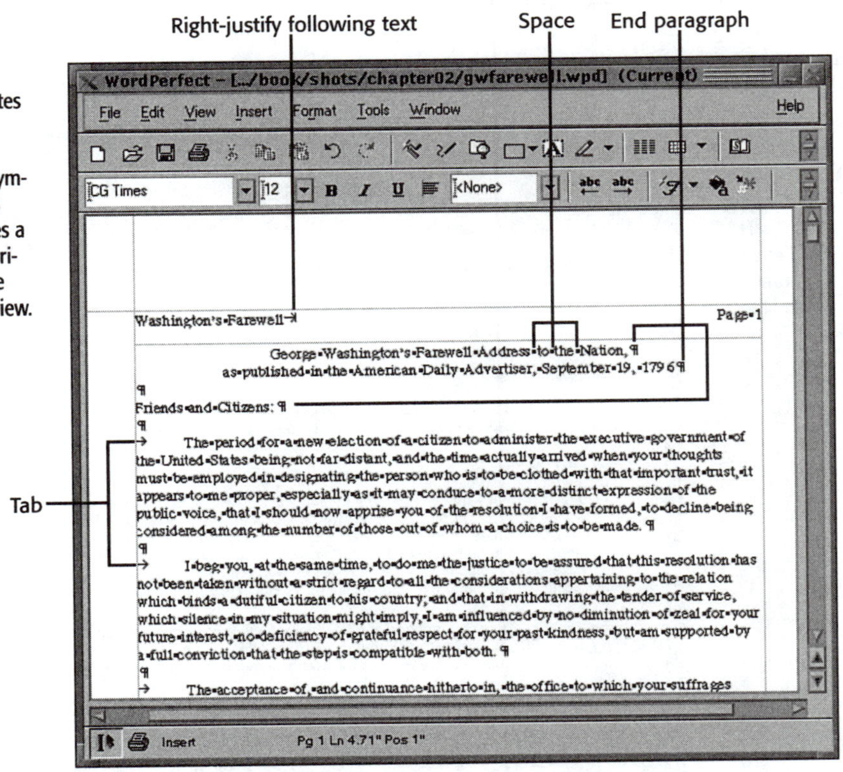

Figure 2.23
When a code creates a large horizontal movement, that code's onscreen symbol is easy to miss because it occupies a relatively small horizontal space in the Show Paragraph view.

Reveal Codes Mode

Whereas the Show Paragraph feature is a small cache of treasure in uncovering how a document was formatted, WordPerfect's Reveal Codes feature is the mother lode. Reveal Codes shows almost everything about how a document is structured and formatted, including not just whitespace characters such as spaces and carriage returns, but also codes for page numbering, graphics, tables, columns, and other advanced elements. As you learn to do more with WordPerfect, you'll find the Reveal Codes feature to be more and more useful in understanding how WordPerfect works and in tracking down the causes of formatting peculiarities.

ENTERING AND LEAVING REVEAL CODES MODE

You can activate Reveal Codes mode by selecting View, Reveal Codes, by pressing Shift+Ctrl+F3, or by selecting Reveal Codes from the pop-up menu when you right-click in an editing area. (If you right-click and don't see Reveal Codes, you may have clicked on an element that imposes its own menu that doesn't include Reveal Codes; try again on another part of the screen.) When active, Reveal Codes splits your Document window, as shown in Figure 2.24. The upper portion contains text as you normally see it, whereas the lower portion displays the underlying WordPerfect codes. These two portions of the window are referred to as *panes* because they're subdivisions of a single window. You can return to normal Editing mode by repeating the steps that activated Reveal Codes mode.

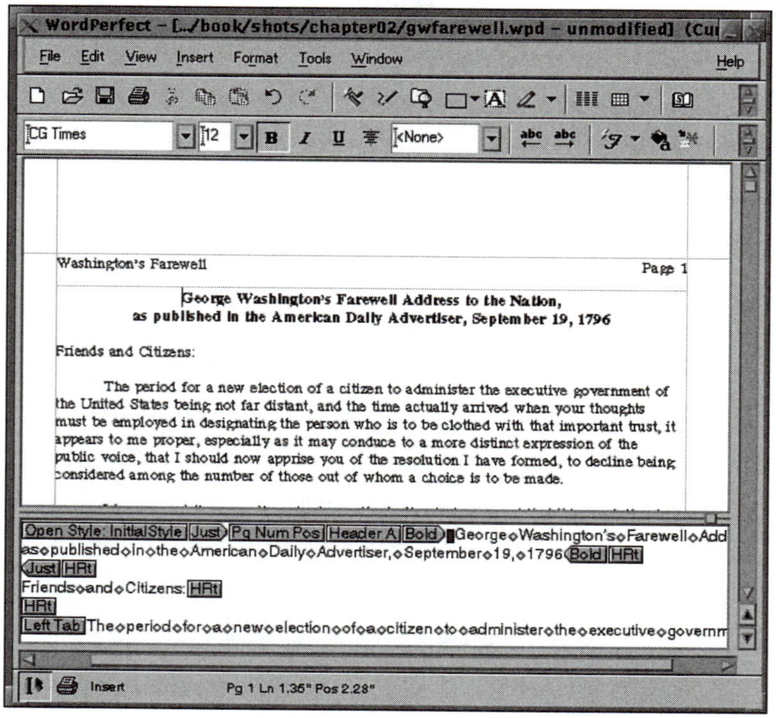

Figure 2.24
You can resize the Reveal Codes pane by clicking and dragging on the small square near the right end of the strip separating the two panes of the window.

UNDERSTANDING THE REVEAL CODES VIEW

When the Reveal Codes pane is open, you can still use the normal editing techniques you use when it is closed; Reveal Codes *augments*, but does not *replace*, those commands. Anything you do in WordPerfect, however, may have an effect on both panes of the window. Any text you enter appears in both the regular view and the Reveal Codes pane. If you enter a formatting command, the change is reflected in a WYSIWYG fashion in the upper editing pane and as a new code or codes in the Reveal Codes pane.

Reveal Codes is obviously not a WYSIWYG view of your document. In Figure 2.24, the centered text is also boldfaced, yet in the Reveal Codes pane, it is not. Instead, there's a gray block with the word Bold printed inside it just to the left of the boldfaced text. This block has a right-facing angle to its side, indicating that it's paired with a code later in the text that deactivates this feature. Its mate, this time with a left-facing angle, appears just after the boldfaced text. Codes with this angling on their edges are known as *paired codes*, and as you might guess by the name, they operate in pairs—one to turn the feature on and another to turn it off. Another paired-code example is centering, which is done by the Just codes in Figure 2.24.

Other codes do not appear in pairs, but rather are codes for discrete features, such as carriage returns, or affect all text or formatting from that point on, such as headers and page numbers. You can see examples of all of these in Figure 2.24. Carriage returns display as [HRt] ("hard return") codes, and the page number and header codes appear on the first line as [Pg Num Pos] and [Header A] markers, respectively. The end of a line of text within a paragraph appears in Reveal Codes with an [SRt] (short for "soft return") code. WordPerfect automatically adds or deletes these codes as needed when you modify your text. A code that affects all subsequent text and doesn't have a matching pair to disable it is referred to as an *ongoing* or *open-ended* code.

In this book, I use brackets to indicate a code, as in [Bold] or [Pg Num Pos]. This convention recalls the days when WordPerfect was a text-mode DOS program, when Reveal Codes used brackets to indicate codes.

The code labels are, in many cases, short abbreviations of the codes' full names. If you can't decipher just what a code is, position your mouse over the problem code and wait a few seconds. A brief description of the code and its function appears in a yellow box.

The insertion point appears in the Reveal Codes pane as an unblinking red rectangle. Like the regular insertion point, the Reveal Codes insertion point falls between two characters but, unlike the regular insertion point, takes up enough space on the display to displace text as it's moved around.

Altering and Deleting Codes

One major benefit of the Reveal Codes mode is that you can double-click on most codes in the Reveal Codes pane to bring up an associated formatting dialog box or window. For instance, double-clicking on the [Pg Num Pos] code brings up the Page Number dialog box. Thus, you can change the *original* command used to create formatting by using Reveal Codes, rather than by trying to apply new formatting over the old. Anybody who has used word processors much knows that the latter doesn't always work as expected, and the reason is simple: The word processor may try to apply the new formatting in a different way or at a different location than was the case for the original. With WordPerfect's Reveal Codes mode, you can control formatting with greater precision.

This technique is not perfect, however. For instance, if you double-click on either of the [Bold] codes, the WordPerfect Font dialog box opens. You might think you could use this dialog box to change the font or style of just the bold text, but that's not so. WordPerfect applies the options you select either to nothing at all or to the entire document from the insertion point on, depending on the options. At least when you use WordPerfect's Reveal Codes mode, you can easily undo this change.

When you type text in regular Editing mode, you can't delete codes (unless you delete all the text to which a code applies). If you press Ctrl+I to italicize text, you've inserted an italics code, but to undo this step, you must either use WordPerfect's undo feature or select the text again and reissue the italics command. With Reveal Codes active, however, you can delete codes as well as text. Position the insertion point just before or after a code and use the Delete or Backspace key to remove it. Another way to remove a code is to click on it and drag it out of the Reveal Codes pane. This technique is useful if you want to undo a change you made to a document but can't quite remember how you made the change. Just fire up Reveal Codes and remove the offending code. If you're not sure of which code to remove, you can try codes at random and use WordPerfect's undo feature to undo the deletion if you get it wrong.

PRECISION POSITIONING

You may occasionally want to position the insertion point precisely in relation to some codes. For instance, if you have a section of bold text and you want to add a word to the beginning or end of that section, your insertion point should be between the [Bold] codes. Without Reveal Codes, getting your insertion point into the correct position is awkward at best. Although the Bold tool on the lower default toolbar indicates whether the text you type will be bold, you cannot actually position your insertion point as desired. In this specific case, of course, you can easily press Ctrl+B to toggle the bold attribute if necessary. If you're trying to edit text with many changes in attributes in a short section, however, you may want to open Reveal Codes and position your insertion point with it rather than try to use the conventional formatting commands to get the result you desire.

If your formatting changes when you delete or edit with Reveal Codes, see "Deleting Text Near Reveal Codes" in the "Troubleshooting" section at the end of this chapter.

→ For a summary of features on the toolbars, **see** "Toolbars," **p. 30**

ALTERING THE REVEAL CODES DISPLAY

If you right-click within the Reveal Codes pane, the context menu gives you the option of setting Preferences for Reveal Codes. This option produces the Display Preference dialog box, shown in Figure 2.25. From this dialog box, you can select the font and font size for the Reveal Codes display, affect the way text wraps, and specify the size of (and thus precisely how much information is displayed in) the Reveal Codes pane. You can even set Reveal Codes mode to start by default on any document you create or open.

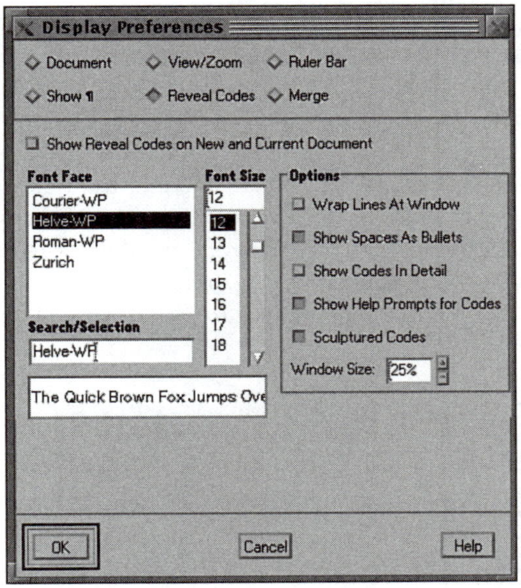

Figure 2.25
You can select any font you like for the Reveal Codes window from the Display Preferences dialog box; the list of fonts available expands if you add fonts to your WordPerfect for Linux installation.

SAVING YOUR DOCUMENT

When you're done editing, you can save your document by choosing either File, Save (Ctrl+S) or File, Save As (F3) from the main menu. If the file has already been saved or was loaded from a preexisting file, the Save option saves the file to the existing filename. If the file is new (that is, it has never been saved), then the Save option enables the Save As dialog box in which you enter the destination filename (Figure 2.26).

Note

The default filename in the Save As dialog box is the name by which WordPerfect knows the document. If you've loaded the file from disk or saved it previously, this name is the document's existing filename. If it's a new and unsaved file, the name is Document#, where # is a number from 1 up, depending on which WordPerfect Document window the file occupies.

The Save As dialog box is the same as the Open dialog box, which in turn is identical to the File Manager dialog box. Whatever its name, this dialog box holds considerable power, and it's discussed in greater detail in Chapter 3, "Managing Files and Folders." For now, just a few features deserve attention.

→ To learn how to use the WordPerfect File Manager, **see** "Linux File Manipulation," **p. 80**

The large scrollable area on the right side of the dialog box contains a listing of directories. The current directory is identified above this list, so you can tell where you are. Double-click on a directory name to enter it. To go up one level, double-click on the Parent directory entry. You may need to scroll the view of this section to the right to see the entire string of Linux permissions associated with each directory. These specify who can access the file or directory and in what way (this topic is discussed in greater depth in Chapter 3). The owner and group of each directory also appear here. If you want to change to a directory that's several levels removed from the current one (for instance, to change to /mnt/floppy from /home/rodsmith/letters/friends), click Files, Other Directory from the Save As dialog box's menu bar. This step opens a small dialog box into which you can directly type the name of the destination directory. However, you must know the exact pathname of that directory for this technique to work. If you want to create a new directory, you can do so with the Files, Create Directory menu item. You can then move into that directory by double-clicking its name in the directory list.

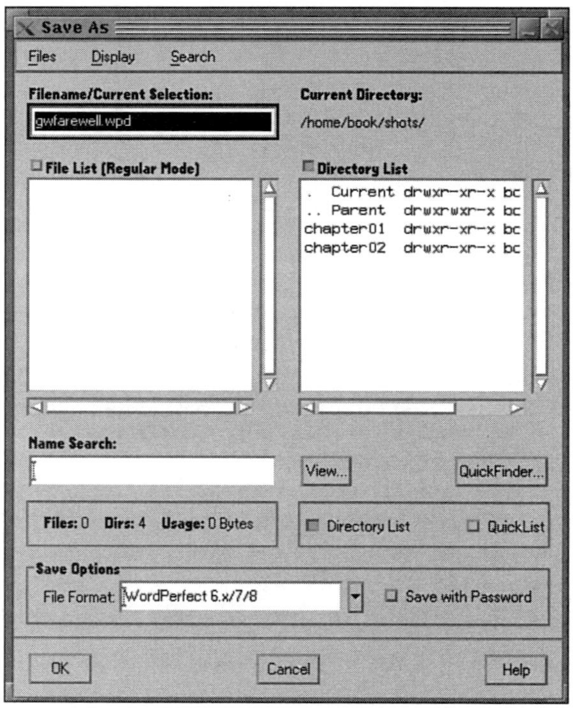

Figure 2.26
The Save As dialog box provides many options for file management in addition to simple filename entry.

To the left of the directory list is the file list, which shows any files in the selected directory. In addition to ownership and permissions information like that in the directory list, the file list shows file sizes and creation times and dates. If you want to replace an existing file, you can click on its filename and then click OK.

To create a file with a new name, type the name into the Filename/Current Selection text edit field near the top of the dialog box and then click OK. WordPerfect saves your file under this name.

If your document contains sensitive data, you may want to encrypt it with a password. You can do so by selecting the Save with Password check button near the bottom of the dialog box, in the Save Options section. When you later click OK, WordPerfect prompts you for a password with the Password Protection dialog box, as shown in Figure 2.27. After you've entered the password once, WordPerfect asks for it again as a safety precaution. WordPerfect saves your document to disk in a scrambled form, so that a casual snoop will see only gibberish. One drawback to this approach is that *you* will not be able to recover your file if you should forget your password.

Figure 2.27
Select the level of security you want for your document encryption based on whether you expect to be exchanging your file with WordPerfect 6 users.

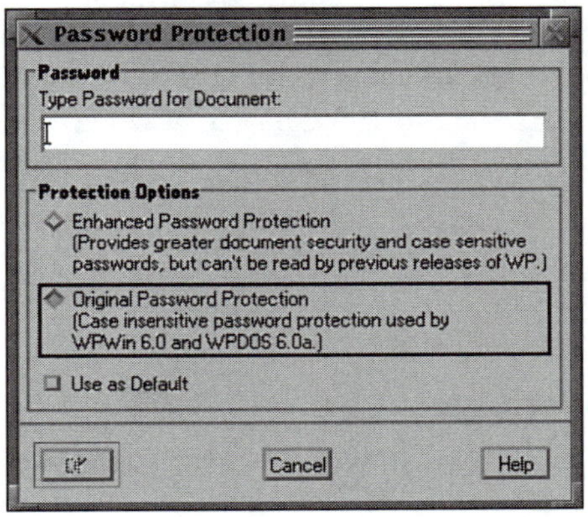

Note

When you type the password, WordPerfect echoes it as pound (#) signs to keep unauthorized people from reading it.

Caution

If you're very security conscious, you should know that WordPerfect for Linux maintains document backup files during editing in the /tmp directory tree and that these files are unencrypted even when the main file is encrypted. These files have permissions set so that other users can't ordinarily access the files. An intruder who has hacked into your account or the root account, however, might be able to read your encrypted file when you're working on it or if WordPerfect crashed while you were working on it.

Linux lets you use long filenames that contain upper- and lowercase letters, digits, and many punctuation marks, such as the ".", "_", and "-". You can also use spaces and various other punctuation marks, but these can sometimes cause problems, especially when dealing with files in an xterm window. Unlike some OSs, Linux is case sensitive, meaning that it differentiates between upper- and lowercase characters in its filenames. Thus, letter.wpd, Letter.wpd, and LETTER.WPD are the names of three different files.

You should probably give your WordPerfect files the filename extension .wpd. Some people use a digit or nothing at all in place of the *d* at the end. The digit in this case is the major version number of the version of WordPerfect that created the file—8 in the case of WordPerfect 8 for Linux. You can, of course, use any other filename extension you like (or none at all), though this change might make finding your documents somewhat difficult. If you don't specify a filename extension, WordPerfect does *not* add one itself.

WordPerfect has used the same file format for versions 6 through 8, and WordPerfect for Linux uses the same file format as WordPerfect for Windows. WordPerfect 8 for Linux can therefore read WordPerfect 6 through 8 for Windows files with no difficulty, and those versions of WordPerfect for Windows can read WordPerfect 8 for Linux files without conversion. Previous versions of WordPerfect used a different file format, but WordPerfect 8 can import these files with no difficulty. You can also export a WordPerfect 8 file to an older format, though you may lose some formatting if you used any features that weren't present in your target version of WordPerfect.

TROUBLESHOOTING

WORD FRAGMENTS AT THE ENDS OF LINES

When WordPerfect wraps lines of text at the bottom of the window, word fragments remain at the ends of lines, but the fragments vanish if you scroll the document somewhere and then return to the original location.

These "ghost words" are a minor screen update bug that sometimes occurs if you resize or move your WordPerfect Document window after opening it. If you don't need to move your WordPerfect window, you can avoid this problem by not doing so. Alternatively, try scrolling your Document window so that your text entry occurs somewhere other than the last line of the window. The glitch disappears, though you have to scroll the window every few lines to keep it at bay. You can also simply try to ignore the problem.

DELETING TEXT NEAR REVEAL CODES

When you delete or modify text, something strange happens to your formatting. Perhaps your margins shift, or your text's font changes.

This sort of behavior typically occurs because you deleted one or more important formatting codes. It most frequently happens when you try to modify text near a formatting code, especially if you're working in Reveal Codes mode, which allows you to delete codes that you can't otherwise delete. Try this: Undo the change, turn off Reveal Codes mode, and do your edit again. Alternatively, open Reveal Codes mode (if it was closed) to see which codes you might have accidentally removed and be sure to avoid them.

CHAPTER 3

Managing Files and Folders

In this chapter

Overview

If you're familiar with operating systems such as Microsoft Windows, OS/2, or MacOS, you may find that Linux's model for disk access takes some practice to master. It's not very different from the others in most respects, but some important differences can confuse you if you're unaware of them. If you're already familiar with the Linux way of handling files and filesystems, you may want to learn a bit about the tools that WordPerfect gives you to manage your files. These tools can help make you more productive in WordPerfect by obviating the need to use other tools for many file functions.

Understanding the Linux Filesystem Model

Most modern operating systems store files in a hierarchical structure of *directories* or *folders*. These two terms are synonymous in this context, and they both refer to listings of files that can be nested, so that one directory can contain several other directories as well as files. Each directory inside the first (sometimes referred to as *subdirectories*) can also contain both directories and files. For instance, you might have separate directories for letters and papers. Your letters directory might contain separate directories for each of your correspondents—your parents, your children, your bank, and so on. If you use this structure efficiently, it can help you to locate your files easily. If you're familiar with this type of structure from using Windows or some other OS, you can apply many of the same techniques you used there in Linux; however, you should also be aware of some important differences.

Directory Separators

One of the more trivial, but easy to overlook, differences is that Linux uses the slash character (/) to separate directories in a hierarchy. Windows, DOS, and OS/2, by contrast, use the backslash (\), whereas MacOS uses the colon (:). Thus, in Windows you might have a file called letters\george.wpd, whereas in Linux that file would be letters/george.wpd. Fortunately, when you use a GUI program, it usually handles the separators automatically; you simply click on directories and filenames to select the ones you want. When you use an xterm or text-mode login to Linux, however, you need to keep this difference in mind.

Filesystem Support

Linux has excellent support for different filesystems. In this context, the word *filesystem* refers to the low-level layout of data on a disk. Most Linux installations use a filesystem called ext2 for their own use. The ext2 filesystem, or "ext2fs" for short, supports various features needed by Linux and other UNIX-like OSs that aren't supported by the filesystems native to Windows. You can also use the FAT filesystem used by Windows, including all its variants such as VFAT and FAT-32. Consequently, you can read and write Windows floppies, removable media such as Zip disks, and hard disk partitions. If you're running a system with both Linux and Windows NT or Windows 2000, you may be interested in Linux's support of the NTFS filesystem that these OSs use (you need a kernel in the 2.2 or above range for this). Similarly, Linux supports read-only access to OS/2's HPFS. Linux kernels numbered 2.2 and above support the Mac's HFS, which means you can access 1.44MB Mac floppies, Zip disks, and hard disks. You can read CD-ROMs intended for Windows by using Linux's ISO-9660 and Joliet support. You can access files on networks of various types using network filesystems such as NFS. Linux also supports a variety of other more exotic filesystems, such as the Amiga FFS and the Minix filesystem.

Most Linux distributions come configured to handle ext2fs, FAT filesystems (under the Linux names msdos or vfat), ISO-9660 and Joliet (as iso9660), and possibly a few others. Check your distribution's documentation for details. To add more filesystems, you may need to recompile your kernel. This scary-sounding task is not difficult if you take your time and read the documentation. A book such as Que's *Special Edition Using Linux* can help you through the kernel recompilation process. Fortunately, you can go a long way with just ext2, FAT, and ISO-9660 support because most desktop computers, even non-Windows ones, can handle FAT disks.

FILENAMING CONVENTIONS

In Linux the filesystem in use determines many of the rules for naming files. In general, ext2fs is the most flexible; you can do with it just about anything you can do in any filesystem in terms of filenames. By contrast, if you use the msdos filesystem, you are restricted to old DOS-style filenames—eight-character names with optional three-character extensions, sometimes referred to as "8.3 filenames." Most other filesystems provide fewer filename restrictions but may have their own idiosyncrasies.

Tip #17 from *Tux Tips*

Linux lets you access any FAT disk using either the msdos or the vfat filesystem. The latter supports long filenames, whereas the former doesn't. If you're exchanging files with Windows users, use the vfat filesystem for the floppies or other removable media you use. If you're exchanging files with people who still use DOS or Windows 3.1, use the msdos filesystem; you'll be stuck with the 8.3 filenames, but you won't make your DOS-using friends or colleagues see strange truncated filenames.

Linux's ext2fs supports upper- and lowercase characters, numbers, spaces, and many punctuation symbols. Using punctuation can be risky, though, because Linux shells may interpret such characters in special ways, making it difficult to enter the filenames. The best approach is to restrict yourself to alphanumerics and the following characters: the period (.), dash (-), and underscore (_). Avoid the following characters at all costs:

```
! @ # $ % ^ & * ( ) { } ' " \ / ¦ ; < > `
```

Sometimes foreign filesystems support characters that give Linux problems; in MacOS, for instance, it's possible to use the slash and backslash characters in filenames. If you try to use a MacOS disk with such filenames from Linux, Linux automatically converts those symbols to others that are legal, though perhaps ugly, in Linux.

CASE SENSITIVITY

Linux is normally case sensitive in its filenaming, meaning that upper- and lowercase versions of the same letter are treated differently. Letter.wpd is not the same as LETTER.WPD or letter.wpd in Linux, for instance, and you can save different files under each of these names. These files can coexist in the same directory and have different contents. Many of the filesystems supported by Linux, however, are not case sensitive, though they are case retentive. That is, the filesystem preserves the case on filenames, but treats filenames that differ only in case as if they were identical. The Windows VFAT and FAT-32 filesystems and MacOS's HFS are case retentive. Case retentiveness doesn't normally present a problem to Linux because Linux simply treats those files in a case-sensitive way. However, if you try to create a file that differs from an existing one only in case, Linux overwrites the existing file. So if you're using a VFAT floppy or disk partition that already has a directory with a file called LETTER.WPD and you try to save another file under the name Letter.wpd, Linux overwrites LETTER.WPD with the contents of Letter.wpd. Some filesystems, such as the old-style FAT with 8.3 filenames, support only single-case filenames. Linux generally converts these to lowercase because the usual practice in Linux and other UNIX-like OSs is to use mostly lowercase filenames.

Tip #18 from *Tux Tips*

Most Linux programs, including WordPerfect, sort filenames so that files starting with uppercase letters appear before those starting with lowercase letters. Thus, if a directory contains the files abra.wpd, Cadabra.wpd, and XYLOPHONE.WPD, the files appear in the sorted order of Cadabra.wpd, XYLOPHONE.WPD, and abra.wpd. Keep this sort order in mind when you're looking for a file!

FILESYSTEM LAYOUT

In many OSs, each disk device or partition gets an identifying letter or volume name, and you access files on that device via that letter. In Windows and related OSs, for instance, floppies are A: (or sometimes B:), the first hard disk is normally C:, and so on. So you might have A:\letters\Sally.wpd referring to the file Sally.wpd in the letters directory on the floppy disk.

Linux, on the other hand, uses a *unified* filesystem structure. In this context, *filesystem* refers to the way directories and files are laid out logically—how you access a file on the hard disk versus the floppy, for instance. In Linux's unified view, there are no drive letters, and volume labels are also irrelevant. Instead, each partition or disk device is *mounted* at some point in the filesystem. That is, the device is grafted onto the filesystem directory structure as if it were an integral part of the whole. For example, you might have a floppy disk mounted at /mnt/floppy. If this floppy has the letters\Sally.wpd file in Windows, that file would be accessible from Linux as /mnt/floppy/letters/Sally.wpd.

I discuss the details of how to mount a device shortly, but for now you should know that mounting tells Linux to look at a particular piece of hardware in a particular way and to give you access to its files at a specific location. Mounting is a way to bridge the gap between the physical disk medium and your view of the files on that disk from within programs such as WordPerfect. Linux's solution is more complex than the simple mapping to drive letters in Windows, but the Linux approach is also more flexible.

The Linux filesystem contains a number of key standard directories, each of which may contain subdirectories. Table 3.1 shows some highlights of this structure.

TABLE 3.1 IMPORTANT DIRECTORIES IN THE LINUX FILESYSTEM

Directory	Importance
/	The root directory—Everything else is mounted off this directory.
/bin	The most vital of the basic Linux commands and utilities, such as `ls`, `cp`, and `tar`, live here.
/boot	Contains boot files, such as the Linux kernel.
/dev	A special part of the filesystem containing device filesthat correspond to low-level hardware devices, such as floppy drives and serial ports.
/etc	System configuration files go here.
/home	This directory houses users' home directories—where individuals store their files, such as WordPerfect documents.
/lib	System libraries live here. These files contain code usable by many programs, similar to DLLs in Windows.
/mnt	You mount media such as floppies and CD-ROMs here, either in /mnt itself or in subdirectories under /mnt, such as /mnt/floppy or /mnt/cdrom.
/opt	Precompiled programs that don't come with your distribution go here. You may want to put WordPerfect itself in /opt/wp8 or something similar.
/proc	/proc is a strange thing indeed; it's not a filesystem in which ordinary files are stored, but it's a way for users and programs to access information about the system.
/root	/root is the system administrator's (root's) home directory. Don't confuse /root with / (the "root directory").
/sbin	Like /bin, but for commands that are normally used only by the system administrator.
/tmp	Temporary scratch space. Programs, including WordPerfect, store temporary files here. Linux normally cleans this directory out every now and then automatically, discarding files that are older than some amount.
/usr	/usr contains most application programs.
/usr/local	Application programs compiled locally go here.
/var	/var is storage for important and frequently changed system information, such as files that are scheduled to be printed and mail that has to be delivered.

Users of WordPerfect are normally concerned with files in the /home and /mnt directories. You should have a subdirectory within /home for your own files. This subdirectory probably has the same name as your login username, and WordPerfect probably resides in a subdirectory under /opt or /usr.

ACCESSING REMOVABLE DEVICES: USING THE `mount` COMMAND

To access files on a removable-media device such as a floppy or CD-ROM, you must first *mount* this device—that is, you must make it available as part of the regular Linux filesystem structure. You do so by issuing the `mount` command at an xterm and specifying both the *device* to be mounted (a file in the /dev directory) and a *mount point* (an empty directory onto which the contents of the disk are grafted). For instance, you might issue the command `mount /dev/fd0 /mnt/floppy` to mount the first floppy drive (A: in Windows) to the /mnt/floppy mount point. Before issuing this command, /mnt/floppy will probably be empty. Afterwards, it will contain the files on the floppy drive, and any files you create in that directory will be created right on the floppy disk. Table 3.2 lists many of the most common devices, typical mount points, and descriptions of what they're for.

TABLE 3.2 COMMON DEVICES AND MOUNT POINTS IN LINUX

Device	Typical Mount Point	Description
/dev/fd0	/mnt/floppy	Floppy disk
/dev/scd0	/mnt/cdrom	SCSI-based CD-ROM drive
/dev/hd?	/mnt/cdrom	IDE/ATAPI-based CD-ROM drive
/dev/cdrom	/mnt/cdrom	Typically a link to the real CD-ROM device file
/dev/sd?4	/mnt/zip	SCSI or parallel port Zip disk formatted with FAT or ext2 filesystem
/dev/sd?	/mnt/zip	SCSI or parallel port Zip or other removable-storage disk formatted with ext2, FAT, or Mac HFS filesystem
/dev/hd?4	/mnt/zip	IDE/ATAPI Zip disk formatted with FAT or ext2 filesystem
/dev/hd?	/mnt/zip	IDE/ATAPI Zip or other removable-storage disk formatted with ext2, FAT, or Mac HFS filesystem

Note

A question mark (?) in the Device column of Table 3.2 indicates that the character may be a lowercase letter from *a* through *d* or *g*, depending on the device and how it's configured.

Note

You may have noticed that some of the device files in Table 3.2 are the same for different devices. For instance, both IDE/ATAPI CD-ROMs and IDE/ATAPI Zip drives can have the identifier /dev/hd?. The details of which devices are associated with which device files vary from system to system. You may have to experiment to find out how your system is configured.

Normally, you need to be root to issue a `mount` command. If you have root access, however, you can configure your system to allow anybody to issue certain types of `mount` commands by editing the file /etc/fstab. Listing 3.1 shows an excerpt from an /etc/fstab file that enables users to mount FAT and ext2-formatted floppies on /mnt/floppy; Mac-formatted floppies on /mnt/macflop; Zip disks on /mnt/zip; and CD-ROMs on /mnt/cdrom.

→ For more information on root access, **see** "Users and the Superuser," **p. 11**

LISTING 3.1 SAMPLE /ETC/FSTAB FILE ENTRIES

```
/dev/scd0       /mnt/cdrom      iso9660     ro,user,noauto                  0 0
/dev/fd0        /mnt/floppy     auto        user,noauto                     0 0
/dev/fd0        /mnt/macflop    hfs         user,noauto,fork=netatalk       0 0
/dev/sdb4       /mnt/zip        auto        user,noauto                     0 0
```


The first column in the /etc/fstab file gives the device name of the device to be mounted, such as /dev/scd0. The second lists the mount point, such as /mnt/cdrom. The third lists either a single filesystem type or auto if Linux is to try to figure out which filesystem to use on its own. The fourth column lists mount options and is where you'll find the key to allowing a non-root user to mount a filesystem, in the form of the user option. Listing 3.1 also shows other options, including `ro` (for creating read-only access to the drive), `noauto` (which prevents Linux from trying to mount the device at boot time), and `fork=netatalk` (an HFS-only option that affects how Linux treats the Mac's resource fork). Note that you may need to alter the device names, mount points, and other options relative to this example to get it to work.

Caution

The /etc/fstab file contains other entries that tell Linux how to mount its standard partitions, including the root (/) partition. *Do not alter these entries unless you know what you're doing!* If you change them, Linux may not boot or may behave strangely in any of a number of ways. Adding new /etc/fstab entries for your removable devices is safe, however, provided you're careful not to modify the existing entries.

When you've altered your /etc/fstab file so that removable devices include the user option, you can (as root) issue the command `mount -a`; then non-root users will be able to issue the `mount` commands, but in a shortened form in which only the mount point is specified. For instance, you could type `mount /mnt/floppy` to mount the floppy drive, or you could type `mount /mnt/macflop` to mount a 1.44MB Mac-formatted floppy. In this case you need separate mount points if you intend to access the same device in different ways, as in the FAT/ext2 floppy versus Mac floppy example.

When you're done using a device, you must *unmount* it before you can eject it. You do this with the `umount` command. (Yes, that is spelled correctly; it's missing the first *n*.) You give the `umount` command the name of the mount point; for instance, `umount /mnt/floppy` to unmount the floppy drive.

Caution

Linux locks storage devices such as CD-ROMs and Zip drives so that you can't eject them before you unmount them. Floppies, however, have manual eject buttons, so it is possible to eject a floppy before unmounting it. Linux might not have written all its data to the floppy when you do so, however, so ejecting a floppy before unmounting it can result in loss of data. Be very careful when you remove floppies!

Caution

Some window or file managers can be configured to provide mount and unmount services for your removable media devices. You may want to look into this topic, especially if you use such devices frequently and don't want to use an xterm for this task.

LINKS

Linux implements a filesystem feature that's common on UNIX-like systems: *links*. A link is a special type of file that points to another file. Links come in two varieties: hard and soft. *Hard links*are essentially duplicate directory entries for a file. They therefore impose no speed penalty but can only link between files on a single partition or disk. *Soft links* point to another file or directory by name. They take a small amount of time to process (not noticeable by humans for a single access), but they can point to files across filesystems. You can create a link from an xterm with the `ln` command, with the `-s` parameter added, if you want to make it a soft link. You give the existing linked-to file first and then the name under which you want the link to exist. Links are supported only under ext2 and other UNIX-like filesystems; you cannot create links on FAT, HFS, or other non-UNIX filesystems, though a soft link may point to a file on such filesystems.

Tip #19 from *Tux Tips*

Because of the Linux filesystem structure, you may find it tedious to access files on a floppy or other removable disk from your home directory. You can reduce this tedium by creating a symbolic link. Suppose you mount floppies on /mnt/floppy. From your home directory, type `ln -s /mnt/floppy floppy` to create a symbolic link. Then, after you've mounted a floppy, you can access its files from your home directory simply by entering the floppy link.

FILE OWNERSHIP AND PERMISSIONS

Because Linux is a multiuser OS, it implements a concept known as *file ownership*. Every file must have an *owner*—a user on the system to whom the file belongs. Files are also associated with a *group*, which is a collection of users. A user can belong to more than one group. This approach allows a system's administrator to arrange his or her users in some logical structure. For instance, in an academic environment, the users would be individual students, and the groups could be individual courses. The user fbrown might belong to the groups cs244 and cs297, corresponding to two computer science courses, and the user tjones might belong only to cs244. Of course, if you're running Linux on your personal desktop computer, you probably have only one or two users, aside from root and the other behind-the-scenes housekeeping users your Linux installation created automatically.

This concept of ownership becomes important when combined with the concept of *permissions*. Each file has associated with it rules specifying whether the owner, other members of the file's group, and anybody else on the system can read the file, write to it, or execute it (that is, run it as a program). These permissions are represented in WordPerfect's file dialog boxes, as well as other places, by a string of characters to the left or right of the filename, shown in Figure 3.1.

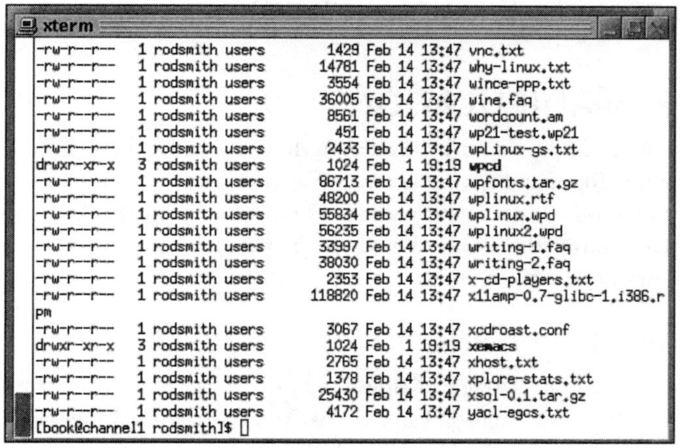

Figure 3.1
Issuing an `ls -l` command in an xterm produces a directory listing showing permissions and ownership for each file.

The permissions string consists of 10 characters. The first of these is a dash (-) for ordinary files or a *d* for a directory. Other values designate other special types of files, such as symbolic links or device files. The remaining nine characters are structured as three triplets, each representing the presence or absence of read, write, and execute permissions. These permissions are present when *r*, *w*, or *x* characters, respectively, are displayed, and are absent when - characters appear in their stead. The first triplet represents the permissions of the file's owner, the second is for the file's group, and the third is for all other users of the system. Thus, a file with a permissions string of `-r--------` represents a file that can only be read and only by the file's owner. A file with a permission string of `-rw-rw-rw` can be read or written by anybody, but cannot be executed.

If you're the only user of your Linux computer, you probably don't need to concern yourself with how your permissions are set, assuming you have both read and write access to your own files, as is probably the default. If you're sharing your Linux system with others and if you want to exchange files with them, you may need to change your files' permissions after you create them. You might also want to change your files' permissions if others use your system and you do *not* want them to see your files. If you have many files to hide, you may want to create a subdirectory with restricted permissions or even alter the permissions on your home directory. Restricting access to a directory means that others won't be able to read any files in that directory. You can alter the permissions of a file from the WordPerfect File Manager dialog box, described shortly.

Because filesystems intended for most non-UNIX OSs don't support permissions or ownership in the way Linux does, Linux has to fake it when it uses those systems. The details vary with the filesystem, but typically you can use the `umask`, `uid`, and `gid` options to the `mount` command, or in the options field in /etc/fstab, to set the default permissions, owner, and group of files, respectively. At an xterm, try typing `man mount` to get lots of details on this topic. Typically, if the user option appears in the /etc/fstab file for a given device, then the user who mounts it will have full read/write access, assuming it's a read/write device like a floppy and not a read-only device like a CD-ROM.

LINUX FILE MANIPULATION

Linux provides many tools for manipulating files, including basic UNIX-style commands at an xterm or text login and a wide array of file managers. It's therefore difficult to summarize these many methods of file manipulation. Fortunately, WordPerfect also provides a number of useful file-manipulation tools. These can be very useful for performing some basic tasks, and even some not-so-basic ones.

THE FILE MANAGER DIALOG BOX

The WordPerfect File Manager dialog box must lie at the heart of any discussion of WordPerfect's file-manipulation capabilities. You encountered this dialog box before when you saved files in Chapter 2, "Working Within a Document," but we only scratched the surface of its capabilities then. Figure 3.2 shows this dialog box. It is used by several WordPerfect file operations, though it's given a different title in its drag bar for each. You can get to the File Manager dialog box by choosing File, Open or File, Save As (Ctrl+S) from the Document window or by choosing Program, File Manager from the WordPerfect Control window, among other ways.

Tip #20 from *Tux Tips*

Choosing Program, File Manager allows you to open a file, just as File, Open does; you just select a file and click the Open button.

The File Manager dialog box includes several buttons, text entry fields, and so on, as well as three menu items. Most of the Files menu items are active only when a file or directory is selected. These items relate to manipulations you can perform on files or directories. The Display menu item contains options to alter what information you have about files in this dialog box. The Search menu item contains options for searching for a file in a number of ways.

FILE OPERATIONS

You can perform common file operations from the Files menu item, such as Move/Rename, Copy, and Delete. Click on a file and then select the operation you want. The

Move/Rename, Copy, and Delete options are all fairly obvious. The Move/Rename and Copy options both pop up a simple dialog box asking for a new name/location for the file, as shown in Figure 3.3. To move a file to another directory, you need to know the exact pathname for that directory because the dialog box doesn't give you the option to select it graphically. Both the Move/Rename and Delete options can be applied to directories as well as individual files.

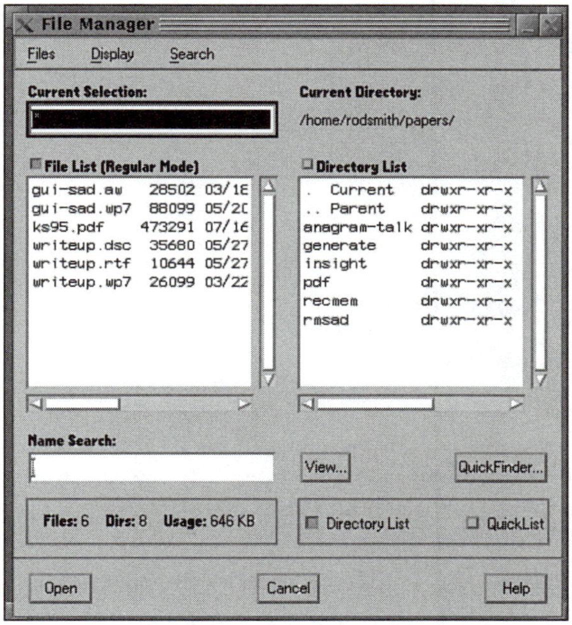

Figure 3.2
The File Manager and Open dialog boxes are identical except for their titles. The Save As dialog box includes extra options for selecting the format of the saved file.

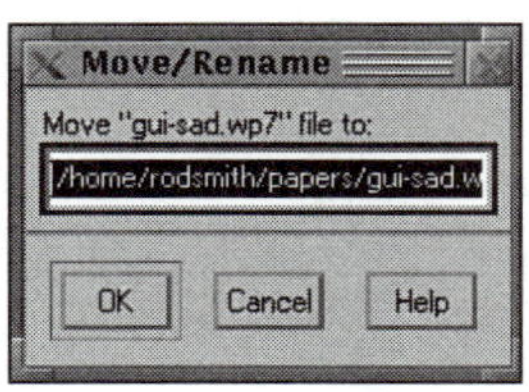

Figure 3.3
The Move/Rename dialog box starts up with the file's existing name and location already present. The Copy dialog box starts with a blank in that space.

Tip #21 from *Tux Tips*

If you want to rename or copy a file within its original directory, you can omit the complete path and just type in the new name. To move or copy a file between directories, you can enter the complete path starting from root (/) or a relative path starting from the current directory, as in ../letter.wpd to copy a file from /home/fjones/letters/home to /home/fjones/letters.

To change the permissions on a file or directory, select Files, Change Mode from the File Manager menu. This step displays the Change Mode dialog box, as shown in Figure 3.4. Click on the appropriate check buttons to select or deselect read, write, or execute permissions for the file. As you do so, the fields labeled Mode (String) and Mode (Octal) also change, showing the permissions in the string form described earlier and in a numeric form that experienced Linux users often use. You cannot edit the permissions by altering either of these fields, though.

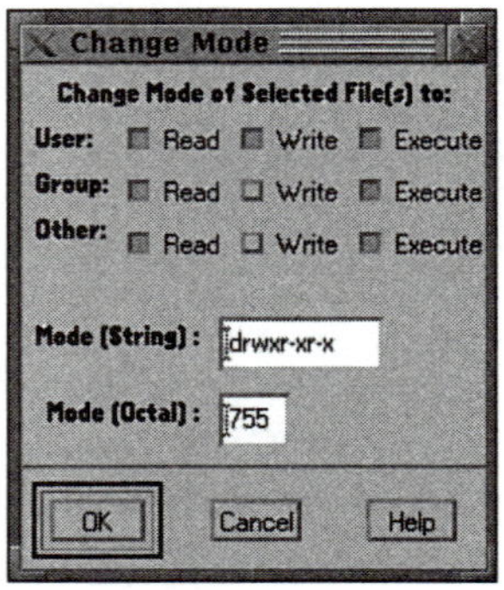

Figure 3.4
Despite the reference to "File(s)," you can only change the permissions of a single file or directory at a time from the Change Mode dialog box.

You can print a file from the File Manager by clicking on the file and choosing Files, Print File(s). WordPerfect *immediately* prints the entire file to the currently selected printer. You do not get a chance to select special print options or to change your printer. If you want to do so, you need to use the Print dialog box.

→ If you want to know more about printing, **see** "Creating Paper Output: Printing," **p. 148**

You can change to a directory by choosing Files, Other Directory, and you can create a new one by choosing Files, Create Directory. Both options show you a dialog box much like that in Figure 3.3 for moving or copying a file. Enter the destination or new directory in that field.

Tip #22 from *Tux Tips*

You can change to your home directory by typing a tilde (~) in the Current Selection field of the File Manager dialog box and pressing the Enter key.

ALTERING THE APPEARANCE OF THE FILE MANAGER DIALOG BOX

You can alter both the appearance of the File Manager dialog box and details of its operation from the Display menu. Three of the items on this menu toggle a feature on or off. The Show File View option turns on a file preview feature; when this feature is active and you click on a file, WordPerfect attempts to display it (Figure 3.5). For some types of files, such as ASCII text and WordPerfect documents, you can see the file's contents immediately. For other file types, you may see a summary of what the file contains; WordPerfect shows a list of the files in a zip archive, for instance. Some file types, such as certain foreign word processor formats or graphics files, may require conversion to be viewed, and you can click a button in the Viewer window to do this conversion. You can still use the File Manager dialog box menus and buttons when the Viewer window is open.

Tip #23 from *Tux Tips*

> The file preview feature is a handy way to clean out a directory of old or unused files or to search for a file with particular contents (but WordPerfect also has file search features, described shortly).

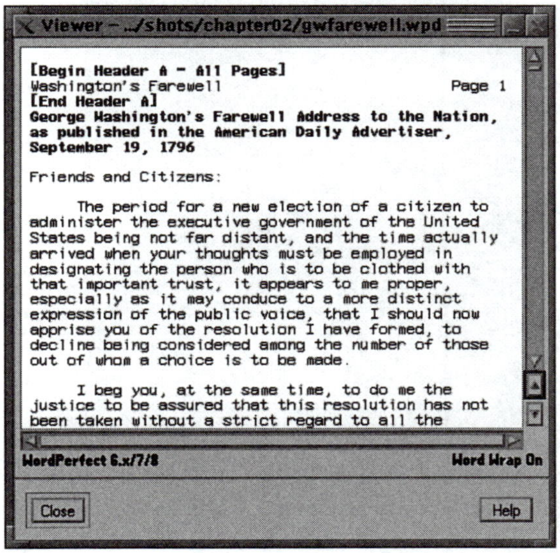

Figure 3.5
The Viewer shows the text in the file, but not the formatting details.

The Display, Show WP Descriptive Names option turns on or off a display of special WordPerfect file information in the dialog box. This information was used in certain DOS versions of WordPerfect to help get around the 8.3 filename limitations of DOS by providing longer descriptive fields within the file. Because Linux supports long filenames, you don't need to use this option unless you're exchanging files with a user of a DOS version of WordPerfect.

Instead of having a "hidden" attribute attached to files, as is supported by Windows, Linux hides files by placing a dot in front of their filenames. Thus, letter.wpd is visible, but .letter.wpd is not. To see these hidden files, turn on the Display, Show Dot Files option; WordPerfect then shows the hidden files in the file and directory listings.

Tip #24 from *Tux Tips*

> When Linux mounts a Windows disk, it loses the hidden file attribute, so you can see all hidden files. If you want to hide a file on such a disk, you can do so from Linux and for Linux by prepending the name with a dot, but in Windows the file is still visible. Some foreign filesystems may not support dot files at all.

To effect the greatest changes in the WordPerfect File Manager display, choose Display, Change File Manager Setup. This sequence opens the File Manager Setup dialog box, as shown in Figure 3.6. This dialog box allows you to change options for file sorting, what information is displayed in the file lists, how to handle passwords, and the default file format when saving files.

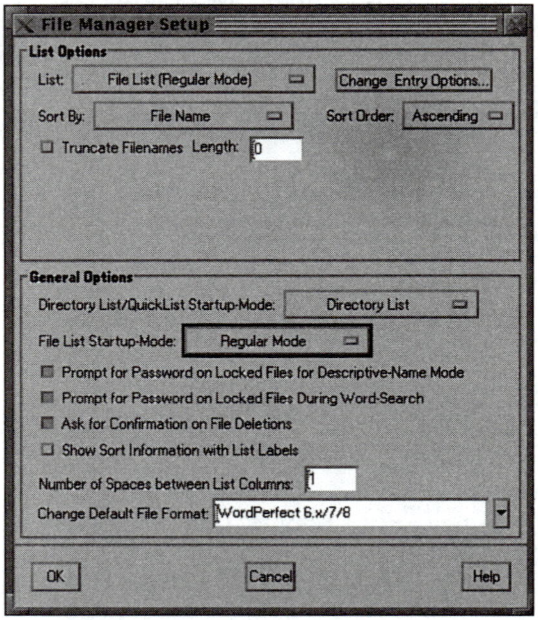

Figure 3.6
The default file format option has no effect on the Open and File Manager dialog boxes, only on the Save As dialog box. This option can be set from any of the three dialog boxes, however.

SEARCH FOR A FILE

You can find a file in several ways with the WordPerfect File Manager dialog box. These search options fall into two broad categories—file characteristic searches and file content searches.

FINDING A FILE BY NAME OR DATE

If you know how a filename begins, you can search for it by typing its first few letters in the Name Search field near the bottom left of the File Manager dialog box. Be sure that you've clicked on either the File List or Directory List buttons above the file and directory lists first, however (you can search for files or directories in this way). As you type the name, the list scrolls down to just above the first best-matching item. If you type the entire name, WordPerfect selects it for you; or you can click on the name to select it.

If you don't know the first few characters of the filename, but you do know that it contains some particular string, you can search by choosing Search, QuickFinder. This sequence opens the QuickFinder dialog box, as shown in Figure 3.7.

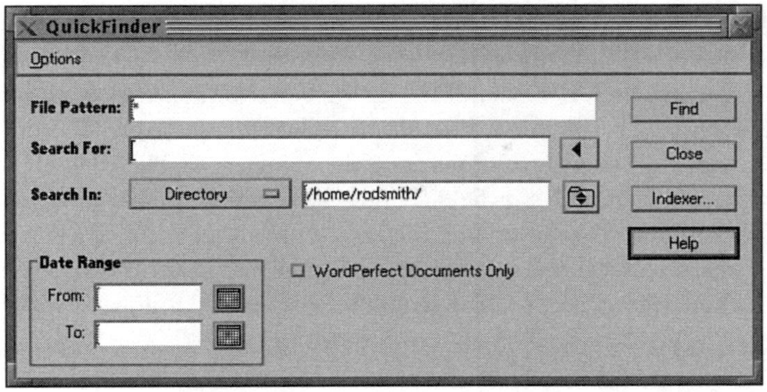

Figure 3.7
QuickFinder lets you search for files by name, date, or contents.

To use QuickFinder to search for a file by name, type in a pattern in the File Pattern box. A pattern can use *wildcards* in much the way they can be used in an xterm. The two wildcards that WordPerfect recognizes here are the question mark (?) and the asterisk (*). A question mark matches any single character. For instance, if you enter `la?.wpd`, WordPerfect finds files called lab.wpd, lad.wpd, and so on, but not late.wpd or la.wpd. The asterisk, on the other hand, matches any character or group of characters (including no characters), so `la*.wpd` finds all of these examples, but not long.wpd or lab.doc. Remember when you type in your pattern that Linux is case sensitive, so if you're not sure about the case of your filename, you may need to use wildcards or enter more than one search.

You can alter the directory in which WordPerfect performs its search by typing its name in the text entry area of the Search In field or by clicking the folder icon next to it and selecting the directory with a file selection dialog box. You can also set WordPerfect to search in a single directory or an entire directory tree by clicking the option button in the Search In line. You can tell WordPerfect to search all files or only WordPerfect documents by using the WordPerfect Documents Only check button.

To find a file by its date, you can enter a range of dates in the Date Range area of the dialog box. Either enter the date in numeric format (such as 2/2/99) or click the calendar next to the date field and select a date from the onscreen calendar (see Figure 3.8). Use the arrow buttons to move backward or forward by a month or a year.

Figure 3.8
Click the Today button to enter today's date in the date search field.

FINDING A FILE BY CONTENT

You can also use QuickFinder to locate a file that contains certain words. Type a word or words into the Search For field of the dialog box and then click Find to start the search. WordPerfect begins snooping through your files for that string. You can further refine the search criteria by using the options from the menus available when you click on the triangle to the right of the Search For field (Figure 3.9). These options allow you to add logical operators and to restrict or expand the search in various ways. These operators insert codes into the Search For field. As an example, you can search for documents that contain the words *cat* and *dog* within the same sentence by using the following search string:

```
/Sentence dog & cat
```

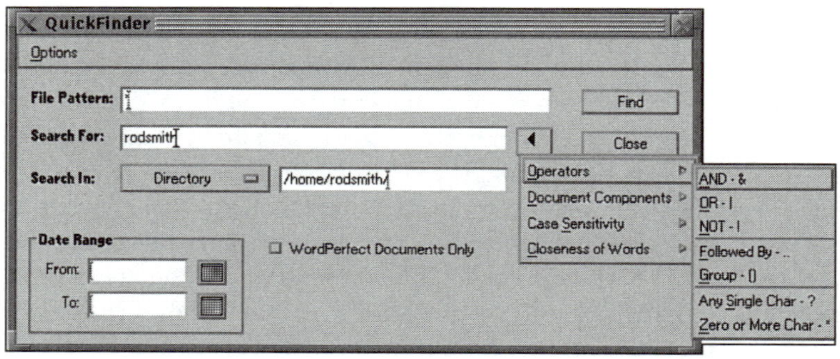

Figure 3.9
You can perform some very complex searches with the operators provided by WordPerfect's search utility.

If you're searching an entire directory tree or a directory with many files, the search may take a while, and if so, WordPerfect displays a progress report from the Search dialog box (Figure 3.10). From this dialog box, you can abort the search by clicking the Stop button.

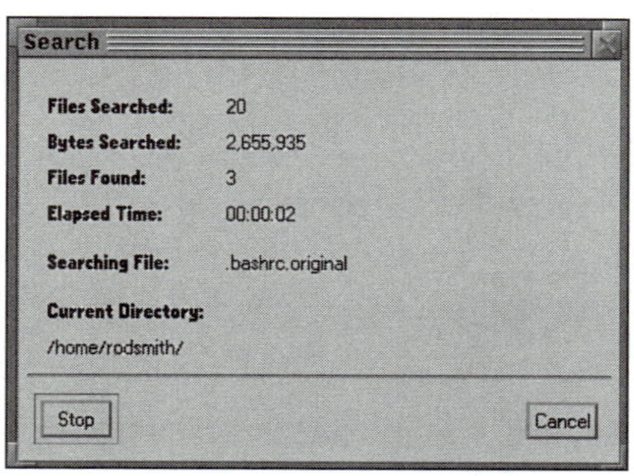

Figure 3.10
WordPerfect keeps you informed of the status of a long search. If the search finds too many files or is taking too long, you can stop it from the Search dialog box.

Tip #25 from
Tux Tips

You can combine any of the search types. For instance, if you're looking for a file that you know has the word *paper* as part of its filename, that was created in 1998, and that contains the word *disgruntled*, you could enter search criteria for filename, date, and content. If the directory you're searching has many files, entering multiple criteria in this way can speed up the search because WordPerfect can then skip files that don't meet the filename and date criteria.

At the end of any QuickFinder search, WordPerfect either displays a simple notice that it's found no files or opens a Search Results dialog box (Figure 3.11).

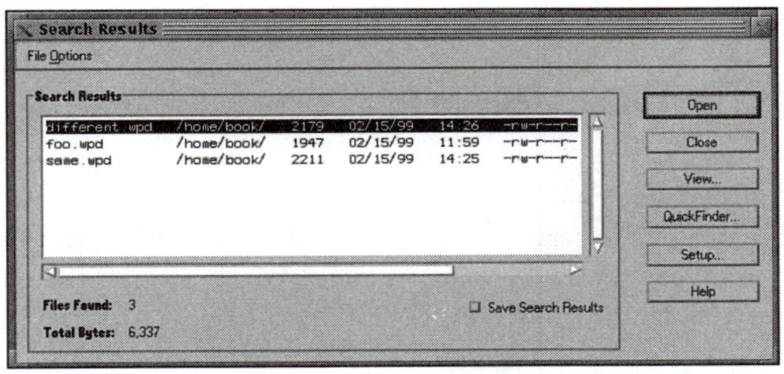

Figure 3.11
You can view or open files directly from the Search Results dialog box.

The Search, First Page and Search, Entire Document menu items provide a more streamlined, but less flexible, means of searching for documents. These options search only the first page or the entire document, as you might guess, and perform only relatively simple keyword searches. If you're just looking for a document that you know contains one specific word, though, these searches can be simpler to initiate than QuickFinder searches.

Opening One or More Files

Opening a file in WordPerfect is simple, assuming you know the filename; just choose Program, Open Window from the WordPerfect Control window or choose File, Open from an existing Document window. Select the file you want to open in the Open dialog box and click the Open button. You can also open files from the Search Results dialog box if you need to search to find the file you want.

When you attempt to open a file into an existing Document window via File, Open, if the current document has been modified since it was last saved, WordPerfect asks whether you want to save the current document first. If you click Yes, WordPerfect saves your changes to the file. If necessary, WordPerfect pops up the Save As dialog box to get a filename. If you click No, you lose any changes to the existing document.

You can load two or more documents into WordPerfect for Linux simultaneously by using the Program, New Window (Ctrl+N) or Program, Open Window items on the program

Control window's menu. The New Window item opens a new, empty WordPerfect window, whereas the Open Window option loads an existing file into a new window. However you do it, each document has its own window, which you can resize, move, and minimize independently. If your window manager supports moving windows from one virtual desktop to another, you can even put your WordPerfect documents on different desktops. These windows are also independent in their WordPerfect controls; you can set up one to show a document in Draft view with no toolbars, status bars, or rulers while another document can be decked out in Page view with as many of these bars as you like. The program Control window's Window menu item contains options to manipulate the Document windows by arranging them on the screen in different ways, minimizing them, closing them, or switching your view between them.

MERGING TWO DOCUMENTS

On occasion you may want to merge two documents together. To do so, go to the file into which you want to insert another file and position the insertion point where you want the new file to appear. Select Insert, File. WordPerfect displays the Insert File dialog box, which is yet another example of the File Manager dialog box. Locate and select the file you want to insert and then click the Insert button. WordPerfect loads the file into the existing one, and when you save the current document, WordPerfect saves the inserted file's contents along with it. This operation does not affect the original version of the inserted file.

Tip #26 from *Tux Tips*

One alternative to merging two or more documents into one large document is to use subdocuments. The subdocument approach enables one WordPerfect document to refer to another one without actually including it. This technique can be useful to keep document sizes manageable or to allow easier collaboration between two or more people on a large document.

→ To learn more about subdocuments, **see** "Master Documents and Subdocuments," **p. 598**

DOCUMENT EXCHANGE

Both WordPerfect and Linux provide mechanisms to help you exchange your documents. *Exchange* can mean moving files between WordPerfect and another word processor, moving files from one computer to another, or both.

READING FOREIGN FILES

WordPerfect includes *filters* that allow it to read foreign files, or files created by other programs. If all goes well, WordPerfect's filters act more or less seamlessly. Nevertheless, WordPerfect asks you to confirm the type of file import to perform in the Convert File dialog box, as shown in Figure 3.12. WordPerfect generally identifies the file type correctly, but occasionally you have to make a correction.

Figure 3.12
If WordPerfect misidentifies a file type, scroll through the list in the Convert File dialog box, select the correct type, and then click OK.

WordPerfect generally does a good job importing foreign files. Figure 3.13 shows WordPerfect displaying an Applix Words file, with Applix displaying the same file alongside for comparison. WordPerfect correctly handled the paragraph formatting, the centering and boldfacing of the headers, and the table. WordPerfect did not correctly translate the fonts, however; the original file used Benguiat Gothic, which was translated to Roman-WP during the importation.

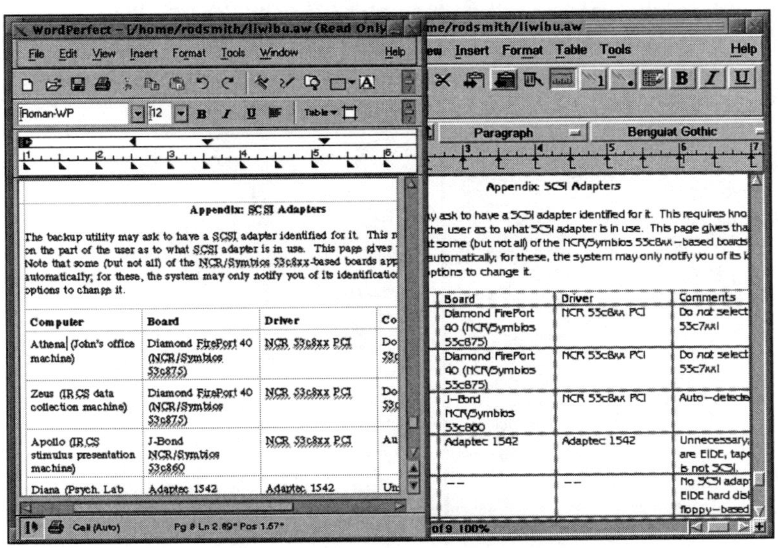

Figure 3.13
WordPerfect's file import/export may create subtle (or not so subtle) differences between the original and the imported version.

WordPerfect can import files from a wide range of sources, including the popular Microsoft Word, earlier versions of WordPerfect, and Applix Words, which is another popular Linux word processor.

Tip #27 from *Tux Tips*

If you need to move a file from Microsoft Word to WordPerfect, try turning *off* the fast save feature of Microsoft Word. From the Microsoft Word Save dialog box, click Options and uncheck the Allow Fast Saves item. This setting makes it easier for WordPerfect (or any other program besides Microsoft Word) to process the file. It also makes the Word file smaller.

Tip #28 from *Tux Tips*

If you have problems loading a file using a program's native format, try saving the file in another format from the other word processor. WordPerfect format is frequently available, but if it's not or if it doesn't work, try Rich Text Format (RTF) or any other format the two programs might have in common.

Writing Foreign Files

If you need to give a file you've created using WordPerfect to somebody who uses another word processor or if you want to use the file in another word processor, you can export the file using WordPerfect's export filters. Choose File, Save As (or press F3) to display the Save As dialog box (Figure 3.14). This dialog box is the same as other File Manager dialog boxes except that it includes a Save Options section near the bottom. Click the arrow next to the file format and select the destination format from the list. You should also be sure to change the filename to reflect the new file type and to prevent overwriting your WordPerfect file with a foreign format. When you click OK, WordPerfect saves your document in the foreign format.

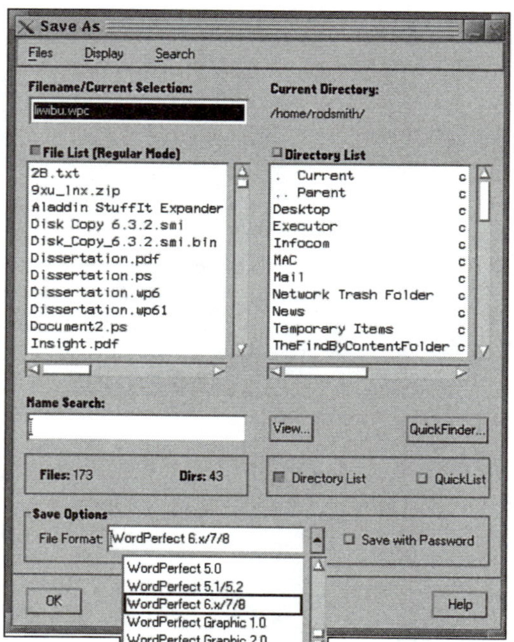

Figure 3.14
Scroll through the list of available file formats in the Save As dialog box as you would scroll through other lists.

Tip #29 from *Tux Tips*

> WordPerfect's file exports, like its imports, are good but not always perfect. To maximize the chances that the file will be readable on the receiving end, try saving it in several different formats, including WordPerfect and RTF. That way, if the destination format poses problems, you or the file's recipient can try other formats.

Getting Your Files from Here to There

If you need to transfer your files to or from another computer, Linux provides several means of doing so. One time-tested method of file exchange is via floppy disks or other removable storage media, such as Zip or magneto-optical disks. As I described earlier in this chapter, Linux can handle such disks in many formats, including disks created by DOS, Windows, OS/2, MacOS, and various forms of UNIX. You simply need to mount the disk and write your files to the appropriate directory inside a mount point. For instance, if you've mounted a floppy disk at /mnt/floppy and you want to save a WordPerfect file to that floppy, you can choose File, Save As to open the Save As dialog box and then choose Files, Other Directory to enter /mnt/floppy as a destination directory. Then enter a filename and click OK. When you're done, unmount the floppy and take or send it to the other computer.

→ To learn more about mounting, **see** "Accessing Removable Devices: Using the `mount` Command," **p. 76**

Tip #30 from *Tux Tips*

> Linux refuses to unmount a disk if any files on that disk are currently open—that is, in use by any program. If you want to unmount a disk and get an error message stating that the device is busy (or something similar), try closing the WordPerfect Document window associated with the floppy-based file. Also, be sure you have no xterm windows open with prompts set to the floppy.

If you want to use FAT floppies, one alternative to mounting them is to use the mtools package, which comes with most Linux distributions. This set of programs lets you access FAT disks without mounting them by issuing commands similar to DOS commands using an xterm or text-mode login. For instance, to copy the file letter.wpd to floppy, you would type `mcopy letter.wpd a:`. Using mtools can be convenient sometimes, but doesn't give WordPerfect direct access to the floppy. You have to save your files to your home directory, use an xterm to copy them to the floppy, and reverse the process when going the other way. Whoever uses mtools also needs full access to the raw floppy device, /dev/fd0, so you may need to change the permissions on that device as root. Changing permissions this way could potentially be a security hole on a multiuser system.

Linux was designed from the ground up as a networking operating system and is very much at home in a wide assortment of networking environments. It has, either built in or readily available in packages, networking options that let it share files with other computers. At the crudest level, you can access your WordPerfect files from another machine using ftp, which is an Internet file transfer protocol. Check your Linux documentation to find out how to configure your machine to function as an ftp server.

Linux can also seamlessly integrate its filesystem with various common office networking protocols. The Linux Samba package allows Windows and OS/2 computers to see Linux disks over a network, netatalk serves a similar function for Apple Macintosh computers, and NFS allows file sharing with other UNIX systems. Each of these packages has a counterpart that allows Linux to mount the foreign disk as if it were a local disk, so you can have a choice of accessing the Linux-based files from a remote location or copying the files from your Linux box to the remote computer. For more information, check the documentation for each package. The Linux HOWTO collection, which comes with most Linux distributions, also has details.

TROUBLESHOOTING

FILES SAVED TO FLOPPY ARE CORRUPT OR NONEXISTENT

You've saved files to floppy disk from Linux, but when you try to read them on another computer, they're either corrupt or are missing.

This phenomenon has two likely explanations. First, you may have ejected the disk before unmounting it. Linux *caches* all its filesystem accesses, which means that it may not actually write the data to disk for several seconds after WordPerfect thinks it's done. Issuing the `umount` command ensures that the disk is in a proper state so you can eject it.

The second possibility is that the disk was mounted with some file translation scheme in place. UNIX uses different characters at the ends of lines in its text files than does DOS or Windows, and MacOS uses yet another scheme. Linux therefore offers mount options to translate these characters. If they're used inappropriately, though, they can corrupt the files. For vfat and msdos filesystems, the default is to use *no* translation, so you won't have a problem unless you've added a mount parameter to do this translation.

FILESYSTEM TYPES

When you mount a floppy, removable disk, or CD-ROM in Linux, you see only 8.3 filenames, but in Windows you see long filenames.

Linux can access DOS/Windows floppies in two ways, using either the vfat or the msdos filesystem drivers. The former supports Windows' VFAT long filename conventions, whereas the latter does not and works only with short 8.3 filenames. Therefore, if you're seeing only 8.3 filenames, chances are you're using the msdos filesystem driver. If you're mounting your files as root, try specifying the vfat filesystem, as in

```
mount /dev/fd0 /mnt/floppy -t vfat
```

(You may need to change your device file or mount point.) If you're mounting as a normal user, you need to change the filesystem type in /etc/fstab to vfat (you need root access to make this change).

With CD-ROMs, Windows uses the Joliet filesystem to store long filenames, and this support was added late in the 2.0.x kernel series. If you're using an older kernel, you may not have Joliet support; or it may not be enabled if your kernel is more recent. You may need to recompile your kernel to add Joliet support.

CHAPTER 4

Making Changes to a Document: Basic Editing

In this chapter

OVERVIEW

The whole point of using a word processor rather than a typewriter is that you can easily alter your text after you've created it. "After" can be a matter of milliseconds or a matter of years, but no matter how long between writing and revision, you use the same techniques to change or expand your document. This chapter covers the basic principles of editing a document, which were introduced in Chapter 2, "Working Within a Document."

→ For basic information on working with WordPerfect documents, **see** "Working Within a Document," **p. 43**

SELECTING TEXT

When you want to change existing text, you have two choices: You can delete the existing text one character at a time by using the Delete or Backspace keys on your keyboard, or you can select the text and delete, replace, or alter it as a whole. If you want to modify more than a few characters, the latter method is usually faster. WordPerfect actually gives you several ways of selecting text. You can

- Position your mouse pointer within your text, click and hold the left mouse button, and drag the pointer across the text you want to select. By default, WordPerfect highlights from the start of the first word to the end of the last word because people usually want to select entire words or groups of words. If you move the mouse out of the editing area, WordPerfect scrolls in the direction you've moved the mouse, allowing you to select more text than can fit in one window.
- Double-click on a word to select it.
- Triple-click on a sentence to select it.
- Quadruple-click on a paragraph to select it.
- Click in the margin to the left of a sentence to select the sentence.
- Double-click in the margin to the left of a paragraph to select the paragraph.
- Position the insertion point at the start or end of the area you want to select, position the mouse at the other end, and then click the left mouse button while holding down the Shift key. As with a click-and-drag operation, this sequence selects the entirety of the words at both ends of the selection.
- Position the insertion point at the start or end of the area you want to select and then hold the Shift key while you use the keyboard commands for moving the insertion point described in Chapter 2. Not all of these keys work with the Shift key for text selection, but most do. This method also allows you to select portions of words, which can sometimes be useful.

→ If you need a rundown of WordPerfect's keyboard commands for moving the insertion point, **see** "Moving Through Text," **p. 51**

- Use the items on the Edit, Select menu to select areas of text.
- Position the insertion point at the start or end of a region; then use Edit, Find and Replace and select its Action, Extend Selection option to select text between the insertion point and some searched-for word.

→ For more information on Find and Replace, **see** "Find and Replace," **p. 105**

- Select a region of text with endpoints that differ both horizontally and vertically; then choose Edit, Select, Rectangle to select only the rectangle marked by those horizontal and vertical end points, as illustrated by Figures 4.1 and 4.2.

Figure 4.1
To select a rectangular region, begin by marking out the top-left and bottom-right points of the area you want to select. Note that more than you intend to ultimately select is highlighted.

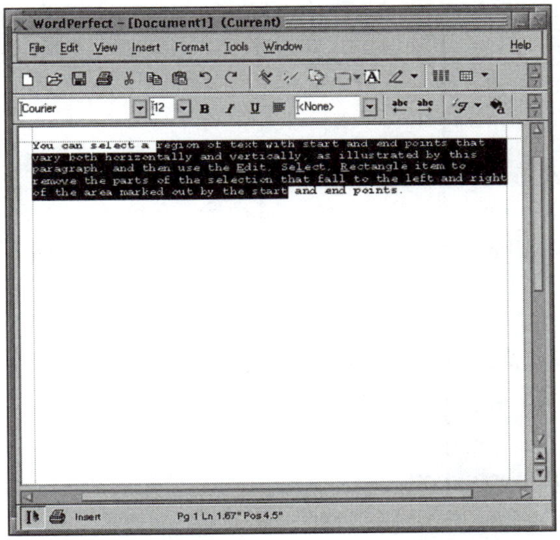

Figure 4.2
After you choose Edit, Select, Rectangle, WordPerfect reduces the selection area to the rectangular area you want to modify.

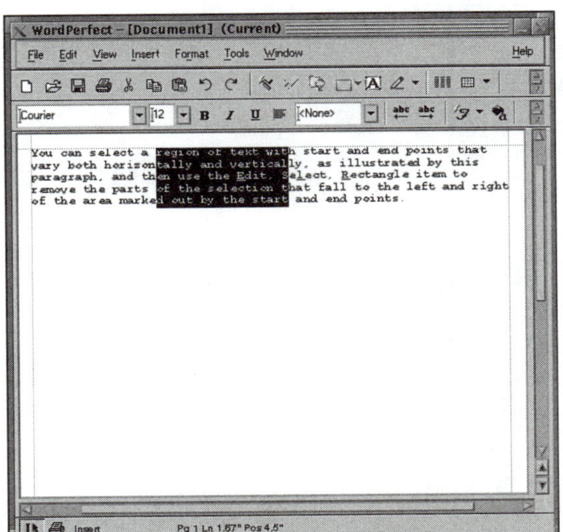

Tip #31 from *Tux Tips*

Rectangular selections generally work best with monospaced fonts like Courier. Such selections allow you to delete columns from tables in ASCII documents, adjust margins in text documents, and so on. WordPerfect has better ways to create this sort of formatting, so you won't often need rectangular selections when you work with WordPerfect files.

Tip #32 from *Tux Tips*

If you want to select text that starts at the left margin, begin your selection inside the first word and allow WordPerfect to expand the selection, start from the end of the selection, or use keyboard selection methods. WordPerfect allows you to adjust the margin with the mouse and might easily mistake a click at the edge of the word as a request to change the margin.

You can perform various operations on the selected text:

- Delete it by hitting the Delete or Backspace keys
- Move it to the WordPerfect Clipboard by choosing Edit, Cut (Ctrl+X)
- Copy it to the WordPerfect Clipboard without removing the original by choosing Edit, Copy (Ctrl+C)
- Replace it by typing new text
- Replace it with text from the WordPerfect Clipboard by choosing Edit, Paste (Ctrl+V)
- Apply various formatting changes to it, such as font changes, tab settings, and so on
- Check the spelling or grammar of the selection by choosing Tools, Spell Check (Ctrl+F1) or Tools, Grammatik
- Print it by selecting File, Print (F5)

Note

A *clipboard* is a data structure that's maintained by WordPerfect or another program for temporary storage of text, graphics, or other data. You use it to move or copy text or other material from one location or application to another. The terms *cut, copy,* and *paste* normally refer to operations on one or more clipboards, whereas the term *delete* refers to removing something without moving it to a clipboard.

Cut and Paste Between Documents or Applications

WordPerfect provides a number of ways for you to include pieces of text from one document in another.

COPYING FROM ONE WORDPERFECT DOCUMENT TO ANOTHER

If you want to copy text from one WordPerfect file to another, try the following procedure:

1. Open both documents. Assuming one is already open, you can open the second by choosing Program, Open Window from the program Control window.
2. Select the text you want to copy in the source document's window.
3. Choose Edit, Copy (Ctrl+C) or, if you want to delete the text from the original, choose Edit, Cut (Ctrl+X) in the source document's window.
4. Switch to the destination document's window and position the insertion point where you want the text to appear.
5. Choose Edit, Paste (Ctrl+V) to insert the text.

This procedure copies text, graphics, tables, and anything else from one document to another. You may not copy precisely what you expect, however, if your document contains any boxed items, because the WordPerfect codes for these elements may not lie precisely where they appear in the text. You may therefore need to copy every box separately, which you can do by following the preceding instructions but selecting the box by clicking on it. When it's selected, eight points mark the box's corners and sides, as in the equation in Figure 4.3.

➔ If you want more information on boxed elements, **see** "Complex Formatting with Boxes for Text, Graphics, and Equations," **p. 355**

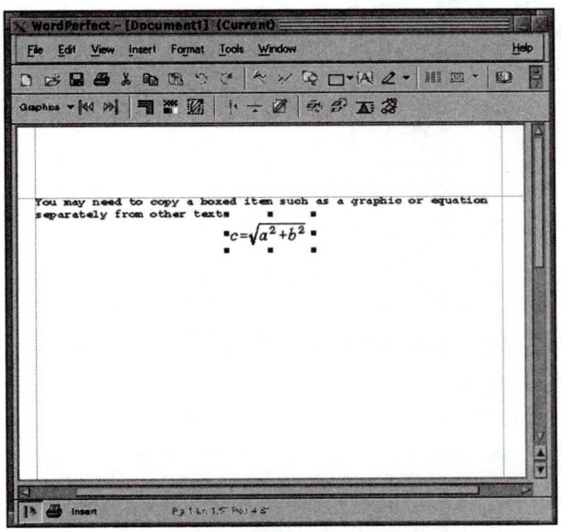

Figure 4.3
Separately selected boxed items are highlighted by points, but boxed items selected with other text will not be so highlighted, making it hard to tell whether they are included in a selection.

You might also find yourself in the opposite situation of having copied a boxed item that you did not intend to copy. If so, try selecting the unwanted copy and removing it by choosing Edit, Cut (Ctrl+X) or by pressing Delete or Backspace.

Tip #33 from *Tux Tips*

Selecting certain boxed items can be tricky because clicking on them might select the item itself or position your insertion point for directly editing the boxed material. If the latter occurs, the deletion procedure I just outlined won't work. If you find that you can't delete a box, try clicking outside the box to deselect it and then select it again by clicking on its border rather than its interior.

If you like, you can copy text using only one WordPerfect window by loading the source and destination documents sequentially. There's seldom much reason to do so, however, because WordPerfect easily supports two or more simultaneously open Document windows.

If you want to remove special formatting when you copy your text, try using the Edit, Paste Special command rather than Edit, Paste. The former sequence brings up the Paste Special dialog box, shown in Figure 4.4. You can select the Unformatted Text Only radio button and click OK to paste the text without special formatting.

Tip #34 from *Tux Tips*

You can paste text with or without its associated formatting by selecting options from the context menu you get by right-clicking the mouse in the editing area, as well as by using the ordinary menus and keystrokes.

Figure 4.4
Selecting the WordPerfect formatted text radio button results in a normal paste operation; the Unformatted Text Only button strips formatting codes from the text when it is pasted into your document.

COPYING FROM FOREIGN APPLICATIONS

You can easily copy a small amount of text from another Linux application into WordPerfect. If it's a text-mode application, you should be running it inside an xterm. To copy text from a Linux application other than WordPerfect, do the following:

1. Select the text in the foreign application by left-clicking and dragging on the source text. Depending on the program, you may be able to double-click to select a word or line, or perhaps you can make your selection in some other way.
2. Switch to the WordPerfect document into which you want to paste the text.
3. Position the insertion point where you want the text to appear.
4. Click with the *middle* mouse button to insert the text.

Tip #35 from *Tux Tips*

Many PC mice lack a middle button. If your mouse does not have a middle button, try *chording,* or pressing simultaneously, the two buttons you do have. XFree86 has an option to interpret a chord of a two-button mouse as equivalent to a click of the (nonexistent) middle button, so this technique should work if X is configured correctly on your system.

Tip #36 from *Tux Tips*

This procedure works for copying between two WordPerfect windows, in addition to copying from a foreign application to WordPerfect. It may not always produce identical results, however, particularly when dealing with text boxes.

You can copy text from foreign applications in this way, but not graphics. Also, you'll lose any special formatting, such as font changes, italics, and so on, when you copy text from other applications. The formatting loss occurs because the X Clipboard is simple compared to WordPerfect's capabilities, and this method relies on the X Clipboard to do its work. WordPerfect maintains one Clipboard for internal use, whereas X maintains a simpler Clipboard for copying text between applications. Selecting text alone copies it to the X Clipboard, and clicking the middle mouse button pastes text from the X Clipboard into the current application. WordPerfect for Linux tries to make these two Clipboards work together seamlessly, and for the most part it succeeds; but if you ever have problems with cut-and-paste operations, you may want to consider which Clipboard you're trying to use. That may help you work around the problem.

INSERTING ONE FILE WITHIN ANOTHER

As described in Chapter 3, "Managing Files and Folders," you can insert an entire file inside a WordPerfect document. You do so by positioning the insertion point where you want the inserted file to appear and choosing Insert, File. This step opens the Insert File dialog box from which you can select a file for insertion. If you insert a foreign application's file in this way, WordPerfect retains the formatting features it understands when it converts the file.

Tip #37 from *Tux Tips*

To copy formatted text from another application to WordPerfect, you can isolate the text you want in the other program, save the text as a separate file, and import it to WordPerfect. Alternatively, you can import the entire foreign file into a separate WordPerfect window and do a cut-and-paste operation between WordPerfect Document windows. Both methods preserve formatting such as italics and centering.

Because the X Clipboard is intended for text, not graphics, you can't directly cut and paste graphics from other applications. You have to use the file insertion method instead. Fortunately, in addition to foreign word processor formats, WordPerfect understands a wide range of graphics file formats, so you can probably find a format that will work for such an import.

→ To learn how to import graphics, **see** "Inserting Graphics from External Sources," **p. 379**

Inserting Special Characters

Chances are that most of the text you will want to enter into WordPerfect will be composed of standard alphanumeric characters and punctuation. You'll have no trouble entering text into WordPerfect, especially if you're a touch typist, because all these characters have keys or simple key sequences dedicated to them on a standard keyboard. You may occasionally find yourself in need of entering characters such as ∞ or ß, however. These characters have no counterparts on a standard U.S. keyboard, so how do you enter them?

The answer is to use WordPerfect's Insert, Symbol (Ctrl+W) feature, as shown in the Symbols dialog box in Figure 4.5. When you open this dialog box, you can insert special symbols from a variety of sets, including ASCII, typographic symbols, Japanese, Cyrillic, and mathematical and scientific. When you select the set you want from the menu button in the Set area of the dialog box, the available symbols change. You can select the symbol you want by clicking it. If you then click Insert, WordPerfect places the symbol at the insertion point in the text and allows you to select another symbol. Double-clicking the symbol has the same effect. If you click OK, WordPerfect inserts the symbol and closes the Symbols dialog box.

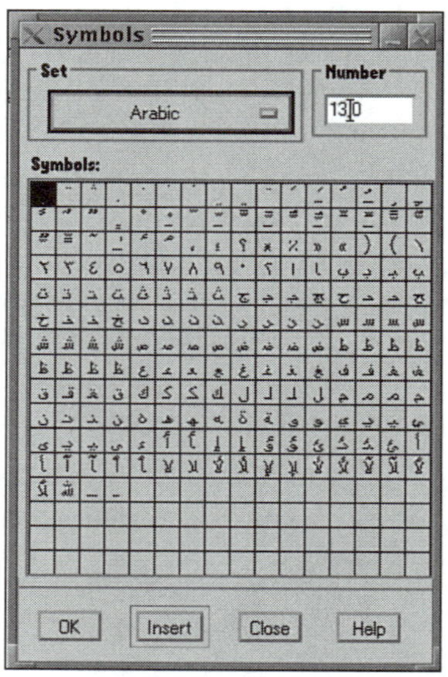

Figure 4.5
In the Symbols dialog box, the Number field gives a numeric code for the symbol. The first number represents the set, and the second, the number of the symbol within the set. You can edit this number by typing it.

Inserting a symbol in this way works, from your document's point of view, very much as if you had an extra-huge keyboard with keys for all these symbols. For instance, if you select text and then insert a symbol, the symbol replaces the selected text, just as if you'd typed an ordinary letter.

If you frequently need certain symbols in your writing, you can tell WordPerfect to enter the symbol you want in either of two ways.

First, you can create a *macro* that inserts the symbol and then assign that macro to a toolbar or key combination. A macro is a recording of procedures that WordPerfect performs automatically whenever you invoke that macro. You record the macro by selecting Tools, Macro, Record (Ctrl+F10); entering a name for the macro; performing some actions that will become the macro; and then selecting Tools, Macro, Record a second time. You assign the macro to a keyboard shortcut by giving the macro a specific name, such as ctrly.wmc if you want it to be played when you press Ctrl+Y. (By default, macro files appear in your home directory.)

→ For more information on macros, **see** "Letting the Computer Do the Work: Macros," **p. 483**

The second way to automate the placement of special symbols is to assign the symbol a QuickCorrect equivalent. To use this method, first enter the symbol and select it in the text as if you were going to copy it. Then click Tools, QuickCorrect to display the QuickCorrect dialog box (Figure 4.6). The symbol you selected should appear in the With: field. Type into the Replace: field some string of ordinary alphanumeric characters that you would like to type instead of the special symbol and then click the Add Entry button. Your addition will appear in the list. When you click OK, WordPerfect saves your changes, and when you subsequently type your special string followed by a space, Tab, or Enter, the symbol automatically replaces the string.

Tip #38 from *Tux Tips*

The QuickCorrect entries are stored in the .wprc subdirectory of your home directory. Therefore, you can safely add and remove QuickCorrect items without affecting QuickCorrect as used by other users of your system. Keeping user-modifiable defaults and settings in a user's home directory is common for Linux programs and helps make Linux a workable multiuser environment.

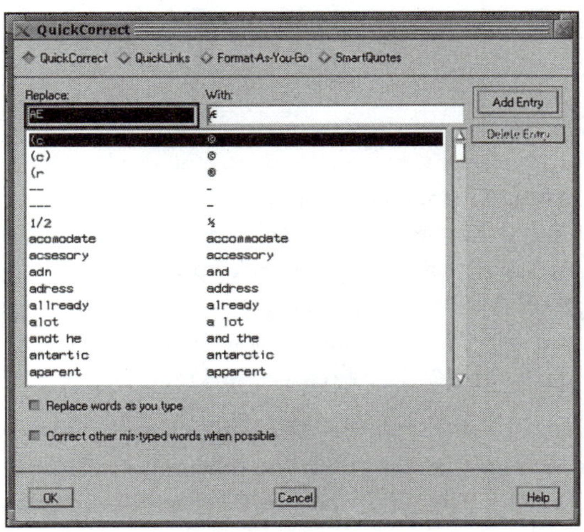

Figure 4.6
WordPerfect comes with quite a few QuickCorrect substitutions already defined. Most of them are designed to correct common typos, but a few are intended to substitute special symbols for common textual equivalents, such as © for (c).

Tip #39 from
Tux Tips

The QuickCorrect feature has many additional applications and features. It's covered in more detail in Chapter 5, "Checking Your Document for Errors."

➔ To learn more about QuickCorrect, **see** "QuickCorrect," **p. 122**

STARTING A NEW PAGE

When you type a document longer than a single page, WordPerfect automatically starts a new page whenever space runs out on the current one. You therefore need not ordinarily concern yourself with the details of starting and ending pages. If you need to know how much you have typed or where you are on a page, the WordPerfect status bar displays this information, as shown in Figure 4.7. You can also get a good idea of how much of a page you've used from the WYSIWYG nature of WordPerfect's display, at least when it's in Page view mode.

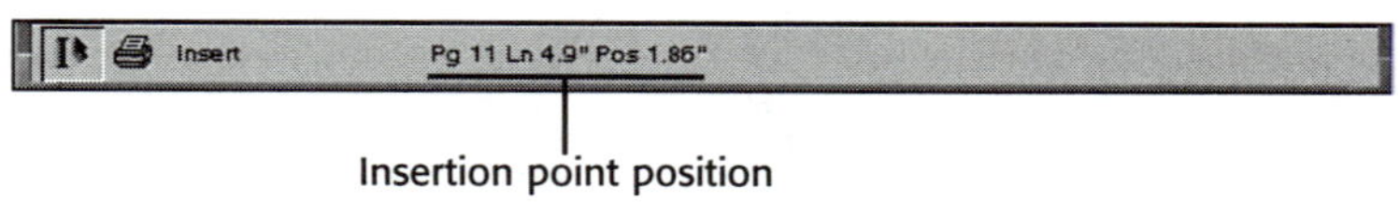

Figure 4.7
The WordPerfect status bar includes a summary of where in a document the insertion point rests in terms of its page, vertical position on the page (Ln), and horizontal position (Pos). These values are relative to the edge of the page, not the margin of the text.

If you need to force WordPerfect to start a new page, you can do so with the Insert, New Page (Ctrl+Enter) command. This sequence inserts what WordPerfect calls a "hard page" code ([HPg] in Reveal Codes). If you're in Page view mode, you can see the insertion point jump to the start of a new visual page; scrolling back up will reveal emptiness from the bottom of the previous page up to wherever you entered the new page command. If you're in Draft view mode, however, you won't see any extra whitespace at the bottom of the page. Instead, a double line represents the page break, as in Figure 4.8.

You may occasionally want to control where WordPerfect places its own page break. For instance, you may want to keep a heading with the paragraph it describes, or prevent WordPerfect from breaking up a poem or computer program listing. You can tell WordPerfect when to break or not break text by using the Format, Keep Text Together feature.

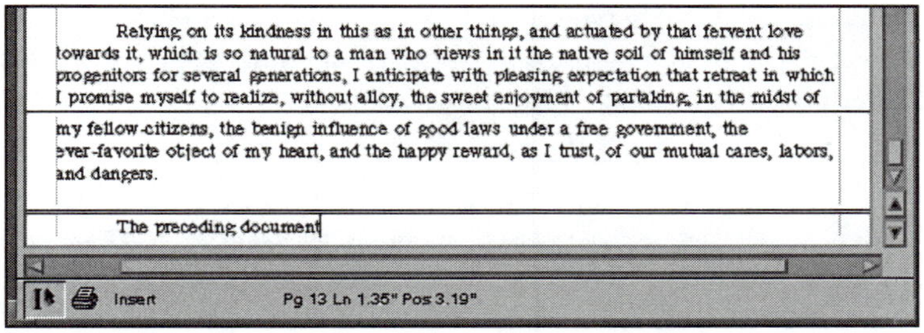

Figure 4.8
Draft view makes more efficient use of onscreen space by depicting page breaks as lines. Single lines denote soft page breaks created by WordPerfect because a page is full, whereas double lines represent hard page breaks inserted by the user.

➔ For more information on controlling page breaks, **see** "Page Breaks," **p. 230**

FIND AND REPLACE

WordPerfect's Find and Replace feature serves two related functions: to find a word, phrase, or character and to optionally replace the found text with new text. You may want to use this feature if you realize you've misspelled a person's name throughout a document, for instance. You can do a find-only operation to quickly locate a segment of text in which you know you use a particular word. If you've used other word processors or text editors, you'll discover that WordPerfect's implementation of this feature is familiar but may even be more powerful in many ways.

BASIC FIND AND REPLACE OPERATION

You start a find or replace operation by choosing Edit, Find and Replace (Ctrl+F2 or Ctrl+F) to open the Find and Replace Text dialog box (see Figure 4.9). Some word processors and text editors use separate dialog boxes for their find and replace operations, but WordPerfect uses one dialog box for both; you simply don't use the replace functions if you only want to do a search. This flexibility allows you to easily change a find operation to a replace operation.

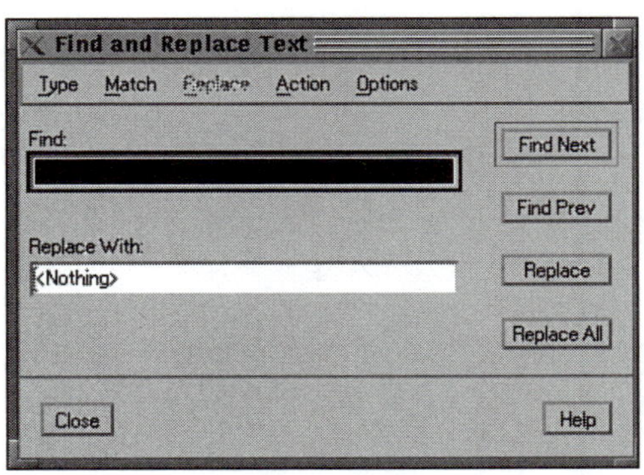

Figure 4.9
The Find and Replace Text dialog box starts with <Nothing> entered in the Replace With field. If you click Replace or Replace All with this default, the Find term is deleted from the text.

To search for text, simply type the text in the Find field and click Find Next or Find Prev. The former searches forward from the insertion point, and the latter searches backward. WordPerfect moves to and selects the searched-for text.

Tip #40 from *Tux Tips*

> WordPerfect sometimes finds text that's located under the Find and Replace Text dialog box. In this case, you won't be able to see the found text, and it's easy to believe that your button click didn't "take." Try moving the Find and Replace Text dialog box to uncover the highlighted text in your Document window.

If you want to replace text, type something in the Find field, click in the Replace With field, and type what you want to use instead. Click Find Next or Find Prev, as in a find operation, to locate and highlight an instance of the word; then click Replace to replace that single instance of the word with the new text and search for the next. If you're certain that you want to replace *all* instances, you can click Replace All. Note that if you've already bypassed some replacements, the Replace All button replaces even those you initially vetoed.

Altering Find Actions

You can alter what WordPerfect does when it finds a word by selecting the desired behavior from the Find and Replace Text dialog box's Action menu. The default action is Select Match, in which the searched-for word becomes selected. If you then close the Find and Replace Text dialog box, the found text remains selected. The Position Before and Position After actions place the insertion point before and after, respectively, the found text. The Extend Selection action creates a large selection from the insertion point to the found text. You can then close the Find and Replace Text dialog box and cut, copy, replace, or otherwise alter the selected text.

Tip #41 from *Tux Tips*

> The find actions apply to *finding* text. When you click Replace, WordPerfect always selects and then replaces the text. If you select the Position After action, perform a search, and then click the Replace button, WordPerfect will search *again*, rather than replace your already-found word.

Defining Find Options

The Match and Options menu items contain options that influence the way WordPerfect performs its search. Specifically:

- If you select Match, Whole Word, WordPerfect matches your Find text only if it is a complete word. This option is handy if you're searching for a short word, such as *is* or *for*, that may appear as part of a larger word, such as *piston* or *format*, which you don't want to find.
- The Match, Case Sensitive option makes WordPerfect match your search only if the text appears with the same capitalization you enter in the Find field. You can use this option to search for acronyms such as ISO or NASA; when you know the text you want

to find appears at the start or in the middle of a sentence; or when you're searching for a word that can be either an ordinary word or a proper name, such as *polish* or *Polish*.

- The Match, Font option lets you specify a font in which the searched-for text must appear (Figure 4.10). Click the Text Only check button to disable this feature or click any of the remaining check buttons to enable it. For instance, you can search for text that's been italicized in any font by checking the Italics button and being sure the Font button is *not* checked; or search for bold text in Helvetica by checking both the Bold button and the Font button and setting the font to Helvetica. Note that the bold, italic, and similar attributes are different from the bold, italic, and similar font styles. The former are what you get when you use Ctrl+B, Ctrl+I, or the toolbar shortcuts to create a bold or italic variant on a font. You specify the latter only from the Font dialog box.

➔ To learn more about fonts and their attributes, **see** "Selecting Fonts and Font Attributes," **p. 46**

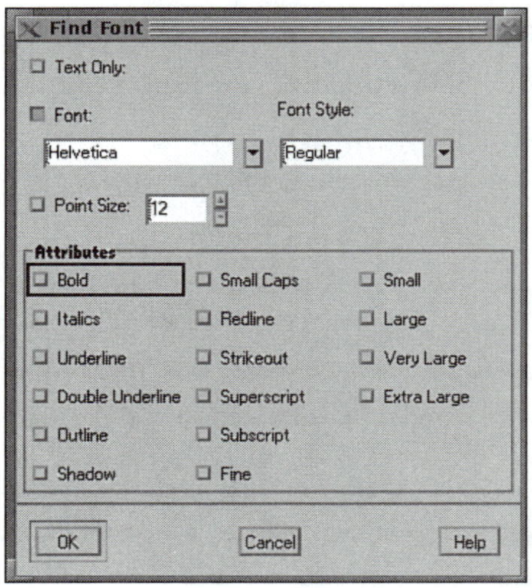

Figure 4.10
Selecting font characteristics in the Find Font dialog box restricts a find or replace to text matching those characteristics that you select. If you don't select a feature, WordPerfect matches any text with or without that feature.

- Select Options, Begin Find at Top of Document to start the search process from the beginning of your file; otherwise, the search starts at the insertion point.
- The Options, Wrap at Beg./End of Document selection causes WordPerfect to "wrap" its search operation from the end of the document to the beginning, or vice versa, so that you search the entire document.
- Select Options, Limit Find Within Selection to do a search or replace only within a region of text that you've already selected.
- Select Options, Include Headers, Footers, etc. in Find to have WordPerfect search through headers, footers, text boxes, and other areas of text that are somehow isolated from the main text.

- The Options, Limit Number of Changes feature opens the Find and Replace Text dialog box and allows you to say how many replacements to make (Figure 4.11). WordPerfect stops after that many replacements. If you've enabled this feature and want to disable it, enter 0 in the Number field and click OK.

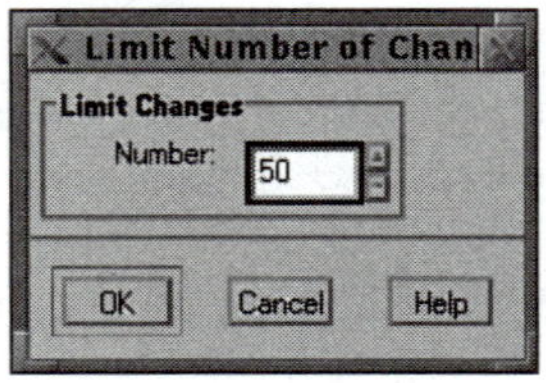

Figure 4.11
You can click the up and down arrows on the spin box to adjust the number of changes, or you can click in the text entry area and enter a value directly.

SEARCHING FOR CODES AND SYMBOLS

One of the more powerful aspects of WordPerfect's Find and Replace feature is its capability to find and replace codes and symbols.

If you want to find or replace a special symbol, do the following:

1. Choose Edit, Find and Replace (Ctrl+F2 or Ctrl+F).
2. If you want to search for a special symbol, click in the Find field; if you want to replace ordinary text with a special symbol, activate the Replace With field.
3. Press Ctrl+W. This step opens the Symbols dialog box, just as in Figure 4.5.
4. Enter your symbols just as you would in ordinary text and dismiss the dialog box.

If you want to replace one special symbol with another, you can do so by pressing Ctrl+W for both fields. You can also do a search or replace with a mixture of normal text and symbols.

You can search for either codes to create specific formatting or general types of codes. The general case is a bit simpler. From the Find and Replace Text dialog box, select Match, Codes. This step brings up the Find Codes dialog box, as shown in Figure 4.12. From the scrollable list in this dialog box, select the code you want to search or replace and either select it and click Insert or double-click the code's name. This step inserts it into the Find or Replace With field. You can select two or more codes to find or replace, and you can mix them with ordinary text.

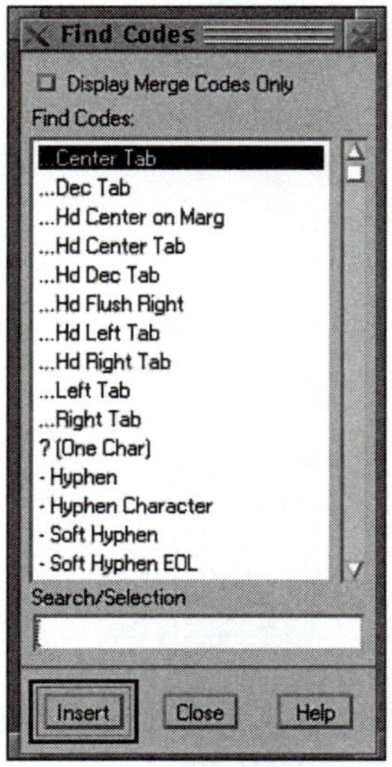

Figure 4.12
The list of codes is shorter when you select it from the Replace With field rather than from the Find field because the find operation needn't match specific values in codes (say, for margin settings), but you do need specific values for replacing such codes.

When you're searching for codes, you might not always be able to identify them in the Find Codes dialog box. These entries are generally abbreviations of meaningful words or phrases, but you may be left guessing at some of them. Unfortunately, WordPerfect's help system may not be of much use in this instance, though you may luck out if you try searching for the code's name or some obvious expansion of it. Of course, if you know you're searching for a code that does something specific, you can probably make some good guesses as to what that code is called if you don't already know its exact name. You could also try creating a similar code and using Reveal Codes mode to see what it's called and then delete your sample.

Tip #42 from *Tux Tips*

If you're searching for two or more codes, the order of the codes is important. For instance, if you search for [Bold On][Italc On], that search won't find [Italc On][Bold On], though the two sequences produce identical results.

Whereas the Match, Codes item lets you search for codes that don't require specific values, the Type, Specific Codes (Shift+Ctrl+O) option lets you search for and replace codes with specific values. Figure 4.13 shows the Specific Code dialog box for selecting the type of code to be found and replaced. After you select a code, the Find and Replace Text dialog box changes to accommodate this new type of search. The details depend on the type of code you're finding, but an example for a find on specific fonts appears in Figure 4.14.

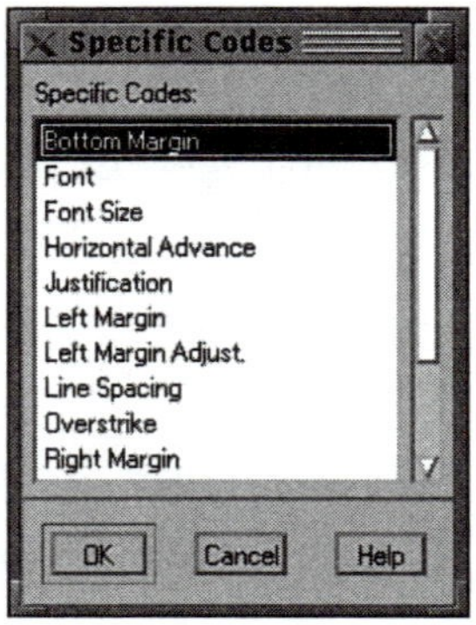

Figure 4.13
The code types in the Specific Codes dialog box require additional data input to include details about the codes to be matched.

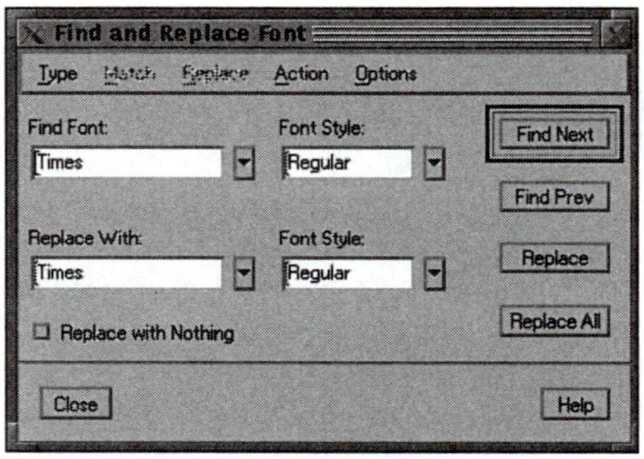

Figure 4.14
The details of the Find and Replace Font dialog box vary with the code type, but all include data entry for the Find and Replace With items, just as in textual find and replace.

Many of the options from the Action and Options menus apply to code searches as well as to textual searches, so you can select where to position the cursor after a find operation completes, whether to search from the beginning of a document, and so on.

Editing with Reveal Codes

Chapter 2 introduced Reveal Codes mode in which you can see and act directly on the codes that underlie WordPerfect's text formatting. For the most part, editing with Reveal Codes active is the same as editing in more conventional ways; however, you should be aware of some differences and extra tricks.

➔ For more information on Reveal Codes, **see** "Reveal Codes Mode," **p. 63**

Entering and Leaving Reveal Codes Mode

You start Reveal Codes mode by choosing View, Reveal Codes (Shift+Ctrl+F3) or by selecting Reveal Codes from the context menu when you right-click in the main text editing area. The same actions close the Reveal Codes pane of the Document window when you're done with Reveal Codes.

SYMBOLS IN REVEAL CODES MODE

Figure 4.15 shows a short document with the Reveal Codes pane open. This document uses the special symbols for 1/2 and 1/4. These symbols appear in the Reveal Codes mode just as they are, unlike formatting codes such as carriage returns. Figure 4.15 shows a feature of Reveal Codes that is sometimes useful, however: When you move the insertion point to the immediate left of the code or symbol, it may reveal more about itself. In Figure 4.15, the insertion point falls just to the left of the second instance of 1/2 in the document. The first occurrence of this symbol, at the start of the text, looks much like its counterpart in the main editing pane, but the second instance shows as a code, complete with information on the symbol set and number for the code. If you double-click on this code, WordPerfect displays the Symbols dialog box and you can enter new symbols.

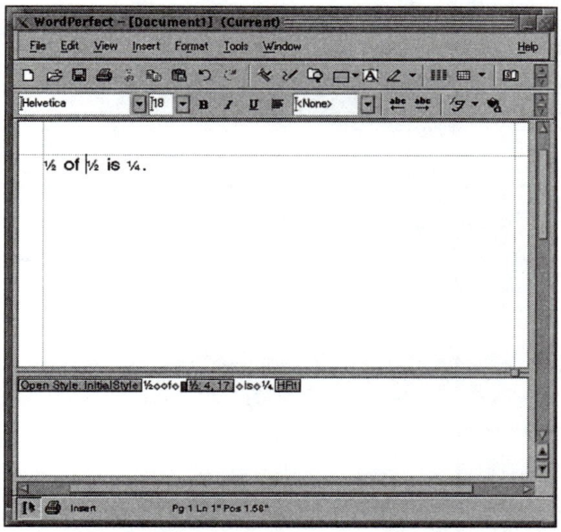

Figure 4.15
The insertion point sometimes, but not always, reveals extra information about codes and symbols when placed to their immediate left in the Reveal Codes pane.

PART I CH 4

Tip #43 from *Tux Tips*

If you double-click on a symbol's code with the intent of changing it, WordPerfect inserts a *new* symbol instead. You can then delete the old code.

USING REVEAL CODES TO SELECT TEXT

You can use Reveal Codes to help you select precisely the text you want—including codes. You select text much as you would in a regular Document window, but because you can see codes, you can select them more precisely. Figure 4.16 illustrates this point. In this figure, I selected the word *precisely*, including the bold attribute but leaving off the italics.

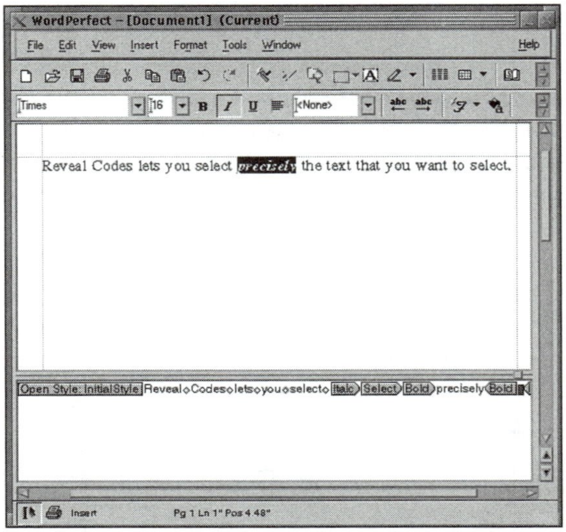

Figure 4.16
A selected area in Reveal Codes is marked out by a Select code on one end and a red insertion point marker on the other.

To select text in Reveal Codes, you click to place the red insertion point marker and then click and drag to select the rest of the desired text. The selection process works much as it does in normal Editing mode, and in fact, you'll see your selection being marked out in the main editing pane.

Tip #44 from *Tux Tips*

To position the insertion point to select codes precisely between codes, click on the code before which you want the insertion point to appear. This step places the insertion point, and you can then click on it and drag to make your selection.

Tip #45 from *Tux Tips*

When you click and drag out of the Reveal Codes pane, it scrolls very quickly. If you want to select text that extends beyond what's shown in the Reveal Codes pane, you may want to use keyboard selection methods rather than the mouse for greater control.

If you select an area that includes specific codes and then copy it to the Clipboard by choosing Edit, Cut (Ctrl+X) or Edit, Copy (Ctrl+C), only the codes you selected are cut or copied. Similarly, if you select text and certain codes and then select Edit, Find and Replace (Ctrl+F2 or Ctrl+F), the selected text and precisely those codes you included in the selection become the default search text. Figure 4.17 shows this effect, based on the selection shown in Figure 4.16.

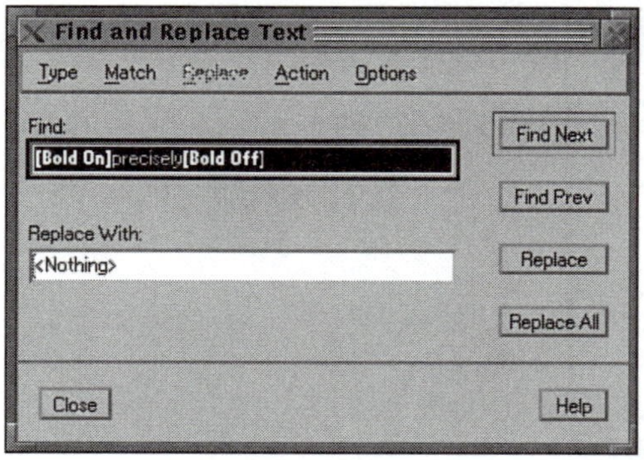

Figure 4.17
The names given to codes in the Find and Replace Text dialog box may not be identical to those in the *Reveal Codes pane* (p. 23), but they identify the same codes that WordPerfect uses internally.

Tip #46 from *Tux Tips*

You may sometimes want to select codes that are difficult or impossible to select from the main Editing window, such as centering codes at the start of a document. In this case, you can use Reveal Codes mode.

PART I
CH 4

USING UNDO AND REDO: TO ERR IS HUMAN

Most people occasionally make mistakes. Fortunately, WordPerfect makes allowances for this tendency and lets you undo your errors—up to 300 of them. If you decide you've made a mistake in performing an undo, WordPerfect even allows you to correct your "correction," with its redo feature.

BASIC UNDO AND REDO

If you've made a mistake, the simplest way to undo it is to choose Edit, Undo. This step reverses the latest change you made to your document. Changes include actions such as altering the font, inserting a header, or typing text. When you undo these changes, they simply disappear from your screen.

What if you want to undo an undo, though? You might want to do so if you change your mind about something or if you accidentally undid too many changes. You can, of course, simply do the changes over again—but that might be difficult if the changes involved typing an entire paragraph of text, for instance. The answer is to choose Edit, Redo (which is only available after you've used the undo feature). Redo yanks the undone operation from the jaws of electronic oblivion.

Both the undo and redo features are available on the default WordPerfect toolbar, as shown in Figure 4.18. Simply click on the counterclockwise arrow to undo a change or on the clockwise arrow to redo it. One or both of these icons are faded and unavailable if there is nothing to be undone or redone.

Figure 4.18
The Undo and Redo toolbar buttons provide quicker access to these features than the menu items, which can be particularly important because these features lack keyboard shortcuts.

UNDO/REDO HISTORY

One of the problems with using undo and redo as just described is that you'll be flying blind. If you forget what your changes have been, you may overshoot your target state. You also have no control over which specific actions you'll be undoing or redoing. You might want to undo some action you took several minutes before, for instance, but you'll be unable to do so without undoing all the intervening actions. That's where the undo/redo history comes in. Choose Edit, Undo/Redo History to display the Undo/Redo History dialog box depicted in Figure 4.19.

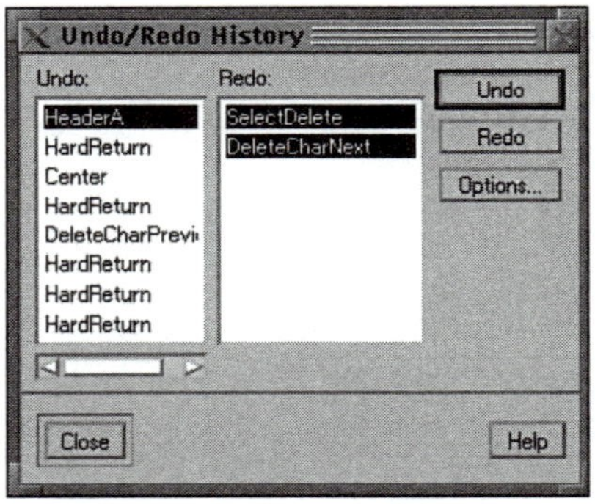

Figure 4.19
If you start the Undo/Redo History dialog box with no actions available to be redone, the buttons along the right edge are smaller, leaving the Options button only partly legible.

You can double-click an action from the Undo list to undo it or from the Redo list to redo it; or you can select the item and then click on Undo or Redo, as appropriate. When you select items that are out of order, however, you'll necessarily undo or redo the items at the tops of the lists, too. Further, if you undo actions and then redo them, they may appear in the Undo list in a new order or form. Overall, this routine can be rather confusing, so I suggest going easy on the undo/redo history and using it only when you must.

Tip #47 from *Tux Tips*

When you undo an action, it moves into the Redo list; but when you perform some new action, WordPerfect clears the Redo list. Therefore, if you think you might want to redo an action, redo it *immediately* or you won't be able to do so at all. This advice applies whether you're using the full undo/redo history feature or are just performing individual undo and redo actions.

CONFIGURING UNDO/REDO OPTIONS

From the Undo/Redo History dialog box, click Options to configure the way undo and redo work (Figure 4.20). This Undo/Redo Options dialog box allows you to set two options. The first is the number of levels of undo/redo allowed. This value can range from `1` to **`300`**; if you enter a value outside of that range, it will be set to either `1` or **`300`**.

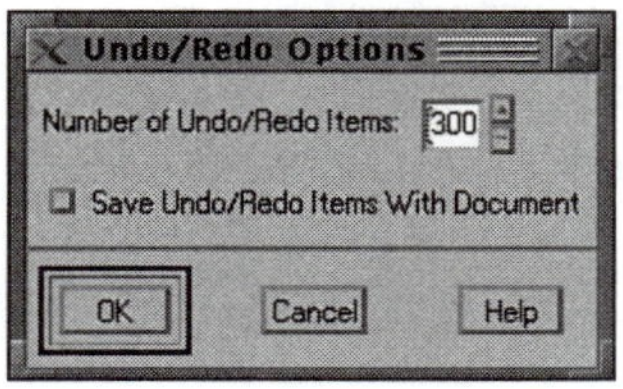

Figure 4.20
The default number of undo or redo actions is 10, but you can increase or decrease this value by clicking the arrows in the spin box of the Undo/Redo Options dialog box or by directly typing in a new value.

The other option is whether to save the undo/redo history with your document. This option is enabled by default and allows you to undo actions you took even in a previous editing session hours, days, or even years in the past.

Tip #48 from *Tux Tips*

If you regularly work with large boxed objects, such as graphics images, saving the undo/redo history can substantially increase the size of your files on disk. You can therefore reduce the size of your WordPerfect document files by disabling this feature or by keeping the number of items in the history small.

WordPerfect remembers the options you set here, so if you intend to change them temporarily, you need to go back and return them to their original values.

TROUBLESHOOTING

WORDPERFECT SLOWS WHEN PASTING TEXT FROM FOREIGN APPLICATIONS

You've been using the X Clipboard to copy lots of text from one or more foreign applications into WordPerfect. As you do so, WordPerfect slows down.

Some people report this problem with WordPerfect for Linux. One possible solution is to use an intermediate WordPerfect window for these operations—paste your text into an empty WordPerfect document and then use WordPerfect's cut-and-paste feature to move the text to the main Document window. If WordPerfect starts to get sluggish, close the temporary Document window. Another workaround is to use WordPerfect's file import feature rather than X's Clipboard to load the desired text.

→ For more information on combining files, **see** "Inserting One File Within Another," **p. 101**

CHAPTER 5

Checking Your Document for Errors

In this chapter

Overview

One reason that people use word processors rather than typewriters is that word processors produce better output on several levels. Visually, the wide range of fonts and crisp text from a laser printer, or even from most inkjets, is far better than what you can get from a typical typewriter. You can also use the computer's capability to handle tedious comparisons to improve your output in another way—by checking your writing for errors in spelling and grammar before you commit the document to paper. Indeed, WordPerfect can even automatically correct common typos as you make them. This chapter covers WordPerfect's document-checking features and shows you how to use them to best effect—or how to disable them if you so choose.

One note: In this chapter, I use the term *error* to refer to any word or language structure identified by WordPerfect as such. In reality, many of these so-called errors are proper names, rare words, unusual but valid grammatical uses, or ambiguous constructions that WordPerfect does not have the intelligence to parse. Using the word *error* instead of a variant of the preceding sentence helps keep this chapter readable.

Proofreading with Spell-as-You-Go and Grammar-as-You-Go

If you've done anything at all with WordPerfect 8 for Linux, you've no doubt already experienced one or both of its on-the-fly error detection technologies, known as *Spell-as-You-Go* and *Grammar-as-You-Go*. These features underline errors in red or blue, respectively, as soon as you type them, as shown in Figure 5.1.

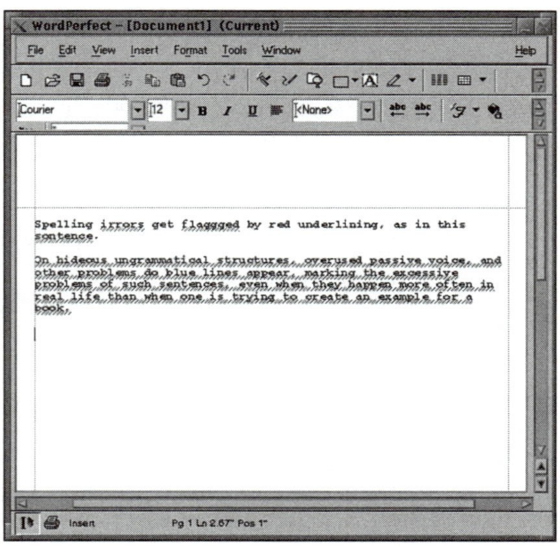

Figure 5.1
You can't see the difference between red and blue underlining in this figure, but the first sentence contains several typos underlined in red by WordPerfect's Spell-As-You-Go. The second is hideously ungrammatical, and so Grammar-As-You-Go has underlined the whole thing.

FINDING YOUR ERRORS RIGHT AWAY

By default, WordPerfect starts with both Spell-as-You-Go and Grammar-as-You-Go enabled, so you'll probably see lots of blue and red underlining if you just start typing. WordPerfect marks any word that it can't find in one of its dictionaries with a red hashed underline. The underline tends to pop out at you as you're typing, so the error is obvious immediately. If you like to correct your errors right away, you'll probably like this feature.

→ For more information on WordPerfect's dictionaries, **see** "Understanding Dictionaries," **p. 136**

Have you spelled a word correctly that WordPerfect flags as being misspelled? See "WordPerfect Flags a Correctly Spelled Word as an Error" in the "Troubleshooting" section at the end of this chapter.

Grammar-as-You-Go tries to do the same for grammatical errors. Computers are pretty good at catching spelling errors—a reasonably complete dictionary is fairly small by today's standards, and so WordPerfect can simply search through its dictionaries each time you type a word, and if it's not there, flag it. Grammar is another matter because it involves complex rules that necessarily intertwine with the meanings of sentences. In years past, few people liked computerized grammar checkers, and with good reason; they missed obvious grammatical errors while they flagged perfectly good sentences for rather dubious reasons. The Grammar-As-You-Go feature of WordPerfect 8 for Linux is, fortunately, much better than the grammar checkers of old; Grammar-As-You-Go is right when it flags errors more often than it's wrong. You might therefore want to leave it activated and heed its warnings.

Even if you don't want to keep these features activated all the time (and I tell you how to deactivate them shortly), you might want to turn them on every now and then. You can use them instead of or in addition to a traditional full-document spell check, for instance, or use them when editing and revising a first draft. WordPerfect's As-You-Go features flag errors in a visually distinctive manner that does not interfere with your normal editing tasks and can be very handy when used in these ways.

Tip #49 from *Tux Tips*

When you use the proofreading features on an existing document, WordPerfect won't necessarily display all the errors on a page just because that page is visible on the screen. To ensure that WordPerfect flags the errors, move the insertion point to the bottom of the Document window by clicking there. This step cues WordPerfect to do its checking and display any errors it finds.

CORRECTIONS AT YOUR FINGERTIPS

When proofreading features are active, you have a shortcut method of finding out how a flagged word or phrase offends WordPerfect. You do so by right-clicking on the underlined word. For instance, Figure 5.2 shows the result of right-clicking on a potential grammatical error; in this case, WordPerfect identifies the error type as well as possible corrections.

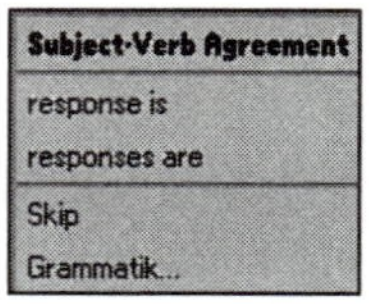

Figure 5.2
You can have WordPerfect make a correction by selecting the alternative you want from the pop-up menu's list.

Tip #50 from *Tux Tips*

> If you're trying to get to the normal context menu for text, say to turn on Reveal Codes mode or to paste text, be sure to not click your mouse when it's over a flagged error. The menus you get over errors and normal text are quite different.

DISABLING PROOFREADING FEATURES

If you find the proofreading features distracting, you can disable them quite easily. To turn off Grammar-as-You-Go but leave Spell-as-You-Go active, choose Tools, Proofread, Spell-As-You-Go. To disable both proofreading features, choose Tools, Proofread, Off. You cannot have grammar checking without spell checking. If you want to enable both features again, select Tools, Proofread, Grammar-As-You-Go. This setting persists between documents and WordPerfect sessions.

PROMPT-AS-YOU-GO

WordPerfect offers a handy adjunct feature to the proofreading features. It's called Prompt-As-You-Go, and it provides you with suggestions relating to both errors and correct text. You may not be able to use it very well immediately, however, since the field in which it displays its information falls far to the right in the property bar. If your Document window is too narrow, the display field disappears onto the second line, which you must access by scrolling the property bar (Figure 5.3).

Figure 5.3
The toolbars are normally so short that their scrollbars are barely recognizable as such, but they do exist.

To make effective use of the Prompt-as-You-Go feature in this situation, you have three choices:

- Widen your WordPerfect document window so that the Prompt-as-You-Go information can display on the first line.

- Use the property bar's scrollbar to view the second line; however, this method makes it more difficult for you to use the items on the first line of the property bar.
- Alter the property bar's defaults to display two lines of information. To do so, right-click on the property bar, select Settings, and then click the Options button. This step brings up the Property Bar Preferences dialog box, as shown in Figure 5.4. Adjust the Maximum number of rows/columns to show spin box to 2, click OK, and then dismiss the Customize Settings dialog box by clicking Close. Your property bar is now taller, and the Prompt-as-You-Go display is visible.

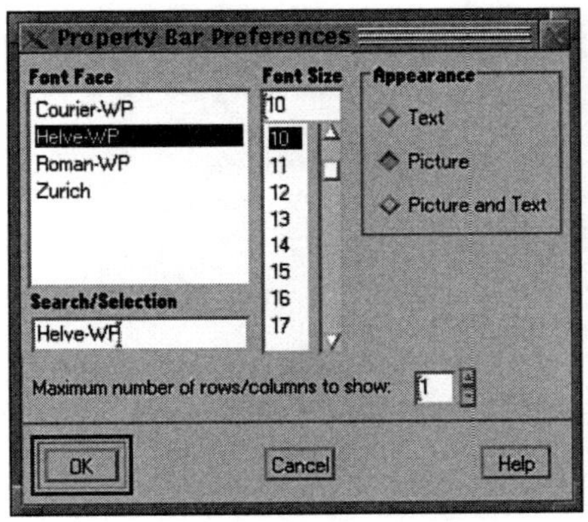

Figure 5.4
Although it is labeled "maximum," the number of rows you set is fixed even if you need fewer rows.

Prompt-As-You-Go is a context-sensitive feature and can be active even when the other As-You-Go features are not. When you place your insertion point on a spelling error, the Prompt-As-You-Go box displays possible corrections, as in Figure 5.5. In this example, the error is really a proper name, but WordPerfect has come up with all sorts of creative possible "corrections." Other times it has few or even no suggestions. Also, notice that WordPerfect's suggestions are based on word sounds as much as they are on spelling; all of the suggested replacements actually begin with the letter *C*, but with a hard sound to match the *K* in the erroneous word.

If you place the insertion point on a word that's spelled correctly, Prompt-As-You-Go provides a list of possible synonyms from the WordPerfect thesaurus. Prompt-As-You-Go also suggests ways to correct grammatical errors, as long as the change is relatively small, such as changing a singular word to a plural. Bigger changes are beyond its ken.

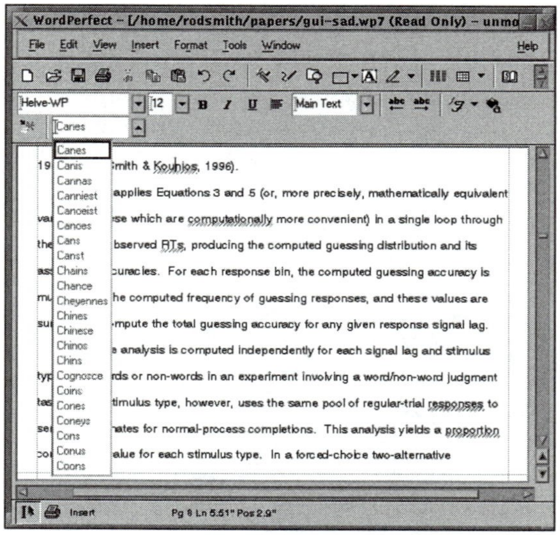

Figure 5.5
The pop-up list of suggestions doesn't appear automatically; you must click the triangle next to the word to display this list.

QUICKCORRECT

Having WordPerfect locate your errors is great, of course, but it can do more than that—it can take an active role in *correcting* errors, through its QuickCorrect feature. To see this feature in action, try typing "teh"; then press the spacebar. If QuickCorrect is active (as it is by default), WordPerfect replaces the typo with the correct word "the." QuickCorrect makes a number of additional miscellaneous corrections, such as helping you keep your sentences capitalized and adjusting spacing between words and between sentences.

Although QuickCorrect couldn't be easier to use, it's actually a fairly complex feature with many configurable options and details to its operation. You can configure QuickCorrect by choosing Tools, QuickCorrect to open the QuickCorrect dialog box shown in Figure 5.6.

CORRECTING TYPOS AND EXPANDING ABBREVIATIONS

The first page of the QuickCorrect dialog box is dominated by its list of words and replacements. The left column contains common misspellings and typos, and the right column lists their corrections. Some things you may want to note about this list include

- You can correct case. For instance, there are entries correcting *april* to *April*, and *cds* to *CDs*.
- You're not restricted to single words, so you can have WordPerfect save you from the embarrassment of having *i snot* appear in your documents by replacing it with what you probably meant to type, *is not*.
- You can include punctuation and nonalphabetic characters. The default list includes corrections for common errors with contractions, for instance, changing *your here* to *you're here*.

- You can include symbols, as evidenced by the first few entries, which include keyboard shortcuts for the copyright symbol (©), among others.

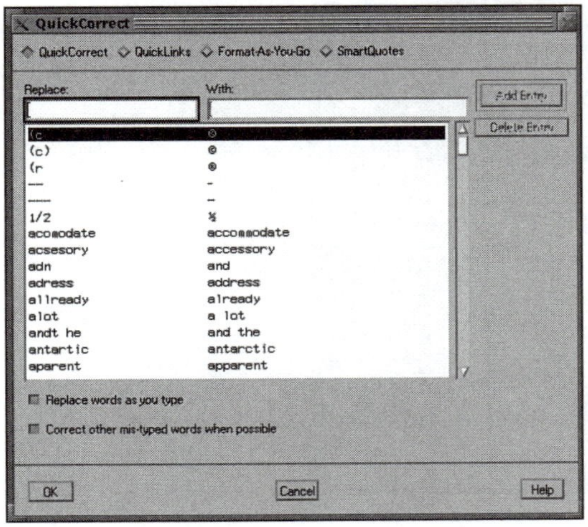

Figure 5.6
When you click the various radio buttons at the top of the QuickCorrect dialog box, you'll get different options to change.

You can add words to the QuickCorrect list by typing the abbreviation or typo into the Replace field, typing the correct word in the With field, and clicking Add Entry. If you want to change an existing entry, you can click on it and make your change in the With field. The Add Entry button changes its text to Replace Entry. Click it to make your change. You can also remove an entry by clicking Delete Entry after you've highlighted a QuickCorrect pair.

Tip #51 from *Tux Tips*

> If you highlight text in the main editing area before choosing Tools, QuickCorrect, WordPerfect places that text in the With field, so you can simply type an abbreviation or common misspelling in the Replace field. If your selected text is particularly long, however, WordPerfect may cut off its end. Just type it back in if this happens.

You can enter special symbols in QuickCorrect by pressing Ctrl+W when the cursor is in the With field. This step brings up the Symbols dialog box, as in Figure 5.7. Select the symbol set you want and click on the symbol; then click Insert to enter that symbol and go on to enter another, or click OK to enter the symbol and close the Symbols dialog box.

→ To learn more about the Symbols dialog box, **see** "Inserting Special Characters," **p. 102**

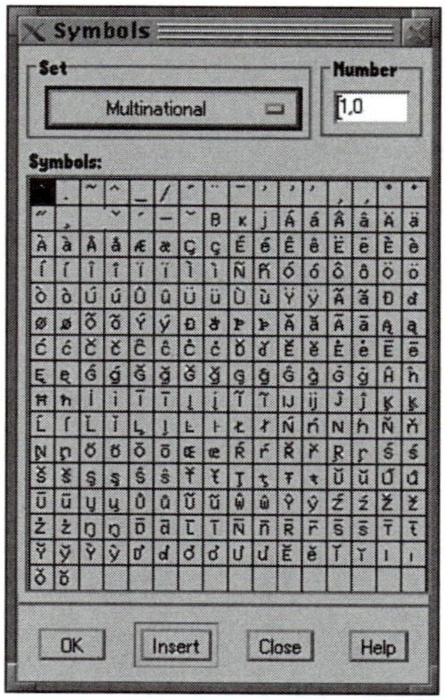

Figure 5.7
Enter special symbols in the With field and an ordinary character string in the Replace field to make your life easier when you need to regularly access particular special symbols.

Note

In theory, the Correct other mis-typed words when possible check button in the main QuickCorrect dialog box controls whether WordPerfect uses its ordinary dictionaries to expand the list of words QuickCorrect uses. With this button selected, WordPerfect is supposed to do a QuickCorrect-style replacement when only one alternative appears for a misspelled word, in addition to when an explicit QuickCorrect dictionary entry is available. I've never seen this happen, however, so I assume this feature is broken. If it's been fixed by the time you read this chapter, you can uncheck the button to prevent WordPerfect from "correcting" errors, which might not be errors, just because it can only "think" of one alternative spelling.

Caution

You might think that you can enter word pairs into the QuickCorrect list and then click Cancel to quit without saving those changes. You can't; as soon as you click Add Entry or Replace Entry, the change is complete. To reverse your change, you have to undo it manually or click on the entry and delete it.

If you want to completely disable WordPerfect's QuickCorrect word replacement feature, you may do so by unchecking the Replace words as you type check button. In this case, WordPerfect keeps the other QuickCorrect features active; they can be disabled from other pages of the QuickCorrect dialog box.

Tip #52 from *Tux Tips*

> If you're typing your document and WordPerfect makes a QuickCorrect change that you don't want, you can reverse the change by pressing Ctrl+Z. This step repositions the insertion point to the start of the changed word, however, so you need to move the insertion point to continue typing.

INSERTING HYPERLINKS

If you're using WordPerfect to prepare Web pages, chances are you'll be including lots of *hyperlinks*. These are the words on Web pages that typically appear in a different color from the rest of the text and on which you can click to load a different Web page, download a file, send email, or read a newsgroup. In the WordPerfect QuickCorrect dialog box, you can click on the QuickLinks radio button to bring up options related to the easy creation of hyperlinks, as shown in Figure 5.8.

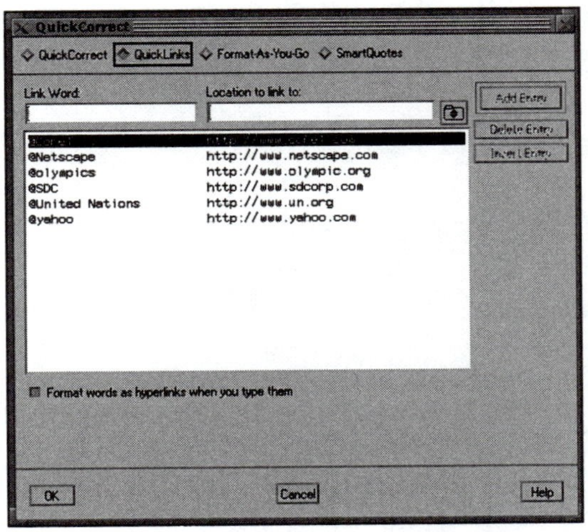

Figure 5.8
You need to know the exact network address of the file or site to create a QuickLinks entry for it.

You can edit the QuickLinks list much the same way as you do the QuickCorrect spelling correction list, by entering a keyword in the Link Word field and the *uniform resource locator (URL)*, or network address, in the Location to link to field. If you enter a Link Word without an @, WordPerfect adds one automatically.

When you type your document, WordPerfect creates a hyperlink in either of two cases: when WordPerfect recognizes the form of a URL or when you type one of the words, complete with @, in the Link Word column of the QuickLinks view of the QuickCorrect dialog box. If you type one of the link words you created with QuickLinks, WordPerfect uses that word, without the @, in the text of the document, but a Web browser uses the URL from the Location to link to column when you click on the link word. If you type a URL that WordPerfect identifies as such, that URL becomes both the text in the document and the network address used by a browser.

Inserting text as a hyperlink won't do you much good unless you're creating a Web page, however. That topic is covered in Part VII, "Web Publishing," so you can check there for more details on this topic. If you're creating documents for paper, you may find the QuickLinks feature to be distracting if you want to include URLs because it changes the appearance of these URLs in a way you may not like. You can therefore disable this feature from the QuickCorrect dialog box by unchecking the Format words as hyperlinks when you type them button.

If you have configured WordPerfect to use a browser, however, clicking on a link inside WordPerfect launches your browser and loads the associated URL. If you need to tell WordPerfect where your browser is, you can do so by choosing File, Internet Publisher in the Document window and then clicking the button labeled Browser Preferences. This step brings up the Browser Preferences dialog box, as shown in Figure 5.9. You can either type in the complete path to your browser in the text entry field near the top of the dialog box or click the folder icon next to that field to locate your Web browser in a WordPerfect File Manager dialog box. You can also tell WordPerfect whether your browser is Netscape, Mosaic, or anything else by making an appropriate selection in the Type of browser option menu.

→ If you'd like to learn more about WordPerfect's support for creating Web documents, **see** "Web Publishing," **p. 659**

Figure 5.9
Common locations for Web browsers are in the /usr/bin, /usr/X11R6/bin, and /usr/local/bin directories.

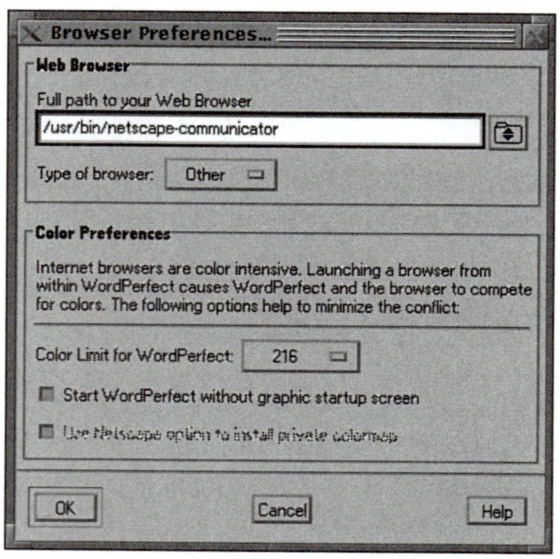

FORMATTING SHORTCUTS AND REGULARIZATION

WordPerfect's QuickCorrect provides a number of automatic corrections and changes that affect miscellaneous typing errors not covered by the normal QuickCorrect list as well as enhancements to the text you type. You can control these items from the panels of the QuickCorrect dialog box revealed by the Format-As-You-Go and SmartQuotes radio buttons. The first of these is depicted in Figure 5.10.

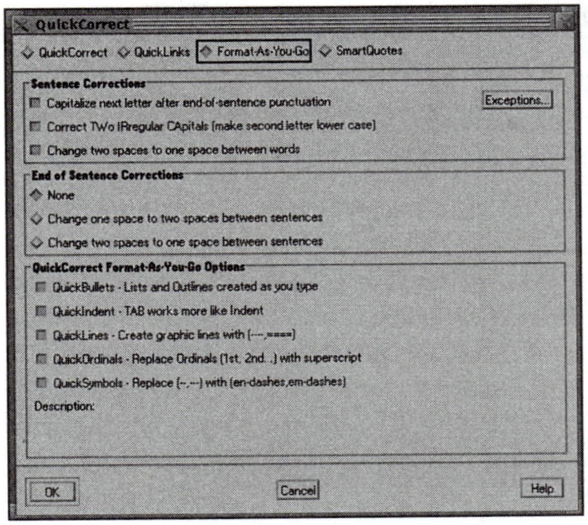

Figure 5.10
The Description line at the bottom of the QuickCorrect Format-As-You-Go area explains the options within that area when you click on them.

CAPITALIZATION

WordPerfect can help you to capitalize your sentences. If you check the Capitalize next letter after end-of-sentence punctuation button, WordPerfect automatically capitalizes your sentences, assuming the previous ones are terminated normally with a period, question mark, or other sentence-ending punctuation.

One potential pitfall with this feature is that if you use some abbreviation that WordPerfect doesn't recognize, it interprets the period in the abbreviation as the end of a sentence and your next word ends up capitalized. To avoid this problem, you can click the Exceptions button in the QuickCorrect dialog box and add your abbreviations to the Exception List, as in Figure 5.11. Simply type your exception and click Add Entry, and you'll be done. You can also tell WordPerfect to automatically add your exceptions when you correct them in the document by clicking the check button at the bottom of the Exception List dialog box. If there's a word on the list that shouldn't be, either because it was added in error or because it's a specialized term in your field in addition to being a common abbreviation, you can click the word and remove it by selecting Delete Entry.

Tip #53 from *Tux Tips*

Although WordPerfect can *help* you to capitalize your sentences, it can't do it all. For instance, if your sentence ends in an abbreviation, WordPerfect will not correctly capitalize the next word. I recommend using this feature only as a typing safety net, not as a way to keep your finger off the Shift key.

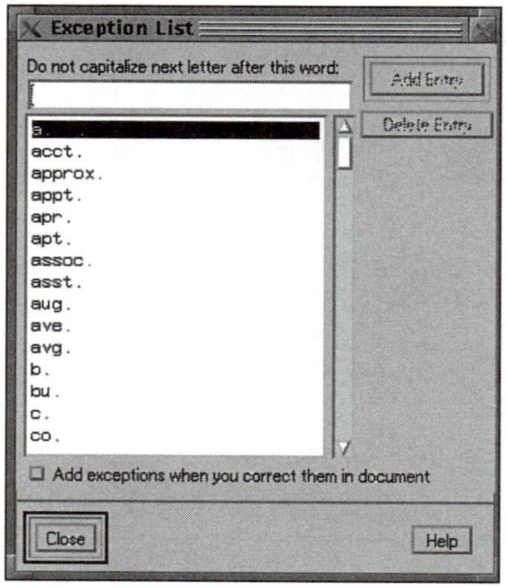

Figure 5.11
WordPerfect ships with a large list of common abbreviations it recognizes.

Some people don't press the Shift key when they should, and others hold it down for too long, resulting in words with two capital letters. WordPerfect can automatically correct this error, too; simply check the Correct TWo IRregular CApitals button in the QuickCorrect dialog box, and WordPerfect will make the appropriate correction for you.

Tip #54 from *Tux Tips*

WordPerfect can sometimes be overzealous with the capitalization-correction feature. For instance, it corrects the word *OSs* (an abbreviation for *operating systems*) to *Oss*. If you run into this situation, you can either disable this correction feature or add the "corrected" form of the word(s) that is giving you trouble to the main QuickCorrect list with the truly correct form as its corresponding replacement. WordPerfect then leaves the word alone, and, in fact, you can type it without any capitalization and WordPerfect will capitalize it correctly.

SPACES

WordPerfect can correct spacing errors both within and between sentences. If you check the Change two spaces to one space between words button in the QuickCorrect dialog box, WordPerfect adjusts your interword spacing within a sentence. If you type more than two spaces, however, WordPerfect does not make any changes.

You can also have WordPerfect adjust your intersentence spacing. If you habitually type two spaces after a sentence but need to use only one for a document, or vice versa, you'll find the End of Sentence Corrections radio buttons to be useful. WordPerfect defaults to making no end-of-sentence corrections.

SYMBOLS, LINES, AND OTHER FORMATTING

The QuickCorrect Format-As-You-Go Options area provides a number of miscellaneous expansions and corrections:

- QuickBullets helps you create short outlines or lists by automating the creation of bullets or outline numbers. If this option is active and you type a bullet, number, or letter at the beginning of a line followed by the Tab key, WordPerfect assumes you intend to create a list or outline and applies the same formatting (with an appropriate change to the number or letter) to the next line.

→ To learn more about creating bulleted or numbered lists, **see** "Creating a Bulleted or Numbered List," **p. 554**

- QuickIndent helps you to create paragraphs that are indented on the second and subsequent lines, as opposed to their first lines. This type of formatting is common in bibliographies, and Figure 5.12, which illustrates this feature, uses fictitious bibliographic entries. To use QuickIndent, position the insertion point at the beginning of the second or subsequent line of a paragraph and press the Tab key. WordPerfect reformats the paragraph in this style, using whatever tab setting is current for the level of indent.

→ For more information on tab stops, **see** "Tabs," **p. 194**

→ For more information on creating indents, **see** "Indents," **p. 208**

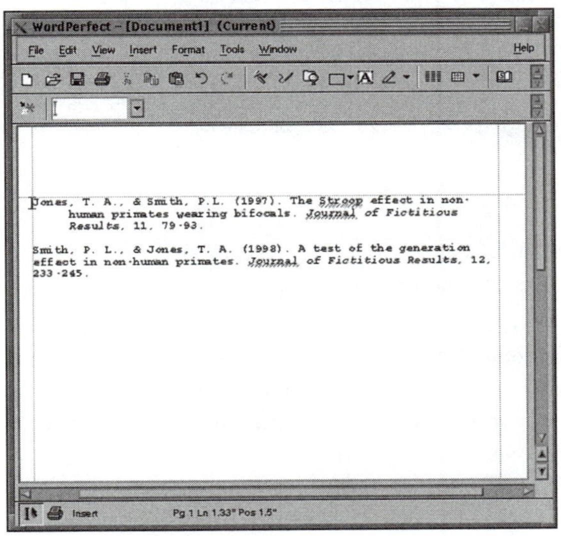

Figure 5.12
I used QuickIndent on the first paragraph, but not the second.

- If you type four or more dashes (-) or equal signs (=) on a line by themselves when the QuickLines feature is active, WordPerfect replaces those characters with single or double lines extending from the left to the right margins, as in Figure 5.13. To remove the line, you can right-click on it and choose Cut or Delete from the pop-up menu, or remove its code in Reveal Codes mode.

→ To learn more about lines in documents, **see** "Graphics Lines," **p. 218**

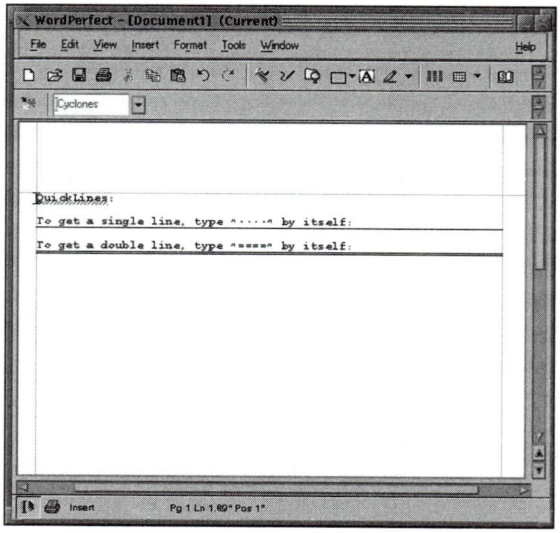

Figure 5.13
QuickLines makes it easy to create simple line formats to separate sections of your text.

Tip #55 from *Tux Tips*

> The lines created by QuickLines can look something like page breaks as viewed in Draft mode. Page break lines extend beyond the margins, however, whereas QuickLines do not.

- When the QuickOrdinals box is selected, WordPerfect makes superscripts from the ordinal indicators in ordinal numbers. For instance, *1st* becomes 1st.
- If you type two or three dashes when QuickSymbols is active, WordPerfect replaces these with en- and em-dashes, respectively. These are wider forms of ordinary dashes used in specific situations. For instance, notice how the dash in a compound word like *em-dash* is shorter than that used to separate clauses of a sentence—the latter is an em-dash and the former is an ordinary dash.

SmartQuotes

If you click the SmartQuotes radio button in the QuickCorrect dialog box, you'll see the options in Figure 5.14. Most conventional typewriters and text editors use a single style of quote mark or apostrophe. If you leave these options active, WordPerfect replaces the quote marks with curly quotes that give a better sense of enclosing the quoted text because they're curved differently for the start and end points. In fact, WordPerfect lets you choose from several different styles of curly quotes for both single and double quote marks. It can also revert to straight quote marks after numbers because that's the standard for the abbreviations for inches and feet.

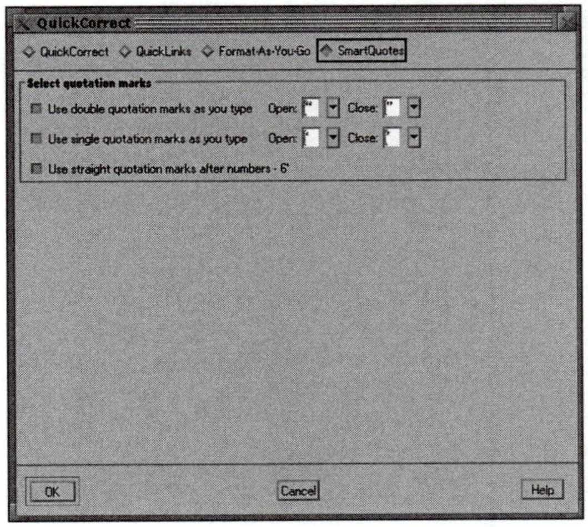

Figure 5.14 SmartQuotes can give your documents a professional appearance.

Tip #56 from *Tux Tips*

If you will be transferring your document to another word processor, you may get better results with SmartQuotes turned off. File import and export filters often distort curly quotes. You'll be fine with it if you expect to read a file in WordPerfect on another operating system, however.

CHECKING AN ENTIRE DOCUMENT AT ONCE

In addition to or instead of checking your document for errors as you type it, WordPerfect can check the entire document in one operation, allowing you to correct those errors in a single pass. You may prefer working this way to having WordPerfect point out errors as you make them, or you may want to do an additional pass in case you didn't notice WordPerfect's highlighting of some small word.

WordPerfect provides two tools for checking an entire document for errors, one for spelling and one for grammar (though the grammar checker also catches spelling errors). These two tools are similar in many ways, but are different in others.

CHECKING FOR SPELLING ERRORS

To check your document for spelling errors, choose Tools, Spell Check (Ctrl+F1). If your document has any errors, WordPerfect responds by scrolling to the first error, highlighting it, and displaying the Spell Checker window, as shown in Figure 5.15.

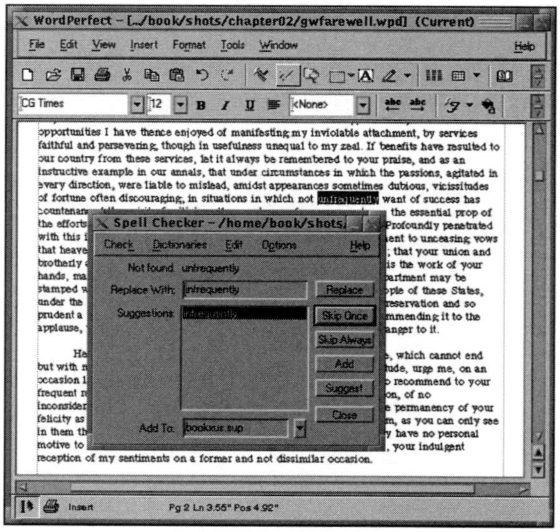

Figure 5.15
You can customize the spell checker's actions from the menus in the Spell Checker window.

When the spell checker highlights an error, you can click any of the buttons along the right edge of the window to deal with the error in one way or another:

- Replace replaces the highlighted word with the first suggestion. If you want to replace it with another suggestion (if there is one), you can select that alternative and then click Replace. You can also manually edit the Replace With field and click Replace to substitute a word that's not on the Suggestions list.
- Skip Once causes WordPerfect to skip over that particular instance of the word. If the same word appears again later in the document or if you start the spell check again, you have to decide about the word again.
- Skip Always skips over that word and any repetition of the word for this and all subsequent checks of the document. Even if you close the file and reopen it, WordPerfect remembers that you don't want to have to see that word flagged as an error again. If you perform a spell check on another document with that word, however, the spell checker asks you about it again.
- Add adds the word to your personal dictionary file, and WordPerfect never considers that word an error again, either in this file or in others. You should always be careful when telling WordPerfect to add a word to its dictionary because you don't want to fill it with misspelled words!
- Suggest doesn't normally have much effect because the spell checker displays its suggestions by default. You can, however, edit the word in the Replace With field and click Suggest to have WordPerfect bring up another set of suggestions. This feature might be handy if WordPerfect has no suggestions initially; you could try an alternative spelling and click Suggest to see whether that shakes any more possibilities out of WordPerfect's dictionaries.
- Close halts the spell check and closes the Spell Checker window.

Tip #57 from
Tux Tips

> You can use the Spell Checker window to look up the spelling of a word that's not in your document. Type your closest guess into the Replace With field and click Suggest. WordPerfect displays the closest matches it can find to your word.

The spell checker hosts a number of options, appropriately under the Options menu. You can check or uncheck these as you see fit. You can tell the spell checker to

- Check words that contain numbers
- Check for duplicate words (as in *the the*)
- Flag irregular capitalization patterns
- Start checking as soon as it's launched
- Include the document's own internal dictionary (used for the Skip Always feature)
- Beep whenever it encounters a misspelled word

All but the last of these are enabled by default.

If you start the spell checker after highlighting some text, WordPerfect checks only the highlighted text, rather than the entire document. This option can be handy if you've disabled the Spell-As-You-Go feature and want to check whether one word is misspelled.

When a spell check is complete, you can leave the spell checker active. If you do so, you can tell it to check a different segment of your document by selecting the range from the Check menu. If you've checked just a word or some selected area, for instance, you can decide to go on to check the entire document or just a page. Click Start after you've selected a range, and the spell checker begins checking that range of your text. (The Start button takes the place of the Replace button when a spell check isn't running.)

CHECKING FOR GRAMMATICAL ERRORS

Performing a grammar check on your document is quite similar to checking it for spelling errors. You start by choosing Tools, Grammatik (Grammatik was once a separate product, and Corel has kept the name; hence the odd spelling). WordPerfect highlights the first error and displays the Grammatik window, as in Figure 5.16.

When Grammatik identifies an error, a brief description of the error appears onscreen. This description may include underlined terms in green; you can click one of these to display a definition of the term. If the error is small, Grammatik may suggest one or more possible corrections, displayed in the text field in the lower left of the window. You can click Replace to substitute the suggested alternative for the original. If you want to leave it alone, you have two choices: You can click Skip to skip that one error, or you can click Next Sentence to go on. The difference is that Grammatik may identify (or misidentify) more than one error in a sentence. If you recognize that Grammatik has made one misidentification and, therefore, is likely to misinterpret something else in the same sentence, you can bypass this process with the Next Sentence button. Otherwise, Skip will do nicely. Similarly, the grayed-out button labeled Ignore Word in Figure 5.16 sometimes changes to Ignore Phrase. In this case, clicking the Ignore Phrase button skips to the next phrase of a sentence.

Figure 5.16
The grammar checker identifies the class of each error it identifies and provides a brief description of the error.

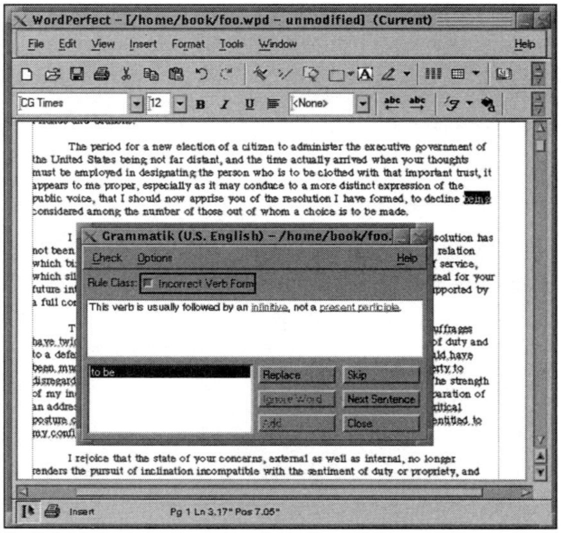

Grammatik checks not only your document's grammar but also its spelling. When Grammatik detects a spelling error, it gives two options in addition to the ones for grammar errors: Ignore Word and Add, which function similarly to the Skip Always and Add functions of the spell checker. There are differences, however; in particular, the spell checker's Skip Always feature keeps WordPerfect from flagging the word again in that document, even on subsequent spell checks. Grammatik flags words that have been marked with the spell checker's Skip Always, however, and Grammatik doesn't remember its own Ignore Word markings if you run a grammar check a second time.

Like the spell checker, Grammatik provides a number of fine-tuning options. Choose Options, Writing Style to open the Writing Style dialog box and customize its checking rules for different types of writing (Figure 5.17). You can select from a number of predefined writing styles or modify one in detail by selecting it and clicking Edit to open the Writing Style Settings dialog box (Figure 5.18). Note that there are three rule classes, each with its own set of rules you can adjust; click the appropriate radio buttons to select the different rule classes.

Figure 5.17
Different writing styles adjust Grammatik's rules in specific areas to conform to the practices in different fields or types of writing.

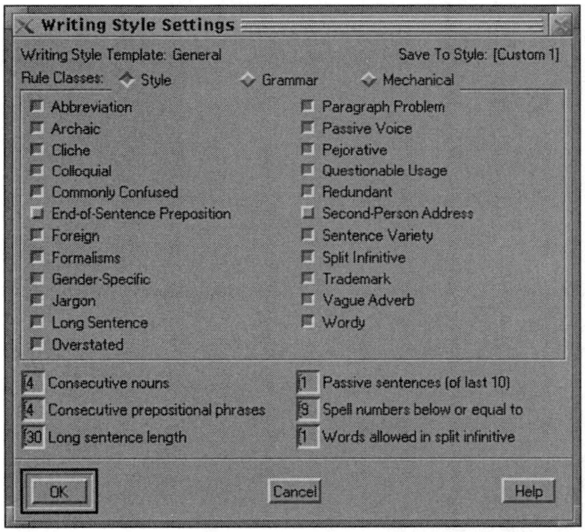

Figure 5.18
You can fine-tune a Grammatik writing style or create a new one to suit your needs.

The Options menu contains a few standalone options, such as the control to Show spelling errors. You can also modify Grammatik's global actions from the Options dialog box (Figure 5.19).

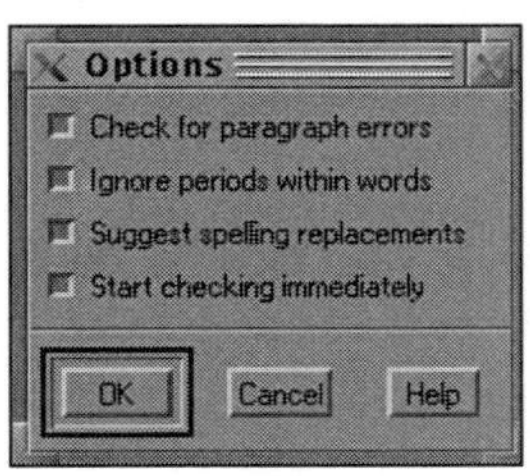

Figure 5.19
These options apply to all Grammatik checks regardless of the writing style.

Tip #58 from *Tux Tips*

If you let a grammar check run its course and then answer No when you're asked whether you want Grammatik to close, you can select various additional items from the Options menu and rerun the check. One of these is Options, Statistics; if you choose Statistics and click Start, Grammatik displays statistics on your document, including average sentence and paragraph lengths and the Flesch Reading Ease measure, in the Document Statistics dialog box (Figure 5.20).

You can get some, but not all, of this information by choosing File, Properties, Document Info from the document's main window.

Note

The Flesch Reading Ease measure is a way for the computer to estimate how easy your text is to read. A score of 100 is the easiest to read, and 0 is the hardest. A typical goal is to get a score in the 60 to 80 range. Like most computerized reading-ease scores, this one relies on metrics such as the average word and sentence lengths and can be thrown off by headings, tables, and so on. You shouldn't take the result of this measure too seriously, though it can be a useful rough guide to readability.

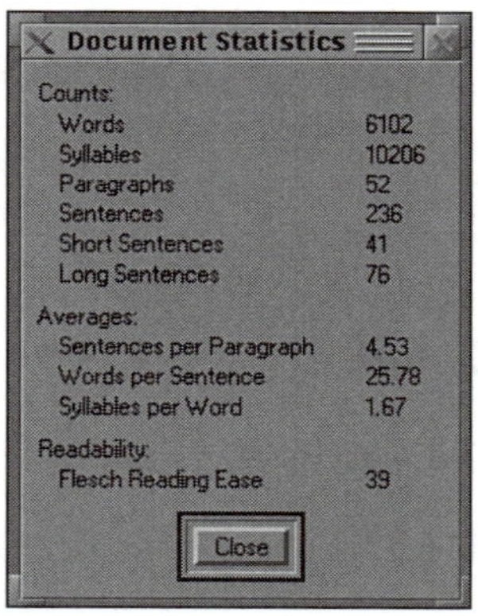

Figure 5.20
You can use document statistics to help determine how difficult or easy your document is to read.

UNDERSTANDING DICTIONARIES

To do its proofreading, WordPerfect maintains several *dictionary files* that contain lists of words and abbreviations for a given language. You'll get the most out of WordPerfect's proofreading capabilities if you understand what these files are and how to manipulate them.

DICTIONARY LOCATIONS AND OPTIONS

WordPerfect maintains three dictionaries with different scopes for spell checking. These files are

- The systemwide spell-check dictionary, located in the shlib10 subdirectory of your WordPerfect installation directory. For American English, this dictionary is called wpus.lex, but it will have a different name if you're using a different language or dialect. All WordPerfect users on a given system normally use the same main dictionary, though you can change this dictionary or add a second one as a user by selecting Dictionaries, Main from the Spell Checker window.

- Your personal "supplementary" dictionary. This dictionary normally resides in the .wprc subdirectory within your home directory and has a filename with a .sup extension (the exact filename depends on your username and language). This dictionary is private for you; if you have co-workers or family members who use your computer under different accounts, they'll have their own supplementary dictionaries. You can modify or add another supplementary dictionary file by selecting Dictionaries, Supplementary from the Spell Checker window.
- The document dictionary associated with the document you're editing. This dictionary is stored in the document file and keeps track of words you've elected to Skip Always.

You won't normally need to do anything about the main dictionary except possibly to change which one you use if you work with several languages. In fact, in most cases you won't be able to modify that file as an ordinary user because Linux's file permissions prevent unauthorized access to systemwide files. If you need to modify that file, then, you need to launch WordPerfect as root (or as whoever installed WordPerfect).

Modifying Dictionaries

You modify dictionary files from within the spell checker by choosing the appropriate dictionary from the Dictionaries menu—Main for the main dictionary (when you're running as root) or Supplementary for either your personal supplementary dictionary or the document's dictionary. After you do so, you'll see a display something like the Supplementary Dictionaries dialog box, as shown in Figure 5.21.

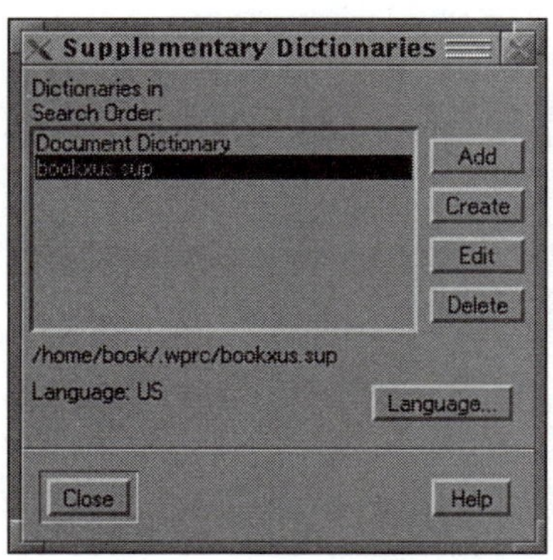

Figure 5.21
WordPerfect searches for words in dictionaries in the order in which they're listed in the Dictionaries dialog box.

From this dialog box, you can add an existing dictionary file, create a new dictionary file, edit one of the files shown in the box, or delete one of those files. If you choose to add or create a dictionary, WordPerfect displays a File Manager dialog box for you to enter an appropriate filename. To modify the contents of a new or existing dictionary, click its name and then click Edit. This step brings up the Edit dialog box, as shown in Figure 5.22.

Figure 5.22
You can tell which type of dictionary entry this is by examining the heading over the second column.

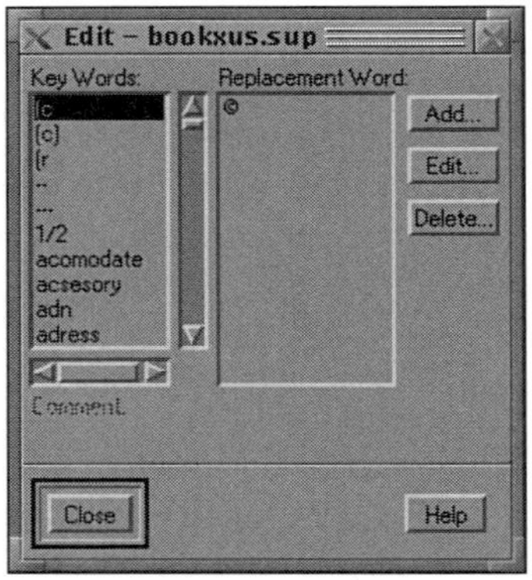

The word list in the Key Words column may look familiar if you read the section earlier in this chapter on WordPerfect's QuickCorrect feature. The reason is that WordPerfect uses one file for both QuickCorrect and spell checks, so you can modify your QuickCorrect entries from the spell checker if you like, though you cannot manage your spell-check dictionary from QuickCorrect.

To modify an entry, first click on it in the Key Words column. You can delete the entry by clicking Delete. If you want to modify it in some way, click Edit. If you want to add an item, click Add. Both the Add and Edit buttons bring up a dialog box similar to that in Figure 5.23, and they function identically except that the Edit button works on an existing entry, whereas the Add button adds a new one.

Figure 5.23
The Alternatives radio button provides the most editable options in the Add or Edit Word/Phrase dialog box.

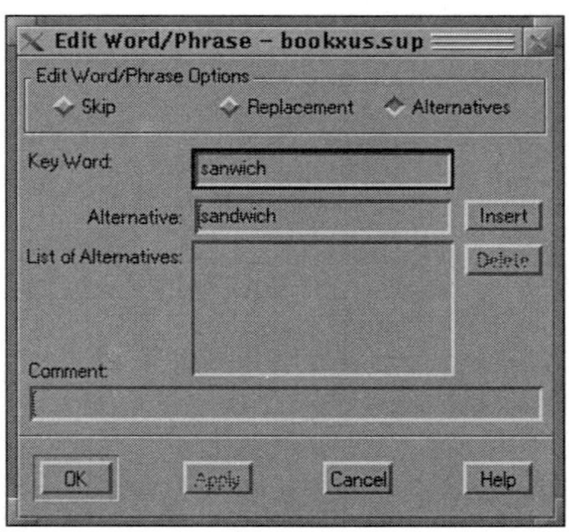

The Add or Edit Word/Phrase dialog box has three radio buttons, each of which produces slightly different options in the rest of the dialog box:

- If you click Skip, you have only one text entry field, and whatever you type there becomes part of the dictionary; it is skipped on all subsequent spell checks that use that dictionary. You use this setting to add a legal word to the dictionary.
- If you select Replacement, you are creating a QuickCorrect entry, and you must enter both the incorrect word or abbreviation in the Key Word field and its replacement in the Replacement field.
- The Alternatives radio button produces the options shown in Figure 5.23. In this case, the Key Word is an incorrect word, and you type, one at a time, one or more possible corrections into the Alternative field. After you type each alternative, click Insert to add it to the list. You can then type another alternative to repeat this process. You can remove an alternative by selecting it and clicking Delete. When a spell check finds the Key Word, it flags it as an error and presents the listed alternatives as possible spellings for you to select. Whatever you type in the Comment field also appears in the Spell Checker dialog box. When you perform a spell check and it finds this word, you can also click Suggest to see WordPerfect's suggestions in addition to those you entered into the dictionary.

Tip #59 from
Tux Tips

Use the WordPerfect spell checker's Alternatives feature if you frequently spell a word incorrectly but in a way that is itself a legal spelling. For instance, if you frequently type *flower* when you should be typing *flour,* you can create an entry for *flower* with both spellings as options. WordPerfect will then flag all instances of *flower* and allow you to choose either spelling. This feature can be annoying, but it's better than distributing a document with errors.

The Thesaurus

Mark Twain wrote, "The difference between the right word and the almost right word is the difference between lightning and the lightning bug." WordPerfect helps you find the lightning with its thesaurus feature. To open the Thesaurus window (Figure 5.24), choose Tools, Thesaurus (Shift+Ctrl+F1). If your cursor is resting on a word when you do so or if you've highlighted a word, the thesaurus automatically looks up that word and displays the results in its left pane. If you've highlighted more than one word, WordPerfect displays results for the first word.

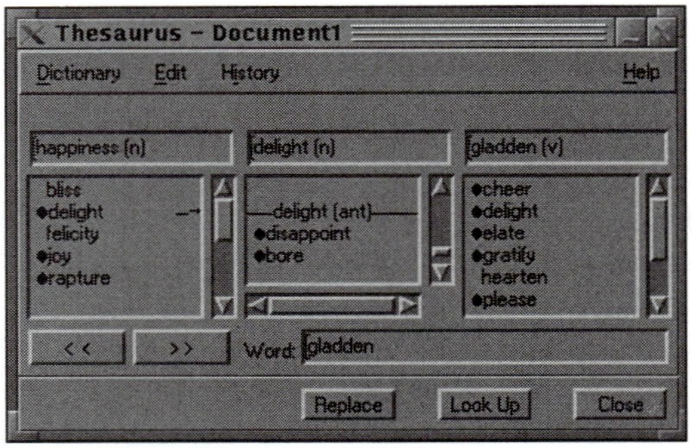

Figure 5.24
The thesaurus lists antonyms at the bottom of its list of words, as shown in the middle pane.

Note

If you're used to using WordPerfect in Windows, you may find the thesaurus feature a bit unfamiliar. The WordPerfect for Linux thesaurus works differently from its Windows counterpart and lacks the word definitions available in the Windows version.

You can easily maneuver from one synonym or antonym to another in the WordPerfect thesaurus by using a number of techniques:

- Word lists may have vertical, and occasionally horizontal, scrollbars. Scroll vertically to see additional entries for a word. Some lists can be quite long. Horizontal scrollbars let you see long words.
- Any word marked with a bullet leads to other thesaurus entries. Double-click on these words to display the additional entries in the next pane to the right. The thesaurus marks the word you used to look up the next with a short arrow to the right of the word.
- If you've looked up more than three words, the << and >> buttons let you move through all the entries by cycling the three panes through the different words. Although these buttons look active initially, they are not until more than three words appear in the thesaurus panes.
- The History menu shows a list of the words you've looked up. Click on a word to return to its entry.
- If you click a word or type one into the Word field and then click Look Up, the thesaurus clears the middle and right panes and displays the word's entry in the left pane. The History menu still lets you get back the previous words' entries, however.

When a word has many synonyms, WordPerfect groups them by parts of speech. Therefore, if you know you're looking for a verb form, you can scroll down to the verb section and look there.

If you've started the thesaurus because you want to replace a word, WordPerfect can do the job for you if the word is highlighted. When you find the word you want, be sure it's displayed in the Word field and then click Replace. WordPerfect makes the substitution for you.

When you're finished with the thesaurus, click Close to dismiss it.

Languages

Because WordPerfect originated in the United States, its support for U.S. English is the most extensive, but it includes varying levels of support for several other languages, including nation-specific variants for some.

Setting Up Language Support

As you might imagine, WordPerfect requires separate dictionary files for different languages. When you install WordPerfect, you get a choice of languages it will support. If you've installed multiple languages, you can tell WordPerfect when to use which language during spell checks. During installation you can also set the default language, which is used both as the default for spell checks and as the language for menus and dialog boxes. The languages supported at install time by WordPerfect for Linux include English (with U.K., U.S., Canadian, and Australian variants), French (with French National and Canadian variants), German, Spanish, Italian, and Dutch.

→ To learn more about installing WordPerfect, **see** "Installing WordPerfect for Linux," **p. 712**

Tip #60 from *Tux Tips*

> If you expect to use any given language in WordPerfect, be sure to include it when you install the software; installing WordPerfect with appropriate language support is simpler than adding that support later.

When you write a document, WordPerfect assumes it's being written in the default language and uses the appropriate spell-check dictionary. If you want to use another language, follow these steps:

1. If you want to use the nondefault language for the entire document, position your insertion point at the beginning of the document.
2. If you want to use the nondefault language for a portion of a document, highlight that portion.
3. Choose Tools, Language, Settings. This step brings up the Language dialog box, as shown in Figure 5.25.
4. Select the language you want to use from the list. If you know the two-letter code for the language, you can type it into the Language box instead.

Figure 5.25
The list of languages includes many for which no dictionaries ship with WordPerfect for Linux, but you can add them in other ways.

Tip #61 from *Tux Tips*

If you regularly create documents in two or more languages, you can create *templates* or *styles* to help set up these options in a consistent way.

→ For more information about styles and templates, **see** "Creating Consistent Documents with Styles and Templates," **p. 303**

If you select a language for which WordPerfect has no installed dictionary files, WordPerfect treats every word as a spelling error.

Tip #62 from *Tux Tips*

If you're using WordPerfect on a system for which you do not have the root password and need to use it with a language for which no dictionary file is installed, you can install your own custom dictionary file in the .wprc subdirectory of your home directory. If you can find a WordPerfect dictionary file, you can use it, but you need to rename it to a name of the form *uuuuxll.sup*, where *uuuu* is your username, *x* is *x*, and *ll* is the two-letter language code. Alternatively, you can set the language as appropriate and perform a spell check. Add every word you know is correctly spelled, and you'll be on your way to creating your own custom language dictionary.

CHARACTER MAPPINGS

If you're using WordPerfect to create documents in a language that uses the Cyrillic or Greek alphabet, you can set up appropriate character mappings to ease your typing and ensure that letters appear correctly in the document. Choose Tools, Language, Character Mapping to open the Document Character Map dialog box as shown in Figure 5.26. Select the desired character mapping, and your keyboard should produce the appropriate mappings of keys to characters.

Figure 5.26
Use US ASCII for most languages that use the Roman alphabet and use Greek or Cyrillic where appropriate.

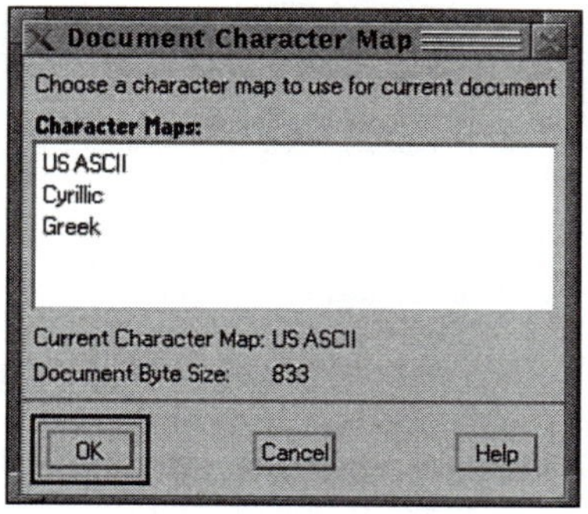

Note

To obtain Cyrillic or Greek characters, you need an appropriate font in addition to the character mapping. The downloadable version of WordPerfect doesn't include these fonts, nor does it let you install additional fonts, so you need to purchase the retail version of WordPerfect to use it effectively with non-Roman alphabets.

Creating a Custom Dictionary

If you have the retail version of WordPerfect, you can use the wpsputl utility to create a custom dictionary for your language. You can also use this procedure to create a specialized dictionary with terms used in your profession, proper names for bibliographies or novels, or other purposes. Otherwise, you needn't bother with this procedure if WordPerfect includes support for your language.

Tip #63 from *Tux Tips*

The downloadable version of WordPerfect includes support only for English, but add-on modules are available for various other languages. Try looking for the add-ons before you go to the effort of creating a custom dictionary.

Here are the steps to follow to create your own custom dictionary file:

1. Locate a raw-text dictionary file for your language. This file should be in plain text (not a word processor or other specialized format). This step is really the hardest part in the process, especially for languages that don't have standard ASCII encodings, because the WordPerfect utilities may react strangely to some of these.
2. If WordPerfect has a code for your language, find it. You can do so by choosing Tools, Language, Settings and selecting your language from the list. The two-letter code in the Language box is the code you want. If your language isn't listed, make up a code but make sure it's not associated with another language by typing it in the Language box. An unused code causes the box to scroll to -Other-. Later, when you want to set your language, you need to select your language from the list or type in your unique code.
3. Start an xterm or use a character-mode login to Linux.

4. If you want to create a main dictionary, you need to be the root user or you need to store your file temporarily in your home directory and then become root to copy the file back to the WordPerfect directory. If you intend to use this dictionary privately, you can do it all as an ordinary user.
5. Go to the shbin10 directory of your WordPerfect installation. For instance, if WordPerfect is installed in /opt/wp8, you would type `cd /opt/wp8/shbin10` to get there.
6. Type `./wpsputl` to start the conversion utility. After a brief pause, a menu of options appears.
7. Select option 1, Change/Create Dictionary.
8. When prompted, enter the name of the dictionary, which should take the form of *wpxx.lex*, where *xx* is the two-letter code you identified in step 2, in lowercase. For instance, if your code is RO, you'd use wpro.lex for the filename. You can enter a pathname here, too. To save the dictionary directly in its ultimate home, add *../shlib10/* before the dictionary's name—for instance, *../shlib10/wpro.lex*. If you're creating a dictionary for your profession or some project, give it a name of the form *uuuuxll.sup*, where *uuuu* is your username, *x* is *x*, and *ll* is your language code. You'll ultimately want to store this file in the .wprc subdirectory of your home directory.
9. The wpsputl program now returns to the original menu. Select option 2, Add Word(s) to Dictionary.
10. At the next menu, select option 2, Add to main word list.
11. At the next menu, select option 2, Add to word list using input file.
12. At the prompt, enter a filename for the input dictionary file. If the file is in a different directory (such as your home directory), you need to provide the complete pathname. If you've forgotten what it is, you may want to open another xterm to find it.
13. The program now returns to the preceding prompt. Select option 3, Exit and add words to dictionary.
14. The wpsputl utility displays a `Please Wait` message while processing the word list and creating the WordPerfect-format dictionary. At the prompt, press Enter to return to the main menu.
15. Select option 0, Exit, to exit.
16. If necessary, move the file you created to the shlib10 directory in the WordPerfect installation directory or to the .wprc subdirectory of your home directory. The file should now be available for use in WordPerfect.

TROUBLESHOOTING

WORDPERFECT FLAGS A CORRECTLY SPELLED WORD AS AN ERROR

WordPerfect has flagged a common word that's spelled correctly as an error, either in Spell-As-You-Go or in a full-document spell check. The word does appear in the spell-check dictionary.

Check to be sure you haven't repeated the word. WordPerfect flags double words as spelling errors, and these are *very* easy to overlook, particularly if one copy of the word appears at the end of a line and the second at the beginning of the next line.

Sharing Personal Dictionaries

You've created a custom spell-check dictionary that you want to use from two or more accounts on the same Linux computer.

Because of Linux's multiuser design, personal spell-check dictionaries aren't shared across accounts. This situation can be an annoyance if you're using two accounts on one system or if you want to share a dictionary with others. There are two ways to get around this problem, though. These examples assume that the user named "alpha" has the desirable dictionary and that the user named "beta" wants to share it. Further, both users are using U.S. English as their primary language; if that's not the case, substitute the appropriate two-letter language code for the *us* prior to the .sup portion of the filenames that follow. To share a dictionary, you can do either of the following:

- From beta's account, start an xterm window. In that window, type `cd .wprc` to change into the .wprc subdirectory. Then type `cp /home/alpha/.wprc/alphaxus.sup betaxus.sup` to copy alpha's desirable custom dictionary into beta's account.
- From beta's account, start an xterm window. In that window, type `cd .wprc` to change into the .wprc subdirectory. Then type `rm betaxus.sup` to remove the existing dictionary. (If that dictionary doesn't exist, you'll get an error message; ignore it.) Now type `ln -s /home/alpha/.wprc/alphaxus.sup betaxus.sup` to create a symbolic link from alpha's desirable dictionary to beta's. Finally, log in as alpha, start an xterm, and type `chmod 666 .wprc/alphaxus.sup` to give beta full read/write access to this dictionary file. Otherwise, beta will be able to read the file, but not write to it. Note that the `chmod` command allows anybody else on the system to write to the file. If alpha and beta are in the same group, you can use `664` instead of `666` to restrict write access to users of that group.

➔ To learn more about Linux's handling of file permissions, **see** "File Ownership and Permissions," **p.78**

Both procedures eliminate any entries beta has already made to his or her personal dictionary. The first procedure creates separate dictionaries for both users, so as they add more words, the contents of their respective dictionaries will diverge. The second procedure allows the users to actually share a common dictionary, so either may add words. If you use the second procedure and alpha's account is subsequently removed from the Linux system, beta loses that dictionary file, as well.

CHAPTER 6

Producing Output

In this chapter

OVERVIEW

After you write a document with WordPerfect, you presumably want to do something with it so that you or others can read it. Usually this process involves printing the file, though there are other ways to distribute a document, such as faxing, emailing, and sending it in a computer-readable format such as PostScript or *PDF (p. 168)*. This chapter covers these topics.

Instead of or in addition to one of these methods of distribution, you might want to use the Web to distribute your document. This option introduces enough new features that this book devotes an entire section to the topic of Web distribution. You should refer to Part VII, "Web Publishing," before preparing a Web document because you need to understand the Web's unique strengths and weaknesses to use it most effectively.

→ To learn more about preparing a Web document, **see** "Web Publishing," **p. 659**

This chapter assumes that you have a working printer configuration, both within Linux and within WordPerfect. Similarly, you need a working fax setup to fax from WordPerfect for Linux.

→ If you need to configure WordPerfect to use your printer, **see** "Fonts and Printing," **p. 739**

If you intend to send your document by email, you need a working email configuration on your system before you can proceed. Most Linux systems set up this configuration automatically, but you may need to do some more to configure your local network or Point-to-Point Protocol (PPP) connection. Consult your distribution's documentation or Que's *Special Edition Using Linux* for details.

Caution

If you need to buy a printer or fax modem for use with Linux, be sure that Linux supports the model you buy. Many low-end printers and internal modems these days work *only* with Windows. These models typically list Windows and no other OS as a requirement on their boxes, and they may even include "Win" or "Windows" in their product names. Some products that are so labeled do work with Linux, however, so check with the manufacturer or ask on the Linux newsgroups to be sure. The Web site `http://www.picante.com/~gtaylor/pht/printer_list.cgi` has a handy listing of printers that are known to work with Linux. External fax modems should be a reasonably safe bet for faxing from Linux.

CREATING PAPER OUTPUT: PRINTING

Producing paper output at or near your computer by using a printer is probably still the dominant way of writing with a word processor. Therefore, WordPerfect includes a wide variety of features for producing exactly the output you want with this medium.

A wide variety of printers are available on the market with varying capabilities and software requirements. WordPerfect lets you fine-tune your document's characteristics to match your printer's capabilities. Understanding these characteristics and capabilities can help you get the most out of WordPerfect and your printer.

PREPARING TO PRINT

Chances are that you'll be printing most of your documents on a standard paper size, such as 8.5×11-inch paper in the United States, and in portrait orientation (with the shorter edge horizontal). If so, you don't need to configure your paper size. If you do need to change your paper size, you do so by choosing File, Page Setup, Page Size, which produces the Paper Size dialog box shown in Figure 6.1. Select the paper size and orientation you want to use and click OK. WordPerfect reformats your document as necessary for the new paper size. You should also be sure that your printer is loaded with the appropriate paper size.

Figure 6.1
The Orientation area shows the dimensions of different page sizes and how text prints on them.

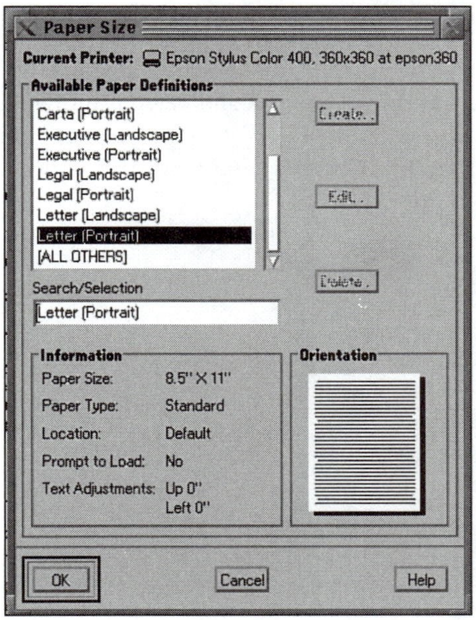

➔ To learn more about WordPerfect's handling of different paper sizes and orientations, **see** "Page and Document Formatting," **p. 225**

If you want to see a full-page view of your document before you print it, choose View, Zoom and select Full Page from the Set Zoom dialog box (Figure 6.2). WordPerfect then displays your document a full page at a time. Unless your screen is quite large, you won't be able to edit effectively at this size, but you can get an overall feel for the layout of your pages.

Tip #64 from *Tux Tips*

If you normally edit in Draft view, you may also want to choose View, Page (Shift+Ctrl+F5) to view headers, footers, and other elements that Draft view hides.

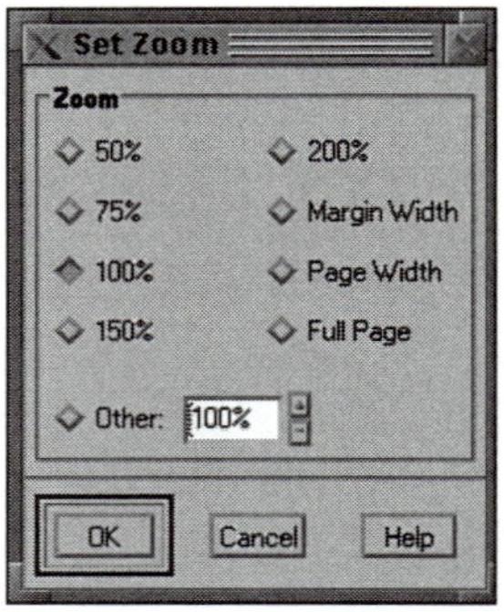

Figure 6.2
Use the Set Zoom feature to get a full-page view of your document.

When you're ready to set up your printer options and begin printing, choose File, Print (F5) to open the Print dialog box, shown in Figure 6.3. You can control most aspects of your printing here, including choosing a printer, selecting how many copies to print, and selecting just how much you want to print.

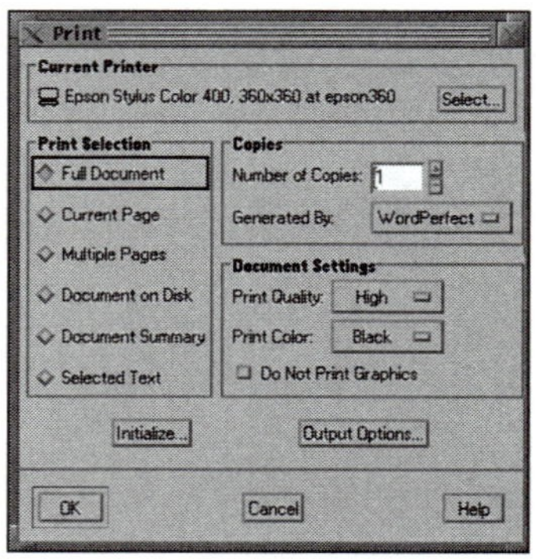

Figure 6.3
You can start the Print dialog box to set your printer or various aspects of your printer's configuration even if you don't intend to print immediately.

PRINTER QUEUES

Linux uses a concept known as a *queue* to handle printing. In computer circles, a queue is a holding area for items that need to be processed by a potentially limited system resource, such as printing to a printer. A multitasking, multiuser system such as Linux may receive two or more print jobs at the same time or a second job before the first has finished printing. Linux therefore places all print jobs in a queue and prints the jobs one at a time from the queue at its leisure. Each queue has its own name, such as lp or okidata, that you assign when you create it.

→ If you want to learn how to set up a Linux printer queue, **see** "Setting Up Linux Printer Queues," **p. 731**

Linux can maintain several print queues that feed one or more physical printers. For instance, if you have two printers, you might have three or more queues serving them, one for one printer and two more to use the second in two different ways (I illustrate this concept shortly).

Linux's printer queues may feed printers that are directly connected to your Linux computer, or printers that are connected to other computers on your local network. You won't see any difference between these different types of printers from within WordPerfect.

WordPerfect adds its own printer drivers atop the Linux printer queues, so WordPerfect can print to the same queue using two or more drivers. Setting up this system can be confusing at times, but *using* queues isn't so hard. For now, just pretend that each queue goes to a separate printer. This idea isn't really inaccurate because one printer can have diverse capabilities, depending on how it's configured in Linux.

→ To learn more about configuring your printers in WordPerfect, **see** "Fonts and Printing," **p. 739**

One more thing: Most UNIX programs assume that they are printing to a PostScript printer. Postscript is a printer "language" that's common on midrange to high-end laser printers; it lets the computer tell the printer where to put text and graphics, what fonts to use, and so on. Because most printers used with desktop PCs are not PostScript models, however, Linux relies heavily on a program called *Ghostscript*. Ghostscript converts PostScript input to any of several printer languages used by non-PostScript printers. Normally, Linux printer queues are configured to automatically send print jobs through Ghostscript when appropriate, so you can pretend that your printer is a PostScript printer in Linux. WordPerfect includes both PostScript drivers and drivers for non-PostScript printers, so you may or may not be printing through Ghostscript when you use WordPerfect. The most important thing to remember here is that Ghostscript *may* be sitting between WordPerfect and your printer and, therefore, may have some effects you should consider when printing.

Have you tried to print a document but find that the fonts come out strangely? See "Unavailable Printer Fonts" in the "Troubleshooting" section at the end of this chapter.

Have you tried printing a document only to have it fail completely (no output or complete gibberish coming from your printer)? See "Printer or Fax Configuration Problems" in the "Troubleshooting" section at the end of this chapter.

SELECTING A PRINTER

WordPerfect displays the printer name and printer queue name in the Current Printer area of the Print dialog box. If this printer is the one you want to use, you need do nothing more to select it. If you need to change the printer, however, you should click Select, which brings up the Select Printer dialog box shown in Figure 6.4. Select the printer you want and click OK to return to the Print dialog box.

Have you reached the Select Printer dialog box and found no selections? See "No Available Printer" in the "Troubleshooting" section at the end of this chapter.

Tip #65 from *Tux Tips*

When you change printers, your document's formatting sometimes changes. Therefore, you may want to check for important formatting changes after you change printers.

Tip #66 from *Tux Tips*

You can go directly to the Select Printer dialog box without entering the Print dialog box by clicking the printer icon in the status bar at the bottom of the WordPerfect window.

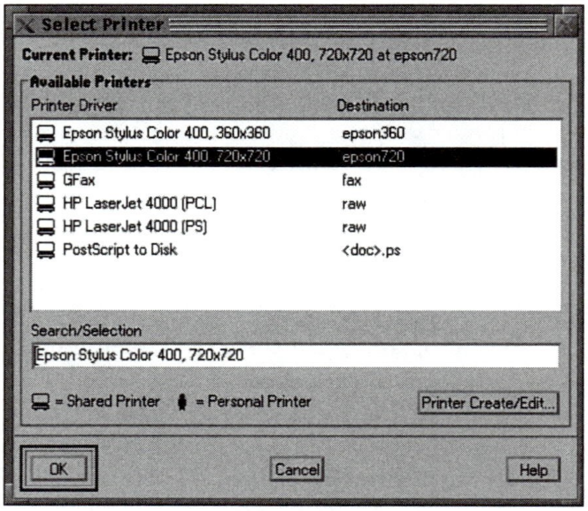

Figure 6.4
Select a printer by clicking its name from the list.

The Select Printer dialog box in Figure 6.4 illustrates some important characteristics of printers and printer handling in Linux and WordPerfect. Some things you may want to take note of and compare to your own setup include

- The Epson Stylus Color 400 in Figure 6.4 has two entries, one for a 360×360-dpi resolution and another for a 720×720-dpi resolution. In Windows, you would select the printer's resolution by setting an option in a printer-specific dialog box, but on my Linux system, I use two printer queues instead. WordPerfect includes a Print Quality setting in its Print dialog box, which may change the resolution if you're printing with a printer-specific driver. When you're printing via Ghostscript, however, this option may not have much effect. The two queues for the Epson Stylus Color 400 on my system simply tell Ghostscript itself to print at differing resolutions.
- The GFax package controls faxing on this system. Like the Epson Stylus Color 400, GFax relies on Ghostscript.

→ To learn more about faxing from WordPerfect, **see** "Faxing," **p. 161**

- My Hewlett-Packard LaserJet 4000 is accessible in two ways: through HP's PCL printer language and through PostScript. The printer itself supports both languages, so I have two WordPerfect queues set up for it. Neither goes through Ghostscript, and in fact, both print to the same Linux queue (named raw).
- The PostScript to Disk printer doesn't print to a conventional queue, but rather creates a PostScript file on disk.

→ If you want to create a PostScript file, **see** "PostScript and PDF Output," **p. 168**

- WordPerfect printers come in two varieties: shared and personal. They're identified by the icons at the bottom of the Available Printers area of the Select Printer dialog box. All the printers in Figure 6.4 are shared printers, meaning that anybody using

WordPerfect on that computer can use the printer. Individuals can create WordPerfect printer definitions for themselves, but other users won't be able to use these definitions. This feature can be handy if you're using WordPerfect on a system for which you don't have root access, but you need to create a new printer definition for, say, creating a PostScript file on disk if the system administrator hasn't provided this capability.

➔ To learn more about setting up a WordPerfect printer, **see** "Fonts and Printing," **p. 739**

PRINT SELECTION

WordPerfect lets you choose how much of your document to print in the Print Selection portion of the Print dialog box. If you start the Print dialog box with no text selected, Full Document is the default selection. As you might expect, this option prints the entire document. You can also, however, elect to print the Current Page, Multiple Pages, or a Document on Disk.

If you select Multiple Pages and then click OK, WordPerfect presents the Multiple Pages dialog box in Figure 6.5. This dialog box gives you assorted options for specifying the page range. For the most part, you'll be dealing only with the Page(s) field, and you can enter values here in various ways:

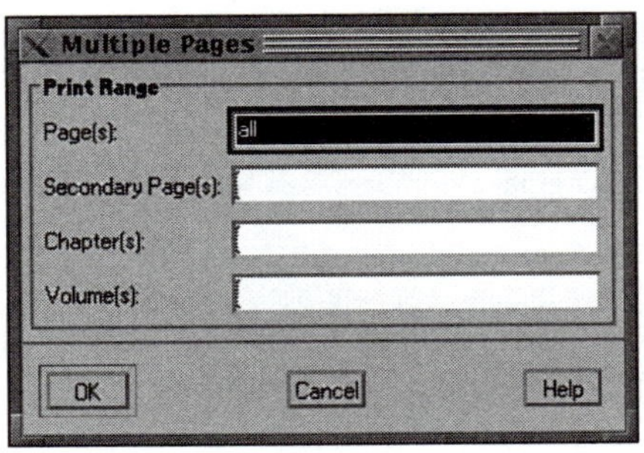

Figure 6.5
The Multiple Pages dialog box defaults to printing all the pages of a document.

- To specify a range of pages, enter the two endpoints separated by a dash (-); for instance, `2-10` prints pages 2 through 10, inclusive.
- To specify a range of pages starting from the first or ending at the last page of the document, you can enter a range but omit the first or last page number. For instance, `-11` prints from the first page through page 11, and `27-` prints from page 27 through to the end of the document.
- If you want to print some selection of individual pages, enter their page numbers separated by commas or spaces, as in `3 8 9` or `3,8,9`.
- Mix and match any of the preceding methods. For instance, `2-7,9,12,15-` prints pages 2 through 7, 9, 12, and 15 through the end of the document.
- If your page numbering system involves Roman numerals, you must use them instead of Arabic numerals when you specify your page ranges.

Caution

WordPerfect won't print an earlier page number after a later one. For instance, if you were to enter `9-12,6` as a page range, WordPerfect would print pages 9 to 12 but would omit page 6. Instead, enter `6,9-12` or do two print jobs.

The Multiple Pages dialog box lets you enter secondary page, chapter, and volume numbers in addition to or instead of page numbers. Briefly, secondary page numbers are an auxiliary way of numbering pages. This approach allows you to keep page numbers rising across files in a multifile document while maintaining separate page numbering within each file or to have separate numbering schemes for physical pages and for textual pages after a table of contents, for instance. Chapter and volume numbers let you divide a large document into chapters and volumes and to refer to these document elements by number.

→ For more information on page and other document numbers, **see** "Headers, Footers, and Page Numbers," **p. 235**

You can reset the page, secondary page, chapter, and volume numbering. If you do so, the page numbers you use when you specify multiple pages to print will correspond to your reset page numbers. If you change your page numbers so that you have multiple pages with one number (for instance, two page 4s), WordPerfect selects the page(s) to print by searching from the start of the document and printing the first page it finds that matches the number you specify. If you give a range separated by a dash, WordPerfect stops when it finds the first page matching the stop page. Consequently, you cannot use the Multiple Pages selection to print the second page 4 by itself.

→ For more information on setting and using page numbers, **see** "Page Numbering," **p. 239**

Tip #67 from *Tux Tips*

To print the second page that has a specific page number, put your insertion point on that page and use the Current Page option in the Print dialog box's Print Selection area.

If you want to print a document that's not currently loaded, select the Document on Disk option from the Print dialog box. When you click OK, you get a dialog box similar to the Multiple Pages dialog box except that you have an extra space in which to enter a filename. Enter the filename there and the page range as well if you don't want to print the entire file. You can also click the folder icon next to the filename to locate the file via the WordPerfect File Manager dialog box.

COPIES

You can tell WordPerfect to print several copies of the document. Simply enter the number you want in the Number of Copies spin box.

You can tell WordPerfect whether it, the printer, or in some cases the spooler should generate the copies. When WordPerfect generates copies, they are printed as one complete document after another, so you won't have to collate them yourself. Generally, when a printer generates copies, it does so a page at a time, so you get all your first pages, then all your second pages, and so on. This method can result in faster printing, especially with laser printers, because the printer receives and processes any page's data only once. The option to have the spooler generate copies is generally unavailable; it is used with WordPerfect's own print spooler, which duplicates the Linux spooler's function and so is not covered in this book.

PRINTER OPTIONS

You can set various print options in the Document Settings area of the Print dialog box or by clicking the Output Options button.

- The Print Quality options are High, Medium, and Draft. Precisely what these options do varies from printer to printer, but in general the High option produces the best quality at the slowest speed, whereas Draft produces speedy but low-quality output.
- If you have a color printer, the Print Color menu button provides two options: Black and Full Color. Use Black if you're printing textual information or a draft for which color is unimportant; you may get faster output that way, depending on your printer. Use Full Color when you're printing a document with color text or graphics.

Caution

Be sure to check the Print Color option before clicking OK when you're printing color. The setting doesn't always default to Full Color when you select a color printer!

- If you select the Do Not Print Graphics check button, WordPerfect omits all graphics from the printout. This option can speed printing on some printers and can also save ink or toner.
- The Initialize button below the Print Selection area sends initialization codes to your printer. You don't normally need this step, though you might want to use it if your printer is doing something strange or if you have another program that puts the printer into a state that causes problems for WordPerfect printouts.

If you click Output Options, you can set several more printer options in the Output Options dialog box (Figure 6.6). You'll probably be most interested in the options at the top of this dialog box.

- Print Document Summary prints summary information about the document, which you can enter by choosing File, Properties, Document Summary from the main document window to open the Document Summary dialog box (Figure 6.7). Here you can enter information on who created the file, keywords for it, and a short abstract. You may find it convenient or useful to print this information along with the document.

Note

WordPerfect actually prints the summary information as a separate print job from the document itself. If you're printing to a networked printer or if you or other users are starting other print jobs at that time, one of these other print jobs may even print between the summary and the document.

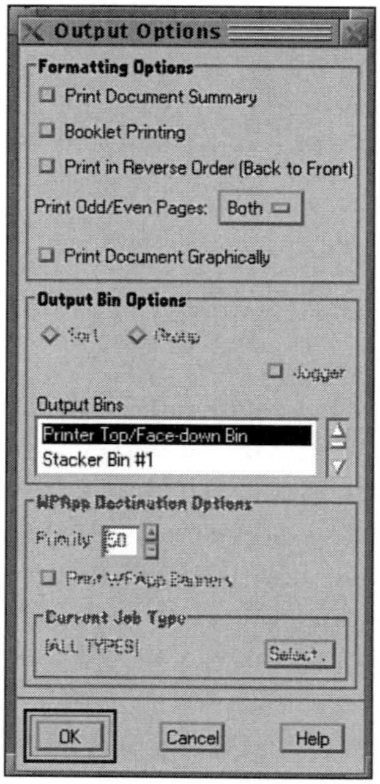

Figure 6.6
Many of the output options may be unavailable (grayed out), depending on which printer and print queue method you're using.

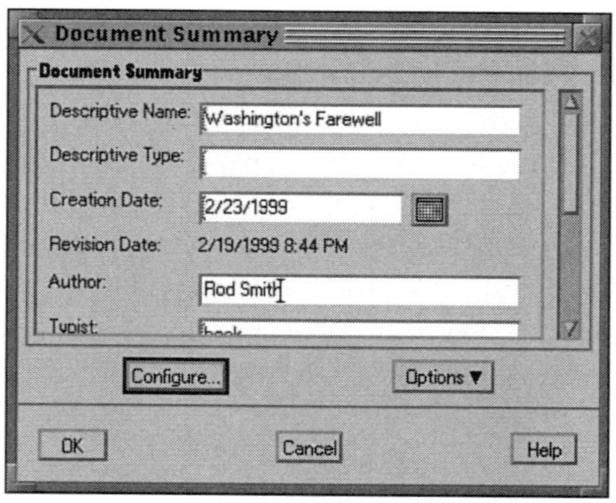

Figure 6.7
You can define what type of information is stored in the document summary by clicking Configure.

- If you select Print in Reverse Order (Back to Front) from the Output Options dialog box, WordPerfect prints the last page first. This option can be handy if your printer's output tray produces a reverse-order printout; WordPerfect reverses the reversal, giving you a correctly ordered printout.
- You can print odd, even, or both types of pages by selecting the appropriate entry in the Print Odd/Even Pages menu button. This setting can be useful if your printer doesn't have a duplexing feature and you want to print a double-sided document. You can select the odd or even pages, print them, select the other option, and feed your pages through again.

Tip #68 from *Tux Tips*

> Because printers vary in how they accept their input pages and orient the output, you may have to experiment to get the page orientation and ordering right. You may have to reverse the ordering of your pages before you feed them through for their second-side printing, too.

- Selecting Print Document Graphically has little effect with certain printer types and a substantial effect with others. With many printers, including PostScript and recent PCL laser printers, WordPerfect normally prints either by using fonts that are built into the printer or by downloading a bitmapped version of the font and telling the printer to use that. WordPerfect can then send the document text to the printer, and the printer forms the text into bitmaps to place on the page. When you select Print Document Graphically, however, WordPerfect forms the entire text of a page into one large graphic and sends that to the printer, assuming that the document doesn't use any built-in printer fonts that would override this feature. Printing a document graphically normally slows down printing, but in some circumstances you might need to use this feature. For instance, if you use a lot of different fonts or create a particularly complex graphic, your printer might run out of memory trying to print, so using this feature allows you to print when you might otherwise not have been able to. On the other hand, with simpler layouts, the memory requirements for handling a document printed graphically from WordPerfect may actually be greater than the requirements of printing in the normal way.

  When you're printing via Ghostscript, the Print Document Graphically option is unlikely to make much difference in terms of your printer's memory capacity because this option simply moves the task that Ghostscript normally performs into WordPerfect. You may see subtle differences in the form of your characters between the two methods, however, and one or the other might print faster.

  If you're printing to a low-end inkjet or dot-matrix printer via WordPerfect's own drivers, the Print Document Graphically option has no effect because it is the way that WordPerfect normally handles these printers. This option also has no effect if you use printer fonts exclusively because WordPerfect can't process them internally—it *must* send text in printer fonts to the printer as text, rather than processed into graphics.

- The Output Bin Options area allows you to specify output bins, as well as sorting and grouping features, for high-end printers that support these features.
- The WPApp Destination Options area is usually grayed out. This area controls options for WordPerfect's print spooler application. These functions replicate Linux's printer queues, so I don't discuss them in this book.

Printing a Booklet

If you select Booklet Printing from the Output Options dialog box, WordPerfect prints your document in booklet format, ready for folding and stapling or binding. This option requires some extra preparation, however, and you need some trial runs to figure out how to feed paper through your printer to get the effect you desire.

Before you open the Print dialog box, you must configure your page for booklet printing:

1. Move your insertion point to the very start of the document.
2. Choose File, Page Setup, Page Size to open the Paper Size dialog box shown in Figure 6.1. To produce a 5.5×8.5-inch booklet (a folded sheet of letter-size paper) with portrait orientation, you need to select letter-size paper with a landscape orientation. (Landscape orientation is a 90-degree rotation of portrait orientation so that the long edges of the page are on the top and bottom.)
3. Choose File, Page Setup, Subdivide Page. This step brings up the Subdivide Page dialog box shown in Figure 6.8. You need to subdivide the page into two parts, either in columns or in rows. For a 5.5×8.5-inch booklet, select two columns, as shown.

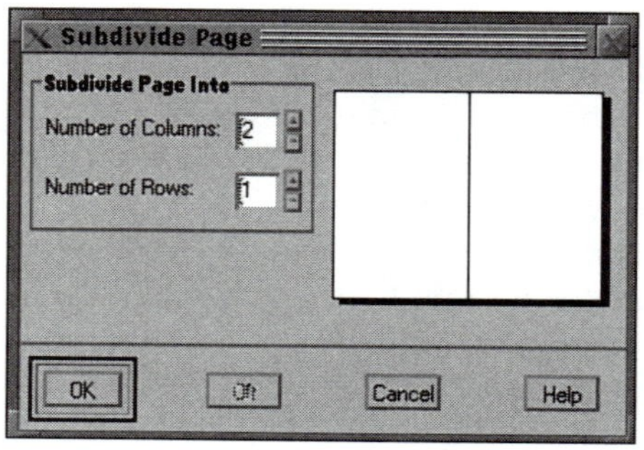

Figure 6.8
Use the Subdivide Page dialog box to prepare for printing a booklet or simply to create columns.

You can now start the Print dialog box, click Output Options, and select Booklet Printing. When you print the document, it prints as a booklet. If your printer doesn't handle duplexing, you need to hand feed your sheets through twice.

Tip #69 from *Tux Tips*

You may want to adjust your page margins, your paragraph indents, and other elements of your document to get it to look good as a booklet.

→ For information on adjusting paragraph formatting, **see** "Paragraph Formatting," **p. 205**

→ For information on adjusting page formatting, **see** "Page and Document Formatting," **p. 225**

The file gw-pamphlet.wpd in the chapter06 directory on this book's companion CD contains the George Washington farewell speech formatted appropriately for booklet printing. If you want to practice printing a pamphlet with this file, I suggest you initially delete all text after page 8 in the file. This modification produces a pamphlet that prints on two sheets of paper for minimal wasted paper while you're figuring out how to do it properly with your printer.

Tip #70 from *Tux Tips*

Some printers, such as many midrange and high-end laser printers, either automatically print both sides of a pamphlet or pause after printing the first side and prompt you to insert the paper again. Other printers, including most printers driven by Ghostscript, don't have this feature and print the booklet straight through, as a single-sided document. To get the desired effect, try loading your printer with precisely the number of sheets you need (the number of pages divided by four and rounded up). After the sheets are printed, insert them again so that the opposite side of page 1 is first to print, oriented so that the tops and bottoms of the front and back sides match, and you're all set.

Tip #71 from *Tux Tips*

If you want to mass-produce a booklet, you may want to print it single sided and then use a photocopier to make multiple copies. If your photocopier can automatically do double-sided copies, be sure to interleave the single-sided printouts so that the first page of the second side follows the first page of the first side, and so on.

Controlling Printing from Linux

Your Linux distribution provides tools to help you control printing. These tools come in the form of several commands you can enter at an *xterm* (p.44). Some of these allow you to do things as an ordinary user; others may require you to be root. Most of these commands take at least the `-P` parameter to specify the printer queue you want to operate on. When you use `-P`, type the name of the printer queue *without an intervening space*; for instance, `lpq -Pcanon` to query the printer queue called *canon*. Without the `-P` parameter, these commands work on the default printer queue, which is generally called *lp*.

`lpq`

Use the `lpq` command to determine what print jobs are present in a print queue and whether that queue is printing or not. Listing 6.1 shows the output of `lpq` when a WordPerfect print job is printing.

Listing 6.1 `lpq`'s Output

```
Rank   Owner     Job  Files                        Total Size
active rodsmith  152  /tmp/_pp_0000012278_58       10832657 bytes
1st    rodsmith  153  /tmp/_pp_0000012335_59       23967 bytes
2nd    rodsmith  154  /tmp/_pp_0000012365_60       22107 bytes
```

This output shows three print jobs in the queue, all started by the user rodsmith. WordPerfect creates print jobs by storing temporary files with obscure names in the /tmp directory, so all these WordPerfect jobs have long and strange names in the Files column. Print jobs from other programs may have more or (believe it or not) less meaningful filenames. The first job is the one that's labeled active in the Rank column and has a job ID of 152. When the `lpq` command was issued, that job was being sent to the printer, hence its "active" status. The second and third jobs have ranks of 1st and 2nd and IDs of 153 and 154, respectively, and are waiting in the queue. The size of the active job is about 1MB, and the next two are each less than 24KB.

Note

The same document may produce print jobs with wildly different sizes when printed to different printers or with different options, such as the Print Document Graphically option.

When you're doing additional print job management, you need to know the job ID listed under the Job column. The job size may or may not be helpful in determining which job is which. If you print two jobs with the same settings to the same printer, the document with more pages is likely to be larger, but that may not be the case if you changed printer options. Even in-document options like your choice of font can affect the print job size.

`lprm`

If you want to stop a print job, either before it begins printing or while it's printing, use the `lprm` command. For instance, `lprm -Plexmark 153` removes the second print job identified by the `lpq` command just discussed, assuming the queue name is lexmark. You need to be the owner of that job or root to use `lprm` on a print job.

Tip #72 from *Tux Tips*

If you issue an `lprm` command on a printing job, printing might stop in the middle of the page. For inkjet and dot-matrix printers, you may need to eject the half-printed page, and possibly reset your printer, to get it to function correctly. You won't need to eject a page from a laser printer, but a reset may be necessary. The Initialize button on the Print dialog box may work, or you may need to use a button or command available on your printer's front panel.

`lpr`

The `lpr` command prints a file from an xterm or text-mode login. If you've created a PostScript file from your WordPerfect document and you have a PostScript printer or a printer queue that uses Ghostscript, you can print it with `lpr`. For instance, to print a PostScript file called file.ps to a PostScript or Ghostscript-driven printer on the queue lexmark, you would type `lpr -Plexmark file.ps`.

→ To learn how to create a PostScript file from WordPerfect, **see** "PostScript and PDF Output," **p. 168**

Depending on the details of how your printer queues are set up, you may also be able to print plain ASCII text and some other file formats directly via `lpr`. Consult your distribution's documentation to see what your printer queues can handle.

Caution

Don't try to print a WordPerfect document file directly via `lpr`; WordPerfect files are not one of the varieties that most `lpr` queues will understand. You need to start WordPerfect to print your WordPerfect files.

`lpc`

The `lpc` command is the workhorse printer control command for Linux, but you need to be root to use many of its features. Also, `lpc` often resides in the /usr/sbin directory, which your shell may not search automatically, so you may need to type the complete pathname to use it (that is, type `/usr/sbin/lpc` rather than just `lpc`). I don't give the complete path in the following discussion, however, for brevity's sake.

→ To learn about the root account, **see** "Users and the Superuser," **p. 11**

You use the `lpc` command to enable or disable queues or to rearrange the order of print jobs. You'll probably use it mostly for this last function. Suppose that you want to rearrange the order of the jobs in the queue in the previous example so that job 154 prints before job 153. To do so, you would type `lpc topq lexmark 154 153`. The `topq` parameter tells `lpc` that you want to place the specified print jobs at the top of the queue in the order you specify. The `lpc` command doesn't interrupt printing on the current job, however, so it remains in place (you can kill it with `lprm` if necessary).

You can also use `lpc` to entirely disable or enable a queue by using the `down` or `up` parameters, as in `lpc down lexmark` or `lpc up lexmark`. Other variants have subtly different effects. If you're interested, type `man lpc` at an xterm to read the Linux documentation on `lpc`. You can use the `man` command to find more information about the other print queue management commands, too.

FAXING

If you have a Linux-compatible fax modem attached to your system, you can use WordPerfect to fax documents directly from your computer. The result is better fax quality on the receiving end than you'd get with a paper printout and fax machine because the computer sends the bitmap directly, rather than relying on an intermediate scan step in the sending fax machine.

LINUX FAXING WITH GFAX

You need a fax package to fax from WordPerfect. In this chapter I describe the use of GFax, which is a GUI front end to the mgetty+sendfax package. The latter comes with most Linux distributions; the former is available from `http://www.gmsys.com/gfax.html` and is also included on the CD that accompanies this book. If you use another fax package, the dialog boxes may be different from the ones shown in this chapter, but the basic principles are the same.

→ To learn how to set up a fax print queue, **see** "Setting Up a Fax Queue," **p. 753**

Note

> Unlike most Linux programs, GFax is shareware. Please read its license before you use it. It's free for most individual users, but if you're using it in a corporate setting, you must pay a small fee.

From WordPerfect's point of view, your fax system is just another printer, and so you select the fax in the same way as you would any other printer. Figure 6.4 shows the Select Printer dialog box, including an entry for GFax on the fax queue. Most Linux fax systems, including the GFax/mgetty+sendfax combination, treat the fax device as a PostScript printer and use Ghostscript to process the PostScript from WordPerfect.

When you select the fax queue and click OK to print, WordPerfect sends the document to GFax, which displays its own dialog box, as shown in Figure 6.9. Enter the fax number and, if necessary, entries for the Name and Attention lines, which print on the optional cover page. You can tell GFax to send the fax in high resolution by selecting the Hi Resolution check button. You can also schedule your fax transmission to go out immediately or at some specific time.

Figure 6.9
The GFax program provides the sorts of fax options with which you're probably familiar if you've used fax software in Windows.

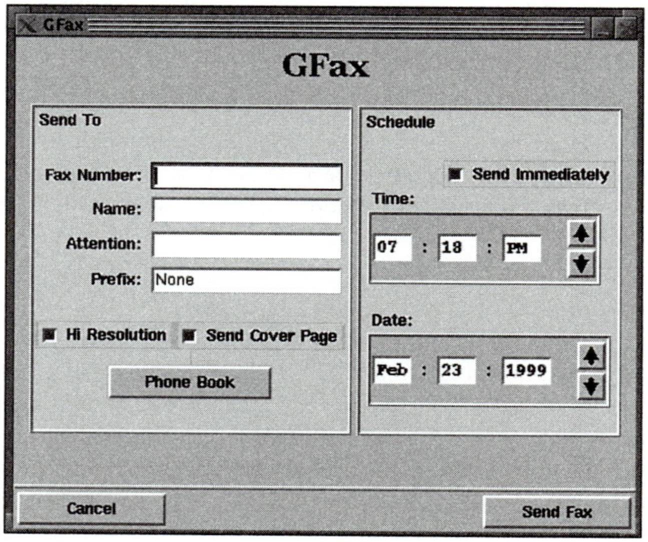

Note

GFax doesn't use the Motif widget set that WordPerfect uses, so GFax's look and feel is slightly different. The differences are just cosmetic, though, such as colored check buttons and larger spin box arrows.

If you regularly send faxes to certain people or companies, you can enter them into the GFax phone book. Instead of entering the name and number each time you send a fax, click Phone Book and select the recipient from the list, as in Figure 6.10. Click the line of the recipient and Ok, and the phone book information appears in the main GFax dialog box. You can also double-click the recipient's line for the same effect.

You can edit a GFax phone book entry by selecting it and clicking Edit, or you can create a new entry by clicking Add New. In either case, you arrive at the Edit Entry dialog box shown in Figure 6.11. Edit the relevant information and click Ok.

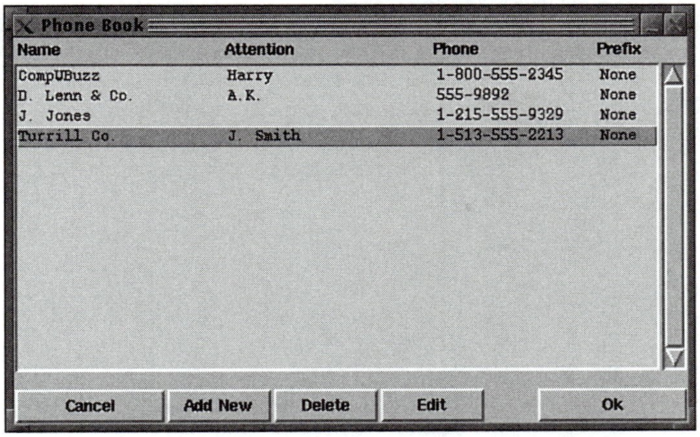

Figure 6.10
You can enter a prefix in the phone book if you need to dial a special number to get an outside line from a business or to use a particular long-distance carrier.

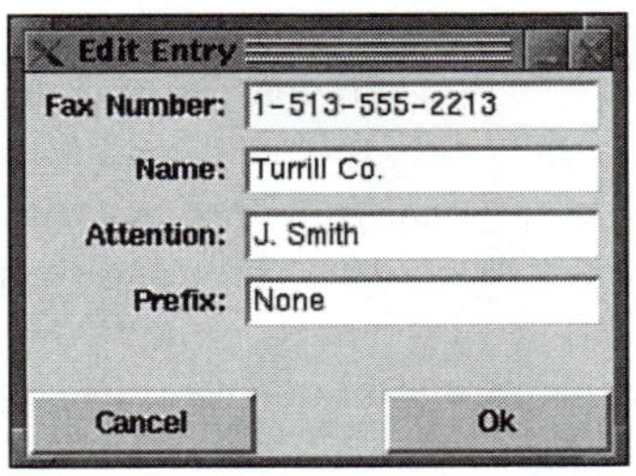

Figure 6.11
GFax's dialog boxes put the Ok button in the lower-right corner, whereas most Linux programs put it in the lower-left corner, so be alert to that difference.

Tip #73 from *Tux Tips*

To use your own personal phone book, you may need to create an appropriate file. At an xterm, first be sure that there is a .gfax subdirectory in your home directory. Then move into that directory by typing `cd ~/.gfax`. Type `touch phbook` to create an empty phone book file. GFax will now use this file for its phone book entries. You can add your own fax phone book entries without disturbing others' entries.

When you click Send Fax, GFax gives you the opportunity to enter some additional cover page information, as shown in Figure 6.12, if you opted to have GFax generate a cover page. (If not, you won't see the Cover Page Information dialog in Figure 6.12.) Edit or enter the information you want to include and click Ok to send your fax.

Note

Linux probably won't send your fax the moment you click Ok to send it for two reasons. First, Ghostscript needs some time to process WordPerfect's PostScript output into an appropriate format for faxing. This delay may be especially noticeable if the document is long or if you're using a slow computer. Second, GFax uses a fax scheduling system that checks for outgoing faxes every minute or so. Unless your fax arrives in the queue just moments before the system checks for faxes, it will sit in the queue for a few seconds before being noticed.

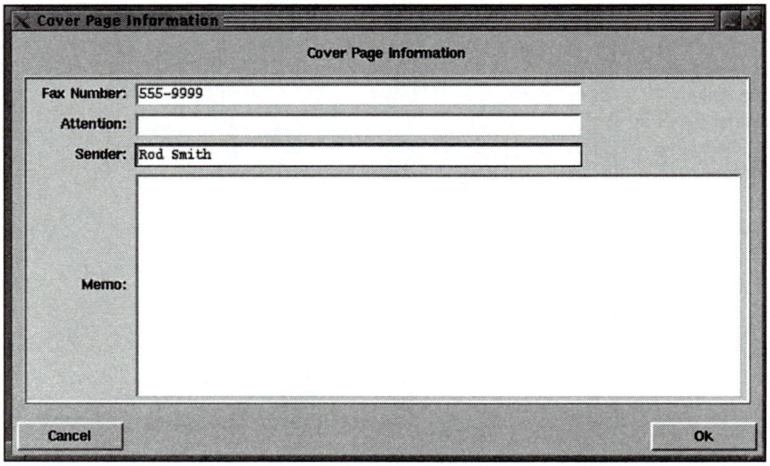

Figure 6.12
You can edit the Sender line and include a cover page memo to help the recipient identify or otherwise interpret your fax.

You can also configure GFax to send email to you when your fax transmission succeeds (or fails, if that's the case). You'll normally configure GFax to send mail to your Linux mail queue. If you use a program such as Netscape to read your Internet mail, chances are it's not set up to read your local Linux mail queue unless your machine is directly connected to the Internet and people use your local machine's name when they send you mail. Therefore, you may need to use pine, elm, or some other mail program to check for your GFax messages.

→ To learn how to set up a fax queue, **see** "Setting Up a Fax Queue," **p. 753**

Have you tried faxing a document only to have the fax transmission fail? See "Printer or Fax Configuration Problems" in the "Troubleshooting" section at the end of this chapter.

Managing Your Fax Queue

GFax, like some other Linux fax packages, relies on the mgetty+sendfax program to do most of the work of sending faxes. This process is largely transparent to you, but you may occasionally want to manage your fax queue in the way you do a print queue. You don't use the `lpq`, `lprm`, and similar commands to manage faxes; instead, you use similar utilities from the mgetty+sendfax package:

- faxq—Type `faxq` at an xterm to see a list of the faxes that are waiting to be sent. The output includes a fax queue number of the form Fxxxxxx/JOB, where *xxxxxx* is a number. You use the Fxxxxxx portion of this ID to delete a fax waiting in the queue if you want to do that. The output of faxq also includes information on the user who sent the fax, the number of pages in the fax, and the telephone number to which it's being sent.
- faxrm—To delete the fax job Fxxxxxx, type `faxrm Fxxxxxx`. If it succeeds, faxrm provides no feedback. You can use faxq to confirm that the fax job was removed if you like.

As with printer queues, you can delete only your own fax jobs from the fax queue unless you're using the root account.

EMAIL

You can send your WordPerfect document directly from WordPerfect via Linux's email system. Choosing File, Send to, Mail, tells WordPerfect to display the Send UNIX Mail dialog box in Figure 6.13. This dialog box functions very much like any other program for sending mail. You enter your recipient's email address in the To field, and you can use the CC or BC fields to enter the addresses of additional recipients. You can enter a Subject for the message and a message text to introduce what you're sending.

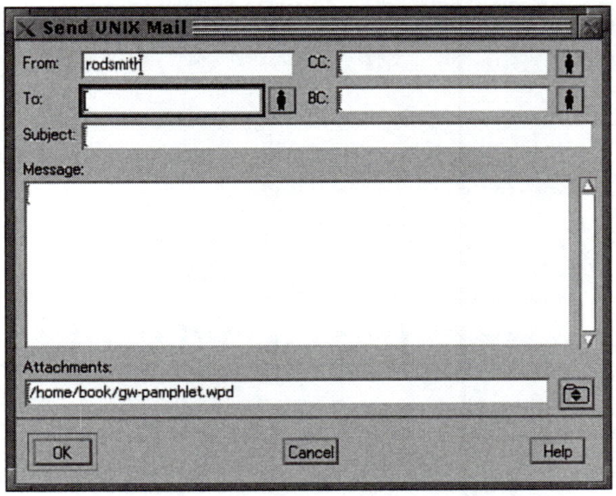

Figure 6.13
The Send UNIX Mail dialog box reveals WordPerfect for Linux's UNIX heritage in its drag bar title.

Note

If you haven't yet saved your file, WordPerfect asks whether you want to do so before sending it. You *must* answer Yes if you want to send the file. If you answer No, the Send UNIX Mail dialog box opens, but the file won't be attached to the outgoing mail message.

Tip #74 from *Tux Tips*

The To and CC fields serve essentially identical functions; they just have different names. Using the BC field results in a *blind copy*, meaning that the other recipients won't know that the individual(s) listed in the BC field also received a copy of the message.

If you click the folder icon next to the Attachments field, you can select other files to attach to the mail message in addition to the WordPerfect file. These files can be WordPerfect files or other types of files, so you can send non-WordPerfect documents that go with a WordPerfect file, such as spreadsheet, graphics, or sound files that are relevant to the topic of your written document.

WordPerfect sends its files in *uuencoded* form, which is a way of sending nontextual data over the fundamentally textual Internet email system. Some email programs automatically decode a uuencoded file on receipt, but others may display what looks like gibberish (Figure 6.14). Such files *have* arrived intact; they're just encoded in a way that requires decoding before they can be used. If your files' recipients are not familiar with this format, you need to advise them to locate uudecoding software for their computers. Under Linux, the program uudecode does the trick, and it's included with most distributions.

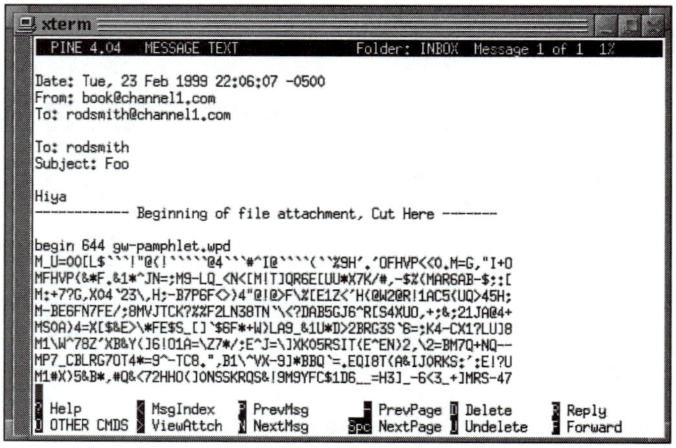

Figure 6.14
Uuencoding ensures that a file can traverse the Internet's myriad mail systems without being corrupted, but without uudecoding the contents are unintelligible.

Note

You should be able to send email to other users of your computer without doing anything special, whether or not you're connected to the Internet. To send email to other systems, you need either a direct connection via an office network, cable modem, or the like or a dialup connection using PPP or something similar. Configuring network connections is beyond the scope of this book, but you can consult Que's *Special Edition Using Linux* for information on this topic.

WordPerfect maintains an address book of email addresses. You can click the person-shaped icon next to any of the email address fields in the Send UNIX Mail dialog box to open this address book (Figure 6.15). Select an address to which you want to send mail from the list and then click the appropriate arrow button to add that address to the To, CC, or BC field.

To add names to your email address book, follow these steps:

1. Click Edit List on the Select Users/Group Names dialog box to open the Create/Edit User List dialog box in Figure 6.16.
2. Click Add to add an address. WordPerfect displays the Add User dialog box shown in Figure 6.17.
3. Type an email address. This entry can be a username of a person who has an account on your Linux computer or a full Internet email address.
4. Click Apply to enter that address into the address book and go on to enter another, or click OK to enter the address and stop there.

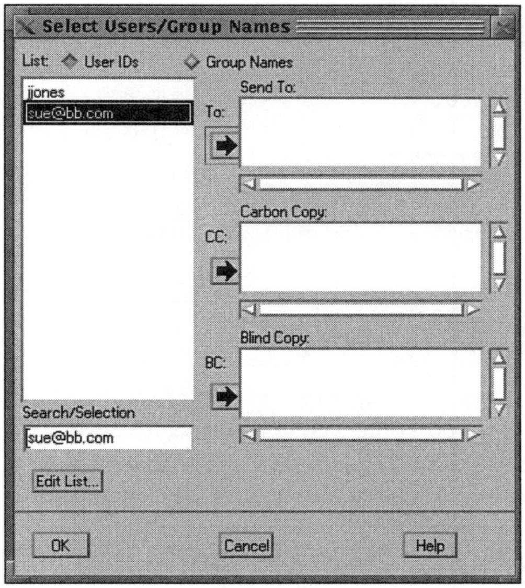

Figure 6.15
If an email address is too long to fit in the list, a scrollbar appears at the top, partially obscuring the first address.

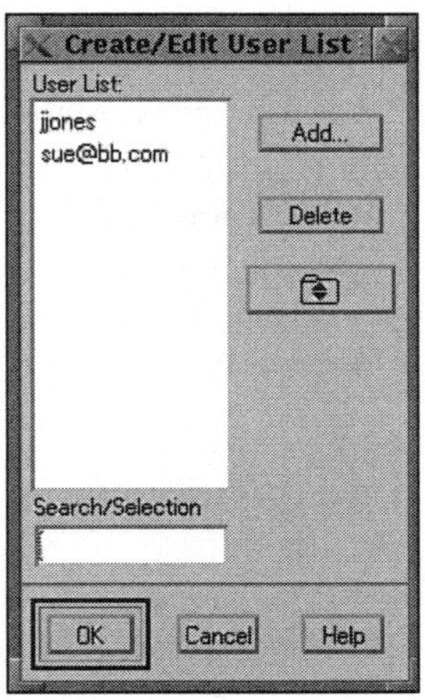

Figure 6.16
Type the first letters of an email address in the Search/Selection field, and WordPerfect highlights that address.

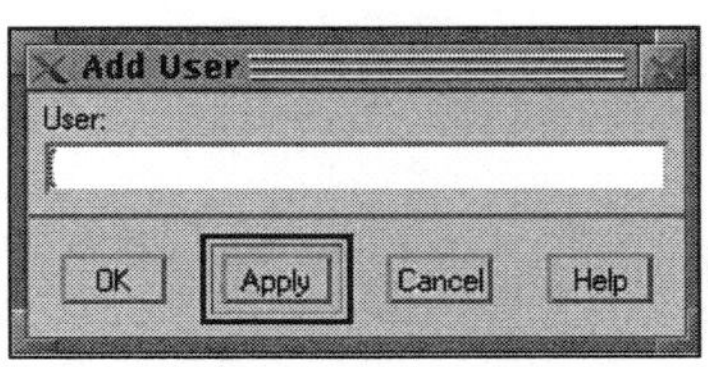

Figure 6.17
WordPerfect refers to email addresses as "users" in this part of the address book.

Note

One thing that's important to remember when you send a WordPerfect document via email is that the recipient must be able to read WordPerfect files. Ideally, the recipient should have WordPerfect 6.0 or greater for any platform (DOS, Windows, or UNIX/Linux). Other word processors may be able to read WordPerfect files, but there's no guarantee that your file will import perfectly.

➔ If you want to learn more about exchanging documents with other word processors, **see** "Document Exchange," **p. 88**

POSTSCRIPT AND PDF OUTPUT

PostScript has long been a common means of document exchange in the UNIX community. Multiuser UNIX systems at corporations and universities have usually been connected to PostScript printers, so a person who produced a PostScript document on one computer could give that file to another person at another UNIX-enabled institution and be reasonably certain that the recipient could print it. Linux, borrowing from this UNIX tradition, provides tools for printing PostScript, even on non-PostScript printers. Indeed, programs built around Ghostscript display PostScript files on the screen without committing them to paper.

Adobe, the originator of PostScript, has also created a file format called *PDF* (short for *portable document format*), which is used by its Acrobat programs. Adobe has made a version of Acrobat Reader available for Linux, and it comes with the retail version of WordPerfect. Like PostScript but more so, PDF has become a standard way of distributing files on computers. Because the Acrobat Reader is freely available, it's practical to give out Acrobat files and be reasonably certain that the recipient can read them. The same isn't necessarily true of word processing formats like WordPerfect's or Microsoft Word's.

Fortunately, WordPerfect can create PostScript output, and Ghostscript can easily convert this output to a PDF file, so you can create these widely interpretable file formats from Linux.

CREATING POSTSCRIPT FILES

All you need to produce a PostScript file is a correctly configured WordPerfect printer. When you select your "print to disk" printer definition and print, WordPerfect asks for a filename (Figure 6.18). PostScript files typically have filenames ending in .ps, so you should give your PostScript output file that extension. Click OK to tell WordPerfect to create your PostScript file.

➔ To learn how to set up a print-to-disk print queue, **see** "Printing to Disk Files," **p. 752**

CHECKING POSTSCRIPT OUTPUT

After WordPerfect creates a PostScript file for you on disk, you may want to check the file before proceeding. You can generate a printed copy by printing the file to your PostScript or Ghostscript-driven printer, using the `lpr` command. You can also view the file onscreen by using any of a number of PostScript-viewing utilities, such as *ghostview*, *gv*, or *kghostview*. These all interface with Ghostscript, which does the actual PostScript processing; the view programs simply display the output. One or more of these programs probably came with your Linux distribution.

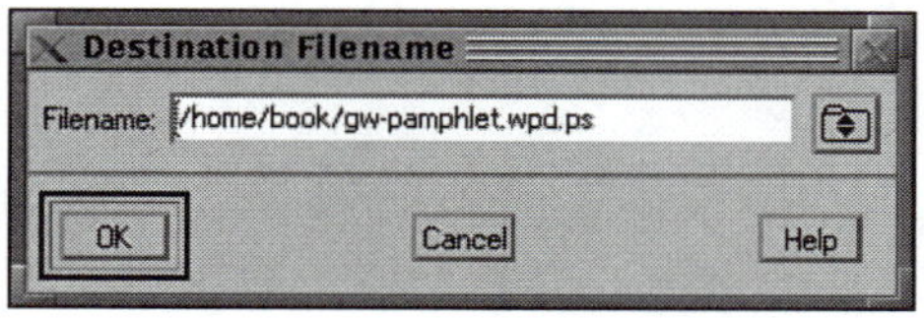

Figure 6.18
WordPerfect creates the default PostScript filename from the document's filename.

To view your PostScript file, open an xterm and type the name of the viewer you intend to use followed by the PostScript filename. For instance, you might type `gv filename.ps` to view the file filename.ps with gv. Figure 6.19 shows gv displaying a PostScript file created by WordPerfect.

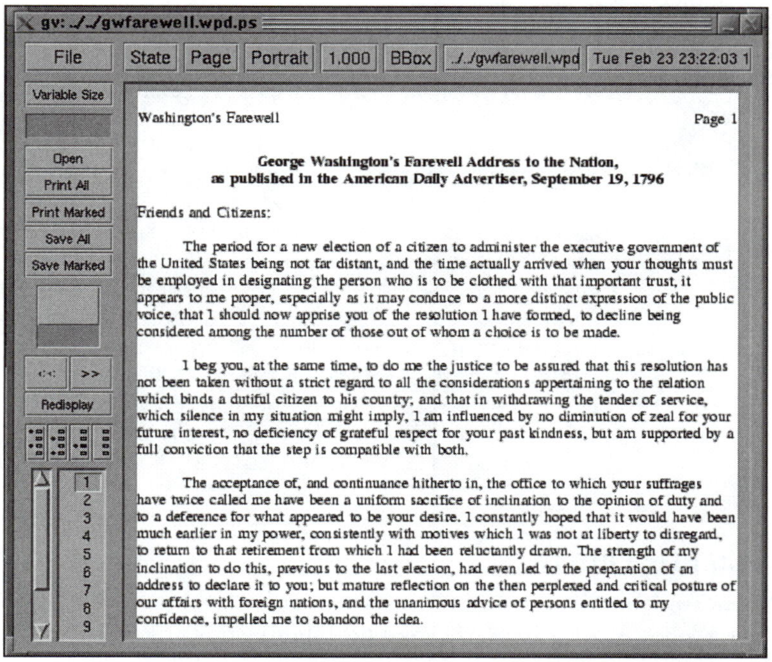

Figure 6.19
The various PostScript-viewing programs have similar basic functionality, though their appearance and details of operation differ.

Note

WordPerfect's PostScript output results in page images with little or nothing in the way of margins. You don't notice the missing margins when the page is printed because the printer can't physically cut away your paper, but this condition does turn up in onscreen previews such as those from gv. WordPerfect occasionally produces a PostScript file that displays nothing in gv but prints fine. Such files also generally produce PDF files that are fine.

CONVERTING POSTSCRIPT TO PDF

If you want to distribute your document electronically to a wide audience, PDF is probably a better format than PostScript. Except for those with PostScript printers, few Windows or Mac users would know what to do with a PostScript file, but Adobe's Acrobat Reader is widely available on these platforms.

Tip #75 from *Tux Tips*

I've found that Windows users generally use the term *PDF file,* whereas Mac users often use *Acrobat file* and may not recognize the term *PDF file*. You might want to use both terms when offering such a file, to be as clear as possible.

Chances are your Linux distribution includes the `ps2pdf` script, which calls Ghostscript with the appropriate parameters to create a PDF file from a PostScript file. To use `ps2pdf`, open an xterm and type `ps2pdf filename.ps` to convert filename.ps to PDF. The result is a file called filename.pdf, and you can read it in Acrobat Reader or in Ghostscript-based viewers such as gv. This operation doesn't delete the original PostScript file, so you can do other things with it.

Tip #76 from *Tux Tips*

Ghostscript is available both in a completely free form ("GNU Ghostscript") and in an almost-completely free form ("Aladdin Ghostscript"). Aladdin Ghostscript is free for most uses as long as you don't distribute it as part of a commercial product. GNU Ghostscript doesn't have this restriction. After a version of Aladdin Ghostscript has been out for a while, it's turned over to become GNU Ghostscript, and a new version of the program becomes Aladdin Ghostscript. The idea is that Aladdin can make money by licensing the latest version of the program, but individuals can still use it without paying for it. The catch is that Linux distributions are often sold on CDs, and so cannot include Aladdin Ghostscript. Linux therefore usually ships with the older GNU Ghostscript. Aladdin has been making major improvements in its PDF output in recent versions, however, so you may want to look into upgrading from whatever came with your distribution to the latest version from Aladdin. You can check the Ghostscript Web site, at `http://www.cs.wisc.edu/~ghost`, or you can check the Web site for your Linux distribution—although that vendor can't include the latest versions on a CD, the company can offer it for download from an ftp or Web site.

CHECKING PDF OUTPUT

When you have a PDF file, you can view it with a Ghostscript-based viewer such as gv or with Adobe's Acrobat Reader. Figure 6.20 shows Acrobat Reader displaying the PDF derivative of the file displayed by gv in Figure 6.19.

You can display this same file in Acrobat Reader for Windows, OS/2, or MacOS with equal ease. You can also print the file from Acrobat Reader on any of these systems.

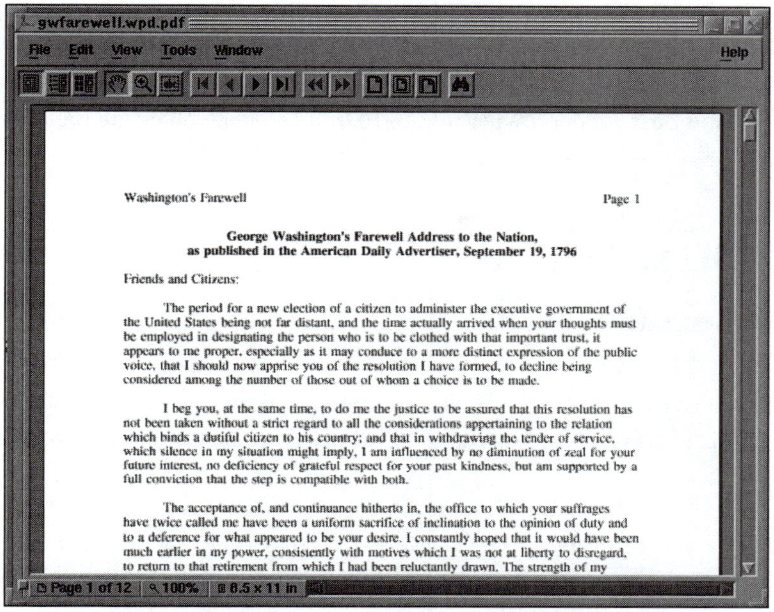

Figure 6.20
Notice that the page margins, which seemed to be missing in the PostScript file, are back in the PDF file.

CONSIDERATIONS WHEN DESIGNING PDF FILES

Producing PDF files using WordPerfect for Linux is relatively straightforward, but there are some points you should keep in mind:

- Because of the way Ghostscript handles fonts, you get the best text in PDF files if you restrict yourself to the most basic PostScript printer fonts: Times, Helvetica, and Courier. Be sure you select a printer font, not a WordPerfect graphics font. The former have printer icons next to their names in most of the WordPerfect font selection menus (Figure 6.21). You can use any font sizes and styles you like, however, without reducing the PDF file's quality.

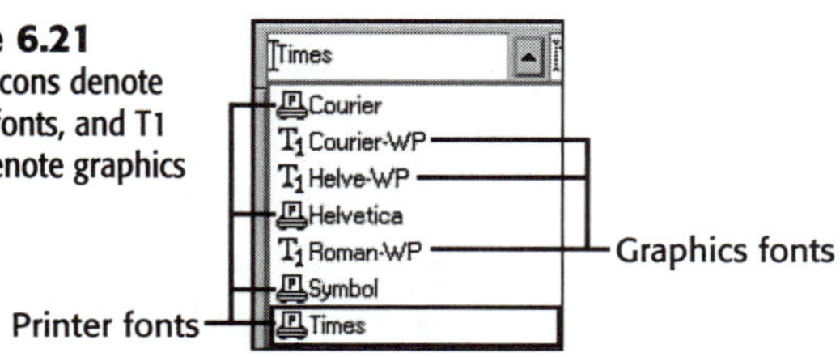

Figure 6.21
Printer icons denote printer fonts, and T1 icons denote graphics fonts.

Are you getting poor font quality on the fonts in your PDF file even if you use Times or Helvetica? See "PDF File Font Problems" in the "Troubleshooting" section at the end of this chapter.

PART I

CH 6

- You get the best results if you use the 8.5×11-inch paper size and portrait orientation. Deviating from this setting may produce sideways text or other oddities.
- One feature of PDF files is that they support *hyperlinks*, much like those on the Web. You can't produce hyperlinks in PDF files when you create them from WordPerfect on Linux, however.

Troubleshooting

No Available Printer

You can't find any printers in the Select Printer dialog box.

You need to create a WordPerfect printer definition first.

→ To learn how to create a WordPerfect printer definition, **see** "Fonts and Printing," **p. 739**

Printer Out of Memory

Your printer only prints part of a page or displays an out-of-memory error message.

Try checking or unchecking the Print Document Graphically check button in the Output Options dialog box, which is accessible from the Print dialog box. You can also try changing to a different font. Fonts built into the printer or downloaded to the printer as font outlines are marked with a printer icon, whereas fonts sent as bitmaps are marked with a T1 icon. Built-in and outline fonts take the least and the most printer memory, respectively, so this information alone may not tell you how much printer memory a font uses. In other words, you may need to experiment.

Unavailable Printer Fonts

The fonts in your document don't appear correctly in the printout.

Chances are you're using a WordPerfect printer driver that doesn't quite match the printer you're using and you've selected printer fonts that aren't present in your printer. Try to find a WordPerfect graphics font that's similar to the printer font you've been using and use it instead.

Printer or Fax Configuration Problems

Your printer or fax setup just doesn't work—you get garbled or bizarre printouts, or faxing doesn't start or fails after sending just one page.

You probably need to modify your WordPerfect printer definition or your Linux printer or fax queue setup.

→ To learn how to create or modify a WordPerfect printer definition, **see** "Fonts and Printing," **p. 739**

PDF File Font Problems

Your PDF file has ugly, chunky-looking fonts.

Try using the PostScript printer fonts Times, Helvetica, or Courier. Note that these are *not* the same as the WordPerfect graphics fonts Roman-WP, Helve-WP, and Courier-WP, though they look the same on the screen and (mostly) when printed. Alternatively, you could try upgrading to a newer version of Ghostscript. As of this writing, Ghostscript doesn't handle font translation for PDF files well, which is the source of the problem, but future versions may improve this situation.

PART II

ENHANCING TEXT BY FORMATTING

CHAPTER 7

Character and Line Formatting

In this chapter

OVERVIEW

This exploration of WordPerfect's formatting capabilities begins by examining the finest-grained formatting commands—those that apply to individual characters and the lines on which they reside. Many of these formatting options have an impact on larger formatting issues, of course, including document length and the overall feel of your document. Indeed, classifying individual WordPerfect commands and features can be tricky sometimes, so check the book's table of contents or index if you can't find a feature in this chapter that you expect to be here.

FONTS

One of the great advantages of word processing with a modern computer rather than using a conventional typewriter is that the word processor allows you to use a wide variety of fonts. You can use one font for the bulk of your document's text and another for headlines or sidebars. You can then start another document and use an entirely different font, including specialty fonts for printing certificates or eye-catching flyers. WordPerfect for Linux gives you access to many fonts, so you can join in the action.

Tip #77 from *Tux Tips*

Many people tend to overuse fonts when they first use a word processor. You'll do best to restrict yourself to just two or three font families per document. More than that tends to make a page look disordered, and one of your goals in creating most documents is presumably to give a very professional and *ordered* impression.

Note

The downloadable version of WordPerfect includes only a handful of fonts and lacks the font installer utility that comes with the retail version of WordPerfect. Therefore, if you want to install more than the basic collection, you need to buy the retail version of the program.

→ For more information on configuring fonts in WordPerfect, **see** "Installing and Managing Fonts," **p. 754**

FONT BASICS

Fonts come in many different styles and families. A *serif* font has small flares or extensions at the ends of its characters' main lines. The text you're reading now is printed in a serif font. In contrast, a *sans serif* font lacks these flares. The major headings in this book are printed in a sans serif font. In fact, this book's use of fonts is typical of printing in the United States: Serif fonts are used for body text, and sans serif fonts for headings. You'll find exceptions to this common practice, of course. Among the fonts that come with WordPerfect, Roman-WP (also called Times or Times New Roman) is a serif font, and Helve-WP (also called Helvetica or Arial) uses a sans serif design.

Most fonts these days use *proportional spacing*, meaning that different characters take up different amounts of space on the page. A lowercase *i* is almost always narrower than an uppercase *W* in such a font, for instance. Most mechanical typewriters, by contrast, use a *monospaced* font, in which all characters are the same width. Among the fonts shipped with WordPerfect, Roman-WP and Helve-WP are both proportionally spaced, whereas Courier-WP (also called Courier or Courier New) is monospaced.

Tip #78 from
Tux Tips

All other things being equal, a proportionally spaced font generally looks better for large amounts of text. You may want to use a monospaced font for text that needs consistent alignment, such as computer code or certain types of tables (though WordPerfect's table features can align tabular information even with proportionally spaced fonts).

→ If you want more information on tables, **see** "Tables," **p. 275**

Fonts typically come in *families*, which are collections of similar designs intended to be used together. For instance, many font families include variant styles for italics or boldface. Depending on how the font is installed in WordPerfect, you may access different members of a font family either by selecting attributes of the *base* font of the family or by selecting an entirely different font name.

→ For more information on configuring WordPerfect to use font families, **see** "Installing and Managing Fonts," **p. 754**

You can also use fonts in various sizes. Depending on the printer you use, many, most, or all of the fonts you can use in WordPerfect will be *scalable*, meaning that the computer or printer can display them at any size within the limits of the technology. When you select a scalable font, you specify its size in points. *Points* is a term from the printing world, and the important thing to know is that most text you read is set somewhere between 9 and 12 points. For most fonts and uses, you'll want to use sizes in this range. You might want to use larger fonts for headlines or flyers, though, or smaller text for putting lots of information on small labels.

Note

Technically, a font is a design of a specific size. For instance, Times 12 point and Times 14 point are different fonts. These fonts, and many others, are examples of Times *typeface*—the overall design of the characters. Because of the way computer font technology evolved, however, the term *font* has come to be synonymous with *typeface* in the computer world. WordPerfect's menus and dialog boxes all refer to fonts rather than typefaces, a convention that this book follows.

Understanding WordPerfect's Font Model

The first thing to know about WordPerfect's handling of fonts is that it's based around two different types of fonts: *printer fonts* and *graphics fonts*. The former are fonts that are *rendered* by the printer—that is, the printer decides precisely which parts of the paper will be black (or some other color, when using a color printer) and which will be white. Printer fonts can be built into the printer, or they can be Adobe Type 1 fonts downloaded to the printer from WordPerfect. Unless you install Type 1 fonts as printer fonts, the only printer fonts on your system are the ones built into your printer.

→ For more information on installing fonts in WordPerfect, **see** "Installing and Managing Fonts," **p. 754**

Note

If you're like most Linux users, you have a non-PostScript printer. Most Linux programs assume they'll be printing to a PostScript printer, though, and so Linux comes with a utility called *Ghostscript (p. 380)* that converts PostScript to the printer's "native tongue." WordPerfect is more flexible than this, but you do have the option of printing PostScript to your non-PostScript printer from WordPerfect. This discussion considers Ghostscript to be part of your printer. "Downloading a font to the printer," in this case, means that Ghostscript processes the font and passes the results on to the printer.

→ For more information on using non-PostScript printers with WordPerfect, **see** "Setting Up Non-PostScript Printers," **p. 742**

Graphics fonts are, like downloadable printer fonts, Adobe Type 1 fonts. WordPerfect renders them, however, and sends either a bitmapped version of the font or a bitmap of the entire page to your printer.

→ To learn more about font rendering options, **see** "Printer Options," **p. 155**

Built-in printer fonts have the advantage of being fast to print. They may not be the same across printers, however. For instance, printers that use the PCL language generally provide fonts called Times New Roman and Arial instead of the Times and Helvetica of PostScript printers. These fonts are similar despite having different names, but they aren't identical. If you write a document using a built-in printer font on one computer or with WordPerfect set up for one printer, your document's formatting may change if you load it on another computer or try printing to another printer.

Downloadable printer fonts print consistently from one printer to another, but only if both printers are PostScript printers. If you try to print using a non-PostScript printer, the downloadable font won't even be available, so WordPerfect has to guess at what font to use instead and may guess something very strange.

Caution

If your PostScript printer doesn't have much memory, it may run out of memory if you try to print with several downloadable fonts.

Graphics fonts work best when you're not sure what printer you'll be using. To make best use of your printer's resolution, however, WordPerfect has to know about your printer's capabilities when it prints. You may be forced to use a WordPerfect driver for another model, though, especially if you're using Ghostscript to print, and WordPerfect may therefore produce bitmaps of a lower resolution than your printer can handle if you use graphics fonts.

➔ To learn more about installing and configuring printer drivers, **see** "Installing and Managing Fonts," **p. 754**

WordPerfect displays different icons in font menus to identify printer and graphics fonts, as in Figure 7.1. Unfortunately, the menus don't distinguish between built-in and downloadable printer fonts, so you have to know which fonts your printer includes, or whether you've installed any downloadable fonts, to know which is which.

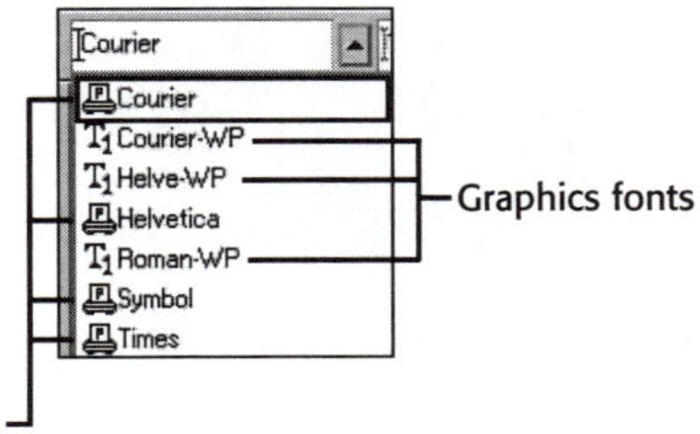

Figure 7.1
WordPerfect identifies printer fonts with a printer icon and graphics fonts with a T1 icon.

Applying Font Codes

You can select a font in any of a number of ways in WordPerfect:

- Selecting the font from the font selector in the default toolbar (shown in Figure 7.1).
- Selecting the font from the QuickFonts selector in the default toolbar (Figure 7.2). QuickFonts shows only the last few fonts you've used, but it shows them graphically and includes font attributes you've applied to the fonts, such as their size. This feature can make it easy to set a number of font characteristics at once.

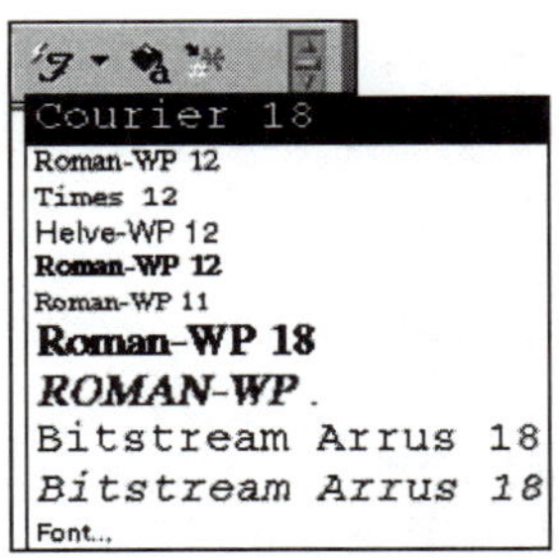

Figure 7.2
If you don't see the font you want, select the Font item from the bottom of the menu to bring up the Font dialog box.

- Choosing Format, Font (F9). This step brings up the Font dialog box (Figure 7.3), which lets you select the font and many font attributes.

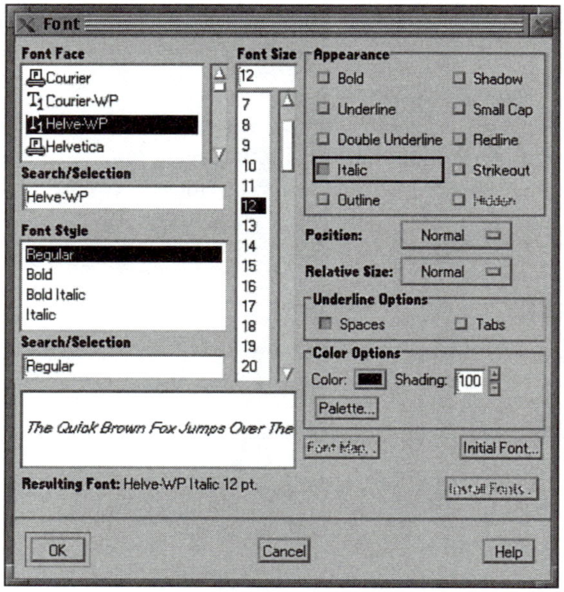

Figure 7.3
You can control just about any aspect of a font you want from the Font dialog box.

- Right-clicking your mouse button and selecting Font from the pop-up menu. This sequence also brings up the Font dialog box in Figure 7.3.

The Font dialog box contains many options, to the point of being overwhelming. Its features include

- Font Face—This box lists the available fonts. Click the font to select it. If you double-click a font name, WordPerfect behaves as if you'd selected the font and clicked OK.
- Font Style—This box lists different font styles from the selected font face's family, usually some combination of Regular, Bold, and Italic or Oblique, though there may be others. You don't usually select a style in this way, however. Instead, you select attributes by choosing the Regular style of the font and then applying bold or italic attributes to it.
- Search/Selection—Both the Font Face and Font Style lists have an associated Search/Selection field. You can type a string to have WordPerfect search for it in the associated list. This feature can be handy if you have lots of fonts installed and know the name of the font you want to use.
- Font Size—Select the point size of the font you want to use. Either click the size you want or enter it into the text entry field.

Tip #79 from *Tux Tips*

You can enter fractional font sizes, such as 12.4 points, by typing the values directly into the field at the top of the Font Size selector.

- Appearance—Click any of the check buttons associated with a font attribute you want, such as Bold or Shadow. WordPerfect either uses the appropriate variant font style for that family or generates something close from the versions it does have.
- Position—Use this button if you want to use superscripted or subscripted text.
- Relative Size—You can enter a size change relative to the main font size by using this menu button. If you enter a size change in this way, the text retains its relative size even if you subsequently change the size of the main text.
- Underline Options—Click the appropriate buttons if you want spaces or tabs to be underlined when you use the underline attribute.
- Color Options—You can select a color for your text by choosing it from the pop-up palette you get when you click the rectangle labeled Color. If that doesn't give you enough options, you can fine-tune a color by selecting Palette, which produces the Define Color Printing Palette dialog box shown in Figure 7.4.

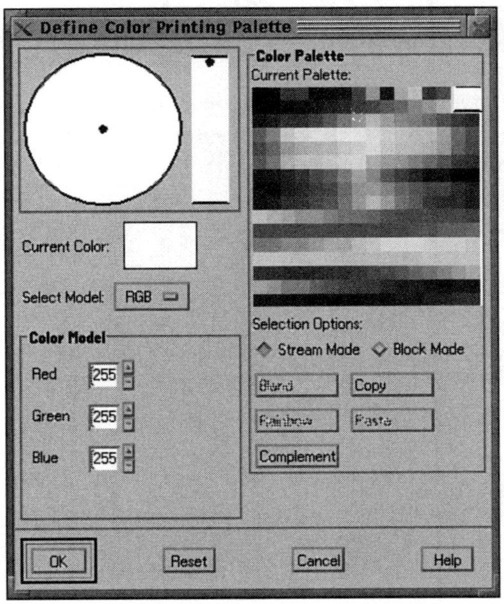

Figure 7.4
You can specify the exact color in any of several color models.

- Initial Font—WordPerfect starts a document with a default font. This font is normally set by your printer driver configuration and is normally Courier or a Courier-like font at 12 points, but you can override this setting from the Document Default Font dialog box you get by clicking Initial Font from the Font dialog box. This subset of the Font dialog box lets you select the font face, font style, and font size.

Tip #80 from *Tux Tips*

Setting the initial font is a good way to change the font for the entire document. You can also change the font by simply inserting a font-change code at the beginning of the document or by selecting the entire text and changing the font that way. These methods, however, may miss fonts in certain embedded objects such as text boxes. Selecting the entire text and setting the font that way also overrides any font changes you've applied to short sections of text.

➔ If you want to know more about text boxes, **see** "Using Text Boxes to Add Variety to Text Layout," **p. 357**

- Resulting Font—WordPerfect displays the name, style, and point size of the font you've selected. The font preview box displays a sample of the chosen font.

Tip #81 from *Tux Tips*

The font preview field is particularly handy if you know what the font you want to use looks like but can't remember its name. Simply select promising-sounding font names and check the preview area until you find the font you want.

WordPerfect provides two types of font-change codes. Figure 7.5 shows the codes in WordPerfect's Reveal Codes pane, as well as the resulting changes in the main editing area. (Choose View, Reveal Codes [Shift+Ctrl+F3] to start Reveal Codes mode.)

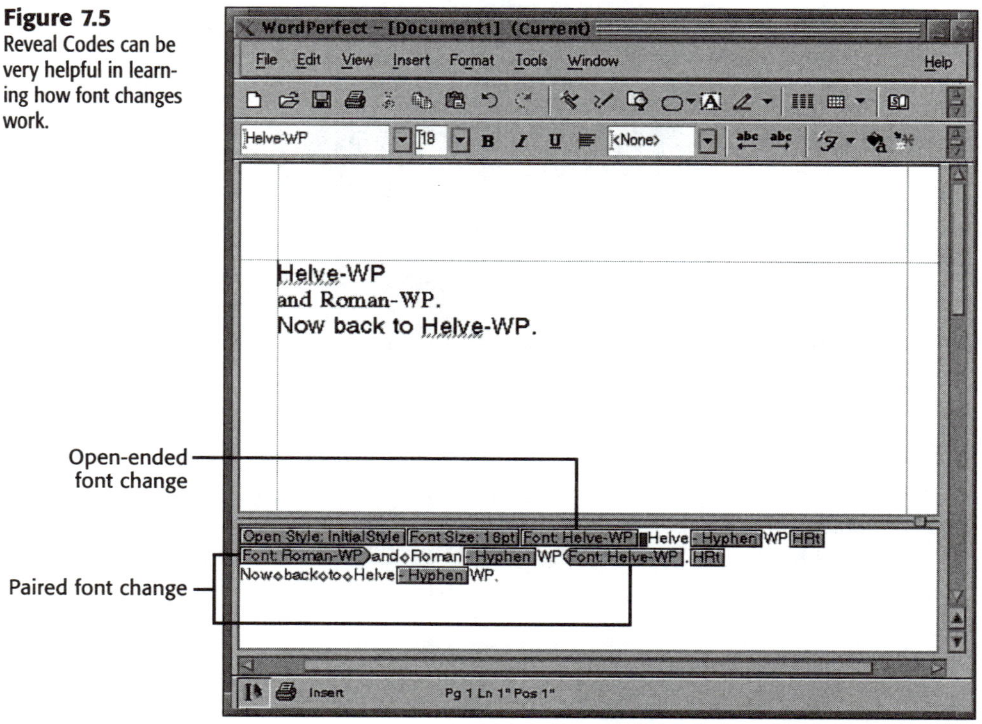

Figure 7.5
Reveal Codes can be very helpful in learning how font changes work.

→ If you want to know more about Reveal Codes, **see** "Reveal Codes Mode," **p. 63**

The first type of font change is an open-ended change, represented by the rectangular font code at the beginning of the text in Figure 7.5. In this case the code sets the font for the document to Helve-WP. Open-ended codes work until another font-change code comes along. WordPerfect inserts a font-change code if you choose a new font when no text is selected and the insertion point is between words or at the end of the document. The code is inserted at the insertion point and takes effect immediately.

The second type of font change is a paired change, as illustrated by the section of text in Figure 7.5 in the Roman-WP font. A paired change, unlike an open-ended change, applies to a limited range of text. WordPerfect uses this type of font-change code when you select some text and then choose a new font. Only the selected text takes on the new font formatting. If you make a font change when the insertion point is in the middle of an unselected word, WordPerfect acts as though that word were selected and applies a paired font change to that word alone.

Note

Unlike most other paired codes, the closing code of a paired font change shows the formatting that applies *after* the closing code. Other paired codes show the formatting in effect between the codes on both sides of the affected area.

FONT ATTRIBUTES

You can apply a number of font attributes to your text to lend it emphasis or otherwise distinguish it. The most complete way to access these attributes is through the Font dialog box shown in Figure 7.3, from which you can set attributes ranging from font size to italics to colors.

For the most part, the various font attributes have their own associated WordPerfect codes. If you want to italicize some text and make it green, for instance, WordPerfect inserts two codes, one for each attribute. You can subsequently delete or modify one attribute alone, though any modification brings up the complete Font dialog box again.

Many font attributes have only paired code versions; you cannot create an open-ended [Bold] font code, for instance, except as part of the main font definition by specifying a bold font style—and that method works only if the font comes with an explicit bold family member. In particular, the attributes in the Appearance section of the Font dialog box, as well as in the Position and Relative Size sections, come only in paired code form. Therefore, to apply such attributes to an entire document or to all of a document past a given point, you must highlight that region and select the code, rather than simply placing the insertion point at the beginning of the affected area and selecting the new format.

Bold, italic, and underline are used so frequently that these attributes have their own keyboard shortcuts and toolbar items. Press Ctrl+B for bold, Ctrl+I for italic, and Ctrl+U for underline. When you're entering new text, these keystrokes serve as toggles, turning the features on and then off.

MODIFYING FONT CODES AFTER THE FACT

If you've entered font codes into a document and then want to modify those font choices, you have a number of options, depending on the nature of the font codes you want to change:

- Choose Format, Font (F9) and click Initial Font to change the default document font. This step doesn't enter any codes you can see in Reveal Codes, but you can see the effect of your change. If you've changed the font by entering font codes, this procedure won't affect these changes.
- Set a font used by the entire document in the initial document style. Go to the top of the document and open Reveal Codes mode by choosing View, Reveal Codes (Shift+Ctrl+F3). Locate the code at the very top of the document called [Open Style: InitialStyle] and double-click it. This sequence brings up the Styles Editor (Figure 7.6). The Contents box may contain a font change, which you can edit by double-clicking it.

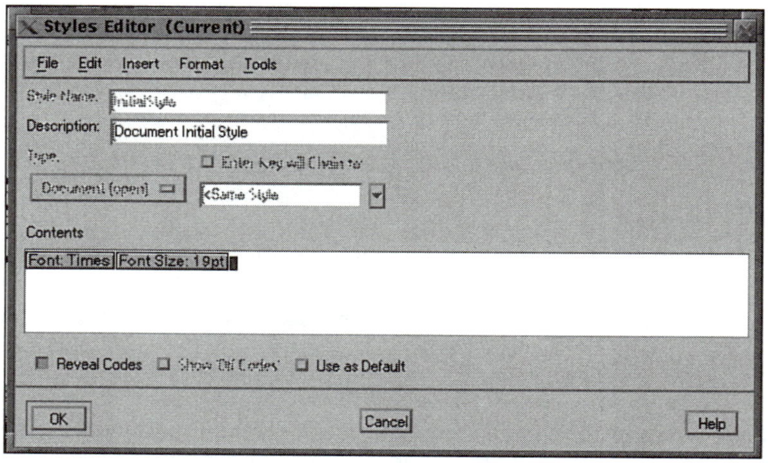

Figure 7.6
You can set up any formatting you like as an initial style.

➔ To learn more about editing styles, **see** "Editing a Style" **p. 321**

- Use Reveal Codes mode to locate the area of text where you want the font change to occur. Find the font code responsible for the change and double-click it. The Font dialog box opens, and you can alter the font from here.

Caution

If you are implementing a paired font change, be sure to select the *opening* font change; otherwise, you'll change the font that takes effect after the text in the font you want to alter.

- Select the text you want to change and apply a new font to it. This option creates a paired font-change code. If you're overriding a previous font change, this step may also delete those existing codes.

Note

Unless you select your text carefully, you may leave some text in the old font with this technique. It's easy to miss spaces and punctuation, for instance. Of course, if you missed it on screen, the difference when printed probably won't be very substantial, but it might be. Even a small inconsistency in your use of fonts can make your document–and you!–look bad in certain cases (resumes, for example), so you should use the Reveal Codes methods of font editing when consistency is important–or better yet, use styles.

TYPESETTING COMMANDS

The Format, Typesetting menu contains several items that can help you fine-tune the appearance of your text by adjusting spacing and precise character placement. Chances are you won't need to use these features very often unless you're doing desktop publishing with WordPerfect for Linux, but when you do need them, they're very helpful.

ADVANCE

Choosing Format, Typesetting, Advance displays the dialog box shown in Figure 7.7. You use this dialog box to position elements on the page precisely. For instance, suppose that you need to insert exactly 2.2 inches of horizontal space between two characters. If you're used to working with typewriters, you might do so by using the spacebar or possibly the Tab key. These methods might not work if other formatting elements, such as the font size, change. So to insert this precise amount of space, you can call up the Advance dialog box, click the Right From Insertion Point radio button, and then enter 2.2 into the Horizontal Distance spin box. The next character you type will be precisely 2.2 inches from the last, and the two will maintain this distance even in the face of other formatting changes.

Figure 7.7
Some of the Advance dialog box items are unavailable until you select an appropriate advance option.

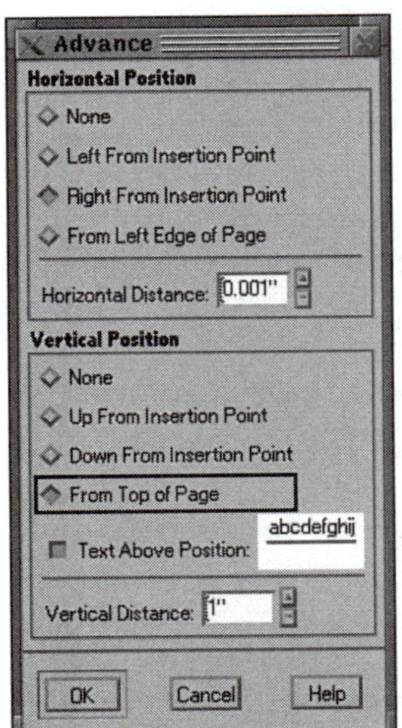

PART II
CH 7

You can use the Advance dialog box to position text changes left, right, up, and down from the current insertion point, as well as absolute positions measured from the top and left edges of the page. When you use the From Top of Page option, WordPerfect must measure the text's position from the top of the page, which introduces an ambiguity: Should the distance measurement be taken from the top or the bottom of the text itself? The Text Above Position check button allows you to toggle these two possibilities, as illustrated graphically to the right of the check button.

Tip #82 from *Tux Tips*

The graphic of text above or below the position is sometimes out of sync with its setting when the dialog box starts. Try toggling the associated check button a couple of times to get it synchronized again.

OVERSTRIKE

The Format, Typesetting, Overstrike command brings up the Overstrike dialog box shown in Figure 7.8. The purpose of this dialog box is to create a compound character by printing one character and printing a second one atop the first. For instance, you could create the not-equal-to mathematical symbol (≠) by overstriking the = and / characters. Simply type the characters you want to overstrike one after another and click OK to see the results. You can also insert various character formatting commands. Click the Format button to see a list of attributes that you can apply to one or more of the characters in the overstrike combination. Options include bold, underlining, and relative font size changes.

Figure 7.8
You can use the overstrike feature to create special mathematical symbols, compound characters, and other effects.

Tip #83 from *Tux Tips*

The Overstrike dialog box includes a memory. Click the Previous and Next buttons to see a history of the overstruck characters you've used. The history is handy when you want to reuse a symbol in a document.

If you repeatedly need to overstrike the same characters, consider creating a macro to do the job.

→ If you want to know more about macros, **see** "Letting the Computer do the Work: Macros," **p. 483**

After you use the overstrike feature to create a character combination, it appears in your document as a single code. You can delete it without Reveal Codes mode open, however, unlike most codes.

Tip #84 from *Tux Tips*

In many cases, you can create an effect similar to or better than an overstrike by using the Symbols dialog box. Overstrike gives greater flexibility, however, and if you need your symbol to match the font of the surrounding text, Overstrike can meet that need, whereas Symbols may not.

➔ To learn more about the Symbols dialog box, **see** "Inserting Special Characters," **p. 102**

PRINTER COMMAND

The Format, Typesetting, Printer Command item is a real "blast from the past"—it's a feature that made sense in the days of text printing on dot-matrix or daisywheel printers when even top-of-the-line word processors didn't support all of a printer's features. In those days, you'd use this option to enter the special printer codes from your printer manual into WordPerfect, and it would send them on to the printer unchanged. If you're using a modern laser printer or inkjet, you'll probably have no need for this feature. You might want it if you have some obscure printer for which no WordPerfect or Ghostscript drivers exist, though.

WORD/LETTER SPACING

If you choose Format, Typesetting, Word/Letter Spacing, you get the Word Spacing and Letterspacing dialog box shown in Figure 7.9. You can use this dialog box to fine-tune the spacing between characters and lines without adjusting the font size.

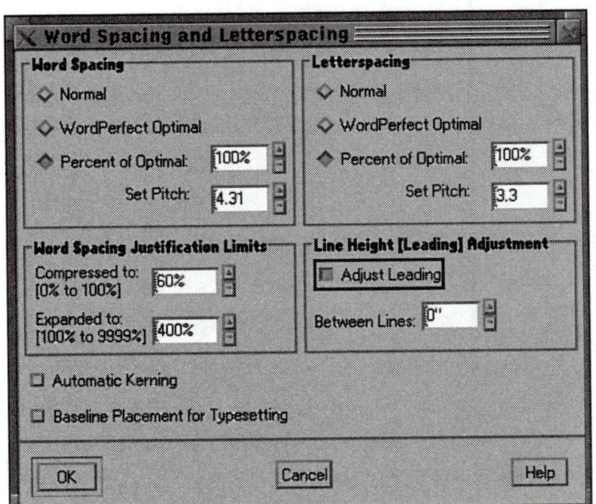

Figure 7.9
Squeeze or stretch your text with the Word Spacing and Letterspacing dialog box.

The Word Spacing and Letterspacing sections of this dialog box provide similar options and do similar things. The difference is that the Word Spacing options apply to the horizontal spaces between words, whereas the Letterspacing options apply to the horizontal distances between letters within a word. In both cases the WordPerfect Optimal setting is the default and specifies that WordPerfect will decide how to space characters. If you select Normal,

WordPerfect uses the letter spacing specified by the font file. To expand or shrink the normal spacing, click Percent of Optimal. You can then enter new spacing settings, expressed either as a percentage of the WordPerfect Optimal spacing or in terms of character pitch (these two values are linked; they're simply two different ways of measuring the same thing). Adjust the percentage up to widen your character spacings or down to narrow them.

Tip #85 from *Tux Tips*

You probably won't see much difference between the Normal and WordPerfect Optimal spacing settings. If you're particular about your spacings, though, I suggest you try both settings with whatever font or fonts you intend to use; neither setting is guaranteed to be the best in all circumstances.

To create fully justified text, WordPerfect can increase or reduce space between words. The Word Spacing Justification Limits section allows you to tell WordPerfect just how far to go to attain this effect. I've not observed this option to have any noticeable effect, however.

The Line Height [Leading] Adjustment area allows you to increase the height of lines of text. This setting doesn't affect the size of your fonts, however; it really adjusts the spacing between lines. You can achieve a similar result by setting the line spacing.

→ To learn more about setting the line spacing, **see** "Line Spacing," **p. 199**

Checking the Baseline Placement for Typesetting button alters the way WordPerfect places the first line of text on the page. Normally, WordPerfect places this line so that the top of the text is just below the top margin of the page. Checking this option, however, places the *baseline* of the text (the bottom portion of most characters) at the top margin.

If you check the Automatic Kerning button, WordPerfect automatically shrinks the spacing between certain pairs of letters. Any letter can be described as fitting inside a rectangular box, and normally these boxes are placed so that they don't overlap. With some letter pairs, however, such as the uppercase *A* and *V*, this placement makes the letters look too far apart. This problem is particularly noticeable at large point sizes. Reducing (or increasing) the space between particular letter pairs is known as *kerning*, and WordPerfect can do this automatically.

Note

You won't see the effect of automatic kerning on the screen, but it will appear in your printout.

MANUAL KERNING

Kerning is such an important issue for certain situations that WordPerfect provides a way to do it manually. Place your insertion point between the two letters you want to kern and choose Format, Typesetting, Manual Kerning to bring up the dialog box in Figure 7.10. You can specify the amount of kerning in several units of measure. As Figure 7.10 shows, manual kerning is visible on the screen, unlike automatic kerning.

Figure 7.10
The unkerned word (top) looks stretched out, but the manually kerned version (bottom) looks better, thanks to the kerning codes inserted between letters.

JUSTIFICATION

Justification and *alignment* refer to where text begins and ends on a line, particularly in relation to the margins and other lines. You can use this feature to create text that's neatly and appropriately formatted for your document. Some subtle interactions occur between these options, however, so be careful!

TYPES OF JUSTIFICATION

WordPerfect offers several types of justification, as shown in Figure 7.11.

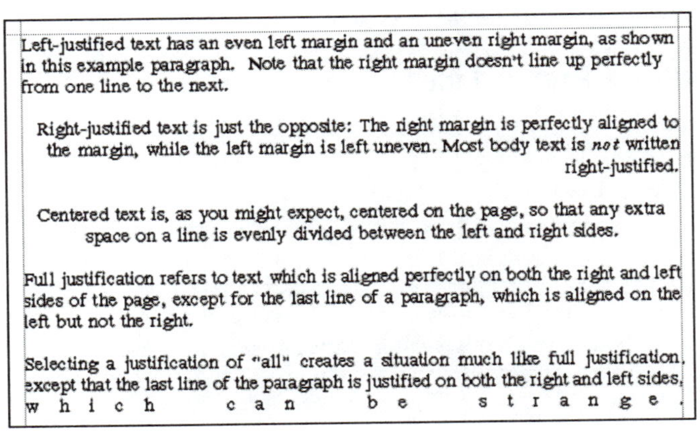
Left-justified text has an even left margin and an uneven right margin, as shown in this example paragraph. Note that the right margin doesn't line up perfectly from one line to the next.

Right-justified text is just the opposite: The right margin is perfectly aligned to the margin, while the left margin is left uneven. Most body text is *not* written right-justified.

Centered text is, as you might expect, centered on the page, so that any extra space on a line is evenly divided between the left and right sides.

Full justification refers to text which is aligned perfectly on both the right and left sides of the page, except for the last line of a paragraph, which is aligned on the left but not the right.

Selecting a justification of "all" creates a situation much like full justification, except that the last line of the paragraph is justified on both the right and left sides, w h i c h c a n b e s t r a n g e .

Figure 7.11
WordPerfect offers a full range of justification options.

- Left—Left-justified text is aligned so that every line begins at the left margin and ends wherever WordPerfect can create the final break between words prior to the right margin. Thus, the right margin in left-justified text is uneven, and this format is sometimes referred to as *ragged right*. In general, left-justified text is the easiest to read. This alignment keeps interword and intercharacter spacing constant, and we're trained from childhood to read from left to right with text aligned on the left (in languages where writing is done from left to right, of course).
- Right—Right-justified text is aligned evenly along the right margin, and the left margin is ragged. You'll probably want to use this setting mainly for short pieces of text and individual lines, such as return addresses on letters or page numbers in page headers.

→ To learn more about page numbers, **see** "Headers, Footers, and Page Numbers," **p. 235**

- Centered—Centered text is aligned so that the text falls equally distant from both left and right margins. You'll probably use it mostly for headings and nontext elements such as figures that you want to center on the page.
- Full justification—This term refers to text that's aligned to both the left and right margins, except for the last line of a paragraph, which is aligned only on the left. This effect is difficult to achieve with a conventional typewriter and makes the page as a whole look good; however, full justification can also make the page more difficult to read because interword and even interletter spacing changes from line to line.
- "All" justification—The justification setting of All is very much like full justification except that the final line of a paragraph is aligned to both the left and right margins. I don't know of any situations in which this effect is used for regular paragraphs because it can be quite ugly. You might want to use it if you're forced to manually insert a page or line break in the middle of a paragraph to control your document's overall layout. WordPerfect does provide other ways to control page breaks and layout, however.

You can achieve most of these formatting effects in two ways from within WordPerfect. Each method has a slightly different result.

Justifying Lines or Partial Lines

First, you can choose Center (Shift+F7), Flush Right (Shift+Ctrl+F7), or Flush Right with Dot Leaders from the Format, Line menu. These options set up centering and right justification. The Flush Right with Dot Leaders option adds a series of dots from the left margin or any text on the left up to the text that's to appear right-justified.

Tip #86 from *Tux Tips*

Dot leaders are handy for directories and similar listings, such as times associated with tracks on a CD or cassette. Leaders provide a visual line to connect the right-justified item with its associated left-justified text.

Tip #87 from *Tux Tips*

You can toggle dot leaders on and off by pressing the keyboard shortcut for the center or flush right commands. For instance, to create dot leaders for centered text, press Shift+F7 twice. The dot leaders appear, and the insertion point moves to the middle of the line so that you can type your centered text.

Setting centering or right justification by using the Format, Line menu items creates an effect that applies to a line or part of a line. The moment you press the Enter key or type too much text to fit on that line, formatting for subsequent text falls back to what it had been before.

Tip #88 from *Tux Tips*

You can have one line with left-justified, centered, and right-justified text. Start with left justification; type some text; choose Format, Line, Center; and then type some text to be centered. Finally, choose Format, Line, Flush Right and type your right-justified text. This method is a convenient way to put a lot of information in a short space at the top of a page, such as home and work addresses on a resume or your name and class on a one-page school paper.

You can create some centered text and then continue without further centering by choosing Format, Line, Other Codes. From the Other Codes dialog box, shown in Figure 7.12, select End Centering/Alignment and click Insert. Any text you type after this point will not be centered *per se*, though it will appear to the right of the existing centered text.

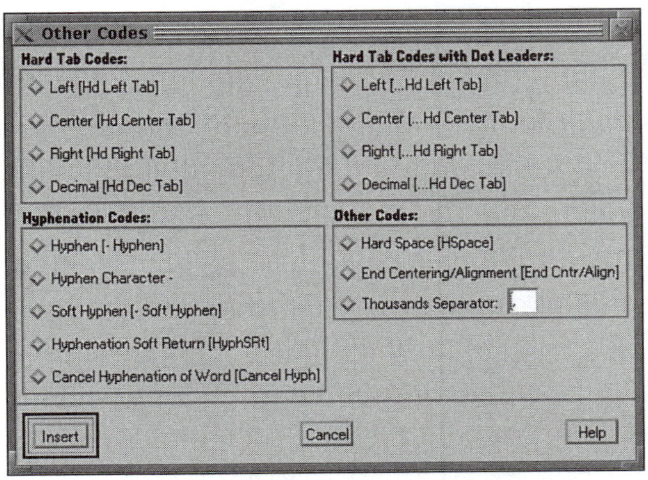

Figure 7.12
You can enter miscellaneous line-oriented formatting codes with the Other Codes dialog box.

Justifying Paragraphs and More

The second means of aligning text uses options from the Format, Justification menu: Left (Ctrl+L), Right (Ctrl+R), Center (Ctrl+E), Full (Ctrl+J), and All. The lower default toolbar contains a tool to select these justification methods, as well. These methods of justification, like font selection, come in two varieties: an open-ended form and a code pair form. If you engage one of these justification options without selecting any text, WordPerfect inserts an open-ended code for that format at the start of the current paragraph and applies that code until the next justification code. If you've selected some text, WordPerfect applies the code in paired form to the current paragraph.

Note

The fact that WordPerfect applies these codes to paragraphs makes them paragraph formatting commands. I discuss them here because of their similarity to the line-oriented justification commands.

Note

If you select more than one line of text and then apply a line-oriented justification command, WordPerfect substitutes a paired paragraph formatting command. Because there's no dot leader option for these commands, you can't apply that style to an entire paragraph or even to several separate lines in one operation.

In many situations, you can use either a line- or a paragraph-oriented justification command. For instance, either will do to center a one-line heading. If your heading is more than one line, you may prefer to use the paragraph-oriented codes. You could also create a centering style to ensure that the text following a temporary justification change isn't inadvertently affected by it. Creating a centering style also allows you to add additional formatting features to whatever you need to center.

→ To learn more about styles, **see** "Ensuring Consistency Within Your Document with Styles," **p. 305**

When formatting your text, keep in mind that the paragraph-oriented commands take precedence over the line-oriented commands. For instance, if you create a right-justified paragraph, you can't then use the line-oriented center command to center a line. Instead, you must revert to the paragraph-oriented right- or full-justification and center, or use the paragraph-oriented center method.

One other difference between the line- and paragraph-oriented alignment commands is that you can delete the line-oriented codes without having Reveal Codes mode open. To delete a paragraph-oriented code, you must use Reveal Codes or override it with another paragraph-oriented code.

TABS

Tabs can be an extremely useful way to position text on a line. You can use tabs to indent a paragraph, to offset a textual element such as a letter's return address, to separate and align components of a list, or to do various other things. WordPerfect for Linux supports a variety of tab types and ways to manipulate them.

Types of Tabs

The word *tab* can actually refer to several different things. For the most part, context will clarify which of these meanings is correct:

- The Tab key on your keyboard.
- A [Tab] code in the document.
- A tab position (or "stop") on the ruler. Setting a tab position determines where a [Tab] code places the insertion point.

WordPerfect supports several varieties of tabs. The different tab types handle text alignment and formatting in different ways:

- Left—Left tabs are the type available on conventional typewriters. When you press the Tab key, the insertion point moves forward to the position of the tab stop. Whatever you type subsequently is left-aligned to the tab stop.
- Center—When you use a center tab, text typed after the tab is centered around the position of the tab stop.
- Right—If you type text after a right tab, the text is right-aligned to the tab stop.
- Decimal—You use a decimal tab to align decimal numbers by their decimal points; the decimal point lies at the tab position.

Tip #89 from *Tux Tips*

> Despite the name, decimal tabs can be used with characters other than a decimal point serving as the alignment symbol, so you can align text around vertical bars, commas, slashes, or anything else that's appropriate.

- Leaders—Each tab type can be used with or without leaders. Leaders are typically dots (periods) but can be other characters.

WordPerfect starts with a left tab every 1/2 inch by default. You can add, remove, or change these tabs easily.

Creating Tab Stops with the Ruler

One of the most intuitive ways to change tabs is with the WordPerfect ruler. Choose View, Ruler to display it. Figure 7.13 shows a ruler with various tab styles. Note that the orientation of the triangles indicates the tabs' alignment, a dot in the center of a triangle indicates a decimal tab, and three dots under the triangle indicate a tab with a leader.

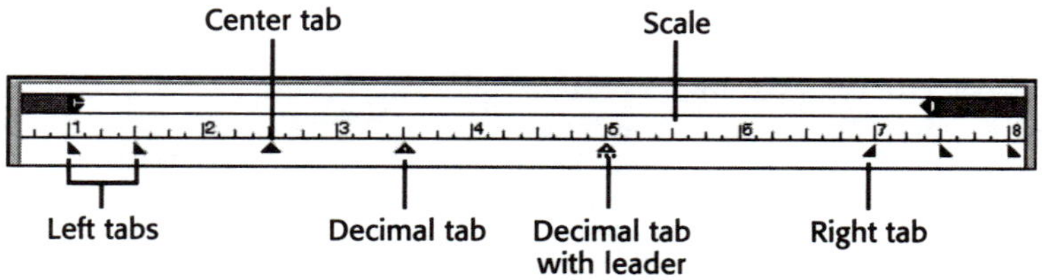

Figure 7.13
The ruler allows you to set tabs and margins graphically.

You can perform a number of operations on your tab stops with the mouse:

- To remove a tab stop, click it and drag it downward. When your mouse pointer dips below the ruler bar and into the text editing area, the triangular tab disappears. At this point, you can release your mouse button, and WordPerfect deletes the tab stop. If you change your mind about deleting the tab before you've released the mouse button, just move the tab back into position and it will remain in place.
- To move a tab, click it and drag it to a new position; then release the mouse button. Be sure to keep your mouse pointer over the tab section of the ruler, just below the ruler's scale but not in the window's text editing area. If your mouse dips below the ruler bar, WordPerfect interprets your action as a request to delete the tab stop.
- To insert a new tab stop, click the left mouse button on an unoccupied part of the ruler. Your new tab will burst into existence there with the same style as the last tab you clicked or created.
- To set a new type of tab for insertion, click the right mouse button anywhere in the tab area of the ruler. You'll get a list of tab types (Figure 7.14). Move your mouse pointer to the tab type you want and click it. When you next left-click in the tab area, WordPerfect creates that type of tab for you.

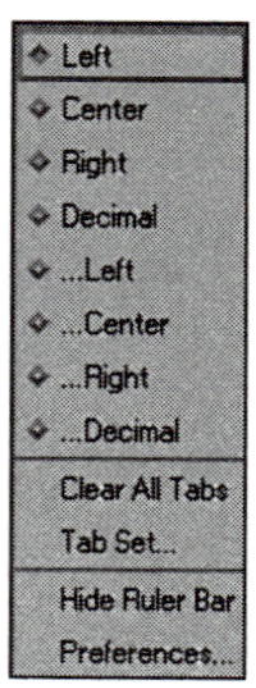

Figure 7.14
You can select the type of tab you want to insert by right-clicking in the tab portion of the ruler bar.

Tip #90 from *Tux Tips*

To change the type of a tab, you must delete it and insert a new tab of the type you want.

Note

When you change a tab stop, WordPerfect inserts a [Tab Set] code at the beginning of the current paragraph. If you've selected any text within that paragraph, the code is a paired code applied to the entire paragraph. Otherwise, it is an open-ended code and will apply to all subsequent paragraphs, as well as the current one.

You can set various options relating to tab setting with the ruler bar by right-clicking the ruler and choosing Preferences from the pop-up menu. This step brings up the Display Preferences dialog box as shown in Figure 7.15. If you use the ruler bar frequently, you can set it to display by default on new documents. If you check the Tabs Snap to Ruler Bar Grid button, any tabs you create or move jump to the nearest ruler grid marker, which can help you create consistent tab markers; however, you cannot fine-tune their positions. Ruler bar guides are vertical lines that extend through the Document window when you position a tab. They can help you align tabs with features in your text.

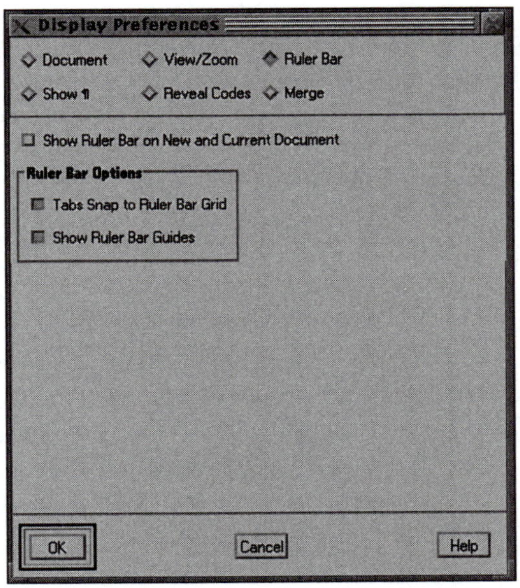

Figure 7.15
Changes you make in the Display Preferences dialog box are in effect until you change them again.

CREATING TAB STOPS WITH DIALOG BOXES

Although working with tabs graphically by using the ruler bar is simple and intuitive, it's relatively imprecise and can be tedious if you want to create a set of tabs at consistent locations. Dialog boxes aren't as simple to use, but they give you more options for setting tabs. You can open the Tabs dialog box by choosing Format, Line, Tab Set (Figure 7.16).

Figure 7.16
You can create finely tuned tab types with the Tabs dialog box.

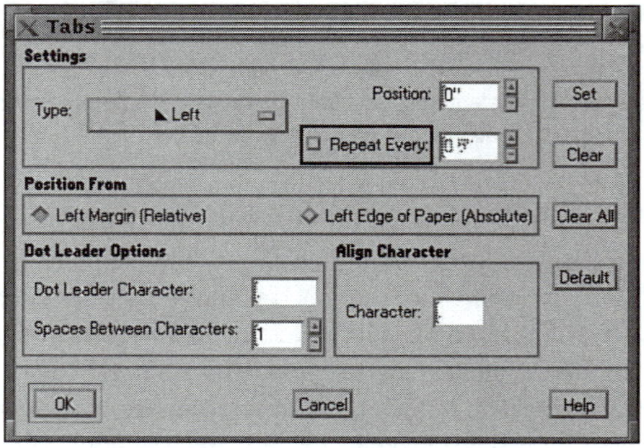

To set a tab with this dialog box:

1. Select the type of tab you want to insert from the Type option button.
2. Enter the position of the tab in the Position spin box. If you want to measure this position relative to the left margin, be sure the Left Margin radio button is active. If you want to measure the tab position relative to the left side of the page, be sure the Left Edge of Paper radio button is selected.
3. If you want to repeat this tab at a regular interval, select the Repeat Every check button and enter a value in the spin box next to it.
4. If your tab stop uses leaders, you can change the character used for the leader by typing it in the Dot Leader Character field. (It's hard to see, but a lone period is in there by default.) You can also adjust the spacing between the leader characters by using the Spaces Between Characters spin box.
5. If your character is a decimal tab, you can select a different alignment character by entering it in the Character field of the Align Character section of the dialog.
6. Click Set.

You can also clear one or all the tabs by clicking Clear or Clear All, respectively. (Clear works only if you enter the correct tab position in the Position box, though you don't need to set the type or other options.) If you click Default, WordPerfect restores the default set of tabs (one left tab every 1/2 inch).

Tip #91 from *Tux Tips*

If you set WordPerfect to display the ruler before you start the Tabs dialog box, your changes are reflected in the ruler as you make them in the dialog box. This visual clue can help you determine whether you're making the changes you intended.

Tip #92 from *Tux Tips*

> You can use any of several units of measure when you set a tab. Replace the quote character (for inches) in the Position box with *c* for centimeters, *m* for millimeters, or *p* for points.

USING TABS

Using tabs is fairly straightforward—simply press the Tab key on your keyboard, and WordPerfect repositions the cursor to the next horizontal tab stop and adds any necessary dot leaders.

Tip #93 from *Tux Tips*

> If you reformat the text prior to the tab, what the "next" tab stop is may change, and so your tabbed text may change, too. To reduce the potential for disruption, I recommend setting your tabs so that you need to use only one to get your desired effect; that way, text reformatting is less likely to disrupt a multiple-tab sequence by eliminating or adding the need for a tab.

You can also insert a tab of a type that's not shown on the ruler by choosing Format, Line, Other Codes and selecting the appropriate tab type from the Other Codes dialog box (Figure 7.12). This technique does not change what's on the ruler and can be useful for making a one-time change to the type of tab.

Tip #94 from *Tux Tips*

> You can cut and paste an open-ended tab setting code by using Reveal Codes (View, Reveal Codes) to display the Tab Set code. Select that code in the Reveal Codes window and choose Edit, Copy (Ctrl+C) to copy the code to the Clipboard. Then reposition the insertion point to where you want to apply the new code and choose Edit, Paste (Ctrl+V). This procedure does not work with paired tab setting codes, however, and you can't use it to apply the open-ended code to a selected portion of text.

LINE SPACING

WordPerfect provides two means for you to adjust line spacing in your document. The first is to choose Format, Line, Spacing, which displays the Line Spacing dialog box shown in Figure 7.17. In this dialog box, you can enter the space between lines in line-height multiples, much as you would with a conventional typewriter. You can, however, enter spacing to increments of 0.1 line. (You may think you can enter greater precision by typing it directly, but WordPerfect rounds your entry to the nearest 0.1-line increment.)

Figure 7.17
The Line Spacing dialog box shows a graphical approximation of the results of your spacing adjustments.

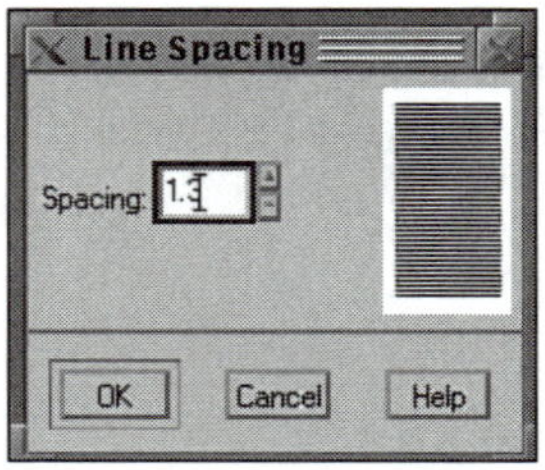

The second means of adjusting line spacing is via the Line Height dialog box, accessed from the Format, Line, Height menu item. This dialog box is shown in Figure 7.18. If you select the Auto radio button, WordPerfect uses the font's point size to determine the height of each line of text. You can fine-tune this value, however, by clicking the Fixed radio button and entering the height you desire.

Figure 7.18
You can enter line heights in inches (i), centimeters (c), millimeters (m), or points (p).

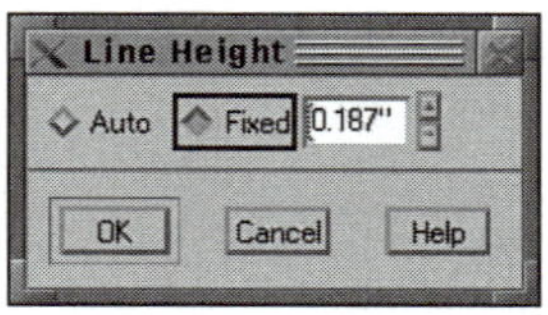

Caution

If you set a fixed line height and then change the font's size, WordPerfect does *not* adjust the height for you; you have to change it again.

Note

WordPerfect applies the line-spacing adjustment to the current line height. Therefore, if you adjust the line height to be larger than usual and then apply a line-spacing value of 2.0, the spacing is greater than it would be by just setting the line spacing to 2.0.

Tip #95 from *Tux Tips*

In general, the Line Spacing dialog box is appropriate for most of your line-spacing needs because it automatically adjusts for font size changes. If you need to adjust your line spacing with greater precision, though, the Line Height dialog box may be what you need.

NUMBERING

Sometimes you may need to refer to specific points in a document. Page numbers can be very helpful but may not be enough. That's why WordPerfect offers an additional level of numbering: line numbers. You can tell WordPerfect to number the individual lines of a document by choosing Format, Line, Numbering to open the Line Numbering dialog box in Figure 7.19. When you start this dialog box, WordPerfect inserts a [Ln Num] code at the start of the paragraph in which the insertion point rests.

→ To learn more about page numbering, **see** "Headers, Footers, and Page Numbers," **p. 235**

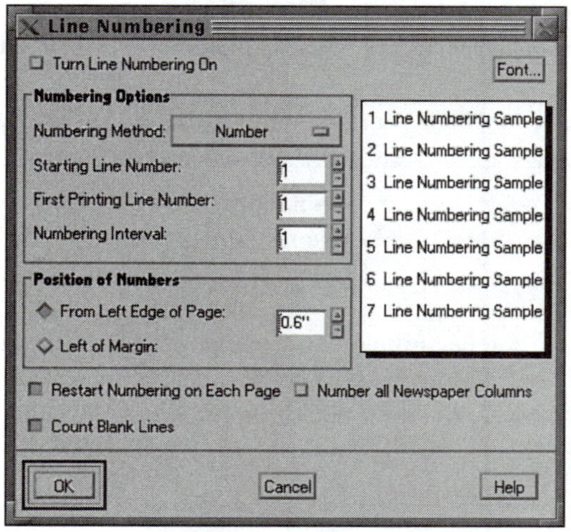

Figure 7.19
Don't forget to select Turn Line Numbering On after you set the other line-numbering options.

WordPerfect labels the lines in your document with a letter or number in the left margin, and you can define many aspects of how this feature works. Specifically:

- Use the Numbering Method option button to tell WordPerfect to label the lines with numbers, letters, or Roman numerals. Letters and Roman numerals may be in either upper- or lowercase.
- You can specify a Starting Line Number. If you enter, say, 200 here, the first line number is 200.
- The First Printing Line Number is the number of the first line to be labeled. This setting can be handy if you want to define line numbering to begin with some line, but don't want to see the line numbers actually begin on your printout for a few lines.

Note

If you increment the Starting Line Number, the First Printing Line Number also increments.

- The Numbering Interval defines whether line numbers appear every line, every other line, or at some other interval.
- You can specify the position of the line numbers, measured either to the right of the left edge of the page or to the left of the left margin.
- If Restart Numbering on Each Page is selected, WordPerfect resets the line number to the Starting Line Number value with each new page. If this option isn't selected, WordPerfect continues to increment line numbers throughout the document.

- If Count Blank Lines is selected, WordPerfect labels and counts blank lines in your document. If not, those lines are skipped. This option applies to both the numbers assigned and the printing of numbers.
- If Number all Newspaper Columns is selected, WordPerfect treats separate newspaper-style columns as if they contained separate lines, giving the second and subsequent columns their own line numbers.

→ If you want to know more about columns, **see** "Creating Professional Layouts with Columns," **p. 255**

- You can select the font in which your line numbers are to appear by clicking the Font button. This step brings up the standard Font dialog box (Figure 7.3).
- Presumably, you want your line numbers to appear in your document, and so you need to be sure that the Turn Line Numbering On box is selected. It defaults to being off, oddly enough.

You can number lines from the beginning of any paragraph by starting the line numbering with no text selected, or you can select a portion of text to number only that part of the text. You can also begin line numbering, select one or more paragraphs, and then change some aspect of the numbering scheme, such as the numbering interval or the number positions.

Caution

Some numbering changes may alter the value of the numbers, either in a changed block or after the changed block. In this case, you need to manually reset your numbering after your temporary changes.

INSERTING THE DATE AND TIME

You can insert the current date and/or time into your document by choosing Insert, Date/Time, which opens the Date/Time dialog box shown in Figure 7.20. Select a date format from the list and click Insert to have WordPerfect insert it for you in your document. If you also select Keep the inserted date current, WordPerfect inserts the date as a WordPerfect code rather than as a string of text. That way, the date is updated whenever you open or print the file.

You can also create your own date formats by clicking New Format to open the Custom Date/Time Format dialog box shown in Figure 7.21. WordPerfect lists the date format in code form in the field labeled Edit Date Format and shows an example of this format in the Date Preview line. You can select and insert date and time codes from the two scrollable lists, and you can type your own explanatory text into the Edit Date Format field. When you're done defining a date format, click OK.

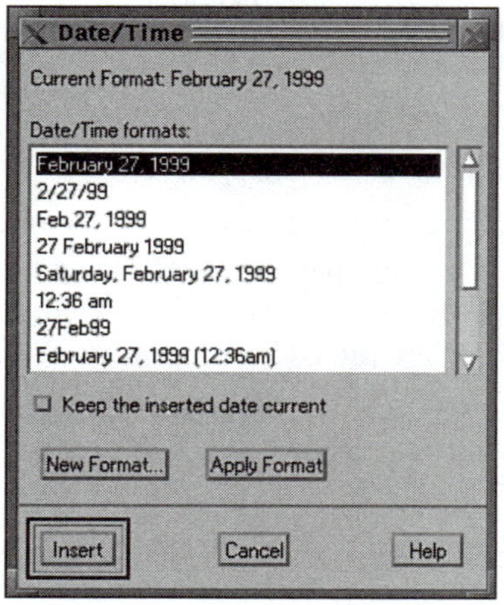

Figure 7.20
Select from any of the standard date formats or define your own with New Format.

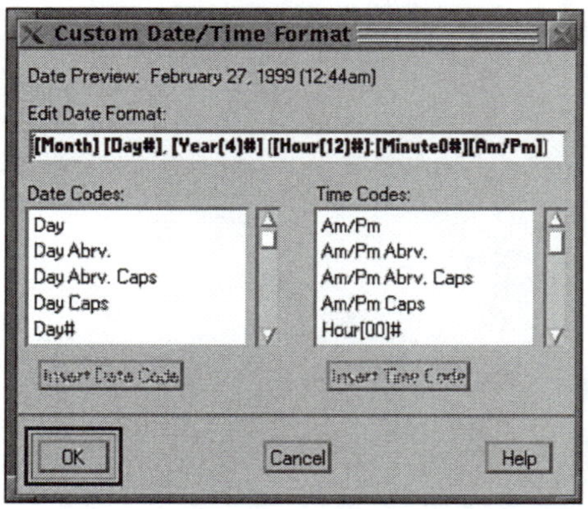

Figure 7.21
Mix and match date codes to create your own custom date format.

TROUBLESHOOTING

DOCUMENT EXCHANGE AND FONTS

You've loaded a document from another computer or word processor, and the fonts aren't what you expect; or you've tried loading your WordPerfect for Linux file on another computer or word processor, and you get the wrong fonts.

This situation is a result of fonts not matching or being imperfectly translated between programs or computers. If you're trying to read a foreign document in WordPerfect, you can change the font by double-clicking its font code in Reveal Codes mode or by setting the initial font for the document. If you need to use a file on another computer or word processor, you might find a solution on that program, or you can try changing the font in WordPerfect to something relatively common, such as Times, which might import better than an unusual font.

→ To learn more about changing fonts, **see** "Modifying Font Codes After the Fact" **p. 186**

JUSTIFICATION PRODUCES OVERLAPPING TEXT

You've tried to align some text, but you ended up with overlapping words.

This problem can result from applying line-oriented justification commands within a paragraph of text. If you intended to align the entire paragraph, try using the commands on the Format, Justification menu instead of those on the Format, Line menu. You may first want to use Reveal Codes to track down and delete the codes that have caused the text to look strange.

→ If you want to know how to use Reveal Codes, **see** "Reveal Codes Mode," **p. 63**

CHAPTER 8

Paragraph Formatting

In this chapter

OVERVIEW

Most of the commands I describe in this chapter are designed to operate on paragraphs of text. As such, WordPerfect inserts the codes for these commands at the beginning of the paragraph in which the insertion point resides. If no text is selected, WordPerfect generally creates an open-ended command, which affects the current paragraph and all subsequent paragraphs. If you have selected text from one or more paragraphs, however, WordPerfect inserts paired codes so that only those paragraphs are affected by your formatting.

Tip #96 from *Tux Tips*

So what *is* a paragraph? From WordPerfect's point of view, a paragraph is any string of text terminated by a hard return code (which appears in Reveal Codes as [HRt]). Therefore, if you enter an address for a letter by typing a line, pressing Enter, typing another line, and so on, WordPerfect considers it to be composed of several paragraphs. You need to keep WordPerfect's view of what constitutes a paragraph in mind when you're working with paragraph-oriented commands.

MARGIN ADJUSTMENTS

The *margins* are the borders outside of which you don't normally place text. WordPerfect defaults to 1-inch margins along the top, bottom, left, and right of the page, meaning that on an 8.5×11-inch sheet of paper, text appears in an area of only 6.5×9 inches. You can adjust WordPerfect's margins, however, in three ways.

First, you can adjust the margins by dragging them directly. Move your mouse pointer to one of the margin guidelines. The mouse pointer changes to a double-headed arrow with a vertical bar (Figure 8.1), and when that happens, you can click and drag your margin to a new position.

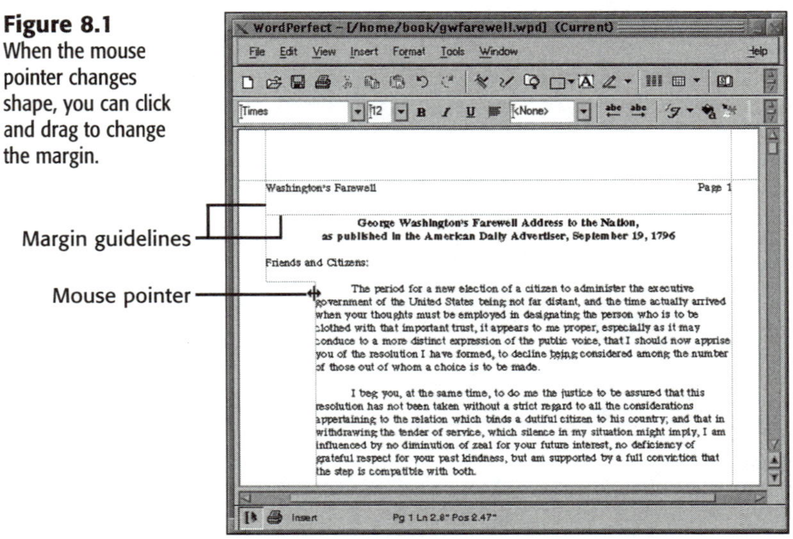

Figure 8.1
When the mouse pointer changes shape, you can click and drag to change the margin.

Second, you can use the Format, Margins (Ctrl+F8) command, which brings up the Margins dialog box in Figure 8.2. Adjust the margins to whatever value you desire. As with other dialog boxes in which you enter measurements, you can enter values in centimeters (c), millimeters (m), points (p), or the default inches (i or the quote character).

Figure 8.2
You can adjust all four margins from a single dialog box.

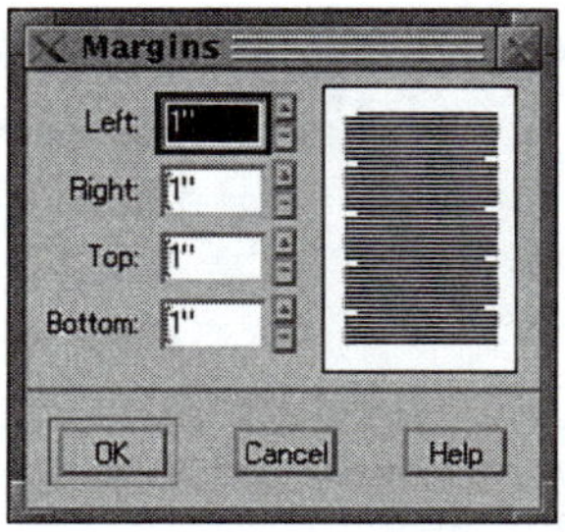

Note

Both of these methods can create either or both paragraph-oriented or page-oriented codes. The top and bottom margin settings are page oriented. WordPerfect inserts them at the start of the current page, and they apply to entire pages. The left and right margin settings, however, are paragraph oriented. They apply from the start of the current paragraph or for however many paragraphs you've selected.

The third method of changing the margins is to use the ruler, which you can display by selecting View, Ruler (Figure 8.3). Click and drag the left margin widget or the right margin widget to modify the margins.

Figure 8.3
The margin adjustment and indent widgets appear here moved inward from their usual locations for easy identification.

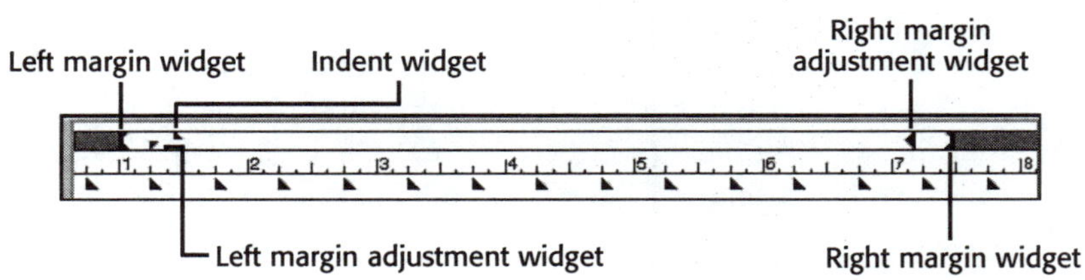

Caution

You can adjust WordPerfect's margins to be quite wide, but your printer may not be able to handle the entire width you can enter in WordPerfect. WordPerfect's printer driver database includes allowable margin width information, but it isn't always accurate, especially if you're using a PostScript driver and are printing through Ghostscript.

→ If you want to learn more about setting up printers, **see** "Fonts and Printing," **p. 739**

INDENTS

Typically, a paragraph of text begins either flush to the left margin or indented 1/2 inch or so (less if you're using multiple columns or narrow paper). WordPerfect provides a number of methods to set these and other indentation effects.

Using WordPerfect's indentation tools can make your life easier in the long run. I've seen people format a single bibliographic entry as multiple WordPerfect paragraphs with all but the first line headed by a tab because they don't know how to use WordPerfect's indentation features. This sort of practice makes subsequent editing harder and can even make the text look bad if it's reformatted for a different font or printer.

These indentation commands, like many of WordPerfect's formatting commands, come in both open-ended and paired code forms. To use an open-ended form, do not select any text prior to entering the code; it then takes effect from the start of the paragraph in which the insertion point resides and continues until the next code that overrides this setting. To use a paired code, select one or more paragraphs and enter the formatting commands; the code pair is in effect from the start of the first paragraph you selected until the end of the last.

INDENTING THE FIRST LINE OF A PARAGRAPH

When you begin typing a new paragraph of text, you're probably accustomed to pressing the Enter key once or twice, possibly pressing the Tab key once, and then typing. With a typewriter or most word processors, these steps achieve the desired effect of separating two paragraphs with a blank line or an indent of 1/2 inch or so on the first line of the paragraph. If you start WordPerfect and use its defaults, these methods work fine. There are other ways to achieve these effects, however.

One way to indent a paragraph is to set an indent on the ruler. Select View, Ruler to display WordPerfect's ruler (Figure 8.3). Near the left margin, you'll find the indent widget, which probably lies immediately above the left margin adjustment widget. You can set a first-line indent by dragging the indent widget to the right. Subsequently, WordPerfect indents any new paragraph by the amount you set, without the need to use the Tab key at the start of a new paragraph.

Tip #97 from *Tux Tips*

> It's easy to grab the left margin adjustment widget or even the left margin widget by mistake when you're trying to use the indent widget. To avoid this problem, don't aim your mouse pointer at the body of the widget itself, but at its upper-right edge.

You can also set a paragraph's indent by choosing Format, Paragraph, Format. This step brings up the Paragraph Format dialog box shown in Figure 8.4. Enter the amount of indentation you want in the First Line Indent spin box and click OK.

Figure 8.4
Enter values in inches (i), centimeters (c), millimeters (m), or points (p).

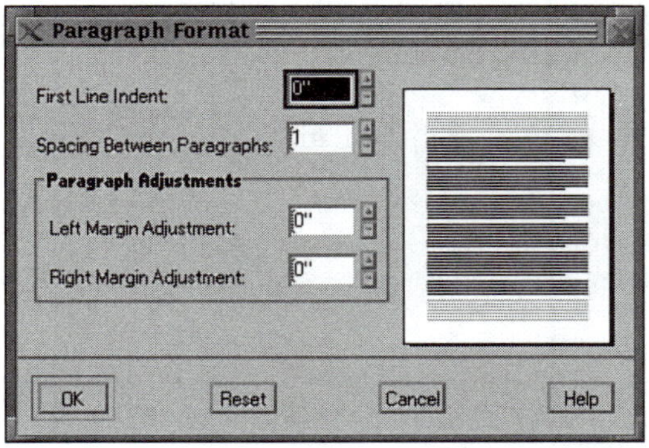

The Paragraph Format dialog box, like most of WordPerfect's formatting dialog boxes, includes a graphical representation of your formatting changes. The first and last of the simulated paragraphs are in gray and show the default settings. The middle paragraphs are in black and show the settings you're creating with the dialog box's controls. This display gives you some idea of what you're doing with the settings, though you need to wait until you apply your changes (and maybe even create a printout) to see their precise effect.

Tip #98 from
Tux Tips

To reset any of the formatting changes you make in the Paragraph Format dialog box, you can use the Reset button. If you want to simply abandon any changes and go back to editing, click Cancel.

Indenting an Entire Paragraph

On occasion, you may want to indent an entire paragraph to set it off from the surrounding text. You might use such a format for a long quotation, for instance, or for a warning or other text you want your reader to notice.

As with a first-line indent, you can indent an entire paragraph either with the ruler or through the Paragraph Format dialog box. On the ruler, click and drag the left margin adjustment widget to change its location. As you do, notice that the paragraph indent widget also moves; the paragraph indent setting is actually set relative to the left paragraph margin. For a more pronounced effect, you can also indent the right margin of the paragraph using the right margin adjustment widget.

You can obtain the same paragraph indentation effects by selecting Format, Paragraph, Format and entering appropriate values in the Paragraph Adjustments section of the dialog box (Figure 8.4). Because the first-line indent is relative to the left paragraph margin, the value doesn't change when you change the Left Margin Adjustment value, but the indentation remains constant relative to that value when you click OK.

There's a shortcut to indenting a paragraph on one or both sides. Choose Format, Paragraph, Indent (F7) or Format, Paragraph, Double Indent to move in the left or both paragraph borders, respectively. This formatting method applies to one and only one paragraph at a time and uses the nearest tab stops in the appropriate direction from the page margins to determine the indentation amount. If you apply both this shortcut and the paragraph margin changes, you get twice the indentation.

Note

The paragraph indent and paragraph double-indent codes can be deleted when Reveal Codes is not active. The paragraph margin adjustment codes cannot be.

Tip #99 from *Tux Tips*

Paragraph indent codes leave tabs in their place, whereas the paragraph-format and ruler-based indentation methods adjust the tabs along with the indent, if the tabs are set relative to the margin.

So how does paragraph indentation differ from adjusting the margin? In many cases, you'll see no practical differences. Both methods adjust the margins, and both the ruler and paragraph-format methods of indentation move tabs that are set relative to the margin. The two methods produce somewhat different looks on the screen, if WordPerfect is set to display margin guidelines, because those change only if you adjust the margin. One practical difference is in paragraph borders and fill, which extend from the left to the right margins regardless of the indentation settings.

→ To learn more about borders and fills, **see** "Borders and Fills," **p. 216**

HANGING INDENTS

A *hanging indent* is a reversal of the usual indentation pattern—in a hanging indent, the first line of a paragraph begins to the left of the remaining lines. This pattern helps make the paragraph beginnings stand out. This formatting method is common in bibliographies and bulleted lists, and WordPerfect makes it easy to create this effect.

The best way to apply hanging indents in an ongoing fashion (to several paragraphs) is usually to choose Format, Paragraph, Format (Figure 8.4) and adjust the First Line Indent spin box to a negative value and the Left Margin Adjustment to a corresponding positive value. With these settings, the paragraphs you type automatically receive hanging indents until you reverse this formatting change.

Note

You cannot perform the corresponding action with the ruler because it doesn't let you drag the left margin indent widget to the left of the left paragraph margin widget.

If you want to apply a hanging indent to just one paragraph, or if you want to apply it to existing text, you can do so by using Format, Paragraph, Hanging Indent (Ctrl+F7). This step applies your formatting change to just one or a group of selected paragraphs. Another way to achieve this effect is to position the insertion point on the second or subsequent lines of the paragraph and press the Tab key. However you do it, this method uses the existing tab stops to format the indent, as opposed to using values you enter specifically for the task.

Tip #100 from *Tux Tips*

> If you apply a hanging indent by using the hanging-indent feature, you also need to apply the same formatting to any intervening paragraphs you type after creating this formatting. If you apply a hanging indent by adjusting the paragraph margins and using a negative first-line indent, you can create more paragraphs with the same formatting just by typing them.

You can also create a hanging indent that obtrudes itself into the left margin in two ways. Do so either by using a negative value for the first-line indent without adjusting the left margin value in the Paragraph Format dialog box, or by moving your insertion point to the start of the paragraph and selecting Format, Paragraph, Back Tab from the document window's menu.

Are you having problems with what appears to be double text on a single line? See "Overlapping Characters in Lines" in the "Troubleshooting" section at the end of this chapter.

WordPerfect can recognize certain patterns that typically signal a desire for hanging-indent formatting and automatically creates a hanging indent. For instance, if you type a small number of characters at the start of a line followed by the Tab key, you're probably typing some sort of listing of information, such as a bulleted list, outline, or a simple table. Hanging indents lend themselves to such formatting. In fact, WordPerfect recognizes the typical starts to a bulleted or numbered list and, when you press the Enter key, begins the next line with an appropriate bullet or number and hanging-indent codes in place.

→ To learn more about creating bulleted or numbered lists, **see** "Creating a Bulleted or Numbered List," **p. 554**

Drop Caps

A drop cap is an extra-large leading letter or word for a paragraph. Novels and magazines frequently use drop caps, and you can, too, with the help of WordPerfect for Linux. Of course, with any word processor you can increase the size of the first character of a paragraph, but this technique may not always produce the effect you desire. WordPerfect provides many options to let you obtain precisely the drop-cap effect you want.

Creating a Drop Cap

To create a drop cap, position the insertion point anywhere in an existing paragraph and choose Format, Paragraph, Drop Cap. This step creates a drop cap of the default style for that paragraph, as in Figure 8.5. Notice that WordPerfect adds another toolbar, the drop-cap toolbar, with which you can fine-tune the drop cap's appearance.

Figure 8.5
The default drop-cap style uses a three-line-high version of the same font used for the rest of the text.

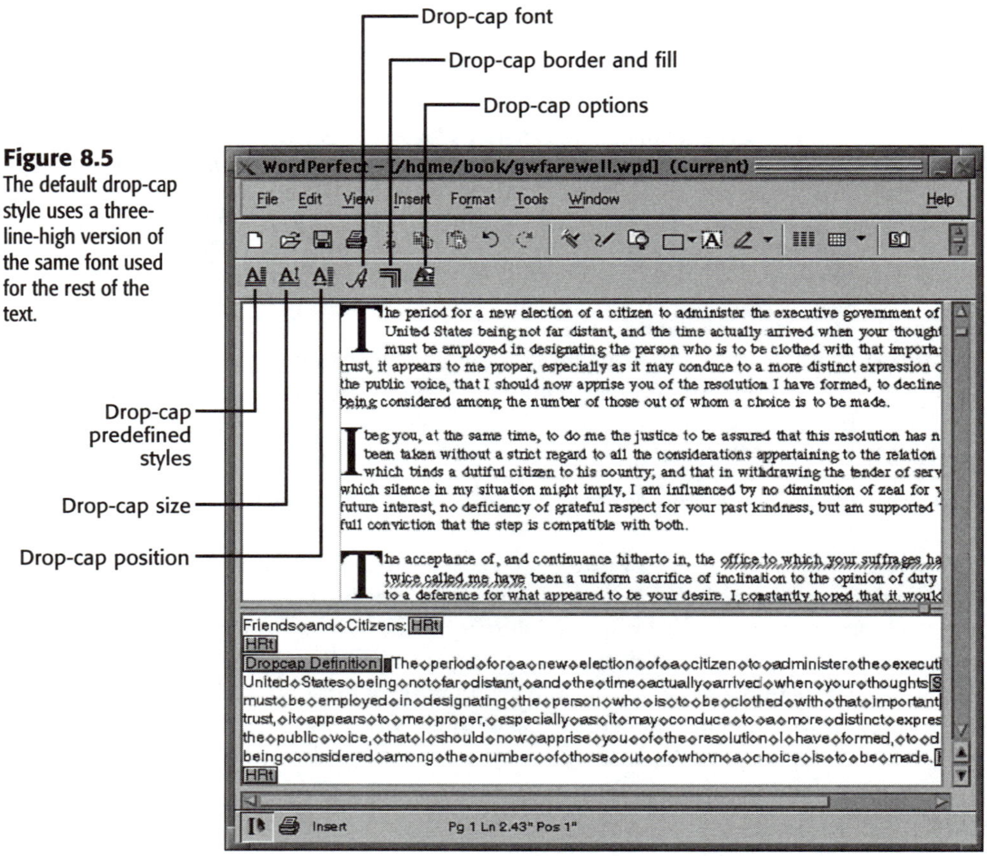

Note

As you might expect by now, you can apply drop caps to a range of paragraphs by selecting parts from all of them and choosing the drop-cap option. Selecting the drop-cap option before typing a paragraph also creates a drop cap when you begin typing.

Tip #101 from
Tux Tips

If the drop-cap toolbar disappears and you want to edit your drop-cap options, try entering Reveal Codes mode (View, Reveal Codes or Shift+Ctrl+F3) and positioning the insertion point by the drop-cap definition.

After you create the drop cap, you can choose its appearance from a number of predefined styles by selecting the style you want from the options that pop up when you click the drop-cap predefined styles toolbar widget (Figure 8.6). Note that these options vary in terms of the position of the horizontal and vertical positioning of the drop cap relative to the first line of the paragraph and in the size of the drop cap. The option in the lower-right corner disables the drop cap.

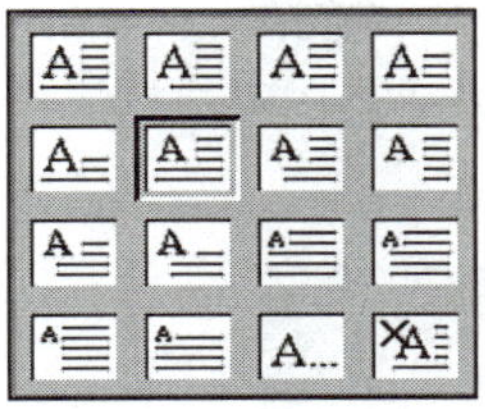

Figure 8.6
You can quickly adjust the positioning and size of the drop cap by selecting from among several predefined alternatives.

ADJUSTING DROP-CAP POSITION AND SIZE

If none of the alternatives available from the drop-cap predefined styles tool appeals to you, you can customize your drop cap further by using the Drop-cap size and Drop-cap position tools.

Clicking the Drop-cap size tool produces a list of sizes ranging from two lines through nine lines, with an extra option of Other in case you want to enter truly huge or fractional drop-cap sizes.

The Drop-cap position tool provides three options: In Text, In Margin, and Other. The first of these places the drop cap horizontally entirely within the paragraph and wraps text around it. The second option places the drop cap entirely to the left of the text, which is usually in the margin. If you select Other, you get the Drop Cap Position dialog box shown in Figure 8.7. This dialog box lets you select how many lines to move surrounding text down relative to the top of the drop cap. You may also enter a percentage value representing how far outside the paragraph the drop cap should fall.

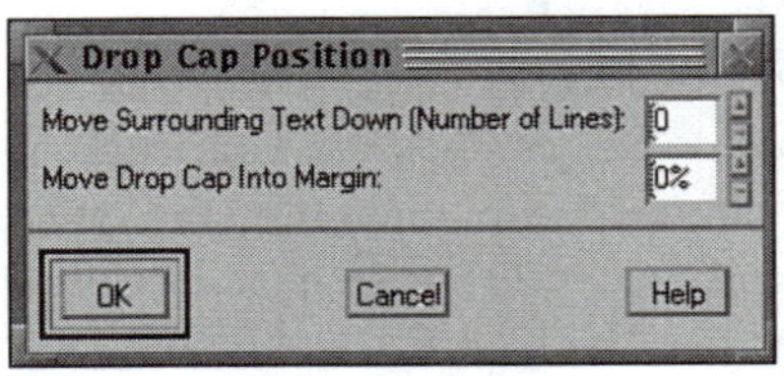

Figure 8.7
The Drop Cap Position dialog box lets you move the drop cap up and to the left of the paragraph to which it belongs.

Note

Although the menus and dialog boxes refer to positioning the drop cap in the margin, this isn't necessarily what these options do. For instance, if you choose an indent option and position a drop cap "in the margin," the drop cap may actually appear to the right of the margin but to the left of the indented paragraph.

CUSTOMIZING DROP-CAP APPEARANCE

Drop caps in books often use a font other than the one used for the bulk of the text. Sometimes the drop-cap font is very ornate; other times it's a *sans serif (p. 178)* font, whereas the body text is serif. You can achieve the same sort of effect by clicking the Drop-cap font tool, which brings up the Drop Cap Font dialog box (Figure 8.8). This dialog box is a variant of the standard WordPerfect Font dialog box. It lets you select the font, font style, appearance attributes, and color.

Figure 8.8
You can't adjust font size from the Drop Cap Font dialog box; use the Drop-cap size tool to do that.

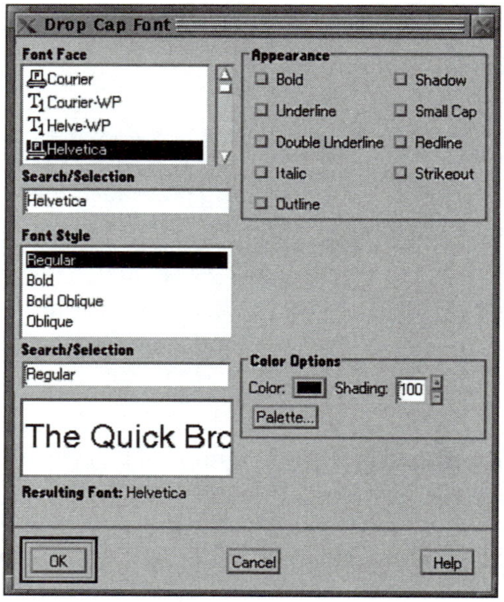

→ For more information on setting fonts, **see** "Fonts," **p. 178**

Clicking the Drop-cap border and fill tool brings up the Drop Cap Border dialog box shown in Figure 8.9, which allows you to select a border and background fill for the drop cap. You can make selections in the Border Options and the Fill Options areas either from a graphic (click the appropriate Graphics selector) or from a description (click the appropriate Listed selector's button). You can disable all border and fill options by clicking the Off button.

Figure 8.9
The drop-cap preview area tends to omit the left and right sides of the border.

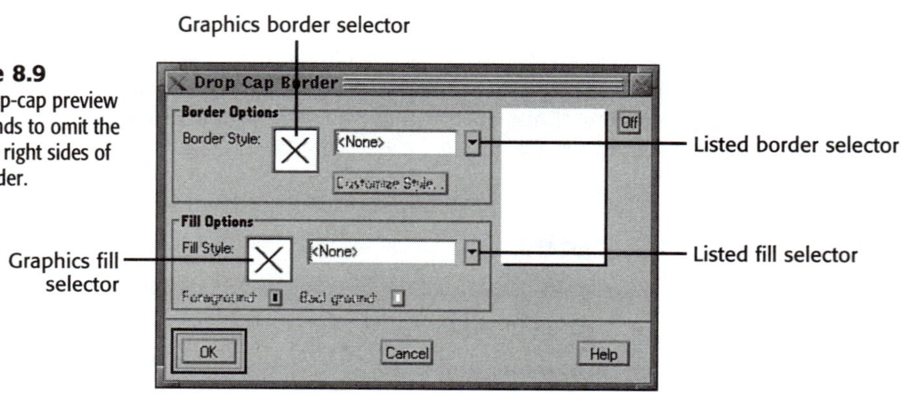

After you select a fill option, you can select foreground and background colors by clicking the appropriate buttons in the Fill Options section of the dialog box. Each button opens a rectangle filled with color selections. After you select a border style, you can tweak it by clicking Customize Style, which brings up the Customize Border dialog box shown in Figure 8.10.

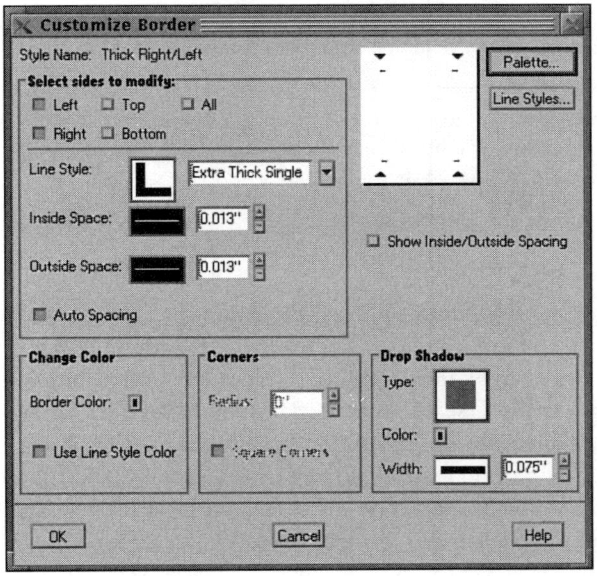

Figure 8.10
You can modify a drop cap's borders to suit your specific needs.

From this dialog box, you can set several options:

- The sides to have borders—These can be left, right, bottom, top, or all.
- The Line Style—Select the visual appearance of the line graphically or by name, as in the Drop Cap Border dialog box.
- Inside and Outside Space—These settings allow you to adjust how much of a gap to leave inside and outside the drop-cap border. If you select Auto Spacing, WordPerfect sets the inside and outside spaces to values it computes—generally `0` for a drop cap.
- Show Inside/Outside Spacing—This option is supposed to show the effect of the inside and outside spacing values in the preview field near the top of the dialog box. That preview field isn't entirely reliable, however, and so neither is this option.
- Change Color—You can set the color of the drop-cap border from a pop-up menu when you click on the color square. If you select Use Line Style Color, WordPerfect uses the color from the Line Style item instead.
- Corners—The default style is for square corners. You can uncheck this option and increment the value in the Radius spin box, however, to have WordPerfect round the corners, even to the point of creating an oval border.

- Drop Shadow—You can create a shadowed effect for the border, as if it were being held slightly above the page, casting a shadow on the paper. Select any of four directions for the shadow to fall from the Type selector and choose a shadow Color and Width from those selectors.
- Palette and Line Styles—These buttons let you customize the color and line style choices used elsewhere in the dialog box.

→ For more information on palettes and line styles, **see** "Graphics," **p. 377**

Note Whenever you select an option that reduces the amount of space available inside the drop-cap border, WordPerfect automatically reduces the size of the drop cap so that it doesn't overlap with the border.

ADDITIONAL DROP-CAP OPTIONS

You can set miscellaneous additional drop-cap options by clicking the Drop-cap options tool to open the Drop Cap Options dialog box shown in Figure 8.11. You can set the number of characters in the drop cap by incrementing the spin box's value; alternatively, you can apply the drop cap to the entire first word by checking the Make First Whole Word as Drop Cap button. You'll normally want to leave the Wrap Text Around Drop Cap button selected; otherwise, text on the second and subsequent lines will overlap the drop cap. The Allow for Descender and Allow for Diacritic options add a bit of space below and above the normal drop-cap space for descenders and diacritic marks, respectively. The drop-cap style is normally applied to uppercase letters without diacritic marks, so these options aren't usually needed, though they might be for some characters and fonts.

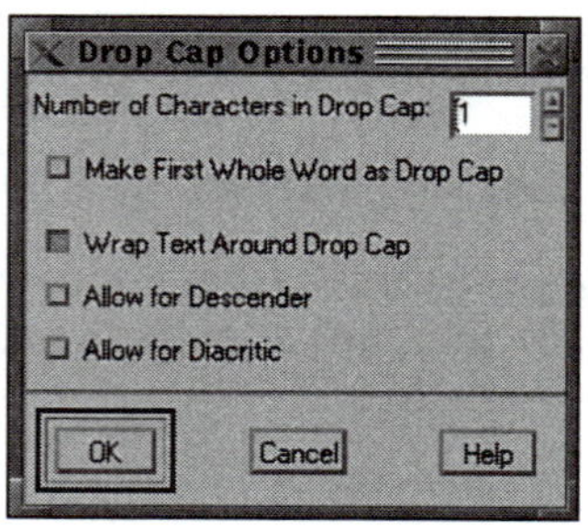

Figure 8.11
Adjust miscellaneous drop-cap options in the Drop Cap Options dialog box.

BORDERS AND FILLS

Borders and fills are lines that surround a paragraph or patterns printed under the paragraph's text, respectively. They're generally used to help make a paragraph stand out from its surroundings, as in a warning or some particularly important or interesting piece of information. WordPerfect lets you easily apply these styles to individual paragraphs or groups of paragraphs. You can create similar effects with text boxes, which have the added advantage of being moveable. You can put text boxes anywhere on a page and have the rest of your text flow around them.

→ If you want to know more about text boxes, **see** "Using Text Boxes to Add Variety to Text Layout," **p. 357**

CREATING BORDERS AND FILLS

You can apply a border or fill pattern to a paragraph by choosing Format, Paragraph, Border/Fill. This sequence brings up the Paragraph Border dialog box, which is almost identical to the dialog box in Figure 8.9, and operates in the same way. Select the border and fill types that you want, and WordPerfect applies them. One addition to the Paragraph Border dialog box compared to the dialog box shown in Figure 8.9 is the Apply border to current paragraph only check button. Select this option if you want to alter only the paragraph in which the insertion point rests or only the highlighted paragraphs. If you uncheck this button, WordPerfect applies your border and fill options to all paragraphs from the beginning of the current one.

If you click Customize Style for a paragraph border, you get a Customize Border dialog box similar to that in Figure 8.10. There are a few differences in how these dialog boxes operate, however. The following items now apply:

- An additional Separator option appears in the Select Sides to Modify area. If checked, WordPerfect creates a separate border around each paragraph. If this option isn't selected, WordPerfect creates one border around all the affected paragraphs.
- The Show Separator Line option shows the separator line, if selected, in the preview area.
- The Corners option isn't available for paragraphs; all paragraph borders must have square corners.

Tip #102 from *Tux Tips*

Borders and fills use the document's left and right margins as their left and right boundaries. Therefore, to place a border or fill around indented text and to have the border or fill closely hug the indented text's extent, you may want to use a margin adjustment rather than an indent command to achieve the indentation effect.

Tip #103 from *Tux Tips*

When creating a shading pattern for text, create one that's on thelight side. Text can be difficult to read if it's printed on too dark a background. Also, shaded text may not photocopy or scan very well. Planning for readability after photocopying can be particularly tricky if you use a color printer and a colored background.

Tip #104 from *Tux Tips*

You can delete a border or fill either by deleting the code that created it in Reveal Codes mode or by selecting the Border/Fill option and specifying no border or fill.

GRAPHICS LINES

In some cases, a full paragraph border or fill may be overkill. If you just want a single line to separate portions of your document, WordPerfect lets you create these, too.

The simplest way to achieve this effect is to use the QuickLines feature, which is part of the QuickCorrect mechanism. Simply type, on a line by itself, four or more dashes (----) or equal signs (====), and WordPerfect replaces these with a single or double line, respectively, extending from the left to the right margin.

→ To learn more about what QuickCorrect can do, **see** "QuickCorrect," **p. 122**

If you want to customize the line you created, you may do so by right-clicking on it and choosing Edit Horizontal Line from the pop-up menu, or by double-clicking the line with the left mouse button. This sequence produces the Edit Graphics Line dialog box shown in Figure 8.12.

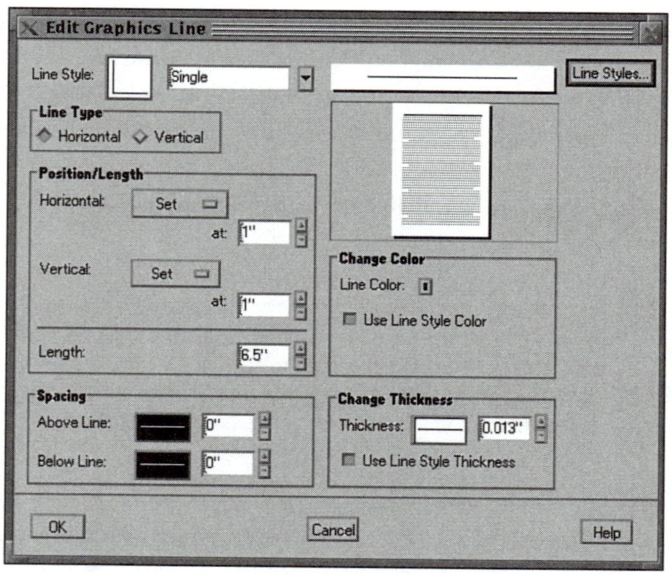

Figure 8.12
Customize a line style with the Edit Graphics Line dialog box.

This dialog box allows you to customize many aspects of your line:

- Change the style of the line with the Line Style selectors, which operate much as other line style selectors do.
- Change the line between horizontal and vertical orientation with the Line Type radio buttons.

- Set the position of the line horizontally and vertically in several ways and adjust its length to whatever you want.
- Specify free space to be given above and below the line in the Spacing area of the dialog box.
- Change the line's color from the Change Color section of the dialog box. Select a line color from the pop-up Line Color dialog box or tell WordPerfect to use the color specified as part of the Line Style by checking the Use Line Style Color button.
- Adjust the thickness of the line in the Change Thickness area. As with the line's color, you can specify a thickness or take it from the Line Style.

In addition to customizing the line in this way, you can click and drag it with the left mouse button to reposition it graphically. This change affects the entries in the Position/Length portion of the Edit Graphics Line dialog box, if you subsequently edit it that way.

In addition to using QuickLines, you can insert lines by choosing various items from the Insert, Shape menu:

- Draw Line—This item lets you draw a line on the page in any orientation. When you choose Draw Line, the mouse pointer changes to a crosshair. Click and hold the left mouse button where you want the line to begin, move the mouse pointer to the endpoint, and release the mouse button. WordPerfect inserts a line between those two points. This line type is more akin to WordPerfect's graphics types than to the other lines discussed here, however.

→ If you want to know more about simple graphics, **see** "Using Lines and Simple Shapes," **p. 380**

- Horizontal Line (Ctrl+F11)—This item is the same as a single QuickLine.
- Vertical Line—This item is like a single QuickLine, except that the line is vertical rather than horizontal. You can use a vertical line to separate columns in a multicolumn layout or as ornamentation in the margins of your page.

→ If you want to know more about columns, **see** "Creating Professional Layouts with Columns," **p. 255**

- Custom Line—This option brings up the dialog box shown in Figure 8.12 except that the dialog box is called Create Graphics Line and allows you to specify the details of the line type before you create it.
- Other items—This menu contains additional choices for inserting circles, rectangles, and other shapes. Like the Draw Line option, these are implementations of WordPerfect's graphics objects.

SPACING

You can adjust the spacing between paragraphs by choosing Format, Paragraph, Format. This step opens the Paragraph Format dialog box, shown in Figure 8.4. Enter a value into the Spacing Between Paragraphs spin box to increase the spacing between paragraphs. You can increment or decrement this box's value by 0.5-line increments by using the spin box controls, or you can enter any other value, in 0.1-line increments, by typing it in directly.

Tip #105 from *Tux Tips*

If you use WordPerfect's spacing control to set the interparagraph spacing in your document, you can easily adjust this spacing for your entire document later simply by modifying this value. By contrast, if you create double-spacing between paragraphs by pressing the Enter key twice, you will have a harder time adjusting interparagraph spacing if you decide to change it later.

Do your paragraphs overlap one another? See "Overlapping Paragraphs" in the "Troubleshooting" section at the end of this chapter.

HYPHENATION

Hyphenation is the breaking of a single word across two lines of text to better distribute the amount of text on multiple lines. For instance, if you have a paragraph in which the word *antidisestablishmentarianism* appears in a position so that it almost, but not quite, fits on a line, WordPerfect for Linux normally places the entire word on the next line. This particularly long word causes a huge gap on the preceding line. If you're using left justification, that gap shows up entirely on the right side of the line, which is visually jarring. If you're using full justification, the gap is spread out across the entire line, which is both visually distracting and makes the line harder to read. Hyphenation exists for cases like these, even for words shorter than *antidisestablishmentarianism*.

→ **See** "Justification," **p. 191**

MANUAL HYPHENATION

One way to perform hyphenation is to do so manually. If you were doing hyphenation with a typewriter or in a raw ASCII text editor (say, for an email message), you'd press the — key followed by the Enter key. You could do the same thing in WordPerfect, or even do a bit better by pressing the spacebar rather than the Enter key. This technique, however, produces a bizarre midline hyphenation if you reformat your paragraph by adding or deleting text, changing margins, or possibly even changing your printer. WordPerfect therefore offers the capability to insert a *soft hyphen* ([—SoftHyphen] code). This special code tells WordPerfect to break the word across lines at that point if and only if there's room on a line before the right margin for the part of the word before the soft hyphen but not for the second part. In this case, WordPerfect hyphenates the word on the screen and in the printout. If not, the soft hyphen lies dormant. If your paragraph formatting changes, making the soft hyphen necessary when it was not, or vice versa, WordPerfect adjusts the text automatically.

To enter a soft hyphen, press Ctrl+Shift+— at the point where you expect to hyphenate the word, if it's needed. Unless you're looking at the word in Reveal Codes mode or it's positioned so that it immediately takes effect, you won't see anything change on the screen, but the code is there.

One hyphenation variant is the hyphenation soft return ([HyphSRt]) code), which is a break in a word *without* the hyphen. If a word has a [HyphSRt] code embedded in it, WordPerfect breaks the word across lines where appropriate. To create a [HyphSRt] code, choose Format, Line, Other Codes and select Hyphenation Soft Return from the resulting Other Codes dialog box (Figure 8.13).

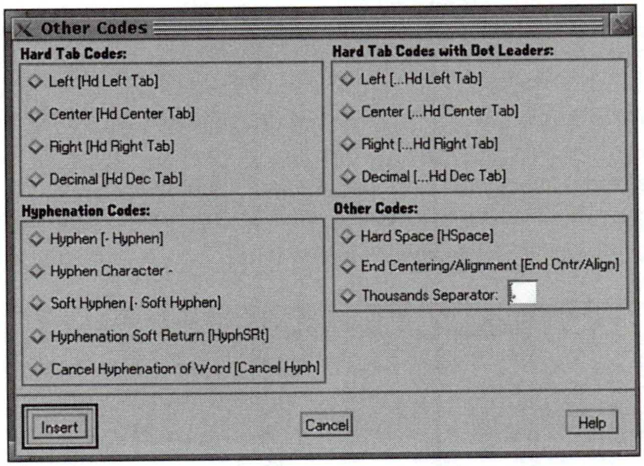

Figure 8.13
You can enter an assortment of miscellaneous codes from the Other Codes dialog box.

Tip #106 from *Tux Tips*

You may want to use [HyphSRt] codes on long Internet addresses embedded in your text.

Tip #107 from *Tux Tips*

When you enter a word that's hyphenated even in the middle of a line, such as *large-grained,* WordPerfect feels free to break that word across lines as if the hyphen were followed by a [HyphSRt] code. To prevent this problem, you must enter a *hard* hyphen by typing the hyphen as Ctrl+– rather than as simply –.

Are hyphenated words in documents you've imported from another word processor not breaking across lines? See "Imported Documents and Hyphens" in the "Troubleshooting" section at the end of this chapter.

Tip #108 from *Tux Tips*

You can enter the hard, soft, and normal hyphen codes from the Other Codes dialog box if you can't remember the keystroke to create the type of hyphen you desire.

You may occasionally need to control line breaks in another way—to *prevent* a line break between words. You can do so by inserting a *hard space* code ([HSpace] in Reveal Codes) into your text. In this case, press Ctrl+spacebar rather than just the spacebar. WordPerfect treats two or more words separated by [HSpace] codes as a single word.

AUTOMATIC HYPHENATION

By default, WordPerfect for Linux does not automatically hyphenate words. You can tell it to do so by choosing Tools, Language, Hyphenation. This step opens the Line Hyphenation dialog box shown in Figure 8.14. With this dialog box, you can specify the parameters WordPerfect uses to decide when to hyphenate. WordPerfect attempts to hyphenate a word if it begins before the region of the line defined by Percent Left and if the second line of the hyphenated result extends past the region defined by Percent Right. Thus, increasing these values reduces the number of automatic hyphenations, and decreasing the values makes for more hyphenations.

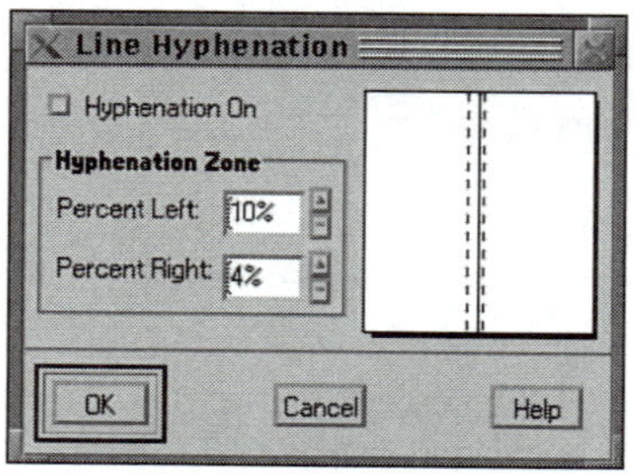

Figure 8.14
Be sure to check the Hyphenation On button before clicking OK, or WordPerfect will not hyphenate.

WordPerfect hyphenates words by consulting hyphenation information in its dictionary. If the program doesn't find a word in its hyphenation dictionary, it won't hyphenate that word. If you want to override WordPerfect's hyphenation decision, you may do so in two ways:

- You may enter a soft hyphen in the word after WordPerfect's chosen hyphenation place. If your soft hyphen is located so that WordPerfect can hyphenate the word there without going over the right margin, your hyphen takes precedence.
- You may select the word and then choose Format, Line, Other Codes. This step brings up the Other Codes dialog box shown in Figure 8.13. Select Cancel Hyphenation of Word and click Insert. WordPerfect no longer hyphenates that word. This setting has no effect if you've manually inserted a soft hyphen or if the word is normally hyphenated even midline. WordPerfect treats words separated by [HSpace] codes as if they'd been given this hyphenation immunity.

HYPHENATION CODES

With so many different types of hyphens floating around, it's often helpful to be able to identify them in WordPerfect's Reveal Codes mode (View, Reveal Codes). Figure 8.15 shows the Reveal Codes pane of a document with several different hyphen and dash characters.

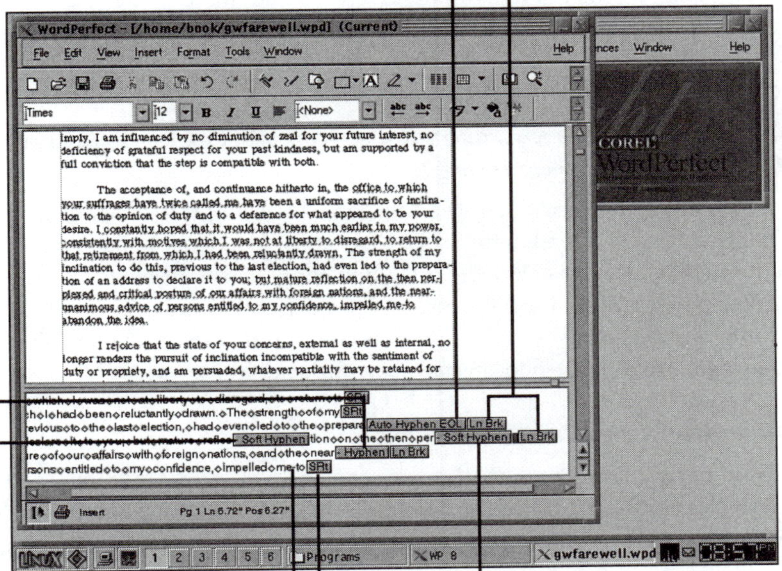

Figure 8.15
Reveal Codes lets you identify the hyphenation and dash characters, even when they're hidden inside a word that is not currently hyphenated.

You should take particular note of a few things about the codes shown in this figure:

- The manually created soft hyphen codes are the same whether the hyphen is visible in a broken word or not.
- The unbreakable hard hyphen code looks like an ordinary hyphen.
- The ordinary hyphen that you create by pressing only the "—" key has a special code ([—Hyphen]).
- Most lines in the paragraph are marked at their ends by soft return ([SRt]) codes, but the hyphenated words are marked instead by line break ([Ln Brk]) codes. These codes are functionally identical.

TROUBLESHOOTING

OVERLAPPING PARAGRAPHS

You've tried adjusting the values on the Paragraph Format dialog box, but now your paragraphs overlap.

The default value for Spacing Between Paragraphs is `1.0`, meaning that the first line of a paragraph falls on the first line after the last line of the previous paragraph. This value does *not* mean that there's one blank line between paragraphs. If you adjust this value to below

1.0, the spacing between paragraphs will be *less than* the spacing between lines within a paragraph, and the last line of one paragraph and the first line of the next will overlap. Set this value back to 1.0 or above to eliminate this problem.

OVERLAPPING CHARACTERS IN LINES

You've tried to create a formatting effect involving tabs, back tabs, or hanging indents, but instead you created a line with overlapping text.

This problem can be caused by a back tab code being embedded in the body of a paragraph. These codes are intended for use at the start of a paragraph, but WordPerfect doesn't enforce that rule. To delete the offending code, start Reveal Codes mode by choosing View, Reveal Codes (Shift+Ctrl+F3) and look for the code within the line that's behaving strangely. You may need to position the insertion point on that line and move through it with the left and right arrow keys to locate the code if it's late in the line.

IMPORTED DOCUMENTS AND HYPHENS

You've imported a file from another word processor, and WordPerfect isn't breaking words with hyphens across lines.

WordPerfect generally converts hyphens in foreign file formats to hard hyphens when it imports a file. This conversion prevents the words from being broken across lines. To correct this problem, you need to replace the hard hyphens with soft hyphens. One quick way to do so for an entire document is to do a find-and-replace operation. Choose Edit, Find and Replace (Ctrl+F2). In the Find field, choose Match, Codes and pick —Hyphen Character. In the Replace With field, press —. The dialog box displays "—" in the Find field and "[— Hyphen]" in the Replace With field, as shown in Figure 8.16. Then click Replace All to replace all the hard hyphens with soft hyphens.

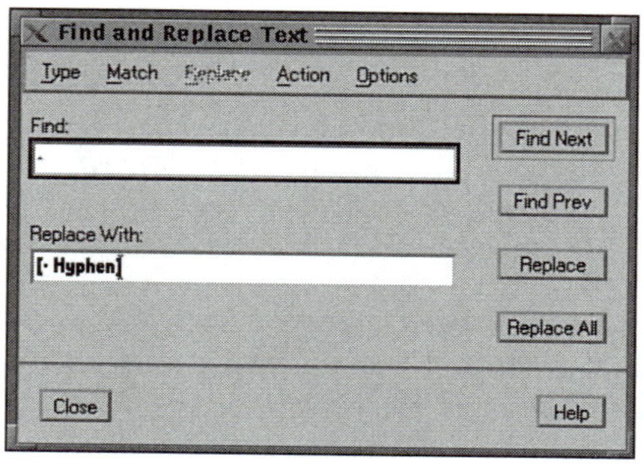

Figure 8.16
You can change one hyphen type to another with Find and Replace.

➔ To learn more about Find and Replace, **see** "Find and Replace," **p. 105**

CHAPTER 9

Page and Document Formatting

In this chapter

OVERVIEW

The commands in this section operate either on whole pages or on the entire document. WordPerfect inserts the codes to control these commands at various points in the file, depending on the code. Some appear at the start of the document, others at the start of the current paragraph, and others at the insertion point. Such diversity can sometimes make it tricky to reverse an edit involving a page-oriented formatting command.

Tip #109 from *Tux Tips*

If you need to delete or alter a page-oriented formatting command, try using WordPerfect's Find and Replace feature to locate the command. From the Find and Replace dialog box, choose Match, Codes and then select the name of the code you want to locate.

→ For more information on finding codes, **see** "Searching for Codes and Symbols," **p. 108**

PAGE SIZE

By default, WordPerfect for Linux starts up with U.S. letter-size (8.5×11 inch) paper in portrait orientation. This setting is fine for most documents in the United States, but you may want to change the default for some specific purpose. You can either select a predefined page size or define your own if you're printing on some particularly exotic paper stock.

SELECTING A PAGE SIZE

You can change WordPerfect's page size by opening the Paper Size dialog box, shown in Figure 9.1. You can get to this dialog box in two ways: choose either File, Page Setup, Page Size or Format, Page, Page Setup, Page Size.

Selecting a page size from one of the predefined types is quite easy; simply find the type in the scrollable list near the top of the Paper Size dialog box, click the name, and then click OK. The page size change takes effect from the start of the current page until the end of the document.

Note

You can't apply a paper size change to a range of pages, though you can apply more than one paper size change per document.

Tip #110 from *Tux Tips*

If you're trying to print a table or some other material that's too wide to fit comfortably in portrait mode (with the wide edge of the page vertical), consider changing the page orientation by selecting your paper size in landscape mode (with the wide edge horizontal) and then switch back to portrait mode.

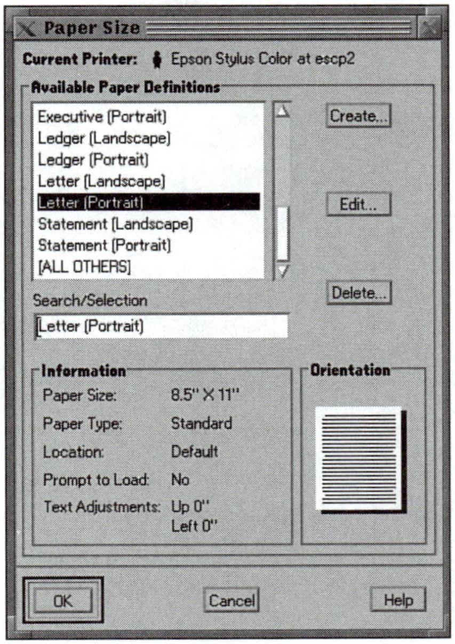

Figure 9.1
You can find a specific paper size if you don't know its name by trying different sizes and watching the Information and Orientation areas of the dialog box.

In general, the page size command sticks with a particular page number, not with any given text on that page. This fact can be frustrating at times if you try to adjust page sizes or orientations when you first create a document because the page to which you want to apply a change may change as you edit your document. Of course, you usually make only one page size change per document, at the start of the document, so this situation isn't normally a major problem.

Tip #111 from *Tux Tips*

If you want to change the page size in the middle of a document, wait until you're ready to print to make the change, if possible. This technique obviates the need for redoing the page size changes as you edit.

Creating a Custom Page Size

If you can't find an appropriate standard paper size in the Paper Size dialog box, you can enter custom values by double-clicking the [ALL OTHERS] definition. This step brings up the Select "[ALL OTHERS]" Paper Size dialog box shown in Figure 9.2 that allows you to enter custom values.

You can enter information on a number of paper attributes in this dialog box.

- Paper Type—You can tell WordPerfect what type of paper you're using (that is, whether it's standard paper, a transparency, labels, or something else). For the most part, this entry has little or no effect on your results, though it might with some printers if you're using WordPerfect printer-specific drivers.

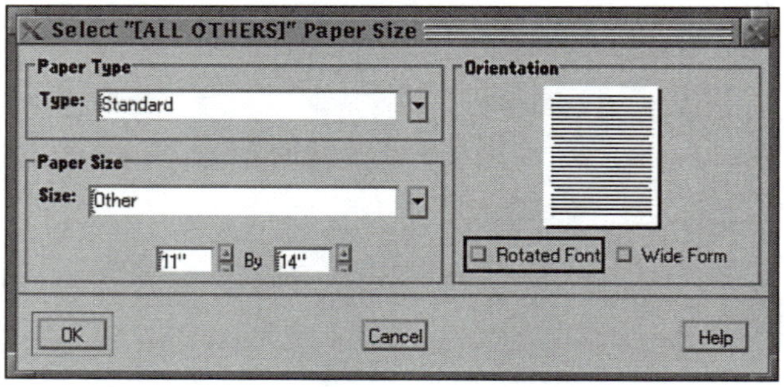

Figure 9.2
Specify paper sizes, types, and orientations for unusual paper sizes in the Select "[ALL OTHERS]" Paper Size dialog box.

→ For more information on WordPerfect printer driver options, **see** "Fonts and Printing," **p. 739**

- Paper Size—In this dialog box, you can usually select from more predefined sizes than you can in the main Paper Size dialog box. If none of the entries in the expanded list suits your needs, you can enter the precise values by selecting a size of Other and entering the dimensions in the spin boxes below the Size entry.
- Orientation—With most printers, you obtain landscape orientation by selecting the Rotated Font option. WordPerfect then tells the printer to print sideways. If you select the Wide Form option, WordPerfect assumes that your printer can print on wider-than-usual paper, as might be the case if you're using a wide-carriage dot-matrix or inkjet printer or a high-end laser printer that can accept extra-large sheets of paper.

Note

WordPerfect quietly rejects an attempt to define a paper size larger than your printer can handle. You might be able to do so if you're printing via Ghostscript and you select a wide-carriage PostScript printer driver, but you won't get the entire width of the printout when you try to print.

Are you having trouble setting a custom paper size, even one that you know your printer can handle? See "When Setting a Custom Paper Size Fails" in the "Troubleshooting" section at the end of this chapter.

If you regularly print to an unusual paper size, WordPerfect gives you two easy ways to select your paper size:

- Create a document template that contains the page size changes you create. When you select the template, you'll automatically use your custom page size.

→ If you want to learn more about templates, **see** "Promoting Cross-Document Consistency with Templates," **p. 333**

- Create a custom paper size definition and then select it from the main options.

The second option actually enters your page size definition into the WordPerfect printer driver on your hard disk. Therefore, you can use this technique only when you're running WordPerfect in Administration mode (by starting it as root with the `-adm` switch) or if you're modifying a personal printer. A regular user cannot create a custom paper size definition for a shared printer.

To permanently create a custom paper size, select the printer you want to modify from the Select Printer dialog box that opens when you click the printer icon in the WordPerfect status bar at the bottom of the document window. To get to the Paper Size dialog box (refer to Figure 9.1), choose File, Page Setup, Page Size; then click Create to create a new paper size definition. At this point WordPerfect greets you with the Create Paper Size dialog box, as shown in Figure 9.3.

Figure 9.3
WordPerfect stores the paper size you create in the printer driver .prs file in the .wprc subdirectory of your home directory or in the shlib10 subdirectory of the WordPerfect installation.

Many items in the Create Paper Size dialog box also appear in the Select "[ALL OTHERS]" Paper Size dialog box (refer to Figure 9.2), and they function the same way in both places. In addition, the Create Paper Size dialog box offers the following options:

- Paper Name—This entry is the name of your paper type definition as it will appear in the Paper Size dialog box.

Tip #112 from
Tux Tips

Most of the standard WordPerfect definitions include (Portrait) or (Landscape) to indicate their orientations, and you may want to do the same for yours.

- Paper Location—You may or may not have many options for paper location. This entry is intended for printers with two or more paper bins, which most low-cost printers don't have.

Caution

Even low-end printers may have an option here for "manual feed." I suggest you not use it, though, because it's likely to not get along well with Linux's printer queue design. Similar comments apply to the Prompt to Load Paper items, which amounts to the same thing for most printers.

- Text Adjustments—When you're printing a small paper size, your printer may feed it through to the extreme left, centered, or maybe even in some other way, relative to a standard-width sheet of paper. You can have WordPerfect adjust the placement of text on the page to account for this behavior by entering appropriate values into the Text Adjustments fields.

Note

If you create a custom paper size in this way and then change printer definitions to one that doesn't support this size, WordPerfect converts the definition to an unnamed custom definition, so you won't lose your document formatting. The results of this conversion will be meaningful and predictable only if your definition is for a paper size that the target printer can handle, however.

Page Breaks

WordPerfect automatically starts a new page when you run out of space on the current one, so for typing most ordinary text, you needn't be concerned with the details of page breaks. There are times, however, when you may want to control just where or how WordPerfect starts a new page, or you may want to force a new one well before the current page is filled. Fortunately, WordPerfect provides options for these situations.

Forcing a New Page

The simplest of WordPerfect's page break control mechanisms is a forced page break. If you know you want to start a new page at some point, say because you're beginning a new section in your document, you can tell WordPerfect to start a new page by choosing Insert, New Page or by pressing Ctrl+Enter. This step creates a hard page break code ([HPg] in Reveal Codes), and whenever WordPerfect encounters this code, it starts a new page.

Tip #113 from *Tux Tips*

If you enter a carriage return followed by a page break, upon reformatting the page break may appear as the first code on a new page, thus creating an entirely blank page. To avoid this possibility, put your page break code at the end of a paragraph; press Ctrl+Enter, rather than the Enter key alone, to end the paragraph and start a new page.

WordPerfect also offers some more sophisticated variants of this option from Format, Page, Force Page, which produces the Force Page dialog box shown in Figure 9.4. Depending upon how you use it, Force Page may or may not actually start a new page. It also inserts its code at the start of the current paragraph, rather than at the insertion point.

Figure 9.4
Use the Force Page feature when you need text to appear on an even- or odd-numbered page.

The Force Page dialog box offers three options (four if you count the None button, which has no effect). The Current Page Odd and Current Page Even buttons ensure that the current paragraph begins on an odd or even page, respectively. If the page number already matches that criterion, WordPerfect does not start a new page, so you may see no changes to the document except in Reveal Codes mode, where WordPerfect inserts a [Force] code at the start of the paragraph. If the paragraph begins on the "wrong" type of page, however, or if reformatting makes that the case, then WordPerfect inserts a page break (code [THPg]) just before the [Force] code.

The third Force Page option, New Page, operates much like a hard page code but appears at the start of the current paragraph. Also, Force Page's New Page option is a little smarter about not creating unnecessary blank pages. If you're creating double-spacing between paragraphs by pressing the Enter key twice, inserting a hard page code can sometimes create a blank page. Force Page's New Page code, however, won't create a blank page unless at least three [HRt] codes precede it.

WIDOW AND ORPHAN CONTROL

In the context of computer word processing, a *widow* is the last line of a paragraph appearing by itself at the top of a page, and an *orphan* is the first line of a paragraph appearing at the bottom of a page. These conditions are generally undesirable, and so WordPerfect provides a means to control them. Choose Format, Keep Text Together to bring up the Keep Text Together dialog box shown in Figure 9.5. Selecting the check button under Widow/Orphan tells WordPerfect to keep at least two lines of each paragraph on any given page. This code affects pages, and so it is inserted at the start of the current page.

Tip #114 from *Tux Tips*

You can create a document template that contains a Widow/Orphan code and then use that template regularly to avoid having to enter the code with each document.

Figure 9.5
The Keep Text Together dialog box lets you specify options to keep certain lines of text together.

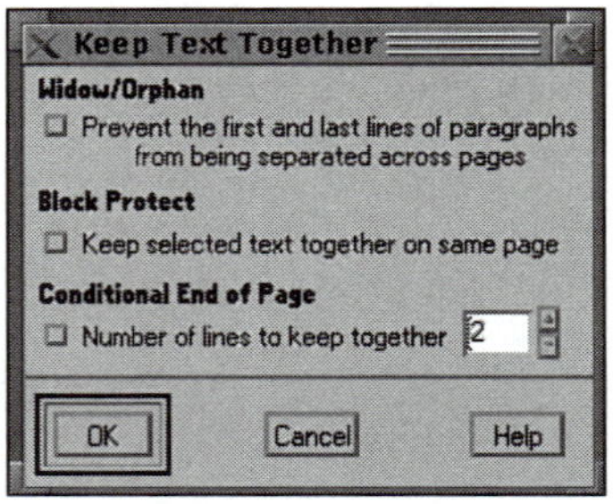

→ For more information on templates, **see** "Promoting Cross-Document Consistency with Templates," **p. 333**

Protecting Blocks of Text

WordPerfect offers two more options for protecting blocks of text on the Keep Text Together dialog box shown in Figure 9.5. The first, Block Protect, is fairly simple. If you highlight a segment of text and select Block Protect, WordPerfect does not break that text across pages.

Block Protect is a paired code feature, so it comes in opening and closing varieties, both shown as [Block Pro] in Reveal Codes. These codes may appear anywhere in text, so in some sense they behave like character-oriented codes. If necessary, WordPerfect inserts a page break several characters or even lines before the beginning of a block protect code. For instance, if the code appears midline, WordPerfect breaks the page at the beginning of that line, or possibly at the beginning of the previous line or earlier if widow/orphan protection or some other formatting code necessitates such a break.

In addition to the Block Protect feature, the Keep Text Together dialog box offers the Conditional End of Page option. Although the conditional end of page is a single code, it influences a specified number of subsequent lines. Selecting this option and using the spin box to enter a number of lines to keep together tells WordPerfect to break the page at or before the resulting [Condl EOP] code if the page break would occur within the specified number of lines. Like Block Protect, Conditional End of Page inserts a page break earlier than the code itself if necessary to accommodate placement within a paragraph or some other code affecting the end of page.

Tip #115 from *Tux Tips*

Block Protect is useful for protecting lists, code samples, and other chunks of text that have variable extent. If you add to or delete from the protected text, WordPerfect still protects the entire block. Conditional End of Page is best suited to keeping headings with their associated text and protecting similar formatting that has a rigidly defined vertical extent.

Tip #116 from *Tux Tips*

These codes can be particularly handy and effective when used in styles. If you incorporate a Conditional End of Page in your heading styles, for example, you can avoid creating headings that get separated from their associated text.

→ For more information on styles, **see** "Ensuring Consistency Within Your Document with Styles," **p. 305**

MEETING PAGE-LENGTH GOALS

Many situations require you to fit your text onto a certain number of pages. For example, your teacher or supervisor may have imposed a page-length limit on your report, or maybe you've written a document and find that the last page has a single line of text—perhaps even your name in the signature field of a letter. To get around these problems, people who use word processors have learned to adjust margins, font sizes, and line and paragraph spacing by hand, nudging one value up and another down to get the desired effect. WordPerfect for Linux has the capability to automate this process, allowing you to get on with things that are more important.

To use Make It Fit, choose Format, Make It Fit. This step opens the Make It Fit dialog box shown in Figure 9.6. From here you specify the desired page count and tell WordPerfect which characteristics it may adjust to meet that page count.

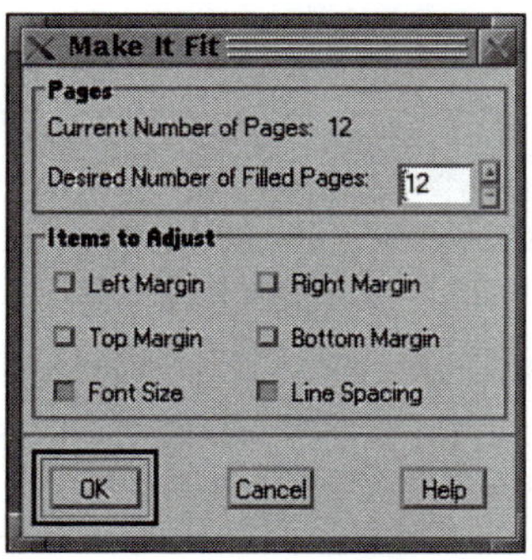

Figure 9.6
Make It Fit adjusts any or all of the items you allow, and it may adjust them in competing ways to fill out the last page.

Some things to note about how WordPerfect accomplishes this task include

- WordPerfect places a limit of a 50% increase or reduction in page length. For instance, if your document currently has 12 pages, the target length must be between 6 and 18 pages. If you enter a value outside this range, WordPerfect beeps and substitutes the closest allowable value.
- Make It Fit doesn't introduce a unified code; rather, it works by altering or creating several other codes. WordPerfect makes global changes in the initial style and also makes changes to formatting alterations you entered outside the initial style.

→ If you want to learn more about the initial style, **see** "The Special Case of the Document Initial Style," **p. 311**

- Make It Fit operates on the entire document; it cannot apply changes to selected pages within a document.

Tip #117 from *Tux Tips*

If you must get around the 50% increase/decrease rule, try applying Make It Fit more than once. Having to change your document's size this much, though, may be a signal that you should add or remove material. If you need to cut, WordPerfect's grammar checker may be able to help you locate verbose passages.

→ If you want to learn more about the grammar checker, **see** "Checking an Entire Document at Once," **p. 131**

Because Make It Fit operates by inserting or altering codes potentially located throughout your document, undoing the effects may be difficult. The primary changes are in the initial style, however, and you can edit these without too much difficulty. To do so, open Reveal Codes either by choosing View, Reveal Codes or by pressing Ctrl+Shift+F3 and go to the top of the document. The first code in your document is called [Open Style: InitialStyle]. Double-click it, and you'll get the Styles Editor (Figure 9.7). Now you can edit or delete the initial codes that Make It Fit entered to adjust your document's size. If your document did not itself change the attributes that you allowed Make It Fit to alter, you'll be able to completely undo its effects from here. Otherwise, you'll have to track down any other Make It Fit changes, or leave them alone.

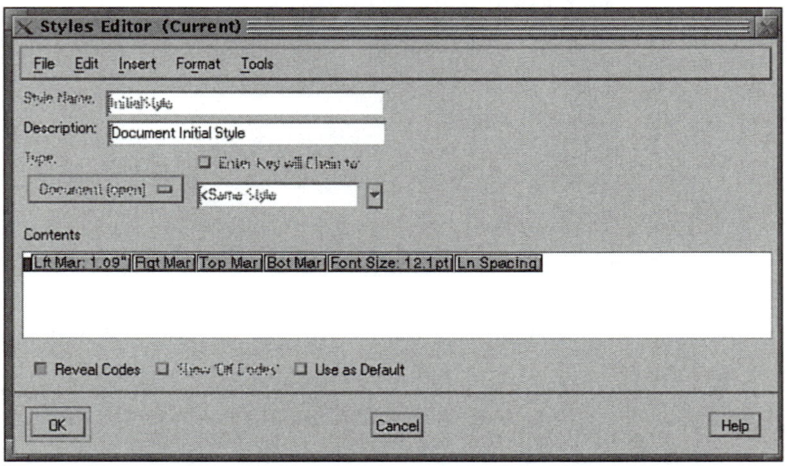

Figure 9.7
The Styles Editor operates much like Reveal Codes mode.

Tip #118 from *Tux Tips*

To avoid possibly having to track down and change Make It Fit's changes, don't apply this feature until you're ready to print.

Tip #119 from *Tux Tips*

> If you make a small change that enlarges your newly fitted document so that it no longer meets your page-length limit, you can apply Make It Fit a second time to create new formatting.

HEADERS, FOOTERS, AND PAGE NUMBERS

Multipage documents often need identifying information on each sheet. This information is typically printed at the top or the bottom of the page as a *header* or *footer*, respectively. Headers and footers can contain fixed information, such as the document's title or author, or variable information, such as the page number. In fact, the page number is so important and so frequently used that WordPerfect offers special commands for quickly formatting a document with page numbers.

CREATING HEADERS AND FOOTERS

To create a header or footer, choose Insert, Header/Footer. This step opens the Headers/Footers dialog box shown in Figure 9.8, which allows you to create, edit, or discontinue a header or footer. WordPerfect supports up to two headers and two footers per document. This feature allows you to have different headers and footers for even- and odd-numbered pages, for instance.

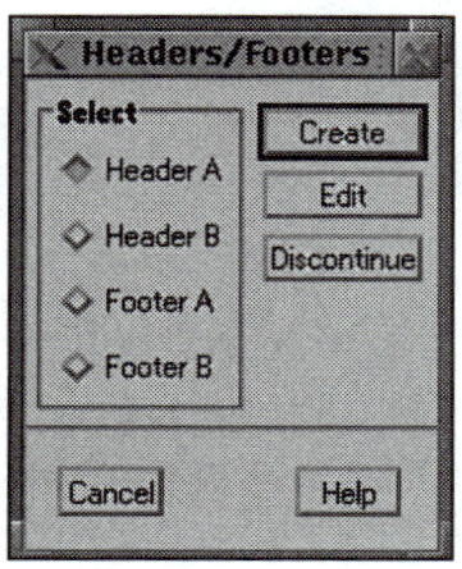

Figure 9.8
Create, edit, or discontinue a header or footer with the Headers/Footers dialog box.

Note

> The only difference between Headers (or Footers) A and B is that they're two different headers/footers that can be used simultaneously in a WordPerfect document. If you only need one header or footer, you can use either A or B and get the same effect. If you need both, it doesn't matter which you use for which purpose. You'll make your life slightly easier if you use them consistently, though—for instance, always using Header A if you need only one header in a document.

When you select Create to create your header or footer, several things happen on the screen, as shown in Figure 9.9. First, the header or footer is actually created and displayed, surrounded by margin guidelines to let you know where the header or footer ends and the main text begins. Second, the lower toolbar changes to incorporate tools specific to headers and footers.

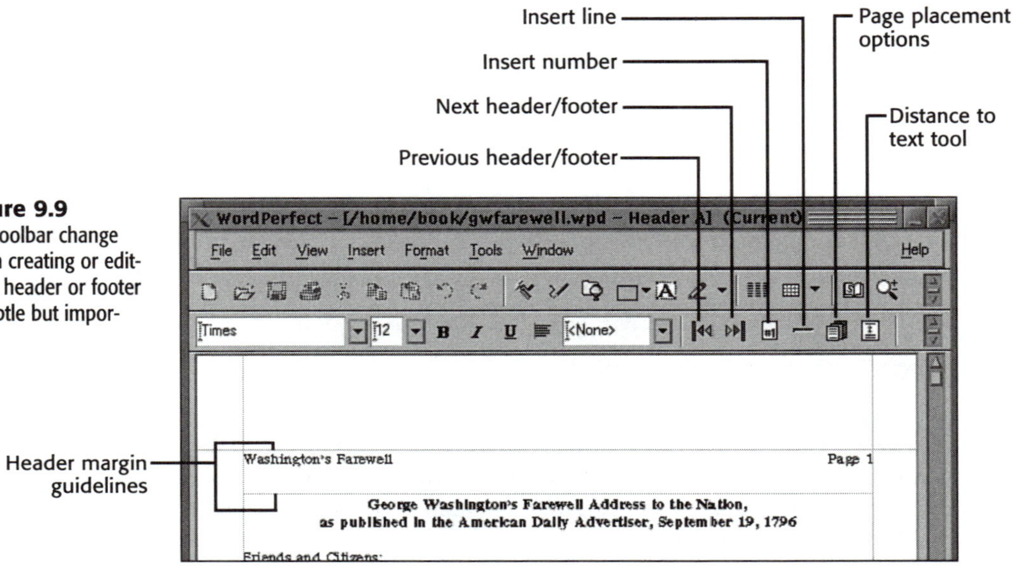

Figure 9.9
The toolbar change when creating or editing a header or footer is subtle but important.

If you're editing in Page view mode, you can simply type your header or footer as you would any other text. You can also alter some special header and footer features, mostly via the header/footer toolbar additions:

- Previous/Next header/footer—Click the Previous or Next header/footer toolbar item to jump to Header B from Header A or to Header A from Header B, respectively. This feature is helpful if you want to create consistent but mirror-image headers and footers for odd and even pages.
- Insert number—You can insert a page, secondary page, chapter, volume, or total number of pages number by selecting the appropriate value from the Insert number toolbar item's pop-up menu. This value appears at the insertion point. You can insert the page number by pressing Ctrl+P, too.

Note

WordPerfect inserts these numbers as codes, not as regular text. That's how it can change them from page to page, which is presumably what you want in a header or footer.

- Insert line—Create a line to separate the header from your body text with the Insert line tool. Creating separator lines is similar to creating graphics lines in the main text, and in fact, you can use the same editing procedures to modify both types of lines.

→ If you want to learn more about graphics lines, **see** "Graphics Lines," **p. 218**

Note

Although *you* can tell where the header or footer and main text are divided because of the margin guidelines, readers do not have this visual clue. Normally the distance between the header or footer and the text is enough of a clue, but you may want to add a line if you've reduced this distance or just to add some visual interest to your document.

- Page placement options—Click the Page placement options item to display the Placement dialog box shown in Figure 9.10. Use this dialog box to tell WordPerfect to place your header or footer on odd, even, or all pages.

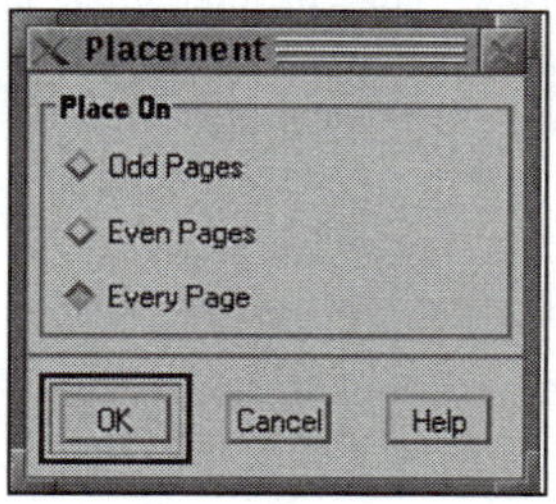

Figure 9.10
Create two headers or footers and assign one to even pages and one to odd pages to produce different headers or footers for two sides of a double-sided document.

- Distance to text—Click the Distance to text tool to display the Distance dialog box in Figure 9.11. WordPerfect automatically maintains the specified distance between the text in your header and your main body text. You cannot edit this space, although it appears onscreen as if it resides in the header or footer.

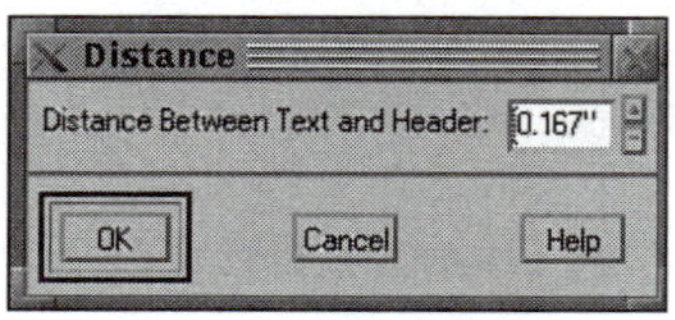

Figure 9.11
A distance of 0 here leaves your header readable but very close to the main text.

When you're done, click your mouse in any portion of the document outside the header or footer or choose File, Close (Ctrl+F4) to return to editing the document itself.

Tip #120 from *Tux Tips*

WordPerfect creates headers and footers so that they fall inside the margins you've set for the document as a whole, thus reducing the space on the page devoted to your document's body text. If you don't want your header or footer to rob space from the text, you'll need to adjust your document's margins yourself. You can use the insertion point position display (near the center of the status bar at the bottom of the window) to help fine-tune this adjustment: Make a margin adjustment; then position the insertion point on the top or bottom line of a page and check its line position. Repeat this procedure until the position is the same as it was on this line before you created the header or footer.

The header/footer code affects the current and all subsequent pages, so you can create a header or footer partway through a document—say, after a table of contents. If you want to change a header or footer, rather than create a new one or discontinue it, you can position your insertion point on the changed page and select Create from the Headers/Footers dialog box (refer to Figure 9.8). Type in your new header or footer, and it replaces the old one. Selecting Edit here simply lets you edit the existing header or footer, affecting its appearance on all pages.

Note

WordPerfect usually inserts the header/footer code at the start of the current paragraph, though sometimes the code ends up at the insertion point. Therefore, your header or footer might drift somewhat from where you intended it to begin unless you're careful about where your insertion point is when you create the header or footer.

EDITING HEADERS AND FOOTERS

If you're working in Page view mode, editing a header or footer is quite simple: Just click in the visible header/footer area and make your changes. These changes affect the headers and footers on all the pages of the document, even if you're making your changes on a page late in the file.

To make a change to a header or footer, you need to choose Insert, Header/Footer and then select Create. Your new header or footer overrides the old one, if you select the same letter choice (A or B), starting from the page on which the insertion point rests. You can also discontinue a header or footer from the same dialog box by selecting Discontinue. If you select Edit from this dialog box, WordPerfect moves the insertion point to the header or footer area to permit editing. You can use this technique to edit a header or footer if you're using Draft mode. Alternatively, you can double-click on the header/footer code in Reveal Codes.

WORKING WITH HEADERS AND FOOTERS IN DRAFT VIEW

When you use Draft view (View, Draft or Ctrl+F5), WordPerfect doesn't display headers or footers on the screen along with the rest of the text. Therefore, you won't be able to judge how they'll appear relative to the rest of your text, and you won't be able to simply click on a header or footer to edit it. When you create or edit a header or footer, WordPerfect opens a separate editing window. This window looks just like other WordPerfect editing windows and has the same menu items and features, though some menu options aren't available. You can even use Reveal Codes in this window.

Tip #121 from *Tux Tips*

To check the appearance of a header or footer, choose View, Page (Shift+Ctrl+F5) to switch to Page view. You can switch back to Draft view after checking the layout of your headers and footers.

Page Numbering

You can enter a page number anywhere in a header or footer by pressing Ctrl+P or selecting the appropriate entry from the Insert number tool in the header/footer toolbar (refer to Figure 9.9). You can even enter a page number in the body of the text itself, if you so desire, by choosing Format, Page, Page Numbering, Insert Page Number, Page Number.

WordPerfect maintains several counters for page and other document element numbers. You can insert these in headers, footers, or other textual elements. These counters are

- Page numbers—Page numbers increment automatically with the page count. By default, page numbers begin at 1 and go up from there.
- Secondary page numbers—Secondary page numbers work just like page numbers, but the two count independently. Therefore, you can set page numbers to count pages from the start of the main text while secondary numbers count pages in a table of contents, for instance.
- Chapter numbers—Chapter numbers are intended to count chapters in a larger document, but you can use them for other things if you like. You must increment chapter numbers explicitly, as described shortly.
- Volume numbers—Like chapter numbers, volume numbers count larger text segments and must be incremented manually.

If you want to create a header or footer with *only* the page number and little or no extra text, WordPerfect offers a shortcut that doesn't involve explicitly creating headers or footers. Choosing Format, Page, Page Numbering, Numbering opens the Page Numbering dialog box shown in Figure 9.12.

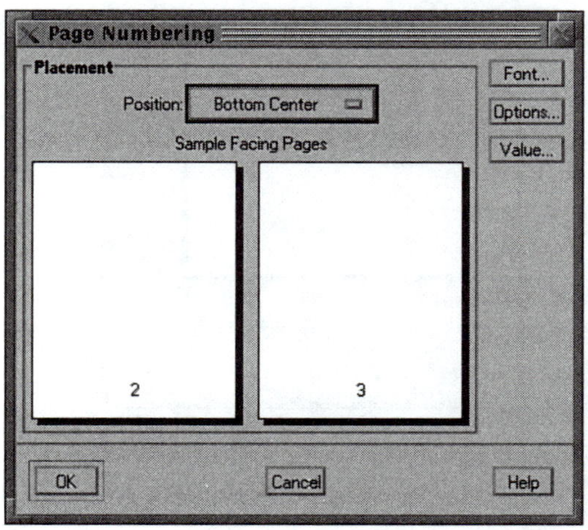

Figure 9.12
The sample pages in the Page Numbering dialog box illustrate how page numbers will appear if you print on both sides of the paper.

This dialog box provides several options for controlling the appearance of the page number.

- Position—The Position button provides assorted options for the position of the page number. The Alternating Top and Alternating Bottom options are intended for double-sided printing because they place the page numbers on the outside corners of the pages.
- Font—Clicking the Font button produces a standard WordPerfect Font dialog box from which you can select the font and font style of the page numbers.

→ To learn about the options in the Font dialog box, **see** "Fonts," **p. 178**

- Options—Clicking the Options button produces the Page Numbering Options dialog box shown in Figure 9.13. You can customize the text that surrounds the page number, as well as how the page number is presented, from this dialog box. For instance, you could add the word *page* before the page number by typing it into the text entry field, or you can add chapter or other numbers. You can also change how each number is displayed—as an Arabic number, a Roman numeral, or a letter. If you click the Insert Format and Accompanying Text at Insertion Point check button, WordPerfect inserts the page number as text in the main document, using the layout you create here, rather than as a header or footer.

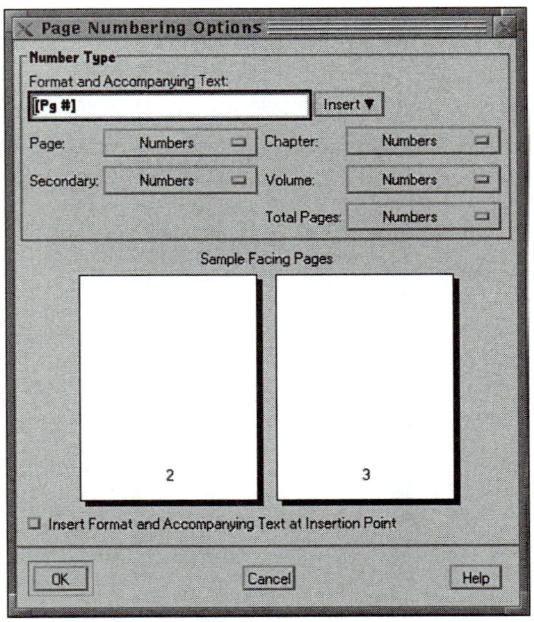

Figure 9.13
The sample pages display a rough guideline of how your changes will appear.

- Value—Clicking the Value button opens the Numbering Value dialog box shown in Figure 9.14. From here you can adjust the page, secondary page, chapter, and volume numbers as you see fit. If you click the Insert and Display at Insertion Point check button, WordPerfect inserts the relevant value into the main body of your text.

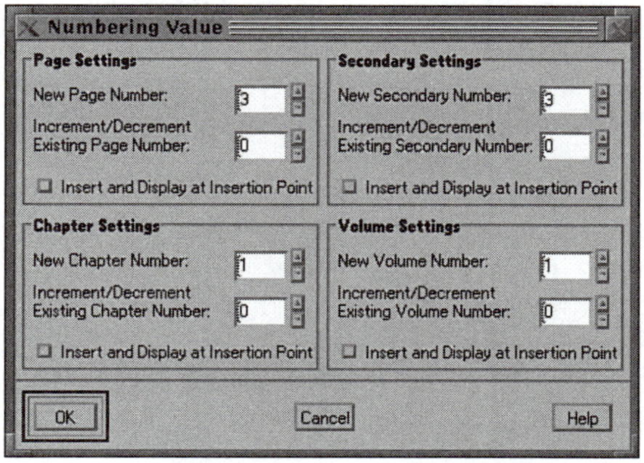

Figure 9.14
You can adjust a number either by entering it directly or by specifying how much to increment it over the existing value.

As you've seen, WordPerfect allows for several types of numbering, including page numbers, secondary page numbers, chapter numbers, and volume numbers. Although you'll probably use page numbers most often, the additional numbering types can be very helpful with larger documents. You can even print pages based on these numbering schemes.

→ To learn how to print a range of pages, **see** "Creating Paper Output: Printing," **p. 148**

WATERMARKS

A watermark is a graphic or piece of text that's printed in faint (typically gray) text underneath the rest of the text in the document. You might use a watermark to print text such as "Confidential" or "Draft Copy" on every page of a report in large but light letters. You could also use this feature to display a company logo or other artwork in similar fashion.

To work with watermarks, choose Insert, Watermark. This sequence opens the Watermark dialog box shown in Figure 9.15. Like headers and footers, watermarks can be created, edited, or discontinued, and you can have two of them in a document to allow for variation across opposite sides in a double-sided document.

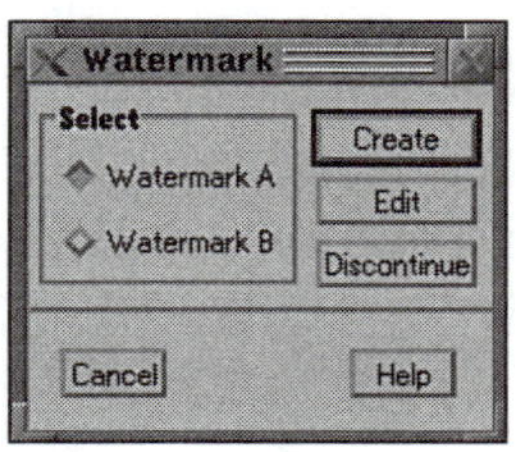

Figure 9.15
You can edit a watermark from anywhere in a document; however, when you create a watermark, it starts to appear from the insertion point's current page.

When you select Create (or Edit, if you want to edit an existing watermark), WordPerfect pops open a new Document window in which you create your watermark much as you would any other document except that any text or graphics you create will be gray, rather than black, by default. As with headers and footers, WordPerfect provides additional toolbar items when you create a watermark (Figure 9.16).

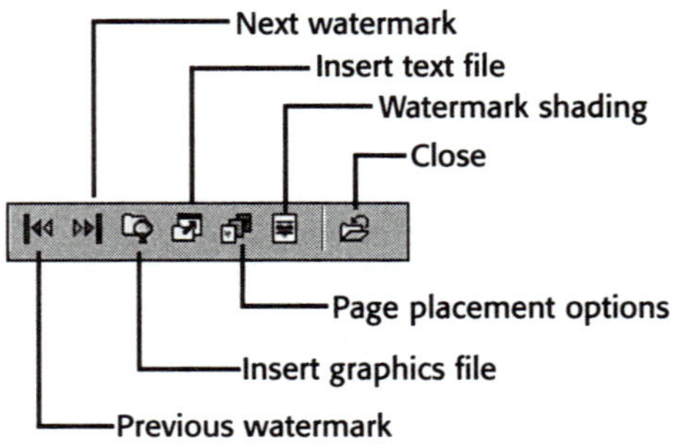

Figure 9.16
You may need to adjust the width of your window to see all the watermark toolbar items.

An explanation of the various watermark tools follows:

- Previous and next watermarks—You can click these toolbar items to bring up Watermark B from Watermark A, or vice versa.

Note

You can use Watermarks A and B much as you would Headers or Footers A and B. You can use them to create mirror-image watermarks for even and odd pages, to create different watermarks for different sections of a document, or to combine two watermarks on one page. If you need only one watermark, you can use either Watermark A or Watermark B.

- Insert files—You can insert a graphics or text file to use as your watermark, or part of it. If you regularly use certain watermarks, you may find this feature particularly helpful—for example, you can save a file with your desired text, formatting, and graphics and load it here. You could also create a template with the watermark predefined.

→ If you want to learn how to create and use templates, **see** "Promoting Cross-Document Consistency with Templates," **p. 333**

- Page placement options—Click on this toolbar item to open a dialog box that allows you to select whether to place the watermark on even, odd, or all pages. This dialog box is also used for headers and footers, as shown in Figure 9.10.
- Watermark shading—This tool opens the Watermark Shading dialog box shown in Figure 9.17. By default, both text and graphics are shaded at 25% of their normal darkness. You can increase this value if you want a darker watermark or decrease it for a lighter watermark. You can adjust the settings for text and graphics independently, so that if you want to combine text and graphics in a single watermark, you can do so and adjust their relative shading to whatever you desire.

Figure 9.17
The graphics shading option is grayed out unless you've inserted a graphic object as a watermark.

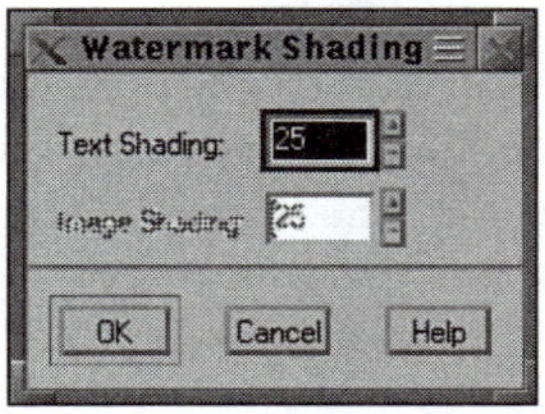

Tip #122 from *Tux Tips*

You can create a watermark with two shades of text either by using both A and B watermarks and adjusting their shading levels independently or by using the color or shading options for the font within a single watermark.

→ If you want to learn about using the font shading options, **see** "Fonts," **p. 178**

Tip #123 from *Tux Tips*

You'll generally want to use a large font for text in your watermark to make it easily visible. If the largest font available via the font list is inadequate, use the Font dialog box (choose Format, Font or press the F9 key) and type in a larger value in the Font Size field.

You can use either text or graphics as a watermark. To use a graphic, you can create it with WordPerfect's graphics editor or insert a graphics file from an external file in various common file formats.

→ If you want to learn about WordPerfect's graphics features, **see** "Graphics," **p. 377**

Tip #124 from *Tux Tips*

If you want to create a watermark with text that's rotated relative to the regular text on the page, you can use the Rotate feature on text in a graphics image.

Does WordPerfect become sluggish after you create a watermark? See "WordPerfect Slows After Creating a Watermark" in the "Troubleshooting" section at the end of this chapter.

TWO-SIDED PRINTING

Actually creating a two-sided printout is something that's very much dependent on your particular model of printer. Some printers have a duplexing option that creates double-sided printouts automatically. With most printers, though, you need to print the even-numbered pages and then pass the same sheets through again to print the odd-numbered pages on the opposite sides of the same sheets. You'll have to experiment with your printer to get the orientation and page order right.

→ To learn about creating a two-sided printout, **see** "Printer Options," **p. 155**

WordPerfect does offer several options relevant to double-sided printing, however. For instance, support for both A and B headers, footers, and watermarks, along with the capability to print these items on even or odd pages only, is precisely what you need to create differences between the headers, footers, and watermarks of even and odd pages (or left and right pages, as they're often oriented when bound in some way). You can also specify alternating placement of page numbers when you use the Format, Page, Page Numbering, Numbering option to create a page number.

Tip #125 from *Tux Tips*

> In double-sided printing, it's common to place page numbers so that, when the pages are bound, the numbers appear on the outside corners of the page (the way this book is printed). When you need to find a page, say, because you're following a cross-reference, you can do so easily by thumbing through the top corners of the pages, without opening the book full to its spine.

WordPerfect has a few more two-sided printing tricks. Choose Format, Page, Page Setup, Two-Sided Printing to display the Two-Sided Printing dialog box shown in Figure 9.18. This dialog box lets you select the *binding width*, a margin adjustment that allows room for binding, which necessarily hides a certain amount of the paper from view. On two-sided documents, you must add the binding width to one side of the text on even-numbered pages and to the opposite side on odd-numbered pages, and the Two-Sided Printing dialog box makes this adjustment automatically. You can add the binding width to any of four sides of the page. This identification applies to the page as you'd view the final printed stack when you begin to read, so for a document such as most English-language books, you'd select Left. Enter the extra space you want to add in the Amount field, and WordPerfect adds this value to the margin for that border. You can see the result in the margins in the Document window.

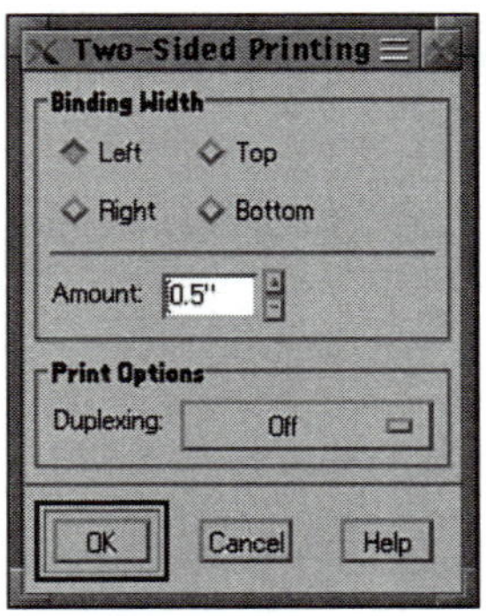

Figure 9.18
Be sure to increment the Amount field in the Binding Width section of the Two-Sided Printing dialog box because it defaults to 0!

Note

> To bind a single-sided document, simply adjust its margins appropriately by choosing Format, Page, Page Setup, Margins (or pressing Ctrl+F8), as described in Chapter 8.

→ If you want to learn about adjusting margins, **see** "Paragraph Formatting," **p. 205**

The Two-Sided Printing dialog box also contains an area for setting duplexing options for printers with built-in duplexing. You can tell WordPerfect whether your printer turns pages along their short or long edges when it duplexes to help get the page orientation right. If you're using a printer without a duplexing option, this feature has no effect. Also, you need WordPerfect drivers that support your printer's duplexing capabilities for WordPerfect to use those capabilities properly.

→ To learn about setting up WordPerfect printer drivers, **see** "Fonts and Printing," **p. 739**

SUPPRESS AND DELAY CODES

WordPerfect can suppress certain codes on any given page. To use this feature, choose Format, Page, Suppress, which brings up the Suppress dialog box (Figure 9.19). You can tell WordPerfect which page-formatting feature(s) to disable for the current page. From the same dialog box, you can also print the page number at the bottom center of the current page, which you might want to do for the first page of a chapter, for instance.

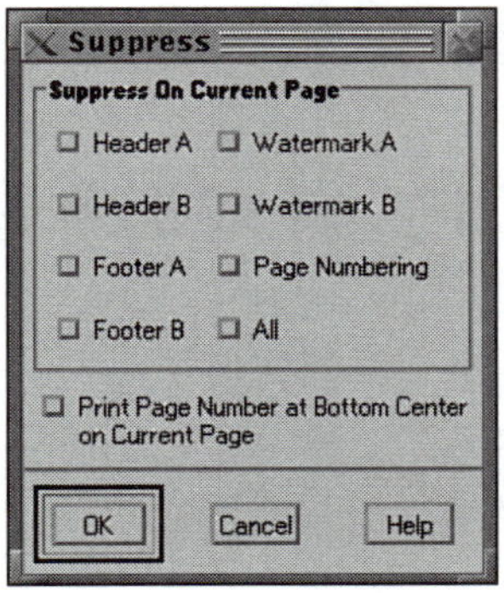

Figure 9.19
Use the Suppress dialog box to turn off headers, footers, page numbers, or watermarks for a single page.

The Suppress Codes feature works only on a single page, and it's the page on which the insertion point rests when you insert the code. WordPerfect inserts the code at the start of that page, even if your insertion point is elsewhere when you select the Suppress Codes feature.

Tip #126 from *Tux Tips*

You cannot apply Suppress Codes to a range of pages, but you can apply it separately to more than one page.

Tip #127 from
Tux Tips

Remember when you suppress a header or footer that WordPerfect formats pages so that these elements appear entirely within the margins. If you want your text to start and end at consistent locations despite the suppression of a header or footer, you have to change your margins for only the page on which you suppress the header or footer.

In contrast to Suppress Codes, Delay Codes allows you to enter a new formatting code and delay its effect by a specified number of pages. To use this feature, choose Format, Page, Delay Codes and enter the number of pages you want the code delayed in the Delay Codes dialog box (Figure 9.20). When you click OK, WordPerfect greets you with a new Document window in which you can enter the codes you want to take effect on a subsequent page (Figure 9.21).

Figure 9.20
WordPerfect accepts only positive values for the Number of Pages to Delay entry.

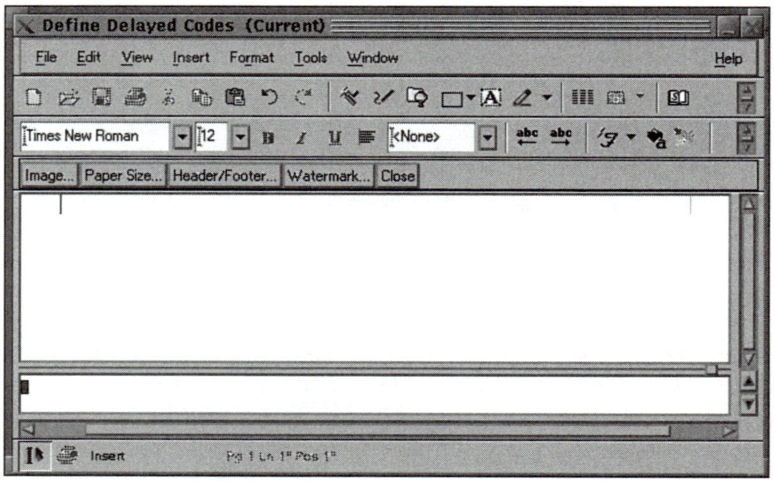

Figure 9.21
WordPerfect ignores text and inappropriate codes you enter in the Define Delayed Codes window.

The Define Delayed Codes window includes a number of extra buttons in a new toolbar for setting common page-formatting features. You can use these or select other formatting options, but WordPerfect rejects some types of codes; for instance, you can issue a font change, but you cannot specify a bold attribute. When you're done creating your code changes, click the Close button in the third toolbar.

When you enter a delayed code, WordPerfect inserts the code at the top of the page you intend it to affect, not at the insertion point. Reveal Codes shows the code as [Delay Codes], and if you position the insertion point just to the left of that code, it expands to show which specific codes it encompasses. Unlike most codes, [Delay Codes] can't be edited by double-clicking it. To change it, you have to move back a page or two and select the Delay Codes option again. This step brings up the Define Delay Codes dialog box that contains the codes you entered already.

Tip #128 from *Tux Tips*

Use the Delay Codes feature when you want formatting to begin a fixed number of pages past some point and continue indefinitely thereafter. Use Suppress Codes to remove a formatting feature from the midst of a range of formatted pages.

Note

WordPerfect actually uses Delay Codes when you select various other page-oriented formatting commands. This approach keeps the code on the page it's supposed to affect, as opposed to drifting around as you edit your text.

Formatting for Special Paper Types

Earlier in this chapter, I described how to set your page size in WordPerfect. This option is very useful if you're printing on something other than a standard sheet of paper. There are some specific types of paper for which WordPerfect has specialized built-in support: envelopes and labels.

Envelopes

If you write a letter in standard letter format, with the recipient's address formatted as a series of lines along the left margin near the top of the page, WordPerfect automatically extracts this address when you choose the Envelope feature, Format, Envelope. The resulting Envelope dialog box (Figure 9.22) lets you enter a return address and print the envelope directly. After you enter a return address, WordPerfect remembers it and uses the same address in the future unless you change it. If you start the Envelope dialog box from a blank page, or from a page that doesn't contain any text WordPerfect can identify as a recipient's address, you must enter the address manually.

Note

If your current printer definition doesn't include any envelope paper sizes, you need to create an envelope paper size to use this feature. Be sure to specify Envelope as the paper type, or WordPerfect won't recognize the definition as an envelope.

→ If you want to set up an envelope paper size, **see** "Creating a Custom Page Size," **p. 227**

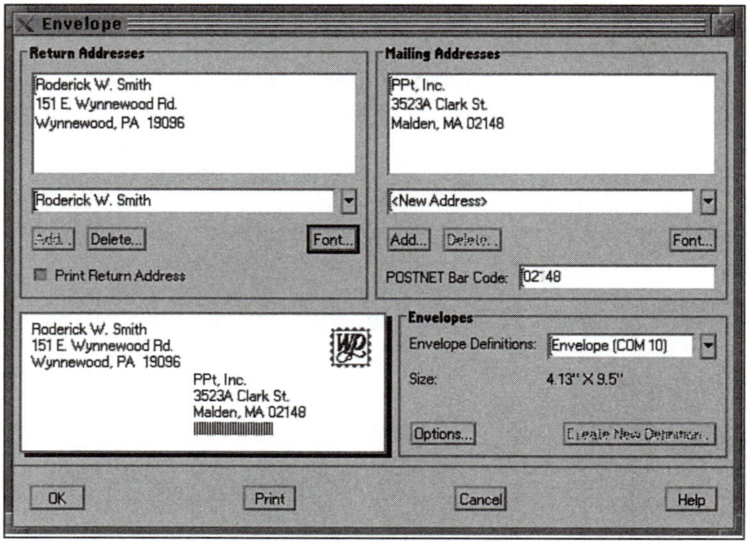

Figure 9.22
As you type a return address in the Return Addresses field, it appears in the envelope preview area.

The Envelope dialog box includes the following features:

- Address books—Both the Return Addresses and Mailing Addresses fields contain simple address books. When you type a new address, or when WordPerfect extracts one from a letter, you can click the Add button to add the address to the address book. You can then retrieve the address by clicking the appropriate address selector button. To delete an address, select it first and then click the Delete button.
- Font selection—If you click the Font button in either address field, you can change the font in which that address is printed.

Note

The printing font bears no relation to the font displayed in the Envelope preview area. The latter font is probably larger than what will print, so don't be concerned if the text runs over the envelope's border; it will probably be fine when printed.

- Envelope selection—You can select from a number of envelope types from the Envelope Definitions selector. If you're printing to a personal printer or if you're running WordPerfect in Administration mode, you can create a definition for a custom envelope size by clicking Create New Definition. This option works the same as the procedure for creating a new paper size definition.
- Envelope options—If you click the Options button, WordPerfect displays the Envelope Options dialog box, shown in Figure 9.23. You can adjust the locations of the addresses from here, and you can tell WordPerfect to include a U.S. Postal Service bar code on the envelope. Such a bar code enables the post office to route your letter to its destination more efficiently.

Figure 9.23
WordPerfect changes the address positions automatically for different envelope sizes.

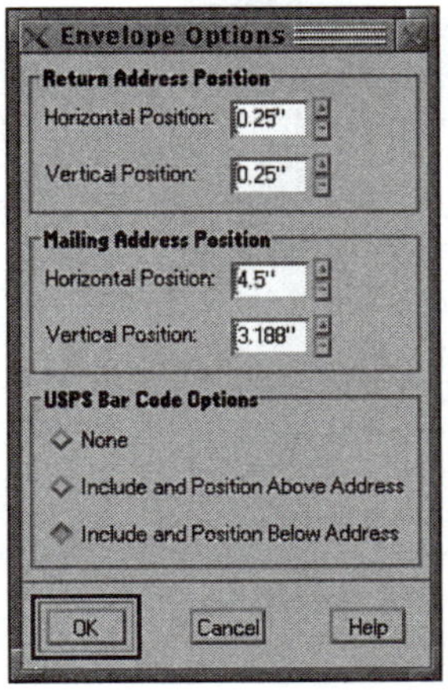

- POSTNET bar code—If you elect to include a postal bar code in the Envelope Options dialog box, the Envelope dialog box contains an extra text-entry field in the Mailing Addresses area. Sometimes WordPerfect correctly identifies the zip code and enters it here; other times you need to enter it manually.

Tip #129 from *Tux Tips*

When you use a Zip+4 zip code, WordPerfect may stumble over the hyphen in both the address field and the bar code field. Delete the corrupted hyphen and retype it to correct this minor glitch. You can avoid the problem by using a hard hyphen (press Ctrl+–), rather than a conventional hyphen, in the Zip+4 code; the hard hyphen enables WordPerfect to read the Zip+4 code correctly.

→ If you want to learn more about hyphenation, **see** "Hyphenation," **p. 220**

When you're done defining your envelope, you can click OK to insert the resulting formatting codes at the end of your document, or you can click Print to immediately print the envelope to the current printer.

Tip #130 from *Tux Tips*

If you intend to print the envelope immediately, be sure to select the correct printer by clicking the printer icon in the status bar before you start this dialog box. WordPerfect does *not* give you the chance to change the printer from the Envelope dialog box.

→ To learn how to select a printer, **see** "Selecting a Printer," **p. 151**

Tip #131 from *Tux Tips*

To format the envelope correctly, WordPerfect relies on printer-specific information. If you're using Ghostscript or a WordPerfect driver for a printer that's similar, but not identical, to the one you're actually using, you may end up printing envelopes in some way that's

continues

continued

> inappropriate for your printer. For instance, some printers feed envelopes narrow edge first and positioned along the middle of the paper path, whereas others may require positioning along the left edge of the paper feed or wide-edge-first orientation. You may be able to adjust the way you feed envelopes to conform to WordPerfect's expectations, or you may want to try other WordPerfect printer drivers. If you normally print via Ghostscript, for instance, you might use drivers for your printer or something similar for envelopes, even if you prefer not to use them for most purposes for some reason. You can also adjust the page size definition for envelopes in the printer driver you're using, as described earlier in this chapter.

→ If you want to know more about WordPerfect's printer drivers, **see** "Fonts and Printing," **p. 739**

→ If you want to know more about setting a page size, **see** "Page Size," **p. 226**

LABELS

Another special type of form for which WordPerfect includes support is labels. You can purchase labels for a variety of purposes in office supply stores. There are labels for use in shipping packages, for labeling videotapes, for name badges, and on and on. Although you can use WordPerfect's labels feature on ordinary paper, you'll probably want to use it with actual labels. With these in hand, then, choose Format, Labels to bring up the Labels dialog box shown in Figure 9.24.

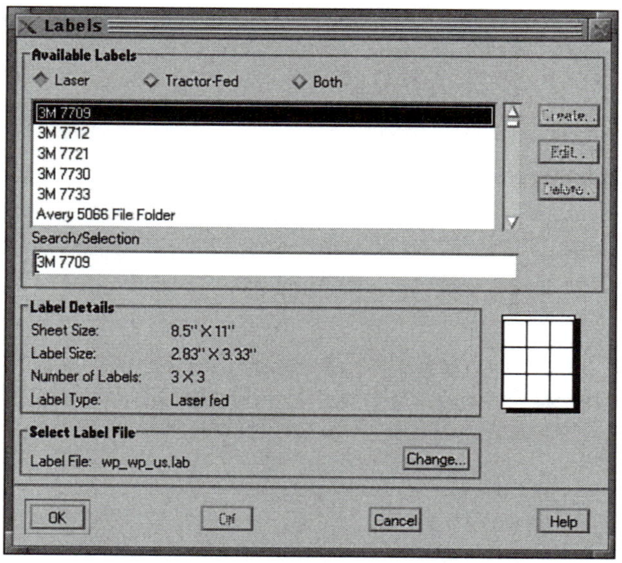

Figure 9.24
You may be able to use a label definition for a label from one manufacturer with a label from another one, but you'll have to check the sizes in Label Details against what you have.

Locate the label type to which you wish to print in the scrollable list at the top of the dialog box. Note that you can elect to see sheets of labels for laser or inkjet printers (identified by the Laser radio button), tractor-fed labels for dot-matrix printers, or both. When you click on a label type, the corresponding information in the Label Details area changes and a preview of the label layout appears.

Tip #132 from *Tux Tips*

Watch the preview and Label Details areas to confirm that you've selected the correct label type.

WordPerfect stores its information on how different labels are laid out in something called a label file. You can change the label file you're using by clicking Change in the Select Label File area. You'll probably want to do this only if you want to define several custom label types and store them in a file in your own directory.

If you're running WordPerfect in Administration mode or if you're using a custom label file, you can create, edit, or delete label definitions by clicking the appropriately named buttons in the Available Labels field of the Labels dialog box. This step produces the Edit Labels dialog box shown in Figure 9.25. Give your label a description you can use to identify it and then tell WordPerfect how the labels are laid out. You specify the size of each label on a page, the position of the top left label, the number of columns and rows, the distance between labels, and the margins you want WordPerfect to use within each label area.

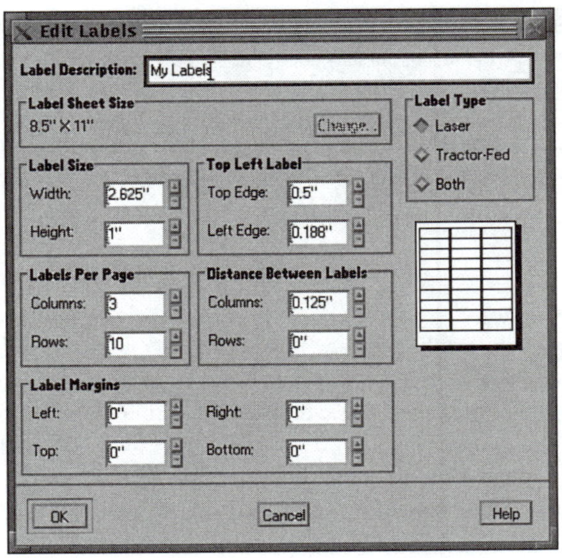

Figure 9.25
WordPerfect assumes that labels are rectangular; if yours aren't, you must specify an appropriate rectangular outline and add appropriate graphics or just make intuitive allowances to get text in the correct area.

When you've selected your label, click OK and WordPerfect returns you to the Document window, but in Label formatting mode. Much of the Document window is now gray, and only the portion corresponding to the upper-left label is white and editable. Enter whatever information you want for that label in this area, and if you want to create another label, choose Insert, New Page (Ctrl+Enter) to go on to the next label. Continue this process for however many labels you need to print, and then print your document as you would any other, but print to your labels.

Is WordPerfect crashing when you try to create a label? See "WordPerfect Crashes When Creating a Label" in the "Troubleshooting" section at the end of this chapter.

Tip #133 from *Tux Tips*

You can print to ordinary paper, too. If you're testing a label layout for a new type of label, you may want to print to paper to check the placement of text on the labels. Create graphics lines in each label area to judge the placement of text within the physical labels. After you print to paper, hold it up in bright light against the label to see whether the areas WordPerfect selected for printing match the areas of the labels themselves.

→ To learn how to use graphics lines, **see** "Graphics Lines," **p. 218**

Tip #134 from *Tux Tips*

If you need to start printing from somewhere other than the upper-left label, you can press Ctrl+Enter as many times as you need to create "empty" labels to bring you to your target label.

Caution

Heed the label and printer manufacturers' recommendations on label use! Using either the wrong type of label or partially used label sheets can cause labels to peel off *inside* your printer—and has the potential to cause expensive damage to your printer. If in doubt, check your printer's manual. Note also that laser and inkjet labels may vary in their glue formulations, and inkjet label glue may not stand up well to the heat in laser printers. You should therefore use *only* laser printer labels in laser printers.

Subdividing a Page

Another WordPerfect feature that is similar to printing labels is subdividing a page. In this case, you tell WordPerfect to split the page into so many rows and columns, and it treats each resulting rectangular area as a single page. You can initiate this feature by choosing Format, Page, Page Setup, Subdivide Page to open the Subdivide Page dialog box shown in Figure 9.26. Specify the number of columns and rows you want and click OK to initiate the subdivision. The Document window shows only the first segment of the subdivided page until you fill it or press Ctrl+Enter to move to the next segment.

Note

Your current margins and ruler settings apply unchanged to *each* subdivided page segment. A standard 1-inch margin is usually too large for subdivided pages, and you'll want to change this setting, either before or after you subdivide the page.

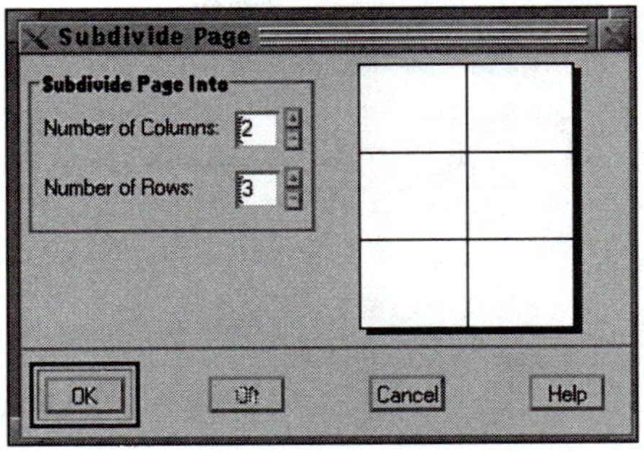

Figure 9.26
The preview area shows how WordPerfect will split the page.

TROUBLESHOOTING

WHEN SETTING A CUSTOM PAPER SIZE FAILS

You're trying to create a custom paper size, but whenever you do, you see no change on screen, though a code does appear in Reveal Codes.

You may be exceeding the paper size for your printer—or what WordPerfect believes your printer to be. If you're printing through Ghostscript to a wide-carriage printer, you may need to select a wide-carriage PostScript model as your WordPerfect printer driver.

→ To learn about WordPerfect printer drivers, **see** "Fonts and Printing," **p. 739**

WORDPERFECT SLOWS AFTER CREATING A WATERMARK

You've created a graphics watermark, and WordPerfect suddenly becomes sluggish.

WordPerfect slows down when displaying graphics, and watermarks can be particularly troublesome this way. If you don't mind losing your onscreen view of the watermark, you can deselect the View, Graphics menu item, which is enabled by default. This change restores WordPerfect's speed by telling it not to display graphics onscreen. They'll still print, however.

WORDPERFECT CRASHES WHEN CREATING A LABEL

You try to create a label, but WordPerfect crashes when you click OK in the Labels dialog box.

This WordPerfect bug is related to ownership of the WordPerfect printer driver (*.prs) files. If you're the only user of WordPerfect on your system, you can eliminate the problem by changing the ownership of these files. For instance, if your username is jiml and if WordPerfect is installed in /usr/wp8, you could do the following from an xterm as *root (p. 229):*

```
cd /usr/wp8/shlib10

chown jiml *.prs
```

Unfortunately, if you have more than one WordPerfect user on the computer or if you don't have root access, this solution won't work. You also won't get the desired results if you give wider ownership to the .prs files because when one user creates a label, the ownership and permissions on that file change, locking out the other users. Another solution, then, is for each user to create a separate *personal* printer definition for printing labels and to select this printer *before* starting to create a label.

→ To learn how to define a personal printer, **see** "Fonts and Printing," **p. 739**

→ To learn how to select a personal printer, **see** "Selecting a Printer," **p. 151**

Another workaround for this bug is to define your own label templates using the page subdivision feature or other formatting and to use these templates rather than WordPerfect's labels feature. You could even run WordPerfect as root to load up the labels you want and save them as templates for other users of the system.

→ If you want to learn to use templates, **see** "Promoting Cross-Document Consistency with Templates," **p. 333**

You may also want to check to see whether Corel has an updated version of the WordPerfect program that fixes this bug.

CHAPTER 10

Creating Professional Layouts with Columns

In this chapter

WHICH TYPE OF COLUMN DO YOU NEED?

WordPerfect for Linux, like most modern word processors, supports formatting text into columns. WordPerfect doesn't stop there, however, because it supports two different *types* of columns: *newspaper* and *parallel*. The former is the type of column you see in newspapers and many magazines—paragraphs of text flow down the left side of the page, then the right (occasionally a page may hold three or more columns). Newspaper columns are very useful on wide paper or for printing in a small font; they make it easier for the reader to find the start of a line of text after finishing the preceding line, a task that can be difficult if the text spans the entire width of a page. Figure 10.1 shows a document that uses newspaper columns.

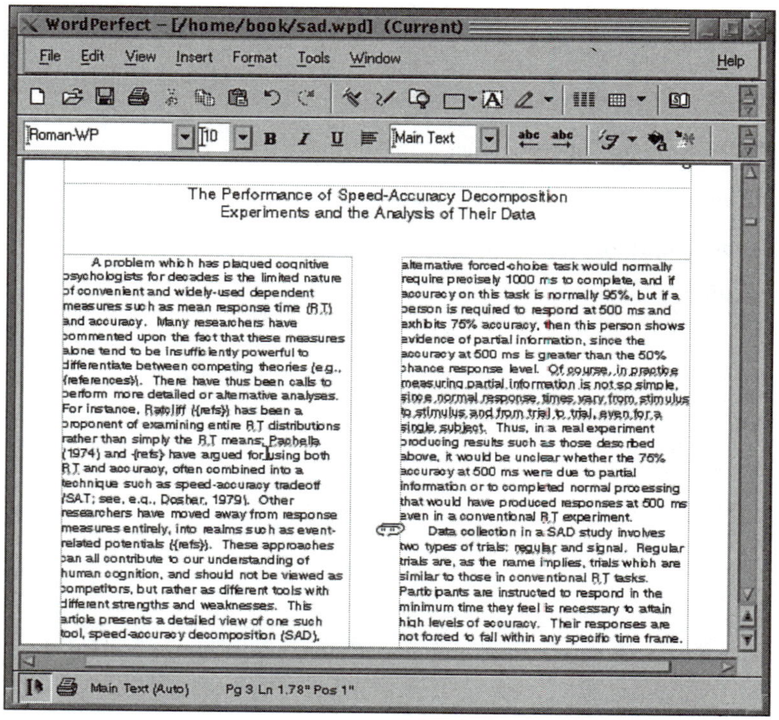

Figure 10.1
Use newspaper columns to format a page of small text into line widths that are easy for the eye to follow.

WordPerfect's parallel columns, on the other hand, are useful for formatting blocks of information, as shown in Figure 10.2. Parallel columns arrange text in rows and columns of text, rather like a table. Parallel columns can more easily span multiple pages, however, and are better suited to heavy text editing than are tables. Parallel columns are therefore the tool of choice for creating scripts that require blocked or columnar layout and for large blocks of tabular information containing lots of text. Tables, on the other hand, make it easier to set up lines separating the individual cells of their design and may export to other word processors better. In many cases, parallel columns and tables are equally well suited to a task, so you can use whichever you prefer.

→ To learn more about tables, **see** "Tables," **p. 275**

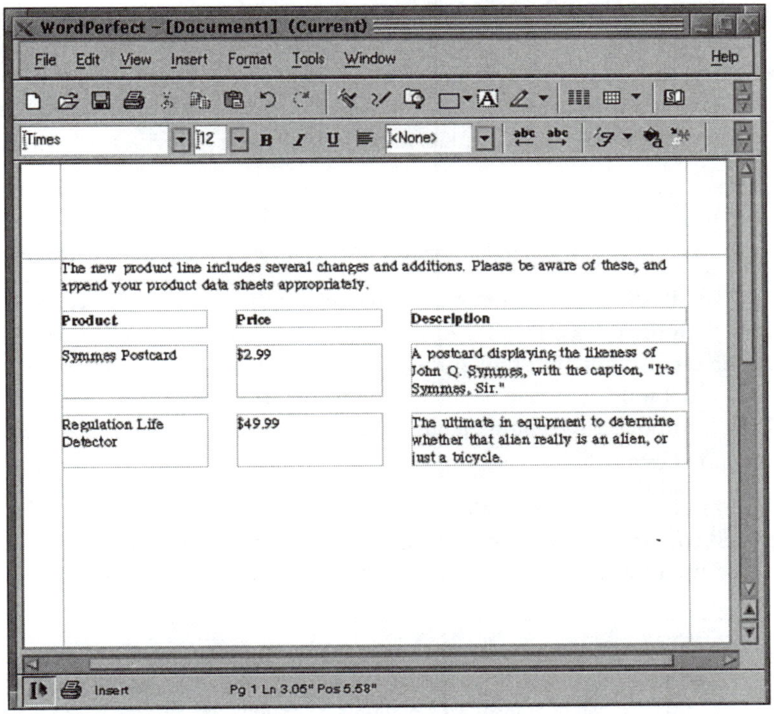

Figure 10.2
Use parallel columns to format lists or other blocks of information.

If you're creating a newsletter or other document with complex formatting, you'll want to read Part IV, "Complex Formatting with Boxes for Text, Graphics, and Equations," which tells you how to insert text, graphics, tables, charts, and equations into your document. You can set WordPerfect to flow your columnar text around these *boxed elements* automatically so you don't need to redefine your columns to accommodate these elements.

→ For more information on boxed elements, **see** "Complex Formatting with Boxes for Text, Graphics, and Equations," **p. 355**

Note

Throughout this chapter, I refer to *Column* mode or *Multicolumn* mode when I mean that WordPerfect is set to display two or more columns. By contrast, the default state for WordPerfect is one in which columns are turned off. This condition can also be thought of as "Single-column mode"—although there's no column definition, text must naturally fall in some sort of vertical organization. That organization is one of a single column.

Understanding Column Codes

Like many of WordPerfect's formatting codes, column codes come in both ongoing and paired types, both of which appear as [Col Def] in Reveal Codes mode. WordPerfect uses an ongoing code if you place your insertion point somewhere in the text without highlighting anything and then select the column definition. All text in the document from that point on appears in column format, unless another code later overrides this setting.

If you select several paragraphs and then turn on a column definition, WordPerfect formats only those paragraphs you've selected in columns, using paired codes. Use this feature if you want to create columns for a portion of text in your document without affecting other text.

You can define a column before or after typing the relevant text. If you define the column first, then the text you type is formatted for columns from the beginning, whereas if you define the column after typing the text, your existing text is reformatted for columns. In general, you'll want to define parallel columns before entering text because they're difficult to format without the column layout in place. Newspaper columns can be handled either way, though you may find it easier to edit text if you define columns after you're finished editing.

Column codes are paragraph oriented. WordPerfect inserts them at the start (and end, for paired codes) of the paragraph in which the insertion point rests or of the highlighted text. You cannot begin or end columns in the middle of a paragraph.

Tip #135 from *Tux Tips*

If you want to begin or end columns inside a paragraph to accommodate some special textual element, consider placing that element in a text box. WordPerfect automatically wraps regular text around a text box, thus obviating the need for a column definition.

→ To learn more about text boxes, **see** "Using Text Boxes to Add Variety to Text Layout," **p. 357**

→ To learn more about wrapping text around boxes, **see** "Text Wrapping," **p. 461**

Attractive Page Layouts with Newspaper Columns

As a general rule, text should average roughly 10 to 12 words per line. If your document has more words per line (because you are using a small font, wide margins, or wide paper), you may want to consider using newspaper-style columns to reduce the count to 10 to 12 words per column. Alternatively, you can increase the size of your font. On the other hand, if you've defined columns and discover that you have fewer than 10 to 12 words per line in each column, you may want to reduce the number of columns in your document, possibly to just one (that is, disable Multicolumn mode). You can also reduce the size of your font in this situation.

These are only rules of thumb, though. If you're in doubt, try printing out two or three pages in several different styles and judge the readability and attractiveness of each style for yourself.

Tip #136 from *Tux Tips*

Don't try to judge legibility or readability from an onscreen display, even if you have a large monitor on which you can view an entire page at once. A computer screen almost always uses lower resolution than a printer, and this factor has a substantial effect on text legibility.

In addition to defining columns, you may want to adjust other elements of your page layout. For instance, you may want to reduce the size of your indents or alter your hyphenation rules to reduce the size of gaps within or at the right edges of your lines.

➔ For more information on indents and hyphenation, **see** "Paragraph Formatting," **p. 205**

METHODS OF DEFINING COLUMNS

The easiest way to start using newspaper columns is by clicking the Define columns toolbar icon on the upper of the two default toolbars. This step produces a pop-up menu of column-creation options, as shown in Figure 10.3. If you select any of the *n* Column items, WordPerfect creates a layout with that many newspaper-style columns evenly spaced across the page. The New Column and Discontinue items have no effect until you define columns. Then these options start a new column and discontinue Column mode, respectively. You can also bring up the Define Columns dialog, described shortly, by selecting Format from this menu.

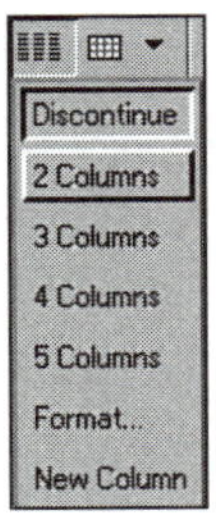

Figure 10.3
Quickly define columns with the Define columns toolbar item.

Tip #137 from *Tux Tips*

Because of the space between columns, turning a single-column document into a multicolumn document usually increases its length unless you also adjust the margins, font size, or some other formatting feature. You can have WordPerfect bring a document to a specified size automatically by using the Make It Fit feature.

➔ To learn how to use Make It Fit, **see** "Meeting Page-Length Goals," **p. 233**

Tip #138 from *Tux Tips*

Editing text in Multicolumn mode can be distracting because the insertion point and the view of the page must shift when moving from one column to another. You might also have to scroll *down* to find text that preceded something you're reading, if that preceding text is in a previous column of the same page. This situation isn't a problem with a big enough screen (or a small enough zoom factor), but I generally prefer to create a document in single-column mode and then define newspaper-style columns, rather than to work with columns from the beginning.

➔ If you want to know more about zoom factors, **see** "Zoom," **p. 55**

When you select one of the predefined column types on the Define column toolbar menu, WordPerfect creates a standard newspaper-column layout with a 1/2-inch *gutter*, or space between columns. If you want to create columns with different details, such as a narrower gutter or uneven column widths, you must use the Define Columns dialog box (see Figure 10.4). You can open this dialog box by selecting Format from the Define columns toolbar icon or by choosing Format, Columns from the WordPerfect main menu. You can use this method of creating columns either to affect all text following the insertion point or to format only a highlighted portion of text; the Define Columns dialog box and the Define columns toolbar entry are the same in this respect.

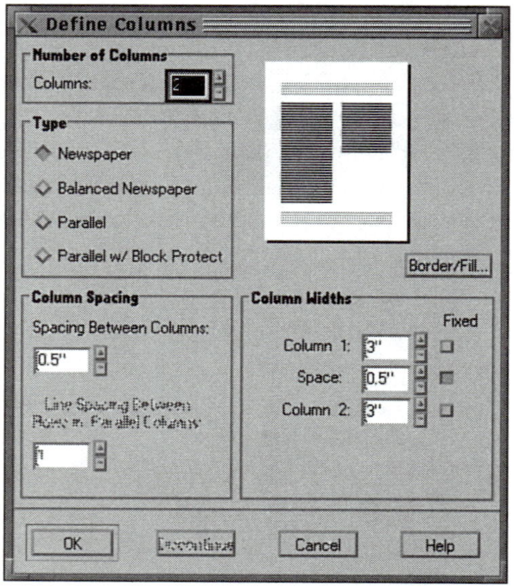

Figure 10.4
The Define Columns dialog box lets you tell WordPerfect how to format your columns.

The default configuration here is for a two-column newspaper layout with a 1/2-inch gutter—the same as you'd get by selecting 2 Columns from the Define columns toolbar menu entry. If that's what you want, you can click OK and your columns will be set. If you want to alter the layout, though, you can do so in a number of ways.

- You can increase the number of columns by incrementing the value in the Columns spin box.

Note

As you change the value in the spin box, both the preview area and the contents of the Column Widths area change to reflect the number of columns you're creating.

- You can adjust the gutter space by modifying the Spacing Between Columns spin box. If you have more than two columns, this value affects all of the Space values in the Column Widths area.

- You can tweak the widths of and spacing between individual columns by modifying the Column Widths values. This technique allows you to create one column that's wider than the others or to put certain pairs of columns closer together than others.

Note

The Define Columns dialog box contains other options that I describe later in this chapter.

The settings in the Column Widths area of the Define Columns dialog box can have important effects both as you define columns in this dialog box and if you subsequently modify your column definition through other means. You should understand that the sum of the values of the columns and spaces listed in these fields must add to the width of the text area on your page (the page width minus the margins). For instance, on a standard 8.5×11-inch sheet of paper with 1-inch margins, the text width is 6.5 inches, so the columns and gutters must add to 6.5 inches. Thus, when you adjust one column or gutter width up or down, WordPerfect must adjust one or more other widths down or up, respectively. The check buttons labeled Fixed next to each column or space value tell WordPerfect which values it can adjust. If a value is selected, WordPerfect won't tamper with it.

Note

If you select all but one check button, WordPerfect beeps if you attempt to adjust the remaining box up beyond the value allowed by your margins. You can adjust this value down, but WordPerfect silently rejects your adjustments when you click OK.

WordPerfect remembers your rules concerning which columns and gutters may be adjusted and applies these rules whenever necessary. For instance, you can adjust the column widths in various ways that don't involve the Define Columns dialog box, but WordPerfect continues to use these settings when you make adjustments by these other means. WordPerfect also remembers which columns and gutters may be adjusted if you change the margins on your page and makes only allowable changes to bring your total text width in line with the new margins.

Tip #139 from *Tux Tips*

If you change a document's margins, WordPerfect adjusts the column widths as allowed by your selections in the Define Columns dialog box. You might not completely approve of whatever method WordPerfect uses, however. For instance, if you increase your margins, hence decreasing column widths, WordPerfect might shrink one column more than you'd like. You should therefore check WordPerfect's changes to be sure they're acceptable for your particular layout.

If you want columns in part of your document only, you can turn off columns by selecting either the Discontinue button from the Define Columns dialog box (Figure 10.4; this button is grayed out until columns are defined) or the Discontinue item from the Define columns toolbar item's menu. When you turn off columns from the second or subsequent column of newspaper-style unbalanced columns, WordPerfect forces a page break before proceeding without columns. If you discontinue columns within the first column, though, WordPerfect starts up single-column operation on the current page.

WORKING WITH TEXT IN COLUMNS

Actual text editing in columns is quite similar to editing without columns. You use the same mouse and keyboard commands to move your insertion point or your view of the page. The presence of columns does add a few twists, however.

→ For basic information on text editing, **see** "Working Within a Document," **p. 43**

MOVING THE INSERTION POINT

The easiest way to move the insertion point from one column to another is usually to click in the desired column with the mouse. You can also move between columns by using Ctrl+Shift+Right Arrow or Ctrl+Shift+Left Arrow.

Note

If you're used to WordPerfect for Windows, you'll note this difference; in Windows, you use the Alt key rather than Ctrl+Shift for column switching. Many Linux window managers, however, use the Alt key for their own purposes, so WordPerfect uses Ctrl+Shift rather than Alt in Linux.

The Go To command (Edit, Go To or Ctrl+G) offers some extra Position options when you're working in Column mode, as shown in Figure 10.5. You can elect to go to the top or bottom of a column, the previous or next column, or to the first or last column. Your insertion point must be located in columnar text before WordPerfect offers these options; if it's in a noncolumnar area of a document that contains columns, you won't get these options.

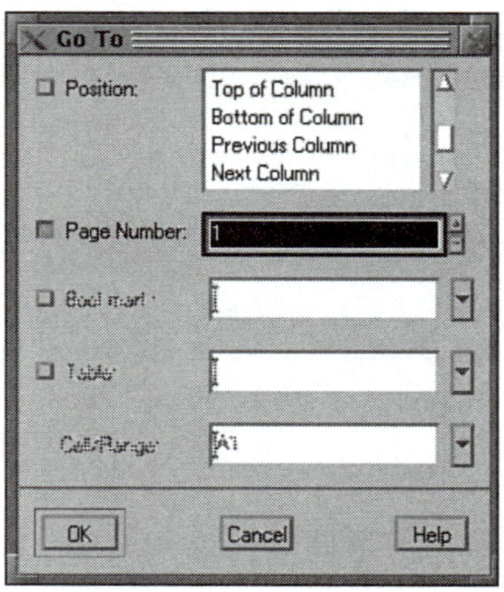

Figure 10.5
The Go To dialog box's Position list adjusts itself to the presence of columns in your document.

Tip #140 from
Tux Tips

If you regularly use keyboard commands to move around columns, you may want to create macros for some of these movement options and assign them to keystrokes.

→ For more information on macros, **see** "Letting the Computer Do the Work: Macros," **p. 483**

PAGE AND COLUMN BREAKS

When you're editing in Column mode, WordPerfect treats page break commands (Insert, New Page or Ctrl+Enter, which normally create [HPg] codes) as column breaks ([HCol] codes). If you've converted a document from single-column to multicolumn format, WordPerfect converts your [HPg] codes to [HCol] codes. When you're editing in Multicolumn mode, a new column command on the last column of a page starts a new page. WordPerfect doesn't accept Force Page commands in Multicolumn mode and ignores any existing Force Page commands in text that's converted into Multicolumn mode.

→ To learn more about page breaks, **see** "Page Breaks," **p. 230**

Tip #141 from
Tux Tips

If you need to force a new page from Multicolumn mode, you can either use multiple column breaks or you can temporarily switch out of Multicolumn mode and enter a page break. The latter will have slightly different effects when working with standard newspaper columns as opposed to balanced newspaper columns, as described shortly.

WORKING WITH THE RULER

Using the WordPerfect ruler (View, Ruler) with columns adds some complexity to the ruler's functionality. Figure 10.6 shows a two-column layout with the ruler. In the figure, the insertion point is in the second column, which affects the ruler's display and action. To understand this behavior, recall that the right and left margin codes, as well as tab and indent codes, are line- or paragraph-oriented codes. They operate on lines and paragraphs and either *are* the margin (or, in this case, column) definitions or operate relative to them. Thus, the iconic representations of these codes shift around on the page along with the columns.

Moving to specifics, you'll note that the margin adjustment and indent widgets appear only in the part of the ruler above the margin in which the insertion point lies. When you highlight text that spans two or more columns, however, these widgets appear over the first column only—even if the text you highlight is in the second and third or later columns.

Tip #142 from
Tux Tips

When the insertion point lies in the second or subsequent column, using the ruler widgets to adjust the margins and indents can be a bit tricky. If you must adjust these items for text past the first column, you may find it easier to do so with the appropriate menu commands rather than via point-and-click ruler operations. Tabs aren't so difficult to adjust via the ruler in Multicolumn mode, though.

→ To learn more about adjusting tabs, indents, and other paragraph formatting features, **see** "Paragraph Formatting," **p. 205**

Figure 10.6
In Multicolumn mode the ruler display is mostly relative to the current paragraph, not to the page as a whole.

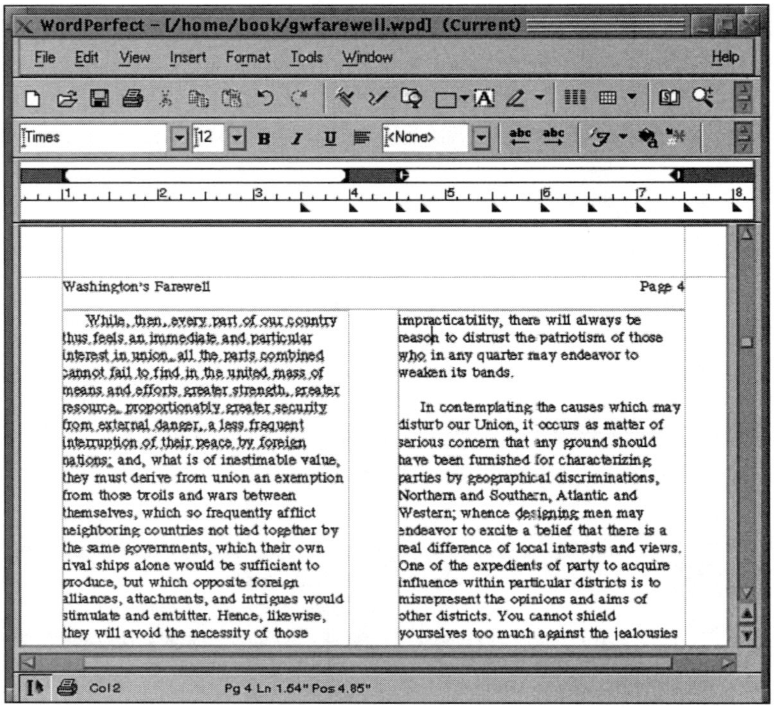

The ruler's tab stops are usually defined relative to the text's left margin, though you may think of them as being fixed on the page. For instance, consider the default tab stops. These appear to start at 0 inches (the left edge of the page), with a new tab stop every 0.5 inch. In reality, though, these tab stops start 1 inch *prior to* the left margin and continue with one new stop every 0.5 inch.

When dealing with columns, therefore, the tab positions in the ruler seem to shift around when you move from one column to another. Further, the ruler's display may not match the effect of tabs in text in a column other than the one in which the insertion point rests. In Figure 10.6, for instance, the insertion point is in the second column, so the first tab stop is at 3.5 inches (the 4.5-inch start of the column minus 1 inch). Although a tab does indent the first line of the paragraph in the first column, a tab stop doesn't *appear* to be there because the ruler is displaying tabs relative to the second column.

You may recall that tabs can be defined relative either to the margin or to the left edge of the page. If you've defined tabs in the latter way, the ruler display and the behavior of tabs in Multicolumn mode are entirely different from the way they were described earlier in this section. When tab stops are defined relative to the left edge of the paper, they are "nailed down" on the page and don't appear to shift as you move your insertion point from one column to another. Although this behavior may seem more intuitive at first, it can actually add to your work because to make consistent tabs across columns, you have to define two or more tabs for every change. For instance, to halve the indent distance from 0.5 inch to 0.25 inch on a two-column layout, such as that in Figure 10.6, you'd need to define tabs at both 1.25 inches and 4.75 inches, as opposed to just one tab stop at 0.25 inch past the column start for relative positioning.

➔ For more information on tabs, **see** "Tabs," **p. 194**

Adjusting Column Guidelines

In theory, you can adjust column guidelines in Multicolumn mode just as you can adjust margin guidelines in single-column mode—just move your mouse pointer over a guideline, click, and drag (Figure 10.7). In practice, this method generally works well with the first column, but sometimes doesn't work with subsequent columns. You can try adjusting your margins this way, but if it doesn't work, choose Edit, Undo and try again using the Define Columns dialog box.

Part II
Ch 10

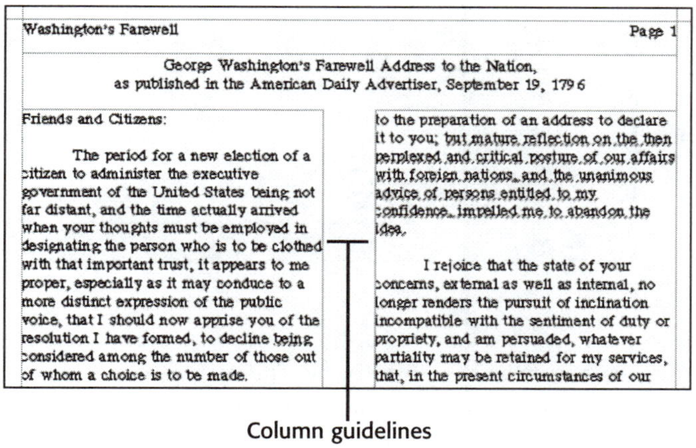
Washington's Farewell Page 1

George Washington's Farewell Address to the Nation,
as published in the American Daily Advertiser, September 19, 1796

Friends and Citizens:

The period for a new election of a citizen to administer the executive government of the United States being not far distant, and the time actually arrived when your thoughts must be employed in designating the person who is to be clothed with that important trust, it appears to me proper, especially as it may conduce to a more distinct expression of the public voice, that I should now apprise you of the resolution I have formed, to decline being considered among the number of those out of whom a choice is to be made.

to the preparation of an address to declare it to you; but mature reflection on the then perplexed and critical posture of our affairs with foreign nations, and the unanimous advice of persons entitled to my confidence, impelled me to abandon the idea.

I rejoice that the state of your concerns, external as well as internal, no longer renders the pursuit of inclination incompatible with the sentiment of duty or propriety, and am persuaded, whatever partiality may be retained for my services, that, in the present circumstances of our

Figure 10.7
Column guidelines work sporadically for adjusting column widths in WordPerfect 8 for Linux.

Are you having troubles adjusting the column guidelines? See "Column Adjustments Don't Work" in the "Troubleshooting" section at the end of this chapter.

Evening Out Column Lengths with Balancing

The preceding discussion covers standard newspaper columns. If you've been using this feature as you've read this chapter, you've probably noticed that the last page of text using standard newspaper columns usually produces uneven columns. The first column may be partly filled, with no text in subsequent columns; or the first column may be completely filled, and the second column may have only a few lines of text (Figure 10.8).

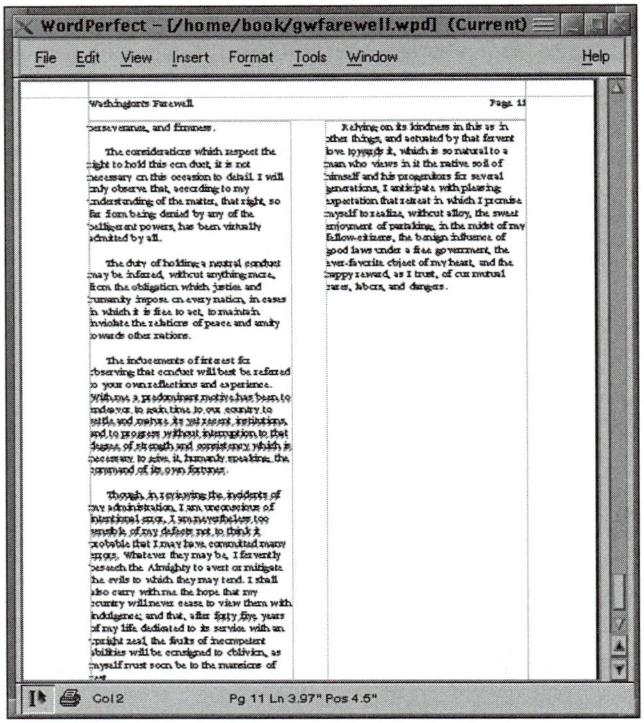

Figure 10.8
Standard newspaper columns can produce an unbalanced appearance on the final page.

To give your columns a more balanced appearance, you can define them as Balanced Newspaper type rather than as ordinary Newspaper type. To do so, choose Format, Columns or select Format from the Define columns toolbar item to bring up the Define Columns dialog box (shown in Figure 10.4); then select Balanced Newspaper in the Type area. This setting produces a fairly even column arrangement on the final page (Figure 10.9).

Note

WordPerfect attempts to balance the column lengths when you use the Balanced Newspaper column setting but may produce a slightly imperfect result. An odd number of lines to be split across two columns, for instance, as in Figure 10.9, necessitates an uneven split. WordPerfect might also be forced to produce a slightly uneven layout because of font changes or other line-height-altering formatting. Finally, stray [HRt] codes at the end of the final column can produce imperfect balancing.

Tip #143 from *Tux Tips*

You can correct imperfect balancing by adjusting line spacing in the final column with WordPerfect's line-height adjustment tools.

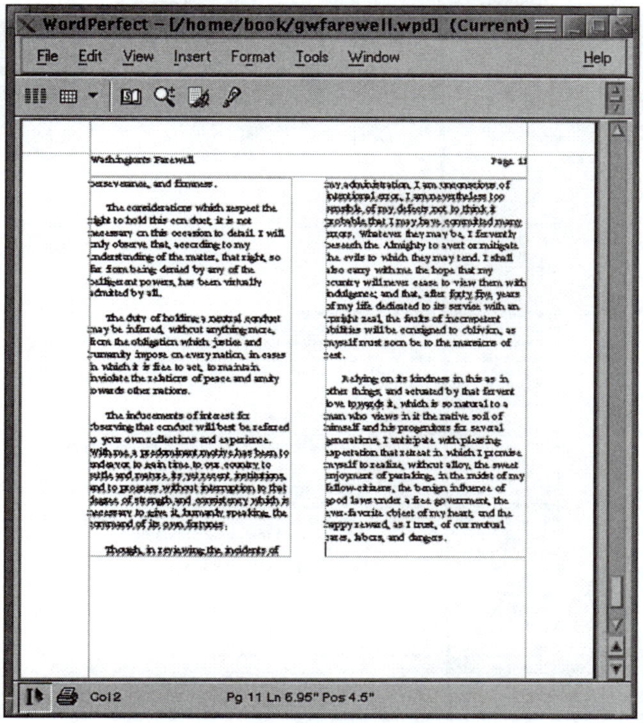

Figure 10.9
Balanced newspaper columns produce a more aesthetically appealing final page for a document.

→ For more information on line-height adjustments, **see** "Word/Letter Spacing," **p. 189**

In addition to looking attractive, balanced columns allow you to switch in and out of Multicolumn mode on a single page. If you turn off conventional columns after beginning a second column on a page, WordPerfect automatically starts another page. With balanced newspaper columns, however, WordPerfect won't do this, so you can switch out of Column mode to, say, insert a headline, and then switch back to Multicolumn mode on the same page (provided that there's still space left on the page, of course).

ORGANIZING INFORMATION WITH PARALLEL COLUMNS

Whereas newspaper columns are useful for presenting text in the most readable manner, parallel columns are useful for organizing information on a page. The resume for Sally Somebody in Figure 10.10 illustrates the use of parallel columns. Parallel columns in the Experience section keep Sally's employers and titles in the left column associated with her accomplishments in bulleted lists in the right column.

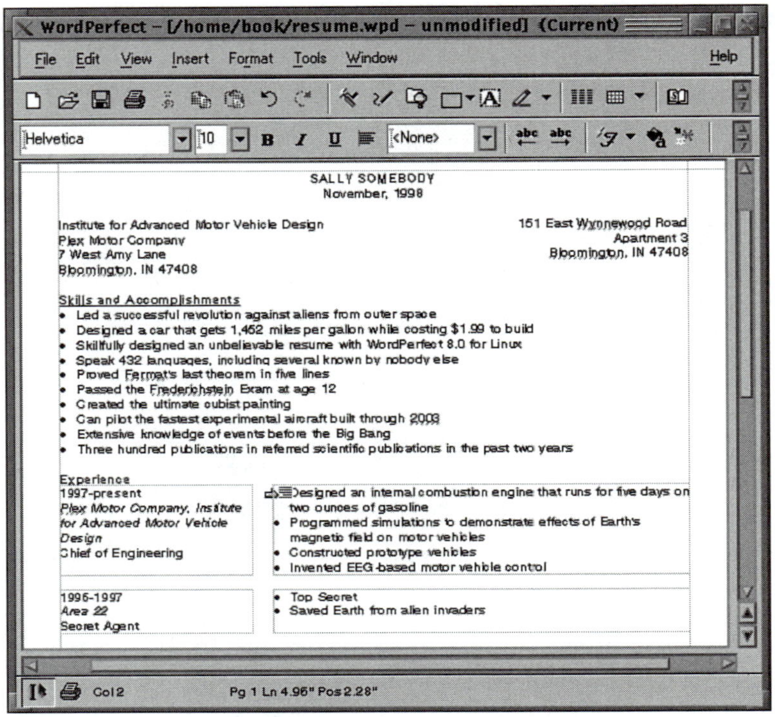

Figure 10.10 Parallel columns keep information in blocks in each column in the correct alignment.

Parallel columns are organized into both columns and blocks. *Blocks* are similar to rows in a table and may sometimes be referred to as rows. Each column in each block is the same length, even if the amount of text is not. For instance, notice in Figure 10.10 how the column guidelines for the left and right columns within each block match in vertical extent. WordPerfect automatically sets these lengths based on the amount of text in each column and block.

DEFINING PARALLEL COLUMNS

Begin your definition of parallel columns as you would other column types, by choosing Format, Columns or by selecting the Format item from the Define columns toolbar icon. When WordPerfect displays the Define Columns dialog box (refer to Figure 10.4), however, you should select Parallel or Parallel w/Block Protect as the Type of column, rather than Newspaper or Balanced Newspaper.

The difference between Parallel and Parallel with Block Protect column types is that the latter does not separate a block across pages. For instance, if Sally Somebody had enough job experience for two pages, a resume with block protect columns would keep a job from being split across two pages. This option functions very much like an automatic block protect on the text in a row.

➔ To learn more about WordPerfect's block protect features, **see** "Protecting Blocks of Text," **p. 232**

The same intercolumn spacing options generally apply to both parallel columns and newspaper columns. In the Define Columns dialog box, you can set the gutter width for all columns in the Column Spacing area and adjust individual column and gutter widths in the Column Widths area. When you select a parallel-column type, however, you get one additional spacing option: Line Spacing Between Rows in Parallel Columns, which sets the distance between blocks of parallel columns.

Note

Because the number of lines of text in two columns of one block need not be the same, the spacing between blocks is likely to vary from one column to another. The Line Spacing Between Rows in Parallel Columns sets the amount of space that's added between the column guidelines.

Entering Text into Parallel Columns

You can enter text into parallel columns much as you do any other text. The same rules concerning the ruler, page breaks, and so on apply to both parallel and newspaper columns.

→ For more information on text manipulation in columns, **see** "Working with Text in Columns," **p. 262**

When you work with parallel columns, pressing Ctrl+Enter or choosing Insert, New Page starts a new column, as is true for newspaper columns. With parallel columns, however, a column exists in one block of text, and when you complete all the columns in one block, the next column is the first one in the next block. If you define a two-column parallel-column layout, for instance, you can type text into the first column; then when you press Ctrl+Enter, your insertion point will jump up and to the right, to the start of the next column, much as it would with newspaper columns. After you enter some text there and press Ctrl+Enter again, however, your insertion point will jump down to a point below the end of the text in the first column of the first block, and you can type additional text. A "new page" command is not guaranteed to start a new page at any time.

Because the exact layout of information is so important with parallel columns, you should probably define them before you begin entering text into them. Although you can convert ordinary text into parallel-column format, it's easy to lose track of your new page commands and get a layout that's not what you expect.

Note

The only way to get text into the second column of a parallel-column layout is to issue a new page command. Thus, converting most ordinary text to parallel-column format results in a single column unless the text contains hard page codes. You can, of course, add column breaks to existing text to format it properly.

Fine-Tuning the Column Layout

You can adjust your parallel-column layout in the same ways you modify a newspaper-column layout—via the ruler, margin guidelines, or the Define Columns dialog box. If you've already defined columns and want to modify their layout throughout the document, or the part of your document that uses them, the best approach is to use Reveal Codes to locate the [Col Def] code that created the columns. Then double-click the code to open the Define Columns dialog box; now you can modify the original column definition—and thereby modify all the columns.

Tip #144 from *Tux Tips*

If you regularly use the same column definition throughout a document, but interspersed with noncolumnar text, you should consider defining the column layout in a style. This way, you can change the column layout later by modifying one style, rather than several different [Col Def] codes.

→ For more information on styles, **see** "Ensuring Consistency Within Your Document with Styles," **p. 305**

Cutting, Pasting, and Moving Parallel Columns

Cutting and pasting text in parallel columns can be tricky, especially if you want to cut or paste more than one column's worth of text. For instance, if you want to rearrange the order of items in a list you've created using parallel columns, you might need to cut and paste several columns at a time. The thing to remember here is WordPerfect's code-based design. At the end of each column in any row, WordPerfect inserts an [HCol] code. To cut or move a row of text, you must select your text so that you preserve the balance of [HCol] codes. If you miss one code when cutting text, WordPerfect shoves subsequent text to the right into the wrong columns.

If you select the text you want to cut or move in standard view, chances are that you'll not get exactly the codes you need. This problem is easily corrected, though; you can position the insertion point at the start of text that's been shifted a column to the right and use the Backspace key to return the text to the correct column. If two columns have become merged together, position the insertion point where they join and press Ctrl+Enter to separate them.

You can avoid these problems altogether by using Reveal Codes mode (View, Reveal Codes or Shift+Ctrl+F3). If you select your text in Reveal Codes, you can include the codes you need and position the insertion point precisely when you want to paste columnar text. If you're selecting a lot of text, though, Reveal Codes can be awkward to use, so you may prefer to just correct any errors that creep in after the fact.

→ For more details on precision code selection, **see** "Using Reveal Codes to Select Text," **p. 111**

Modifying Columns

After you define columns, be they newspaper or parallel, you can modify the column definitions in various ways to enhance their appearance. For the most part, the same principles apply to both newspaper and parallel columns, though there are differences.

No matter where your insertion point is located within a newspaper-column layout, when you change that layout, you change the entire column definition. The only way to change *part* of that layout is to discontinue the columns and then start them up again with a different definition. With parallel columns, however, if you make a change to the column definition (say, altering the relative widths of the columns), WordPerfect applies that change from the insertion point on, leaving earlier blocks of columns unaffected.

CHANGING COLUMN WIDTHS

Particularly when you're using parallel columns, you may decide you need to change the relative widths of your columns. The easiest way to do so is to position the mouse in the middle of the gutter between two columns. When the mouse pointer changes shape, as shown in Figure 10.11, click and drag left or right. This step changes the relative widths of the two columns, making one wider at the other's expense and keeping the gutter width constant. The change affects the entire extent of the columns if you're working with newspaper columns—thus if you make this change on the last page of a document with a single column definition, the change still affects the entire document. With parallel columns, however, this change affects only the columns from the mouse pointer's location forward.

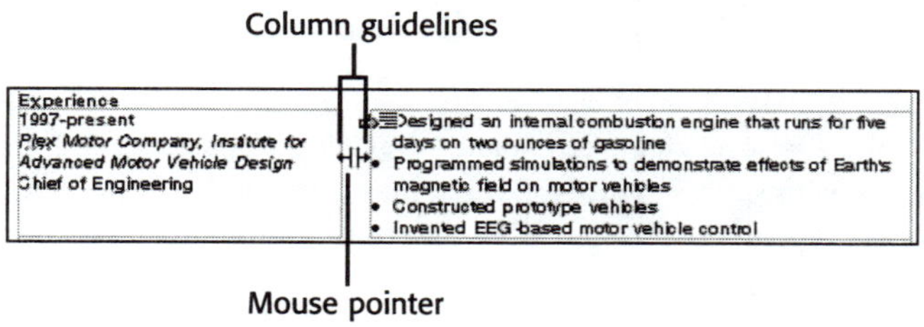

Figure 10.11
The shape of the mouse pointer indicates that WordPerfect will interpret a mouse click-and-drag operation as a command to change the column widths.

You can also change column widths by clicking and dragging on the column guidelines, though this technique may not work well for all columns. Choosing Format, Columns to open the Define Columns dialog box is another way to achieve this change. Either way, a change to parallel columns affects only the current and subsequent column blocks, whereas a change to newspaper columns affects the entire column definition.

Note

If you adjust a column via its column guidelines, only the left and right guidelines have any effect. You cannot adjust a column's top or bottom guideline because they are defined by the location of text. If you need to adjust the vertical placement of your columns, you can insert horizontal advance codes or carriage returns or you can adjust the Line Spacing Between Rows in Parallel Columns item in the Define Columns dialog box.

→ To learn more about advance and line spacing options, **see** "Typesetting Commands," **p. 187**

ADDING BORDERS AND FILLS

You can add impact to columns by applying a border or fill pattern to them—either when you create the column definition or afterwards. Clicking Border/Fill from the Define Columns dialog box produces the Column Border dialog box, shown in Figure 10.12.

Figure 10.12
The Column Border dialog box works very much like other border dialog boxes.

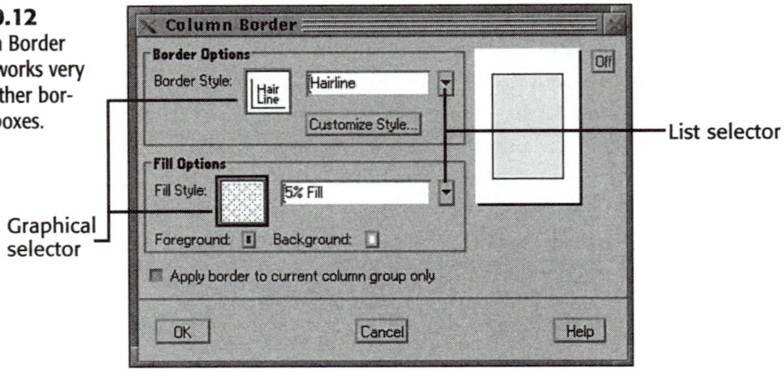

From the Border Options area of the dialog box, you can select the following border styles: single border lines in several widths, double lines, dashed lines, or borders along various sides in pairs or singly. You can also customize a border style by clicking Customize Style and setting the options for each component individually.

WordPerfect also gives you many different selections in the Fill Options area. Select from among solid fills of various darkness or assorted patterns. You can also set the foreground and background colors for the fill pattern.

Figure 10.13 shows one of the many possible border/fill combinations.

Figure 10.13
Most column border/fill styles work best to highlight a limited section of a document.

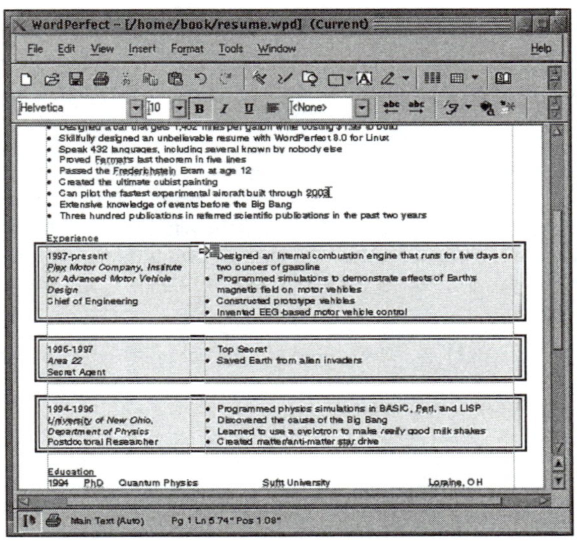

Tip #145 from *Tux Tips*

> Be sure your fill pattern isn't too dark, particularly if you're using colored fills. What's legible on screen may look poor when printed, and photocopying may further reduce legibility, particularly if you use a colored fill.

If you select the Apply border to current column group only check button, WordPerfect uses the border selection only with the parallel block group in which the insertion point rests. This option has no effect with newspaper columns.

Note

> The border and fill options are nearly identical to their paragraph-formatting counterparts.

For information on paragraph borders and fills, **see** "Borders and Fills," **p. 216**

Tip #146 from *Tux Tips*

> You can apply both a paragraph and a column border/fill. WordPerfect applies the paragraph border/fill to the paragraph(s) within a column, so the paragraph border appears entirely within the confines of the column border. If you use fills in both, the paragraph fill takes precedence, completely overriding the column fill.

Are you trying to format your border/fill so that each column has its own border, or there's a gap in the fill pattern between columns? See "Separate Borders for Each Column" in the "Troubleshooting" section at the end of this chapter.

TROUBLESHOOTING

COLUMN ADJUSTMENTS DON'T WORK

You try to adjust column spacing with the column guidelines or ruler, but the cursor jumps around erratically.

This behavior happens sometimes, particularly with two-column newspaper layouts. Try adjusting the columns by using the Define Columns dialog box (Format, Columns or double-click on the [Col Def] code that initiates the column layout).

SEPARATE BORDERS FOR EACH COLUMN

You want to create separate borders for each column, create a line between columns, or apply a fill pattern so that there's a gap to the fill in the gutter area, but you can't get the Column Border option to do this.

WordPerfect gives you two ways to achieve these effects. One is to use paragraph borders/fills rather than column borders/fills. Move your insertion point to the start of the columns you want to be so bordered. Then choose Format, Paragraph, Border/Fill and apply your desired style from the resulting Paragraph Border dialog box. Be sure to uncheck the Apply border to current paragraph only check button. The second technique is to use the List selector rather than the Graphical selector for the Border Style (refer to Figure 10.12).

At the bottom of the list, you'll find options for creating a line between columns (Column Between) and for surrounding all sides of the columns (Column All). These two methods produce slightly different results, however, so you may want to experiment a bit to get the effect you desire.

→ For information on paragraph borders and fills, **see** "Borders and Fills," **p. 216**

CHAPTER 11

Tables

In this chapter

OVERVIEW

WordPerfect's table feature allows you to format textual or numeric information in neat rows and columns, optionally with borders around some or all cells of this layout. The table feature is very similar in many ways to the parallel-column feature, though they each have specific strengths. In general, tables are better suited to containing numbers within cells or to creating complex layouts, whereas parallel columns excel at handling large amounts of text. In many cases, you'll be able to use either formatting feature with equal ease and similar results, however.

→ For more information on parallel columns, **see** "Organizing Information with Parallel Columns," **p. 267**

Any given table is organized in horizontal rows and vertical columns. The basic unit of the table is the *cell*, which is the area defined by the intersection of a row with a column. Text you enter in a cell remains there and does not flow into other areas of the table.

CREATING A TABLE

WordPerfect offers several means of creating a table, each with some unique features. You can build a table from scratch, convert existing text into a table, or build a table inside other elements, such as text boxes.

→ To learn more about text boxes, **see** "Using Text Boxes to Add Variety to Text Layout," **p. 357**

CREATING A BASIC TABLE

Creating a basic table is easy, particularly if you use the appropriate tool on the WordPerfect toolbar. If you click and hold the Create table toolbar icon (Figure 11.1), WordPerfect displays a special pop-up widget that allows you to select the dimensions of your table. Drag down and to the right to increase the number of rows and columns, respectively, in the table and then release the mouse button. WordPerfect creates a table with those dimensions and standard formatting.

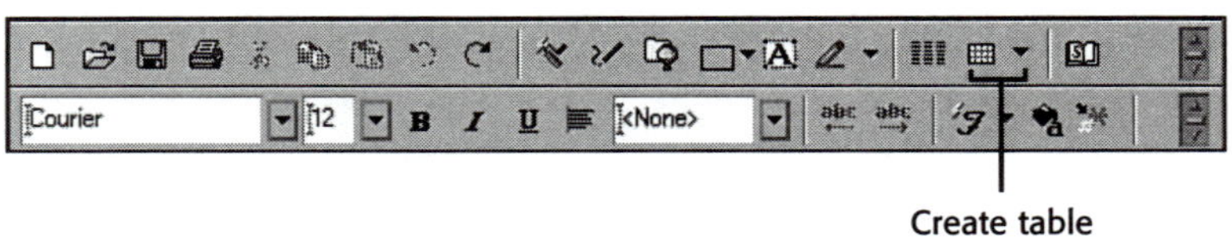

Figure 11.1
The Create table toolbar icon works like a specialized menu when you click it.

If you want to exercise greater control over the initial table options or if you've turned off the toolbar with the Create table icon, you can create a table by choosing Insert, Table (F12), which produces the Create Table dialog box shown in Figure 11.2. You can also open this dialog box by double-clicking on the Create table toolbar icon.

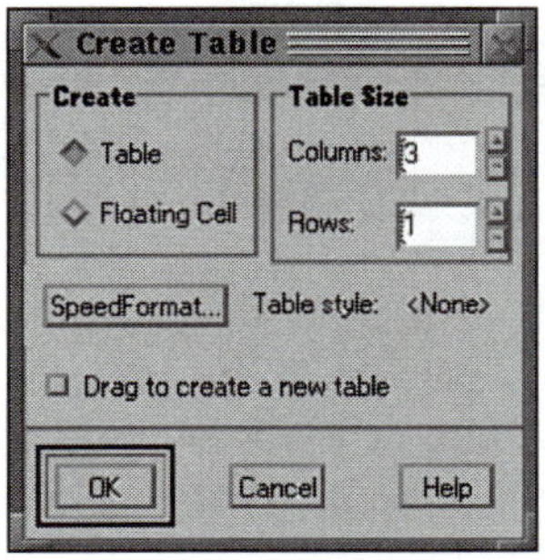

Figure 11.2
The Create Table dialog box gives you added control over the table's initial formatting.

When you use the Create Table dialog box, you select the number of columns and rows you want from the appropriate spin boxes in the Table Size area of the dialog box. For the moment, you'll want to leave the Table radio button checked in the Create area; the Floating Cell radio button is used for creating references to tabular data inside text and is described later in this chapter. Be sure the Drag to create a new table check button is not selected; it's used to create movable tables, a feature that I also describe later.

→ For information on floating cells, **see** "Referencing Table Contents in Text: Floating Cells," **p. 300**
→ For information on movable tables, **see** "Creating Movable Tables," **p. 282**

If you want to create a table with special initial formatting (say, something other than the default line styles), click the SpeedFormat button in the Create Table dialog box. This step opens the Table SpeedFormat dialog box, as shown in Figure 11.3. The default format of <None> produces a spare table with only single lines separating the cells of the table. You can select from among several additional predefined styles, however, such as the Fancy Fills style shown in Figure 11.3. This style uses various levels of shading and an assortment of line types to create a more visually interesting table.

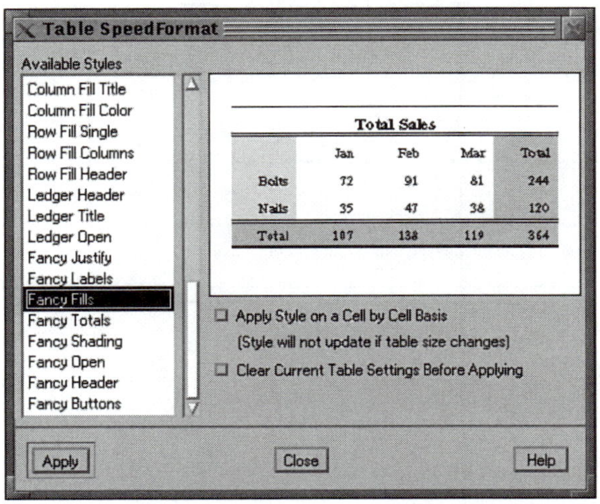

Figure 11.3
The Table SpeedFormat dialog box lets you choose from several appearances for your table by setting line types, shading, and font options.

Note

Several of the predefined table formats include a single title line above the table as a whole. This line is actually part of the table, and you should consider it when you set your table's size.

In addition to the selection of a style, the Table SpeedFormat dialog box gives two options. Neither of these options has any effect when you're creating a new table. You can use them when you use this dialog box to change an existing table's formatting, as described later in this chapter.

→ To learn more about reformatting a table, **see** "Modifying a Table's Format," **p. 284**

Whether you create a table via the Create Table dialog box or by using the Create table toolbar icon, the result is a table inserted in the main body of the text, as in Figure 11.4. When you work with a table, the lower toolbar changes to provide easy access to table-related features.

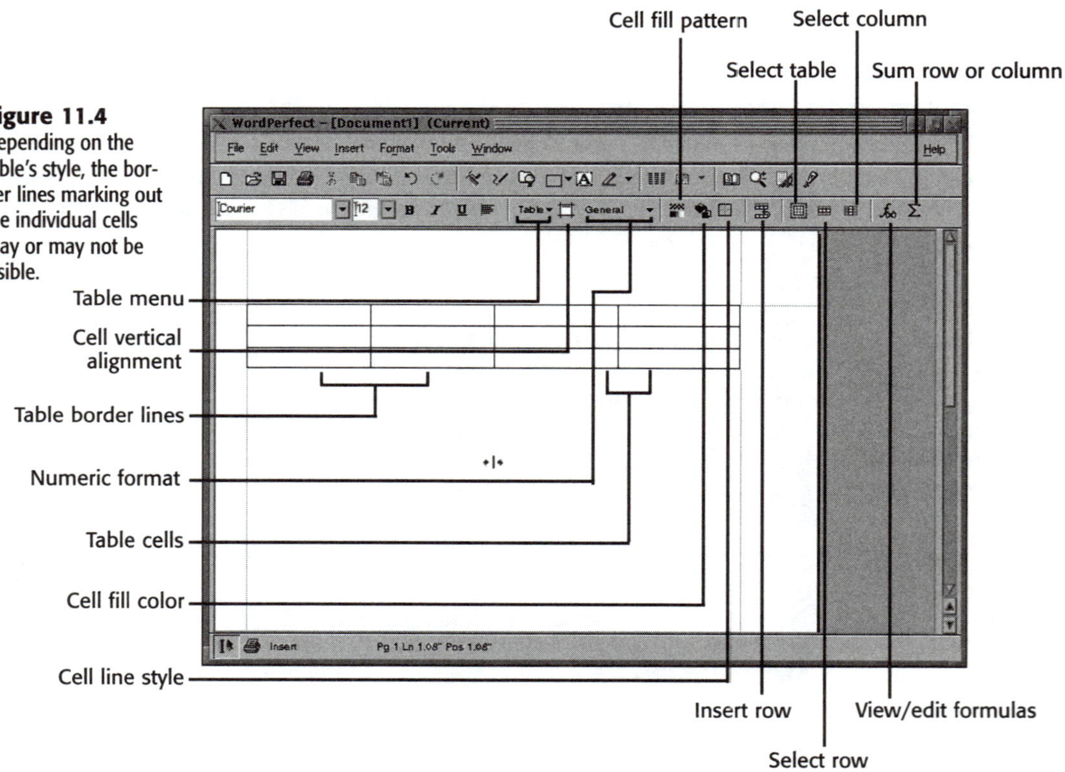

Figure 11.4
Depending on the table's style, the border lines marking out the individual cells may or may not be visible.

You may need to expand your document window to see all these toolbar icons at once. The additional toolbar items are

- Table menu—When clicked, this icon produces a menu of table-related functions. Many of these functions also appear if you right-click your mouse within the table.
- Cell vertical alignment—This icon also produces a menu, which allows you to select whether to align text to the top, center, or bottom of each cell.
- Numeric format—This icon produces a menu that lets you select how to format numeric entries. This feature is particularly handy if you're using a table as a simple spreadsheet in a document.
- Cell fill pattern—Clicking this icon yields a pop-up menu that displays various fill patterns. Select one of these to change the fill pattern of the current cell or cells, or click More to get more options.

→ For more information on table fill patterns, **see** "Modifying a Table's Format," **p. 284**

- Cell fill color—This tool effectively serves as an adjunct to the Cell fill pattern tool. It sets the color used by the fill pattern.
- Set line style—This tool lets you set the style of the line surrounding a cell.

Note

The Set line style tool does *not* affect the line styles of the cells that surround the current cell, so you may need to change their line styles, too, depending on the effect you want to achieve.

- Insert row—This tool quickly inserts a row in the current table, just above the current row.
- Select table—Clicking this tool selects the entire table, which allows you to set tablewide options, as described shortly.
- Select row—You can select one row of a table by clicking this tool.
- Select column—Click this tool to select the column in which the insertion point rests.
- View/edit formulas—This tool opens the formula-editing toolbar described later in this chapter. Use it to perform spreadsheetlike operations within a table.

→ To learn more about doing spreadsheet operations in tables, **see** "Table Mathematics," **p. 296**

- Sum row or column—If your table contains numeric data, you can position the insertion point in a cell to the right of or underneath a column of values and click this button to have WordPerfect calculate the sum of the values in that row or column, respectively. WordPerfect will do this by inserting a table formula, so you can update this value by recalculating the table after changing other cells.

→ For more information on updating table mathematics, **see** "Table Recalculation," **p. 301**

CONVERTING EXISTING TEXT TO A TABLE

WordPerfect allows you to convert text into tables. WordPerfect can handle two different types of text for such a conversion. The first is a tab-formatted table—that is, a table that's been laid out using tabs to move from column to column and carriage returns ([HRt] codes) to create new rows. The second possible source for a conversion is WordPerfect's parallel-column formatting.

→ If you want more information on parallel columns, **see** "Organizing Information with Parallel Columns," **p. 267**

TAB-FORMATTED TABLE CONVERSION

You're most likely to encounter tab-formatted tables either in imports from other programs (including raw ASCII text created in a text editor) or in WordPerfect documents you or somebody else created before learning about WordPerfect's table features. Whatever the source, the conversion process is simple:

1. Select the entire table you want to convert, using the mouse or other text-selection method.

→ To learn more about selecting text, **see** "Selecting Text," **p. 96**

2. Choose Insert, Table (F12). Rather than the Create Table dialog box you normally get when you choose this menu item, in this case you get the Convert Table dialog box shown in Figure 11.5.

Figure 11.5
The Convert Table dialog box's options are quite simple.

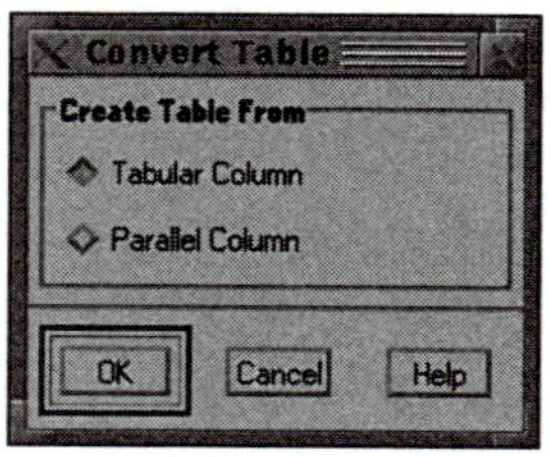

3. Select the Tabular Column radio button and click OK.

WordPerfect converts the tab-formatted table into a standard WordPerfect table. The results are likely to be imperfect, but cleaning them up should be relatively simple with the help of WordPerfect's table editing features. Figure 11.6 shows the results of converting a simple tab-formatted table to a WordPerfect table. This figure shows both the original table and what happened to it when WordPerfect did its conversion.

→ To learn how to reformat a table, **see** "Modifying a Table's Format," **p. 284**

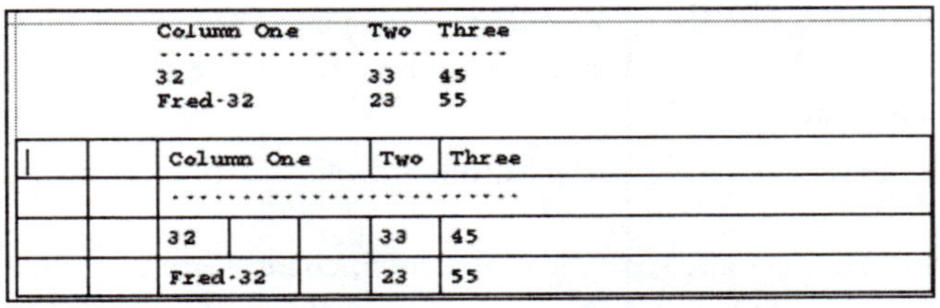

Figure 11.6
A converted table usually requires some reformatting to look good.

Note some of the characteristics of the conversion:

- Like other WordPerfect tables, the converted table extends from the left margin to the right margin, even though much of that space is empty.
- The original table used a pair of tabs at the beginning of each line to offset the table from the left margin. The converted table has turned these into table cells. You could eliminate these tabs before doing the conversion by using WordPerfect's rectangular block selection feature to prevent the extra cells from appearing in the conversion.

→ To learn how to select a rectangular block of text, **see** "Selecting Text," **p. 96**

- The original table used a row of dashes to separate the table headings from the data in the table. This row has been converted into a separate row of cells. Again, you can edit out this row of dashes before the conversion to keep it from being converted inappropriately.
- The original used varying numbers of tabs to align text properly, which results in a varying number of columns per row. It's easiest to remove these extraneous cells after conversion by joining them together.

→ For more information on joining cells, **see** "Joining and Splitting Cells," **p. 285**

- WordPerfect has preserved text alignment in the conversion, so the various columns line up correctly, despite differing numbers of tabs per line. WordPerfect also preserves alignment within cells. For instance, if the original table used center tabs, the converted text is centered in its cells.

→ For information on different types of tabs, **see** "Tabs," **p. 194**

Are you having trouble converting a tab-formatted table to a WordPerfect-style table? See "Table Conversion Doesn't Work" in the "Troubleshooting" section at the end of this chapter.

Parallel-Column Conversion

Converting text from a parallel-column format to a table format works very much like converting a tabular table:

1. Select all the text in the columns you want to convert to a table. If you miss some text, it won't be included in the conversion. If you select too much text, the extraneous text will not be converted, but WordPerfect will create an extra row in the table.

2. Choose Insert, Table (F12). WordPerfect displays the Convert Table dialog box (shown in Figure 11.5). You use the same dialog box to convert a tab-formatted table into a WordPerfect table.
3. Select the Parallel Column radio button and click OK to perform the conversion.

When you perform a parallel-column conversion, chances are good that your text's formatting will change slightly because the exact widths of the columns change. Tables don't use gutters, as parallel columns do, but tables do use margins within each cell, which parallel columns don't.

Tip #147 from *Tux Tips*

If you want to export a document containing parallel columns to another format, including HTML for creating Web pages, you'll probably get better results if you convert any parallel columns to tables first.

→ For information on exporting WordPerfect files in foreign word processors' file formats, **see** "Document Exchange," **p. 88**

→ For information on creating Web pages from WordPerfect documents, **see** "Exporting Existing Documents as Web Pages," **p. 673**

Creating Movable Tables

For greater flexibility in page layout, you may want to create a table that you can move around on the page. The trick here is to insert the table inside a text box, which is a WordPerfect feature designed to allow you to place text most flexibly on the page. To create a table in this way, follow these steps:

1. Select Insert, Table (F12) or double-click on the Create table toolbar icon to open the Create Table dialog box (refer to Figure 11.2).
2. In the Create Table dialog box, set up your table size and, if desired, use SpeedFormat to specify a style for the table.
3. Select the Drag to create a new table check button.
4. Click OK.

WordPerfect removes the Create Table dialog box and returns you to the main document window. When you click and drag your mouse pointer, though, it will create a rectangular box. Release the mouse button when the box is the size you want, and WordPerfect will create a text box and place the table you specified inside it. This formatting allows you to move the table around, adjust the size of the text box in which it resides, edit the table as you would any other table, and otherwise perform any action you can perform on either a table or a text box.

→ To learn about text boxes, **see** "Using Text Boxes to Add Variety to Text Layout," **p. 357**

Tip #148 from
Tux Tips

The way your mouse pointer works depends on whether it's over the table itself, the text box, or the main body of your document. Unfortunately, two or more of these might seem to be the case at any given time, and WordPerfect can seem to behave inconsistently. In reality, what's happening is that different components of these formatting structures may be selected or deselected at different times. You can watch the shape of the mouse pointer as you move it across a movable table to get clues about what's going on. If you're in doubt, try clicking your mouse pointer well outside the table's area and then select whatever you want to work with again. Alternatively, Draft view may be easier to use in cases like this.

Note

If you're working in Draft view (View, Draft or Ctrl+F5), WordPerfect opens a second Document window whenever you edit a text box. Therefore, when you create a movable table, WordPerfect opens such a Document window for you. Select File, Close (Ctrl+F4) or click the Close button in the third toolbar WordPerfect displays in this window to close it and return to editing the document as a whole.

A Word About Selections in Tables

When you're working with most ordinary text in WordPerfect, selections are fairly unambiguous. When you select something, it's text, and that's that. With tables, though, it's a bit more complex because you might be selecting the text, a cell, a group of cells, or perhaps the entire table. Precisely how a table formatting option behaves, or whether it's available in the menus, sometimes depends on which item you've selected. Here's how to control what you're selecting:

- To select a row, a column, or the entire table, the easiest path is to use the Select table, Select row, and Select column toolbar icons (Figure 11.4). Position your insertion point in the table, row, or column you want to select and then click the appropriate icon.
- Alternatively, try positioning your mouse pointer near the top or left edge of a cell. The pointer will become extra-large and point at the nearest cell edge. When the mouse pointer looks like this, it selects the entire cell. You can click and drag to select more than one cell, or you can double-click to select the entire row or column, depending on which way the pointer is facing.
- Most of the rest of the time, when you select text normally, you are selecting, and affecting with any formatting changes, the text itself.

When you select the text in the table, WordPerfect highlights the text but not the lines composing the cells of the table. In contrast, when you select entire cells, rows, or columns, WordPerfect highlights entire cells or groups of cells, including their borders.

Tip #149 from
Tux Tips

Pay attention to precisely what you are selecting! Your extra care here can head off confusion and frustration later because WordPerfect's table features depend on the types of elements you select.

MODIFYING A TABLE'S FORMAT

One of the great advantages of using WordPerfect's table feature instead of tabs to create a table is that the table feature gives you much more flexibility in formatting your table. For example, the table feature enables you to place shading and borders in and around cells, rows, and columns.

ADJUSTING COLUMN AND TABLE WIDTHS

WordPerfect normally creates a table that spans from the left margin to the right margin, with columns of equal width. This layout may not be what you want, however. Fortunately, it's easy to adjust the widths of either the entire table or of individual columns.

One way to adjust these widths is with the mouse pointer. If you move the mouse pointer over one of the vertical lines that separate columns, the pointer changes to a double arrow, as in Figure 11.7. Simply click and drag the pointer to adjust the column widths.

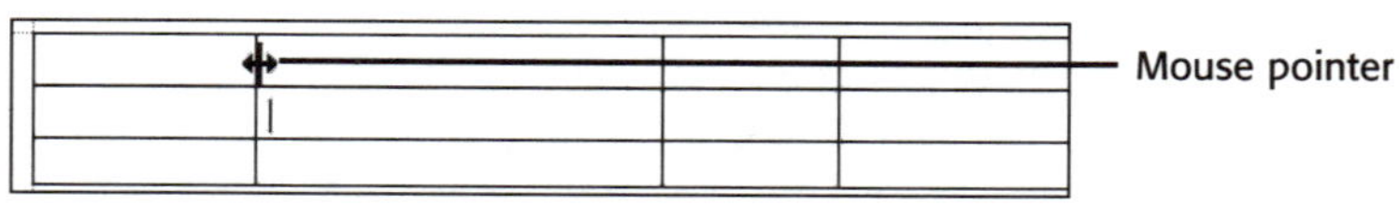

Figure 11.7
The double-arrow mouse pointer indicates that you can adjust a column's width.

Note

This method enables you to adjust *column widths* only; you cannot click and drag to adjust *row height*. WordPerfect automatically adjusts row height for the amount of text in the cells of any given row. If you want to increase the amount of space in a specific cell or row, you can adjust the row margins, as described shortly.

When you adjust the column's width, all cells in that column are affected. You cannot adjust the relative sizes of columns on just one row, though you can adjust the *number* of cells on any given row and possibly work something out from there.

Another way to adjust column widths is from the Column or Table portion of the Format dialog box. Open this dialog box by choosing Format from the context menu or from the Table menu toolbar item. This feature is described in detail later in this chapter.

→ To learn how to format a table, **see** "Table Text Formatting Options," **p. 289**

JOINING AND SPLITTING CELLS

You may occasionally need to merge two or more cells or to split one cell into two or more parts. For example, merging cells is a convenient way to create headings that span multiple columns, as shown in Figure 11.8. You may also need to join or split cells if you need more or fewer columns than you originally anticipated.

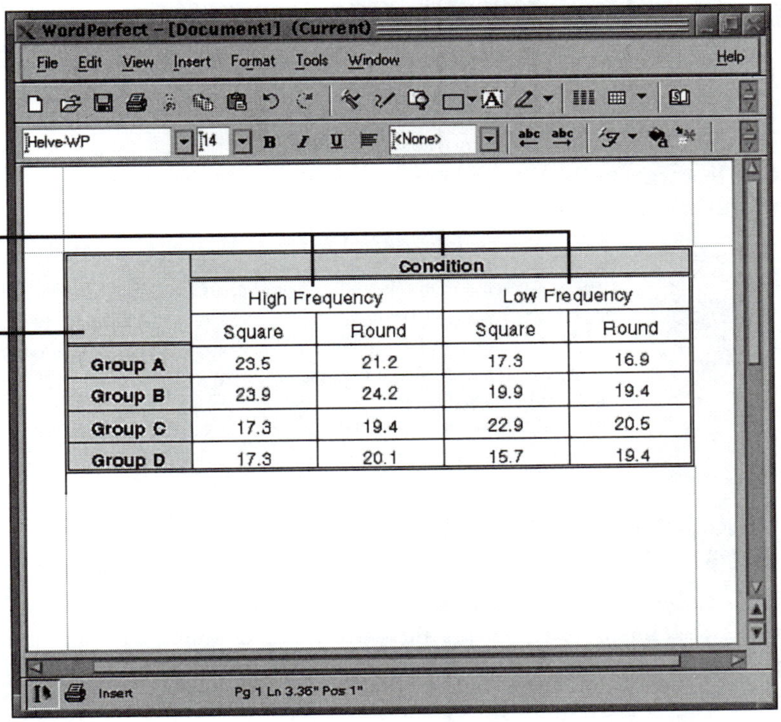

Figure 11.8
You can merge or split cells to create uneven numbers of cells in different columns or rows.

To split a cell, position the insertion point in the cell you want to split and then choose Split, Cell from the Table menu on the table toolbar. Alternatively, you can right-click and choose Split Cell from the context menu. In either case, you'll see the Split Cell dialog box in Figure 11.9. You can elect to split the cell either vertically or horizontally, but not both ways. Select the desired split direction and adjust the value in the associated spin box to tell WordPerfect how many cells to create from the original; then click OK.

If you select multiple cells before you bring up the Split Cell dialog box, WordPerfect splits each of them in the way you specify. You can add an entirely new row or column in this way, in fact.

To join two or more cells, you must first select them. Then you can either choose Join, Cell from the Table menu on the table toolbar or choose Join Cells from the context menu when you right-click on the highlighted cells. In either case, WordPerfect immediately merges the selected cells into one.

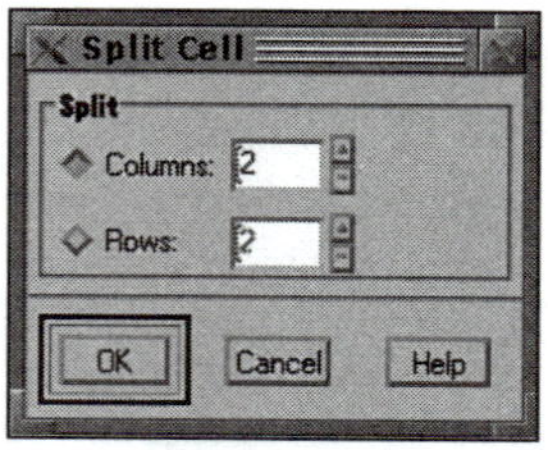

Figure 11.9
You must select the appropriate radio button before you can adjust the spin box to select how many cells you want to create from each original one.

When you split or merge cells that contain text, WordPerfect has to do something with that text. In the case of joining, WordPerfect combines the text from the merged cells. Text entries from cells that were in the same row appear side by side, whereas text entries from cells that were in the same column appear one above the other. When you split cells, WordPerfect puts the original contents in only one of the cells.

Tip #150 from
Tux Tips

You can use multiple split and join operations to create the effect of an altered size for a cell in one row relative to others. For instance, if you have a three-column layout, you can make the middle cell in the first row wider than the middle cells in the other rows. Here's how: Split the end cells of the first row; then join the middle cell with the inner portion of each of the two split cells. You now have three cells, and the middle one is as wide as two of the originals.

Adjusting Table Placement

WordPerfect treats tables as a series of codes. This series begins with a [Tbl Def] code, is followed by a series of [Row] and [Cell] codes, and ends with a [Tbl Off] code. WordPerfect places the table in text wherever this series of codes appears. If you want to adjust the table's placement, there are a number of ways to do this.

Centering or Right-Justifying a Table

By default, WordPerfect creates tables that span from the left margin to the right margin. Therefore, there's little point in trying to adjust such a table's horizontal placement unless the table is first slimmed down. You can do so by adjusting the column widths. Reducing the width of the column on the far right shrinks the total width of the table.

Now you might think that you can use ordinary text-justification commands to center the table or place it flush against the right margin. Unfortunately, that approach won't work. Instead, you need to choose the Format (Ctrl+F12) command from the table toolbar's Table menu or choose Format from the pop-up menu you get when you right-click over text in the table. This step produces the Format dialog box. Click the Table radio button at the top of this dialog box, and you'll see the options shown in Figure 11.10. You can adjust the table's horizontal position by selecting the alignment you want from the option button in the Table Position area. You can select from Left, Right, Center, Full, and a specified distance from the left edge of the page.

Note

The Full option restores the table to spanning the entire width of the page margins. The From Left Edge option measures distance from the left edge of the page, *not* from the margin.

Figure 11.10
You can enter a value in the spin box in the Table Position area only if you select From Left Edge; this entry places the table at the distance you specify from the left edge of the page.

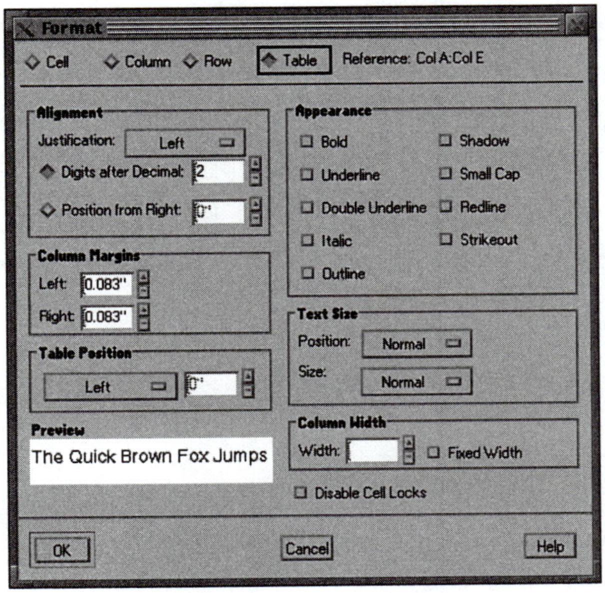

NARROW TABLES WITH TEXT WRAPPING

Sometimes a table needs to be quite wide; for example, it may have lots of columns, or perhaps each cell contains a lot of information. Other times, though, you may need to produce a very narrow table with only two or three skinny columns. In cases like this, rather than create a table with lots of blank space on one or both sides, you may want to run your main text alongside the table, as in Figure 11.11.

One of the simplest ways to achieve this effect is to use tables in conjunction with columns. Create a column definition by choosing Format, Columns and designing your layout as described in Chapter 10. You can use either newspaper-style or parallel columns for this purpose. Place text in one column and your table in the other. When you've filled the horizontal space, discontinue columns and continue with your text.

→ For information on columns, **see** "Creating Professional Layouts with Columns," **p. 255**

The tricky part in designing a layout in this way is switching into and out of Multicolumn mode. Because column formatting is paragraph oriented, you can't simply switch it on and off within a paragraph. Therefore, you may not have a block of text that's of the suitable length to match your table. You may get lucky, though, and you can tweak something that's almost right by adjusting the relative widths of the columns.

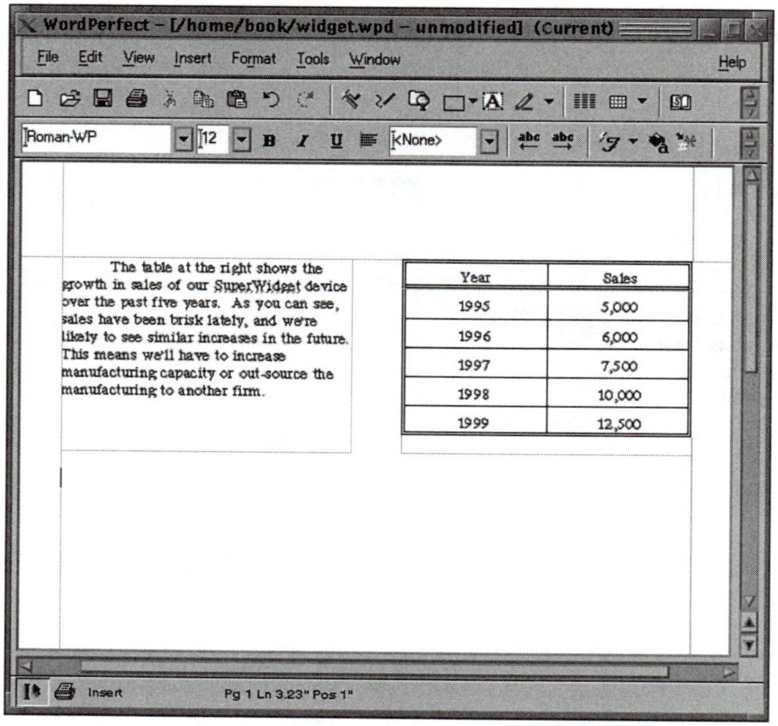

Figure 11.11
Narrow tables may look better with text alongside them on the page.

Boxed Tables

Another option is to use a boxed table. Create your text as you normally would. Then, when you want to create your table, create a movable table, as described earlier in this chapter. By default, text *flows* around a boxed table, as if the table were a physical obstruction in a river of text.

→ For information on tables in boxes, **see** "Creating Movable Tables," **p. 282**

→ For information on wrapping text around boxed objects, **see** "Text Wrapping," **p. 461**

The text box method produces results that look very much like the results when you combine columns with a table. The text box is more flexible, however. For instance, if your layout already uses columns, you can use a text box to place your table *between* the columns, as shown in Figure 11.12.

Of course, with either method, you may need to adjust spacing between the box and your main text. Use whichever methods are appropriate for columns or for boxed text.

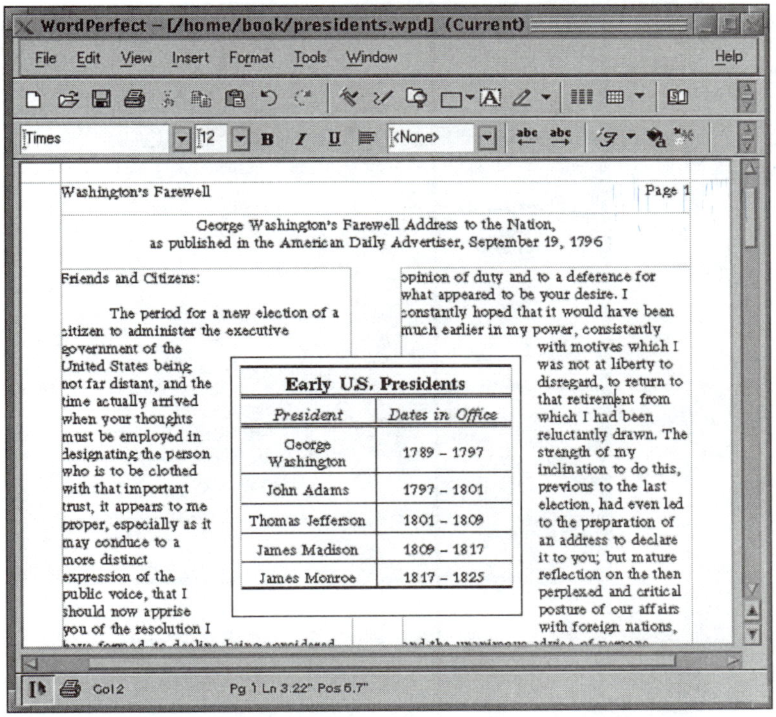

Figure 11.12
By default, when you place a boxed table between columns, the text in both columns wraps around the table.

TABLE TEXT FORMATTING OPTIONS

Whenever you work with a table, you're dealing not only with the formatting of your document's text but also with the formatting of text within the cells of the table. In many cases, you can apply the usual WordPerfect formatting operations to this text. You can highlight entire cells or text within them and apply formatting commands to change the font, justification, and so on. Sometimes these commands won't take across more than one column or row, however; and columns have specialized formatting needs. WordPerfect therefore provides table-text formatting options. You access these by positioning your insertion point in a cell, or by selecting a range of cells, and then choosing Format (Ctrl+F12) from the Table menu toolbar item. Alternatively, you can right-click on your cell or highlighted group of cells and choose Format from the resulting pop-up menu. Either way, WordPerfect opens the Format dialog box shown in Figure 11.13. This dialog box has four sets of options; the set you see depends on which of the four radio buttons at the top of the dialog box is selected.

→ To learn more about selecting text in tables, **see** "A Word About Selections in Tables," **p. 283**

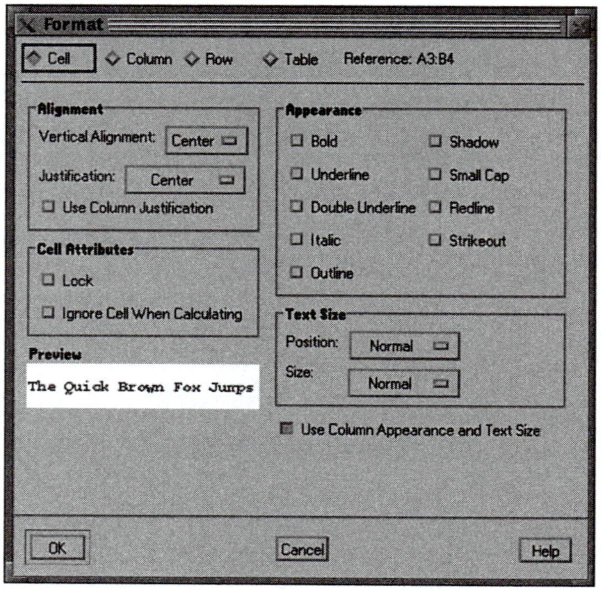

Figure 11.13 The Format dialog box allows you to set options related to individual cells, columns, rows, or the entire table.

Cell

The Cell option of the Format dialog box lets you set options for the display of text within individual cells. If you've not highlighted any cells, or only one, then the options you apply here affect only the cell in which the insertion point rests; otherwise they affect all the cells you've highlighted. You can set the following items:

- Text alignment (both horizontal and vertical). Alternatively, you can tell WordPerfect to use the justification as set in the Column portion of this dialog box.
- Various font appearance attributes.
- Adjustments to the text's size, including whether it's superscripted or subscripted and whether it should be the same size as, smaller than, or larger than surrounding text.
- A lock to prevent changes in the cell; if you click the Lock option in the Cell Attributes section, WordPerfect doesn't allow changes to the cell's text. To undo this setting, select Disable Cell Locks from the Table portion of the Format dialog box.
- Ignore Cell When Calculating to remove the cell from consideration in spreadsheet-like calculations in tables.

Column

If you click on the Column radio button in the Format dialog box, the options change to match those shown in Figure 11.14.

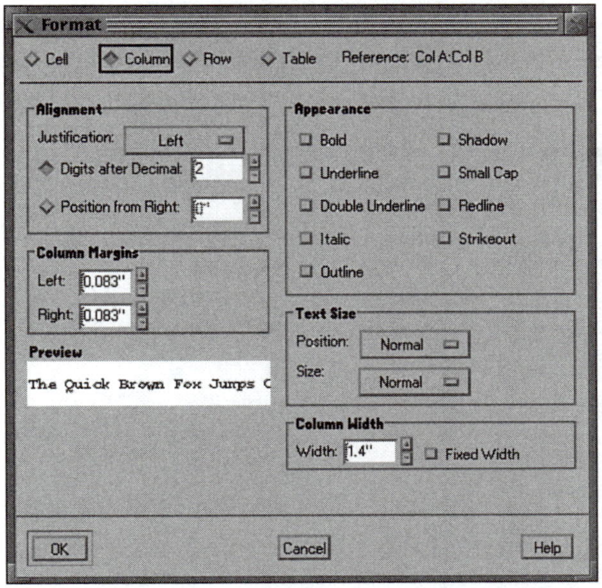

Figure 11.14
The Column options include many of the same text formatting features as the Cell options.

From here, you may adjust the following items:

- Various font attributes similar to those you can set when you're adjusting cell formatting.
- Left and right margins for the text within a column. This entry is the minimum space between the column lines and the text within the cell.
- Additional options in the Alignment field. These options are active only if you select Decimal Align for the Justification. If you then select Digits after Decimal, WordPerfect aligns the text so that as many digits as you specify appear after the decimal point before the right edge of the cell. If there are more than this number of digits, WordPerfect wraps them to the next line in the cell. If you select Position from Right, WordPerfect aligns the decimal point so that it falls the distance you specify from the right edge of the cell.

Note

The Alignment options in the Column portion of the Format dialog box work only if you've checked the Use Column Justification option in the Cell portion of this dialog box.

- The Column Width setting specifies the column's width in whatever units you prefer. If you select Fixed Width, WordPerfect keeps the column width set at the value you specify if you subsequently change another column's width; otherwise, WordPerfect may adjust other column widths to keep the total table width less than the margin width.

Row

When you select Row, you can set options relating to the formatting of rows of cells (Figure 11.15).

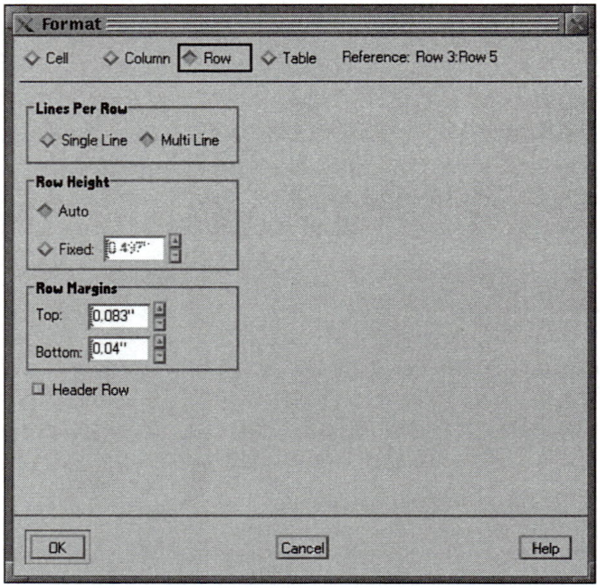

Figure 11.15 The Row portion of the Format dialog box is relatively sparse because it lacks font attribute options.

From this part of the dialog box, you can tell WordPerfect to

- Restrict cells to a single line of text or allow multiline cells.
- Adjust the row height automatically or leave it fixed at a value you specify.
- Set the top and bottom margins for text in a row.
- Set the Header Row option for the first row in a table. WordPerfect then reproduces that row at the top of each page if your table spans more than one page.

Table

The Table option (refer to Figure 11.10) lets you set formatting features that apply to the entire table. These adjustments include

- Alignment options similar to those in the Column portion of the dialog box
- Left and right column margin settings that are, again, similar to those in the Column portion of the dialog box
- A Table Position setting option that can be used to align the table to the left or right margin, to center the table, to stretch it full against both margins, or to specify a distance between the left edge of the paper and the table

- Appearance and Text Size options similar to those in other portions of this dialog box
- Column Width settings for the entire table
- The Disable Cell Locks setting that turns off all cell locks set from the Cell portion of this dialog box

Numeric Formatting Options

In addition to the options in the Format dialog box, you can set options specific to numeric table entries from the Number Type dialog box (Figure 11.16). You can display this dialog box by choosing Numeric Format (Shift+Ctrl+F12) from the Table menu toolbar icon or from the pop-up menu you get when you right-click in a table.

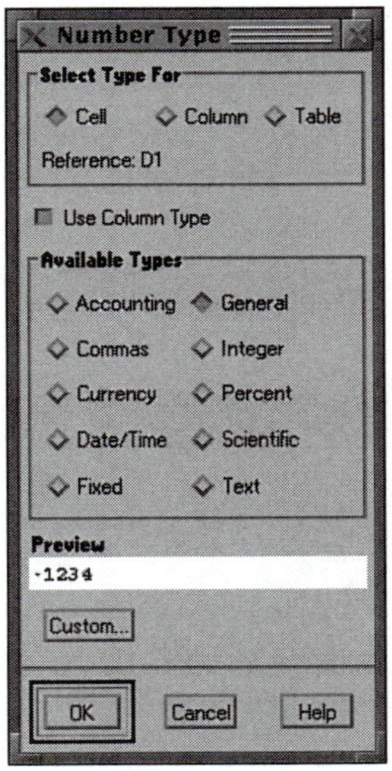

Figure 11.16
Fine-tune your column's presentation of numerical data by using the Number Type dialog box.

As in the Format dialog box, you can apply the options in the Number Type dialog box to individual cells, full columns, or the entire table (but not to individual rows). You can select from various predefined types or click Custom to design your own numeric presentation format (Figure 11.17).

Tip #151 from
Tux Tips

If a field you want to modify is grayed out in the Customize Number Type dialog box, click Cancel and then select another option from the Available Types in the Number Type dialog box. When you click Custom again, your option may now be available. For instance, the Decimal Amounts options for controlling the number of digits displayed after a decimal point are unavailable if you select General as the base type, but these options become available when you select Fixed instead.

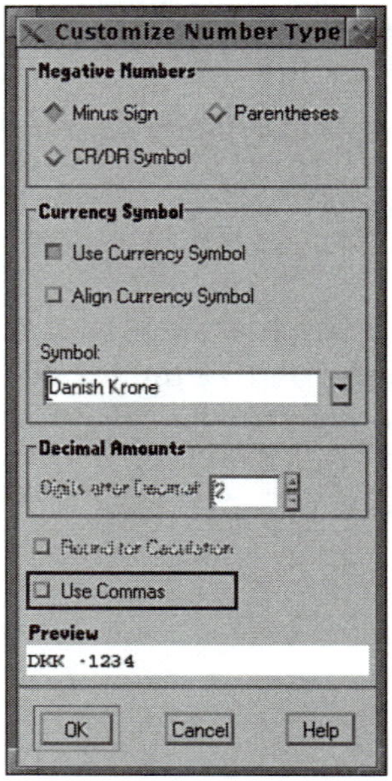

Figure 11.17
Create special-purpose number formats from the Customize Number Type dialog box.

TABLE CELL FORMATTING OPTIONS

A WordPerfect table has two parts: text and the lines and background that surround the text. So naturally, WordPerfect lets you adjust the appearance of these lines and backgrounds, as well as the text itself. If you want to apply these options to a range of cells, be sure to select the cells themselves first.

→ To learn how to select cells rather than the text inside them, **see** "A Word About Selections in Tables," **p. 283**

SPEEDFORMAT

If you select SpeedFormat from the Table menu toolbar or from the table context menu, WordPerfect displays the Table SpeedFormat dialog box (also accessible when you create a table, as shown in Figure 11.3). You can select from any of the Available Styles. When you apply a style to an existing table, the two check buttons also have an effect:

- Apply Style on a Cell by Cell Basis affects what happens if you insert or delete rows or columns. With this option checked, WordPerfect applies the current row or column's formatting to a new row or column. If you leave this box unchecked, WordPerfect recomputes the styles to be applied to all rows and columns. This step is most important if your style involves alternating rows or columns of shading or color.
- Clear Current Table Settings Before Applying, if checked, causes WordPerfect to reset any customized settings: For instance, if you've applied an extra-dark border around a cell, WordPerfect clears the border before applying the new style. If this box is not checked, WordPerfect doesn't touch the customized border.

Note

The SpeedFormat options apply to an entire table, even if you selected only a portion of that table beforehand.

BORDERS AND FILL

If you choose Borders/Fill (Shift+F12) from the Table menu toolbar item or from the table context menu, WordPerfect opens the Table Lines/Fill dialog box in Figure 11.18. This dialog box provides two slightly different sets of options, depending on whether you select the Current Cell or Selection radio button or the Table radio button. Both options work in much the same way, though: You select the style of border for any of the specific sides of the cell or table. You can also select the line color and a fill option. When you select Table, your selections apply to the table as a whole, as opposed to the individual cells within the table. The details of setting the line styles and fills are very much like setting these options for paragraphs in normal text.

→ To learn more about borders and fills, **see** "Borders and Fills," **p. 216**

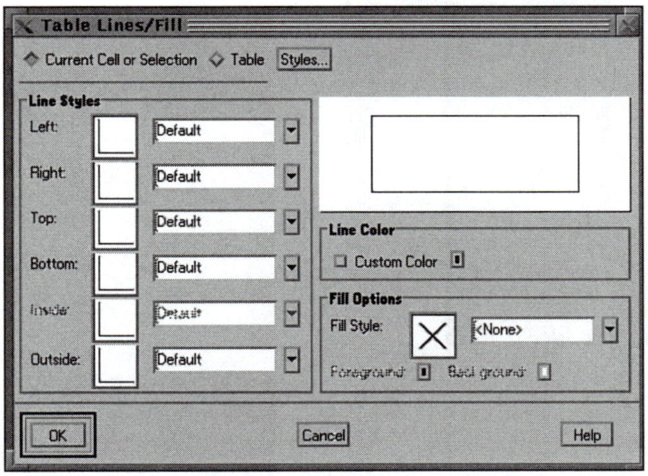

Figure 11.18
Fine-tune the appearance of cells with the Table Lines/Fill dialog box.

EDITING A TABLE'S CONTENTS

Fortunately, editing a table's contents is relatively straightforward, at least if those contents are ordinary text and numbers. You can enter and edit text in a table very much as you would anywhere else. There are a few exceptions, though:

- The Tab key doesn't work in the usual way. Rather, you use Tab to jump between cells in the table.
- Use Ctrl+Shift with the arrow keys to move between cells.

Note

Strangely, when used with Ctrl+Shift, the Up Arrow key moves the text cursor *down* one row, whereas the Down Arrow key moves it *up* one row. The Left Arrow and Right Arrow keys operate in a more predictable fashion, though.

- Don't attempt to enter a page break or other page-related formatting. Most of these features are disabled from the menus, but a few simply have no effect or produce error messages.
- If you select an entire row or column and perform a conventional delete function (say, pressing the Delete key), WordPerfect displays the Delete dialog box shown in Figure 11.19. You can use this dialog box to tell WordPerfect to delete the entire row or column, or just its contents. If you perform a cut operation (Edit, Cut or Ctrl+X), the resulting dialog box is slightly different but serves the same function.

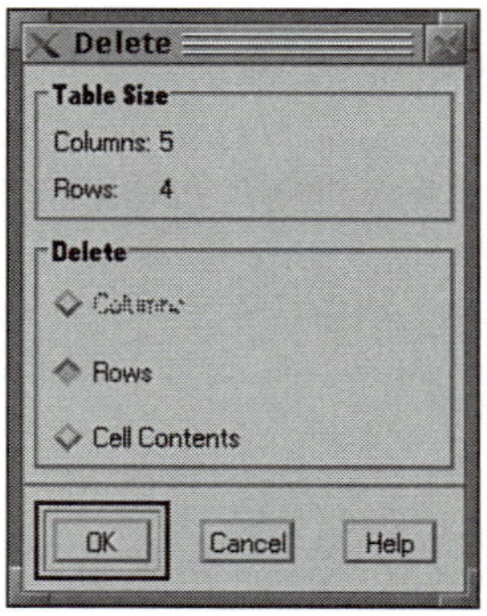

Figure 11.19
Tell WordPerfect precisely what you want to delete with the Delete dialog box.

Table Mathematics

WordPerfect for Linux's tables features may look like just another way to format static text, but they're much more than that. You can actually perform many spreadsheet operations with WordPerfect's tables and even include references to table cell values in your text. That way, if you change a value in the spreadsheet, WordPerfect can update the in-text references to that value and its dependent values.

Basic Principles

When you perform table mathematics, you must start with numeric values. These could be monthly personal budget figures, annual sales and expenses, observations from an experiment, or anything else you can express numerically. You can include text in tables in which you want to perform mathematical operations, too; textual cells just won't affect the mathematical results.

Each cell in the table is identified by a name. This name consists of a *table name* (such as TABLE A) followed by a period and a *cell name*. The cell name consists of a *column letter* followed by a *row number*. For instance, in an unmodified 3×3 table that's the first one in the document, the upper-left cell is TABLE A.A1, or A1 for references within the table. The cell to A1's right is B1, and the cell below that is B2.

Note

If you join or split cells, this numbering scheme may be disrupted. For instance, if you join cells A1 and A2, the combination cell is called A1, whereas the new second cell is A3, and there is *no* A2.

When you need to refer to cells for calculations, you refer to them using their cell names. This system works very much the way spreadsheets work, so if you're familiar with a spreadsheet, you should find WordPerfect's table labels quite familiar.

As an example of table mathematics, let's consider a modification of the table shown in Figure 11.8. That table presented data from a fictitious experiment. Suppose you want to find the means of certain columns or groups of columns, though. You could do this manually, of course, but you can also have WordPerfect do the work. The first step is to create new rows for the averages, as shown in Figure 11.20. This figure also shows the result of selecting the View/edit formulas toolbar button: a new toolbar used to manipulate WordPerfect table formulas.

Figure 11.20
You can display the Formulas toolbar by pressing the View/edit formulas toolbar button or by choosing Formula Toolbar from the Table context menu.

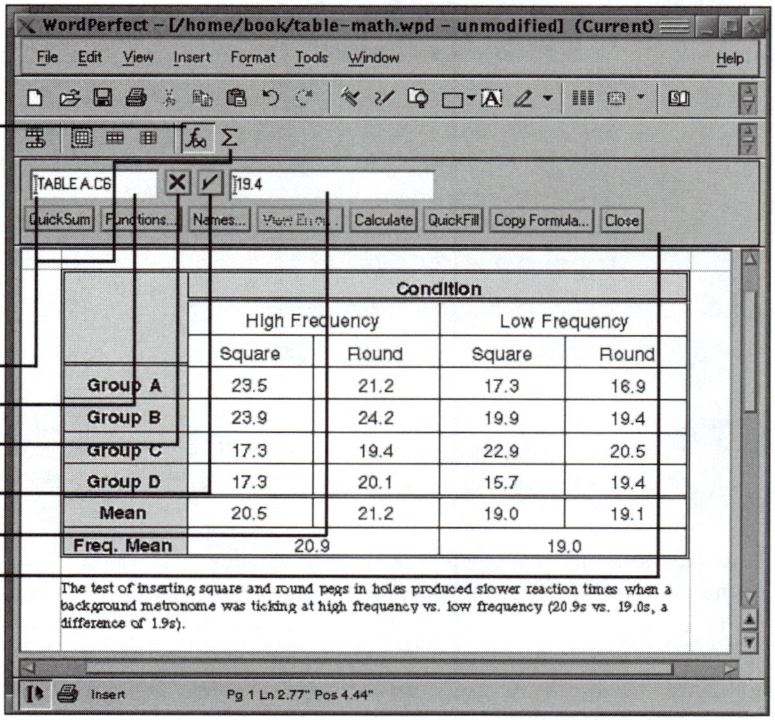

Note

You may find it helpful to examine a sample of a table that uses math functions, either as you read this text or separately. The CD that accompanies this book includes a file called table-math.wpd in the chapter11 directory. This file produced Figure 11.20 and illustrates the principles described here.

INSERTING FUNCTIONS INTO A TABLE

One of the easiest functions to insert is a sum. You can do so by placing the insertion point in a cell and pressing either of the QuickSum buttons. This step inserts the sum of the values in the cells immediately above or to the left of the current cell. If the Formulas toolbar is active when you insert the sum, a formula for computing the sum appears in the Cell contents field of the toolbar; however, this formula bears little resemblance to what actually appears in the table. The formula reads something like `SUM(B4:B7)`, which means that WordPerfect is to take the sum of the values of the cells from B4 through B7.

You can select from many more functions by clicking the Function toolbar button, which opens the Table Functions dialog box shown in Figure 11.21. If you have trouble finding the function you want from the original list, you can restrict it by selecting an appropriate subset from the Type of functions to list button.

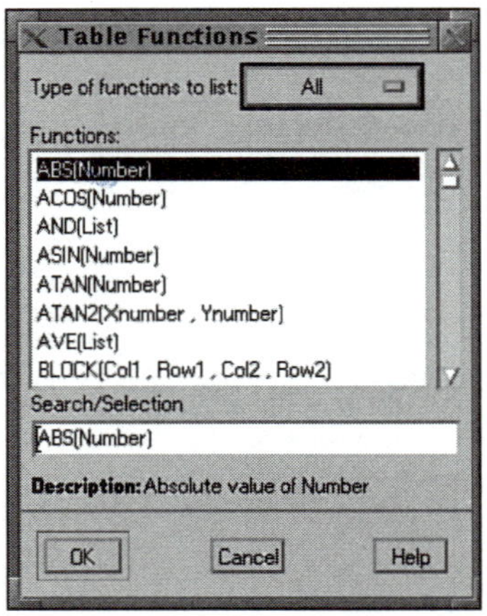

Figure 11.21
WordPerfect supports many common mathematical functions for various types of operations.

Note

When you select a function, WordPerfect displays a description of it near the bottom of the dialog box. You can use this information to determine what each function does.

Most functions take one or more *arguments*, which are the values that go into computing the function. For instance, the `AVE` (average) function takes a list of values. The argument can be a range, such as B4:B7 or several values separated by commas. When you select a function, WordPerfect inserts it into the Cell contents field of the Formulas toolbar and highlights the arguments. You must then replace these "dummy" arguments with the ones you actually want in this formula.

Tip #152 from *Tux Tips*

> You can provide a second function as the argument to a function, which can be useful when you're performing complex calculations that WordPerfect does not support directly. For instance, suppose you need to find the average of the square roots of some values in the table. You could create a series of cells in which you compute the square roots and then find the average of those. Alternatively, though, if you don't need the square root values for other things, you could feed the `AVE` (average) function the `SQRT` (square root) functions as parameters directly.

It's generally possible to perform a calculation in more than one way with WordPerfect's tables. For instance, each of the following functions finds the average of cells B4 through B7:

```
AVE(B4:B7)
AVE(B4,B5,B6,B7)
SUM(B4:B7)/4
(B4+B5+B6+B7)/4
```

The first of these is the easiest to modify if you should add more values that need to be averaged, though.

CELL NAMING

You probably don't think in terms of alphanumeric codes to represent the cells in your table. Instead, you probably think of a value by what it represents, such as "February's sales" or "test totals." You can tell WordPerfect about your more meaningful labels and then use them as references in your table calculations. To do so, click the Names toolbar button. This step opens the Table Names in Current Document dialog box shown in Figure 11.22.

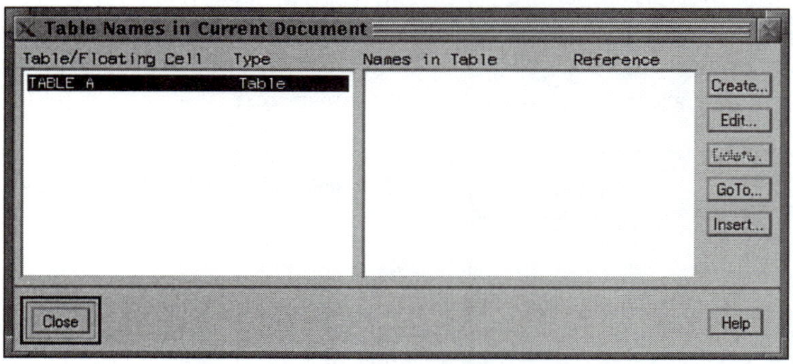

Figure 11.22 The Table Names in Current Document dialog box lets you edit, insert, or delete cells, as well as give them meaningful names.

To assign a name to a cell, be sure that the correct table is selected in the Table/Floating Cell field and then click Create. WordPerfect now opens the Create Name dialog box shown in Figure 11.23. Either type a name into the Enter Name Manually field or click Use Text from Current Cell to Name radio button to take the name from whatever text is in the current cell. You can name an individual cell, a row, or a column in this way and then use that name, rather than the usual spreadsheet-style references, in your formulas.

Figure 11.23
The Create Name dialog box lets you use meaningful names in your spreadsheets.

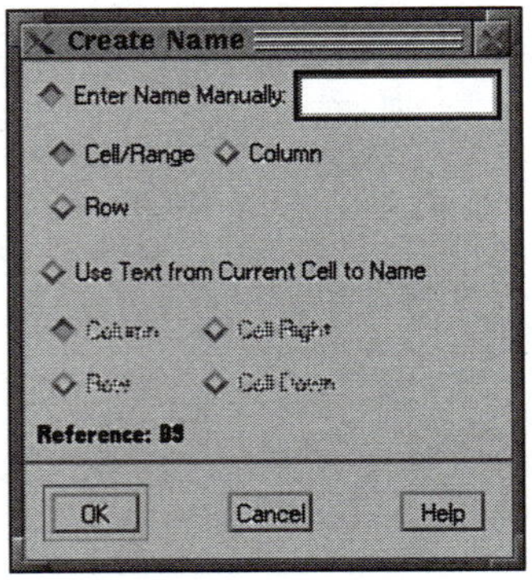

REFERENCING TABLE CONTENTS IN TEXT: FLOATING CELLS

Letting WordPerfect do the math for you can be great, but if you need to refer to the results of this math in your text, keeping that text in sync with the table's values can be tricky. That's why WordPerfect lets you create *floating cells* in your text. These are a way to reference table values in the main body of your text. To create a floating cell, your insertion point must *not* be in a table, so click outside of the table and type some introductory text. Then choose Insert, Table (F12), which opens the Create Table dialog box (refer to Figure 11.2). Select Floating Cell, rather than Table, in the Create area of the dialog box and then click OK. WordPerfect turns on the Formulas toolbar. You must enter a reference to the cell you want to use in the Cell contents field, and this reference must include the table name, not just the cell reference. For instance, `TABLE A.B7` puts the value from cell B7 in Table A into your text. Now click the Accept formula button and check to be sure that WordPerfect has processed your entry correctly; finally, click Close to remove the Formulas toolbar from the screen. Because they're part of your text, floating cells take on whatever formatting is applied to your text. Unless you look at them in Reveal Codes or try to edit them, they are indistinguishable from ordinary text.

Tip #153 from *Tux Tips*

> **You can enter regular table formulas for a floating cell to provide further data analysis within the text of your document.**

If you want to change the appearance of the floating cell's number, you can double-click on the opening [Flt Cell] code in Reveal Codes. From the Number Type dialog box (Figure 11.16) that now appears, you can specify the numeric format for the floating cell. You can set other formatting aspects as well, such as font and font attributes, as you would for any other text.

TABLE RECALCULATION

When you change a value in a table, WordPerfect doesn't automatically recalculate the entire spreadsheet to match that change. You tell WordPerfect to do so by right-clicking over a table and choosing Calculate from the pop-up menu, or by clicking the Calculate button from the Formulas toolbar. In response, WordPerfect recalculates all changed values in the table, as well as references to those values in floating cells in the rest of your text.

Caution

Remember to recalculate your table values if you change anything; otherwise, you'll lose the benefits of table math. Even saving and reloading a file won't redo these calculations; you *must* do it yourself!

IMPORTING A SPREADSHEET

WordPerfect for Linux can read Lotus 1-2-3 spreadsheet files and ASCII delimited text files into a table. To read such files, you can use a regular File, Open or Insert, File operation. Alternatively, to ensure that WordPerfect at least tries to interpret the data as a spreadsheet, use Insert, Spreadsheet/Database, Import. This sequence opens the Import Data dialog box shown in Figure 11.24.

Figure 11.24 Use the Import Data dialog box to tell WordPerfect how to import data into a table.

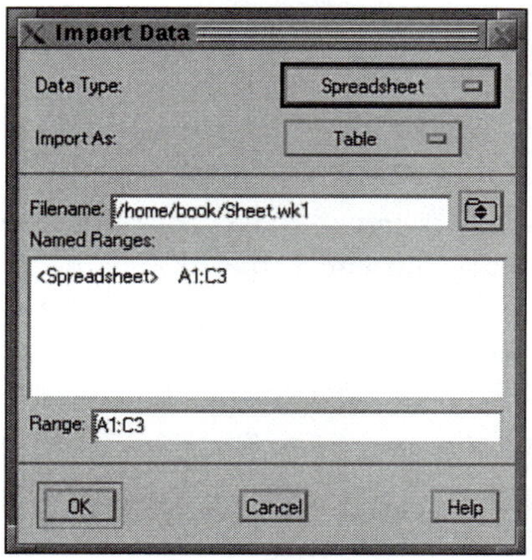

Type in the original filename in the Filename field; alternatively, click the folder icon next to that field to select the file from a WordPerfect File Manager dialog box. When you've selected a file, WordPerfect displays some basic information in the Named Ranges and Range fields.

You can control several options from this dialog box:

- Edit the range in the Range field to have WordPerfect import only a subset of the spreadsheet.
- Select Text or Merge Data File from the Import As button if you want to import the data as raw text or as data for a mail merge rather than as a table.

→ To learn how to use WordPerfect's mail merge feature, **see** "The Same Letter, Different Information: Mail Merge," **p. 515**

- Select different import file formats to force WordPerfect to try to apply different filtering techniques to the file. The ASCII Delimited Text and ANSI Delimited Text options may be useful if you're trying to import ASCII text as a table, for instance.

Caution

WordPerfect occasionally crashes when importing unfamiliar spreadsheet formats. You would therefore do well to save any open file(s) before attempting such an operation, and be especially cautious when importing new file formats.

Are you having trouble importing your spreadsheet as a table? See "Spreadsheet Import Troubles" in the "Troubleshooting" section at the end of this chapter.

TROUBLESHOOTING

TABLE CONVERSION DOESN'T WORK

You're trying to convert text to a table, but it's not working. You're getting tables with only a single column, or WordPerfect is complaining about an incorrect format.

WordPerfect requires specific types of formatting when converting existing text to table form. If you're trying to convert tabular text into a table, the existing columns *must* be separated by tabs, not spaces or other formatting. You can manually replace spaces with tabs and do the conversion again. If you're having problems with a conversion from columnar text, be sure the existing text uses *parallel* columns, not *newspaper* columns. The latter won't work.

→ If you want to know more about parallel and newspaper columns, **see** "Creating Professional Layouts with Columns," **p. 255**

SPREADSHEET IMPORT TROUBLES

When you try to import a spreadsheet, you get a lot of text, rather than a table, or WordPerfect reports that the import failed or even crashes.

Any of these things can happen if you try to import a spreadsheet from a format that WordPerfect does not support or if WordPerfect tries to apply the wrong spreadsheet import filter. Formats supported include Lotus 1-2-3 and ASCII text files. If you were unsuccessful with a normal file load or import operation, try Insert, Spreadsheet/Database, Import and experiment with the different Data Type options instead. If that alternative doesn't help, try saving the original file as a Lotus 1-2-3 (*.wk? file extension) file or as an ASCII-delimited text format.

PART III

Creating Consistent Documents with Styles and Templates

CHAPTER 12

Ensuring Consistency Within Your Document with Styles

In this chapter

STYLES: PACKAGES OF FORMATTING FEATURES

If you've read much of this book, you've probably encountered advice to place certain formatting features in a "style" to make it easy to create, modify, or use that feature to its fullest. Just what does this advice mean, though? What are styles?

Styles are a way to combine arbitrary sets of formatting codes, and even text, into a single text formatting code. You can then select the style, rather than the individual formatting commands, to create the format you desire. WordPerfect's style feature enables you to

- Include just about any formatting features you desire in one style
- Include one style within another
- Link styles so that when you finish a paragraph in one style, another will automatically begin
- Save styles to disk so that you can use them in different documents
- Create styles that apply to sets of characters, to entire paragraphs, or to all text subsequent to the style code

To use the style feature properly, you need an understanding of its basic principles, so I spend some time at the outset of this chapter providing this background information. Please be patient as you read this, then, because the payoff in terms of increased efficiency at using WordPerfect is well worth the time it will take you to read the whole chapter.

USING STYLES FOR CONSISTENCY AND EFFICIENCY

Why use styles? Styles can make your documents more consistent, and they can make your use of WordPerfect more efficient.

Consider the following scenario: You're writing a report for your boss or for a class, and you must follow certain formatting rules. For instance, the section headings must be boldfaced in 14-point Helvetica, the main body text must be in 12-point Times, and block quotes must be formatted with a 1/2-inch paragraph indent on the right and a 1-inch indent on the left. If you don't use styles, you must manually create these formats every time you use them. This process can be tedious and error prone, especially because some of these formats don't match WordPerfect's defaults exactly. You can use styles to create shortcuts for each of these formats, though, and simply select the appropriate style from the Styles toolbar item (Figure 12.1) as you work. If you set up linking appropriately, you won't even need to explicitly select a style change if you always use one style after another; for instance, you're likely to use the main text style after a section heading.

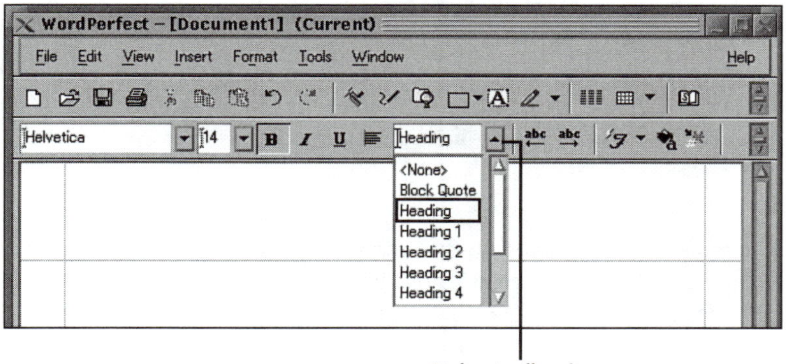

Figure 12.1
When you click the triangle next to the style display, you get a scrolling list of current styles and can select a new one.

→ To learn how to set fonts, **see** "Fonts," **p. 178**

→ To learn how to set indents, **see** "Indents," **p. 208**

Now imagine that you're almost finished with your report and discover you've been using the *wrong* formatting. Instead of Helvetica 14-point bold for headings, you must use Helvetica 16-point italic; and instead of a 1-inch left indent on block quotes, you need a 3/4-inch indent. If you've been entering your formatting on a case-by-case basis, you'll need to go back and change *every* instance by hand. Alternatively, you could use a find-and-replace operation, but that might miss something because of inconsistencies in the order in which you entered your formatting codes, or it might change codes it shouldn't. Having to make format changes at this point not only takes a lot of time and effort but also increases the likelihood that the final product will contain errors or inconsistencies. If you use styles, though, you can change the formatting in just a few seconds because you can change every instance of text that's formatted in a given style by editing the original style definition.

Tip #154 from *Tux Tips*

Styles can be helpful if you want to experiment with different formatting options for a document. Create your document using styles, but create rough formatting at first. When you're done, change the styles and observe the results; then stop whenever you're satisfied.

Styles can also increase your efficiency as you work across documents. You can create a style and then save it for use in multiple documents. You can, for instance, create a letterhead style that includes a company logo and return address and sets up proper formatting for the rest of the letter. Select this style, and your document will suddenly have the letterhead. This use of styles is particularly effective when combined with WordPerfect's templates feature.

→ To learn about templates, **see** "Promoting Cross-Document Consistency with Templates," **p. 333**

TYPES OF STYLES

WordPerfect's styles can be classified in two major ways, each of which has some influence on how the style is created and used. Each classification dimension is important for different reasons, although they sometimes interact in important ways.

SCOPE

The scope of a style refers to the amount of text to which the style applies. WordPerfect's dialog boxes sometimes refer to this characteristic as the style's *type*, but that's a rather imprecise term. WordPerfect supports three possible scopes for its styles, and these should be familiar if you've used WordPerfect's other formatting features:

- Character—Character styles apply to individual characters or groups of characters in text. You could, for instance, create a style called Emphasis, containing the bold attribute. You could then use the Emphasis style whenever you want to emphasize something in your text, be it a single character or several paragraphs.
- Paragraph—Paragraph styles apply to a minimum text unit of one paragraph. As such, if you select a paragraph style when the insertion point rests in a paragraph, the entire paragraph takes on that style.
- Document—Document styles apply in an open-ended manner to all text from the insertion point forward. Whatever formatting they set up is undone only when you tell WordPerfect to use another style or when you override the individual formatting in the style with its counteracting codes.

Caution

You can't turn off a document style after you enable it. If you want to use a style for only a portion of your document, you need to use a paragraph or character style or you must create another document style to counteract the effects of the first.

PAIRED AND ONGOING CODES

Character and paragraph style codes come in pairs and are applied as *paired codes (p. 65)*. Therefore, you can apply them to a range of existing text by selecting the text and then picking the appropriate style from the style list. You can also apply them while entering text by selecting the appropriate code and then entering text. When you press the Enter key while typing text in a paired style, WordPerfect either discontinues the style or *chains* to another, meaning that it starts the new style after the paragraph containing the first is done. There's more information on the chaining of styles shortly.

The document style codes are ongoing codes, also known as *open-ended* or *document* codes. These don't come in pairs, so when you select a document style, it remains in place until you select a new style.

Competing Scopes

WordPerfect's style types mirror the types of basic codes available in the program. If you use basic codes with a scope that's different from the scope for the style in which they're used, the result depends on the details of the basic codes' and the style's scopes. In general, the higher scope takes precedence.

For instance, if you create a document style that consists of character-oriented formatting codes such as [Bold], WordPerfect simply applies these codes as if they were open-ended codes. If you put the [Bold] code in a paragraph style and then position your insertion point in a paragraph and select this paragraph style, the entire paragraph is formatted as bold. In either case, you'll be able to deactivate the bold code for individual characters, words, or paragraphs within the range of the style.

If you create a character style that includes paragraph- or document-oriented formatting, however, any attempt to apply that style forces WordPerfect to reformat text to accommodate the individual code's requirements. For instance, Figure 12.2 shows the effect of applying a character style that contains paragraph margin-setting and paragraph-spacing codes to a single word within a paragraph. Note that Reveal Codes doesn't show any [HRt] codes around the selected text, but the formatting effect is the same as if they were there. Those codes are, in effect, implicit in the style itself. WordPerfect still treats the entire unit as a single paragraph, however. If you apply further paragraph-oriented formatting, it works on the entire original paragraph and is not be interrupted by the [Char Style: Char-w/Parag] code.

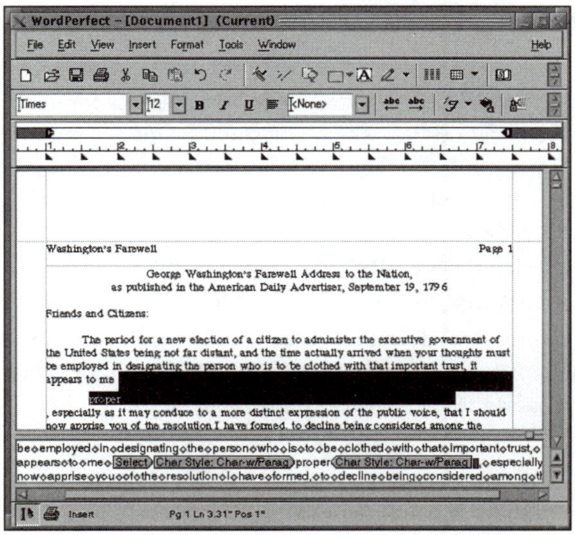

Figure 12.2
The selected word, *proper,* was formatted with the character style Char-w/Parag, which contains paragraph-oriented formatting codes. Hence WordPerfect created an implicit paragraph just for that word.

LOCATION

When applied to a classification of WordPerfect styles, the word *location* refers to where the style information is stored on your system, not to where the style is used inside a document. The possible locations for a style are

- System styles—WordPerfect includes several styles as part of its basic operation. If you start WordPerfect and immediately click the Styles toolbar item (refer to Figure 12.1), you'll see some predefined styles. More important, however, are the styles that are (mostly) transparent to you. For instance, if you create a header, WordPerfect uses its default Header style. You can see this style by using Reveal Codes when editing a header, as in Figure 12.3. WordPerfect maintains system styles for other standard formatting features, as well. It is possible to edit these styles, but when you do, your edits affect only the current document; WordPerfect restores its defaults for the next document you edit.

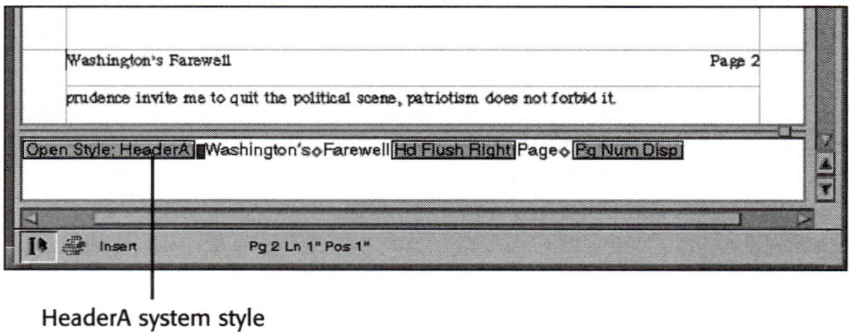

Figure 12.3
Many of WordPerfect's formatting features use styles internally, although you might not notice this fact.

- Shared library—WordPerfect maintains a library of styles for all users. These are separate from the system styles, which you should consider immutable except on a document-by-document basis. The WordPerfect administrator (generally root) can permanently change shared styles. WordPerfect for Linux doesn't ship with any standard shared library styles.

→ If you need more information about the root account, **see** "Users and the Superuser," **p. 11**

- Personal library—WordPerfect for Linux maintains a library of styles for each individual who uses the program. These styles are stored in the .wprc/.wp8styles file in each user's home directory. You can change or add to your own personal library styles (WordPerfect creates the file, but it's empty by default).

- Template styles—You can store a collection of styles in a template and use that when you create new documents. Template styles help you maintain a consistent look for all the documents you create based on these styles.

→ For more information about templates, **see** "Promoting Cross-Document Consistency with Templates," **p. 333**

- Document styles—If you create a style in a document and don't save the document as a template, that style will be saved with the document and will be accessible only from that document. Whenever you use a template, personal library, or shared library style, its styles effectively become document styles, and whenever you modify a system style, it is essentially converted into a document style. Thus, document styles are the type you use most; the others are ways to package styles for easy use.

Caution

The term *document style* can refer to either the scope of the style or its storage location. I've disambiguated these terms by using them in fuller phrases, such as *document-scope style* or *styles stored in a document* throughout this chapter.

Where a style is stored influences when you can use the style. For instance, if you create a style in a document, you won't be able to use it in other documents unless you import that style into another document. If you use a standard system style, though, it will be accessible from all your documents.

→ For information on sharing styles across documents, **see** "Retrieving Styles from Disk," **p. 324**

The Special Case of the Document Initial Style

One style in particular deserves special attention: the Document Initial Style. If you create a new WordPerfect document and immediately open Reveal Codes with View, Reveal Codes (Shift+Ctrl+F3), you'll see your document already has one code, [Open Style: InitialStyle] and you cannot delete it. [InitialStyle] sets the initial codes that you do not want to delete accidentally. Other than the fact that the [InitialStyle] code can't be deleted, it's very much like a regular document-scope style placed at the beginning of the file. Because it's the first style in the document, it applies to all the text in the document unless subsequent styles or formatting changes override it.

Tip #155 from *Tux Tips*

Place changes you want to apply to the entire document, such as margins for the entire document or font changes, in the [InitialStyle] code to ensure they apply to everything and aren't accidentally deleted. Changing a default in this way also ensures that the change applies to headers, footers, text boxes, and other special formatting elements.

You can edit the [InitialStyle] much as you would any other style. This process is described later in the chapter.

→ If you want to know how to edit a style, **see** "Editing a Style," **p. 321**

Additional Specialty Styles

Many of WordPerfect's formatting features, such as headers and footers, graphics, and text boxes, are somewhat like miniature self-contained WordPerfect documents embedded within the original document. As such, each feature contains its own equivalent of the [Open Style: InitialStyle] code. For instance, a header's initial style code is called [Open Style: HeaderA] or [Open Style: HeaderB]. By default, these styles include the main document's [InitialStyle] code, which is why changing the document's [InitialStyle] code automatically changes these elements.

→ For information on headers and footers, **see** "Headers, Footers, and Page Numbers," **p. 235**

→ For information on graphics, **see** "Graphics," **p. 377**

→ For information on text boxes, **see** "Using Text Boxes to Add Variety to Text Layout," **p. 357**

Caution

Although I've described these special document elements as being like a document embedded within another document, you should not confuse these items with a document that you merge with another via the Insert, File menu item or with WordPerfect's Subdocument feature. They are both entirely different features.

→ To learn how to combine two documents, **see** "Merging Two Documents," **p. 88**

→ To learn about subdocuments, **see** "Master Documents and Subdocuments," **p. 598**

Are you having trouble understanding how the Style choice from the context menu for a graphic relates to the contents of this chapter? See "Graphics Styles" in the "Troubleshooting" section at the end of this chapter.

The document-like nature of these elements is manifested in another way: When you're working in Draft view (if you've chosen View, Draft or pressed Ctrl+F5), WordPerfect opens a new document window when you edit these elements. For the most part, you can treat these elements as entirely separate documents. You will, however, have access to all the main document's styles when you edit one of these elements. You won't be able to create new styles when you're editing a specialty element, so if you expect to need one when working with such an element, create it before you begin editing. This rule applies even if you're using Page view (View, Page, or Shift+Ctrl+F5).

When you create a specialty element, WordPerfect inserts a single code in your document for that element. Double-clicking on that code in Reveal Codes or clicking within the element in your main document's window (if it's visible that way) lets you edit the element. At this point, WordPerfect's Reveal Codes pane shows the codes only for the specialty element, including its initial style code. You can find more details on how to work with these elements in the chapters devoted to them.

Creating a Style

Unless you're using a template with predefined styles, WordPerfect provides only a handful of default styles. Aside from the document initial style and styles for specialty document elements, these styles are intended for different types of headings in a document; they're called Heading 1 through Heading 5, and you can choose them from the Styles menu on the lower default toolbar (refer to Figure 12.1). Chances are you'll want a richer assortment of styles, and you can create these quite easily.

USING QUICKSTYLES TO CAPTURE A STYLE

One of the simplest ways to create a style is to use WordPerfect's QuickStyles feature to take a snapshot of the formatting of text at some point in the document and to use that as a style. To use QuickStyles, follow these steps:

1. Enter some text and format it in the way you want the style to appear. For instance, you might apply a first-line indent and interparagraph spacing change to allow you to quickly enter paragraphs of text without pressing the Enter or Tab keys unnecessarily.

➔ If you want to know how to apply indent and inter-paragraph spacing options to paragraphs, **see** "Paragraph Formatting," **p. 205**

2. Position your insertion point within your formatted text.
3. Select the Styles item in the lower default toolbar and select the QuickStyles item. This step opens the Styles QuickCreate dialog box shown in Figure 12.4. Alternatively, you can choose the Format, Styles (Shift+Ctrl+F8) menu item, which brings up the Style List dialog box shown in Figure 12.5. Then click QuickStyle to display the Styles QuickCreate dialog box.

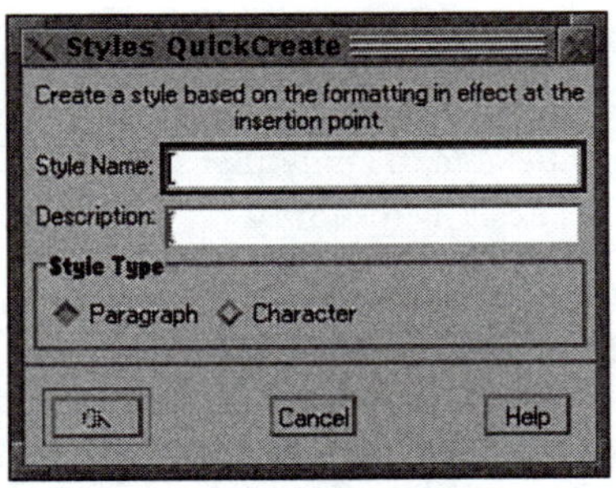

Figure 12.4
You won't be able to click OK until you've typed in an appropriate name for the style.

4. Type a name for the style in the Style Name field of the Styles QuickCreate dialog box and optionally type a description of the purpose or effect of the style in the Description field.
5. Tell WordPerfect whether the style is a paragraph or character style by clicking the appropriate radio button. Recall that paragraph styles apply to entire paragraphs of text, whereas character styles can be applied to smaller units of text. In the example of a style to automatically create appropriate indentation and spacing for a paragraph, you'd select Paragraph here.

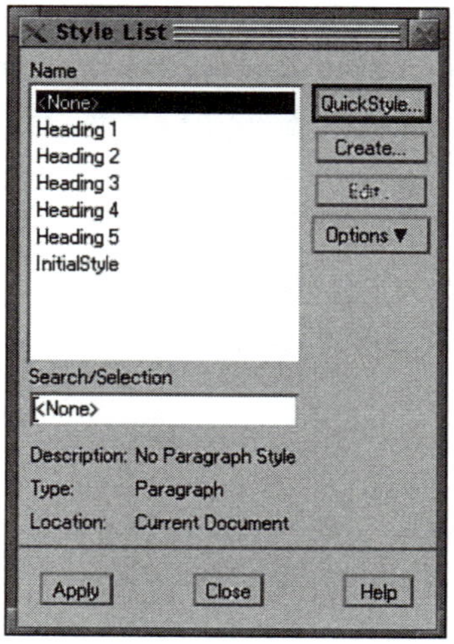

Figure 12.5
The Description area of the Style List dialog box provides information on the type and location of the selected style.

Caution

You cannot select Document in the Styles QuickCreate dialog box. If you want to create a document style, you have to do it from within the Style Editor, described shortly.

6. Click OK. WordPerfect creates your style based on the formatting at the insertion point and makes it available for selection via the Styles toolbar item or the Style List dialog box.

When you're done, the original formatting on the text you used as a model for the QuickStyle remains in place. You'll now be able to apply that same formatting to any other text by selecting the appropriate style, as described shortly.

Tip #156 from
Tux Tips

As one of the great benefits of styles is the capability to alter styles later and have the alteration apply to all text, leaving the original formatting for a style's model in place isn't optimal. You may want to delete the original formatting codes and apply the style instead. This step allows you to take the best advantage of that style if you decide to change it in the future.

DEFINING A NEW STYLE

Using QuickStyles can be a convenient way to define a style; however, it does have its drawbacks. Creating a format and then deleting it and applying a style in its stead, as I've just suggested, can be a bit awkward. In addition (and even more important), using QuickStyles prevents you from using some of WordPerfect's styles capabilities. To use these, you need to define your styles in another way:

1. Choose Format, Styles (Shift+Ctrl+F8) to open the Style List dialog box, shown in Figure 12.5.
2. Click Create to define a new style. WordPerfect now displays the Styles Editor dialog box, shown in Figure 12.6.

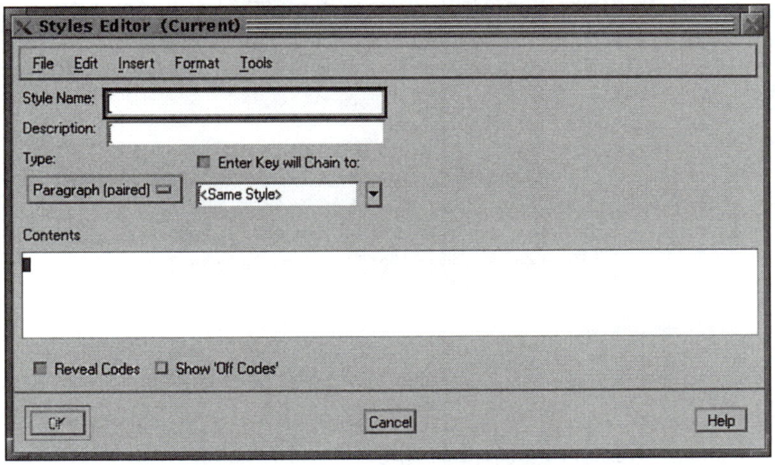

Figure 12.6
The Styles Editor dialog box contains a data entry area that works very much like either the Reveal Codes or the normal view of a normal Document window.

3. Type a name for the style in the Style Name text entry field and type a description in the Description field.
4. Select the scope of style by using the Type option button. You can choose from Character, Paragraph, and Document types.

→ If you want to know about different style scopes, **see** "Scope," **p. 308**

5. Select your chaining options by selecting or deselecting the Enter Key will Chain to check button and choosing a chaining action from the drop-down menu list immediately below that button. These options are complex, and I describe them in detail shortly.
6. Click in the large Contents area of the dialog box and enter your formatting commands. You can use keyboard shortcuts or the Styles Editor menus, which give you access to WordPerfect's editing features, much as the Document window menus do. You'll find that a few keys, such as Tab, don't work. The Insert menu has entries for these keys.

Caution

By default, the Styles Editor shows codes in a manner similar to that used by the Reveal Codes pane. If you like, you can disable this feature by unchecking the Reveal Codes check button near the bottom left of the window. In this case, WordPerfect displays what you type as it will appear in WordPerfect. Because you normally enter codes, not text, into a style, this approach isn't generally very useful. You might want to use this technique briefly if you're creating a style that includes text, however, to see how that text will appear.

Caution

Not all menu commands insert codes. Some, such as Tools, Spell Check (Ctrl+F1) and Edit, Find and Replace (Ctrl+F2), perform their usual functions but on the text and codes in the Styles Editor window, rather than on the text in the document.

7. If you want WordPerfect to do something at the end of every style, select Show 'Off Codes.' This option displays a special code [Codes to the left are ON—Codes to the right are OFF] (referred to hereafter simply as [OFF]). WordPerfect inserts anything you type after this code at the end of the style.

Tip #157 from *Tux Tips*

With character styles, WordPerfect uses the Enter key as a command to stop using that style, but Enter won't then insert an [HRt] code. Therefore, if you know this style will always end with a paragraph, you can use the Off Codes feature to have WordPerfect automatically insert an [HRt] code at the end of the style.

8. Click OK to finalize your style. WordPerfect adds it to the list of available styles.

STYLE CHAINING

Style chaining is an important feature of paragraph and character styles, and its effects can be both substantial and subtle. The idea is fairly simple: When one style ends, WordPerfect can automatically begin another. This feature has an effect only when you're typing in new text; if you apply a style to existing text, the chaining options for the style don't do anything. Style chaining is also meaningless with respect to document styles because they don't really end—they can only be overridden by another style.

The Styles Editor window always gives you at least two options for chaining of paragraph or character styles: <None> and <Same Style>. The first reverts to the default style for that part of the document (the InitialStyle or a previous document style) after the current one has run its course. The <Same Style> option starts the same style again after its first run ends. Although this behavior may seem odd, it's useful in the case of paragraph styles, given WordPerfect's method of handling them as you type. Specifically, when you press the Enter key to end a paragraph, the current paragraph style also ends. With style chaining to <Same Style>, WordPerfect reenables the previous paragraph style, so you can continue creating text in this style indefinitely.

If you've created more than one style, you can also select your previously defined styles as chaining options. This feature is particularly useful if you regularly use one style immediately after another. For instance, a heading in a document is normally followed by one or more paragraphs of text, so you can chain from a heading style to a style for text paragraphs and save yourself having to select (and possibly forget to select) the appropriate paragraph style.

Tip #158 from *Tux Tips*

If you want to have two or more styles chain to each other (say, for a question-and-answer format for an interview), you can. Create the first style without chaining. Then create the second style and chain it to the first. Finally, *edit* the first style, as described shortly, and chain it to the second style.

→ To learn how to edit a style, **see** "Editing a Style," **p. 321**

When you create a character style, you can elect to have the style chain or not chain, at your discretion. If you enable chaining for a character style, then pressing the Enter key when you're typing in text does the chaining, but does *not* insert an [HRt] code in the text (unless you include this code after the [OFF] code). If you disable chaining for a character style, then the only way to get out of the style is to select <None> or another style. Pressing the Enter key inserts an [HRt] code in the text, and the style remains active.

When you create a paragraph style, you *must* use some chaining option. If you opt to create a paragraph style and then deselect the Enter Key will Chain to option, WordPerfect automatically changes the style's type to Character. To keep using the current style from paragraph to paragraph, either define it as a Document style or chain it to itself.

When you create a document style, you *cannot* use chaining. Because document styles never really end, the concept of chaining is meaningless when applied to them. WordPerfect disables the chaining features in the Styles Editor when you opt to create a document style.

INCLUDING ONE STYLE WITHIN ANOTHER

You probably want to design your styles in a way that can be described hierarchically. For instance, you might have a document in mind with headings, subheadings, paragraphs of text, indented paragraphs for block quotes, and paragraphs with hanging indents for bibliography entries. All of these elements will share certain formatting features, such as margins and perhaps the font. The headings might then add justification options and perhaps font size or style changes. The paragraphs will use first-line indent features, and the block quotes may use this as well as paragraph indents. Bibliography entries will require different indent options entirely. In sum, the styles in this document might be organized as shown in Figure 12.7.

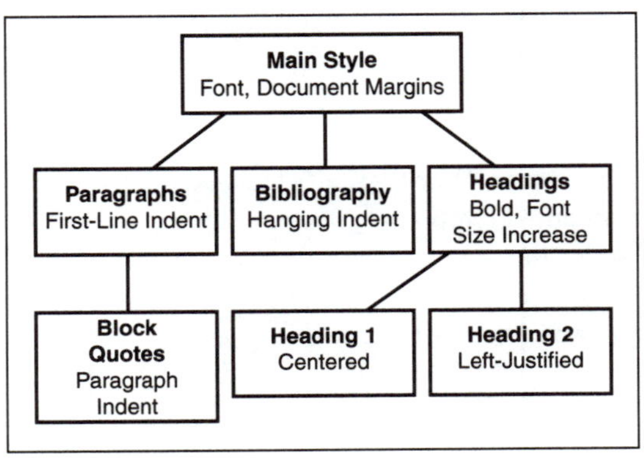

Figure 12.7 Styles can be organized hierarchically, so that those far down in the hierarchy inherit the characteristics of those above them.

At each level, the style takes on the characteristics of the style at the preceding level (its *parent* style). Ideally, you want to structure your styles so that if you need to change some high-level attribute, such as the font, that change will automatically propagate down to all its descendant styles. Some word processors structure their styles in an explicitly hierarchical manner. WordPerfect doesn't, but you can achieve the same effect by including one style within another.

To use this technique, choose Format, Styles (Shift+Ctrl+F8) in the Styles Editor window and locate a style to serve as this one's parent. This step brings up the Style List dialog box, as in Figure 12.5. Select the style on which the one you're creating should be based and click Apply. Depending on the style type, WordPerfect inserts either an opening or paired opening and closing codes for the style you've selected. Create whatever changes or additions you want for your new style, within the confines of the paired codes if that's what WordPerfect inserted, and click OK.

Tip #159 from *Tux Tips*

Because WordPerfect doesn't show the hierarchical structure of your styles, you might want to provide this information in the Style Name. For instance, you could call your main text style MainText and then use MainText-Quote as the name of an indented style for block quotes. This naming pattern will help you to understand the structure of your styles and will keep related styles grouped together in the style lists.

When you create a child style, you can override the parent style's settings. For instance, if you set a font and set margins in InitialStyle, and then discover that you need to create a style that uses a different font but the same margins as InitialStyle, you can override the InitialStyle font setting in your new style. Simply include the new font code as an ordinary part of your new style.

Tip #160 from *Tux Tips*

Occasionally when you select a WordPerfect feature, it will insert more than one code. This condition is particularly common with certain font attributes. For instance, you may want to change the size of a font. When you open the Font dialog box by choosing Format, Font (F9) and select a new font size, WordPerfect inserts both the font size change and the name of the font. You can—and probably should—delete any extraneous codes that WordPerfect inserted. You certainly don't want the extra [Font] code if you want the font itself to be the same as that used in a parent style.

Saving Styles

As described earlier in this chapter, WordPerfect's styles can be stored in any of several places, including the document, a template, a personal library, or a shared library. By default, WordPerfect stores the styles that you create in the current document. If you want to use the same style in a variety of documents, though, you'll want to store it elsewhere in order to access it easily without having to re-create it.

Creating a Styles Template File

One option for storing styles is to store them in a template file. Whenever you want to create a file using the styles you've created, simply load that template file.

→ For information on templates, **see** "Promoting Cross-Document Consistency with Templates," **p. 333**

Tip #161 from *Tux Tips*

> Storing your styles in a template file is particularly useful when your styles are related to a particular type of document. For instance, if you've created styles specifically for reports that follow a format required by your company or profession, you can store those styles as a template and then load that template whenever you need to use that format. When you need to write something else, such as a business letter, your styles list won't be cluttered with the unnecessary report styles.

One of the simplest methods of saving a template with the styles you've created is to save a document with the styles but no text. Give the document the filename extension .wpt, though, rather than the usual .wpd, to flag it as a template file. You can then load the file as an ordinary file or set it up as a template using WordPerfect's ExpressDocs feature.

Caution

> If you elect to save and load your template file normally, be careful not to select File, Save (Ctrl+S) until you've saved a new document under another name; otherwise, you'll overwrite the template file. Overwriting isn't an issue if you use an ExpressDocs template, though.

You can also save *only* the styles in a document, even if the document contains text. To do so, choose Format, Styles (Shift+Ctrl+F8) to open the Style List dialog box (refer to Figure 12.5). Click the Options button and select Save As, which produces the Save styles to... dialog box, as shown in Figure 12.8. From the Save styles to... dialog box, you can elect to save the styles you've created (User Styles), your modifications of the default system styles (System Styles), or both. If you type a filename into the Filename field, WordPerfect stores your file under that name in the .wprc subdirectory of your home directory. You can also click the folder icon to get a standard WordPerfect File Manager dialog box with which to specify a filename.

→ For more details on using the File Manager dialog box, **see** "Linux File Manipulation," **p. 80**

Tip #162 from *Tux Tips*

> Even if you have no text in your document, using the Save styles to... dialog box to save your styles may result in a smaller file than using the ordinary file save operation, which saves additional information that's not needed for the styles alone.

Figure 12.8
Specify a file into which your styles will be stored.

STORING A STYLE IN A LIBRARY

If you regularly use a style or set of styles across multiple document types, you may want to save the style or styles in a library. To do so, follow these steps:

1. Create a style or load a document that contains the style you want to save.
2. Choose Format, Styles (Shift+Ctrl+F8) to bring up the Style List dialog box (refer to Figure 12.5).
3. Select the style you want to copy to the library.
4. Click the Options button and select the Copy item. This step opens the Styles Copy dialog box shown in Figure 12.9.

Figure 12.9
The Shared Library option is available only when you're running WordPerfect in Administrative mode.

5. Select the library to which you want to copy the style, personal or shared, by clicking the appropriate radio button. Because you're copying from the current document, there's little point in leaving that button selected, despite the fact that it's the default.

Caution

To copy a style to a shared library, you need to run WordPerfect in Administrative mode, which typically means logging in as root and running the `xwp` binary using the `-adm` switch. If you then copy a style to the shared library, the style will be available to all users of WordPerfect on your computer. If you copy the style to the personal library, only you will have access to the styles.

➔ To learn more about the root account, **see** "Users and the Superuser," **p. 11**

6. Click OK to copy the style to the specified library.

Tip #163 from *Tux Tips*

If you want several users of one computer to have access to a style, you can repeat this procedure for each person, or you can copy the .wprc/.wp8styles file from the directory of the person who has the styles in his or her library to the equivalent directories of the other users. This latter option wipes out any other styles those users might have defined, however.

After you store a file in a library, it will always be available, either to the person who owns the library, if it was a personal library, or to everybody who uses that computer, if it was the system library.

In case of conflicting style names, the document's style always takes precedence. Because WordPerfect converts template and library styles to document styles when you use the styles, after you save a document any changes you make to the style stored in the template or library have no effect on the documents already created with that style. When you load those documents, you get the old style's formatting.

EDITING A STYLE

One of the great advantages of styles is that you can alter a style after you've created and used it. When you do so, all the text that's been formatted with that style within your document automatically changes to match the new formatting. This feature can be a great timesaver if you need to make formatting changes in a long or complex document.

ALTERING AN EXISTING STYLE

One of the easiest ways to change a style that already exists is to double-click on the style's code in the Reveal Codes pane. This step opens the Styles Editor (refer to Figure 12.6) with the relevant style definition in place. You can then make any changes you desire to the style, and they'll take effect as soon as you click OK. Editing a style works just like creating one except that the style already contains formatting codes when you begin to edit.

→ For information on creating a style, **see** "Defining a New Style," **p. 314**

You can also edit an existing style by choosing Format, Styles (Shift+Ctrl+F8) to open the Style List dialog box (refer to Figure 12.5). Click on the name of the style you want to edit and then click Edit to open the Styles Editor (refer to Figure 12.6). Again, editing the style works just like creating one except that the style already has formatting codes in place when you begin editing.

Have you edited a style and now can't see the change? See "Edited Style Changes Don't Work Immediately" in the "Troubleshooting" section at the end of this chapter.

Resetting a Style

A few styles—namely, the InitialStyle and the five default Heading styles—can be altered by you and then reset to their original states if you decide you don't like the changes you've made. To reset a style, first choose Format, Styles (Shift+Ctrl+F8) to bring up the Style List dialog box (refer to Figure 12.5). Click on the name of the style you want to reset and click the Options button in the Style List dialog box. Select the Reset option. When WordPerfect asks for confirmation that you want to reset this style, if you're sure, click Yes.

Deleting a Style

You may occasionally want to delete a style. Perhaps you're no longer using it or perhaps WordPerfect created extraneous styles in a file import operation or perhaps the style just isn't working out the way you'd hoped. Whatever the case, the procedure for deleting a style is simple:

1. Choose Format, Styles (Shift+Ctrl+F8) to bring up the Style List dialog box (refer to Figure 12.5).
2. Click on the name of the style you want to delete.
3. Click the Options button. From the list of actions, choose Delete.
4. WordPerfect displays the Delete style dialog box (Figure 12.10). This dialog box gives you two choices: Include Codes and Leave Codes. If you select the former, WordPerfect removes all traces of the style from your document. Therefore, if you delete a style that, say, centers text, selecting this option returns any text that's formatted in that style to its "native" state. The Leave Codes option, by contrast, converts any codes for the style in the text into the raw formatting for the style. Therefore, the centered text just described remains centered, but you can no longer use the style to modify all instances of that centering; instead, you have to modify them individually.

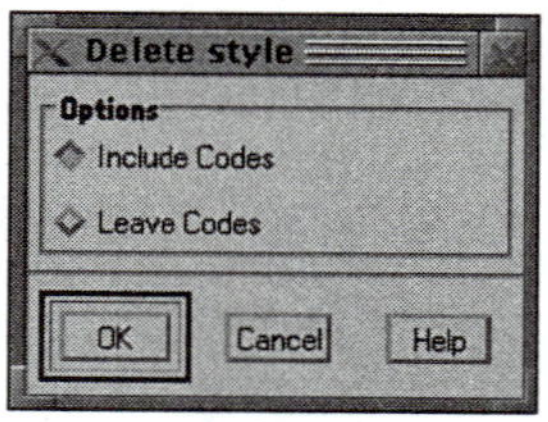

Figure 12.10
In this case, *include* and *leave* refer to the deletion process–you'll be including the codes in the deletion or leaving them behind in the document.

5. Click OK to remove the style.

Caution

Unlike most operations in WordPerfect, you cannot undo a code deletion. Therefore, you should exercise extreme caution when performing this operation. If this style resides in your document, I recommend you save the document immediately before deleting the style. This way, if you make a mistake, you can close your document window and load the undamaged version of the file. If you're deleting a style from a library, consider moving the style to an empty file first so you can recover it if you need to.

Tip #164 from *Tux Tips*

You can use a code deletion while leaving the codes behind to convert a document from a style-based form to a conventionally formatted form. This technique might make it easier for some foreign word processors to load the WordPerfect file.

→ For information on exchanging files with other word processors, **see** "Document Exchange," **p. 88**

Caution

You can only delete styles that exist in the document you're currently editing or that are stored in the style libraries to which you have access.

USING STYLES

After all this discussion of creating styles, editing styles, saving styles to disk, and so on, you may feel that you know everything you need to about styles, and yet we've not covered one very important topic: *using* them! Fortunately, basic use of styles is pretty simple. The main trick is really in using them regularly in your day-to-day work so that you become familiar with them and know how best to apply them. Using styles effectively is a skill that you must learn through practice, though at the end of this section I've included some examples to help you.

APPLYING STYLES TO TEXT

As with most of WordPerfect's formatting features, you can apply styles in one of three ways:

- Select a section of text and apply the style to it. WordPerfect automatically expands the selection, if necessary, to include the entire paragraph(s) you've selected if you are working with a paragraph style.

Caution

If you apply a document (open ended) style in this way, WordPerfect inserts the code at the beginning of your selection, and the style will apply from that point on, even past the end of your selected text.

- Place your insertion point somewhere in your text and select a style. If the style is a character style, nothing seems to happen because WordPerfect places both open and close codes in the text at that point. If the style is a paragraph style, WordPerfect applies the style to that paragraph alone, even if you've defined a style to which the one you selected chains. If the style is a document style, WordPerfect applies the style from that point on in your document.
- As you're typing text at the end of a document, select a style. WordPerfect applies that style to any text you subsequently type. WordPerfect may automatically switch to another style if you've selected a chaining option and press the Enter key, or the style may continue indefinitely if you've chosen a document style or a character style with no chaining options.

WordPerfect provides two ways for you to select a style using any of the preceding three models of style application:

- Select the style from the Style toolbar item (refer to Figure 12.1). Assuming that this toolbar is displayed, using the toolbar is generally the easiest way to select a style, but this method also provides the least information about the style you're about to apply.
- Choose Format, Styles (Shift+Ctrl+F8) to open the Style List dialog box (refer to Figure 12.5). Choose the style you want from the list and then click Apply. This method takes a bit more effort than the first, but the Description area of the Style List dialog box provides information about the style, including its type and location. This information is vital for understanding just how your style will interact with your text. You can also click Edit from this dialog box to see more information about the style, including to what, if any, other style it chains.

Have you applied a style that includes text and found that you cannot edit that text? See "Can't Edit Text in a Style" in the "Troubleshooting" section at the end of this chapter.

RETRIEVING STYLES FROM DISK

WordPerfect allows you to import styles from a WordPerfect document (be it a template or an ordinary document) without affecting the current document's text. This technique is useful if you want to apply an existing document's styles to a new document or if you want to update the styles in your current document with new ones from another document. To perform such a retrieval, follow these steps:

1. Choose Format, Styles (Shift+Ctrl+F8) to open the Style List dialog box (refer to Figure 12.5).
2. From the Style List dialog box, select Retrieve from the Options button. This step opens the Retrieve styles from... dialog box (Figure 12.11).
3. Enter a filename in the Filename field. You can do so by typing the name directly or by clicking the folder icon and selecting the file in a WordPerfect File Manager dialog box.

Figure 12.11
The Retrieve styles from… dialog box is very similar to the Save styles to dialog box in Figure 12.8.

Caution

If you type a filename directly and without a complete path, WordPerfect assumes that file is in the .wprc subdirectory of your home directory. Therefore, if you want to load a file from elsewhere, be sure to provide a complete path to it (for instance, /home/rodsmith/letter.wpd).

4. Select the type(s) of style you want to retrieve from the Style Type radio buttons. If you select User Styles, WordPerfect retrieves only conventional document-scope styles. If you select System Styles, WordPerfect retrieves only your modifications to WordPerfect's System styles, such as the InitialStyle. If you select Both, WordPerfect retrieves all styles.
5. Click OK.
6. WordPerfect usually asks whether you want to overwrite the current styles. If you select Yes, then WordPerfect overwrites any styles in your current document with the same names as those in the source file. If you select No, WordPerfect does not import *any* styles.

Tip #165 from *Tux Tips*

If you want to import only a few styles, try using an empty document to load the styles from the source; then delete the extraneous styles and save the remainder to another file. Import this third file into the document in which you need the styles.

→ To learn how to save styles, **see** "Saving Styles," **p. 318**

After you retrieve the styles from the source, they should be available in your current document as document-scope styles. You can use them and modify them as you see fit.

You can retrieve styles from regular WordPerfect documents, from templates, and from any file to which you've previously saved styles. In any of these cases, WordPerfect extracts the styles and the styles alone.

SOME EXAMPLES OF STYLES IN ACTION

WordPerfect's styles feature is the sort of feature that tends to grow on you. Initially, you may not see much use for it, but as you apply it here and there, you begin to see how useful it can be. Here are some examples of how you can use WordPerfect's styles to improve your productivity.

LETTERHEAD

Most businesses have special preprinted stationery with their name and address at the top. This *letterhead* paper often requires special formatting in a word processor, even if that's just a few extra carriage returns at the start of the first page of a letter. WordPerfect offers an alternative: You can create a letterhead style that contains your company's name and address and even graphics to go with it. You can use the same feature for your personal correspondence so that you don't have to enter your name and address over and over again.

Tip #166 from *Tux Tips*

Even if you use preprinted letterhead paper for normal correspondence, when you fax directly from WordPerfect for Linux, you can't use this paper. Thus, a letterhead style that mimics your normal letterhead stationery can be a real boon.

Figure 12.12 shows a sample letterhead style in the Styles Editor window, and Figure 12.13 shows the result (including the start of a letter). You can examine this document yourself; it's on the CD accompanying this book in the chapter12 directory, under the filename QuadroBean.wpd. For information on accessing files on the CD, see Chapter 3 or your Linux distribution's documentation.

→ If you need to know how to access a CD in Linux, **see** "Accessing Removable Devices: Using the `mount` Command," **p. 76**

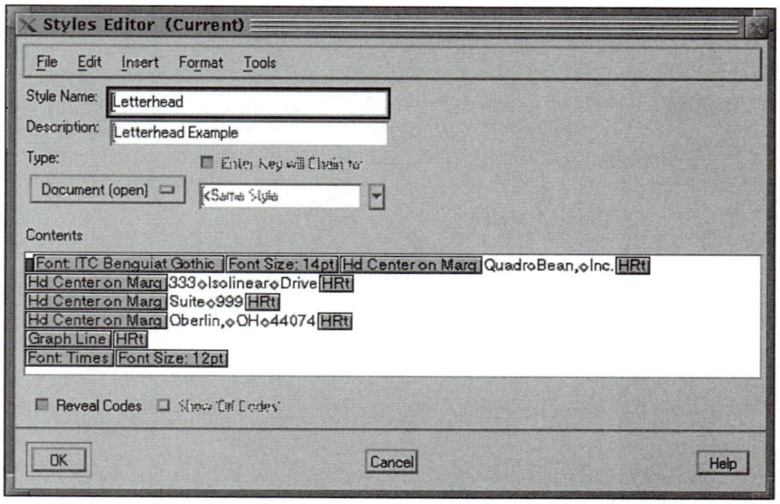

Figure 12.12
A letterhead style can be quite simple, or it can include lots of special formatting and even graphics.

Figure 12.13
The result of the letterhead style in Figure 12.12 is an easy-to-create letterhead.

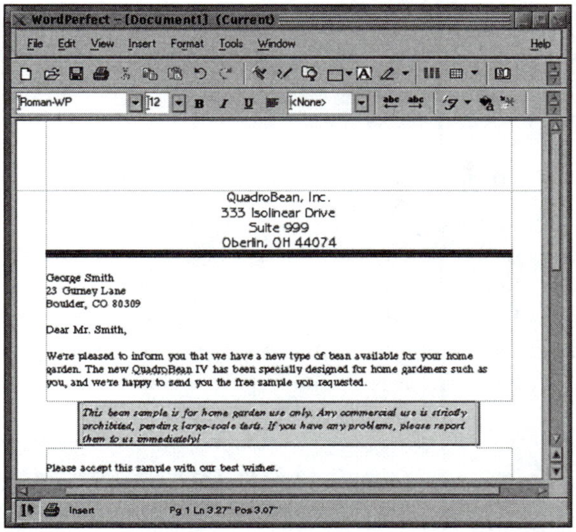

This style is done as a document (open ended) style. It is intended to be the first style on the page. The letterhead style inserts the return address for the company and a graphics line and then changes the font to 12-point Times. When you select this style, WordPerfect displays the return address and you can begin typing the text of the letter without worrying about fonts or other formatting. You can easily expand this type of style to include page numbers, margin adjustments, or other formatting features.

→ For information on setting fonts, **see** "Fonts," **p. 178**
→ For information on creating a graphics line, **see** "Graphics Lines," **p. 218**
→ For information on adjusting margins, **see** "Margin Adjustments," **p. 206**
→ For information on page numbering, **see** "Headers, Footers, and Page Numbers," **p. 235**

Tip #167 from *Tux Tips*

Styles similar in principle to this letterhead style can be used in other contexts, such as for first-page headings for in-house forms or for watermarks to help identify drafts or confidential documents. You can even delete or change a style as your document goes through revisions. For instance, you could have different watermark styles for first, second, and subsequent drafts and simply replace one watermark style with another as you revise your document.

→ To learn about watermarks, **see** "Watermarks," **p. 241**

Offset Text Box

For some types of documents, you may have a need for a paragraph that's set off from the rest of the text in some manner. In WordPerfect you can create this effect in a couple of ways. One is by creating a border or fill around an ordinary paragraph, and another is by creating a separate text box. If you regularly do this sort of formatting, you may want to create a style for it, especially if you regularly modify WordPerfect's defaults to something that's ideally suited to your needs.

→ If you want to learn how to create a text box, **see** "Using Text Boxes to Add Variety to Text Layout," **p. 357**

→ If you want to learn how to create a border around a paragraph, **see** "Borders and Fills," **p. 216**

Figure 12.14 shows style codes for creating an inline offset text box for a warning or other important text. When you're ready to use this style, just select it and anything you type will be properly formatted with margin adjustments, a surrounding text box, and italicized text. The text box in Figure 12.13 was created with this style and no other special formatting. When you press the Enter key while this paragraph style is active, the style ends and your text reverts to its preceding state because the style chains to <None>.

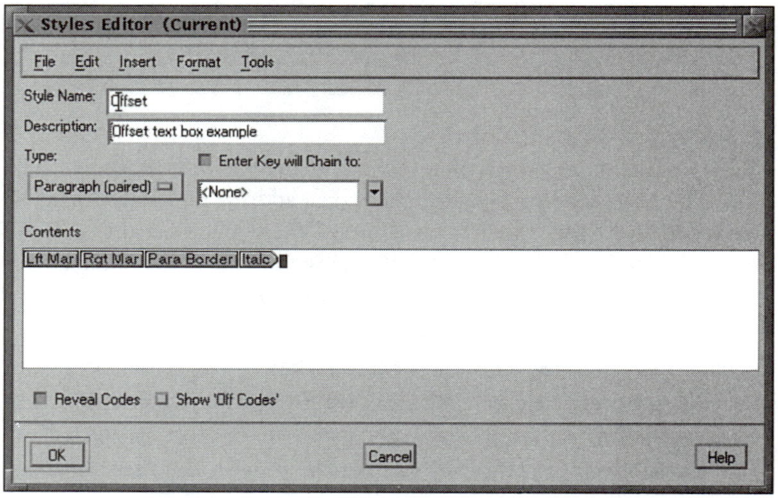

Figure 12.14
A paragraph style can include several formatting options, which can greatly reduce the effort required to produce special formatting.

QUESTION/ANSWER FORMATTING

Suppose you're writing a problem-and-solution section for a product manual or perhaps you're transcribing an interview. You might find something like the pair of styles shown in Figures 12.15 and 12.16 useful. These chain to each other, so that you don't need to manually enter the headings for the formatting on these lines or enter the text formatting for the different fonts. Each style sets a tab and creates a hanging indent so that the text flows naturally for this type of formatting. Each also sets fonts for the questions and answers, as well as their labels. Finally, the Solution style places an [HRt] code after the [OFF] code to separate the problem/solution pairs.

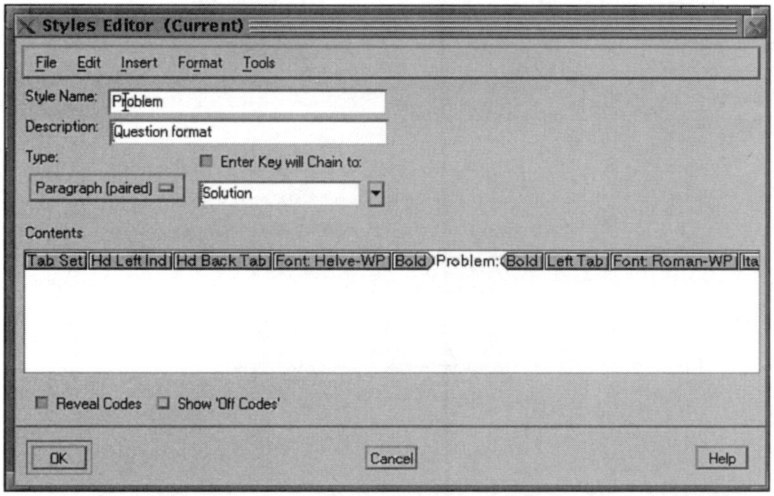

Figure 12.15
The Problem style uses Helve-WP bold for the label and Roman-WP italic for the answer.

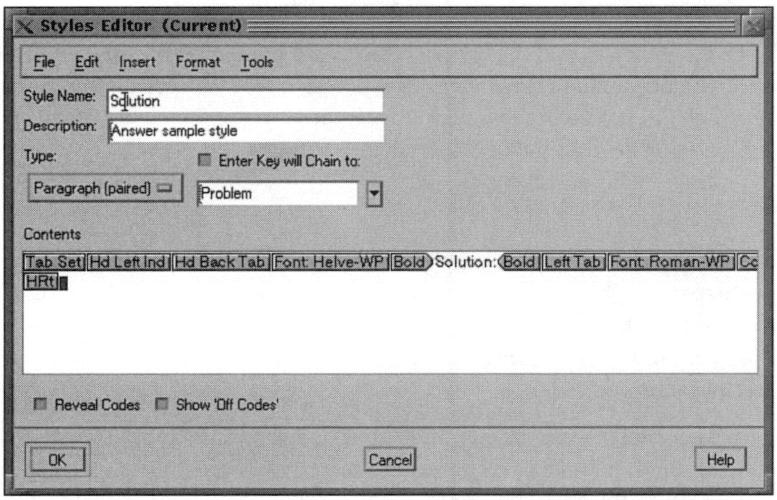

Figure 12.16
The Solution style uses Helve-WP bold for the label and Roman-WP for the solution, giving extra visual clues as to which is which.

These styles and a sample of their use are in the file troubleshoot.wpd on the CD accompanying this book. To use these matched problem/solution styles, simply select the Problem style. WordPerfect displays the word *Problem* and places your insertion point appropriately for you to begin typing a problem description, which appears in italics. When you're done typing the problem, press the Enter key and WordPerfect starts the Solution line, complete with the Solution label. When you're done with a solution, press the Enter key again. WordPerfect spaces down a line and begins the next Problem line. When you're done with all your problem/solution pairs, select any other style and the leading "Problem" text disappears. The result of using these styles is shown in Figure 12.17.

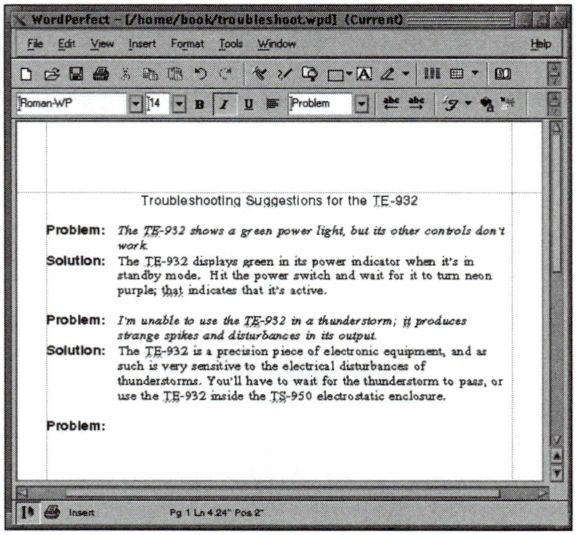

Figure 12.17 One style may contain different formatting codes applied to different types of text.

Both the Problem and the Solution styles are paragraph styles, so they apply to entire lines of text. They include text themselves, and that text isn't formatted in the same way as the text that you'll type in conjunction with these styles. These are also characteristics of the letterhead style described earlier.

Inline Heading

The final style example is an inline heading—a heading for a paragraph that's actually part of the paragraph, as opposed to one that appears on the line above the paragraph. This sort of style must be done with a character style because the style must end midline. A style to produce this effect is shown in Figure 12.18, and the file in which this style appears is on the CD as game.wpd.

To reduce unnecessary typing, the Inline-Head style includes two spaces after the [OFF] code to separate the heading from the text in the rest of the paragraph. This style also relies on a chain to the MainText style, which simply sets the main paragraph's font and a first-line indent. One consequence of this layout is that, as you type the first line with Inline-Head, the heading doesn't appear to be indented. When you press the Enter key to end the Inline-Head style and begin the MainText style, though, the paragraph suddenly indents itself.

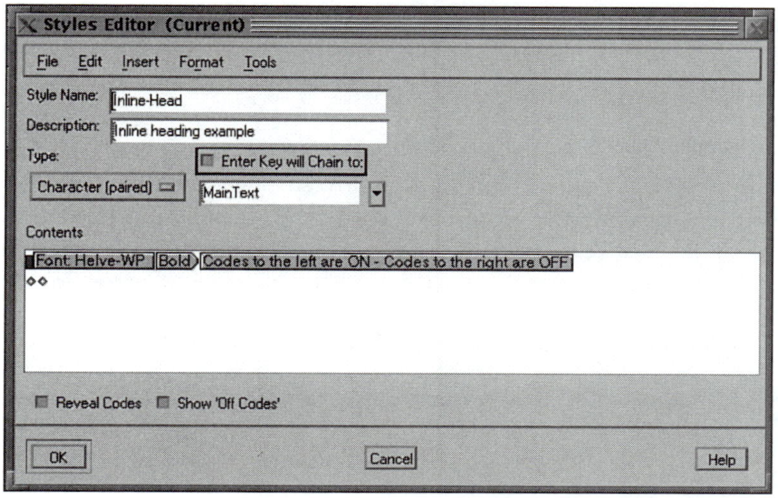

Figure 12.18
The Inline-Head style chains to the MainText style, which sets a font and paragraph indent.

TROUBLESHOOTING

EDITED STYLE CHANGES DON'T WORK IMMEDIATELY

You've edited a style and clicked OK in the Styles Editor window (refer to Figure 12.6), but your change doesn't seem to have taken effect.

The cause of this problem is a minor window redraw bug. If you scroll your document down or up and then return to the original location, you'll see that the style actually has been updated. WordPerfect simply neglected to show the update onscreen immediately.

GRAPHICS STYLES

You're trying to create or edit a graphics style, but it seems quite different from what's described in this chapter.

Graphics styles, while similar to regular styles in many ways, are also different. They're described in Chapter 15.

→ To learn about graphics styles, **see** "Graphics Styles," **p. 408**

CAN'T EDIT TEXT IN A STYLE

You've started a style that inserts some text, such as text in a return address or as part of a question-and-answer style. You want to edit this text, but when you click on it, nothing happens.

This behavior is normal. Text in a style is part of the style itself, and you can't change it using the normal WordPerfect editing commands. You have two choices for how to proceed:

- Choose Format, Styles (Shift+Ctrl+F8) and select the style from the Style List dialog box (refer to Figure 12.5). Click Edit to edit the style using the Styles Editor dialog box (refer to Figure 12.6), and change the text there. This change affects all instances of the style in the current document, but not in other documents.

→ To learn how to edit a style, **see** "Editing a Style," **p. 321**

- Don't use a style. If you regularly need to alter text that's part of a style, you might want to use a template or a macro instead. You'll then be able to edit the text, but the text won't change everywhere it appears in your document. You can still use styles for other things in both templates and macros, though; these features are not mutually exclusive.

→ For more information on templates, **see** "Promoting Cross-Document Consistency with Templates," **p. 333**

→ For more information on macros, **see** "Letting the Computer Do the Work: Macros," **p. 483**

CHAPTER 13

Promoting Cross-Document Consistency with Templates

In this chapter

Templates: Starting Points for Often-Used Document Types

In word processing parlance, a *template* is a structure for a document. A template file contains many of the elements necessary for a particular type of document but has no content. For instance, a template for a scientific research report might contain headings for the various sections of a report along with styles appropriate for each of these sections and table formats. A template for an invoice might contain a formatted layout with spaces for a description of the product or service rendered, the billing address, the invoice number, and so on. In both cases, the person using the template would fill in the content—paragraphs of text and tables of data for the research report or addresses, service information, or product listings for the invoice.

The advantage of a template is that you can reuse it. You can write one research report or fill out one invoice, save the file as a regular document, and then load the same template and create an entirely new document with it. You need not re-create the common formatting elements, however, only the content that changes.

Note

WordPerfect's implementation of templates is somewhat tied into its ExpressDocs feature. This feature is not available in the downloadable version of the program, only in the personal retail and server versions. You can certainly still create files to be used as templates with the downloadable version, but you won't be able to use the library of ExpressDocs templates or add your own templates to that library.

Using Templates for Consistency and Efficiency

Why use templates? For the same reason you use styles: Both can help you produce professional, consistent-looking documents, and both can improve your efficiency. Styles, however, improve consistency mainly *within* a single document, whereas templates increase consistency *between* documents—that is, templates help you to produce many documents that follow the same overall format.

→ If you want to learn more about styles, **see** "Ensuring Consistency Within Your Document with Styles," **p. 305**

Suppose you regularly create some type of document. This could be a memo, a legal brief, a book chapter, or anything else. Your document probably has certain types of text and formatting requirements that diverge from WordPerfect's defaults. For instance, if you're creating a memo, you probably want to plaster the word *memo* or *memorandum* across the top of the first page, and you'll need Subject, From, and To fields. If you have to enter these elements each time you create a memo, you can waste a lot of time—and you'll probably make a mistake at some point. Depending on your attention to detail, you may produce memos that vary in formatting minutiae like the tab spacing or font size. You'll create a more professional image for yourself if you produce consistent documents, and templates can help.

If you just finished reading Chapter 12, "Ensuring Consistency Within Your Document with Styles," you may be thinking, "Ah, but I can create a style to handle these elements and then insert the style for a consistent look." This is very true, and a style may be all you need to set up initial formatting for many documents; but a single style isn't well suited to creating documents in which the constant elements are interspersed with text you need to change routinely. For instance, Figure 13.1 shows a typical memo. Notice that information such as the recipients and subject are interspersed with unchanging elements, such as the text "Memorandum" and the graphics lines. You could certainly use several styles to achieve this effect, but after a certain point, that approach becomes awkward and error prone. A template creates the unchanging elements for a document, leaving you to fill in only those elements that are unique, regardless of their location.

Figure 13.1
This memo was created from a WordPerfect ExpressDocs template called Memo - Traditional Style.

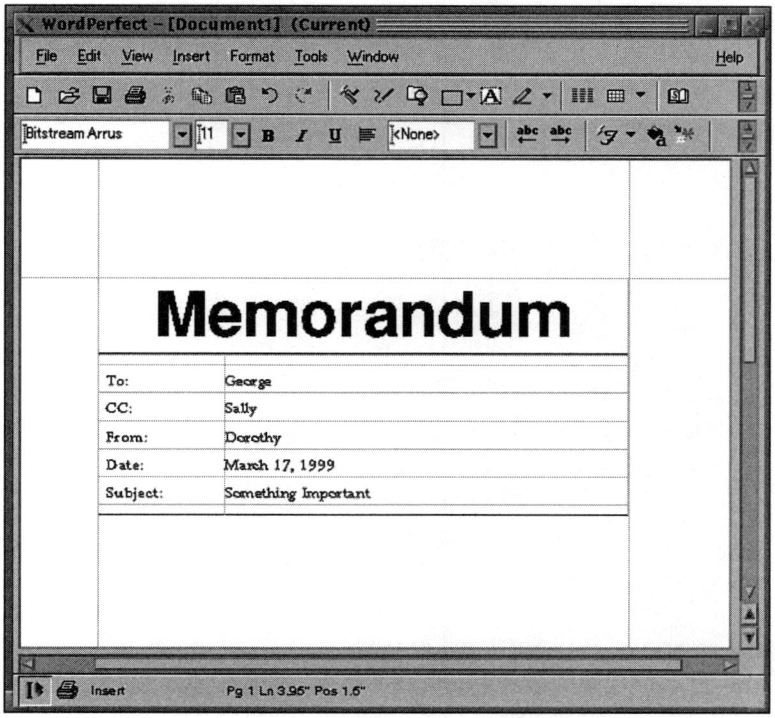

One additional feature of templates bears mentioning: Templates can automatically prompt you for information to go in certain parts of the document. When you load an ExpressDocs memo template, for instance, WordPerfect prompts you for information on the memo's recipient, the subject, and so on. This feature ensures that you don't accidentally miss one of these fields. WordPerfect can also automatically fill in certain personalized pieces of information, such as your name, from information stored in your Linux home directory.

USING TEMPLATES

The easiest way to use templates is via WordPerfect's ExpressDocs feature. There are other ways to use templates, though, and if you have the downloadable version of WordPerfect, you may be particularly interested in these because ExpressDocs is available only in the retail versions of the product.

USING EXPRESSDOCS

The personal retail and server versions of WordPerfect come with about 90 templates, ranging from attendance logs through weekly planners. Accessing these is simple:

1. Choose File, ExpressDocs.
2. If you are using this feature for the first time, WordPerfect may display the Personal Info dialog box shown in Figure 13.2. WordPerfect automatically inserts the identifying information you type into this box into any ExpressDocs templates that require it.

Figure 13.2
If you leave all the fields in the Personal Info dialog box blank, WordPerfect displays it again the next time you start the ExpressDocs feature.

Note

WordPerfect stores the personal information you provide it in the .wprc/.wp8x.set file off your home directory. Therefore, different people using the same computer can have different personal information.

Is your personal information outdated or incorrect? If so, you may find yourself constantly correcting WordPerfect's templates, but there's a better way. See "Editing Personal Info" in the "Troubleshooting" section at the end of this chapter.

3. Select the document template you want from the Document Template list in the ExpressDocs dialog box (Figure 13.3).

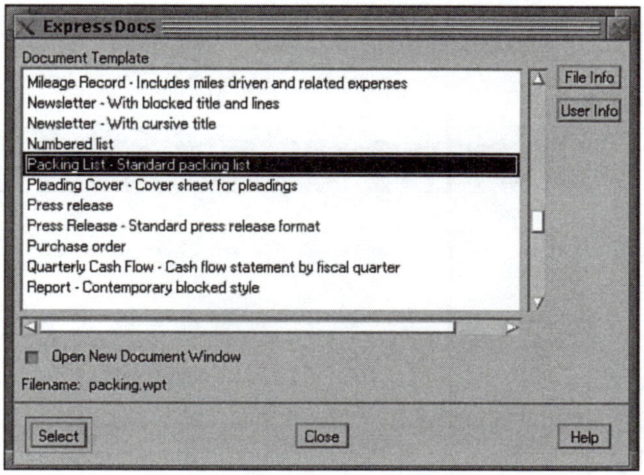

Figure 13.3
The Document Template list displays descriptive text about the templates, rather than their filenames, to give you some idea of their purpose.

4. If you want to open the template in a new document window, be sure the Open New Document Window check button is selected. If you want to open the template in the current document window, be sure that button is not selected.

Tip #168 from *Tux Tips*

Open a template in the current window if you want to merge the template with existing text, for example, to use both a fax cover sheet template and a letter template. In this case, you would use the fax cover sheet template first and then open a letter template into the same document. Some templates work best when they're opened into their own windows, though.

5. Click Select to open the template.

What happens next depends on the template. Most of the templates that come with WordPerfect for Linux prompt you for various pieces of information using a prompt dialog box, as in Figure 13.4. In this figure, the prompt dialog box is called Express Docs, but other prompt dialog boxes may have different names. This particular dialog box is for a packing list template.

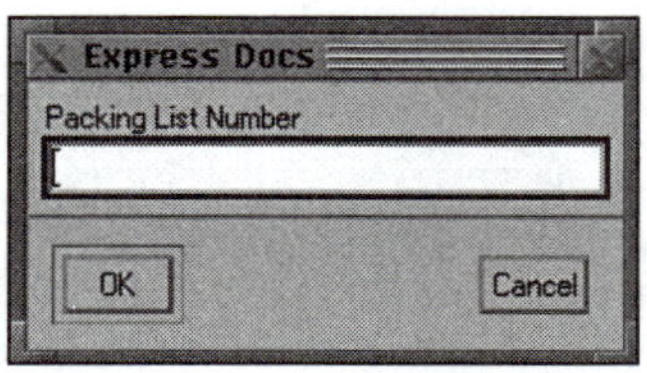

Figure 13.4
WordPerfect displays a different prompt in the Express Docs dialog box depending on the information you need to enter for a template.

After you fill out a few prompts, WordPerfect displays the document, as in Figure 13.5.

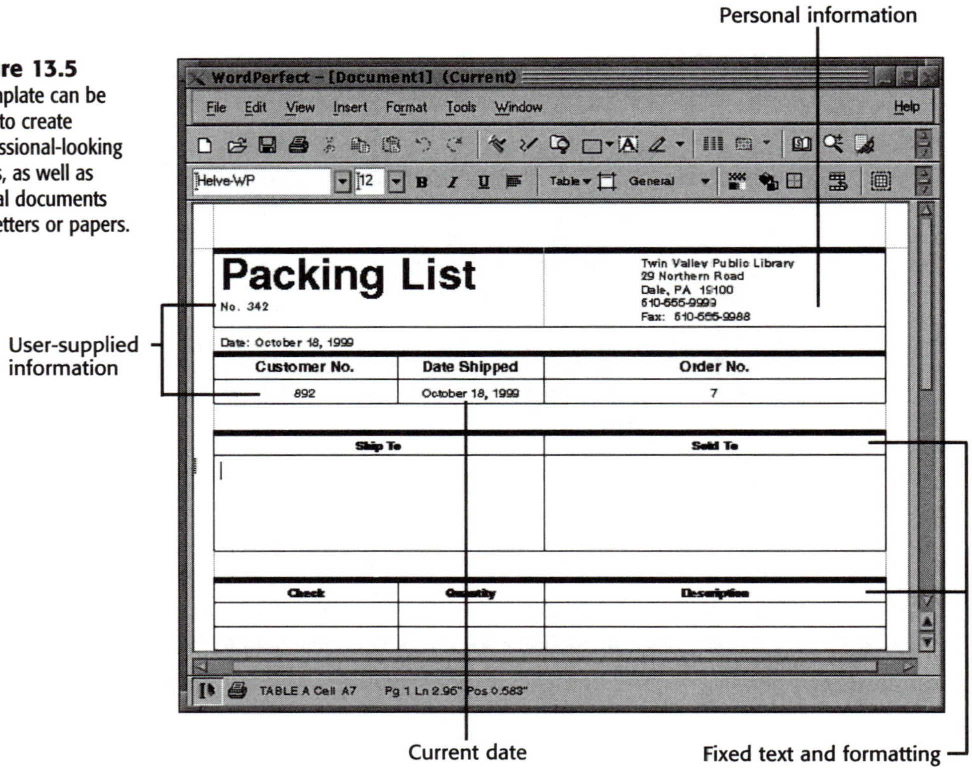

Figure 13.5
A template can be used to create professional-looking forms, as well as textual documents like letters or papers.

The packing list template is typical in its mixture of information and formatting from various sources. In particular:

- WordPerfect places the information for which it prompted you in the template at the points where the template designer intended.
- WordPerfect can automatically place some or all of the personal information it maintains in the file. Although this information is grouped in one area in the packing list template shown in Figure 13.5, that need not always be the case.

Note

WordPerfect inserts personal information as regular text. Therefore, if you create a file using an ExpressDocs template, save the file, and then modify your personal information, the original information remains unchanged in this document if you load it again. Any documents you create subsequently use the new information, however.

- WordPerfect can extract the current date and insert it into the text. The resulting text is actually just an ordinary WordPerfect [Date] code, but it is particularly useful in templates.

➔ To learn how to insert a date code into a document, **see** "Inserting the Date and Time," **p. 202**

- A template can contain any formatting or text an ordinary document can contain. In the case of the packing list template, these items include several tables, the large Packing List heading, and smaller headings in each of the tables. You insert your own text in and around the existing formatting.

The packing list template prompts the user for certain vital pieces of information, such as the customer and order numbers. Of course, a packing list must also contain other information. In principle, the template could prompt for those, too, but there are too many variables involved to do that efficiently. For instance, WordPerfect can't know how many lines you'll need for the recipient's address or just what format you'll want to use for reporting the items to be packed. It therefore leaves these items for you to fill in and format as you see fit using conventional text entry and editing techniques. If you regularly fill out a form in a particular way, though, you might consider modifying the template to prompt you for this information.

➔ If you want to know how to customize an existing template, **see** "Modifying an Existing Template," **p. 347**

USING TEMPLATES WITHOUT EXPRESSDOCS

One important thing to remember about WordPerfect for Linux's templates is that the prompting you get for filling out information in templates normally works only if you load the template via the ExpressDocs feature. If you load the template file normally, WordPerfect displays awkward-looking *merge codes* in place of the items for which it prompts. Figure 13.6 shows the packing slip template when loaded "raw." These merge codes tell WordPerfect how to obtain information to display in a given location. In fact, they're closely related to WordPerfect's macro feature.

➔ To learn more about macros, **see** "Letting the Computer Do the Work: Macros," **p. 483**

If you have the downloadable version of WordPerfect, which doesn't support ExpressDocs, you can still have WordPerfect prompt you for the fill-in-the-blank items in a template. You can use the j-card.wpt template file from the chapter13 directory of the CD that accompanies this book to see how this process works. Follow these steps:

1. Load the template file. You'll notice that the template toolbar (Figure 13.6) appears along with your file.
2. Click the Merge button on the template toolbar to display the Merge dialog box (Figure 13.7).

Figure 13.6
The template merge codes may not be visible when you first load a template file; use the Options, Display Codes template toolbar selection to be sure they're visible.

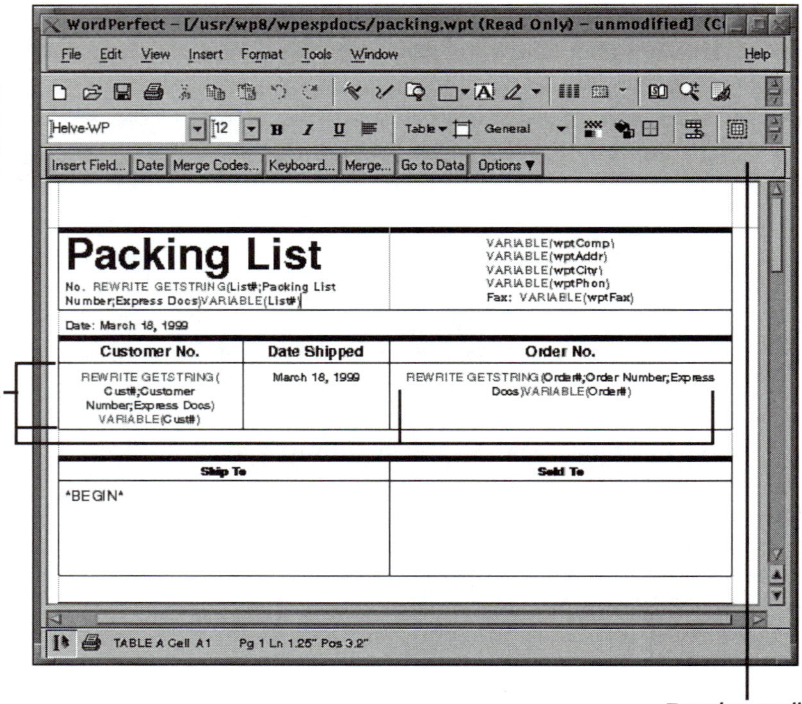

→ To learn how to read files from a CD in Linux, **see** "Accessing Removable Devices: Using the `mount` Command," **p. 76**

Figure 13.7
The Merge dialog box lets you set up or perform a merge of information from two different sources.

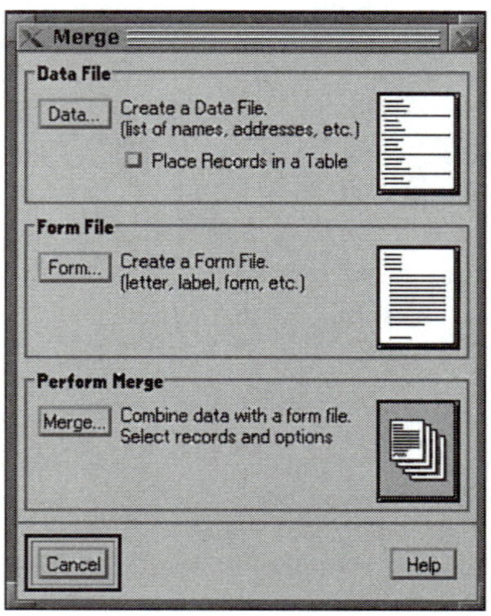

3. Click the Merge button in the Merge dialog box to open the Perform Merge dialog box (Figure 13.8).

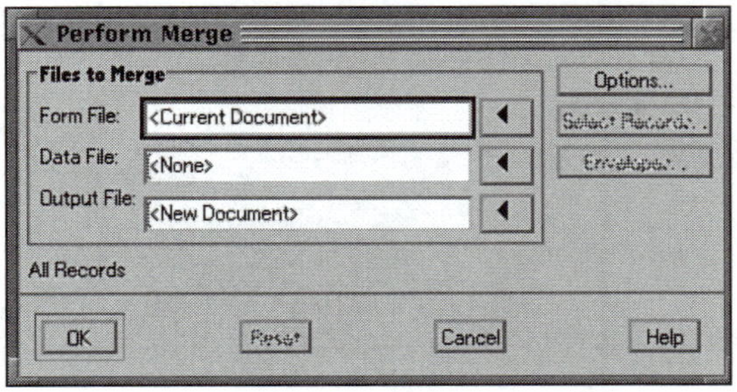

Figure 13.8
You should be able to leave the defaults in place when acquiring data for a template.

4. Make sure that the Form File field reads <Current Document>, that the Data File field reads <None>, and that the Output File field reads <New Document>; then click OK. WordPerfect opens a new document window and prompts you for the information to begin working in the template.

This procedure won't get you anything, of course, unless you already have templates that require prompting, and no such templates come with the downloadable version of WordPerfect. You might have obtained such files from somebody else, however, or you might have created them yourself.

Note

As you might guess by the frequency with which the word *merge* appears in the preceding discussion, not to mention the template toolbar and associated dialog boxes, WordPerfect uses its mail merge capabilities when prompting you for information to place in templates. This situation is an example of a WordPerfect feature expanding beyond its initial applications.

→ For more information on mail merge, **see** "The Same Letter, Different Information: Mail Merge," **p. 515**

AUTOMATING TEMPLATE USE WITH MACROS

Because the preceding procedure is a bit awkward for regular use, this section jumps ahead and explains how to use macros to automatically load a template and perform the merge operation so that you can enter prompted information into the template.

→ To learn more about macros, **see** "Letting the Computer Do the Work: Macros," **p. 483**

First, you need an appropriate template file. You can create one yourself, as described later in this chapter, or you can use the sample template file on the CD that accompanies this book. It's called j-card.wpt and is in the chapter13 directory on the CD. Copy the file to your home directory or anywhere else that's convenient. (These instructions assume that the file is in your home directory.) The j-card.wpt template is designed for creating a "J-card" for a music CD (Figure 13.9). If you create audio CDs using a CD-R drive, you'll find this template useful. After you create a J-card, you fold and cut it, and then insert it in the CD jewel case.

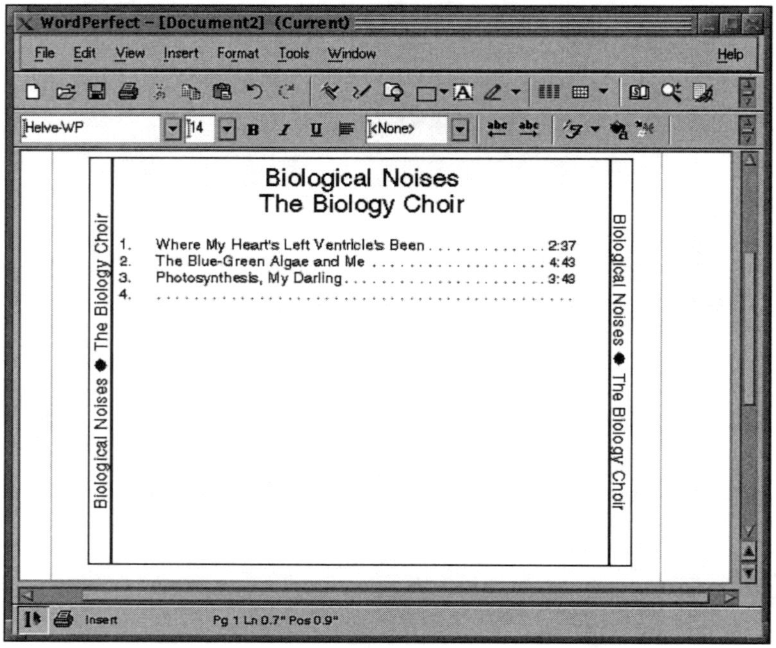

Figure 13.9
The j-card.wpt template saves you time by collecting the title and performer information once but using it in three places.

To set up a macro to automatically load the j-card.wpt template and have it prompt you for the CD's title and performer, do the following:

1. Start a clean WordPerfect document by choosing Program, New Window from the WordPerfect Control window.
2. In the new document window, choose Tools, Macro, Record (Ctrl+F10).
3. Type a name for the macro in the Record Macro dialog box (Figure 13.10). If you don't give the file an extension of .wcm, WordPerfect does so automatically. For example, you could call it j-card or j-card.wcm, and in either case it is saved as j-card.wcm. You don't need to give the macro file a name that's related to the template file's name, though. By default, WordPerfect stores the macro file in your home directory.
4. Click Record in the Record Macro dialog box. You'll be returned to your empty WordPerfect document window.

Figure 13.10
Enter a name for your macro directly or click the folder icon to use a WordPerfect file selector.

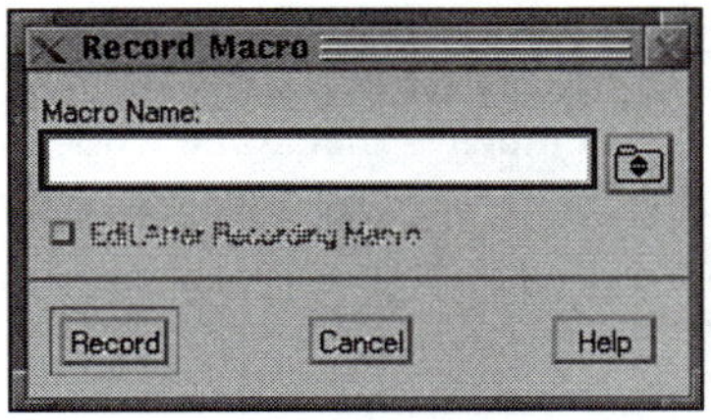

5. From that window, choose File, Open to open a WordPerfect file selector dialog box. Locate the j-card.wpt file and load it.
6. Choose Tools, Merge (Shift+F9) to open the Merge dialog box (refer to Figure 13.7).
7. In the Merge dialog box, click Merge in the Perform Merge area. This step opens the Perform Merge dialog box (refer to Figure 13.8).
8. In the Perform Merge dialog box, click OK. WordPerfect opens a new document window and presents the template prompts. Click OK to each one without entering any text.
9. When you finish entering the template variables, switch back to the j-card.wpt window.
10. Choose Tools, Macro, Record (Ctrl+F10) again to halt the recording.

You've now recorded a macro for loading and using a template. To use this macro, choose Tools, Macro, Play (Shift+Ctrl+F10). This sequence displays the Play Macro dialog box, which looks very much like the Record Macro dialog box (refer to Figure 13.9). Enter the name of the macro and click Play to run the macro's steps.

Tip #169 from *Tux Tips*

You can assign the macro to a single keystroke by giving it a special name, such as `ctrly.wcm`, to launch it by pressing Ctrl+Y. Many of the possible keyboard commands are already used by WordPerfect's standard features, though, so you may have to hunt a bit to find an available keystroke.

→ If you want to know more about assigning macros to keystrokes, **see** "One-Key Access to Macros," **p. 492**

Note

The j-card.wpt template uses many of the formatting features discussed in this book. Of particular note, the template uses several styles to create its formatting. If you want to adjust the font for all text in the file, do it from the document Initial Style. To enter text for tracks, position the cursor just after the 1 in the central text box, type the track title, and then press the Enter key. The cursor will jump over the dot leaders to the end of the line, where you can type a track length. Press the Enter key again, and WordPerfect will begin a new line with a new track number. When you're done, use the Backspace key to erase the extra track number and, if you like, enter any additional text you want after the track listing.

CREATING AND EDITING TEMPLATES

You can go a long way with just the default templates that come with WordPerfect. Past a certain point, though, you may need to create entirely new templates, or at least modify those that ship with WordPerfect. Fortunately, you can do so, and you'll find that considerable power lurks in the template feature.

ADDING A TEMPLATE TO EXPRESSDOCS

Most WordPerfect for Linux features let you add personal customizations by modifying files in the .wprc subdirectory of your home directory. For instance, WordPerfect maintains a personal dictionary there, as well as assorted preferences and defaults for the program's operation. ExpressDocs is one of the few exceptions to this rule; the *only* files to appear in the ExpressDocs dialog box (refer to Figure 13.3) are those in the wpexpdocs subdirectory of your WordPerfect for Linux installation. Therefore, if you create your own template, you have four choices for how to use it:

- Load it as an ordinary document. This option is viable if your template (unlike most ExpressDocs templates) doesn't involve prompts for input to specific fields. The drawback is that it's easy to mistakenly choose File, Save (Ctrl+S) after you've modified the template, thus overwriting the template itself. This option also becomes awkward if your template prompts for information.

Tip #170 from *Tux Tips*

> You can change the permissions on a template file so that you can't overwrite it in this way. From an xterm window, type `chmod a-w foo.wpt` to write-protect the foo.wpt template file. You can also write-protect the file from the WordPerfect File Manager dialog box.

→ To learn more about using the File Manager to manipulate files, **see** "File Operations," **p. 80**

- Create a macro to load your template and perform the merge operation to get it to perform its prompting, as described earlier in this chapter.
- Copy your template into the wpexpdocs subdirectory. It will then appear in the ExpressDocs dialog box. You may need root access to copy a template to this directory.

→ For more information on root access, **see** "Users and the Superuser," **p. 11**

- Create your own ExpressDocs directory and change the default ExpressDocs directory, using WordPerfect's Files Preferences, described in Appendix C, "Customizing Preferences." If you do change this default directory, you lose access to the standard ExpressDocs files unless you copy them or create symbolic links to them from your new directory. In any event, you can copy your own templates into this new directory for easy access.

→ To learn how to change the default ExpressDocs directory, **see** "Files," **p. 774**

This fourth option is probably the best, overall. You can selectively or comprehensively link existing templates to that directory. Because symbolic links take almost no space on your disk, you won't be wasting any appreciable amount of disk space, even if many people are using WordPerfect on the same computer. On the other hand, if you want everybody who uses WordPerfect on your computer to have access to the new templates, you should copy them to WordPerfect's wpexpdocs subdirectory.

→ For more information on symbolic links, **see** "Links," **p. 78**

Before you copy a template file to the wpexpdocs subdirectory, you should give the file a descriptive name in WordPerfect's Document Summary feature. If you don't, the template will appear in the ExpressDocs dialog box as *****, which isn't terribly descriptive. To give the file a descriptive name, choose File, Properties, Document Summary from the Document window. This step brings up the Document Summary dialog box shown in Figure 13.11. Type the name you want to see in the ExpressDocs dialog box in the Descriptive Name text field and click OK. Be sure to save this change to your template before copying it to the wpexpdocs directory, too!

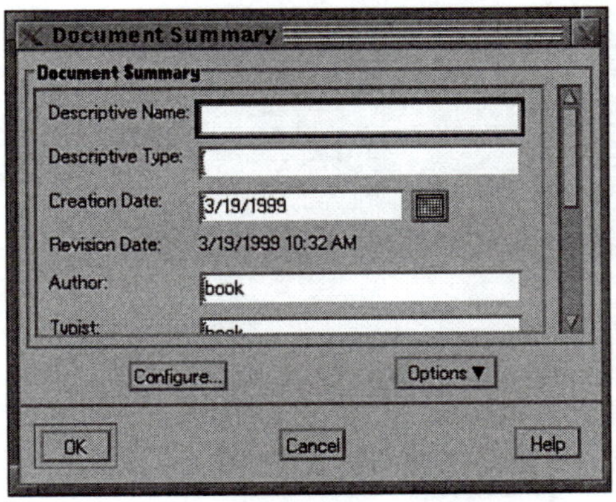

Figure 13.11
If the Descriptive Name field isn't an option when you open the Document Summary dialog box, click on Configure to open a dialog box that allows you to add the field.

To copy a file into the wpexpdocs directory so that all WordPerfect users can access it, you can do one of three things. For any of these actions, you may need to be root, so first log on as root or use `su` to obtain root privileges. If you don't have access to the root account, you'll need to contact your system administrator to copy the file for you. These instructions also assume that WordPerfect is installed in the /usr/wp8 directory. If you've put WordPerfect somewhere else, make appropriate changes to the pathnames.

→ If you want to know more about root privileges, **see** "Users and the Superuser," **p. 11**

- Use an xterm window. Use the `cp` command as root to copy the file. For instance, `cp /home/jennie/j-card.wpt /usr/wp8/wpexpdocs` copies the j-card.wpt file from the jennie home directory to the wpexpdocs directory.

Tip #171 from *Tux Tips*

You can place symbolic links in the wpexpdocs subdirectory in place of complete files. For instance, if you place the j-card.wpt file in the /home/jennie/.wprc directory, you can create a symbolic link by using the command `ln -s /home/jennie/.wprc/j-card.wpt /usr/wp8/wpexpdocs` instead of using `cp`. Placing symbolic links in the wpexpdocs subdirectory has the advantage of allowing the user jennie to edit the template from her own directory without root access even after the template is placed in the wpexpdocs subdirectory. This approach is particularly useful if you want to give all users of the system access to a template that's occasionally updated by one person.

- Use WordPerfect's File Manager dialog box (Figure 13.12) to do the copy. Launch WordPerfect as root and then choose Program, File Manager to launch the WordPerfect File Manager dialog box. Locate the template you want to copy, highlight it, and choose Files, Copy from the File Manager's menu. Type in the path to the wpexpdocs subdirectory and then click OK.

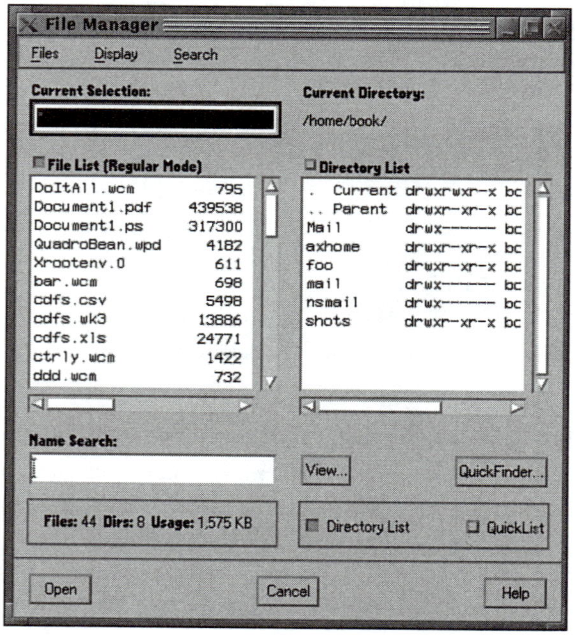

Figure 13.12
You can use the File Manager dialog box to copy files even if they aren't WordPerfect document files.

→ For more information on using the File Manager to manipulate files, **see** "File Operations," **p. 80**

- Launch WordPerfect as root and then load the template file you want to copy. Immediately choose File, Save As (F3) and locate the wpexpdocs subdirectory. Type in a name for the file (ending in .wpt) and click OK.

MODIFYING AN EXISTING TEMPLATE

Given the wide variety of templates with which the retail version of WordPerfect for Linux ships, chances are you'll be able to create any custom template you need by modifying an existing one. This technique can certainly save a lot of time, especially if your modifications are small, like adding a colored background to a table or changing some text.

→ To learn how to give a table a background, **see** "Table Cell Formatting Options," **p. 294**

STARTING THE MODIFICATIONS

To modify a template, you should begin by copying the original file to your home directory. If you've been using a template via ExpressDocs, you can find the template's name by checking near the bottom of the ExpressDocs dialog box. Just above the Select button, WordPerfect lists the filename of whatever template you've highlighted. At an xterm, type `cp /usr/wp8/wpexpdocs/foo.wpt ~/` to copy the template file foo.wpt from the wpexpdocs directory to your current directory (assuming WordPerfect is installed in /usr/wp8).

MERGE CODES AND THEIR PARAMETERS

You can then load the template into WordPerfect as you would any other file. Figure 13.6 shows the results of loading the packing.wpt file. There are two basic types of components in this template: ordinary text and template merge codes. The ordinary text is the text and codes that appear in the final document. In the case of packing.wpt, the ordinary text includes the tables and their fixed text, such as Ship To and Order No. The template merge codes appear in red. Figure 13.6 shows three merge codes, each used in several places: *REWRITE*, *GETSTRING*, and *VARIABLE*. Others are available, too.

Tip #172 from *Tux Tips*

> WordPerfect may not display the merge codes when you initially load a template. Select Options, Display Codes from the template toolbar to display the codes.

Some merge codes take *parameters*, which are pieces of information to refine what WordPerfect does with the merge code. For instance, the GETSTRING merge code takes three parameters: a variable name to reference the information the user types into a prompt dialog box, the prompt to be displayed in that dialog box, and a title for the prompt dialog box. The VARIABLE merge code, by contrast, takes just one parameter (a variable name), and the REWRITE code takes none. The parameters to the merge code, unlike the merge codes themselves, appear in black text, surrounded by red parentheses.

If you want to modify a template so that it doesn't display some prompted piece of information, you need to remove one or more merge codes. For instance, suppose you want to modify the packing list template so that it doesn't display (or prompt you for) a customer number. To do so, you need to remove the REWRITE, GETSTRING, and VARIABLE merge codes from the Customer No. box in the template. You can remove merge codes just as you would ordinary text except that the code itself disappears with one press of the Delete or Backspace key if you're doing a character-by-character deletion. If you press either key to delete a merge code, the associated parameters turn into ordinary text, so be sure to delete those, too. If you stop here, you'll still be able to enter a customer number manually when you use the template, just as you can enter the shipping and billing information, the package contents, and so on. If you prefer not to have a field for customer number on the form at all, you can delete the Customer No. text and either merge those cells with the ones to their right or reformat the document in some way to make it look good.

Note

The merge codes appear in WordPerfect in whatever format will apply to the text they will ultimately insert. For instance, if you format the merge codes for Helvetica 12-point bold, the text they create will appear in that font.

Additional Template Changes

You can make simple textual or formatting changes to a template just as you can make changes in any other document. You should be careful not to change any of the merge codes, though. Also, keep in mind that merge codes may take more or less space on the screen than the text you enter in their place when you use the template. If your formatting depends on information fitting into a certain amount of space, you have to consider this factor.

You may have noticed a piece of text in the document that isn't there when you use the template: the word **BEGIN** in the Ship To field. If you scroll to the bottom of the packing.wpt template, you'll see a merge code that reads CHAINMACRO(edocbeg), which launches a WordPerfect macro called edocbeg.wcm. This macro searches for the word **BEGIN** in the text, deletes it, and places the insertion point there, as illustrated by Figure 13.13. Thus, when you use the template, your insertion point appears automatically at this point. To change that location, delete the existing **BEGIN** and place a new one wherever you like.

→ To learn more about macros, **see** "Letting the Computer Do the Work: Macros," **p. 483**

Note

The asterisks surrounding the word *BEGIN* are part of what's searched for, so they're very important. Be sure to include them if you want to change the beginning point of the template, or your change won't work. You may, however, change the case of the word *BEGIN* and write it as **begin** or some other variant.

Completing the Modifications

When you've finished modifying the template, you should save it and test it. Save the template by choosing File, Save (Ctrl+S) or File, Save As (F3). Test it by choosing Tools, Merge (Shift+F9). This step opens the Merge dialog box (refer to Figure 13.7) from which you should click on Merge to open the Perform Merge dialog box (refer to Figure 13.8). Click OK to perform a test. You should get the prompts from the original template, plus or minus any changes you've made to them, and then see your modified template with any prompted text included in a new document window. If you're dissatisfied with your results, go back to the original template window and try again. You can safely close the merged template document window whenever you like, but you might want to keep it open for reference if you're dealing with tricky formatting.

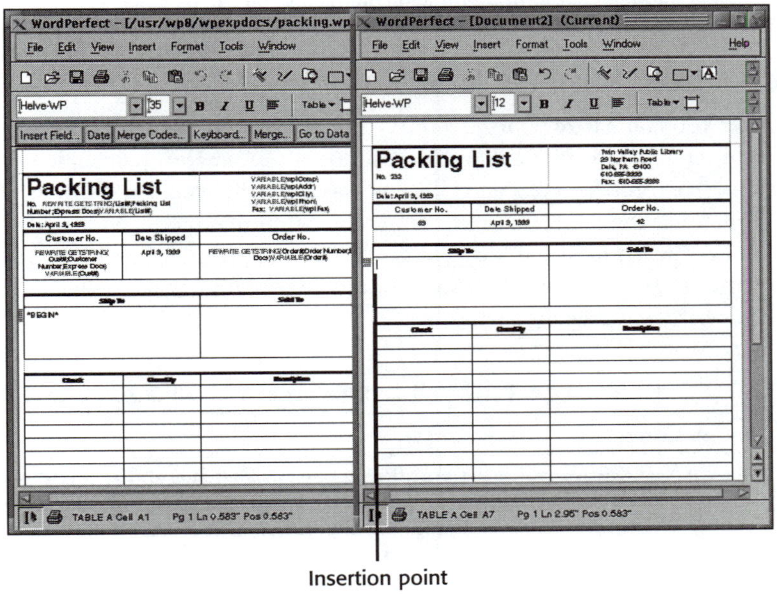

Figure 13.13 The "*BEGIN*" in the template (left window) marks where the insertion point appears (right window).

Tip #173 from *Tux Tips*

> If the relative sizes of the merge codes and the text that's likely to replace them is giving you trouble, you can adjust the formatting in the merged template and then duplicate that formatting in the original template.

When you're satisfied with your changes, you have the same choices outlined earlier for how to use the template. You can use it as-is by loading it as a normal file, you can assign it to a macro, you can copy it back to the wpexpdocs directory, or you can create a new ExpressDocs directory and store the template there. If you choose to copy the template back to the wpexpdocs directory, you can either overwrite the existing file or copy it under another name (by using, for instance, `cp foo.wpt /usr/wp8/wpexpdocs/foo2.wpt` as the copy command from an xterm window). If you copy the template under another name, be sure to change the Descriptive Name from the Document Summary dialog box. If you don't change this entry, you'll get two documents in the ExpressDocs dialog box with the same descriptive name.

→ For more information on adding templates to the ExpressDocs list, **see** "Adding a Template to ExpressDocs," **p. 344**

Caution

It's wise to check that you're not about to overwrite a file of the same name before you perform a copy operation, especially when you're operating as root. Use the `ls` command to check that no files exist with the filename you intend to use before using `cp` to copy the file. If you skip this step, you could accidentally wipe out some important template.

→ For more information on the xterm and commands you can use in it, **see** "Running the xterm: Your Key to Linux from X," **p. 18**

CREATING NEW TEMPLATES WITH PROMPTS

After you familiarize yourself with WordPerfect's templates, you may decide to write one from scratch. Creating a template that doesn't automatically include any variable information is easy: Just create a document as you normally would and save it with a .wpt extension. If you want the template to prompt you for information or to include data you enter into the Personal Info dialog box (Figure 13.2), you need to learn at least a bit about how to use merge codes. Table 13.1 lists some of the merge codes you'll be most likely to use and describes what they do.

TABLE 13.1 SOME OF THE MOST CRITICAL MERGE CODES FOR TEMPLATE USE

Merge Code	Description
CHAINMACRO	Runs a WordPerfect macro file. This allows you to modify text in the template and perform other operations. WordPerfect's macro facility is essentially a programming language, and you can do with it many of the things you could do in C, Pascal, or Perl. WordPerfect's macros apply most readily to the contents of a WordPerfect document, though. For more information, see Chapter 19, "Letting the Computer Do the Work—Macros."
DATE	Inserts the current date. This merge code differs from using a [Date] code because the [Date] code changes every time you load the document, whereas the date merge code inserts the date as ordinary text. This feature makes the DATE merge code a better choice than [Date]when you want a record of the document's creation date. The DATE merge code takes no parameters.
GETSTRING	Places a prompt dialog box on the screen and assigns whatever the user types to the variable name parameter. This merge code takes three parameters: a variable name to hold the information the user types, a prompt to appear in the dialog box, and a name for the dialog box. For instance, you might use GETSTRING(title; Enter the CD's title; CD Title). You could then incorporate the user's response by passing the variable title to future merge codes.
POSTNET	Inserts U.S. Postal Service bar codes corresponding to the string that you provide. Use this code if your template is for an envelope or mailing label and you want to include these codes on it.
REWRITE	Causes WordPerfect to draw the contents of the screen up to that point. If you omit REWRITE codes, you won't be able to see how your document is progressing until you've answered all the prompts.

Merge Code	Description
VARIABLE	Inserts a variable's contents into the document. As such, the code takes one parameter, the name of the variable.

Table 13.1 is by no means exhaustive; it contains only the most frequently used merge codes. Because WordPerfect uses its mail merge facility to perform these template prompts, you can use other mail merge features here if you like.

→ To learn more about the mail merge feature, **see** "The Same Letter, Different Information: Mail Merge," **p. 515**

To insert a merge code in your template, do the following:

1. Position your insertion point where you want the merge code to appear
2. Click the Merge Codes button from the template toolbar. This step produces the Insert Merge Codes dialog box (Figure 13.14).

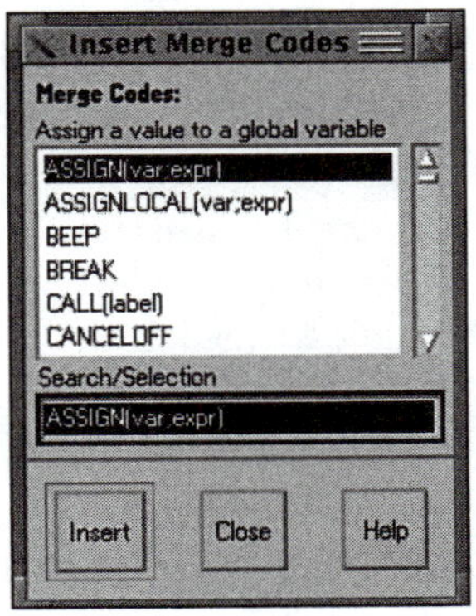

Figure 13.14
The Insert Merge Codes dialog box lets you select the merge codes you want from a scrollable list.

3. Select the code you want from the scrollable list and click Insert. If the code you've selected takes any parameters, WordPerfect displays the Insert Merge Code dialog box (Figure 13.15).

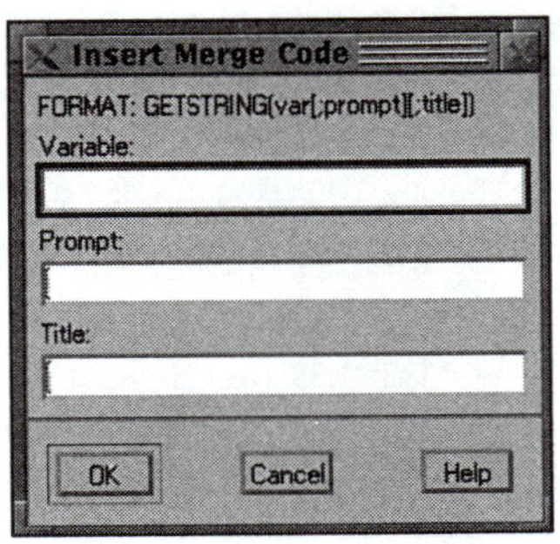

Figure 13.15
The Insert Merge Code dialog box may have a different number of fields than shown here, depending on the code you're inserting.

4. Type the parameters into the associated fields of the Insert Merge Code dialog box and then click OK. WordPerfect inserts the merge code in the text.

Tip #174 from *Tux Tips*

> If you need to insert several codes, you can leave the Insert Merge Codes dialog box open. Try moving it to the side of the main document window for easy access. You can then edit in the main document window and, when you need to insert another code, just click on the Insert Merge Codes dialog box to use it again.

→ *Are you having trouble getting your prompted text to appear correctly in your text? See "Merge Code Ordering" in the "Troubleshooting" section at the end of this chapter.*

TEMPLATES IN ACTION

Templates are great for any type of document that you create often. You can get a good idea of what you can do with templates by checking out the templates available from ExpressDocs if you have the retail version of WordPerfect. If not, or if you'd just like a quick rundown of what's possible, here are some good applications for templates:

- Reports and papers—You can create a template with appropriate predefined formatting and styles for different sections of a report. A template like this would consist largely of its styles and formatting, but you could have it prompt for the report's title and insert your name from the Personal Info that WordPerfect maintains.
- Letters—You can create a single style that inserts a letterhead and configures your margins, font, and so on, which may be enough for you. Templates let you do a bit more, however, because you can include the salutation, closing, and even text that's likely to appear in certain types of letters, such as directions to your location (complete with a map, if you like).

→ To learn more about styles, **see** "Ensuring Consistency Within Your Document with Styles," **p. 305**

Tip #175 from *Tux Tips*

> If you regularly fax letters, you may want to create a template that includes your scanned signature. Be aware of the security implications if others have access to your computer, however. Be sure to restrict access to this template to you alone! You may also want to protect this template with a password.

→ To learn how to give a document a password, **see** "Saving Your Document," **p. 67**

- Certificates, labels, or special forms—With today's high-resolution color printers, you can produce visually appealing special forms, and a template can be just the ticket if you need to create these on a regular basis. WordPerfect includes its own special feature for labels, but some irregular labels may require special formatting best handled by templates. For instance, if you have circular labels, you could lay out a watermark to display the label location and then adjust your margins to put text within the label most easily.

→ For information on using labels, **see** "Labels," **p. 250**

→ For information on using watermarks, **see** "Watermarks," **p. 241**

- Business Forms—The packing list template used as an example through much of this chapter is just the beginning. You can use templates for employment applications, invoices, tax forms, or any other business form. You can even use a WordPerfect table inside such a template and have it perform calculations to help fill out the form. Between that and WordPerfect's prompts, you may not have to type anything into the document window!

→ To learn about having WordPerfect perform calculations inside tables, **see** "Table Mathematics," **p. 296**

- Presentations—If you need to create overhead transparencies, slides, or posters, you can create an appropriate style in WordPerfect to set up fonts and other formatting features, such as a bulleted list, your organization's logo, and so on.
- Cover sheets—You can create a cover sheet template for inclusion with printed matter for identification or routing. Although GFax has its own cover sheet mechanism, you can bypass that and design your own using a WordPerfect template for faxes. WordPerfect's prompting can help ensure that you don't forget some important information for the cover sheet.

→ To learn how to send a fax from WordPerfect, **see** "Faxing," **p. 161**

- Newsletters and brochures—These types of forms typically require careful layout. Why re-create it all each time you write one? Create a template once, and you'll at least have the basics in place, though of course, you'll still need to work at fitting in all the elements.
- Web pages—If you're trying to maintain a Web site with lots of pages, you can use a template to set up your main formatting features and then add the content page by page. This approach can greatly enhance the consistency of a Web site. Consistency is particularly important on the Web because your readers are likely to view several pages in rapid succession, making any inconsistencies stand out.

→ To learn more about creating and manipulating Web pages, **see** "Web Publishing," **p. 659**

- Calendars and agendas—Use a WordPerfect table with its numbering and date/time manipulation capabilities to create a custom calendar or agenda and then fill in the details. This technique is great for distributing a calendar of events that influence many people, such as a project timeline. You could also create a customized wall or desk calendar with scanned photos or artwork from the Web or a clip art CD.

Templates are like styles in that the more you use them, the more uses you'll discover. I therefore recommend that you dig in! Create a template or two and check out the ones that ship with WordPerfect. If you use WordPerfect much at all, you won't regret the time you invest in learning to use templates.

TROUBLESHOOTING

EDITING PERSONAL INFO

You've moved or you've discovered a typo in the Personal Info that WordPerfect maintains, making templates that use this information less than perfect.

You can easily correct your Personal Info. Choose File, ExpressDocs to open the ExpressDocs dialog box (Figure 13.3) and then click User Info to open the Personal Info dialog box (Figure 13.2). You can edit any of the information there and then click OK to save it.

MERGE CODE ORDERING

You've created a template that uses prompts to obtain information from the user; however, the first time you use the template, you get either nothing or just gibberish, and the second time it displays the data you input the first time.

WordPerfect performs the actions associated with the template's merge codes in the order in which they appear in the Reveal Codes window. The GETSTRING merge code is used to assign a value to a variable (its first parameter). If you try to access that variable by using the VARIABLE merge code before the GETSTRING merge code appears, you won't succeed. The ordering is usually pretty obvious, but if you're using text boxes, headers, or other formatting, the order of the codes might not be entirely clear until you open Reveal Codes and check where the actual codes reside. Remember that many codes that themselves contain codes may display in Reveal Codes as single codes, so you may not be able to see the VARIABLE merge code unless you edit the feature in which it rests, but the code is there.

PART IV

Complex Formatting with Boxes for Text, Graphics, and Equations

CHAPTER 14

Using Text Boxes to Add Variety to Text Layout

In this chapter

PLACING TEXT ARBITRARILY ON THE PAGE WITH TEXT BOXES

WordPerfect was originally designed for editing fairly straightforward text documents, such as letters and reports. WordPerfect's core functionality therefore emphasizes the capabilities that are most important for this type of document: creating and editing lines of text. Manipulating the minutiae of the appearance and placement of individual lines wasn't a priority in the 1980s, in part because most printers weren't capable of any great feats back then. As WordPerfect evolved, however, it gained new layout capabilities, such as support for an arbitrary selection of fonts and increasingly complex column formatting. One of the features added to WordPerfect builds on this strength by allowing you to create a sort of document within a document. This feature is the text box.

The basic idea behind the text box is to allow you to place text anywhere you like on the page. The emphasis on text manipulation in early versions of WordPerfect made it great for *creating* and *editing* text, but not so great for producing newsletters and other complex text layouts. Figure 14.1 shows an example of a text box in action, used for a pull quote, and if you're familiar with WordPerfect's basic text editing features, you can probably tell why this effect might be difficult to produce using only those features. The text in the center of the window is in a larger font than the rest of the text, and yet that other text flows around the large text. Producing this effect without text boxes to place the text where you desire would be difficult at best, and very likely impossible.

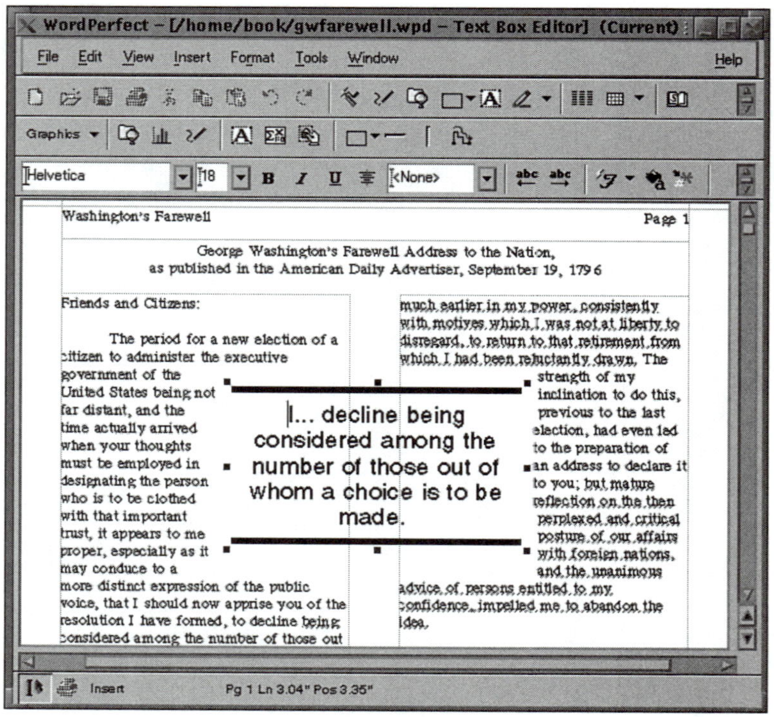

Figure 14.1
You can use text boxes to produce an offset segment of text to catch the reader's attention.

Note

Some other word processors rely on text boxes much more heavily than does WordPerfect. For them, *all* text resides inside text boxes; it's just that the program comes up with one predefined text box for the size of a page, and when it's filled, another is created. These programs typically refer to their text boxes as *frames,* so if you've ever used a so-called frame-based word processor, you can think of WordPerfect's text boxes in much the same way. WordPerfect achieves certain of its effects, however, such as headers and footers, without using text boxes, whereas some frame-based word processors use frames for their equivalent features.

→ To learn how to create headers and footers, **see** "Headers, Footers, and Page Numbers," **p. 235**

If you're used to dealing with a page layout program or frame-based word processor, the appeal of text boxes should be obvious. You may be tempted, however, to use text boxes to solve a formatting problem when other, and better, solutions are available in WordPerfect. If you're unfamiliar with text boxes or frames, you may be wondering just what they're good for. To help out in both cases, here are some text box *dos* and *don'ts*:

- *Do* use text boxes to place a small amount of text somewhere arbitrary on the page, as in the example in Figure 14.1.
- *Do* use text boxes to place a table on the page when you'll need to move it or wrap text around it.

→ For information on creating tables you can move around on the page, **see** "Creating Movable Tables," **p. 282**

- *Don't* use text boxes for tables that will span most or all of a page and that you don't expect to move; simply place the table directly in the text. Placing a table inside a text box unnecessarily increases the complexity of the document and makes editing the table a bit more difficult.

→ For information on creating a fixed table, **see** "Creating a Basic Table," **p. 276**

- *Do* use text boxes for newsletter sidebars or similar short text segments when you want surrounding text to flow around a fixed sidebar.
- *Don't* use text boxes simply to highlight text that's to remain fixed relative to its surrounding paragraphs. You can place a border around or shading within one or more paragraphs of text with WordPerfect's border and fill features.

→ To learn how to create a border around or fill within a paragraph, **see** "Borders and Fills," **p. 216**

- *Don't* use text boxes to create watermarks. Although text boxes can be used in this way, doing so degrades WordPerfect's performance unnecessarily because WordPerfect has a special watermark function.

→ For information on creating and using watermarks, **see** "Watermarks," **p. 241**

Part of the reason I've given you this advice on when to use text boxes and when not to is that, for all their power, they make editing more complex. Text boxes won't massively increase WordPerfect's complexity, at least not for just a box or two; but if you overuse text boxes when other WordPerfect features could do as well, your editing may become unnecessarily complex and slow.

Did WordPerfect become sluggish after you added a few text boxes? See "Speed Problems When Using Text Boxes" in the "Troubleshooting" section at the end of this chapter.

Text boxes are one of several different types of boxes that WordPerfect supports. Other types are graphics boxes and equations. There's no need to create a text box only to fill it with one of these box types. If you want to create a large box that contains both text and graphics or an equation, however, you can embed a graphics or equation box within a text box. In fact, you can even embed one text box within another.

→ To learn how to use graphics, **see** "Graphics," **p. 377**

→ To learn how to create charts, **see** "Creating a Chart," **p. 411**

→ To learn how to create mathematical equations, **see** "Equations," **p. 433**

→ To learn how to manipulate any boxed element, **see** "Box-Related Functions," **p. 459**

Creating Text Boxes

To create a text box, choose Insert, Text Box. WordPerfect creates a default text box with a hairline border along each side (see Figure 14.2), selects the text box, and places your insertion point inside it. You can tell the text box is selected by the presence of control points at its corners and midway along each side. You use these control points to resize the text box.

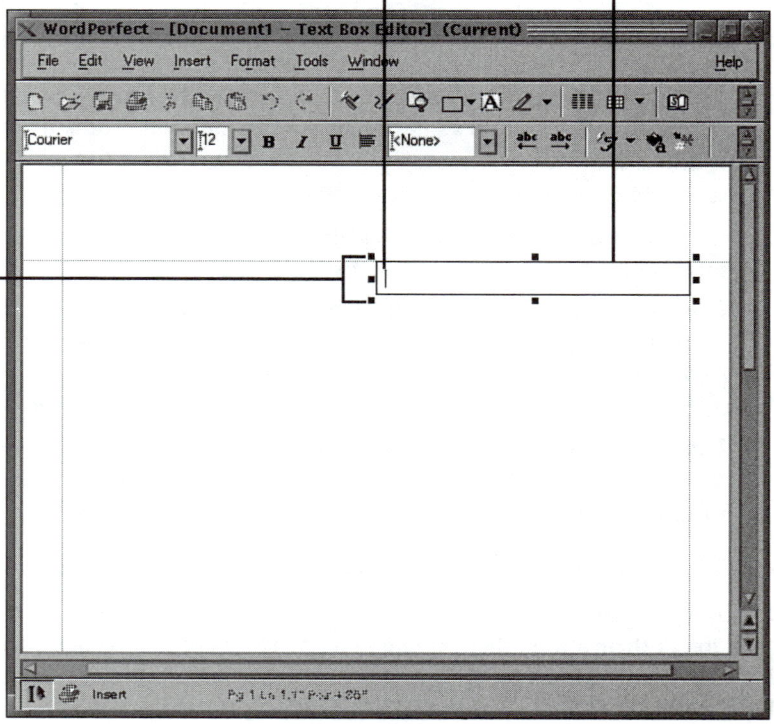

Figure 14.2
WordPerfect assumes that when you create a text box, you'll immediately want to create text inside it.

When you first create a text box, WordPerfect inserts a [Box] code in your text. This code actually contains the box text and formatting, though you don't see this in Reveal Codes. Figure 14.3 shows a document with a text box, and the Reveal Codes pane of that window shows the [Box] code, expanded by placing the insertion point to the immediate left of the code.

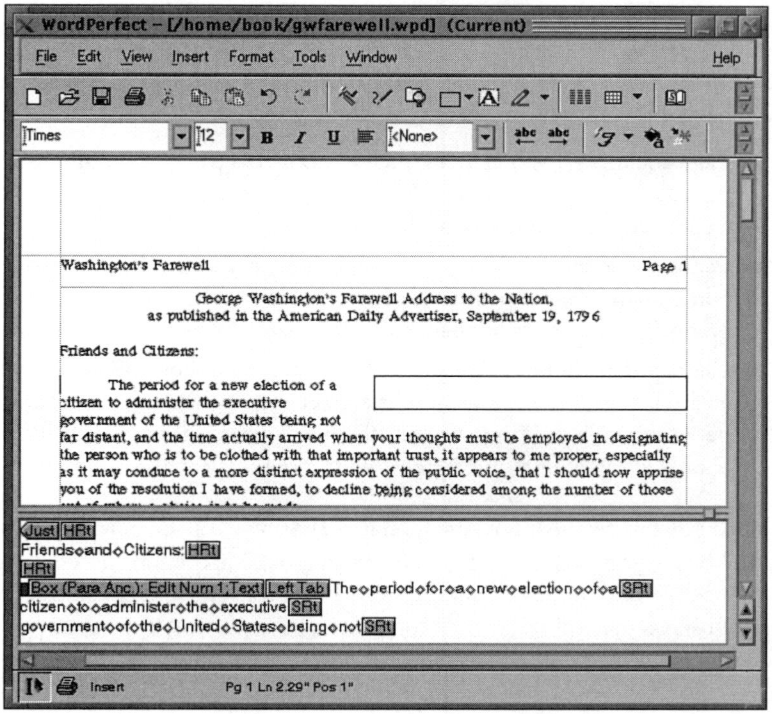

Figure 14.3
WordPerfect places the text box on the right side of the page by default.

WordPerfect provides many options for the formatting and placement of text boxes, so when it creates a text box, it must select certain default values. You should be aware of the following points about newly created text boxes:

- The codes function much like paragraph codes; WordPerfect inserts them at the start of the current paragraph and keeps them there.
- By default, WordPerfect uses a *paragraph anchor* on its text boxes (which is what (Para Anc) refers to inside the [Box] code in Figure 14.3). *Paragraph anchoring* means that the text box will be associated with the paragraph in which its code resides. If that paragraph moves to another page, the text box also moves.
- The box is initially tall enough for one line of text and occupies the right half of a line of text. If you type text into the box, its height expands to match the number of lines of text you enter. The box has a hairline border and no fill. You can, of course, change any of these defaults.

Moving and Resizing Text Boxes

Chances are you'll need to make some changes to WordPerfect's default text box. For starters, you'll want to move it into the position you desire and possibly resize it. You can perform these operations visually, by manipulating the text box as it appears on the page, or you can set its size and position precisely by entering values into dialog boxes.

Understanding Text Box Anchors

Text boxes (and other boxed elements) are positioned relative to other items on the page. You have three options for text box anchors:

- Page anchor—If you elect to anchor a box to a page, you can tell WordPerfect to position the box relative to the edge of the page, the page margin, or occasionally relative to a column or the center of a paragraph.
- Paragraph anchor—With a paragraph anchor, you specify a horizontal location much as you do with a page anchor, relative to a margin, the page edge, or the center of the paragraph. The box is positioned vertically relative to the top of the paragraph, however.
- Character anchor—If you select the character anchor option, WordPerfect treats the entire box as if it were a single character. For instance, if you place a [Box] code between two words in your text, the box appears there as well. If the box is large, then WordPerfect adjusts line heights appropriately to make room for the box.

When you create a text box, paragraph anchoring is the default. To change the anchoring on a text box, right-click on it (first left-clicking outside the box, if necessary, to deselect it) to produce the box context menu (Figure 14.4). From this menu, choose Position to get the Box Position dialog box (Figure 14.5). Select the placement method you want from the radio buttons in the Box Placement section of the dialog box.

Figure 14.4
The box context menu is a convenient way to control many box-related features.

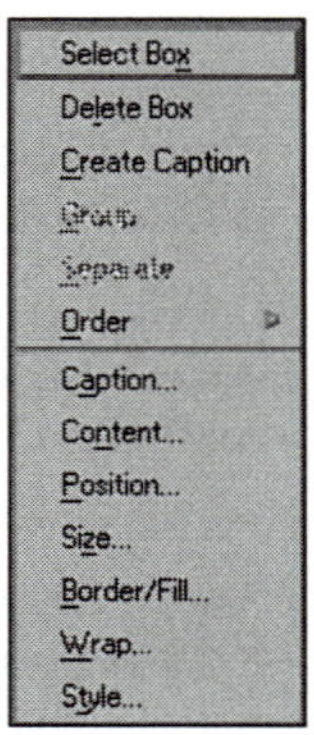

Are you having trouble getting the text box context menu? See "Selecting Text Boxes and Normal Text" in the "Troubleshooting" section at the end of this chapter.

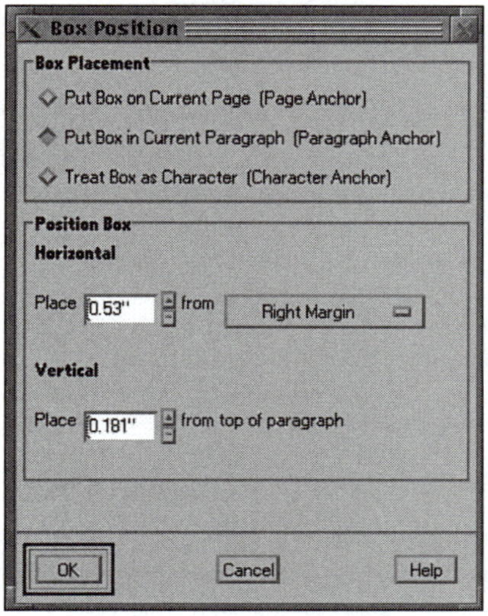

Figure 14.5
The options in the Position Box area depend on the Box Placement options you select.

Note

If you change the anchor option on your box, chances are the box will shift its location, so you may need to reposition it.

The anchor option you select has a substantial impact on what happens to your box as you edit the surrounding text. With a page anchor, the box maintains its position on the page where its code lies, no matter what happens to the surrounding text. If you insert or delete enough text to shift the code onto another page, however, the box jumps to the same relative position on that page. If you use a paragraph anchor, the box maintains its horizontal position but moves up or down the page along with its paragraph. Such boxes always remain on the same page as their codes, which reside at the start of the paragraph. If you select character anchor, the box moves both horizontally and vertically with its surrounding text.

Tip #176 from *Tux Tips*

Chances are you'll want to use page or paragraph anchors most of the time for text boxes because you're presumably using this feature to create text that's related to nearby text, but doesn't flow directly with it. You're more likely to use character anchors with small graphics or equations.

→ For more information on graphics, **see** "Graphics," **p. 377**
→ For more information on equations, **see** "Equations," **p. 433**

Making Adjustments Visually

The easiest, but least precise, way of adjusting a text box's size and position is to drag the box with the mouse. To make either type of change, you need to select the text box by clicking in it. Then try moving the mouse pointer around inside the box and near its edges.

You'll notice the mouse pointer change into various shapes. These shapes are visual indicators of what will happen if you click and drag the mouse.

Figure 14.6 shows a text box with a four-arrow mouse pointer. The mouse pointer takes this shape when it's near any edge of the text box, but not over one of the control points. When you see a four-arrow pointer, you can click and drag to move the text box to a new location.

Figure 14.6
A four-arrow mouse pointer indicates that you can move the text box.

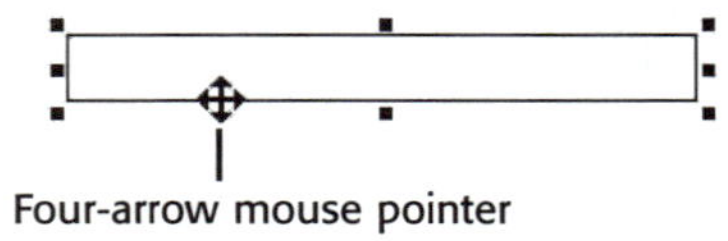

When you move a text box in this way, the corresponding [Box] code may or may not shift position. If you move the box across whatever unit you've set for the anchor point, the code also moves, but otherwise the code is changed rather than moved. For instance, with paragraph anchoring, if you shift a box from the top line of the paragraph to something closer to the bottom, the [Box] code remains at the start of the paragraph. If you move the box to reside within another paragraph, however, WordPerfect shifts the [Box] code to that paragraph.

When you move the mouse pointer over one of a box's eight control points, it takes on the appearance depicted in Figure 14.7. This shape indicates that you can click and drag either in the direction of the arrow to increase the size of the text box in that direction or in the opposite direction to reduce the size. The rules for resizing a box work much like the rules for resizing a window with most window managers: Clicking on a corner control point adjusts the box's size both horizontally and vertically, whereas clicking on a control point on a box edge results in horizontal or vertical adjustment, but not both.

Figure 14.7
A thin arrow pointing in one direction indicates that you can perform a resize operation.

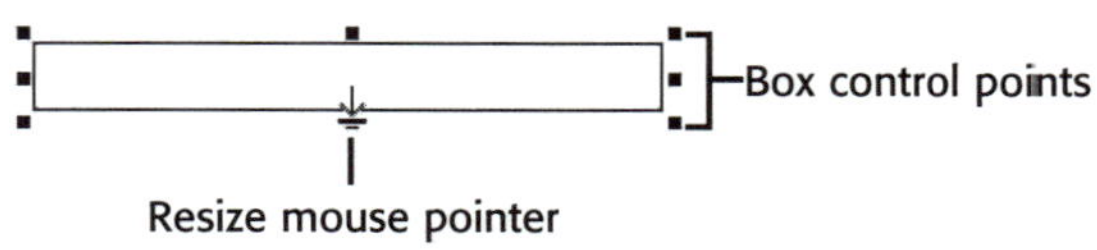

Tip #177 from *Tux Tips*

You can attain greater precision of control over your text box positioning if you use WordPerfect's zoom feature. Choose View, Zoom to bring up the Set Zoom dialog box (Figure 14.8). Set the zoom level to something higher than you would normally use, such as 200%. Because everything onscreen is now larger than usual, positioning text boxes as you please should be easier. For still greater precision, though, you need to enter position values directly from a dialog box, as described shortly.

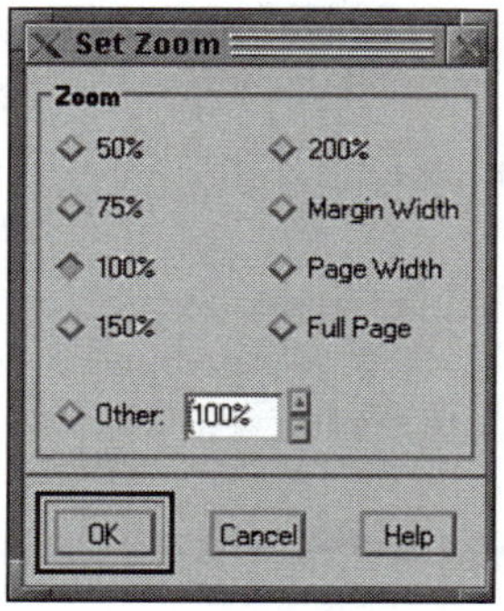

Figure 14.8
Use WordPerfect's Zoom feature to position a text box precisely with the mouse.

Making Position Adjustments from a Dialog Box

Adjusting a text box's position via point-and-click operations can be convenient, but it's also often imprecise. To align two text boxes precisely or to position one exactly halfway between the margins, you need to use dialog boxes.

To adjust the box's position in this way, right-click the text box and select Position. This step opens the Box Position dialog box, as in Figure 14.5. The box positioning options vary substantially, depending on the box placement options in effect. In general, you select the horizontal and vertical positions for the text box. Sometimes you have to tell WordPerfect relative to what you're measuring, whereas with other anchor options the position is fixed.

Are you having trouble getting the text box context menu? See "Selecting Text Boxes and Normal Text" in the "Troubleshooting" section at the end of this chapter.

Page Anchor

The page anchor option has the greatest number of options for box positioning, as shown in the Box Position dialog box with Page Anchor selected in Figure 14.9.

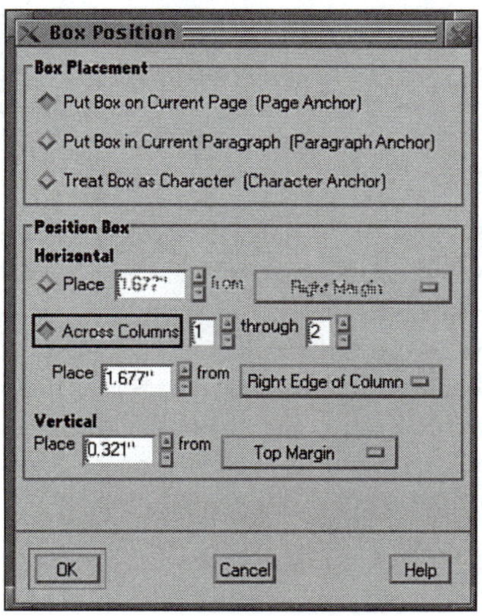

Figure 14.9
Position a box with a page anchor for precise control over the box's position on the page.

When you use a page anchor, the Position Box area has the following options:

- Horizontal via place—You can specify the horizontal location of the box by specifying its placement relative to either margin, the left edge of the paper, or the center of the paragraph.≥
- Horizontal across columns—If you have a multicolumn layout, you may want to tell WordPerfect to place a text box relative to a column. This option keeps the text box positioned appropriately even if you change the column widths—though the text box's own size may need to be changed.

→ To learn how to use a dialog box to adjust a text box's size, **see** "Making Adjustments from a Dialog Box," **p. 368**

- Vertical—Specify the vertical position of the box relative to the top or bottom margin, the top edge of the page, or the center of the top and bottom margins.

Tip #178 from *Tux Tips*

The horizontal and vertical center options can be particularly helpful if you want to center a text box on the page. They can also be helpful if you want to precisely align text boxes relative to *each others'* centers. For instance, if you have two text boxes and you want to center one over the other, you can give them both the same alignment relative to the center of the paragraph. This value need not be `0`; you can shift the boxes' positions left or right by entering negative or positive values, respectively.

PARAGRAPH ANCHOR

If you elect to anchor your box to a paragraph, your options are relatively simple, as shown in Figure 14.5. You can specify horizontal position relative to the left edge of the page, the left or right margin, or the center of the paragraph. You can specify vertical position relative to the top of the paragraph only.

Note

You must enter positive values for the position relative to the top of the paragraph. Thus, you can't position a box above the paragraph in which its code resides. You can, however, position a box *below* the paragraph in which its code resides by entering a sufficiently large positive value here.

Boxes anchored to a paragraph appear on the same page as the start of their paragraphs even if you specify additional vertical space. For instance, suppose a paragraph begins 3 inches above the bottom margin. If you create a text box and specify that it should appear 4 inches below the first line of the paragraph, WordPerfect places the box so that its bottom edge falls at the bottom margin, even though this placement means that the box starts less than 4 inches below the [Box] code. If a text box must appear at a specific location relative to a paragraph, consider using a character anchor, or use a block protect code to force a page break if space runs out at the end of the page.

→ For more information on WordPerfect's handling of page breaks, **see** "Page Breaks," **p. 230**

CHARACTER ANCHOR

If you select the Character Anchor option, the Box Position dialog box contains the options shown in Figure 14.10. Because anchoring a box as a character keeps it positioned where the [Box] code resides in the text, you can't set the box's position relative to margins, the page edge, or a paragraph. Instead, the positioning options available to you relate to the box's location relative to the surrounding text.

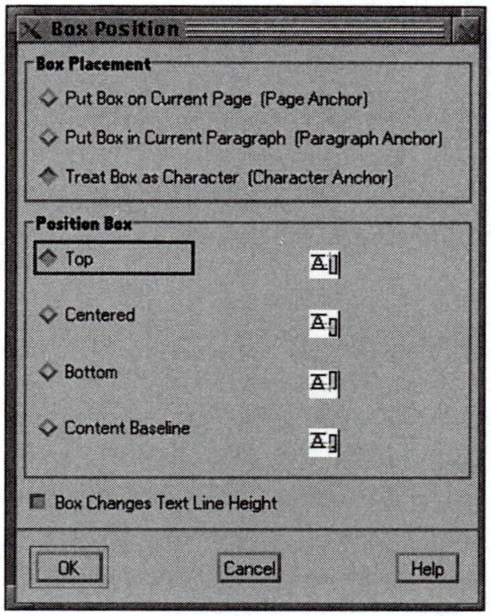

Figure 14.10 Control where a box lies relative to the surrounding text when you select Character Anchor in the Box Position dialog box.

You should think of these options as describing the position of the surrounding text relative to the box. Given this viewpoint, the Top, Centered, and Bottom options are fairly self-explanatory. The Content Baseline option aligns the surrounding text with the text inside the text box, which may be desirable in some cases. All four options are illustrated in Figure 14.11.

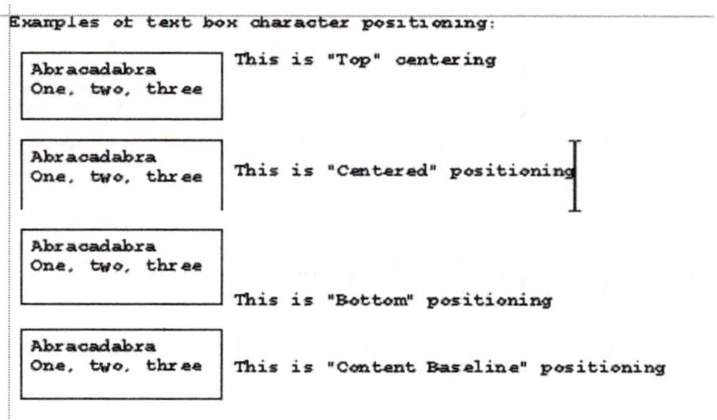

Figure 14.11 When you use character positioning, you can control where the surrounding text falls relative to your text box.

When you select the Box Changes Text Line Height check button in the Box Position dialog box, WordPerfect adjusts the line height for surrounding text to keep text from overlapping with the text box. If the text box is very tall, the result is a very tall line of text, which can be visually jarring. You can uncheck this option, but doing so is likely to cause ugly formatting problems because non-boxed text will overlap the text box; so uncheck the Box Changes Text Line Height option only if you have some specific formatting goal in mind.

MAKING SIZE ADJUSTMENTS FROM A DIALOG BOX

Just as you can adjust the position of a text box by using the Box Position dialog box, you can adjust the size of a text box by using the Box Size dialog box (Figure 14.12). You produce this dialog box by choosing Size from the box context menu.

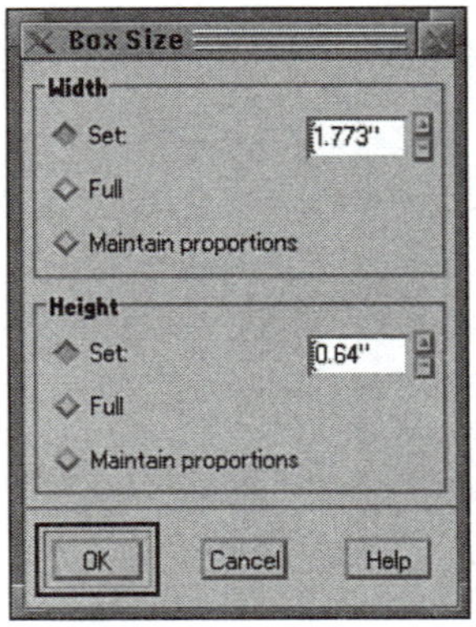

Figure 14.12 The Box Size dialog box lets you set the size of a text box with precision.

You can set the horizontal or vertical size of the text box by specifying an exact value for the Set field in the Width or Height section, respectively. If you select Maintain proportions for either or both of these fields, WordPerfect adjusts the text box's dimensions to match that of any text you've entered. The Full option sets the box's dimensions to the entire margin width or height. In addition, the Full option converts the text box into a page-anchored box if that's not already the case.

Note

The default text box height is Maintain proportions. This setting means that when you add or delete text in a text box, WordPerfect automatically adjusts the box's height to match.

EDITING TEXT WITHIN TEXT BOXES

Working with text within a text box is very much like working with text in an ordinary document. Although you can use most of the same editing techniques in both cases, some exceptions and special formatting capabilities are associated with text boxes.

→ If you want to know more about WordPerfect's editing features, **see** "Making Changes to a Document: Basic Editing," **p. 95**

Exceptions to Normal Text Editing Practices

Although most text editing techniques work fine within text boxes, there are a few exceptions:

- Text boxes are limited to one page and therefore are good only for short runs of text. If you try to insert too much text, say, by using Insert, File and loading a large file, WordPerfect complains that you've tried to insert too much text. It stores the entire extent of the text but displays only the amount that fits in the dialog box. WordPerfect reminds you that there's too much text in the text box whenever you select it to edit the text that is there. You get a similar complaint if you try to enter a page break [HPg] code.

Note

Some frame-based word processors let frames flow from one page to another to accommodate large amounts of text or elements repeated across pages. WordPerfect doesn't use this approach, although you may be able to achieve certain effects with another feature, such as watermarks or headers/footers.

- The ruler tends to be a bit unreliable when used with text boxes. You can use the corresponding menu commands to modify tabs, margins, and so on from dialog boxes, however.

Tip #179 from *Tux Tips*

If you prefer to use the ruler, try switching to Draft view (View, Draft or Ctrl+F5). You can then edit text boxes in a separate window, and the ruler is more likely to behave itself.

- The main document begins with an [Open Style: InitialStyle] code. A text box begins with an [Open Style: BoxText] code instead (Figure 14.13). You can therefore set the default *style (p. 142, 306)* of boxed text independently of that for the main text. The BoxText style does inherit the InitialStyle style, however. Therefore, unless you override them in the BoxText style, changes you make to the InitialStyle apply to boxed text, as well.

→ For general information on styles, **see** "Ensuring Consistency Within Your Document with Styles," **p. 305**

- When you're done editing text in a text box, click outside the text box area to finish. Alternatively, you can choose File, Close (Ctrl+F4).

Note

When you edit text inside a text box, Reveal Codes shows only the codes within that text box. To see codes outside the text box, you need to click outside the text box. Now instead of seeing the codes inside the text box, you see only the [Box] code corresponding to the text box itself.

Part IV
Ch 14

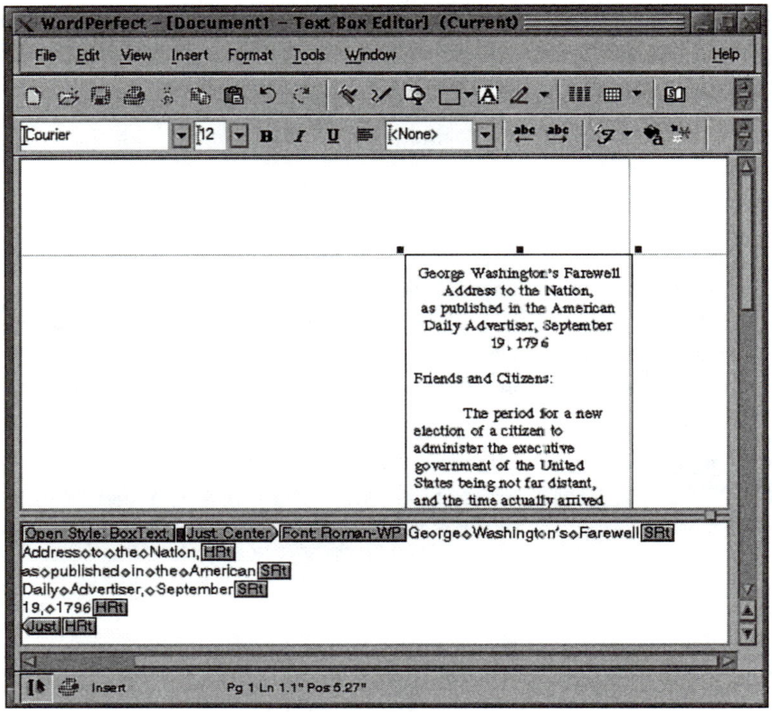

Figure 14.13
Use the BoxText style to set formatting options you want to apply to all boxed text, but not to other textual elements.

SETTING MARGINS

When you place a text box within other text, chances are you don't want that text butting up against the outline of the text box itself; you probably want to keep some distance between the outside text and the text box. Similarly, you probably want to keep the text *inside* the text box from touching the border lines of the text box. You can, of course, control the left and right margins for the text inside the box by using paragraph-oriented margin adjustment commands, but what about the other margins? WordPerfect has several options for controlling them, and in fact, you may want to use these options for the left and right margins of inside text, as well.

→ To learn more about margins, **see** "Margin Adjustments," **p. 206**

To adjust the spacing both inside and outside the text box's borders, right-click on the text box and select the Border/Fill option. This step opens the Box Border/Fill dialog box in Figure 14.14. Click the Customize Style button to get the Customize Border dialog box (Figure 14.15). To adjust the spacing inside and outside the text box, modify the Inside Space and Outside Space items, respectively. You can also click Auto Spacing to have WordPerfect produce spacing automatically.

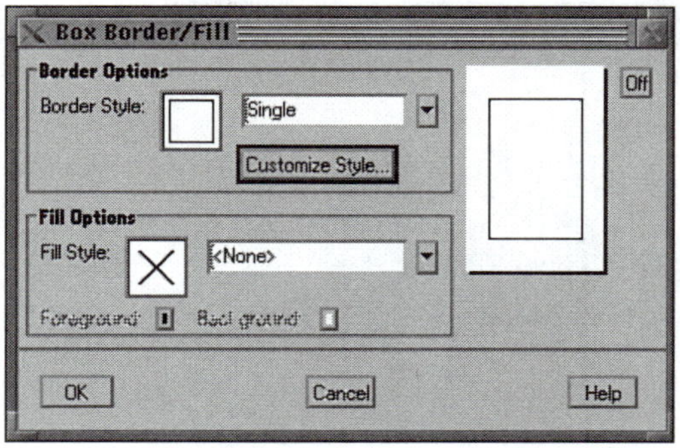

Figure 14.14
You can set the text box's border and fill options from the Box Border/Fill dialog box.

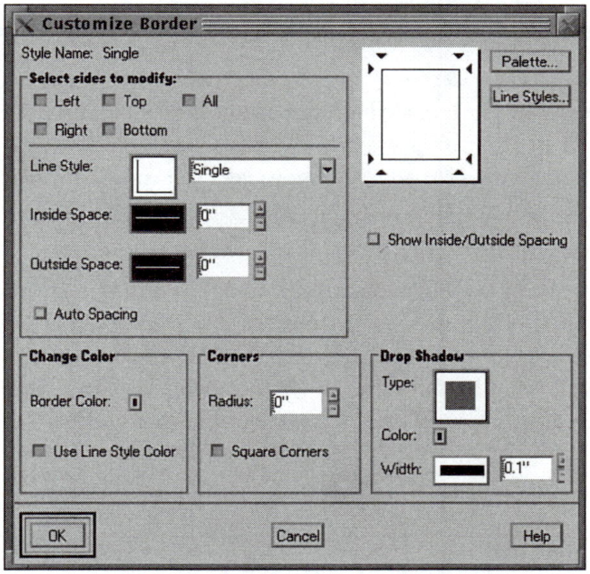

Figure 14.15
You can adjust much more than the inside and outside spacing from the Customize Border dialog box.

Are you having trouble getting the text box context menu? See "Selecting Text Boxes and Normal Text" in the "Troubleshooting" section at the end of this chapter.

The Box Border/Fill and Customize Border dialog boxes can be used to set many additional options. These options are discussed in detail in Chapter 18, "Box-Related Functions."

→ To learn more about box borders and fills, **see** "Borders and Fills," **p. 475**

SPECIAL BOXED TEXT OPTIONS

Boxed text allows for a few text-formatting tricks that are difficult or impossible to attain in other ways with WordPerfect. You can explore these possibilities by right-clicking the text box and choosing Content from the context menu. This step opens the Box Content dialog box, shown in Figure 14.16.

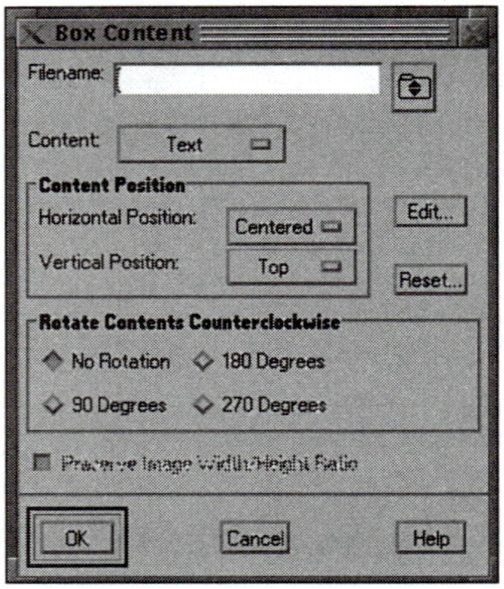

Figure 14.16
Use the Box Content dialog box to adjust details of text positioning within a text box.

You can align text in the text box to the top, bottom, or center vertically by using the options in the Content Position portion of the Box Content dialog box. The Horizontal Position options frequently have no effect, however. You can set these with conventional text justification commands if you like.

→ For information on conventional text justification commands, **see** "Justification," **p. 191**

The most unique trick of the Box Content dialog box is setting text rotation. You can set text to No Rotation, 90 Degrees, 180 Degrees, or 270 Degrees. You can use this option to create a piece of text that's rotated relative to the rest of the text on a page. The J-card template discussed in Chapter 13, "Promoting Cross-Document Consistency with Templates" uses this feature to print properly oriented text on the spines of the J-card insert, as shown in Figure 14.17.

→ For information on the J-card template, **see** "Automating Template Use with Macros," **p. 341**

Tip #180 from *Tux Tips*

If you want to rotate an entire page of text, say, for a wide table, you'd be better off using the page size formatting options from File, Page Setup, Page Size instead of the text rotation feature in the Box Content dialog box. The page size formatting options enable you to change the page orientation for one or a range of pages and change it back immediately thereafter.

→ For more information on setting the page orientation, **see** "Page Size," **p. 226**

Note

If you use rotated text, WordPerfect opens a new document window when you try to edit it, as if you were using Draft view mode, as described shortly.

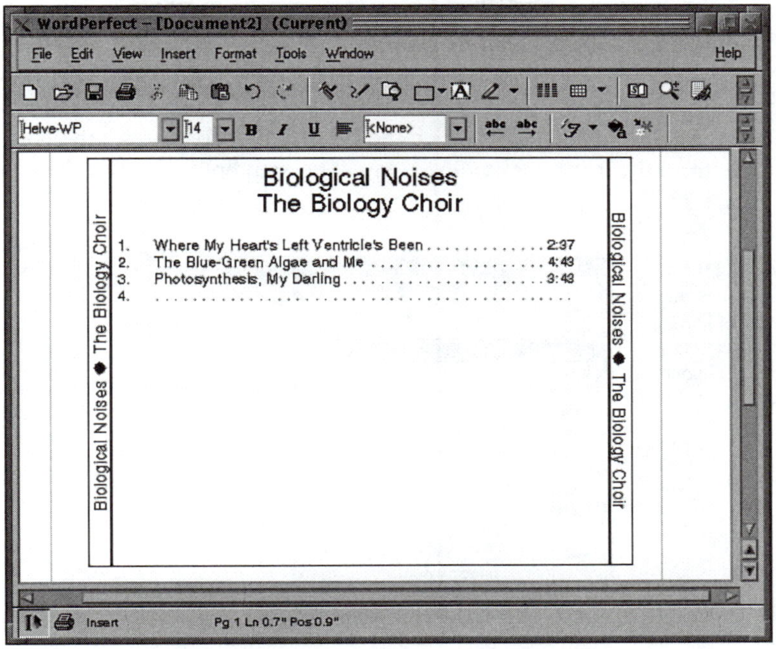

Figure 14.17
The CD J-card is composed of three text boxes, two of which use rotated text.

EDITING BOXED TEXT IN DRAFT VIEW MODE

If you're using WordPerfect's Draft view mode (View, Draft or Ctrl+F5), you won't be able to edit boxed text directly. Instead, when you select a text box for editing, WordPerfect opens a new document window (Figure 14.18). This window doesn't show the box's margin gridlines, so you may have a hard time judging just how your text will fit in the box until you finish editing and click the Close button to finish the job. On the other hand, WordPerfect's ruler is more reliable for text boxes in Draft view than in Page view.

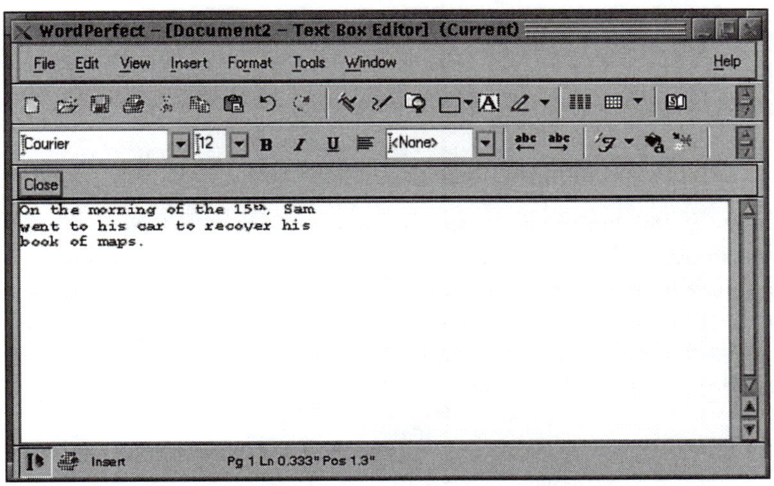

Figure 14.18
Click the Close button or choose File, Close (Ctrl+F4) when you're done editing a text box in Draft view.

TEXT BOX STYLES

WordPerfect lets you define a style for the text box's appearance, but it isn't a full-fledged style in the sense of WordPerfect's ordinary style feature. Rather, you can define a set of formatting rules for the box's border and fill patterns. You can set the box's style by selecting Style from the text box's context menu. This step opens the Box Style dialog box shown in Figure 14.19

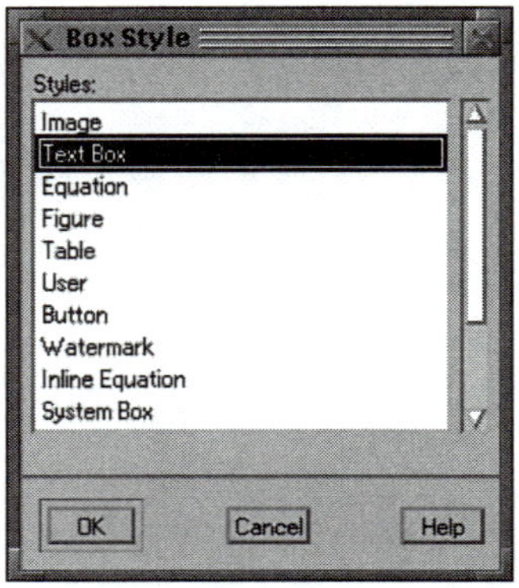

Figure 14.19
Select from styles to adjust a set of line types, shading, and line spacing for your text box.

→ For information on box borders and fills, **see** "Borders and Fills," **p. 475**

Are you having trouble getting the text box context menu? See "Selecting Text Boxes and Normal Text" in the "Troubleshooting" section at the end of this chapter.

Tip #181 from *Tux Tips*

Most of the box styles in the Box Style dialog box are quite similar, so you may prefer to customize the available options.

TROUBLESHOOTING

SPEED PROBLEMS WHEN USING TEXT BOXES

You've created a document with several text boxes and now WordPerfect has slowed down considerably.

Unfortunately, WordPerfect does exhibit speed problems when your document contains text boxes. If you don't need to view the contents of the text boxes, you can tell WordPerfect not to show graphics by using the View, Graphics menu item. (This item is a toggle and shows a square to its left, much like a check button in a dialog box, when it's active, which is the default. To turn it off, you want to remove that square.) You can still edit the contents of your text boxes when you deselect this option. When the text box isn't selected, the text inside the text box is invisible.

An alternative is to find a way to achieve your desired effect without using text boxes. For instance, you might be able to use a table, paragraph border, or columns to accomplish your goal.

→ To learn how to use tables, **see** "Tables," **p. 275**

→ To learn how to use paragraph borders and fills, **see** "Borders and Fills," **p. 216**

→ To learn how to use columns, **see** "Creating Professional Layouts with Columns," **p. 255**

SELECTING TEXT BOXES AND NORMAL TEXT

You try to access a menu item, either on WordPerfect's normal menus or from a context menu, but the item you want is either not there or is grayed out.

WordPerfect is very context sensitive, meaning that its menu options change according to the text that is selected and the location of the insertion point and mouse pointer. This behavior can make for some confusion when you're dealing with text boxes because you may not be able to choose common menu commands like File, Save (Ctrl+S) or View, Draft (Ctrl+F5) when you're editing in a text box. In addition, the context menu you get when you right-click over a text box depends on whether or not you're editing it. If you're not getting the menu options you want, look for the insertion point. Is it in the text box or in ordinary text? Try moving the insertion point and then try the menu option you want again.

CHAPTER 15

Graphics

In this chapter

OVERVIEW

Early word processors, including many earlier versions of WordPerfect, lacked support for graphics. If you needed to include graphics in a document created with such a program, either you had to prepare the graphics separately and insert them on separate sheets from your text or you had to do a literal cut-and-paste operation with the printouts. Today, of course, word processors allow you to embed graphics within your document. WordPerfect even incorporates the ability to *create* graphics. This chapter covers WordPerfect's capability to use graphics from external sources and to create drawings. This capability is useful for creating maps, diagrams, clip art, and so on. The next chapter covers WordPerfect's capability to create graphics from data. Use this feature if you want to create a bar chart from data in a WordPerfect table or from an external source.

→ For information on charts, **see** "Creating a Chart," **p. 411**

→ For information on using tables to store and manipulate numerical data, **see** "Table Mathematics," **p. 296**

WordPerfect treats graphics as boxed elements, similar to text boxes. Many of the box manipulation functions are common to both text and graphics boxes, so if you know how to manipulate text boxes, you already know a lot about handling graphics boxes.

→ For more information on manipulating boxes in general, **see** "Box-Related Functions," **p. 459**

Note

Some features described in this chapter are available only in the retail versions of WordPerfect for Linux. Specifically, WordPerfect's Draw module is unavailable in the downloadable version of the program. To create graphics, then, you need to either buy the retail version of WordPerfect or use another external program.

UNDERSTANDING GRAPHICS FILE FORMATS

Before proceeding, it's important to clarify one key distinction in computer graphics: that between *bitmap* and *vector* graphics formats. A bitmap graphic describes an image in terms of a fixed number of *pixels* (short for *picture elements*), which are individual points of color in a larger image. Given enough pixels in a small enough space (that is, at a high enough *resolution*), bitmap graphics can produce very realistic images. One drawback to bitmapped graphics is that they don't *scale* very well. That is, if you want to increase or decrease the size of such an image, it may start to look chunky or uneven. This factor is a major problem because printing an image necessarily involves scaling it because most printers have much higher resolutions than most monitors have. (Modern monitors have a greater capacity to handle lots of colors, though, so printers don't have a perfect advantage here.) Another drawback to bitmap graphics is that it's hard to build an image from smaller component parts with them. For instance, if you want to create an image of a house, you must insert the appropriate components correctly the first time. If you create a house and then decide that the trees in front of the house are too big, you can't just resize the trees and expect the rest of the house to appear from obscurity. Instead, you have to redraw those components yourself.

A vector representation, on the other hand, describes an image in terms of lines, curves, and other shapes. For instance, you could create a vector representation of a house by using a rectangle for the doorframe, more rectangles for the windows, freehand shapes for trees, and so on. Vector images are very good for creating diagrams and simple line art. In addition, vector images scale well because the image is *rasterized* (turned into a bitmap format) separately for devices of differing resolutions.

If you have a graphic that you've scanned using a scanner or pulled off the Web, chances are it is in a bitmapped file format; likewise, for files you create using most "paint" programs. A graphics file that you create with a "draw" program, however, is likely to be in a vector format. WordPerfect favors vector images; when you use the WordPerfect Draw module, it creates vector files. You can incorporate a bitmapped image inside a vector file with WordPerfect, however, or use a bitmapped image created with another program.

Inserting Graphics from External Sources

One of the easiest ways to incorporate graphics into a WordPerfect document is to import a graphic that already exists by choosing Insert, Graphics, From File. This sequence produces a standard WordPerfect File Manager dialog box, from which you can select your graphics file. WordPerfect inserts the graphic into the current document (Figure 15.1).

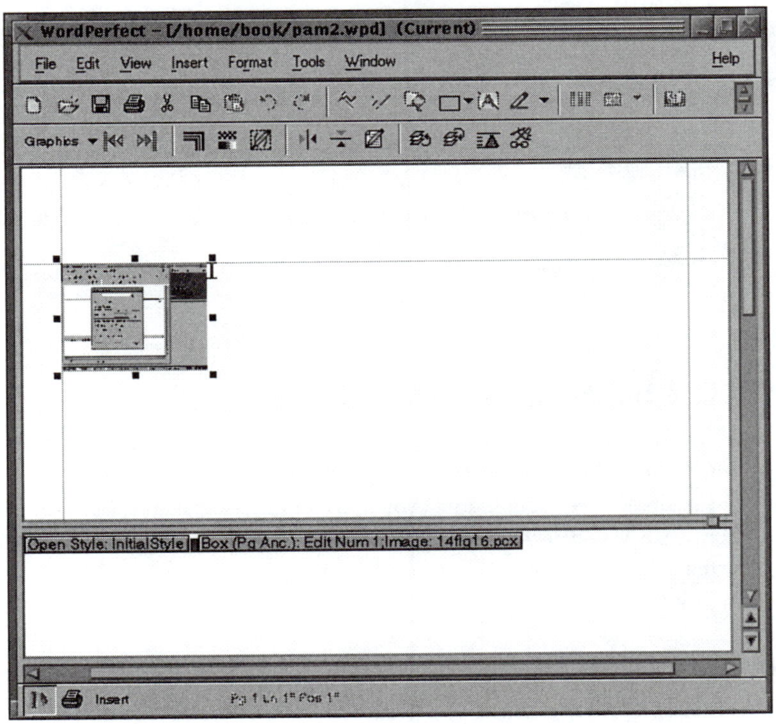

Figure 15.1
Graphics boxes, like text boxes and equations, appear in WordPerfect's Reveal Codes pane as individual codes.

If you click the Image on Disk check box in the Retrieve Options area of the File Manager dialog box when retrieving a file, WordPerfect won't actually store it in your WordPerfect file. Instead, it stores a reference to the file on disk and retrieves the file only as needed to display it onscreen or print it. This storage technique can make WordPerfect documents considerably smaller than they would be if you imported the images completely, but if you delete the original image or move your WordPerfect file to another computer, you lose the graphic.

WordPerfect understands a number of common graphics file formats and converts these into its native internal format. Supported formats include Windows bitmap (.bmp), CGM, HPGL, EPS, PCX, GIF, JPEG (.jpg), Sun Rasterfile, TIFF (.tif or .tiff), X Bitmap (.xbm), X Window Dump, and of course WordPerfect Graphics (.wpg). Most of these are bitmapped formats, but some are vector formats. In general, the bitmapped formats import without difficulty, but sometimes the nonnative vector formats, such as HPGL and EPS, cause problems.

Note

When you import an Encapsulated PostScript (EPS) file, WordPerfect does *not* display the graphic itself; instead, it displays the words `File contains data for PostScript printers only` inside a graphics box. You can still resize and manipulate this graphics box, but you won't be able to see what's in it until you print, and the graphic will print only on a PostScript printer or one driven by *Ghostscript (p. 732)*. You also won't be able to edit this graphic.

→ For information on configuring both PostScript and non-PostScript printers, **see** "Fonts and Printing," **p. 739**

Tip #182 from *Tux Tips*

If you have the downloadable version of WordPerfect and want to create drawings for inclusion in WordPerfect, you can use the `xfig` program. This package comes with most Linux distributions, and it can export files in EPS and other formats that WordPerfect can use.

USING LINES AND SIMPLE SHAPES

Whether you're using the downloadable or the retail version of WordPerfect 8 for Linux, you can quickly create an assortment of lines and simple shapes for use in your document. These shapes are included in WordPerfect graphics boxes, but you won't be able to edit them with WordPerfect's Draw module, even if you have the retail version of the program. Instead, WordPerfect includes tools to help you edit these objects directly in the main document window.

CREATING SHAPES

When you create a shape, you generally have to position your mouse pointer in the document editing area and perform some sequence of mouse clicks and drags to insert the shape.

The details differ from shape to shape. To create a shape, choose an appropriate option from the Insert, Shape menu. The available shapes are

- Draw Line—This option draws a line at an arbitrary orientation. To draw a line, click and hold the mouse pointer, drag it to a new location, and release. If you hold the Shift key when creating a line, WordPerfect restricts it to horizontal, vertical, or diagonal orientations.
- Horizontal Line (Ctrl+F11)—Unlike most shapes you can create, the horizontal line doesn't require any further intervention; it inserts a single horizontal line at the insertion point, extending from the left margin to the right margin.
- Vertical Line—Like the horizontal line, the vertical line option creates a line at the current insertion point position but running from the top margin to the bottom margin.
- Custom Line—This option works like the Horizontal Line and Vertical Line options except that WordPerfect first displays the Create Graphics Line dialog box (Figure 15.2). You use this dialog box to set various line attributes, such as the line's thickness and color.

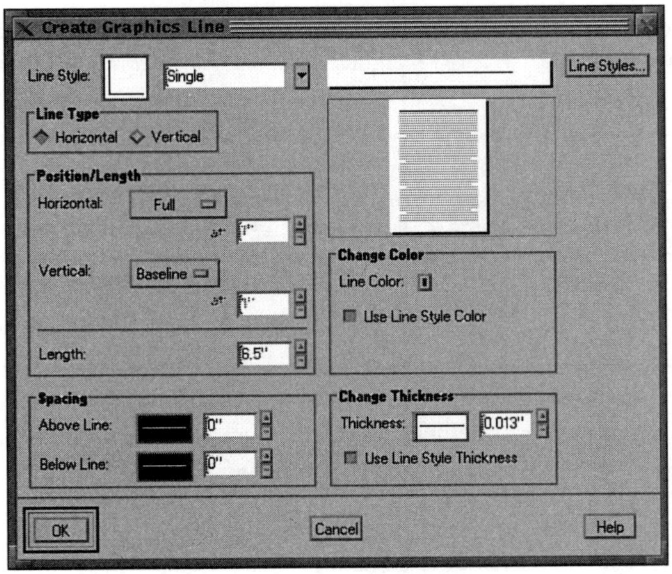

Figure 15.2
The Create Graphics Line dialog box lets you set various line attributes before the line is created.

- Polyline—You can create a shape composed of multiple joined line segments with this option. When you select Polyline, position your mouse pointer where you want the line to begin and then click and release the mouse button. Move to another position and click again. The first segment of the shape runs between those two positions. Click again in a third position to create the second segment running from the second to the third positions. When you're finished, double-click to create the final line segment.

- Polygon—A polygon is a shape with an arbitrary number of straight sides joined at points called *vertices* (singular: *vertex*). When you select the Polygon option, click in the document window, move your mouse, click again, and so on. Each time you click, WordPerfect adds a vertex to the polygon. The polygon is a complete shape at all times. When you're done, double-click to create the final vertex.
- Rectangle—To create a rectangle, click and hold the left mouse button, drag, and release when the rectangle is the size you want.
- Rounded Rectangle—This option works just like the Rectangle option, except that this rectangle has rounded corners.

Have you noticed prompts or directions for creating squares? Are they not working? See "Constrain to Square Not Working" in the "Troubleshooting" section at the end of this chapter.

- Ellipse—Click, drag, and release to create an ellipse. Positioning an ellipse can be tricky, though because the points where you click and release are *outside* the ellipse. Imagine an ellipse enclosed by a rectangle. The first point you click is one corner of this rectangle, and the second is the opposite corner. WordPerfect doesn't display the rectangle around your ellipse, though; this image is just a way to describe how the ellipse is defined. The best way to understand it is to try it.
- Circle—You create a circle much as you do an ellipse. The difference is that WordPerfect extends the imaginary rectangle so that it becomes a square, thus defining a circle.

Tip #183 from
Tux Tips

If you want to position a circle or ellipse so that it rests against another object or along a margin, the way WordPerfect provides for creating these shapes works well. If you want to center a circle or ellipse in an area, though, WordPerfect's method can be confusing. In this case, try creating the object first and then moving it into position.

→ To learn how to move graphics, **see** "Moving and Resizing Graphics," **p. 408**

- Arrow—WordPerfect can create an arrow that points left or right. It can be long and squat or short and tall, depending on how you size it. You create an arrow just as you do a rectangle, circle, or ellipse, and the arrow is contained within the invisible rectangle described by your mouse movements. If you drag left after clicking, WordPerfect creates a right-facing arrow, and if you drag right, the arrow faces left.

Figure 15.3 shows many of these graphics elements. The shapes in Figure 15.3 use the WordPerfect defaults for line thickness, color, and so on. You can, of course, change any of these attributes.

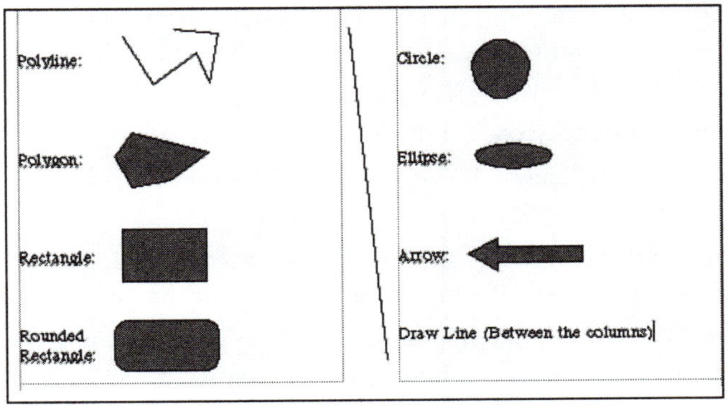

Figure 15.3
WordPerfect's stable of simple graphics elements is enough to let you add visual interest to many documents if you use these elements creatively.

ALTERING SHAPE ATTRIBUTES

If you click on a shape, WordPerfect displays a toolbar with relevant tools for altering attributes such as the shape's color and line style. The exact attributes available, and therefore the toolbars that WordPerfect presents, vary from shape to shape.

DRAW LINE AND POLYLINE ATTRIBUTES

When you select a draw line or a polyline, WordPerfect displays the line toolbar shown in Figure 15.4.

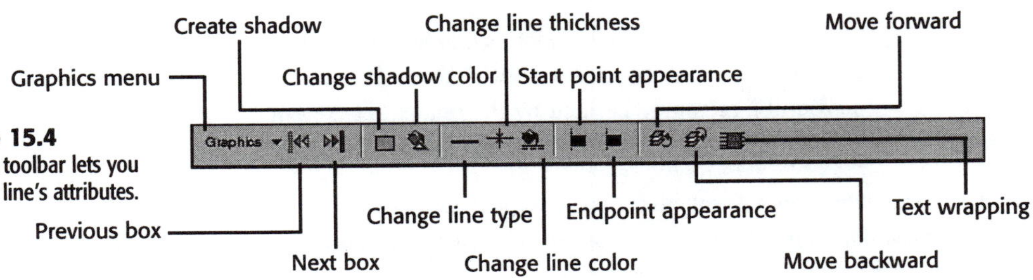

Figure 15.4
The line toolbar lets you adjust a line's attributes.

The toolbar contains the following items:

- Graphics menu—This item produces a menu of options for further manipulating the graphics box.
- Previous box and Next box—These icons move the selection to the previous and next boxes, respectively.
- Create shadow—You can create a shadow effect by selecting an option from the pop-up menu you get when you click this icon. Set the shadow to fall in any of four directions, or none at all (the default, which disables the shadow effect).

- Change shadow color—You get a pop-up color menu when you click this icon. Select the color you want for the shadow you create with the Create shadow tool.
- Change line type—You can change the line from its default solid state to any of several dashed or dotted styles with this tool.
- Change line thickness—Adjust the thickness of the line, from the default thin value on up to some very thick selections.
- Change line color—Like the Change shadow color tool, this one produces a pop-up color square from which you can pick a color for the line.
- Start point and endpoint appearance—You can alter the appearance of the start points and endpoints of the line with this tool. Select from a square, rounded, or arrow-shaped point.
- Move forward and Move backward—Graphics objects can overlap one another on the page, and when that happens, WordPerfect treats one object as if it falls in front of another. You can use the Move forward and Move backward tools to adjust the relative positioning of graphics boxes.

Note

Each graphics object can be thought of as being drawn on a sheet of invisible plastic that fills the entire page. Thus, if you have three or more objects, some of which don't overlap and some of which do, you may need to move an object forward more than once to move it in front of the object that's obscuring it. In this case, you may need to move the object "in front of" the objects that aren't actually obscuring it.

- Text wrapping—You can tell WordPerfect how to wrap text around the line. You can have text flow behind the object, flow in front of the object, hug the object closely, or wrap around a rectangle defined by the object. These options are discussed in more detail in Chapter 18, "Box-Related Functions."

→ To learn more about WordPerfect's text wrapping options, **see** "Text Wrapping," **p. 461**

Are you having problems getting WordPerfect to display text behind or in front of a graphics object? See "Graphics Not Printing in Front of or Behind Text" in the "Troubleshooting" section at the end of this chapter.

HORIZONTAL AND VERTICAL LINE ATTRIBUTES

When you create or modify a horizontal or vertical line, WordPerfect displays the horizontal/vertical line toolbar, as shown in Figure 15.5.

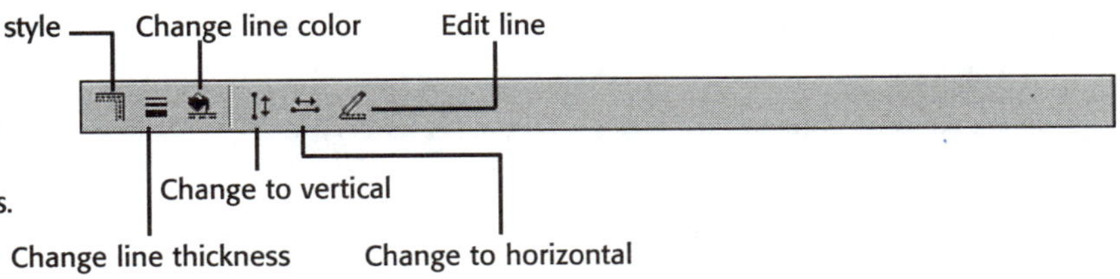

Figure 15.5
The horizontal/vertical line toolbar lets you adjust a line's attributes.

The following options are available on the horizontal/vertical line toolbar:

- Change line style—You can change the style of the line with this tool. When you click it, you get a pop-up list of different line styles, including double lines, triple lines, thick lines, and dotted lines.
- Change line thickness—Use this option if you want to change the thickness of the line.
- Change line color—As you might expect, this option allows you to change the color of the line.

Note

If you select a line color or thickness, that selection overrides any line style settings you may have made.

- Change orientation—Use the Change to vertical and Change to horizontal tools to convert a line from horizontal to vertical orientation or vice versa.
- Edit line—Use this tool to obtain finer control over the line's characteristics by opening the Edit Graphics Line dialog box, which is virtually identical to the Create Graphics Line dialog box (refer to Figure 15.2). You can set the line style, orientation, color, thickness, position, length, and spacing around the line from this dialog box. You can also open the dialog box by right-clicking on the line and choosing Edit Vertical Line from the context menu.

Tip #184 from *Tux Tips*

If you accidentally set a line's color or thickness and want to undo the change to use a pre-defined line style, you can do so from the Edit Graphics Line dialog box. Simply be sure that the Use Line Style Color and Use Line Style Thickness check boxes are selected.

You'll note that fewer options are available for modifying the appearance of horizontal and vertical lines than for modifying a draw line. If you want one of the draw line options, such as control over the appearance of the line's endpoints, you can simply create a draw line and use the Shift key to ensure that the line is strictly horizontal or vertical in orientation.

FILLED SHAPE ATTRIBUTES

Filled shapes are the rest of the shape options from the Insert, Shape menu—polygons, rectangles, rounded rectangles, circles, ellipses, and arrows. These shapes all consist of an outline that's filled with some color, hence the term *filled shape*.

When you create or edit any filled shape, WordPerfect displays the filled shape toolbar, depicted in Figure 15.6

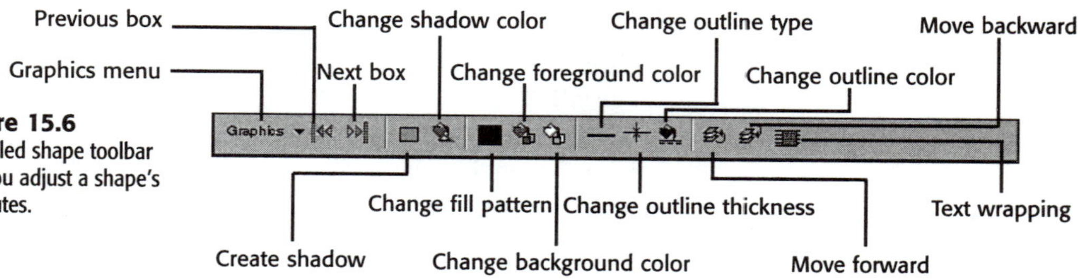

Figure 15.6
The filled shape toolbar lets you adjust a shape's attributes.

Many of the tools on the filled shape toolbar are the same as those on the line toolbar, and you should refer to that section for a description of those items. The rest of the filled shape toolbar items are

- Change fill pattern—By default, WordPerfect uses a solid fill pattern, meaning that the entire shape is filled with a single color. You can select from several different fill patterns, however, such as checkerboards, lines, and an assortment of gradients. You can create an outline of a shape with no fill pattern by selecting the X in the upper-left corner of the fill pattern menu.
- Change foreground/background color—Each of these tools creates a pop-up color selector from which you can select a color. The foreground color is the only one that's used by the default solid fill pattern. If you select another pattern, though, the background color comes into play. Using both colors, you could create, say, a red-and-green checkerboard fill.
- Change outline tools—These tools operate just like the Change line tools for lines but on the outline of the filled shape.

CHANGING A POLY'S SHAPE

You can, of course, click on any shape and use its control points to alter its size, or move it when the mouse pointer takes on a four-arrow appearance. These operations work just like they do for text boxes.

→ If you want to know how to move and resize a box, **see** "Moving and Resizing Text Boxes," **p. 362**

Tip #185 from *Tux Tips*

If you hold down the Shift key when selecting a shape, WordPerfect selects multiple shapes, much as it does when you hold down the Shift key when you're selecting text. You can use this feature to move or resize several shapes at once.

You can also alter the shapes of polylines and polygons. Start by double-clicking the shape. WordPerfect then places a control point at each of the shape's vertices, as shown in Figure 15.7. Click on a control point and drag it to change the shape's appearance.

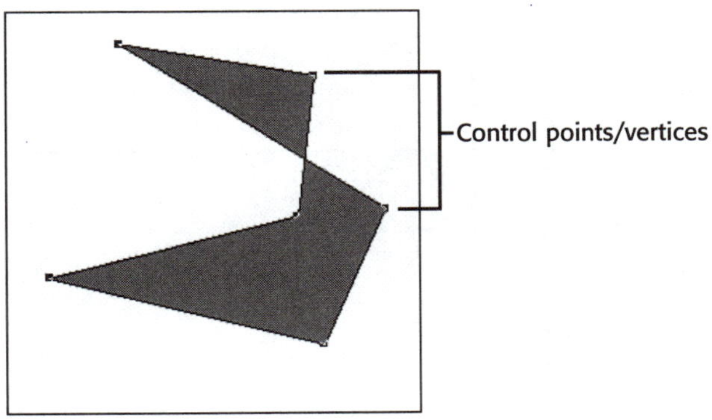

Figure 15.7
The vertices define the shape of a polyline or polygon, though you may think of them in terms of their lines.

Tip #186 from *Tux Tips*

If your shape is particularly small, you can use WordPerfect's Zoom feature (View, Zoom) to enlarge the view to up to 400%. Making precision adjustments to a shape is much easier when the shape is large.

Using WP Draw

The ability to insert shapes in your document can be very helpful when you need to create a quick and simple graphics effect. If you want to create a more complex image, however, such as a logo or a line art figure, you'll probably need a full-fledged graphics editor. WordPerfect's Draw module is just such a program.

Note

The Draw module *is* a separate program. Its executable file is called xwpdraw and is located in the wpbin subdirectory of your WordPerfect installation. You can use it as a standalone program, but you won't be able to insert text in your image unless you launch the Draw module from within WordPerfect.

Note

The Draw module is included only in the retail versions of WordPerfect. If you're using the downloadable version, you won't be able to use Draw. You can use some other drawing program, however, and import the results into WordPerfect using a file format such as EPS.

→ To learn how to import an existing graphics file into WordPerfect, **see** "Inserting Graphics from External Sources," **p. 379**

BASIC PRINCIPLES

You start WP Draw by choosing Insert, Graphics, Draw Picture. This step produces a pair of windows, shown in Figure 15.8, with which you create or edit your graphics. The window to the left in Figure 15.8 is the *tool palette*, and it contains shortcuts to perform graphics editing actions. In many ways, it's like a toolbar in WordPerfect's normal document windows. The window on the right in Figure 15.8 is the main Drawing window where you create or edit graphics.

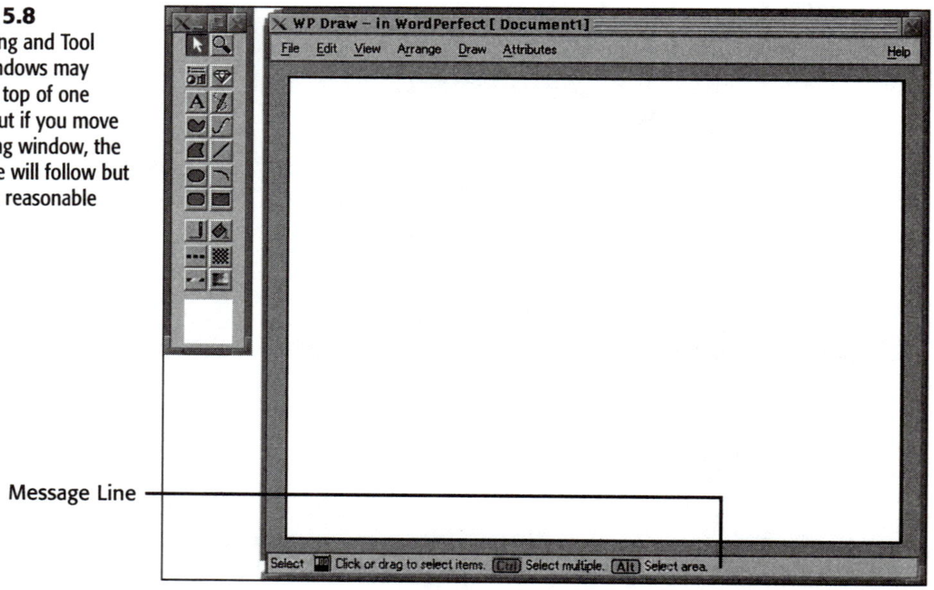

Figure 15.8
The Drawing and Tool palette windows may appear on top of one another, but if you move the Drawing window, the tool palette will follow but maintain a reasonable distance.

WP Draw is primarily a vector-based graphics utility. You therefore work with it by creating and manipulating shapes, rather than individual pixels. You can import a bitmapped graphic into WP Draw, but you won't be able to manipulate it in any substantive way; WP Draw treats the graphic as a single graphics element. Therefore, if you want to do something like crop or adjust the color balance of a photograph you've scanned, you'll have more success with a separate bitmapped graphics utility, such as the GIMP (`http://www.gimp.org`).

When you create an image with WP Draw, you have a fixed area in which to work. When you're finished, you can scale the image you create up or down in size to fit whatever area on the page you need it to occupy. If you want to create an image with proportions that are different from what you see in the WP Draw window, you can adjust these proportions. *Before* you begin to create your graphic, choose File, Page Layout from the WP Draw window's menu to open the Page Layout dialog box shown in Figure 15.9. Adjust the width or height of the age by a proportion to meet your requirements. For instance, if you want something that's half as wide as the default, change the width from its default of 11 inches to 5.5 inches. When you click OK, you'll find that your editing area's proportions have changed appropriately.

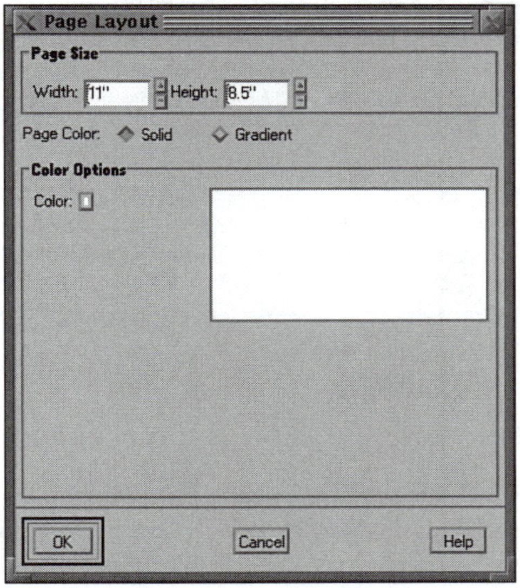

Figure 15.9
You can set the background color as well as the size for your graphic from the Page Layout dialog box.

> **Note**
> The sizes you enter in the Page Layout dialog box should not be interpreted literally. By default, your drawings will not occupy an actual printed size of 11∞8.5 inches. Remember that a vector graphic is fully scalable. Therefore, the units used to describe it initially are somewhat arbitrary. You may, of course, scale the resulting image up to a size of 11∞8.5 inches if you like, but by default it will actually be smaller.

The Tool palette window provides quick access to many of WP Draw's features. You can click the Draw ellipse tool, for instance, and then immediately click in the main WP Draw window to draw an ellipse. Other tools, such as the Set line color tool, display pop-up menus with selections from which you may choose. Figure 15.10 identifies the functions of the buttons in the Tool palette window.

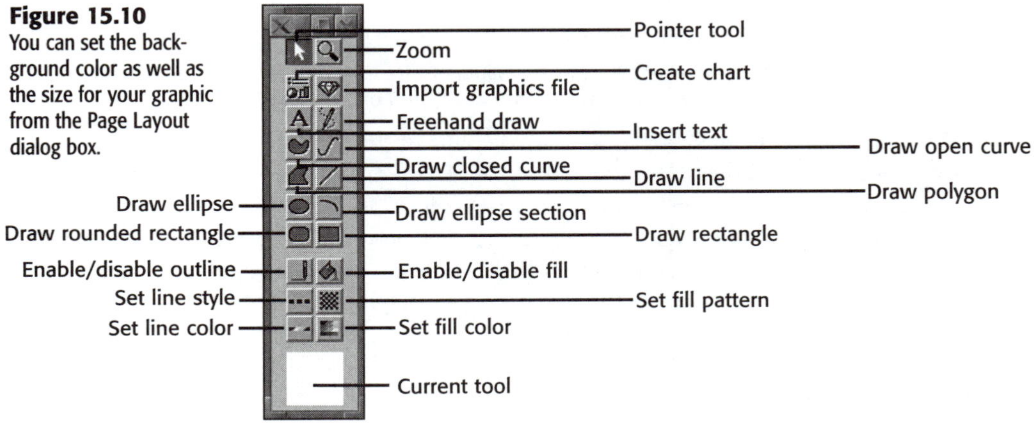

Figure 15.10
You can set the background color as well as the size for your graphic from the Page Layout dialog box.

Note

When you select a tool from the tool palette or select a graphics object, the Current tool area of the tool palette shows the current selection. Checking this area before you click in the Draw window can help ensure that you don't perform some unintended action.

Note

When your mouse pointer rests atop one of the tools in the Tool palette window, the title of the main WP Draw window changes to reflect the function of the tool over which the mouse pointer lies.

DRAWING CURVES, LINES, AND FIGURES

You create most lines and figures much as you do with WordPerfect's Shape feature:

1. Click on the shape or line type you want to create in the Tool palette window.
2. Click in the Document window when you want to create closed or open curves, polygons, or lines. With these object types, you can click more than once to alter the appearance of the item, by adding vertices to polygons, for instance. Double-click when you're done creating the shape you want.

Note

The WP Draw line type is similar to the polyline option of WordPerfect's Shape feature in that you can create multiple joined line segments. If you want just one line, click once at the start point, move the mouse pointer to the endpoint, and double-click.

3. Click and drag to create a freehand drawing, ellipses, arcs, and both rounded and regular rectangles.

Tip #188 from *Tux Tips*

WP Draw has no separate options for creating squares. If you want to create a square, choose View, Snap to Grid to constrain an object's start point and endpoint to a grid pattern. This option makes it easier to create a square.

→ To learn more about the snap to grid feature, **see** "View Options," **p. 401**

Have you noticed prompts or directions for creating squares? Are they not working? See "Constrain to Square Not Working" in the "Troubleshooting" section at the end of this chapter.

4. Pay attention to the messages in the Message line of the Drawing window. You can use keystrokes to alter WordPerfect's handling of certain objects. For instance, pressing the Alt key while you create an ellipse or rectangle causes WordPerfect to create the shape from the center. That is, the point where you click becomes the center, and as you drag the mouse you define an outer point of the rectangle (or of an imaginary rectangle surrounding the ellipse).

In addition to selecting the object you want to draw from the tool palette, you can choose an object from the Draw menu. All the tool palette shapes have entries on this menu.

Tip #189 from
Tux Tips

The tool palette does not have a circle tool, but you can draw a circle by choosing Draw, Circle from the Drawing window's menu.

Inserting Text

You can insert text into a Draw object by clicking the Insert text tool or by selecting Draw, Text from the Drawing window. Then you have two choices:

- Click once in the editing area to create a single line of text. Even if you type more than one line of text, WP Draw puts it all on one line, which usually results in the text not fitting within your drawing area. You can resize the text box in the Text Editor, as described shortly, to correct this situation.
- Click and drag in the Editing window to place text in the area specified by your drag operation. If you type too much text to fit in the vertical space provided, WP Draw expands that space.

Either way, WP Draw displays the WP Draw Text Editor window shown in Figure 15.11. This window presents an image very much like a text box, and you can treat it in a similar way. In particular, you can enter text into this box, and you can scale it horizontally by dragging its control points.

➔ For more information on text boxes, **see** "Using Text Boxes to Add Variety to Text Layout," **p. 357**

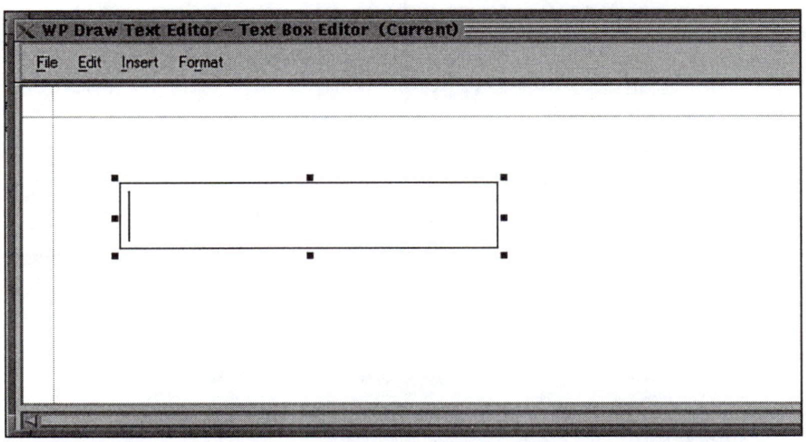

Figure 15.11
The Text Editor window gives you access to a subset of WordPerfect's usual editing features.

Compared to WordPerfect's normal assortment of text manipulation features, text entry within a WP Draw object is crude; you can't create tables, create columns, or use Reveal Codes, for instance. What you can do with text inside a WP Draw object is

- Enter ordinary text. Like a text box, the text entry field expands vertically as you type text into it, though this expansion is mirrored in the final text object only if you used a click-and-drag operation to define that area.

- Enter special symbols from the Symbols dialog box (shown in Figure 15.12) by choosing Insert, Symbol (Ctrl+W). Select a symbol and click Insert to insert it into the Text Editor's text entry field, or click OK to insert it and remove the Symbols dialog box.

Figure 15.12
The Symbols dialog box lets you select many special symbols.

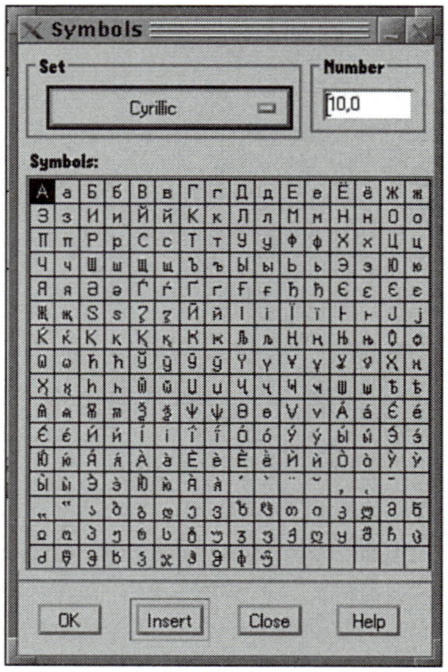

- Cut, copy, and paste text by using the Edit, Cut (Ctrl+X), Edit, Copy (Ctrl+C), and Edit, Paste (Ctrl+V) options, respectively. You can paste text from elsewhere in the WP Draw Text Editor window or from a regular WordPerfect document window.
- Set text justification options by using the Format, Justification menu options. You can select from Left (Ctrl+L), Right (Ctrl+R), Center (Ctrl+E), Full (Ctrl+J), and All justification. These options are similar to those you apply to text in a regular WordPerfect document window.

→ To learn more about text justification, **see** "Justification," **p. 191**

- Set a font and font size by choosing Format, Font (F9) to open the Font dialog box shown in Figure 15.13. This dialog box is a subset of the standard WordPerfect Font dialog box and functions in much the same way.

→ To learn how to use the Font dialog box, **see** "Fonts," **p. 178**

Note

Because WP Draw uses a "virtual" page size that's much larger than the image is likely to be when printed, the font sizes you select here need to be correspondingly larger. The default font size in WP Draw is 36 points, but if you don't adjust your image size, that font is closer to 12 points when printed.

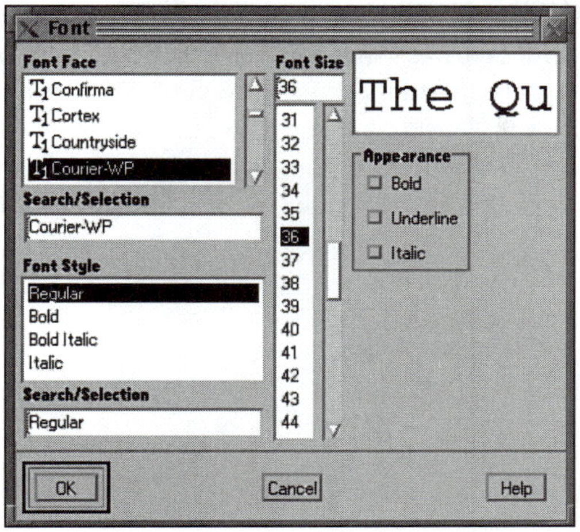

Figure 15.13
The Fonts dialog box is a subset of its usual WordPerfect document window counterpart.

When you're done entering text, choose File, Close to close the WP Draw Text Editor window. When the text appears in the main WP Draw editing area, you can click on the control points to scale your entry. You can stretch it vertically or horizontally or resize it to make it larger or smaller overall.

Tip #190 from *Tux Tips*

If you enter only a short amount of text that doesn't reach the right edge of the text entry area, you may want to resize that text entry area while you're still in the WP Draw Text Editor window. Otherwise, you'll have a text object that's wider than the text it contains. This condition doesn't cause any real harm but could get confusing if you click on a part of the object that has no text, thus selecting the text object when you meant to select something else.

Inserting a Graphic

You can insert a graphics file from an external source within a WP Draw image just as you can insert such a file directly into a WordPerfect document. Inserting a graphic in this way allows you to modify it by adding graphics elements. If the graphic is a WP Draw file (with a .wpg filename extension), then you can delete or modify existing elements, too.

To insert an existing graphics file, choose File, Insert. This step opens a standard WordPerfect File Manager dialog box from which you can select a graphics file. If the WP Draw document on which you're working is currently empty, Draw resets your document's size to match the file you're loading. Therefore, if you want to add elements that require more space, you have to choose File, Page Layout to get the Page Layout dialog box, as shown in Figure 15.9, and adjust the page size.

Are you having troubles loading an EPS file in this way? See "Inserting an EPS File into a WP Draw Object" in the "Troubleshooting" section at the end of this chapter.

You can also insert a WordPerfect Chart into your Draw object. To insert a chart, click the Create chart tool in the Tool palette window (refer to Figure 15.10) or choose Draw, Chart. You can then either click once in the editing area to devote the entire draw area to the chart or click and drag to place the chart in only a portion of that area. WP Draw displays the Create Chart dialog box (Figure 15.14) from which you select the type of chart you want to create. The next step, creating a chart, is covered in Chapter 16, "Creating a Chart."

➔ For more information on creating a chart, **see** "Creating a Chart," **p. 411**

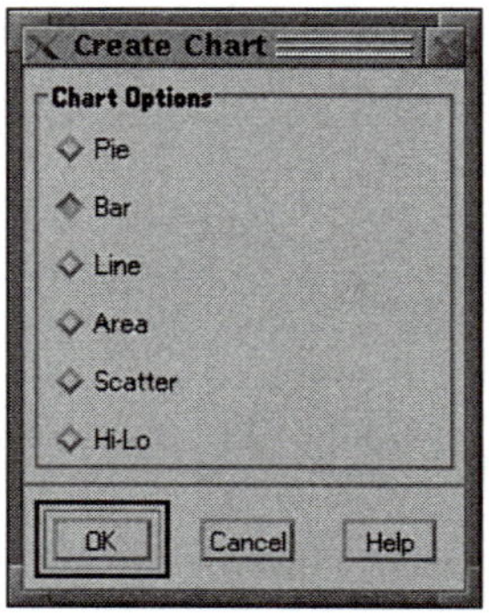

Figure 15.14
Specify the type of chart you want to create from the Create Chart dialog box.

SETTING OBJECT ATTRIBUTES

Creating objects with black lines and solid blue fills, as is the default in WP Draw, can be a good start in creating a drawing, but chances are you'll want to modify these defaults. You may want dotted lines to show a route on a map, or different colors or fill patterns to lend an image a greater sense of realism, not to mention variety.

You can set various attributes for the objects from the Attributes menu on the main WP Draw window. You can set any of these values either before you create an object or after the fact, by selecting the object and then setting the options you want.

COLOR

Set color-related options by choosing Attributes, Color from the WP Draw menu to open the Color Attributes dialog box shown in Figure 15.15. You can set three colors from this dialog box by selecting the Fill Foreground, Fill Background, and Line Color radio buttons. Just click one of these buttons, select a color from the selections to the left, and you're done.

If the colors in the color square don't meet your needs, you can click the Define button to open the Define Color Palette dialog box (Figure 15.16). This dialog box allows you to create a customized color for your every need.

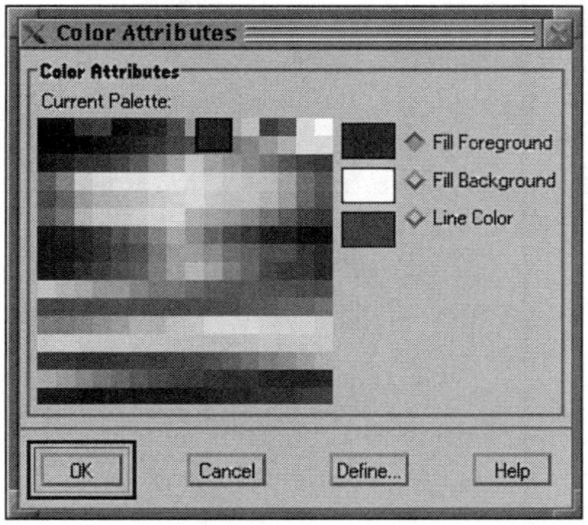

Figure 15.15
The squares next to each radio button preview the color used for that feature.

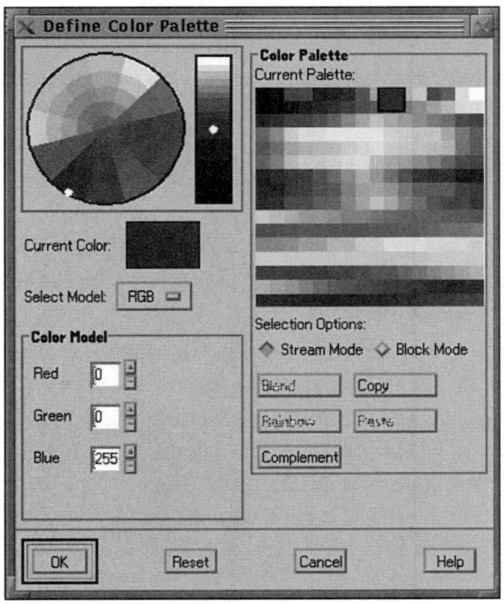

Figure 15.16
Define a color precisely with the Define Color Palette dialog box.

To create a custom color, you must first select a color from the color palette in the upper right of the dialog box. After you modify a color, it's available from the color palette elsewhere in WP Draw. Whenever you're editing, WP Draw displays a sample patch of the current color in the Current Color area, just under the color wheel in the dialog box. You can modify a color in a number of ways:

- Color wheel—You can select a color from the circular color wheel in the upper-left corner of the dialog box. Select a color from the wheel and then make the color darker or lighter by selecting an appropriate level in the bar to the wheel's right.

- Color model—You can specify a color numerically by using one of three color models: red-green-blue (RGB), cyan-magenta-yellow-black (CMYK), or hue-lightness-saturation (HLS). The RGB and CMYK models are both in common use in the computer world. RGB is a way to describe how colors are mixed in a computer's monitor. Selecting `255` for each of the red, green, and blue components describes white, whereas selecting `0` for each of these elements describes black. The CMYK model is used in many color printers, but here the correlation between darkness and numeric values is reversed—you must have values of `0` in each of the four cells to produce white, whereas black is produced by a value of `255` for black or by `255` for each of cyan, magenta, and yellow. The HLS scheme is less common in the computer world. It describes colors in terms of their hue (a value describing the color's position around the edge of the color wheel), their lightness (the position of the color on the vertical bar next to the color wheel), and saturation (the distance from the center of the color wheel).
- Color complement—Click on the Complement button, and the color changes to its complement. A color's complement is its opposite on the hue and lightness portions of the HLS scale. When you click this button, the points identifying the color on the color circle move 180 degrees while the lightness point jumps up or down the scale as if mirrored around a point at the scale's center.
- Color copies—You can select a color and perform a copy-and-paste operation on it. You might use this technique to make both a color and its complement available or if you want to use variants of some original color. First select a color and click Copy. Then select another color that you don't intend to use and click Paste. The first color now occupies both squares on the color palette. You can perform the complement operation or otherwise modify one of these colors.
- Blend and Rainbow—The Blend and Rainbow options work on multicolor selections. If you click and drag in the color palette display, you'll select a range of colors. Precisely how this selection occurs depends on the Selection Options radio buttons; Block Mode selects a square area, whereas Stream Mode selects a range including complete lines between the start point and endpoint (similar to selecting text in a WordPerfect text window). The Rainbow option creates a rainbow-style selection, ranging from the first color to the last. The Blend function creates a color blend, ranging from the first to the last.
- Reset—If you ever feel that you've messed up your color palette beyond redemption, you can click the Reset button to restore the default values.

Note

> The color you select onscreen may not appear exactly the same when printed. You may need to adjust your color use in WordPerfect to account for this discrepancy. Also, if you intend to give your file to somebody else or print it on another printer, the results may vary. If exact color matches are important to you, you need to look into color calibration techniques. Such topics are beyond the scope of this book.

WP Draw has one more color option. This option applies to an entire object, so to try it, first select an object in WP Draw's Editing window. If you choose Attributes, Invert, WP Draw immediately inverts the color selection. The effect is similar to what you find in color negative film—black becomes white, blue becomes magenta, and so on.

Note

If you want to adjust the background color for your image, choose File, Page Layout to get the Page Layout dialog box (refer to Figure 15.9). Click the Color button to get a pop-up selection of colors and then pick the one you want.

Line

When you click Attributes, Line, WP Draw displays the Line Attributes dialog box (Figure 15.17).

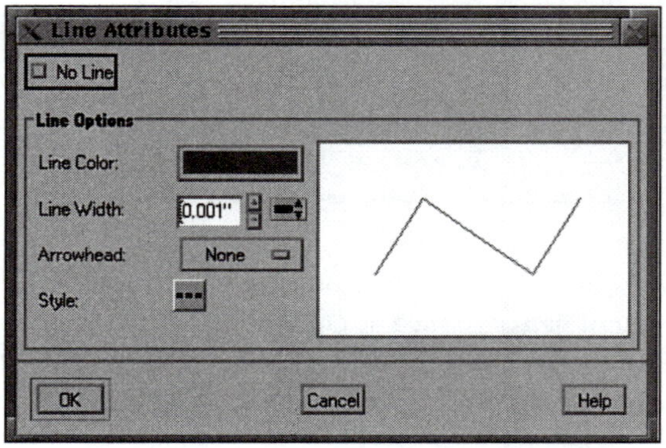

Figure 15.17
Watch the preview area in the right of the Line Attributes dialog box to see what sort of line style you're creating.

From the Line Attributes dialog box, you can set various attributes for the line or object outline:

- Line Color—Click the Line Color button for a pop-up menu of colors. (It's a button, though it just looks like a solid color.) Click on a color to set it.

Note

To achieve greater precision, you can set the line color from the Attributes, Color menu, as just described.

- Line Width—You can either set the width in inches (or other units by specifying *c* for centimeters, *m* for millimeters, or *p* for points) or select from several predefined widths by clicking the pop-up list of options to the right of the spin box.
- Arrowhead—You can set an arrowhead at the beginning, end, both ends, or neither end of a line from the Arrowhead option button. This feature is only meaningful for open lines. With closed shapes such as circles and polygons, it has no effect.
- Style—Select from several predefined line styles by clicking on the button. You can set solid, dashed, dotted, and several variants.

FILL

Choose Attributes, Fill from the WP Draw menu to open the Fill Attributes dialog box, shown in Figure 15.18. This dialog box has three components, selectable from the Fill Type radio buttons. Figure 15.18 shows the first of these, the Pattern fill.

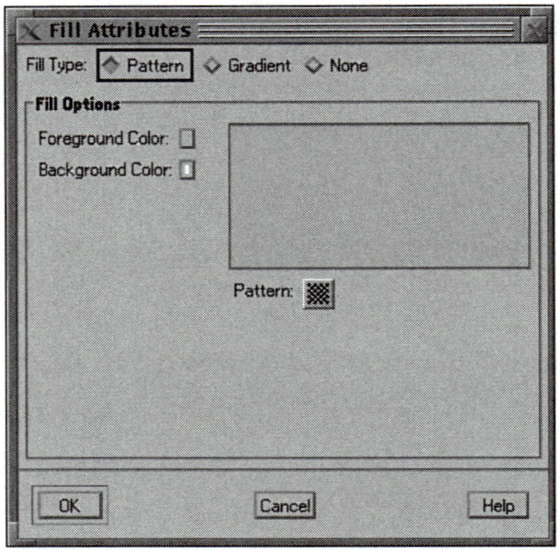

Figure 15.18
You can create an interesting fill pattern from the Fill Attributes dialog box.

Use the Pattern fill to produce a fill based on a particular pattern. Click the Pattern button to see a pop-up menu of patterns, such as checkerboards and diamonds. You can then select colors for the foreground and background components of this fill pattern from the Foreground Color and Background Color buttons.

Caution

The size of the objects that compose the pattern you see on screen may bear little resemblance to the size of these patterns in the printed output. In general, the printed pattern will have a much finer grain than what you see on the screen.

If you select a Gradient fill type, the dialog box's contents change to those depicted in Figure 15.19.

From here you can create a fill pattern that blends in various ways from one color to another:

- Color—You must select two colors. The terms Center Color and Outer Color used in the dialog box apply most readily to Circular and Rectangular gradients with 50% offsets (these terms are described shortly). For other types of gradients, don't be too concerned with these terms—just know that these two colors are blended in some way.
- Type—You can select from Linear, Circular, and Rectangular gradients. Figure 15.19 shows a Circular gradient. The Rectangular gradient is similar except that its pattern of color change follows a squared-off shape rather than a circular one. A Linear gradient shows the Center Color along a single line going through the object and blends to the Outer Color at the points furthest from this line.

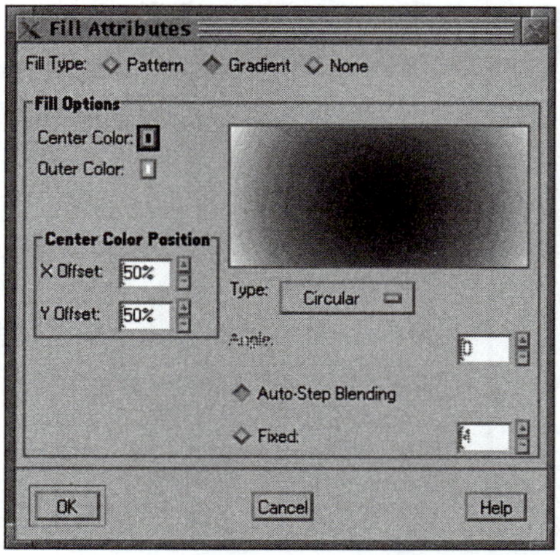

Figure 15.19
Create an object with a color that blends from one color to another with the Gradient Fill Attributes option.

- Center Color Position—Set the position in the object of the center color. If both the X and the Y Offsets are set to 0%, the center color appears in the upper-left corner. Increasing these values moves the offset along the X and Y axes, respectively.
- Angle—This value applies to the Linear and Rectangular types only. It describes the angle of the linear or rectangular pattern of the gradient. You can adjust the Angle value to turn a linear pattern from a horizontal to a vertical orientation, for instance.
- Blending—If you select Auto-Step Blending, WP Draw creates a smooth blend from the Center Color to the Outer Color. If you select Fixed, the blend occurs in the number of steps you specify. You can use Fixed blending to create a bull's-eye effect, for instance, or to create an image of several darkening concentric rectangles.

If you select a Fill Type of None, WP Draw doesn't use a fill in the object, and all the fill options in the dialog box are grayed out.

Shadow

When you choose Attributes, Shadow, the Shadow Attributes dialog box opens (Figure 15.20). From here you can set the color of the shadow and its X and Y Offsets from this dialog box. To set the Offsets, use either the sliders to the right and bottom of the preview area or the spin boxes. If you select Transparency On, WP Draw dims the shadow relative to the color you select.

Tool Palette Shortcuts

The lowest cluster of six tools in the WP Draw tool palette window (refer to Figure 15.10) contains shortcuts to set various color and fill attributes. You can turn on and off lines and fills, set line and fill patterns, and set line and fill colors from these tools. In each case, however, you have only a pop-up menu with a handful of predefined selections. These shortcuts are adequate for many situations, but not for more advanced uses. For instance, you can't set a fill gradient or define a custom color from the tool palette.

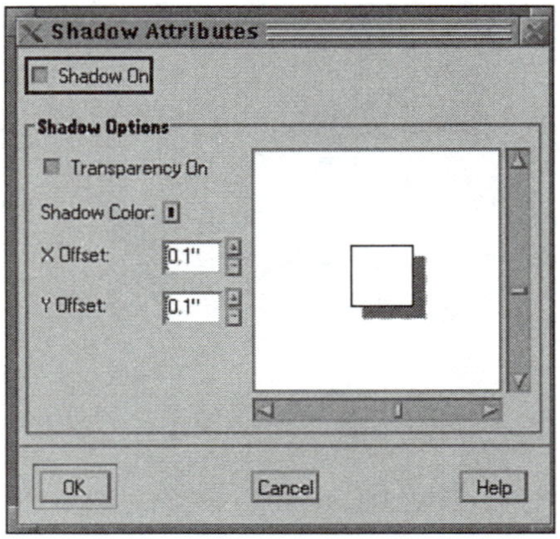

Figure 15.20
Fine-tune a shadow effect from the Shadow Attributes dialog box.

Additional Line and Fill Information

WordPerfect treats text inside a Draw object as a filled object. Therefore, you can use many of the line and fill options to achieve unusual text effects, as in Figure 15.21. Not all of these effects work with text, however. For instance, gradients don't work, though fills do.

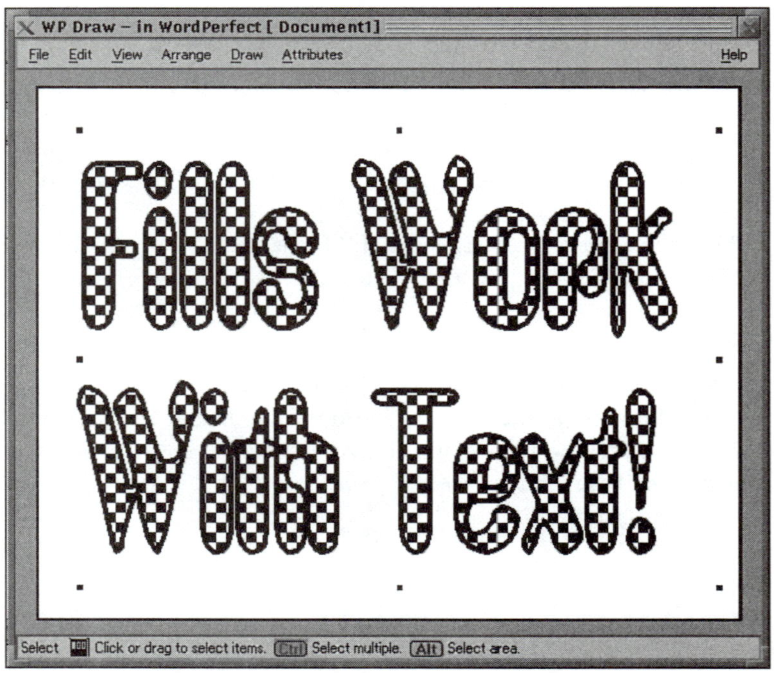

Figure 15.21
Set outline and fill styles for text with WP Draw.

You might think that in the case of line-only objects such as arcs and lines, WP Draw wouldn't let you use the fill-related options. WP Draw is more resourceful, however, and can find a way to *close* a shape—to connect the start point and endpoint to create a shape that can be filled. You tell WP Draw to create such a shape by selecting Fill from the Enable/disable fill tool palette button. When you enable this option, WP Draw may not actually form a line between the endpoints, but the program does fill the intervening area as if there had been such a line. Figure 15.22 illustrates this situation with a filled polyline. This figure uses an extra-thick line style and a light fill so that you can easily differentiate the two.

Figure 15.22
You can fill an open shape in WP Draw.

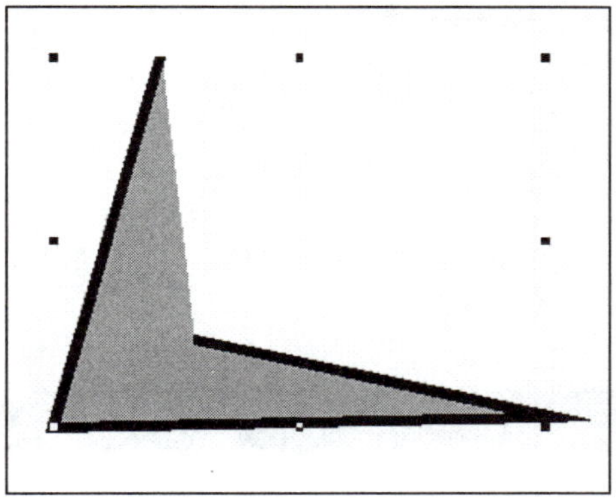

VIEW OPTIONS

If you want to position your objects with improved precision, you can use the options under the View menu. The Ruler option turns on rulers along both the horizontal and vertical axes of the image, and the Grid option displays gridlines at 1-inch intervals. Figure 15.23 shows the WP Draw window with rulers and gridlines displayed. If you select Snap to Grid, WP Draw restricts your control points to 0.25-inch intervals, which can help if you need to position your objects with precision. These three options are all toggles, and they show small squares to their left in the menu. These squares appear indented when the options are selected, much like check buttons in dialog boxes.

When you enable the rulers, WP Draw displays markers showing the cursor's current position, as well as a shadow cursor similar to the one in WordPerfect's main Document window except that this one is X-shaped. Figure 15.23 shows these items (the main cursor has been removed to better highlight the shadow cursor).

→ If you want to know more about the insertion point or shadow cursor, **see** "The Insertion Point and Shadow Cursor," **p. 29**

Figure 15.23
Use gridlines and rulers to position objects consistently and precisely.

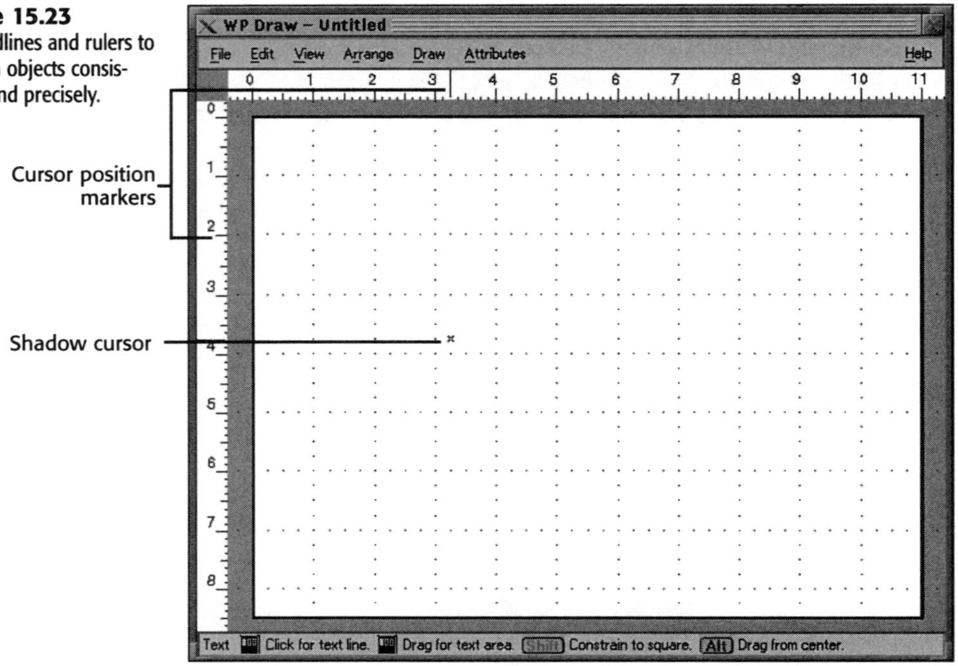

You can further refine the grid options by choosing View, Grid/Snap Options. This step opens the Grid/Snap Options dialog box, shown in Figure 15.24, from which you can adjust the grid's spacing intervals and the units to which grid snap occurs. You can also set special snap lines to align objects to a position that's not normally snapped to by clicking in the ruler. You can remove an existing snap line by selecting it and dragging it out of the ruler.

Figure 15.24
Adjust the details of the gridlines and grid snap options from the Grid/Snap Options dialog box.

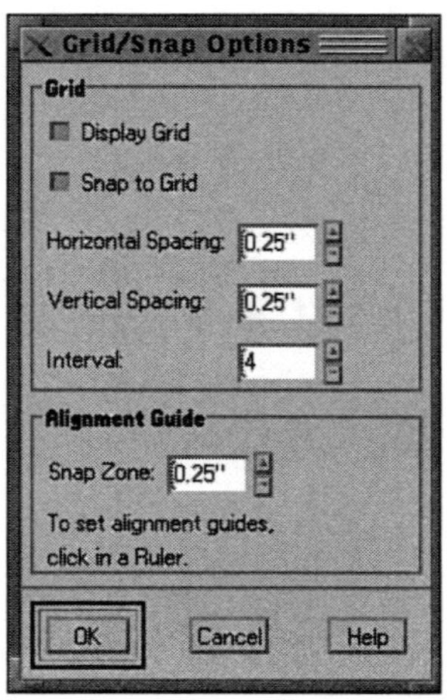

Finally, you can zoom in and out to perform more detailed operations. Choose View, Zoom, Zoom Area to zoom in. Click and drag to select an area you want to examine in more detail, and WP Draw expands its view to show you that area. To see the entire drawing area again, choose View, Zoom, Drawing Area, and WP Draw zooms all the way out. The Zoom item on the tool palette provides similar functionality, but the options are iconic: a magnifying glass to zoom in and a page to zoom out.

Editing Existing Graphics

After you create an object, you can manipulate it in various ways to alter its appearance or the way it interacts with other objects. One of the more obvious types of editing you can do is to alter any of the usual creation options, such as the object's fill color or outline style. To make these changes, simply select the object and then alter the options as previously described. You can do other things, though, to further alter an object's appearance.

Note

If you click an object once, you select it. If you double-click an object, you can edit it. Sometimes an object you're editing seems to move in front of other objects that partially obscure it, but this state is only temporary. This feature lets you access the object's control points when they would otherwise be obscured.

Arrangements

WP Draw provides several options for altering the way objects arrange themselves on the page. When considering these options, remember that WP Draw treats the objects *as* objects, not as pixels on the screen, as a paint program would. You can think of these WP Draw objects as being drawn on transparent plastic sheets that you can stack one atop another and move around to achieve the effect you desire.

- Front or back—You can use the Arrange, Front and Arrange, Back options to move an object to the front or the back of the stack of objects. Each of these options effectively removes the object from the stack and places it at the top or bottom of that stack. If you want to shift an object's position just slightly, you may need to shift several objects to achieve the effect you desire.
- Groupings—If you select several objects by holding down the Ctrl key when selecting them in sequence, you can group them together by selecting the Arrange, Group option. Then whatever you do to one object affects all the objects. This action could be moving them, resizing them, or changing their attributes. If you want to undo this grouping, you can, by selecting Arrange, Ungroup.
- Flips—You can flip an object horizontally or vertically by selecting the Arrange, Flip Left/Right or Arrange, Flip Top/Bottom options, respectively.

CHANGING AN OBJECT'S SHAPE

You can, of course, change an object's size and proportions by clicking and dragging its control points. Depending on the object, though, you may also be able to alter its appearance in more fundamental ways, sometimes even in ways that aren't options when you first create the object.

Tip #191 from *Tux Tips*

If possible, you may want to avoid adjusting a text object's size in this way, especially if you have several text objects in one drawing and want to maintain the same size for all of them. Instead, edit the text objects and increase or decrease their font sizes. This will result in font sizes that are precisely the same across all font objects.

→ For information on inserting text in a graphic, **see** "Inserting Text," **p. 391**

To edit an object, you can either double-click on it or right-click and choose Edit Points from the context menu. This option enables you to edit the shape's control points—that is, the points that define the shape. You can click and drag these control points to redefine the shape's appearance. For instance, Figure 15.25 shows a closed curve that is being redrawn.

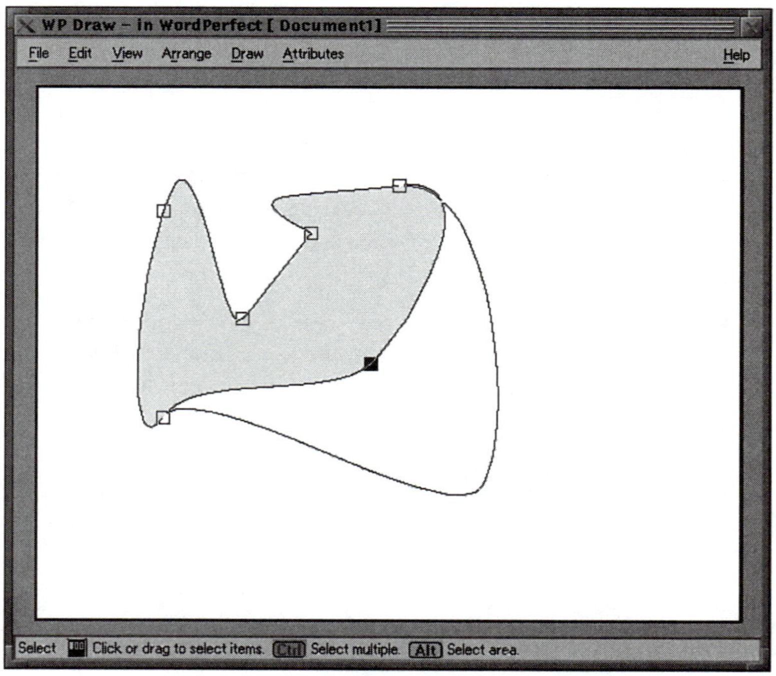

Figure 15.25
If your shape didn't turn out quite the way you wanted, you can redefine it after the fact.

Note

Text objects are a special case. When you double-click on a text object, WP Draw opens its Text Editor window (refer to Figure 15.11) from which you can edit the text in the object. Imported bitmap graphics are another special case because the only control points they have to be edited are the ones for resizing the image.

In addition to editing existing points, you can often add or delete them. Click on a point to highlight it, then right-click on the same point, and you'll see a pop-up menu. The options available differ from object to object but usually include selections to add a new point or delete an old one. An option for many objects is to open or close them. For instance, you can turn a polygon into a polyline, or vice versa.

With open and closed curves, double-clicking a control point produces a pair of control points associated with the first one. These modify how the line curves near the first control point. You can produce a tighter or looser curve around the control point by modifying these secondary control points.

Object Distortions: Skew and Rotate

If you select an object and then choose Edit, Skew or Edit, Rotate, you'll be able to skew or rotate the object, respectively. When you skew an object, you get four control points on which you can click and drag to twist an object so that one edge moves in one direction and another edge moves in another. When you rotate an object, you get five control points. The center one shows the point around which rotation will occur, and you can use any of the other four to rotate the object. Figure 15.26 shows the effect of skew, rotation, and both on text objects.

Figure 15.26
You can create unusual textual effects by skewing or rotating text in a WP Draw image.

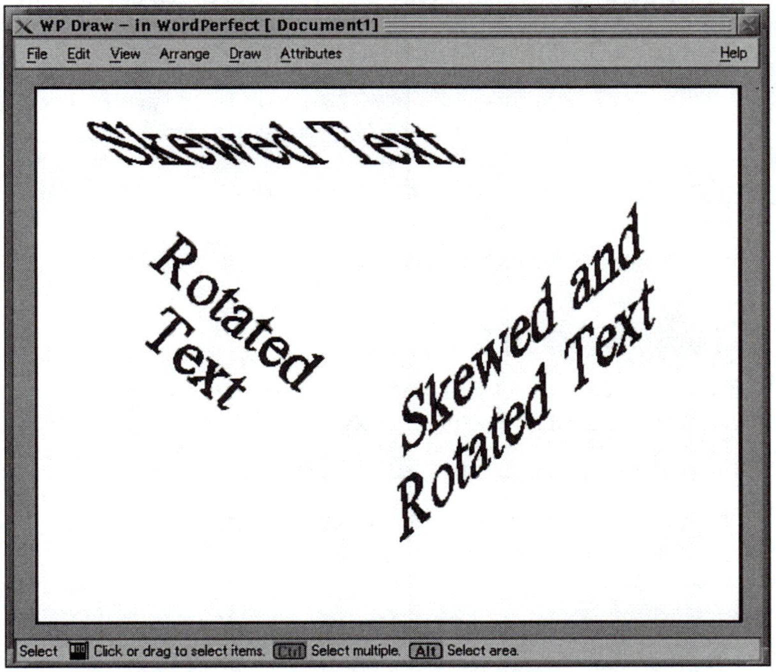

Tip #192 from *Tux Tips*

You can create parallelograms and diamonds from rectangles and squares by using the skew and rotate commands, respectively.

SAVING YOUR WORK

If you've been using WP Draw solely to create an image for one document, you can simply choose File, Exit and Return to [Document1] to close WP Draw and return to your document. (If you've already named your document, the menu option contains the document's filename rather than [Document1].)

Tip #193 from *Tux Tips*

> When you return to WordPerfect, the main Document window may not immediately reflect any changes you made to the Draw object. Try clicking it or scrolling your document to get WordPerfect to show you the changes.

If you want to save your WP Draw document for use in other documents, you may want to use the File, Save Copy As command. This command opens a standard WordPerfect File Manager dialog box, pointed initially at the wpgraphics subdirectory of your WordPerfect installation directory. You probably can't save files there because the permissions will be wrong, so enter an appropriate path and filename for your file and click Save As to save the file. You'll typically want to save the file with a .wpg (WordPerfect Graphics) filename extension so that you can easily identify it as a WP Draw document. In the future, you'll be able to load this file into other documents.

→ To learn how to use a graphic you've saved to an external file, **see** "Inserting Graphics from External Sources," **p. 379**

Tip #194 from *Tux Tips*

> You can use a saved WP Draw document as a watermark, in addition to using it as ornamentation or clip art. This technique is particularly useful if you want to create a watermark with rotated text, such as "Confidential" running sideways down the page, or if you want to place your company's logo or some other graphic in your watermark.

→ For more information on watermarks, **see** "Watermarks," **p. 241**

THE IMAGE TOOLS PALETTE

If you right-click on a WP Draw object or an imported graphic in a Document window, WordPerfect displays the image tools palette, shown in Figure 15.27.

You can use the tools on the image tool palette to adjust various image characteristics:

- Rotate—Use the Rotate tool to rotate an image. The rotation works much as it does within WP Draw itself.

→ To learn more about skewing and rotating images, **see** "Object Distortions: Skew and Rotate," **p. 405**

- Move within box—Use this tool to shift the image within the box. For instance, if you discover that your WP Draw image isn't centered properly, you could use this tool to center the image.

Figure 15.27
The image tools palette lets you manipulate color, brightness, and assorted other image attributes.

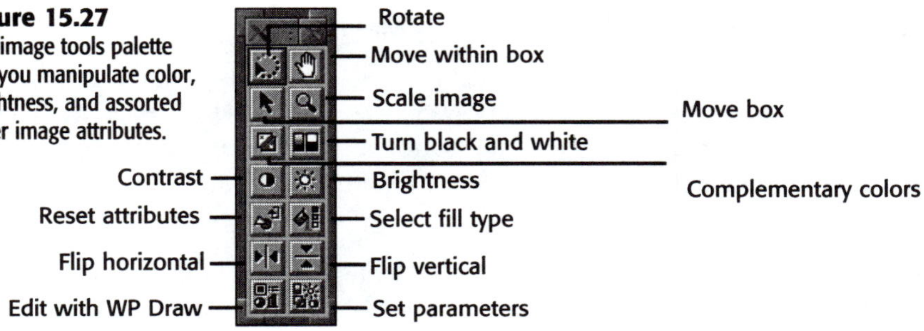

- Move box—Click this tool to return to normal editing after using the Rotate or Move within box tools.
- Scale image—If you click the Scale image tool, you get three options: a magnifying glass icon with which you can select an area to be magnified, a pair of vertical arrows that results in a sliding scale with which you can increase or decrease the magnification, and a 1:1 symbol that restores the image to its normal size.
- Complementary colors—Click this icon to reverse the image's colors, much as with color negative film.
- Turn black and white—Click this button to reduce the color map to two colors: black and white. You can use this tool to create a "line art" effect from a scanned photograph.
- Contrast—You can adjust the contrast of the image, that is, the difference in brightness between the brightest and the darkest areas.
- Brightness—You can make an image brighter or dimmer with this control.
- Reset attributes—Click here to restore the original settings of all the image attributes you set from the image tools palette.
- Select fill type—Select this tool to choose between color and black-and-white fills for WP Draw objects.
- Flip horizontal and Flip vertical—These tools reverse the image horizontally and vertically, respectively.
- Edit with WP Draw—Click here to edit the image with WP Draw. This step has the same effect as double-clicking the image.
- Set parameters—If you click here, you get a dialog box that allows you to set all the attributes controllable from the image tools palette.

Note

The Complementary colors, Brightness, and Contrast tools work only with imported color bitmapped graphics, not with WP Draw objects. The Select fill type tool works only with WP Draw objects.

MOVING AND RESIZING GRAPHICS

Whether it's a simple graphics Shape, an imported bitmap, or a very complex WP Draw document, you can move and resize graphics objects in the regular WordPerfect document window much as you would any other boxed object. You can click on the object and drag it when the mouse pointer resembles a four-headed arrow, or click and drag one of the control points to resize the image.

As with text boxes and other box types, graphics images can have one of three anchors: page, paragraph, or character. In each case, WordPerfect describes the position of the object relative to the object after which the anchor is named. Thus, you should use a page anchor to fix the graphic in one location on a page, a paragraph anchor to associate the graphic with one paragraph, or a character anchor to have the graphic flow with a specific word or character within a paragraph of text.

→ To learn more about different anchoring options, **see** "Moving and Resizing Text Boxes," **p. 362**

GRAPHICS STYLES

If you right-click a graphics object and choose Style from the context menu, the Box Style dialog box opens (Figure 15.28). You can select predefined styles for the graphics box's outline and background from the list in this dialog box.

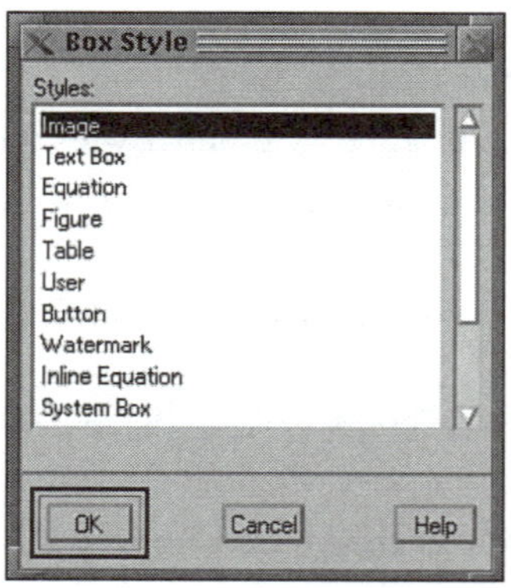

Figure 15.28
Many of the styles in the Box Style dialog box are very similar to one another, though they aren't identical.

If you want to further customize the border around the graphics box, you can right-click on it and choose Border/Fill from the context menu. When the Box Border/Fill dialog box opens (Figure 15.29), you can select from a wider range of border and fill options.

→ To learn more about box borders and fills, **see** "Borders and Fills," **p. 475**

Figure 15.29
Adjust the border around and fill within a graphic from the Box Border/Fill dialog box.

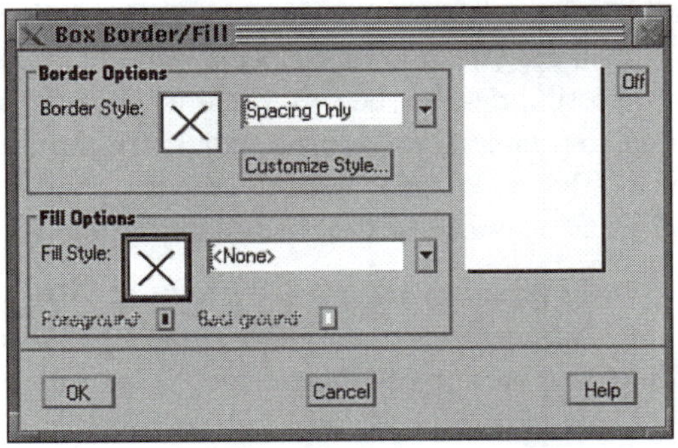

TROUBLESHOOTING

GRAPHICS NOT PRINTING IN FRONT OF OR BEHIND TEXT

You've selected a text wrapping option to place a graphic in front of or behind regular text, but this option isn't working correctly. The onscreen display and the printout may not show the same thing, and one or the other isn't what WordPerfect claims it is.

This problem is a known bug in the way WordPerfect handles layers of text and graphics. Try checking with Corel to see if an updated version of the program is available.

CONSTRAIN TO SQUARE NOT WORKING

According to the directions in WP Draw or in the help files for WordPerfect's Shape feature, to create a square, you should hold down the Shift key when creating a rectangle to constrain it to be a square. You followed the directions, but you keep getting rectangles.

The onscreen prompts are simply wrong in this case. If you're using WP Draw, you can choose View, Snap to Grid to help you create a square, but this method doesn't guarantee a perfectly square shape. Try adjusting the grid snap options to be particularly coarse to make it easier to get a square.

→ To learn more about graphics display options, **see** "View Options," **p. 401**

INSERTING AN EPS FILE INTO A WP DRAW OBJECT

You're trying to include an EPS file into a WP Draw document, but the file doesn't seem to import properly.

WP Draw can't handle an EPS file, though WordPerfect can use one as a graphics object when printing to a PostScript printer. If you must use this file, though, you may be able to convert it to a bitmapped format with *Ghostscript (p. 732)* and then import the bitmap. If your EPS file is called *graphic.eps*, you can convert it at an xterm with the following command:

```
gs -dNOPAUSE -dBATCH -sDEVICE=tiffg4 -r300 -sOutputFile=graphic.tif graphic.eps
```

This command converts the file to TIFF format at 300-dpi resolution, thus changing the raster graphic to a bitmapped graphic. You can select another file format if you like (try `gs —help` for a listing of possibilities, but be aware that most of the ones you see are intended for direct output to printers). You can also adjust the resolution up or down by altering the value following the `-r`. You may need to crop the graphics image using a program such as the GIMP (`http://www.gimp.org`) before importing into WP Draw.

If Ghostscript balks at your EPS file, try using a GUI front end like `gv`, `ghostview`, or `kghostview`. It may be able to coax Ghostscript into producing an appropriate output file. You can also try upgrading your version of Ghostscript. Check with your Linux distribution's Web site, or `http://www.cs.wisc.edu/~ghost`, to see whether a more recent version is available.

CHAPTER 16

Creating a Chart

In this chapter

Overview

Humans are very visually oriented creatures. Give us an image, and we quickly get an intuitive feel for its content and meaning, including sizes, distances between components, and so on. Our visual orientation explains why graphs are such powerful tools for presenting data: We can glance at a graph and immediately grasp at least something of the underlying relationships between its elements. By contrast, a table with perhaps dozens of numbers generally takes us longer to process. The graph and the table may present exactly the same data, but we handle them differently.

WordPerfect 8 for Linux includes the capability to convert numeric data into any of several different types of graphs (WordPerfect calls them *charts*). You may want to use this feature if you're preparing a report or paper that must present much numeric data.

Note

WordPerfect uses the term *chart* almost exclusively, but as a generic term, *graph* does as well, and I use both throughout this chapter. Outside of WordPerfect, one term or the other may be more frequently applied to certain types of figures. The term *graphic*, however, is more general, and can include WP Draw objects, imported bitmaps, and so on.

Tip #195 from *Tux Tips*

You can also import a graph created in another program, using WordPerfect's graphics import capabilities. Creating the chart within WordPerfect allows you to more easily alter the graph within WordPerfect, however. In addition, compared to a graphic you import, a chart created by WordPerfect may look better when printed, depending on the graphics export options available in the other data-graphing package.

→ For more information on importing graphics, **see** "Inserting Graphics from External Sources," **p. 379**

As a type of graphic, a completed chart in a WordPerfect document window behaves just like any other boxed graphic. You may therefore want to read the chapters on graphics and on box-related functions for information on manipulating such objects. In addition, you can move and resize charts much as you do text boxes, so those discussions are relevant to this topic. This chapter focuses on operations *within* WordPerfect's Chart utility.

→ To learn about graphics in WordPerfect, **see** "Graphics," **p. 377**
→ To learn about functions for all boxed objects, **see** "Box-Related Functions," **p. 459**
→ To learn how to move and resize boxed objects, **see** "Moving and Resizing Text Boxes," **p. 362**

Note

As part of WP Draw, the Chart feature is available only in the retail versions of WordPerfect 8 for Linux. If you're using the downloadable version of the program, *none* of the information in this chapter is relevant to what you have. If you need any of these features, you must either use an external program and import the graphic or buy the retail version of WordPerfect.

STARTING A CHART

WordPerfect provides three ways to create a chart:

1. You can create a chart directly from a WordPerfect document window by choosing Insert, Graphics, Chart. This step opens WP Draw in its Chart mode, displaying a default chart (Figure 16.1). Creating a chart this way is fine if all you want is a simple chart (that is, if you don't want to add extra elements to the chart with WP Draw) and if the data to be charted doesn't already exist in a WordPerfect table.

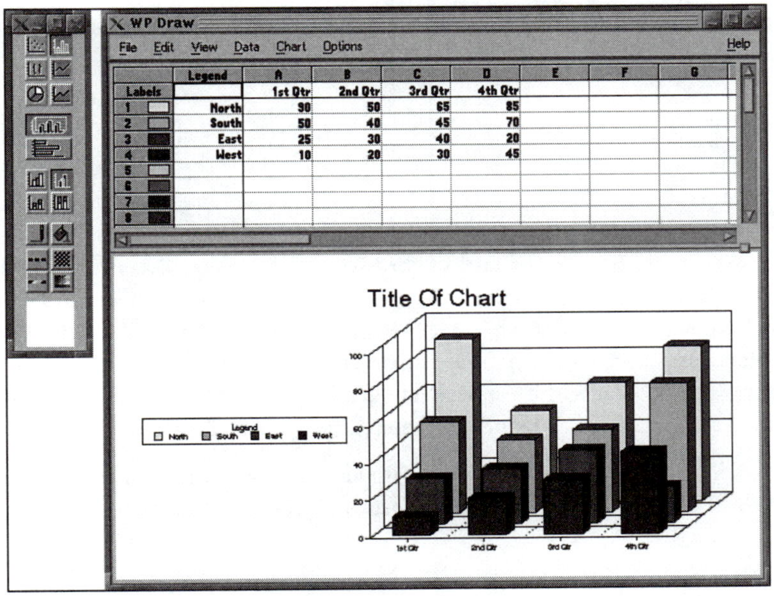

Figure 16.1
The upper pane of the WP Draw Chart window works much like a spreadsheet, whereas the lower pane displays the data in graphical form.

2. You can create a chart from within a regular WP Draw session by choosing Draw, Chart and then either clicking once in the draw area to devote all of it to the chart or performing a click-and-drag to set aside a particular area for the chart. In either case, WP Draw displays the Create Chart dialog box (Figure 16.2). Select the type of chart you want to create and click OK to open a new WP Draw editor for creating your chart.

Note

Whether you create the chart directly from a WordPerfect document window or from WP Draw, you can change the chart type later.

→ For more information on the available chart types, **see** "Types of Charts," **p. 419**

3. If you have data in a WordPerfect table that you'd like to turn into a chart, you can right-click on the table and choose Chart from the context menu. WP Draw opens in Chart mode, using the data contained in the table.

Figure 16.2
You can select the type of chart you want to create when you start one from within WP Draw, but you have to set this option later if you create the chart directly from a WordPerfect document window.

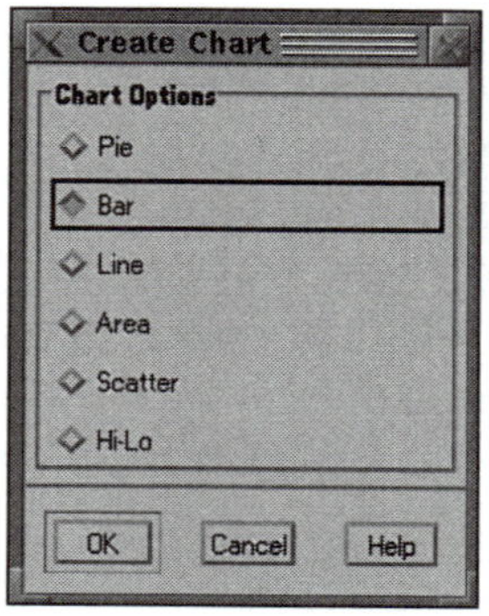

Tip #196 from *Tux Tips*

Right-clicking on the table to get the correct context menu can be tricky. If you have this problem, try clicking over one of the lines between cells in the table or along the table's edge.

Regardless of how you create a WordPerfect chart, you always use WP Draw to edit it, but in a special Chart mode, as shown in Figure 16.1. The small window to the left is the Tool palette window (as it is when you use WP Draw to create a drawing) from which you can select various chart-related functions. The larger window is the main WP Draw window. It's split into two parts: The upper pane displays the data in numeric format, and the lower pane shows the chart that's being constructed from those data.

The Tool palette window (Figure 16.3) is functionally similar to a toolbar in the main WordPerfect document window except that the tool palette is a separate window.

Figure 16.3
The tools in the Tool palette window enable you to create and edit charts quickly.

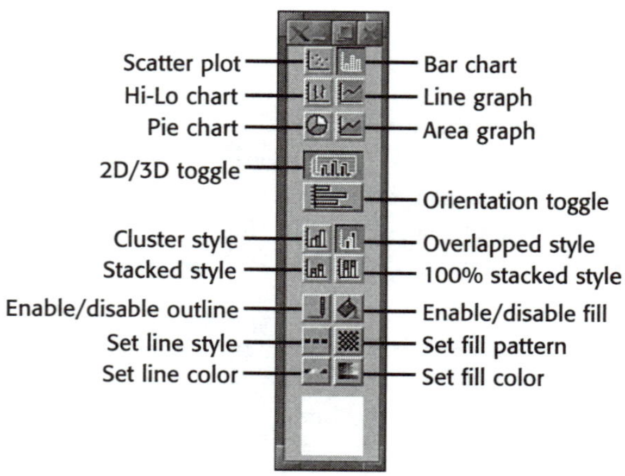

The Tool palette window contains tools that help you create and edit charts:

- Chart type—The cluster of six tools near the top of the Tool palette window allows you to select the type of chart you want to create. You can create bar charts, line charts, pie charts, and others. These options are discussed in greater detail shortly.
- 2D/3D and Orientation toggles—These tools let you turn on or off a 3D appearance to the chart and change its orientation.
- Chart styles—The Cluster, Overlapped, Stacked, and 100% stacked style buttons alter the appearance of your chart. These options are available only with some chart types. They're meaningless, for instance, with pie charts.
- Line and fill options—The cluster of six tools at the bottom of the tool palette let you select various options for line and fill patterns and colors for charted data. These options are active only if you select a row or column from the data pane by clicking to the left of or above the row or column. They're also inactive for certain types of 2D charts

Caution

If you modify an attribute after selecting a column, that change may affect the *entire* chart, depending upon the chart type.

IMPORTING DATA TO BE CHARTED

To do any good, Chart needs data. If you started Chart by right-clicking a table and choosing Chart from the context menu, then the table's data is already loaded and you need do nothing more. Otherwise, you need to enter data somehow. You can do so manually if there's not much data or if it's not in a machine-readable form, but importing the data from an external file is generally easier than entering data manually.

When you start Chart without any data, Chart automatically loads some default sample data (refer to Figure 16.1). You can use File, Clear to remove the sample data if you like. If you import data from another source, WP Draw removes the existing data automatically.

To import data from an external spreadsheet, you should first ensure that the data is in a form that WordPerfect's Chart utility can read. Comma-separated value (CSV) files work well. Lotus 1-2-3 .wk1 and .wk3 files also work. If the data is in another file format, such as a Microsoft Excel .xls file, you may need to convert this file first. You can perform the conversion in the original spreadsheet program or in another program that can read the format for the original data.

Note

CSV files are simple ASCII data files in which columns of values are separated by commas, and each row has its own line. You can create a CSV file (which typically has a .csv filename extension) in an ASCII text editor. Most spreadsheets also have the capability to create CSV files.

Assuming you have a file with data you want to graph, you can import the file by choosing File, Import from the WP Draw main menu. This step opens a standard WordPerfect file selector dialog box. When you click the Import button, you will probably see the Text Import dialog box (Figure 16.4).

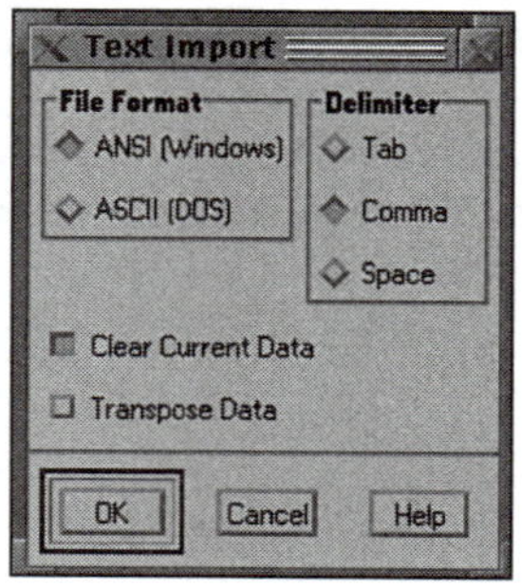

Figure 16.4
Fine-tune your data import options in the Text Import dialog box.

You can use the Text Import dialog box to set options that influence how the data is imported.

- File Format—You can select ANSI (Windows) or ASCII (DOS). For most files, this option makes no difference. If you have problems with certain characters, though, you might want to try both options.
- Delimiter—You can select Tab, Comma, or Space. The delimiter character is what's used to separate columns. In a CSV file, it will be a comma. If you have a data file that uses tabs or spaces, though, you can import it as if it were a CSV file but change this option to handle the file appropriately.
- Clear Current Data—If you check this option (it's checked by default), Chart deletes the current data before importing your file.
- Transpose Data—If you check this button, Chart swaps the roles of rows and columns. This setting determines whether rows are arrayed along the X-axis of the graph with columns being displayed as separate bars or lines, or vice versa. (The data formats for some other chart types are more complex; see the appropriate section on these chart types for details.)

Note

To alter the final chart so that the values in the table are displayed horizontally rather than vertically, use the Horizontal option in the Layout Options dialog box.

→ To learn how to change the orientation of the chart, **see** "Layout Options," **p. 427**

Are you having trouble loading small data files in formats that Chart should understand? See "Importing Short Data Files" in the "Troubleshooting" section at the end of this chapter.

When you import data, WordPerfect always interprets the first row and the first column as labels. For instance, if you're displaying household budget information for the first three months of the year, you might have columns labeled January, February, and March, and rows labeled Housing, Car, Food, Taxes, Utilities, and Misc. Your actual data comprises a 3×6 matrix, and that's how Chart displays it. Your CSV file has *seven* rows and *four* columns, though. If your input data file doesn't have these labels, you may need to edit the data after importing it.

Tip #197 from *Tux Tips*

In the preceding example, a bar chart would show groups of bars, one group for each month. Each group would contain separate bars for each budget category. If you prefer to group budget categories and have separate bars for each month within a group, you can either use the Transpose Data option when you import the data or perform a cut-and-paste operation using transpose, as described shortly, on the data you've loaded.

EDITING THE RAW DATA

You may need to edit the data you've imported to fix typos or correct problems during the importation process. The Chart data editor is straightforward, and you shouldn't have too much difficulty getting the hang of it. Like a WordPerfect table or most spreadsheets, the imported data is laid out in a simple two-dimensional matrix. The individual rows are arranged horizontally, and columns vertically. The intersection of a given row with a given column is called a *cell*, and you can edit an individual cell by clicking it and typing in a new value. After you click a cell, you can't edit just part of its contents; *anything* you type immediately replaces what was in the cell. Therefore, to replace the value `10347` with `10348`, you can't just replace the final digit; you must type the entire replacement value. You can get around this rule, however, by choosing Edit, Edit Cell to open the Edit Current Cell dialog box, as shown in Figure 16.5. From this dialog box, you can use the keyboard arrow keys and the mouse pointer to position the insertion point and edit only part of the cell's contents.

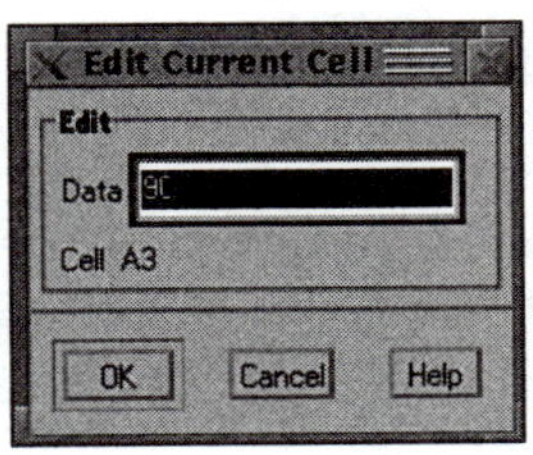

Figure 16.5
The Edit Current Cell dialog box starts with the entire contents selected, so be sure to use the mouse or an arrow key before typing if you want to edit only part of the cell's contents.

You can perform cut, copy, and paste operations on data in a cell by choosing Edit, Cut (Ctrl+X), Edit, Copy (Ctrl+C), and Edit, Paste (Ctrl+V), respectively. If you want to edit a specific cell, you can specify one by choosing Edit, Go To Cell to open the Go To dialog box in Figure 16.6. Type the number-letter identifier for the cell you want to edit, such as `C4` or `E10`.

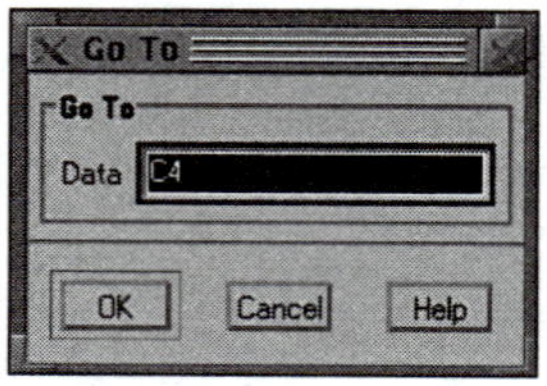

Figure 16.6
The Go To dialog box defaults to the current cell.

In addition to editing the contents of individual cells, you can operate on entire rows, columns, or ranges of cells. Click the label above or to the left of the column or row to select it all or click and drag to select a rectangular region of cells. You won't be able to type values for the entire range, but you can perform cut, copy, and paste operations on it. You can also choose Edit, Paste Transposed to paste a selected region with row and column roles reversed.

If you need to insert a row or column, you can do so with Edit, Insert. If you choose this sequence after selecting a row or column, Chart inserts that item in the selected position, moving the selected row or column down or to the right, respectively. If you choose Edit, Insert after selecting a range of cells, or just one cell, Chart displays the Insert dialog box (Figure 16.7) where you specify whether to insert rows or columns, and how many.

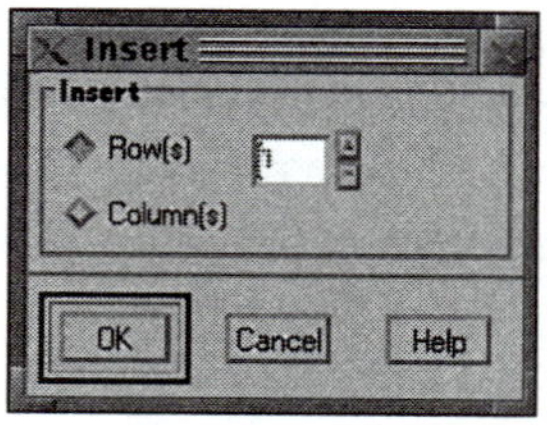

Figure 16.7
Tell WP Draw how many rows or columns to insert with the Insert dialog box.

The Edit, Delete option works much like Edit, Insert but has the opposite effect. If you choose Edit, Delete when an entire row or column is selected, Chart confirms the deletion with a pop-up dialog box and then deletes the entire row or column if you give your approval. If you choose Edit, Delete when you've only selected one cell or a small range of cells, Chart displays a Delete dialog box that looks just like the Insert dialog box (Figure 16.7) except for the title. From here you can tell Chart whether to delete rows or columns, and how many.

Tip #198 from *Tux Tips*

Use the insert and delete options to clean up an imported data file that lacks label rows or columns or that has blank rows or columns.

Types of Charts

WordPerfect's Chart module enables you to display data with various chart types. Which chart type is best depends on the type of data you're presenting. Although you *can* display any data using any chart type, the different types are best suited to different sorts of data. You select a chart type by clicking the appropriate button in the tool palette or by choosing the type from the Chart menu.

Scatter Plots

A scatter plot presents the individual data points as that—individual points (Figure 16.8). In a scatter plot, the values in the data file's columns correspond to points on the plot's X-axis, and each row gets a separate entry in the legend along with a specific symbol or color.

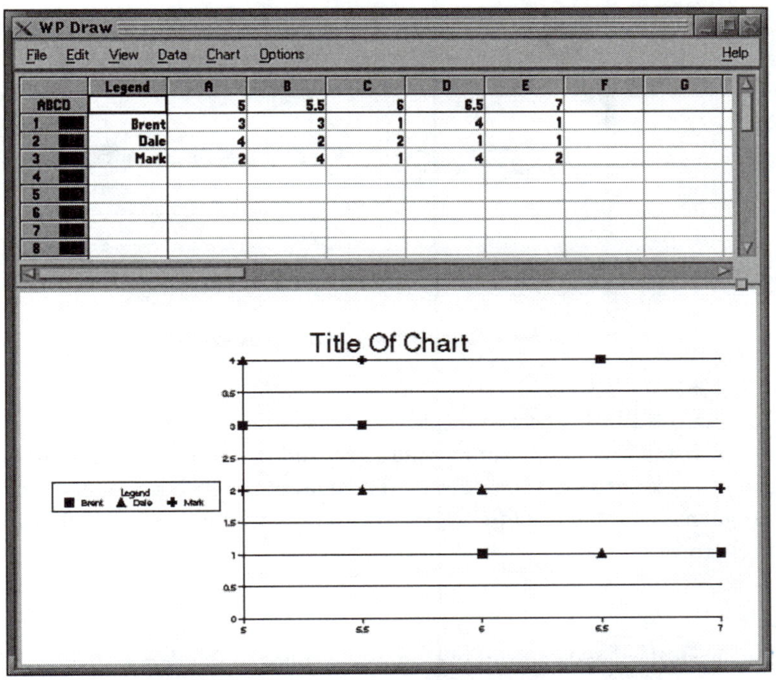

Figure 16.8
Use scatter plots to examine raw data.

Scatter plots are useful for examining raw data before averaging or performing other analyses. They're also good for finding clusters of data points and *outliers* (data points that are radically different from others in a category).

Hi-Lo Charts

A hi-lo chart takes clusters of four values along the rows in the input data spreadsheet and presents these values as individual shapes depicting the data. The first two values are used to plot high and low points for a range, and the next two values show two intermediate values that may be of interest. The result is depicted in Figure 16.9.

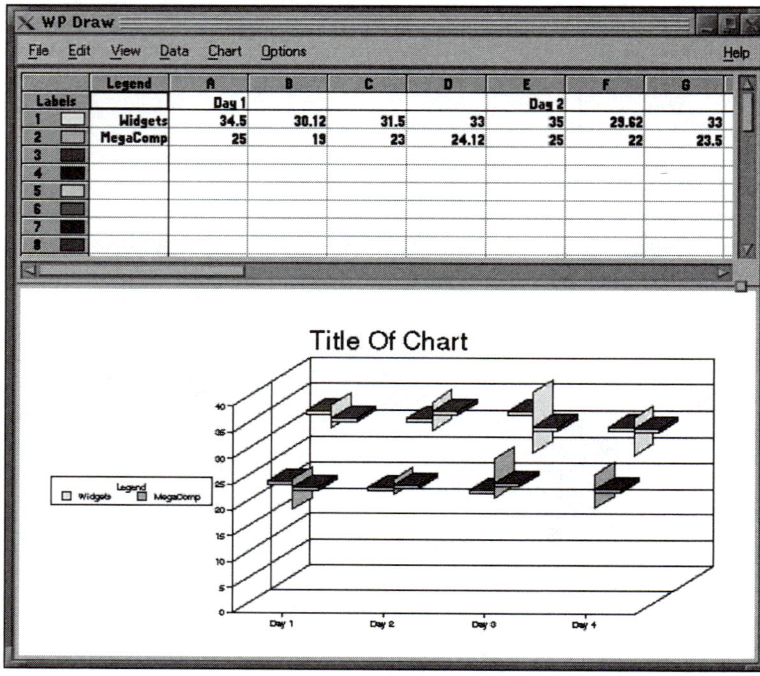

Figure 16.9
Use hi-lo charts to present statistical information in a compact form.

You can use a hi-lo chart to display statistical information in very little space. For instance, the high and low values can be the largest and smallest recorded values during a period, and the two intermediate values can be the edges of the 95% confidence interval for the measurement (a confidence interval is a measure of the region in which the true average is most likely to lie).

PIE CHARTS

Pie charts display data from two-column groups in the data display. Within each group, the first column contains a label and the second a value. In addition, the first column of the data set contains a legend label. The Chart module places each two-column group within a single circular area, segmented into wedges proportional to the values in the second column of the group. The result looks like Figure 16.10.

Pie charts are particularly useful for representing how some fixed resource is allocated or how a population is divided. You might use these in your home budget to see where your money is going. Pie charts don't cope well with displaying more than one dimension of data, however, because each dimension requires its own pie and these can become difficult to compare.

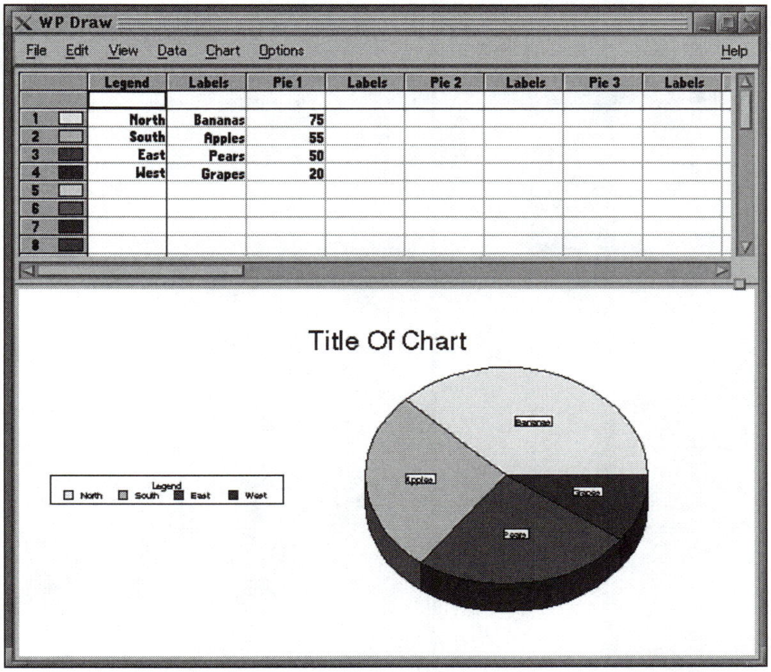

Figure 16.10
Pie charts show what proportion of the total each of several categories represents.

Tip #199 from *Tux Tips*

Consider using bar charts with 100% stacked style if you want to compare allocations in two or more conditions, such as changes in your budget over several months or years.

→ To learn how to set the 100% stacked style, **see** "Layout Options," **p. 427**

Bar Charts

A bar chart presents each data point as a bar rising from the X-axis (Figure 16.11). As with scatter plots, each column of data corresponds to one location on the X-axis, and each row gets its own identifying color in the legend.

Bar charts are frequently used to present comparative data on different groups. For instance, you could show grade point averages across different college majors for several years, with the majors listed across the X-axis, each major holding a clustering of years.

Line Graphs

A line graph uses the same sort of data as a bar chart, but rather than present this data in discrete bars, the line graph displays connected lines, as shown in Figure 16.12.

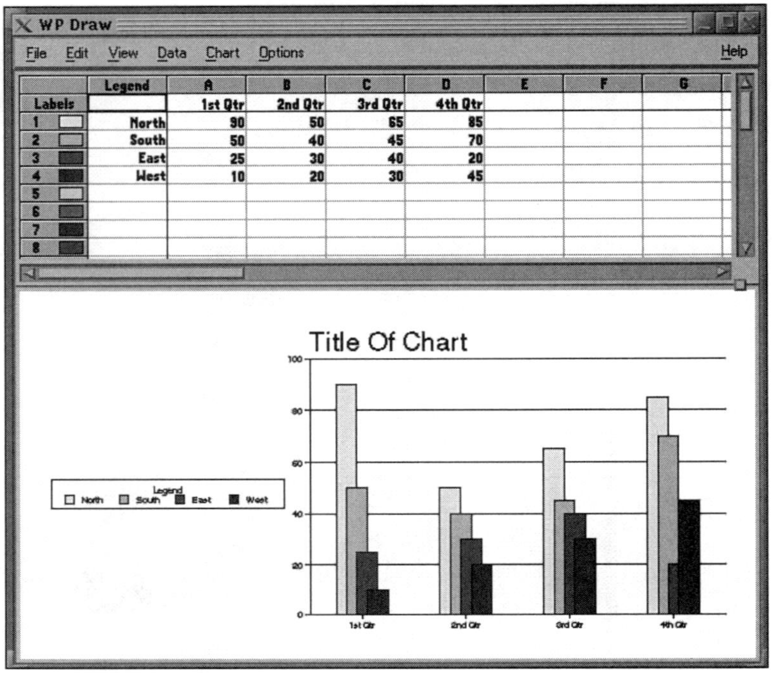

Figure 16.11
Bar charts let you compare discrete groups, as well as groups of groups.

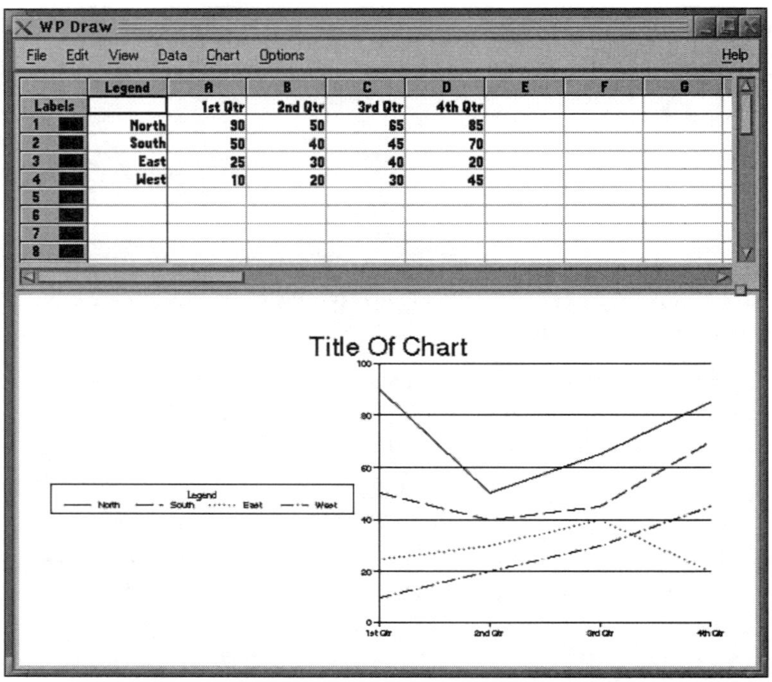

Figure 16.12
Line graphs are useful for displaying continuous variables.

One implication of a line graph is that the points between values on the X-axis have meaning. After all, when you use a line graph, you're displaying values at these points, even if you didn't actually collect data from them. Thus, one rule of thumb is to use bar charts for discrete characteristics (male versus female, different continents, Linux versus Microsoft Windows versus MacOS, and so on) and to use line graphs for continuous variables (age, oxygen content, altitude, and so on). Not all people follow this rule, however, and in some cases following it may make your graph look unnecessarily cluttered, so use your discretion when deciding between these two graph types.

Area Graphs

An area graph is a variant on a line graph in which the area below the line being plotted is filled in (Figure 16.13). An area graph may add some visual interest; however, if your lines cross (as they do in Figure 16.13), you lose some of the data.

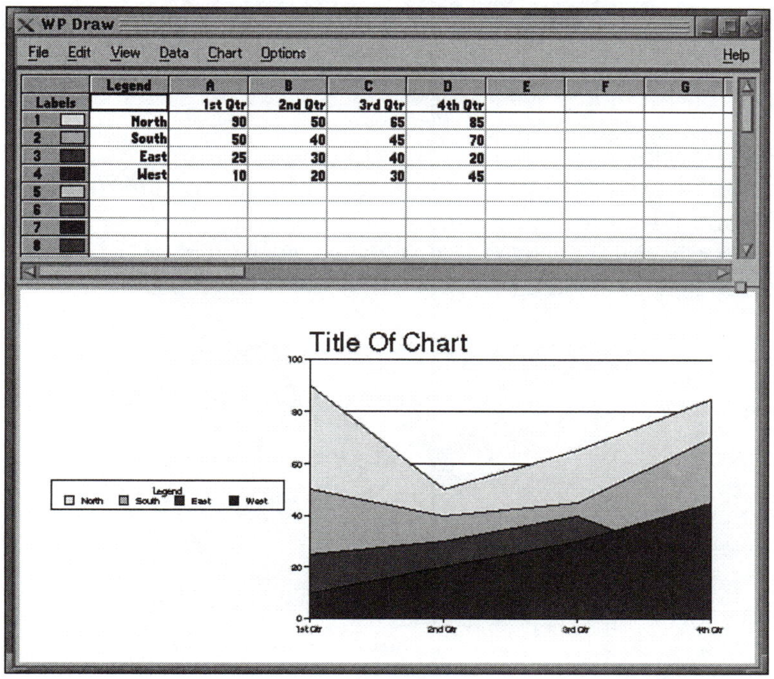

Figure 16.13
Area graphs can add color but at the potential cost of lost data points.

Tip #200 from *Tux Tips*

If you want to use an area graph but have crossed lines, using a 3D style may help to restore some of the "lost" data.

ADDING IDENTIFYING ELEMENTS

Presenting data in a chart is good, but unless your reader knows what the data represent, the chart won't help you convey your ideas. That's why WordPerfect's Chart module lets you identify the information in a chart in various ways.

Note

You can give your chart a caption to describe it in overall terms. Applying captions is a part of WordPerfect's box feature in general, and in fact, you can apply captions to text boxes, graphics, and even equations.

➔ To learn more about captions, **see** "Captions," **p. 466**

Note

Many of the dialog boxes I describe here have a Preview button. Click this button to have Chart redraw the contents of the window with the attributes you've just set. If you subsequently click Cancel, these changes will be undone. Using Preview is a handy way to check what an option will do without having to set it and then go back if you don't like the result.

TITLES

You can give your chart several titles. The most obvious is the title for the chart as a whole. You can also enter a subtitle for the chart, as well as labels for the X- and Y-axes, which Chart classifies as titles. Start by choosing Options, Titles from the main WP Draw menu to open the Titles dialog box, shown in Figure 16.14.

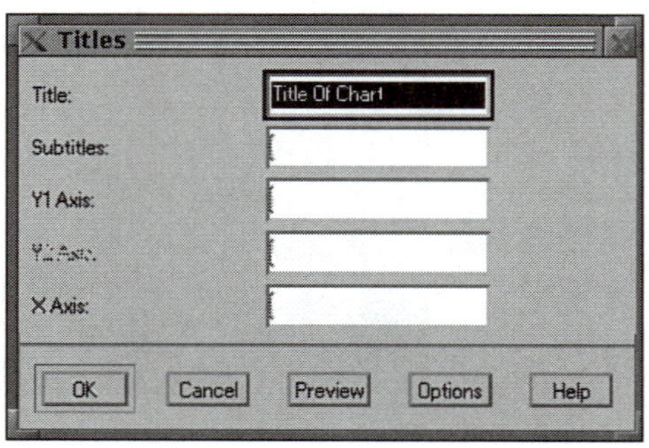

Figure 16.14
You can set several titles from one dialog box.

You can set additional title options by clicking the Options button in the Titles dialog box to open the Title Options dialog box (Figure 16.15). Select the title you want to adjust from the radio buttons at the left.

From the Title Options dialog box, you can adjust the following attributes:

- Position—Set the title's position to the left, center, or right of the chart for the title or subtitle. You can set the axes to display horizontally or vertically.

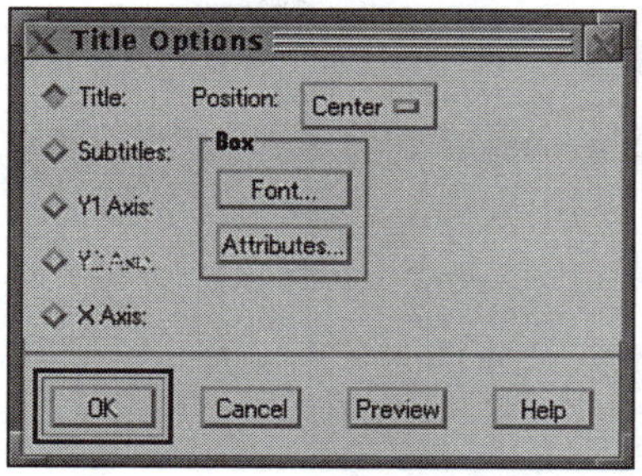

Figure 16.15
Additional title options can be created in the Title Options dialog box.

- Font—Click Font to bring up a subset of the standard WordPerfect Font dialog box and choose a font for the title in question. You can select the font, font size, font color, and font attributes (bold, italic, or underlined) for the title.

➔ If you want more information on selecting fonts and font attributes, **see** "Fonts," **p. 178**

- Attributes—Click Attributes to open the Box Attributes dialog box (Figure 16.16). From this dialog box, you can create a box to surround the title to offset it and provide visual interest. You can set the shape of the box, as well as the border and fill styles and colors. The available fill attributes are similar to the options available for fill attributes for objects in WP Draw.

➔ To learn about fills in WP Draw objects, **see** "Fill," **p. 398**

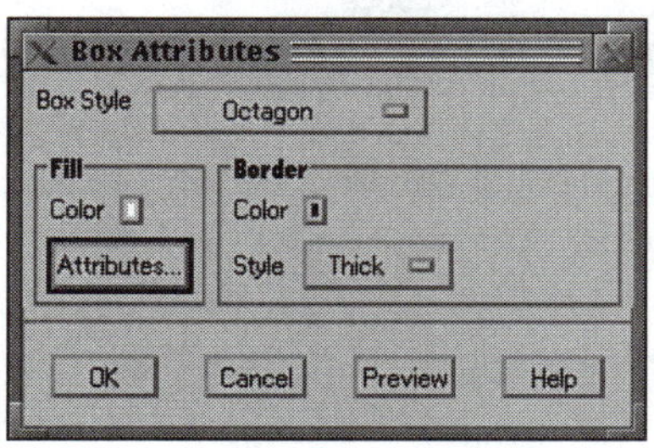

Figure 16.16
Create a border around a title with the Box Attributes dialog box.

LEGENDS

A *legend* (or *key*) is a set of labels to show the reader what each bar, line, or point in a chart represents. By default, WordPerfect's Chart module places a legend to the left of a chart, but you can turn off the legend, place it elsewhere, or modify its appearance in various ways. To change the placement or appearance of the legend, choose Options, Legend to open the Legend dialog box shown in Figure 16.17.

From this dialog box, you can adjust the following aspects of the legend:

- Display Legend—If you want to completely disable the legend, deselect the Display Legend check button.
- Placement—You can place the legend inside or outside the chart area. In either case, you can also specify the location relative to the chart using the Position option button.

Figure 16.17
Create the legend you want from the Legend dialog box.

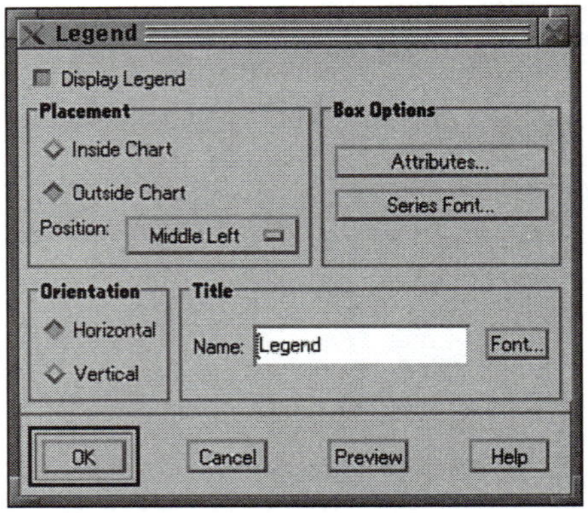

Note

The Inside Chart option frequently obscures the data. This option can work well if your data produces a chart that leaves one corner untouched, however.

- Orientation—You can display the items in the legend next to each other (Horizontal orientation) or above one another (Vertical orientation).

Tip #201 from *Tux Tips*

In general, if you place the legend to either side of the chart, Vertical orientation is more space efficient. If you place the legend above or below the chart, Horizontal orientation often works well, though not if the labels in the legend are particularly long.

- Title—You can set the title for the legend box. It defaults to Legend, but you might want to call it Key or something else. You can also set the font for this title by clicking the Font button.
- Attributes—If you click the Attributes button in the Box Options area, WP Draw opens a Box Attributes dialog box like that in Figure 16.16. Use this dialog box to adjust the appearance of the box that surrounds the legend.
- Series Font—If you click the Series Font button in the Box Options area, WP Draw opens a font selector dialog box. This option controls the font for the individual labels within the legend box (as opposed to the legend box title).

LABELS

In most chart types, the axes have labels. These labels identify the groups or values along their axes, and of course, Chart lets you modify the appearance of these labels. To do so, choose Options, Labels to open the Labels dialog box (Figure 16.18). Select the labels you want to modify from the radio buttons on the left side of the dialog box and then set the options in the remaining areas.

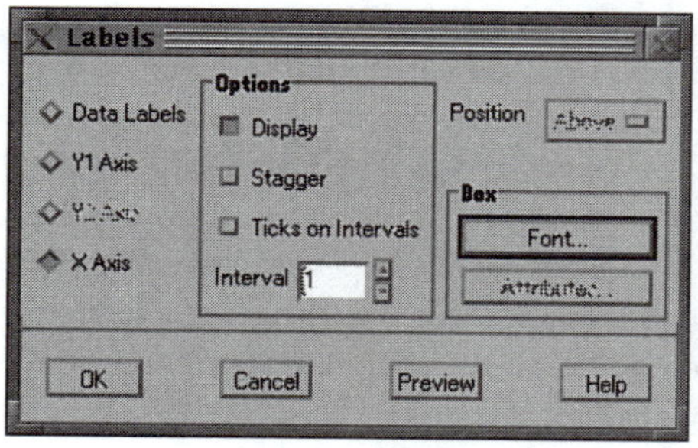

Figure 16.18 The options in the Labels dialog box depend on which label type you're modifying.

Note

In WordPerfect's default layout, only pie charts use the Data Labels. You can enable these for other chart types, however. A small box then displays the exact value of the plotted point, which may be desirable if you want your readers to see this information easily. It can clutter the display if the chart has a lot of points, however.

FINE-TUNING THE CHART'S APPEARANCE

WordPerfect provides a number of options that alter the appearance of a chart in various ways. You can use these to make a chart conform to the standards of your field or simply to make a chart more readable or interesting.

LAYOUT OPTIONS

Choose Options, Layout to open the Layout dialog box (Figure 16.19).

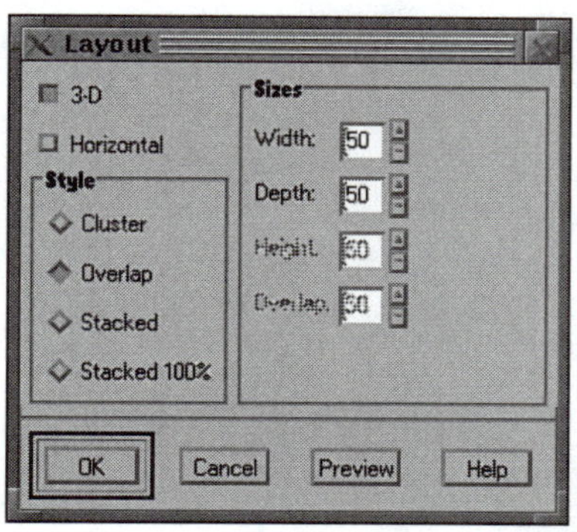

Figure 16.19 The precise layout options available in the Layout dialog box vary somewhat with the type of chart you're creating.

You can adjust the following options from the Layout dialog box:

- Style—The style of the graph affects how Chart plots the data. The styles available vary somewhat from one type of chart to another, but the selection shown in Figure 16.19 is typical. Cluster and overlap are variants on the same type of bar graph display. A stacked bar graph stacks one condition atop another, which might be good for displaying a total and its components. The Stacked 100% option is like the Stacked option except that every value is now converted to a percentage of the total value. With pie charts, you can "explode" a slice—that is, move it away from the rest of the pie for emphasis. Hi-lo charts provide options for displaying data as different styles of error bars or as area graphs, which are rather like line graph conversions of the hi-lo data. The four style tools in the middle of the Tool palette window (refer to Figure 16.3) serve as shortcuts to many of these settings.

- Sizes—You can adjust the sizes of the bars, lines, or other graph elements from the spin boxes in the Sizes portion of the Layout dialog box. As with the Style options, the exact features available vary from chart type to chart type.
- 3D—If you select the 3-D check button, Chart uses perspective to create the illusion of a three-dimensional representation. This effect can be visually interesting, and in some cases it can make your data easier to see. More frequently, however, 3D tends to obscure the data because the depth cues are confusing. For instance, Figure 16.20 shows the same data plotted in 2D and in 3D. The 2D chart on the right clearly shows that the two lines converge on the same point. The perspective cues in the 3D chart on the left, however, obscure the fact that these values are identical (you might even think that the two lines cross). The Tool palette window (refer to Figure 16.3) contains a button to toggle between 2D and 3D representations.

Tip #202 from *Tux Tips*

For some chart types, WordPerfect disables color options when switching to a 2D representation, and instead uses different types of dashes in lines and fills patterns in bar and pie charts (this is illustrated in Figure 16.20). Turning off color can actually be beneficial if you're using a black-and-white printer because colors may come out as identical shades of gray on such devices.

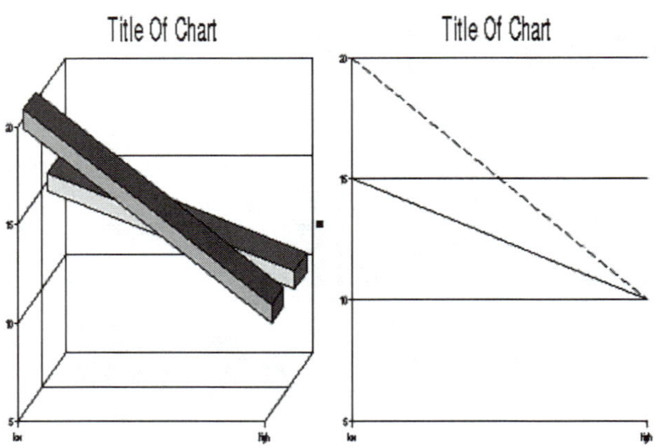

Figure 16.20
Graphs in 3D format can look good, but 3D also tends to make it more difficult to interpret the data.

- Horizontal—The default orientation places the Y-axis vertically. In some disciplines or for some data, however, you may want to place the X-axis horizontally. To make this change, click either the Horizontal check button or the Orientation toggle tool in the Tool palette window (refer to Figure 16.3).

AXIS OPTIONS

Choose Options, Axis to open the Axis Options dialog box (Figure 16.21). From this dialog box, you first choose the axis whose values you want to adjust. Then you can set various options.

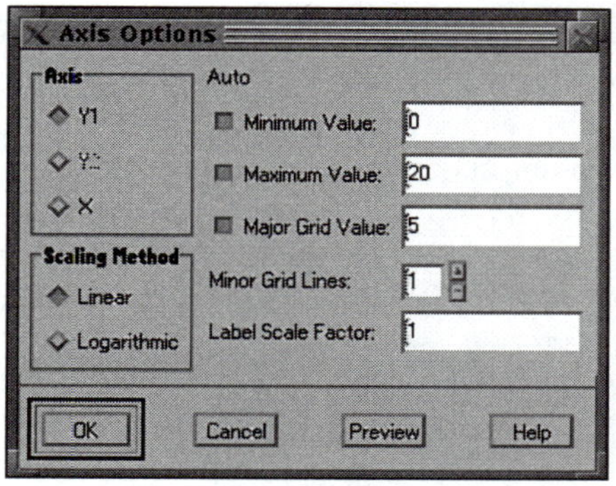

Figure 16.21
Use the Axis Options to adjust the scaling of your chart.

From the Axis Options dialog box, you can set the following options:

- Scaling Method—You can choose linear or logarithmic scaling for the chart's axis. Linear is the default. You probably already know whether you need to use logarithmic scaling. It compresses data as the values get higher and higher.
- Minimum and Maximum Values—You can tell Chart to display a particular range of values if you like. You might use this feature for data that's restricted to a small range at high values. For instance, if you're charting the value of new jetliners over the past few years, you won't have many values of $100 on the chart, so you can cut off the lower part of the scale to better display the actual differences. If the check button to the left of these values is selected, Chart chooses values automatically.

Tip #203 from *Tux Tips*

Another option is to place the legend in the unused portion of the chart, which could also result in a larger graph area, thus producing the same effect. This approach has the added benefit of not overplaying differences, which may look bigger than they are if you restrict the range of a scale.

- Major Grid Value—This field controls how frequently Chart places a line inside the figure marking a major grid point. As with the Minimum and Maximum Values, if you select the corresponding check button, Chart sets this value automatically.

Are you having trouble setting a custom grid value? See "Can't Set Major Grid Values" in the "Troubleshooting" section at the end of this chapter.

- Minor Grid Lines—This field creates unlabeled grid lines between the major grid lines. For this setting to have any effect, you need to select an appropriate grid option, as described shortly.
- Label Scale Factor—WordPerfect adjusts your scale by the factor you specify here. For instance, suppose that your data records distances in meters, but you want to display kilometers in the chart. Enter a value of `1000` here, and Chart will alter the grid values appropriately.

Grid/Tick Options

Choose Options, Grid/Tick to open the Grid and Tick Options dialog box (Figure 16.22). A *grid* line is a line running along a labeled or unlabeled point on the axis. A *tick* is a marker on the axis corresponding to the point in question. If used in moderation, grid lines can help the reader interpret a chart, particularly if precise values on the chart are important. If overused, however, grid lines clutter the chart and make it difficult to interpret.

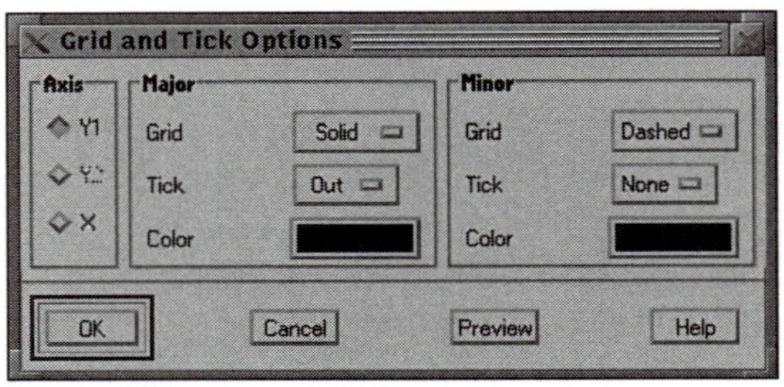

Figure 16.22
Minor grid and tick marks are visible only if you select Minor Grid Lines in the Axis Options dialog box.

You can use the Grid and Tick Options dialog box to set options for the grid line type (solid, dashed, and so on); for whether the tick mark appears inward, outward, or not at all; and for the color of the grid and tick lines.

Tip #204 from *Tux Tips*

To keep from overwhelming your data, use a light color or a thinly dashed pattern for your grid lines, particularly if you use a 2D display. Chart defaults to solid black lines, which can compete with the data lines or points for your attention.

Frame Options

The *frame* is the box that surrounds a chart. You can use the Frame dialog box (Figure 16.23), accessed by choosing Options, Frame, to set options for the chart's frame. You can turn on or off the lines or planes that compose a chart's frame, and you can set various attributes for its colors and patterns. If you want to give the chart a raised appearance, you can enter a value into the Base Height spin box. This option extends the Y-axis downward without applying grid or tick marks to the extension.

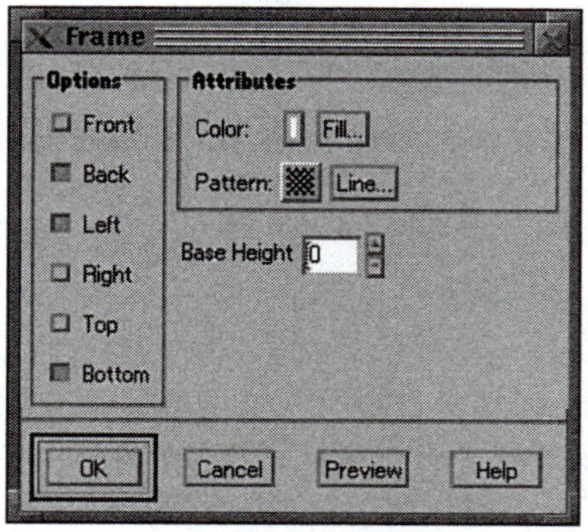

Figure 16.23
Some frame options have no effect on a 2D chart style.

PERSPECTIVE OPTIONS

If you're using a 3D representation, you can choose Options, Perspective to control the position from which you view the chart, using the Perspective dialog box (Figure 16.24).

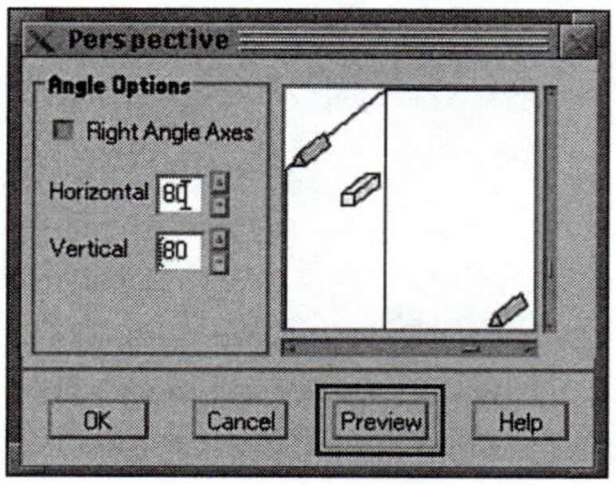

Figure 16.24
The field in the right of the Perspective dialog box gives a restricted preview of the chart with the settings you specify.

If you select Right Angle Axes, then Chart plots the X- and Y-axis at right angles to one another on the two-dimensional computer screen and paper printout. If you deselect this option, Chart provides a more varied range of viewing angles. In either case, these angles are controlled through the Horizontal and Vertical spin boxes. These boxes accept values from `0` through `100`. If you enter values of `0` in both boxes, the view is head-on, producing a result that's very much like a two-dimensional view. Adjusting these values rotates your virtual viewing position to the right and up.

TROUBLESHOOTING

IMPORTING SHORT DATA FILES

You're trying to load a short data file, but Chart keeps reporting that the import failed, especially when it's a simple CSV file.

Chart's import feature requires a CSV file that is at least 100 bytes long. You can easily pad a CSV file with a text editor (including WordPerfect—just be sure to save the file as ASCII text). Enter as many carriage returns or spaces as you need at the end of the file, and Chart should be able to handle it.

➔ For information on creating an ASCII text file from WordPerfect, **see** "Writing Foreign Files," **p. 90**

Can't Set Major Grid Values

You've tried to set the axis grid values for a chart, but when you click OK, you don't get the grid values you selected, even when the Auto check button is not selected.

This situation is a minor problem with WordPerfect. Go back into the Axis Options dialog box. You'll probably find gibberish in the Major Grid Value field, and the Auto button may be selected. Delete the gibberish, make sure the Auto button is deselected, and type in your value a second time. It will be accepted on this go-around.

CHAPTER 17

EQUATIONS

In this chapter

OVERVIEW

Mathematics has been called the universal language because the basic principles of mathematics are invariant across culture, geography, and time. As WordPerfect for Linux includes support for many different languages, it's only fitting that mathematics be one of them. WordPerfect's support for the language of mathematics resides largely in its Equation Editor feature, which is the topic of this chapter.

Considered as a language, mathematics is very precise, and mathematical equations have very strict requirements for formatting. In addition, mathematical equations frequently contain special mathematical symbols. Some of these are derived from the Greek alphabet, and others are unique to mathematics. For these reasons, the ordinary methods of text entry and formatting for WordPerfect are inadequate for equations. WordPerfect therefore uses a separate program module, known as the Equation Editor, to handle equations. You use the Equation Editor to enter an equation by using the keyboard, mouse, or some combination of the two, and WordPerfect inserts an equation into your document.

Note

Although it's used primarily for mathematical equations, WordPerfect's Equation Editor can be used for displaying other material, such as chemical reactions.

The inserted equation functions as a single boxed element within the WordPerfect document as a whole. You can position an equation at a fixed point on a page, relative to a paragraph, or within a paragraph of text. You can perform operations on an equation similar to the operations you typically perform on other boxed elements. By default, equations use character anchoring, meaning that they behave as if they were single characters in the document as a whole and can therefore be placed within a paragraph.

→ To learn about functions that apply to all boxed elements, **see** "Box-Related Functions," **p. 459**
→ To learn how to move and resize boxes, **see** "Moving and Resizing Text Boxes," **p. 362**

Note

The downloadable version of WordPerfect 8 for Linux does not include the Equation Editor. If you're using that version of WordPerfect, you'll still be able to load and edit documents that contain equations, but you won't be able to modify the equations or create new ones. If you must use equations, then, you have three choices: You can purchase the retail version of WordPerfect; you can construct an equation with WordPerfect's regular text-editing features, including its capability to insert special symbols into ordinary text; or you can create your equations in another program and import them in a graphics file format.

→ For information on placing symbols in ordinary text, **see** "Inserting Special Characters," **p. 102**
→ For information on importing graphics files, **see** "Inserting Graphics from External Sources," **p. 379**

If you're familiar with TeX or LaTeX, you'll find WordPerfect's Equation Editor similar in principle to these packages' equation-handling features but substantially different in detail.

UNDERSTANDING EQUATION PRINCIPLES

Because mathematical equations present many unique formatting requirements, WordPerfect's Equation Editor works very differently from a regular document window. In fact, working with the Equation Editor involves describing an equation in words similar to what you might use when reading an equation to another person. It helps to understand the rules for this description before attempting to create equations.

THE EQUATION EDITOR WINDOW

To start the WordPerfect Equation Editor, choose Insert, Equation from a document window. This step opens the Equation Editor window, shown in Figure 17.1 displaying a simple equation.

Figure 17.1 WordPerfect's Equation Editor is split between a non-WYSIWYG editing pane and a WYSIWYG display pane.

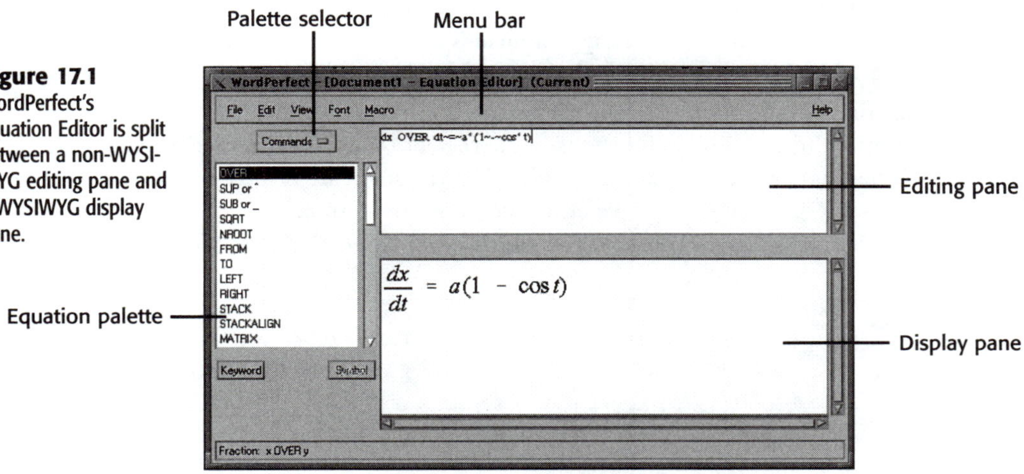

The Equation Editor window has several important components:

- Menu bar—This menu bar is like that in other WordPerfect windows and dialog boxes.
- Editing pane—You type or otherwise enter a description of the equation into the editing pane. Many of the usual editing rules, such as cursor movement and text selection, apply in this pane. Other normal WordPerfect features, such as the use of centering or italics, have no effect here. (There are other ways to achieve many of these effects in an equation, however.)
- Display pane—The display pane shows you how your equation will appear in your document. WordPerfect doesn't update the display pane automatically, however; you must choose View, Redisplay (Ctrl+F3) to see the effects of any change you make in the editing pane.

- Equation palette and Palette selector—WordPerfect provides you with a number of commonly used equation commands, symbols, and functions in the Equation palette. If you're unsure of the name for one of these, you can search the appropriate list to find it. You can use the Palette selector to select from among eight different Equation palettes.

KEY EQUATION EDITOR CONCEPTS

When you enter text into the editing pane, you're telling WordPerfect what sort of equation to create. WordPerfect tries to break this input into smaller chunks and to work with those chunks, just as a human breaks down a sentence into smaller chunks (words). Just as words come in different varieties (verbs, nouns, adjectives, and so on), equation components come in several different varieties:

- Commands—Commands tell the Equation Editor to perform a specific formatting action on the following text. For instance, in Figure 17.1, the word OVER in the editing pane is a command that tells WordPerfect to place the preceding text above the following text, separated by a line, as in a fraction or differential. You can select from the available commands by selecting Commands from the Palette selector option button. Table 17.1 summarizes the commands available in the WordPerfect Equation Editor.
- Functions—Functions are standard mathematical operators, such as *log* or *cos*. The Equation Editor displays functions in the regular (nonbold, nonitalic) version of the selected font, as in "cos" in Figure 17.1. You can select from the available options by selecting Function from the Palette selector option button.

➔ To learn how to set a font for your equations, **see** "Setting the Equation's Font," **p. 451**

Note

If you use a function in your equation that WordPerfect doesn't realize is a function, you can force it to appear without italics by using the FUNC command, described later in this chapter.

- Symbols—A symbol is any character that's not a regular letter or number. For example, letters from the Greek alphabet are common symbols in mathematical equations. WordPerfect provides six sets of symbols from which you can choose in the Equation palette, by selecting Large, Symbols, Greek, Arrows, Sets, or Other from the Palette selector option button. You can also insert additional symbols by choosing Font, Symbol (Ctrl+W) to get to the regular WordPerfect Symbols dialog box (Figure 17.2).

Are you trying to insert a special symbol from a particular font, but without success? See "Using Special Math Fonts" in the "Troubleshooting" section at the end of this chapter.

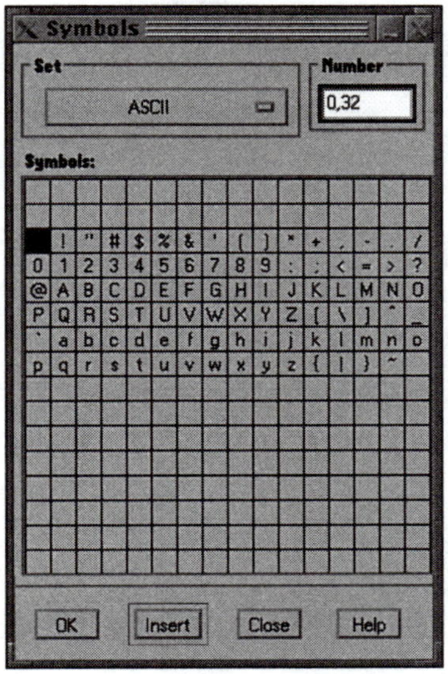

Figure 17.2
You can insert any symbol known to WordPerfect with the Symbols dialog box.

- Keywords—A keyword is a word that describes a command, function, or symbol. For instance, *beta* is a keyword to produce the Greek letter β. With most symbols, you can use either a keyword or the symbol directly in the editing pane. When you select a symbol in the Equation palette, both the Keyword and the Symbols buttons become available to you. If you click the former, you get a keyword, and if you click the latter, you get the symbol. Using the symbol in the editing pane takes less space, but because you can't type it easily from the keyboard, probably takes more time to produce than by using the keyword.
- Variables—A variable is a sequence of letters and numbers that the Equation Editor doesn't recognize as a keyword. For instance, in Figure 17.1, WordPerfect interprets *dx*, *dt*, *a*, and *t* as variables. The Equation Editor displays variables in italics.
- Numbers—Numbers are non-negative integers followed by a symbol, period, alphabetic character, or space. The only number in Figure 17.1 is *1*. Because of WordPerfect's rules for what constitutes a number, you have to perform special formatting to get the Equation Editor to treat negative and decimal numbers in the way you probably intend. More on this topic shortly.

Table 17.1 Equation Editor Commands

Command	Description
OVER	Fraction
OVERSM	Small fraction

continues

TABLE 17.1 CONTINUED

Command	Description
SUP or ^	Superscript
SUB or _	Subscript
SQRT	Square root
NROOT	Nth root
FROM	Limits (displays below previous symbol)
TO	Limits (displays above previous symbol)
LEFT	Left delimiter
RIGHT	Right delimiter
STACK	Vertical stack (multiline equations)
STACKALIGN	Vertical stack with character alignment
MATRIX	Matrix
FUNC	Format without italics
UNDERLINE	Underline
OVERLINE	Overline
{	Start group
}	End group
HORZ	Horizontal move
VERT	Vertical move
~	Normal space
`	Thin space (1/4 of normal space)
BINOM	Binomial
BINOMSM	Small binomial
&	Column separator
#	Row separator
MATFORM	Matrix column format
ALIGNL	Align left
ALIGNC	Align center
ALIGNR	Align right
PHANTOM	Insert empty space
.	No delimiter
\	Literal
BOLD	Bold attribute

Command	Description
ITAL	Italic attribute
LINESPACE	Vertical (line) spacing
LONGDIV	Long division
LONGDIVS	Long division (square symbol)
SCALESYM	Scale symbol

EQUATION EDITOR SYNTAX

Syntax is the set of rules that specify a legal combination of words in a language. The language can be a natural human language, such as English or Russian, or it can be a computer language, such as Pascal or C. WordPerfect's Equation Editor uses its own syntax, and understanding how the Equation Editor interprets entries in the editing pane will help you to create the equations you want.

Note It's possible to create a syntactically correct construction that's nonetheless meaningless or that conveys an unintended meaning. For instance, "the carrot ate the man" is a syntactically correct English sentence, but it's probably not what the writer or speaker intended (except possibly in a fantasy or science fiction story). Similarly, you can create syntactically correct WordPerfect equations that produce unintended results, or that are mathematically nonsensical. WordPerfect won't complain if you create an equation that reads 3 = 2, for instance. On the other hand, you can create syntactically *incorrect* sentences, such as "carrot the man ate the." If you create a syntactically incorrect equation and then choose View, Redisplay (Ctrl+F3), WordPerfect informs you that the equation has incorrect syntax, but won't tell you where the problem is. You'll have to track it down yourself.

The following syntactical rules apply to WordPerfect's equations:

- Commands are not case sensitive. For instance, `OVER`, `Over`, and `over` are all equivalent to WordPerfect's Equation Editor. When you select a command from the Equation palette, WordPerfect inserts it into the equation entirely in uppercase, but you need not follow this convention if you type the command in directly.
- Commands entered as keywords need separation from variables. For instance, if you want to create a fraction, a/b, you would enter `a OVER b`. Typing `aOVERb` instead won't work; WordPerfect interprets this entry as a single variable. A few commands in the form of special characters, such as ~, don't need to be surrounded by spaces.

Note Commands don't need to be separated from numbers that immediately precede them, but they do need to be separated from numbers that follow them. For instance, `7OVER 2` produces the fraction 7/2, but `7 OVER2` produces the number 7 followed by the variable `OVER2`. It's therefore wise to use spaces around *all* your commands, and in fact, when you insert a command with the Equation palette, WordPerfect does just that.

- Commands act on individual equation elements. If you want a command to work on two or more elements, you must group them by surrounding the elements with brace characters ({ and }). For instance, if you want to produce the equation

$$\frac{a+b}{c}$$

you need to include the superscript in braces, as in

```
{a`+`b} OVER c
```

Tip #204 from *Tux Tips*

> If in doubt, use braces. It may be quicker to use braces than to not use them and then go back to edit your equation to correct the problem after you're previewed it or inserted it into your document.

- You can nest groups of braces—that is, you can have braces within braces.
- Braces *must* appear in pairs. If you have a different number of open braces ({) than close braces (}), WordPerfect flags it as a syntax error.

When you create an equation, WordPerfect applies formatting and commands according to *precedence rules*. These rules are similar to the rules you learned in school for the precedence of mathematical operators, such as multiplication coming before addition, but they apply to whatever formatting commands are present in the equation. WordPerfect's order of precedence is

1. Items within braces
2. Diacritical marks (such as tildes, caps, and overbars)
3. Primes, `SUP`, `SUB`, `FROM`, and `TO`
4. Roots
5. `OVER`, `BINOM`
6. Other elements from left to right

CREATING AN EQUATION

WordPerfect supports many equation features and components. These range from simple symbols such as Greek letters to complex combinations of several multiline formulas, such as complex fractions, integrals, matrices, and so on. Some examples of equations follow.

These should help you understand just what can be done with the WordPerfect Equation Editor, and how to do it, though I don't give examples of every possible command.

If you haven't already done so, you can start the Equation Editor by choosing Insert, Equation from a WordPerfect document window. When you're done creating your equation, it will appear in your document where the insertion point rests.

As you use the Equation Editor, remember to choose View, Redisplay (Ctrl+F3) from time to time. This step updates the display pane to show you what the equation you've entered will look like. If you choose View, Redisplay before you've completed an equation, though, WordPerfect may complain about a syntax error, depending on whether the portion you've entered so far is itself a syntactical equation expression.

INSERTING HORIZONTAL AND VERTICAL SPACE

Because the Equation Editor restructures the text you enter in the editing pane to produce the equation in the display pane, you can't count on ordinary formatting tools to work as you might normally expect. This applies even to the most basic of formatting tools, spaces. The Equation Editor, in fact, ignores spaces, tabs, and even carriage returns except for their role in separating different keywords. For instance, if you enter the equation `a + b = c`, WordPerfect removes the spaces and produces a much more cramped-looking equation.

Tip #205 from
Tux Tips

You can use this feature to your advantage when editing large equations. Because the Equation Editor ignores carriage returns, you can break a single lengthy equation line across two lines in the editing pane, but it remains on one line in the display pane and in the finished equation.

If you want to insert horizontal space, use the tilde (~) key or the grave accent (`) key. These two symbols generally appear on the same key on the keyboard. They're often, but not always, located in the upper-left corner of the keyboard, next to the 1 key. The grave accent character is *not* the same as an apostrophe. The Equation Editor inserts a full-sized space for a tilde and a space one-fourth that size for a grave accent. You can use these symbols to control spacing in an equation, and in fact, the equation in Figure 17.1 demonstrates this technique. It's common to use a full space around equal signs and other major equation elements, and a quarter space to separate multiplicands. For instance, you might use ``x~=~2`a`b`` to produce the following equation:

$$x ~=~ 2\,a\,b$$

Vertical spacing in an equation is handled in a number of ways, depending on the equation type. Some commands, such as `OVER` and `MATRIX`, create appropriate vertical space in their formatting. If you want to create two or more lines yourself, for instance, if you're creating a mathematical proof on several lines, you can use the STACK or STACKALIGN commands, as described shortly. You can also use the pound sign (#) to separate lines. For instance,

```
x~=~2`a`b # y~=~3`c`d
```

produces the following equations:

$$x = 2ab$$
$$y = 3cd$$

→ For more information on the OVER command, **see** "Creating Fractions," **p. 443**

→ For more information on the MATRIX command, **see** "Using Matrices," **p. 449**

→ For more information on the STACK and STACKALIGN commands, **see** "Creating Multiline Equations," **p. 452**

Inserting Symbols

WordPerfect's Equation Editor provides easy access to the most frequently used mathematical symbols through the Equation palette. Use the Palette selector to select from one of the symbol palettes (Large, Symbols, Greek, Arrows, Sets, or Other). You can then click on the symbol you want to insert and click the Symbol or, if it is available, the Keyword button to insert the symbol into the equation at the insertion point.

Many, but not all, of these symbols have keywords associated with them. When you select one of these symbols in the Equation palette, the Keyword button becomes active. If that button is grayed out, the associated symbol generally has no keyword. If you type in a symbol's keyword, it displays in the equation just as if you'd selected it from the Equation palette, which can often save some time and effort, if you know the keyword.

Note

Usually, when a keyword is available, you can select it from the Equation palette, as just described. A few symbols, however, have keyword-style abbreviations that do not appear in this way. For instance, -> produces a right-facing arrow, and <-- produces a left-facing arrow.

Note

Most keywords are not case sensitive. For instance, NOT and not both produce the same symbol in a finished equation. The Greek alphabet symbols, however, *are* case sensitive. If entered in uppercase, they produce the uppercase Greek letter, and if entered in lowercase, they produce the corresponding lowercase Greek letter.

If you want to insert a symbol that's not present in one of the symbol palettes, you can select from any of WordPerfect's usual array of symbols by choosing Font, Symbol (Ctrl+W). This step opens the Symbols dialog box (refer to Figure 17.2) from which you can insert any symbol you like into an equation.

→ To learn how to use the Symbols dialog box, **see** "Inserting Special Characters," **p. 102**

Are you trying to insert a special symbol from a particular font, but without success? See "Using Special Math Fonts" in the "Troubleshooting" section at the end of this chapter.

CREATING SUPERSCRIPTS AND SUBSCRIPTS

Many mathematical equations contain superscripts and subscripts. These can be created with the caret (^) and underscore (_) characters, respectively, or by the `SUP` and `SUB` keywords.

These commands are easy to use. For instance, to create the expression x^2, you would enter `x SUP 2` or `x^2`. To create a_4, you'd enter `a SUB 4` or `a_4`.

Note

You should use spaces around SUP and SUB to ensure that they're interpreted as you intend, and not as variables. You can omit these spaces when you use the caret and underscore characters, however, because WordPerfect interprets them specially, even without surrounding spaces.

Like other Equation Editor commands, the superscript and subscript operators work on individual groups, variables, or numbers. Therefore, you must use braces if you want to enter a superscript or subscript that's at all complex. For instance, to create x^{a+1}, you'd enter `x SUP {a ~ + ~ 1}`. If you omit these braces, you get $x^a + 1$ instead of your intended expression.

It's possible to nest multiple levels of superscripts and subscripts and to combine both within a nesting level. For instance, if you enter ``x^{b`a_2}^3``, you get the following equation:

$$x^{b\,a_2^{\,3}}$$

CREATING FRACTIONS

Fractions represent another staple of equations, and they're one of the equation types that can be difficult to do properly without an Equation Editor, especially as the fractions themselves become complex and nested. For starters, consider the following equation:

$$a = \frac{b}{c}$$

This equation can be created by entering `a~=~b OVER c` into the Equation Editor's editing pane. The `OVER` command itself is quite simple: It just places the preceding variable, number, or group in the numerator of a fraction and the following variable, number, or group in the denominator. The only tricky thing is in the grouping. For instance, here's a more complex equation:

$$a = \frac{x + \frac{1}{x + 2}}{\frac{2}{x + 1}}$$

This equation contains three nested fractions, and it can be created using the `OVER` command along with some braces:

```
a~=~{x~+~{1 OVER {x~+~2}}} OVER {2 OVER {x~+~1}}
```

This expression is quite an eyeful, and it's easy to miss a brace when constructing an equation such as this. Once you're used to WordPerfect's Equation Editor, though, you won't have much trouble with such constructions.

Tip #206 from *Tux Tips*

> Make frequent use of the display pane when you're constructing a complex equation. Choose View, Redisplay (Ctrl+F3) whenever you finish a section that should be legal to display, such as just after you finish the numerator of the complete expression in the preceding example. This practice makes it easier to track down problems when they occur. Of course, if you redisplay your equation when it has unbalanced braces or other problems, WordPerfect will complain.

CREATING SUMS AND INTEGRALS

Sums and integrals, those favorites of calculus textbooks, are very difficult to produce in a conventional document window. They're easy with WordPerfect's Equation Editor, though. The general syntax for creating such equations is

```
operator FROM x TO y
```

In this case, `operator` is `SUM` or `INT` for a sum or an integral, respectively. The `x` and `y` values represent the range over which the summation or integration is to occur, and they're displayed below and above, respectively, the summation or integration sign. These two values are optional, though if you have a `TO` value you must also have a `FROM` value.

Tip #207 from *Tux Tips*

> If you want to dispose of the `FROM` value, you can use a tilde (~) to display a space there.

As an example, suppose you want to create the following equation:

$$t = \int_{0}^{100} 6(1 + x)^{-\frac{1}{3}} dx$$

You could create the equation in WordPerfect's Equation Editor by entering the following expression:

```
t~=~INT FROM 0 TO 100 6`(1~+~x)^{-{1 OVER 3}}dx
```

Note

> When you use an integral or sum, *only* the integral or sum symbol and the associated `FROM` and `TO` values are included in that symbol. The equation you're integrating or summing is handled separately, though it will display on the same line, assuming you don't do anything to change that. In the preceding example, then, even the trailing *dx* is separate from the integral sign, despite the fact that the *dx* is a necessary part of the integration.

Tip #208 from *Tux Tips*

> The FROM and TO values appear directly below and above the summation or integration symbols. If you prefer to have these values appear slightly to the right of the symbols, use SUB and SUP instead of FROM and TO, respectively.

When you're dealing with a particularly tall equation to be summed or integrated, you may want to make the summation or integration symbol larger than it normally appears. You can do this with the `SCALESYM` command, which has the following syntax:

```
SCALESYM (% of normal) operator
```

For instance, `SCALESYM 200 SUM` creates a summation symbol that's twice as large as usual. If for some reason you want an extra-small symbol, you can do it by using a value of less than 100.

You can use the `SCALESYM`, `FROM`, and `TO` values with symbols and functions other than the summation and integration marks, as well. For instance, if you need to use limits, you could type something like the following:

```
lim FROM {t -> inf}~x(t)
```

This entry produces the following equation:

$$\lim_{t \to \infty} \; x(t)$$

Creating Roots

You can create square roots in your equations by using the `SQRT` command. The syntax for this command is

```
SQRT {expression}
```

Of course, `{expression}` can be anything from a simple digit to a very complex equation surrounded by braces. For instance, `SQRT 2` creates

$$\sqrt{2}$$

and `SQRT {{x`+`y^2} OVER z}` produces

$$\sqrt{\frac{x+y^2}{z}}$$

The square root is only one of an infinite number of possible roots, however, and WordPerfect's Equation Editor doesn't restrict you to just square roots. You use the NROOT command to create a root other than a square:

```
NROOT {expression 1} {expression 2}
```


The first expression is the root number, and the second is what appears under the radical. As with square roots, either of these expressions can range from a very simple number or variable up to a complete equation in and of itself. You can also nest roots within each other. For instance, entering

```
x~=~NROOT {3.2} {a`+`SQRT {{c OVER d}`+`NROOT 41 b}}
```

produces

$$x = \sqrt[3.2]{a + \sqrt{\frac{c}{d} + \sqrt[41]{b}}}$$

Note

In the preceding equation, the outermost root requires braces around the root number (3.2) because the Equation Editor interprets a period as the end of a number. The NROOT within the square root, however, has braces around neither of its expressions because they are both simple numbers or variables.

USING DELIMITERS

It's common for equations to use parentheses, brackets, lines, and so on to delimit parts of an equation. If you try to enter such characters around extra-tall equation components such as fractions, however, the Equation Editor won't automatically adjust the heights of the delimiters. This condition can produce some awkward-looking equations. To get around this problem, you can use the LEFT and RIGHT commands just before the appropriate delimiter character to expand its height. For instance, you can enter

```
x~=~LEFT ({y`+`1 OVER y} OVER z RIGHT )`a
```

to produce this equation:

$$x = \left(\frac{y + \frac{1}{y}}{z} \right) a$$

The delimiters with which LEFT and RIGHT work are outlined in Table 17.2.

TABLE 17.2 SCALABLE DELIMITERS

Delimiter	Description
(	Left parenthesis
)	Right parenthesis
[	Left bracket
]	Right bracket
< or LANGLE	Left angle bracket
> or RANGLE	Right angle bracket
\{	Left brace
\}	Right brace
¦ or LINE	Single vertical line
DLINE	Double vertical line

USING MATRICES

The Equation Editor's `MATRIX` command formats numbers, variables, or even entire equations in neat rows and columns, much like a WordPerfect table. The syntax for this command is

MATRIX {e_{11} & e_{12} & … # e_{1n} & e_{21} & … e_{mn}}

Each of the `es` in this command is an expression. The values that fall along each row of the matrix are separated by ampersands (`&`), whereas you separate rows by using a pound sign (`#`). Take the following matrix command as an example:

```
MATRIX {a1 & a2 & a3 # b1 & b2 & b3 # c1 & c2 & c3}
```

This command produces the following output:

$$\begin{matrix} a1 & a2 & a3 \\ b1 & b2 & b3 \\ c1 & c2 & c3 \end{matrix}$$

Matrices are usually bounded by straight lines on either side. You add the lines by using the `LEFT LINE` and `RIGHT LINE` commands before and after the `MATRIX` command; for instance:

```
LEFT LINE `MATRIX {a1 & a2 & a3 # b1 & b2 & b3 # c1 & c2 & c3} `RIGHT LINE
```

➔ For more information on the `LEFT LINE` and `RIGHT LINE` commands, **see** "Using Delimiters," **p. 447**

By default, the Equation Editor centers the contents of each matrix column. If you want to create a different alignment for one or more columns, you can include the `MATFORM` command within the matrix. This command itself takes three keywords to specify alignment: `ALIGNL`, `ALIGNC`, and `ALIGNR`, for left, center, and right alignment, respectively. Its syntax is

```
MATFORM {align1 & align2 & … & alignn}
```

The use of the `MATFORM` command will become clearer with an example:

```
LEFT LINE `MATRIX {MATFORM {ALIGNR & ALIGNC & ALIGNL} AAA & BBB & CCC # 23 & 34
& 45 # x & y & z} `RIGHT LINE
```

This input produces the following equation as output:

$$\left| \begin{matrix} AAA & BBB & CCC \\ 23 & 34 & 45 \\ x & y & z \end{matrix} \right|$$

EQUATION FORMATTING

Once you've created your equation, or even while you're creating it, you may want to modify the way it's formatted. WordPerfect lets you set fonts, create multiline equations, and otherwise modify the way it displays an equation.

OVERRIDING FORMATTING WITH THE BACKSLASH

Certain characters, such as the caret (^), braces ({}), pound sign (#), and tilde (~), have special meaning to the Equation Editor, as previously described. You're not out of luck if you need to use one of these characters, though; you can use the backslash character (\) as a way to tell WordPerfect to include the subsequent character in the equation as-is. For instance, suppose you want to use braces in an equation like this:

$$a \;=\; w(x[\sqrt{y}\{z+1\}-2]+3)$$

The following sequence produces the preceding equation:

```
a~=~w`(x`[SQRT y`\{z`+`1\}`-`2]`+`3)
```

You can even use the backslash character on itself to include a backslash in your equation.

SETTING TEXT STYLES

When you're working with the WordPerfect Equation Editor, you can't use the normal WordPerfect commands to create variant text attributes, such as bold or italic. WordPerfect automatically formats variables in italic and all other characters in normal text. If you want to override this, you must use the ITAL, BOLD, and FUNC commands. These commands produce italic, bold, and normal formatting of the equation text. For instance, if you want to create an equation in which subscripts attached to constants and variables are not italicized, you might do something like this:

```
P(t_FUNC g`<=`FUNC C)~=~1`-`{P(t_FUNC s`>`FUNC C} OVER {P(t_FUNC n`>`FUNC C)}
```

This line produces an equation in which the constant `C` and the variables' subscripts are not italicized:

$$P(t_{\mathrm{g}} \leq \mathrm{C}) = 1 - \frac{P(t_{\mathrm{s}} > \mathrm{C})}{P(t_{\mathrm{n}} > \mathrm{C})}$$

Note

The `ITAL` and `BOLD` commands each apply their one attribute to the text. Thus, if you want to format a variable to be regular (not italic) bold text, you must apply *both* the `FUNC` *and* the `BOLD` commands. It doesn't matter in which order you apply them, though.

As commands, `ITAL`, `BOLD`, and `FUNC` apply to individual variables, numbers, or functions, unless you group more than one item with braces. In fact, you can group an entire equation and apply a style to the whole group if you like.

SETTING THE EQUATION'S FONT

By default, an equation displays in the document's default font. If you don't select any fonts in your document, this setting is the default font specified by the printer driver. You can override this setting by choosing Format, Font... (F9) from the document's window and clicking Initial Font from the Font dialog box, as described in Chapter 2.

→ To learn more about setting fonts, **see** "Selecting Fonts and Font Attributes," **p. 46**

Note

You can't use printer fonts for equations. Thus, if your document initial font is a printer font, WordPerfect changes the font to its closest matching graphics font.

Setting your fonts by inserting a font change code in the main document or by using a style (including the document's InitialStyle) does not affect your WordPerfect equations' fonts. Thus, if you want your equations to appear in the same font as the rest of the document, you should set the font via the Initial Font option from the Font dialog box.

→ For more information on setting fonts, **see** "Applying Font Codes," **p. 181**

→ For more information on the document's InitialStyle, **see** "Ensuring Consistency Within Your Document with Styles," **p. 305**

If you want to set an equation's font independently of the fonts used by the rest of the document or by other equations, you can do so by choosing Font, Equation Font from the Equation Editor's menu. This step opens the Equation Font dialog box shown in Figure 17.3. This dialog box is a subset of the usual Font dialog box and allows you to select the font face, style, size, and color.

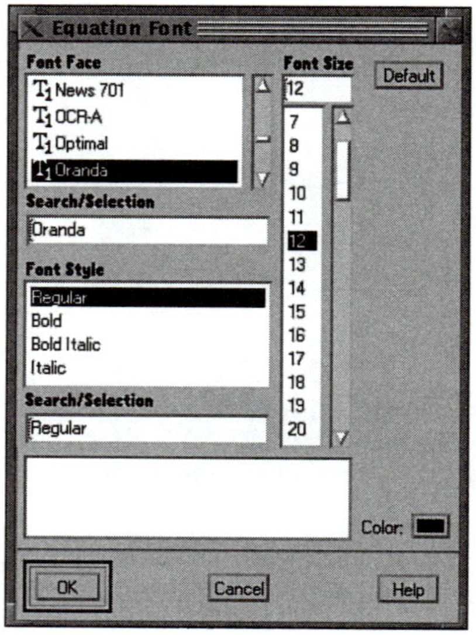

Figure 17.3
The Equation Font dialog box is a subset of the main WordPerfect Font dialog box.

Are you having trouble setting the font for an equation? See "Can't Change the Equation's Font" in the "Troubleshooting" section at the end of this chapter.

Creating Multiline Equations

You may want to create an equation that occupies more than one line. For example, the equation might be particularly complex, or it's really a mathematical proof with each line representing one step in the proof. Whatever the case, you can create multiline equations using the `STACK` and `STACKALIGN` commands. The syntax for the `STACK` command is as follows:

```
STACK {line1 # line2 # … # linen}
```

Each pound sign (#) separates one line of the equation, and each line can contain whatever formatting is needed to enter a complete equation or a portion of one. For instance, suppose you enter this expression for an equation:

```
STACK {x~=~a`LEFT (1 OVER a`+`a RIGHT ) # x~=~1`+`a^2}
```

This entry produces the following equations as output:

$$x = a\left(\frac{1}{a} + a\right)$$
$$x = 1 + a^2$$

By default, WordPerfect centers two or more equation lines within the same equation box. Centering may be good for some situations, but it's not for others. If you want to change the alignment of one or more lines, you can do so by using the `ALIGNL` or `ALIGNR` commands to align to the left or right, respectively. To do so, insert the command at the beginning of the line you want to affect.

You may want to align your equations along a fixed point within each line, such as the equal sign. To do this, you use the `STACKALIGN` command. This command works very much like the `STACK` command, except that you use an ampersand (`&`) immediately after the symbol around which you want to align the different lines. For instance, you might redo the previous equation in this manner:

```
STACKALIGN {x~=&~a`LEFT (1 OVER a`+`a RIGHT ) # =&~1`+`a^2}
```

This variant produces neatly aligned equations, as in a derivation:

$$\begin{aligned} x &= a\left(\frac{1}{a} + a\right) \\ &= 1 + a^2 \end{aligned}$$

Tip #209 from *Tux Tips*

If you're dealing with several lines or more complex equations per line than in these examples, you may find it useful to center each line of your equation on a separate line in the Equation Editor's editing pane. Because the Equation Editor ignores carriage returns in the editing pane, you can split a single equation across multiple editing pane lines quite easily. Simply press the Enter key either before or after each # sign, and you'll be set. This technique will make it easier for you to edit your equation.

One further command is particularly relevant to alignment, and that is `PHANTOM`. This command places space in an equation equivalent to the size of the following group, variable, or number, but it won't actually print the item. For instance, if you include `PHANTOM {x`+`y}` in an equation, WordPerfect inserts a space exactly as wide as `x + y` would be. This command is a handy way to align multiple equation components or to leave space in an equation for something that will be filled in a subsequent line or equation, as you might if you're writing an example of how to solve a math problem.

Equation Editor Display Options

In addition to the many commands that affect how the equation is printed, the Equation Editor provides features that influence how you view and interact with an equation on the screen. These features don't affect the printed appearance of the equation, but can make the Equation Editor itself easier to use.

- Redisplay—As discussed earlier in this chapter, you must choose View, Redisplay (Ctrl+F3) to have WordPerfect interpret the contents of the editing pane and display the results in the display pane. Use this preview to check that your equation is syntactically correct and produces the results you expect before finishing with the equation.
- Zoom—If you choose View, Zoom Display, WordPerfect displays the Zoom Equation Display dialog box (Figure 17.4). You specify the size of the text in the display pane by using this dialog box. The default value is 200%, so you can better make out superscripts, subscripts, and other small features that might get lost on screen but still be legible when printed. You can change the zoom factor to anything you like, however. The Equation Height, Equation Width, and Full Equation options may be of particular interest because they allow your equation to be as large as possible in the space available in the display pane. If you set a large enough zoom factor, your equation might not fit in the pane, but you can scroll through it with the scrollbars.

Tip #210 from *Tux Tips*

If you have a large enough screen, you can expand the size of the Equation Editor's window to get a larger view on the equation.

Figure 17.4
You can manually enter any zoom factor from 10% to 1000%, or select from a predefined or scalable value.

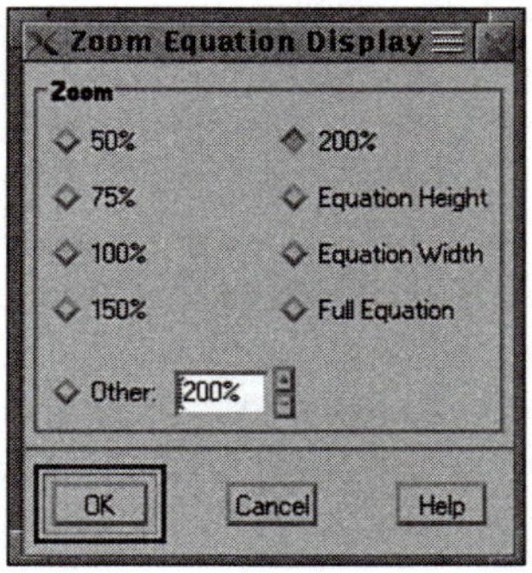

- Palette display—Choosing View, Palette toggles the display of the Equation palette and its associated buttons on or off. When you remove these features, both the editing and display panes expand horizontally to fill that space, giving you a wider view of the equation.

Saving or Retrieving an Equation

WordPerfect provides you with several means of saving and retrieving equations. Some of these techniques operate only on equations within a document, whereas others allow you to store equations on disk.

Inserting an Equation in Your Document

When you're done editing an equation, choose File, Close from the Equation Editor's menu, and WordPerfect inserts your creation into your document.

Caution

If you close your equation and it contains syntax errors, WordPerfect *will* close the Equation Editor, but it *will not* display an equation. That is, you end up with a [Box] code in your document, but it displays only a small and invisible box. So it's best to use View, Redisplay (Ctrl+F3) prior to closing the Equation Editor window to be sure your equation will work as expected.

Tip #211 from
Tux Tips

If you close the Equation Editor and find that your equation doesn't display, the (empty) equation will still take up a small amount of space. Double-click on that space to open the Equation Editor again. If you can't find the equation, use Reveal Codes mode to find the [Box] code corresponding to the equation. Double-click that, and WordPerfect will highlight the empty equation with eight squares to outline its position.

By default, WordPerfect creates equations with character anchoring, meaning that they behave as if they were characters in a paragraph. Such equations move with text. This sort of placement is perfect for equations that reside within a paragraph and may also work well if you want to place an equation on a line by itself. Equations are, however, boxed objects, just like text boxes or graphics. You can therefore replace the character anchors with either paragraph or page anchors, if you like, and position equations just as you do other boxed items. In fact, if you want your equation to reside on a line by itself with a number to identify it, you'll want to change its style from Inline Equation to Equation and give it a caption. This setting changes the equation to use a paragraph anchor, and numbers the equation automatically.

➔ For information on positioning boxes on the page, **see** "Moving and Resizing Text Boxes," **p. 362**

➔ For information on changing a box's style, **see** "Box Styles," **p. 465**

ALTERING AN EQUATION IN YOUR DOCUMENT

To edit an existing equation within a document, double-click on the equation to load it into the Equation Editor. You can then apply any changes just as if you had typed the equation, and when you close the Equation Editor, WordPerfect applies those changes to the existing equation.

When an equation uses character anchor, you can delete it by using the Delete or Backspace keys on the keyboard. No matter which anchor the equation uses, you can select it and use the usual cut, copy, and paste features to delete, duplicate, or move the equation. If you duplicate an equation, any changes you make to one copy will apply only to that copy. You can use this feature to make several variants of an original equation.

SAVING AND RETRIEVING EQUATIONS FROM DISK

If you choose File, Save As (F3) from the Equation Editor, you'll get a standard Save As dialog box and you'll be able to save your equation in WordPerfect's normal file format. You can select other formats, as well. The filename selector suggests an extension of .eqn to identify Equation Editor files, but you can enter any extension you like.

If you choose File, Insert File, WordPerfect displays a File Manager dialog box from which you can select a file to import into your equation. WordPerfect inserts the file you select into any existing equation text; it won't replace what's already there. You may select any file that WordPerfect can import, but if you select a non-equation file, WordPerfect strips much of the formatting and ignores anything that's not applicable to equations. For instance, if the file you import contains bold text, although that text imports as bold, when you choose View, Redisplay or close the Equation Editor, the text is not bold unless a `BOLD` command is applied to it. Figure 17.5 illustrates this situation; the editing pane includes italicized, bold, and underlined text, as well as text in different fonts. The display pane, however, shows an equation in which these different attributes have been stripped and replaced by the attributes determined by the Equation Editor's parser (namely, italic for the words because the Equation Editor interprets the words as variables).

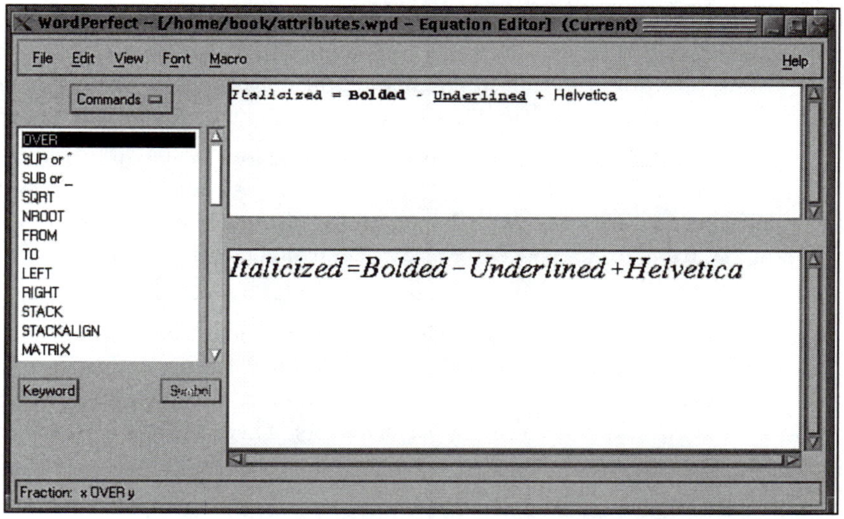

Figure 17.5 The Equation Editor uses its own rules, not WordPerfect font attributes, to determine what attributes to give to equation elements.

You can use the save and insert features to build up a library of equations. If you regularly use certain equations in your documents, just insert them in this way and you'll be set. You may want to consider using this technique even if you regularly create equations that resemble, but are not identical to, one another. It's often easier to modify an existing equation than to create one from the start, especially if it's a complex equation.

TROUBLESHOOTING

CAN'T CHANGE THE EQUATION'S FONT

You've tried changing the font for the equation, but the change just doesn't seem to work properly. Either there's no change you can see, or the font changes to something you didn't want.

Occasionally, WordPerfect for Linux silently refuses to change a font in an equation because the font you've requested lacks the appropriate variant styles. Check to be sure that the font you're using has normal and italic variants, at least. If it lacks one of these, you may be unable to do anything about it. Sometimes WordPerfect will be able to "fake it," though, and create a suitable approximation of the font styles it needs.

This problem could also be a manifestation of a corrupt mapping between font variants. When you install a font in WordPerfect, it's important to check that these mappings are set up correctly, else you may get strange font changes when you select font attributes like bold or italic. Because equation variables appear in italic text, a bad font mapping will be obvious in equations, though it might not be in normal text if you don't use italics much.

→ To learn more about font mappings, **see** "Installing and Managing Fonts," **p. 754**

USING SPECIAL MATH FONTS

You're trying to use a specialty font that contains exotic characters you need for your equations, but with little luck. You either get the normal character set, or you get nothing but the exotic characters from the specialty font. You want to be able to mix and match.

Unfortunately, WordPerfect supports only one font in any given equation, so your options are limited. The best solution is to find a font that contains *both* the exotic characters and the conventional ones you need. You could also check to see whether the symbols you need are available from WordPerfect's Symbol dialog box (Font, Symbol or Ctrl+W).

One other possibility is to take advantage of WordPerfect's font mapping and its limitations, as outlined in the previous Troubleshooting tip. The idea is this: Install your font as a regular style font and do not install any italic or bold variants. Instead, set up the font mapping such that the italic and bold variants map to some other font that you would find acceptable in your equation for conventional characters. Then, when you create your equation, set it to the specialty font. You can then use the `FUNC`, `BOLD`, and `ITAL` commands to control which font is used for which characters.

→ For more information on font mapping, **see** "Installing and Managing Fonts," **p. 754**

CHAPTER 18

Box-Related Functions

In this chapter

OVERVIEW

In the preceding four chapters, I described WordPerfect's *box* features—text boxes, graphics, charts, and equations. Each of these is a type of object that's inserted into a WordPerfect document. These objects can be moved, resized, and ordered when they overlap one another. They can have borders and captions, and ordinary text can wrap around them in various ways. Despite the fact that the different types of boxes are used for very different purposes, they all share these characteristics, and this chapter is devoted to describing most of these. Moving and resizing boxes, however, are described in Chapter 14, "Moving and Resizing Text Boxes," so check there for information on these functions.

When working with boxes, you should keep in mind WordPerfect's method of dealing with them: They're handled as individual [Box] codes in the document, but these codes can be *anchored* to a character, paragraph, or page. Both text and graphics boxes (including charts) default to paragraph anchoring, whereas equations default to character anchoring. You can change any of these anchors, as described in Chapter 14.

Many of the features you need for working on boxes are accessible from the box context menu (Figure 18.1). You open this menu by right-clicking on the box itself. Some of the menu's items change or become inaccessible depending on what type of box is selected or what features the box already sports. For instance, the Edit Image item isn't available for text boxes.

Figure 18.1
The box context menu lets you manipulate many of a box's features.

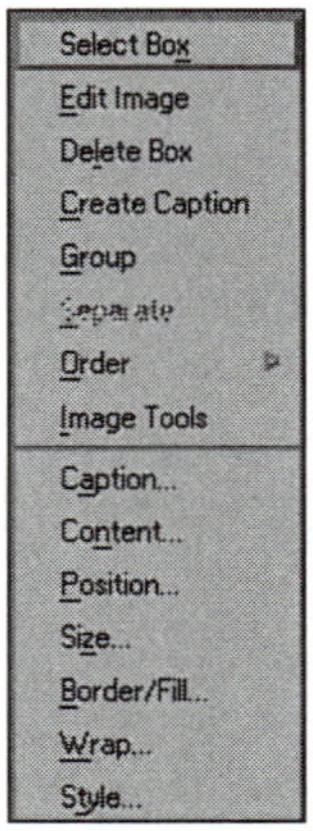

Note

With graphics and equation boxes, you can right-click anywhere in the box at any time to get the context menu. When a text box is already selected, however, you'll get the text-editing context menu if you right-click in the text box. To get the box context menu, first click outside of the box area to deselect the text box and then right-click the box.

Although the downloadable version of WordPerfect doesn't let you create many types of boxes, such as equations and charts, you *can* use the features described in this chapter with this version of the program. Thus, if you have the full version of WordPerfect on one computer and the downloadable version on another, you can still use the latter to modify the formatting of boxed items created on the former computer.

TEXT WRAPPING

Imagine that the text of your document is like a river, flowing from the top of the page downward. A box is then like an object dropped into that river. WordPerfect lets you control how the river will flow around the object. You control the flow by choosing Wrap from the box context menu or by clicking the Text wrapping tool from the box toolbar (Figure 18.2).

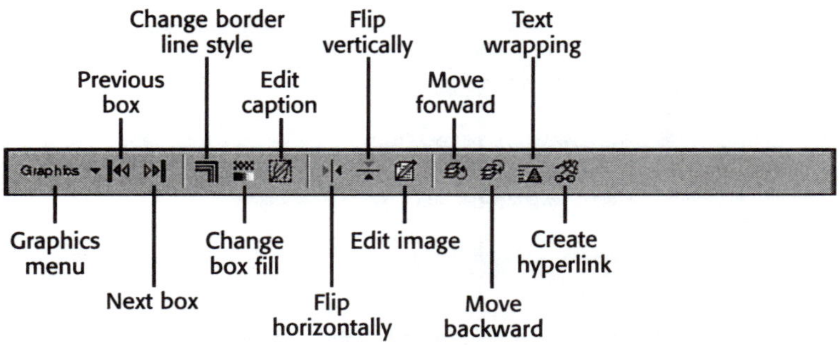

Figure 18.2 The Graphics menu item on the box toolbar produces a menu that duplicates most of the context menu's options.

When you choose Wrap from the context menu, you get the Wrap Text dialog box (Figure 18.3). This dialog box allows you to select one option from each of two text-wrapping characteristics. When you select the Text wrapping toolbar item, you get a single menu in which all these characteristics appear.

Figure 18.3 The Wrap Text dialog box includes icons demonstrating the effect of each option.

WRAPPING TYPE

The Wrapping Type area of the dialog box presents five options that control the main wrapping strategy. The effects of these options are shown in Figure 18.4.

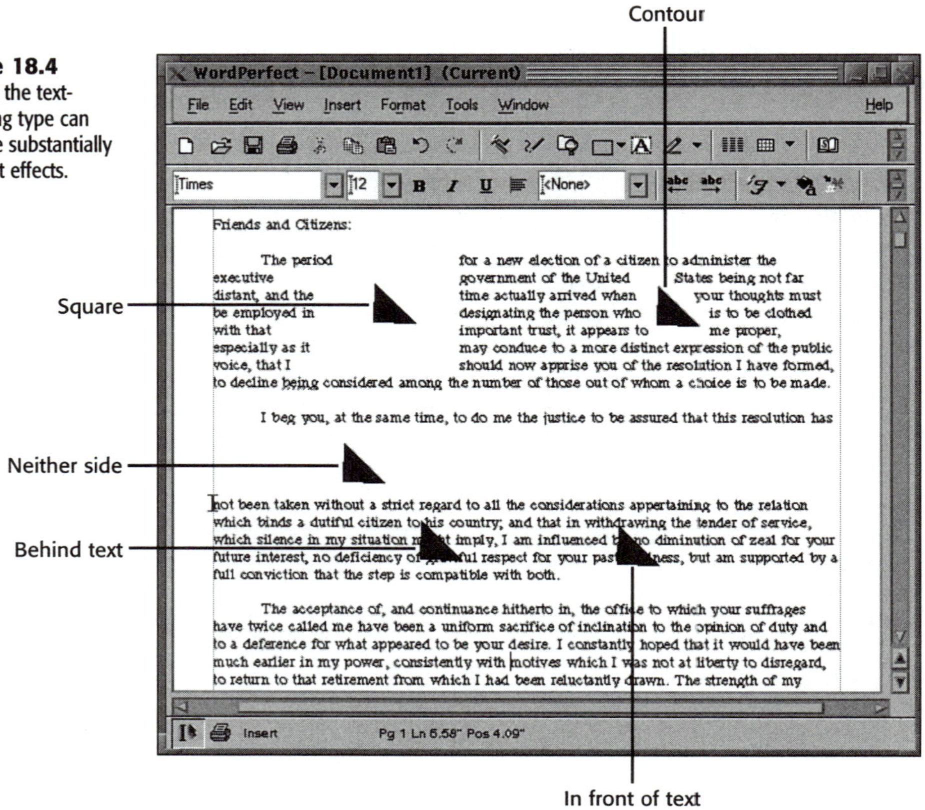

Figure 18.4
Altering the text-wrapping type can produce substantially different effects.

The following options are available in the Wrapping Type area of the Wrap Text dialog box:

- Square—When this option is selected, text wraps around the rectangular outlines of the box, even if the contents of the box are much smaller than the box. For instance, if you have a graphics box that contains a small triangle, WordPerfect wraps text around the entire box, despite the fact that doing so results in a large amount of empty space.
- Contour—When you select the Contour option, WordPerfect moves text in on the area occupied but unused by a graphics box. Thus, if your image is small compared to the graphics box, a Contour wrapping style masks this discrepancy. The Contour wrapping style also causes text to wrap more closely around a shape, mimicking its outline. For instance, notice how in Figure 18.4 the text to the right of the contoured shape starts at different horizontal positions, mirroring the angle of the triangle in the graphic.

Note

The Contour option works just like the Square option for text boxes, equations, and any box that has a border or background fill pattern.

Tip #212 from *Tux Tips*

Turn on hyphenation or right justification to have text to the left of a box follow the contour more closely.

→ To learn how to enable hyphenation, **see** "Hyphenation," **p. 220**

→ To learn how to enable right justification, **see** "Justifying Paragraphs and More," **p. 193**

Note

The Contour option won't work with Encapsulated PostScript EPS files you've imported. If you must use contour wrapping around such an image, you need to convert it to some other format before you import it or re-create the image with WordPerfect's graphics editor.

→ For more information on file imports, **see** "Inserting Graphics from External Sources," **p. 379**

- Neither Side—The Neither Side option causes text to skip over the box entirely, as if it extended from the left to the right margin. You may want to use this option if your box nearly fills the width of the page because a narrow slice of text next to a wide graphic can look rather out of place.
- Behind Text—This option is supposed to place the box "behind" the text—that is, the box won't displace the text, and you should be able to read the text as if it were on a transparent sheet in front of the box.
- In Front of Text—This option, like Behind Text, is supposed to leave the text formatting undisturbed. In this case, though, the boxed item is supposed to block out the text that occupies the same space.

Are you having trouble getting the Behind Text or In Front of Text options to work? See "Behind Text and In Front of Text Wrapping Not Working" in the "Troubleshooting" section at the end of this chapter.

Wrapping Side Options

In addition to the type of wrapping, when you select Square or Contour wrapping, you can select around which side(s) WordPerfect will wrap your text. You set this option with the Wrap Text Around radio buttons in the Wrap Text dialog box, and your options are

- Largest Side—WordPerfect wraps text around whichever side provides the greatest space for text. Using this option usually fits the greatest amount of text into a given amount of vertical space, without having the boxed item interrupt a line of text.

Tip #213 from *Tux Tips*

Use Largest Side if you want to position a boxed item near the left or right side of the page, but don't know which side it will be until after you've done some additional editing. If you need to change the side later, you won't then need to change your wrapping option.

- Left Side—Text appears to the left of the boxed element, and *only* to the left.
- Right Side—Text appears to the right of the boxed element, and *only* to the right.

Tip #214 from *Tux Tips*

Use the Left Side and Right Side options when you want text to wrap around only one side of the text, but when there's space between the box and the opposite border.

- Both Sides—Text wraps around both sides of the box. You'll continue to read text in single lines—this option does *not* create a multicolumn layout—but the boxed element interrupts lines of text. (You can use this option with a multicolumn layout to place a box between columns, however.)

→ To learn more about columns, **see** "Creating Professional Layouts with Columns," **p. 255**

Tip #215 from *Tux Tips*

When you select a wrapping option that creates a very narrow region for text, there may not be enough room for a complete word, especially if the word is long. WordPerfect will then break the word, which can make your text quite difficult to read. The Largest Side and Neither Side options will reduce or eliminate this problem. If you use text wrapping, especially with large font sizes or narrow wrapping margins, you should check your text to be sure you don't create truncated words in this way.

BOX ORDERING

If you create boxes that sit alone on the page, separated from other boxes, you need not concern yourself with box ordering. If your boxes overlap one another, however, you need to consider which ones will appear in front of which other ones.

You can think of boxes as being printed on sheets of transparent plastic, which you can shuffle forward and backward in a stack by using the Move forward and Move backward toolbar buttons (refer to Figure 18.2). In addition, the Order context menu item has four subitems that allow you to change a box's ordering: To Front, To Back, Forward one, and Back one. The last two of these do the same as the Move forward and Move backward toolbar buttons, respectively. The first two move an item to the top or bottom of the "stack" of boxes.

Note

You'll be able to adjust boxes forward and backward even when they don't overlap one another. This adjustment doesn't ordinarily have any effect, but it could be important if two boxes both overlap a third box or if you subsequently move the boxes so that they do overlap one another.

⚠ *Are you trying to place a text box behind some other boxed object? Is it not working when you print? See "Text Box Ordering Doesn't Work" in the "Troubleshooting" section at the end of this chapter.*

If you want to move, resize, or reorder several boxes at once, you can combine them into a single *group*. When you use the group feature, WordPerfect links the boxes to one another and treats them as if they were a single object whenever you move, resize, or reorder them. You still need to treat the boxes as separate objects for most other operations, however, such as adding captions or borders.

To form two or more boxes into a group, first select them all by clicking on them while holding down the Shift key. You'll notice that all the boxes you select become highlighted simultaneously. You can then right-click on any of the boxes and choose Group from the context menu. To undo the grouping operation, you can right-click on the grouped boxes and choose Separate from the context menu. Any changes you applied to the grouped objects still apply, but you can now manipulate each box individually.

Box Styles

If you choose Style from the box context menu, you get the Box Style dialog box (Figure 18.5). You can use this dialog box to set various combinations of formatting options, including border, fill, size, text flow, and default caption type. Select the style you want and then click OK to have these changes take effect. Each style provides a different combination of formatting options intended for specific purposes. You can later override any or all of these settings, of course.

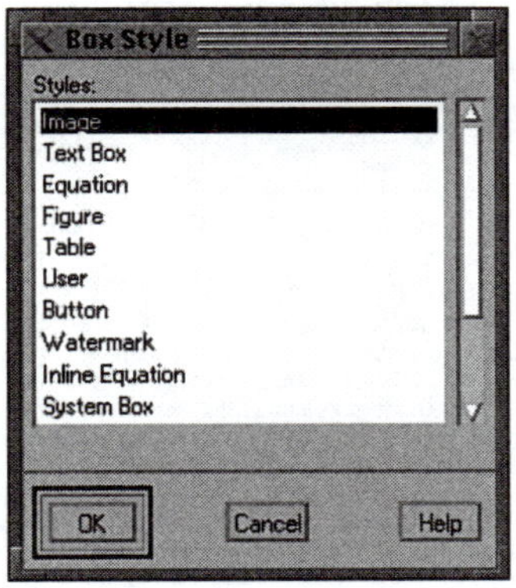

Figure 18.5
Adjust your box's appearance with the Box Style dialog box.

WordPerfect sets a default box type when you create your boxed item. For instance, if you insert a graphic, it takes on the Image style. Text boxes normally use the Text Box style unless you created the text box by using the Drag to create a new table option when creating a table, in which case the text box will have the Table style. Equations normally acquire the Inline Equation style.

→ To learn how to create a movable table, **see** "Creating Movable Tables," **p. 282**

Tip #216 from *Tux Tips*

If you want your equation to occupy a line of its own and have an equation number, you should change its style from Inline Equation to Equation immediately after creating it. The default Inline Equation style is good for placing an equation in a paragraph of text, but to make an equation stand out and be numbered, the Equation box style is much more appropriate. To automatically number an equation, you need to give it a caption.

CAPTIONS

A *caption* is a short text segment that's associated with a boxed item. In printed documents, it's not uncommon for charts, figures, and tables to have captions. In documents that make heavy use of equations, they're often numbered, and these equation numbers can be considered captions, as well. You probably don't want to place captions in the regular text flow, especially if you use paragraph or page anchors because then you'll be constantly struggling to keep your captions near the box with which they're associated. That's why WordPerfect provides a caption option for all box types, to keep caption text associated with the box.

CREATING A CAPTION

You can create a caption in either of two ways:

- Right-click on the box to open the box context menu and choose Create Caption.
- Select the box and click on the Edit caption toolbar item (refer to Figure 18.2).

In either of these cases, WordPerfect creates a caption and lets you edit it (Figure 18.6).

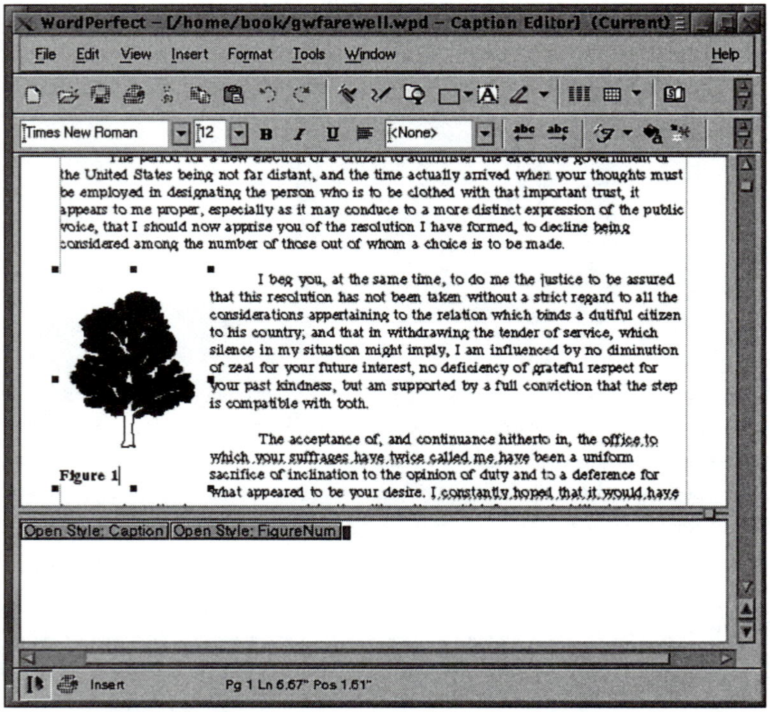

Figure 18.6
Many text boxes have associated caption styles that automatically insert an appropriate label and number.

Note

If you're using Draft view (View, Draft or Ctrl+F5), WordPerfect opens a new document window in which you can edit the caption's text. If you're using the default Page view (View, Page or Shift+Ctrl+F5), you edit the caption in the same document window as the rest of the document. Figure 18.6 shows a caption in Page view.

Many box styles insert an appropriate label and number for the boxed element. For instance, if your box has the Image style when you create the caption, you get a caption that begins with the boldfaced words *Figure n* (where *n* is the figure number, from 1 up). The predefined caption text, if any, varies from style to style, as does its location—for tables, it's above the box, for equations, it's to the right, and for graphics and nontable text boxes, it's below the box.

Editing Caption Text

You can start editing caption text in any of three ways:

- Double-click in the caption area of the boxed item. This procedure can be tricky, however; WordPerfect will sometimes start the editor for the boxed item itself, rather than for the caption.
- Right-click on the box and choose Edit Caption from the context menu.
- Select the box and then click on the Edit caption toolbar item.

If you're in Page view, you can edit the caption text directly. If you're using Draft view, you get a new document window in which to edit the caption's text. In either case, editing a caption is very much like editing a text box. You can use most of WordPerfect's formatting features, including font changes, font attributes, justification, and even columns. The caption area expands or contracts vertically as necessary to accommodate your caption text.

→ For information on editing text boxes, **see** "Editing Text Within Text Boxes," **p. 368**

Caption Styles

The insertion point in Figure 18.6 is in the graphic's caption. Therefore, the Reveal Codes pane in that figure shows the styles used in the caption. Captions use one or possibly two styles by default:

- Caption—The first style code in Figure 18.6 is [Open Style: Caption], and this same style is used, by default, in all captions.
- Numbering style—The second style code in Figure 18.6 is [Open Style: FigureNum]. This style controls the caption numbering, as described shortly. Different caption types may use distinct numbering styles, which may format the text in quite different ways. This style may not be present at all, in fact, for some box types.

You can edit a caption style by double-clicking on the style's code in Reveal Codes. This step opens the Styles Editor dialog box, shown in Figure 18.7. You can then alter the style as you see fit by entering the appropriate formatting codes from the Styles Editor's menus or the keyboard. For instance, if you want your caption text to be centered in the caption field, choose Format, Justification, Center (Ctrl+E).

➔ To learn about styles, **see** "Ensuring Consistency Within Your Document with Styles," **p. 305**

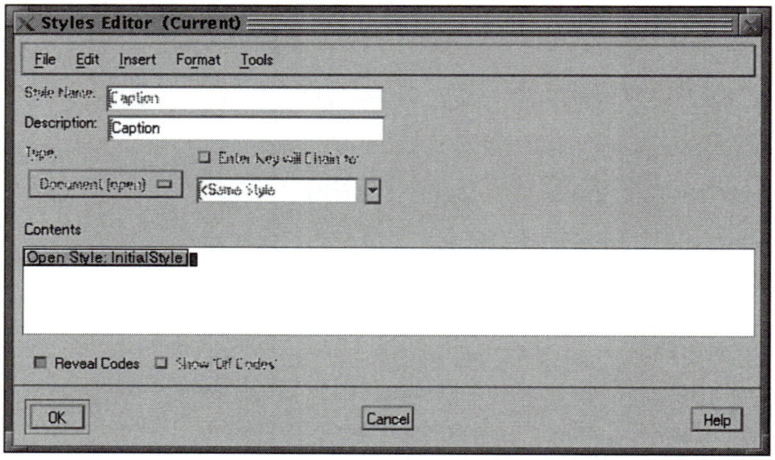

Figure 18.7
The caption styles can be edited just like any other WordPerfect style.

Note

Some style features are not available for caption and numbering styles. For instance, you can't chain the style to another one.

When you alter a caption style, you alter all caption text that uses that style. Thus, whatever changes you make after double-clicking the [Open Style: Caption] code affect *all* the captions. If you want your changes to affect only some captions, either make the changes to the appropriate numbering style (say, to apply the changes to all table or equation captions) or create a new style and use it in selected captions. If you want a formatting change to apply to only one caption, apply the change directly, rather than via a style.

Caption Numbering

WordPerfect can automatically number your captioned boxes. With graphics box captions, table captions, and equation captions, WordPerfect inserts an appropriate numbering style, as described in the preceding section. Figure 18.8 shows the Styles Editor dialog box opened and displaying the FigureNum style.

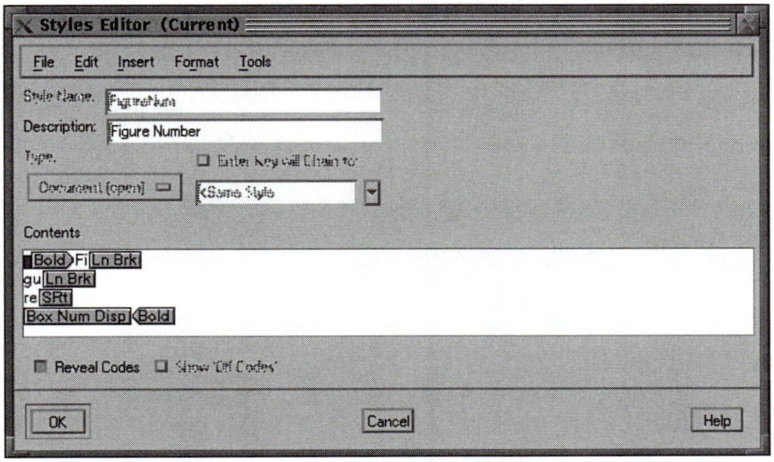

Figure 18.8
The numbering styles typically include formatting, text, and a [Box Num Disp] code to display the box number.

MODIFYING CAPTION STYLES

If you're satisfied with how WordPerfect does the numbering, you can leave the numbering style alone. If not, you can edit it and have the change apply to all the captions for that type of box. For instance, to remove the bold attribute from the figure number, you can remove the [Bold] codes that surround the FigureNum style.

Note

The numbering style may include spurious [Ln Brk] or [SRt] codes. You can safely ignore these.

ALTERING COUNTER SEQUENCES

If you want to change the type or sequence of the number itself, you need to delete the [Box Num Disp] code and replace it by choosing Insert, Other, Counter. This step opens the Counter Numbering dialog box (Figure 18.9). WordPerfect maintains several counters of its own for various box types. You can select one of these and insert it in the document by selecting Display in Document. WordPerfect keeps track of the numbering, so don't use Increase and Display or Decrease and Display unless you want to modify the numbering sequence—for instance, to skip a number.

Tip #217 from *Tux Tips*

If you want to alter the numbering in the middle of a sequence, choose Insert, Other, Counter from the document window, with the insertion point located appropriately to effect your change. Making such a change within a style is likely to be ineffective or may produce peculiar results.

If you only want to alter the value of a counter, you can click Increase, Decrease, or Value. The Increase and Decrease options increment or decrement the counter by one, respectively. The Value button pops up the Set Value/Number Method dialog box (Figure 18.10) in which you can set any value you want and select a numbering style from among numbers, letters, or Roman numerals.

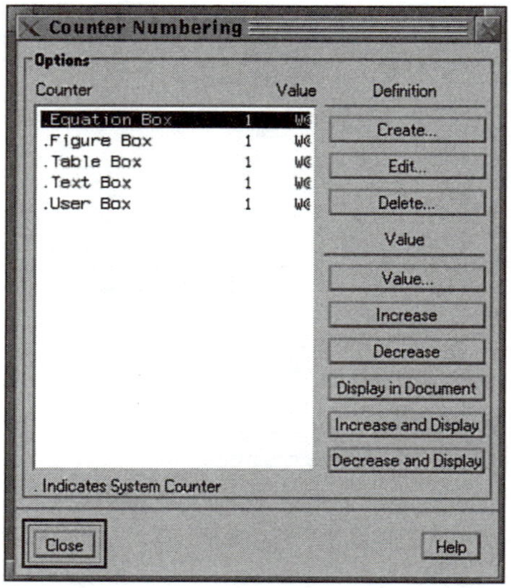

Figure 18.9
Customize a caption's numbering options from the Counter Numbering dialog box.

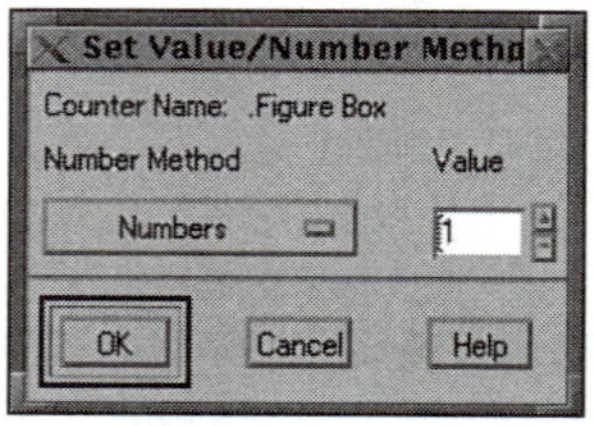

Figure 18.10
You can set a counter to any value and to use any numbering style from the Set Value/Number Method dialog box.

Are your boxed item numbers appearing out of sequence (such as Figure 5 appearing below Figure 6 on a page)? See "Keeping Boxes in Sequence" in the "Troubleshooting" section at the end of this chapter.

Creating Custom Counter Definitions

To create a custom numbering style or to make massive changes to one that already exists, you can use the Create or Edit buttons, respectively, from the Counter Numbering dialog box (Figure 18.9). These buttons open the Create Counter Definition or Edit Counter Definition dialog boxes, which are virtually identical (Figure 18.11).

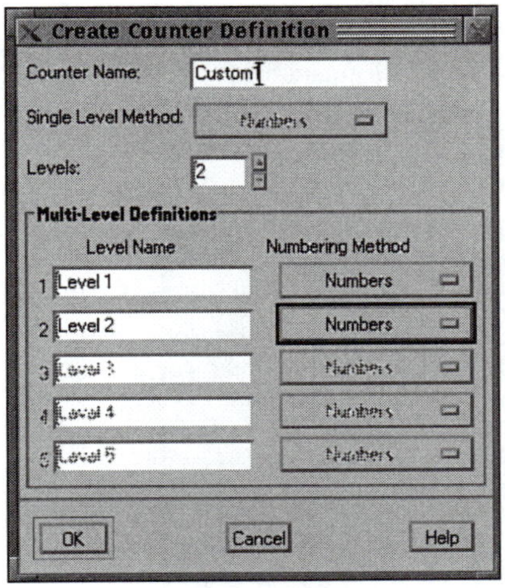

Figure 18.11
The Create Counter Definition dialog box lets you create box counters that include more than one level, such as the chapter and figure numbers used for figures in this book.

When you create a custom counter definition, you can select the number of *levels* in the number. The default counters all use one-level definitions, which consist of a single number. Counters with more levels use more numbers. For instance, the figures in this book follow a two-level system because the figure numbers count both chapters and figures within a chapter. To use more than one part of a multilevel definition, however, you need to choose Insert, Other, Counter, select the appropriate level, and select Display in Document once for each level. WordPerfect also does not automatically increment these counter levels for you, and using Increase and Display in a style doesn't do the trick, either. To get your counters to increment, you need to choose Insert, Other, Counter from the document window and click Increase between each box.

Tip #218 from *Tux Tips*

> To get the benefits of automatic incrementing with a multilevel display, you may prefer to create a counter for all but one level of the number and use the default box counter for the last level. For instance, to number figures as this book does, you could create a chapter counter that you increment manually between chapters and use the .Figure Box counter for the figures. Adjust your FigureNum style so that it includes the chapter number counter, a period, and the figure number counter in succession.

When you create a single-level counter definition, you select the numbering method from the Single Level Method option button near the top of the dialog box. When you create a multilevel definition, however, that option button is disabled (as in Figure 18.11) and you have to select the numbering method for each level separately in the Multi-Level Definitions area of the dialog box. You can elect to use numbers, letters, or Roman numerals for each of these numbering methods.

CAPTION OPTIONS

You can specify various options for the caption, such as where it's located and how it's oriented, by right-clicking the box and choosing Caption from the context menu. This step produces the Box Caption dialog box, shown in Figure 18.12.

Figure 18.12
Set the position, orientation, and size of a caption from the Box Caption dialog box.

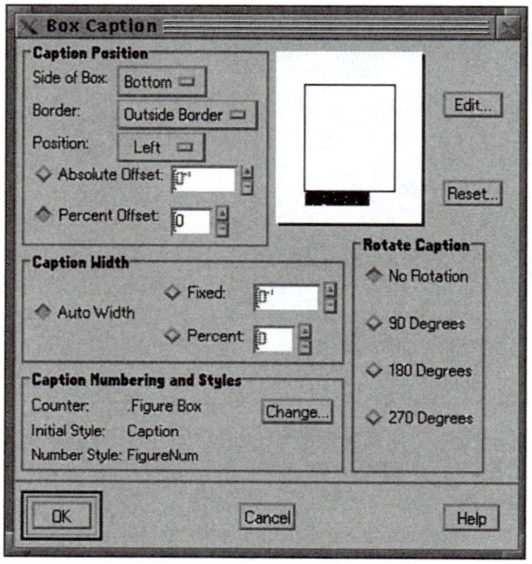

You can use the Box Caption dialog box to set the following options:

- Caption Position—You can tell WordPerfect to place the caption along any of the four sides of a box; inside, outside, or on any border that may be present; and aligned to the left, center, or right of the box (or top, center, or bottom, as appropriate). If the caption's width or height is less than the width or height of the box, you can also specify an offset value to shift the caption relative to this positioning. For instance, you could have the caption's left edge begin 0.5 inches from the left edge of the box. The default values for these options depend on the type of box for which this item is a caption. For instance, figure captions fall below their graphics boxes, whereas equation captions appear to the right of their equation boxes.

Note

The Border option won't appear to have any effect unless there's a visible border around the box.

→ To learn about borders, **see** "Borders and Fills," **p. 475**

- Caption Width—You can adjust the width of the caption area in the Caption Width part of the dialog box. This option usually defaults to Auto, but you can specify a width, either fixed in inches (or another unit of measure by using *c* for centimeters, *m* for millimeters, or *p* for points), or expressed as a percentage of the box's width.

- Rotate Caption—No Rotation, the default, produces normal text. You can rotate the text to lie on its side or to be upside-down, however.
- Caption Numbering and Styles—If you click the Change button in the Caption Numbering and Styles area of the Box Caption dialog box, WordPerfect displays the Select Counter dialog box (Figure 18.13). Use this dialog box to select a new numbering counter *before you create a caption*. If you change the numbering counter after you create the caption, this option has no effect. Also, this option affects only the counter, not the rest of the style. For instance, if you use the Select Counter dialog box to change the numbering counter of a figure, you get a change in the number, but the text still reads *Figure*, followed by the number.

Figure 18.13
Change the numbering style for a caption before you create it.

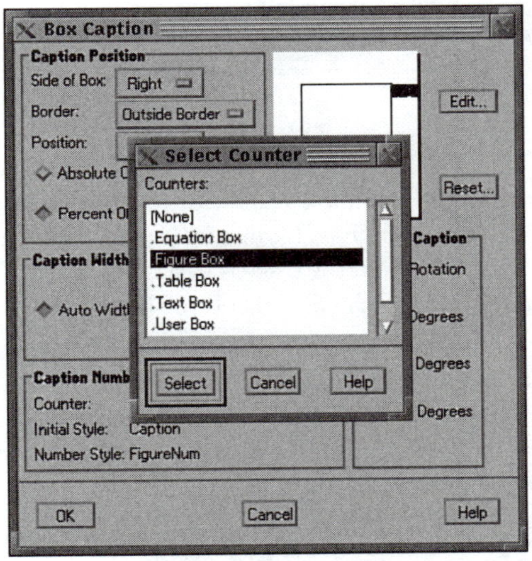

Tip #219 from *Tux Tips*

If you want to change the text of the counter, use the box style feature to adjust the style of the box. You need to change an equation from the Inline Equation style to the Equation style to give it an automatically numbered equation number caption.

→ For information on box styles, **see** "Box Styles," **p. 465**

- Edit and Reset—If you click the Edit button, WordPerfect lets you edit the text of the caption. If you click Reset, WordPerfect removes the caption completely. You can create it again later if you like.

Box Contents

You can alter certain aspects of box formatting by choosing Content from the box's context menu. This step produces the Box Content dialog box, shown in Figure 18.14.

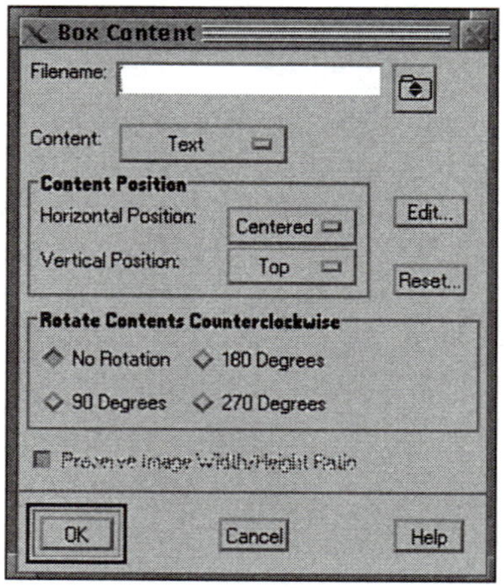

Figure 18.14
Set formatting of information inside the box using the Box Content dialog box.

You can use the Box Content dialog box to adjust the following items:

- Filename—You can enter a filename for a graphic or text file here. If you specify a text file, WordPerfect loads it into a text box. If you give a graphics filename, WordPerfect either loads it or uses it directly from the disk, as described shortly. You can change the image in a graphics box by altering the filename listed here.
- Content—Use this option button to specify what type of data the box contains. The Empty option produces an empty box, and the Image, Text, and Equation options format the box to contain their respective types of information. If you select Image on Disk, WordPerfect loads the image from the disk file and does not store the image inside the WordPerfect file.

Tip #220 from *Tux Tips*

Use Image on Disk to minimize the size of the WordPerfect file, especially if you're using lots of clip art, perhaps from a library you maintain on your hard disk, or bitmapped graphics that tend to be large. Using this option reduces the portability of your files, however; the graphics may not appear if you give the file to someone who doesn't have the appropriate file on disk or whose computer system does not store that file in the same directory as your system does.

Are files you've received from another person missing their graphics? See "Missing Graphics" in the "Troubleshooting" section at the end of this chapter.

- Content Position—If the box is larger than its content, you can tell WordPerfect to position the contents in the center, top, or bottom vertically and in the center, left, or right horizontally.
- Rotate Contents Counterclockwise—For text boxes, you can rotate the contents of the box in 90-degree increments. These options are unavailable for equation and image contents, though you can use the options in WP Draw to rotate a graphic.

→ To learn how to rotate a graphic, **see** "Object Distortions: Skew and Rotate," **p. 405**

- Preserve Image Width/Height Ratio—If you select this check button, WordPerfect adjusts the image's width in proportion to any height changes you make, and vice versa. If you don't check this option, you can stretch the image in one direction more than in another, thus introducing distortions.
- Edit and Reset—If you click the Edit button, WordPerfect lets you edit the box's contents, using whichever editor is appropriate for the image type. If you click Reset, WordPerfect clears the box.

Note

If the box contains an Image on Disk and you click Edit, WordPerfect converts the graphic to a conventional image, thus moving it into your document. You'll therefore lose the benefits of having it on disk.

BORDERS AND FILLS

WordPerfect boxes can optionally have borders composed of various line styles surrounding them or fills composed of various patterns underneath the text or graphics they contain. The easiest way to create a border or fill is to use the box toolbar (refer to Figure 18.2). When clicked, the Change border line style produces a pop-up menu of options for borders, and you can pick one to apply it to the selected box. The Change box fill tool produces a similar menu of fill options.

For greater flexibility in setting your borders and fills, you can right-click on the box and choose Border/Fill from the context menu. This sequence opens the Box Border/Fill dialog box (Figure 18.15). You can select both borders and fills using either a graphical selector or a text-mode selector.

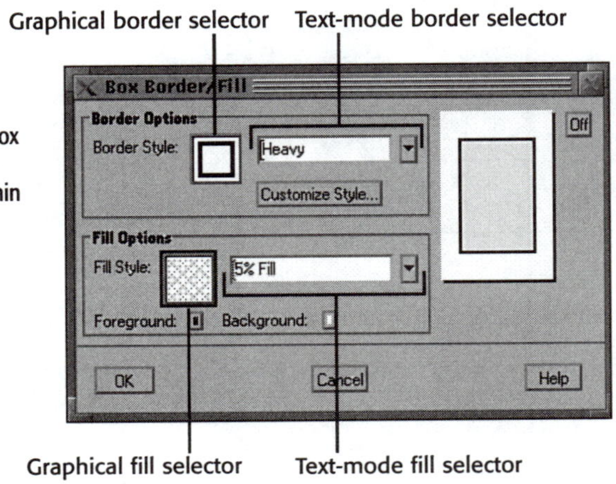

Figure 18.15
Use the Box Border/Fill dialog box to create a border around or a fill within a box.

Fills are composed of foreground and background colors, which default to black and white, respectively. You can select other colors for either of these, however, so that you can produce, for instance, a text box that prints on a light green background if you're using a color printer. To change the fill colors, click on the color selectors to the right of Foreground and Background in the Fill Options area of the dialog box and then select the colors you want.

Tip #221 from *Tux Tips*

In general, light fills work best on text and line art. Darker fills tend to make text difficult to read. This effect can be compounded by photocopying or scanning, so if you expect to copy or scan your document, be particularly cautious when you use fills, or do a test printout and photocopy the page to see how it looks. Darker fills may be appropriate for some types of graphics or if you're using light text against a dark background.

If the selection of border styles is inadequate for your needs, you can further customize the border by clicking Customize Style in the Border Options area of the Box Border/Fill dialog box. This step produces the Customize Border dialog box, shown in Figure 18.16.

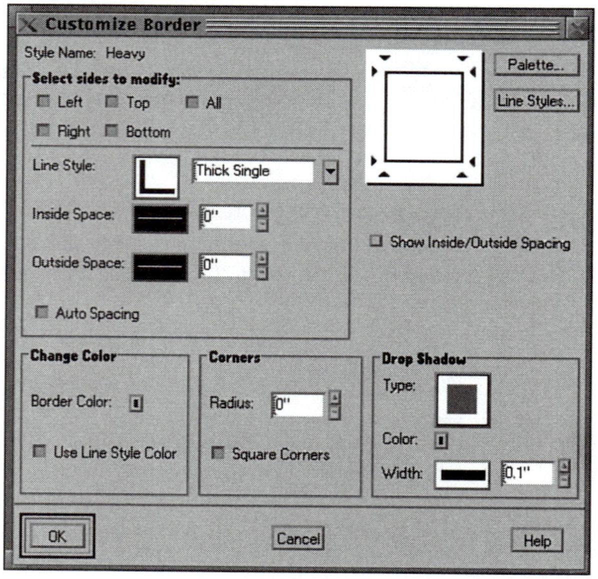

Figure 18.16
The Customize Border dialog box lets you create a customized border type.

From the Customize Border dialog box, you can define all the aspects of a border:

- Sides—You can elect to alter the line style and spacing for any combination of the four box sides.

Note

Your side selections don't influence the color, corners, or drop shadow features; these features apply to all sides of the border.

Tip #222 from *Tux Tips*

> You can set different sides of the border to use different styles by selecting Customize Style from the Box Border/Fill dialog box multiple times, selecting a different side or sides each time.

- Line Style—You can select the line style by using either a graphical or a text-mode selector, similar to the selectors in the Box Border/Fill dialog box. You have more choices for line styles in the Customize Border dialog box, however.
- Line Styles—You can create a customized line style by clicking the Line Styles button near the upper-right corner of the dialog box. This step produces the Line Styles dialog box, shown in Figure 18.17. You can select a style you like and click Select, or you can create a new style or edit an existing one by clicking Create or Edit, respectively. Both buttons produce a dialog box very much like the Edit Line Style dialog box shown in Figure 18.18. From here, you can set the color, pattern, width, and spacing between lines. You can add or delete lines to create a multiline border. You'll then be able to select your style from the text-mode Line Style selector in the Customize Border dialog box, though you won't be able to use the graphical selector unless you modified a style that was already on that selector.

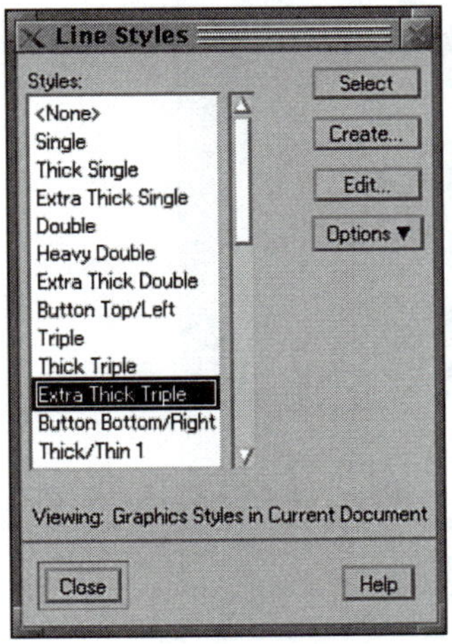

Figure 18.17
The Line Styles dialog box only lets you select styles that already exist, but you can use this dialog box to start another that lets you modify or create a new line style.

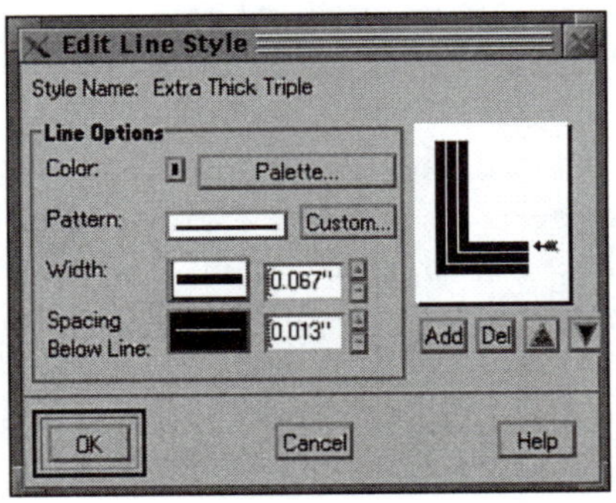

Figure 18.18
The Edit Line Style dialog box is identical to the Create Line Style dialog box (not shown) except for the title and the fact that the Create Line Style dialog box starts with a blank Style Name, which you must fill in.

- Inside Space and Outside Space—You can create space either inside or outside the border. Use the Inside Space option to create a "frame" around a graphic or to keep text inside a text box from appearing too cramped. Use the Outside Space option to keep text outside the box at a distance from the box. If you select the Auto Spacing check button, WordPerfect sets these values automatically, based on the box type. If you select Show Inside/Outside Spacing, WordPerfect shows your spacing selections in the preview area in the upper-right portion of the dialog box.
- Color—You can select a color for the lines of the border with the Border Color option button. If you select the Use Line Style Color check button, WordPerfect ignores this selection and instead uses whatever color is used in the Line Style definition.
- Palette—If you find the default color selections too constraining, try clicking the Palette button. WordPerfect then displays the Define Color Printing Palette dialog box (Figure 18.19). You can then select a color to redefine from the color square in the Color Palette area and modify it by using either the spin boxes in the Color Model area or the color wheel above that.

→ For more information on the color selection options, **see** "Color," **p. 394**

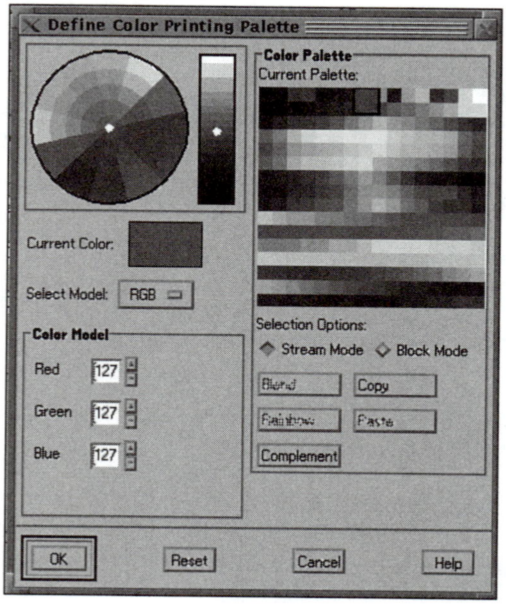

Figure 18.19
The Define Color Printing Palette dialog box provides a plethora of options for precisely defining a color.

- Corners—With most border styles, you can set the corners to be rounded rather than square. To do this, adjust the contents of the Radius spin box in the Corners section of the Customize Border dialog box. The larger the Radius value, the more curved the border. If you later decide you want square corners after all, you can select the Square Corners check button, which overrides the value in the Radius spin box.

Tip #223 from *Tux Tips*

If you set up rounded corners, the corners will encroach on the box's contents. This can be a problem if the contents themselves have square corners because those corners will then extend outside of the border. You may need to edit the graphic itself, either using WP Draw or an external graphics editor, to achieve the best results with rounded corners. At the very least, you can use the Inside Space spin box to fit squared-off box contents within the rounded border you create.

→ For information on editing and importing graphics, **see** "Graphics," **p. 377**

- Drop Shadow—You can give the box a drop shadow—a shaded area to give the document a somewhat three-dimensional appearance, as if the box were levitated slightly above the page. The Type selector lets you select the angle of the shadow, the Color option button lets you specify the color, and the Width selector lets you specify the shadow's size, either from a predefined set using a graphical selector or more precisely with the spin box.

TROUBLESHOOTING

BEHIND TEXT AND IN FRONT OF TEXT WRAPPING NOT WORKING

You've tried using Behind Text or In Front of Text wrapping, but they don't behave as expected.

WordPerfect for Linux, version 8.0.0076, contains a bug that prevents the Behind Text and In Front of Text options from working. Onscreen both work like In Front of Text is supposed to, but in a printout, both work like Behind Text is supposed to. Check with Corel for an updated version of WordPerfect to fix this problem. Alternatively, you might consider using a watermark to get both onscreen and printed output that shows the Behind Text formatting. If you need In Front of Text formatting, you could try reformatting the text. If a completely black graphic is acceptable, you could use it to blot out the text; even with the graphic "behind" the text, a black graphic will effectively cover the text.

TEXT BOX ORDERING DOESN'T WORK

You're trying to adjust the ordering of text boxes, but no matter what you do, the text prints atop graphics.

Like the overlapping of ordinary text with graphics boxes, overlapping a text box with another box doesn't follow the rules it should. The onscreen display may be correct, but when you try to print, the text inside the box overlaps all other objects. The best solution is to check with Corel to see whether it has fixed this problem in an update to the program. If Corel hasn't or if this solution isn't practical, you might be able to get by using an equation box instead of a text box. You won't have the same formatting options for equations as you do for text, and whatever text you want to display must conform to WordPerfect's equation syntax, but equations do at least follow the correct ordering rules.

Keeping Boxes in Sequence

You've inserted numbers for boxed items, but the numbers don't always appear in a logical order. For instance, Figure 5 may fall below Figure 6 on a page.

WordPerfect numbers boxes according to the order in which their *codes* appear in the document. If you use character anchoring, this order is same as the order in which the contents of the boxes appear, but this might not be the case if you use paragraph or page anchoring. In both of these cases, the exact location of the box may drift, relative to the [Box] code, and this drift can be greater for page anchors than for paragraph anchors. To correct this problem, use Reveal Codes mode to locate the [Box] codes for the out-of-order boxes and move them so that they appear in the correct order in the text. You need to select one of the two [Box] codes; cut it with Edit, Cut (Ctrl+X); move the insertion point, and paste the box back with Edit, Paste (Ctrl+V).

Missing Graphics

You've received a file from another person or have moved a file from one computer to another; now the graphics have disappeared, or instead of the graphic, you get a box that reads `Image Not Available.`

The likely cause of this problem is graphics that have been inserted as on-disk images, meaning that WordPerfect doesn't store them internally, but looks for them on the disk. If you right-click the graphics box and choose Content from the context menu, you get the Box Content dialog box (Figure 18.14). If the Content option button reads Image on Disk, then this is likely the problem. Check the Filename and see whether it reflects a real file. If the file was originally created with WordPerfect for Windows, the filenaming conventions may follow Windows patterns, rather than Linux patterns, so you may need to change the name.

If the graphic doesn't appear, but you get a message reading `file contains data for PostScript printers only`, the graphic is an imported Encapsulated PostScript (EPS) file. This graphic prints fine if you're using a PostScript printer or a printer that's being driven by Ghostscript. The graphic doesn't print at all on other printers, however, and doesn't display on the screen.

→ To learn about importing graphics, **see** "Inserting Graphics from External Sources," **p. 379**

→ To learn about printer drivers and Ghostscript, **see** "Fonts and Printing," **p. 739**

PART V

Automating Repetitive Tasks

CHAPTER 19

Letting the Computer Do the Work: Macros

In this chapter

WHAT ARE MACROS?

One of the fundamental purposes of a word processor is to make your life easier. WordPerfect for Linux sports many features to do this, ranging from its automatic reformatting of text as you add or delete words to sophisticated tools for performing complex operations on documents with a single command. One of these tools is the *macro* facility.

At its simplest, a macro is a way to record a series of actions and play them back at a later time, much as you can record and play back sounds on an audiocassette tape. These actions can be anything from typing a few characters to performing complex formatting tasks. For instance, suppose you regularly write business letters, which you close with

```
Yours truly,

George K. Balst
Director, Space Sciences Division
```

Those words don't take long to type, but if you do this task repeatedly, the short time it takes to type them can add up. There's also the lurking possibility of your making a typo, especially if the text includes numbers or if you've disabled WordPerfect's Spell-As-You-Go feature. Macros help you to produce your documents more quickly and with greater consistency and accuracy.

→ For more information on the As-You-Go error checking features, **see** "Proofreading with Spell-as-You-Go and Grammar-as-You-Go," **p. 118**

Improving document creation speed, consistency, and accuracy are also the advantages of using styles and templates. Macros differ from these features in that macros are *dynamic*—they can operate on existing text and formatting, rather than serve as a basis for new documents (templates) or lay down fixed formatting and text (styles). Of course, in some cases, two or even all three features could be used to accomplish the same goal. You could create a letter template that includes the preceding closing lines, for instance, or create a style that does nothing but enter that text. Indeed, flexibility is one of the characteristics of a sophisticated tool such as WordPerfect: It often provides multiple solutions to a problem. Table 19.1 gives an overview of the strengths and weaknesses of styles, templates, and macros.

→ To learn more about styles, **see** "Ensuring Consistency Within Your Document with Styles," **p. 305**

→ To learn more about macros, **see** "Promoting Cross-Document Consistency with Templates," **p. 333**

TABLE 19.1 A COMPARISON OF STYLES, TEMPLATES, AND MACROS

Feature	Strengths	Weaknesses
Styles	Change all instances of a format at once, to maintain internal consistency.	Inserted text is not editable (though formatting may be overridden). Difficult to mix inserted text with your own text.

Feature	Strengths	Weaknesses
Templates	Include formatting for many parts of a document. Can include styles or run macros.	Apply to entire documents, not sections of text. ExpressDocs missing from downloadable version of WordPerfect for Linux.
Macros	Dynamic; can operate on existing text. Can perform complex operations.	Text and formatting are not centrally controlled; result of running a macro is the same as typing in text directly.

Of course, you can combine styles, templates, and macros in a single document. As Table 19.1 indicates, for instance, a template can include various styles or run macros. You can, of course, run macros in a document that you created from a template or that includes styles.

WordPerfect's macro facility can do more than simply record and play back your actions; it can use the features of a computer programming language to customize the macro's effects. For instance, you can create a macro that looks for key document features and applies appropriate formatting to them after the fact. This technique is particularly useful if you find it difficult to remember to apply appropriate styles to parts of your document, to give just one of a myriad possible examples. Because WordPerfect's macro facility is, in effect, a programming language, fully describing its use and applications is well beyond the scope of this book, but I do give some pointers at the end of this chapter. Fortunately, you don't need to know anything about programming to record a basic macro or to use the macros that come with WordPerfect.

WHY USE MACROS?

Ultimately, macros save you time and allow you to create documents that are more consistent and accurate than they might otherwise be. Of course, as with any sophisticated feature, you'll have to invest some time up front to get these benefits, but this investment is very worthwhile if you use WordPerfect very much. Here are some examples of applications for macros:

- Insert "boilerplate" text—You can insert addresses, phone numbers, letter closings, directions, even entire paragraphs or pages of text.
- Apply text formatting—Instead of entering half a dozen or more commands to format text in some way, create a macro to do it. In many cases, styles are a more appropriate solution to this problem, but you may want to use a macro for some reason—for instance, if you need to export the document to another word processor and if styles don't make that transition properly.

→ To learn more about styles, **see** "Ensuring Consistency Within Your Document with Styles," **p. 305**

- Aid in text entry—If you're like me, you probably make occasional typos that involve *transposing* two characters, or even words. For instance, you might type *adn* instead of *and*. WordPerfect comes with a transpose character macro to which you can assign a keystroke and then use if you realize your typo as you make it. You can also create a macro to set several text attributes, activate features, and so on when you're entering text.
- Alter text formatting—If you regularly need to change existing documents in certain ways, such as removing tabs from the beginnings of paragraphs and changing font definitions, a macro can save you considerable time. Enter appropriate find-and-replace functions, format changes, and so on and then run the macro on as many documents as you need to alter.
- Set up a box or table—If you regularly create boxed items or tables that require particular types of formatting, you can create a macro to do this job. For instance, the default format for an equation is inline. If you regularly want to create equations on their own lines with numbering in a caption, however, you need to go through several steps to do this. A macro eases this task considerably.

→ For more information on equations, **see** "Equations," **p. 433**
→ For more information on captions, **see** "Captions," **p. 466**

These are just a few examples of how macros can be applied. As you learn about them, you'll be better able to determine for yourself how you can use them for the type of writing *you* do with WordPerfect.

Creating Macros

Creating a macro is a fairly straightforward task, but if you're not careful you can produce a useless macro or even damage your existing document. Therefore, you should follow some safety precautions prior to recording a macro, and you should test your macro on an unimportant or saved file before using it on anything important.

Precautions and Preliminary Steps

If the macro you're creating is simple, such as one that merely inserts some text, it's unlikely that you'll do any damage to your file when you're recording it, so you need not be as diligent in following these recommendations as when you create a more complex macro. Macros can perform tasks like find-and-replace operations, text deletions, and other potentially destructive operations. If you modify the macro and include more sophisticated programming features, you can run into bugs, just as if you were using C, Java, PERL, or any other programming language. Therefore, with more sophisticated macros, you should be cautious about how you create and test them. Sensible precautions include

- Save your work before starting—If you try to enter a complex macro but make a mistake, you'll be able to return to your saved and undamaged file. You should also save your work before you run an existing macro, especially if it's intended to make many or sophisticated changes to your document.

Caution

WordPerfect's undo feature may not undo an entire macro run. You can undo individual commands executed by the macro, such as a deletion, text insertion, or find-and-replace operation, but a complex macro can easily overrun your undo history's capacity.

→ To learn more about WordPerfect's undo feature, **see** "Using Undo and Redo: To Err Is Human," **p. 113**

- Create your macro in a blank document window—Choose Program, New Window (Ctrl+N) in the WordPerfect program Control window to create a blank document window for creating your macro. This way, any error will only affect an empty document. Of course, if your macro is intended to operate on existing text, this approach may not be practical.
- Pay attention to WordPerfect's state when you create the macro—Because the macro will perform whatever actions you do when you create it, you need to be in the same state when you run the macro. For instance, if you create a macro that operates on an existing header, your file needs to have a header for that macro to work. Further, if your insertion point is in the header when you begin recording the macro, the insertion point needs to be in the header when you play the macro, too; else it will work on the regular document text rather than on the header's text!

Part V
Ch 19

Tip #224 from *Tux Tips*

As you build more and more macros, you may forget the starting state for different macros. Therefore, it's best to build them from a consistent starting state, such as with the insertion point in ordinary text in a document window. If you need to modify a header, then, the macro will include the codes for switching into that header and then back into ordinary text when you're done.

- Don't make unnecessary assumptions—For instance, if you want to modify text at the start of a line, you might logically place the cursor at the start of the line and *then* begin recording the macro. This approach, however, requires you to position the insertion point every time you run the macro. If you forget, the results may not be what you want. It's better to use the Home key *inside* the macro to position the insertion point. Of course, some assumptions may be necessary. If you write a macro to transpose words, for instance, you have to make assumptions about the words to be transposed.
- Run through the steps involved in the macro before recording it—This practice will ensure that you don't run into any surprises when you record it, so you won't need to fumble around and correct mistakes while recording the macro.

- Take your time—WordPerfect records your key presses, menu selections, and so on, but does *not* record your times to perform these actions. You could take a week to record a macro to insert half a dozen characters, and it would still play back in a fraction of a second.

Recording Macros

To record a macro, follow these steps:

1. Choose Tools, Macro, Record (Ctrl+F10). WordPerfect prompts you for a macro name with the Record Macro dialog box (Figure 19.1). If you type a filename without an extension, WordPerfect adds .wcm (the default WordPerfect macro filename extension) automatically.

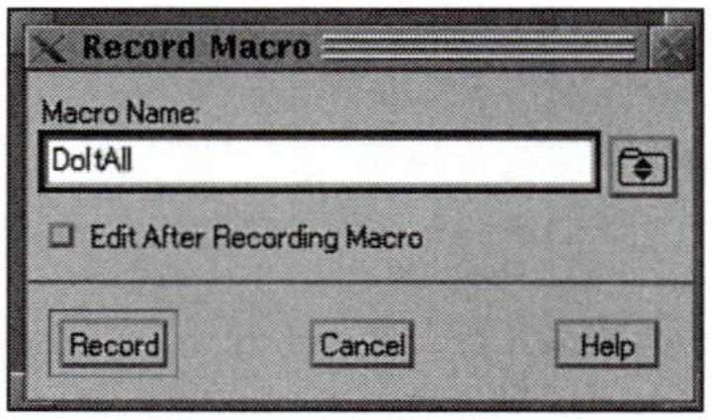

Figure 19.1
You can click the folder icon if you'd like to use a WordPerfect File Manager dialog box to select a filename and directory for the macro.

Note

By default, WordPerfect stores macros in your home directory. You can store them elsewhere by typing an absolute or relative path. For instance, if your username is fred, you can store a macro called DoIt.wcm in the macros subdirectory under your home directory by typing `macros/DoIt.wcm` or `/home/fred/macros/DoIt.wcm` in the Record Macro dialog box. You can change the default macro directory, as described shortly.

→ For information on changing your macro directory, **see** "Maintaining a Macro Directory," **p. 490**

2. If you want to edit your macro immediately after creating it, check the Edit After Recording Macro button in the Record Macro dialog box. If not, leave this box unselected.

→ For information on editing a macro, **see** "Modifying and Programming Macros," **p. 505**

3. Click Record. You'll notice a couple of changes in the display. First, the words Macro Record appear in the status bar at the bottom of the document window. Second, your mouse pointer changes shape to indicate that you're recording a macro. Both these changes are shown in Figure 19.2.
4. Perform whatever actions you want the macro to do—insert text, change formatting, do a find and replace, or what have you.

Figure 19.2
The shape of the mouse pointer indicates its reduced capabilities in macro recording mode.

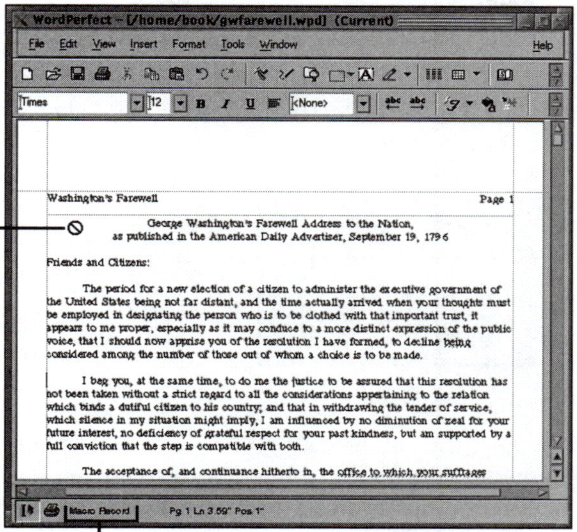

Note

You can't use the mouse to move the insertion point, select text, or open a context menu when you're recording a macro. You can use it to select toolbar and menu items, and you can use keyboard commands to move the insertion point and select text.

5. Choose Tools, Macro, Record (Ctrl+F10) a second time. This step halts the recording of the macro.

If you elected to edit the macro immediately after recording it, WordPerfect now displays a Macro editing window, as described later in this chapter. Otherwise, you're done, and your macro is recorded.

Note

If, during recording, you want to do something that shouldn't be recorded as part of the macro, you can choose Tools, Macro, Pause. Any actions you take from now until you choose Tools, Macro, Pause again won't be recorded as part of the macro.

PLAYING BACK AND TESTING MACROS

Recording a macro is only half the story, of course. You must also be able to *play* or *run* the macro—that is, to use it. When you're done recording a macro, it's a good idea to test it by playing it once, but be sure to use a test file or to save your work, just in case there's a problem with the macro.

To run a macro, choose Tools, Macro, Play (Shift+Ctrl+F10). WordPerfect then displays the Play Macro dialog box (Figure 19.3). If you know the name of the macro you want to play, type it in (with or without the .wcm extension) and click Play. If the macro is in something other than your home directory, enter the correct path in this dialog box. WordPerfect then runs the macro, and you can check its behavior.

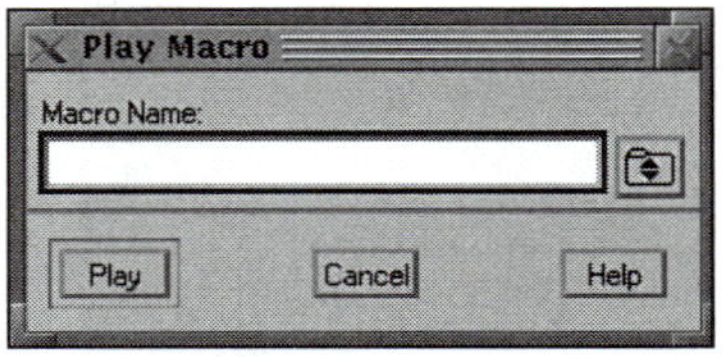

Figure 19.3
Use the Play Macro dialog box to specify which macro to play.

If you don't recall the exact name of the macro, you can click the folder icon next to the Macro Name field in the Play Macro dialog box. This step produces a standard WordPerfect File Manager dialog box from which you can select a macro.

Note

The File Manager dialog box defaults to showing the macros in the wpmacros subdirectory of the WordPerfect installation. To see the macros in your own home directory, type `~/*.wcm` in the Current Selection field and press the Enter key.

Are you trying to create a macro that involves other programs? Is it not working? See "Macros Won't Work with Other Programs" in the "Troubleshooting" section at the end of this chapter.

MACRO MANAGEMENT

To get the most out of macros, it's helpful to know how to manage them. Macro management is largely a matter of configuring WordPerfect and the macros so that they're easier to run. After all, choosing Tools, Macro, Play (Shift+Ctrl+F10) and typing in a macro name every time you want to run a macro isn't the easiest way to access a feature that's intended to save you effort.

MAINTAINING A MACRO DIRECTORY

One of the first things you may want to do when you begin creating macros is to create a subdirectory especially for them. WordPerfect's default of your home directory is a bit suboptimal because if you create a lot of macros, you will have a very cluttered home directory. To change this default, you first need to create a directory in which to place your macros. For instance, you could type, at an xterm, `mkdir ~/.wprc/macros` to create a directory called *macros* within the .wprc subdirectory off of your home directory. This directory is a logical place to store WordPerfect macros because other WordPerfect configuration files are stored in the ~/.wprc subdirectory.

Caution

If your WordPerfect configuration files ever become corrupt, you may be tempted to delete the entire ~/.wprc directory tree and start anew. Although this action fixes certain WordPerfect problems, you should be sure to rescue your macros before taking this step, if you store them in the ~/.wprc directory.

To tell WordPerfect to use your macros directory, select Preferences on the WordPerfect program Control window. This step produces the Preferences dialog box shown in Figure 19.4. Click Files to open the Files Preferences dialog box. Select the Macros/Toolbar/Keyboard radio button, and the Files Preferences dialog box takes on the appearance shown in Figure 19.5.

Figure 19.4
You can set many of WordPerfect's defaults from the Preferences dialog box.

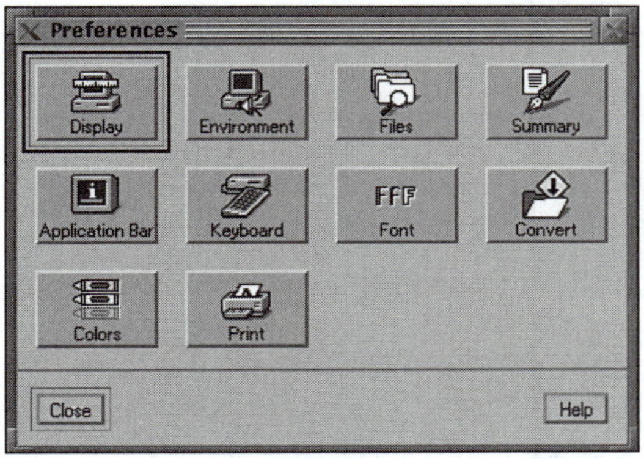

Figure 19.5
WordPerfect searches in two directories for macros, plus your home directory.

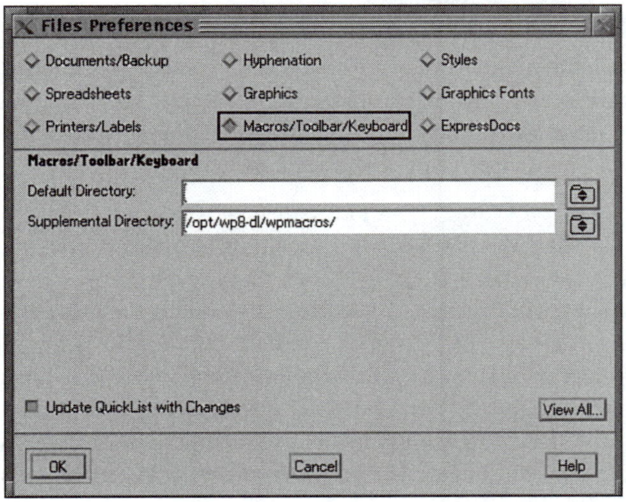

Type the path to your macros directory in the Default Directory field of the Macros/Toolbar/Keyboard area of the Files Preferences dialog box. As with most Linux file specifications, you can use the ~ as a placeholder for your home directory. Thus, type `~/.wprc/macros` to have WordPerfect use the .wprc/macros subdirectory off your home directory for its macros path. You can then click OK in the Files Preferences dialog box and Close in the Preferences dialog box to dismiss these dialog boxes.

Changing the macros path in this way also alters the default directory in which WordPerfect starts a File Manager when you use this option to select a macro to play or record. WordPerfect still searches the wpmacros subdirectory of the Linux installation and your home directory, in that order, for macros if they're not found in the Default Directory you specify in the Files Preferences dialog box. The total set of directories WordPerfect searches for macros is known as the *search path* for macros.

Note

You can change the Supplemental Directory in the Files Preferences dialog box, as well as the Default Directory. Make this change if you don't want WordPerfect to search its own wpmacros directory or if you want to search some other directory instead.

After you set the Default Directory for macros, any new macros you create appear in that directory. You can also move macros you created elsewhere by using the `mv` command in an xterm window or by using WordPerfect's File Manager dialog box. You can rename, delete, and copy macro files just as you would any other Linux file.

→ To learn how to use the File Manager dialog box to manipulate files, **see** "Linux File Manipulation," **p. 80**

ONE-KEY ACCESS TO MACROS

To get the maximum effect from macros, you need quick ways to access them. One quick method of macro access is to assign them to keys on your keyboard. This method is particularly useful for macros that you'd normally use when typing, such as a transpose character macro or one to enter superscript or subscript formatting.

When you press a key, WordPerfect looks up what to do in a special internal *key-binding* table. This table tells WordPerfect which actions to associate with each keystroke—it *binds* the keys to their associated actions. If a given Ctrl+key or Shift+Ctrl+key combination doesn't have an entry in this table, WordPerfect searches for a macro file using the name corresponding to that key: ctrl*x*.wcm or ctrlsft*x*.wcm, where *x* is the key. The first of these plays when you press Ctrl+*x*, and the second runs when you press Shift+Ctrl+*x*. Therefore, giving your macro an appropriate name is an easy way to get it to run by pressing a key combination. The trick is figuring out which keystrokes are available. WordPerfect uses most of the Ctrl+key combinations, and many of the Shift+Ctrl+key combinations, as shown in Table 19.2.

Tip #225 from *Tux Tips*

You can use a symbolic link to set up macros using both descriptive filenames and key binding names. For instance, suppose that you create a macro to transpose words, called transpose-words.wcm. Assuming this macro is in a directory on the macro search path, you can start an xterm and, in this directory, type `ln -s transpose-words.wcm ctrly.wcm` to create an appropriate link to have Ctrl+Y activate the macro.

TABLE 19.2 KEY COMBINATIONS USED BY WORDPERFECT 8 FOR LINUX

Key	Ctrl+Key	Shift+Ctrl+Key
a	Select all	Expand abbreviation
b	Bold	Insert bullet
c	Copy text	*<None>*
d	Date text	Date code
e	Center justification	*<None>*
f	Find and replace	FROM-TO macro
g	Go to	*<None>*
h	Outline body text	*<None>*
i	Italic	*<None>*
j	Full justification	*<None>*
k	Case toggle	*<None>*
l	Left justification	*<None>*
m	Macro command inserter	*<None>*
n	New blank document	*<None>*
o	*<None>*	*<None>*
p	*<None>*	*<None>*
q	Find QuickMark	Set QuickMark
r	Right justification	Redo
s	Save document	Save all
t	New blank document	*<None>*
u	Underline	*<None>*
v	Paste	Paste simple
w	WP symbols	*<None>*
x	Cut	*<None>*
y	*<None>*	*<None>*
z	Undo	Undelete

WordPerfect actually provides several key-binding tables, and you can define others. Therefore, if you very much want to assign a macro to a particular key that's already in use, you can do so by altering the key-binding table.

➔ For information on binding keys, **see** "Keyboard," **p. 778**

Note

Your X server, such as XFree86, also uses its own key-binding table, and your window manager may intercept certain keys. Keystrokes can be intercepted and reinterpreted at each of these stages. Alt+key combinations are particularly likely to get lost between the keyboard and WordPerfect. Therefore, some WordPerfect key-binding tables may be more appropriate than others for your system. If a key does something you don't expect, it could be because of WordPerfect's key bindings, your window manager intercepting keystrokes, or X's key bindings.

Macros in Toolbars

You can add a macro to a WordPerfect toolbar (but not to the property bar, which is the lower of the two default toolbars). You can then access the macro by clicking its icon in the toolbar. To add a macro to a toolbar, follow these steps:

1. Right-click in the toolbar to get its context menu.
2. Choose Edit. This step opens the Toolbar Editor dialog box (Figure 19.6).

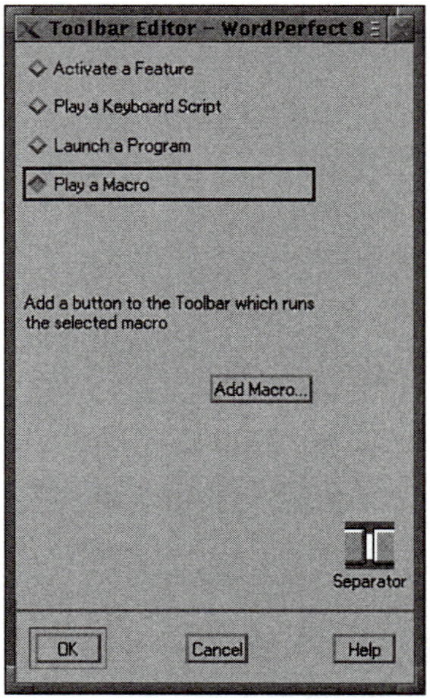

Figure 19.6
You can edit various aspects of the toolbar from the Toolbar Editor dialog box.

3. Select the Play a Macro radio button and then click the Add Macro button. This step opens the Select Macro dialog box, which looks and works just like the Play Macro dialog box (refer to Figure 19.3).
4. Type the name of the macro into the Macro Name field of the Select Macro dialog box or click the folder icon to get a WordPerfect File Manager dialog box with which to select the macro. When you've entered the macro's filename, click Select.
5. WordPerfect asks whether you want to save the full pathname to the macro. If you select Yes, then WordPerfect remembers the full pathname, which can be useful if you've selected a macro that's not in a usual macro directory. If you select No, then WordPerfect stores only the macro's filename, not its complete path, so if you move your macros around, you needn't reconfigure any altered toolbars, just change the default macro path, as described earlier.
6. Click OK in the Toolbar Editor to finish editing the toolbar.

At this point, a cassette icon appears in the toolbar, as shown in Figure 19.7. You can use this tool to run your macro—click on it, and your macro will play.

Figure 19.7
Macro tools look like audiocassettes.

Note

If you place more than one macro tool on a toolbar, they all look identical. If you leave your mouse pointer over one of them for more than a second or two, though, WordPerfect displays the macro's filename in a yellow box.

Examples of Macro Use

Throughout this chapter, I refer to examples of what macros can do, such as transposing words and inserting text. Now that you know how to create a macro, I can go into greater detail on what you can do with macros, including step-by-step procedures for creating some useful macros.

Some Useful Macros You Can Create Yourself

It doesn't take a lot of effort to create a few macros to make your life easier. Is there some set of functions you routinely apply to your documents? If so, consider creating a macro for these functions and assigning that macro to a keystroke or toolbar item. This technique can save you a great deal of time in the long run.

Note

Although using macro key mapping is very useful, in the examples that follow I use more descriptive macro filenames, because your system may already have functions assigned to any key I might select.

The macros I describe here are all available on the CD that accompanies this book, in the directory called chapter19.

→ To learn how to access a CD in Linux, **see** "Accessing Removable Devices: Using the `mount` Command," **p. 76**

SETTING UP COLUMNS WITH BLOCK PROTECT

Suppose that you routinely create documents that present lists of items with associated descriptions, similar to the document shown in Figure 19.8. If you routinely use the same formatting options (number of columns, gutter size, and so on), you'll probably want to use some method to automate the creation of this format. You can use styles for this task, but macros are another option, and one with certain advantages, as I describe shortly.

→ To learn more about styles, **see** "Ensuring Consistency Within Your Document with Styles," **p. 305**

→ To learn more about parallel columns, **see** "Organizing Information with Parallel Columns," **p. 267**

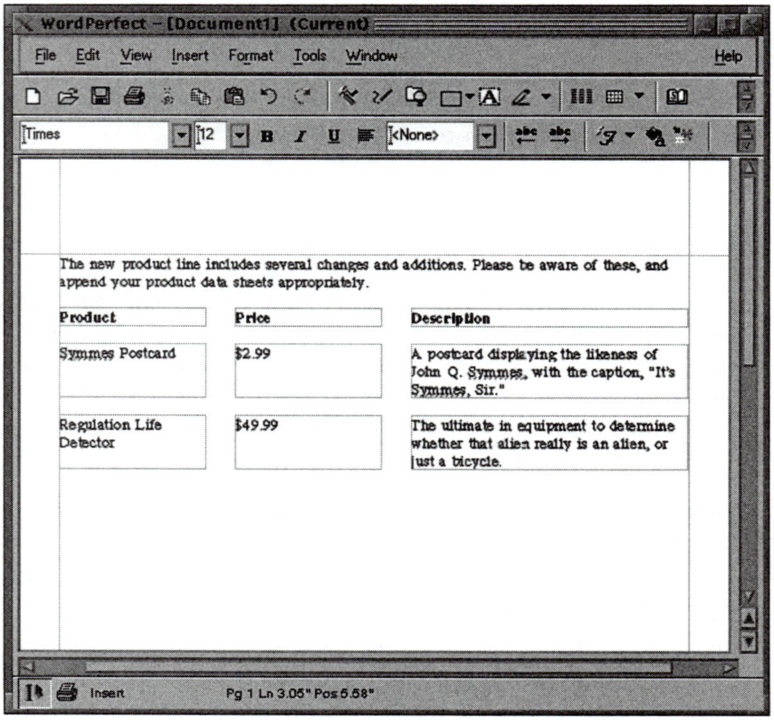

Figure 19.8
Parallel columns are useful for displaying lists of items with associated descriptions or other information.

To create a macro to begin formatting similar to that in Figure 19.8, follow these steps:

1. Open a new document window by choosing Program, New Window (Ctrl+N) from the WordPerfect Control window.
2. Choose Tools, Macro, Record (Ctrl+F10) from the document window. WordPerfect displays the Record Macro dialog box (refer to Figure 19.1).
3. Enter the macro name, such as `product-columns`, and click Record.
4. Choose Format, Columns to open the Define Columns dialog box (Figure 19.9).

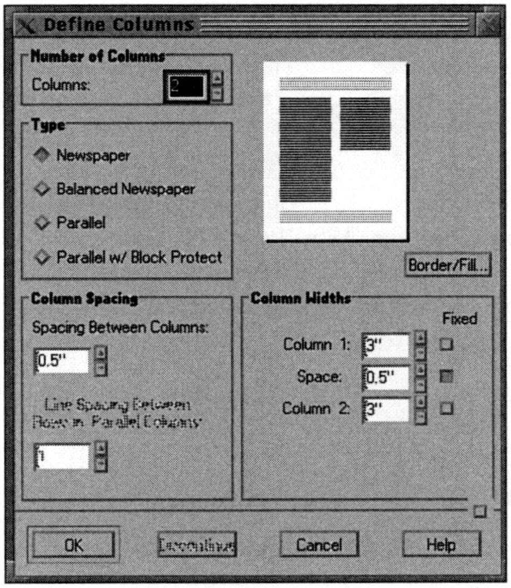

Figure 19.9
When you run the macro, you won't see the Define Columns dialog box; instead, WordPerfect uses the options you entered when you created the macro.

5. Click the Parallel w/Block Protect radio button in the Type area of the Define Columns dialog box.
6. Increase the Columns spin box to read 3.
7. Reduce the Spacing Between Columns from 0.5" to 0.3".
8. Reduce the width of Column 1 to 1.4" and then select the Fixed button next to the Column 1 spin box.
9. Reduce the width of Column 2 to 1.4" and click in another spin box to be sure that change is accepted.
10. Click OK to define the columns.
11. Press Ctrl+B to turn on the bold attribute; then type `Product`, press Ctrl+Enter, type `Price`, press Ctrl+Enter, type `Description`, press Ctrl+B, and press Ctrl+Enter.
12. Choose Tools, Macro, Record (Ctrl+F10) to stop recording the macro.

To test your macro, choose Tools, Macro, Play (Ctrl+Shift+F10) and type `product-columns` for the macro name. When you press the Enter key or click Play, you should see your column definitions appear, complete with the column headings.

As I mentioned earlier, you can achieve much the same effect by using styles, and if you want your column definitions to be identical throughout your document even if you need to change them, styles are a preferable solution. Macros may be better if you want to be able to change the definition for one price list, however, or if you sometimes need to change the heading text. You can also use more advanced macro programming techniques to customize your column definitions. For instance, you could create one macro that lets you select from among several predefined column definitions.

→ For information on macro programming, **see** "Modifying and Programming Macros," **p. 505**

Deleting a Word

WordPerfect provides a number of useful keyboard controls for entering, selecting, deleting, and copying text. No program can provide absolutely every keyboard editing feature conceivable, however, and you may find that WordPerfect lacks one that you'd like to have. For instance, you can delete the word nearest the insertion point by pressing Ctrl+Backspace, but what if you want to delete the current *sentence*? To create a macro to perform this task, follow these steps:

1. Open a new document window by choosing Program, New Window (Ctrl+N) from the WordPerfect Control window.
2. Type at least one sentence into this window. You can use short one- or two-word sentences, just so long as they end in sentence punctuation, such as periods.
3. Position the insertion point in the middle of one of these sentences.
4. Choose Tools, Macro, Record (Ctrl+F10) from the document window. WordPerfect will display the Record Macro dialog box (refer to Figure 19.1).
5. Enter the macro name, such as `delete-sentence`, and click Record.
6. Choose Edit, Select, Sentence. WordPerfect should highlight one sentence.
7. Press the Backspace or Delete key to delete the sentence.
8. Choose Tools, Macro, Record (Ctrl+F10) to stop recording.

You can now run the delete-sentence.wcm macro whenever you want to delete an entire sentence. To make this task even easier, you could give it a keyboard shortcut by renaming the macro or creating a symbolic link to it, as described earlier.

→ To learn how to assign a macro to a keystroke, **see** "One-Key Access to Macros," **p. 492**

Caution

Be careful about assigning keyboard shortcuts to potentially destructive macros. You don't want to create a situation in which you'll be accidentally deleting sentences, paragraphs, or larger amounts of text, by having your finger stray from a common keystroke. For instance, although Ctrl+O is available for a macro, this might not be a good choice for the delete-sentence.wcm macro because of the proximity of Ctrl+O to Ctrl+I, the keyboard shortcut for italicizing text. Better to use a Shift+Ctrl+key combination, such as Shift+Ctrl+O or (perhaps easier to remember via the mnemonic word *kill*, Shift+Ctrl+K).

Document Reformatting

One common need in many offices is the need to reformat a document to meet certain criteria. For instance, you may receive a document that uses two [HRt] codes between paragraphs and two spaces between sentences, but you may need to format this so that it uses only one [HRt] code between paragraphs and a single space between sentences. You can do this job with a pair of find-and-replace operations, which isn't too difficult. If you receive several of these documents each week, however, and if they each contain a dozen such necessary changes, entering the multitude of find-and-replace operations will become quite tedious and error prone.

→ For more information on find and replace operations, **see** "Find and Replace," **p. 105**

Macros are a perfect solution to this problem. You can create a macro that performs all the necessary find-and-replace operations and run it on each file. Several minutes' tedium suddenly becomes a task of a few seconds! To create a macro such as this, follow these steps:

1. Open a document that's representative of those you need to modify. Be sure there's a copy saved on disk because you'll be modifying this document as you create the macro.
2. Choose Tools, Macro, Record (Ctrl+F10) from the document window. WordPerfect displays the Record Macro dialog box (refer to Figure 19.1).
3. Enter the macro name, such as `reformat`, and click Record to begin recording.
4. Press Ctrl+Home to move the insertion point to the top of the file.

Note

You should explicitly position the insertion point at the start of the document even if the insertion point is already at the top of the file because you may run this macro after you've moved the insertion point elsewhere.

5. Choose Edit, Find and Replace (Ctrl+F2) to open the Find and Replace Text dialog box (Figure 19.10).
6. Choose Match, Codes from the Find and Replace Text dialog box's menu. This step opens the Find Codes dialog box (Figure 19.11).

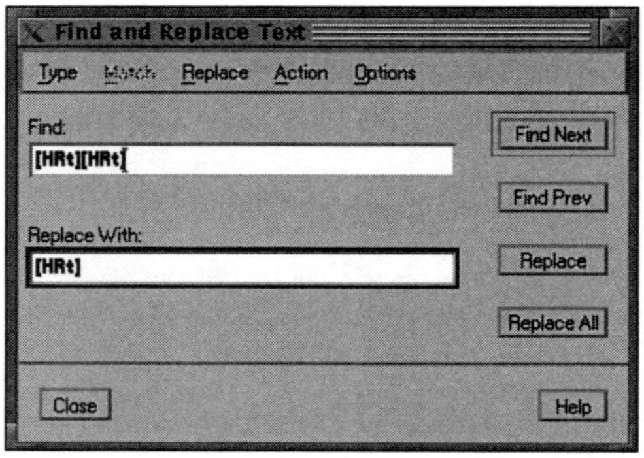

Figure 19.10
When you run the macro, you won't see the Find and Replace Text dialog box.

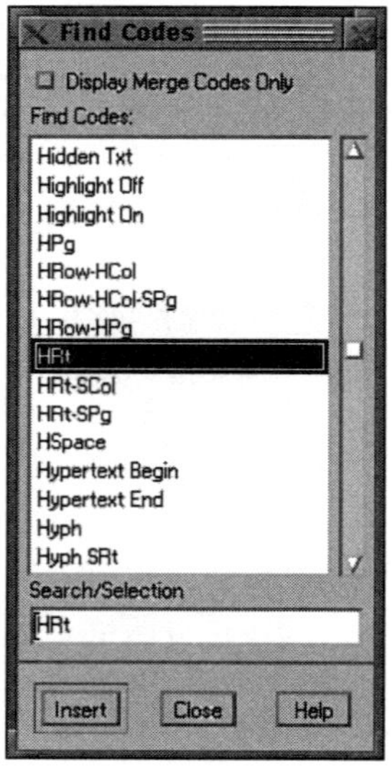

Figure 19.11
If you're uncertain of a code's name, use Reveal Codes to find it before you begin the find-and-replace operation.

7. Locate the [HRt] code, select it, and click Insert.
8. Click Insert in the Find Codes dialog box again.
9. Click Close in the Find Codes dialog box. It will disappear, and you should see two [HRt] codes in the Find field of the Find and Replace Text dialog box.
10. Click in the Replace With field of the Find and Replace Text dialog box.
11. Choose Replace, Codes to open the Replace Codes dialog box, which is very similar to the Find Codes dialog box (Figure 19.11).
12. Locate and select the [HRt] code.
13. Click Insert once, followed by Close. The Replace With field will now contain one [HRt] code.
14. Click Replace All. WordPerfect tells you how many occurrences it replaces. Dismiss this dialog box.

15. Click Close in the Find and Replace Text dialog box.
16. Press Ctrl+Home to go to the start of the document.
17. Choose Edit, Find and Replace (Ctrl+F2) to reopen the Find and Replace Text dialog box.
18. Double-click in the Find field and type a period and two spaces there.
19. Double-click in the Replace With field and type a period and a single space there.
20. Click Replace All. WordPerfect again tells you how many occurrences it's replaced; dismiss this dialog box.
21. Click Close in the Find and Replace Text dialog box.
22. Choose Tools, Macro, Record (Ctrl+F10) to stop recording the macro.

Now, whenever you run the reformat.wcm macro, your document will be reformatted by those find-and-replace operations. Of course, you can change the operations from those specified here or add more. For instance, you probably want to add operations to locate double spaces that follow question marks and exclamation marks, as well as periods. You could also search for margin and font change codes and then insert your own standard margins and font. The possibilities go on and on.

Note

Many macros, such as the delete-sentence.wcm macro, run so quickly that you'll not notice anything but the result. A complex reformatting macro, especially when coupled with a large document or a slow computer, may take more than an eye blink to finish. You may notice the text Macro Play appear in the status bar, in the Macro status indicator area (refer to Figure 19.2).

Using Macros Included with WordPerfect

WordPerfect ships with several macros in its wpmacros subdirectory. Because WordPerfect searches this subdirectory when you enter a macro's name, you can type the name in the Play Macro dialog box (refer to Figure 19.3) and WordPerfect will find it. A description of these macros, and what they do, follows.

- abbrev.wcm—Expands abbreviations in a document based on the contents of the abbrev.data file in the macro search path. To create your own abbreviation file, copy the abbrev.data file from the WordPerfect wpmacros subdirectory to your own macros directory. You can then select Create/Edit when you run the abbrev macro to add abbreviations to the file.
- calfill.wcm—Fills in the dates in a calendar layout for a given month and year.

Caution

Use the calfill.wcm macro only if you've already created a calendar layout. This macro is used internally by the calendar templates provided with the retail versions of WordPerfect.

➔ For more information on the calendar templates, **see** "Using ExpressDocs," **p. 336**

- capital.wcm—Capitalizes the first letter of a word or selection.
- checkbox.wcm—Places a square box in the text. You can use this macro if you're designing a questionnaire in which you expect people to mark their answers with a check mark or *X*.
- closeall.wcm—This macro closes all open WordPerfect document windows.
- ctrlm.wcm—Starts the WordPerfect Macro Command Inserter, which is used in programming macros. Because of its name, this macro runs when you press Ctrl+M.

➔ To learn about the Macro Command Inserter, **see** "Entering New Commands," **p. 507**

- ctrlsftf.wcm—Prompts for FROM and TO values for the WordPerfect equation editor. This macro runs when you press Shift+Ctrl+F, whether or not you're in the equation editor.

➔ For information on FROM and TO values in the equation editor, **see** "Creating Sums and Integrals," **p. 445**

- edocbeg.wcm—Searches for the string **BEGIN** in a document, removes it, and places the insertion point at that location. This macro is used by many of the ExpressDocs templates included with the retail versions of WordPerfect 8 for Linux.

➔ For more information on ExpressDocs, **see** "Using ExpressDocs," **p. 336**

- endfoot.wcm—Converts endnotes to footnotes.

➔ For more information on footnotes and endnotes, **see** "Footnotes and Endnotes," **p. 607**

- exec.wcm—Starts an xterm window.
- filestmp.wcm—Places the document's filename in a header or footer. You specify where you want it by selecting options in the File Stamp Options dialog box (Figure 19.12).

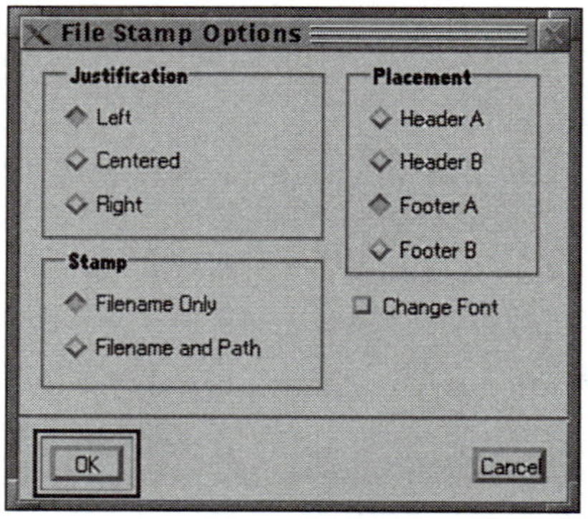

Figure 19.12
Macros can create their own custom dialog boxes, as illustrated by the File Stamp Options dialog box.

- fontdn.wcm—Reduces the size of the font by 2 points.
- fontup.wcm—Increases the size of the font by 2 points.
- footend.wcm—Converts footnotes to endnotes.

➔ For more information on footnotes and endnotes, **see** "Footnotes and Endnotes," **p. 607**

- graph.wcm—Creates a graph paper style—a set of horizontal and vertical lines covering the page. This macro is intended to be run from the graph paper template included with retail versions of WordPerfect.

→ To learn more about the graph paper templates, **see** "Using ExpressDocs," **p. 336**

- lawfax.wcm—Creates a cover sheet for law offices. Intended to be called from the legal fax cover sheet template in ExpressDocs.
- linenum.wcm—Moves the insertion point to a specified line and character number. For instance, if you enter 15 and 20 as the line and character numbers in the Macro Line Finder dialog box (Figure 19.13), WordPerfect shifts the insertion point to the 20th character of the 15th line of the document.

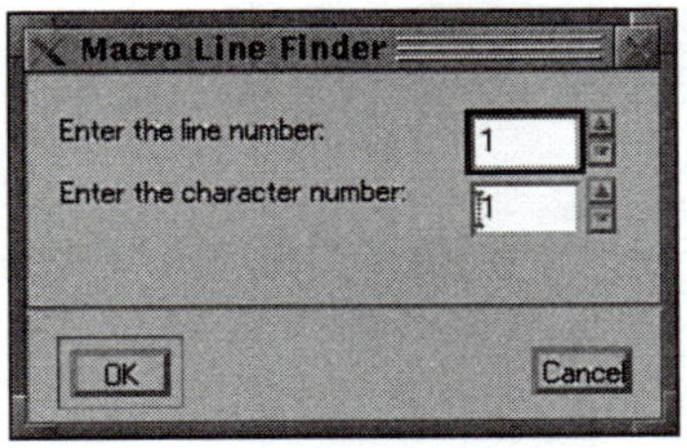

Figure 19.13
Enter the line number and the character on that line to move the insertion point there.

- macroconv.wcm—Converts WordPerfect 5.1 macros for use with WordPerfect 8 for Linux.
- pagexofy.wcm—Creates page numbers, including the total number of pages, in "page *x* of *y*" format, where *x* is the current page number and *y* is the total number of pages in the document.
- parabrk.wcm—Creates a break between paragraphs based on any of a number of special characters. Figure 19.14 shows the Paragraph Breaks dialog box in which you enter your choice of break type, along with several different breaks in the document window.
- pgborder.wcm—Creates any of several ornate borders around a page.

Tip #226 from *Tux Tips*

WordPerfect may become slow when you create these borders. To speed it up again, toggle off the View, Graphics menu option (so that a hollowed square does *not* appear next to the option in the menu). This step eliminates the border from your onscreen display, though the border will still print.

- pleading.wcm—Creates a legal pleading paper format.
- resume.wcm—Selects the résumé toolbar.
- reverse.wcm—Reverses the colors in a selected area. For instance, this macro turns black text on a white background into white text on a black background.

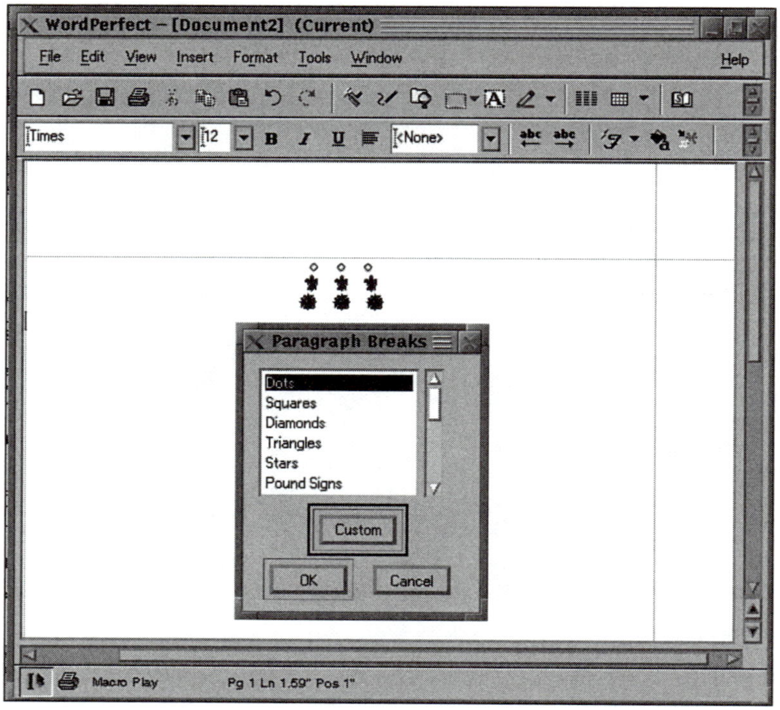

Figure 19.14
Use the paragraph break macro to create a distinctive break between sections or at the end of a document.

- saveall.wcm—Saves all open WordPerfect documents.
- softfonts.wcm—Displays all the graphics fonts with which the retail versions of WordPerfect 8 for Linux ship.

Do most of the fonts you get when you run the softfonts.wcm macro look like Courier? See "Fonts Don't Display" in the "Troubleshooting" section at the end of this chapter.

- transpos.wcm—Transposes the two characters to the immediate left of the insertion point. For instance, if you place the insertion point at the end of the typo *kye* and run this macro, the word *key* results.
- watermrk.wcm—Takes a WordPerfect graphics file from the wpgraphics subdirectory and formats the file as a watermark.

→ To learn more about watermarks, **see** "Watermarks," **p. 241**

Note

Some of these macros are intended to be used from templates. In general, these won't produce the expected results when run in other ways.

Tip #227 from *Tux Tips*

> If you want to use any of these macros on a regular basis, you can assign them to keys by placing an appropriately named symbolic link in your macros directory. For instance, at an xterm in your macros directory, type `ln -s /opt/wp8/wpmacros/transpos.wcm ctrlsftt.wcm` to set up the transpose macro to run when you press Shift+Ctrl+T. (This command assumes that WordPerfect is installed in /opt/wp8; adjust as necessary for your system.)

You may have noticed that some of these macros do some pretty sophisticated things, including displaying their own dialog boxes, getting user input, and altering their actions based on that input. Such actions are achieved by programming the macro, much as you can program in other computer languages, such as C or Pascal.

MODIFYING AND PROGRAMMING MACROS

As I indicated earlier in this chapter, WordPerfect's macro facility is really a full-featured computer programming language. The fact that you can record macros by performing ordinary WordPerfect actions lets you use some of the features of this language without knowing anything about programming. To make the fullest use of the macro facility, though, you need to actually *program* the macros, rather than simply *record* them. When you program a macro, you use WordPerfect as a text editor and you type in commands or modify existing ones with words rather than with mouse clicks or ordinary WordPerfect actions.

Even if you don't have any interest in learning to program macros, you may want to use this capability to make slight modifications to an existing macro. For instance, if you've created a complex search-and-replace macro but find you typed one of the replace terms incorrectly, it might be quite a nuisance to re-create the entire macro. Instead, you can load it into the macro editor, correct your typo, and save the macro in much less time—and without the risk of introducing a different error during a second recording session.

STARTING THE MACRO EDITOR

To edit an existing macro, choose File, Open and select a macro file from the WordPerfect Open dialog box. When you load the macro, WordPerfect automatically goes into Macro editing mode. Figure 19.15 shows WordPerfect displaying the reformat.wcm macro for reformatting a document. Notice that the macro is only eight lines long, though it took 22 steps to describe how to create the macro using the macro recorder. WordPerfect's macro language is much more efficient than unadorned English for conveying information on WordPerfect's functions.

If, instead of modifying an existing macro, you want to create a new one, you can choose Tools, Macro, Macro Toolbar. This step produces the macro toolbar, and you can begin typing in the code for your macro. Alternatively, you might want to copy an existing macro that does something similar to what you want and see whether you can modify it.

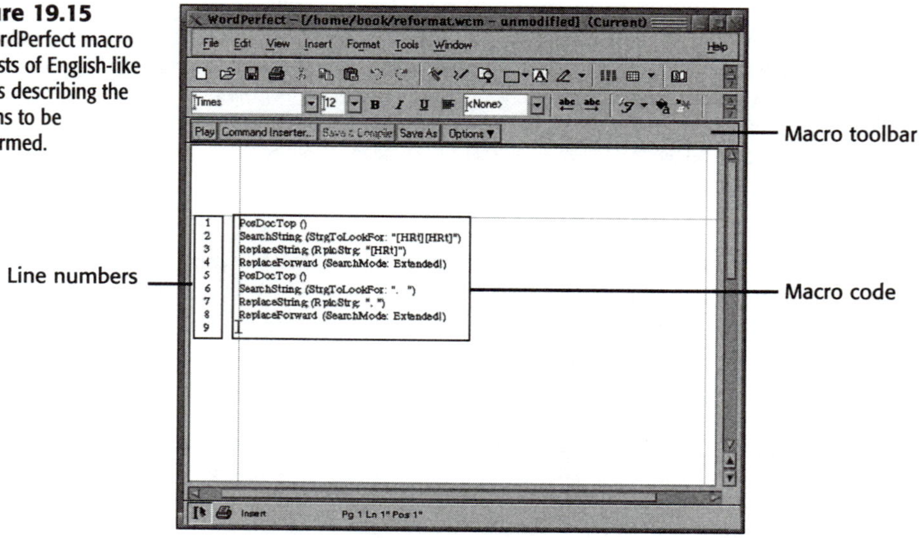

Figure 19.15
A WordPerfect macro consists of English-like words describing the actions to be performed.

MACRO EDITOR BASICS

The macro editor contains many tools for creating and modifying macros. You can do a lot with macros using just a few tools because most of what macros do is handled by the text you enter into the macro editor.

MODIFYING AN EXISTING MACRO

You can modify macro commands as you would any other WordPerfect text by using ordinary mouse movements, keyboard entry, and so on. For instance, suppose you want to modify the replace.wcm macro so that it replaces two spaces with one after question marks as well as after periods. You can do so by following these steps:

1. Select lines 5 through 8 by moving the insertion point to the start of line 5 and holding the Shift key while you press the Down Arrow key four times.
2. Choose Edit, Copy (Ctrl+C).
3. Position the insertion point on line 9.
4. Choose Edit, Paste (Ctrl+V) to duplicate that text.
5. On the new lines 10 and 11, replace the periods (.) in the strings with question marks (?). Be sure not to disturb the spaces that also fall between the quotation marks, or the quotation marks themselves.
6. Click Save & Compile in the macro toolbar. The macro will now shrink two spaces after question marks, as well as after periods.

Tip #228 from *Tux Tips*

You may find it easier to create a short version of a macro such as reformat.wcm, which performs several very similar tasks, and expand it with the macro editor, rather than to create the macro entirely by recording it.

Entering New Commands

If you want to add to a macro by entering new commands that aren't simply variants of existing commands, you have two choices: You can type the command directly, or you can click the Command Inserter button in the macro toolbar. This step opens the WordPerfect Macro Command Inserter dialog box (Figure 19.16).

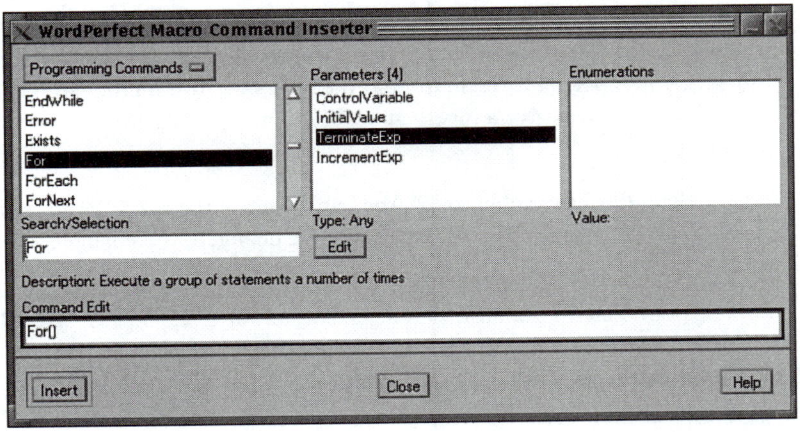

Figure 19.16
You can use the WordPerfect Macro Command Inserter to locate commands and insert them into a macro.

You can select commands from any of three classes by using the option button in the upper-left corner of the dialog box: Programming Commands, System Variables, and WordPerfect Commands. The first of these is the set of commands you use to control your macro and to interact with the user. The WordPerfect Commands each correspond to one command you can enter into WordPerfect, such as a find operation. The System Variables are values that are maintained by WordPerfect, such as the size of the font.

When you've selected the class of command or variable you want to use, you locate the specific one you want from the scrollable list on the left side of the dialog box. You can either double-click on the name of the command or variable or select it and click Edit. In either case, the command then appears in the Command Edit field near the bottom of the dialog box. With some commands, you get a list of parameters the command takes. These are values you must provide to the macro to make it work. For instance, you may need to provide the name for a prompt, or a variable into which you'll store a user's response. Double-clicking on a parameter won't accomplish anything; parameters are listed mainly to let you know what they are and that you can (and sometimes must) use them. Some parameters produce a set of enumerations on the right side of the dialog box. These are possible values that a parameter can take. When you've finished constructing the command with the WordPerfect Macro Command Inserter dialog box, you click the Insert button to enter the command into the macro.

There are a large number of commands, and they vary substantially in how they're used, so it's difficult to be more precise about where and how you'd use commands without getting into details for specific commands. The main thing to know at this point is that the WordPerfect Macro Command Inserter dialog box can help you to find commands. It won't do you much good unless you have some idea of how to program a macro, but if you know the basics and just can't recall precisely what a command is called, this dialog box is helpful.

VARIABLES AND PARAMETERS

Most programming languages, including WordPerfect's macro language, rely heavily on *variables*. These are simply a way to identify a value that can vary as the macro runs. For instance, if you were to write a macro to prompt the user for the font size and then set that font size, you'd need a variable to hold the font size. You might call it something like SizeOfFont. You can give a macro an initial value by using the := operator. For instance,

```
SizeOfFont:=12
```

sets the SizeOfFont variable to 12. You can then use the SizeOfFont variable in place of the constant value 12, and if the variable's value is changed, the changed value will be used instead.

Variables can contain numeric values, such as 12, or text, such as "WordPerfect." Text that can go in a variable is often referred to as a *string*. You specify a string by enclosing it in quotation marks. For instance, you might use

```
ThePrompt:="What size font?"
```

WordPerfect provides you with quite a few *system variables*. You can see what these are by selecting the System Variables item in the WordPerfect Macro Command Inserter dialog box. These variables allow you to access certain variables relating to WordPerfect's current state, such as the font size and text alignment. System variables are all preceded by a question mark, so they're easy to identify in a macro.

To tell WordPerfect to use a variable, you use it as a *parameter* in a command. Parameters appear in parentheses after the command's name. For instance, the command

```
FontSize (SizeOfFont)
```

sets the size of the font to the value of the parameter, which is the variable SizeOfFont. You can also often pass *constants* to commands instead of variables. Constants can't be changed. For instance, the number 12 is a constant, as is the string "font size." The WordPerfect Macro Command Inserter tells you what type of value to pass for each of the parameters.

Note

The WordPerfect command inserter uses terms such as *word* and *float* for parameter types. The term *word*, in this context, is not a word in a human language, but rather it's a numeric value. A *float* isn't something that appears in a parade, but is a floating-point number (one that can have a decimal point). WordPerfect uses other programming terms to describe parameters, such as *Boolean* (a value of 1 or 0) and *enumeration* (a set of possible values, such as *Digit* or *Letter*). These are all possibilities for the type of parameter needed by a command.

When you pass a parameter to a command, the command may use the value of the parameter or, if the parameter is a variable, the command may alter the value of the parameter. You can generally tell which will be the case based on the type of command. For instance, a command to acquire input from the user takes string constants for prompts and alters a variable.

If a command takes more than one parameter, you separate them with semicolons, as in `Assign(VariableName; Value)`. If you omit the semicolon or use another character, such as a comma, WordPerfect complains when you try to use your macro.

Prompting for Input

One of the major reasons for writing a macro, rather than recording one, is to make the macro *interactive*—to get information from the user and incorporate it into a document. For instance, you might want your macro to prompt you to enter a telephone number, which will then be entered into the text the macro creates.

WordPerfect macros support quite a few methods of prompting for input, and going into them all is beyond the scope of this book. One of the more useful of these methods, however, is the `GetString` command. This is one of the Programming Commands in the WordPerfect Macro Command Inserter, and its syntax is


```
GetString(VariableName, Prompt, Title, Length)
```

The first of the four parameters, `VariableName`, is the variable in which you'll record the user's input. You needn't call it VariableName, of course; you use whatever name you've given the variable, such as *PhoneNumber* or *ZipCode*. The `Prompt` and `Title` parameters are both strings. The first is a string that appears inside the dialog box as a prompt, and the second is the dialog box's title. Length is a number (an "unsigned word," according to the WordPerfect Command Inserter) that is the maximum length of the variable to be returned.

As an example, the command

```
GetString(PhoneNumber, "Enter your phone number:", "Phone", 20)
```

produces the Phone dialog box shown in Figure 19.17. If this command appeared in a macro, the user could type a phone number into the dialog box and you'd be able to access it later in the macro by using the PhoneNumber variable.

Figure 19.17
A dialog box created by `GetString` stops accepting input after accepting the number of characters specified by the `Length` parameter.

The `GetNumber` and `GetUnits` commands work much like `GetString`, except that these commands don't take the `Length` parameter and are intended to accept numeric input. `GetNumber` takes any numeric value, whereas `GetUnits` is designed to get a measurement, such as 2c for 2 centimeters. You can use `GetUnits` to prompt for changes to margins, tab stops, and so on.

If you want to insert the value of a variable into a document, you can use the `Type` command, which takes a single parameter—the variable to be entered into the text. For instance, the following two-line macro prompts the user for a telephone number and inserts it into the text:

```
GetString(PhoneNumber, "Enter your phone number:", "Phone", 20)
Type(PhoneNumber)
```

Plenty of commands other than `Type` can take variables as parameters, and you can use similar pairings to collect data to be used to set formatting features in WordPerfect, to search for or replace text, and so on.

Loops

One of the strengths of any programming language is its capability to do the same thing over and over again, with little effort on the programmer's part. This capability is handled by a type of programming structure called a *loop*. WordPerfect's macro language provides several loop structures to give you flexibility in how to handle a loop.

One of the most basic—and useful—loop commands is `For`. This command takes four parameters: a control variable it uses for counting, the initial value of the control variable, a means to test when to stop the loop, and some means of incrementing the control variable. For instance, the following command creates a loop that runs 10 times:

```
For(Loop; 1; Loop<=10; Loop+1)
```

Any commands that follow the `For` command are part of the loop until an `EndFor` command appears. For instance, if you placed the preceding `GetString` and `Type` commands between the above `For` command and an `EndFor` command, WordPerfect would collect 10 telephone numbers and insert them into the document.

Conditional Statements

You may want your macro to do different things depending on certain conditions, such as the value of a user's input or the size of a font or margin. You can handle this situation with the `If` and `Switch` commands, which are two examples of *conditional* statements. Both of these commands take a single parameter containing a test condition, and they both end with a matching end statement (`EndIf` or `EndSwitch`). The `If` statement can also use an `Else` statement between the `If` and `Endif` statements.

In the case of `If`, the test condition must be a Boolean—that is, it must be describable in terms of true or false. For instance, if the test were 4=5, it would always be false and the commands between the `If` and `Else` (or `Endif` if there's no `Else` clause) would never execute. In the case of a test of 4=5, the commands between `Else` and `Endif`, however, would always execute. For instance, consider the following macro code:

```
If(OldMargin<NewMargin)
   Type("Wider")
Else
   Type("Narrower")
EndIf
```

This code enters the string Wider into the document if the value of `OldMargin` is less than the value of `NewMargin` but enters Narrower if this isn't the case.

Note

> Indenting the lines affected by a conditional or loop statement is a tradition in programming, and it helps you to see the structure of your program.

A `Switch` command allows several values in the test statement, as opposed to the two in an `If` statement, and executes a single command for each value that matches. For instance:

```
switch(TheStyle)
   CaseOf 1: Type("Style 1")
   CaseOf 2: Type("Style 2")
   CaseOf 3: Type("Style 3")
endswitch
```


Saving and Compiling Macros

When you've entered your macro, you must *compile* it to make it work. When you compile, WordPerfect converts the macro text into a format that it can more easily understand, to speed the running of the macro. You'll still be able to edit the macro, though; the compiling step is not destructive. You use the Save & Compile button on the macro toolbar to save your work and simultaneously compile the macro, or the Save As button to save the macro under a new name and compile it. If you select the second option, WordPerfect displays the Macro Save As dialog box, which works just like the Play Macro dialog box (refer to Figure 19.3) except that you enter a new filename for the macro.

Does WordPerfect complain about errors when you try to compile your macro? See "Problems Compiling a Macro" in the "Troubleshooting" section at the end of this chapter.

If you like, you can also click on the Play button in the macro toolbar to immediately run the macro. You will, however, be running the macro inside the Macro editor window, which may damage the macro if it is designed to alter text. It's probably a better idea to open a new document window and run the macro from there, using the normal macro-running commands.

LEARNING MORE ABOUT MACRO PROGRAMMING

Programming WordPerfect macros, like programming any computer language, is something that you learn best by doing. The preceding discussion can help you to understand the basics, but WordPerfect's macro language is quite complex. One good way to learn more about it is to examine the macros that come with WordPerfect. Some of these do some sophisticated things, including constructing and displaying dialog boxes, using loops and conditional statements to do more than enter different pieces of text in the document, and so on. Start by modifying existing macros, even if you don't need to change them. You can then try cutting and pasting pieces of these macros into macros you recorded. It won't be long before you can write sophisticated macros yourself.

TROUBLESHOOTING

MACROS WON'T WORK WITH OTHER PROGRAMS

You're trying to create a macro that exits from WordPerfect or that interacts with another program (say, to cut and paste text from another application), but you can't get the macro to work.

WordPerfect's macro facility is largely self-contained; it affects *only* WordPerfect. You can't use it to control other Linux programs, at least not when you're recording a macro (as opposed to programming one). It's also not possible to create a macro that will exit from WordPerfect; although you can choose Program, Exit (Shift+Ctrl+F4) from the program Control window when recording a macro, doing so will result in an unusable macro.

If you want to create a macro that uses other Linux programs' features, you may be able to program the macro and use the AppExecute macro command to run another program. For instance, `AppExecute("emacs")` runs the emacs program. Your control of the external program ends with that program's command-line parameters, however, and getting this program's output back into WordPerfect may be awkward.

You might also want to look into other Linux scripting and programming languages, like sed, awk, and perl. You can accomplish a lot with these, though you can't use them to control WordPerfect and its features.

FONTS DON'T DISPLAY

You've tried running the softfonts.wcm macro, but most of the fonts turn up looking just like Courier.

The downloadable version of WordPerfect 8 for Linux ships with only a handful of fonts, and you can't install more because there's no font installer for this version of the program. Therefore, this result is normal if you have this version of the program. Early shipments of the personal retail version of the program also lacked the fonts, so you'll see the same result with these versions of the program. If you have the retail version and are missing the fonts, you can ask Corel to send them.

PROBLEMS COMPILING A MACRO

You're trying to save or compile your macro, but you get an error message, such as that in Figure 19.18.

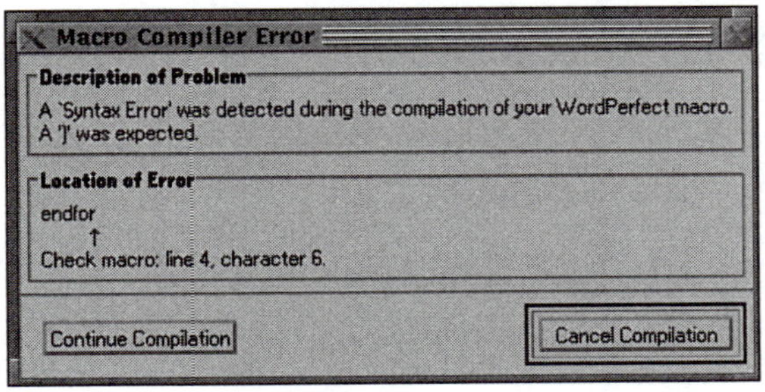

Figure 19.18
If WordPerfect can't understand your macro, it complains.

WordPerfect is very fussy about its macro language. If you make a slight mistake in the syntax, such as omitting a parenthesis or inserting a stray character, WordPerfect complains. As Figure 19.18 shows, WordPerfect tells you where it thinks the error is. Be wary, though; sometimes the real cause of the error is several lines earlier in the macro.

CHAPTER 20

THE SAME LETTER, DIFFERENT INFORMATION: MAIL MERGE

In this chapter

OVERVIEW

One of the common problems faced by many businesses is the need to produce large numbers of customized documents. Typically, these documents are letters, and the letters need to have the recipients' names and addresses printed on them. You might also want to insert other customizations, such as account numbers, product names, or even entire paragraphs into some documents but not others.

Even individuals sometimes need to produce customized letters or other documents. Have you ever wanted to send holiday letters to friends and family? Have you ever moved and needed to notify all the magazines, banks, miscellaneous businesses, and friends of your new address?

This need is the one for which WordPerfect's mail merge feature was designed. It allows you to create two documents and, from those two, print out half a dozen, a hundred, or even thousands of copies, each of them customized for the recipient. The first document contains only the customized information, such as names and addresses. This document is known as the *data file*. The second file you create is the letter or other document, and it uses special codes instead of the information that's to be customized. This document is referred to as a *form file*. When you perform the merge, WordPerfect combines the two files into one file containing as many customized copies of the form file as there are data entries in the data file.

Note

Creating customized letters is probably the most popular application of the mail merge feature and is the one that inspires the name. You could use this feature for other purposes, however, such as printing certificates with recipients' names or price lists for a chain of stores customized with the stores' addresses and phone numbers. If you find yourself changing a few details in a document to produce minor variants of that document, it's a candidate for use with mail merge.

WordPerfect's mail merge feature is quite powerful, and it can do a lot more than just drop text from one file into another. You can use it to include a certain piece of text for some people but not others or to create a merge using only a subset of the entries in the data file, for instance. Chances are you won't need all these features for every merge. Therefore, I first present the basics of using the mail merge feature. This information is probably enough for most of your mail merge needs. Later in this chapter, I delve into some of the more advanced features of mail merge.

Note

WordPerfect uses its merge feature in its ExpressDocs templates. Therefore, if you've used this feature, you may already be familiar with some important merge concepts. Further, after you read this chapter, you'll be able to design more complex ExpressDocs templates.

→ To learn more about ExpressDocs and templates, **see** "Promoting Cross-Document Consistency with Templates," **p. 333**

A Basic Mail Merge Example

Suppose you want to thank previous clients for their business while telling them of your latest offerings or specials. You can write one letter and, with WordPerfect's mail merge feature, create customized letters and envelopes for each client.

Creating a Data File

The first step in performing a merge is to create an appropriate data file. Of course, if you already have a data file, you can skip this step and use what you have. This feature is one of the keys to the value of mail merge: You can use the same data file for different mail merges. In fact, WordPerfect 8 for Linux lets you use data files that were created outside of WordPerfect, so you may not need to reenter this information if you already have it in some other program.

→ If you want to use a foreign file for a merge's data, **see** "Using Foreign Data Files," **p. 532**

A WordPerfect mail merge data file stores the unique data you want to put into a letter. The file is divided into *records* and *fields*. A record is the data associated with one person or company. In the end, you'll get one letter per record in the data file, unless you use advanced merge techniques to limit the number of records you use, as described later in this chapter. Each record is composed of one or more fields, each of which corresponds to one piece of information to be inserted into the merge form file. For instance, in a typical mail merge, you'll have fields for name, address, city, state, and zip code. You can include additional fields that may be relevant, such as account numbers, birthdays, telephone numbers, professions, and so on. You don't need to merge every field in your data file into any given letter, so don't be shy about creating fields you don't immediately need, if you think you *will* need them in the future.

→ To learn how to limit the records you use in a merge, **see** "Merging from a Restricted Set of Records," **p. 542**

Tip #229 from *Tux Tips*

In general, if you can think of a way to break up a field into two or more fields, you should do so because this technique imparts greater flexibility on your design of the form file. For instance, you *could* use one field for the city, state, and zip code; but if you use multiple fields, you can refer to the recipient's city in the text of the letter. Similarly, you may want to split the name into first and last names and perhaps include a field for a title or honorific. This approach allows you to use a salutation of Catherine for informal letters or Dr. King for more formal occasions.

To define the structure of your data file, follow these steps:

1. Choose Tools, Merge (Shift+F9) to open the Merge dialog box (Figure 20.1).
2. Click the Data button in the Data File section of the Merge dialog box. WordPerfect responds by displaying the Create Data File dialog box (Figure 20.2).

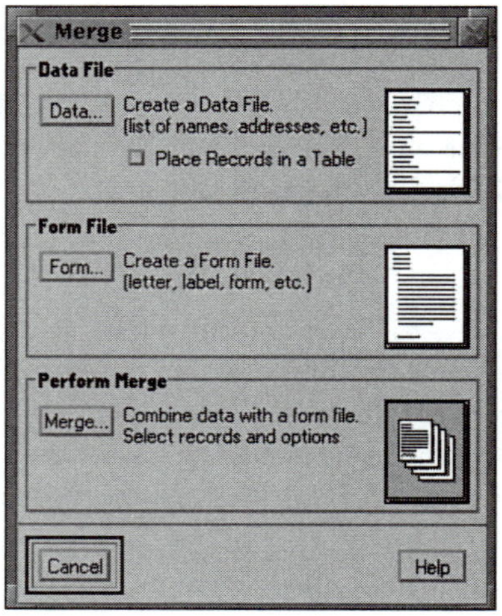

Figure 20.1
You create both data files and form files from the Merge dialog box, as well as perform the merge itself.

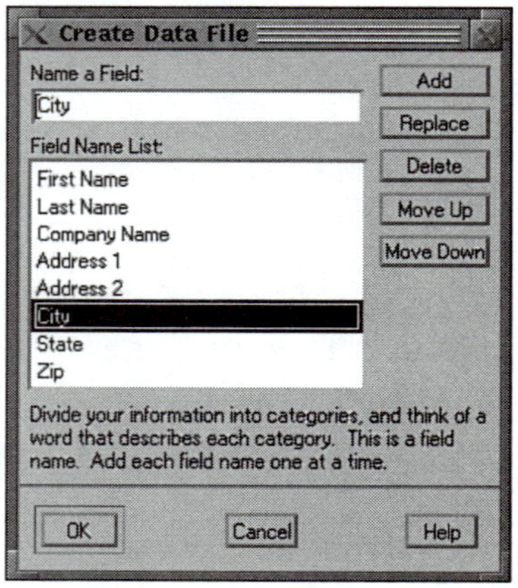

Figure 20.2
The Field Name List starts out empty, but as you enter field names, they appear here.

3. Type in the name of a field and press the Enter key or click Add. You'll later identify the field with this name, so it should be descriptive, and you shouldn't use the same field name for more than one field.
4. Repeat the previous step for as many fields as you intend to create.

Note

Figure 20.2 shows the fields I use for the rest of this example.

5. Click OK in the Create Data File dialog box to finish defining the fields.

When you've defined the fields in your data file, WordPerfect automatically proceeds to the second stage of creating the data file: creating records. It uses the Quick Data Entry dialog box (Figure 20.3).

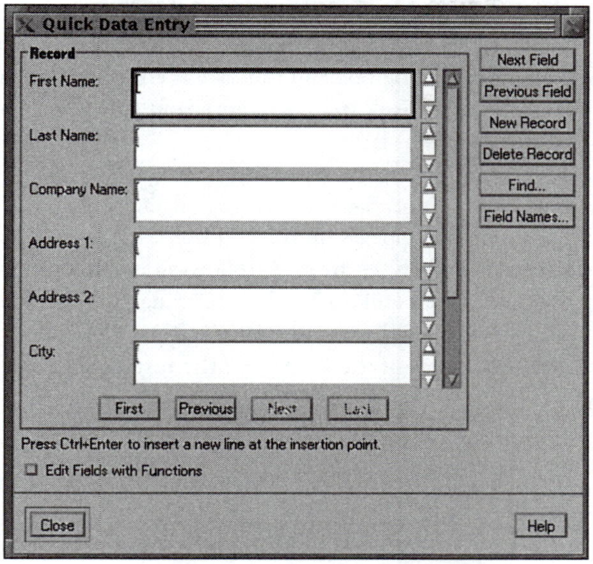

Figure 20.3
The record names in the Quick Data Entry dialog box are the same as the names you assigned in the Create Data File dialog box.

To use the Quick Data Entry dialog box, follow these steps:

1. Click in the first field's area (called First Name in Figure 20.3, though it may be different for your file). You can then enter text into this field. When you're done, press the Enter key, the Tab key, or click Next Field to move to the next field.

Tip #230 from *Tux Tips*

If you need to enter two lines of data in one field, you can use Ctrl+Enter to separate the lines. You could use this technique if an address uses more lines than you have address fields in your data file format. When you merge a form file that uses this field, WordPerfect inserts an [HRt] code at that point in the file.

2. Enter data into all the remaining fields in the same way you did the first.

Note

You can leave a field blank. For instance, the definitions in Figures 20.2 and 20.3 give two lines for an address, but both lines may not be necessary. Leaving a field blank does no harm, though of course, if your form file calls for information from that field, it will be blank.

3. When you've entered information into all the fields, click New Record or press the Enter or Tab key. WordPerfect enters the data you just typed into the data file and clears the fields in the Quick Data Entry dialog box. You can then create a new data record.

4. When you're done entering information for all the data records, click Close. WordPerfect asks whether you want to save the file to disk. If you select Yes, you get a standard File Manager dialog box and can save the file (the filename extension .dat is standard for data files, but isn't strictly required). If you select No, you can further edit the data file and then save it later.

Whether you save the file or not, the result of your data entry should look something like Figure 20.4. (This sample file appears on the CD in the chapter20 directory under the name merge.dat.) Each record is separated from its following neighbor by an ENDRECORD code and a pair of double lines, and each field is terminated by an ENDFIELD code and a carriage return. The ENDFIELD and ENDRECORD codes appear in red. The first record serves only to define the names of the fields, and the rest of the records contain the actual data for a merge.

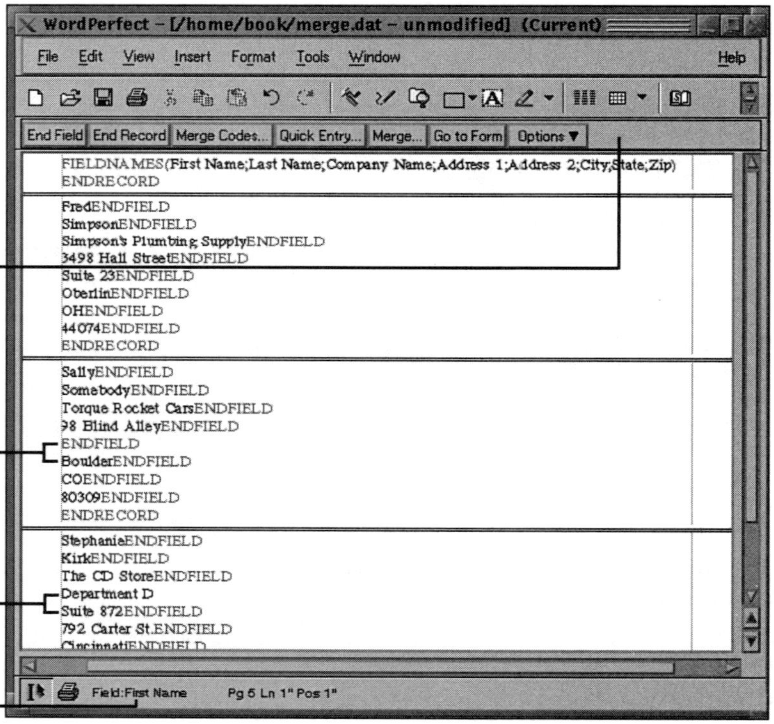

Figure 20.4
A merge data file normally displays in Draft view, with double lines indicating [HPg] codes.

If you don't want to make any changes or additions to the merge data file, you can close the window or choose File, New to clear it. When you perform the merge, WordPerfect accesses the file from disk, not from the document window (though you can change this behavior, as explained later in the chapter). For this reason, though, you should be sure that you've saved the file to disk before you proceed with creating the form file.

Tip #231 from *Tux Tips*

A merge data file may contain sensitive information. You can prevent others with access to your Linux computer from reading the file by setting appropriate file permissions on the file. For even greater protection, you can save the file with encryption enabled. If you do enable encryption, WordPerfect prompts you for the password at merge time.

→ To learn about Linux file ownership and permissions, **see** "File Ownership and Permissions," **p. 78**

→ To learn about file encryption options, **see** "Saving Your Document," **p. 67**

Creating a Form File

The merge data file is only half of what's needed to do a merge. You presumably want to write a letter to the people whose names appear in the merge file. You begin such a letter somewhat differently than you do an ordinary WordPerfect document, but much of what you do with the letter once you've started it is the same:

1. Start with a blank WordPerfect document window, by choosing Program, New Window (Ctrl+N) if necessary.
2. In the new document window, choose Tools, Merge (Shift+F9). This step opens the Merge dialog box (refer to Figure 20.1).
3. Click Form in the Merge dialog box. WordPerfect then displays the Create Form File dialog box, as shown in Figure 20.5.

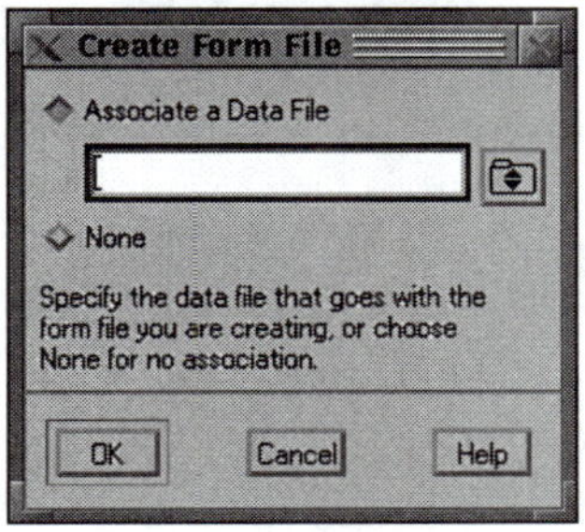

Figure 20.5
You can associate a merge data file with a form file when you create the latter by using the Create Form File dialog box.

4. Type the name of the merge data file or click the folder icon in the Create Form File dialog box to select a data file using a File Manager dialog box. Alternatively, click None if you want to specify the merge data file at the time of merge. You might use this technique if you intend to use the same letter with more than one merge data file. (You need to specify your merge fields by number rather than by name if you do so, however.) After you specify the name of the merge data file, click OK; WordPerfect then returns you to the document window, which now has a Merge form toolbar (Figure 20.6).

Figure 20.6
When you edit a merge form file, WordPerfect displays the Merge form toolbar to give you easy access to merge-related features.

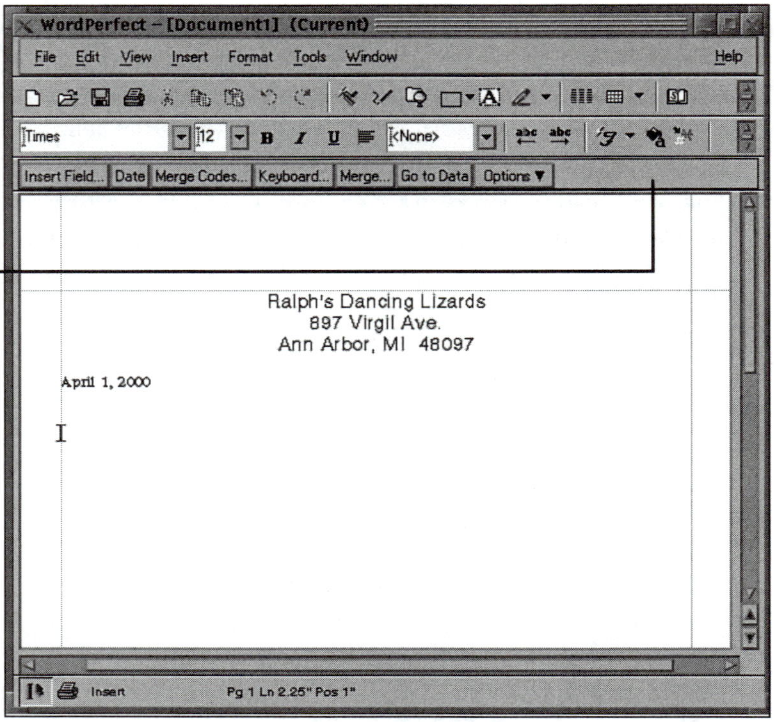

5. If there's text you want to appear before any merged information, enter the text now. You may use styles or macros, if appropriate, as well as any other formatting features you see fit to include. You can even use templates, though if they include their own prompts for recipients' names and addresses, you have to delete that information to make way for the data from the merge file. Figure 20.6 shows a return address and date already entered in the merge form file.

→ For more information on styles, **see** "Ensuring Consistency Within Your Document with Styles," **p. 305**

→ For more information on macros, **see** "Letting the Computer Do the Work: Macros," **p. 483**

→ For more information on templates, **see** "Promoting Cross-Document Consistency with Templates," **p. 333**

Tip #232 from *Tux Tips*

If you expect to be reusing this merge form (for instance, if it's for a billing notice that goes out to different clients weekly), you may want to use a [Date] code rather than a manually entered date. To insert a [Date] code, choose Insert, Date/Time and check the Keep the inserted date current check button, or click Date on the merge form toolbar. The former option gives you more formatting choices. Either way, whenever you perform a merge, WordPerfect places the current date in that location.

→ To learn more about WordPerfect's date and time features, **see** "Inserting the Date and Time," **p. 202**

6. To enter a merge field, click the Insert Field button in the merge form toolbar. This step opens the Insert Field Name or Number dialog box, as shown in Figure 20.7. Select the field you want to insert and click the Insert button. Assuming the document window isn't obscured by another window, you'll see an appropriate code in the document window.

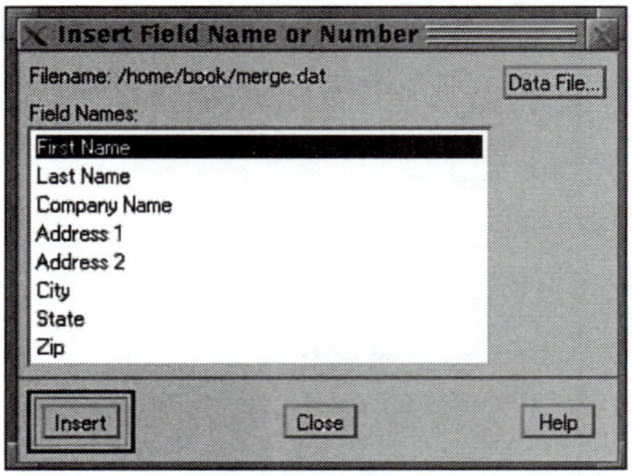

Figure 20.7
You can insert a field without closing the Insert Field Name or Number dialog box by clicking Insert, or remove the dialog box by clicking Close.

Do you get a warning that there are no field names or numbers when you click on Insert Field? See "No Field Names or Numbers" in the "Troubleshooting" section at the end of this chapter.

Tip #233 from *Tux Tips*

You can also double-click on the name of the field you want to insert to insert it.

7. Chances are you don't want all the fields mashed together, so you can click in the document window to enter whatever formatting you need to separate the fields, such as spaces, commas, carriage returns, or even entire paragraphs of text.

Note

Different window managers have different rules for when windows acquire *focus (p. 15)* or move in front of or behind other windows. Depending on your window manager, then, the Insert Field Name or Number dialog box will probably remain in front of the WordPerfect document window, but the document window will be active after you click in it or simply move your mouse pointer over it.

→ For more information on window manager focus, **see** "Understanding Window Managers," **p. 12**

8. You can switch back and forth between the document window and the Insert Field Name or Number dialog box until you're done entering the fields.

9. Finish writing your letter. The result might look something like the letter shown in Figure 20.8.

Figure 20.8
The merge codes appear in red, with the names of the fields you're inserting in black, between red parentheses. Note the reference to the recipient's city in the letter's main paragraph, as well as in the address.

Merge codes

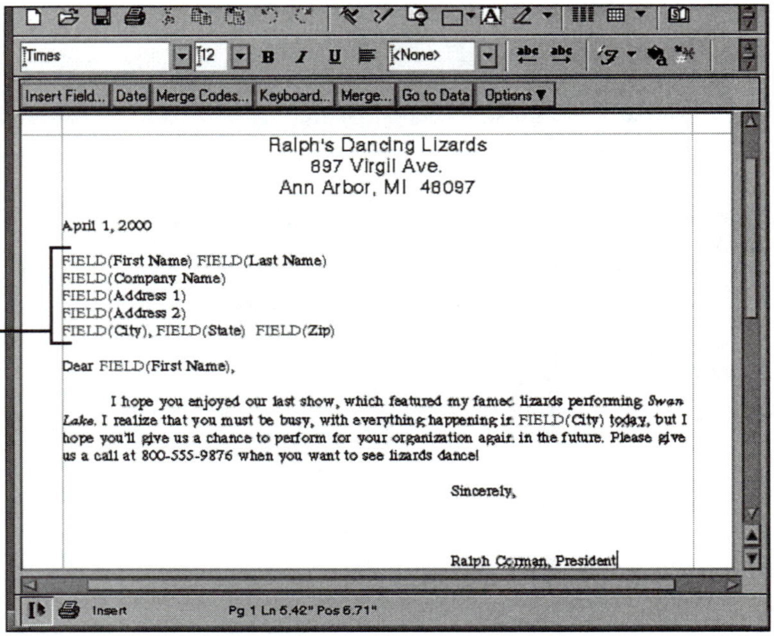

10. Save the letter file by choosing File, Save As (F3). You should typically give WordPerfect merge form files a filename extension of .frm, though it isn't a requirement.

Note

The file depicted in Figure 20.8 is on the CD that accompanies this book, in the chapter20 folder under the filename merge.frm.

When you design a merge form file, you should remember that the field codes may take up more or less space than the text that's to replace them. Indeed, any given field may be larger than its replacement text for one record but smaller for another. Therefore, if you need any text that includes merged codes to be formatted or aligned in any particular way, you should be careful to do this formatting in as flexible a way as possible. For instance, you should use a WordPerfect table rather than one you create the old-fashioned way with tabs or spaces. If you insert text into a paragraph that contains large words, you may want to enable automatic hyphenation to reduce the likelihood of getting an ugly gap at the end of a line from one of these words being wrapped down to the next line.

→ To learn how to create a table, **see** "Tables," **p. 275**
→ To learn how to turn on automatic hyphenation, **see** "Hyphenation," **p. 220**

Performing the Merge

With both merge form and merge data files in hand (or on disk, more precisely), you're ready to do the merge. When you perform a merge, you get a new file that contains the combination of the merge form file with each of the records from the merge data file. You'll then be able to print this file. You can also optionally modify it further. For instance, you may find that the merge has done unfortunate things to the formatting of your text for one or more records, and you might want to correct this problem manually.

To do a merge, follow these steps:

1. Click Merge on the merge form toolbar or choose Tools, Merge (Shift+F9) to open the Merge dialog box (refer to Figure 20.1).
2. Click the Merge button in the Merge dialog box to open the Perform Merge dialog box (Figure 20.9).

Figure 20.9
You can control many aspects of the merge process from the Perform Merge dialog box.

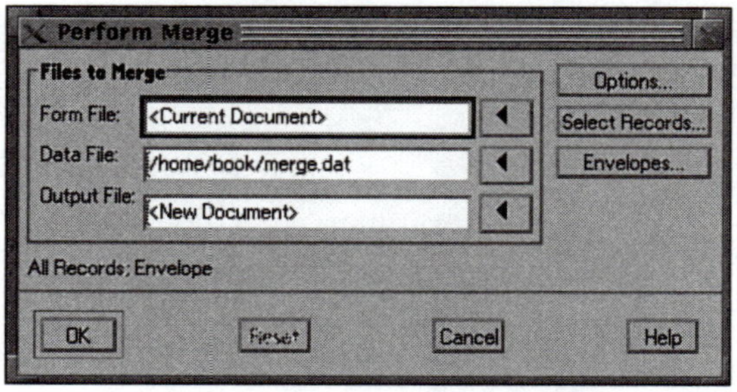

3. If you want to produce envelopes for your mail merge at the same time you produce the letters, click Envelopes. This step opens the Envelope dialog box (Figure 20.10). Type your return address into the Return Addresses field and then click in the Mailing Addresses field. Click the Field button in the Envelopes area of the dialog box to open the Insert Field Name or Number dialog box (refer Figure 20.7) and build the recipient's mailing address from the fields there, much as you did when you constructed the merge form file. You can also change the envelope type, tell WordPerfect to use a POSTNET bar code, change the envelope fonts, and so on. When you're done, click OK.

→ To learn more about the Envelope dialog box, **see** "Envelopes," **p. 247**

Do you get a warning that there are no field names or numbers when you click on Field in the Envelope dialog box? See "No Field Names or Numbers" in the "Troubleshooting" section at the end of this chapter.

Tip #233 from *Tux Tips*

If you'd like to produce mailing labels rather than envelopes, you need to do a second merge operation. Before you choose Tools, Merge (Shift+F9) for this second merge, choose Format, Labels and follow the instructions in Chapter 9, "Page and Document Formatting," for creating a label layout. Place your merge codes in the first label space, and when you perform the merge, WordPerfect automatically places the text for each record in its own label field.

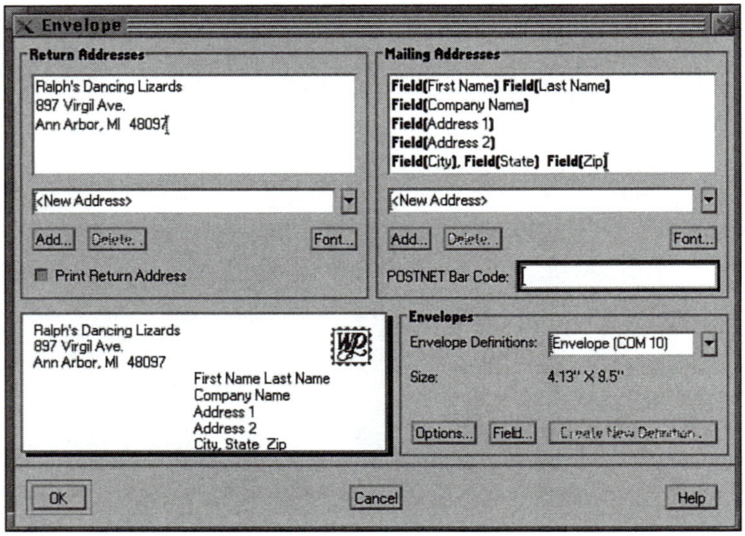

Figure 20.10
The preview area shows text in a font that's larger than the default, so if text seems squeezed or even doesn't fit, don't be concerned; it will probably print fine.

→ For more information on creating labels, **see** "Labels," **p. 250**

4. Click OK to perform the merge. WordPerfect opens a new document window that contains the merged data file. If you elected to print envelopes, they'll appear at the end of this file.

After you merge your file, you can save it or print it. Before you do that, however, you may want to check the file for any obvious problems. You can correct many problems, such as formatting that was disrupted because of unusually large or small merge field contents, on a case-by-case basis. If a problem is more systematic, you may want to close the merged data file, correct the problem in the merge form file, and perform the merge again.

If you elected to produce envelopes, you may want to copy them to a new file and print them separately because many printers don't have a separate envelope feeder and won't provide you with an easy way to specify a change from normal paper to envelopes. Alternatively, you can use the Multiple Pages option in WordPerfect's Print dialog box to print the envelopes separately from the letters.

→ To learn how to copy text between WordPerfect documents, **see** "Copying from One WordPerfect Document to Another," **p. 99**

→ To learn how to use the Multiple Pages print option, **see** "Print Selection," **p. 153**

Are you trying to create a merge document with page numbers? You may be running into problems getting the page numbers to start at 1 for each copy of the merged document. For a solution to this problem, see "Page Numbering and Mail Merge" in the "Troubleshooting" section at the end of this chapter.

Merge Data Files

The flexibility of the mail merge feature is only as great as the quality of the merge data files you use. For instance, if you have a data file that includes a single record with a person's complete address, you will be able to use it to address an envelope or letter. You won't, however, be able to include the recipient's name in a salutation or refer to any other information from this address in the body of a letter.

Fortunately, you can manipulate merge data files in various ways. WordPerfect also provides options and features that allow you to use different types of merge data files and to import data from external sources for use in a mail merge.

Data Files as Records Versus Tables

If you look back at Figure 20.1, you'll see a check button called Place Records in a Table in the Data File portion of the Merge dialog box. If you select this option, WordPerfect stores and displays merge records in a tabular format, rather than in the textual format I've described up to now. You still enter data into a file in the same way, and you perform the merge operation identically; the difference is largely one of display. The data text format does allow for more text in each field, as well as more fields in total, however—data tables support a maximum of 64 fields, whereas data text files support 255 named fields and an unlimited number of numbered fields. Figure 20.11 shows the same data displayed in both text and table formats.

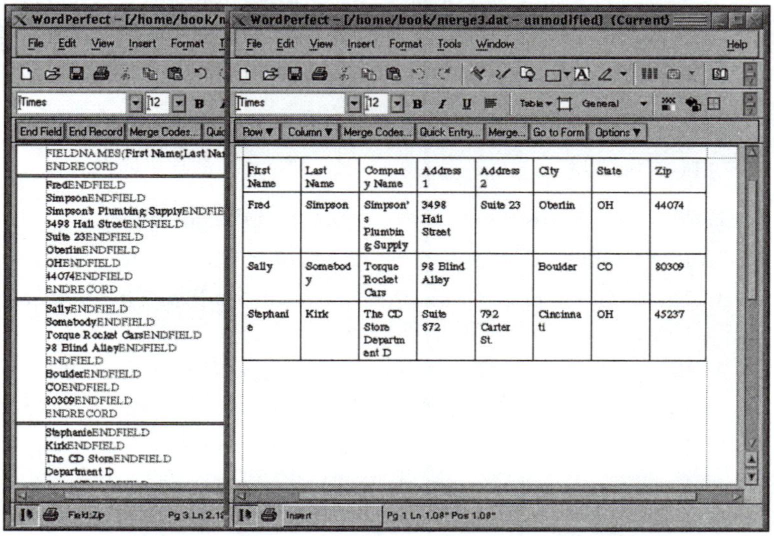

Figure 20.11
The window on the left shows data in a textual format; the same data appears in tabular format on the right.

Tip #234 from *Tux Tips*

Use the tabular data format if you need few fields and if you want to print the merge data file in a more easily read format. If you have more than a few fields or if the fields contain lots of text, the table form may become difficult to read. You can improve a table's practical field capacity by adjusting the page margins and changing from portrait to landsca pe orientation.

➔ For information on adjusting margins, **see** "Margin Adjustments," **p. 206**

➔ For information on setting the page orientation, **see** "Page Size," **p. 226**

You can convert an existing table merge data file to textual form by selecting the entire table and choosing Delete from the context menu or pressing the Delete or Backspace keys. This step produces the Delete Table dialog box (Figure 20.12). To convert the table, select Convert to Merge Data File. You can also use this procedure on any other table, even if you didn't create it as a merge data file initially.

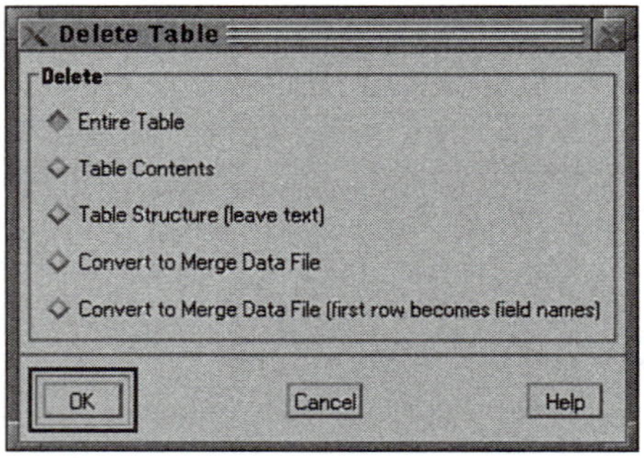

Figure 20.12
The Delete Table dialog box lets you convert the table to a textual merge data file.

Tip #235 from *Tux Tips*

The Convert to Merge Data File (first row becomes field names) option may not work correctly with tables that weren't originally created as table-formatted merge data files, so use it with caution.

When you've converted the table to a text-format merge data file, you may need to enter field names. Follow these steps:

1. Position the insertion point at the top of the document.
2. If there's a record corresponding to column labels, delete this record.
3. Select Merge Codes from the merge data toolbar. This step produces the Insert Merge Codes dialog box (Figure 20.13).
4. Select the FIELDNAMES(name1;...;nameN) item from the scrollable list and click Insert. WordPerfect opens the Create Data File dialog box (refer to Figure 20.2).
5. Enter the field names as described earlier and then click OK.
6. Click Close in the Insert Merge Codes dialog box.

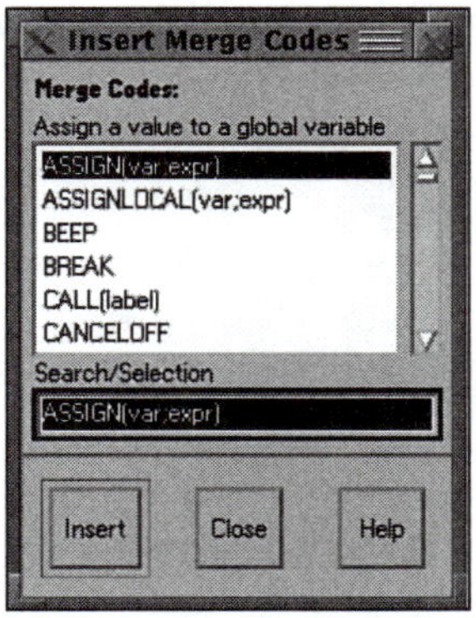

Figure 20.13
You can insert any of a number of advanced merge codes from the Insert Merge Codes dialog box.

7. If necessary, remove any extraneous merge codes and [HPg] codes. The first record should contain the FIELDNAMES code and an ENDRECORD code, followed by an [HPg] (indicated by a double line). The next record should contain the first data record. Check Figure 20.4 for an example.

EDITING MERGE DATA FILES

You can edit a merge data file in one of two ways: directly in the WordPerfect document window or indirectly, using the Quick Data Entry dialog box (refer to Figure 20.3).

DIRECT DATA EDITING

For many operations, editing data directly is the simplest way to accomplish your goal. You can use this method to correct typos, change addresses, and even create entirely new records. You can use most of WordPerfect's usual editing features when you edit your data in this way, including cut and paste, find and replace, and so on. You can also use direct data editing to enter text attributes such as bold or centering. Finally, you can enter special merge codes such as the FIELDNAMES code described earlier with this method.

Caution

> Do not apply a text attribute across more than one field! WordPerfect will enter a start code for that attribute at the start of the selection and a close code at the end of the selection, just as it does in normal text editing. The problem is that if you merge the first field but not the second or if you merge them out of order, your document will contain a code to turn on this attribute but not one to turn it off. You should apply the attribute you want separately to each field. Be careful not to include the ENDFIELD code when you apply the attribute; if you do, WordPerfect may insert an [HRt] code along with the field.

If you enter new records using direct data editing, you need to insert ENDFIELD and ENDRECORD codes. You do so by clicking the End Field and End Record buttons, respectively, in the merge data toolbar. When you press End Field, WordPerfect inserts the

ENDFIELD code and a carriage return. When you press End Record, WordPerfect inserts an ENDRECORD code and an [HPg], which shows up as a pair of horizontal lines in the Draft view that's used by default for merge data files.

Tip #236 from *Tux Tips*

If you examine WordPerfect's status bar when you're editing a merge data file, you'll notice the presence of a field name indicator (refer to Figure 20.4). This indicator tells you the name of the field in which the insertion point rests and can be very helpful when you're creating new records.

If your data file is in table format, you can use the usual WordPerfect table editing commands to edit the merge data. You won't need to explicitly insert ENDFIELD or ENDRECORD codes, as these functions are handled by the table formatting codes.

→ To learn how to edit a table, **see** "Editing a Table's Contents," **p. 295**

Tip #237 from *Tux Tips*

Because a table structures its data with each field corresponding to a column, you won't be likely to mistakenly enter too many or too few fields for any given record. Therefore, the table data format offers some extra safety when you're editing data directly, assuming that you have few enough fields to edit the table comfortably.

QUICK DATA ENTRY DATA EDITING

If you click the Quick Entry button on the merge data toolbar, WordPerfect displays the Quick Data Entry dialog box (Figure 20.3) from which you can edit existing entries and create new ones. Doing your data entry in this way has the advantage of ensuring that you won't miss a field or produce too many fields, which could cause a merge to place incorrect information in the merged file.

Editing data in an existing file works much like entering new data; you click in the field you want to modify, edit the text, and click Close when you're done. In addition, the Quick Data Entry dialog box supports a number of features you may want to exploit when editing data:

- Record selection—The First, Previous, Next, and Last buttons in the Record area of the dialog box open different records in the data file. You can use the buttons to move through the file and make several changes.
- Find text—If you click Find, WordPerfect displays the Find Text dialog box, shown in Figure 20.14. You can use this dialog box to find text and even perform find-and-replace operations on the data file. When you perform a find operation, WordPerfect searches *all* the fields of *all* the records, so you can use this operation to locate a record by any unique information you know it to contain—a person's name, a city, or anything else.

→ For more information on the find and replace feature, **see** "Find and Replace," **p. 105**

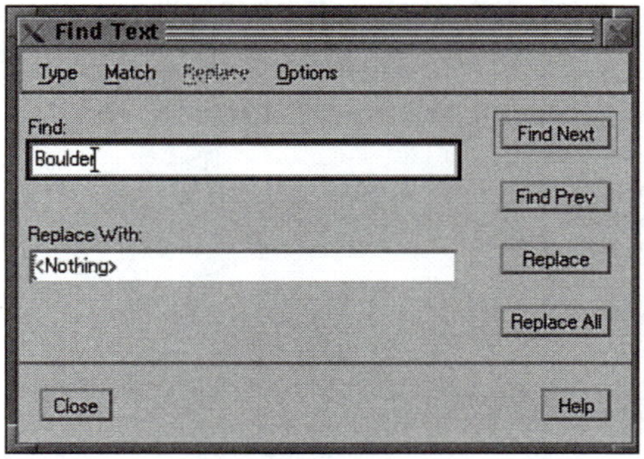

Figure 20.14
The Find Text dialog box lets you locate key words in a data file.

- New Record—If you click New Record, WordPerfect creates a new record immediately after the record that's currently displayed. All the Record fields clear, and you can enter new information into them.
- Delete Record—If you click Delete Record, WordPerfect removes the record whose contents are currently displayed in the dialog box.
- Alter field names—If you click Field Names, you can add, delete, or change the names of fields in the data file, as described shortly.

When you're done editing your data file with the Quick Data Entry dialog box, click Close. WordPerfect asks whether you want to save the changes to disk and, if you select Yes, prompts you for a filename with a File Manager dialog box.

Tip #238 from *Tux Tips*

If you want to save the changes to the original file, click No when WordPerfect asks whether you want to save the file and then choose File, Save (Ctrl+S) at the document window. WordPerfect then saves your changes to the original data file without prompting you for the filename.

Adding or Deleting Fields

If you want to add or remove fields from your data file, you should click on Quick Entry to get the Quick Data Entry dialog box (Figure 20.3) and then click the Field Names button. WordPerfect responds by displaying the Edit Field Names dialog box, as shown in Figure 20.15.

The following options are available in the Edit Field Names dialog box:

- Add—Type the name of a new field in the Name a Field text field and then click Add to insert that field. It appears either after the selected field in the Field Name List or, if no field is selected, after all the field names.
- Add Before—This option works like Add except that the field is added before the selected field or at the start of the list if no field is currently selected.

- Replace—Select a field, alter its name, and click Replace. WordPerfect renames the field.
- Delete—This option removes a field from the database.

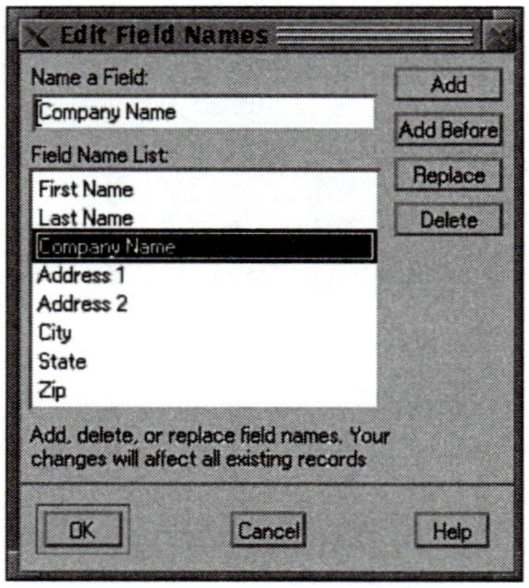

Figure 20.15
You can add, delete, or change the names of fields with the Edit Field Names dialog box.

The Add, Add Before, and Delete options in the Edit Field Names dialog box cause fields to be changed in every record of the data file. For instance, if you delete a field, WordPerfect removes that field, including its contents, from every record. WordPerfect therefore asks for confirmation before performing a delete operation.

The Replace option alters only the field names, which are stored in the FIELDNAMES code of the first record. You can alter these names directly by editing them in the document window if you like.

Using Foreign Data Files

WordPerfect allows you to import data files from other programs for use in a mail merge operation. To do so, follow these steps:

1. Create a blank WordPerfect document window by choosing Program, New Window (Ctrl+N) from the WordPerfect program Control window if necessary.
2. In the new document window, choose Insert, Spreadsheet/Database, Import to open the Import Data dialog box (Figure 20.16).
3. Select the source data type—Spreadsheet (Lotus 1-2-3 format) or ASCII or ANSI delimited text.
4. Select the destination format—Table or Merge Data File.
5. Type the filename or click the folder icon button to select the file using a File Manager dialog box.
6. If you know that the original file contains extraneous rows or columns, you can specify the range of rows and columns you want to import by using standard spreadsheet notation in the Range field of the Import Data dialog box.

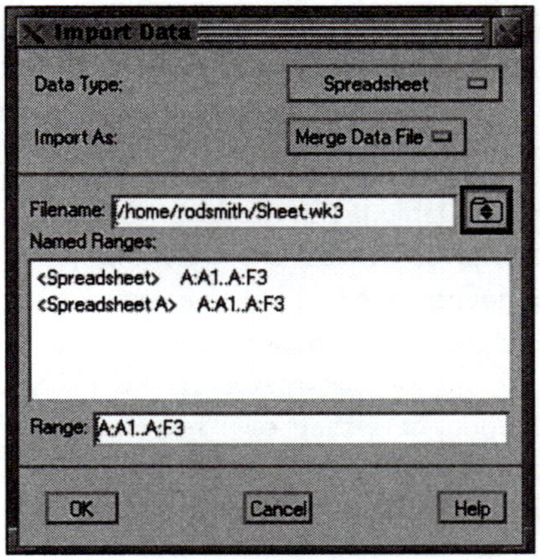

Figure 20.16
You can import a foreign data file for use in a mail merge much as you would to place it in a table.

7. Click OK.
8. If necessary, edit the resulting file. You may want to insert record names, for instance, as described in the earlier discussion of converting tables to text-mode merge files.

→ To learn about converting from tables to text-mode merge files, **see** "Data Files as Records Versus Tables," **p. 527**

9. Save the newly converted merge data file to disk.

Merge Form Files

As you've seen, you can customize a merge form file with data from a merge data file in order to create letters with customized addresses, salutations, and even references to arbitrary information in the merge data file. WordPerfect's merge feature is more sophisticated than this, however. You can prompt the user for input at the time of merge in case you need to include information that's not in the merge data file. You can set up a merge form file with words, sentences, or even entire paragraphs of text that will be included only with certain records.

This level of customization isn't the end of it, either; WordPerfect's merge codes include many features that are similar to the macro programming facility. Although going into detail about all these features is beyond the scope of this book, the sampling provided here should help you solve many mail merge problems and give you the grounding you need to experiment further.

→ For more information on macro programming, **see** "Modifying and Programming Macros," **p. 505**

Inserting Merge Codes

To insert a merge code, you must first open or create a merge data file, as described earlier in this chapter. You can then click the Merge Codes button on the merge form toolbar, which produces the Insert Merge Codes dialog box (refer to Figure 20.13).

→ To learn how to create a merge data file, **see** "Creating a Data File," **p. 517**

Many merge codes require *parameters (p. 347, 508)*, just like macros. These appear in parentheses after the name of the merge code. Parameters can be *constants (p. 508)*, which never change, or *variables (p. 508)*, which can take on different values. You might use a variable to pass merge-time input from the user into the document, for instance.

When you click on a code in the scrollable list in the Insert Merge Codes dialog box, WordPerfect provides a brief description of what the code does just above that list. You click Insert to insert the code into the document's text at the insertion point. If the code takes parameters, WordPerfect prompts you for the parameters, as in the Insert Merge Code dialog box shown in Figure 20.17. Type in the parameters in the boxes and click OK. Some merge codes, such as the FIELD code, have their own specialized dialog boxes to prompt you for information (when you insert a FIELD code, for instance, you get the Insert Field Name or Number dialog box shown in Figure 20.7).

Figure 20.17
The Insert Merge Code dialog box varies for each merge code; some have fewer than three text entry fields, and some bear little resemblance to the dialog box shown here.

Inserted merge codes ordinarily appear in the merge form data file in red so that you can easily spot them and so that you know what codes go where. You can alter this behavior by selecting Display as Markers or Hide Codes from the Options option button in the merge form toolbar, however. You might want to apply this technique if your merge codes are too large relative to the data they'll be inserting and thus cause formatting problems when you're editing the file.

WordPerfect ordinarily formats text you acquire from the merge data file according to whatever font, font attributes, and other options are active at that point in your file. You can override the formatting for any given field in a record by editing the merge data file, however, as described earlier.

The most important merge code for most merges is the FIELD code. It inserts the data from a specified field in the merge data file. When you insert this merge code, you get the Insert Field Name or Number dialog box, and you select the field name. Each field also has a field *number*, however, ranging from 1 for the first field up to the number of fields present in the data file. You can refer to a field by the number rather than the name, which may sometimes be useful if you need to use the same merge form file with two different merge data files that use the same ordering of information, but different field labels. In the document window, you can use WordPerfect's normal editing features to remove the names of the fields from between the parentheses in the FIELD codes and replace them with their corresponding numbers.

Do you get a warning that there are no field names or numbers when you select the FIELD code? See "No Field Names or Numbers" in the "Troubleshooting" section at the end of this chapter.

PROMPTING FOR INPUT AT MERGE TIME

You may sometimes need to collect information at merge time, rather than from a merge data file. For instance, you might want to add a paragraph of text to personalize a letter, or add information that's not available in a data file. In this case, you need to prompt the user for input at merge time. Many of the WordPerfect ExpressDocs templates make heavy use of this capability.

➔ To learn more about ExpressDocs templates, **see** "Using ExpressDocs," **p. 336**

USING THE KEYBOARD MERGE CODE

One quick way to obtain merge-time input is to use the KEYBOARD merge code. When you enter a KEYBOARD merge code, WordPerfect asks you for a prompt. Enter some text here to tell you (or whoever runs the macro) what to enter. For instance, you might type `Enter personalized note:` as the prompt.

When you perform the merge, WordPerfect pauses the merge and displays a dialog box with no buttons and the prompt you entered. Your insertion point is available in the main document window, and the merge toolbar changes, as shown in Figure 20.18.

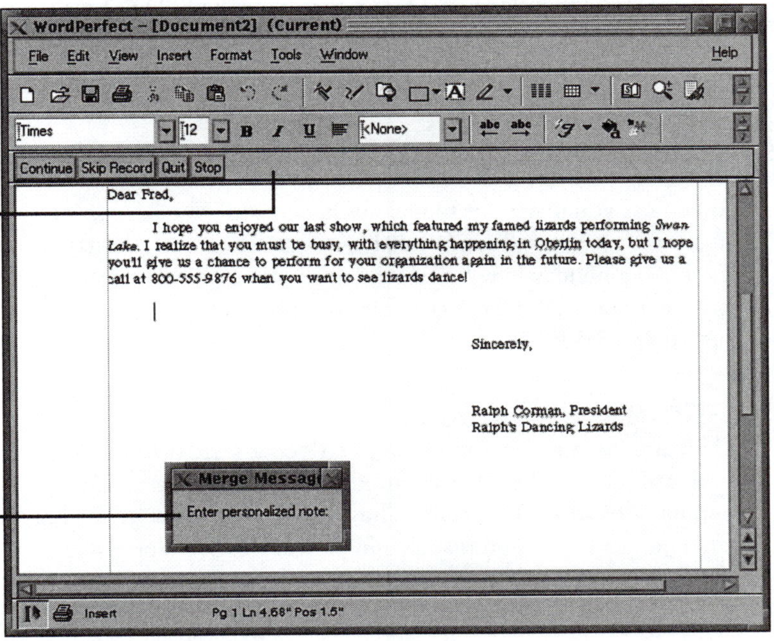

Figure 20.18
The KEYBOARD merge code lets you enter an arbitrary amount of text in each merged document.

Merge toolbar

KEYBOARD merge code prompt

When you select the document window, you'll be able to enter any text you like, including WordPerfect formatting codes. How you select the document window varies with your window manager. With some, you need to click in the window or on the scrollbar; with others, you may just need to move your mouse pointer over the window. When you're done editing your input, click the Continue button in the merge toolbar. You won't be able to do anything directly with the Merge Message dialog box, but it will go away when you click the Continue button in the toolbar.

The KEYBOARD merge code is easy to use when you design your merge form but can be a bit awkward to use at merge time because of the switching between the Merge Message dialog box and the document window—especially if you need to enter just one or two words per record.

Using the GETSTRING and VARIABLE Merge Codes

You can reduce your merge-time effort somewhat by using a pair of merge codes, GETSTRING and VARIABLE, in place of KEYBOARD. Insert GETSTRING *before* VARIABLE in the merge form file. GETSTRING requires three parameters: a variable name, a prompt to go within the prompt dialog box, and a title for the prompt dialog box. For instance, you might use `Animal`, `Animal used in show:`, and `Extra Info`, respectively, for these parameters. The VARIABLE merge code takes only one parameter, the variable name.

When you've finished entering these codes into your merge form file and start a merge, WordPerfect prompts you for this information for each record, as shown in Figure 20.19.

This method generally works better than the KEYBOARD merge code if you need to insert only a few words of text because you don't need to switch between the prompt dialog box and the document window. You might find the opposite is true with your particular window manager and its options, however, so try both methods.

Conditionally Including Text

Sometimes you may want to include some text if some condition is met. For instance, you might want to remind residents of certain states about sales tax regulations or include information in a holiday letter that you think will be of interest to your family members, but not to your friends. WordPerfect offers two classes of merge codes to help you in situations like this: IF and SWITCH.

The IF Codes

There are three major variants on the IF code—IF, IFBLANK, and IFNOTBLANK. All three codes cause WordPerfect to include anything after their code only if their condition is met, up until an ENDIF code. The IF code is the most general of the four. It takes as a parameter an expression that should be true or false. For instance:

```
IF(FIELD(State)=OH) Remember to pay your sales tax! ENDIF
```

Figure 20.19
You can use a given variable more than once, but be sure to use it only *after* you've collected its name with the GETSTRING merge code.

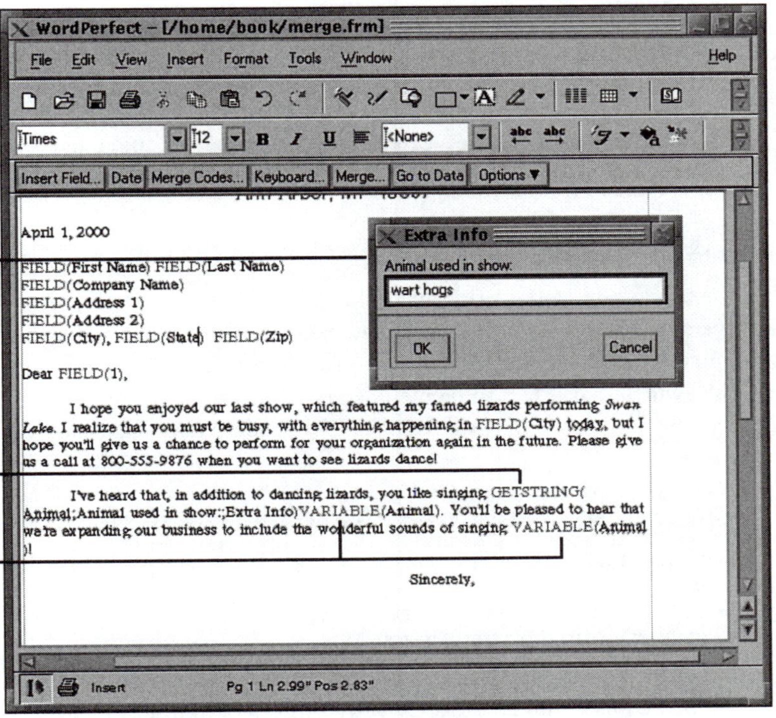

Dialog box produced by GETSTRING merge code

GETSTRING merge code

VARIABLE merge codes

Note

Remember that spaces and other formatting codes are included with other text controlled by conditional codes. (To improve readability, the examples shown here include spaces that you might not want in your actual data files.) After you've used these features a bit, you'll get a good feel for how best to use the codes to produce a well-formatted document.

The preceding example includes the sentence `Remember to pay your sales tax` only in merge forms where `State` is `OH`. You can use the vertical bar (¦) and ampersand (&) symbols within the IF code as shorthand for *or* and *and*, respectively. For instance, `IF(FIELD(State=OH) ¦ FIELD(State=NV))` causes the subsequent text to be included when `State` is either `OH` or `NV`.

The IFBLANK and IFNOTBLANK codes work much like IF, but they test whether the field is empty. For instance, if you have a database in which the product a customer purchased sometimes, but not always, appears, you could use something like the following:

```
IFNOTBLANK(Purchase) Your FIELD(Purchase) is a fine product, so let us service it
for you! ENDIF
```

This statement inserts the sentence only when you know what product you're asking to service.

All of the IF variants can use the ELSE clause. This merge code appears between the IF and the ENDIF and allows you to set up something to print for either outcome of the test. For instance:

```
IF(FIELD(State=MA)) We offer free shipping within the state of Massachusetts. ELSE
Shipping is based on the weight of your shipment. Call us for a quote. ENDIF
```

This statement inserts the first sentence if `State` is `MA`, and the second sentence otherwise.

THE SWITCH CODE

The SWITCH code works much like the IF codes, except that it offers a result for each of several possible outcomes. It takes an expression as its parameter and uses the CASEOF code to mark each possible outcome. In addition, the SWITCH code offers a DEFAULT code, which marks text to be included in all unspecified situations. The SWITCH code as a whole is terminated by an ENDSWITCH code. Here's an example of SWITCH in action:

```
SWITCH(FIELD(City))
CASEOF(Boston) Please call 555-3428 for the representative nearest you.
CASEOF(New York) This service will be available in your area within 30 days. Check
our web site for more details.
CASEOF(Philadelphia) Our central office at 555-9923 will be happy to serve you
immediately.
DEFAULT FIELD(City) is not yet in our service area, but we hope to have nationwide
coverage within the next 18 months.
ENDSWITCH
```

The preceding codes and text cause WordPerfect to insert a unique sentence for letters destined for Boston, New York, and Philadelphia and a generic text (which nonetheless includes the name of the city) for all others.

Caution

> Both the IF and the SWITCH codes are extremely sensitive to the exact spelling of field contents (if that's what you're using for their conditions). For instance, if instead of `New York` a City field read `NewYork` (without the space), the recipient would receive the less optimistic DEFAULT sentence. Further, in the SWITCH example, only the City field is tested, so this example might fail if a recipient were in a small hamlet with the same name as one of the big cities, but in a different state. You can make a more complex merge form file to compensate, say, by using nested CASE and IF codes, but nesting becomes awkward after a point.

MERGING INTO A TABLE

You can merge data into a table for various reasons. For instance, you might want to print some or all of the information from a text-mode merge data file in a format that's more readable than the file's native format. Or you might want to convert a text-mode data file to a table format. When you perform this merge, you can put more than one field's data in each column if you like.

To merge data into a table, follow these steps:

1. Create a blank document window by choosing Program, New Window (Ctrl+N) from the WordPerfect Control window if necessary.
2. In the blank document window, choose Tools, Merge (Shift+F9). WordPerfect displays the Merge dialog box (refer to Figure 20.1).
3. Click Form in the Form File area of the Merge dialog box. WordPerfect then displays the Create Form File dialog box (refer to Figure 20.5).
4. Type the name of the merge data file, or click the folder icon to select the data file from a File Manager dialog box.
5. Click OK. The merge form toolbar appears in the document window.
6. Define your table layout with as many columns as you want and two rows. You can either use the column toolbar tool or choose Insert, Table (F12).

➔ For more information on ways to create a table, **see** "Creating a Table," **p. 276**

7. Enter appropriate headers in the first row of the table—for instance, you might use *Name*, *Company*, and *Address*. These headers need not correspond to the labels by which the existing file's merge fields are known.
8. With the insertion point in the final position of the last cell of the first row, click the Merge Codes toolbar button. WordPerfect displays the Insert Merge Codes dialog box (refer to Figure 20.13).
9. Select the LABEL merge code from the Insert Merge Codes dialog box and insert the code. When you're prompted for a label, type something to identify this location, such as *loop*. You'll later enter another code to cause the merge process to loop back to this point, and this label tells WordPerfect where to go.
10. Move the insertion point to the first column of the second row.
11. Click Insert Field on the merge form toolbar to get the Insert Field Name or Number dialog box (refer to Figure 20.7).
12. Select and insert all the field names you want in the first column. Be sure to click in the document window and insert any punctuation, spaces, or other fixed text you want in that column, as well.
13. Click in the subsequent cells in the second row and repeat the previous step for them.
14. When you're done with the final cell, locate the Insert Merge Codes dialog box and insert the NEXTRECORD and GO codes. When WordPerfect asks for a label for the GO code, respond with the label you entered in step 9. When the merge is performed, these two codes move the merge to the next record and insert a new row into the table by jumping execution back to the end of the previous row in the merge form file. The result of your activity to this point should look something like the table displayed in Figure 20.20.

Name	Company	Address LABEL(loop)
FIELD(First Name) FIELD(Last Name)	FIELD(Company Name)	FIELD(Address 1) FIELD(Address 2) FIELD(City), FIELD(State) FIELD(Zip) NEXTRECORD GO(loop)

Figure 20.20
When you run the merge, WordPerfect treats the area between the *LABEL(loop)* code and *GO(loop)* code as instructions for creating a table of multiple rows.

15. Save the merge form file.
16. Perform the merge as described earlier in this chapter.

ADVANCED MERGE

Some mail merge options are controlled when you perform the merge operation, rather than from the data or form file. You can use these options to select data and form files and select the specific records you want to include in the final merge operation.

When you click Merge on the merge toolbar and then select Merge from the Merge dialog box (refer to Figure 20.1), WordPerfect displays the Perform Merge dialog box (refer to Figure 20.9), which controls the options described in this section.

SELECTING ALTERNATE MERGE FILES

The Perform Merge dialog box lets you specify the two source files for the merge, as well as the destination file. You specify the files in the Files to Merge area of the dialog box either by clicking on the option button to the right of each field or by directly editing the field.

The Form File's option button presents two options: Select File and <Current Document>. If you select the former, you get a File Manager dialog box from which to select a file. If you select the latter, WordPerfect uses the current document window as the source for the merge form file. If you've started the merge from a form file, <Current Document> is the default option.

When you create a merge form file, WordPerfect prompts you for the data file with which to associate the merge form file or gives you the option to not specify a file (refer to Figure 20.5). If you give the name of an associated data file, it becomes the default option for the Data File field in the Perform Merge dialog box. You can change this entry by typing a new filename or by selecting Select File from the option button to the right of the Data File field. If you select <None> here, WordPerfect creates one instance of the final document with each field empty.

Caution

If you change the data file to one that uses different field labels than those you used when you created the form file, WordPerfect treats each reference to a field that doesn't have a matching name as if it were blank. If none of the fields in the new file have the same names as those used in the form file, your merge results in identical copies of the file without any merged data.

The Output File options are the most numerous. The default is <New Document>, which creates a new WordPerfect document window containing the entire new file. You can also select <Current Document>, which causes the new document to merge into whatever document window was active when you started the merge operation. If you're merging a particularly large job, you may prefer to send it directly to the printer by selecting <Printer> from the Output File option button. If you select the Select File option, WordPerfect lets you enter a filename to which to write a WordPerfect (.wpd) document containing the final merged file. You might want to use this option if you plan to merge the file on one machine and transfer it to another to print.

Caution

Don't send a large merge file directly to the printer without first testing it with a subset of the merge data file. It's easy to create a merge file that needs some minor reformatting before it prints acceptably, and there's no point to wasting reams of paper on such a mistake.

SETTING MISCELLANEOUS MERGE OPTIONS

If you click the Options button in the Perform Merge dialog box, WordPerfect displays the Perform Merge Options dialog box (Figure 20.21).

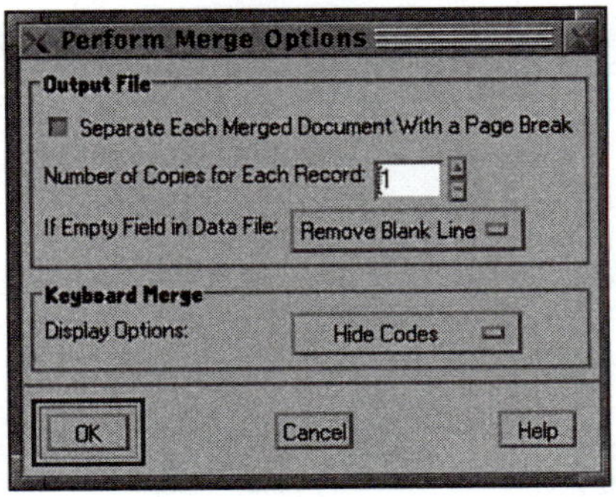

Figure 20.21
The Perform Merge Options dialog box lets you set assorted merge-time options.

You can use the Perform Merge Options dialog box to do the following:

- Tell WordPerfect whether or not to use separate pages for the final merged files by using the Separate Each Merged Document With a Page Break check button.
- Create multiple copies of each merged document by using the Number of Copies for Each Record spin box. If you set this spin box to 2, for instance, WordPerfect creates two copies of each letter. You can also use print options to create multiple copies when you print.

→ For information on setting the number of copies, **see** "Copies," **p. 154**

- Tell WordPerfect how to handle blank fields. If there's a blank field that merges into a code that resides by itself in the document, you may want to remove that line. If so, set the If Empty Field in Data File option button to Remove Blank Line (the default). To leave it in, set the option button to Leave Blank Line.
- Display or not display codes when performing a keyboard merge. Use the Display Options option button to set how remaining merge codes display when WordPerfect encounters a KEYBOARD merge code.

MERGING FROM A RESTRICTED SET OF RECORDS

You can tell WordPerfect to merge data from only some records of the merge data file by clicking the Select Records button in the Perform Merge dialog box. This step opens the Select Records dialog box, shown in Figure 20.22.

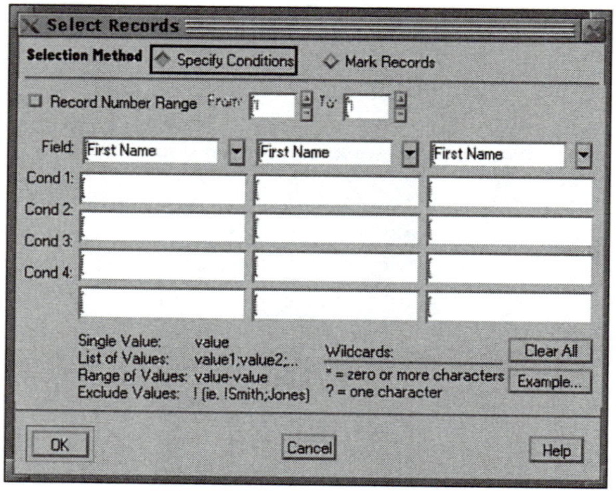

Figure 20.22
Use the Specify Conditions option to set rules for which records to include in a merge.

If you click the Record Number Range check button, you can select a range of records. For instance, if your merge data file contains 400 records, you can limit the merge to records between 50 and 100 by entering those values in the appropriate spin boxes.

In addition to or instead of this selection, you can specify which records to merge based on field values. Each of the four rows represents one set of conditions that may satisfy the requirements for inclusion of a record. Each column within each row represents a set of conditions that must all be satisfied to include the record. If you click Example, WordPerfect displays an example and its interpretation, as shown in Figure 20.23.

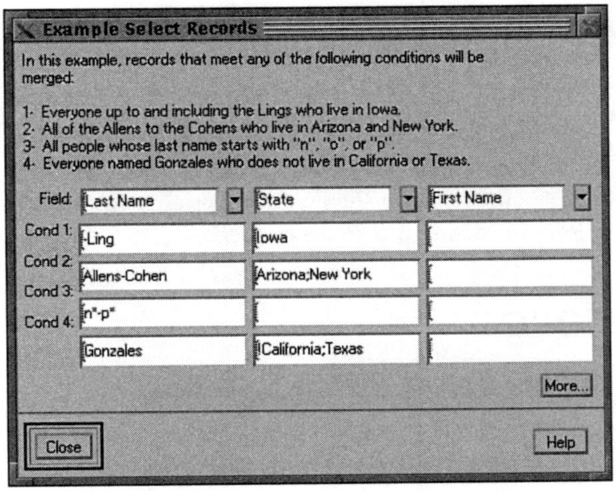

Figure 20.23
The Example Select Records dialog box demonstrates the rules governing record selection in the Select Records dialog box.

If you click the Mark Records radio button in the Select Records dialog box, WordPerfect changes the dialog box's contents to allow you to select records on a case-by-case basis, as shown in Figure 20.24.

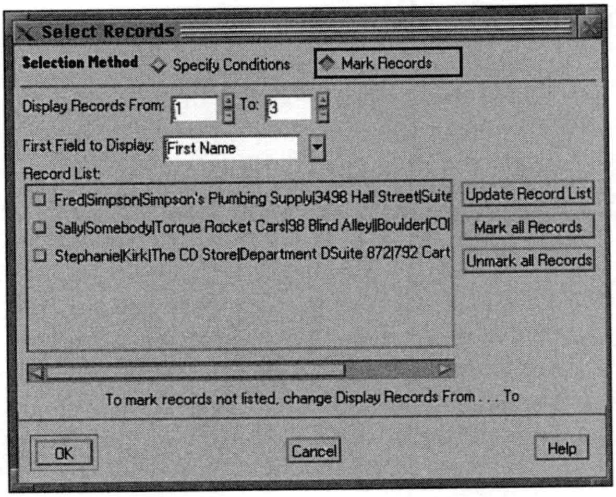

Figure 20.24
Select records to include in a merge on a case-by-case basis by using the Mark Records selection in the Select Records dialog box.

Tip #239 from *Tux Tips*

Use the Mark Records method of record selection when you have few records in your merge data file; use the Specify Conditions method when you have a very large merge data file.

To select a record for inclusion in the merge, check its box in the Record List (by default, no records are selected). You can mark or unmark all the records by clicking the Mark all Records and Unmark all Records buttons, respectively. Clicking Update Record List causes WordPerfect to check the file's contents again, which can be useful if you want to change it just prior to the merge. You can also alter the order in which fields display by using the First Field to Display option button.

However you do it, when you've finished entering criteria for which records to select, click OK in the Select Records dialog box to dismiss it.

Sorting Records

You may sometimes need to sort a list of records prior to a mail merge. For instance, you might want to sort by zip code to get lower rates from the post office, or you might want to sort two merge data files in the same way so that you can produce two mail merge documents and then combine them easily for mailing. WordPerfect allows you to do these sorts with its sort command.

WordPerfect's sort capability goes beyond mail merge, however. You can sort lists, tables, paragraphs, and columns in addition to merge records. The procedures are the same, but the details of how WordPerfect performs the sort may differ.

Understanding Sort Records and Keys

When you sort data, WordPerfect uses *records* and *keys* to define how to sort. When you sort a textual merge data file, the records correspond to the merge data file's records, but when you sort an ordinary text file, the records are other units—for instance, lines or paragraphs. WordPerfect moves records as a whole, so all the information associated with the record moves to its new location after the sort. Sort keys are the units you use for sorting. For example, if you want to sort a merge data file by zip code, the zip code merge field would be the key. In case you want to sort by only one word in a field that contains several words, or even lines, WordPerfect subdivides Sort records into several pieces. For instance, with merge data text files, the subdivisions are Fields, Lines, and Words. For example, suppose you have a data file that includes a single field for the address, including the street address, city, state, and zip code, on multiple lines. This merge field corresponds to the sort field. WordPerfect's sort feature breaks down the sort field into lines ("3489 Hall St.", "Suite 23", and "Oberlin, OH 44074", for instance), and it further breaks down each line into words ("Oberlin", "OH", and "44074"). You can specify to sort by a specific field, line, and word. In this case, you could sort by zip code by specifying the final word of the final line of the address field. (More on precisely how to do this shortly.) Table 20.1 summarizes the hierarchy of data files as handled by Sort.

TABLE 20.1 SORT RECORD SUBDIVISIONS

Record Type	Record Termination	Record Subdivisions
Line	[HRt]	Fields Words
Paragraph	[HRt][HRt]	Lines Fields Words
Merge text	ENDRECORD code	Fields Lines Words
Parallel	Column block	Columns Lines Words
Table	[Row]	Cells Lines Words

When you perform a sort, WordPerfect moves data records—those blocks of text terminated by the code or unit specified in the Record Termination column of Table 20.1. You tell WordPerfect *how* to sort by specifying a unit of text using the numbered subdivisions.

In the case of Line records and Line subdivisions in Paragraph records, the Field subdivision is defined by tabs or indents. If a line doesn't have any tabs or indents, the Line subdivision has a single Field. In either case, the Field is considered a subdivision of the Line. By contrast, with Merge text data files, a Line is a subdivision of a Field because you can enter an [HRt] code in a Field to break it into two or more lines. It's important to keep this distinction in mind, especially if you regularly work with both types of records.

Tip #240 from *Tux Tips*

To have WordPerfect treat two-word names, such as van Horn, as a single word, use an [HSpace] character, created by pressing Ctrl+Enter. By contrast, an ordinary hyphen character [–Hyphen] is not a word separator, so to treat a hyphen as such, you must enter a hard hyphen by pressing Ctrl+–. This treatment of hyphens is just the opposite of the way it works with line breaks.

→ To learn more about the various types of hyphens and spaces, **see** "Hyphenation," **p. 220**

PERFORMING A SORT

To perform a sort, choose Tools, Sort (Shift+Ctrl+F9) to open the Sort dialog box (Figure 20.25).

From the Sort dialog box, you can set various options to control the sort:

- Input File and Output File—You can type in the name of a document file, click the option button to the right of the field name and select Select File to select a file from a File Manager dialog box, or select <Current Document> from the option button. (This last is usually the default).

Tip #241 from *Tux Tips*

For safety, either save your work before you perform a sort or sort to a separate Output File from the original file. You might want to create a new file for the sort, even if you think it sorted properly, in case you need to revert to the unsorted ordering in the future.

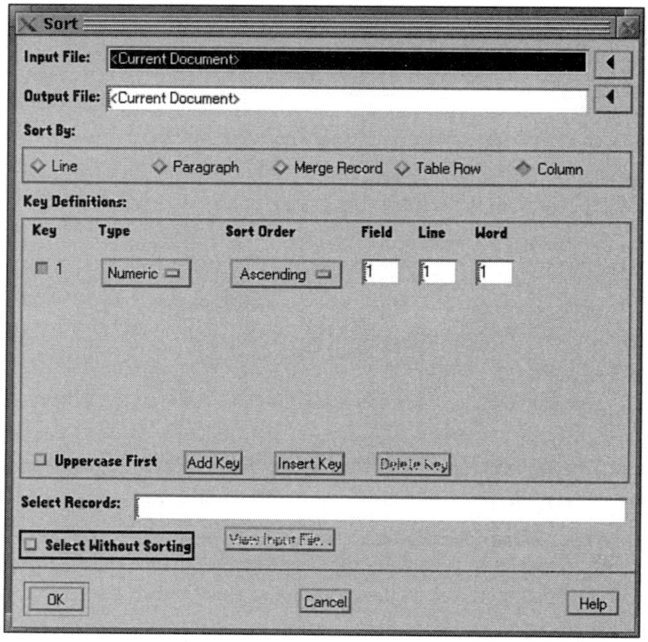

Figure 20.25
You can add additional key definitions by clicking Add Key or Insert Key.

- Sort By—You can tell WordPerfect which record type, as outlined in Table 20.1, to use when sorting. With some files, WordPerfect selects the correct record type when you start Sort, but other times you have to select the type manually.
- Uppercase First—Normally, WordPerfect doesn't distinguish between upper- and lowercase letters when sorting. If you select this check button, however, WordPerfect places uppercase letters before lowercase letters when you sort in ascending order.
- Add Key, Insert Key, and Delete Key—These three buttons allow you to add and delete keys to the sort procedure. The dialog box starts with only one key, but you can add more if you want to sort by more than one key. For instance, suppose you have a data file with 1,000 records. You might want to sort this listing by last name and then first name, making it easier for you to find individuals with common names like Smith or Jones. The active key is indicated by the selected radio button (which looks like a check button) in the Key column of the Key Definitions area. This key is the one that's deleted if you click Delete Key and after which a new key appears if you select Insert Key.
- Key Definitions—For each key, you can select whether to sort numerically or in alphabetic order; in ascending or descending order; and the numbers for the record subdivisions, as described earlier and outlined in Table 20.1. If you want to sort by the third line of the second field, for instance, you would enter `2` under Field and `3` under Line. If you prefix a number with a `-`, WordPerfect counts from the *end* of the record subdivision. For instance, if a record contains city, state, and zip code, you could specify zip code by entering `-1`. This entry is less likely to fail than a `3` because some city names (such as New York and San Diego) contain two words, so `3` would sort by the state name, rather than the zip code.

As an example, suppose you want to sort a table by last and first names and suppose the table provides separate columns for these two names. You could do this sort by following these steps:

1. Place the insertion point in the table you want to sort.
2. Choose Tools, Sort (Shift+Ctrl+F9) to open the Sort dialog box. Make sure that the Table Row radio button is selected in the Sort By area of the dialog box.
3. Click Add Key to add a second key. WordPerfect ultimately sorts the list by whatever you specify for the first key, and in the event of ties, it uses the second key to break the tie.
4. Select Alpha for both keys' Type and select Ascending for their Sort Orders.
5. For the first key, type the number of the column in which the last name appears in the Cell text entry field.
6. For the second key, type the number of the column in which the first name appears in the Cell text entry field.
7. Click OK to sort the table.

Figure 20.26 shows a before-and-after example of following this procedure with a three-record table. The upper table is the original, and below that is the sorted version.

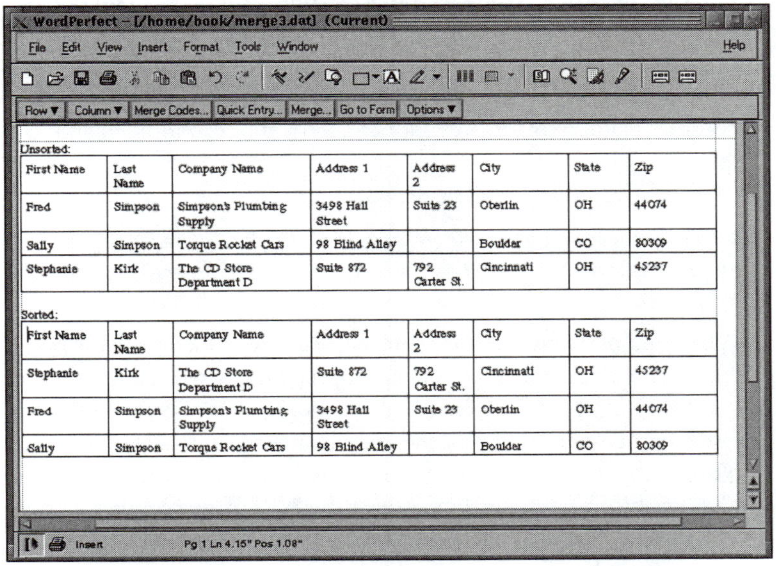

Figure 20.26
Sorting a table created for data merge won't sort the header line, but will sort the table contents.

Are you following these directions, but find that WordPerfect doesn't sort your records? See "Sort Doesn't Work" in the "Troubleshooting" section at the end of this chapter.

TROUBLESHOOTING

NO FIELD NAMES OR RECORDS

When you click on Insert Field in the merge toolbar or select the FIELD code from the Insert Merge Codes dialog box, you get a warning that no field names or records were found, followed by the Insert Field Name or Number dialog box, as shown in Figure 20.27.

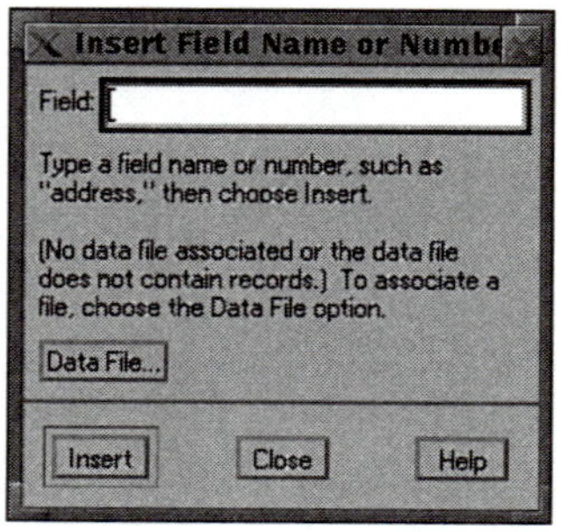

Figure 20.27
You may need to enter field numbers rather than names if you use a data file without field names.

If you specified None in the Create Form File dialog box (refer to Figure 20.5) or if you're using a data file that doesn't specify field names with a FIELDNAMES code, WordPerfect won't be able to present a list of field names. You have to do one of the following: Enter the names manually (if you'll ultimately be using a data file that contains field names), click the Data File button to tell WordPerfect what data file you intend to use with the form file, or refer to the fields by number rather than by name. Type 1 in the Insert Field Name or Number dialog box for the first field, 2 for the second field, and so on.

PAGE NUMBERING AND MAIL MERGE

You're trying to perform a merge on a document that has page numbers. Unfortunately, the page number counter doesn't reset for each new copy of the document, so only the first one's page numbers are correct.

When you perform a merge, you create a single document for printing, so unless you correct page numbers by inserting appropriate codes, the result is continuously increasing page numbers through all the copies of the merge form. To correct this problem, position your insertion point at the top of the merge form file and select Format, Page, Page Numbering, Numbering to get the Page Numbering dialog box. Click the Value button to open the Numbering Value dialog box. Adjust the New Page Number setting in the Page Settings area up to 2 and then click OK twice to remove both dialog boxes. Then perform that procedure again but this time adjust the New Page Number value down to 1. You must *change* the page number in the Numbering Value dialog box, or WordPerfect won't actually insert a page numbering change code; hence the need to go to this dialog box twice.

→ For more information on setting and adjusting page numbers, **see** "Headers, Footers, and Page Numbers," **p. 235**

SORT DOESN'T WORK

You've followed the instructions in this chapter, but Sort doesn't do anything.

One possibility is that your file is already sorted. If that's the case, invoking a WordPerfect Sort won't change the file in any way. Another possibility is that you've entered information for the sort incorrectly—say, by specifying the wrong field on which to sort. Sometimes, though, WordPerfect just won't sort a file when it should. To correct this problem, save your file, exit from WordPerfect, and start it again. When you try the sort now, it should work.

PART VI

Managing Large Documents

CHAPTER 21

Organizing a Document with Outlining

In this chapter

OVERVIEW

Every high school English teacher advises students to write an outline before writing a document of any size, and with good reason: Creating an outline helps you to organize your thoughts and create structure in your document. Without an outline, it's all too easy to create a rambling mess that drifts back and forth between topics. It's much easier to correct this sort of problem in an outline than it is in the final document.

Of course, word processors do help you to rescue a rambling or drifting document with their cut-and-paste features and capability to excise and edit text before committing it to paper. Creating an outline is still a good idea, though, because it's still easier to catch structural problems at the outline stage than when editing a draft. To sweeten the deal a bit, WordPerfect includes an Outline editing mode that helps you to create an outline most effectively. You can have WordPerfect take care of numbering your outline according to one of several styles, and you can use its editing features to move outline elements around effectively. You can also expand and collapse the outline, so that you can look at the broad structure of the document or "zoom in" on the minutiae. Finally, when you're done with the outline, you can incorporate it into your finished document, by using the outline headings as text headings for the final product.

In addition to supporting a full outline, WordPerfect allows you to quickly create lists of items, handling the renumbering of the list when you insert or delete items from it. This feature is very useful when you need to provide a list of short examples or present a numbered series of instructions, to name just two applications.

WordPerfect uses *styles (p. 31, 142, 306)* to implement its outlining features, including simple lists. Therefore, you may want to read the chapter on styles, if you haven't already, before proceeding with this chapter. You'll still be able to use lists and outlines without knowing much about styles, but understanding styles will help you understand how and why the features described here work as they do.

→ To learn more about styles, **see** "Ensuring Consistency Within Your Document with Styles," **p. 305**

CREATING A BULLETED OR NUMBERED LIST

If you don't need to create an entire outline, but only a simple list, WordPerfect provides a way to do so rapidly and easily. Position your insertion point where you want the list to begin and choose Insert, Bullets & Numbering. You'll see the Bullets & Numbers dialog box appear, as shown in Figure 21.1. This dialog box presents several types of *bullets* (symbols used to indicate a new item in an unnumbered list) and numbering schemes.

Figure 21.1
Select the bullet or numbering type you want from the Bullets & Numbers dialog box.

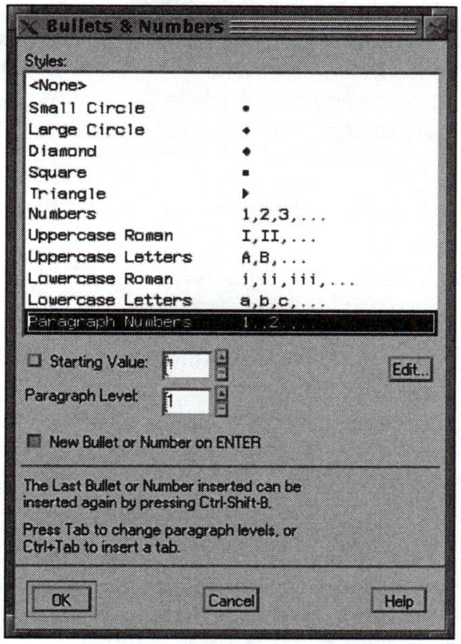

SELECTING A BULLET/NUMBER TYPE

You can choose a style for your bulleted or numbered list by selecting an appropriate style in the Bullets & Numbers dialog box and then clicking OK. The dialog box disappears, and WordPerfect inserts an appropriate bullet or number. As you type your text, you'll notice that WordPerfect includes a hanging indent for most of the styles, as in the bulleted items in Figure 21.2. This format provides your list with an appealing appearance. When you press the Enter key, WordPerfect creates a new bullet or number with a hanging indent and following tab, to minimize the keystrokes you need to enter to continue the list. If you prefer not to have WordPerfect do this work for you, you can deselect the New Bullet or Number on ENTER item in the Bullets & Numbers dialog box when you create your list. When you've entered all the items you want, press the Enter key, backspace over the extra bullet or number, and then continue with your text entry as usual.

Figure 21.2
You can use more than one bullet or list type in a single document.

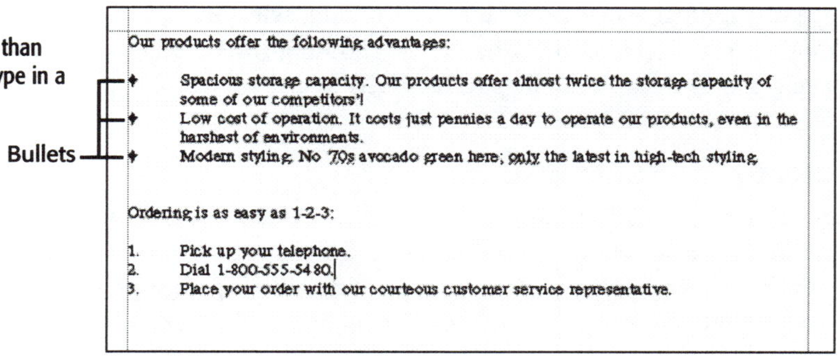

PART VI CH 21

Note

If this list is the second or later numbered list you're creating of a given style, you may need to click the Starting Value check button in the Bullets & Numbers dialog box and set the spin box to 1 to get the list numbering to begin at one again. Alternatively, you can alter the starting number for a list subsequent to its creation, as described shortly.

→ If you want to learn more about bulleted or numbered lists, **see** "Editing Bulleted or Numbered Text," **p. 556**

In addition to using the Bullets & Numbers dialog box, WordPerfect provides several alternative ways to create a bulleted or numbered list:

- At the start of a line, type a number or letter, followed by a period or close parenthesis, and press the Tab key. WordPerfect creates a numbering style based on your initial entry and applies it as just described.
- Select an appropriate bullet character from the Symbols dialog box (Insert, Symbol or Ctrl+W) at the start of a line and press the Tab key. WordPerfect creates a bulleted list using that symbol.

Note

The styles that WordPerfect uses for these two types of formatting are not the same styles you select from the Bullets & Numbering dialog box, but both styles work in the much same way.

- If you've already created or used an appropriate numbering style, you can choose it from the Styles selector in the property bar (Figure 21.3). Bullet and numbered list styles you've not yet used don't appear here, however, so you need to use them some other way first.

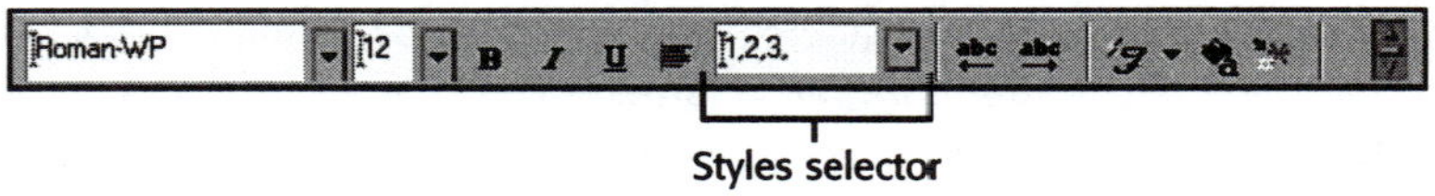

Figure 21.3
Bullet and numbering systems are instances of styles, and after you use one, it appears in the Styles selector for your document.

- Type text normally, one paragraph per list item, select it all, and then apply the style by choosing it from the property bar, or by choosing Insert, Bullets & Numbering and picking a style from the Bullets & Numbering dialog box.

EDITING BULLETED OR NUMBERED TEXT

You can edit text in a bulleted or numbered list much as you would any other text. You can use the normal text selection, cut and paste, formatting, and other techniques on it. Of course, the text is already formatted with a specific style (a *paragraph* style, to be precise), which means that you probably won't want to apply certain types of formatting to it.

Note

You'll be able to apply character styles to the text in a list without harm. If you attempt to apply a paragraph or document style, however, you'll override the list style and lose the numbering.

→ If you want to know more about how styles of different scope interact, **see** "Scope," **p. 308**

Because the number or bullet characters are part of the style, you won't be able to edit them directly. If you try, you'll remove the style altogether. Of course, you can add the style back again or undo your action if you attempt this by mistake. If you want to change the numbering system, say, to skip a number or because you're starting a new list, you can do so by following these steps:

1. Position the insertion point before the number you want to alter and choose Insert, Other, Counter. WordPerfect displays the Counter Numbering dialog box, as shown in Figure 21.4.

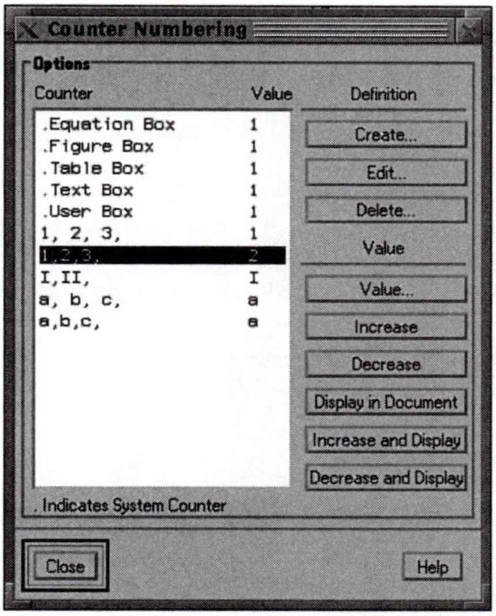

Figure 21.4
The exact list of counters in the list box on the left side of the Counter Numbering dialog box depends on which numbering and outline styles you're using and which other counters may be in use in your document.

2. Select the counter you're using in the current list.

Tip #242 from *Tux Tips*

If the Counter Numbering dialog box shows several counters and you're not sure which one to select, check the listings in the Value column for a clue. The value of the list number following the insertion point appears in this column for the counter you want to change.

3. Click Increase or Decrease to change the value by one.
4. If you want to change the value by more than one or if you want to change the numbering scheme (say, from Arabic numbers to letters or Roman numerals), click Value to open the Set Value/Number Method dialog box (Figure 21.5).

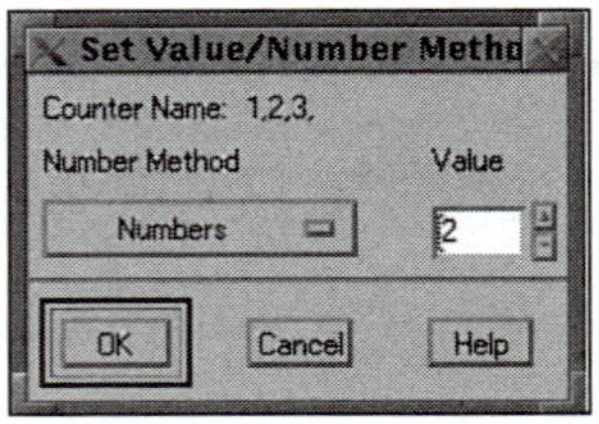

Figure 21.5
You can set any number in the Value spin box and choose a numbering system (Arabic numbers, Roman numerals, or letters) from the Number Method option button.

5. Set the value you want for the counter and change the numbering method if you like.

If you insert a new line within a numbered list or delete an existing line, WordPerfect automatically renumbers any affected list items. This feature means that you don't need to worry about numbering your lists, so you can move text around and concentrate on other aspects of your writing.

EDITING A BULLET OR NUMBER STYLE

You can alter the appearance of all text in a bullet or number style by using the WordPerfect Styles Editor. There are three ways to start the Styles Editor:

- Choose View, Reveal Codes (Shift+Ctrl+F3) to open a Reveal Codes pane on your document window. Use this pane to locate a style code for the style you want to alter. Double-click on this style code to edit it.
- Choose Format, Styles (Shift+Ctrl+F8) to open the Style List dialog box shown in Figure 21.6. Select the style you want to edit and click the Edit button.

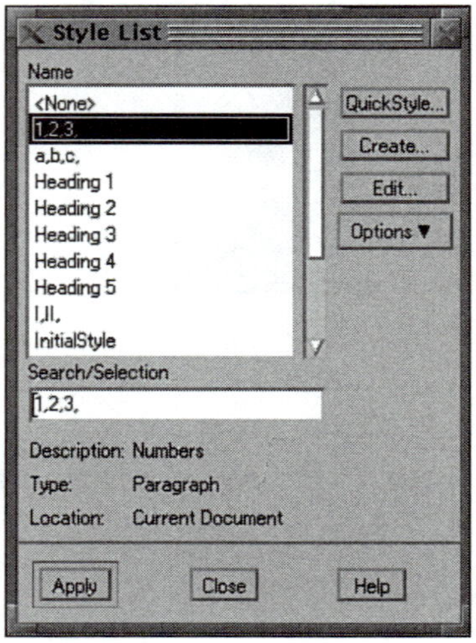

Figure 21.6
You can edit any style, whether it's a list style or not, by selecting it from the Style List dialog box.

- Choose Bullets & Numbering to open the Bullets & Numbers dialog box (refer to Figure 21.1). Select the style you want to alter and click the Edit button.

Note

The last option works only for styles you first selected from this same dialog box. If you created a bullet or number style by typing a bullet or number followed by a tab, you can't select that style for editing from the Bullets & Numbers dialog box.

Whichever way you start it, the WordPerfect Styles Editor, shown in Figure 21.7, lets you change the formatting of the style. You could, for instance, enter text attributes such as bold or italic for the number or text, alter fonts, or add or change text such as the period following a number. For the most part, editing text in the Styles Editor works much like editing text in Reveal Codes mode.

→ To learn more about editing a style, **see** "Editing a Style," **p. 321**

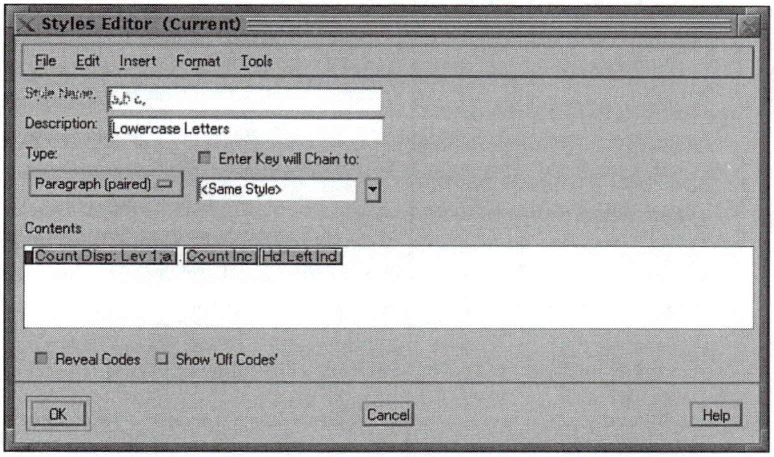

Figure 21.7
The WordPerfect Styles Editor functions much like WordPerfect with a Reveal Codes pane open.

You may want to change one particular formatting detail for a list: the location of the first tab stop. The default in WordPerfect places tab stops every 0.5", which is a bit wide for most lists, unless you're using Roman numerals or very large fonts. You can change tabs from the Styles Editor by positioning the red insertion point at the start of the style and choosing Format, Line, Tab Set. This step opens the Tabs dialog box (Figure 21.8). You can set a tab at, say, 0.3" by entering `0.3"` in the Position spin box and clicking Set, followed by OK.

→ If you want to learn more about setting tab stops using the Tabs dialog box, **see** "Creating Tab Stops with Dialog Boxes," **p. 197**

Note

You can also set the tabs on a list-by-list basis by selecting the entire list and setting the tab as described here (but choosing Format, Line, Tab Set from the document's window) or by using the WordPerfect ruler. This method sets the tab for only the selected list, not for all the lists using the same style.

Figure 21.8
If you want to set a wider tab than the default, you must clear the tab at 0.5", as well as create the new one.

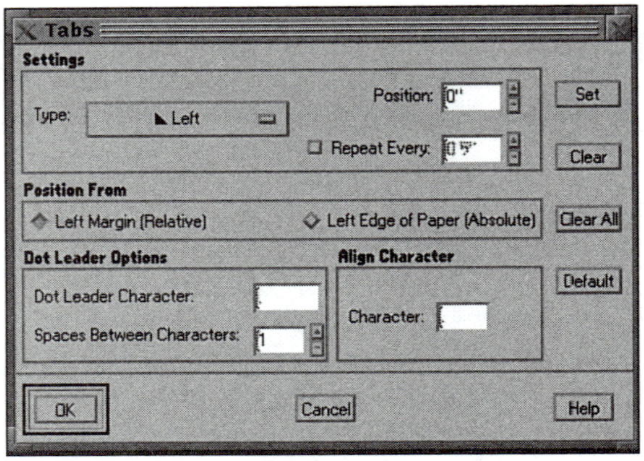

Have you made changes to a list style but find that they aren't applying to all the lists you thought they should? See "Changes to a List's Style Are Only Sporadically Effective" in the "Troubleshooting" section at the end of this chapter.

THE SPECIAL CASE OF PARAGRAPH NUMBERS

The final entry in the Bullets & Numbers dialog box (refer to Figure 21.1) is called *Paragraph Numbers*, and this numbering style works somewhat differently from the others. The other bullet and number styles are intended for creating lists with hanging indent formatting, and probably only a single level. Paragraph numbers, however, are used with paragraphs and don't use hanging indents. In addition, they borrow many characteristics of full-fledged outlines, including different outline *levels*—that is, how deep in the outline structure the paragraph falls. Figure 21.9 shows the effect of using the Paragraph Numbers style without additional formatting. You can, of course, add formatting to create a hanging indent style or other effects if you like—but WordPerfect's full outlining features might be more suitable if you want that sort of appearance.

When you begin this numbering style, you can press the Tab key on the keyboard to select the type of number you want at the top level. WordPerfect cycles through the choices and eventually begins repeating them. For subsequent paragraphs, however, pressing the Tab key causes an indent relative to the previous level and a change in the number type, as illustrated in Figure 21.9. For instance, if you press Enter and get the number *6*, pressing Tab indents this number and changes it to *a*. If you want to move back to the previous level, you must press Shift+Tab or select the Previous level item from the outline toolbar.

Does the Shift+Tab key combination have no effect? See "Shift+Tab Doesn't Work" in the "Troubleshooting" section at the end of this chapter.

Tip #243 from *Tux Tips*

If you want to place a conventional tab at the beginning of a line with Paragraph Numbering enabled, you need to choose Format, Line, Other Codes to get the Other Codes dialog box (Figure 21.10). Select the type of tab you want to use and click Insert. This tab code appears after the line number code.

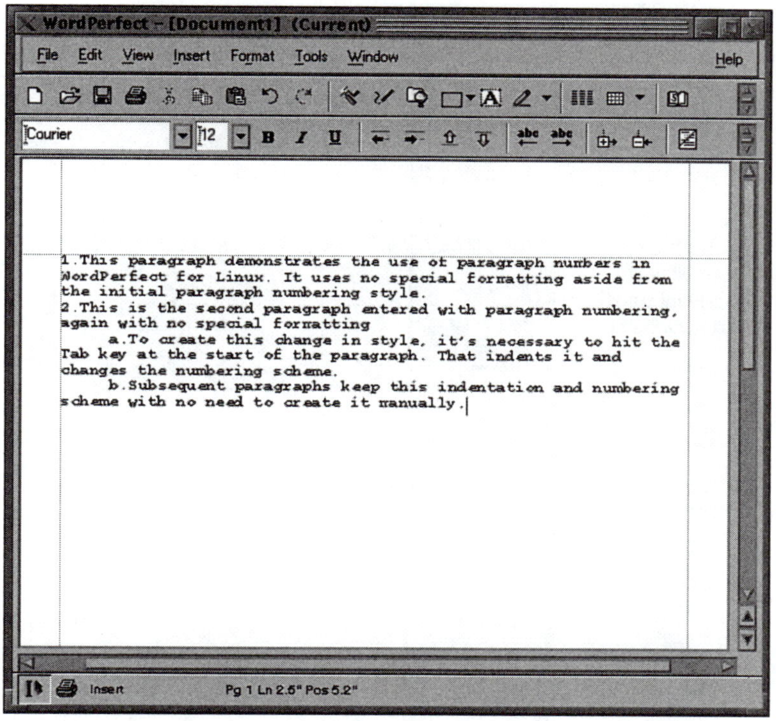

Figure 21.9
When you use the Paragraph Number feature, WordPerfect displays the outline toolbar.

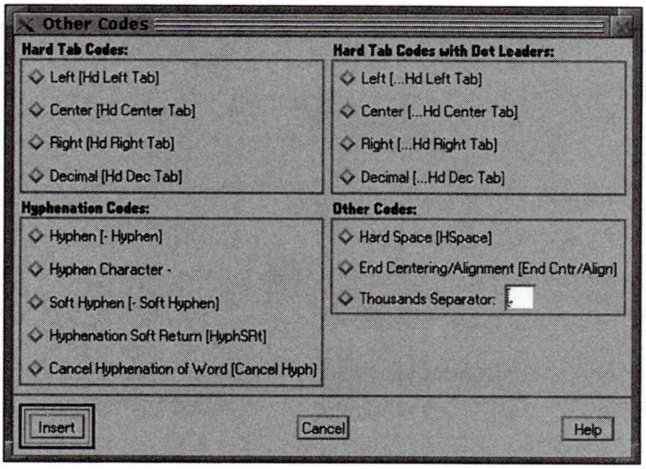

Figure 21.10
Use the Other Codes dialog box to insert codes that have become inaccessible because of special features in Paragraph Numbering or Outline modes.

CREATING A NEW OUTLINE

Simple lists are very useful for entering information in the body of a document, but they can't do much when it comes to actual outlining. An outline uses indentation, numbering, different bullet styles, text attributes, or some combination of these features to differentiate topics in the text to show you how your document will be structured. Many styles of outlining are possible, and WordPerfect offers great flexibility in this regard. It supports nine predefined outline styles, and it includes the capability to create more. Figure 21.11 shows four of the outline styles that are included with WordPerfect. As that figure shows, the same outline can look quite different when formatted in different styles.

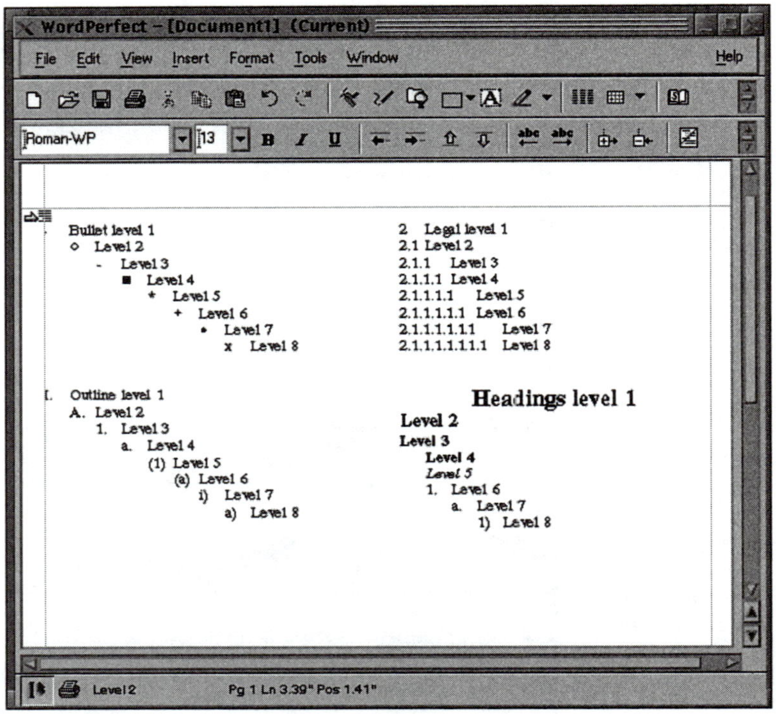

Figure 21.11
Outline types differ in indentation, numbering, and character attributes.

STARTING AN OUTLINE

To start a new outline, choose Tools, Define Outline, and WordPerfect displays the Outline Define dialog box shown in Figure 21.12. Select the outline style you want from the Name list box. WordPerfect shows a summary of the style's effects just below the Search/Selection text entry field, so you can verify that you've selected the style you want. Make sure that the Start New Outline check button is selected and then click OK to start the outline.

ENTERING TEXT

In many ways, entering text into an outline is no different from entering text in an ordinary WordPerfect document. You can use most of the same formatting commands to apply text attributes, cut and paste text, and so on. Outlining does present some unique characteristics, though, and these often influence how you enter text into an outline.

Figure 21.12
You can use the Search/Selection field in the Outline Define dialog box to type in the name of an outline style if you know it.

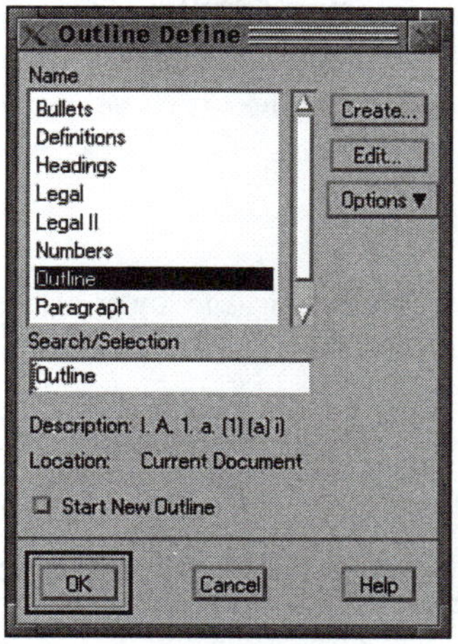

UNDERSTANDING OUTLINE LEVELS AND FAMILIES

WordPerfect supports up to eight *levels* of text—that is, different formatting in the outline. Typically, general topics come at very *high* levels in the outline, whereas *lower* levels cover more specific information. For instance, in an outline of this book, each chapter would get a very high-level entry in an outline, whereas headings such as *Creating a New Outline* would get a lower-level entry, *Entering Text* would get a lower entry than this, and *Understanding Outline Levels and Families* would get a still-lower entry. I describe how you create these levels shortly.

→ If you want to learn more about entering text in an outline, **see** "Entering Text," **p. 562**

Each outline entry is associated with a *family*. This family consists of an entry and the outline levels immediately under it. For instance, consider the following outline:

I. The Novels of Susan Blaine
 A. *Neutral Star*
  1. Characterization
  2. Metaphor
 B. *The Huts of Titanium*
  1. Analogy
  2. Characterization
II. The Poetry of Samuel Jones
 A. Meter and Rhyme
 B. Anaphor

The A and B items under "The Novels of Susan Blaine," as well as the A and B items' subsidiary entries, constitute the family for the first heading in this outline. Similarly, the 1 and 2 items under "*Neutral Star*" represent the family for that heading.

You'll often want to manipulate a family of outline entries, and WordPerfect provides tools for doing so. Because WordPerfect creates the outlines from individual styles, however, nothing intrinsic to the outline keeps families together. It is therefore possible to accidentally separate sections of families. You should be aware of this factor as you edit your outline.

USING THE OUTLINE TOOLBAR

When you create an outline, WordPerfect transforms the lower of the two default toolbars into the outline toolbar, as shown in Figure 21.13.

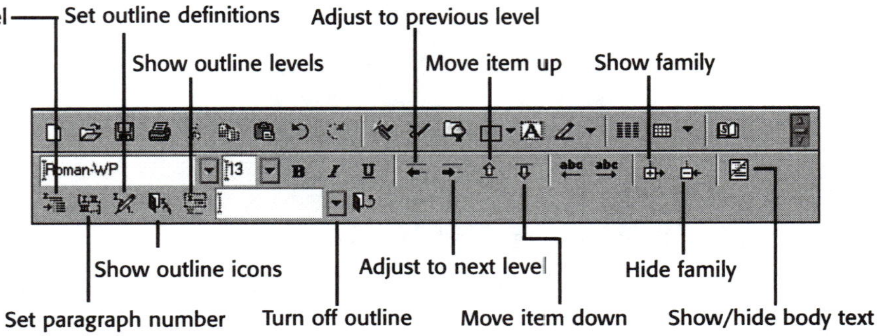

Figure 21.13
By default, the outline toolbar occupies one line, and you must scroll to see items on the second level, but this figure shows a two-line toolbar so that you can see all the tools at once.

I discuss the outline tools in greater detail through the rest of this chapter, but here's a brief summary of them:

- Adjust to previous level—This tool moves the outline item up one level in the outline hierarchy. Clicking this item is equivalent to pressing Shift+Tab with the insertion point at the start of the entry.

Does the Shift+Tab key combination have no effect? See "Shift+Tab Doesn't Work" in the "Troubleshooting" section at the end of this chapter.

- Adjust to next level—This tool moves the outline item one level down in the level hierarchy. You can achieve the same effect by pressing Tab when the insertion point is at the start of the entry.
- Move item up—This tool moves a line or selection up in the outline without changing the item's level.
- Move item down—This tool moves a line or selection down in the outline without changing the item's level.

Note

The *Adjust* toolbar items shift the *level* of the item, without changing its position in the outline. The *Move* items, on the other hand, shift the physical location of the outline item, without altering its level in the outline hierarchy.

- Hide family—When you select this tool, WordPerfect hides all lower levels in that family. They'll still be present, but hidden.

- Show family—When you select this tool, WordPerfect displays all lower levels in the family if any are hidden.
- Show/hide body text—If you expand your outline to include regular paragraphs of text, this toolbar item allows you to hide or display them.
- Adjust to specified level—Click here to move an outline item up or down more than one level in the outline hierarchy.
- Set paragraph number—You can change the paragraph number (the number that applies to the highest-level outline element) with this tool.
- Set outline definitions—Open the Outline Define dialog box (refer to Figure 21.12) by clicking this tool.
- Show outline icons—Click here to show icons representing the level of each outline item, 1 through 8 or T for text, as shown in Figure 20.14.

Figure 21.14
The outline icons can help you understand your outline's structure, and you can manipulate the outline with them.

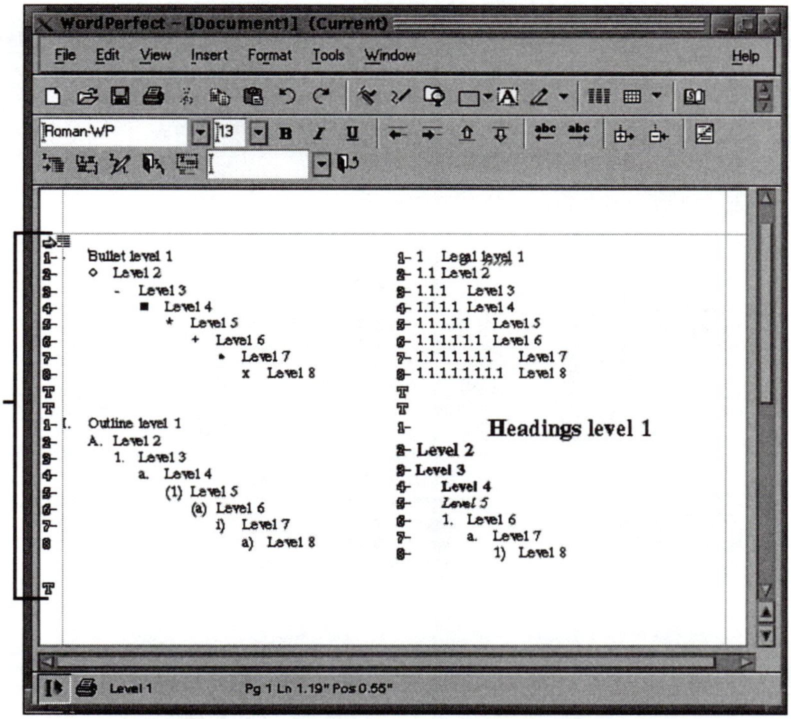

Tip #244 from *Tux Tips*

Outline icons are extremely useful. I recommend leaving them enabled whenever you're working with an outline.

PART VI
CH 21

- Show outline levels—Click here to get a drop-down menu with which you can select how many levels of the outline to display. Set this item to *One* to show only the top-level item, *Two* to show that and the first subsequent level, and so on.
- Turn off outline—If you've finished your outline and want to begin entering normal text or if you want to insert normal text within the outline proper, select this tool.

BASIC TEXT ENTRY

When you start to create your outline, WordPerfect uses whatever numbering is appropriate for the first level of that outline and you can begin your text. With most outline styles, you get full text indentation or a hanging indent style with the number to the left of any paragraph you enter and a straight left margin on the paragraph text itself. Both styles are illustrated in Figure 21.15.

Figure 21.15
Different outline styles often provide different types of indentation, numbering, and other formatting for any given outline level.

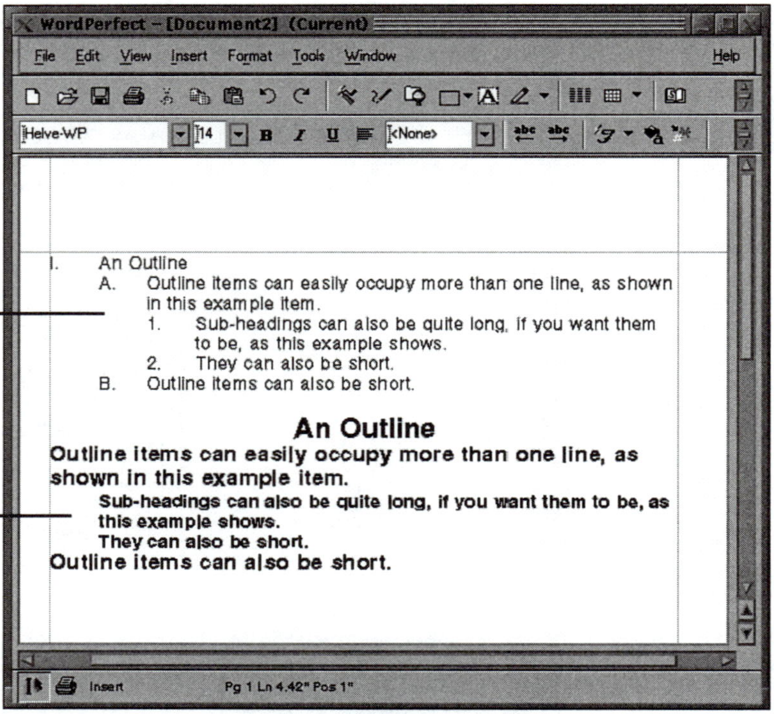

With most outline styles, whenever you press the Enter key, WordPerfect assumes that you want to create a new outline entry at the same level as the previous entry. For instance, after you type a fourth-level entry, WordPerfect enters any appropriate bullet or numbering and creates the appropriate formatting for another fourth-level entry. To change the level of an outline entry before you type it, press the Tab key or click the Adjust to next level toolbar item to move the entry down one level; alternatively, press Shift+Tab or click the Adjust to previous level toolbar item to move the entry up one level.

Does the Shift+Tab key combination have no effect? See "Shift+Tab Doesn't Work" in the "Troubleshooting" section at the end of this chapter.

INCORPORATING NORMAL TEXT IN AN OUTLINE

You can build your document around your outline with WordPerfect. You might want your outline entries to serve as headings in the document. If so, you can change the outline style, as described later in this chapter, and build your text around the outline. If you later discover that your document's structure needs modification, you can hide the text (click the Show/hide body text toolbar item), modify the outline's structure, show the body text (click the Show/hide body text toolbar item again), and continue writing.

→ If you want to learn how to change the style of an outline, **see** "Changing the Outline Style," **p. 572**

The simplest way to incorporate normal text in an outline is to press the Enter key at the end of an outline item and use the Backspace key to remove the new outline heading. You can now enter ordinary paragraphs of text, apply styles to them, and so on. Text you create in this way is hidden when you select the Show/hide body text toolbar item.

Caution

If you use the Backspace key to remove the final item in an outline, you'll have difficulty restarting the outline. Therefore, I recommend using this technique only to enter text into an existing outline. If you want to restart an outline after entering text in this way, you can choose Tools, Define Outline to get the Outline Define dialog box (refer to Figure 21.12) and start the outline again. (Be sure the Start New Outline check box *is* selected.) Then use Reveal Codes mode to locate the [Para Num Set] code and remove it.

Tip #245 from *Tux Tips*

If you use the Headings outline style, you won't need to delete extraneous numbers or style codes because when you press the Enter key after one of these heading styles, WordPerfect doesn't chain to the same style. Instead, you get whatever the current document style is. Therefore, switching to Headings outline style, or a similar style you create yourself, prior to entering normal text can speed the process somewhat.

A second way to include normal text in an outline is to select the Turn off outline toolbar item. Text you enter in this way is still part of the outline, however, and the Show/hide body text toolbar item won't hide this text. You might want to use this method for normally formatted text that you don't want to hide.

EDITING OUTLINE TEXT

After you create your outline, chances are you'll want to alter it. After all, that's why you create an outline in the first place: to make alterations quickly and easily, before you commit substantial amounts of text to paper (or to hard disk, as the case may be).

You can, of course, alter the text itself much as you would any other WordPerfect text. You can add or remove words, apply formatting such as underlining or font changes, and so on.

As with lists, applying paragraph or document-scope styles is inappropriate for outlines, however, because these styles would override the styles that define the outline.

Much of the editing you do with outlines is likely to involve moving text around, including shifting its position within the outline and altering the levels of items and families to subordinate one family to another or to move a family up a level.

ADJUSTING TEXT LEVELS

It's not unusual to write an outline and then discover that you need to subordinate one or more headings to another topic, either a new one or an existing one. Similarly, you might edit or move enough material out of a family so that it has only one remaining member, in which case you might want to eliminate the head of that family and promote its single member (along with *its* family). You can perform these tasks with the Adjust to next level and Adjust to previous level toolbar items, respectively.

With either of these tools, you can either position the insertion point anywhere in the outline entry you want to move and select the tool to shift that one entry or select a range of entries and apply the tool to all of them. When you use the second technique, the entries you select need not all be of the same level; you can select entries on several levels to adjust them all at once. In this case, the entries retain their relationships relative to one another. For instance, if you select one Level 3 item and its two Level 4 children and promote them all, then the first item becomes a Level 2 item and its children become Level 3 items.

You can adjust a single line via the keyboard alone, as well. Place the insertion point at the start of the item to be moved (that is, between the item's number, if any, and the text you typed); then press the Tab key to promote the item to the next level or press Shift+Tab to demote the item to the previous level. These keystrokes work only when the insertion point is at the *start* of the line, however. If the insertion point rests midline, these keys insert a [Tab] code or a [Hd Back Tab] (backwards tab) code, respectively.

MOVING OUTLINE FAMILIES

When you edit an outline, you'll probably want to move items from one location to another, in addition to shifting the level of items. WordPerfect provides three means of moving outline families.

DRAGGING OUTLINE ICONS

One of the quickest ways to move outline families is to select them via their outline icons, which appear to the left of the outline as a whole. If you click on the number corresponding to an outline level, WordPerfect selects that item and its descendants, as shown in Figure 21.16. You can then click and drag in the outline icons region to move the family, and a horizontal line indicates where the family will reside when you release the mouse button.

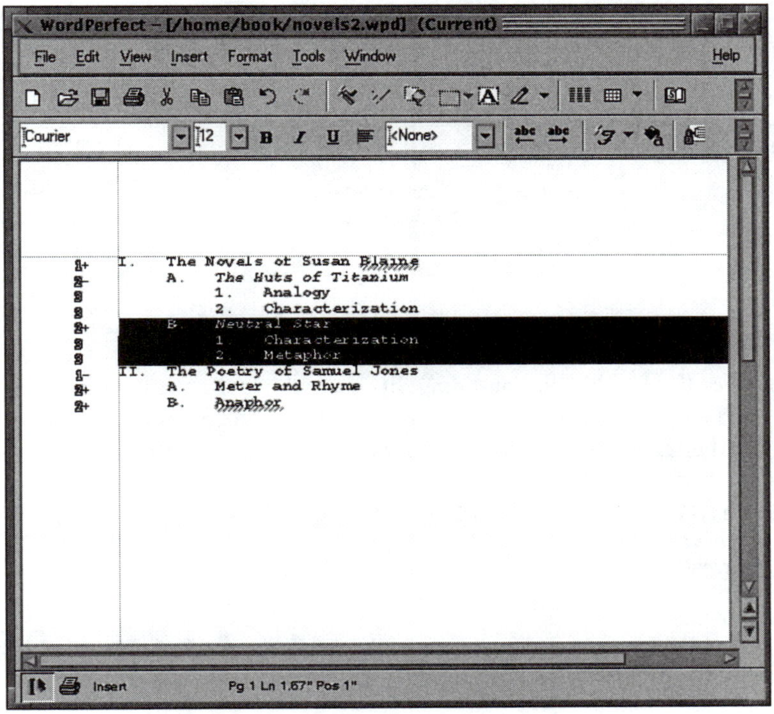

Figure 21.16
Clicking the 2 next to the Neutral Star entry selected the entire family.

> **Note**
>
> The plus (+) and minus (-) symbols next to some of the outline icons in Figure 21.16 indicate whether that outline item has hidden children. I discuss this topic in greater detail shortly.

Cut and Paste

A second method for moving outline families is to use conventional cut-and-paste operations. Select the family (either by clicking its outline icon number or by using a click-and-drag operation in the text itself); use Edit, Cut (Ctrl+X) to remove the family from one location; and then use Edit, Paste (Ctrl+V) to insert it elsewhere. The trouble with this method is that it can be quite tricky to select exactly the text you need and then paste it back where it belongs. Chances are good that you'll need to insert an extra line, delete a blank line, or adjust the level of one or more items if you do it this way.

The Move Item Toolbar Tools

The final method of moving a family is to use the Move item up and Move item down toolbar options. These, as you might expect, move a selection up or down in the outline. The best way to use these to move an entire family is to use the Hide family toolbar item to collapse a family and then move the family's head item up or down. When you're done, use the Show family item to restore it. Collapsing the family causes this toolbar item to operate on the entire family, whereas it operates on only one outline entry if it's not collapsed.

Caution

Using the Move item up or Move item down tools on a selection of text works but also causes the outline toolbar items to disappear. They'll reappear when you click in the outline, but this behavior makes using the Move item tools impractical for moving a selection of text.

WATCH THOSE CODES!

WordPerfect handles outlining, like everything else, with codes. An outline is, in fact, quite a mess of codes, as shown in Figure 21.17. Each line of the outline has style codes at its start and end. There may also be codes for hidden text, text formatting, and so on.

Figure 21.17
Each item in an outline has paragraph open and close codes, and the outline may contain codes for hidden text and normal formatting, as well.

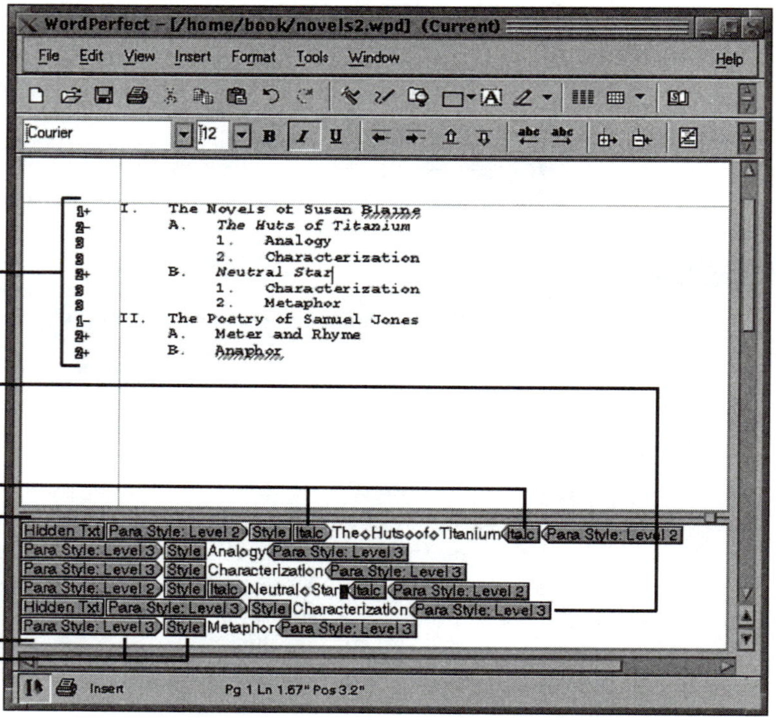

One of the dangers in moving outline families is that you might miss or mishandle codes. Thus, ordinary cut-and-paste operations on multiple lines of an outline frequently produce stray lines and altered outline levels. When you hide text, as described shortly, it's also possible to miss the hidden text, which can result in text not moving with the outline level with which it was associated.

The general rule here, then, is to be aware of WordPerfect's code-based nature and to check the structure of your outline after making changes. Reuniting "orphaned" sections of an outline with their associated sections can be tedious.

→ To learn more about Reveal Codes, **see** "Reveal Codes Mode," **p. 63** and "Editing with Reveal Codes," **p. 110**

Expanding and Collapsing Outlines

One of the great capabilities of WordPerfect's outlining feature is that it can expand and collapse outlines. I've already mentioned one aspect of this: WordPerfect's capacity to hide ordinary text. When you click the Show/hide body text toolbar icon, WordPerfect hides (or shows, if it was previously hidden) text that's not entered in an outline style. This capability can make continued use of WordPerfect's outlining features feasible even after you've moved on to regular text entry.

WordPerfect can also hide or show entire levels of an outline, however. The first pair of relevant tools is the Show family and Hide family toolbar icons. If you place the insertion point on an outline item and click Hide family, WordPerfect collapses that family into its first line. This technique is useful when you're moving families, or just to unclutter the display a bit if one family is particularly elaborated and you want to see the "big picture." You can undo this collapse by clicking the Show family icon in the outline toolbar.

The second way to control expansion and collapse is on a full-document level, by selecting a level from the Show outline levels toolbar item. If you choose Three from this menu, for instance, WordPerfect shows all menu items up to and including level 3, but no levels deeper than this.

Tip #246 from *Tux Tips*

A documentwide adjustment like this is particularly helpful when you want to see different levels of structure for a large document. If you're outlining an entire book, for instance, it's easy to lose track of the overall structure when viewing the entire outline with full detail. Collapse the outline to a level of two or three, and the document becomes much more manageable. Expand the outline when you need to work on an individual chapter or a section of a chapter.

No matter how you do an expand or collapse, the outline icons indicate whether something is hidden or not, as shown in Figure 21.17. An outline icon that's followed by a plus sign (+) indicates that it's hiding something—either normal text or a family of outline levels. A minus sign (-) next to an outline icon indicates that it has subordinates, but that they're all visible. An outline icon that is accompanied by neither a plus sign nor a minus sign indicates that the item is the terminus of its family—it can be neither expanded nor collapsed. Reveal Codes is another way to determine where hidden text resides. Such text appears as a [Hidden Txt] code.

Caution

Don't try to enter a new line after a line that's hiding something (bears a plus sign next to its outline icon). If you do, you'll insert the item *before* the text it's hiding, which is probably not what you intended. Expand the item and then insert the new item after the formerly hidden text.

Are some of the levels of your outline missing? See "Missing Outline Levels" in the "Troubleshooting" section at the end of this chapter.

Changing the Outline Style

You may want to alter the style of the outline after creating it for any of a number of reasons. You might discover that the style is inappropriate for your needs, for instance, or you might want to alter it after initially outlining your document to use the outline as headings in the body of your text.

To change the outline style, click the Set outline definitions toolbar icon to open the Outline Define dialog box (refer to Figure 21.12). Select the new outline style, check to be sure that the Start New Outline check button is *not* selected, and click OK. WordPerfect immediately alters the outline's style. You can change it back, if you like, by following the same steps.

Tip #247 from *Tux Tips*

If your document contains two outlines in the same style, but you want to change the style of only one outline, select the entire outline that you want to change before following these instructions. Otherwise, the styles of *both* outlines will change.

Customizing Outline Styles

As is true of many other features, WordPerfect's outlining relies on styles to work. In fact, outlining uses *several* styles—one for each level of the outline. If none of the nine predefined outline styles suits your needs, you can create your own. To do so, first choose Tools, Define Outline or click the Set outline definitions toolbar icon to open the Outline Define dialog box (refer to Figure 21.12).

The outline definitions that format individual levels of an outline consist of a number code, a regular WordPerfect style, or both. For instance, a style could contain number codes and associated text only, in which case it would not use hanging indents or any special text formatting, though it could use different number types, parentheses, periods, and so on at each level. Another outline style might use conventional WordPerfect styles only, in which case it wouldn't use numbering at all, only text formatting such as different font sizes, indenting, and so on. Most outline styles use both components, however.

Note

The word *style*, when applied to outlines, really refers to a *group* of styles. Each outline level is formatted according to its own style, and when you select an "outline style," you're really selecting a group of styles that go together. When necessary, I use the terms *outline style* and *individual style* to differentiate these two uses of the word *style*.

Creating a New Outline Style

If one of the existing outline styles is close to what you need, but not exactly right, you can save yourself some effort by selecting it and choosing Copy from the Outline Define dialog box's Options option menu button. This step produces the Outline Definition Copy dialog box (Figure 21.18). Select Current Document if you want to use this definition for one document only; otherwise, select Personal Library. The Shared Library option is available only if you start WordPerfect with the `-adm` switch and you have write access to the WordPerfect directory tree. Because outline styles are just a collection of ordinary individual styles, the consequences of storing them in any given location are the same as for other types of styles. In any event, when you've copied the style, you can edit the copy rather than create an entirely new style.

→ To learn more about managing style collections, **see** "Saving Styles," **p. 318**

Figure 21.18
You can copy an existing outline definition to modify it or to make it more widely available.

To create a new outline style, click Create in the Outline Define dialog box. This step produces the Edit Outline Definition dialog box, shown in Figure 21.19.

You use the options in the Outline Define dialog box to create specific individual styles and associate them with the various outline levels.

- Name—Enter a name for the style in the Name field of the dialog box.
- Description—You can give the style a description to help further identify it.
- Levels—The Levels field shows what each of the eight levels of the outline style will look like, assuming the individual style doesn't add any text. Click on an item in this list to edit that style level.
- Associated Style—This field displays a list of available individual styles you can use in your outline style definition. This field and the Levels field are linked; when you click a specific style level in the Levels field, its matched Associated Style entry is selected.

Figure 21.19
The Edit Outline Definition dialog box is used both for editing existing outline styles and creating new ones.

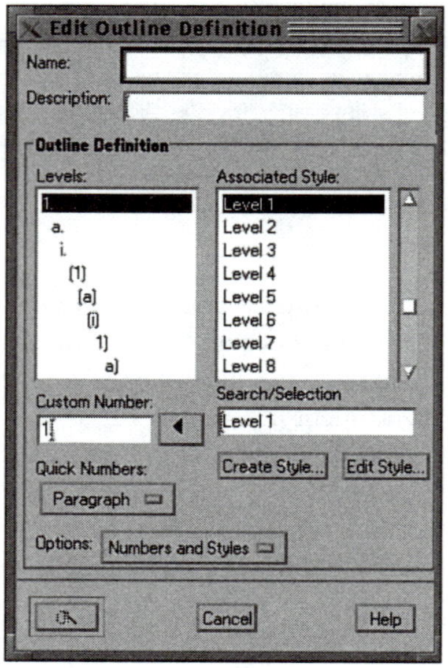

- Custom Number—Click the option button next to this field to get a list of "numbers" for outline level, including Arabic numbers, Roman numerals, letters, and assorted bullets. You can also edit the text to incorporate any textual formatting, such as periods, parentheses, or even whole words, that you want to include in this level.
- Search/Selection—You can type a style name in this field to have WordPerfect select it, rather than scrolling to it in the Associated Style field.
- Quick Numbers—You can select from any of several predefined numbering schemes with this option button.
- Create Style and Edit Style—You create or edit, respectively, an individual style with these buttons. Clicking either of these buttons produces the Styles Editor dialog box (refer to Figure 21.7), into which you can enter any special text formatting you want to use with this style.

→ To learn more about the Styles Editor, **see** "Creating a Style," **p. 312**

- Options—You can tell WordPerfect to use numbers alone, styles alone, or both for this outline style definition.

The general procedure for creating a new style is to select a numbering scheme that's at least close to what you want and then to customize it and the Associated Styles you need to implement the style.

Tip #248 from
Tux Tips

> If you edit an existing Associated Style, you'll replace it within the document or style library you're editing and, if you save the style in other places, in those locations as well. This situation could cause problems if, for instance, another outline style or template relies on the style you've changed. Therefore, I recommend that you create a new individual style rather than edit an existing one. You can base a new style on an old one by selecting the old one before you click Create Style. WordPerfect uses the old style's settings as a basis for the new style.

EDITING AN EXISTING OUTLINE STYLE

You can edit an existing outline style much as you would create a new one except that you click Edit, rather than Create, in the Outline Define dialog box (refer to Figure 21.12). If you choose to edit one of the nine default outline styles, WordPerfect won't let you change the name of the style, though you can change the name of a style you create.

As with the individual styles within an outline style, if you intend to reuse your edited style, I recommend copying the style (by selecting Copy from the Options button in the Outline Define dialog box) and editing the copy rather than the original style. If you do change a default style and decide you prefer the original, you can select Reset from the Options button to reset the selected style to its original state.

SAVING AND RETRIEVING OUTLINE STYLES

Saving and retrieving outline styles works much like saving and retrieving individual styles. You can copy styles from one location to another, such as from the current document to your personal style library, by selecting a style and choosing Copy from the Outline Define dialog box's Options option button. This step opens the Outline Definition Copy dialog box (refer to Figure 21.18). If you want your style to be readily available whenever you use WordPerfect, you should copy the style to a Personal Library. If you run WordPerfect in administrative mode (by typing `xwp -adm` at an xterm prompt) and you have write access to the WordPerfect directory tree, you can also copy the styles to the systemwide Shared Library. This way, others who use the computer can use the outline style you've created.

→ If you want to learn more about saving and retrieving styles, **see** "Saving Styles," **p. 318** and "Retrieving Styles from Disk," **p. 324**

To copy the style to a separate file so that you can move it between systems or retrieve it only when needed, select Copy from the Options button. This step opens the Save Outline Definitions dialog box shown in Figure 21.20. You can save the regular system styles, your own user styles, or both to the filename you specify.

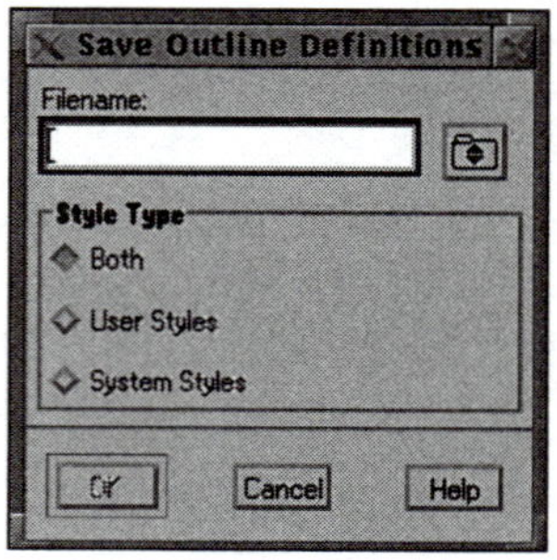

Figure 21.20
You generally need to save only the User Styles when you create a new outline style.

If you want to retrieve an outline style, you can select Retrieve from the Options button in the Outline Define dialog box. This step opens the Retrieve Outline Definitions dialog box, which is identical to the Save Outline Definitions dialog box shown in Figure 21.20 except for the title. Both dialog boxes function in much the same way.

Troubleshooting

Changes to a List's Style Are Only Sporadically Effective

You've changed the style for a simple list, but only some lists in the document use your changes.

The most likely cause of this problem is that the different lists don't use the same styles. Because you can create styles that work the same but are distinct in different ways, it's easy to create two or more list styles when you need only one. One way to fix this problem is to locate the style that's not been affected by the change and edit it. You can either duplicate your style change, or you can delete everything in the Contents area of the Styles Editor dialog box (refer to Figure 21.7) and replace it with a reference to the style that you've changed. To do so, choose Format, Styles (Shift+Ctrl+F8) from the Styles Editor's menu and apply the target style from the Style List dialog box (refer to Figure 21.6).

Shift+Tab Doesn't Work

Nothing happens when you press Shift+Tab to move back an outline level.

This problem occurs because of a flaw in some X keyboard maps. If you're experiencing this problem, try adding the following line to the .xinitrc file in your home directory:

```
xmodmap -e 'keycode 23 = Tab'
```

Then, at an xterm prompt, type `~/.xinitrc` to rerun the .xinitrc script; alternatively, log out and log in again.

MISSING OUTLINE LEVELS

You're certain you created more levels to an outline than WordPerfect is displaying.

If you haven't already, you should display the outline icons by clicking the Show outline icons toolbar item. You can then check whether the last heading that appears has a plus sign (+) next to it. If so, the levels below that are simply hidden, and you can display them by clicking the Show family toolbar icon.

If you're having problems with several families, you may want to use the Show outline levels tool to adjust the number of levels WordPerfect displays overall. Occasionally, WordPerfect gets "stuck," and you need to adjust the number of levels all the way up to Eight, and then down again, if you don't want to display that many levels.

A final possibility is that you accidentally moved part of the family. Try adjusting the number of levels displayed all the way up to Eight and then searching for some text you know is in the part of the family that's missing. This technique should turn up the missing text, and then you can move it back to where it belongs.

CHAPTER 22

Document Management

In this chapter

OVERVIEW

When your document becomes large, or when you revise a document frequently, you'll probably want some way to control, track, and otherwise manage your revisions. WordPerfect 8 for Linux provides a number of tools to help do this, including some that are new to version 8 of the program. You can add nonprinting comments to the document, highlight areas of text, maintain a history of document versions, and keep bookmarks so you can quickly find key sections of your document. All of these features can help you work more efficiently with the document, including collaborating on it with one or more other people.

DOCUMENT COMMENTS

When you write a document that you intend to revise, you may want to include nonprinting notes for yourself or your collaborators. For instance, you may want to remind yourself to add information on a particular topic, or you might want to flag some text for your co-authors to examine with particular care. You can, of course, embed such notes in the text itself, using special formatting to distinguish the comments from ordinary text. If you use this technique, however, your comments will print when you print the document, which may not be what you want to do, especially if you give preliminary printed copies to others. Therefore, WordPerfect supports a feature that's known as *comments*. A WordPerfect comment does not print, but you can read it from within WordPerfect with little effort. In addition, when WordPerfect is properly configured, each comment bears a code to identify the individual who created it, which is useful if you're collaborating on a document.

CREATING COMMENTS

To create a comment, place your insertion point where you want the comment to appear and then choose Insert, Comment, Create. WordPerfect responds by opening a new document window in which you edit your comment, as shown in Figure 22.1.

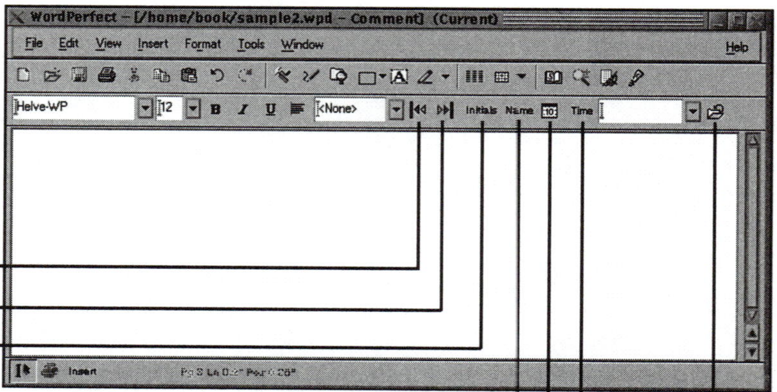

Figure 22.1
To use all the features in the customized toolbar in the Comment editing window, you may need to widen the window (as shown here) or scroll the toolbar.

You can use most of the usual WordPerfect editing features when you create your comment, including text attributes such as italics, fonts, columns, and so on. You can even create or includes graphics in a comment. When you later view your comment, all these elements will be present.

→ To learn more about fonts, **see** "Fonts," **p. 178**

→ To learn more about columns, **see** "Creating Professional Layouts with Columns," **p. 255**

→ To learn more about graphics, **see** "Graphics," **p. 377**

Note

It's possible to create a comment that's larger than WordPerfect can display on screen at one time. You'd therefore do well to restrict the size of your comments to a few lines. If your comment must be large, consider altering the font size or spacing to reduce the amount of screen space it consumes. If you're collaborating with others, remember that your collaborators may have smaller monitors, or prefer to use WordPerfect with smaller windows, than you do. A comment you can see in its entirety may be too large for your colleagues to see.

The Comment editing window provides several toolbar items specific to comments. These are

- Previous comment and Next comment—Click these tools to edit the previous and next comments, respectively. You normally use these when editing comments, but you might want to refer to a previous comment when creating a new one.
- Insert initials—As described shortly, WordPerfect maintains identifying information for you, and you can insert your initials into the comment by clicking this tool.

→ If you want to know more about using your initials in comment codes, **see** "Individualizing Your Comment Codes," **p. 585**

- Insert name—Just as you can insert your initials, you can insert your name with this tool.
- Insert date and Insert time—If you want to keep track of *when* you made your comment, you can do so by inserting the date and time with these two tools.
- Close Comment editing window—When you're done editing your comments, you must close the Comment editing window. You can do so by choosing File, Close (Ctrl+F4) from the Comment editing window's menu or by clicking on the Close Comment editing window tool. WordPerfect then closes the window and returns you to the normal document editing window.

As an alternative to typing a comment from scratch, you can convert existing text into a comment by selecting the text and choosing Insert, Comment, Create. WordPerfect doesn't display the Comment editing window, but instead converts the text directly into a comment, deleting it from the main body of the document.

Reading Comments

When you insert a comment, WordPerfect places a comment code where your insertion point had been. This code causes WordPerfect to display a comment icon in the document's left margin, as shown in Figure 22.2. If you've not defined your personal information, as described shortly, the comment icon looks like the Anonymous comment icon in Figure 22.2. If you have specified your personal information, however, your comment looks like the Signed comment icon in Figure 22.2. Although it's difficult to make this out even onscreen, the Signed comment icons include your initials. They're also shaded with the color you specify in your preferences, as described later in this chapter. If you have more than one comment on a line of text, you get the Multiple comments icon.

→ For more information on setting your initials and comment color, **see** "Individualizing Your Comment Codes," **p. 585**

Figure 22.2
WordPerfect comment icons can take on any of several appearances.

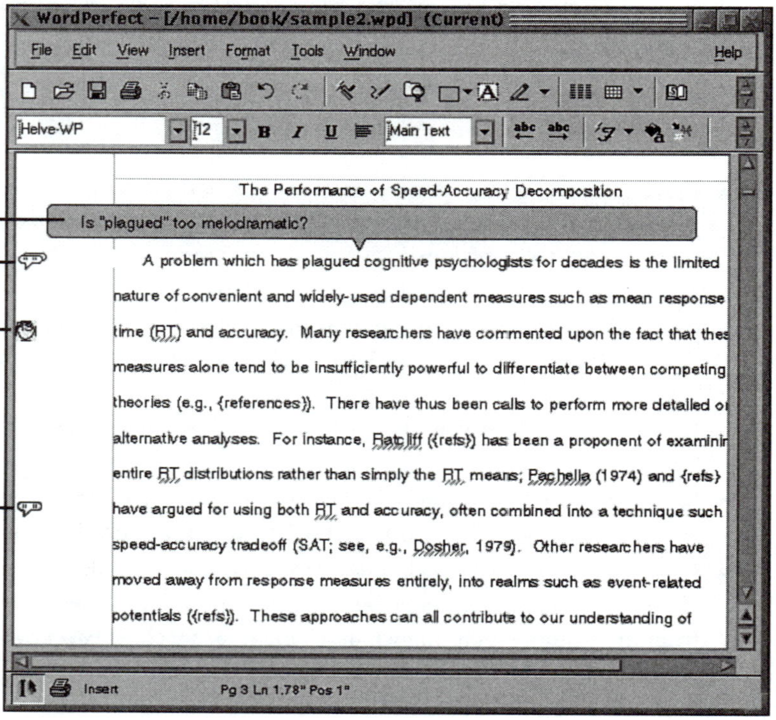

Tip #249 from *Tux Tips*

If you use comments, don't use the Margin Width Zoom option because it hides the comment icons outside of the window's view.

→ If you'd like to know more about the zoom options, **see** "Zoom," **p. 55**

To view a signed or anonymous comment, click on the comment's icon. WordPerfect displays the comment "bubble," as shown in Figure 22.2. This bubble includes a pointer that identifies the location of the comment code in the text. You can use this feature to draw your or your collaborator's attention to a specific word on the line. The background shading for the bubble matches the color assigned to the person who created the comment, so if you know those colors, you can instantly tell who created the comment.

If two or more comments appear on one line, WordPerfect displays the Multiple comments icon. When you click on this icon, WordPerfect displays a menu of available comments (Figure 22.3), listed in their order on the line. When you click on one of these identifiers, WordPerfect displays its matching comment. WordPerfect identifies the comments in the menu by their creators' names (or "Comment" for anonymous comments).

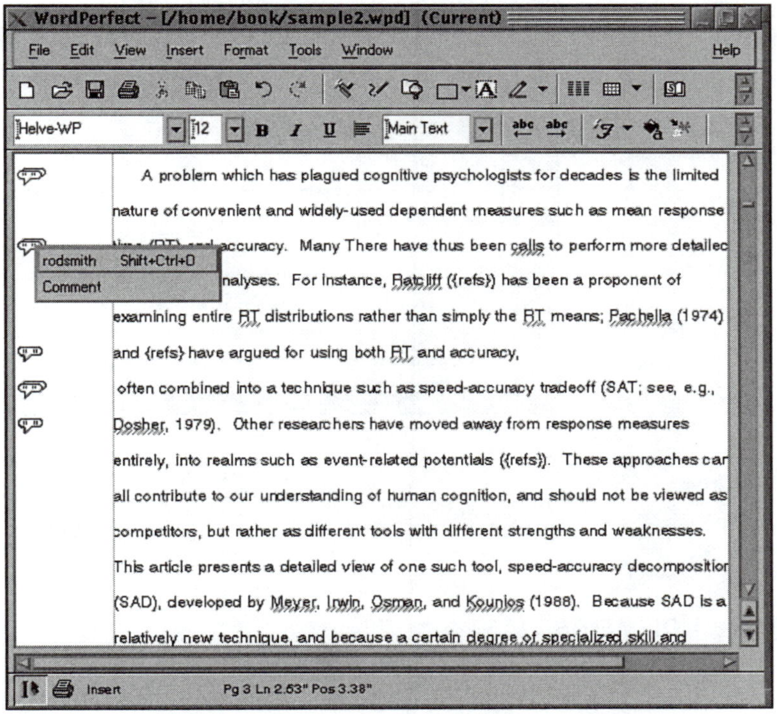

Figure 22.3
You can select the comment you want to view based on the creator's name.

Note

The first entry in the menu shows a shortcut key of Shift+Ctrl+O, but it doesn't work.

Note

WordPerfect also sometimes displays the Multiple comments icon when a comment appears on the same line as some other formatting codes. In this case, you see only one entry in the menu when you click on the icon.

If you print a file that contains comments to a PostScript file and create from that PostScript file an Adobe Acrobat PDF file, Acrobat Reader provides access to the comments, as shown in Figure 22.4. The comments lose their precise association with text, however; their icons appear next to the lines with which they're associated, but not at the exact point in the text in which they occurred.

➔ To learn how to create a PDF file from a WordPerfect document, **see** "PostScript and PDF Output," **p. 168**

Figure 22.4
Adobe Acrobat can display comments in PDF files you've created by using WordPerfect and Ghostscript.

Open comment

Closed comment icons

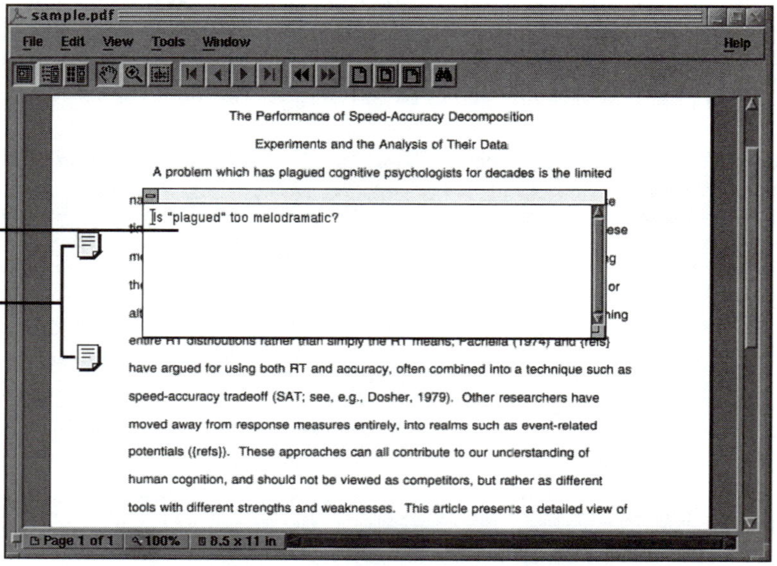

Tip #250 from *Tux Tips*

If you intend to distribute your document via PDF file, you might consider using WordPerfect comments in place of or in addition to footnotes or endnotes.

Editing Comments

You can edit a comment in any of several ways:

- Double-click on a Signed or Anonymous comment icon in the margin.
- Right-click on a Signed or Anonymous comment icon and choose the Edit item from its context menu.

Note

Double-clicking or right-clicking on a Multiple comments icon won't do anything.

- Place the insertion point after the location of the comment you want to edit, but before any subsequent comment, and choose Insert, Comment, Edit.

- Double-click on the [Comment] code when you view the document in Reveal Codes mode (View, Reveal Codes or Shift+Ctrl+F3).

In any of these cases, WordPerfect opens a new window in which you can edit your comment. Editing a comment works just like creating a new comment, as described previously, except that any existing comment text is already present in the window.

Tip #251 from
Tux Tips

> You can edit both your comments and the comments of others. When you do so, the original author's identification codes remain intact. Therefore, if you're using comments to exchange ideas with collaborators, you may want to use new comments or at least identify yourself when you modify existing ones.

Converting Comments to Text

If you've created a comment that you think you *might* want to include in the text or if you want to print a document with its comments, you can convert one or more comments into normal text. To do so, position your insertion point after the comment you want to convert, but before any subsequent comment, and choose Insert, Comment, Convert to Text. WordPerfect removes the comment and places its contents in the document's text.

Note

> In this conversion, WordPerfect also places the [Open Style: Comment] code in the text, so if you're using that code to modify the formatting of your comments, you need to delete the code after you convert the comment.

Tip #252 from
Tux Tips

> If you want to convert many comments to text, you can write a macro to do so by repeatedly searching for [Comment] codes and then performing the conversion. If your macro then selects the newly converted comment text (by using the Extend Selection option in Find and searching for the [Open Style: Comment] code), you can apply a special style to the comment code. This approach allows you to format your comments in some special way and to write a macro to reverse the comment conversion. With these two macros, you could convert all your comments back and forth repeatedly, say, to print your document with all its comments and then restore it to its original state.

→ For more information on macros, **see** "Letting the Computer Do the Work: Macros," **p. 483**

→ For more information on finding and replacing text, **see** "Find and Replace," **p. 105**

Individualizing Your Comment Codes

If you're collaborating on a document, you may want to use your comments to communicate with your co-authors about the document without cluttering the text with material that you'll have to remove before printing the final copy. To use your comments most effectively, you should personalize your comment codes. You can do so by choosing Preferences from the WordPerfect control window and then clicking on the Environment button. WordPerfect then presents the Environment Preferences dialog box, shown in Figure 22.5. In the upper-left corner of this dialog box, you'll find the User Info for Comments/Summary area. The Name field in this area defaults to your username, which is what WordPerfect inserts when you click the Insert name toolbar button (refer to Figure 22.1). You can change this entry to whatever you like, however. Similarly, you can adjust the initials you get when you click the Insert initials toolbar button, and that WordPerfect displays in the Signed comment icon, by adjusting the Initials text entry field. Finally, you can choose a color by selecting it from the pop-up color selector when you click on the colored area next to Use Color. WordPerfect uses this color for both the Signed comment icon and as the background color for comment text when you display a comment.

Figure 22.5
You can set the name, initials, and color used for comments you create.

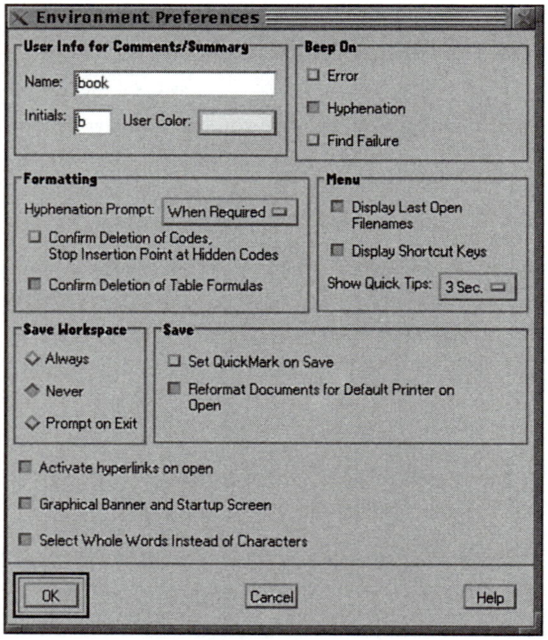

Tip #253 from *Tux Tips*

If you leave the Initials field blank, WordPerfect creates an anonymous comment—it displays no color or initials in its onscreen icon—and the background when you view a comment is gray.

Text Highlighting

You may want to draw your or a co-author's attention to an incomplete or awkward passage or otherwise make some text stand out. Of course, you could use text formatting, such as boldface, for this task; but this type of formatting is frequently used for other purposes and so isn't usually the best choice for this purpose. WordPerfect therefore lets you *highlight* text. Highlighted text uses a background color other than the usual to make it stand out.

Highlighting Text

WordPerfect provides several means for you to highlight text:

- Select a portion of text that you want to highlight and then choose Tools, Highlight, On. WordPerfect immediately highlights that text.
- Before typing the text you want to be highlighted, choose Tools, Highlight, On. WordPerfect highlights any text you type from that point on until you choose Tools, Highlight, On again.
- Choose Tools, Highlight, On and then select some text. WordPerfect immediately highlights the text you select.

If you choose Tools, Highlight, On and type some text, you'll also be able to select words to have WordPerfect immediately highlight them. You won't be able to type new text and have WordPerfect automatically highlight it after selecting words for highlighting, however.

Setting a Highlight Color

By default, WordPerfect uses bright yellow as the highlight color, but you can change this color if you like. The easiest time to change the highlight color is before you select text. To change the color of the highlighting later, you need to remove the existing highlighting and reapply it in the new color.

To change the highlighting color, choose Tools, Highlight, Color to get the Highlight Color dialog box (Figure 22.6). You can pick a color by clicking the Color option button to produce a palette of colors. You can also lighten colors by reducing the value in the Shading spin box. For instance, to get 20% gray, select black as the color and enter `20` in the Shading spin box. After you select a color, subsequent highlighted selections take on that new color, but existing highlighted selections retain their original colors. You can therefore mix different highlight colors in your document if you like.

Figure 22.6
Select a unique highlight color from the Highlight Color dialog box.

HIDING, SHOWING, AND REMOVING HIGHLIGHTING

If you're using highlighting to help your own editing efforts, you presumably don't want to print the highlighting and you may not want to display it at all times on the screen. WordPerfect therefore allows you to control whether highlighting displays and prints or not and to remove it altogether.

You can toggle the display of highlighting by choosing Tools, Highlight, Print/Show. By default, this option is selected, meaning that WordPerfect both prints and displays highlighting. If you choose this menu item, WordPerfect deselects the option, thus hiding all instances of highlighting, both on screen and when printed. Choosing it again redisplays all highlighting.

Are you trying to print a document with highlighting enabled but find that the printout doesn't highlight the same text you see highlighted on the screen? See "Highlighting Prints Strangely" in the "Troubleshooting" section at the end of this chapter.

If you want to remove highlighting from a word or larger body of text, you can do so with the Tools, Highlight, Remove selection. If your insertion point rests in a highlighted word when you choose this option, WordPerfect removes highlighting from that word. If you've selected a section of text when you choose this option, WordPerfect removes highlighting from the entire selected region. You can also remove highlighting by deleting the [Highlight] codes that surround highlighted text in the Reveal Codes pane (View, Reveal Codes or Shift+Ctrl+F3).

COMPARING DOCUMENTS

You may occasionally have two copies of a document that are different, and you need to know in what ways they're different. For instance, if you're collaborating with somebody, you might want to know what changes your co-author has made to the document; or you might simply discover two versions of a file in different locations and not be sure which one you should be using. In situations like these, WordPerfect's document comparison feature comes in handy.

PERFORMING A COMPARISON

To compare two documents, follow these steps:

1. Load one version of the document. If you know which one is more recent, load it.
2. Choose File, Document, Add Compare Markings. WordPerfect responds by displaying the Add Compare Markings dialog box shown in Figure 22.7. You specify the filename of the second (earlier) document in the Compare Current Document to field, and you tell WordPerfect whether to compare by Word, Phrase, Sentence, or Paragraph. If you've changed only one word in a document, the latter options mark the entire phrase, sentence, or paragraph in which that word appears.
3. Click OK. WordPerfect compares the files and displays changed or added text in red and text that's been deleted from the current document in strikeout (with a line drawn through it). Figure 22.8 illustrates these effects.

Figure 22.7
Specify the file for and granularity of the comparison from the Add Compare Markings dialog box.

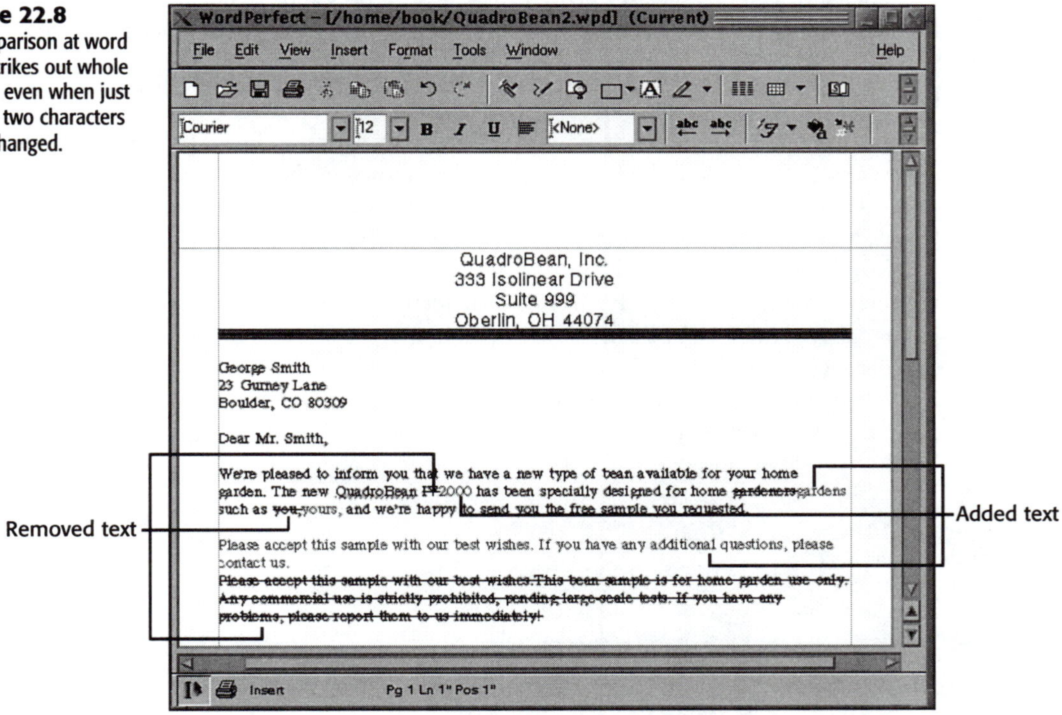

Figure 22.8
A comparison at word level strikes out whole words, even when just one or two characters have changed.

If the comparison markings have served their purpose, you can remove them by choosing File, Document, Remove Compare Markings, which produces the Remove Markings dialog box shown in Figure 22.9. You can choose to remove only the struck-out text or both that and the redline markings.

Tip #254 from *Tux Tips*

You can redline text yourself, without comparing documents, by selecting the text and choosing Format, Font (F9). Select Redline in the Font dialog box's Appearance area and then click OK.

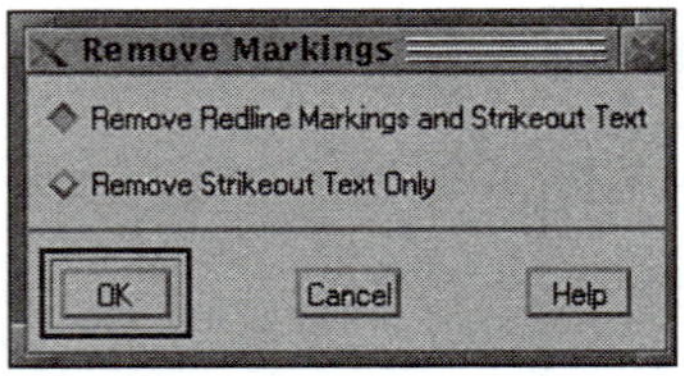

Figure 22.9
If you select Remove Redline Markings and Strikeout Text, WordPerfect keeps red-lined text but leaves it un-redlined while completely removing struck-out text.

Redline Method

By default, WordPerfect displays redlining using red text on the screen. How this text prints varies from printer to printer, however. If you're using a PostScript printer (or a printer that's being driven by Ghostscript), WordPerfect outputs appropriate color codes to turn the text red if the printer is color capable. If it's not, the text prints gray rather than black. If you're using a PCL printer with WordPerfect PCL drivers, the text prints with a gray background. Other printers may use other redlining methods.

You can alter how redlined text appears, however, by choosing File, Document, Redline Method, which produces the Redline dialog box shown in Figure 22.10. The default selection is Printer Dependent, which works as I've just described. You can choose to mark the left, right, or alternating margins with the Redline Character you specify in the dialog box, however. When you select one of these options, WordPerfect marks redlined text in red on the screen, but does not use the printer-dependent redlining method. Instead, WordPerfect places the Redline Character in the left or right margin of a line that contains altered text. It also displays a red version of the Redline Character in this position on the screen. When you select one of these options, therefore, your printouts look the same no matter which printer type you use, but you'll not be able to tell precisely which text within a line is redlined, only which lines are redlined.

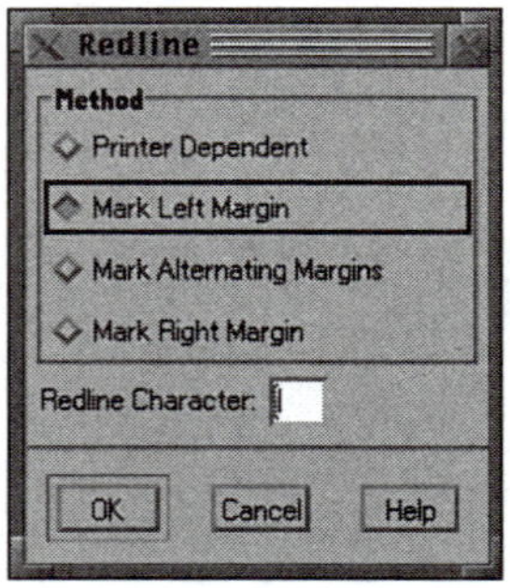

Figure 22.10
Specify how you want WordPerfect to print redlining with the Redline dialog box.

Tip #255 from *Tux Tips*

If you click in the Redline Character field of the Redline dialog box and then press Ctrl+W, WordPerfect displays the Symbols dialog box and lets you select a redline character from any symbol WordPerfect supports.

→ To learn how to use the Symbols dialog box, **see** "Inserting Special Characters," **p. 102**

Document Revision Tracking

Have you ever saved an updated version of your document, only to realize that you've blundered? Perhaps you accidentally deleted an important section of text, or maybe you found that your painstaking reformatting needs to be undone. Whatever the case, this situation calls out for a system to track and archive different revisions of your document. One of the additions from WordPerfect 7 to WordPerfect 8 is just such a system, known as Version Control.

Basic Principles of Version Control

The basic idea behind Version Control is to save several different versions of the document. That way, if you need to restore text you've removed during a revision, you can load up the previous version, copy the text, and paste it into a later version. You can even revert entirely to a previous version if necessary.

WordPerfect maintains several versions of the document, located within a single file whose name begins with the $ character. Rather than accessing this *versions file* directly, you load a file with a normal WordPerfect name and then use special dialog boxes to switch between the various archived versions in the versions file.

WordPerfect maintains a set number of versions of a file, each of which is associated with some identifying text, which you can alter when you save the file. When you've saved the number of versions you've specified, WordPerfect starts removing the oldest versions. You can tell WordPerfect to make one or more specific versions of the file "permanent," however, meaning that it won't be deleted when its time would otherwise be up.

To begin using Version Control, you start a document just as you would any other. Use whatever formatting, styles, and so on, you want. The WordPerfect document is essentially the same as any other; there are no restrictions on the features you can include in a version-controlled document.

Saving a Document with Version Control

To save your document using Version Control, follow these steps:

1. Choose File, Version Control, Save Current. If you've not saved the document before, WordPerfect asks whether you want to name it. Select Yes and then select a name using the File Browser dialog box shown in Figure 22.11.
2. WordPerfect displays the Version Control Save Current dialog box shown in Figure 22.12. If you select Make first version permanent, WordPerfect does not automatically delete this version of the file when you've saved more than the specified maximum number of versions. If you select Save file to single location, WordPerfect saves all version files to one directory, which may be different from the directory in which the main file resides. You can specify the number of versions WordPerfect maintains by setting the number you want in the Max number of Temp versions spin box.

→ For more information on setting the version options, **see** "Setting Versioning Options," **p. 593**

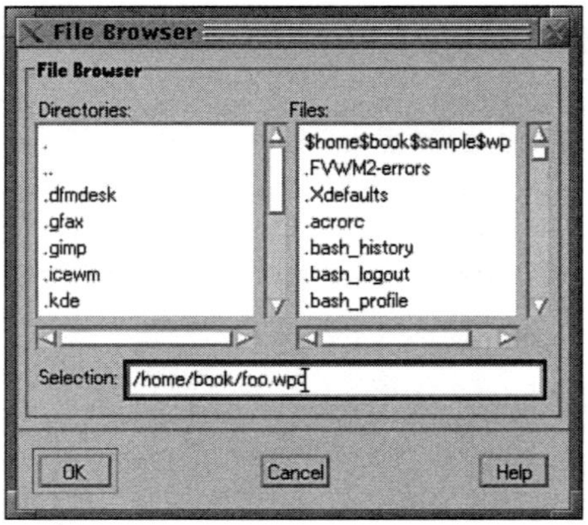

Figure 22.11
The File Browser dialog box is a subset of the normal WordPerfect File Manager dialog box.

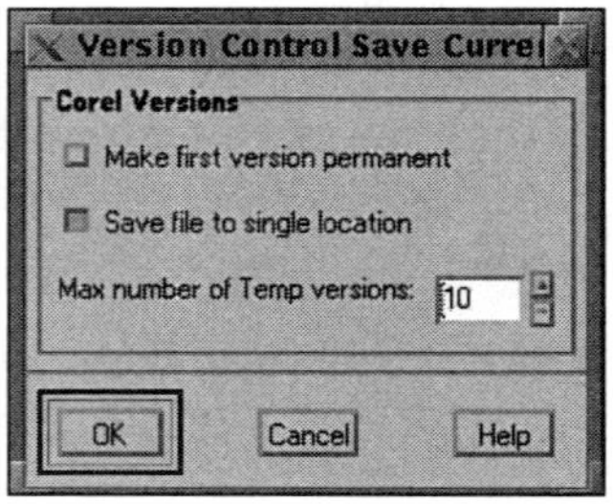

Figure 22.12
Set Version Control options with the Version Control Save Current dialog box.

3. When you've set all the options you want, click OK. WordPerfect saves the file.

After you've entered more text or otherwise edited your document, you can choose File, Version Control, Save Current again to save your document, which produces the Version Control Save dialog box shown in Figure 22.13. You can specify that you want to make this version permanent by clicking the Permanent check button, and you can enter text to identify this version in the Comment field. *Do not* use the normal File, Save (Ctrl+S) or File, Save As (F3) commands to save your file if you want to take advantage of versioning.

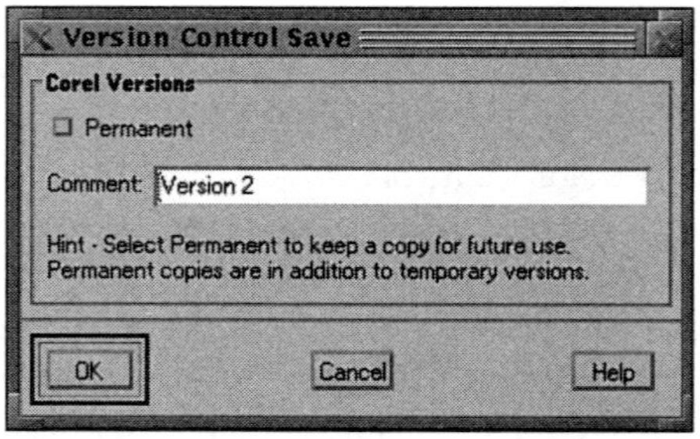

Figure 22.13
You can replace the Comment field's text with something more descriptive to help you identify the version.

In addition to saving your document in the version file, when you select the version save feature, WordPerfect saves the file under the filename you specified earlier. This file always contains the most recent version of the document, which makes it easy to retrieve the document or to pass it on to others without including all your revisions.

Note

Because WordPerfect saves several copies of your document, using the versioning feature can substantially increase the amount of disk space needed for the documents, especially if your documents are large because of included graphics or other size-increasing features. You may want to occasionally remove unneeded version files from your hard disk.

Retrieving Previous Versions

When you've saved two or more versions of a document, you can switch between them by using the File, Version Control, Retrieve Current and File, Version Control, Retrieve Document commands. These work alike except that the latter prompts you to select a new file from the File Browser dialog box (refer to Figure 22.11) before prompting for a version to retrieve. You should select the WordPerfect document file you specified when you created the file; don't go looking for the version file itself (the one with a filename that starts with $). Both commands display the Version Control Retrieve dialog box, as shown in Figure 22.14. You select the version you want to use and then click Retrieve. WordPerfect asks you to confirm that you want to replace the current version with the one you've selected and, if you click Yes, does so.

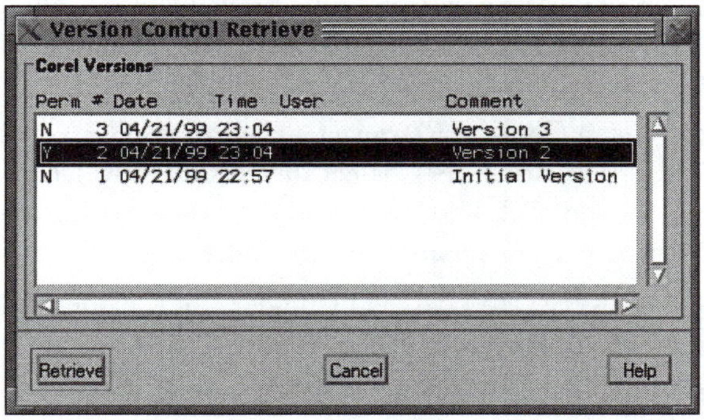

Figure 22.14
You can use the version creation date and time, as well as any comment you entered, to help you identify a version you want to use.

Setting Versioning Options

You can set a few options that affect how Version Control works by choosing File, Version Control, Preferences, which produces the Version Control Preferences dialog box shown in Figure 22.15. You can set the number of versions WordPerfect will store by default, and you can tell WordPerfect where to store versions. If you click Default versioning at single location and enter a directory in the Path text entry field, WordPerfect uses that directory for version files if you subsequently select Save file to single location in the Version Control Save Current dialog box (refer to Figure 22.12). If you leave this field blank, WordPerfect stores version files in the same directory as normal files.

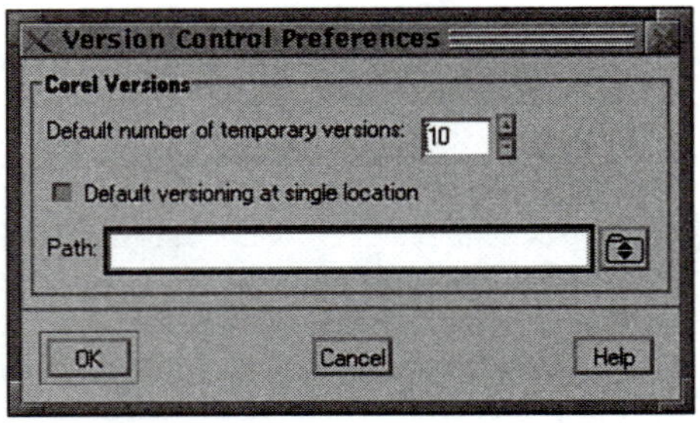

Figure 22.15
Clicking the folder icon next to the Path field produces a File Browser dialog box that you can use to select a directory in which to store version files.

Tip #256 from *Tux Tips*

Because you never have to select version files yourself, you may want to create some out-of-the-way directory in which to store them. This practice can save you from having to view their rather ugly names, which are related to the complete pathname to the file, but with slash (/) directory separators replaced by dollar signs ($), and .cv appended to the file-names.

Comparing Versions

If you want to use WordPerfect's comparison feature, as described earlier in this chapter, to compare two versions of your document, you need to save the earlier version in a separate file and then compare it to the latest version. Follow these directions:

1. If necessary, save the current version of your document by choosing File, Version Control, Save Current and by entering any options you like in the Version Control Save dialog box (refer to Figure 22.12).
2. Choose File, Version Control, Retrieve Current to get the Version Control Retrieve dialog box (refer to Figure 22.14). Select the version to which you want to compare the current document and click Retrieve. Select Yes when WordPerfect asks whether you want to replace the current version with the older one.
3. Choose File, Save As (F3) to save the file under a new name, without revision features. Use any temporary filename you like.
4. Choose File, Version Control, Retrieve Document to get back to your latest document version. WordPerfect displays the File Browser dialog box (refer to Figure 22.11) and then, when you select a file, displays the Version Control Retrieve dialog box (refer to Figure 22.14). Select the version you want to compare to the file you just saved.
5. Perform the document comparison as described earlier. Specify the file you've saved as the one to which you want to compare the current file.

→ If you want to review how to compare two documents, **see** "Comparing Documents," **p. 588**

When you're done with a comparison within document versions, you can delete the file you created in step 3. If you plan to do more comparisons against that version of the file, of course, you may want to leave it until you finish.

Bookmarks

When you work with a document of more than a few pages, you may find yourself frequently looking for a particular part, or specific parts, of the document. For instance, if you're writing a paper that includes a bibliography, you may want to jump quickly to the bibliography to add an entry or check on the spelling of an author's name when you refer to that entry in the body of your text. Jumping around this way can be awkward, particularly as your document grows. That's why WordPerfect includes a *bookmark* feature. You can flag several different points in the text as bookmarks and then jump to them quickly.

Note

WordPerfect has the capability to automatically generate a bibliography (known in WordPerfect as a *table of authorities*) if you mark your references appropriately in the text. You may want to use this feature rather than maintain your bibliography the "old-fashioned" way.

→ For more information on having WordPerfect maintain your bibliography, **see** "Tables of Authorities," **p. 644**

Setting Bookmarks

To set a bookmark, place your insertion point at the bookmark location and choose Tools, Bookmark. WordPerfect presents the Bookmark dialog box (Figure 22.16). Click Create to create the bookmark and enter its name in the Create Bookmark dialog box (Figure 22.17). WordPerfect uses the text you enter here as a label for the bookmark in the Bookmark dialog box, so use something descriptive like Bibliography or Chapter 2. When you click OK in the Create Bookmark dialog box, WordPerfect removes both it and the Bookmark dialog box from the screen. You can then go on with editing as usual.

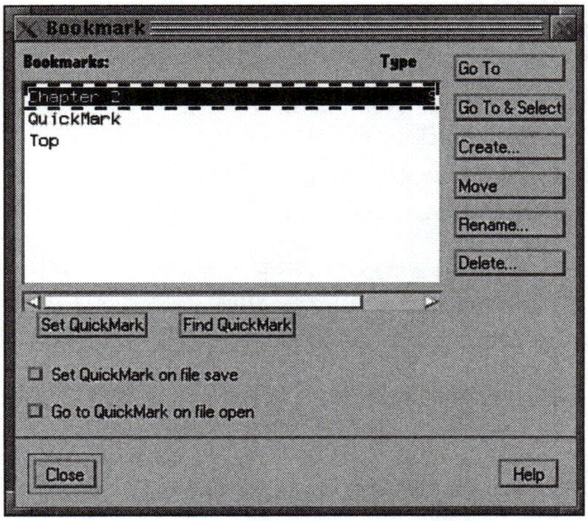

Figure 22.16
Many of the buttons to the right of the bookmark list aren't available until you create at least one bookmark.

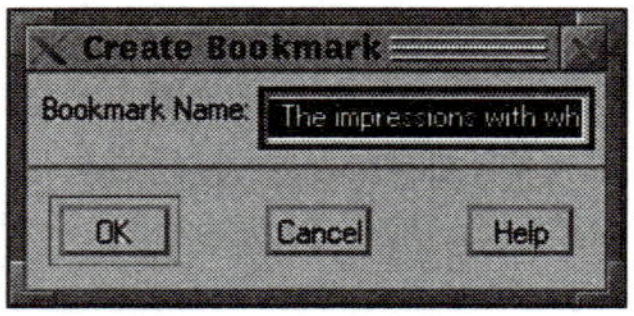

Figure 22.17
WordPerfect uses the first few words following the insertion point as the default bookmark name, but you can change this entry by typing something new.

One variant method of creating a bookmark is to select an area of text; choose Tools, Bookmark; and then create the bookmark as I've just described. When you use this method, the bookmark is that entire region of text, rather than simply a point in the text, and you can select that region of text when you later go to it. This item is known as a *selected bookmark*.

Tip #257 from *Tux Tips*

You might want to place a bookmark in order to select and copy text to insert it elsewhere. You can use this technique to copy a reference that you cite frequently, for instance. If so, a selected bookmark is just what you need because it will save you from having to select the region you want when you go to that bookmark. You might also want to consider using a macro or style, however, if you're copying the same sort of text again and again, especially if you do so in several documents.

→ To learn about macros, **see** "Letting the Computer do the Work: Macros," **p. 483**
→ To learn about styles, **see** "Ensuring Consistency Within Your Document with Styles," **p. 305**

When you insert a bookmark, WordPerfect places a [Bookmark] code at the point where the bookmark resides. When you create a selected bookmark, WordPerfect places a pair of [Bookmark] codes, one opening and one closing, much as it does for paired codes for bold text or paragraph styles.

JUMPING TO BOOKMARKS

To use a bookmark, you choose Tools, Bookmark to open the Bookmark dialog box (refer to Figure 22.16). Select the bookmark to which you want to go and click Go To. WordPerfect immediately moves the view in the document window, places the insertion point at that location, and removes the Bookmark dialog box.

In the case of selected bookmarks, you can jump to their starting [Bookmark] codes as if they were conventional bookmarks, or you can click Go To & Select in the Bookmark dialog box. WordPerfect then selects the text between the two [Bookmark] codes when it jumps to the bookmark location.

THE QUICKMARK

WordPerfect supports a special bookmark, known as the QuickMark. It is simply a regular bookmark with the name *QuickMark*, but WordPerfect provides some extra options for working with the QuickMark. If you examine Figure 22.16, you'll see two buttons below the bookmark list, Set QuickMark and Find QuickMark. You can use these buttons to create and go to the QuickMark, respectively. Because you don't need to enter the QuickMark's name or click on it in the bookmark list, the QuickMark is quicker to use than other bookmarks; hence its name.

Tip #258 from *Tux Tips*

> You can go to the QuickMark without opening the Bookmark dialog box by pressing Ctrl+Q, and you can set a QuickMark by pressing Shift+Ctrl+Q.

Note

> WordPerfect supports only one QuickMark per document. If you create one, move the insertion point, and set another, WordPerfect removes the first QuickMark when it creates the new one.

Tip #259 from *Tux Tips*

> If you've set bookmarks other than the QuickMark, you may want to set the QuickMark before jumping to one of the other bookmarks. This step lets you quickly return to your original editing location by pressing Ctrl+Q after you finish with whatever you want to do at the other bookmark's location.

In addition to being fast, the QuickMark can help you edit your document across multiple sessions. If you select the Set QuickMark on file save check button in the Bookmark dialog box, WordPerfect automatically places the QuickMark at the insertion point's location whenever you save your document. The next time you load it, you can go directly to the QuickMark to resume editing where you left off. You can even automate this last step by selecting Go to QuickMark on file open in the Bookmark dialog box. These options are independent of one another, so you can set either one without the other if you like.

MODIFYING BOOKMARKS

WordPerfect lets you manage your bookmarks in various ways from the Bookmark dialog box (refer to Figure 22.16). The following options are available from buttons to the right of the list of bookmarks:

- Move—If you want to move a bookmark from one location to another, place your insertion point at the desired new location; choose Tools, Bookmarks; and select the bookmark you want to move. When you click Move, WordPerfect moves the bookmark to the current insertion point position and dismisses the Bookmark dialog box.
- Rename—If you want to change a bookmark's name, use the Rename button. WordPerfect responds by displaying the Rename Bookmark dialog box, which looks (and works) just like the Create Bookmark dialog box (refer to Figure 22.17).
- Delete—If you're done with a bookmark, you can remove it by selecting its name in the bookmark list and clicking Delete. WordPerfect asks for confirmation. If you click Yes and more bookmarks are present, WordPerfect highlights the next bookmark in the list and asks whether you want to delete it. Click No to stop deleting bookmarks.

In addition to managing bookmarks in this way, you can use Reveal Codes (View, Reveal Codes or Shift+Ctrl+F3). You can delete and move [Bookmark] codes just as you would any other codes. You cannot copy a bookmark code, however, because WordPerfect doesn't allow two bookmarks to have the same name.

→ For more information on Reveal Codes, **see** "Reveal Codes Mode," **p. 63**

MASTER DOCUMENTS AND SUBDOCUMENTS

When you create a document that has several sections, you may want to use WordPerfect's subdocument feature. This feature allows you to create a single document from several different files. One document, known as the *master document*, serves as an overall framework for the document. It specifies the order in which the parts appear and may include formatting commands and text to link the parts together. The remaining documents, called *subdocuments*, typically contain the bulk of the text for the document as a whole. They can be edited independently of one another and then linked together when the document is complete and ready for printing. This linking process is called *expansion*; when you're ready to print the document, you expand the master document, thus producing one large document that you can print in one operation. When you're done, you can, if you like, *condense* the master document into its component documents. There's more on these operations later.

Note

To print a document that contains subdocuments, you first have to expand the document; otherwise, you'll get nothing but the contents of the master document.

A SAMPLE DOCUMENT

WordPerfect's subdocument capabilities can be particularly useful when you're collaborating with others and different individuals are responsible for different parts of a document. Each individual can write his or her portion of the document, and you can then link them together when all the subdocuments are finished. Another application for this feature is when the document as a whole needs to contain text elements that might be mixed and matched between different documents. For instance, you might have a price list and warranty statement that you routinely include in letters to customers and in brochures. You can put the price list and warranty text in two subdocuments and include these in letters and brochures that you write later. Finally, subdocuments can be very useful when you're writing large documents such as books or user manuals. Create each chapter as a subdocument and then create a master document with a table of contents and formatting between the individual chapters.

→ To learn more about creating a table of contents, **see** "Table of Contents," **p. 638**

Throughout this section, I refer to a sample document set based on the second of these applications. The CD accompanying this book contains, in its chapter22 subdirectory, the files prices.wpd, warranty.wpd, and letter.wpd. The first two files are subdocuments, and the last is a master document that uses the subdocuments.

CREATING SUBDOCUMENTS

You create subdocuments just as you would any other document. They needn't be small documents, though the sample subdocuments on the CD are. If you're writing a book, for instance, each subdocument could be a chapter with dozens of pages.

Figure 22.18 shows the prices.wpd subdocument, which consists mostly of a single table. Note that you can use any formatting or other document features in a subdocument; you're not restricted to simple text. You can even include graphics and use styles in subdocuments.

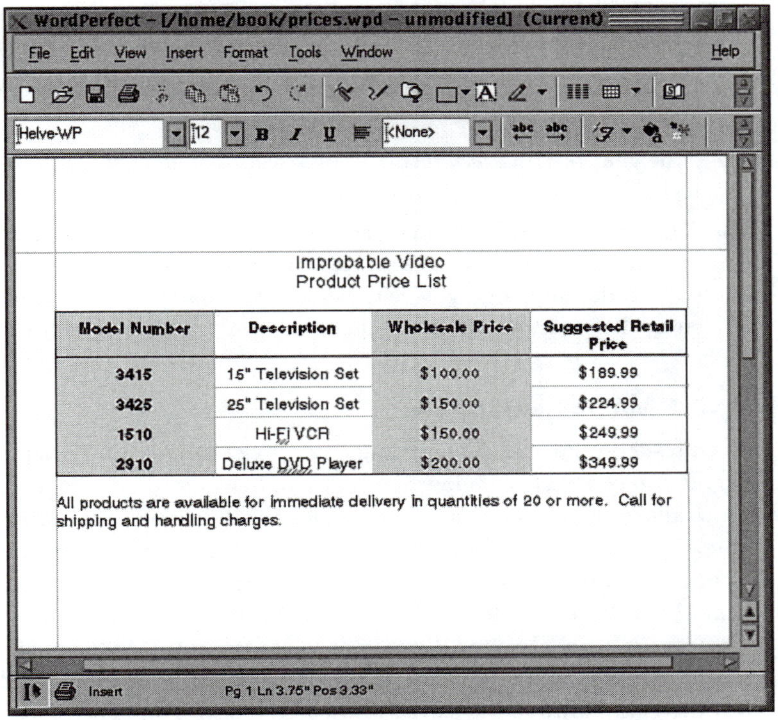

Figure 22.18
A subdocument is just like any other WordPerfect document.

One important consideration in creating subdocuments is that, when WordPerfect combines the subdocuments into the master document, the program discards the subdocuments' initial style codes. Therefore, you shouldn't place any vital formatting codes in the subdocuments' initial styles. Instead, use other style codes or set the formatting you want without styles. Similarly, if the master document and subdocument use styles with the same name, the master document's style takes precedence. When you expand the document, WordPerfect adds styles from the subdocuments into the master document, so if two subdocuments use styles of the same name, the first one listed in the master document takes precedence.

➔ For more information on the document initial style code, **see** "The Special Case of the Document Initial Style," **p. 311**

Tip #260 from *Tux Tips*

When you design your documents, use a consistent set of styles for all of them. You can also use a style library if you like. If you want to create a style for just one subdocument, give it a unique name, even if what you want is just a minor variant of a style from the library. Alternatively, try modifying the formatting in the text *after* you've applied the style.

→ To learn about different locations for styles, **see** "Types of Styles," **p. 308**

Another factor to consider when you create your subdocuments is that any formatting you leave in place at the end of one document remains in effect at the beginning of the next. For instance, if you change the font in an open-ended way, this change carries over into the next document. To avoid this sort of problem, you can do one of three things:

- Do all your formatting using paragraph or character styles, which don't apply *open-ended codes (p. 65)*.
- Type your text and *then* format it by selecting it and applying the new formatting. This technique ensures that *paired codes (p. 65)*, rather than open-ended codes, are used.
- Adjust the formatting in the master document immediately after the point where you include the subdocument. This step can be awkward, however, and you might forget some codes until after you've expanded the document. You could use a special "default style," however, which sets all the codes you're likely to change in the subdocuments.

Creating a Master Document

As with subdocuments, the master document can contain any text or formatting codes you like. It also contains one important addition, however: the [Subdoc] codes. Figure 22.19 shows the letter.wpd master document along with its [Subdoc] codes in both Draft view (View, Draft or Ctrl+F5) and Page view (View, Page or Shift+Ctrl+F5). When you use Page view, the only indication of the presence of a [Subdoc] code outside of the Reveal Codes pane (View, Reveal Codes or Shift+Ctrl+F3) is the Subdocument icon in the left margin. This icon is easily obscured if you use a zoom factor that hides much of the margins. In Draft view, the subdocument's location is more obvious because WordPerfect places marker text there, as shown in the partially obscured left window. Therefore, you may want to use Draft view when you create a master document, even if you normally prefer Page view.

→ If you want more information on zoom factors, **see** "Zoom," **p. 55**

When you expand the document, WordPerfect replaces the [Subdoc] code with the contents of the subdocument to which it points. WordPerfect does not insert any additional page breaks or codes to change the formatting of the document. Therefore, if you want each subdocument to start on a new page, you have to press Ctrl+Enter to insert an [HPg] code, or use WordPerfect's Force Page features. As described earlier, you should be careful to create subdocuments that won't alter the formatting of text after their [Subdoc] codes in the master document. If you must include such a document, you should enter appropriate codes to reset your formatting to the desired state after the [Subdoc] code.

→ For more information on controlling page breaks, **see** "Page Breaks," **p. 230**

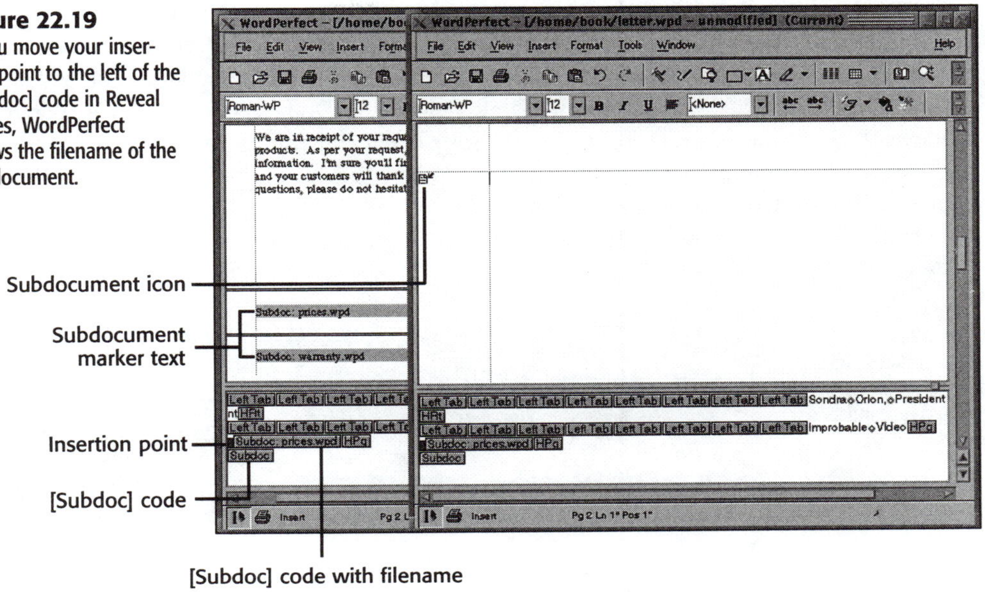

Figure 22.19
If you move your insertion point to the left of the [Subdoc] code in Reveal Codes, WordPerfect shows the filename of the subdocument.

DOCUMENT OPERATIONS

After you create your documents, you can expand the master document by choosing File, Document, Expand Master in the master document's window. WordPerfect displays the Expand Master Document dialog box, shown in Figure 22.20.

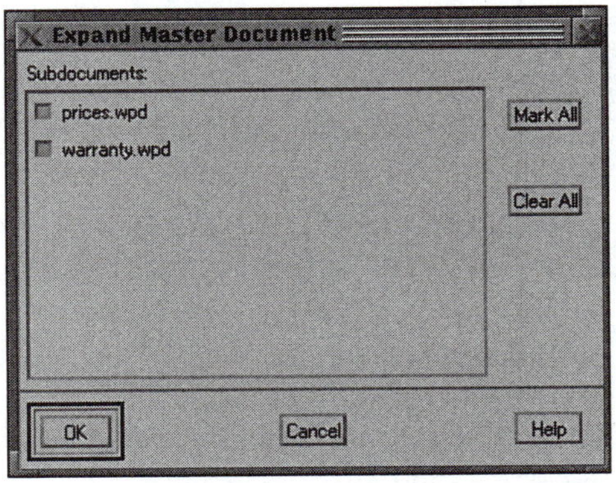

Figure 22.20
The Expand Master Document dialog box lets you select which subdocuments to include in the master document.

You can select individual subdocuments for inclusion in the expansion operation. If you want to include all of them, click Mark All. You can also click Clear All to remove all of them, though you'll probably want to follow this step by manually selecting a small number of subdocuments. When you've selected all the subdocuments you want to include, click OK and WordPerfect performs the expansion.

When WordPerfect finishes with the expansion, you have a single document that contains the contents of the master document and all of the subdocuments you selected. Each subdocument is preceded by a [Subdoc Begin] code and followed by a [Subdoc End] code, as shown in Figure 22.21. If you're using Draft view (View, Draft or Ctrl+F5), WordPerfect shows the locations of these codes in the regular document editing window as well, as Figure 22.21 shows. If you're using Page view (View, Page or Shift+Ctrl+F5), WordPerfect displays the start and end points of a subdocument with icons similar to the one used to mark the subdocument location when the master document is condensed (refer to Figure 22.19).

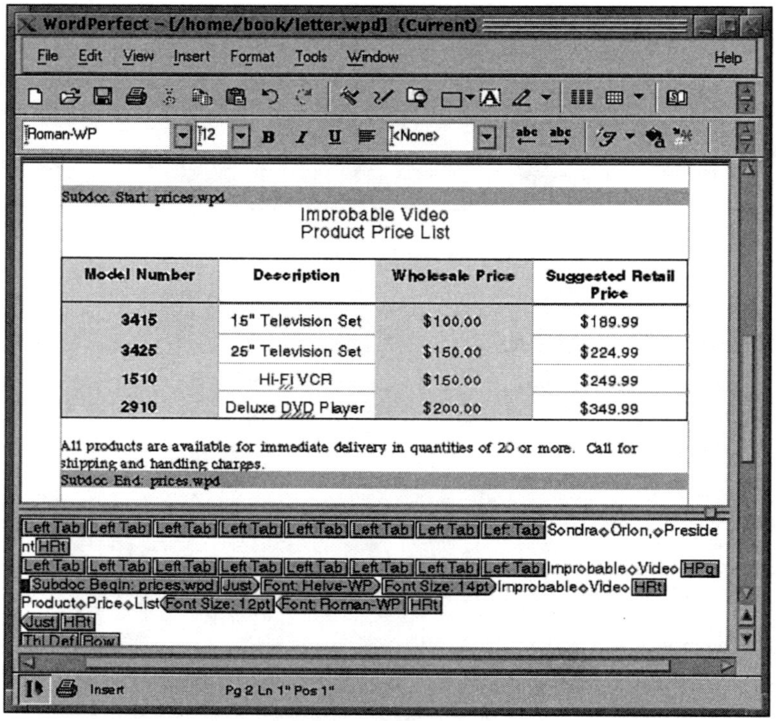

Figure 22.21
WordPerfect keeps track of subdocuments even after you've expanded a master document.

You can edit the master document and all its component subdocuments in the expanded master document mode. You might need to edit this way to make the various documents better match each other in terms of style or if one subdocument has stray codes that affect another's layout. When you're done editing, you can collapse the document back into its component parts by choosing File, Document, Condense Master. WordPerfect presents the Condense/Save Subdocuments dialog box shown in Figure 22.22.

Figure 22.22
Choose whether to condense and save each subdocument with the Condense/Save Subdocuments dialog box.

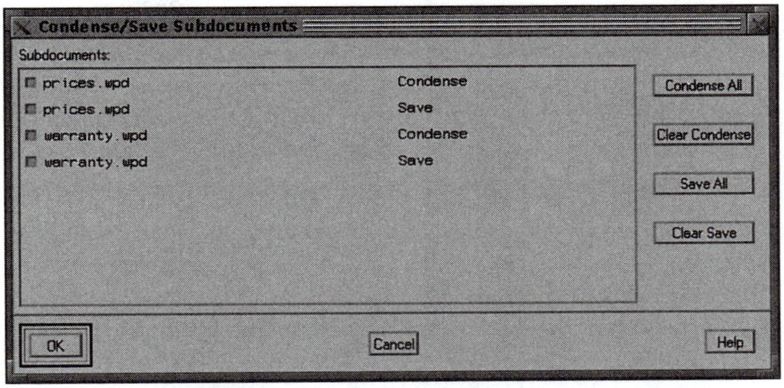

You can set various options from the Condense/Save Subdocuments dialog box:

- Condense—Each subdocument is listed twice in the Subdocuments area of the Condense/Save Subdocuments dialog box, once with the word Condense, and once with the word Save. Select the check button on the line that bears the word Condense to remove the contents of the subdocument from the master document and replace them with a [Subdoc] code.
- Save—If you select the check button next to the filename in the line that contains the word "Save," WordPerfect writes the contents of the text between the [Begin Subdocument] and [End Subdocument] Codes back to the file that originated this text.
- Condense and Save buttons—The buttons to the right of the Subdocuments section of the dialog box allow you to select and deselect groups of documents.

Are the buttons to the right of the Subdocuments portion of the Condense/Save Subdocuments dialog box behaving strangely? See "Condense/Save Subdocuments Button Troubles" in the "Troubleshooting" section at the end of this chapter.

Caution

When you edit the expanded document, be careful not to remove any [Begin Subdocument] or [End Subdocument] codes. Removing those codes will confuse WordPerfect, and it may not perform the condense operation as you'd expect.

Note

If you've elected to save a subdocument that's open in another document window, WordPerfect won't be able to comply, so some or all of the condensation operation will fail. Your document will still be intact, and you can complete the operation by closing the open subdocument window and doing the condensation again.

You can save the master document in either its condensed or its expanded form. You do this save just as you would save any other document; just save it before you expand it (or after you condense it) to save it in condensed form or after you expand it to save it in the expanded form. Saving the document in its expanded form consumes roughly as much disk space as the master document in its condensed form and all its subdocuments combined, so this approach can chew up a lot of disk space if the subdocuments are themselves large.

WordPerfect maintains the path to your subdocuments in the master document, generally relative to your home directory. Thus, if you move your master and subdocuments to a new directory, WordPerfect may have trouble locating the documents. If WordPerfect has this problem, it displays the Subdocument Error dialog box (Figure 22.23). You can type in the new path to the subdocument here or click on the folder button to the right of the Subdocument Name text entry field to use a File Manager dialog box to locate the subdocument. You may also get this dialog box if you try to expand a master document that was created with WordPerfect for Windows because Windows doesn't use the same directory structure as Linux uses.

Figure 22.23
WordPerfect presents the Subdocument Error dialog box when it can't find the subdocument.

Note

If you place the sample master and subdocument files from the CD in a directory other than your home directory or if you try to use them directly from the CD, you'll encounter the Subdocument Error dialog box.

CONSIDERATIONS IN DESIGNING MASTER DOCUMENTS

Typically, a master document serves as a framework for the bulk of the text in other files. At its simplest, it could be nothing more than a series of [Subdoc] codes, perhaps with [HRt] codes between to get the subdocuments to print on multiple pages.

Tip #261 from *Tux Tips*

You can use WordPerfect's subdocument feature to fax several documents at one time. Create a framework document that does nothing but place the subdocuments, expand it, and "print" to your fax print queue. This method can be simpler and quicker than faxing the documents one at a time.

→ To learn how to send faxes from WordPerfect for Linux, **see** "Faxing," **p. 161**

Of course, there's nothing wrong with including text in a master document, and you may find a need to place substantial amounts of text in one. You can certainly use WordPerfect's subdocument feature merely to include some existing file as, say, an appendix to a larger document.

The main thing to remember when you create a master document, whether it contains little or substantial amounts of text, is that the subdocuments *may* modify your master document's formatting, including the formatting of subsequent subdocuments. Similarly, any formatting you enter in the master document has the potential to affect subdocuments you include in that master document. As discussed earlier, it's best to use a common set of styles for all the documents you'll be including and to deviate from these styles by creating new styles in the subdocuments or by selecting the text to be modified and applying formatting to that text. This technique ensures that WordPerfect will use paired on and off codes, rather than create an open-ended code that will affect subsequent text.

If you're creating a large document, you may want to create a table of contents, cross-references, or other lists or references. You need to place the codes to generate the list or table in the master document and then place the tags on individual words in the subdocument. For instance, you can tag headings within a subdocument for inclusion in a table of contents and place the table of contents marker in the master document. You'll expand the master document and then generate the table of contents. You can then print the document, condense it, or otherwise modify it.

→ For more information on creating tables of contents or other lists, **see** "Organizing Lists and References," **p. 627**

TROUBLESHOOTING

HIGHLIGHTING PRINTS STRANGELY

When you print a document with highlighting, you get a printout with more than the highlighted text actually highlighted. The entire line from the left margin up to the highlighting, a rectangular area of the page, or some other section is highlighted, not just the highlighted word or words.

A bug in WordPerfect's printer system causes this problem. The exact results depend on the printer driver, but most or all drivers are affected. You have to disable highlighting to get a reasonable printout; you can also check with Corel to see whether it has an updated version of the program that fixes this bug. If you're trying to use highlighting in a printout as an alternative means of emphasizing text, you might consider applying color to the text itself, rather than to its background, as highlighting does. With some printers, you may be able to obtain the result you desire by using the redline font attribute. Finally, you can apply a background fill effect to paragraphs and larger sections of text to highlight them.

→ To learn how to set the font's color or the redline font attribute, **see** "Fonts," **p. 178**

→ To learn about redlining as a means of comparing documents, **see** "Performing a Comparison," **p. 588**

→ To learn how to set a border or fill, **see** "Borders and Fills," **p. 216**

CONDENSE/SAVE SUBDOCUMENTS BUTTON TROUBLES

The Clear Condense and Save All buttons in the Condense/Save Subdocuments dialog box (refer to Figure 22.22) don't work the way you expect. Clear Condense clears all selections in the Subdocuments area of the dialog box, whereas Save All selects all the check buttons, not just the Save buttons.

Some versions of WordPerfect for Linux contain this bug. You can contact Corel to see whether it has an updated version of the program, or you can select and deselect the items you want individually.

CHAPTER 23

Footnotes and Endnotes

In this chapter

Uses for Footnotes and Endnotes

When you're writing a document, and especially one that's more than a page or two long, you may occasionally need to provide information that's somewhat peripheral to the main discussion but that's still relevant and doesn't belong elsewhere. Examples include a citation of a specific book or paper, an elaboration of a point in the main text that would be distracting if placed there, or an interesting tidbit or piece of trivia. You can handle this type of situation in four ways:

- Place the information in parentheses—This approach can be effective if the text is short, and in fact, some academic disciplines require this method for providing references to other works. Just use the parentheses keys on your keyboard to use this approach.
- Create a sidebar—A sidebar places text in a WordPerfect text box near the text to which it's related. You might use this approach in a newsletter or similar document for material that's longer than a paragraph. Sidebars are overkill for shorter text segments, however.

➔ If you want to use a text box, **see** "Using Text Boxes to Add Variety to Text Layout," **p. 357**

- Create a footnote—A footnote places text at the bottom of the page, frequently separated from the main body of text by a short dividing line. Footnotes generally use a superscripted number or symbol in the body text and a duplicate of that number in the footnote itself to help the reader to locate the correct footnote from the text. Figure 23.1 shows an example of a footnote created with WordPerfect 8 for Linux.
- Create an endnote—The reference to an endnote in the text of a document looks just like a footnote reference, but endnotes are printed at the end of a document, in sequence, typically on one or more pages devoted entirely to the endnotes.

This chapter is devoted to WordPerfect's support for these last two options, footnotes and endnotes (or "notes" to refer to either type). They're very similar in operation and in the types of uses to which they're put, so many techniques for using them in WordPerfect apply to both types of note.

Which you'll use depends largely upon the formatting requirements for the document you're writing. For instance, the American Psychological Association (APA) guidelines for papers submitted to its journals specifies the use of endnotes. Therefore, if you're writing a paper for a journal that uses APA style, you need to use endnotes. For other types of writing, however, you may need to use footnotes.

Tip #262 from
Tux Tips

Footnotes are generally easier on the reader than endnotes, so if you have a choice, you should probably use footnotes. Many requirements for the use of endnotes date back to the days before word processors, when endnotes were easier for the writer to create. With WordPerfect, both note types are easy to create, so there's no advantage to using endnotes from a writing point of view. WordPerfect even allows you to convert note types easily, so you can use endnotes for some copies of a document and footnotes for another.

Figure 23.1
By default, WordPerfect numbers footnotes and separates them from the body text with a short line.

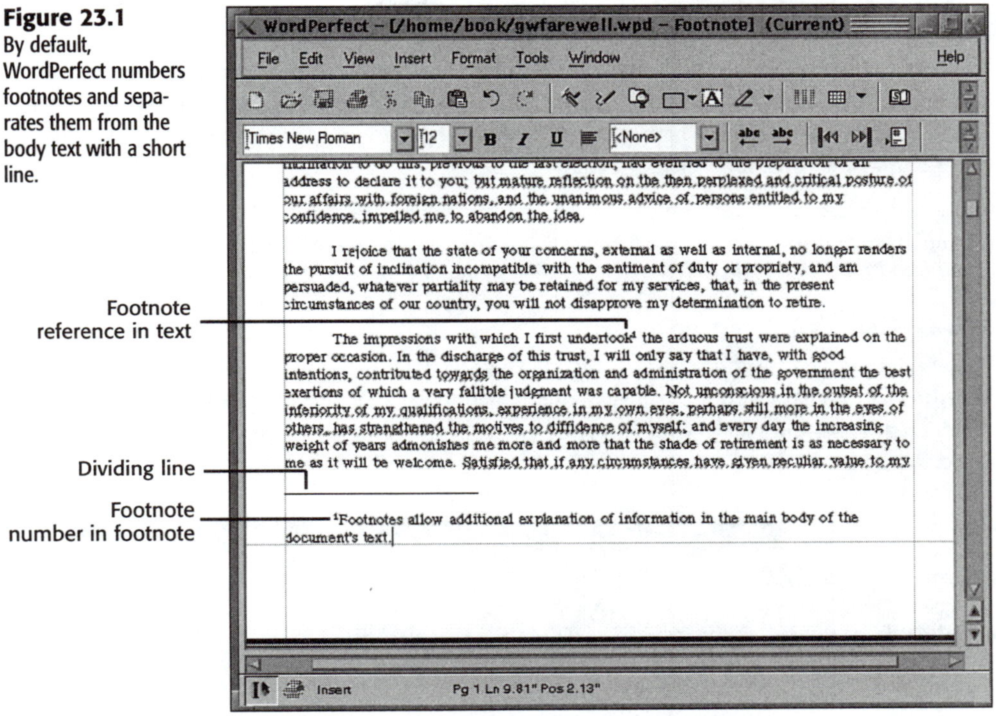

➔ To learn how to convert a note from one type to another, **see** "Converting Notes from Type to Type," **p. 625**

When you place text in a footnote or endnote, you're implying that it's not as important as the main text. Therefore, you shouldn't put anything vitally important in a note. Further, both footnotes and endnotes, but especially endnotes, impose a cost on the reader to read them. Therefore, you should not overuse notes. If you find yourself regularly writing paragraph-long notes, try to work them into the main body of your text or dispose of them entirely; likewise, if you create several notes per paragraph. Overuse of notes may indicate a need for better document organization, so you might want to use WordPerfect's outlining features before you begin actual writing. Of course, these cautions don't apply to the use of notes for references to other documents if that's your use for them.

➔ To learn how to use WordPerfect's outlining features, **see** "Organizing a Document with Outlining," **p. 553**

Whether you use footnotes or endnotes, WordPerfect automatically maintains their numbering, so when you insert a new note between two other notes, you needn't manually adjust the note numbers. Similarly, when you use footnotes, WordPerfect automatically adjusts the point at which the page breaks to leave enough room for the footnotes you create.

CREATING A NOTE

To create a note, place your insertion point where you want the note's in-text reference to appear and choose Insert, Footnote, Create or Insert, Endnote, Create for a footnote or an endnote, respectively. What happens next depends on the type of note and whether you're using Page view or Draft view.

➔ For more information on Draft and Page views, **see** "Draft Versus Page Views," **p. 54**

Are you trying to use some feature while creating a note, but find that it's unavailable? See "Restricted Features in Notes" in the "Troubleshooting" section at the end of this chapter.

FOOTNOTES, PAGE VIEW

If you're creating a footnote using Page view (View, Page or Shift+Ctrl+F5), which is WordPerfect's default mode for displaying a document, WordPerfect creates a footnote on the current page and places your insertion point in the footnote so that you can enter text directly (Figure 23.2).

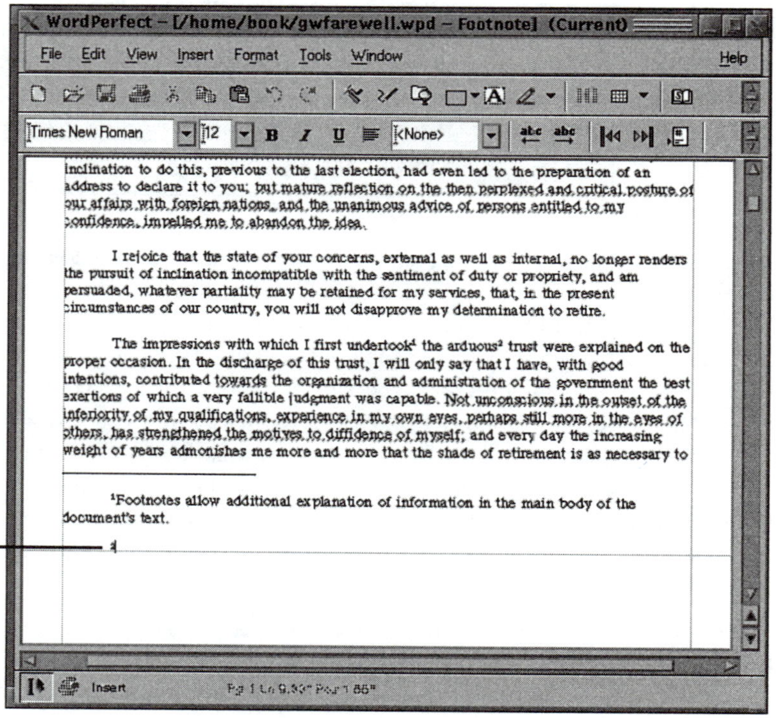

Figure 23.2
In Page view mode, you edit your footnote directly on the page on which it will appear.

The default formatting for a footnote indents it and places a superscripted footnote number immediately preceding the note text itself. You can change this setting, as described later in this chapter, or keep the default formatting. Either way, you can type in your footnote text as you would most any other text, including the use of character attributes such as bold or italic, font changes, and so on. You can even insert a graphic, equation, or other boxed element within a note. When you're done, click back in the main text area to continue editing.

➔ To learn more about boxed elements, **see** "Complex Formatting with Boxes for Text, Graphics, and Equations," **p. 355**

Tip #263 from *Tux Tips*

Instead of clicking in the main text area, you can choose File, Close (Ctrl+F4). This step ends the Footnote editing mode and returns the insertion point to the footnote reference in the text.

ENDNOTES, PAGE VIEW

Creating an endnote using Page view mode works much like creating a footnote in Page view except that now WordPerfect shifts your insertion point to the end of the document or to the end of the endnote area to create the endnote, as shown in Figure 23.3.

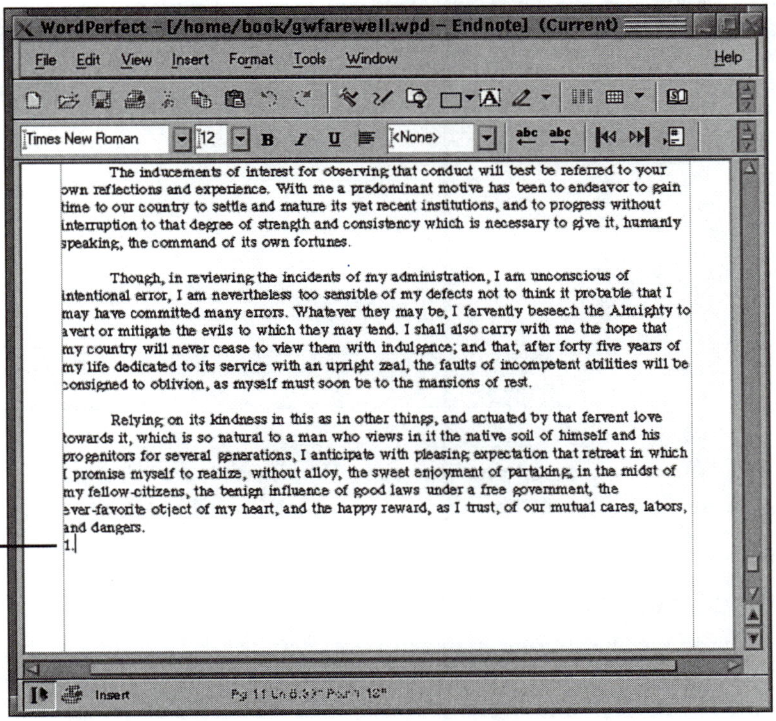

Insertion point

Figure 23.3
In Page view mode, you edit your endnote directly on the page on which it will appear.

Endnotes use slightly different formatting from footnotes. Whereas the text of a footnote includes a tab and a superscripted footnote number, an endnote includes no tab, and the endnote number is followed by a period and is not superscripted, as you can see in Figure 23.3. As with footnotes, you can change this formatting if you like.

➔ If you want to learn how to change footnote formatting, **see** "Setting Note Options," **p. 618**

To return to your normal text, you can either scroll to your previous location and click there, or choose File, Close (Ctrl+F4). The latter option returns your insertion point to the endnote reference in the text, so you probably want to use it, especially if you're adding an endnote to the middle of a document that's already large.

DRAFT VIEW

In Draft view, WordPerfect doesn't display certain document features, such as headers, footers, or notes. Therefore, if you're using Draft view and create a footnote or endnote, WordPerfect opens a second document window in which you can create the note, as shown in Figure 23.4. You use this window much as you would other document windows, entering text, adding text attributes, and so on. When you're done, choose File, Close (Ctrl+F4) to close the window and return to the location of the note's text reference.

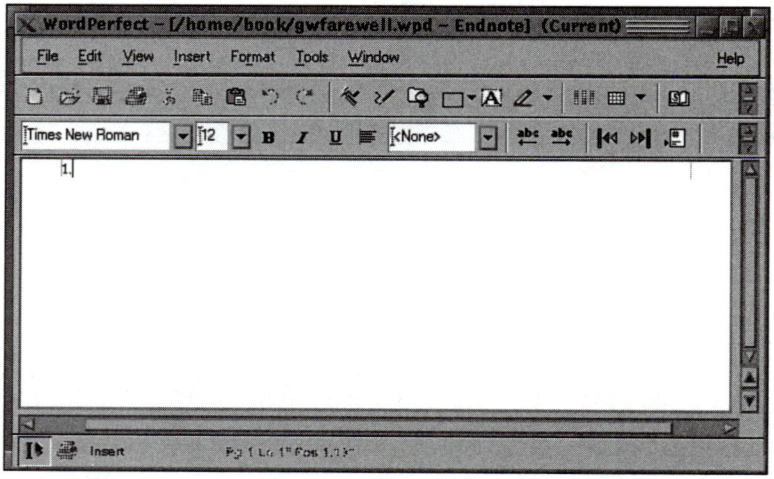

Figure 23.4 When you use Draft view, you edit your note in a separate window.

Even after you finish editing your note, WordPerfect doesn't display it. To see where your note will appear in the document, choose View, Page (Shift+Ctrl+F5) to switch to Page view.

CREATING SPACE FOR ENDNOTES

By default, endnotes appear immediately after the last of the text in the document, as shown in Figure 23.3. This location is probably closer to the end of your text than you'd like, so you may want to shift the position of the endnotes, perhaps moving them onto their own page.

To move the endnotes, place your insertion point at the end of the document and enter whatever formatting you'd like to distance the endnotes from that text. For instance, you could press Ctrl+Enter to create a hard page break to place the endnotes on their own page, or you could choose Format, Typesetting, Advance to insert a fixed amount of vertical space. If you'd like to include a heading or other text to identify the endnotes, you can do so, as well. The rule to remember is that the endnotes appear after the final code or text in the document, wherever that is.

→ To learn more about placing text where you want it on a page, **see** "Advance," **p. 187**)

There is an exception to this rule, however: You can insert an [Endnote Placement] code to place your endnotes wherever you like in your document. To do so, position your insertion point where you want the endnotes to appear and choose Insert, Endnote, Placement. WordPerfect responds by displaying the Endnote Placement dialog box, shown in Figure 23.5.

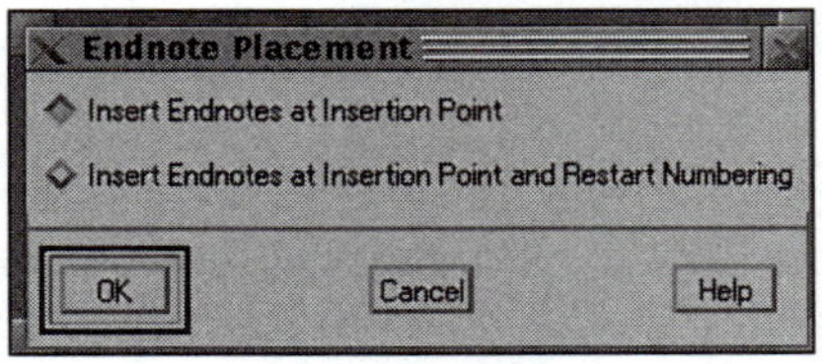

Figure 23.5
You can reset endnote numbering after placing endnotes by using the Insert Endnotes at Insertion Point and Restart Numbering option.

When you click OK in this dialog box, WordPerfect inserts an [Endnote Placement] code followed by a [HPg] code to start a new page. The [Endnote Placement] code tells WordPerfect to compile all the endnotes up through that point in the document and display them. You can therefore use this feature to place endnotes prior to some other element of your document, such as a bibliography.

If you enter any endnotes subsequent to the [Endnote Placement] code, WordPerfect compiles and displays them at the next [Endnote Placement] code or at the end of the document, whichever comes first. Therefore, the Endnote Placement dialog box provides you with the option to restart endnote numbering. If you click the Insert Endnotes at Insertion Point and Restart Numbering radio button, any endnotes you create after the first endnote listing are numbered starting with 1. If you accept the default of Insert Endnotes at Insertion Point, WordPerfect numbers subsequent endnotes in sequence with the earlier ones.

When you use Page view, WordPerfect displays alterations you make to influence the position of the endnotes on the screen as they will appear in the final document. In Draft view, however, when you insert an [Endnote Placement] code, WordPerfect displays the text *Endnote Placement* on a gray background where it will ultimately place the endnotes, as shown in Figure 23.6. If you don't use an [Endnote Placement] code, however, or if you place endnotes after the last such code, WordPerfect won't display where the final endnotes will go—that's always at the end of the document.

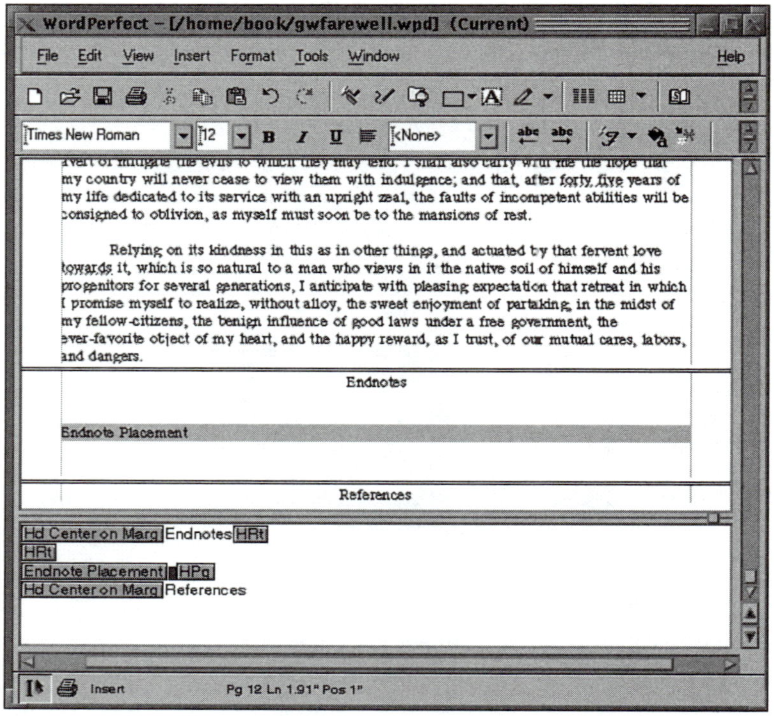

Figure 23.6
The *Endnote Placement* text shows the location of the [Endnote Placement] code when you are working in Draft view.

EDITING NOTES

WordPerfect provides many tools to help you edit notes. You can edit their text, use Reveal Codes on them, move them, and more.

BASIC NOTE EDITING

To edit a note, choose Insert, Footnote, Edit or Insert, Endnote, Edit, whichever is appropriate. WordPerfect opens either the Edit Footnote dialog box or its identical endnote namesake, as shown in Figure 23.7. Type in the number of the note you want to edit and click OK. WordPerfect positions the insertion point in the note and lets you edit it.

Note

If you're in Draft view, WordPerfect opens a new window in which you can edit your note. If you're in Page view, WordPerfect repositions the display to the page on which the note rests and places the insertion point in the note. In either case, if you choose File, Close (Ctrl+F4) when you're done editing, WordPerfect returns the insertion point to its previous location. This behavior allows you to edit a note even when a different part of the main document is visible in the document window, and you won't lose your place.

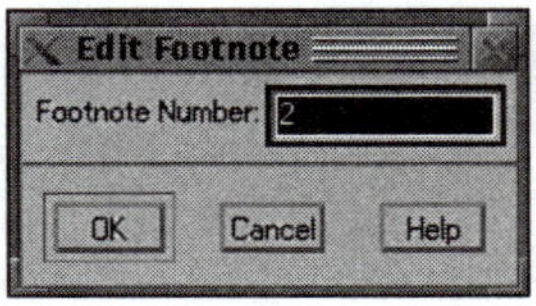

Figure 23.7
The default number for editing a note is the last note before the insertion point, or 1, whichever is higher.

If you're using Page view, you can also edit a note by scrolling to its location and clicking in it. Editing a note is very much like editing regular text; you can switch back to regular text by choosing File, Close (Ctrl+F4).

One further means of editing a note is to double-click its code in WordPerfect's Reveal Codes pane (View, Reveal Codes or Shift+Ctrl+F3). This step opens the Edit Footnote or Edit Endnote dialog box (Figure 23.7) with the appropriate note number already entered. Click OK to edit that note.

Are you trying to use some feature while editing a note, but find that the feature isn't available? See "Restricted Features in Notes" in the "Troubleshooting" section at the end of this chapter.

If you're editing one note and want to edit the next one, you can use the Next note item in the note toolbar (Figure 23.8) that WordPerfect displays whenever you edit a note. Similarly, you can click on the Previous note item to edit the previous note.

Figure 23.8
WordPerfect provides a few note-related shortcuts in the note toolbar whenever you edit a note. You may need to widen your window or scroll the toolbar to see all of these items, however.

Tip #264 from *Tux Tips*

Use Next note and Previous note if you want to proofread notes when you're in Draft view.

Using Reveal Codes with Notes

WordPerfect treats notes very much as if they were self-contained documents. This is similar to its treatment of text boxes, in fact, though WordPerfect doesn't use boxes to process notes. When you edit a note, WordPerfect's Reveal Codes pane shows *only* the contents of that note, as shown in Figure 23.9. This is true whether you're editing your document in Page view or Draft view. WordPerfect won't even show other notes on the same page in any given note's Reveal Codes pane, as Figure 23.9 demonstrates.

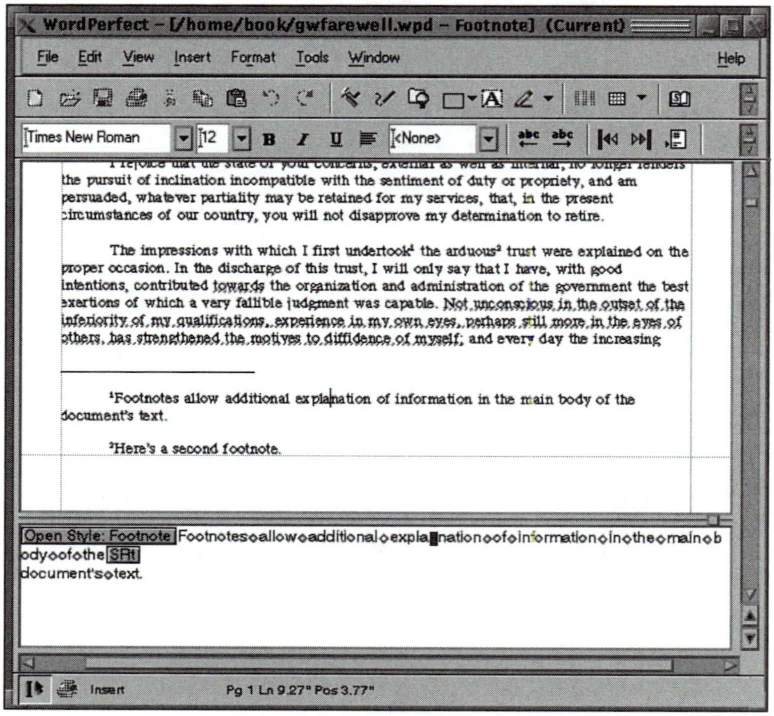

Figure 23.9
Reveal Codes shows that WordPerfect treats a note almost as if it were a self-contained document.

You can control many aspects of a note's formatting by editing the note's opening style (the [Open Style: Footnote] code in Figure 23.8). You can do so by double-clicking the opening style code in Reveal Codes or by editing the in-note numbering style from the Footnote Options or Endnote Options dialog boxes, as described later in this chapter.

→ If you want to learn more about how to edit styles, **see** "Editing a Style," **p. 321**

Moving, Copying, or Deleting a Note

To move a note from one location to another within your document, you need to work outside the note editor and in the main document. Essentially, you move a note just as you would any other text: by selecting it and performing a cut-and-paste operation on it. You can use Edit, Cut (Ctrl+X) to cut the note, then reposition the insertion point and use Edit, Paste (Ctrl+V) to restore it in a new location. Alternatively, to move the note a short distance, you can select it and then click and drag the selected note to reposition it.

Tip #265 from *Tux Tips*

Because WordPerfect displays note reference numbers in a smaller font than normal, you may find it difficult to select only the note with your mouse. You can use the Shift key with the arrow keys to make this selection more easily.

If you have two footnotes that are identical or almost so, you may want to copy the first one rather than create a new note. Copying a note is similar to moving one except that you use Edit, Copy (Ctrl+C) rather than Edit, Cut (Ctrl+X) on the original note.

If you decide that you no longer need a note, you can delete it. To do so, treat the note reference in the text as if it were an ordinary character—just select it and delete it, or position your insertion point before it or after it and press the Delete or Backspace keys, respectively.

Note

The in-text note reference is composed of *one* code, even when the note itself is more than one character wide, as in a two-digit number. You therefore need to press Delete or Backspace just once to remove the note. More than this, and you'll remove surrounding text.

When you reposition, delete, or insert a note, WordPerfect automatically renumbers it and all subsequent note references. In the case of footnotes, WordPerfect also adjusts the size of the footnote area at the bottom of the page. Therefore, moving or deleting a footnote can cause a moderate amount of text reformatting on a page.

Caution

When you cut, copy, delete, or move text, any notes in that text are affected as well. Thus, if you want to preserve a note inside text you're about to delete, be sure to move or copy that note elsewhere before you delete the host text. This step is easy to overlook when, say, you decide to completely rewrite a paragraph.

Renumbering Notes

If you're creating a large multisection document, you may want to start numbering your notes anew for each section. WordPerfect lets you do this by choosing Insert, Footnote, New Number or Insert, Endnote, New Number, as appropriate. This step opens the Footnote Number or Endnote Number dialog box, which are identical except for their titles (Figure 23.10).

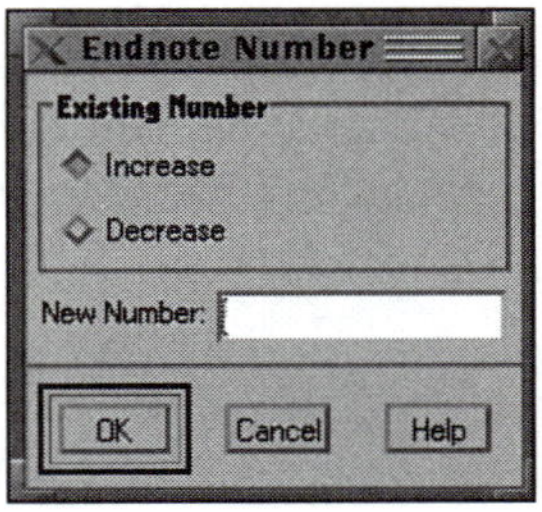

Figure 23.10
Adjust note numbering with the Footnote Number or Endnote Number dialog box.

If you select Increase or Decrease, WordPerfect adjusts the note number up or down by one, respectively. If you type a number into the New Number field, you'll notice that both Increase and Decrease deselect themselves, and when you click OK, the next footnote or endnote, as appropriate, takes on the number you typed.

These numbering options work by inserting WordPerfect codes in the document's text. These codes embed themselves wherever the insertion point rests when you create them, so you should be careful about where you place these codes.

Tip #266 from *Tux Tips*

> If your note numbers are changing mysteriously, you may have one or more stray set codes in your text. Try using WordPerfect's Find feature to locate them. Search for the codes [Footnote Num Set], [Footnote Num Inc], [Footnote Num Dec], or their similarly named endnote equivalents.

→ If you want to learn more about locating codes in text, **see** "Searching for Codes and Symbols," **p. 108**

Tip #267 from *Tux Tips*

> If you're creating a document that resets note numbering regularly, say, with each new section, and if you're using a style that's not used anywhere else to start new sections, you can embed the note numbering code in that style. For instance, you might use a style to display a section title, and you can place the note numbering code in that style. Doing so will ensure that you don't forget to reset the note numbering scheme.

Setting Note Options

WordPerfect's default note formatting works well for many, but not all, situations. For instance, by default notes use any options you set in the document's initial style. You might want single-spaced footnotes with double-spaced text, however, or you might want to use a different font or font size for your notes. Fortunately, WordPerfect provides many options for note formatting. These options vary somewhat for footnotes as opposed to endnotes, though there are many commonalities, as well.

Setting Footnote Options

To set footnote options, you must choose Insert, Footnote, Options to open the Footnote Options dialog box (Figure 23.11). This dialog box contains several sections and options for controlling different aspects of footnote display.

Note

> With the exception of numbering options, changes you make in the Footnote Options dialog box affect *all* footnotes, regardless of where the insertion point rests when you make the change.

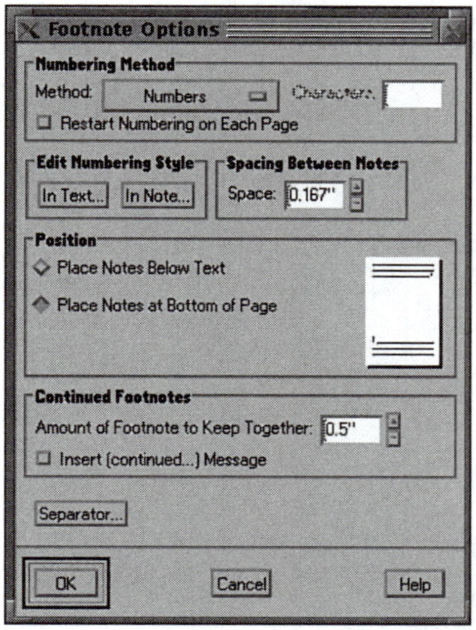

Figure 23.11
The Footnote Options dialog box lets you adjust many aspects of footnote formatting.

Adjusting Footnote Numbering

WordPerfect defaults to numbering footnotes sequentially, starting with the Arabic number 1 and continuing to increment these values through the entire document. WordPerfect does provide other options, however. You can elect to use Arabic numbers, Roman numerals, letters, or special characters by choosing one of these from the Method option button in the Numbering Method area of the Footnote Options dialog box.

Normally, WordPerfect starts footnote numbering at 1 and continues incrementing those values throughout the document. As a result, page 1 might have footnotes 1 and 2, page 2 might have footnote 3, page 3 might have footnotes 4, 5, and 6, and so on. You can alter this numbering scheme so that each page's footnotes begin with the number 1 by selecting the Restart Numbering on Each Page check button.

If you select Characters from the Method option button in the Footnote Options dialog box, WordPerfect uses the characters you enter into the Characters text field, cycling through them and then doing so again with double characters, then triple, and so on. For instance, if you enter `*+` in this field, WordPerfect uses a single asterisk for the first footnote, a single plus sign for the second, a double asterisk for the third, and so on. If you don't type anything in the Characters field, WordPerfect works as if you'd typed a single asterisk, so you get footnotes "numbered" in asterisks—* for 1, ** for 2, and so on.

If you type Ctrl+W when your insertion point is in the Characters field, WordPerfect displays the Symbols dialog box (Figure 23.12). You can select symbols here for use in footnote labels. Select a Set (the Typographic Symbols and Iconic Symbols sets are most likely to have symbols useful for footnotes), select a symbol, and click Insert to insert it into the

Characters field. Click OK to insert a symbol and dismiss the Symbols dialog box. Be careful not to select symbols that look too much alike, however; especially when reduced in size in a superscript, certain symbols may be too similar to one another to be used effectively as footnote labels.

→ To learn more about using special symbols, **see** "Inserting Special Characters," **p. 102**

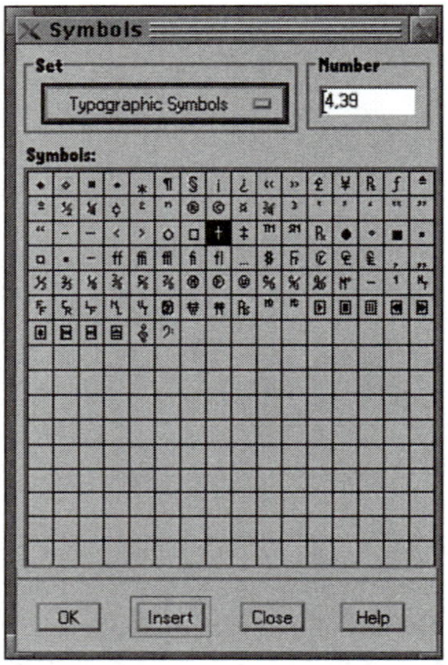

Figure 23.12
You can use the Symbols dialog box to insert special symbols for use in footnote labeling.

Tip #268 from *Tux Tips*

If you use the Characters method of numbering, it's best to also use the Restart Numbering on Each Page option, at least when a document has more than a handful of footnotes. A footnote number in the tens or even hundreds is quite manageable when printed as an Arabic number, but dozens or hundreds of asterisks would look quite silly.

Note

When you change the numbering method, WordPerfect places a code in your document at the insertion point. Therefore, if you want to change the numbering method for an entire document, you should place this code at the start of the document, or in the document's initial style or document template. The fact that ordinary code handles this option means, however, that you can intentionally change the numbering style mid-document. You might, for instance, want to use a different numbering system for appendixes as opposed to the rest of your text.

→ To learn more about setting options for the document, **see** "The Special Case of the Document Initial Style," **p. 311**

→ To learn more about using templates to set options for documents, **see** "Promoting Cross-Document Consistency with Templates," **p. 333**

Editing Styles

WordPerfect uses styles to format both the footnote reference in the text and the footnote itself. You can edit both of these styles from the Footnote Options dialog box by clicking In Text to edit the in-text footnote reference or In Note to edit the style used for the footnote. In either case, WordPerfect opens the Styles Editor, as shown in Figure 23.13. You can think of the Styles Editor as working as a sort of stripped-down Reveal Codes mode. You can select many of WordPerfect's formatting features from the Styles Editor window's menus or by using keyboard shortcuts such as Ctrl+B to select the bold attribute.

→ To learn more about editing a style, **see** "Editing a Style," **p. 321**

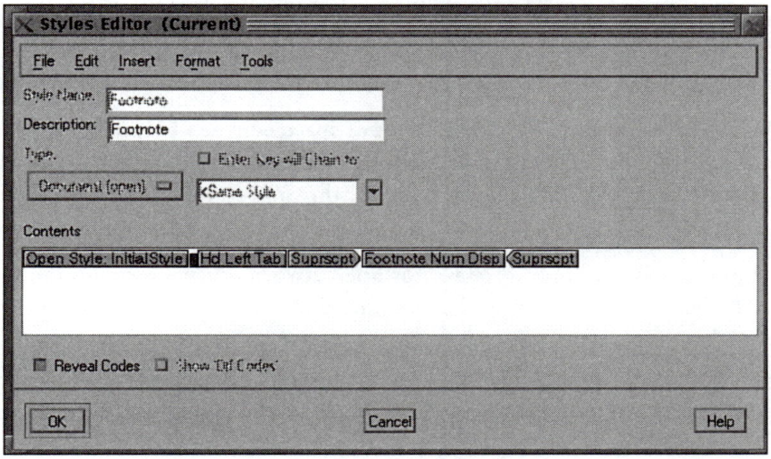

Figure 23.13
When the Reveal Codes check button is selected, the Contents area of the Styles Editor works much like a WordPerfect Reveal Codes pane.

Tip #269 from *Tux Tips*

Instead of selecting In Note from the Footnote Options dialog box, you can double-click the [Open Style: Footnote] code at the start of any footnote in Reveal Codes mode to get to the Styles Editor. The effect will be the same as if you'd selected In Note.

Figure 23.13 shows the default codes used for footnotes. The codes for in-text footnote numbers are somewhat simpler. In both cases, however, the [Footnote Num Disp] code is surrounded by [Suprscpt] codes to superscript it. In the footnote itself, there's also an [Hd Left Tab] code to indent the first line of the footnote. You can delete or change this tab code, remove the superscript code, or otherwise alter the style for the footnote number. You can also add text and codes after the footnote number and its closing [Suprscpt] code to affect only the footnote text. For instance, if you want to change the footnote's font, you could choose Format, Font (F9). These changes affect *only* the footnote text.

→ If you want to know more about setting fonts, **see** "Fonts," **p. 178**

Caution

You should avoid entering codes after the closing [Suprscpt] code in the in-text footnote number style because they will affect your document text *after* the footnote number.

Have you edited a footnote style but find that your change doesn't appear to have taken effect? See "Edited Style Changes Don't Work Immediately" in the "Troubleshooting" section at the end of this chapter.

SETTING FOOTNOTE POSITION AND SPACING

By default, WordPerfect places footnotes at the bottom of the page (via the Place Notes at Bottom of Page option in the Position area of the Footnote Options dialog box). If you want your footnotes to begin immediately below the text, you can click the Place Notes Below Text radio button instead. This option has a substantial effect only on pages that have relatively little text. On pages that are full, these two options produce similar results.

WordPerfect places a certain amount of space between footnote entries to help visually separate them. The effect is something like double-spacing between paragraphs of text. You can adjust this amount up or down by using the Space spin box in the Spacing Between Notes area.

To set spacing around the separator line between normal text and footnotes, click Separator and adjust this option from the resulting dialog box. I discuss the separator line options shortly.

You can adjust the spacing of lines within your footnotes by clicking In Note and setting the appropriate spacing options in the Styles Editor dialog box (Figure 23.13).

Note

Because the default style for footnotes incorporates the document initial style, footnotes reflect any options you enter there. If you set double-spacing in the document's initial style, therefore, your footnotes are also double-spaced. You can override these codes by entering new ones in the footnote's style as described earlier, however.

FOOTNOTE CONTINUATION

If you have a particularly large footnote that's referenced in text very close to the bottom of the page, WordPerfect is forced to do one of two things:

- Place a page break early in the text to move the complete footnote and its reference onto the next page
- Break the long footnote across two pages

WordPerfect uses a numeric criterion to determine which of these two actions to take, and you can adjust this by changing the Amount of Footnote to Keep Together option in the Continued Footnotes area of the Footnote Option dialog box. Adjusting this value up from the default value of 0.5" makes it more likely that WordPerfect will insert an early page break and keep the entire footnote on one page. Reducing this value makes it more likely that WordPerfect will split the footnote across two pages.

When WordPerfect does split a footnote, by default it simply starts the footnote area on the next page with the split note, without a footnote reference number or identifying text. If you click the Insert (continued) Message check button, however, WordPerfect inserts a message that reads `continued...` at the end of the first page's footnote area, and another `continued...` note, along with footnote number, preceding the continued portion of the note. You can adjust the language in which the message appears by setting an appropriate language in the document's initial style.

➔ To learn how to set WordPerfect's default language, **see** "Languages," **p. 141**

➔ To learn how to use the document initial style, **see** "The Special Case of the Document Initial Style," **p. 311**

Setting a Separator

By default, WordPerfect places a 2-inch separator line between the regular document text and footnotes, as shown in Figures 23.1 and 23.9. You can change this line's length, position, appearance, and the spacing above and below it by clicking Separator in the Footnote Options dialog box. This step opens the Line Separator dialog box, shown in Figure 23.14.

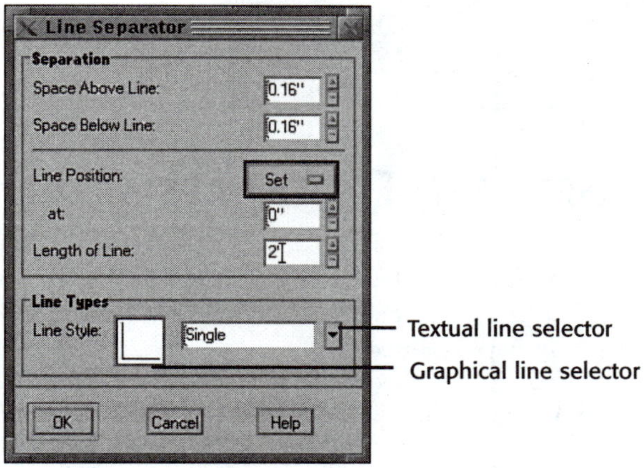

Figure 23.14
Adjust the appearance of the footnote separator line from the Line Separator dialog box.

You can set the amount of blank space above and below the line by using the Space Above Line and Space Below Line spin boxes, respectively. The default line is positioned at the left and is 2 inches long, but you can choose Left, Center, Right, Full (across the entire width of the page, minus the margins), or Set (you specify a starting location and length). The at spin box below the Line Position option button is active only for the Set option.

Tip #270 from *Tux Tips*

To eliminate the line, set a length of 0. You can still use the Space Above Line and Space Below Line items to insert space between your ordinary text and footnotes.

You can change the appearance of the line by choosing from the graphical or textual line selectors in the Line Types area. Choose from a thin single line (the default), various dashed lines, assorted line lengths, multiple lines, and various colored lines.

Setting Endnote Options

You set endnote options by choosing Insert, Endnote, Options, which produces the Endnote Options dialog box shown in Figure 23.15. The options here work very much like the options in the Footnote Options dialog box, though you have fewer of them for endnotes. Because endnotes go at the end of your document, there are no positioning options or a separator line; instead, you can control where WordPerfect places endnotes with ordinary text formatting and the Insert, Endnote, Placement feature, as described earlier in this chapter.

➔ To learn how to place your endnotes where you want them, **see** "Creating Space for Endnotes," **p. 612**

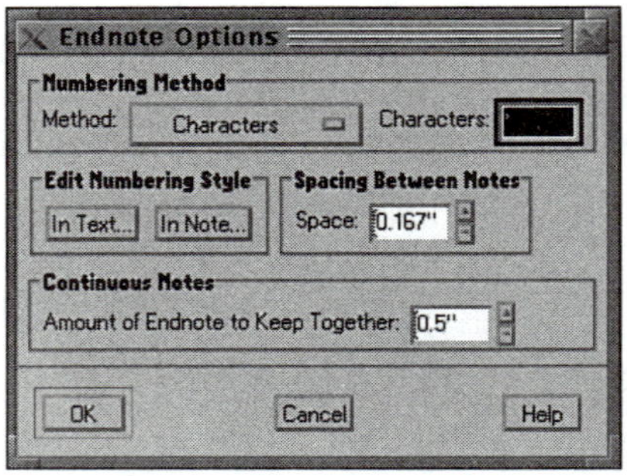

Figure 23.15
The Endnote Options dialog box contains a subset of the options in the Footnote Options dialog box.

Interactions Between Footnotes and Endnotes

Most documents need only footnotes or endnotes, but in some cases you may want to use both, or you may need to reformat a document that uses one type of note to use the other. Fortunately, WordPerfect lets you mix and convert between the two types of notes, so you can accomplish these goals with minimal fuss.

Mixing Footnotes and Endnotes

Despite their similarities, WordPerfect's footnote and endnote features are distinct, so you can include both of them in a single document. You might do so, for instance, to place citations to other works in endnotes and tangential discussions or elaborations in footnotes. You'll need to remember which type of note you're using for which purpose, of course.

Your reader may have more difficulty with a document that contains both note types, however. By default, WordPerfect uses identical in-text note references for footnotes and endnotes (namely, a superscripted Arabic number). To disambiguate your note numbers, you should change the numbering scheme for one or both note types so that they use distinct systems. For instance, you could use Arabic numbers for endnotes and characters for footnotes.

➔ To learn how to set note numbering, **see** "Renumbering Notes," **p. 617**

CONVERTING NOTES FROM TYPE TO TYPE

WordPerfect 8 for Linux comes with two *macros (p. 484)*, endfoot.wcm and footend.wcm, which convert endnotes to footnotes' and vice versa. If you ever need to convert from one note type to another, you can use these macros to do the job. To do so, follow these steps:

1. Make sure your insertion point is positioned *outside* the notes in ordinary text.
2. Choose Tools, Macro, Play (Shift+Ctrl+F10) to open the Play Macro dialog box (Figure 23.16).

Figure 23.16
You can use the Play Macro dialog box to play macros other than the note conversion macros.

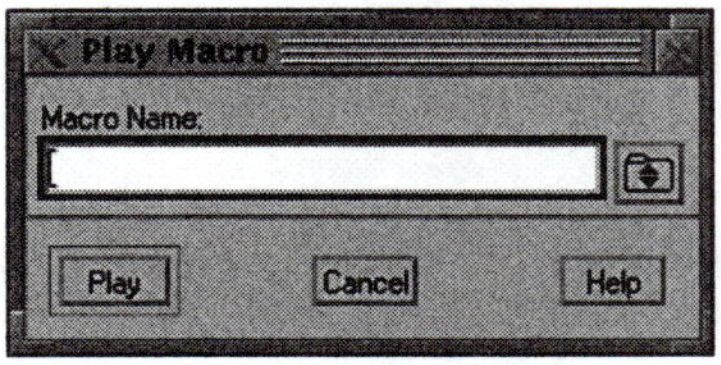

3. To convert footnotes to endnotes, type `footend` into the Macro Name field. To convert endnotes to footnotes, type **`endfoot`** into that field.

Note

WordPerfect automatically adds the .wcm filename extensions. You can include them if you want to, but you don't need to do so.

4. Click Play to run the macro. WordPerfect tells you that it's converting the notes. On a short document, this message appears for a fraction of a second, but on a long document or a slow computer, WordPerfect may take a while to finish the job.

WordPerfect tries to preserve any formatting changes you've made to your footnote styles, such as changes in font or line spacing. It also tries to restore the original settings if you then change your notes back to another type. WordPerfect is usually, but not always, successful, so you should check the formatting of the notes after this change.

Note

If your document contains both footnotes and endnotes, the conversion from one type to the other is not easy to reverse because you get a single ordering of the target note type. You should therefore save your document before doing such a conversion in case you change your mind.

Are you getting error messages when you try to run one of the note conversion macros? See "Note Conversion Macro Won't Run" in the "Troubleshooting" section at the end of this chapter.

If you want to convert only some of the notes in your document, you have to do work on a case-by-case basis: Select all the text in the note, copy it to the Clipboard, and delete the note. Then create a new note of the type you want and paste the text from the old note into it. (The endfoot.wcm and footend.wcm macros perform this task repeatedly, using Find to locate all the notes in the text.)

→ To learn how to search for note (and other) codes, **see** "Find and Replace," **p. 105**

TROUBLESHOOTING

RESTRICTED FEATURES IN NOTES

You're creating or editing a note, and you can't perform some action or use some feature—saving your file, creating columns, changing between Draft and Page views, or various others.

WordPerfect doesn't provide absolutely unrestricted access to all WordPerfect features when you're editing a note. Some features, such as creating another note or setting up columns, are of dubious utility within a note. Others, such as saving files or printing, are simply unsupported. To gain access to these features, choose File, Close (Ctrl+F4) to exit from Note editing mode and try again.

EDITED STYLE CHANGES DON'T WORK IMMEDIATELY

You've edited a note style and clicked OK in the Styles Editor window (Figure 23.13), but your change doesn't seem to have taken effect.

This problem is a minor window redraw bug. If you scroll your document down or up and then return to the original location, you'll see that the style actually has been updated. WordPerfect simply neglected to show this change onscreen immediately.

NOTE CONVERSION MACRO WON'T RUN

You're trying to convert footnotes to endnotes, or vice versa, but WordPerfect refuses to run the macro.

The cause of this problem is likely to be one of three things:

- Your insertion point is placed in the note text itself, or possibly in another special text element like a header or footer. WordPerfect runs the conversion macros only when the insertion point is placed in ordinary text.
- You've changed your macro execution path to exclude the wpmacros subdirectory of your WordPerfect installation directory.

→ To learn about the macro path, **see** "Maintaining a Macro Directory," **p. 490**

→ To learn about setting WordPerfect's file preferences, **see** "Files," **p. 770**

- You never installed, or you deleted after installing, the necessary macros. You might need to rerun the WordPerfect installation procedure to get these macros. Be sure to do a full install, not a minimal install.

→ If you want to know about the WordPerfect installation process, **see** "Installing WordPerfect for Linux," **p. 712**

CHAPTER 24

ORGANIZING LISTS AND REFERENCES

In this chapter

BASIC PRINCIPLES OF LISTS AND REFERENCES

No idea is an island. The ideas in any document relate to one another, and you may need to refer to one idea from another's location. This situation can pose a problem with large documents because you want to be able to refer your reader to an idea's exact location, but you may not know what that location will be after you change formatting, add text, delete text, and so on. That's where WordPerfect's reference features come in handy; they can refer your reader to one place in the text from another.

Although no single idea is an island, your document as a whole might be considered an island, and if it's large enough, you may need a roadmap to find your way around. In fact, you may need several, just as you may want both street maps and subway maps to navigate the island of Manhattan. WordPerfect can help you to generate these maps, in the form of a table of contents, table of authorities (otherwise known as a bibliography), index, and other lists. As with references, WordPerfect keeps track of exact page numbers and other minutiae, allowing you to concentrate on the ideas, not the formatting and numeric details. I refer to all of these document "maps" as lists.

In general, the various list and reference features work as follows:

1. As you type your text, or after you've done it, you *mark* certain words and phrases for inclusion in a list or as a *target* for a reference.
2. You place a special code somewhere in your document to mark where the list is to appear or where you want to refer to a previously marked target.
3. You select a special WordPerfect feature to *generate* the lists and references.

Caution

All lists and references are *static* after they are generated. Consequently, if you generate a list or reference and then alter your text, the list or reference may become inaccurate. You may face a similar problem if you change the printer for the document because WordPerfect may reformat your document for the new printer. Therefore, it's good practice to regenerate all your lists and references just prior to printing to ensure that they're accurate.

The details of how you mark your text, what text you should mark, and how to generate the associated list or reference vary from type to type. The rest of this chapter is devoted to these details. Even if you want to generate a specific type of list or use the reference feature, you should read the next section, titled "Lists," because it introduces many of the techniques used by the other list types, including tips you'll want to know about for generating any list type.

LISTS

WordPerfect provides specific list types to handle tables of contents, tables of authorities (bibliographies), and indexes. If you don't want to create one of these specific types of lists, however, you can use the more general-purpose list type, known in WordPerfect simply as a list. Tables of contents and the rest are also lists, but WordPerfect provides tools specifically for them.

Marking Text

The first step in generating a list is generally to mark the text you want to include in the list. For instance, suppose you're writing a book on computer programming, and this book includes many source code listings from sample programs. You might want to include a listing of the program titles so that your readers can find them easily. The first step to generating such a list is to mark some appropriate identifying text either in the source code or in the surrounding text. Follow these steps:

1. Select the text that you want to reproduce in the list. Be careful to include only the exact text you want and nothing more.

Tip #271 from *Tux Tips*

> If you want to select only a single word, you can place the insertion point in the word. If you don't actively select text, WordPerfect will mark the word in which the insertion point rests.

Tip #272 from *Tux Tips*

> It's often helpful to use Reveal Codes (View, Reveal Codes or Shift+Ctrl+F3) when you select text, especially when the word or words you want to select are formatted differently from the surrounding text. If you don't use Reveal Codes, it's easy to grab the formatting codes, which you may not want, or omit them if you do want them.

→ For more information on selecting text and codes in Reveal Codes, **see** "Using Reveal Codes to Select Text," **p. 111**

2. Choose Tools, Reference, List. WordPerfect places the List toolbar in the document window, as shown in Figure 24.1
3. Type a name for the list into the List selector's text field. If you've already marked one or more items for this list, you can instead select it by using the button to the right of the selector's text field.
4. Click the Mark button in the List toolbar. WordPerfect marks the text you selected for inclusion in the list.

The result is a pair of [Mark Txt List] codes surrounding the marked text, as in the Reveal Codes pane in Figure 24.1. Outside of Reveal Codes, there won't be any indication that the text is marked for inclusion in a list (though Figure 24.1 also shows the text as being selected so that you can find it more easily).

Tip #273 from *Tux Tips*

> If you want to create a list of figures or other boxed items and if these items have captions, you can have WordPerfect use the captions in place of marking text. I describe this process shortly.

→ To learn how to create a list of captions, **see** "Creating and Editing List Types," **p. 631**

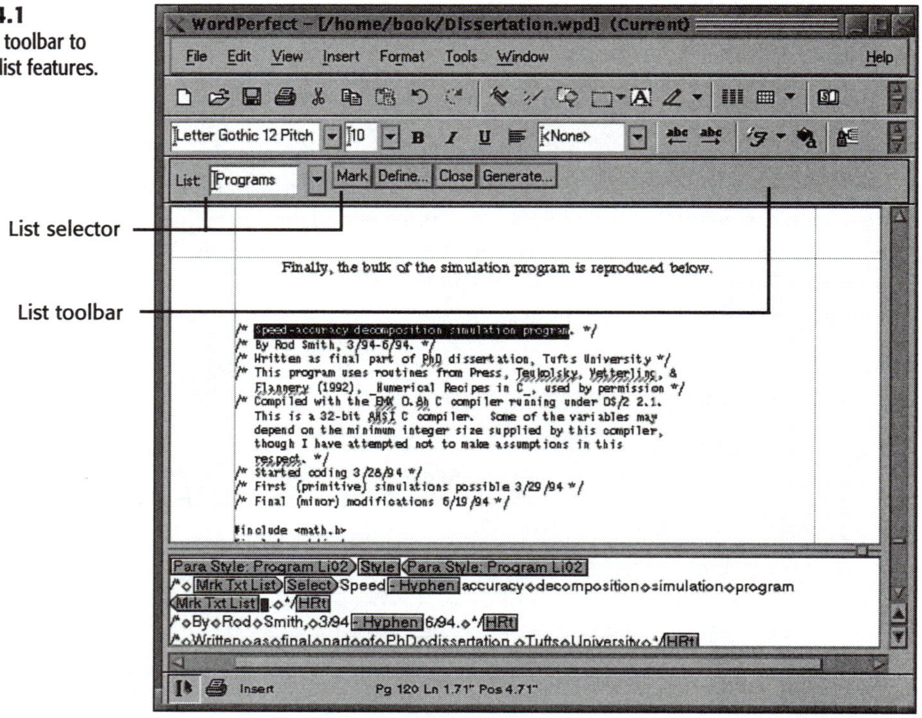

Figure 24.1
Use the List toolbar to control the list features.

DEFINING A LIST

You determine the details of the final list's formatting when you *define* the list. To do so, you should position your insertion point where you want the list to appear and then click the Define button in the List toolbar (Figure 24.1). (You may need to select Tools, Reference, List if you don't see this toolbar in your document window.) This step produces the Define List dialog box, shown in Figure 24.2. From here, you can perform several actions to control the formatting of the list.

Figure 24.2
Click Insert for the default list formatting or customize it by clicking Edit.

Creating and Editing List Types

If you've already marked text to reside in the list, you may want to edit the list type by clicking Edit. If you've not yet marked text for the list, you can create a new list type by clicking Create. Either way, WordPerfect displays a dialog box similar to the Edit List dialog box, shown in Figure 24.3. If you're creating a new list, you should first type in a List Name. You won't need to do this step if you're editing an existing list.

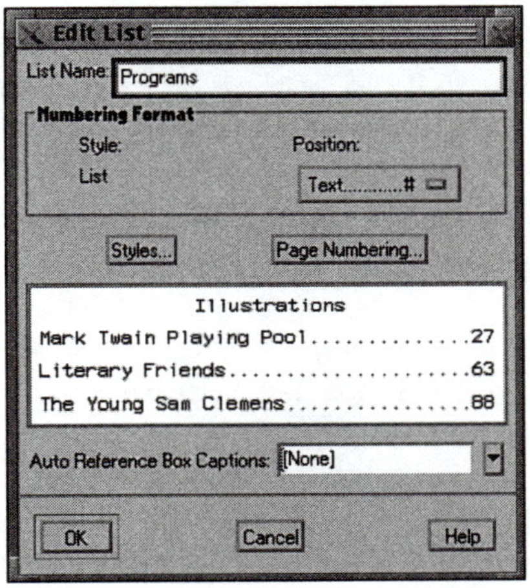

Figure 24.3
The Create List dialog box looks and works just like the Edit List dialog box.

The options in the Create List and Edit List dialog boxes are centered around formatting and numbering the list items. Unlike the more specialized list types, there's only one "level" of the basic list item, so you can't have some list items indented and not others except by manually formatting the list items after creation. The specific options you can control include the following:

- Numbering Format—If you click the Position option button, several numbering options appear. The default option uses a right-justified page number with dot leaders, but you can select from several other styles. When you do so, the preview area in the middle of the dialog box changes to illustrate the numbering option you've chosen.
- Page Numbering—If you want to add text to the page number, you can click the Page Numbering button. WordPerfect then displays the Page Number Format dialog box, shown in Figure 24.4. The default selection here is Document Page Number Format, which sets WordPerfect to use the formatting details you've set elsewhere for the page number. If you click User-Defined Page Number Format, however, you can add text to the page number, or if you click on the Insert option button, you can select alternative numbers, such as chapter numbers. You could, for instance, use this feature to create both chapter and page numbers in the numbering reference.

→ For more information on page numbering options, **see** "Page Numbering," **p. 239**

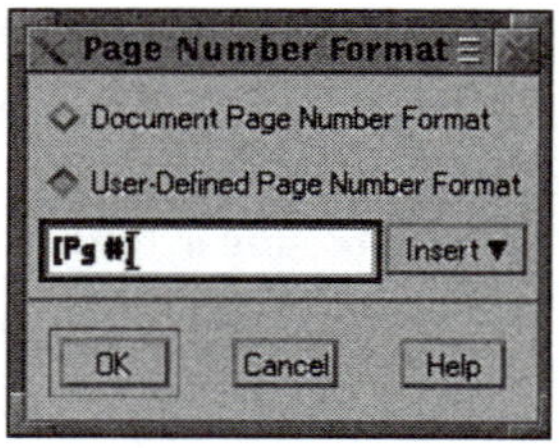

Figure 24.4
You can add text such as the word "page" to the page number with the Page Number Format dialog box.

Tip #274 from *Tux Tips*

If you click in the text entry field in the Page Number Format dialog box and press Ctrl+W, WordPerfect displays the Symbols dialog box with which you can enter special symbols.

→ To learn how to use the Symbols dialog box, **see** "Inserting Special Characters," **p. 102**

- Styles—WordPerfect formats text in lists by using styles. You can modify these styles by clicking the Styles button in the Edit List or Create List dialog box. This step produces the List Style dialog box, shown in Figure 24.5. Click on a style to change the formatting of the list entry to that style and then click Select to confirm this choice. You can create a new or modify an existing style by clicking the Create or Edit buttons, respectively, in the List Style dialog box. Either way, WordPerfect displays the Styles Editor dialog box, as shown in Figure 24.6. You can edit the style by entering codes and even text, much as you would in a regular WordPerfect document window.

→ For more information on editing a style, **see** "Editing a Style," **p. 321**

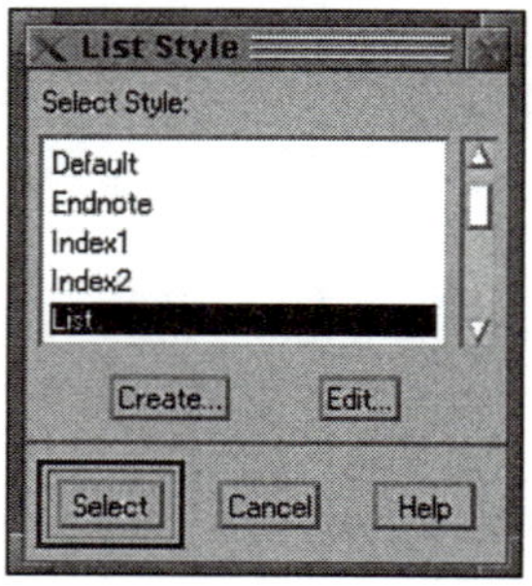

Figure 24.5
Double-click on a style to select it.

Note

WordPerfect does not reflect the changes you make in a style in the sample of list formatting in the Edit List dialog box.

- Box Captions—If you're creating a list of figures, text boxes, or other boxed items with captions, you can use the Auto Reference Box Captions selector in the Edit List dialog box to configure the list. Click on the button to the right of this field, and WordPerfect displays the different box types that it supports, such as .Figure Box and .Text Box. Click on the box type for which you want to create a list. When you later generate your list, WordPerfect uses the caption text from the matching boxed items as the list text.

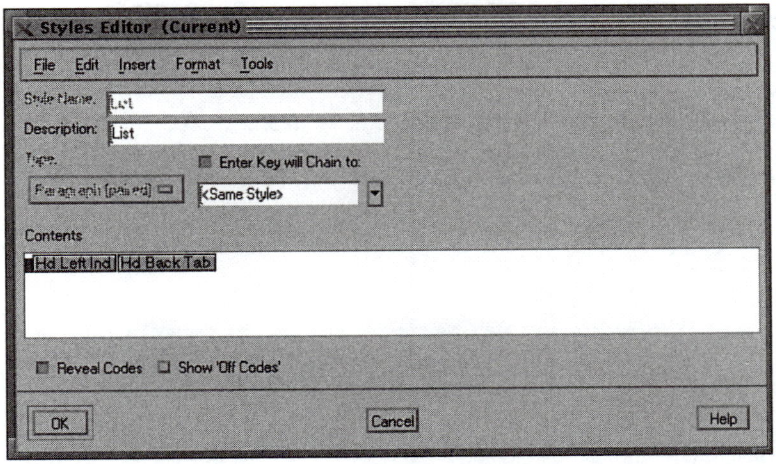

Figure 24.6
The default style for regular lists creates a hanging indent.

➔ To learn how to create captions for your boxed items, **see** "Captions," **p. 466**

When you're finished changing your options in the Create List or Edit List dialog box, click OK to set them.

List Management

In a particularly large document with several lists, you may need to manage your lists in various ways. For instance, you might find that you've defined several lists, but don't need them all. To avoid confusion, you can delete a list by selecting it in the Define List dialog box and then clicking Delete. WordPerfect asks you to confirm this deletion and then, if you do confirm it, deletes that list type.

Note

Deleting the list does *not* delete the markings around text that's been marked for this list type. These markings won't do any harm, except for taking up some disk space, so you might as well leave them in place. That way, if you subsequently decide to restore the list, you can do so by creating a list of the same name as the original.

Another type of list management you can perform is to import list types from another document. For instance, if you've created a document with list formatting, including styles and other customizations, and you want to reuse that list formatting, you can do so. Click Retrieve in the Define List dialog box to get the Retrieve List Definitions dialog box (Figure 24.7). Type the name of the file that contains the list types you want to retrieve in the Filename field or click the folder icon to its right to select the file from a WordPerfect File Manager dialog box. In this case, WordPerfect displays the list types used in that file in the List Name list box. You can select one or more of the list types by clicking with the mouse. (To select more than one, use the Ctrl key along with a mouse click to select individual separated list types or use the Shift key to select a range of list types.) When you've selected the list types you want to include, click OK to include them.

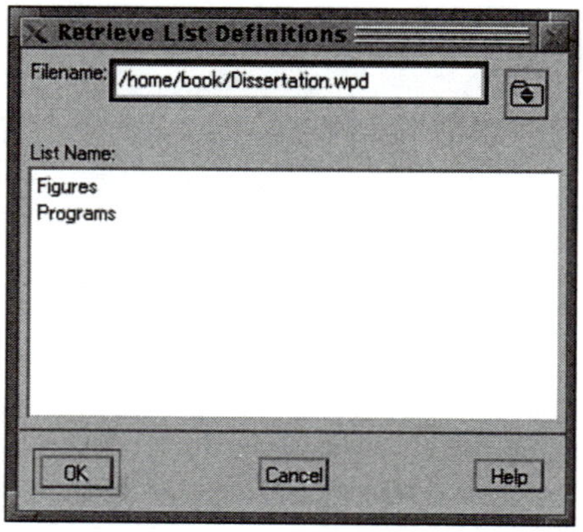

Figure 24.7
You can select one or more list types to import from another WordPerfect document by using the Retrieve List Definitions dialog box.

Tip #275 from *Tux Tips*

If you regularly create documents that use the same customized lists, consider creating a template file to automate this and other formatting options.

→ For information on templates, **see** "Promoting Cross-Document Consistency with Templates," **p. 333**

INSERTING YOUR LIST

When you finish customizing your list's styles and other features, you need to insert a list code. You do so by clicking Insert in the Define List dialog box. WordPerfect inserts three codes and some text at the location of the insertion point, as shown in Figure 24.8.

The first code that WordPerfect inserts when you define your list is a [Def Mark] code. This code tells WordPerfect what sort of text to generate at that point in the document. In Figure 24.8, the insertion point is to the left of this code, revealing that a list called Programs is to be generated.

The second and third codes that WordPerfect inserts when you define a list are a matched pair of [Gen Txt] codes. These tell WordPerfect where to generate the list. When you generate the list, WordPerfect deletes any text that falls between these codes and replaces it with the list itself. Because WordPerfect leaves the [Gen Txt] codes intact when it's finished with the generation, subsequent generations won't produce duplicate lists; instead, WordPerfect will *replace* the outdated list with the new one. The fact that the text between the [Gen Txt] codes will be replaced also means that WordPerfect can place some text between those codes to identify where the list will appear, and WordPerfect does just that, as Figure 24.8 illustrates.

Caution

You should be careful not to delete the [Def Mark] or [Gen Txt] codes because if you do, WordPerfect won't be able to generate its lists properly.

Figure 24.8
The [Def Mark] code shows the list with which it's associated when the insertion point is to its immediate left.

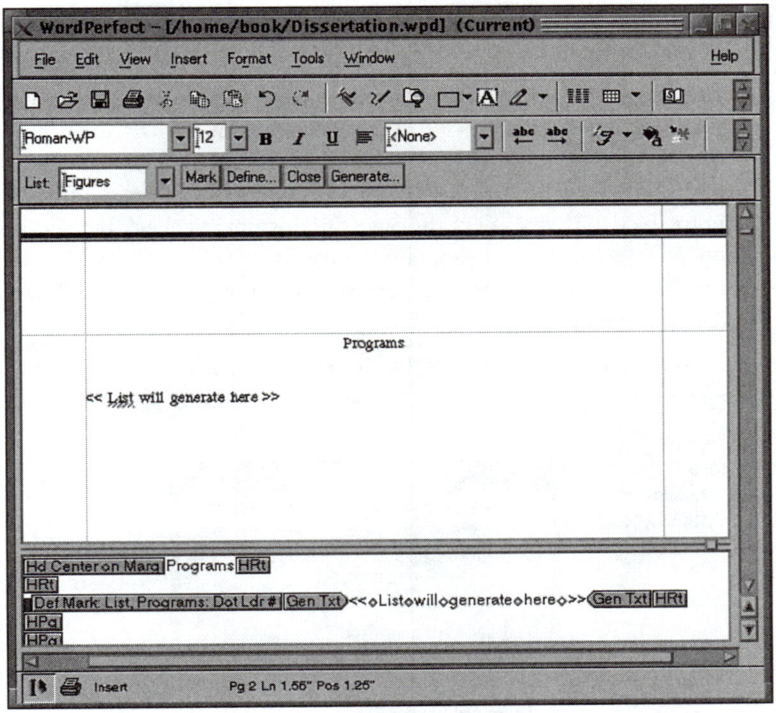

Chances are you want to create some text and formatting around your list. For instance, you might want to give it a label, such as Programs or Figures, and perhaps place the list on its own page. You perform this task with ordinary WordPerfect editing procedures. The document in Figure 24.8 places the list of programs on its own page using [HPg] codes and centers the word *Programs* at the top of the page.

→ To learn about options for starting a new page, **see** "Page Breaks," **p. 230**

Note

Because WordPerfect replaces *any* text between the [Gen Txt] codes with the list contents, you should be careful not to place your intended table formatting and headers in that area. If in doubt, use Reveal Codes to be sure you're placing your text and formatting correctly.

Tip #276 from *Tux Tips*

If you define a list, generate it, and then decide to change something about the definition, you'll need to define the list again. Try placing your insertion point just before the [Def Mark] code and define a new list. You'll then have a [Def Mark] code, two [Gen Txt] codes with WordPerfect's intervening blurb, a second [Def Mark] code, a third [Gen Txt] code, the unacceptable list, and a fourth [Gen Txt] code. Delete the second [Def Mark] code and the first and second [Gen Txt] codes along with their intervening text. This step will associate the new [Def Mark] code with the old [Gen Txt] codes, so WordPerfect will replace the bad list automatically the next time you generate. This procedure may sound complicated, but it sounds worse than it is. After you've done it a couple of times, it will be easy.

GENERATING A LIST

When you've placed your list definition, it's time to actually generate the list. You do so by clicking Generate from the List toolbar (refer to Figure 24.1) or by choosing Tools, Reference, Generate (Ctrl+F9). WordPerfect then displays the Generate dialog box (Figure 24.9). Normally, you can immediately click OK to generate the list. You can set a few options, however, by clicking Options. This step produces the Generate Options dialog box (Figure 24.10). If you select the Save Subdocuments check button, WordPerfect saves any subdocuments when you generate the list. If you check Build Hypertext Links, WordPerfect turns the page numbers into hypertext links to those sections, which may be helpful if you're creating a Web page.

→ For information on subdocuments, **see** "Master Documents and Subdocuments," **p. 598**

→ For information on links in Web pages, **see** "Creating a Link," **p. 689**

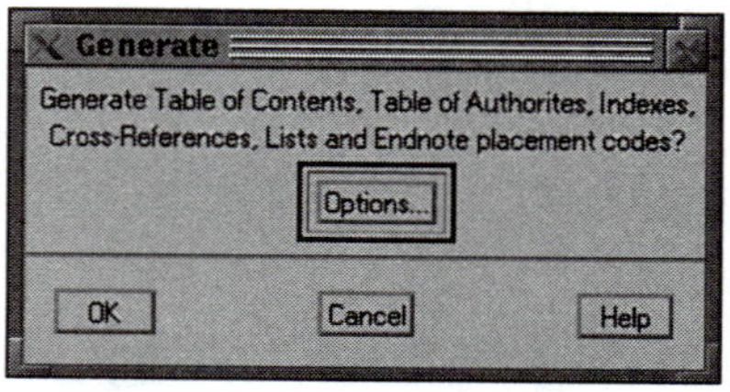

Figure 24.9
The Generate dialog box exists mainly to let you set some last-minute options.

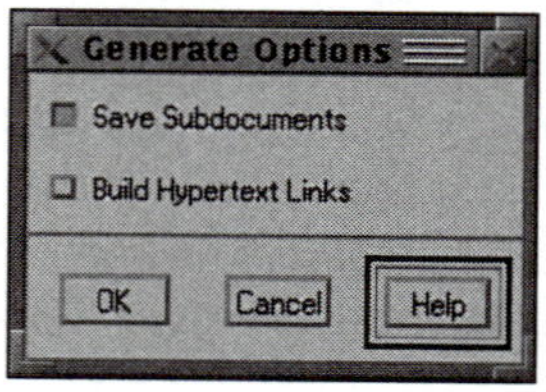

Figure 24.10
Set miscellaneous options in the Generate Options dialog box.

When you click OK in the Generate dialog box, WordPerfect generates *all* your lists and references for this document. Although you can do this step before you define all your lists and references, you should refrain from doing so if you know that a definition is in poor shape (say, one of the [Gen Txt] codes is missing from a pair). WordPerfect displays a dialog box telling you that it's performing the generation. How long this dialog box remains onscreen depends on the size of your document and the speed of your computer. When the generation is done, WordPerfect places your insertion point at the end of the last list or reference it's generated, so you may need to move your insertion point to get back to where you want to be.

Tip #277 from *Tux Tips*

You can use WordPerfect's QuickMark feature to get back to your original location after executing a generation. Press Shift+Ctrl+Q to place the QuickMark in the location to which you want to return, perform the generation, and then press Ctrl+Q to return to the QuickMark.

➔ To learn how to use the QuickMark, **see** "The QuickMark," **p. 596**

Figure 24.11 shows the result of a list generation. As you can see in the Reveal Codes pane, the [Def Mark] and [Gen Txt] codes remain intact, but the placeholder text WordPerfect inserted between the two [Gen Txt] codes has been replaced by the list contents.

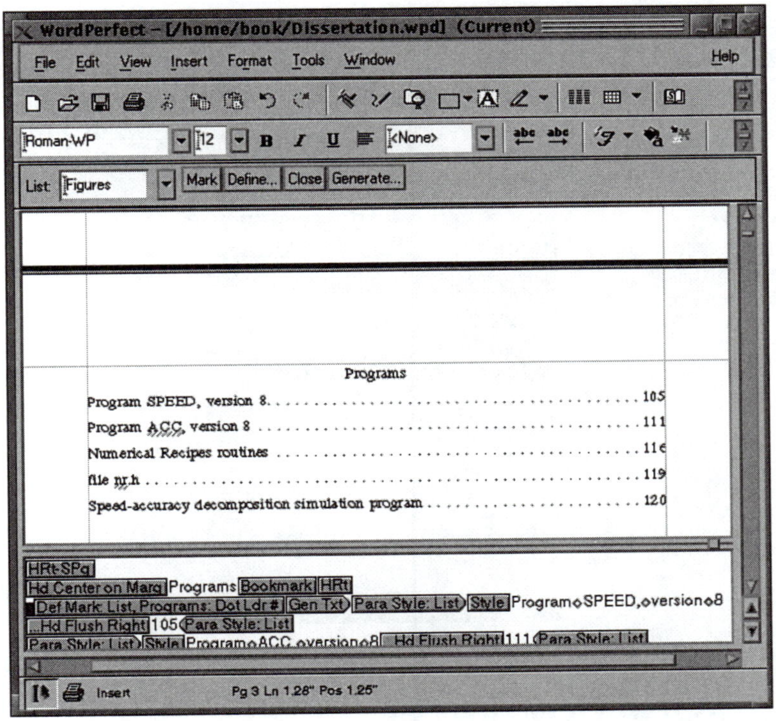

Figure 24.11
The generated list is ordinary text between the [Gen Txt] codes.

Part VI Ch 24

You may find that your generated list isn't quite what you want. For instance, in Figure 24.11, all five lines in the list except for the `file nr.h` line begin with uppercase characters. You might want to correct this sort of inconsistency. You might also find that some text appears italicized or with other undesirable formatting. In any of these cases, you have two choices:

- Edit the text directly in the list, leaving the original intact. Doing so may be your only option in some cases. For instance, you might have included text that needs to fall on two lines in the body of the text but that you want to have on a single line in the list. WordPerfect grabs the [HRt] code in the main text and incorporates it in the list, but you'll want to remove that code from the list yourself. This option does have the disadvantage, however, of requiring you to repeat this editing the next time you generate this list. If the editing alters page breaks, you could even invalidate the page numbers in the list.

- Edit the original text and regenerate the list. If you find the location of the original text, you can edit it to eliminate the undesirable formatting. If you do so, the list will be correct the next time you generate it. All other things being equal, this approach is definitely the preferable solution when you expect to edit your document and generate your lists again.

Tip #278 from *Tux Tips*

> Select Edit, Go To (Ctrl+G) to display the Go To dialog box shown in Figure 24.12 if you want to edit the original text. You can then type the page number as displayed in the table and click OK to jump to that page. Locating the original text should then be easy.

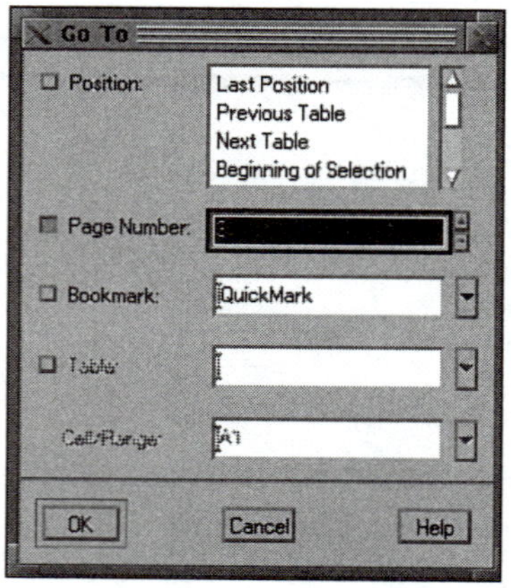

Figure 24.12
Use WordPerfect's Go To feature to quickly locate the source for text in a list.

TABLES OF CONTENTS

Although the general-purpose list type can be very useful for many applications, you'll occasionally need to create a list that's too complex for this feature to produce. WordPerfect therefore provides several specific list types to meet these needs. The first of these is the table of contents. Conceptually, creating a table of contents works much like creating a general-purpose list: You mark the items you want to appear in the table of contents, define the table's location and exact appearance, and then generate it. Some of the details do differ, however.

You control most of the table of contents operations from the table of contents toolbar. Figure 24.13 shows this toolbar, which you activate by selecting Tools, Reference, Table of Contents.

Figure 24.13
You can quickly specify the level of a table of contents entry from the table of contents toolbar.

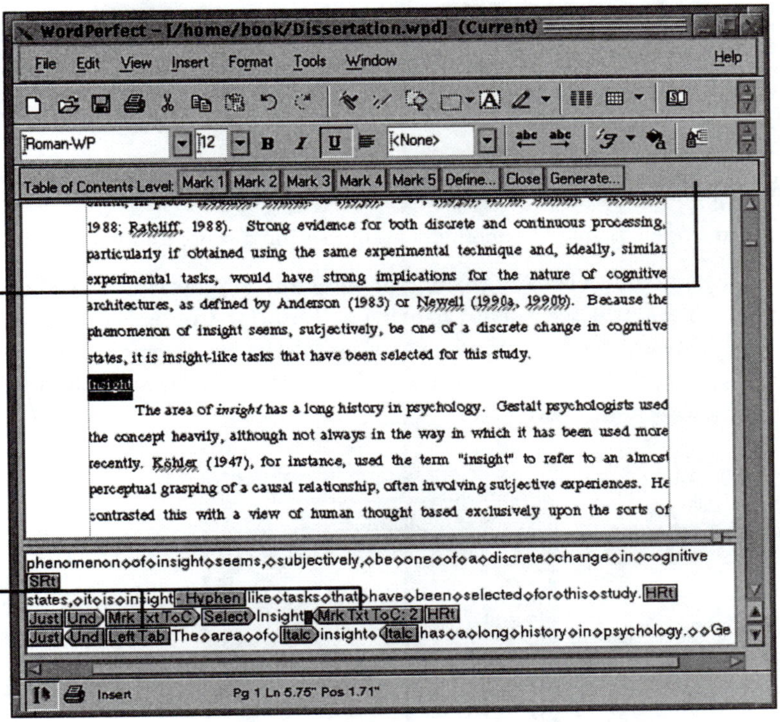

Marking Text

WordPerfect's table of contents feature lets you create a table of contents with several different *levels (p. 638)*. Table of contents levels are very similar to levels in outlining. They're used to indicate the hierarchical structure of your document, with broad topics taking up high levels and specific ones occupying lower levels.

Marking Individual Words and Phrases

When you mark text for inclusion in a table of contents, you must specify its level by selecting from one of the five Mark buttons in the table of contents toolbar. Use Mark 1 for the top-level item in the table of contents. Use Mark 2 for the level immediately below the first level. Use Mark 3, Mark 4, and Mark 5 to mark still lower divisions.

WordPerfect records the level with which the text is associated when you mark it for the table of contents, much as it records the name of the list when you mark it for a generalized list. You can see this information in the closing [Mrk Txt ToC] code in Figure 24.13, which is marked as belonging to level 2. When you subsequently build your table of contents, WordPerfect may format the table's items differently depending upon their levels.

USING STYLES TO MARK TEXT

If you're using styles to create your document headings, you can simplify your task considerably by including the table of contents marking codes inside the style itself. You can do the same thing with other list and reference types, but you're most likely to want to use this feature with tables of contents.

➔ To learn how to use styles, **see** "Ensuring Consistency Within Your Document with Styles," **p. 305**

To embed a table of contents marking within a style, follow these directions:

1. You must begin by creating or editing an appropriate style. You can do so in several ways, as described in Chapter 12, "Ensuring Consistency Within Your Document with Styles," but let's suppose you want to start from scratch and create a new heading style. You should choose Format, Styles (Shift+Ctrl+F8) to display the Style List dialog box shown in Figure 23.14.

Figure 24.14
Select and edit WordPerfect styles from the Style List dialog box.

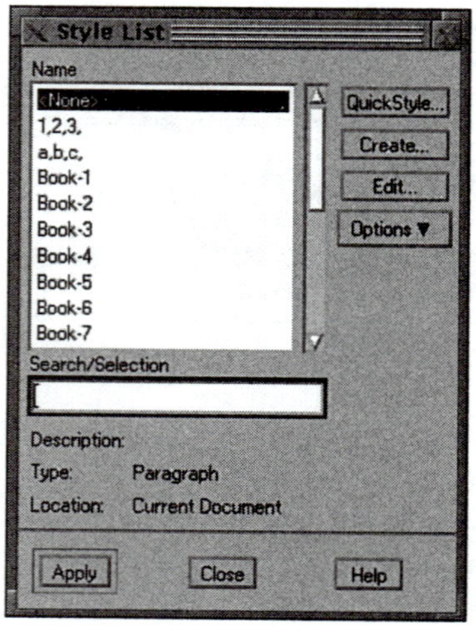

2. Click Create to create a new style with the WordPerfect Styles Editor dialog box (refer to Figure 24.6). Enter whatever formatting codes you want for the heading in the text.
3. Select the Show 'Off Codes' check button. WordPerfect displays a code with the lengthy name [Codes to the left are ON - Codes to the right are OFF] (hereafter simply [OFF]).
4. Choose Tools, Reference, Table of Contents in the Styles Editor's menu. WordPerfect displays a miniature table of contents toolbar inside this dialog box.
5. Position the red insertion point marker in the Styles Editor dialog box just to the left of the [OFF] code.

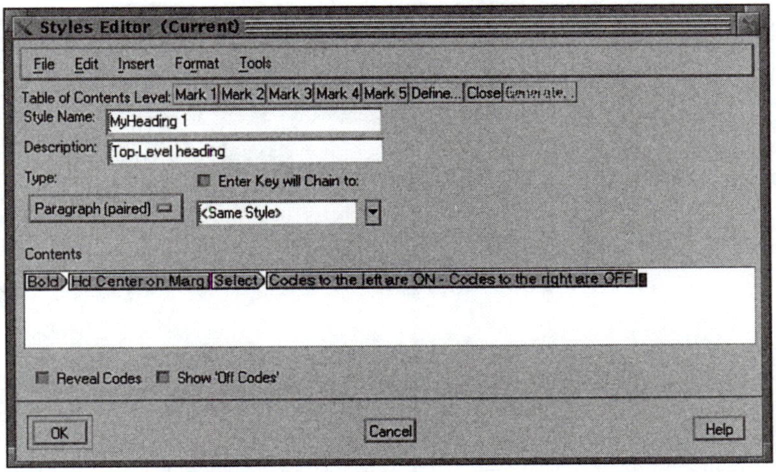

Figure 24.15
To mark other reference types using styles, choose an appropriate option from the Tools, Reference menu and use the toolbar item for that reference type.

6. Press Shift+Right Arrow. WordPerfect places a [Select] code to the left of the [OFF] code and moves the insertion point to the right of that code. Your Styles Editor dialog box may now look something like the one in Figure 24.15, though the formatting codes you entered may be different from the ones shown here.
7. Click the appropriate Mark button from the miniature toolbar in the Styles Editor dialog box. The [Select] code vanishes and you should see a pair of [Mark Txt ToC] codes, one before the [Off] code and one after it.

Performing these operations as I've just outlined results in WordPerfect extracting the text and *only* the text for the table of contents. If you select formatting codes as well as the [Off] code, you'll include those in the table of contents as well, so with some careful planning, you can automatically generate a table of contents that contains some or all of the formatting used in your headings.

Note

WordPerfect's default heading styles (Heading 1 through Heading 5) include appropriate markings to add them to the table of contents, so if you use those headings, they *will* be added to any table of contents you generate. If you don't want the headings to appear or if you want to change the table of contents level for some headings, you need to edit the standard heading styles.

DEFINING A TABLE OF CONTENTS

As with a generic list, you must define certain characteristics for a table of contents before you can generate one. You do so by clicking the Define button on the table of contents toolbar (you can choose Tools, Reference, Table of Contents to display this toolbar if it's not already on the screen). Be sure your insertion point rests where you want the table of contents to appear in your text, however, before you click. When you do click Define, WordPerfect displays the Define Table of Contents dialog box shown in Figure 24.16.

Figure 24.16
The preview area in the right of the Define Table of Contents dialog box shows only those levels you've elected to use in your table of contents.

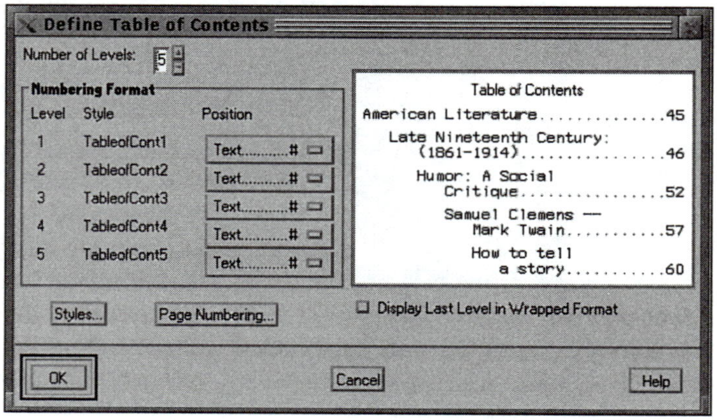

From this dialog box, you can control many aspects of the table of contents:

- Number of Levels—You can define how many levels you want in your table of contents by using the Number of Levels spin box. If you select fewer levels than you used when marking your table of contents entries, WordPerfect omits the extra levels. If you specify more levels, WordPerfect doesn't actually use the additional levels, but they'll be available should you later mark additional entries. The default number of levels is one, but Figure 24.16 shows five levels.
- Numbering Format—For each level, you can specify a unique numbering format, similar to the numbering format for generic lists.
- Wrapped Numbering—If you select the Display Last Level in Wrapped Format check button, WordPerfect alters its formatting of the last level. Normally, WordPerfect places each table of contents entry on its own line. When you select this option, however, WordPerfect wraps all the entries in a single paragraph. This format makes for a more compact table of contents, but it may not be as readable as it might otherwise be.

Note

If you select wrapped format, WordPerfect automatically changes the Numbering Format for the final level, if necessary, to remove right justification and dot leaders for the page number. If you subsequently deselect this option or increase the number of levels, you may need to reset these options.

- Styles—If you click the Styles button, WordPerfect displays the Table of Contents Styles dialog box, shown in Figure 24.17. Use the radio buttons across the top of this dialog box to select a table level and then select the style you want from the style list field. You can click the Create or Edit buttons to create or edit a style, as described earlier for general lists.

→ To learn more about list styles, **see** "Creating and Editing List Types," **p. 631**

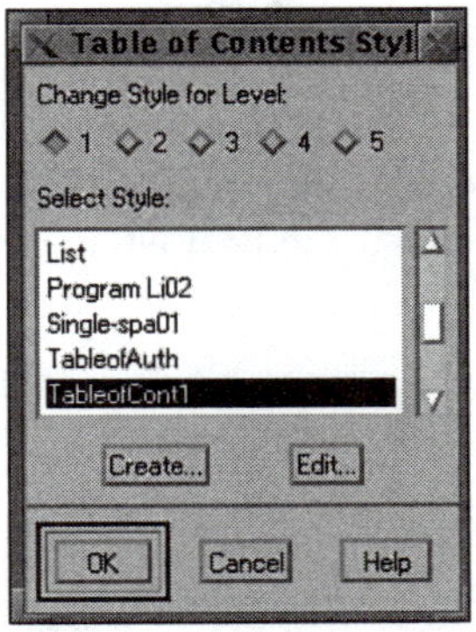

Figure 24.17
Select the style you want for the table of contents entries from the Table of Contents Styles dialog box.

- Page Numbering—If you click the Page Numbering button, WordPerfect displays the Page Number Format dialog box (refer to Figure 24.4). You can use this dialog box to modify the nature of the page number, including adding text, chapter numbers, and so on. I discuss these features in relation to general-purpose lists.

→ To learn more about list styles, **see** "Creating and Editing List Types," **p. 631**

When you've defined the table of contents options to your satisfaction, click OK in the Define Table of Contents dialog box, and WordPerfect inserts three codes: a [Def Mark] code and a pair of [Gen Txt] codes. WordPerfect also places the text << `Table of Contents will generate here` >> between the [Gen Txt] codes. These codes, and the associated text, work just as they do for other list types.

Tip #279 from *Tux Tips*

You can define more than one table of contents for a document. You might use this feature, for instance, to provide a broad overview of a large document containing, say, just chapter titles, in one table of contents. Another might then contain a more in-depth listing of the contents of the document.

GENERATING A TABLE OF CONTENTS

Before you generate a table of contents, you may want to create appropriate formatting around the future table of contents location. For instance, you might want to place [HPg] codes on either side of it, to put it on its own page or pages; and you might want to add text, such as `Contents`, to identify the table of contents for what it is. As with generic tables, you should be careful to add such formatting outside of the paired [Gen Txt] codes. If you don't, your formatting will disappear when you generate your table of contents.

To do the generation, either click the Generate button in the table of contents toolbar or choose Tools, Reference, Generate (Ctrl+F9). This step produces the Generate dialog box (refer to Figure 24.9), which works in the same way as it does when you're generating other lists. In fact, when you click OK, you generate or regenerate all your lists and references, not just your table of contents.

TABLES OF AUTHORITIES

A *table of authorities* is a type of bibliography, especially common in legal documents. This feature is specifically designed for the legal profession, which has somewhat more complex bibliographic needs than most other professions. It can be used, however, in other fields and for other types of documents.

As with other WordPerfect lists and references, you control most of the table of authorities features from a special table of authorities toolbar, shown in Figure 24.18. You display this toolbar by choosing Tools, Reference, Table of Authorities.

Figure 24.18
The table of authorities toolbar provides many of the same features as the other list and reference toolbars.

MARKING TEXT

When you mark the text for the table of authorities, WordPerfect creates two different marking codes. The first, known as the *full form*, contains the complete reference, including information such as all the authors, the full title of the work, and where it appeared. This text is what WordPerfect places in the table of authorities. The second type of marking code is the *short form*, and it's a shorthand you use for marking subsequent references to the same work. Using these two forms, WordPerfect maintains a list of all references to the work and, if you so desire, prints their page numbers in the table of authorities.

WordPerfect's table of authorities feature supports multiple lists. For instance, you might want to have separate lists for cases and statutes. When you define the full forms, you indicate to which list the references belong.

CREATING FULL FORMS

You normally create a full form as a table of authorities marker for the first reference to a document in your text. When you find such a reference, click Create Full Form from the table of authorities toolbar, and WordPerfect responds by displaying the Create Full Form dialog box, shown in Figure 24.19.

You need to do two things in the Create Full Form dialog box:

1. Specify the section in which the reference is to appear. For instance, if you want separate tables of authority for cases and statutes, you would either enter one or the other of these in the Section Name field or select one from the drop-down menu by pressing the option button to the right of that field. With documents that have only one monolithic bibliography, you'll want only one option here.

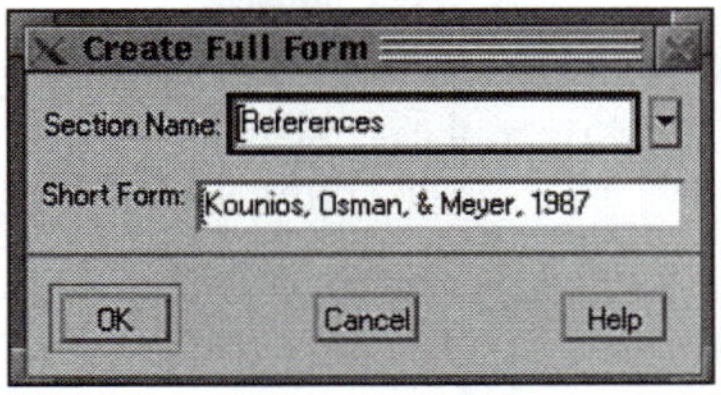

Figure 24.19
The first step in creating a full form is to label a short form.

2. Enter a Short Form name. If you select the text you want to use for the short form reference before clicking Create Full Form, WordPerfect will automatically place it in the Short Form area of the Create Full Form dialog box. If not, or if you want to modify that text, you can type your text or changes here. Be sure to select something that you can remember, and preferably something short, because you need to select this short form item from a list of references for each subsequent occurrence, at least if you want to have page numbers for each reference in your table of authorities. For instance, rather than the `Kounios, Osman, & Meyer, 1987`, shown in Figure 24.19, you might use `Kounios et al., 1987` or `KOM, 1987`.

Note

WordPerfect does not place the short form identifier in your text, and it doesn't need to be present in the text. It's used solely to link future references to the document to that document's full form.

When you click OK, WordPerfect opens a document editing window with which to create the full form, as shown in Figure 24.20. You should format the full form in whatever way you see fit, including attributes such as italics, boldface, and underlining as appropriate for the type of bibliography or table of authorities you need to create.

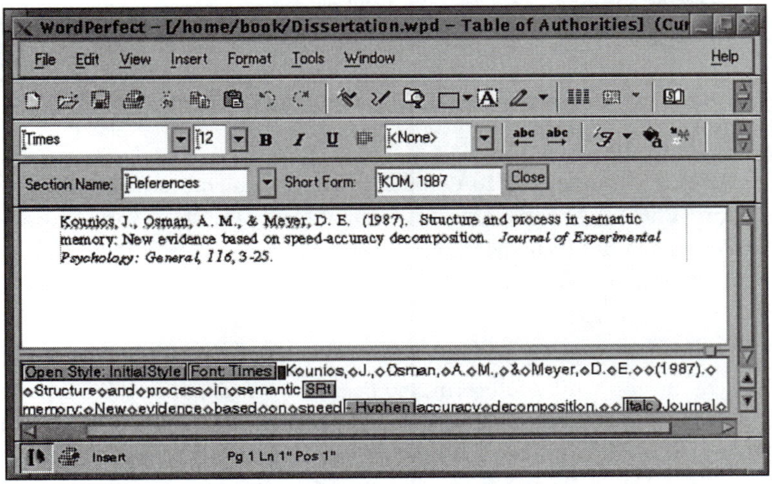

Figure 24.20
You can use typical WordPerfect formatting, including font changes and font attributes, in the full form of the table of authorities entry.

Note

> You can set many formatting features later, when you define the table of authorities. For instance, you might want to leave the individual table of authority entries formatted flush left with no indentation and then set a hanging indent for all entries when you define the table.

When you've created the table of authorities full form entry to your satisfaction, press the Close button in the lowest toolbar of that window (Figure 24.20) or choose File, Close (Ctrl+F4). This step closes the window and places a [ToA] code in the text where the insertion point rests. If you start Reveal Codes mode (View, Reveal Codes or Shift+Ctrl+F3) and move the insertion point to the immediate left of the [ToA] code, you'll see additional information about it, including its section name, the short form name, and an indication that it's a full form entry.

Marking Short Forms

If you don't want page number references in your table of authorities, you needn't mark subsequent references. If you do want a comprehensive list of page numbers, however, you need to mark all the references to the work to have them all listed in the table. This is the function of the short form. To mark a short form, follow these steps:

1. Place the insertion point in the text that's to be referenced.
2. From the table of authorities toolbar (refer to Figure 24.18), select the reference you want from the Short Form list.
3. Click Mark in the table of authorities toolbar.

Tip #280 from *Tux Tips*

> When you mark your references for inclusion in a table of authorities, you may want to do so with the insertion point *inside* the reference. This way, if you subsequently select and delete the reference, you'll also delete the code and you won't be left with references that no longer exist in your document. (This method works only if you *select* the word to be deleted; if you just use the Backspace or Delete key several times to delete the reference, the short form mark remains intact.)

WordPerfect inserts a [ToA] code in your document, just as it did when you marked the full form. This [ToA] code, however, contains only the short form reference, and you don't need to enter the full form. Marking the short forms is therefore much less time-consuming than creating the full forms.

Editing Forms

If you need to edit a full form, say because you've discovered a typo, you should click Edit Full Form from the table of authorities toolbar (refer to Figure 24.18). WordPerfect displays the Edit Full Form dialog box, shown in Figure 24.21. Select the form you want to edit and click OK. WordPerfect then presents a document window (shown in Figure 24.20) in which you can edit the full form.

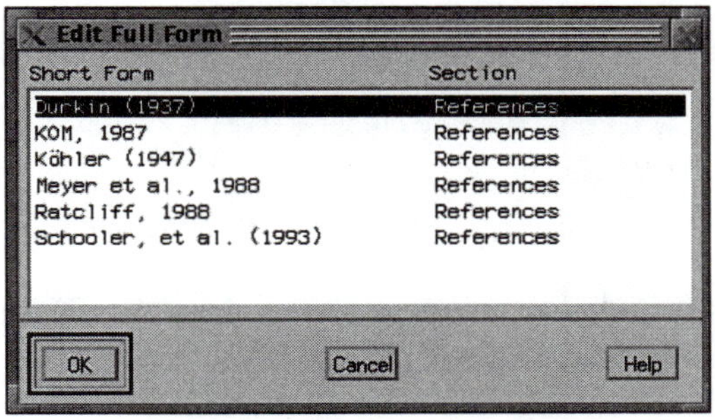

Figure 24.21
Use the short form name to select a full form to edit.

When you're editing a full form, you can alter either the full form or the short form name. When you do so, WordPerfect adjusts all the references to that form as appropriate.

If you want to delete a reference, you need to locate the [ToA] codes in the document itself and delete them. You can do so most easily by generating a table of authorities with page references and then using Edit, Go To (Ctrl+G) to open the Go To dialog box (refer to Figure 24.12) to jump to the page on which the [ToA] code exists. You can then locate it with Reveal Codes turned on.

Defining a Table of Authorities

After you mark your table of authority entries, it's time to define the table. As with other lists, you start with your insertion point where you want the table to appear in the text. Then choose Define from the table of authorities menu bar. This step opens the Define Table of Authorities dialog box, shown in Figure 24.22.

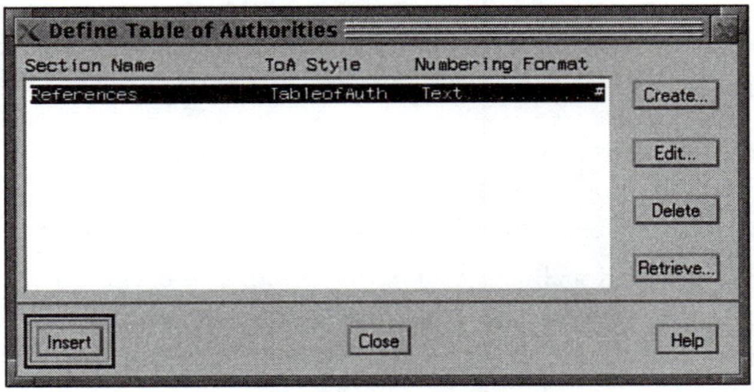

Figure 24.22
You can adjust formatting and other options for a table of authorities in the Define Table of Authorities dialog box.

The Define Table of Authorities dialog box works very much like the Define List dialog box shown in Figure 24.2. To summarize:

- You select the name of the section you want to generate by choosing it in the list that dominates the dialog box.
- You can delete a section's definition by clicking Delete after you select it. This step leaves the codes in your document intact, but you won't be able to generate this section unless you create it anew with the Create button.

- You can retrieve table of authorities styles from other documents by using the Retrieve button.
- You can create a new section style by clicking Create, or you can edit an existing section style by selecting it and clicking Edit.
- When you're done, click Insert to insert the codes to create the table of authorities.

When you create or edit the table of authorities definition, WordPerfect presents the Edit Table of Authorities dialog box, shown in Figure 24.23.

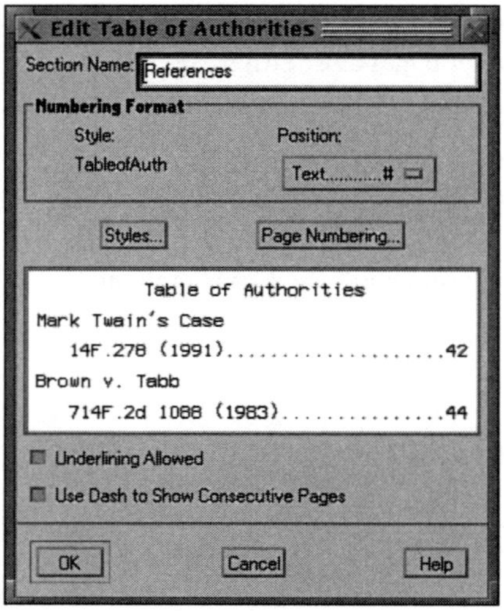

Figure 24.23
The Edit Table of Authorities dialog box allows you to set numbering options and styles for your table of authorities entries.

The Edit Table of Authorities dialog box works very much like the Edit List dialog box shown in Figure 24.3 but with a few differences. To summarize:

- You can set different styles for displaying line numbers by using the Position option button in the Numbering Format area of the dialog box. If a reference occurs more than once in the text, WordPerfect separates the specific occurrences with commas.
- You can set additional page numbering options by clicking Page Numbering, which produces the Page Number Format dialog box (refer to Figure 24.4).
- You can select a new style or modify the default one for table of authority entries by clicking Styles. This step produces the Table of Authorities Style dialog box, which is identical in function to the List Style dialog box (refer to Figure 24.5).
- If you select the Underlining Allowed option button, WordPerfect permits underlining in table of authority entries. If this button is not selected, WordPerfect strips underline attributes from table of authority text when it generates the table. (The underlining codes remain intact in the full form entries, however.)

- If you reference the same entry on consecutive pages, WordPerfect can use a dash to specify the first and last pages, rather than list each page in the sequence separated by commas. To enable this feature, be sure that the Use Dash to Show Consecutive Pages check button is selected.

As with other lists, when you click Insert in the Define Table of Authorities dialog box, WordPerfect inserts a [Def Mark] code defining the table of authorities and a pair of [Gen Txt] codes specifying the location for the table. Whenever you generate the table of authorities, WordPerfect deletes any text between these two [Gen Txt] codes.

If you want to separate your references into different sections, you need to define a separate table for each section. You can, of course, include appropriate formatting to separate and label these sections but be sure you do so outside the region between the [Gen Txt] codes; otherwise, WordPerfect deletes this formatting.

GENERATING A TABLE OF AUTHORITIES

To generate a table of authorities, click Generate in the table of authorities toolbar. WordPerfect produces the Generate dialog box (refer to Figure 24.9), which works just as it does when you generate other list types. Click OK to generate the table of authorities, along with other lists and references. WordPerfect churns for a bit, then places your insertion point at the last list or reference in your document.

You can edit the text in your table of authorities after you generate that text if you like. If you find an error, you'll normally want to correct it by clicking Edit Full Form in the table of authorities toolbar, however, which opens the Edit Full Form dialog box (refer to Figure 24.21). You can then select the entry you'd like to edit and modify the text to that entry. When you're done, you can regenerate the table of authorities. If you decide to edit the text directly, on the other hand, you lose those changes the next time you generate your lists and references.

Are you trying to generate a table of authorities, but get an error message about missing full forms? See "Missing Full Forms" in the "Troubleshooting" section at the end of this chapter.

INDEXES

An index is an alphabetized list of words or concepts that are particularly important, including references to their locations within the document. WordPerfect's index feature supports two levels. This flexibility allows you to create a major index heading, such as "trees," and list specific instances or types as subheadings, such as "larch" and "pine."

Note

Some books may include more than one index. For instance, a book might contain both a concept index and an index of authors. In this particular case, the index of authors may simply be another term for a bibliography, or a table of authorities as WordPerfect calls it. In any event, WordPerfect supports only one actual index, so if you want to generate two lists that you refer to as indexes, you need to use another type of list, be it a table of authorities, a generic list, or even a table of contents, for one of the indexes.

You control most of WordPerfect's indexing features from the index toolbar shown in Figure 24.24. You display this toolbar by choosing Tools, Reference, Index.

Figure 24.24
The index toolbar provides text entry fields for specifying the heading and, if necessary, subheading for an index entry.

MARKING TEXT

WordPerfect provides two ways for you to mark text for an index. First, you can mark text manually, as you can with other list types. When you use this method, you specify a heading and a subheading and place an index mark code in or next to the words you want to include in the index.

Tip #281 from *Tux Tips*

If you place the index mark code *within* the term to be included in the index, you'll be less likely to accidentally leave that code behind if you later decide to remove the term from the document.

The second way to mark indexed terms is to use a *concordance* file. Such a file contains a list of words that you want to include in the index. When you tell WordPerfect to use this file, the software searches through your document and includes in the index the words that match the ones in the concordance file.

Marking text manually has the advantage of allowing you to catch minor variant terms, such as plurals, and placing them under a single index entry. Using a concordance file is typically much faster, however, particularly on large documents and if you already have a concordance file, say, for a similar document you've already indexed. You can get some of the advantages (and drawbacks) of both methods by using a concordance file to do a quick pass and then manually searching for and marking additional items.

PLACING INDIVIDUAL MARKS

To mark a word or phrase as a standalone index entry, follow these steps:

1. Locate and highlight the word or phrase you want to include in the index.
2. Click in the Heading text entry field of the index toolbar. WordPerfect places the word or phrase you highlighted there.
3. If necessary, edit the word or phrase in the Heading text entry field. For instance, you might want to convert a plural form to the singular or remove capitalization.
4. Click the Mark button in the index toolbar. WordPerfect places an [Index] code in the text.

Alternatively, you can place the insertion point where you want the [Index] code to appear and type in the heading or select it from the list that appears when you click the option button to the right of the Heading text entry field in the index toolbar. The key point is that WordPerfect enters the contents of the Heading field as the word to be indexed; it does not use the text in the document. This feature allows you to index synonymous words, both singular and plural forms, and other variants in the same index entry.

If you want to create an index entry with subheadings, you can enter text into the Subheading text entry field just as you do the Heading field. For instance, if you're writing a bread cookbook, you might want to index the various recipes that use specific ingredients, such as

Pesto
Apple Pesto Bread (38)
Spicy Wheat Bread (23)

You can accomplish this effect by creating two index marks, both with `Pesto` in the Heading field. One mark will have `Apple Pesto Bread` in the Subheading field, and the other will have `Spicy Wheat Bread` there.

Tip #282 from *Tux Tips*

If you want page numbers on the main heading (`Pesto` in the example), you need to create index entries for that Heading with blank Subheadings, one for each instance you want to index. If all the [Index] codes you place for a given Heading contain Subheading text, the main heading in the index will not have any page numbers associated with it.

CREATING A CONCORDANCE FILE

A concordance file is an ordinary WordPerfect document file. Each line contains one word or phrase that you want to be indexed. So to create a concordance file to index the words *Free Software Foundation* and *Linux*, you would open a new WordPerfect document window, type `Free Software Foundation`, press the Enter key, and then type `Linux`. Then save the file under any filename you like.

When you define your index, you tell WordPerfect where to look for the concordance file, and WordPerfect adds all the words it finds in that file to the index. If you like, you can specify a file that's in a format other than a WordPerfect document file, as long as WordPerfect can load that file. For instance, you can use an ASCII text file. If you do so, however, WordPerfect asks for confirmation of the file's type so that it can import the file when you generate the index. If you want to use such a file, you might do well to convert it to WordPerfect format before specifying the file as a concordance file.

→ For information on importing foreign files, **see** "Reading Foreign Files," **p. 88**

WordPerfect must alphabetize the entries in the concordance file before using them. Therefore, if your concordance file is large, you might want to sort it before you use it to make the generation go more quickly. You won't save any time for just one generation, but if you generate your lists and references more than once, you will save time this way.

→ To learn how to sort a concordance file (or any file), **see** "Sorting Records," **p. 544**

When you use a concordance file, WordPerfect does *not* mark the individual words from the concordance file in your document. You can therefore remove those words from your index by removing the concordance file from the index definition and generating the index again. If you've placed an index mark manually and added a word to the concordance file, WordPerfect creates one entry for the word showing all its occurrences, plus the mark location if it's on a page that doesn't contain the literal word. You can use this feature to manually add index entries for synonyms of words in the concordance file.

DEFINING AN INDEX

As with other list types, you must define your index before you can generate it. To do so, you must first position your insertion point where you want the index to appear and then select Define from the index toolbar. WordPerfect responds by displaying the Define Index dialog box (Figure 24.25), which works much like its counterparts for other list types.

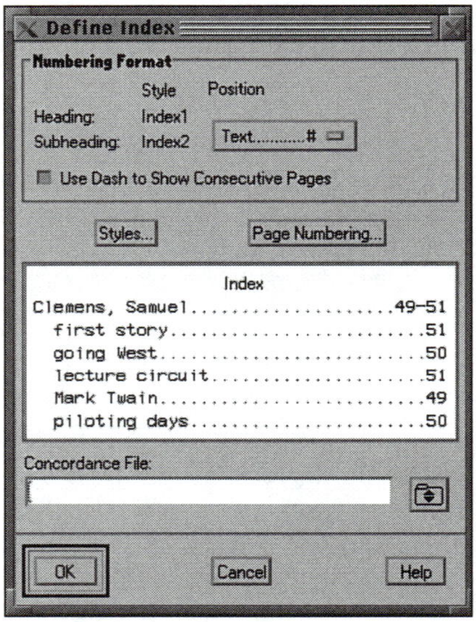

Figure 24.25
The sample index in the Define Index dialog box shows a main heading with page numbers, but a main heading only shows page numbers if you've marked a heading with no defined subheading in the text.

Some key points and differences between the Define Index dialog box and its counterparts include the following:

- You set the page number position and leader options for both headings and subheadings with the one Position option button in the Numbering Format area.
- You can select the Use Dash to Show Consecutive Pages check button to reduce the length of some index entries.
- If you click Page Numbering, WordPerfect displays the Page Number Format dialog box (refer to Figure 24.4) with which you can adjust the details of how WordPerfect displays the page numbers.

- If you click the Styles button, WordPerfect presents the Index Styles dialog box shown in Figure 24.26. You can use this dialog box to select the styles to be used for headings and subheadings. As with similar dialog boxes for other list types, you can use this one to create a new or edit an existing style, as well.

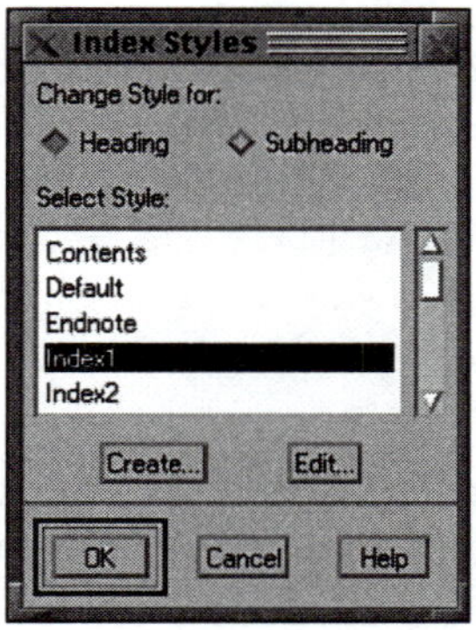

Figure 24.26
When you click Heading or Subheading, WordPerfect displays the style used in the Select Style area. You can change the style by using that list, as well.

- You can specify the concordance file by typing its name in the Concordance File text entry field of the Define Index dialog box or by clicking the folder icon to its right and using a WordPerfect File Manager dialog box to locate the file.

When you click OK in the Define Index dialog box, WordPerfect places a [Def Mark] code and a matched pair of [Gen Txt] codes, along with some text between the [Gen Txt] codes showing where it will create the index. As with other list types, you can place formatting and text around these codes to make a suitable context for the index. For instance, you might place an [HPg] code before these codes to put the index on its own page and add a suitable centered label. Be sure to add such formatting *outside* of the [Gen Txt] codes because WordPerfect deletes any text between these codes when it generates the index.

GENERATING AN INDEX

To generate an index, click Generate in the index toolbar or choose Tools, Reference, Generate (Ctrl+F9) from the document's menu. WordPerfect responds by displaying the Generate dialog box (refer to Figure 24.9), just as with other list and reference generations, and you can click OK to generate them all. WordPerfect automatically alphabetizes the items in the index.

When the generation is complete, you can edit the text within the index to suit your needs. For instance, you might want to capitalize some terms or remove capitalization from others.

CROSS-REFERENCES

Cross-references, unlike the topics I've covered so far in this chapter, don't produce entire lists of page number references. Instead, you use a cross-reference to refer to an item or page number in the normal flow of text. For instance, you can use a cross-reference to refer to a figure in your text or to the location of a definition elsewhere in your document. When you use this feature, you don't need to concern yourself with the numbering of your figures or with how such page numbers will change as you reformat your document because WordPerfect keeps track of those details. WordPerfect calls the location to which you're referring the *target*, and the location where you make the referral is the *reference*. You can mark either the targets or the references first, or you can mix them up as you enter your text. Just be sure you're consistent in labeling these elements and that a target exists for every reference.

To use the cross-reference features, you must first choose Tools, Reference, Cross-Reference, which produces the cross-reference toolbar shown in Figure 24.27.

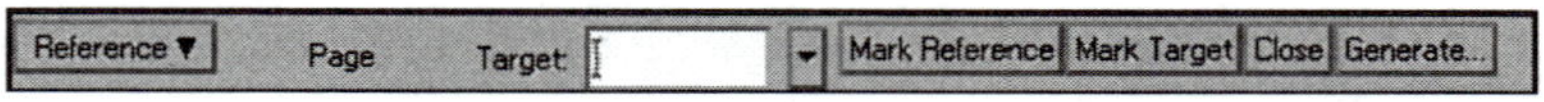

Figure 24.27
Control the target and reference placement with the cross-reference toolbar.

> **Note**
>
> Cross-references can be either *forward* or *backward,* depending on whether the target comes after or before the reference, respectively. WordPerfect allows both types, and you define them in precisely the same way.

MARKING TARGETS

To mark a target, follow these steps:

1. Place your insertion point in the target text.
2. Type a name for the target item in the Target field of the cross-reference toolbar. If you want to mark more than one location as a target for the same reference (say, if you discuss related information in several places), you can select the target name by clicking the option button to the right of the Target field and selecting from the options.
3. Click Mark Target in the cross-reference toolbar.

WordPerfect places a [Target] code at the insertion point's position. When you later refer to that target by page number, WordPerfect uses the page number on which the [Target] code appears.

If you want to refer to a boxed item such as a figure or equation, you should place your insertion point on the same page as the boxed item, after the [Box] code but before any subsequent [Box] codes. The best way to do so is to use Reveal Codes (View, Reveal Codes or Shift+Ctrl+F3) and, if necessary, WordPerfect's Find feature (Edit, Find and Replace or Ctrl+F2) to locate the box's code. You can then place the [Target] code immediately after the [Box] code.

→ For information on boxed elements, **see** "Complex Formatting with Boxes for Text, Graphics, and Equations," **p. 355**

→ For information on using Reveal Codes, **see** "Reveal Codes Mode," **p. 63**

→ For information on finding codes, **see** "Searching for Codes and Symbols," **p. 108**

If you want to refer to text in a footnote or endnote or to such a note by number, you should place the [Target] code inside the note's text itself.

You don't need to concern yourself with the nature of the reference when you're marking your targets. For instance, it doesn't matter whether you want to refer to a figure by its page number or its figure number at this point. You just need to place the [Target] code, which WordPerfect uses to extract the reference information you specify later.

MARKING REFERENCES

To mark a reference, follow these steps:

1. Enter any text you need to introduce the reference. For instance, you might want your text to read "this is discussed further on page 26." If so, you'd type `this is discussed further on page` and a space.
2. Type in the target name, using the Target field in the cross-reference toolbar, or select it by using the option button to the right of that field.
3. Select the type of reference from the list in the Reference option button in the cross-reference toolbar. Some of these items (such as Page) result in the insertion of a plain number, whereas others (such as Caption Number) produce more identifying text (such as "Figure *n*" for a figure caption, where *n* is the number).
4. Click Mark Reference in the cross-reference toolbar. WordPerfect inserts a [Ref] code at the insertion point. This code indicates what type of reference it is, as in [Ref Pg] for a page number reference or [Ref Box] for a box caption reference. Initially, the reference displays in the document as a question mark (?) character, but WordPerfect expands this notation when you generate the references.

You can enter more than one reference to a single target. For instance, you might want your text to read "see Figure 4 on page 23." You'd do this by entering one [Ref Box] code in place of "Figure 4" and a [Ref Pg] code in place of "23." Figure 24.28 illustrates these codes before generating the references.

Figure 24.28
When you first enter references, they have undefined values; generating is the step that defines the references.

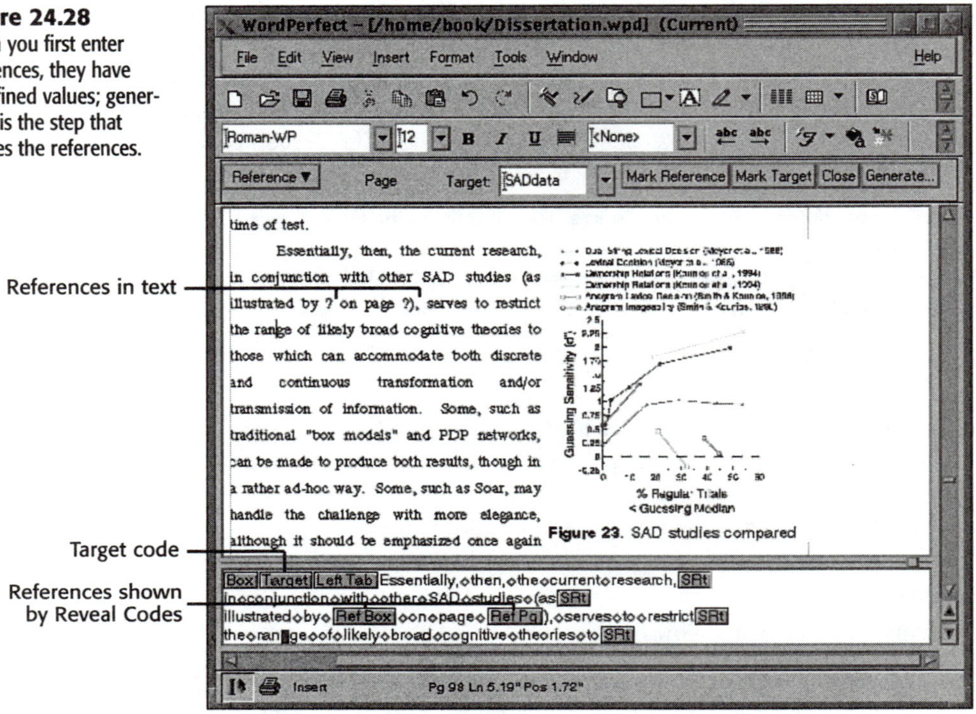

Caution

Be sure that all your references have targets. If a reference lacks a target, WordPerfect silently fails to create reference text, instead leaving the reference text as a question mark (?) in the text. Because WordPerfect doesn't display an error message, you may not realize what has happened unless you carefully proofread your document.

GENERATING REFERENCES

You generate cross-references in the same way you generate lists: by clicking Generate in the cross-reference toolbar or by choosing Tools, Reference, Generate (Ctrl+F9). WordPerfect presents the Generate dialog box (Figure 24.9), and you can click OK to have it generate all your references and lists. At this time, WordPerfect replaces the "?" placeholder for the reference with the number and, if applicable, text. Figure 24.29 illustrates the effect of generating references; it shows the same text as Figure 24.28, only after a generation operation.

Figure 24.29
The generated text sometimes acquires attributes from its source, for example, the bold attribute for "Figure 23."

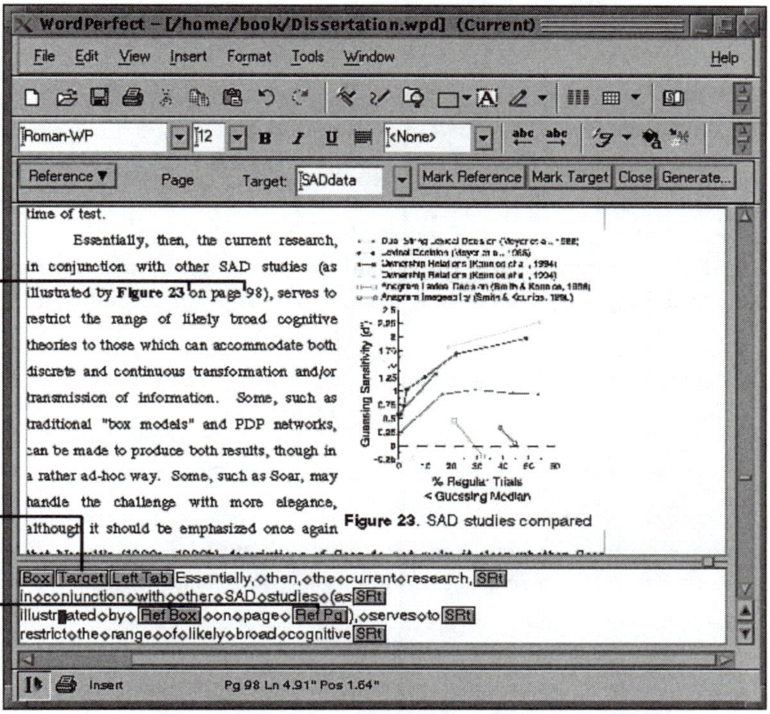

Tip #283 from *Tux Tips*

If you want to generate references and see the change to the one you just created, you may want to place a QuickMark by pressing Ctrl+Shift+Q. When WordPerfect generates lists and references, it places your insertion point at the end of the last item it generates. If this item isn't the one you want to view, you can then press Ctrl+Q to jump back to the QuickMark location.

→ To learn more about the QuickMark, **see** "The QuickMark," **p. 596**

Have you generated references but found that they include text attributes or other formatting you don't want? See "Generated References Contain Undesirable Formatting" in the "Troubleshooting" section at the end of this chapter.

TROUBLESHOOTING

MISSING FULL FORMS

You've created entries for a table of authorities, but when you try to generate your table, you get a message complaining that there are short forms for which no matching full forms have been found.

There are two possible causes of this error. One is that you've accidentally deleted the full form in editing your text. Try using WordPerfect's Find feature (Edit, Find and Replace or Ctrl+F2) to locate the text that would contain either full or short forms for the offending entry and replace the first short form with an appropriate full form.

It's also possible that you mistyped a short form name, in which case you can locate and delete the "typo" [ToA] code and replace it with a [ToA] code using the correct short form name.

GENERATED REFERENCES CONTAIN UNDESIRABLE FORMATTING

You've generated references, particularly references to boxed items' captions, and find that the in-text references contain formatting used in the source material. For instance, Figure 24.29 shows that the reference to the figure in the document takes on the boldface attribute from the figure's caption.

Here are two solutions to this problem:

- Eliminate the undesirable formatting from the source material. In the case of box captions, you must edit the style in which this formatting appears, for instance. This solution may be awkward if you want to apply this formatting in the caption, but not in the text, however, because you have to apply the formatting manually in the box captions.

→ To learn how to modify caption formatting styles, **see** "Caption Styles," **p. 467**

- Place the [Target] code for the boxed item *before* the [Box] code, rather than after it, and use the Counter reference type when you mark the reference. WordPerfect displays the Counter dialog box (Figure 24.30) from which you can select the counter type you want to use, such as .Figure Box for a figure counter. Because these [Ref Count] codes produce only numbers in the final generated text, you need to precede the codes with appropriate identifying text (such as Figure).

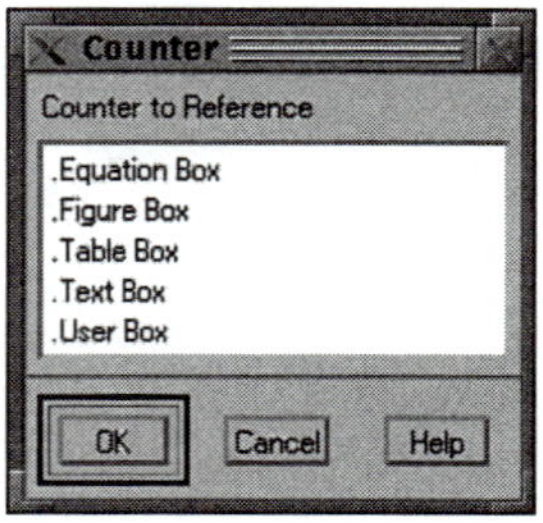

Figure 24.30
You can insert a reference to any of several counters from the Counter dialog box.

PART VII

WEB PUBLISHING

CHAPTER 25

An Introduction to the Web

In this chapter

AN INTRODUCTION TO THE WEB

Writing is a means of communication that's been used by humans for thousands of years. Through much of that time, the written word has appeared primarily on paper or on paper-like materials. Empires have come and gone, languages have mutated, and even the alphabets we use have changed over the centuries, but the utility of words on paper has remained steady, and even increased with the rise in literacy rates in the last few centuries. Computers, however, are now nipping at the heels of the time-tested medium of words on paper, displacing the words to computer screens. Fortunately, you can use WordPerfect's many capabilities to help bridge the gap and produce electronic documents of various forms. Chapter 6, "Producing Output," covers some of these electronic media, including the use of email, PDF (Adobe Acrobat) files, and fax output. This chapter and the next cover one of the most recent new platforms for the written word: the World Wide Web (WWW, or just "the Web" for short).

→ If you want to learn more about producing both paper and electronic output, **see** "Producing Output," **p. 147**

This chapter covers what might be called the theory of the Web, including its basic principles and the unique challenges and opportunities in designing a Web-based document. The chapter also discusses importing existing Web *pages* (that is, its basic documents) and exporting existing WordPerfect documents as Web pages. Chapter 26, "Designing a New Web Page," discusses using WordPerfect to design a new Web page or to edit an existing document to make better use of the Web's features. The present chapter also provides an overview of WordPerfect's Web page editing features. The next chapter goes into greater detail about how to use WordPerfect to create and edit Web pages.

UNDERSTANDING WEB PAGES

The Web is a medium that's new and that is rapidly changing. It's based on several key principles that don't apply to paper-and-ink documents, and many Web page designers fail to recognize these principles and their implications. Before you embark upon a Web page design journey, you should understand something about the medium with which you'll be working.

WEB SERVERS AND CLIENTS

The Web is a high-tech medium that requires the cooperation of several computers for one person to view just one Web page on a remote site. Of most concern to you are two computers and especially the software they run for handling Web pages: the *server* and the *client*. A server is a computer that makes Web pages available to other computers on the Internet. To serve Web pages, a server must run some special software, such as the *Apache* program that comes with many Linux distributions. The server must also have a reliable connection to the Internet. If left to its own devices, the server machine just sits there. It doesn't actively distribute its Web pages; it waits until it receives a request from another computer before doing anything.

That other computer is known as a *client* computer. It could be your own Linux box, for instance, and the requests it sends normally go out at your direct command. The software you need to use a computer as a client on the Web is known as a *Web browser* (or just *browser* for short), and you're probably already familiar with these. On Linux, the programs Netscape, Mosaic, and Lynx are three popular Web browsers. To locate a server with files you want to read, you need the server's Web *address*, known technically as the *uniform resource locator* or *URL*. The URL is the Internet machine name, typically preceded by `http:` and often followed by some directory and filename, such as `http://linux.corel.com` for Corel's main Linux site. You can either type a URL into a Web browser (for instance, in the Location field in Netscape, as shown in Figure 25.1) or click on a *link* in a Web page, which causes the browser to grab the URL from the Web page itself.

Figure 25.1
Netscape for Linux is one of several Web browsers available for Linux.

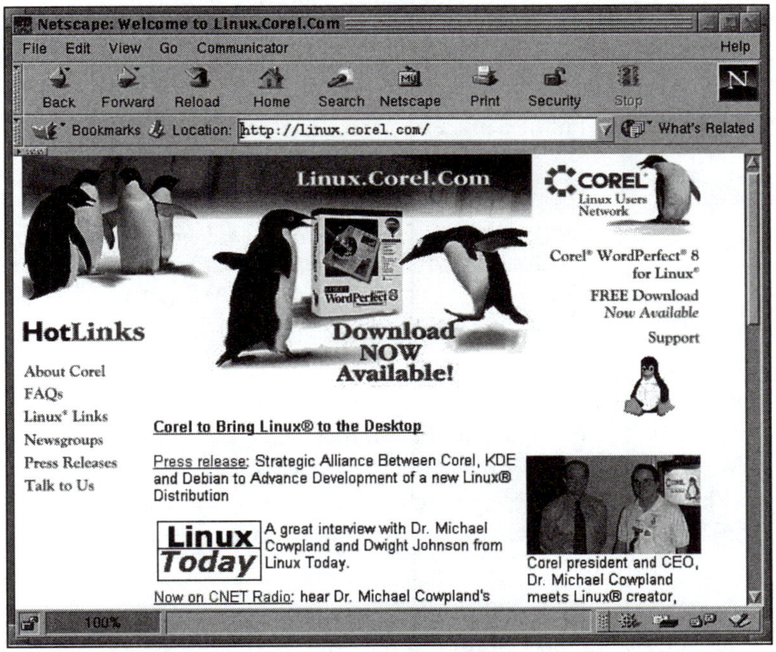

One of the most important things to understand about Web pages is that different Web servers, and most especially different Web browsers, have different capabilities and may respond differently to the same Web page. Even a single browser can work differently depending upon how it's configured. When you design a Web page, you need to consider this fact or risk losing a potentially large portion of your audience.

HTML: The Currency of the Web

At the moment, the Web is built mostly around a document format called *HTML*, which stands for *hypertext markup language*. You may occasionally hear people refer to *programming HTML*, but this phrase is inaccurate. HTML is not a computer programming language like C, Pascal, or WordPerfect's macro facility. It is, rather, a document format, in principle similar to a WordPerfect document file. In fact, if you're familiar with WordPerfect's Reveal Codes feature (View, Reveal Codes or Shift+Ctrl+F3 from a WordPerfect document window), you'll understand many of the key features of HTML, because it's based on *tags* that work very much like WordPerfect's document codes. For instance, when you italicize a word in WordPerfect, it places [Ital] codes around the word. In HTML, you'd use an `<EM>` code (short for *emphasize*) before an italicized word and an `</EM>` code after it. HTML is, really, just ordinary ASCII text and HTML tags, most of which reside within angle brackets (<>).

→ If you want to learn more about Reveal Codes, **see** "Reveal Codes Mode," **p. 63**

Typically, HTML files have filename extensions of .htm or .html. Some Web servers use special scripts to generate HTML files "on-the-fly," and these pages may bear other extensions, such as .cgi, though when you read them in a Web browser, you're just reading ordinary HTML. When you create a Web page with WordPerfect, you'll probably want to give it an .htm or .html filename extension.

Note

Web pages may refer to non-HTML documents. For instance, a Web page might contain links to PDF files, sound files, compressed programs, and so on. Depending on how your Web browser is configured, it may be able to handle and display these files directly, download the file to your hard disk, or pass the file on to another program. Thus, HTML is not the *only* document type on the Web, but it's the one that holds all the rest of it together.

In the future, HTML may be supplanted by other document types, such as the up-and-coming XML. When and if HTML is replaced as the dominant file format on the Web, WordPerfect will have to adapt, or you may eventually have to use another tool to create Web pages. Chances are, though, that HTML will remain viable for years to come, even if XML or some other format quickly dominates the Web.

Because HTML is a simple document format, it's possible for a human to directly generate HTML code using nothing but a text editor and an HTML reference. In addition, many Web browsers allow you to view the original HTML for a Web page (for instance, Netscape's View, Page Source menu option). If you'd like to create HTML directly, you can use existing Web pages as a guide to creating your own. WordPerfect does support HTML creation, however, much as it supports exporting files to Microsoft Word or ASCII text formats. You can therefore create a Web page quickly by using WordPerfect. Alternatively, you can create the Web page with WordPerfect and then modify the page by hand to better achieve the results you desire, using an ASCII text editor (WordPerfect can serve this function, too).

Web Document Structure

A typical Web document consists of several individual pages, each of which is a single HTML file. A reader can navigate between these pages using links that you provide. In addition, each page may have links to other Web documents entirely, to downloadable files, and to graphics. Typically, a graphical Web browser displays a graphic when it loads the first page, though there are text-only Web browsers, and many graphical browsers allow the reader to disable the graphics loading.

Note

The term *Web page* can apply either to an individual HTML file and its associated graphics or to a group of HTML files and their graphics. I use *Web page* in the former way in this chapter and the next, using *document* or *site* instead of *page* for an entire collection of HTML files.

Note

The term *Web page* shouldn't be taken to imply anything about the size of the page. Unlike a printed page, which is limited in size by the paper, a Web page can be any size, from a single line to thousands of lines.

Consult with your *Internet service provider (ISP)* or with whomever is in charge of your Web site to determine how to place files on it and what to call them (some Web sites limit you to DOS-style 8.3 filenames, and all have a special name reserved for your main Web page, such as index.html). By default, WordPerfect creates a subdirectory for graphics you place on each Web page. You need to create an identically named subdirectory on the computer that hosts your Web page, or override WordPerfect's default, when you place your Web document on the Web server.

Note

Linux can function as a Web server, and if you're using WordPerfect on the computer that's playing this role, you can create your Web documents in the directory that hosts your Web site.

Using Links Effectively

Much of the power of the Web comes in the form of links. As I mentioned earlier, links allow a Web page to integrate seamlessly with other documents that are hosted on machines halfway around the globe. Links are also useful for structuring your own document. If you have a document of any size that you want to present on the Web, chances are you'll want to break it up into several smaller pages. Such a breakup has two primary benefits:

- It can help you to structure your document so that your reader can quickly locate the information he or she needs. For instance, Figure 25.2 shows a Web page with several links in a table of contents, as well as a section showing downloadable programs as links.

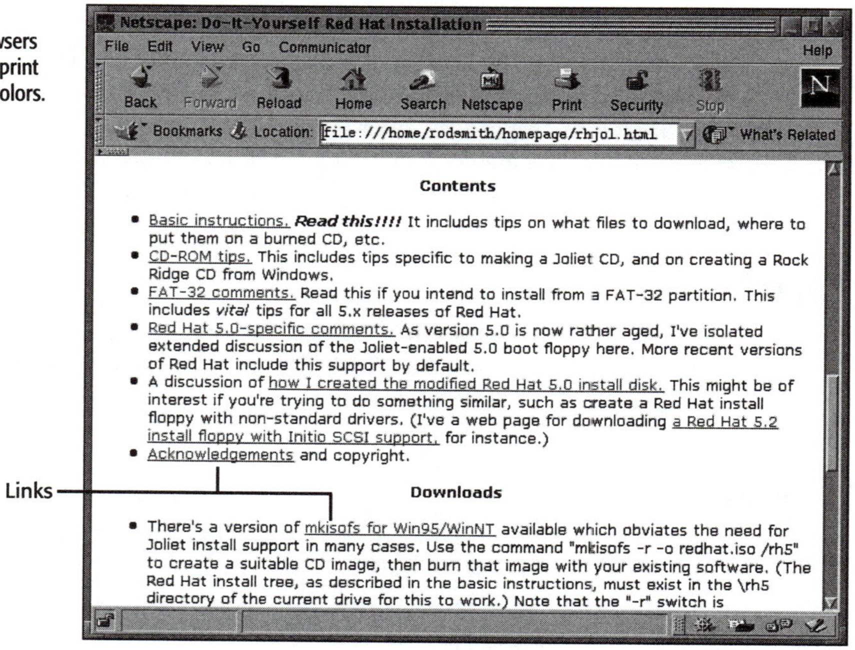

Figure 25.2
Most GUI Web browsers underline links and print them in distinctive colors.

- It can reduce the size of files your reader must download. When you create your Web page on your own system, it's easy to forget how large it is. A reader who tries to access a 50,000-word tome, however, is likely to run into speed problems in accessing such a big page. (Remember that the Web page text is a file that must be downloaded, just like any other file.) Breaking a Web page into chunks and linking them together appropriately can make downloading faster.

As a general rule of thumb, you should structure your Web documents so that no individual page contains more than a few screens of text. More than that, and the page starts to become intimidating to a casual reader. Of course, you may need to violate this rule of thumb—say, for a large table or for a particularly complex topic.

In general, you'll probably want to use an introductory Web page containing a table of contents and perhaps some introductory text. Each element in the table of contents will contain a section of the text—say, the contents of a Level 1 heading in a document outline. If your document is particularly large, you might want to break down each of these Level 1 pages in a similar way.

→ f you want to learn more about creating outlines, **see** "Organizing a Document with Outlining," **p. 553**

Each Web page should contain a link to the previous page and to the main page of your document. Including links to your main page on all document pages makes it easier for your reader to navigate through your document. Most Web browsers do include buttons to go to the previous page, but given the nature of the Web, your reader might have come to some subpage of your document directly from some other site. If so, your reader's "back" button won't find your document's main page.

Note

Keep in mind that the Web allows others to link to *any* page of your document. This fact can be troublesome at times because of the need for additional links within your own document. One more subtle issue is that a reader linking to a particular page of your document may not understand the context of the document as a whole. For instance, if you're creating a Web document devoted to sound card support in Linux, a reader linking to a particular page of that document may not realize it's for Linux if you don't make that information explicit on each and every page. You may then get swamped with requests for information on Windows drivers for sound cards.

You can, of course, create links between the pages of your document, not just from a table of contents page to its subpages. For instance, you might have a Web document devoted to cameras, with individual pages devoted to 35mm SLR models, 35mm point-and-shoot cameras, and APS models. Suppose you discuss the various limitations of 35mm point-and-shoot cameras in its page. A reader might encounter this discussion, realize that the limitations are too restricting, and want information on something without those limitations. You can then provide a link directly from that page to the 35mm SLR page, thus making your reader's life easier. In addition, you can link two or more pages to a single subpage. For instance, you could create a page that contains a table summarizing the differences between the various camera types and link to it from each of the three camera type pages. Figure 25.3 illustrates the Web document structure I've just described.

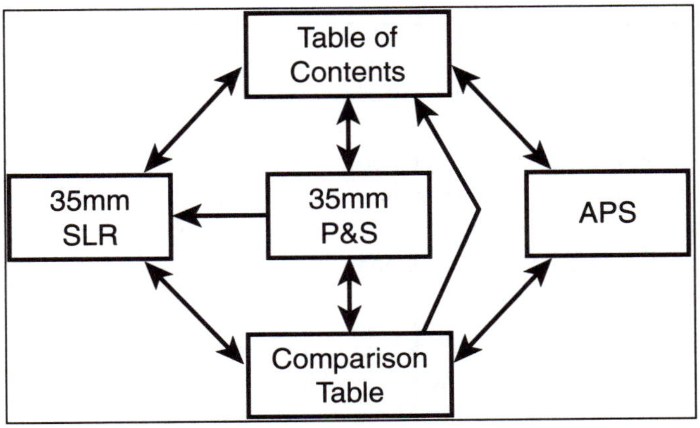

Figure 25.3
A Web page can link to another page that doesn't have a direct reciprocal link, as in the links from the 35mm P&S page to the 35mm SLR page, or from the Comparison Table page to the Table of Contents page.

In addition to linking the pages of your own document, you can create links to documents outside of your own—that is, after all, half the point of the Web. You'll typically scatter these links throughout your document, but you can cluster them within a small area if you like.

Tip #283 from *Tux Tips*

> Keep in mind that the Web is a very dynamic structure. Web sites can and do change their organization. Therefore, your links will be safest if you link to the highest-level page that's practical. For instance, if you wanted to create a link to the camera Web site whose structure is outlined in Figure 25.3, you'd be safest to link to the Table of Contents page, even if the context of your reference concerns only APS cameras. That way, if the site's maintainer renames the APS page or breaks it down into separate subpages, your link will still work.

If you check back to Figure 25.2, you'll see that the links appear as ordinary text, not URLs. The URLs are "hidden" in the HTML document, just as are the codes for creating the bulleted lists, centering, and other formatting. You can also create a graphic that serves as a link. I describe the mechanics of using WordPerfect to place a URL in a Web page in Chapter 26.

→ If you want to learn more about how to create links, **see** "Managing Links," **p. 689**

Browser Capabilities

One of the most important things to remember when you design your Web page is that not all Web browsers have the same features. The latest versions of Netscape Navigator and Microsoft Internet Explorer generally contain similar sets of sophisticated features. These currently include support for style sheets, fonts, security, and so on. WordPerfect itself doesn't support some of these features. If you want to use a feature that WordPerfect *does* support, though, you should try to create a design that will *degrade gracefully*—that is, that will still work when your reader is using a browser that doesn't support that feature.

Let's take fonts as an example. Web pages often specify the fonts to be used on the screen. You can specify the fonts with WordPerfect by selecting fonts much as you do in a conventional document—by choosing Format, Font (F9) and picking the font you want from the Font dialog box (Figure 25.4). Not everybody has a Web browser that can accept the HTML font change codes, however. Such browsers should ignore the font change codes, so you don't need to be concerned with their doing strange things in response to the font change code. Even if a browser supports font change codes, the system on which it's running might not have the appropriate font installed, in which case the browser substitutes something else. You should therefore be sure that your Web page doesn't rely upon the font in some way. For instance, you shouldn't format text and assume that your reader will see lines of the same length as you see.

Tables are another feature that can cause problems. Most browsers today do support tables, but a few older ones don't. These browsers may do unpredictable things when faced with tables. For instance, Figure 25.5 shows the same Web page displayed in Netscape Navigator with table support and in NCSA Mosaic with table support disabled. In this particular case, Mosaic displays all the table's text, and because the table is simple, it's still understandable; a more complex table might become unintelligible. Fortunately, most browsers today do support tables, so if you need to use tables, you needn't be scared away by these problems; however, you shouldn't use tables unnecessarily.

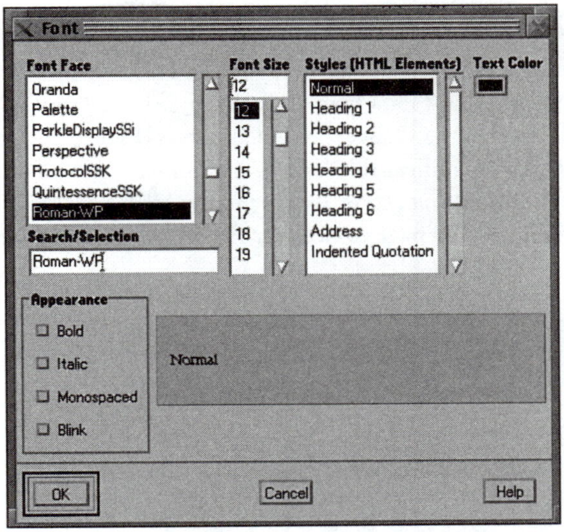

Figure 25.4
You specify fonts in a Web page much as you do in ordinary WordPerfect documents.

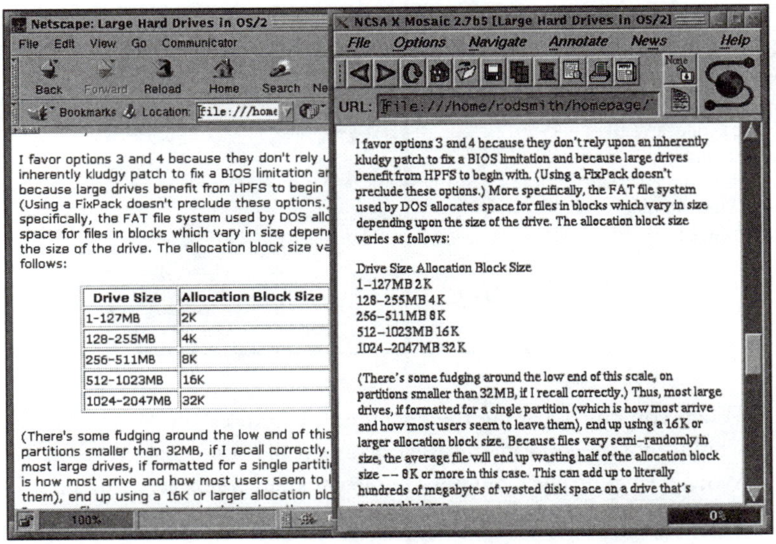

Figure 25.5
When a browser doesn't support tables (NCSA Mosaic, on the right), it may make the text in a table difficult or impossible to interpret.

Tip #284 from *Tux Tips*

You can create tablelike formatting by specifying monospaced text and formatting the text "by hand." You can also create alternative versions of a page that uses tables or some other feature that's not universally supported. If you create links between the pages, the reader can decide which version to use.

→ If you want to learn more about how to use monospaced text in Web pages, **see** "Formatting for the Web," **p. 700**

Importing Existing Web Pages

If you want to modify an existing Web page, you can load it into WordPerfect just like any other document, by choosing File, Open and selecting the file in the Open dialog box. WordPerfect then presents the Convert File dialog box (Figure 25.6) to confirm that the file is a Web page. When you confirm this fact, WordPerfect, oddly enough, warns you that you're entering Web formatting mode, in which only Web features work (Figure 25.7). Because you're loading a Web page to begin with, you presumably want to click Yes.

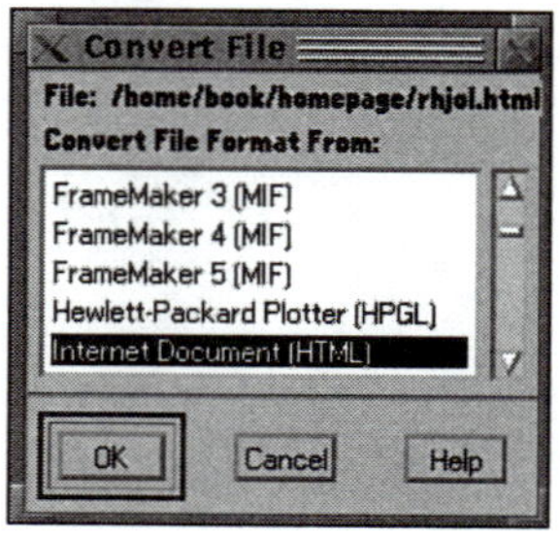

Figure 25.6
When you import a Web page, WordPerfect refers to the page as an Internet Document (HTML).

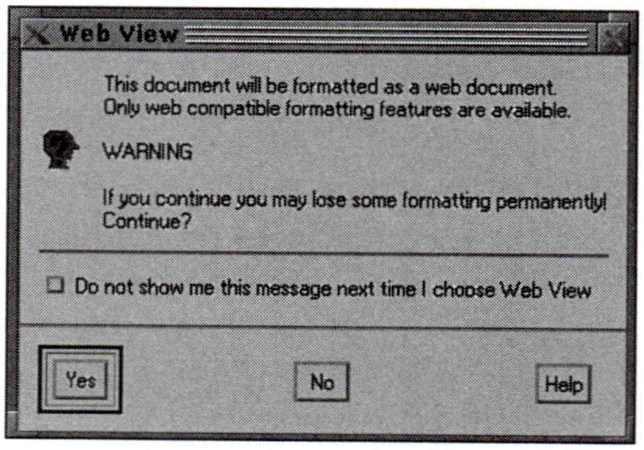

Figure 25.7
If you select the Do not show me this message next time I choose Web View check button, WordPerfect won't display the Web View warning dialog box again.

When you finish loading the document, WordPerfect displays it much as a Web browser would, though you can edit the document like a normal WordPerfect document. WordPerfect displays modified toolbars when you're working on a Web page, as shown in Figure 25.8.

The tools available to you in Web page mode include

- Insert graphic—This option opens a File Manager dialog box in which you can select a graphic image to include in your Web page.

→ If you want to learn more about how to insert graphics in your Web page, **see** "Inserting Graphics," **p. 695**

- Insert text box—Use this option to create a text box.

→ If you want to learn more about creating and using text boxes, **see** "Using Text Boxes to Add Variety to Text Layout," **p. 357**

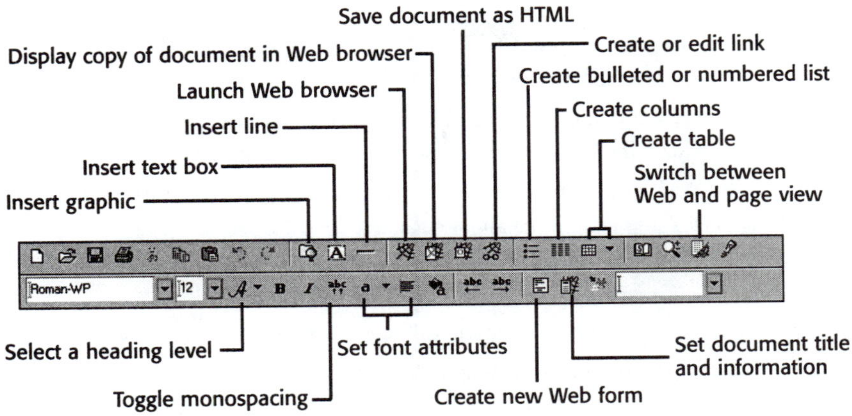

Figure 25.8
The Web page toolbars provide quick access to many of WordPerfect's Web page features.

Caution

WordPerfect converts any text boxes into graphics elements, and the text may be colored strangely and be a bit blurry in that form. I suggest avoiding text boxes in Web documents. If you must use one, consider replacing the graphic with one you create in an external graphics package, such as the GIMP.

- Insert line—This option creates a horizontal line in the Web document.

→ If you want to learn more about how to create and use lines in your document, **see** "Using Lines and Simple Shapes," **p. 380**

- Launch Web browser—This option starts the Web browser you've defined for WordPerfect. If you haven't defined one, WordPerfect instead displays the Browser Preferences dialog box (Figure 25.9). You can enter the path to your Web browser in the Web Browser area of this dialog box or search for the browser by clicking the folder icon. Common locations for a Web browser are /usr/bin/, /usr/local/bin/, and /usr/X11R6/bin/.

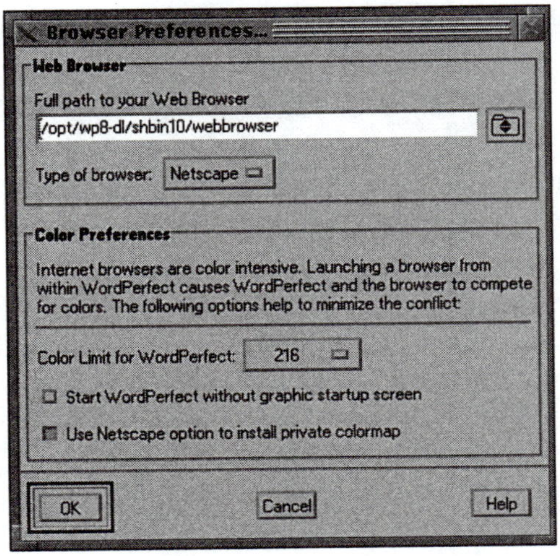

Figure 25.9
If you tell WordPerfect which Web browser you're using, WordPerfect can pass the browser parameters to help keep it from competing for colors with WordPerfect.

- Display copy of document in Web browser—When you select this option, WordPerfect creates a temporary copy of the document as a Web page and launches your Web browser to view it.
- Save document as HTML—Click this tool to save your document in HTML format. I describe this option in more detail shortly.
- Create or edit link—You can create and edit links by selecting this option.

➔ If you want to learn more about creating Web links, **see** "Managing Links," **p. 689**

- Create bulleted or numbered list—This option starts a new bulleted or numbered list, using whatever style of list you last used.

➔ If you want to learn more about lists, **see** "Creating a Bulleted or Numbered List," **p. 554**

- Switch between Web and Page view—If you've loaded a Web page with the intention of converting it to a paper document, you can click here to gain access to all of WordPerfect's formatting features. If you use non-Web features, however, you'll lose them if you subsequently save the document as a Web page.
- Select a heading level—You can select heading and list types from the drop-down list you get when you select this toolbar item. Different browsers may display these styles differently.
- Toggle monospacing—Because each Web browser may use different fonts, you can't be sure of which font your readers will see. You may occasionally want to use a *monospaced (p. 179)* font, however. For instance, you might use monospacing for samples from program listings. You can specify a monospaced font by clicking this toggle, much as you would select boldface or italic attributes. WordPerfect uses Courier as its monospaced font, but your readers may see another font. The monospaced attribute usually works even on Web browsers that don't support setting specific fonts.
- Set font attributes—You can set a number of font attributes, such as italics and superscript, by choosing them from the menu you get when you click this toolbar item.

Caution

"Blink" is one of the attributes you can select. Don't. Blinking text is almost universally regarded as obnoxious on Web pages.

- Create new Web form—If you've seen many Web pages, particularly on commercial sites, you've probably encountered Web forms. Forms let you enter information and submit it back to the originating site to place orders, participate in a survey, or register software, to name just three possible applications. If you click this tool, you'll be able to create a Web form of your own.

➔ If you want to learn more about Web forms, **see** "Creating Web Forms," **p. 702**

- Set document title and information—Web pages include special codes within them, such as titles and information for search engines. You can specify some of this information by clicking this tool, which opens the Title dialog box shown in Figure 25.10.

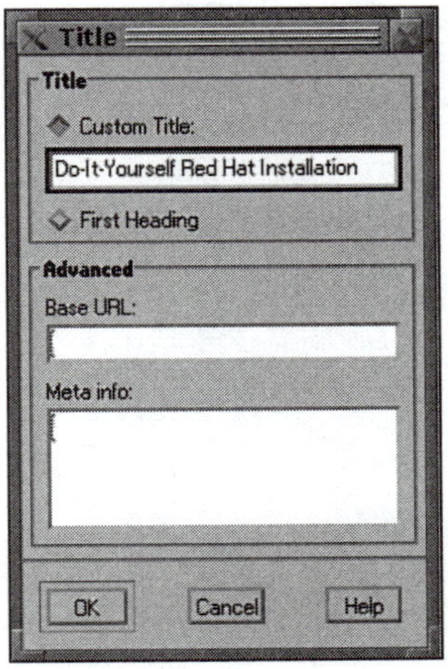

Figure 25.10
Most GUI browsers display the title of your Web page in their title bars.

In addition to the unique features related to Web pages, WordPerfect provides many of its normal formatting features when you're editing Web pages, such as font selections, columns, tables, and so on. You can also spell check your document, undo and redo changes, and otherwise use most of WordPerfect's editing features.

EXPORTING EXISTING DOCUMENTS AS WEB PAGES

If you have an existing document that you want to make available on the Web, you have several options. You could place the WordPerfect file directly on the Web as a link from some other Web page. This approach would allow others to download, print, and edit your file, but only if they have WordPerfect or can load WordPerfect files. Another option is to create a PDF file from the document and make that available. This approach makes for a more universally readable file but with strict formatting that may be inconvenient for some readers. If you want to take advantage of the Web's unique features, you'll want to convert the document into one or more Web pages. Fortunately, WordPerfect makes this conversion easy. Follow these steps:

1. Load the existing document.
2. Choose File, Internet Publisher. WordPerfect presents the Internet Publisher dialog box, shown in Figure 25.11.
3. Click Publish to HTML. WordPerfect displays the Publish to HTML dialog box (Figure 25.12).
4. Type in a filename for the Web page or click the folder icon to enter a filename using a File Manager dialog box; then click OK. WordPerfect exports a copy of the file as a Web page.

At this point, you can load up a Web browser and view the result. You can also load the Web page into a text editor or some other Web page editor if you prefer to handle your Web pages outside of WordPerfect.

Figure 25.11
You can select various Web-related functions from the Internet Publisher dialog box.

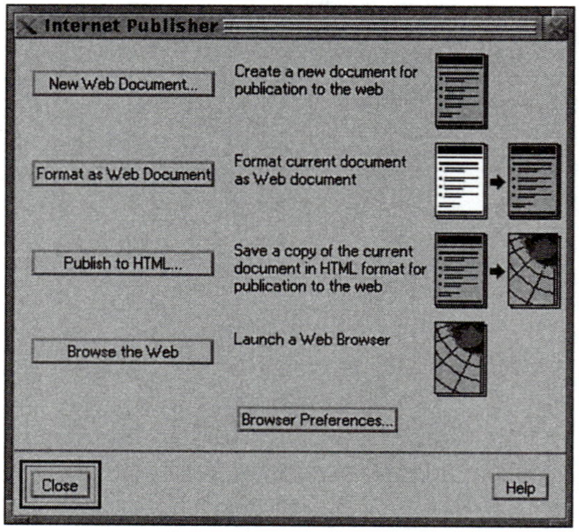

Figure 25.12
WordPerfect tells you where it will save graphics associated with the Web page when you elect to save the page.

Tip #285 from *Tux Tips*

You can break one existing document into several separate Web pages by deleting all the text except that which you want to appear in one page and then saving that page. Reload the original document and repeat this process, each time deleting a different section of text.

Tip #286 from *Tux Tips*

If you know when you create a conventional document that you'll also want to create a Web-based version of it, you may want to use WordPerfect's subdocuments feature. This approach allows you to create one continuous document for printing to paper while maintaining separate subdocuments to serve as the Web document.

➔ If you want to learn more about creating subdocuments, **see** "Master Documents and Subdocuments," **p. 598**

If you're converting a document to a Web page, chances are you'll want to modify it after the conversion. For instance, you might want to break it into multiple pages or tweak the formatting to better conform to Web standards. If so, you may want to convert the file, edit it, and then save it. To do so, you can select Format as Web Document rather than Publish to HTML from the Internet Publisher dialog box (refer to Figure 25.11) in step 3 above.

In response, WordPerfect displays the Web View dialog box (refer to Figure 25.7) as confirmation that you want to reformat the document as a Web page. You can then edit and save your document as a Web page.

Whether you loaded an existing Web page, created one from scratch, or converted an existing document to Web format, when it comes time to save the document, you should not use the usual File, Save (Ctrl+S) or File, Save As (F3) options. Instead, you should either use the Save document as HTML toolbar item or select File, Internet Publisher to get the Internet Publisher dialog box (refer to Figure 25.11) from which you can select Publish to HTML.

FORMATTING FOR THE WEB

Throughout history, the printed word has been *static*. After being committed to paper, words have stayed put; the only way to rearrange them or change their appearance has been to copy the words to another sheet of paper, typically by writing, typing, or using a printing press. Even the invention of the photocopier hasn't changed the static nature of text. The Web, however, does. When you create a Web document, *you don't know how it will look* to any given reader. As I discussed earlier in this chapter, different browsers support different features, and so you can't be sure that, say, the fonts you use will be available on the reader's system or that your reader will be able to read tables. You can't even assume that your reader will be able to see your graphics. There's nothing wrong with including advanced features in Web pages, but unless those features are vital to your page, such as graphics in a discussion of Picasso's paintings, you should design your Web pages so that they don't *rely upon* these formatting features.

The lack of knowledge about the state of the reader's display can be very unsettling to Web page designers who are used to designing layouts for the printed page, and many pages on the Web display a profound lack of understanding of this fact. To maximize the readability of your pages, the first rule, and the one from which all others derive, is to keep your design *simple*.

Tip #287 from *Tux Tips*

If you must have control over exact formatting, consider using the Adobe Acrobat PDF file format for distributing your file. You can create a simple Web page with a link to the PDF file and another to Adobe's site in case a reader doesn't have the PDF software. You can use WordPerfect, along with Ghostscript, to produce PDF files of your documents.

→ If you want to learn more about creating PDF files, **see** "PostScript and PDF Output," **p. 168**

DOCUMENT FORMATTING INCONSTANCY

Figure 25.5, in addition to showing the differences in features between Web browsers, demonstrates how different browsers can display the same Web page differently. This Web page didn't specify any specific fonts, and so each browser used the font that was set by the user—Microsoft's Verdana for the Netscape window on the left and Times for the Mosaic window on the right. The different fonts result in differences in line breaks and in other formatting details. I needn't have used two different browsers for this demonstration either; most modern browsers let you select the font you use by default, and any windowed browser lets you adjust the width and height of the browser, thus affecting the reader's view of the Web page.

Some Web pages claim to be designed for 800×600 displays. What this claim really means is that the designers assume you have your Web browser maximized on an 800×600 display. Although making assumptions about the reader's hardware and software can make your life easier if you're using a graphic for navigation or other sophisticated formatting, doing so is inconsiderate to the reader and may actually make your layout look bad on displays that are larger or smaller than the one you had in mind.

FONTS IN WEB PAGES

As I mentioned earlier, you can set fonts in a Web page just as you do in other WordPerfect documents, by choosing Format, Font (F9) in the WordPerfect document window. You can't assume that your font will be seen in the way you intend, however, because a user might not be using a browser that supports the HTML font codes or might have disabled font changes.

Note

When you create a Web page that uses specific fonts, you are *not* distributing the fonts in the Web page itself. You're simply telling the reader's browser to use the fonts of the name you specify if they're available on the reader's system. The only way to use text in a font that's not on the reader's system is to use a graphic that contains the text in the font you want, and this technique will slow the Web page's loading.

The availability of fonts on the target system is especially an issue if you're designing a page that you expect to be viewable from Windows or MacOS because Linux doesn't always use the same fonts as those OSs use. In general, your safest bets for fonts are, in this order:

1. No font specification—If you don't change your font, WordPerfect won't insert any HTML font change code and your reader will get his or her default fonts.

Note

WordPerfect normally displays Web pages without any specified font in its Roman-WP font.

2. Times, Helvetica, and Courier—These are present on most systems.

Note

You may need to set your WordPerfect printer driver to a PostScript model to gain access to these fonts. If you're using a PCL printer, you can use Times New Roman and Arial in place of Times and Helvetica. *Do not* use Roman-WP, Helve-WP, or Courier-WP.

3. The fonts available from Microsoft's Web site at `http://www.microsoft.com/typography/fontpack/default.htm`—Microsoft has made these freely available, and they're now installed with many Microsoft products, so many systems have them. You can even use these fonts with Netscape in Linux.

Note

Microsoft has made these fonts available only as TrueType fonts, but WordPerfect requires fonts in Type 1 format. You can convert them to Type 1 format, but WordPerfect may not display them as you'd expect. This limitation can make working with these fonts using WordPerfect awkward.

→ If you want to learn more about how to install fonts in WordPerfect, **see** "Installing and Managing Fonts," **p. 754**

4. Other fonts—Fonts you've installed from font CDs or other sources may work on a few systems, but you can't assume they'll work everywhere. This category includes the fonts that come with WordPerfect for Linux.

Have you specified fonts in WordPerfect but find that Netscape or some other browser isn't using them? See "Fonts Don't Show Up in Browser" in the "Troubleshooting" section at the end of this chapter.

When you specify a font, the worst thing that's likely to happen is that the reader's system will substitute an unviewable font. For instance, if you use a font with a strange name from a small font foundry and a *different* font with the same name exists, then a reader with the second font may get it when the Web page specifies the font change. A more likely scenario is that the font won't exist on the reader's system, and the browser will substitute a standard font, such as Times, Helvetica, or Courier. Your reader will probably consider this substitution a step down from his or her default font.

Rules for Using Graphics

The Web has become a very graphics-intensive medium, and you may want to include graphics in your Web pages. I discuss the actual mechanics of placing them in your Web page in Chapter 26. For now, though, I'd simply like to present some tips and rules of thumb for including them:

→ If you want to learn more about placing graphics in Web pages, **see** "Inserting Graphics," **p. 695**

- Make your graphics as small as possible. You can use an external image-editing program, such as the GIMP (`http://www.gimp.org`), to shrink the size of your graphics. You can reduce the size in bytes of any graphic by reducing its size in pixels. For GIF-format graphics, you can reduce the file size by reducing the number of colors used. For JPEG graphics, you can reduce the image "quality." Graphics that are too large can substantially increase the time it takes for a page to load.
- Use external JPEG or GIF graphics. WordPerfect's WPG-format graphics tend not to convert well to the GIF and JPEG formats that Web browsers support.

- Don't assume your graphic will appear exactly where you see it in the WordPerfect document window, either relative to text or relative to other graphics.
- Set the Alternate Text HTML attribute for the graphic. You set this attribute by right-clicking the graphic and choosing HTML Properties from the context menu. This action produces the HTML Properties dialog box (Figure 25.13) in which you can enter the Alternate Text. Browsers that don't support graphics, or which have graphics disabled, display the Alternate Text in place of the graphic.

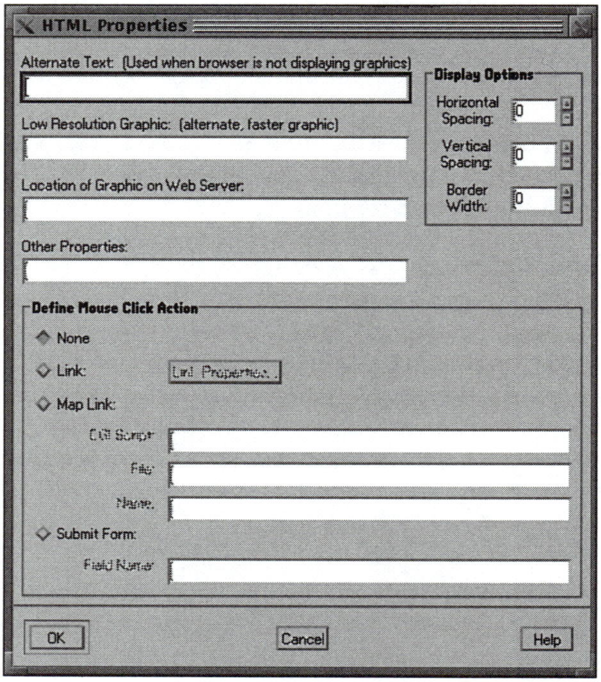

Figure 25.13
You can define how a browser will treat a graphics object by setting the items in the HTML Properties dialog box.

CONSIDERATE WEB PAGE DESIGN

When you create a Web page on one system, it's easy to forget that not everybody uses the same system as you do. It's also easy to get carried away with some of the Web's flashier features, such as graphics, to the detriment of accessibility. A Web page that uses graphics to produce fancy logos and menus may *look* good, but to somebody who has a slow Internet connection or who's visually impaired, such a design will be very frustrating. You can minimize frustrations by following some simple rules of thumb:

- Keep Web pages and graphics small. It's easy to lose sight of the time it takes to load your Web pages if you're testing them locally, or if you have a fast Internet connection. Try testing your Web page on a 28kbps modem connection. In general, you shouldn't have any pages or graphics that are larger than about 50KB.

- Don't use wide graphics or tables. You shouldn't assume that your reader has a Web browser that's wider than *any* amount. After all, there are Web browsers for palmtop computers, people running them on 640×480 displays, and so on.
- Use alternate text tags for graphics, as described earlier, and include text-only links if you must use a graphics map for links.
- If you want to use a background bitmap, use a background that's low in contrast but that affords high contrast to the text. For instance, you might use a background that's very light in color with black text. It's best not to use background bitmaps at all, however.
- Be cautious about assigning background colors and text colors. If you set a background color, always set a text color. Some people set their Web page backgrounds to white; but some readers set their browsers' *text* color to white, with a black background. If a Web page sets only a white background and relies on the default black text color, the reader will end up with white-on-white text. Link colors can also be washed out in this way. Keep in mind that some readers suffer from various forms of color blindness and may not be able to distinguish between some colors, such as red and green. Overall, it's safest not to set any text or background colors at all. Although most browsers provide ugly defaults, users can override these defaults to suit their own tastes.
- Avoid the use of *frames*. Frames are subdivisions in a Web page that allow you to present a constant set of options over several different Web pages on a site. Frames have several drawbacks, however, including taking up space, having different appearances in different browsers, not being universally supported in all browsers, and being confusing to many readers.
- Include contact information. Provide an email link so that readers can contact you with questions and suggestions about your site.

Figure 25.14 illustrates many of these Web page design principles, with both good (left) and bad (right) Web page design examples. Netscape displays both pages in identically sized (391×536) windows. The example of poor design is truly horrific. In addition to the inconsiderate design elements, it conveys little information and uses HTML features needlessly. For example, although you probably can't tell from Figure 25.14, given the page's other flaws, its table lists "went sightseeing" for most days of the itinerary, which is hardly an effective use of a table. This poor example also includes no email links or links to other pages on the author's site, so nobody can inform the author of his poor design (fortunately, this action is unnecessary, as I made up this example—you won't find it on the Web, though you may find others just as bad).

Of course, if taken to extremes, this advice can lead to some pretty dull-looking Web sites. In the end, you'll be faced with a trade-off between making a Web site that's accessible to all and one that has visual appeal to people with the most popular Web browsers and modern computer hardware. In my opinion, it's possible to design a Web page that's reasonably appealing and still accessible simply by avoiding the "latest and greatest" browser features and not making assumptions about the size of the fonts and browser window.

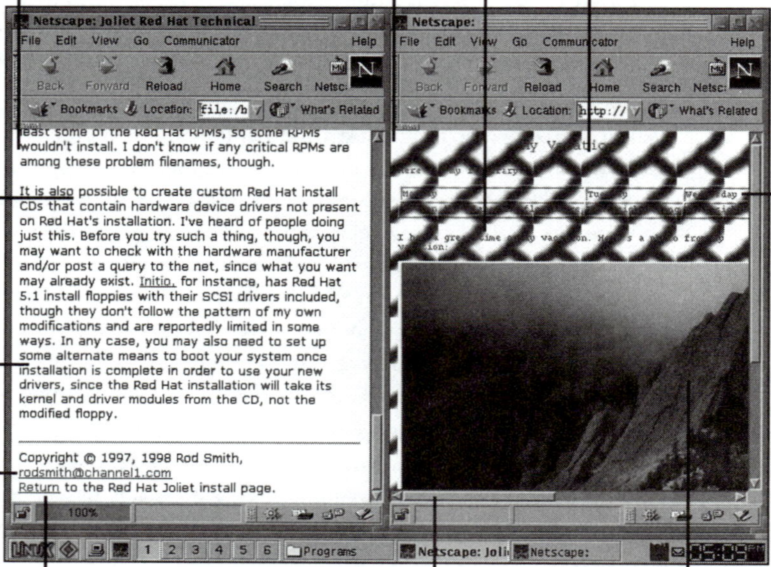

Figure 25.14
A well-designed Web page (left) is easy to read; a poorly-designed one (right) can be painful to contemplate.

If you'd like more information on creating the most accessible Web pages possible, check the Web site `http://www.anybrowser.org/campaign/`. This site hosts a great deal of information on making Web pages that can be read with any browser, along with useful links to other sites on Web page design.

CHECKING WEB PAGES

After you create a Web page, or several pages that constitute an entire Web document, you should check it. There are two main ways to check your Web pages: by using various browsers to view it and by using an HTML syntax checker.

VIEWING YOUR DOCUMENT IN BROWSERS

You can use the Display copy of document in Web browser tool in the WordPerfect toolbar (refer to Figure 25.8) to automatically launch your chosen Web browser from WordPerfect and check your Web page. This option is extremely useful, and you should definitely avail yourself of it. You shouldn't stop there, however. As I've said, Web browsers vary in their capabilities and features. You may run across problems with a page in some Web browsers but not in others. I therefore recommend checking your Web page in several browsers. Here are some specific suggestions:

- Lynx—The Lynx browser is a text-only browser that comes with most Linux distributions. If you want to check a Web page (let's call it *page.html*) with Lynx, you can start an xterm, go to the location where the page is stored, and type `lynx page.html`. You'll then get something like Figure 25.15, which shows the same page as does Figure 25.2, displayed by Lynx. One of the advantages of Lynx is that, if Lynx can display it, people who browse the Web using computerized speech synthesizers can also use your page.

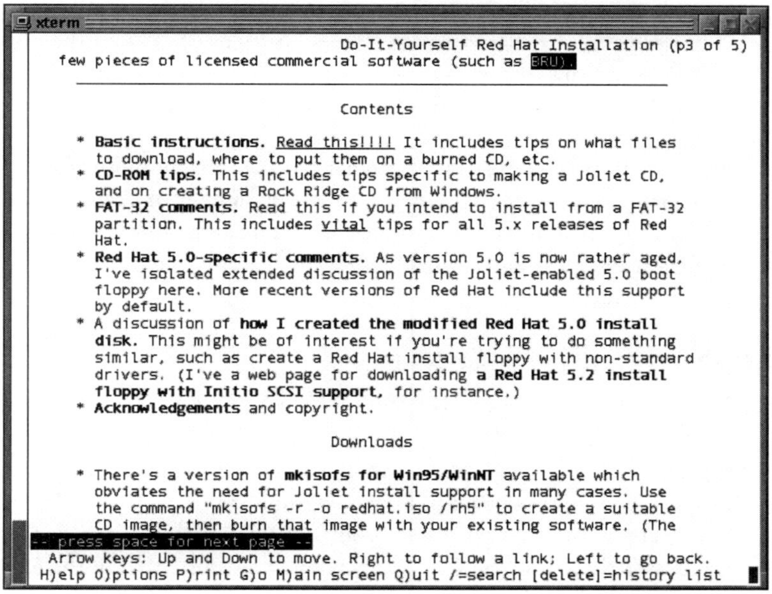

Figure 25.15
Use Lynx to check a Web page's utility on a minimal browser.

- Netscape for Linux—Netscape is one of the most popular browsers for Linux, and if you're creating a Linux-oriented page, you should definitely be sure that the page works well in this browser.
- NCSA Mosaic—Mosaic is the progenitor of all Web browsers. It's no longer being maintained, but is still useful as a minimal GUI Web browser.
- Other Linux browsers—Other browsers for Linux include Arena and the KDE browser. Some of these products may come with your Linux distribution, and all are useful for testing Web page compatibility.
- Microsoft Internet Explorer—This browser and Netscape make up the largest share of the browser market. You should therefore check that your page works well with Internet Explorer. Internet Explorer isn't available for Linux, however.

- Test on various OSs—Netscape and Internet Explorer are the most popular browsers on Windows and MacOS, but there are other browsers for these OSs, such as Mosaic and Opera. Less common OSs, like IBM's OS/2 or the EPOC system used on Psion palmtops, often include their own browsers. Even browsers that are common across platforms may display their information differently in the different OSs. For instance, if you've set fonts in your Web page and tested them in Linux, they may not appear as you expect in Windows or MacOS, because font names may be different.
- Test at low resolution—You should check your Web page at low screen resolution or with a small browser window to simulate this situation.
- Test at low color depth—Graphics images typically consume a lot of colors, so Web pages can change appearance radically if a system has few colors. Background bitmaps are particularly likely to cause problems at low color depth because the colors the computer uses for the background bitmap may interfere with the text. Keep in mind that different browsers and OSs may handle low-color situations differently. Netscape in Linux with a low-color display is a pretty brutal test of color handling.

Of course, testing *all* possible combinations is probably more than you care to do—indeed, it's probably more than you *can* do, given the number of exotic operating systems like EPOC. I do recommend that you test a Web page in at least two browsers in each of at least two OSs if such testing is possible. Further, I recommend that Lynx be one of the browsers you use for testing.

USING AN HTML SYNTAX CHECKER

Although HTML isn't a programming language *per se*, it does have certain rules of syntax, and you can check the HTML files that WordPerfect produces to be sure that they follow proper HTML conventions. Syntax checking is the sort of tedious task at which humans are not very good, because we can easily overlook the little details. Therefore, assorted computer programs exist to let you check your HTML files for accuracy. One of these, known as *weblint*, is included on the CD that accompanies this book, in the chapter25 directory.

To install weblint, follow these instructions:

1. Log in as root or use the `su` command to gain root access.
2. Mount the CD that accompanies this book, using a command such as `mount /dev/cdrom /mnt/cdrom`. You may need to use a different device file and mount point, depending upon how your system is configured.

→ If you want to learn more about using CD-ROMs, **see** "Accessing Removable Devices: Using the `mount` Command," **p. 76**

3. Change to the chapter25 directory on the CD, by typing `cd /mnt/cdrom/chapter25` (or something similar, depending upon your mount point).
4. Run the weblint install routine: `make install`.

5. You'll see a summary of the files that have been copied and information as to where they are. The most important is the *weblint* program itself, which goes in /usr/local/bin.
6. Move out of the CD mount point and unmount the CD, by using the commands cd ~ and umount /mnt/cdrom.

You should now be able to use the weblint command to check your HTML files. Type weblint followed by the name of the file(s) you want to check. You can optionally include the -x netscape or -x microsoft parameters immediately after weblint to get weblint to recognize some common HTML extensions that are used by Netscape and Internet Explorer, respectively. So to check a file called mypage.html, you would type weblint -x netscape mypage.html. This command might produce output something like this:

```
mypage.html(4): empty container element <TITLE>.[sr]
mypage.html(12): unknown attribute "WP" for element <BR>.[sr]
mypage.html(12): unknown attribute "WP" for element <BR>.[sr]
mypage.html(15): unknown attribute "WP" for element <BR>.[sr]
mypage.html(15): unknown attribute "WP" for element <BR>.[sr]
mypage.html(20): unmatched </P> (no matching <P> seen).
```


Note

If you run weblint and find that it produces no output, your Web page contains no errors that weblint could find.

Weblint begins every line of its output with the filename it's checking and, in parentheses, the line number of the problem. This line number may not be the line number as you'd see the file within WordPerfect; it's the line number in the ASCII HTML file itself. Not everything that weblint prints is necessarily an error. In the preceding example, for instance, the first line reports a potential problem on line 4 involving an empty container element. In other words, the <TITLE> element is empty, so the Web page has no title. You should probably correct this situation, but it won't cause real problems for any Web browser. The unknown attribute errors flag codes that WordPerfect inserts in the HTML file for its own edification and won't cause any problems for regular browsers. The unmatched </P> error, however, is potentially more serious. In HTML, each paragraph is supposed to begin with a <P> tag and end with a </P> tag, but weblint has found a paragraph that has the closing </P> but no opening <P>. In fact, sometimes weblint itself gets a bit confused, which can result in errors in its reporting. The file that I used to produce the preceding output actually did have an opening <P>, but the ordering of other elements confused weblint. Specifically, the HTML looked something like this:

```
<FONT FACE="Comic Sans MS"><P>Text</FONT></P>
```

In absolutely perfect HTML, when you apply two tags to text (such as a <FONT> tag and a <P> tag), the "off" tags must appear in the opposite order of the "on" tags. Thus, the </P> should come before the </FONT> in this case. Therefore, weblint squawked, but it complained about the wrong error.

Note

WordPerfect does tend to produce improperly nested HTML codes. Fortunately, most Web browsers cope well with this error. If you want to create the best code possible, you can edit the HTML file that WordPerfect produces to correct this error and any other errors that weblint finds. Some errors, such as the missing title, you can correct within WordPerfect.

Are you having trouble getting weblint to run? See "Weblint Won't Run" in the "Troubleshooting" section at the end of this chapter.

Weblint tends to be pretty fussy about Web pages. If you save a random selection of Web pages to your disk and run them through weblint, you'll probably find that most of them contain several errors, yet they display fine in most Web browsers. The same is true of WordPerfect's output. Still, weblint can be a very useful tool if you want to modify WordPerfect's HTML output by hand and to catch things like missing Web page titles.

Conversely, a page that checks out fine in weblint may not look the way you want it to look. Weblint can only find syntax errors in the HTML tags; it cannot find poor Web page layout, spelling errors in your text, or bad links. To find these problems, you need to check your page in one or more browsers or use tools like WordPerfect's spelling and grammar checkers.

→ If you want to learn more about WordPerfect's spelling and grammar checking features, **see** "Checking Your Document for Errors," **p. 117**

TROUBLESHOOTING

FONTS DON'T SHOW UP IN BROWSER

You've designed a Web page that uses certain fonts, but when you try to view it with Netscape or some other browser, you don't see the fonts you've specified.

If you're viewing the Web page on a system other than the one you used to create the page, one possibility is that the fonts simply don't exist on that system. Even with one system, however, this problem can occur for several reasons:

- The browser you're using doesn't support the HTML font change codes.
- You've disabled font changes in the browser's preferences.
- WordPerfect and X aren't using the same fonts. Most X-based Linux Web browsers, including Netscape, use fonts that are installed in X. WordPerfect, however, has its own font-handling system. You must install a font with the correct name in X to use that font from Netscape. Consult your X documentation for information on how to install fonts in X.
- WordPerfect and X don't agree on the font's name. Even if you install the same font in both X and WordPerfect, the font must use the *same name* in both environments for WordPerfect-created Web pages to use the correct font in X.

If you're having problems getting WordPerfect and a Linux browser to agree on fonts, however, you should think about whether you should be using that font at all because there's a good chance that your readers won't have access to it.

WEBLINT WON'T RUN

You've tried running the Weblint program, but it doesn't work. Instead, you get a message along the lines of `exec: perl: not found`.

The Weblint program is written in Perl, which is one of several common programming languages in Linux. If Perl isn't installed on your system, Linux won't be able to run Perl programs, such as Weblint. Perl comes with most Linux distributions, so installing it shouldn't be difficult, but the installation details vary from one distribution to another. You should consult your distribution's documentation on how to install packages, locate the Perl package, and then install it.

CHAPTER 26

Designing a New Web Page

In this chapter

STARTING A NEW WEB PAGE

Chapter 25, "An Introduction to the Web," introduces basic principles of Web page design and summarizes the tools WordPerfect provides to ease Web page creation. This chapter goes into greater detail on how to use these features to create a Web page or a group of Web pages (a Web document or Web site). You should definitely read Chapter 25 in conjunction with this one to learn how to create a Web page that's flexible and that won't frustrate your readers. Chapter 25 also includes instructions on importing existing Web pages and on converting existing WordPerfect documents into Web pages.

If you want to start a new Web page, you should follow these directions:

1. Start a new document by choosing File, New from the document window or by choosing Program, New Window from the WordPerfect program Control window.
2. In the new document window, choose File, Internet Publisher to open the Internet Publisher dialog box shown in Figure 26.1.

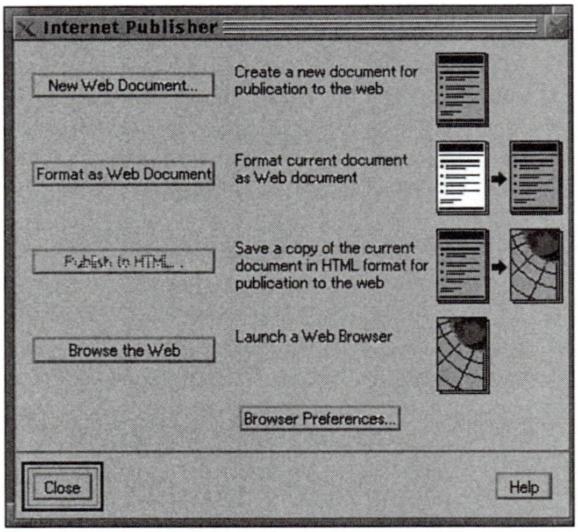

Figure 26.1
You can perform assorted Web page conversions from the Internet Publisher dialog box.

3. Click Format as Web Document. WordPerfect changes your menus and toolbars to support Web page functions rather than conventional WordPerfect functions. By default, WordPerfect displays a gray background, rather than the usual white when you're editing a Web page, so it's easy to tell that you're in this mode.

When you start a new Web page, WordPerfect's toolbars change, giving you quick access to Web page features and no access to some features that don't apply to Web pages. Figure 25.8 (p. 671) shows the Web page toolbars.

When you're editing a Web page, many of WordPerfect's formatting and editing features work just as they do for conventional documents. For instance, you can press the Enter key to end a paragraph, and you can use WordPerfect's spell checker in the normal way.

Other features work almost, but not quite, the same. Although you have access to bold and italic attributes, for example, the browser your reader uses interprets these attributes, and not all browsers display them as such. The text-mode Lynx browser, for instance, underlines text that you code as being either bold or italic, so these attributes are indistinguishable.

MANAGING LINKS

Links (p. 78) are at the heart of the Web. They allow you as a Web page author to direct your reader to helpful information, either on your Web site or on a site you don't control. As a reader, links let you quickly jump around the Web in search of information or simply "browse" for anything that you might find interesting.

In most Web browsers, links appear as underlined text in a different color than the rest of the text. Typically, links that the reader has already followed and links that the reader has not followed appear in different colors. Usually, the URL isn't part of the text that your reader sees; it's hidden in the *HTML (p. 663)* codes. When the reader clicks on the link, though, the Web browser extracts the URL and loads that Web page or performs some other appropriate action, as described shortly.

The Web supports several different types of links—links to other Web pages, links to send email, links to download files, and so on. Although the basic procedures to create each of these link types are the same, each has unique characteristics.

CREATING A LINK

In some circumstances WordPerfect creates a link for you automatically or semiautomatically. Two of the most important such situations are

- Reference lists—Mark text you want to include in a table of contents or other reference list; then choose Tools, Reference, Generate (Ctrl+F9) to open the Generate dialog box, and from that the Generate Options dialog box (Figure 26.2). If you select the Build Hypertext Links option from the Generate Options dialog box, WordPerfect creates hyperlinks from the entries.

→ If you want to learn more about creating a table of contents or other reference list, **see** "Organizing Lists and References," **p. 627**

- QuickCorrect—You can configure WordPerfect's QuickCorrect feature to create a hyperlink whenever you type something that WordPerfect identifies as a URL (say, something that's preceded by `http://`). You set this option from the QuickCorrect dialog box's QuickLinks field, which you get by choosing Tools, QuickCorrect (Figure 26.3). You can also set QuickCorrect to automatically create a link when you type certain words. In either case, you need to click the Format words as hyperlinks when you type them button.

→ If you want to learn more about configuring QuickLinks to insert links, **see** "Inserting Hyperlinks," **p. 125**

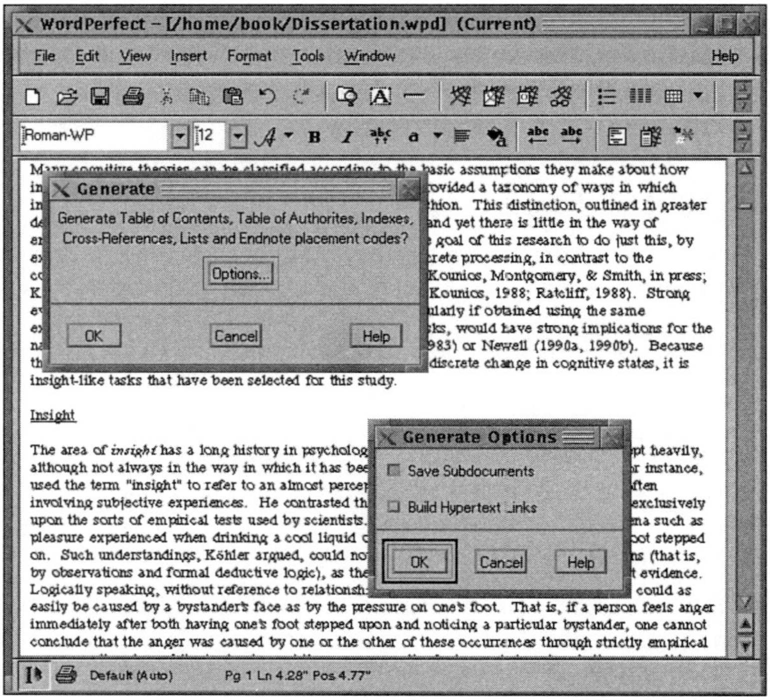

Figure 26.2
Click Options in the Generate dialog box to open the Generate Options dialog box.

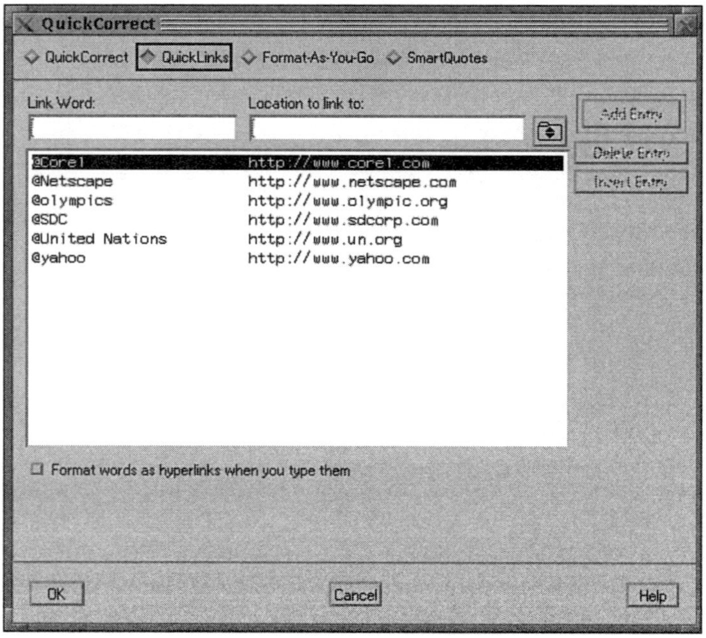

Figure 26.3
Using QuickCorrect can make creating common hyperlinks very easy.

In many cases, you'll want more control over your link creation process than you can get with automatic link creation. To create a link in a more "manual" way, follow these steps:

1. Place the insertion point in the word you want to serve as the link, select multiple words if you want the entire group to serve as the link, or click on a graphic if you want it to serve as the link.
2. Choose Tools, Hyperlink from the document window's menu or choose Create Link from the menu you get when you click on the Create or edit link toolbar icon. This step produces the Create Hypertext Link dialog box shown in Figure 26.4.

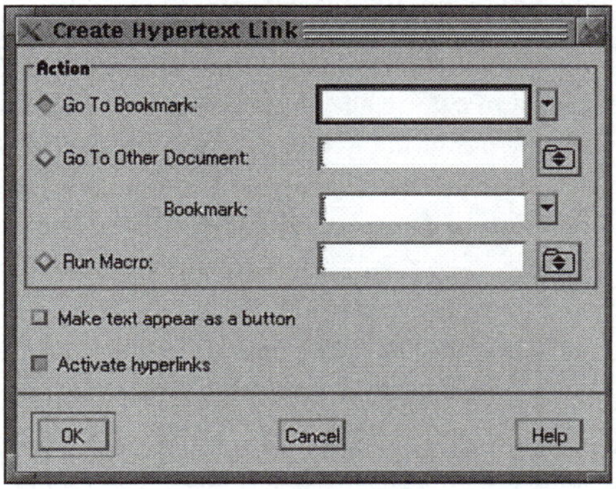

Figure 26.4
The Create Hypertext Link dialog box lets you create several types of links.

3. Enter the URL for the destination file or other link type. You normally use the Go To Other Document field (with or without the Bookmark field to link to a bookmark within that file, as described shortly) or the Go To Bookmark field to link to a bookmark within the current file. The Run Macro field is useful mainly for creating links to be used within WordPerfect; when you click such a link from within WordPerfect, it runs a WordPerfect macro, but this type of link is useless in regular Web browsers.
4. If you want to create a "button" out of a link, you can check the Make text appear as a button check button.

Tip #288 from *Tux Tips*

WordPerfect uses a graphic image to create the "button" effect. Unfortunately, the graphics WordPerfect generates for this effect can be very difficult to read. You may want to use a program like the GIMP (`http://www.gimp.org`) to create new, more-legible button graphics to replace the ones WordPerfect generates.

5. Click OK. WordPerfect now formats the word or words you selected as links.

Note

If you leave the Activate hyperlinks button selected in the Create Hypertext Link dialog box, WordPerfect lets you load additional Web pages by clicking on the links in your Web document. This feature can be handy but can also make editing those links more difficult because you need to use cursor commands to move into the link text. This setting affects *all* the links on a Web page, so you can change it back and forth by editing any link.

TYPES OF LINKS

As I've mentioned, Web links come in a number of different forms. These forms, and what you enter in the Create Hypertext Link dialog box to create them, are as follows:

- Within-page link—If you set bookmarks in your page, you can create a link to them by typing the name of the bookmark in the Go To Bookmark field or by selecting the name from the list you get when you click the option button to the right of that field. You can use this type of link if you have a large Web page and want to refer readers to specific parts of it. This link type is useful only for shifting the view of your reader's Web browser within a single Web page, not for creating links between the separate pages that may make up a complete Web site.

→ If you want to learn more about bookmarks, including how to set them, **see** "Bookmarks," **p. 595**

- Link to another page of the same document—If you have a multipage Web document, you can link to other pages of that document by typing the page's filename in the Go To Other Document field. For instance, if you have a Web page about cameras, the main page might be called cameras.html, and you could have additional pages under the filenames cameras-slr.html, cameras-aps.html, and so on. If you want to create a link to the cameras-slr.html file, you would enter its filename in the Go To Other Document field. In addition, you can go directly to a bookmark in the other document by typing the bookmark's name in the Bookmark field.
- Link to another page—If you want to link to a page that's located on another Web server entirely, you must entire its entire URL in the Go To Other Document field. For instance, to link to Corel's main Linux Web site, you would enter `http://linux.corel.com`. You can also use the complete link for Web pages that are part of your own document, but this approach can sometimes slow down transfers slightly, and you'll have to change the links if your Web site's address ever changes.
- Download link: http—Many Web servers let you place downloadable files and Web pages in the same directory. You can use this feature to make program files, PDF documents, or other files available. In this case, you specify the URL precisely as you would for another Web page, by typing its filename alone or the filename preceded by `http://` and the machine name and directory path. Depending upon how the reader's browser is configured and the type of the file, the file might appear in the browser window, be loaded into an external program, or be saved to disk.

Caution

Web servers try to notify Web browsers of what type of file they're transferring. If they guess incorrectly, the result can be a corrupted download. This can happen sometimes with files that have odd filename extensions. If you think your Web server might guess your files' types incorrectly, try the download from several different browsers in different OSs; if you do have problems, contact your Web server administrator or place a warning message on your Web page. In my experience, Netscape Navigator under Windows is particularly prone to corrupting files for which the server doesn't know the correct file type.

Tip #289 from *Tux Tips*

For any download of more than a few tens of kilobytes, include a statement of the file size on any download link. That way, your readers will have some idea before starting a download of how long it will take.

- Download link: ftp—Before the Web came into existence, *ftp* (short for *file transfer protocol*) was used for transferring files across the Internet. You can create a link to a file that's on an ftp site by specifying the file's complete URL, such as `ftp://bogus.machine.com/pub/download.tgz`.

Note

An ftp transfer normally works only if the file is on an *anonymous* ftp site, meaning that it allows anybody to log on with any password merely by using the username *anonymous*. If an ftp site requires a real password, most Web browsers will fail to download a file that's linked in this way.

- Mail link—You'll probably want to place your email address on your Web page to allow others to send you email about your page. You do so by typing `mailto` and a colon before the address in the Go To Other Document field, as in `mailto:jennie@turrill.com`. When your reader clicks on such a link, the browser normally opens a mail program with the email address already entered. Some browsers do not support email, however, in which case the link does nothing. You may therefore want to be sure that your email address appears in plain text in the Web page (the plain text can itself be the link).

Caution

Some companies market products by sending *spam* to lists of email addresses. Spam is an unsolicited mail message that's sent to a large number of email addresses, and many people find spam annoying. One of the ways that so-called *spammers* get their addresses is by looking for mailto: links on Web pages. Placing such a link on your Web page may therefore increase the amount of spam you get. If your Internet service provider (*ISP [p. 665]*) or Web site supports it, you might be able to use a Web form, described later in this chapter, instead of a mailto: link to get feedback from your Web page's readers.

- News link—*Usenet news* is a massive collection of cyberspace "meeting places" for discussion of topics ranging from computers to employment to gardening. These newsgroups are invaluable sources of information and help, and you can create a link to newsgroups much as you do mail links, using the keyword `news` rather than `mail`, along with the name of the newsgroup. For instance, typing `news:comp.os.linux.misc` into the Go To Other Document field causes the reader's Web browser to open a newsreader on to the `comp.os.linux.misc` newsgroup, assuming that browser is configured to support this feature.

Many, but not all, Web browsers give some indication of what type a link is, often by displaying the complete URL (including `http:`, `mail:`, or similar header) when the user's mouse pointer passes over the link. Still, you should try to make the type of link clear to the reader in your own text. For instance, you could include text such as "download the Sock Eaters game from `ftp://ftp.sockeaters.com`" or "send me mail at `amym@harvard.edu`" and embed the link in that text. These two examples also show the relevant information (the ftp site's address and the email address) in the text, which makes it easier for your reader to extract that information and use it in another program, if desired.

Deleting Links

There are two ways to remove a link from the text of a Web page:

1. Position the insertion point inside a link (you may need to use keyboard cursor movements) and choose Remove Link from the menu that appears when you click on the Create or edit link toolbar button.
2. Use Reveal Codes mode (View, Reveal Codes or Shift+Ctrl+F3) to locate the [Hypertext] codes in the WordPerfect document and delete them.

Setting Text and Link Colors

You can set the colors that your reader will see for links, as well as for ordinary text, by choosing Format, Text/Background Colors to open the Text/Background Colors dialog box shown in Figure 26.5.

The Regular Text color is, as you might expect, the color for ordinary text in the Web page. Links get three colors: one for an unvisited link, one for a visited link, and one that's only used for a moment when the reader clicks on the link.

Do you want to let your reader decide on a color scheme for the Web page? See "Recovering Web Browser Default Colors" in the "Troubleshooting" section at the end of this chapter.

Figure 26.5
You can control the colors assigned to ordinary text, various links, and the background from the Text/Background Colors dialog box.

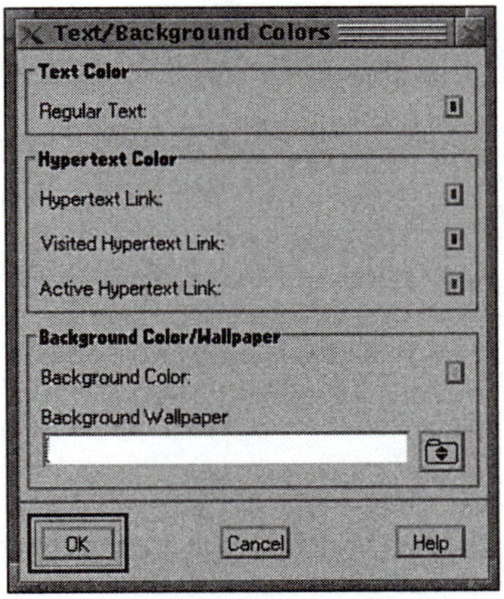

INSERTING GRAPHICS

Graphics can be a critical part of many document types, and Web pages are no exception to this rule. Whether you want to display a photo of your family dog or an image showing scientific data, WordPerfect lets you insert graphics into your Web page to help liven it up.

GRAPHICS FORMATS FOR THE WEB

Most Web browsers support two types of *bitmapped (p. 378)* graphics format: GIF and JPEG. These two formats have different characteristics, and understanding them is important if you are to pick the format that's best for any given image.

Note

Web browsers don't support WordPerfect's internal graphics formats, so WordPerfect must export any graphic you create with it. Generally, it does a poor job, so you're better off creating your Web page graphics outside of WordPerfect.

Tip #290 from *Tux Tips*

One of the keys to producing a Web page that readers like is to keep the size of your graphics files small to reduce the load time for the page. You can do so in different ways for the different graphics formats, as I describe shortly. In general, no graphic image should be larger than 50KB.

You can mix both of these graphics formats on a single Web page. There's no reason to restrict yourself to only one format unless that's all your graphics software supports or your images are all of a type that's best in only one format.

JPEG

JPEG files usually have a filename extension of .jpg. JPEG files support 24 bits per pixel, meaning a theoretical maximum of more than 16 million colors. This feature makes JPEG files very good for color photographs or images with fine gradations of color. Both JPEG and GIF support built-in compression of their graphics images, but JPEG's compression method is *lossy*, meaning that it loses some information. Just how much information it loses depends on the settings of the graphics software used to save the image. With high "quality" settings, JPEG images can be virtually indistinguishable from uncompressed originals, but with low settings, JPEG images can be quite ugly. Figures 26.6 and 26.7 illustrate this point; 26.6 is a JPEG file saved at high quality, and 26.7 is the same file saved at low quality. Compared to Figure 26.6, Figure 26.7 is blocky and difficult to interpret.

Figure 26.6
A high-quality JPEG file looks good but can take up a lot of space (67KB for this one).

Most software that supports JPEG allows you to specify the quality of the image you save. Different software uses different scales, so you need to check your software's documentation for what to use, but a scale of 1 to 100 is common for JPEG quality. I used the Linux program xv with a quality setting of 82 for Figure 26.6, and I saved Figure 26.7 at a setting of 4. In practice, many Web images can be saved at settings of 20 to 50 with acceptable results.

Another way to reduce the size of the files is to reduce the size of the images. Use your image-editing software to shrink the size of the image, say, from 1024×768 to 512×384.

Figure 26.7
A low-quality JPEG file looks bad but is small (7KB for this one).

Tip #291 from *Tux Tips*

If you have high-quality images you want to display, but you don't want to bloat your Web page with them, you can create lower-quality or smaller versions to place on the page and use these images as links to the high-quality images. Put text on the page saying to click on the picture to get a larger and better version, and you're set.

GIF

GIF is a format that supports up to 256 colors. GIF files, like JPEG files, support compression, but the GIF compression method is *lossless*. This means that what you get out when you view the image in a Web browser is the same as what went into it when you created the image. This characteristic makes GIF better than JPEG for displaying graphics that include large areas of the same color because GIF reproduces this static swath of color perfectly, whereas JPEG can introduce distortions, making the image look blurry or muddy.

Tip #292 from *Tux Tips*

Use JPEG for photographs, and use GIF for line drawings, cartoons, and screen shots.

As with JPEG, you can reduce the size of the file by reducing its resolution. Alternatively, you can often shrink the file's size by reducing the number of colors it contains. Various image-editing programs provide this option. Reducing the number of colors may introduce color shifts or a speckled appearance to certain colors. You can't reduce the "quality" of a GIF file in the way you can a JPEG file because the JPEG quality setting relates to its lossy compression method, and the GIF compression is lossless.

Images for Illustration

One purpose for placing graphics in a Web page is to illustrate some point. You might want to put a photo of your pets or an image of a product you're selling on your Web page. You can insert such a graphic in your Web page just as you would if you were creating a paper document: Choose Insert, Graphics, From File to open a WordPerfect File Manager dialog box and select the file you want to insert as a graphic using this dialog box.

→ If you want to learn more about using graphics in WordPerfect, **see** "Graphics," **p. 377**

Do your graphics not appear when you check your Web page in a browser? See "Graphics Don't Display Properly in Web Pages" in the "Troubleshooting" section at the end of this chapter.

Background Images

If you want to give your Web page a distinctive background, you can do so by choosing Format, Text/Background Colors, which opens the Text/Background Colors dialog box shown in Figure 26.5. You can set the background in two ways from this dialog box:

- Set a constant background color by selecting one from the Background Color button. This color will cover the entire background of the page.
- Set a background image by entering the filename for the image in the Background Wallpaper text entry field or by clicking the folder icon to select a file from a WordPerfect File Manager dialog box. A background image, if set, overrides a background color.

After you set this option, you won't see any change in the file as displayed by WordPerfect because it doesn't display the background image. WordPerfect inserts appropriate codes in the HTML file it generates, however, so that Web browsers that support background bitmaps will load the image and display it when they read the file.

Tip #293 from *Tux Tips*

Most Web browsers *tile* a background image, meaning that they repeat it over and over. Therefore, you can create effects such as a background that's yellow on the left and green on the right by creating an image that's just a very short stripe of yellow, followed by a short stripe of green. The Web browser fills out the rest.

Setting Image Properties

If you right-click on an image and choose HTML Properties from the context menu, WordPerfect displays the HTML Properties dialog box shown in Figure 26.8.

Figure 26.8
You can alter the appearance of a graphic, as well as what happens when you click on it, by using the HTML Properties dialog box.

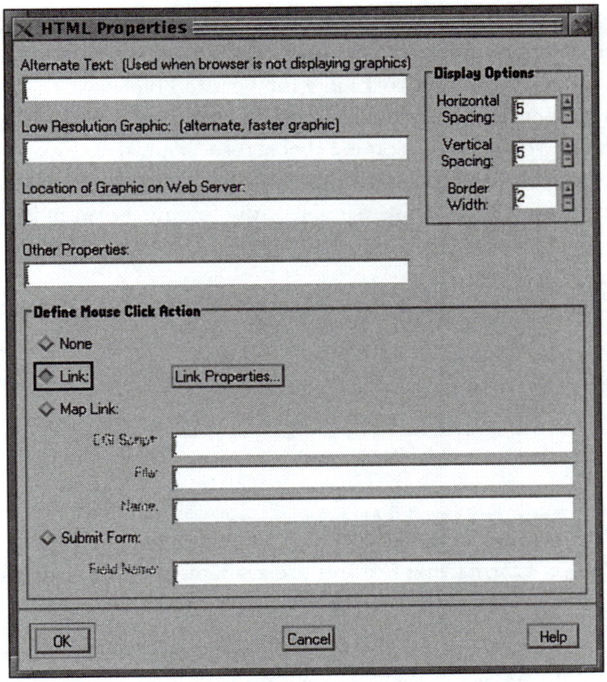

From this dialog box, you can set several characteristics for the graphic:

- Alternate Text—You should always set the Alternate Text field in the HTML Properties dialog box. This text displays on text-only Web browsers, and on graphical Web browsers if the user has disabled graphics loading. You can use this field to describe the graphic to let your reader know whether it's worth getting.
- Low Resolution Graphic—If you enter the name of a graphic file here, some Web browsers will get it before the main graphic in order to display something while the larger main graphic is downloading.
- Location of Graphic on Web Server—If you know where you'll be putting graphics on your Web server, you can enter that directory here, and some Web browsers will look in that directory for the graphics.
- Display Options—You can set a border around the graphic, as well as horizontal and vertical distance from other elements by using the spin boxes in this area.

Tip #294 from *Tux Tips*

If you're using your graphic as a link, it's useful to put a border around it. With most Web browsers, the border takes on the text link colors, which makes it more obvious that the graphic is a link.

- Define Mouse Click Action—None is the default selection and results in a static image. If you select Link, you can then click the Link Properties button to get the Edit Hypertext dialog box, which works like the Create Hypertext Link dialog box shown in Figure 26.4, and enter a link; the graphic will then function as a Web link. The Map Link button lets you create a graphic that serves as multiple links, depending on where you click it; but you need to create a special script for this feature to work, and doing so is beyond the scope of this book. Similarly, the Submit Form button lets you set the graphic to submit a Web form, but you need Web form support on the Web server.

→ If you want to learn more about creating Web forms, **see** "Creating Web Forms," **p. 702**

Caution

Don't use the Other Properties field, because entering anything there will result in an incorrect HTML entry.

FORMATTING FOR THE WEB

Many of WordPerfect's formatting features are available for Web pages but sometimes these features work in subtly different ways. Here are some pointers on these differences.

- Bold and italic text—You can enter bold and italic text in a Web page just as you can in a conventional document. There's no guarantee that your reader's Web browser will display the text using these specific attributes, however. The Lynx Web browser, for instance, displays both of these using underlining.
- Columns—If you choose Format, Columns, Format, you'll see the Web Columns dialog box shown in Figure 26.9, which presents fewer options than you get for columns on a printed page. You can set the number of columns and the width of the columns, expressed as either a percentage of the browser's window or the total number of pixels. Again, you can't assume that a browser will actually display your text in columns.

→ If you want to learn more about using columns for conventional documents, **see** "Creating Professional Layouts with Columns," **p. 255**

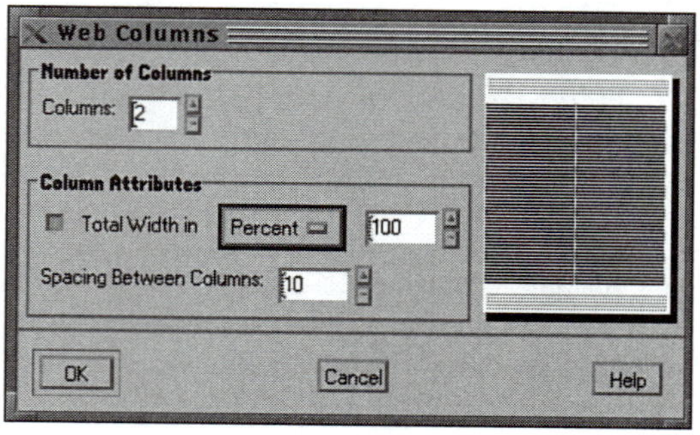

Figure 26.9
The Web Columns dialog box lets you set up a multicolumn layout for a Web page.

Are you trying to set a column width in pixels, but find that WordPerfect keeps changing the value you enter? See "Web Column Width in Pixels Is Incorrect" in the "Troubleshooting" section at the end of this chapter.

- Monospaced text—You can select a monospaced text attribute that works much like bold or italic. This attribute tells the reader's Web browser to display text in a monospaced font, though not what specific font to use. This feature can be useful if you want to display something that should be in a monospaced font, such as a program listing.

Tip #295 from *Tux Tips*

In a conventional document, you might use a specific font, such as Courier, for this purpose. Using the monospaced attribute is better for Web pages, however, because it works even on Web browsers that don't accept font change codes.

- Document title—Choose Format, Title or click the Set document title and information toolbar item to display the Title dialog box shown in Figure 26.10. You can enter a title for the document, which usually appears in a Web browser's title bar. You can also enter a base URL, which is the location of the main page of a document, and meta information, which is a set of special HTML codes to help index or classify a Web page.

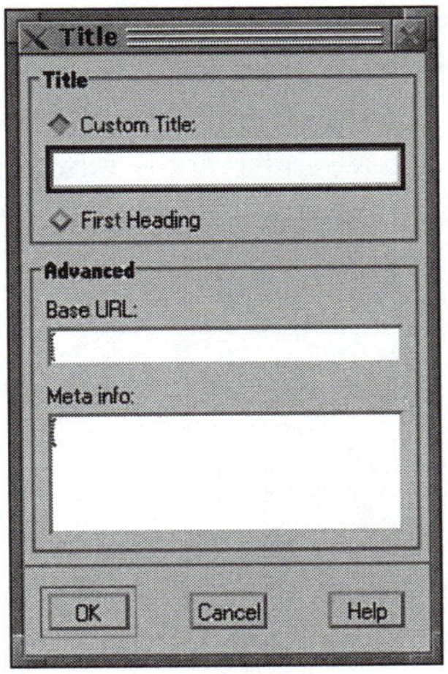

Figure 26.10
You'll probably want to enter only a document title in the Title dialog box; the Advanced information is usually unnecessary.

- Fonts—If you choose Format, Font (F9), WordPerfect displays the Font dialog box shown in Figure 26.11. This variant on the usual Font dialog box allows you to set the font, the font size, assorted HTML styles, the text color, and a few appearance attributes. As with many formatting features, you can't be absolutely certain that your reader will see what you intend when you specify any of these font characteristics.

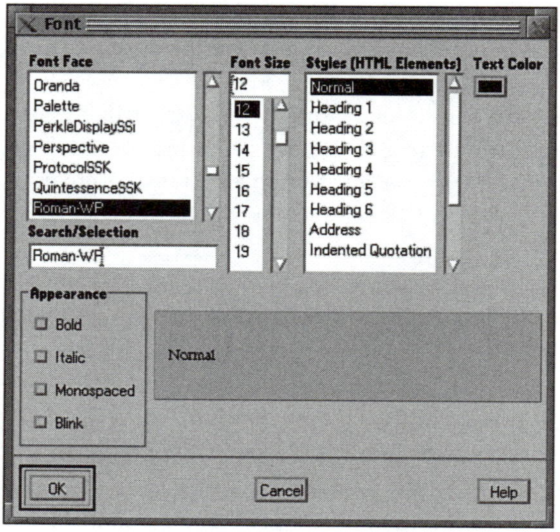

Figure 26.11
The Font dialog box displays all the fonts available to WordPerfect, but you can't know which of these will be available to your reader.

- HTML styles—If you click the Select heading level toolbar item, you'll see a menu of headings and other HTML styles. These are similar to WordPerfect's normal styles, but not as flexible. You can't change the formatting of an HTML style, for instance. Using these styles, you can create numbered or bulleted lists, various headings, and a few other formats. The Preformatted Text style is similar to the monospaced text feature, but you generally apply monospaced text to a few words and Preformatted Text to several lines of text.

CREATING WEB FORMS

A Web form is a type of interactive Web page. It allows a reader to enter information that is then submitted to the Web server and processed in some way. You might use forms to collect information for a purchase, to gather bug reports, or to present a survey. To create a Web form, select the Create new Web form tool from the Web toolbar. WordPerfect inserts appropriate codes, as well as onscreen markers of the form's location and a new Web form toolbar, as shown in Figure 26.12.

Note

To use a Web form, your Web server needs to support this feature. Contact your ISP or Web site administrator for details.

Figure 26.12
The Web form begin and end markers let you know where the Web form exists, even when the Reveal Codes pane is closed.

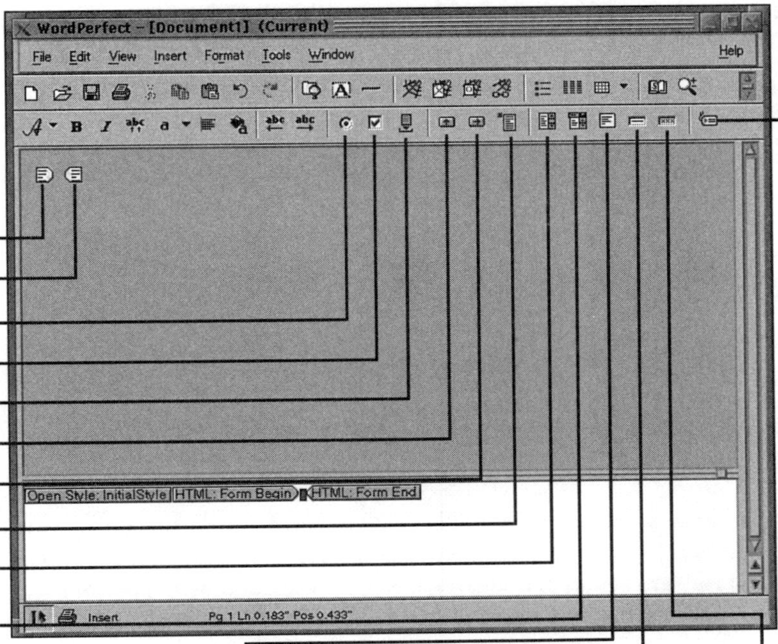

The major elements available for constructing a Web form include

- Insert radio button—A Web form radio button works like a WordPerfect radio button; only one in any given set can be active at a time.
- Insert check box—A Web form check box works like a WordPerfect check button; the user can select any combination of check boxes.
- Insert hidden field—Click here to place a hidden input field in the form. This field won't be immediately obvious to readers, but will be present.
- Insert reset button—Place a reset button on your form to allow users to reset all the values to their defaults.
- Insert submit button—Users click on the submit button to send their entries to you.
- Insert submit image—A submit image works just like a submit button except that the former contains a graphic image rather than the text `Submit`.
- Insert listbox selection list—Use a listbox selection list to present your users with a set of onscreen choices displayed in list form. After you insert a list, you double-click on it to open the Listbox/Combobox Properties dialog box shown in Figure 26.13, in which you enter the options to appear in the list.

Figure 26.13
Use the Add button to add items to the list of options for a listbox or combobox.

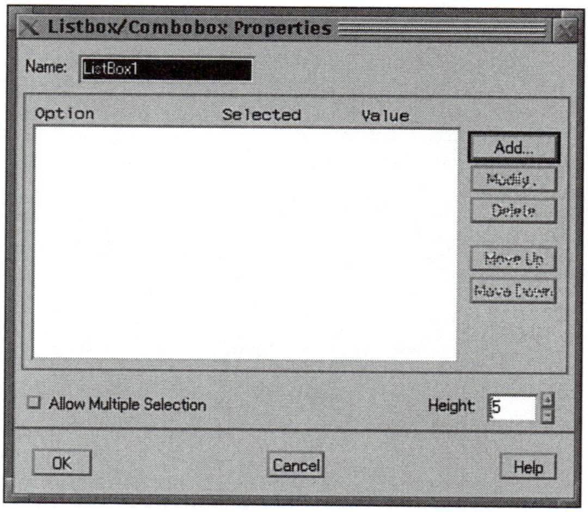

- Insert combobox selection list—This feature works much like a listbox selection list except that rather than a box that displays several or all of the choices at all times, the browser uses an option button, which displays only the current selection at any one time.
- Insert text area—Your users can type information into the text area. You can use this area to let users send you messages, bug reports, or other information that can't be easily broken into multiple-choice selections.
- Insert text line—A text line is much like a text area except that it supports only a single line of text.
- Insert password field—A password field is just like a text line except that the user sees asterisks instead of the words he or she types. Thus the field is ideal for entering confidential information such as passwords.
- Edit form properties—Click on an existing form and then click this tool to open a dialog box to modify the form properties. The Listbox/Combobox Properties dialog box (Figure 26.13) is an example of the dialog box you see when you use this tool, but the properties vary for the different form elements. Double-clicking on a form element has the same effect as clicking once and selecting this tool.

You can intersperse ordinary text and other HTML elements within a form to label the form elements and to make the form useful; otherwise, it's just a collection of buttons and text entry fields.

Tip #296 from *Tux Tips*

If you think you'll want to communicate with whomever submits information via a Web form, be sure to collect the user's email address or some other method of contact.

As you might expect, forms can look quite different on different browsers and platforms. Therefore, testing your forms on an assortment of Web browsers before you make them generally available on your Web site is especially important.

INSERTING YOUR OWN HTML CODES

If you know how to use HTML codes directly, you may want to insert codes that WordPerfect can't produce into your document. You can do so by choosing Format, Custom HTML. WordPerfect won't place any obvious indication that anything's different on the screen, but the next characters you type appear in red and continue to be red until you choose Format, Custom HTML again. WordPerfect passes these red characters unchanged to the HTML file.

Note

If you use custom HTML to create a feature that WordPerfect *does* support, WordPerfect interprets your custom HTML and displays the feature as such if you subsequently load the HTML file.

TROUBLESHOOTING

RECOVERING WEB BROWSER DEFAULT COLORS

You've created a Web page and want to let the user specify the color scheme for text, links, and backgrounds, but the Web page WordPerfect produces overrides the browser defaults.

WordPerfect always saves the HTML tags to set text and background colors, even if you don't change any of them. If you want to let your reader's browser determine the color scheme, you need to delete some HTML tags from the files that WordPerfect produces. To do so, load the file into a text editor (WordPerfect itself will do, but be sure to specify ASCII Text as the file format when WordPerfect asks about it and be sure to save the document as ASCII text when you're done). You then need to look for a line near the top of the file that reads something like this:

```
<BODY TEXT="#000000" LINK="#0000FF" VLINK="#551A8B" ALINK="#FF0000"
BGCOLOR="#C0C0C0">
```

This line is an HTML `BODY` tag that includes several fields. The `TEXT` field sets the color for the body text, `LINK` sets the color for an unvisited link, `VLINK` sets the color for a visited link, `ALINK` sets the color for an active link, and `BGCOLOR` sets the color of the background. If you've set a background bitmap, a `BACKGROUND` field also appears within this code, pointing to the filename of the background bitmap. You need to delete whichever of these fields (including the color values in quotation marks) you want the reader's browser to set from its defaults. For instance, to give complete control to the reader's browser, the line should read

```
<BODY>
```

When you are done, save the file. You need to repeat this step each time you save the file from WordPerfect.

GRAPHICS DON'T DISPLAY PROPERLY ON WEB PAGES

You've created a Web page with graphics, but when you test it, you get blue boxes or nothing in place of your graphics.

WordPerfect often does a poor job of converting graphics for use on the Web. If the graphic you're trying to use was originally in GIF or JPEG format, you can try using that image directly. WordPerfect creates a directory named after the Web page, but with .htg appended—for instance, floorwax.html.htg if your Web page is floorwax.html. Copy the original JPEG or GIF file into that directory and try again. If the graphic wasn't originally in GIF or JPEG format, you should try to find some other way to convert it to one of these formats. The Linux programs the GIMP and xv are both very useful for this purpose. They come with most Linux distributions.

Note

If you resized your graphic in WordPerfect, you may need to resize it from your graphics editor before moving it to the Web graphics directory. If you don't, you can end up with a graphics file that's far too large to use comfortably on the Web. Even if the graphic appears as the correct size in your browser, the graphics file might be quite large because HTML supports codes to resize graphics *within* the Web browser. This feature can produce the image size you want, but won't affect the size of the graphics file or the time it takes to download.

Another possibility is that WordPerfect has placed an *absolute* reference to the graphic in the HTML file, which may happen if you chose the Image on Disk option when you imported the graphic or if the graphic is a background bitmap. To correct this problem, you have two options:

- Right-click on the image and choose Content from the context menu to get the Box Content dialog box shown in Figure 26.14. Set the Content option button to read Image instead of Image on Disk; then click OK and save the Web page. You're likely to find that you have a distorted image, however, in which case you need to copy the original file into the image subdirectory, as I've just described.

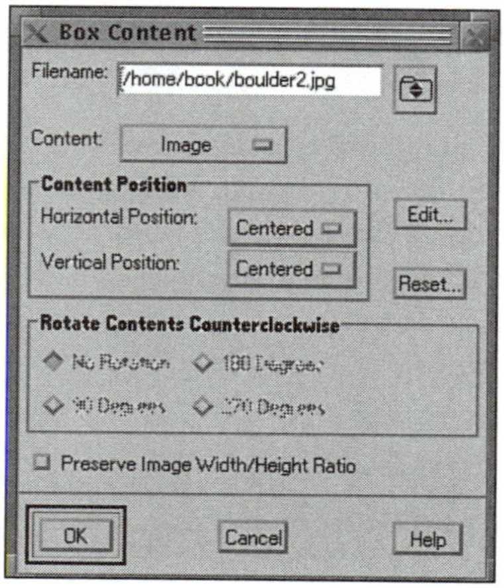

Figure 26.14 WordPerfect saves Content of Image and Image on Disk image references differently.

- Edit the HTML file that WordPerfect produces and locate the image code, which looks something like`<img src=/home/yourname/file.jpg>`. Remove the leading directory names, leaving only the `file.jpg` part. (There may be other components to this code, including height, width, and other specifications. Leave them alone.) If you subsequently save the Web page from WordPerfect, you have to redo this change.

One final possible cause of the problem is that many Web browsers don't support background images or support only GIFs as background images. Try the Web page from Netscape Navigator or Microsoft Internet Explorer. If it turns up there and if the image is a GIF, then the original browser may not support background images.

WEB COLUMN WIDTH IN PIXELS IS INCORRECT

You're trying to set the total width of your Web columns in pixels, but WordPerfect creates much narrower columns than you want.

A bug in WordPerfect for Linux causes this problem. You can work around it by entering a substantially larger column width than you really want. You may want to reconsider whether you really need to specify a column width, however, because a layout like this will look bad on any Web browser that's not adjusted to something close to the window size you're assuming. In addition, some Web browsers just don't understand the column-width command.

PART VIII

APPENDIXES

APPENDIX A

Installation and System Configuration

In this appendix

Installing WordPerfect for Linux

Before you can actually *use* WordPerfect for Linux, it must be installed on your system. Perhaps it already is, in which case you can skip this section, or this entire appendix. If not, though, you need to get WordPerfect set up on your computer.

Preliminary Steps

You normally install software in Linux as root, which guarantees that you have full write access to normal software-installation directories such as /opt and /usr. If you don't have root access to the computer on which you want to run Linux, you may need to consult with your system administrator to get WordPerfect installed. Alternatively, you can install it in your home directory, though if you do and if your system administrator has implemented *disk quotas* to limit the amount of disk space given to each user, you may run out of disk space.

➔ To learn more about the root account, **see** "Users and the Superuser," **p. 11**

If you're installing the full version of WordPerfect for Linux, you need to mount the distribution CD. You do so from an xterm window with a command such as `mount /dev/cdrom /mnt/cdrom`. You may need to use a different device file or mount point; consult your distribution's documentation or a book such as Que's *Special Edition Using Linux* for help on this topic. If you move to the CD-ROM root directory with a command such as `cd /mnt/cdrom` and then issue an `ls` command, you should see several subdirectories and the script files install.wp and remove.wp.

➔ To learn more about mounting CD-ROMs, **see** "Accessing Removable Devices: Using the `mount` Command," **p. 76**

The downloadable version of WordPerfect comes as one or more tarred gzip files. Ordinarily, files of this type use .tgz or .tar.gz extensions, but the main site Corel uses for distributing WordPerfect assigns the files .GZ extensions alone, which can cause confusion, particularly because Netscape automatically un-gzips these files, but doesn't remove the .GZ extension. The result is a tar file with a .GZ extension, which is rather confusing because tar files usually have .tar extensions.

The CD that accompanies this book includes a file called guilg00.tar.gz that contains the downloadable version of WordPerfect, so you're spared this confusion. To use this file, you should

1. Mount the CD (using a command such as `mount /dev/cdrom /mnt/cdrom`).
2. Create and move into a temporary directory for the WordPerfect install files, for instance, `mkdir install-wp; cd install-wp`.

3. Uncompress the install files: `tar xvzf /mnt/cdrom/wp-files/guilg00.tar.gz`. This command creates a series of install files, including the install script, called Runme.

The installation instructions presented here work for both the personal retail and the downloadable versions of WordPerfect. For the most part, the installation process is identical for both varieties. I point out the differences.

RUNNING THE INSTALLER

Caution

You normally install Linux software as root. Corel has announced a potential security problem involving WordPerfect for Linux when it's installed this way, however. The safest way to proceed is to create, as root if necessary, a directory in which to install WordPerfect (using, say, `mkdir /opt/wp8` from an xterm prompt) and to change the ownership of this directory to your ordinary user account (for instance, `chown mchugh /opt/wp8`). You can then run the installation procedure as the ordinary user. If you install directly as root and if you're running a networked or multiuser computer, a malicious individual could gain access to your WordPerfect documents or cause WordPerfect to overwrite important files.

To begin the installation, you must run the install program, by typing either `./install.wp` or `./Runme` from the retail version's installation CD or the temporary install directory for the downloadable version, respectively.

Do you get a `Permission denied` *message when you try to run the installer? See "CD-ROM Mount Permissions" in the "Troubleshooting" section at the end of this chapter.*

You may be asked to answer one or two questions in the xterm in which you run the installer. For instance, if you're installing the downloadable version, you are asked to confirm that you've extracted all the files from their archives; and if you're running a recent Linux kernel, you may see a query such as

```
Linux 2.2.3 not certified. Continue? (y/n)
```

You should type `y` here, and WordPerfect eventually presents its Install Configuration dialog box, as shown in Figure A.1.

Figure A.1
WordPerfect comes with its license terms in several different languages.

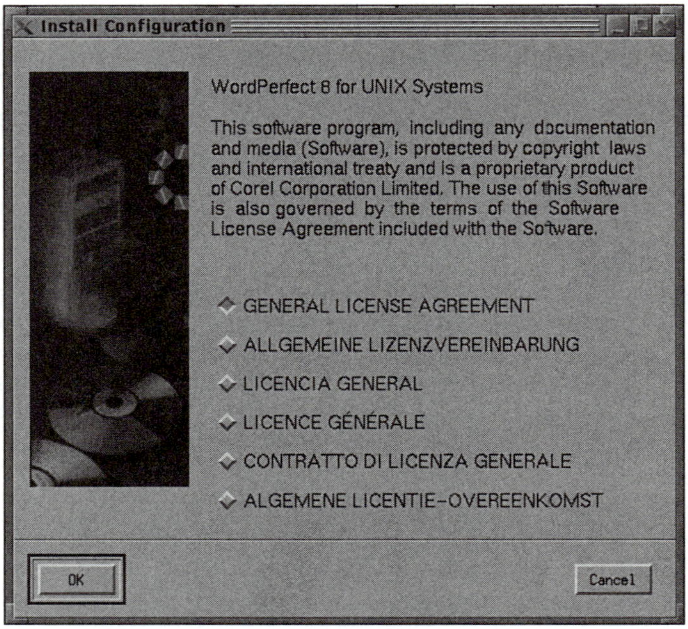

From here, you're in the install program and should follow these steps:

1. Select the radio button bearing the text in the language in which you want to read WordPerfect's license agreement and click OK.
2. The Install Configuration dialog box disappears and then reappears with the text of the WordPerfect license, as shown in Figure A.2. Read the license and, if you agree with the terms, click Accept to continue.
3. Assuming you clicked Accept, the Install Configuration dialog box changes again (Figure A.3) to ask for an installation directory. Good locations for WordPerfect 8 include /opt/wp8 and /usr/wp8. Precisely where you place it is a matter of personal preference and available disk space. For the rest of this procedure, I assume that you are selecting /opt/wp8 here.

Tip #297 from *Tux Tips*

To find out how much free disk space you have, try using the `df` command. This command lists, for each mounted partition, the total partition size, the used space, and the available space (all in kilobytes). A full WordPerfect installation takes close to 90MB of disk space, but you'll need less if you're installing the downloadable version or cut back on the features to be installed.

4. Click OK and the Install Configuration dialog box changes again, as shown in Figure A.4. You can select how much of WordPerfect you want to install from this dialog box.

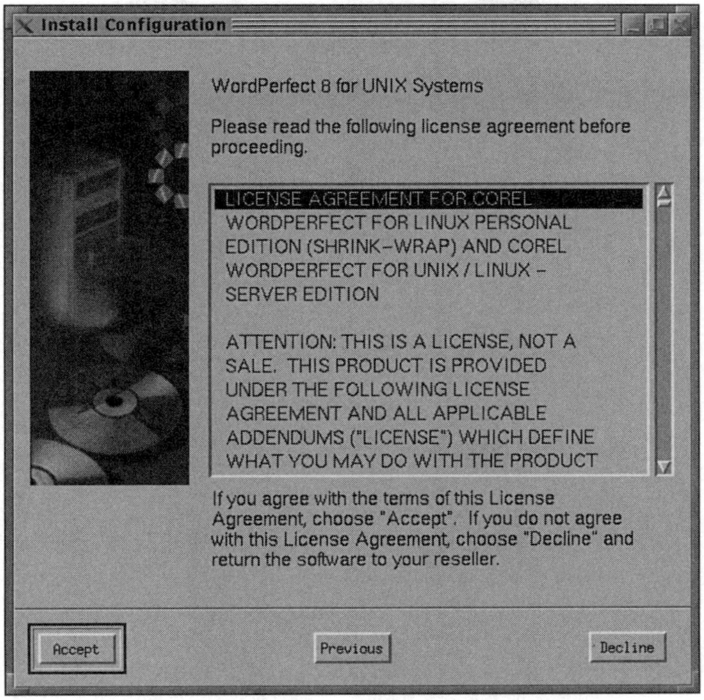

Figure A.2
You can use the scroll-bar to view the entire text of the license agreement.

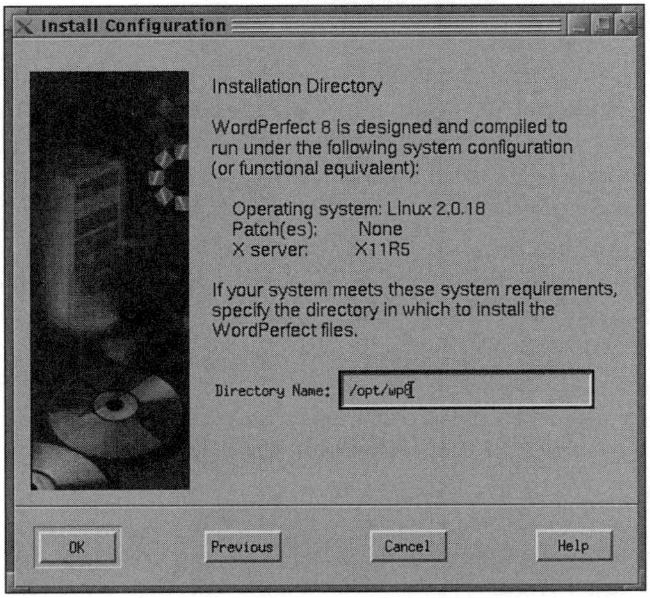

Figure A.3
WordPerfect for Linux was compiled for Linux kernel 2.0.18, but it works fine with both older and newer kernels, including kernels in the 2.2.x series.

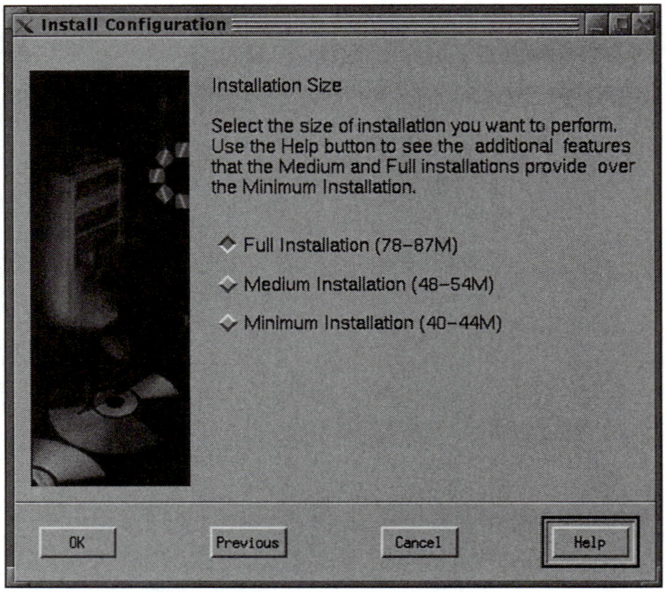

Figure A.4
Click Help to see what extra features you get for each installation size.

Note

This book is written with the assumption that you are selecting a Full Installation at this point. If you select a Medium or Minimum Installation, some of the features described in this book may not be available.

5. When you click OK, the Install Configuration dialog box changes again (Figure A.5), this time giving you the opportunity to pattern your installation after an existing WordPerfect installation. You can use this feature to copy some configuration files from a previous installation of WordPerfect, either a WordPerfect 7 installation or an installation of the downloadable version of the program if you're upgrading to the retail version.
6. If you're installing the personal retail version, the installer asks for your WordPerfect registration number after you click OK again. This number is printed on the registration card that came with your package. If you're installing the downloadable version, you won't get this request.
7. When you click OK, the installer asks whether you want to update the /etc/magic file. Some utilities use this file to help identify file types. You can check to see whether you have this file and, if it's present, elect to update it.
8. The Install Configuration dialog box changes yet again (Figure A.6), this time prompting for the languages you want to install. These languages can be used in dialogs, menus, and spell checkers. It's best to select the languages you may want to use when you first install WordPerfect because adding them later may require that you reinstall the program.

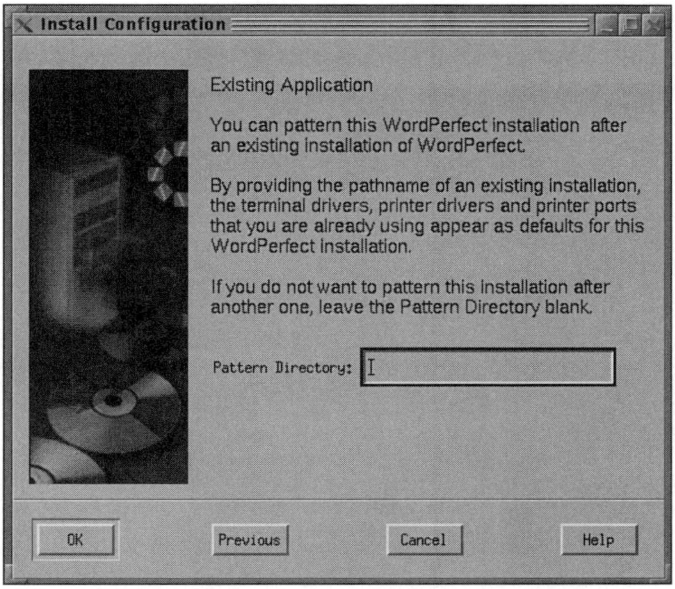

Figure A.5
If you've installed a previous version of WordPerfect, type the path to its directory in the Pattern Directory field.

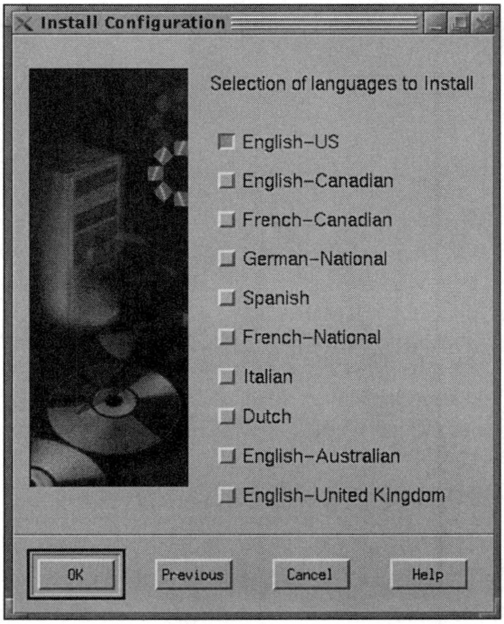

Figure A.6
You can select any or all of the languages and then later select a specific language as the default.

9. When you click OK, the Install Configuration dialog box changes again, this time presenting a list of printers, as shown in Figure A.7. You can click a printer name from the list and then click Select to install it. Repeat this action for as many printer definitions as you want to install. It's best to select all your printer definitions here; if you don't, you may need to get your printer drivers from the Web.

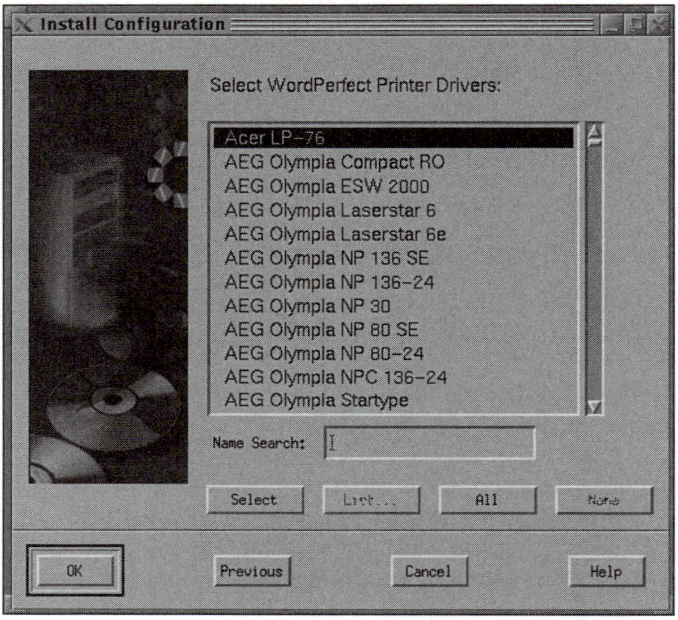

Figure A.7
Click List to see all the printer drivers you've selected so far.

Tip #298 from *Tux Tips*

As a general rule, Linux programs assume that they're printing to PostScript printers. Linux includes software to convert PostScript into formats that non-PostScript printers can understand. Thus, even if your printer is listed in WordPerfect's printer list, you may want to install a PostScript printer driver, such as the Apple LaserWriter IINT for 300-dpi monochrome printing, the Apple LaserWriter Pro 630 for 600-dpi monochrome printing, or the QMS magicolor for color printing at up to 600 dpi. You can install more drivers than you think you'll need and then delete the extras.

→ If you want to learn about installing printer drivers, **see** "Fonts and Printing," **p. 739**

10. The installer now gives you the opportunity to link the printer drivers you just selected to Linux printer queues (Figure A.8). Click the name of the printer driver in the Install Configuration dialog box and then click Assign to get the Destinations dialog box in Figure A.9. Select the Linux printer queue associated with that printer and click OK to make the assignment. You can also click Setup to get the Setup Information dialog box shown in Figure A.10. You should probably leave the Lpr radio button checked, but you can change the name by which the printer will be known in WordPerfect, and add any `lpr` spool options you want in the Spool Options field.

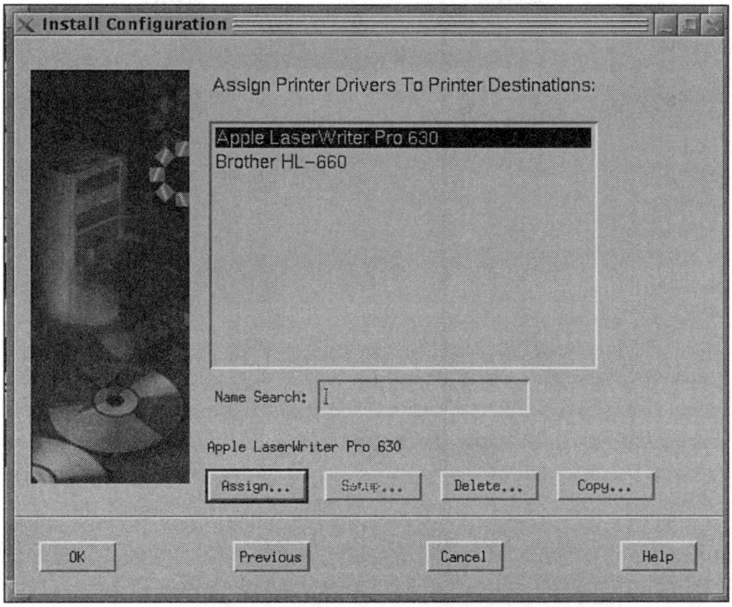

Figure A.8
You can customize your printer configuration before finishing the WordPerfect installation.

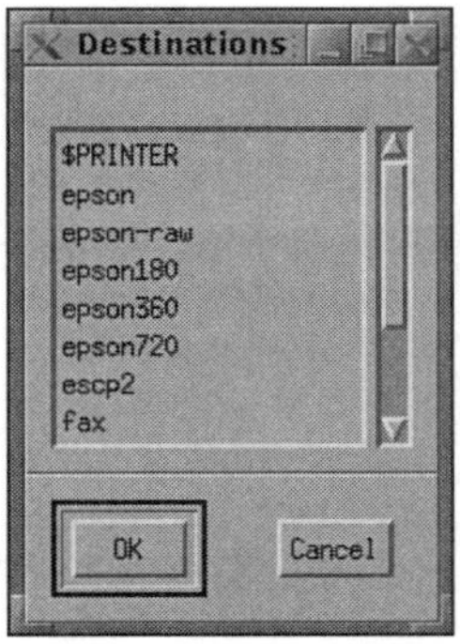

Figure A.9
The Destinations dialog box shows the printers listed in your /etc/printcap file.

11. The installer now gives you the option of installing a Common Desktop Environment (CDE; a commercial GUI for UNIX and Linux) desktop icon, the WordPerfect online manual, and Netscape 3.04. (Only the CDE Desktop icon is available with the downloadable version of WordPerfect.) If you're using CDE, you should definitely select the first option, and you might find it useful even in other environments, as described later in this chapter. The online manual can be handy if you like to use online documentation. Netscape 3.04 is quite old, however, and this book's CD comes with Netscape 4.6, so you may want to opt out of the Netscape option.
12. After you click OK again, the installer reviews the steps it has just taken. This is your last chance to change an option; you can do so by clicking Previous as many times as you must to return to the option you want to change. If you're satisfied with the configuration you've chosen, click OK.

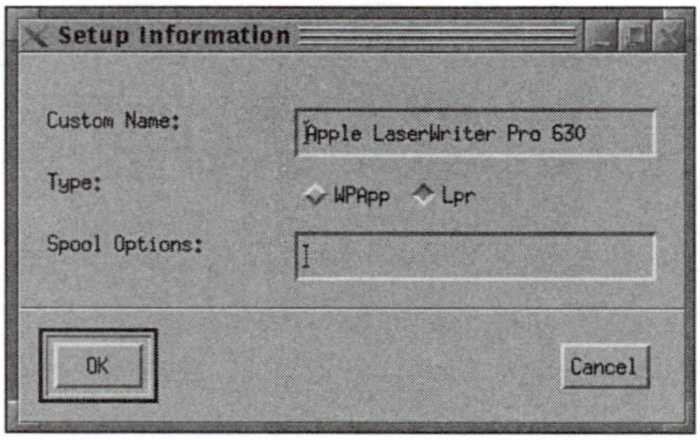

Figure A.10
If you're using a printer driver for a printer that's similar, but not identical, to the one you own, you may want to change the Custom Name field to reflect your actual printer's name.

The installer presents a series of installation progress bars, each of which counts up from 0% to 100% completion. When these dialog boxes finish, the installer displays the README Review dialog box (Figure A.11). If you'd like information on any of the topics displayed in this list, select the topic and click View. Otherwise, click Done.

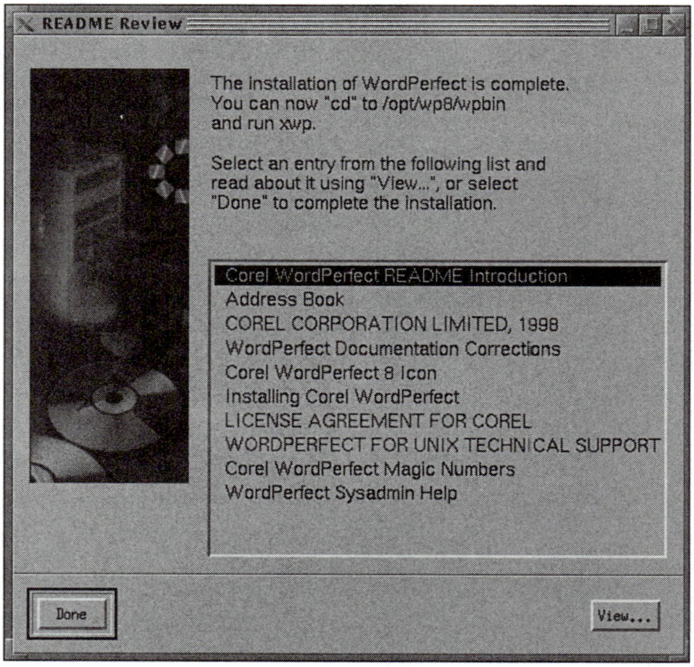

Figure A.11
You can get information about various WordPerfect topics from the README Review dialog box.

POSTINSTALLATION CLEANUP AND TESTING

If you're installing the personal retail version of WordPerfect, you should now move out of the CD directory and unmount the CD. The commands `cd ~; umount /mnt/cdrom`, or something similar using the correct mount point, should do the trick.

If you're installing the downloadable version, you'll find that the installer has renamed the install files, so they now have .bk extensions. You no longer need these, so you can delete them.

Tip #299 from *Tux Tips*

If you try to run the installation again after the downloadable version has finished one install, the installer sees the .bk extensions and reports that there's nothing to install. Therefore, if you aborted the installation, you should use the `mv` command to rename all the .bk files to remove that extension. Alternatively, delete them all and use `tar` to extract the files from the original archive a second time.

To test your WordPerfect installation, type `/opt/wp8/wpbin/xwp` (or something similar using whatever install path you specified in place of /opt/wp8). WordPerfect should start, and you can begin using it. If you installed the downloadable version, WordPerfect immediately begins asking you for a registration number, using the Warning dialog box in Figure A.12, and continues to ask every time you start the program. The simplest way around this request is to register with Corel, at `http://nas.corel.com/nasapps/wp8linuxreg/register.html`. Registration costs nothing but the time it takes to fill out the forms.

→ For a quick introduction to WordPerfect and its features, **see** "A Window and Menu Quick Tour," **p. 28**

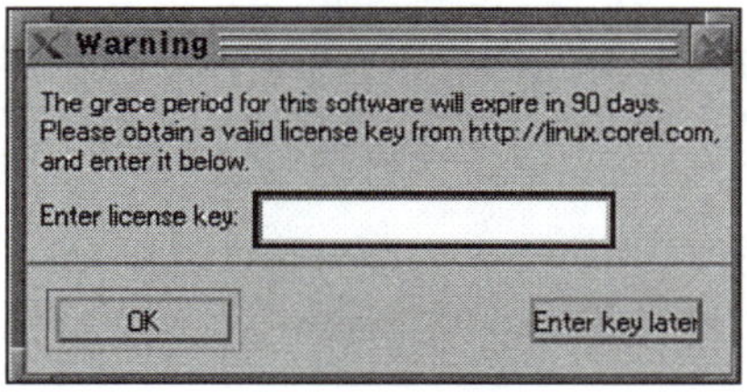

Figure A.12
You can use the downloadable version of WordPerfect for up to 90 days by clicking the Enter key later button whenever you start the program, or you can obtain a free registration number from Corel to use it indefinitely without the need to dismiss this dialog box.

Note

The downloadable version of WordPerfect is free for most *noncommercial* uses. If you intend to use WordPerfect in a business setting, you can use the downloadable version for evaluation purposes, but you're legally obligated to purchase the retail version to continue using it. The WordPerfect license has the details.

Note

If you've previously tried any version of WordPerfect for Linux, the downloadable version may complain that your time has expired. You can obtain a registration number from Corel's Web site to get around this problem.

Note

You may need to run WordPerfect the first time as root, or as the user who installed the software, for WordPerfect to update its registration information.

Are you unable to run WordPerfect after you've installed it? See "WordPerfect Won't Run" in the "Troubleshooting" section at the end of this chapter.

CONTROLLING ACCESS TO WORDPERFECT

Starting WordPerfect by typing `/opt/wp8/wpbin/xwp` at an xterm every time you want to run it can get awkward. If you're an experienced Linux user, you may have tried creating a *link (p. 19)* to the WordPerfect binary in a directory that's on your path. This procedure won't work, however (WordPerfect is just peculiar in this respect). Instead, try this:

1. Log in as root or use `su` to gain root access.
2. In an xterm, change to a directory that's on your regular search path. For instance, `cd /usr/local/bin` will probably work.
3. Use your favorite text editor to create a text file called xwp. That file should contain one line: `/opt/wp8/wpbin/xwp $1 $2`. (Change the path to the xwp file, if necessary, to match your system.)
4. Save the file and exit from the text editor.
5. Type `chmod a+x xwp` to make the text file executable.

These steps create a simple *shell script* that does nothing but run WordPerfect. The `$1 $2` part of the script passes up to two filenames to WordPerfect, so you can load up to two files by typing their names after you type `xwp`. If you want to be able to load more files, you can include `$3, $4`, and so on. Extra parameters won't do any harm.

You can add any of the startup options that WordPerfect recognizes to your shell script, if you like. These options alter the way WordPerfect works with Linux fonts, its color handling, and so on.

TABLE A.1 WORDPERFECT STARTUP OPTIONS

Option	Description
filename	Add a filename to have WordPerfect load it. The `$1 $2` part tells the shell script to pass any filenames the script gets on to WordPerfect.
filename -bk bookmark	Load the file and move to the named bookmark.
-adm	Start WordPerfect in administrative mode. Use this mode to add or edit printers or fonts or to perform other administrative tasks.
-docGeometry	Specify a size for the WordPerfect Document window. For instance, `-docGeometry 600x400` forces WordPerfect to start with a 600×400 document window.

Option	Description
-fastGraph	Use faster but less accurate graphics rendering.
-fontSize	Display fonts in menus and dialog boxes in the specified size. For instance, `-fontSize 50` displays fonts at 50 points, which may be useful if you have a small screen or are visually impaired.
-help	Displays the available startup options.
-ignore	Ignores the options in your `~/.wprc/.wp8x.set` file.
-lang language	Starts using the specified language (`language` is a 2-letter language code).
-macro macroname	Runs the macro called `macroname` as soon as WordPerfect starts.
-maxColors xx	Limits the number of colors WordPerfect uses. `xx` can be 2, 16, 125, or 216.
-noauto	Turns off auto code placement. This feature lets you place paragraph- or page-oriented codes anywhere you want—a risky proposition.
-nobackup	Disables WordPerfect's automatic backup features.
-onewp	Tells WordPerfect to open a new file in an existing instance of WordPerfect, if one is already open, and if that instance also used the `-onewp` option.
-printer "name"	Sets the default printer to the one called `name`. Quotation marks are required if the printer name contains a space.
-recover	Attempts to fix a document that's not working properly, especially if the problem is related to tables.
-restore	Deletes your ~/.wprc/.wp8x.set file, allowing you to restore various WordPerfect settings to their default values.
-setup path	Sets the directory from which WordPerfect reads the .wp8x.set file to `path`.
-temp path	Sets the directory in which WordPerfect stores temporary files to `path`.

→ To learn how to install printers and fonts in WordPerfect for Linux, **see** "Fonts and Printing," **p. 739**

→ To learn how to set assorted WordPerfect defaults, **see** "Customizing Preferences," **p. 767**

→ To learn more about WordPerfect's language handling, **see** "Languages," **p. 141**

→ To learn more about macros, **see** "Letting the Computer Do the Work: Macros," **p. 483**

→ To learn how to select a printer, **see** "Selecting a Printer," **p. 151**

If you're managing a multiuser system and you want to configure WordPerfect so that not all your users can run it, do the following:

1. Log in as root or use `su` to gain root access to your system.
2. Create a new group for WordPerfect access by editing /etc/group or by using whatever user and group management tools your Linux distribution provides. Be sure to add all the users to whom you want to give access to WordPerfect to this group. (In the following steps, I refer to this group as the *wp* group, but you can call it something else.)

3. Change the group for the WordPerfect files. For instance, `chgrp --recursive wp /opt/wp8` changes all the files in the /opt/wp8 directory to the wp group.
4. Remove world read and execute permissions on the WordPerfect files. For instance, `chmod o-rx /opt/wp8 --recursive` removes these permissions for the files in /opt/wp8.

Now, only users in the wp group can run WordPerfect or access the files in the WordPerfect directory tree.

CONFIGURING GUI ENVIRONMENTS TO WORK WITH WORDPERFECT

You may want to run WordPerfect directly from a menu of some sort in X, rather than by typing `xwp` in an xterm window. You can do so, but you need to know something about configuring your window manager or file manager to get this method to work. As such environments for Linux number in the dozens, it's not practical to present a complete enumeration here. I can, however, present instructions for some of the more popular window managers and GUI environments.

Note

If you elected to install the CDE icon when you installed WordPerfect, you can find it in the wplib/dt/appconfig/icons/C subdirectory under the WordPerfect install directory. This subdirectory contains two files, wp8App.m.pm and wp8Run.m.pm, both of which contain the same 32×32 image. You can copy either file to an appropriate icons directory for your GUI environment and, if that environment can use a 32×32 XPM image, use the file to provide a WordPerfect icon for that environment.

KDE

K Desktop Environment (KDE) is an increasingly popular GUI desktop environment for Linux. It includes a window manager, a file manager, and assorted utilities to help make a pleasant working environment. Configuring it to launch WordPerfect for Linux involves creating a couple of files. The CD that accompanies this book includes the files you need to create, so you can copy them from the CD to the appropriate directories. If you want to configure your system so that all KDE users have access to WordPerfect through KDE, you need to copy these files to your KDE directory tree (probably /opt/kde). To configure just your own individual login, create the necessary configuration files in your ~/.kde directory:

1. Copy the x-wp.kdelnk file to the share/mimelnk/application directory (under either ~/.kde or your KDE install directory). This file contains information on a KDE desktop entry for WordPerfect so that you can double-click WordPerfect document files to launch the program.
2. Copy the xWP.kdelnk file to the share/applnk/Applications directory. This file contains an entry to place WordPerfect in the KDE menu structure.

3. Copy the CDE icon or some other icon you'd like to use for WordPerfect, to the share/icons directory. For instance, `cp /opt/wp8/wplib/dt/appconfig/icons/C/wp8App.m.pm /opt/kde/share/icons/wp.xpm` makes the icon globally accessible.

Note

The KDE configuration files on the CD assume that WordPerfect is on your search path, so you need to follow the directions earlier in this chapter to set up a shell script to launch WordPerfect from a location on your path. Alternatively, you can edit both the x-wp.kdelnk and xWP.kdelnk files to provide them with the complete path to the WordPerfect executable on your system.

You can customize the KDE configuration files further for WordPerfect. For instance, the ones on the CD recognize files with extensions of .wpd, .wp8, .wp7, and .wp6 (and uppercase versions of these) as WordPerfect files. You can add other document types so that, for instance, KDE launches WordPerfect whenever you click on a Microsoft Word .doc file. You can also edit the share/mimelnk/magic file to let KDE detect WordPerfect files even when they have odd filename extensions. If you like, you can provide a different icon for the KDE menu icon than for the KDE file manager. Consult your KDE documentation for help on these advanced configuration topics.

When you're done editing the configuration files, you may need to restart KDE to get it to recognize the changes. Quit your open applications and log out or exit from X; then log back in or restart X. You should now see WordPerfect in the KDE menu under the Applications heading, and you should be able to load WordPerfect documents into WordPerfect from the KDE file manager.

GNOME

GNU Network Object Model Environment (GNOME) is KDE's main competitor as an integrated desktop environment for Linux. As such, it has capabilities similar to those of KDE, including the capability of launching WordPerfect from its menus or by double-clicking a WordPerfect document in its file manager.

To add a WordPerfect icon to the main GNOME panel at the bottom of the screen (Figure A.13), right-click on a blank area of the panel and select Add new launcher from the context menu. This action opens the Create launcher applet dialog box shown in Figure A.14.

Figure A.13
GNOME includes a panel that normally resides at the bottom of the screen. This panel includes quick-launch tools, a clock, and other conveniences.

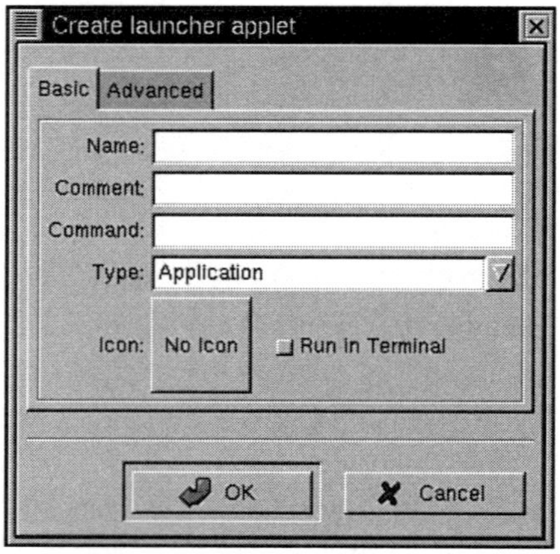

Figure A.14
The GNOME Create launcher applet dialog box lets you create a new panel launcher for a program such as WordPerfect.

In the Create launcher applet dialog box, you can enter information on WordPerfect:

- Name—The name as you want it to appear in a menu. For instance, you could use *WordPerfect* or *WP for Linux*.
- Comment—A comment to be associated with the program. You might specify whether you are using the personal retail or downloadable version, for instance.
- Command—The command to launch WordPerfect. If the program or a shell script to launch it is on your path, you can type `xwp` here. Otherwise, type the complete path to the program, such as `/opt/wp8/wpbin/xwp`.
- Type—Leave the default value: Application.
- Icon—Click on the Icon button to get the Choose an icon dialog box shown in Figure A.15. If you previously copied the CDE icon from wplib/dt/appconfig/icons/C/ to /usr/share/pixmaps and gave it an .xpm extension, GNOME presents it as one of the options in this window. Select it and click OK.

Figure A.15
You can select any icon you like for WordPerfect from the Choose an icon dialog box.

When you've entered the information in the Create launcher applet dialog box, click OK and GNOME creates a new WordPerfect launcher in its panel. You can manipulate it as you would other GNOME launchers—consult the GNOME documentation for details.

If you want to be able to automatically launch WordPerfect files when you double-click them in a GNOME file manager window, select Settings, GNOME Control Center from the main gnome panel menu to open the Control Center window shown in Figure A.16. Select Mime Types in the panel on the left side of the Control Center window to display and edit the Mime entries, with which you associate document types with applications.

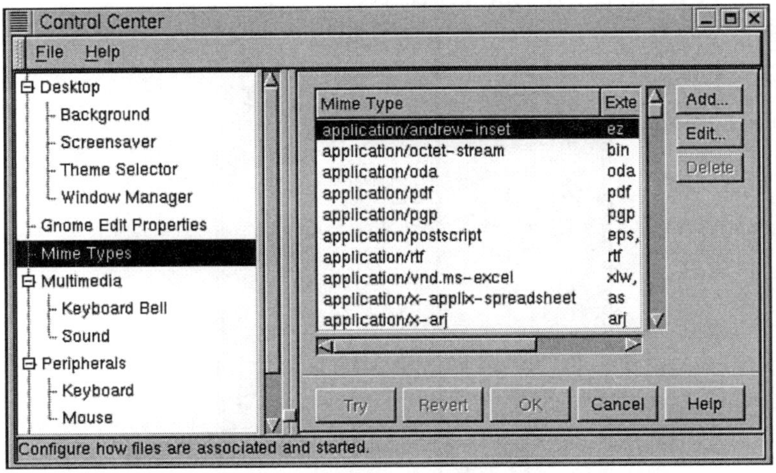

Figure A.16
You can create a Mime type to link WordPerfect documents to WordPerfect.

To add an appropriate Mime type:

1. Click Add in the Control Center window to open the Add Mime Type dialog box shown in Figure A.17.

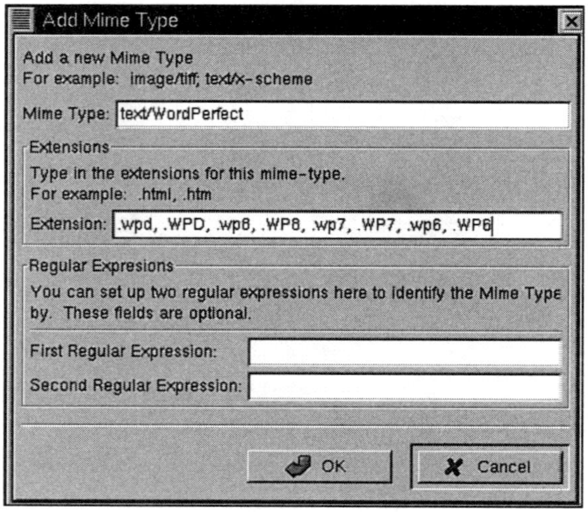

Figure A.17
The Add Mime Type dialog box lets you associate filename extensions with a Mime type.

2. Enter `text/WordPerfect` in the Mime Type field of the Add Mime Type dialog box, as shown in Figure A.17.
3. Enter the filename extensions you want to associate with WordPerfect in the Extension field, as shown in Figure A.17.
4. Click OK in the Add Mime Type dialog box. The text/WordPerfect type should now appear in the Control Center dialog box.
5. Be sure that text/WordPerfect is selected and click Edit to bring up the Set actions for text/WordPerfect dialog box shown in Figure A.18.

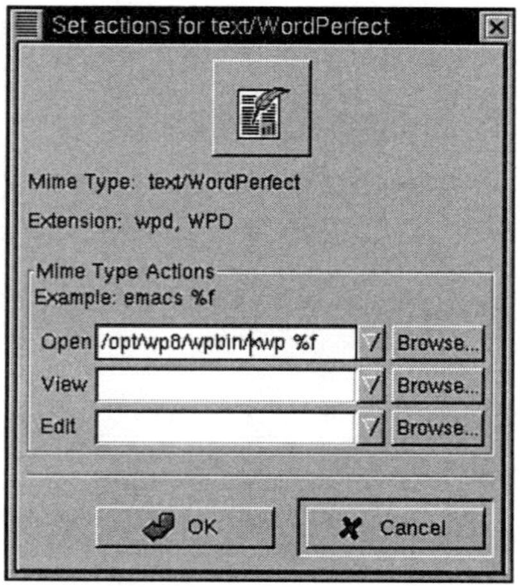

Figure A.18
You can tell GNOME what program to launch and what icon to display for WordPerfect files with the Set actions for text/WordPerfect dialog box.

6. Type the name of the WordPerfect executable and, if necessary, its complete path, in the Open field, followed by a space and `%f`, as shown in Figure A.18.
7. Click the icon in the upper portion of the dialog box to get the Select an icon dialog box, which looks and works just like the Choose an icon dialog box (refer to Figure A.15). Pick an icon for your WordPerfect files and click OK.
8. Click OK in the Set actions for text/WordPerfect dialog box; then click OK in the Control Center dialog box and choose File, Exit (Ctrl+Q) from the Control Center dialog box's menu.

Now your WordPerfect files appear in GNOME file manager windows with their unique icons, and when you double-click on one, WordPerfect starts and loads that file.

THE FVWM FAMILY

One of the older of the popular window managers in Linux is fvwm. In fact, fvwm has "forked" into several related window managers, such as fvwm2 and fvwm95. Fortunately, configuring these programs is similar. Unlike KDE and GNOME, fvwm is only a window manager; it doesn't include a file manager or extra utilities. You can, however, set up pop-up menus and, in some varieties, a toolbar at the bottom of the screen, and place entries to WordPerfect in these menus and toolbars.

When you left-click on the desktop when you're running fvwm, it presents a context menu (Figure A.19). Typically, this menu contains a number of submenus, indicated by the entries with arrowheads next to them. Clicking on a submenu heading opens the submenu.

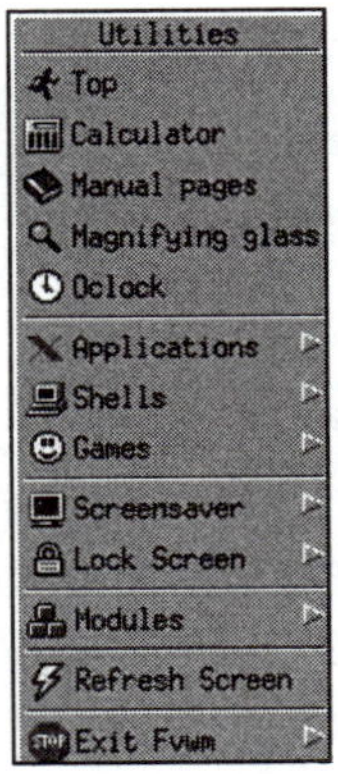

Figure A.19
The fvwm pop-up menu looks different for the different varieties of fvwm. This one is an fvwm2 context menu.

Note

With some varieties of fvwm, you need to click and hold the mouse button to keep the context menu on the screen. With other varieties, you click once, and the menu stays up until you make a selection from it.

Some varieties of fvwm include menu bars at the bottom of the window. With these, you can click on an item, typically labeled Start, in the lower-left corner of this menu bar, to bring up a menu similar to the context menu.

You perform basic fvwm configuration by editing a text file in your home directory. This file is called .fvwm or some variant of that, such as .fvwm2rc95. If you don't know what yours is called, try using the command `ls ~/.fvwm*` to see all the possibilities. After you locate the fvwm configuration file, load it into a text editor.

> **Note**
>
> If you don't see any files starting with .fvwm in your home directory, fvwm must be loading its configuration file from a systemwide default file. This file may be located in /usr/X11R6/lib/X11/fvwm or a similarly named directory. You can use the command `find /usr -name "*fvwm*"` to locate it. The file itself may be called system.fvwmrc or a similar variant. You can edit the systemwide file directly, or you can copy it to your home directory and rename the file by removing the *system* portion of the filename so that it begins with a period.

If your variety of fvwm includes a Start menu, you can edit it by editing text in the various AddToMenu items in the fvwm configuration file. Search for `AddToMenu`, and you'll probably find something that looks like this:

```
AddToMenu "StartMenu"
+ "New Shell      %mini-sh1.xpm%"   Exec nxterm &
+ "Manual Pages   %mini-book1.xpm%" Exec xman &
+ "Applications   %mini-x2.xpm%"   Popup Applications
+ "Games          %mini-happy.xpm%"  Popup Games
+ ""
+ "Exit Fvwm      %mini-stop.xpm%"   Popup Quit-Verify
```

You can add WordPerfect to the main Start menu by inserting a line where you want the WordPerfect item to appear similar to the New Shell or Manual Pages entries, thus:

```
+ "WordPerfect   %wp-mini.xpm%"   Exec xwp &
```

> **Note**
>
> If fvwm finds the file wp-mini.xpm and if it's a 16×16 bitmap, fvwm uses that icon in the menu. If you installed the CDE icon, you can load it into an image editor such as the GIMP, reduce the size, and save the icon to the /usr/share/icons/mini directory to have fvwm use it.
>
> The preceding entry assumes that either the WordPerfect executable or a shell script such as I described earlier in this chapter exists on the regular Linux search path. If it doesn't, you need to give the complete path to the WordPerfect executable file, rather than just `xwp`, for this example to work.

If you prefer to have WordPerfect appear on a submenu, say under *Applications*, you'll need to locate the section of the configuration file beginning with `AddToMenu "Applications" "Applications" Title`. It contains a series of lines preceded by plus signs similar to those for the main Start menu, and you can add WordPerfect to the submenu just as you do to the Start menu itself.

Some versions of fvwm don't use `AddToMenu`, but instead use `Popup`. They may also use a somewhat different arrangement of information in their pop-up entries. For these fvwm varieties, you might want to use the following line to add WordPerfect to their menus:

```
Exec  "WordPerfect"  exec xwp &
```

In general, you'll be able to get an appropriate result by mimicking what you see in other entries in the fvwm configuration file.

When you finish editing the fvwm configuration file, you need to restart the window manager. Left-click on the desktop to get the context menu. You'll probably see a menu item called Exit Fvwm, Quit, or something similar. Select it. Fvwm may log you out immediately, but more likely it shows you another menu with options for starting other window managers, quitting, or restarting. If this option is present, select the restart option. This action causes fvwm to reload its configuration files without actually logging you out or exiting from X. If there's no restart option, you need to quit from fvwm and then log back in again or restart X.

→ If you want more information on logging in and starting X, **see** "If You Are New to Linux," **p. 10**

OTHER WINDOW MANAGERS AND FILE MANAGERS

Most Linux window managers use configuration files that are similar in basic principle to the fvwm configuration files, although some window managers have several configuration files inside a subdirectory whose name starts with a period (.). Whatever their names and locations, you need to edit the configuration files with a text editor to add one or more lines pointing the program to your WordPerfect executable file. You can generally model the WordPerfect entry on one for an existing program. You can also consult the documentation for the window manager to find out how to configure it in the way you want.

In addition to window managers, quite a few file managers are available for Linux. These programs let you manipulate files on your hard disk in a GUI environment. Both KDE and GNOME include file managers, and in fact, some of the configuration I outlined for them involved setting up their file managers. You may want to do the same if you're using a Linux file manager other than KDE or GNOME. As with window managers, the details differ, but you generally need to edit a configuration text file and add information to it to let the file manager know how to identify WordPerfect files and what to do when it finds them. Try modeling your entries after those for existing file types, such as .gif or .txt files.

SETTING UP LINUX PRINTER QUEUES

Linux's printing system varies substantially from one distribution to another, so it's not possible to give directions on setting up printer queues that are guaranteed to be both concise and accurate for all systems. All Linux printing systems do share certain characteristics, however, and this section discusses some of them.

Note

You need root access to do most of the things I discuss in this section, so you need to either log in as root or use `su` at an xterm to gain root privileges.

Understanding Linux Printing

Most Linux systems use a utility called `lpr` to handle printing. You can invoke this program at an xterm or text-mode login to print a file, as in `lpr output.ps`. This command sends the file output.ps to the default printer (you can specify another printer by using the `-P` parameters, as in `lpr -Plw630 output.ps`—note that there's no space between the `-P` and its value).

The `lpr` utility does not itself send the file to the printer, however. Instead, `lpr` places a copy of the file in a *spool directory* and lets other programs do much of the real work. A spool directory is a location where files reside temporarily before they're processed and deleted. The printer spool directory is closely related to the concept of a *printer queue (p. 150)*.

➔ To learn more about the `lpr` command, printer queues, and manipulating files in printer queues, **see** "Controlling Printing from Linux," **p. 159**

Once `lpr` places a file in the printer spool, a program called `lpd` takes over. This program is also known as the *printer daemon*, and in a properly configured Linux system, it's always running. It checks to see whether any files are in the printer spool and, if there are, processes them for printing. `lpd` follows the instructions it finds in the /etc/printcap file. This file tells `lpd` what other programs to call to modify the file before sending it to the printer and what sorts of options are available for that printer. Therefore, most of the task of configuring a Linux printer queue involves editing the /etc/printcap file.

Note

The /etc/printcap file may refer to other configuration files, and it may invoke programs that have their own configuration files. Therefore, you may need to edit more than just /etc/printcap to get printing to work the way you like it. If you're not printing to a PostScript printer, you're especially likely to need to configure the *Ghostscript (p. 380)* program, which converts PostScript into formats that non-PostScript printers can understand.

One feature that many Linux printer queues use is a smart filter. A *smart filter* is a program that `lpd` calls to determine the type of file being printed—whether it's a PostScript file, ASCII text, a JPEG graphic, or something else. The smart filter then converts the file into a common format (typically PostScript). With a smart filter, you can print many file types without being concerned with what they are.

Caution

Most smart filters don't know how to process WordPerfect files. Therefore, chances are you won't be able to print a WordPerfect file by using `lpr` directly; you must load the file into WordPerfect before you can print it.

USING A DISTRIBUTION'S GUI PRINTER CONFIGURATION TOOLS

Many Linux distributions include utilities to help you configure your printer. These take care of most or all of the details of setting up configuration files, including the use of smart filters. Examples of such configuration tools include SuSE's YAST and Red Hat's printtool and linuxconf. Figure A.20 shows printtool's main window, called RHS Linux Print System Manager. If you're running Red Hat Linux, you can start it by typing `printtool` at an xterm window.

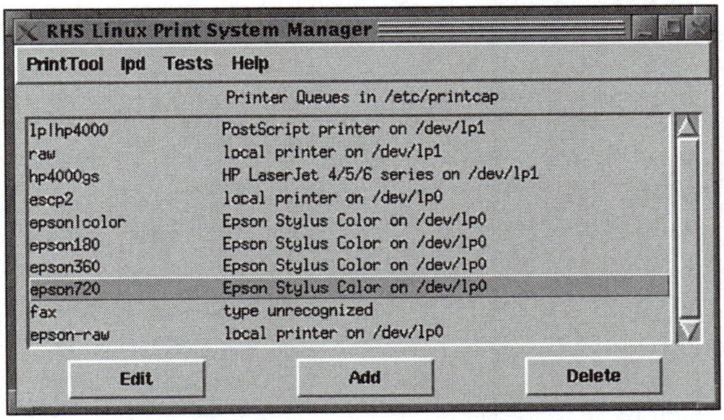

Figure A.20
Red Hat's printtool makes setting up printer queues easy if Red Hat supports your printer.

The details of how to set up a printer with a GUI tool vary from one tool to another, of course, but Red Hat's printtool is typical. You add a printer queue by clicking the Add button in the RHS Linux Print System Manager window. At this point, you tell it whether the printer is *local* (connected directly to the computer) or one of several *networked* types (connected via a network). The printtool utility then shows the Edit Local Printer Entry dialog box (Figure A.21) in which you enter information such as the printer's name, spool directory, printer device, and input filter. The name and spool device are largely arbitrary; you can set them to whatever you like, though it's generally best to keep the printer name short and the spool directory inside /var/spool/lpd. The printer device is how `lpd` communicates with the printer hardware. For a parallel printer, this device is normally /dev/lp0 or /dev/lp1, but it could be /dev/ttyS0 for the first serial port, or even more exotic devices. If you're defining a queue for a network printer, printtool asks for the remote printer's host and queue name instead of a printer device.

Much of the trick in configuring a printer is in choosing the correct Ghostscript driver and options. If you're using printtool, you do this configuration from the Configure Filter dialog box (Figure A.22), which you obtain by clicking Select in the Edit Local Printer Entry dialog box. If you're lucky, you can select your printer from the Printer Type list, set the resolution and paper size, and be done with it. It's more likely, however, that you'll need to select a printer that's compatible with yours. For instance, if your printer uses Hewlett-Packard's PCL 5e language, you need to select HP LaserJet 4/5/6 series, which is the HP series that uses PCL 5e. The Configure Filter dialog box provides only a single entry for all PostScript printers, so select that if you're using *any* PostScript printer.

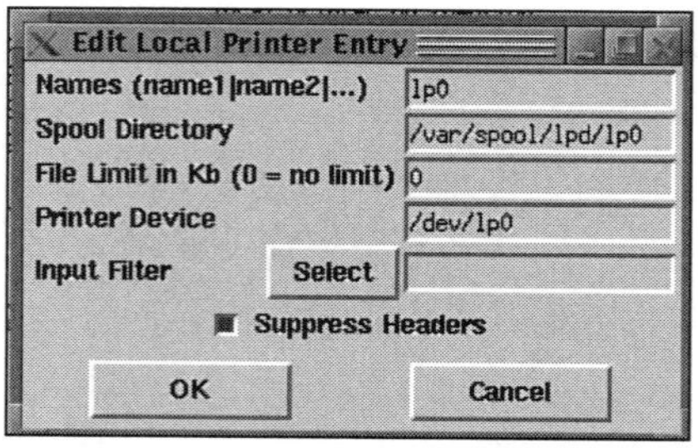

Figure A.21
If you want to limit the size of files Linux will print, you can adjust the File Limit in Kb field.

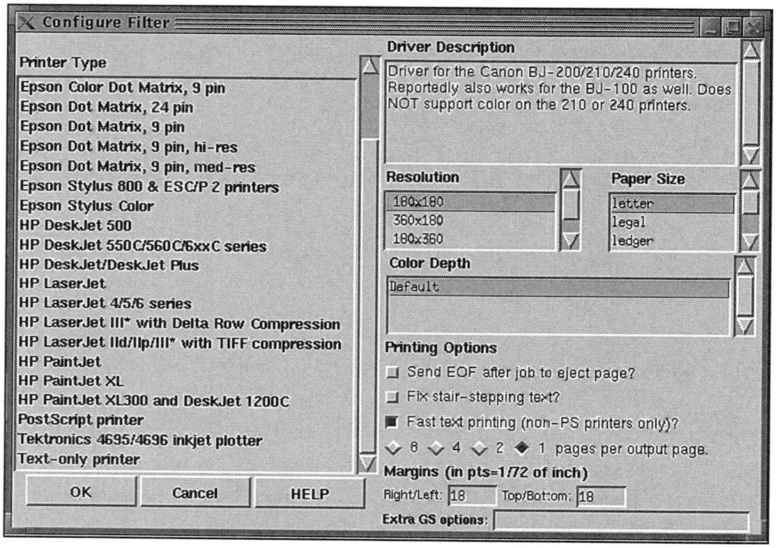

Figure A.22
You can set many printer options in the Configure Filter dialog box.

In addition to setting the printer model and obvious options such as the resolution, a GUI utility such as printtool may let you modify other printing options:

- Send EOF—You can configure a Linux printer queue to send an end of file (EOF) code after a print job to eject the page. This option isn't normally necessary if you use Ghostscript to print.
- Fast text printing—When `lpd` encounters a text file destined for a non–PostScript printer, `lpd` has two options: either send the file directly as text or convert the file to PostScript and run the result through Ghostscript. The first option generally prints faster using the printer's font, but you might prefer the latter to get better-looking text on some printers. This option might be called something other than Fast text printing, depending upon your print configuration utility.
- Fix stair-stepping text—If you elect to print ASCII as plain text, some printers may misinterpret the UNIX-style end-of-line codes. In this case, you get a line of printout, then the next line begins one line down but to the right of the first line, and so on for subsequent lines. You might therefore get only a couple of lines of text on such a printout

because the rest of the lines begin beyond the page's right edge. Most printer configuration utilities provide an option to fix this problem.

Tip #300 from *Tux Tips*

If you want to create a printer queue for use with a WordPerfect driver for a non-PostScript printer, *do not* click Select to set up a filter in the Edit Local Printer Entry dialog box (refer to Figure A.21). Red Hat's printtool then creates a printer entry that doesn't try to process the output through its smart filter, and the results get passed to the printer unchanged, which is what's needed if you're using a WordPerfect driver for a specific model printer.

When you're done configuring your new printer queue's options, you can test the queue. Many GUI configuration utilities provide options for doing such tests, such as printtool's Tests, Print ASCII test page; Tests, Print Postscript test page; and Tests, Print ASCII directly to port features. If you get acceptable results from these tests, you're set. If not, you may need to go back and edit your configuration in one way or another.

Tip #301 from *Tux Tips*

If you get a printout when you try something like printtool's Tests, Print ASCII directly to port, your printer port *is* configured correctly, so problems with other printout types relate to the filter configuration. If Tests, Print ASCII directly to port produces no output and no blinking lights on the printer, then you should try other printer port options before anything else. If your GUI utility provides no "direct to port" test option, you can run the test with a short ASCII text file by issuing a command such as `cat test.txt >/dev/lp0` from an xterm. If you're testing a laser printer, you may need to use its front-panel controls to eject a page after doing a test printout in this way.

Configuring Ghostscript

As I've said, picking the correct Ghostscript driver for your printer can be one of the challenges in configuring Linux to work with your printer. If you're not sure what to use, try these Web sites for advice:

`http://www.cs.wisc.edu/~ghost`

`http://www.picante.com/~gtaylor/pht/printer_list.cgi`

The first site is the main Ghostscript Web site, and the second is a printer compatibility database. Both have information on the availability of Ghostscript drivers for specific models of printer. Most printers use a Ghostscript driver for a broad *class* of printers, such as printers that use the PCL 5e language. You therefore don't generally use a driver that's named after your specific model of printer.

Even if Ghostscript includes a driver for your model of printer, the specific Ghostscript binary file you have installed may not include that driver. This is because the person who compiled Ghostscript had to choose which drivers to include and which to omit. You can find out which you have by typing `gs --help` at an xterm. If you discover that a Ghostscript driver does exist for your printer, but is not included in your binary, and if you're using a

Linux distribution based on the *Red Hat Package Manager (RPM)* distribution format, you can try the Ghostscript binary from `http://www.users.dircon.co.uk/~typhoon/`. This RPM includes more drivers than most Ghostscript binaries. As a last resort, you can compile Ghostscript yourself. The directions for compiling the program come with the Ghostscript files from the program's Web site. This process is a bit tedious, so be sure to read the directions carefully.

Because of Ghostscript's licensing terms, most distributions ship with a version of Ghostscript that's out of date. You can upgrade to a more recent version from many sources, including the main Ghostscript Web site, which I mentioned in the preceding paragraph, and your distribution's Web site. If you can print from WordPerfect without such an upgrade, you may not want to upgrade; but if you're having problems, a Ghostscript upgrade might be what you need to fix them.

You can configure Ghostscript to use different sets of fonts by editing its Fontmap file. This file typically resides in the Ghostscript directory at /usr/share/ghostscript/x.yy, /usr/local/ghostscript/x.yy, or a similar location (*x.yy*, in this case, is the Ghostscript version number, such as *5.50* for version 5.50). Editing the Fontmap file lets you add fonts to Ghostscript much as you could install them on a hard disk attached to a PostScript printer. Adding fonts to Ghostscript's Fontmap file isn't required for using fonts with WordPerfect, but you might want to do so to reduce the size of your PostScript files or to make available fonts that a particular WordPerfect printer driver assumes are installed in the printer.

→ For more information on configuring fonts in WordPerfect, **see** "Installing and Managing Fonts," **p. 754**

EDITING THE CRITICAL CONFIGURATION FILES

The most critical Linux printer file is /etc/printcap. This file defines the capabilities of printers, hence its name. It contains one entry for each printer. Each entry looks something like this:

```
lp¦canon:\
    :sd=/var/spool/lpd/canon:\
    :mx#0:\
    :sh:\
    :lp=/dev/lp0:\
    :if=/var/spool/lpd/canon/filter:
```

Each /etc/printcap entry is supposed to reside on a single line; the backslash (\) character at the end of all but the final line in this example tells `lpd` to treat the next line as part of the preceding line in order to get around this single-line rule. The fields of the /etc/printcap entry are separated by colon (:) characters. Most of the fields contain assignments to *variables*; the variable falls on the left of an equal sign (=) or pound sign (#), and the value it's assigned falls to the right. Here's what each field does:

- `lp¦canon`—The first line identifies the name for the queue. The vertical bar (¦) separates two names for the same queue. In this example, the queue can be accessed under the names *lp* and *canon*.

- `sd=/var/spool/lpd/canon`—This field identifies the spool directory (hence, `sd`). This directory is where `lpr` places files destined for this printer and where `lpd` looks for them.
- `mx#0`—This field specifies the maximum file size that Linux will accept. The value of `0` indicates that this option is disabled, so there's no file size limit.
- `sh`—This field indicates that the system won't print a banner page. If you remove this line, Linux prints a page identifying the user who created the printout before each job prints.
- `lp=/dev/lp0`—The `lp` variable identifies the printer device.
- `if=/var/spool/lpd/canon/filter`—The `if` variable indicates the name of the *input filter* file. This filter is a program that processes the input file in some way, such as converting ASCII to PostScript or PostScript to PCL 5e. On most systems, you'll want to set this variable to point to a smart filter, but you can omit this field entirely if you want to print to the printer with a WordPerfect driver for that specific model.

Setting up the input filter is probably the trickiest part of configuring a Linux printer queue. Most distributions do include tools to help with this job, but if yours doesn't or if you want to do it yourself, you may want to look for a package such as APS Filter, which is one of the smart filter programs. Consult the filter's documentation for how to configure the filter and then pass its name in the `if` variable.

There are many more possible variables in an /etc/printcap file; the preceding example just scratches the surface. If you want to know more, try typing `man printcap` to get your system's documentation on the /etc/printcap file, check the Linux Printing HOWTO, or consult a book such as Que's *Special Edition Using Linux*.

TROUBLESHOOTING

CD-ROM MOUNT PERMISSIONS

You're trying to run a WordPerfect install program from a CD, but you get a Permission denied *error even when you run the installer as root.*

This problem is a result of a safety precaution. If you have an /etc/fstab entry for your CD-ROM and mount the CD with a shortened `mount` command (such as `mount /mnt/cdrom` rather than `mount /dev/cdrom /mnt/cdrom`) or if you use a tool from a GUI environment such as GNOME to mount the CD, you won't be able to run programs from the CD. This limitation exists to reduce the likelihood of damage by malicious individuals who mount a CD and use programs on the CD to damage a system. To get around this problem, you need only unmount the CD and remount it *as root* using the full `mount` syntax, as in `mount /dev/cdrom /mnt/cdrom`.

WORDPERFECT WON'T RUN

You've installed WordPerfect, but it won't run when you type its path and filename, as described earlier in this appendix.

This problem has a number of possible causes, and the text of any error message in the xterm you use to launch WordPerfect may provide helpful diagnostic clues. One potential problem is a mismatched or missing *libc library*. This library is a set of routines common to many programs, including WordPerfect. WordPerfect 8 for Linux was compiled with version 5 of the Linux libc libraries (or *libc5* for short), but most new Linux distributions use libc6, more commonly known as *glibc*. Distributions based on glibc usually include libc5 for compatibility with older programs, but it's possible that libc5 isn't installed on your system or that it has an incompatible version of libc5. Try consulting your distribution's documentation to determine (1) which version of libc is included and (2) how to install support for libc5 programs if libc5 is not installed already.

You can't run WordPerfect by setting a symbolic link to it in a directory that's on your regular search path and then typing the name of the symbolic link. This procedure does work for most other programs, but not for WordPerfect.

➔ To learn how to configure WordPerfect to run when you type `xwp`, **see** "Controlling Access to WordPerfect," **p. 722**

It also bears mentioning that both the downloadable and personal retail versions of WordPerfect for Linux require X; they cannot be run from a text-mode login. If you want to run the text-mode version of WordPerfect for Linux, you'll need to buy the retail server edition of the program.

You can run WordPerfect for Linux from a computer other than the one on which it's installed. For instance, if you have a network with a Macintosh running LinuxPPC and an Intel-based computer running Linux for x86, you can install WordPerfect on the x86 computer and then use `telnet` from the Mac to log into the x86 computer. If you then type `xwp`, WordPerfect will run on the x86 machine, but display on the Mac. You'll need to set your DISPLAY environment variable correctly for this remote access to work, however. You can set the DISPLAY environment variable by typing either `export DISPLAY=`*`your.machine`*`:0.0` or `setenv DISPLAY=`*`your.machine`*`:0.0` where *`your.machine`* is the name of the machine from which you want to run WordPerfect. (Whether you use `export` or `setenv` depends on the Linux shell you're using.) You can do the same thing from a computer running Windows, MacOS, OS/2, or various other OSs if you're running an X package for those OSs.

APPENDIX B

Fonts and Printing

In this appendix

Understanding the Font and Printing System

WordPerfect for Linux is a powerful word processor, but if you want to get output from it, you need to be able to configure it to work with your printer. Unlike many OSs such as Windows, MacOS, and OS/2, Linux doesn't have a centralized printer driver system or an easy way to use the same fonts both on the screen and in printed output. Therefore, WordPerfect itself provides much of this "glue." Understanding how the WordPerfect and Linux printing systems work together will help you to decide how to configure your printer and fonts in WordPerfect for Linux.

A Brief History of Font and Printer Handling in Word Processors

To fully understand WordPerfect for Linux's current printing system, it's necessary to know something about how WordPerfect, and other word processors, have historically handled printers and fonts. This section therefore presents a little history.

In the Days Before GUIs

WordPerfect originated in the days when top-of-the-line printers were daisywheel impact devices, requiring manual intervention to change fonts, and when onscreen displays were restricted to a single monospaced font. Particularly flexible dot-matrix printers included half a dozen fonts.

In this environment, it made sense for a word processor to give the user a limited choice of fonts, which would be printed but not displayed in anything resembling a what-you-see-is-what-you-get, or *WYSIWYG* *(p. 10)*, fashion. WordPerfect adopted a strategy, consistent with its code-based design, of embedding font change codes in the document. WordPerfect maintained a database of character widths for various fonts, which it used to compute the ends of lines. WordPerfect also maintained an extensive set of printer drivers that, in terms of font support, served as a means of telling the printer to switch from one font to another. This system was simple and effective for any individual printer but was crude by today's standards. This approach also required a massive library of printer drivers.

Non-X GUI Font Handling

With the arrival of popular GUIs (mainly the Mac and Windows), as well as more flexible printer font handling, things changed. The demand for WYSIWYG word processing led to the emergence of a new font and printer strategy, where the word processor relied on the underlying OS for font handling, both for the onscreen display and the printer. In the conceptually simplest case, the word processor goes to the OS with a font request, and the OS then displays text in the desired font. When it comes time to print, the OS uses the *same* font file to print. Because MacOS and Windows include printer driver support in the OS, it's possible in these environments to strip away the extensive collection of printer drivers from the word processing application. This simplification makes coordinating onscreen and printer fonts relatively simple, at least from the word processor's point of view.

Of course, the details of such an arrangement can become tedious, especially when mixing font types (say, TrueType fonts in the OS, but a PostScript printer requiring Type 1 fonts). The word processor, however, needn't concern itself with most of these details; the tedium is handled by the OS and its drivers. Unfortunately, X doesn't directly provide printer support, so this approach doesn't work in UNIX or Linux.

WordPerfect's GUI Adaptations

Presumably in order to maintain compatibility with its older and non-GUI versions, WordPerfect adapted to the demands for WYSIWYG word processing in its 5.x versions for Windows and OS/2 by taking a kitchen-sink approach: Retain the existing base of printer drivers and *add* support for OS-driven fonts and printer drivers. (Mac versions of this era were more OS-centric, and UNIX versions owed more to the old text-mode heritage, as I describe shortly.) With version 7 for Windows, WordPerfect switched to a more OS-centric view, but that's not the case with versions 7 or 8 for UNIX or Linux. This mix makes for a confusing hodge-podge of options, and adapting the old text-mode approach to a WYSIWYG product makes for yet more complexity.

Basically, WordPerfect adds both screen fonts and downloadable printer fonts and attempts to match the two. As described shortly, it's possible to add two types of fonts to WordPerfect: *graphics* fonts and *soft* or *printer* fonts. The former are PostScript Type 1 (also known as Adobe Type Manager, or ATM) fonts used for display on the screen; these fonts can also be used to produce font bitmaps for printing. Soft fonts are downloaded to and processed by the printer and are not used for screen display purposes. In the case of a PostScript printer, soft fonts are also Type 1 fonts, and so you can have the same font installed twice, once for screen display and once for printing.

In the end, you must install a graphics font in WordPerfect for it to display text in that font (it doesn't use the fonts installed in X itself). WordPerfect can use these fonts for printing, or you can install the same fonts again or install different fonts to be used for printing. In the latter case, WordPerfect manages the mapping of display and printer fonts in an attempt to keep the two matched. It doesn't always succeed, though.

Linux Printer Queues

Although Linux doesn't host printer drivers in quite the sense that they exist in Windows, OS/2, or MacOS, Linux does maintain a system for handling printing. Normally, each physical printer will have one or more Linux printer *queues (p. 150, 732)* associated with it. When you print a document from WordPerfect or some other application, Linux keeps a temporary copy of the file on disk until the printer to which it's going signals its availability. Linux then sends the file to the printer. As a user, you can manipulate the printer queue by deleting jobs, rearranging job orders, and so on.

→ If you want to learn more about managing Linux printer queues, **see** "Controlling Printing from Linux," **p. 159**

You can configure a printer queue to process the input in some way before passing it on to the printer. Such processing is known as *filtering*, and the program that does the filtering is known as a *print filter*. For instance, you can set up a printer queue to add a cover page with information such as the user who created the print job. Cover pages can be very useful in an office environment with shared printers. Other filters are possible, too, such as filters that create two pages of output on a single printed page (a *two-up* filter) and so on. Ordinarily, you'll create two or more printer queues if you want to apply a filter to some print jobs but not others.

In Linux, one particularly common type of filter is to use *Ghostscript (p. 151, 180, 380, 409)* to process PostScript input into a format that's suitable for printing on a non-PostScript printer. This is because most Linux programs expect to print to a PostScript printer, so using Ghostscript as a print filter lets these programs work with the inexpensive inkjet printers and PCL laser printers that Linux users are likely to own.

→ To learn how to set up a Linux printer queue, **see** "Setting Up Printer Queues," **p. 731**

WordPerfect can print using a PostScript printer driver and a printer queue that's configured to use Ghostscript. Because of its library of drivers for many printers, however, you may be able to print using a queue that does *not* pass its data through Ghostscript. Which method works best for you depends very much upon your specific model of printer and perhaps also on the printing features that are important to you.

Setting Up Non-PostScript Printers

One way to print from WordPerfect for Linux is to use WordPerfect's driver for your specific printer or for a compatible model. This technique avoids the use of Ghostscript, which is the usual way of printing to non-PostScript printers from Linux. Because Ghostscript *rasterizes (p. 379)* an entire page and prints it as a large graphic, whereas WordPerfect's drivers can use the printer's built-in fonts or will (when possible) download a font to the printer, printing with the WordPerfect driver may be faster than printing with a PostScript driver through Ghostscript. Printing with a WordPerfect driver may or may not produce results that are as good as printing with Ghostscript, however. You'll have to try both ways with your specific printer to be sure.

To print with a WordPerfect driver for your specific non-PostScript printer model, you must have some way to pass "raw" data through to the printer. Depending on your distribution and how you've configured your printer, you may or may not currently be able to do so. If not, adding an /etc/printcap entry for the task isn't too difficult. Here's what I use on my system:

```
raw:\
   :sd=/var/spool/lpd/raw:\
   :mx#0:\
   :sh:\
   :sf:\
   :lp=/dev/lp1:
```

→ For details on setting up a Linux printer queue, **see** "Setting Up Printer Queues," **p. 731**

DRIVER FILES

Two types of driver files are important in setting up a WordPerfect printer driver:

- .all files—The driver files you download from your printer manufacturer's or from Corel's Web site will have .all extensions. These files contain a family of drivers for printers that use similar printer languages. For instance, one .all file might contain drivers for several different Hewlett-Packard LaserJet printers.
- .prs files—Files with .prs extensions contain drivers for only a single model of printer. When you install a printer driver, WordPerfect creates the .prs file for your printer based on information in the .all file. For instance, the .prs file might contain the information necessary for WordPerfect to control the Hewlett-Packard LaserJet 4000 printer.

To install a printer driver in Linux, therefore, you normally need an .all file for your model of printer or for one that's closely related to your printer. You *can* install a printer driver using only a .prs file, but neither Corel nor printer manufacturers distribute these files directly. There are three places you might find an .all file:

- With WordPerfect for Linux—WordPerfect comes with an assortment of printer driver files. When you installed WordPerfect, you could tell it which drivers to include. If you did so, your printer driver is already installed and you just need to link it to the appropriate Linux printer queue. If you didn't select your printer when you installed WordPerfect, the installer may not have placed the appropriate .all file on your disk, so you may need to look elsewhere for the file.

➔ To learn about installing WordPerfect for Linux, **see** "Installing WordPerfect for Linux," **p. 712**

- On Corel's Web site—You can search for the latest WordPerfect printer drivers at `http://www.corel.com/support/printerdrivers/index.htm`.
- From your printer manufacturer—Many, but not all, printer manufacturers have WordPerfect printer drivers for their printers.

WordPerfect 8 for Linux uses the same printer drivers as WordPerfect 6 for DOS, so you may need to look for DOS drivers for your printer to find what you want. When you find them, they may come as DOS self-extracting zip files. These files have .EXE extensions and can be decompressed with InfoZip on Linux, using a command such as `unzip DRIVER.EXE`. When you extract the files, they have uppercase filenames that aren't of quite the correct format for WordPerfect for Linux; they must be converted to lowercase and have .us added before the .all extension. For instance, the file WP60DM01.ALL must be renamed to `wp60dm01.us.all` to work in WordPerfect for Linux.

When you have an .all file, you may want to copy it to the shlib10 subdirectory of your WordPerfect installation. This step isn't strictly required, but WordPerfect checks there first for the driver files, and you need to redirect it if the .all file isn't there.

Setting Up the WordPerfect Printer Driver

WordPerfect allows printers to be set up in either of two ways: as a *shared* printer that's accessible to all users of the system or as a *personal* printer that's created and used by only one user. You set up a shared printer by invoking WordPerfect as root with the `-adm` flag, as in `xwp -adm`. To set up a personal printer, run WordPerfect normally as the user who wants to use this printer definition.

Note

You need to have full read/write access to the WordPerfect installation directory to install a shared printer. Therefore, you'll probably need to issue the `xwp -adm` command as root for the installation to work.

→ If you want to know more about the root account, **see** "Users and the Superuser," **p. 11**

When you create a printer definition, WordPerfect creates appropriate .prs files. In the case of shared printers, WordPerfect places these files in the shlib10 subdirectory of its installation. If you install a personal printer, WordPerfect places the .prs files in your own .wprc subdirectory.

To install a printer driver, follow these steps:

1. Click the select printer icon on the status bar at the bottom of the WordPerfect Document window (Figure B.1) to get the Select Printer dialog box (Figure B.2).

Figure B.1 The WordPerfect status bar provides a shortcut to the Select Printer dialog box.

2. Click Printer Create/Edit in the Select Printer dialog box to get the Printer Create/Edit dialog box shown in Figure B.3.
3. Click Add to display a list of printers in the Add Printer Driver dialog box (Figure B.4).
4. If the .all file containing the driver you want to use isn't in the WordPerfect shlib10 subdirectory, click the folder button in the Printer Driver Directories box to locate the directory in which the .all file exists.

Tip #302 from *Tux Tips*

If you're creating a second definition using the same driver file, as you might if you have two similar printers or to print to the same printer using different Linux printer queues, you can click the Printer Driver Files (*.prs) radio button to see a shorter list of installed .prs drivers, rather than the longer list of .all driver files.

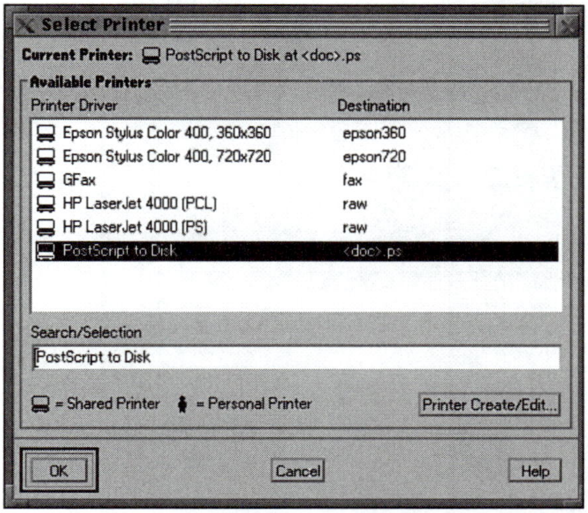

Figure B.2
The Select Printer dialog box lists all the printers you currently have defined, and lets you choose which one to use.

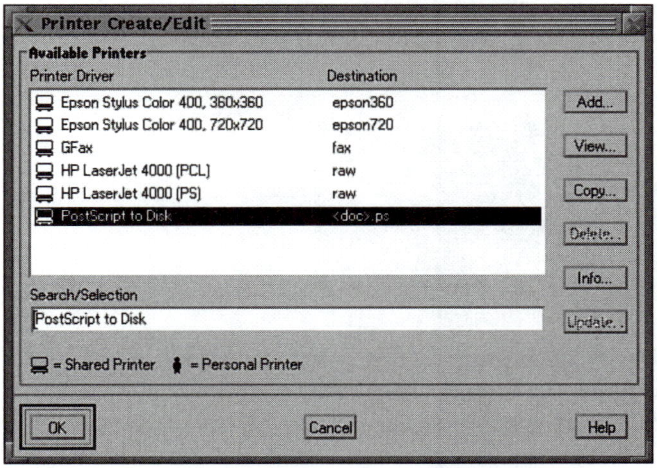

Figure B.3
The Printer Create/Edit dialog box lets you add a WordPerfect printer driver.

5. Select the driver you want to install and click OK. WordPerfect displays the Create Printer dialog box shown in Figure B.5, which lets you know what .prs file it will create. You can edit the filename or leave it alone. Either way, click OK in the Create Printer dialog box to dismiss it.

Tip #303 from *Tux Tips*

WordPerfect drivers sometimes include information about how they work, their limitations, and so on. Click the Info button in the Add Printer Driver dialog box before you click OK in that dialog box to read this information.

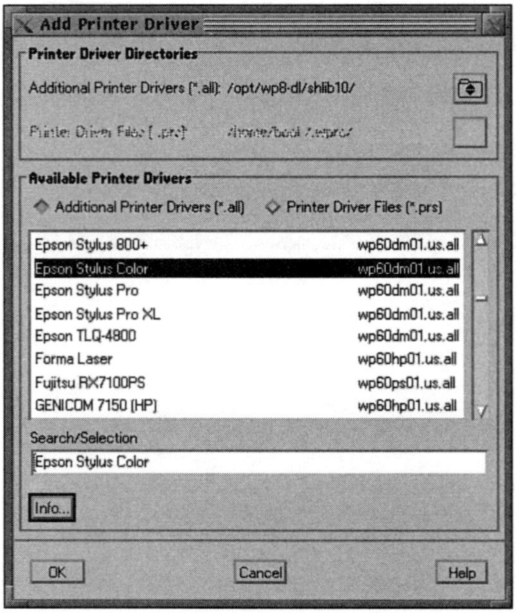

Figure B.4
The Add Printer Driver dialog box shows all the printer drivers in all the .all files in the printer drivers directory.

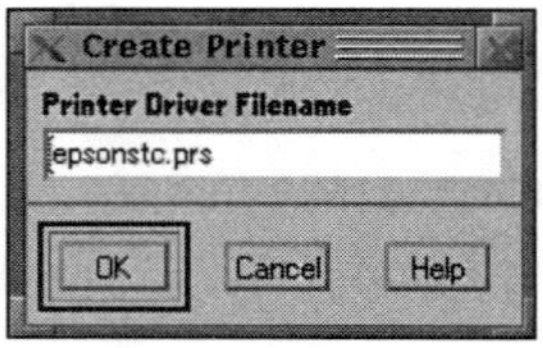

Figure B.5
You can alter the name of the .prs file to prevent a filename collision or to create a file with a more meaningful name than the WordPerfect default.

6. WordPerfect removes the Create Printer and Add Printer Driver dialog boxes and returns you to the Printer Create/Edit dialog box (refer to Figure B.3). Your new driver should be listed there with a destination of None. Be sure the printer is selected and then click the Setup button (which replaces the View button) to open the Printer Setup dialog box shown in Figure B.6.
7. You can use the Printer Setup dialog box to change the initial font of the printer, the driver name, the sheet feeder, and whether or not WordPerfect will attempt to print color to the printer. The most important thing to configure is the destination of the printer. To set the destination, click Destination to get the Select Destination dialog box (Figure B.7).
8. The Select Destination dialog box lists your existing printer queues. Choose whichever one is appropriate. You can also select Disk to have the driver "print" to a disk file, though an easier way to print to disk is described shortly.
9. Click OK on each open dialog box to dismiss them all and finish the installation process.

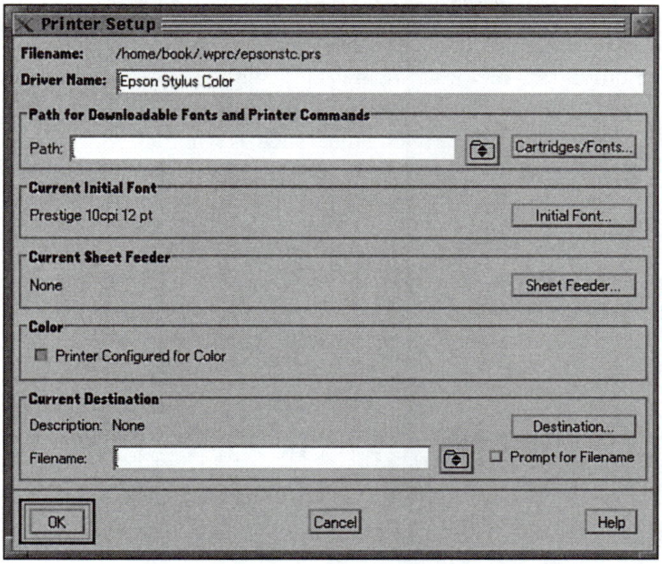

Figure B.6
The Printer Setup dialog box lets you set printer-specific options such as the initial font and sheet feeder.

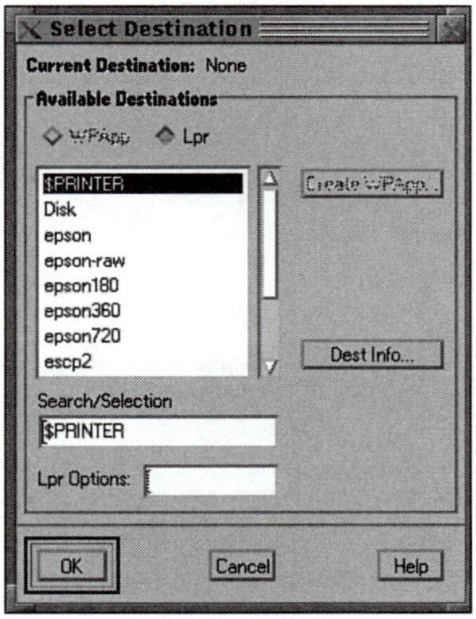

Figure B.7
The Select Destination dialog box shows the printer queues you've created in Linux.

When you finish defining your printer, you should probably do a test printout. I recommend printing something very simple first, such as a single line of text. That may minimize the wasted paper if you picked the wrong printer or if there's some other problem—when printers get confused, they sometimes spew reams of paper containing a few characters per sheet.

➔ For instructions on printing from WordPerfect, **see** "Creating Paper Output: Printing," **p. 148**.

If you can't find a driver for your specific model, try to find a driver for a model that's as similar in capabilities as possible to the one you're using. Unfortunately, doing so is easier said than done unless you're a walking encyclopedia of printer specifications. Your printer manufacturer may be able to offer advice on this issue, so try checking its Web site or contacting its technical support line. Aside from issues such as whether your printer will respond correctly to commands intended for another model, there's the issue of resolution. If you select a 300-dpi printer driver but have a 600-dpi model, for instance, WordPerfect is likely to print at 300 dpi.

Are you getting poor print resolution when you print? See "Poor Print Resolution" in the "Troubleshooting" section at the end of this chapter.

Are you getting one blank page at the end of each print job? See "Stray Form Feeds" in the "Troubleshooting" section at the end of this chapter.

Setting Up PostScript or Ghostscript-Driven Printers

Setting up a PostScript or Ghostscript-driven printer is actually very much like setting up a non-PostScript printer, and you should read the preceding section for details on where to get drivers and how to install them. There are a few additional concerns, however, especially if you're using Ghostscript.

Tip #304 from *Tux Tips*

> If you have a choice of using a WordPerfect driver for your model of printer or using a PostScript driver along with Ghostscript, try both approaches. Sometimes the native driver works better; other times using Ghostscript is best.

The Linux Printer Queue

To create a PostScript-capable Linux printer queue using a non-PostScript printer, the queue must pipe the output through Ghostscript. This process can be done in any of a number of ways, and many distributions help to automate it. If you have a genuine PostScript printer, of course, you need not involve Ghostscript; your queue needn't do any processing, though many distributions do create printer queues with "smart filters" that will try to interpret the file type and, if necessary, convert it to PostScript before sending it on to the printer.

→ To learn how to configure a Linux printer queue, **see** "Setting Up Printer Queues," **p. 731**

For the PostScript-enabled Linux printer queue to work on a system printing to a non-PostScript printer, you also need Ghostscript. All major Linux distributions ship with Ghostscript, but they often include older versions that may not include the best drivers or may not even have the driver for your printer. You may therefore want to look into getting a more recent version. The main Ghostscript Web site is at `http://www.cs.wisc.edu/~ghost`. This site includes the latest Ghostscript source code, documentation, and so on. If your system uses Red Hat's *Red Hat Package Manager (RPM)* system for package distribution, the Web site `http://www.users.dircon.co.uk/~typhoon/` has RPMs for Ghostscript, including the most commonly requested drivers.

WHICH WORDPERFECT DRIVER?

WordPerfect provides a huge selection of PostScript printer drivers (searchable on Corel's Web site at `http://www.corel.com/support/printerdrivers/index.htm`), and they don't all work equally well. Some drivers, like most or all drivers for Hewlett-Packard LaserJets with PostScript capabilities, prepend special codes that are intended to switch the printer into PostScript mode. Unfortunately, these characters may confuse a print filter, either on your Linux box or on a remote print server if that's how the printer is connected, into thinking that the file is ASCII text. The result is page after page of PostScript commands printed in (typically) Courier font, which can be quite a waste of paper. Other WordPerfect drivers include commands for features specific to certain printers, and Ghostscript or PostScript printers without these features might choke on these commands. I recommend that you try any WordPerfect PostScript driver first in print-to-file mode; then examine the file. If it begins with `%!PS-Adobe-3.0`, you at least won't print reams of PostScript code; but if this is the second or later line of the file, it may or may not print correctly. If it doesn't print properly, you can either try another printer driver or try to modify the printer filter so that it won't try to modify the output from WordPerfect.

On the other hand, you might have precisely the opposite problem—that is, your printer might not process a raw PostScript file correctly, but might require the sort of mode-switching code that WordPerfect inserts with some of its drivers. If you set up your printer to print from other Linux PostScript-producing programs, though, you have this problem licked, or at least your printer should work with a WordPerfect driver that doesn't generate any mode-switching code.

When WordPerfect prints with graphics fonts, it creates a bitmap version of the font at whatever resolution it believes the printer will be using. Therefore, for best results, the driver you use should support the resolution at which you expect to print. If you use a driver that's too low in resolution for your printer, you will effectively lose some of your printer's resolution. Here are some suggestions for PostScript drivers to try:

- If you're using a true PostScript printer, you might as well try a WordPerfect driver for that specific model, observing the caveats above about mode-switching code.
- If you're using a lower-resolution printer, something that produces 300-dpi output, like the Apple LaserWriter II NT (in wp60ps01.us.all), might be in order.
- The LaserWriter Pro 630 driver (in wp60ps02.us.all) works well with Ghostscript to produce monochrome 600-dpi output.
- Drivers for the Lexmark Optra R+ (in wp60ps01.us.all) produce 1200-dpi output and works with a Hewlett-Packard LaserJet 4000, but Ghostscript chokes on these PostScript files. This driver might be a good place to start for other 1200-dpi laser printers.
- For color printers, I've found that the QMS magicolor (in wp60ps01.us.all) driver produces good results at 600 dpi. This resolution is close enough to the 720 dpi of many color inkjet printers that the difference is difficult to detect even on high-quality paper.
- The NEC Colormate PostScript driver (in wp60ps01.us.all) produces 300-dpi color output, but it has a bug that causes it to drop back to black text if you select any text attribute (such as italics or bold) in a line of colored text.

ADJUSTING PRINTER MEMORY

WordPerfect includes information on the amount of memory installed in different models of printers. If you're using Ghostscript, a PostScript printer other than the one whose driver you've installed, or a PostScript printer with additional memory installed, this memory amount may be inaccurate. Unfortunately, if you use Type 1 soft fonts and put enough of them on a page to overwhelm the amount of memory that WordPerfect *thinks* is installed, Ghostscript crashes when you try to print the file. It's therefore a good idea to increase the amount of memory that WordPerfect thinks is installed. Follow these directions:

1. Click the select printer icon on the WordPerfect status bar (refer to Figure B.1) to get the Select Printer dialog box (refer to Figure B.2)
2. Click the printer in question and click Printer Create/Edit to get the Printer Create/Edit dialog box (refer to Figure B.3).
3. Click Setup to get the Printer Setup dialog box (refer to Figure B.6).
4. Click Cartridges/Fonts to get the Select Fonts dialog box shown in Figure B.8.

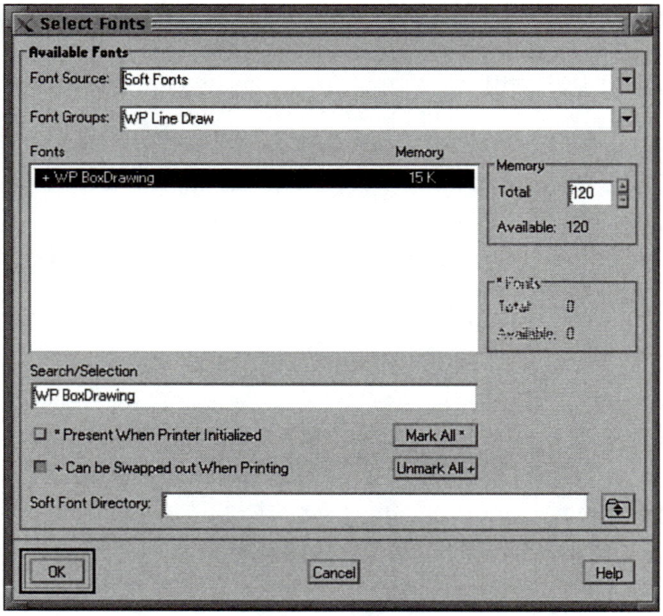

Figure B.8
You can adjust assorted font-related features of WordPerfect's printer support with the Select Fonts dialog box.

5. Change the Font Source option button from Built In to Soft Fonts, and the Memory box becomes editable.
6. Adjust the Total field in the Memory box upwards. Values of up to 32,767 seem to work well; beyond that and WordPerfect sometimes behaves strangely.
7. Click OK on each dialog box to dismiss them all and accept the new printer memory settings.

BUILT-IN FONTS

WordPerfect's printer drivers include information on the fonts that are standard on their respective printers. Therefore, using a driver for a printer that's different from the one you have may result in an inability to access fonts that are present in the printer or in the presence of fonts in WordPerfect's menus that don't exist in your printer. The same thing can happen with Ghostscript. With the latter, if you see printer font names in WordPerfect's menus that aren't present in Ghostscript, it may be possible to add them by editing Ghostscript's Fontmap file, assuming you have the fonts themselves.

To find a Type 1 version of a font, you can try these approaches:

- Check out `http://www.ssifonts.com`, which has a handy font name cross-reference database. This information may point you to the originator of the font or to a clone of the font.
- Hewlett-Packard has an archive of the fonts that are standard on its laser printers in its OS/2 software archive. It's a 2MB self-extracting zip archive (with an .EXE extension, but InfoZip can process it—use the `unzip` command).
- You may already have the font either in WordPerfect's shlib10 directory or in one of the font archives that are optional with Ghostscript.
- Check the appendix-b directory on the CD that accompanies this book. It contains a number of freeware fonts in various subdirectories.

When you have the font, you can modify Ghostscript to use it by following these steps:

1. "Print" a file to disk using the WordPerfect driver and font in question. Try creating just a single line of text with the font, and nothing more.
2. Examine the PostScript file to determine which font name to use. Search for the text you created. Immediately before it you'll likely find the font name, which with any luck will be similar to the name used in the WordPerfect font menu. For instance, Palatino appears as Palatino-RomanR.
3. Locate the exact filename for the font you want to use. You can use `grep` here, as in `grep Palatino *.afm` to locate the .afm file(s) for Palatino. You'll actually use the matching .pfb file when you edit the Fontmap file.
4. Edit the Ghostscript Fontmap file, which is located in different places on different systems, but most likely somewhere in the /usr/share/ghostscript or /usr/local/ghostscript directory trees.
5. Copy an existing Fontmap entry and modify it so that it points to the Type 1 file you located in step 3, including the complete path (or create a link from this file to your Ghostscript fonts directory). You should use the name you determined in step 2 as the font's name.
6. If necessary, install the font in WordPerfect so that you can use it for onscreen display.

It's possible to tell WordPerfect about fonts you've added to your Ghostscript Fontmap file or installed in a PostScript printer. To do so, follow these directions:

1. Click the select printer icon on the WordPerfect status bar (refer to Figure B.1) to get the Select Printer dialog box (refer to Figure B.2).
2. Click the printer in question and click Printer Create/Edit to get the Printer Create/Edit dialog box (refer to Figure B.3).
3. Click Setup to get the Printer Setup dialog box (refer to Figure B.4).
4. Click Cartridges/Fonts to get the Select Fonts dialog box shown in Figure B.8.
5. Change the Font Source option button from Built In to Additional Fonts.
6. Change the Font Groups to WPFI Type 1 (Soft Fonts). The soft fonts you installed in the driver appear in the Fonts area of the dialog box.

Note

The WPFI Type 1 (Soft Fonts) option is available only if you already installed soft fonts in your printer driver.

→ To learn how to install soft fonts, **see** "Installing and Managing Fonts," **p. 754**

7. You can select whether each individual font is always present in the printer (as it would be with Ghostscript if you created an appropriate Fontmap entry) or should be downloaded (the default). You do so by checking or unchecking the *Present When Printer Initialized check button in the Select Fonts dialog box for each of the fonts in the Fonts list.

Note

To gain access to fonts built into a PostScript printer, you need access to Type 1 font files with the correct font names. Try contacting the printer's manufacturer if you don't have these files.

8. When you finish configuring the list of fonts available in the printer, click OK on all the open dialog boxes to dismiss them.

Note

You need to install the font twice—once as a soft font and once as a graphics font—to be able to see the font onscreen as well as on the printed page. If it's not installed as a graphics font, WordPerfect substitutes some other font on the screen, though the font will print fine.

Printing to Disk Files

Configuring WordPerfect to print to a disk file is no more difficult than configuring it to print to a regular PostScript or Ghostscript-driven printer. To do so, follow these directions earlier in this chapter for installing a WordPerfect printer driver. You'll presumably want to select a driver that works well with Ghostscript. If you have a printer you're using via

Ghostscript, you may want to use the same driver to simplify tasks such as installing soft fonts. When you're installing the WordPerfect driver, however, you need to deviate from the procedure in these ways:

- You don't need to click the Destination button in the Printer Setup dialog box because you won't be sending the output to a regular queue.
- At the Printer Setup dialog box, click the Prompt for Filename check button in the Current Destination box. WordPerfect displays <doc>.dsk in the Filename field. You should probably change this entry to <doc>.ps, to give the file a .ps filename extension by default, which is standard for PostScript files.

When you use WordPerfect regularly, you can then select the queue you created in this way to print the document to a PostScript file.

Note

You can perform these steps to create disk files containing files that can be printed to non-PostScript printers, too. For instance, if you have a Canon inkjet printer on a computer other than the one on which WordPerfect is installed and if these computers aren't networked so that you can print to the inkjet from WordPerfect, you might want to create a disk file with data in Canon's printer language. You can then move the file via floppy disk or some other means to the computer that hosts the Canon printer and send the file to the printer to obtain a printout.

SETTING UP A FAX QUEUE

If you have fax software configured in Linux and if this software takes PostScript input (as does most Linux fax software), configuring WordPerfect to send faxes works just like configuring the program to print to any other PostScript printer. As most Linux fax software uses Ghostscript to create a faxable bitmap, you should use a WordPerfect printer driver that works well with Ghostscript. Again, if you choose to install soft fonts, you can save yourself some effort by using the same .prs file you use for other Ghostscript-driven printers.

The GFax software that's discussed in Chapter 6, "Producing Output," is a GUI front end to the mgetty+sendfax package. Both these packages have their own installation instructions, and you should consult their documentation for details on setting them up. As a summary, however, you will need to

1. Install mgetty+sendfax. This package comes with most Linux distributions, so consult your distribution's documentation for how to install a package; then locate and install it. The plus sign (+) in the mgetty+sendfax package name is sometimes changed to a minus sign (-) in the package's filename.
2. Edit various files in the /etc/mgetty+sendfax directory to reflect your needs. For instance, you'll want to add the usernames of those users who are allowed to use the fax system to the fax.allow file, alter the faxheader file to show your fax number, and edit sendfax.config to reflect your phone number and fax modem hardware. You can find more information on configuring mgetty+sendfax at `http://www.leo.org/~doering/mgetty/`.

3. Install the GFax package by running the `gfax_install` installation script as root.
4. Edit the /etc/printcap file to add an entry for the fax queue; otherwise, WordPerfect won't know that the queue is available. Here's an entry that works if you accept the defaults when you install GFax:

```
gfax:\
    :sd=/var/spool/gfax/outgoing:\
    :mx#0:\
    :sh:\
    :lp=/dev/ttyS1:
```

INSTALLING AND MANAGING FONTS

Note

The downloadable version of WordPerfect 8 for Linux lacks the font installer utility. Therefore, you can't install new fonts in this version of the program. If font support is important to you, you should consider purchasing the personal retail version of the program.

Fonts come in *families* of related designs. For instance, the Times font that's installed in PostScript printers has four common variants: Roman (the *base* font), italic, bold, and bold-italic. Each of these is technically a separate font and comes in a different file. When you install a font in WordPerfect, you normally install several font files for these common variants. WordPerfect can fake it and generate these variants from a base font, but using specific variant fonts generally produces better-looking results.

WordPerfect supports two basic types of fonts: *graphics* fonts and *soft* or *printer* fonts. WordPerfect uses the former type of font for displaying text in its windows, so you *must* have a graphics font installed if you expect to see your text as it will appear on the printed page. WordPerfect can also *rasterize (p. 379)* a graphics font when printing, so you can use graphics fonts for printing, as well. Soft fonts, on the other hand, are rasterized by the printer. Rather than install these fonts in WordPerfect per se, you install them in the printer driver.

Tip #305 from *Tux Tips*

Because the printer rasterizes soft fonts, they print at the printer's best possible resolution. This feature can be useful if you're using Ghostscript to print to a printer that uses a resolution that's mismatched to the resolution specified in the WordPerfect PostScript driver you're using. In this situation, Ghostscript can print at the best possible printer resolution if you use soft fonts, but not if you use graphics fonts. Thus you may want to use soft fonts whenever possible if you're using a printer with an odd resolution, such as 720×1440 dpi.

Graphics fonts are PostScript Type 1 fonts—a standard format for Linux fonts in general. If you're using a PostScript or Ghostscript-driven printer, soft fonts are also Type 1 fonts, so you can install the same font twice. WordPerfect also supports soft fonts for printers using Hewlett-Packard's PCL language, but these fonts are not Type 1 fonts. PCL fonts are quite rare compared to Type 1 fonts, so I don't discuss them further here.

OBTAINING FONTS

Because WordPerfect uses Type 1 fonts, you need appropriate Type 1 font files to install fonts in WordPerfect. I recommend you look for a font CD that's advertised as containing Type 1 (also known as ATM) fonts. You need two files, one with a .pfb extension and one with an .afm extension, to install a font in WordPerfect. Unfortunately, some Type 1 font collections lack the .afm files, and the package labels generally omit this information. Here are some font collections I've used successfully with WordPerfect for Linux:

- Bitstream 500 Font CD—`http://www.bitstream.com`.
- CorelDRAW fonts—`http://www.corel.com`. This collection comes with CorelDRAW for Windows and includes many of the fonts from Bitstream's CD.
- Southern Software Just My Type CD—`http://www.ssifonts.com`. SoftKey also markets this collection, but without the .afm files.
- Expert Software 2000 Fonts CD—`http://www.expertsoftware.com`.

Note

You should purchase a font CD marketed for Windows, not for MacOS. Although the Type 1 fonts themselves are not platform dependant, Mac fonts come in special "suitcase" files that require decoding to use anywhere but on a Mac. Although programs such as the `t1utils` package on the CD that accompanies this book can do this decoding, you'll save yourself some effort by purchasing fonts that come in the correct format to begin with.

Many Web sites have collections of freeware and shareware fonts. Some of these are good; others are of poor quality. Some good fonts have been released as freeware and come with Ghostscript. The CD that accompanies this book includes these fonts, as well as several other freeware fonts. You can even use some of these to improve the quality of the display of standard Times, Helvetica, and Courier fonts in WordPerfect.

→ For information on improving WordPerfect's display of fonts, **see** "Optimizing Onscreen Font Display," **p. 761**

Although Type 1 font collections are common, TrueType font collections are more common because TrueType is the native format for Microsoft Windows. WordPerfect for Linux can't use TrueType fonts, unfortunately, but the companion CD includes a font conversion utility that converts TrueType fonts to Type 1 format. The conversion is likely to degrade the quality, however, and WordPerfect won't use some of these fonts very well at all, so you should use this utility only as a last resort.

Tip #306 from *Tux Tips*

If you're not concerned with WYSIWYG display, you may still be able to use a "problem" converted TrueType font by installing it as a printer font; then your printer's PostScript rasterizer, rather than WordPerfect, will have to deal with it. If you're printing via Ghostscript, you can install the original TrueType font directly in Ghostscript.

STARTING THE FONT INSTALL PROCESS

To install fonts, you must have write access to the WordPerfect files, so you'll probably need to be root. There are two ways to do the installation; either use the `xwpfi` utility in the shbin10 subdirectory of the Linux installation or launch WordPerfect with the `-adm` switch and then select any of a number of menu sequences to get you into the font installation utility. One such sequence is choosing Format, Font (F9) to get the Font dialog box (Figure B.9) and then clicking Install Fonts to get the Select Font Type dialog box (Figure B.10).

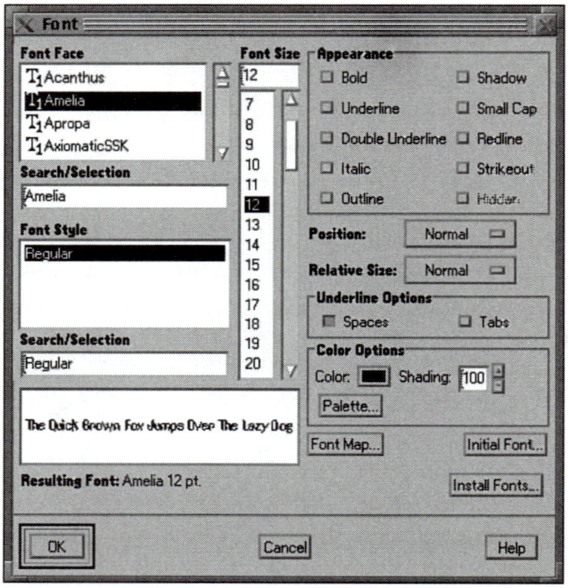

Figure B.9
The Install Fonts button is available only if you started WordPerfect by using the *-adm* switch.

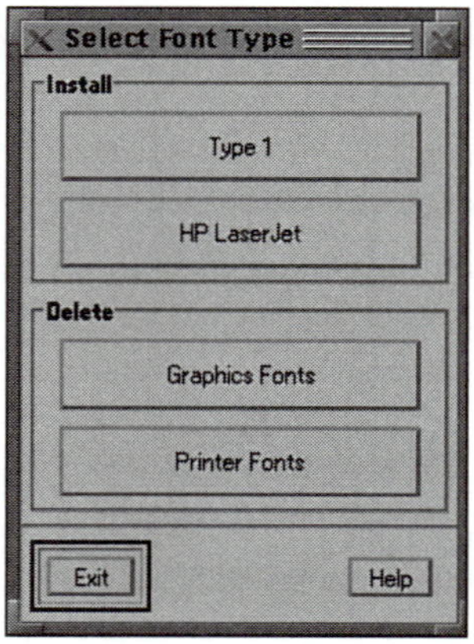

Figure B.10
Use the Select Font Type dialog box to choose the type of font you want to install or delete.

If you choose to use xwpfi, you'll first see the Font Installer window shown in Figure B.11. You can set the directory from which you'll install fonts by choosing Preferences, Graphics Fonts, or you can install or delete fonts by choosing an appropriate option from the File menu.

Figure B.11
The Font Installer window lets you install or delete fonts without launching WordPerfect.

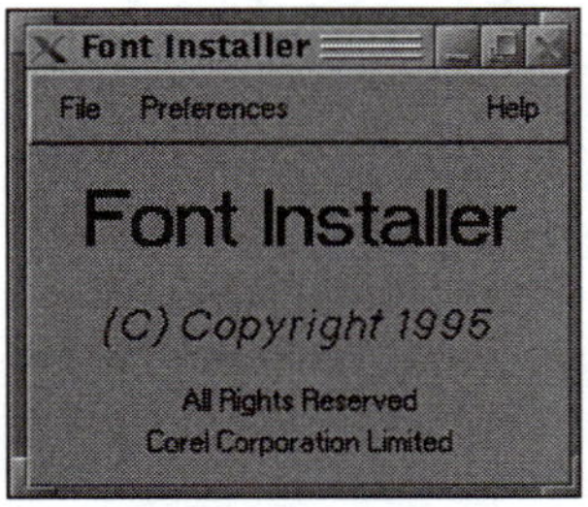

If you want to install soft fonts and if you start the font installer from within WordPerfect, you should first select the printer to which you want to install them. If you use the xwpfi program, you need to tell it whether to install both printer and graphics fonts or graphics fonts only when you opt to install Type 1 fonts. If you choose both types, you'll see the Printer Driver Select dialog box (Figure B.12) where you select the printer driver to which you want to install the fonts.

Figure B.12
Select the driver into which you want to install soft fonts with the Printer Driver Select dialog box.

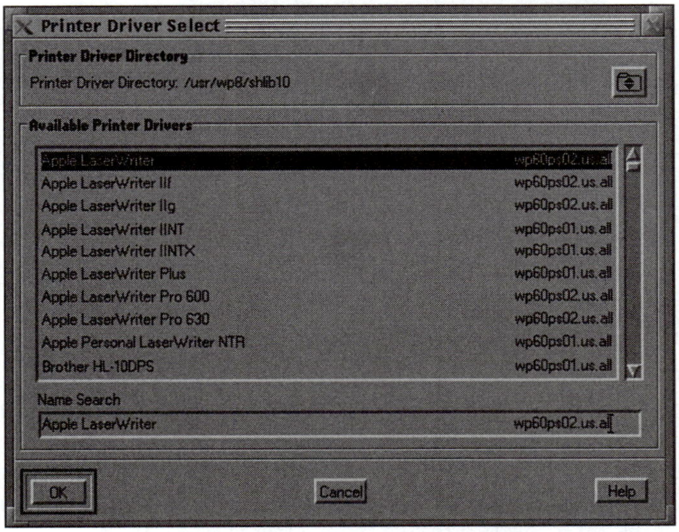

Note

Prior to installing fonts, you should copy the font files, or create links to the font files, to the shlib10 directory. WordPerfect requires both .pfb and .afm files to install a Type 1 font.

Tip #307 from *Tux Tips*

If you install soft fonts directly from WordPerfect, they go into the .prs printer driver file. If you use the `xwpfi` utility, the fonts are installed in the .all file. Therefore, you should use WordPerfect itself if you don't want to risk damaging the original .all file, and use `xwpfi` if you want to take the same .all file and install the modified printer drivers on several computers. If you install the fonts to the .all file, you need to reinstall the printer driver to get access to the fonts.

When you finish with the preliminaries, both the `xwpfi` utility and the font installer within WordPerfect show the Select Fonts dialog box shown in Figure B.13.

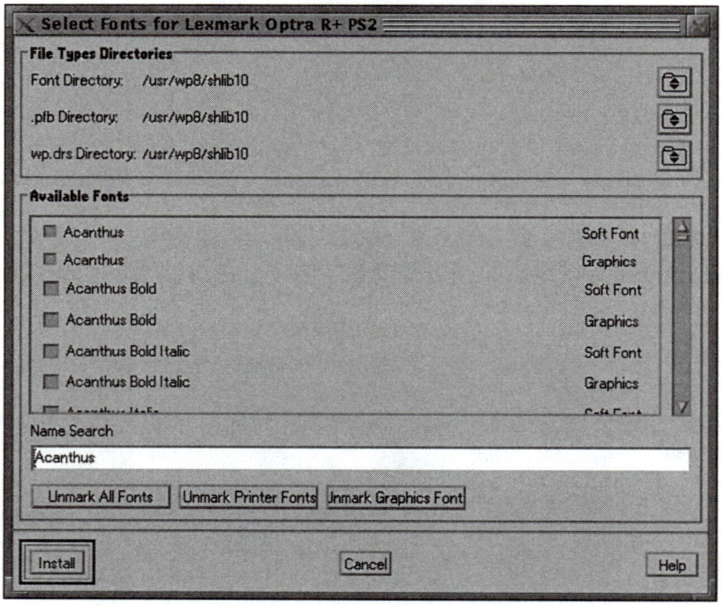

Figure B.13
The title of the Select Fonts dialog box depends on the printer driver you selected.

INSTALLING GRAPHICS FONTS

Graphics fonts are necessary to display a font onscreen. They can also be used to print to most printers (only daisywheel printers can't use graphics fonts). To print graphics fonts, WordPerfect rasterizes the font into a bitmap and then either prints the entire page as a huge graphic or downloads the bitmap font to the printer. In either case, graphics fonts print at whatever resolution WordPerfect believes the printer can handle. This process is the Achilles' heel of graphics fonts because it means that you may not get the maximum benefit of your printer's resolution if it's mismatched to the driver you're using.

When the Select Fonts dialog box comes up, all the fonts are selected. You should either deselect all the fonts with the Unmark All Fonts button and then select the ones you want or deselect only those fonts you don't want. When you've selected only the fonts you want, click on the Install button to finish. Installing graphics fonts doesn't actually copy any files if you leave the File Types Directories defaults; rather, it creates entries in the shlib10/wp.drs file reflecting the availability of the fonts.

Note

Depending on how you started it, the Select Fonts dialog box may show soft fonts as well as graphics fonts. Thus each font may be shown *twice*, once as a graphics font and once as a soft font. Be careful about which fonts you select in this case.

INSTALLING SOFT FONTS

You install soft fonts in much the same way as you install graphics fonts. If you want to be able to see the font you're installing on the screen as well as on the printed page, though, you must install the same font as both a graphics font and a soft font.

Type 1 soft fonts print at whatever resolution the printer can handle. Thus, if you're printing to a 600-dpi PostScript printer, soft fonts will print at 600 dpi, even if you're using a 300-dpi PostScript driver.

As with graphics fonts, installing soft fonts doesn't copy any font files if the File Types Directories settings are left alone. The contents of the .all and .prs files associated with the printer in question do change, however.

WordPerfect shows a soft font in a WYSIWYG manner in the Document window only if you also install a matching graphics font. Therefore, you may have the same font *twice* in your font menus, once as a soft font and once as a graphics font. To use the soft font, be sure to select the soft font from the menu. The soft font is the one with the printer icon next to its name. If you select the graphics font (the one with T1 next to its name), you won't get the benefits of using a soft font, such as having a PostScript printer print at its maximum resolution no matter which WordPerfect driver you use.

CORRECTING THE AFC FONT MAPS

As I described earlier, many fonts come in families of related designs. To use these families more easily, WordPerfect provides an *automatic font change (AFC)* feature to tell the program when to use variant fonts. For example, AFC tells WordPerfect to use Times Italic when you press Ctrl+I to select the italic attribute when you're using the Times base font. Unfortunately, WordPerfect sometimes doesn't set up the AFC map very effectively. It is possible for you to intervene in this matter and correct WordPerfect's mistakes, however. As with installing the fonts initially, you should perform this task as root.

Choose Format, Font to get the Font dialog box (refer to Figure B.9) and then click the Font Map button to get the Edit Printer/Document Font Mapping dialog box shown in Figure B.14. Click the Map Individual Styles check button in the Printer Font area of the dialog box for full access to the font styles.

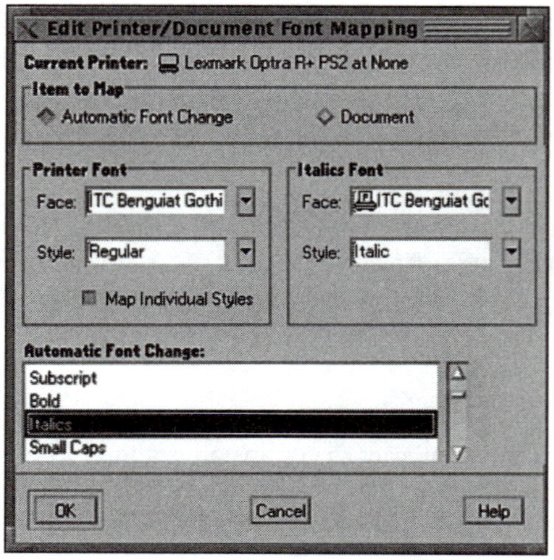

Figure B.14
Set up the font mapping to get the correct font variants when you use font attributes like bold or italic.

To correct a font mapping, follow these steps:

1. Select a font in the Printer Font area, with a base Style (such as Regular).
2. In the Automatic Font Change area, select an attribute, such as Italics. If the font entry displayed in the Italics Font area (which changes name with the attribute you select) doesn't show the correct font and style, you need to edit it. For instance, if you chose the Printer Font called Paradise Sans but the Italics Font shows up as Times Bold, you need to make a correction.
3. Change the Face in the Italics Font box to the appropriate font.
4. Change the Style to an appropriate attribute, such as Italics.
5. Repeat steps 2 through 4 for additional attributes, such as bold.
6. WordPerfect creates multi-attribute font effects (bold-italic, for instance) by chaining the effects. For instance, to produce a bold-italic effect, WordPerfect looks up the font to use for a bold variant and then looks up the font to use to get an italic variant of the bold font it just found. Therefore, you need to repeat steps 1 through 5 for each style that's likely to be used in such a way. For instance, you need to select the Bold Style in the Printer Font box, then select the italic attribute in the Automatic Font Change selector, and pick the Bold Italic Style in the Italics Font box. You need to do the same thing with the Italic Style in the Printer Font box and a Bold style in the Automatic Font Change box.

Note

You may notice that some fonts map to blank fonts. For instance, if you select an Italic Printer Font and click Italics in the Automatic Font Change field, the Italics Font field will most likely be blank. In this case, you should leave it blank because you won't normally be selecting an italic variant on an already-italicized font. If a font doesn't have a matching variant, this field should also be blank in which case WordPerfect will try to fake it by applying a slant to italicize a font, thickening the lines to create a bold effect, and so on.

If you have multiple printer drivers installed, you must repeat this process for each printer driver's soft fonts, but the graphics fonts should stay the same from printer to printer.

Some font families don't use the most common names for attributes. For instance, oblique fonts are often used where italics are, but WordPerfect doesn't always correctly map font families that use the term *oblique* rather than *italic*. Specifically, the standard, bold, and oblique versions of such a font will appear under one name (say, *Benquad*), but the bold-oblique variant will get another name (*Benquad-*, in this case). If you configure the AFC correctly, you can still get Benquad bold-italic; you just need to map the bold attribute on Benquad Italic to Benquad- bold-italic and the italic attribute of Benquad Bold to the same. This situation can also happen with fonts from certain foundries because of their font-naming conventions. Similarly, some fonts may use *black* rather than *bold* or have other naming oddities, and such fonts are likely to appear twice in your font menus.

Tip #308 from
Tux Tips

WordPerfect stores its font mappings in the wp.drs file for graphics fonts and in the .all and .prs files associated with the printer drivers for soft fonts. Remember to back up these files after changing the font mapping. This precaution can save you a lot of time if WordPerfect decides to regenerate the font mapping for some reason.

DELETING FONTS

If you want to remove fonts from your system, the procedure works very much as it does when you install them. From the Select Font Type dialog box (refer to Figure B.10), however, you select an appropriate option from the Delete box rather than from the Install box. If you use the `xwpfi` utility, you select an appropriate delete option from the File menu. You then select the font(s) you want to delete in a dialog box that's very much like the Select Fonts dialog box (refer to Figure B.13).

If you had to adjust the AFC mappings for fonts other than the ones you deleted, you may need to do so again after you delete a font.

OPTIMIZING ONSCREEN FONT DISPLAY

WordPerfect includes its own font rasterizer, which it uses both for printing and for displaying fonts on the screen. The program also includes its own selection of Type 1 fonts. Unfortunately, many people find the quality of fonts displayed in WordPerfect for Linux to be inferior to fonts in Windows. Fortunately, you can take steps to improve WordPerfect's font display.

THE PRIME COMPARISON THE PRIME COMPARISON

Figure B.15 shows a sample of WordPerfect 8 for Linux's font display as delivered, using standard fonts at 12 points. Note that several characters, such as the Helvetica bold `m` and the non-bold Courier `w` extend below the baseline when they shouldn't. Others, such as the Helvetica bold `a`, are difficult to read because they're poorly formed. It's hard to tell that the Times bold-italic is actually a bold variant.

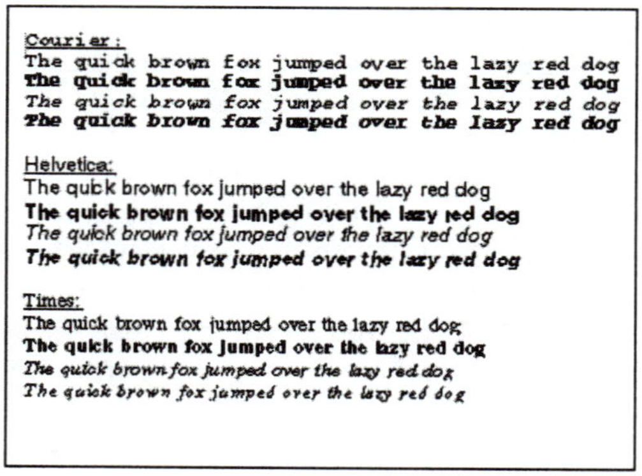

Figure B.15
WordPerfect's default font display is less than perfect.

WordPerfect for Windows delivers a much clearer display, as shown in Figure B.16. There are several reasons for this, including the fact that the Windows text is larger, delivering more pixels per character. Other possible factors include the use of different font files (Corel's fonts versus Microsoft's), different font technologies (Type 1 versus TrueType), and different font rasterizers (the one in WordPerfect versus the one in Windows).

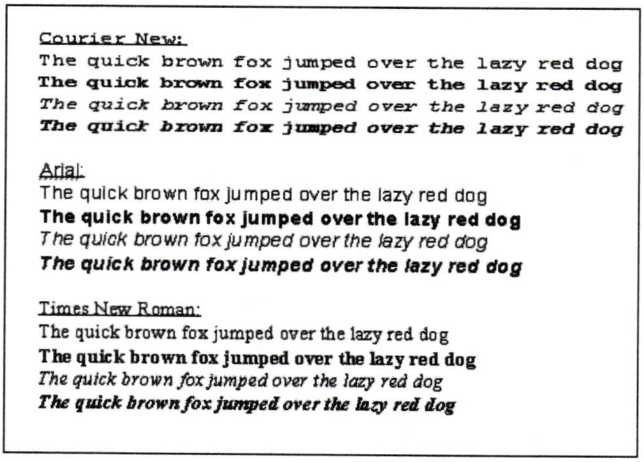

Figure B.16
I used the Windows equivalents of Courier, Helvetica and Times for this display: Courier New, Arial, and Times New Roman.

CHANGING WORDPERFECT'S STANDARD FONTS

WordPerfect's standard fonts are located in the shlib10 subdirectory of the WordPerfect installation. Table B.1 lists the files that are of interest, along with equivalent font files that come with Ghostscript.

TABLE B.1 WORDPERFECT'S BASIC FONT FILES

WordPerfect Font File	Ghostscript Font File	Description
wpro____.pfb	n021003l.pfb	Times (aka Roman-WP)
wproi___.pfb	n021023l.pfb	Times Italic (aka Roman-WP Italic)
wprob___.pfb	n021004l.pfb	Times Bold (aka Roman-WP Bold)
wprobi__.pfb	n021024l.pfb	Times Bold-Italic (aka Roman-WP Bold-Italic)
wphv____.pfb	n019003l.pfb	Helvetica (aka Helve-WP)
wphvi___.pfb	n019023l.pfb	Helvetica Italic (aka Helve-WP Italic)
wphvb___.pfb	n019004l.pfb	Helvetica Bold (aka Helve-WP Bold)
wphvbi__.pfb	n019024l.pfb	Helvetica Bold-Italic (aka Helve-WP Bold-Italic)
wpco____.pfb	n022003l.pfb	Courier (aka WP-Courier)
wpcoi___.pfb	n022023l.pfb	Courier Italic (aka WP-Courier Italic)
wpcob___.pfb	n022004l.pfb	Courier Bold (aka WP-Courier Bold)
wpcobi__.pfb	n022024l.pfb	Courier Bold-Italic (aka WP-Courier Bold-Italic)

These fonts may be known under other names when specific printer drivers are used; for instance, the Helvetica fonts are used onscreen when Arial is selected for a PCL printer. You may want to look into replacing other onscreen fonts. Try using the Linux `grep` command to locate the fonts that are of interest. Use `grep` on the .afm files only, to reduce the amount of gibberish that's returned from binary files.

Basically, all you need do is replace these files with other files that contain equivalent fonts. For instance, you can use the font files that come with Ghostscript or OS/2. The safest approach is to back up the existing file and then copy the new one in its place; for instance:

```
mv wpro____.pfb wpro____-orig.pfb
cp /usr/share/ghostscript/fonts/n021003l.pfb wpro____.pfb
```

The preceding commands replace the main Times font with the one that comes with Ghostscript, assuming Ghostscript is located in /usr/share/ghostscript. (You can use the Ghostscript Fontmap file to locate the specific fonts you need from Ghostscript if you want to try this step with more fonts than I listed in Table B.1.) If you like, you can use symbolic links instead of copying the font file—just substitute `ln -s` for `cp` in the preceding command.

What do you gain? Figure B.17 shows the results of using Ghostscript's fonts in place of those shipped with WordPerfect.

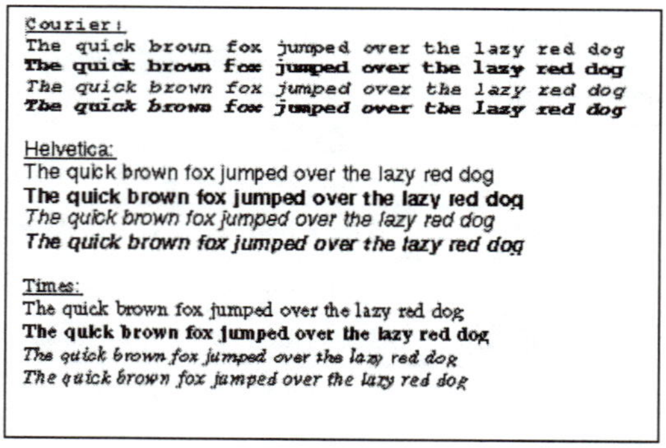

Figure B.17
For the most part, the Ghostscript basic fonts are superior to those that come with WordPerfect.

In general, the Ghostscript fonts are more even in height than are their WordPerfect counterparts, but the Ghostscript Times bold-italic doesn't look any bolder than the italic version (a flaw shared by the WordPerfect font at this size). Some of the Courier bold letters in both implementations are rather too dense.

Tip #309 from *Tux Tips*

> If you want to find Type 1 fonts to replace the standard set and are not satisfied with those that come with Ghostscript, try looking for a copy of Adobe Type Manager (ATM) for Windows. This program comes with Times, Helvetica, and Courier. Also, check with the manufacturer of your printer for standard fonts in Type 1 format. Note that not all fonts are completely equivalent; some font files may lack certain characters, for instance, which may produce unpredictable results when you try to display or print text that contains special symbols or other effects.

Changing your standard fonts in this way also changes the fonts you get when you print. At typical printer resolutions, you aren't likely to notice the difference, but you might, especially if the font you're using has an oblique rather than an italic variant or is missing certain important characters. This problem affects you only if you select the font as a screen font; if you select a printer font, it displays onscreen using your modified font but prints using the printer's (or Ghostscript's) font.

Overall, replacing the font files can help a bit in terms of legibility, but this technique won't improve the text display to the level of text display in Windows. What more can be done?

Font Size Font Size

One of the more obvious differences between WordPerfect for Linux and WordPerfect for Windows is that the latter produces larger fonts than the former. This disparity is caused by the programs' making different assumptions about screen size, and of course, your system might be configured in a way that produces different effects. Still, it's only fair to compare fonts of the same size. In WordPerfect, onscreen font size can be manipulated without affecting the size of fonts on paper by choosing View, Zoom to open the Set Zoom dialog box. For my system, 120% brings the Linux and Windows displays into rough parity. Figure B.18 shows the result, using the Ghostscript fonts.

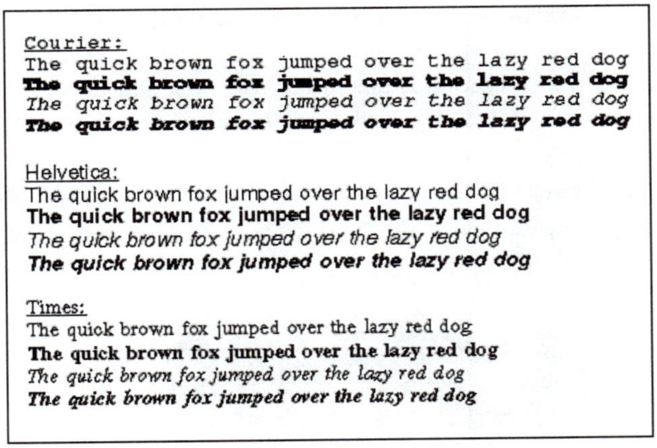

Figure B.18
Increasing the size of the display by using WordPerfect's Zoom feature improves legibility at the cost of reduced amounts of text on the screen.

This change in font size improves legibility substantially—the result begins to approach the quality of the Windows rendering, though a few characters here and there can still be improved.

These principles also apply to fonts that you add. You can easily find several implementations of common fonts like Benguiat, Optima, and others, and one may look far better on the screen than another. In this case, it's a matter of installing one rather than another. Increasing the display size is certainly one means of producing better-looking text no matter which font you're using.

TROUBLESHOOTING

POOR PRINT RESOLUTION

You selected what you believe to be the correct printer driver, but you find that you're getting poor resolution from the printouts—300 dpi on a 600-dpi printer, for instance.

There are four likely causes of this problem:

- The driver you selected actually is *not* correct. Sometimes printer manufacturers change their printers radically and change the model numbers only very slightly, so double-check the driver or try another one for another printer.
- Despite the fact that PostScript printers are perfectly capable of handling the Type 1 fonts that WordPerfect uses for its graphics fonts, WordPerfect converts them to bitmaps when it prints to PostScript printers. Therefore, you are likely to get chunky text if you select a printer driver that's mismatched to your printer. You can get around this problem by choosing a different driver or by using soft fonts, which WordPerfect *does* download to the printer.

- If you're printing to a non-PostScript printer using Ghostscript, your Linux printer queue may need to be modified. In particular, you may need to change the printer options relating to resolution. Try printing from an application other than WordPerfect, such as Netscape. If the problem persists, then it is likely to be in your printer queue configuration.

→ To learn more about configuring a Linux printer queue, **see** "Setting Up Printer Queues," **p. 731**

- You may need to change the Print Quality setting in the Print dialog box (Figure B.19). For best quality, set this option to High.

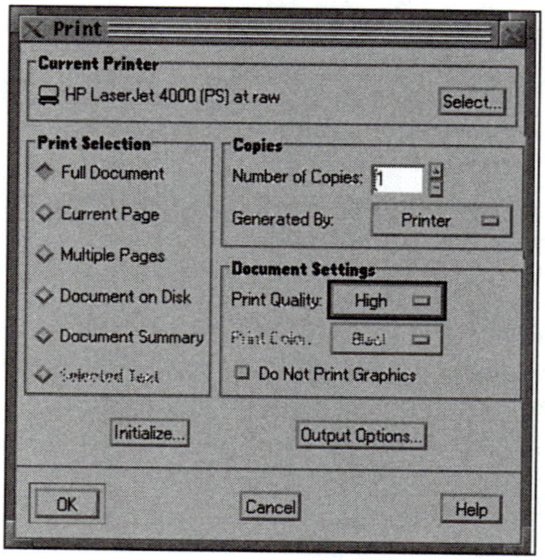

Figure B.19
The Print dialog box allows you to set some quality-related print options.

STRAY FORM FEEDS

You're printing to a non-PostScript printer and get a stray form feed after each print job.

This problem is easily corrected by adding the `sf` option to the /etc/printcap entry for the printer. This change may cause the last page of jobs from other applications to not print, however, so you may need to create two printer queues, one for WordPerfect and one for everything else.

→ To see a sample printer queue with the `sf` option, **see** "Setting Up Non-PostScript Printers," **p. 742**

APPENDIX C

Customizing Preferences

In this appendix

The Preferences Options

A good tool conforms to its user rather than requiring the user to conform to the tool. Throughout this book, I've referred to *default* values for assorted WordPerfect features. Sometimes WordPerfect remembers a change you make in a default value and uses your new value when you next start WordPerfect. For example, the Spell-As-You-Go and Grammar-As-You-Go features, which you can turn on and off from the Tools, Proofread menu, retain the settings you established the last time you used WordPerfect.

➔ To learn more about the As-You-Go error-checking features, **see** "Proofreading with Spell-as-You-Go and Grammar-as-You-Go," **p. 118**

With other options, you need to tell WordPerfect to change its behavior in a different way, by using WordPerfect Preferences. With most of the options described in this appendix, you begin by choosing Preferences from the WordPerfect Program control window. This step produces the Preferences dialog box shown in Figure C.1. You select the class of options you want to modify by clicking the appropriate button in this dialog box.

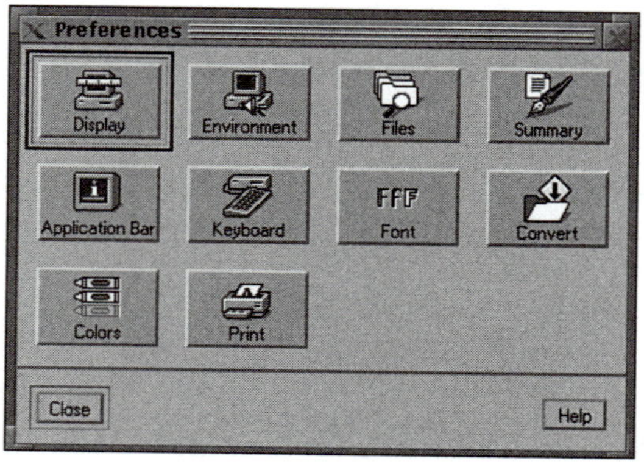

Figure C.1
The Preferences dialog box is your starting point for configuring many of WordPerfect's default settings.

WordPerfect stores these options in the .wprc subdirectory of your home directory. This way, two or more people using the same computer can have entirely different Preferences settings.

Display

When you click Display in the Preferences dialog box, WordPerfect responds by showing the Display Preferences dialog box (Figure C.2). This dialog box has six settings, selectable from the radio buttons near its top. Each setting provides a new set of options for the display of information in WordPerfect.

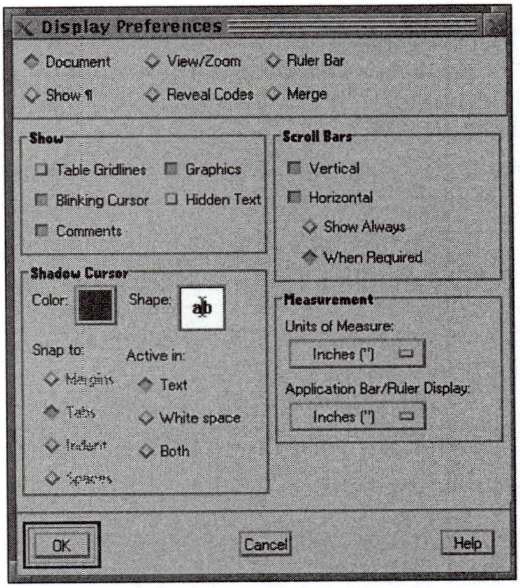

Figure C.2
The Display Preferences dialog box lets you control various aspects of how WordPerfect displays information.

DOCUMENT

The first set of options relates to major document display features:

- Show—You can elect to show or not show a number of features, including table gridlines, a blinking cursor (insertion point), comments, graphics, and hidden text. You can set some of these options on a temporary basis from the Document window, as in using View, Graphics to toggle the display of graphics in a document.
- Scroll Bars—You can enable or disable both horizontal and vertical scrollbars. For the horizontal scrollbar, you can set WordPerfect to display the scrollbar only when WordPerfect is displaying more text than will fit horizontally in the window.
- Shadow Cursor—You can set the shadow cursor to be active only in text, only in whitespace, or both. If you choose White space or Both, WordPerfect gives you assorted options for where it will "snap" your insertion point if you click in whitespace. You can also use the Color and Shape items to change the color and shape, respectively, of the shadow cursor.

→ If you want to know more about the shadow cursor, **see** "The Insertion Point and Shadow Cursor," **p. 29**

- Measurement—By default, WordPerfect uses inches as its default unit of measurement, but you can change this setting to centimeters, millimeters, or points. In addition, you can set different units for dialog boxes and for the ruler. In dialog boxes that ask for measurements, you can override the default by using *i* or double quotes (") for inches, *c* for centimeters, *m* for millimeters, or *p* for points. For instance, *2c* sets a measurement to 2 centimeters, even if the default is inches.

Show ¶

You can tell WordPerfect to display a number of symbols in the main editing area of the Document window from the Show ¶ area of the Display Preferences dialog box (Figure C.3). You can activate the show symbols feature on a temporary basis from the Document window, by choosing View, Show Paragraph, but you can set the specific symbols to display only from the Display Preferences dialog box.

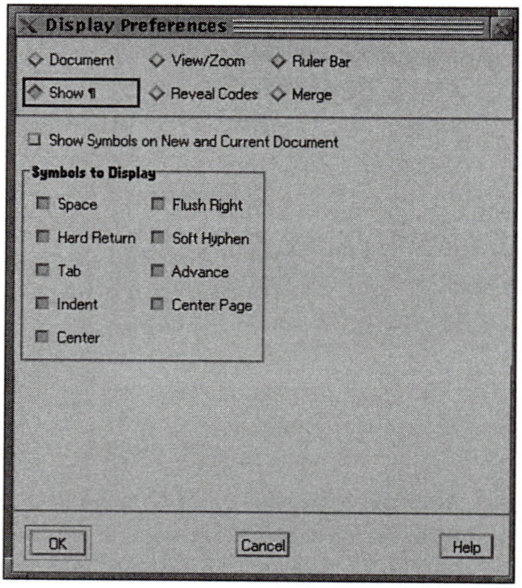

Figure C.3
Select the symbols you want to display and then click Show Symbols on New and Current Document to display them.

View/Zoom

When you first use it, WordPerfect defaults to a 100% zoom factor and Page View. You can alter these settings, however, to maximize the amount of text you see on the screen, improve font legibility, or otherwise customize the display (Figure C.4).

→ To learn about the view modes and zoom, **see** "Document Views," **p. 54**

→ To learn about improving font legibility, **see** "Optimizing Onscreen Font Display," **p. 761**

Reveal Codes

You can tell WordPerfect to open the Reveal Codes pane on all documents from the Reveal Codes portion of the Display Preferences dialog box (Figure C.5). You can also change the font used in the Reveal Codes pane and change various options for how WordPerfect displays information in that pane.

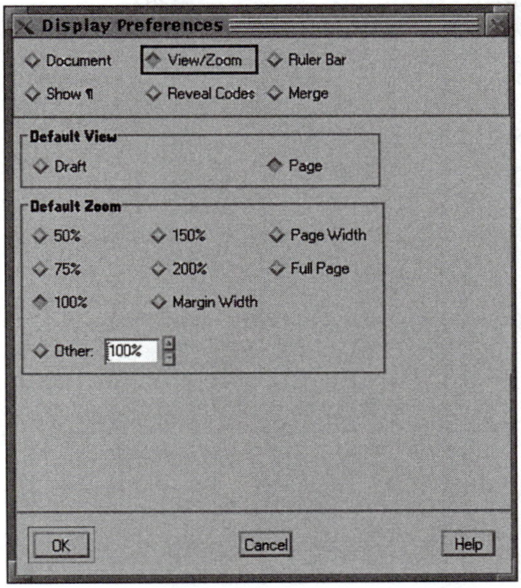

Figure C.4
You may want to adjust the default zoom factor if you run a smallish monitor at high resolution.

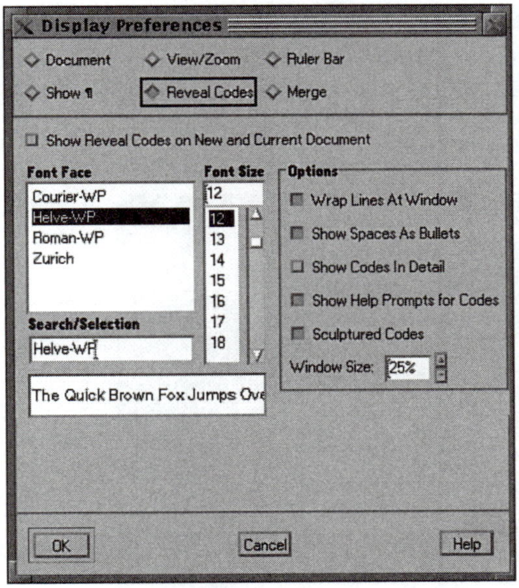

Figure C.5
You can use any font you can install as a graphics font in WordPerfect for use in the Reveal Codes pane.

RULER BAR

As with Reveal Codes, you can set WordPerfect to display the ruler whenever you open a document. You can also set options relating to the display of the ruler from the Ruler Bar Options portion of the Display Preferences dialog box (Figure C.6).

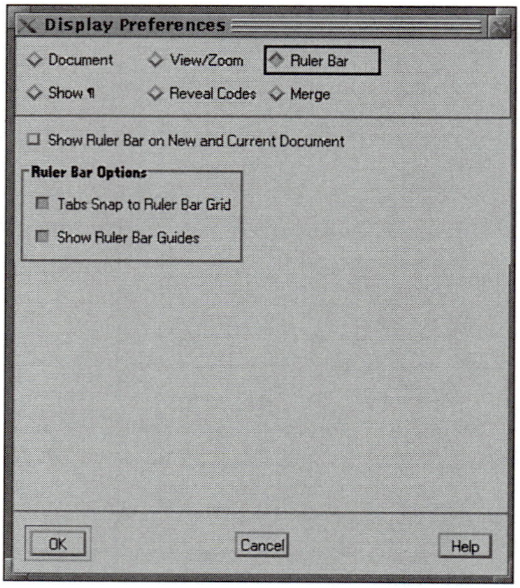

Figure C.6
Use the Tabs Snap to Ruler Bar Grid option to make placement of tabs more precise.

MERGE

When you perform a mail merge, WordPerfect can display the merge codes in the form file, show only markers, or show nothing at all. You set the default in the Merge Codes section of the dialog box (Figure C.7).

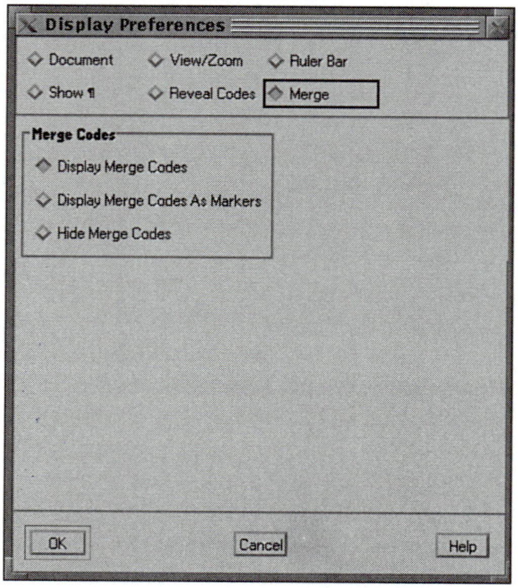

Figure C.7
If you merge short pieces of information, choosing Display Merge Codes As Markers or Hide Merge codes can help you lay out your form file.

→ If you want to learn how to perform a mail merge, **see** "The Same Letter, Different Information: Mail Merge," **p. 515**

Environment

The Environment Preferences dialog box, shown in Figure C.8, lets you set assorted options related to how WordPerfect interacts with you when you're editing a document.

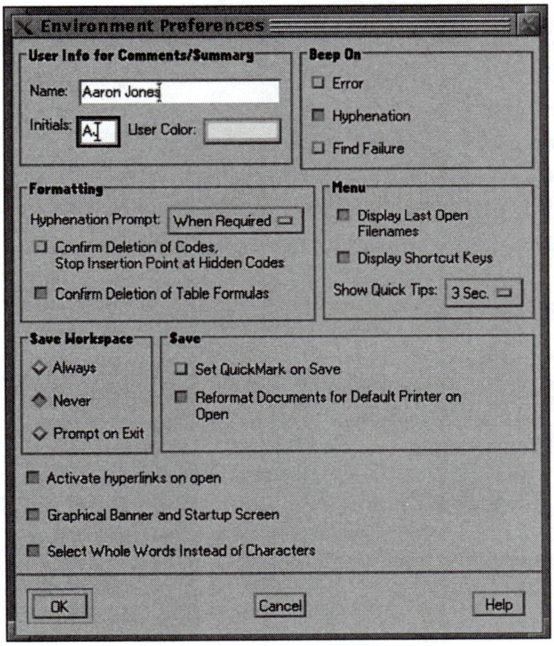

Figure C.8
The Environment Preferences dialog box is crowded, but it doesn't change its contents in the way the Display Preferences dialog box does.

You can set the following options in the Environment Preferences dialog box:

- User Info for Comments/Summary—You should enter your full name and initials here to use them in comments and document summary information. You can also specify a color, which WordPerfect uses as a background for comment markers and comments.

→ To learn more about comments, **see** "Document Comments," **p. 580**

- Beep On—WordPerfect can beep at you under the circumstances you specify. Disable all of these if you prefer to work in silence.
- Formatting—You can set options related to when WordPerfect prompts you for hyphenation information, whether or not to confirm the deletion of codes, and whether or not to confirm the deletion of table formulas.

→ If you want to learn more about hyphenation, **see** "Hyphenation," **p. 220**
→ If you want to learn more about editing with Reveal Codes, **see** "Editing with Reveal Codes," **p. 110**
→ If you want to learn more about table formulas, **see** "Table Mathematics," **p. 296**

- Menu—By default, WordPerfect shows a list of the last several files you've opened in the Program menu in the Program control window, but you can disable this setting here. Similarly, you can turn off the display of shortcut keys in the WordPerfect menus. Finally, you can tell WordPerfect how long to wait before displaying a hint about the function of an item when you place the mouse pointer over a button or toolbar icon.
- Save Workspace—If you save the workspace, WordPerfect automatically loads the last document you edited whenever it loads.
- Save—If you check Set QuickMark on Save, WordPerfect sets the QuickMark whenever you save a document. If you click Reformat Documents for Default Printer on Open, you may find that your document changes formatting if you move it back and forth between computers.

→ To learn more about the QuickMark, **see** "Bookmarks," **p. 595**

- Activate hyperlinks on open—If you select this option, WordPerfect launches a browser or tries to load a file if you click on a link in a Web page you open in WordPerfect.
- Graphical Banner and Startup Screen—This option enables the graphical WordPerfect "splash page" image when you launch the program, as well as the similar graphic in the Program control window.

Tip #310 from *Tux Tips*

If you run X at a low color depth, you may find that all the colors on your screen change when you move your mouse in or out of certain programs' windows, or some programs may display with very few available colors, which can look quite bad. Eliminating the color-hungry graphical banner and graphic in the Program control window can reduce the severity of these problems. Increasing the color depth can also help, but that slows down your display, your video card may not have enough RAM to handle increased color depth, and WordPerfect doesn't work well at a depth of 24 bits with some X servers (although it's fine at both 16 and 32 bits).

- Select Whole Words Instead of Characters—If you select this option, WordPerfect selects the entire word rather than just the part on which you've clicked when you select text with the mouse. This feature can help you to select text quickly but can get in the way if you want to select part of a word.

→ If you want to know more about selecting text in WordPerfect, **see** "Selecting Text," **p. 96**

Files

The Files Preferences dialog box (Figure C.9), like the Display Preferences dialog box, has several sections, each accessible by clicking on its radio button in the top of the dialog box. These sections all look very much alike, so I only show the first one, which also has the most options.

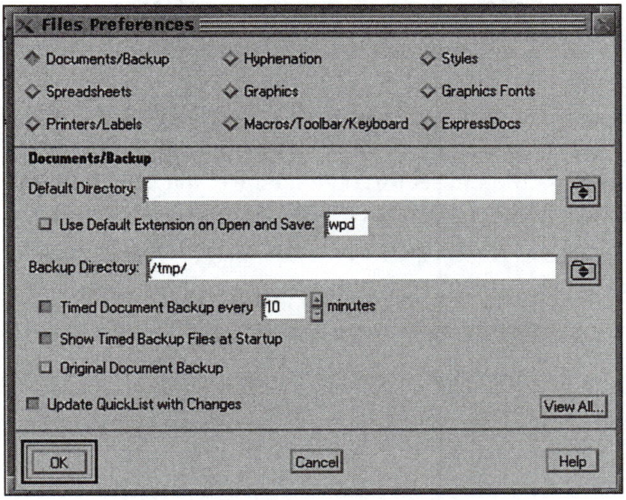

Figure C.9
Most of the selections in the Files Preferences dialog box contain between one and three fields for entering directory names.

Each of the nine selections in the top portion of this dialog box lets you enter directories relating to the location of files WordPerfect uses. If a field is blank, WordPerfect normally uses your home directory for that function.

- Documents/Backup—You can specify a default document directory, or WordPerfect will use your home directory. You can also set WordPerfect to automatically append the .wpd extension (or another one you specify) to all files. The Backup Directory field gives the location of the backup files that WordPerfect creates. You can also set assorted backup options here.
- Spreadsheets—When you import a spreadsheet, WordPerfect looks in up to two directories to find a spreadsheet if you don't specify a full path.

→ If you want to learn how to import a spreadsheet, **see** "Importing a Spreadsheet," **p. 301**

- Printers/Labels—You can specify the directory in which WordPerfect stores customized labels, as well as what type they are (laser, tractor-fed, or both).

→ To learn more about creating labels, **see** "Labels," **p. 250**

- Hyphenation—You can specify the location of hyphenation dictionaries here.

→ To learn more about WordPerfect's hyphenation features, **see** "Hyphenation," **p. 220**

- Graphics—WordPerfect searches for graphics files in up to two directories that you specify. The second of these defaults to the wpgraphics directory within the WordPerfect installation tree.

→ If you want to know more about importing graphics, **see** "Inserting Graphics from External Sources," **p. 379**

- Macros/Toolbar/Keyboard—WordPerfect searches for macros, toolbars, and keyboard layout files in the directories you specify here.

→ To learn more about macros, **see** "Letting the Computer Do the Work: Macros," **p. 483**

- Styles—WordPerfect keeps its styles in a file and directory you specify here. If you launch WordPerfect as root with the `-adm` switch, you can also specify the systemwide shared style file from here.

→ If you want to know more about styles, **see** "Ensuring Consistency Within Your Document With Styles," **p. 305**

- Graphics Fonts—You can set this option only if you launch WordPerfect as root with the `-adm` switch. This entry tells WordPerfect where to look for graphics font files.

→ To learn how to install and manage fonts, **see** "Installing and Managing Fonts," **p. 754**

- ExpressDocs—WordPerfect looks in the directory you specify here for the ExpressDocs templates.

→ To learn more about templates, **see** "Promoting Cross-Document Consistency with Templates," **p. 333**

Note

Although you can set this directory in the downloadable version of WordPerfect, the ExpressDocs feature isn't available in this version of the program.

SUMMARY

The Document Summary Preferences dialog box shown in Figure C.10 lets you give greater emphasis to WordPerfect's document summary features.

Figure C.10
Use the Document Summary Preferences dialog box to ease document exchange with WordPerfect for DOS users.

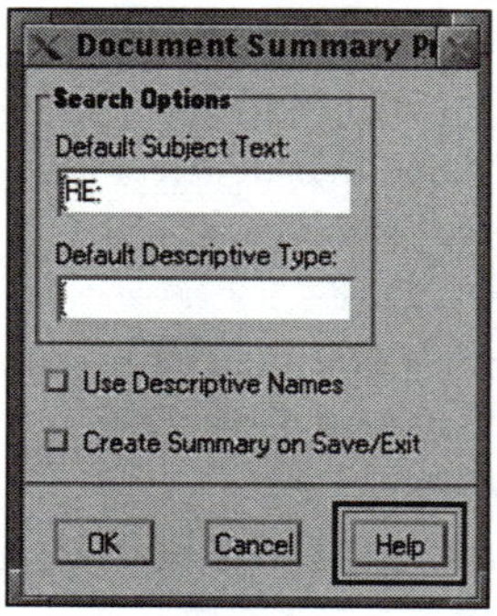

- Search Options—Document summary information includes assorted text fields, and you can specify defaults for two of them here.
- Use Descriptive Names—If you select this option, WordPerfect displays the WordPerfect descriptive names whenever it presents a File Manager dialog box. This option can be convenient if you regularly load files that were created with a version of WordPerfect for DOS that uses these fields to get around the DOS 8.3 filename limits.

- Create Summary on Save/Exit—The WordPerfect document summary features normally sit largely unnoticed, but you can tell WordPerfect to display the Document Summary dialog box (Figure C.11) whenever you save a file. This option may be useful if you regularly save files that will be loaded in WordPerfect for DOS, since you can then give the file a descriptive name to get around DOS's 8.3 filename limits.

Figure C.11
The Document Summary dialog box lets you enter information about who you are and the document's contents.

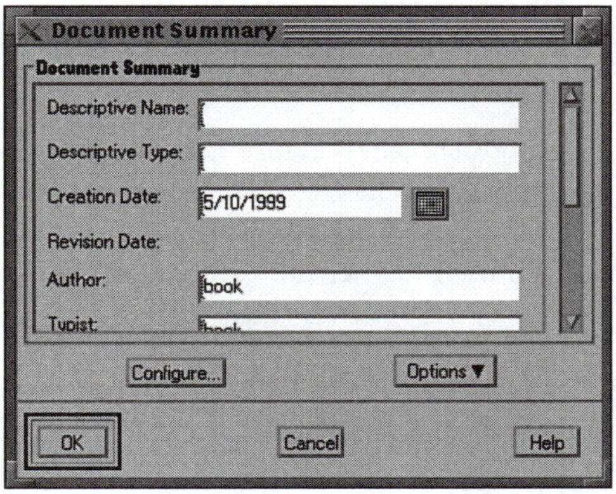

APPLICATION BAR

The WordPerfect application bar (also known as the status bar), at the bottom of the WordPerfect Document window, contains buttons and information relating to the current state of WordPerfect features. You can customize the application bar from the Application Bar Preferences dialog box shown in Figure C.12.

Figure C.12
When you click an item in the Select items to appear on the bar list, the Item description box describes the item.

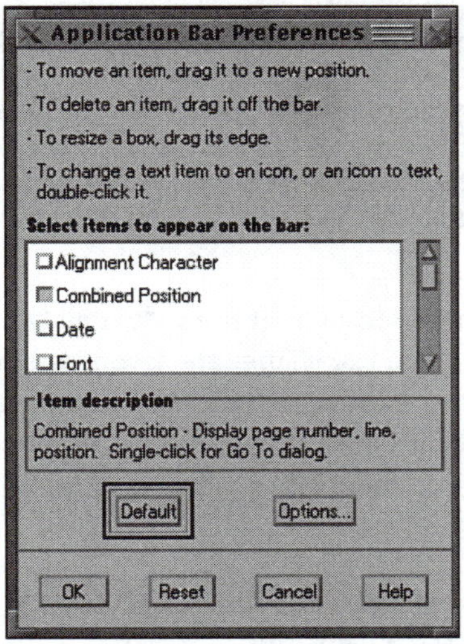

You can activate and deactivate features by clicking the check buttons in the Select items to appear on the bar list. In addition, when the Application Bar Preferences dialog box is open, you can move and resize the application bar items by clicking and dragging them. (Normally, clicking on an application bar item either has no effect or activates a feature associated with the item.) You can restore the application bar to its normal settings by clicking Default, and you can set the font used in the application bar by clicking Options.

KEYBOARD

The Keyboard Preferences dialog box shown in Figure C.13 lets you select the keyboard layout you want to use. You might want to change the layout if you're familiar with a previous version of WordPerfect, with WordPerfect on some other platform, or with a different word processor altogether, and you want to alter the keyboard mapping to match your expectations.

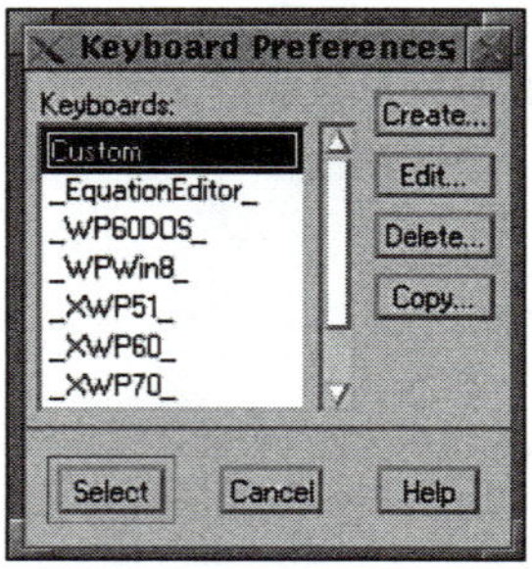

Figure C.13
The Edit and Delete buttons are only active for keyboard definitions you create yourself.

Caution

Your X server and window manager may intercept certain keystrokes for their own uses. If you try a new keyboard layout and find that certain keystrokes produce a window manager action, such as changing from one virtual display to another, when you try to activate a WordPerfect feature, then you've run into this problem. You may be able to reconfigure your window manager to use keystrokes that WordPerfect doesn't use, or you may need to customize the WordPerfect keyboard layout to avoid your window manager's reserved keys. This problem can even occur with the default _XWP8_ keyboard layout.

WordPerfect lets you do more than choose from a set of predefined keyboard layouts; you can create one or more custom layouts. You can change the layout either to work around a conflict with your window manager or to create a set of keystrokes that's simply more to your liking. You can even create a keyboard layout to emulate a completely different word processor. To modify the keyboard layout, follow these steps:

1. Select a keyboard layout that's close to what you want.
2. Click Copy to create a copy of this layout. WordPerfect prompts you for a name for the new layout. When you've entered a name, the new layout name appears in the Keyboard Preferences dialog box.

3. Be sure your new custom layout is selected and then click Edit. WordPerfect displays the Keyboard Editor dialog box (Figure C.14).

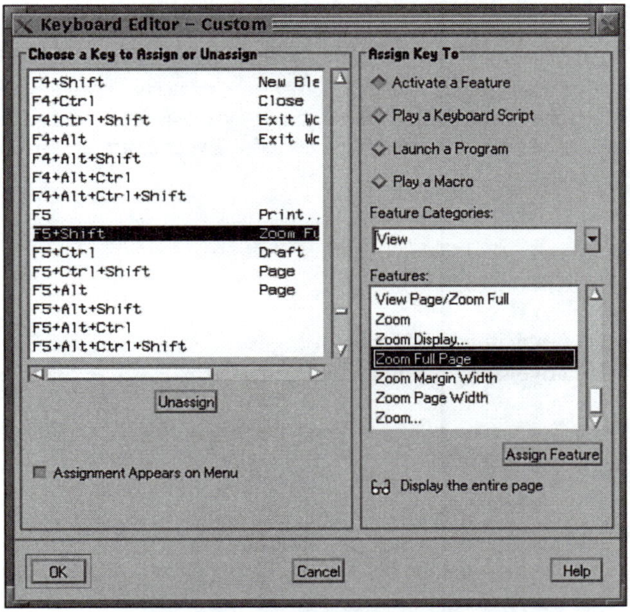

Figure C.14
The Keyboard Editor dialog box's title bar includes the name of the keyboard layout being modified—Custom, in this case.

4. Select a key you want to reassign in the Choose a Key to Assign or Unassign list box. Note that each key is available with or without various modifiers, such as Ctrl and Shift.
5. Select the class of actions you want to perform from the radio buttons in the Assign Key To box. Activate a Feature links the key to a WordPerfect feature; Play a Keyboard Script lets you type at the keyboard and link what you type to the key; and Launch a Program runs an external program. The portion of the dialog box immediately below the radio buttons changes slightly with each radio button selection.

Are you trying to assign a macro to a keystroke? Is it not working? See "Can't Assign a Macro to a Keystroke" in the "Troubleshooting" section at the end of this chapter.

6. Fill out the specific information for the action you want to perform. For the Launch a Program option, for instance, you must select a program file. The Activate a Feature option requires you to select a WordPerfect feature. These features are categorized according to their locations in the Document window's menus.
7. Click Assign Feature.
8. If you want the keyboard shortcut to appear on the menu item for this feature, select the Assignment Appears on Menu check button. (This button won't be available for all features, however.)
9. If you want to undefine a keystroke so that it does nothing, click Unassign, rather than Assign Feature.

10. Repeat steps 4 through 9 for each keystroke you want to define or redefine.
11. Click OK in the Keyboard Editor dialog box and Select in the Keyboard Preferences dialog box to use your new keyboard definition.

Note

This book includes references to keyboard shortcuts for features based on the _XWP8_ keyboard layout. If you change your keyboard layout, either by selecting a different layout that comes with WordPerfect or by defining your own, these references may no longer be accurate.

FONT

The Font Preferences dialog box shown in Figure C.15 lets you adjust the default redlining method and the relative sizes of fonts when you use the Relative Size font attribute.

Figure C.15
You set preferences for redlining and relative font size adjustments in the Font Preferences dialog box.

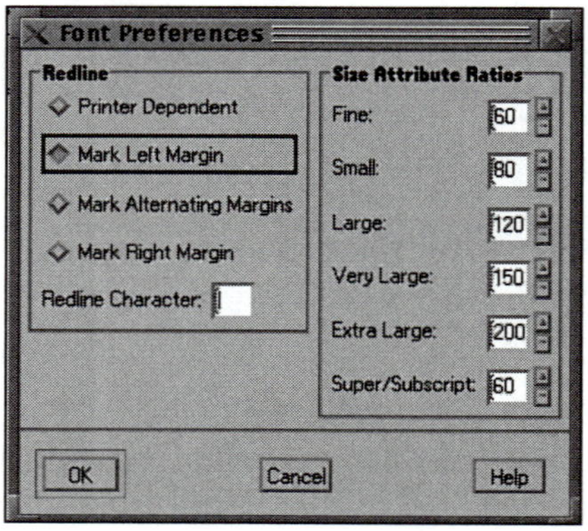

If you select the Printer Dependent redline method, WordPerfect uses a redlining method defined in the printer driver file. This method varies from printer to printer; for instance, it creates red text with color for PostScript printers, but a gray background with PCL printers. The various Mark methods place the character you type in the Redline Character box at the left, right, or alternating left and right sides of the page regardless of the printer type you use. You can override the default setting on a document-by-document basis by choosing File, Document, Redline Method from the Document window's menus.

→ If you'd like to learn more about redlining, **see** "Comparing Documents," **p. 588**

The spin boxes in the Size Attribute Ratios box let you tell WordPerfect how much larger or smaller you want to make the text with each of these characteristics compared to ordinary text. For instance, WordPerfect normally creates superscripts and subscripts that are

60% the size of ordinary text—that is, if the main text is at 12 points, WordPerfect uses 7.2-point text for superscripts and subscripts. You can change these relative text sizes here.

→ To learn more about font size attributes, **see** "Fonts," **p. 178**

CONVERT

The Convert Preferences dialog box, shown in Figure C.16, lets you give WordPerfect some pointers on how to import files. The main Convert Preferences dialog box is useful mainly in handling imported spreadsheets. The Field and Record fields let you specify which characters delimit fields and records, respectively, in the foreign file. The Encapsulated field gives a character that tells WordPerfect to treat the enclosed text as a single field, even if the Field code is present between two instances of the Encapsulated code. Finally, use Strip to specify characters you want removed from the file as it's imported.

→ To learn how to import a spreadsheet into a table, **see** "Importing a Spreadsheet," **p. 301**

→ To learn how to import a spreadsheet for a mail merge, **see** "Using Foreign Data Files," **p. 532**

Figure C.16
Use the Convert Preferences dialog box to control import options for spreadsheets.

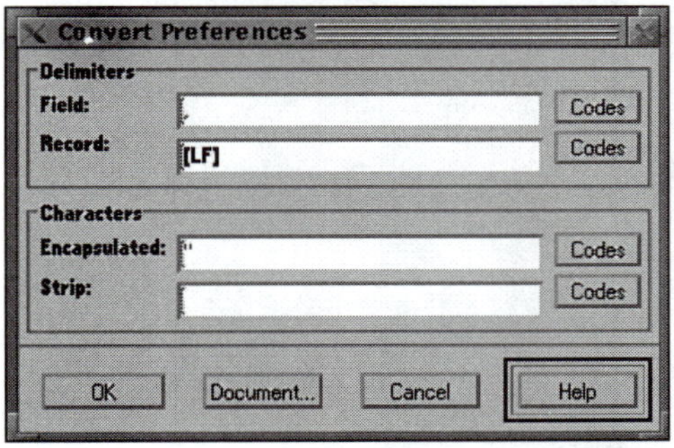

If you click Document in the Convert Preferences dialog box, WordPerfect presents the Conversion Preferences dialog box shown in Figure C.17. You use this dialog box to set preferences for importing text files, including the margins, paper size, and whether to underline spaces and tabs when converting underlined text. If you click Language, you can specify the language code for imported files, which can be useful if you expect to spell-check imported documents in WordPerfect. If the file you're importing sets these values, those settings override the ones you specify here.

→ To learn how to import a foreign word processor file, **see** "Reading Foreign Files," **p. 88**

Figure C.17
Use the Conversion Preferences dialog box to control import options for spreadsheets.

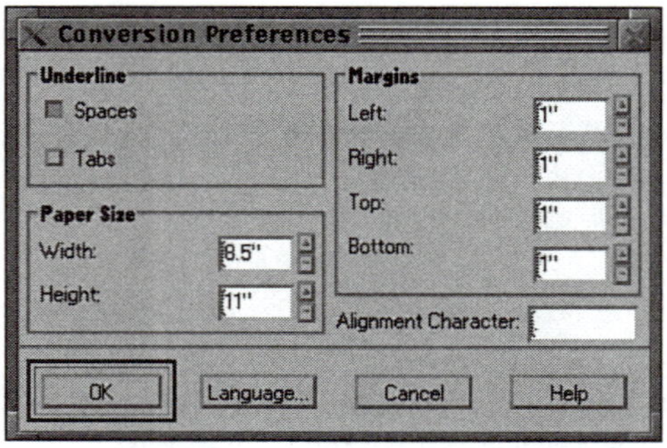

COLORS

You can customize WordPerfect's color scheme by using the Color Preferences dialog box shown in Figure C.18.

Figure C.18
Customize the colors WordPerfect uses for its windows and dialog boxes with the Color Preference dialog box.

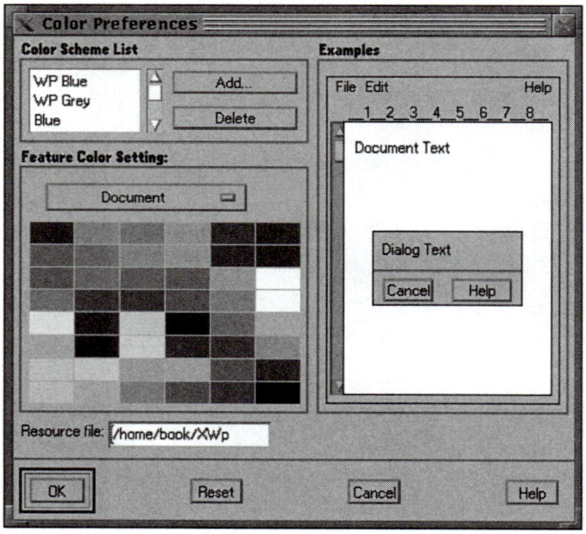

This dialog box has several components:

- Examples—The Examples area on the right of the dialog box shows a mock-up of a WordPerfect window and dialog box so that you can see the effect of your changes before committing them.
- Color Scheme List—You can select from any of several predefined color schemes in the Color Scheme List box in the upper-left portion of the dialog box. If none of the existing color schemes meet your needs, click Add to create a new one. You can also delete an existing scheme by selecting it and clicking Delete.

- Feature Color Setting—Use the option button in the Feature Color Setting box to choose which of nine window elements, such as the document text or the menu bar, you want to adjust. Then click on one of the colors in the color rectangle immediately below that option button to set the color.
- Resource file—WordPerfect stores its color options in a file, normally XWp in your home directory. You can change the location of this file, however.
- Reset—You can reset the color settings for a color scheme to its original values by clicking the Reset button.

Note

When you're finished editing your color scheme and click OK, WordPerfect doesn't immediately apply the changes. You have to exit from WordPerfect and restart the program to see the new color scheme in the actual windows.

PRINT

The Print Preferences dialog box (Figure C.19) lets you set various defaults for printing. You can override most of these from the Print dialog box (Figure C.20) you get when you choose File, Print to print a document.

→ If you want to know how to print from WordPerfect, **see** "Creating Paper Output: Printing," **p. 148**
→ If you want to know how to set up a WordPerfect printer queue, **see** "Fonts and Printing," **p. 739**

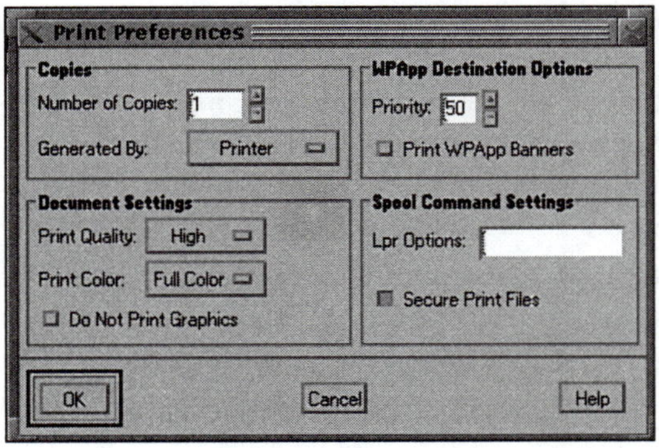

Figure C.19
If you usually print multiple copies or use special lpr queue options, you can set them from the Print Preferences dialog box.

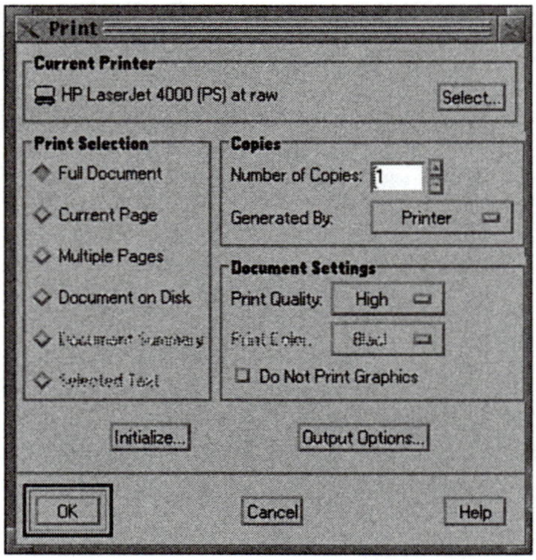

Figure C.20
You can override many of the Print Preferences defaults when you print an individual document.

The following options are available in the Print Preferences dialog box:

- Copies—WordPerfect normally produces only one copy of a document when you print, but you can change this default. You can then use the Generated By option button to tell WordPerfect to generate the copies or to have the printer do it.
- WPApp Destination Options—This box controls options for WordPerfect's own spooler, which merely duplicates the standard Linux spooler, so I don't discuss it in this book.
- Document Settings—You can change the default document resolution with the Print Quality option button, and you can use the Print Color option button to tell WordPerfect to print in color or only black. You can also check the Do Not Print Graphics check button to omit graphics from your printouts. These options depend, to a greater or lesser extent, upon the printer's and printer driver's capabilities. For instance, a Print Quality setting of High means one thing for a modern 1200-dpi laser printer and another for a 15-year-old, nine-pin dot-matrix printer.
- Spool Command Settings—If you want to include a special `lpr` spool command option with your printouts, you can enter one here. Type `man lpr` at an xterm to learn about these options. If you select the Secure Print Files check button, WordPerfect generates print files with permissions that prevent other users from reading them. This setting is useful if you sometimes print sensitive data on a shared system, but on some systems this option may interfere with the spooler's ability to print.

Are you unable to print from WordPerfect? See "Can't Print from WordPerfect" in the "Troubleshooting" section at the end of this chapter.

Note

If you have more than one printer queue defined in WordPerfect, the options you enter in the Print Preferences dialog box apply to all the queues. If you want to change a default for some printers but not others, you may be out of luck. You can set some printer options on a printer-by-printer basis when you configure the printer queue, however.

➔ If you want to know how to set up a WordPerfect printer queue, **see** "Fonts and Printing," **p. 739**

TOOLBARS

You access most of the features I discuss in this appendix through the Preferences dialog box (refer to Figure C.1). You can configure WordPerfect's toolbars from the toolbars themselves, however. To do so, right-click on a toolbar in a WordPerfect Document window to get a context menu for that toolbar, as in Figure C.21. This menu includes a list of available toolbars, and you can click on a toolbar name to activate it (or to remove it if it's already active).

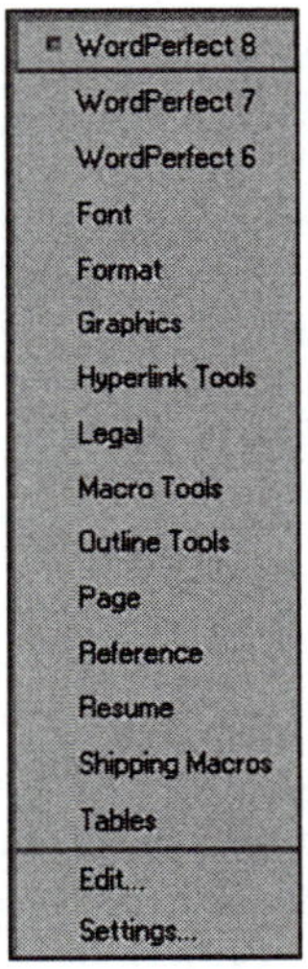

Figure C.21
WordPerfect places a 3D square next to each active toolbar (only WordPerfect 8 in this figure), similar to an active check button in a dialog box.

If you want to customize a toolbar, activate it, right-click on the toolbar you want to edit, and choose Edit from the toolbar context menu to open the Toolbar Editor dialog box shown in Figure C.22. You can add a button to Activate a Feature; Play a Keyboard Script; Launch a Program; or Play a Macro. You select these features much as you do when designing a new keyboard map. You can also drag the Separator icon from the Toolbar Editor dialog box to the toolbar you're editing to add a vertical line to separate groups of toolbar tools. To remove a tool from a toolbar, click the tool and drag it off the toolbar when the Toolbar Editor is open.

Figure C.22
The Toolbar Editor dialog box's name changes to include the name of the toolbar you're editing.

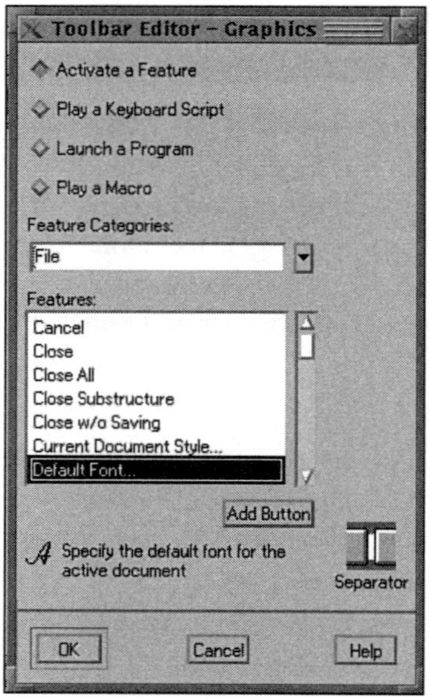

If you choose Settings from the toolbar context menu (refer to Figure C.21), WordPerfect displays the Customize Setting dialog box shown in Figure C.23. You can use this dialog box to turn on and off any of WordPerfect's toolbars or property bars. You can also click Reset to reset a toolbar to its original state, click Edit to open the Toolbar Editor dialog box for a toolbar, or click Create to create an entirely new toolbar to suit your specific needs. Finally, you can click Options to get the Toolbar Options dialog box (Figure C.24), in which you can set the toolbar's font and specify whether to display the toolbar's items as text, as pictures only, or as pictures and text together.

Figure C.23
You can work with either Toolbars or Property Bars in the Customize Settings dialog box, but you can't create a new property bar.

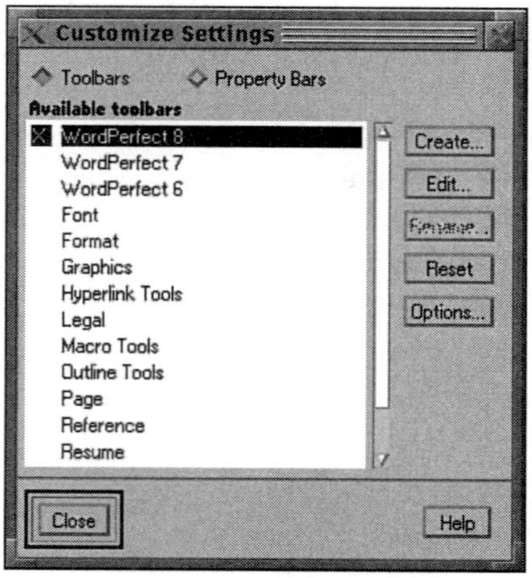

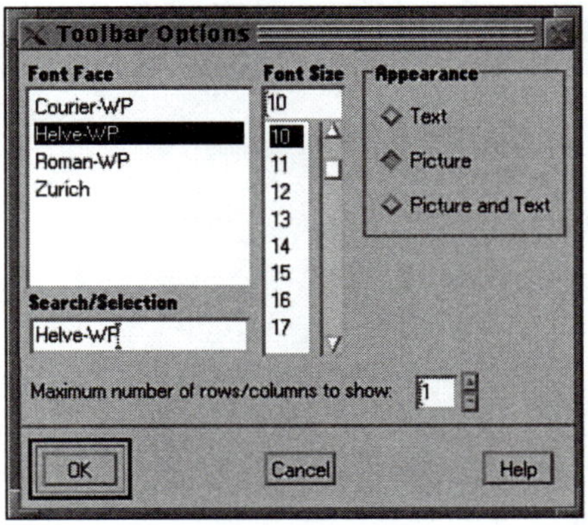

Figure C.24
The font options have an effect only if you've chosen Text or Picture and Text for the toolbar's Appearance.

➔ For more information on using toolbars, **see** "Toolbars," **p. 56**

TROUBLESHOOTING

CAN'T ASSIGN A MACRO TO A KEYSTROKE

You're trying to assign a macro to a keystroke from the Keyboard Editor, but nothing happens when you type the appropriate keystroke.

Not being able to assign a macro to a keystroke this way is a bug in WordPerfect. Fortunately, you can work around it for Ctrl+key and Shift+Ctrl+key combinations by naming your macro appropriately. Call your macro ctrl*x*.wcm or ctrlsft*x*.wcm where *x* is the key you want to use with Ctrl or Shift+Ctrl, respectively. Your macro file needs to exist in the macro search path for this method to work.

➔ To learn more about assigning macros to keystrokes, **see** "One-Key Access to Macros," **p. 492**
➔ To learn how to set the macro search path, **see** "Files," **p. 774**

CAN'T PRINT FROM WORDPERFECT

You can print from most applications, but not from WordPerfect. When you try to print, you simply do not receive a printout.

One possibility is that you've specified the wrong printer driver. Try configuring your driver to print to disk and then use `lpr` to print the file yourself, as in `lpr test.ps`. If this procedure doesn't work and if the file size is greater than 0, try another printer driver. If printing a WordPerfect-generated PostScript file with `lpr` works, disabling the Secure Print Files option in the Print Preferences dialog box may fix your printing problem. If doing so eliminates the problem but you still want to be able to use the Secure Print Files option, you may need to change the ownership or permissions on one or more of the files Linux uses for printing, such as /usr/bin/lpr or /usr/sbin/lpd. These files should both be owned by root and have permissions strings of `-rwxr--r--` on most systems. Be cautious about changing these settings; however, there are several subtle variant methods of configuring a Linux printer queue, and some of these may rely on having some other set of permissions or ownership to these files.

INDEX

Symbols

A

B

C

E

F

H

I

J

K

L

M

N

O

P

S

T

U

V

W

X

Y-Z

USEFUL WEB SITES FOR WORDPERFECT FOR LINUX

The following Web sites contain useful information and files for users of WordPerfect for Linux. Some of these sites relate directly to WordPerfect or Linux, whereas others contain general information of interest to WordPerfect for Linux users.

- `http://linux.corel.com`—Corel's main Web site for Linux. Check here for information on updates to WordPerfect for Linux and for the latest information on the program.
- `https://livewire.corel.com/wp8LinuxReg/register.html`—The site at which you register your downloaded version of WordPerfect for Linux or the version of WordPerfect that comes with this book. Note that this site is a *secure* site, so you need to type `https` at the start of the URL, and you need a browser that supports security features.
- `http://www.corel.com/support/printerdrivers/index.htm`—Corel's printer driver site. Go here (or to your printer manufacturer's Web site) to search for drivers for WordPerfect 8 for Linux. Be sure to check Appendix B, "Fonts and Printing," to learn what to do with the files you find.
- `http://www.cs.wisc.edu/~ghost/`—The main Ghostscript Web site. Linux uses Ghostscript for printing to non-PostScript printers, and WordPerfect can print using a PostScript driver and Ghostscript. See Chapter 6, "Producing Output," and Appendix B for more details.
- `http://www.users.dircon.co.uk/~typhoon/`—The site for RPMs (Red Hat Package Manager files) of recent versions of Ghostscript for easy installation in Linux distributions from Red Hat, Caldera, and others. These binaries include more printer drivers than many other Ghostscript binaries.
- `http://www.gmsys.com/gfax.html`—The homepage for GFax, an X-based front end to fax software for Linux.
- `http://www.math.utah.edu/~beebe/fonts/postscript-type-1-fonts.html`—The site for information on the PostScript Type 1 fonts used by WordPerfect 8 for Linux, including links to font utilities and fonts.
- `http://www.plig.org/xwinman/`—The site for summary information about and links to dozens of window managers. As described in Chapter 1, "Introduction to WordPerfect for Linux," window managers provide much of the look and feel of an X-based Linux desktop.
- `http://www.kde.org`—The home site for the K Desktop Environment (KDE) project, which aims to put an easy-to-use desktop environment on Linux.
- `http://www.gnome.org`—The home site for the GNU Network Object Model Environment (GNOME) project, which is another project to create an easy-to-use Linux desktop environment.
- `http://www.linux.org`—One of the major Linux sites on the Web, with links to lots more Linux-related information.
- `http://www.netscape.com`—The home site of the Netscape Web browser. You should check Web pages you create using WordPerfect with Netscape.
- `http://n9vst.com/pem/wplinux.html`—A site with an extraordinary number of additional links related to WordPerfect for Linux.

What's on the CD-ROM?

This CD-ROM contains a Trial Version of Corel WordPerfect 8 for Linux—Personal Edition courtesy of the Corel Corporation. No registration is required to install the product. Once installed, the product will be fully functional for a period of 90 days. To use the product beyond the 90-day period, you will need to visit `linux.corel.com` to register. This program can be found in the Programs folder.

This Special Edition Using CD-ROM also contains the associated files from selected chapters of the text. The files are named and located in individual folders on the CD with their corresponding chapter numbers.

Also on this CD-ROM, Netscape Communicator 4.6 for Linux and Adobe Acrobat Reader 3.0. These programs can be found in the Programs folder.

Note

See Corel's End User License Agreement on the CD-ROM.

Note

This CD-ROM uses long and mixed-case filenames requiring the use of a protected-mode CD-ROM driver.